T0206406

CAMBRIDGE LIBRARY COLLECTION

Books of enduring scholarly value

Botany and Horticulture

Until the nineteenth century, the investigation of natural phenomena, plants and animals was considered either the preserve of elite scholars or a pastime for the leisured upper classes. As increasing academic rigour and systematisation was brought to the study of 'natural history', its subdisciplines were adopted into university curricula, and learned societies (such as the Royal Horticultural Society, founded in 1804) were established to support research in these areas. A related development was strong enthusiasm for exotic garden plants, which resulted in plant collecting expeditions to every corner of the globe, some-times with tragic consequences. This series includes accounts of some of those expeditions, detailed reference works on the flora of different regions, and practical advice for amateur and professional gardeners.

Flora Capensis

This seminal publication began life as a collaborative effort between the Irish botanist William Henry Harvey (1811–66) and his German counterpart Otto Wilhelm Sonder (1812–81). Relying on many contributors of specimens and descriptions from colonial South Africa – and building on the foundations laid by Carl Peter Thunberg, whose *Flora Capensis* (1823) is also reissued in this series – they published the first three volumes between 1860 and 1865. These were reprinted unchanged in 1894, and from 1896 the project was supervised by William Thiselton-Dyer (1843–1928), director of the Royal Botanic Gardens at Kew. A final supplement appeared in 1933. Reissued now in ten parts, this significant reference work catalogues more than 11,500 species of plant found in South Africa. Volume 7 comprises sections that were published individually between 1897 and 1900, covering Pontederiaceae to Gramineae.

Cambridge University Press has long been a pioneer in the reissuing of out-of-print titles from its own backlist, producing digital reprints of books that are still sought after by scholars and students but could not be reprinted economically using traditional technology. The Cambridge Library Collection extends this activity to a wider range of books which are still of importance to researchers and professionals, either for the source material they contain, or as landmarks in the history of their academic discipline.

Drawing from the world-renowned collections in the Cambridge University Library and other partner libraries, and guided by the advice of experts in each subject area, Cambridge University Press is using state-of-the-art scanning machines in its own Printing House to capture the content of each book selected for inclusion. The files are processed to give a consistently clear, crisp image, and the books finished to the high quality standard for which the Press is recognised around the world. The latest print-on-demand technology ensures that the books will remain available indefinitely, and that orders for single or multiple copies can quickly be supplied.

The Cambridge Library Collection brings back to life books of enduring scholarly value (including out-of-copyright works originally issued by other publishers) across a wide range of disciplines in the humanities and social sciences and in science and technology.

Flora Capensis

Being a Systematic Description of the Plants of the Cape Colony, Caffraria & Port Natal, and Neighbouring Territories

VOLUME 7:
PONTEDERIACEAE TO GRAMINEAE

WILLIAM H. HARVEY *ET AL.*

CAMBRIDGE
UNIVERSITY PRESS

University Printing House, Cambridge, CB2 8BS, United Kingdom

Cambridge University Press is part of the University of Cambridge.
It furthers the University's mission by disseminating knowledge in the pursuit of education, learning and research at the highest international levels of excellence.

www.cambridge.org
Information on this title: www.cambridge.org/9781108068154

This edition first published 1897–1900
This digitally printed version 2014

ISBN 978-1-108-06815-4 Paperback

FLORA CAPENSIS.

VOL. VII.

DATES OF PUBLICATION OF THE SEVERAL PARTS OF THIS VOLUME.

PART I., pp. 1–192, was published *December*, 1897.

PART II., pp. 193–384, was published *July*, 1898.

PART III., pp. 385–576, was published *August*, 1899.

PART IV., pp. 577–791, was published *May*, 1900.

FLORA CAPENSIS:

BEING A

Systematic Description of the Plants

OF THE

CAPE COLONY, CAFFRARIA, & PORT NATAL

(AND NEIGHBOURING TERRITORIES)

BY

VARIOUS BOTANISTS.

EDITED BY

SIR WILLIAM T. THISELTON-DYER, K.C.M.G.,
C.I.E., LL.D., F.R.S.

HONORARY STUDENT OF CHRIST CHURCH, OXFORD.
DIRECTOR, ROYAL BOTANIC GARDENS, KEW.

Published under the authority of the Governments of the Cape of Good Hope and Natal.

VOLUME VII.
PONTEDERIACEÆ TO GRAMINEÆ.

LONDON
LOVELL REEVE & CO., LTD.,
6, HENRIETTA STREET, COVENT GARDEN.
Publishers to the Home, Colonial, & Indian Governments
1897-1900.

LONDON:
PRINTED BY GILBERT AND RIVINGTON, LD.,
ST. JOHN'S HOUSE, CLERKENWELL ROAD, E.C.

PREFACE.

It was considered advisable to commence the continuation of the *Flora Capensis* with Volume VI., which was almost exclusively devoted to the orders which furnish what are familiarly known as "Cape Bulbs." These are perhaps now more largely cultivated in Europe than any other South African plants, and a systematic description of the species it was felt would meet a long-acknowledged want.

Several considerations suggested the desirability of next attacking the seventh and concluding volume of the work in advance of the fourth and fifth, which are still unpublished. What perhaps more weighed with me is the fact that the orders contained in it admittedly present more difficulties than are likely to be encountered in any other part of the work. It has, however, been my good fortune to be able to enlist the aid of contributors who, in each case, have had the advantage of a special previous study of the groups they have undertaken.

Amongst these I must enumerate :—Arthur Bennett, Esq., F.L.S., who has particularly devoted his attention to the *Naiadaceæ;* Dr. Masters, F.R.S., who is an acknowledged authority on the *Restiaceæ;* C. B. Clarke, Esq., F.R.S., who has long been occupied with a comprehensive memoir on the *Cyperaceæ;* and finally Dr. Stapf, A.L.S., who had previously collaborated with Sir Joseph Hooker in preparing the *Gramineæ* for the Flora of British India. The latter order occupies more than half the volume, and this portion of it at least will, I trust, be found of especial usefulness in a country which is largely pastoral.

For the limits of the regions under which the localities in which the species have been found to occur are cited, reference may be made to the preface to Volume VI.

I have again to acknowledge the assistance I have received from Mr. C. H. Wright, A.L.S., and Mr. N. E. Brown, A.L.S., Assistants in the Herbarium of the Royal Botanic Gardens, the former in reading the proofs, and the latter in working out the geographical distribution.

Besides the maps already cited in the preface to Volume VI., the following have also been used :—

Natal. By Alexander Mair, 1875.

Kaffraria und die östlichen Grenz-Distrikte die Cap-Kolonie. By H. C. Schunke in Dr. A. Petermann's Mitteilungen, 1885, t. 9.

Spezial-Karte von Afrika. Gotha : Justus Perthes, 1893.

To the South African correspondents enumerated in the preface to Volume VI., I have again to tender my acknowledgments for the contribution of specimens.

I must further record my obligations to others, and especially to those whose aid in various ways has been of the greatest value in the preparation of the volume :—

Dr. Wilhelm Brehmer, of Lübeck, has lent the study set of Drège's *Restiaceæ* and *Gramineæ*, without which it would have been impossible to arrive at a correct knowledge of some of the species.

Dr. A. Fischer von Waldheim, Director of the Imperial Botanic Gardens, St. Petersburg, has lent the specimens of *Danthonia* and *Pentaschistis* collected by Ecklon and Zeyher.

Dr. Theodor Magnus Fries, Professor of Botany in the University of Upsala, has lent Thunberg's collection of grasses, which have afforded valuable help in the correct identification of some of the species of the earlier authors.

Leo Hartley Grindon, Esq., has contributed a small collection from the Orange Free State.

Professor Eduard Hackel, of St. Pölten, Austria, has kindly lent some type-specimens of grasses.

Major Wolley Dod, R.A., has contributed a very large collection of plants made by himself in the Cape Peninsula, which is unexpectedly rich in new species.

It only remains again to add that the expense of preparation and publication of the present volume has been aided by grants from the Governments of Cape Colony and Natal.

W. T. T. D.

Kew, March, 1900.

SEQUENCE OF ORDERS CONTAINED IN VOL. VII. WITH BRIEF CHARACTERS.

Continuation of Series III. Coronarieæ. Ord. CXXXVIII.—CXL.

[*Liliaceæ*, the first of this series, was by an oversight included under Series II., Epigynæ, on p. xi. of the preceding volume.]

CXXXVIII. PONTEDERIACEÆ (page 1) *Flowers* hermaphrodite. *Perianth* petaloid; lobes 6, biseriate. *Stamens* 6 or 3, rarely 1. *Ovary* in the South African genus 1-celled with 3 parietal placentas, or imperfectly 3-celled. (*Aquatic or marsh herbs. Leaves with long sheathing petioles. Flowers subtended by a membranous spathe.*)

CXXXIX. XYRIDEÆ (page 2). *Flowers* hermaphrodite. Outer *perianth-segments* glumaceous; inner petaloid. *Stamens* 3; staminodes 3 or 0. *Ovary* 1-celled, with 3 parietal or basal placentas. (*Rush-like herbs. Leaves radical. Flowers in peduncled heads or spikes.*)

CXL. COMMELINACEÆ (page 7). *Flowers* hermaphrodite, rarely some male. Outer *perianth-segments* herbaceous; inner petaloid. *Stamens* 6–2; staminodes 1–4, or 0. *Ovary* 3–2-celled. (*Succulent herbs. Leaves cauline, alternate. Inflorescence cymose, usually subtended by boat-shaped bracts.*)

Series IV. Calycinæ. Ord. CXLI.—CXLIII.

CXLI. FLAGELLARIEÆ (page 15). *Flowers* hermaphrodite in the South African genus. *Perianth-segments* 6, subpetaloid or scarious, biseriate. *Stamens* 6. *Ovary* 3-celled; ovules solitary. *Fruit* a 1–3-seeded berry. (*Robust sarmentose herbs. Leaves prolonged into apical tendrils. Panicle terminal.*)

CXLII. JUNCACEÆ (page 16). *Flowers* hermaphrodite in the South African genera. *Perianth-segments* 6, scarious, biseriate. *Stamens* usually 6. *Ovary* 1- or 3-celled; ovules numerous or rarely few. *Fruit* capsular. (*Perennial or annual herbs* (*Rushes*). *Leaves terete or linear. Flowers small, clustered.*)

CXLIII. PALMÆ (page 28). *Flowers* diœcious in the South African genera. *Perianth-segments* 6, scarious or fleshy, biseriate. *Stamens* 6 in the South African genera. *Ovary* 3-celled, or carpels 3 and distinct; ovules solitary. *Fruit* drupaceous or baccate. (*Trees with erect simple stems. Leaves pinnate or flabellate. Spadix axillary.*)

Series V. Nudifloræ. Ord. CXLIV.—CXLVI.

CXLIV. TYPHACEÆ (page 30). *Flowers* unisexual. *Perianth* of 3–6 scales or hairs or none. *Stamens* with free or connate filaments. *Ovary* 1-celled; ovule 1, pendulous. (*Perennial, aquatic or marsh herbs. Leaves linear. Flowers densely crowded in cylindric spikes.*)

CXLV. AROIDEÆ (page 32). *Flowers* unisexual in the South African genera. *Perianth* none, or of 4–9 free or united segments. *Stamens* 4–6. *Ovary* 1–4-celled. (*Floating or tuberous-rooted herbs. Leaves net-veined. Inflorescence enclosed in a spathe.*)

CXLVI. LEMNACEÆ (page 39). *Flowers* very minute, naked or enclosed in a spathe. *Anthers* 1–2, on filaments or sessile. *Ovary* 1-celled; ovules 1–7, erect. (*Minute floating plants. Flowers seated in a cavity of the frond.*)

Series VI. Apocarpæ. Ord. CXLVII.

CXLVII. NAIADACEÆ (page 41). *Flowers* uni- or bi-sexual. *Perianth* green (white in *Aponogeton*), tubular or of 2–6 segments or absent. *Stamens* 2–6, hypogynous. *Carpels* 1–4, distinct, 1-ovuled. (*Aquatic or marsh herbs of various habit.*)

Series VII. Glumaceæ. Ord. CXLVIII.—CLI.

CXLVIII. ERIOCAULEÆ (page 51). *Flowers* minute, unisexual, crowded into heads surrounded by short bracts. *Perianth-segments* 4–6, free or connate in two series which are usually separated by a distinct stipes, scarious or hyaline. *Stamens* 3, inserted on the inner perianth-segments, or 6, opposite and alternating. *Ovary* 2–3-celled; ovules solitary, pendulous. (*Perennial or annual often marsh herbs. Leaves linear or subulate, radical or cauline. Heads usually monœcious.*)

CXLIX. RESTIACEÆ (page 59). *Flowers* unisexual, disposed in spikelets with scarious bracts. *Perianth-segments* 6 (or 4), glumaceous, biseriate; outer keeled. *Stamens* 3. *Ovary* 3–1-celled; ovule solitary, pendulous. (*Rush- or sedge-like herbs, usually perennial. Stems solid or fistular, with* (*rarely without*) *leaf-sheaths. Inflorescence cymose or reduced to a single spikelet.*)

CL. CYPERACEÆ (page 149). *Flowers* uni- or bi-sexual in spikelets with scarious bracts. *Perianth* of 6 or fewer scales or bristles or 0. *Stamens* 3–1, rarely more. *Ovary* 1-celled; ovule 1, erect. *Embryo* in the base of the albumen. (*Herbs* (*Sedges*). *Stems solid. Leaves narrow, with closed sheaths. Spikelets arranged in corymbs or apparent umbels.*)

CLI. GRAMINEÆ (page 310). *Flowers* uni- or bi-sexual in spikelets with scarious bracts (glumes and valves). *Perianth* reduced to 2 (rarely 3) minute scales (lodicules). *Stamens* usually 3. *Ovary* 1-celled; ovule 1, ascending. *Embryo* outside the albumen. (*Herbs or* (*in Bambuseæ*) *tall shrubs. Stems branched at the base. Leaves with split or closed sheaths. Spikelets spicate, racemose, or panicled.*)

FLORA CAPENSIS.

ORDER CXXXVIII. PONTEDERIACEÆ.

(By N. E. BROWN.)

Flowers hermaphrodite, all alike or some cleistogamous, regular or slightly irregular. *Perianth* inferior, petaloid; segments 6, biseriate, united into a tube in the lower part, rarely free, equal or the outer series smaller. *Stamens* 3 or 6, rarely 1, usually of two sizes and often of different colours; filaments free; anthers basifixed or dorsifixed, 2-celled, introrse, opening by longitudinal slits, or rarely by terminal pores. *Ovary* superior, 3-celled with axile placentas, or 1-celled with 3 parietal or more or less prominent placentas; style filiform; stigma entire, thickened or subcapitate, or shortly 3-lobed; ovules numerous and biseriate in each cell, or rarely solitary, anatropous. *Fruit* a many-seeded capsule, or rarely a 1-seeded achene. *Seeds* small, ovoid, ribbed; embryo cylindrical, straight, embedded in the centre of a copious albumen.

Aquatic or marsh herbs, with the lower part of the stem or rhizome creeping and rooting in the mud, or entirely floating; leaves alternate, hastate, cordate, orbicular, ovate, lanceolate or linear, entire, usually with a sheathing petiole, the submersed leaves sometimes without blades, or different in form; inflorescence terminal, either fascicled in the sheath of the flowering-leaf, or spicate, with the flowers solitary or fascicled along the axis of the spike, rarely solitary and axillary.

DISTRIB. Species about 22, chiefly natives of Tropical Africa and Tropical America, about 4 occurring in North America, 2 or 3 in Tropical and Temperate Asia, and 1 in Australia. Only one species has yet been found within the confines of the South African Flora.

I. HETERANTHERA, Ruiz. & Pav.

Perianth with a distinct tube; segments equal, spreading, oblong. *Stamens* 3, affixed to the throat of the perianth-tube, exserted, more or less unequal; filaments filiform; anthers oblong. *Ovary* 1-celled with 3 parietal placentas, or imperfectly 3-celled with very prominent placentas; style filiform; stigma thickened; ovules numerous, biseriate. *Capsule* oblong or linear, with a thin pericarp. *Seeds* numerous, ovoid, ribbed.

Lower part of the stem creeping and rooting in the mud; leaves with long petioles sheathing at the base, and cordate, ovate, or reniform blades, or all linear and submerged; flowering-shoots bearing one leaf, whose sheath embraces the membranous spathe which subtends the terminal flower-spike or includes 1–3 flowers; flowers spicate, small, blue, whitish, or yellow, all alike and expanding, or one or more cleistogamous and enclosed in the sheath of the leaf at the base of the spike.

DISTRIB. Species 9, Tropical Africa and Tropical America, three extending into North America.

1. **H. kotschyana** (Fenzl ex Solms in Schweinf. Beitr. Fl. Æthiop. 205); plant 5–12 in. high; stem below the leaves creeping and rooting at the nodes, herbaceous, glabrous; petiole 2½–8 in. long, erect or ascending, sheathing at the base, glabrous; blade 1–2¾ in. long, ½–1¾ in. broad, cordate, obtusely pointed, glabrous; flowering-stems 1½–4 in. long, bearing one leaf, whose sheath embraces a membranous spathe 1–1¼ in. long; flower-spike terminal, 2–4 in. long, several-flowered; basal flower enclosed in the sheath of the leaf, never opening; stamen 1; all the other flowers expanding, sessile, ebracteate; perianth-tube 2½ lin. long, cylindric, very slightly curved, and a little oblique at the mouth, lobes 2 lin. long, ¾ line broad, oblong, obtuse, spreading; stamens 3, shortly exserted; ovary oblong, trigonous; style filiform, as long as the stamens, with a simple slightly thickened stigma; capsule oblong, trigonous. *Solms in DC. Monog. Phanerog.* iv. 522. *Monochoria vaginalis, Kirk in Journ. Linn. Soc.* viii. 147.

KALAHARI REGION: Transvaal; Bosch Veld, *Rehmann*, 5148! 5149!

Also in Eastern Tropical Africa.

ORDER CXXXIX. XYRIDEÆ.

(BY N. E. BROWN.)

Flowers hermaphrodite. *Calyx* inferior, irregular; sepals 3, or rarely 2, the two lateral exterior, boat-shaped, keeled, glumaceous, the third interior, membranous, convolute, forming a closed, obtuse, spathe-like hood, or obtusely calyptriform, circumscissile at the base and split open on one side, closely enveloping the corolla when in bud, and pushed off as the corolla grows out and expands, rarely absent. *Corolla* regular, petaloid, with a slender tube usually split into claw-like segments in the lower part by the growth of the ovary, and three spreading, cuneate-obovate lobes, marcescent. *Stamens* 3, affixed at the mouth of the tube opposite the corolla-lobes; filaments short, flattened; anthers basifixed, oblong, 2-celled, cells parallel or slightly divergent at the base, contiguous, or more or less separated by a broad connective, extrorse, opening by longitudinal slits. *Staminodia* 3 or 0, alternating with the corolla-lobes at the mouth of the tube, simple or bifid at the apex and hairy or glabrous, or divided into two entire or bifid, glabrous or hairy arms, or more or less deeply divided into brush-like tufts of hairs. *Ovary* superior, 1-celled or imperfectly 3-celled; placentas 3, parietal or shortly united at the centre, or free and erect from the base; ovules indefinite, 2-seriate, orthotropous; style filiform, sometimes with tubercles or processes at or below the middle, trifid or 3-armed at the apex, or entire; stigmas dilated or subcapitate, rarely simple. *Capsule* dividing into 3 valves between the placentas. *Seeds* minute, ovoid, apiculate; hilum basilar; testa thin, usually ridged; albumen copious, transparent; embryo minute, broadly depressed-conical, seated at the apex of the albumen under the terminal apiculus.

Perennial or rarely annual herbs, of tufted habit, growing in damp places. Leaves all radical, linear, terete or filiform, sheathing at the base. Peduncles erect, simple, terminated by a solitary dense head or spike, leafless or bearing one or more pairs of convolute sheaths, and embraced at the base by a leafless or leaf-bearing sheath. Flower-heads or spikes globose, ovoid, or elongate; bracts glumaceous, somewhat rigid, imbricate, concave, or rarely somewhat convolute, the lower ones in a few species elongated and leafy, forming an involucre. Flowers solitary and sessile in the axils of the bracts, yellow or blue, usually of small size.

DISTRIB. Genera 2; species about 130, dispersed throughout the Tropics and warmer regions of the earth.

I. XYRIS, Linn.

Sepals 3; two lateral and exterior, boat-shaped, keeled; one interior, membranous, convolute or obtusely calyptriform, closely enveloping the corolla when in the bud, circumscissile at the base, and deciduous as the corolla develops. *Corolla* with a slender tube and three spreading cuneate-obovate lobes. *Staminodia* 3, simple, bifid, or two-armed, glabrous or penicillate. *Style* filiform, without tubercles or processes, 3-branched above; stigmas dilated or subcapitate.

Peduncles with a basal sheath, naked above. The rest as in the Order.

DISTRIB. The same as for the Order. Species about 120.

Spikes dark brown:
  Bracts with a distinct, oblong-lanceolate, grey dorsal area ... ... (1) **decipiens.**
  Bracts without a grey dorsal area:
    Leaves terete: ... ... (2) **natalensis.**
    Leaves flat:
      Keel of lateral sepals ciliate or scabrid from the base to the apex... ... ... (3) **Umbilonis.**
      Keel not ciliate to the apex:
        Bracts entire; lateral sepals 3–3½ lin. long, pale straw-coloured ... ... (4) **Rehmanni**
        Bracts with lacerated margins; lateral sepals 2½–2¾ lin. long, brown... ... (5) **Gerrardi.**
      Keel of lateral sepals entire, not ciliate ... (6) **capensis.**
Spikes straw-coloured (perhaps greenish when alive), or pale brownish; keel of lateral sepals not ciliate:
  Flowering-spike globose or ovoid, 3–5 lin. thick... ... (7) **anceps.**
  Flowering-spike lanceolate, about 1 line thick ... ... (8) **filiformis.**

1. **X. decipiens** (N. E. Brown); leaves 5–16 in. long, 1½–2½ lin. broad, laterally flattened, linear, acuminate at the apex, glabrous, with concolorous sheaths 1¼–3 in. long, the sheath embracing the peduncle with a distinct blade like the other leaves, but flexuose a little above the sheath; peduncle 2–2½ ft. high, 1–1⅓ lin. thick, terete, distinctly sulcate-striate, glabrous; spike 6–9 lin. long, 5–6 lin. thick, ovoid, acute or subacute; bracts 3–3½ lin. long, 2½–2¾ lin. broad, oblong-obovate, very obtuse, minutely subdenticulate, shining chestnut-brown, with a very distinct, oblong-lanceolate, greiysh dorsal area at the apex about 1–1¼ lin. long and ½ lin. broad, 7–9-nerved; nerves reticulate at the apex; lateral sepals 2½ lin. long, ½ lin. broad, pale-brown, linear-falcate or linear-lanceolate,

acute, keeled; keel narrowly wing-like, serrulate, pubescent with rust-coloured appressed hairs; corolla only seen in bud; arms of staminodia (in the bud) 1 line long, divided into tufts of fine hairs; anthers 1 line long, linear; stigmas very slightly dilated; capsule $2\frac{1}{2}$ lin. long, oblong-lanceolate, subacute, trigonous, with a thin, membranous pericarp.

WESTERN REGION: West Africa, south of the Tropic, *Curror*!

This is very similar to *X. communis*, Kth., from Tropical America, but the dorsal area of the bracts is narrower, and the lateral sepals are serrulate on the keel, not fimbriate with fine hairs as in *X. communis*.

2. **X. natalensis** (Nilss. in Öfvers. Vet. Akad. Förhandl. Stockh. 1891, 157); leaves 10–20 in. long, $\frac{1}{2}$–1 line thick, terete, acute, glabrous, with large, dark brown sheaths 2–$3\frac{1}{2}$ in. long; peduncular sheath leafless, $4\frac{1}{2}$–$5\frac{1}{2}$ in. long; peduncle 16–24 in. long, $\frac{2}{3}$–1 line thick, terete, glabrous; spike 4–6 lin. long, 3–$3\frac{1}{2}$ lin. thick; ovoid or oblong; bracts 2–3 lin. long, $1\frac{1}{2}$–2 lin. broad, elliptic or elliptic-oblong, very obtuse, entire, not ciliate, opaque dark brown, glabrous, 7–9-nerved; nerves more or less reticulate in the upper part; lateral sepals $2\frac{1}{2}$–$2\frac{3}{4}$ lin. long, falcate-lanceolate, brown, with a darker brown, ciliated, wing-like keel, produced into an acute point at the apex; corolla yellow, with a tube about 3 lin. long, and broadly cuneate-obovate, denticulate lobes, $2\frac{1}{2}$ lin. long, 2 lin. broad; arms of staminodia divided into tufts of fine hairs; anthers oblong, 1 line long. *Nilss. in Svensk. Vet. Akad. Handl.* xxiv. *no.* 14, 28.

EASTERN REGION: Natal; Flats between Umlazi River and Natal Bay, *Krauss*, 141! and without precise locality, *Sanderson*, 455! (456 according to *Nilsson*). Zululand, *Gerrard*, 1524!

The large dark brown sheaths and terete leaves readily distinguish this from all the other South African species.

3. **X. Umbilonis** (Nilss. in Svensk. Vet. Akad. Handl. xxiv. no. 14, 30); leaves 9–20 in. long, $\frac{3}{4}$–1 line broad, laterally flattened, linear, tapering to a very fine point at the apex, straight or twisted, glabrous, with brown sheaths $1\frac{1}{2}$–$4\frac{1}{2}$ in. long; peduncular sheath $3\frac{1}{2}$–8 in. long, leafless, acutely pointed, dark brown in the lower part; peduncle 18–28 in. high, $\frac{2}{3}$–1 line thick, subterete or slightly compressed, sometimes slightly 2-edged in the apical part, glabrous; spike $3\frac{1}{2}$–7 lin. long 3–$3\frac{1}{2}$ lin. thick, ovate or oblong-ovate; bracts 2–$2\frac{1}{4}$ lin. long, $1\frac{1}{2}$–$1\frac{3}{4}$ lin. broad, elliptic-obovate, or suborbicular, very obtusely rounded at the apex, dark brown, slightly shining, 7–9-nerved; the nerves scarcely or but very slightly reticulate at the apex; lateral sepals $1\frac{3}{4}$–2 lin. long, $\frac{1}{2}$ line broad, falcate-lanceolate, rather dark brown, the margins a little paler than the rest, narrowly keeled; keel produced at the apex into a short and often more or less flexuose point, and finely and densely ciliate from base to apex with short hairs; corolla only seen in the bud, yellow, lobes broadly cuneate-obovate, denticulate; arms of staminodia divided into tufts of fine hairs; anthers oblong.

KALAHARI REGION: Transvaal; Magalies Berg, *Burke*, 128! *Zeyher*, 1727! Lomati Valley, near Barberton, *Galpin*, 1223!
EASTERN REGION: Natal; Umbilo Waterfall, *Rehmann*, 8139! Inanda, *Wood*, 155! and without precise locality, *Buchanan*, 304!

This is very nearly allied to X. *Bakeri*, Nilss., from Madagascar, but differs in its much narrower and more distinctly awned lateral sepals.

4. X. Rehmanni (Nilss. in Svensk. Vet. Akad. Handl. xxiv. no. 14, 28); leaves 1–2 ft. long, $\frac{1}{2}$–$1\frac{1}{4}$ line broad, laterally flattened, linear, tapering to a fine hair-like point at the apex, straight or twisted, glabrous, with sheaths about 3 in. long, brown; peduncular sheath 6–7 in. long, leafless, but with a short flat point $1\frac{1}{2}$–$2\frac{1}{2}$ lin. long, dark brown below, pale or greenish at the apex; peduncle 2–$2\frac{1}{4}$ ft. long, $\frac{3}{4}$–1 line thick, somewhat compressed, not two-edged, glabrous; spike globose, about $4\frac{1}{2}$ lin. in diam.; bracts $2\frac{3}{4}$–3 lin. long, $1\frac{1}{2}$–2 lin. broad, the outer orbicular, the inner elliptic, obtuse, mucronate, entire, not ciliate, glabrous, dark olive-brown with a slightly paler apical area at the base of the mucro, 9-nerved; nerves not reticulate in the upper part; lateral sepals shortly exserted beyond the bracts, 3–$3\frac{1}{4}$ lin. long, $\frac{2}{3}$ line broad, lanceolate, pale straw-coloured with a brown keel, which is produced into a short mucro at the apex, and shortly ciliate-scabrid from near the base to about $\frac{1}{2}$–$\frac{3}{4}$ line below the apex; corolla yellow, lobes broadly cuneate-obovate, denticulate; arms of the staminodia divided into tufts of long fine hairs; anthers oblong, $1\frac{2}{3}$ lin. long.

KALAHARI REGION: Transvaal; Houtbosch, *Rehmann*, 5764!

5. X. Gerrardi (N. E. Brown); leaves not seen; peduncular sheath over $3\frac{1}{2}$ in. long, dark brown in the lower part, leafless, mucronate; peduncle 9–16 in. long, $\frac{1}{2}$ line thick, somewhat compressed, especially in the upper part, two-edged, glabrous; spike ovoid or subglobose, 3–$3\frac{1}{2}$ lin. long, about $2\frac{1}{2}$ lin. thick; bracts $2\frac{1}{2}$–$2\frac{3}{4}$ lin. long, 1–$1\frac{1}{2}$ lin. broad, ovate-oblong or elliptic-oblong, acute, carinate, and somewhat complicate at the apex, membranous and slightly lacerated at the margin, three-nerved, dark brown, glabrous; lateral sepals $2\frac{1}{2}$–$2\frac{3}{4}$ lin. long, $\frac{2}{3}$ line broad, lanceolate (not falcate), acute or obtuse, keeled, dark brown, paler towards the margins; keel narrowly wing-like in the lower $\frac{1}{2}$–$\frac{2}{3}$ only, and serrulate-scabrid or ciliate to half or three-fourths the way up, its apex very shortly, or not at all produced into a point; corolla yellow, only seen in bud, with broadly cuneate-obovate, denticulate lobes; arms of staminodia divided into tufts of fine hairs; anthers $\frac{3}{4}$ line long, oblong; capsule $1\frac{1}{2}$ line long, about 1 line diam., obovoid, dorsally flattened, trigonous, apiculate.

EASTERN REGION: Zululand, *Gerrard*, 1526!

This species is readily distinguished from all the other South African ones, by the slightly lacerate, 3-nerved bracts, and from all but *X. Rehmanni* by the nearly straight (viewed sideways) lateral sepals being ciliate or scabrid to only about $\frac{1}{2}$ or $\frac{3}{4}$ the way up; in all the other ciliated species, the ciliation is more pronounced and extends from the base to the apex. Viewed dorsally, the lateral sepals are more or less curved, as in all the other species. The only specimens I have seen are without leaves.

6. X. capensis (Thunb. Prod. 12); leaves 1–7 in. long, $\frac{1}{2}$–$1\frac{1}{2}$ lin. broad, laterally flattened, linear, glabrous; peduncle 6–18 in. high, $\frac{1}{3}$–$\frac{3}{4}$ line thick, terete, sometimes distinctly striate, sometimes the striations scarcely visible, often with one or two slightly raised lines or angles, glabrous, with concolorous or pale brown sheaths $\frac{3}{4}$–$2\frac{1}{2}$ in. long, peduncular sheath 1–5 in. long, leafless or produced into a leafy point $\frac{1}{2}$–6 in. long; spike at first ovoid, in fruit subglobose, 2–$3\frac{1}{2}$ lin. long; bracts 2–$2\frac{3}{4}$ lin. long, $1\frac{1}{2}$–$2\frac{1}{2}$ lin. broad, orbicular, elliptic, or oblong, obtuse, or the inner subacute and often more or less carinate at the apex, entire, 3-nerved, glabrous, not ciliate, brown; lateral sepals 2–$2\frac{1}{2}$ lin. long, $\frac{1}{2}$ line broad, falcate-lanceolate, acute, not mucronate, pallid, transparent, with a pale brown, narrow wing-like keel, which is neither ciliate nor produced at the apex; corolla yellow; tube $2\frac{1}{2}$–3 lin. long; lobes $1\frac{1}{2}$–$1\frac{3}{4}$ lin. long, and about the same in breadth, broadly cuneate-obovate, denticulate; arms of the staminodia divided into tufts of fine hairs; anthers oblong, about as long as the filaments; capsule $1\frac{3}{4}$ lin. long, trigonous, oblong, obtuse. *Thunb. Fl. Cap. edit. Schult.* 81; *Vahl, Enum.* ii. 206; *Willd. Sp. Pl.* i. 255; *Roem. et Schultes, Syst. Veg.* i. 552; *Kunth, Enum.* iv. 24; *Nilss. in Öfvers. Vet. Akad. Förhandl. Stockh.* 1891, 154; *and in Svensk. Vet. Akad. Handl.* xxiv. *no.* 14, 40.

SOUTH AFRICA: without locality, *Zeyher*, 1725!

COAST REGION: Cape Div.; Wynberg, *Harvey!* Tulbagh Div.; banks of streams near Ceres, *Bolus*, 5498! *MacOwan and Bolus, Herb. Norm. Aust. Afr.*, 1100! Queenstown Div.; marshy ground on the summit of N'Qebanya Mountain, 5600 ft., *Galpin*, 1907!

CENTRAL REGION: Somerset Div.; summit of Bosch Berg, 4800 ft., *MacOwan*, 1749!

KALAHARI REGION: Transvaal, Magalies Berg, *Burke*, 170!

EASTERN REGION: Tembuland; Bazeia Mountain, 4000 ft., *Baur*, 598! Natal; Kar Kloof, *Rehmann*, 7364! Inanda, *Wood*, 103!

7. X. anceps (Lam. Tabl. Encycl. i. 132); leaves 2–14 in. long, $1\frac{1}{4}$–$4\frac{1}{2}$ lin. broad, laterally flattened, linear, obtuse or acute, glabrous, with concolorous sheaths $\frac{3}{4}$–$4\frac{1}{2}$ in. long; peduncular sheath $1\frac{1}{2}$–$6\frac{1}{2}$ in. long, leafless, obtuse, or with a leafy obtuse point $\frac{1}{4}$–$\frac{1}{2}$ in. long; peduncle $\frac{1}{2}$–$2\frac{1}{2}$ ft. long, $\frac{1}{2}$–$1\frac{1}{3}$ line broad, compressed and acutely two-edged, especially towards the apex, glabrous; spike globose or ovoid, 3–5 lin. in diam.; bracts 2–3 lin. long, $1\frac{3}{4}$–2 lin. broad, elliptic, very obtuse, or the inner ones subacute, entire, not ciliate, pale yellowish-brown, or pale straw-coloured, with a narrowly ovate or lanceolate, acute, green area at the slightly keeled apex; nerves about 7–9, very slender and obscure; lateral sepals $2\frac{1}{4}$–$2\frac{1}{2}$ lin. long, $\frac{1}{2}$ line broad, subfalcate-lanceolate, acute, with a rather broad, wing-like keel, neither ciliate nor produced at the apex, entirely pale straw-colour; corolla yellow, with a tube $2\frac{1}{2}$ lin. long, and broadly cuneate-obovate, denticulate lobes, about $1\frac{1}{2}$ lin. long, and $\frac{3}{4}$–1 line broad; arms of staminodia divided into tufts of fine hairs; anthers oblong, $\frac{2}{3}$ line long; capsule $1\frac{1}{2}$ lin. long, trigonous, oblong, obtuse, not apiculate. *Nilss. in Svensk. Vet. Akad. Handl.* xxiv. *no.* 14, 37. *X. platycaulis, Poir. Encycl.* viii. 820; *Roem. et Schultes, Syst.*

Veg. i. 551; *Kunth, Enum.* iv. 18; *Nilss. in Öfvers. Vet. Akad. Förhandl. Stockh.* 1891, 153. *X. nitida, Willd. ex Dietr. Sp. Pl.* ii. 372.

EASTERN REGION: Natal; Durban, *Rehmann*, 8560!

Also in Tropical Africa and Madagascar.

This is a totally different plant from that described by Poiret and nearly all subsequent authors as *X. anceps.* Poiret redescribed Lamark's plant (which was from Madagascar) as *X. platycaulis*, and wrongly identified an Indian species (*X. complanata*, R. Br., syn. *X. Walkeri*, Wight) with *X. anceps*, Lam.

8. **X. filiformis** (Lam. Tabl. Encycl. i. 132); leaves $\frac{1}{3}$–5 in. long, $\frac{1}{3}$–1 line broad, laterally flattened, linear, obtuse or acute, spirally twisted or straight, glabrous, with sheaths $\frac{1}{4}$–$1\frac{1}{2}$ in. long, concolorous or pale; peduncular sheath $\frac{1}{2}$–2 in. long, leafless and mucronate, or with a leafy point $\frac{1}{4}$–$\frac{1}{2}$ in. long; peduncle $1\frac{1}{2}$–12 in. long, very slender, $\frac{1}{5}$–$\frac{1}{4}$ line thick, terete or subcompressed, striate, glabrous; spike 2 lin. long, lanceolate, acute when in flower, 1–3-flowered; bracts few, $1\frac{1}{2}$–2 lin. long, $\frac{3}{4}$–1 line broad, oblong or elliptic-oblong, boat-shaped or convolute, apparently acute, but obtuse when flattened out, entire, not ciliate, 3-nerved, membranous, very pale brownish or straw-coloured; lateral sepals about 2 lin. long, rather more than $\frac{1}{3}$ line broad, straight, lanceolate, acute, keeled; keel neither winged nor ciliate, membranous, straw-coloured; corolla yellow, only seen in very young bud; capsule $1\frac{1}{2}$–$1\frac{3}{4}$ lin. long, $\frac{2}{3}$ line diam., oblong, obtuse, apiculate, trigonous (or triangular ?) in cross section. *Poir. Encycl.* viii. 821; *Vahl, Enum.* ii. 207; *Kunth, Enum.* iv. 24. *X. straminea, Nilss. in Öfvers. Vet. Akad. Förhandl. Stockh.* 1891, 153; *and in Svensk. Vet. Akad. Handl.* xxiv. *no.* 14 40.

KALAHARI REGION: Transvaal; Apies River, *Burke*!

Also in Tropical Africa.

I have only seen flowers of this species in very young bud, either in the Tropic or South African specimens, and have failed to discover the nature of the staminodia; the anthers in that state are very small and ovate.

ORDER CXL. COMMELINACEÆ, A. Rich.

(By C. B. CLARKE.)

Flowers small, bisexual, or the upper part of the cyme male. *Sepals* 3; one (dorsal) entirely without, hooded. *Petals* 3, blue, yellow, or white, free or claws united into a linear tube in *Cyanotis* and *Coleotrype*. *Stamens* 6, whereof 1–4 are often sterile, reduced to staminodes or wanting; filaments often bearded by beaded hairs. *Ovary* free, 3- (sometimes 2-) celled; style simple; ovules one or several in each cell, attached to the inner angle. *Capsule* loculicidal; embryo small, shortly cylindrical, far from the hilum, loose in the floury albumen; radicle close to the foramen covered externally by the embryostega.

Herbs; leaves alternate, ovate lanceolate or linear; bases sheathing.

DISTRIB. All warm countries. Genera 26; species 330.

Tribe 1. **COMMELINEÆ.** Fertile *stamens* 3–2, sterile 0–4. Flowers zygomorphic.

I. **Commelina.**—*Inflorescence* of 2–1 cymes included (or nearly so) within a folded or funnel-shaped bract (spathe).

II. **Aneilema.**—*Flowers* panicled or clustered, not included in a spathe.

Tribe 2. **TRADESCANTIEÆ.**—Fertile *stamens* 6–5. Flowers actinomorphic.

III. **Coleotrype.**—*Flowers* axillary clustered. *Stamens* inserted high up on the linear corolla-tube. Capsule 3-celled.

IV. **Cyanotis.**—*Flowers* axillary clustered. *Stamens* hypogynous. *Capsule* 3-celled.

V. **Floscopa.**—Racemes panicled. Capsule 2-celled.

I. COMMELINA, Linn. *partly.*

Inflorescence of 2–1 cymes included (or nearly so) within a folded or funnel-shaped spathe. *Petals* clawed; 2 equal, the third exterior, smaller. *Stamens* 3–2 anterior perfect, 3–2 dorsal sterile. *Ovary* cells 2 anterior equal, 2–1-ovulate, dehiscent; the third dorsal 1-ovulate, or empty or suppressed. *Seeds* 5–1; hilum linear vertical.

Annual succulent weeds; flowers evanescent; the spathe is an ovate leaf-like bract, either simply folded boat-like, or its lower margins connate, so as to form an oblique funnel; the flowers of the lower cyme (and the upper flowers of the upper cyme) are usually male.

DISTRIB. Species 100, in all warm countries.

Subgenus I. DIDYMOON. Two anterior cells of ovary 2-ovulate; dorsal cell 1-ovulate or empty or suppressed.

EU-COMMELINA. Capsule 5-seeded; petals blue.

Spathe simply folded, boat-shaped ... ... (1) **nudiflora.**
Spathe simply folded, small, falcate ... ... (2) **subulata.**
Spathe an oblique funnel ... ... (3) **benghalensis.**

HETEROCARPUS. Capsule 5–1-seeded, i.e. 2 or 4 ovules of the two anterior cells generally infertile; petals yellow or orange.

Capsule usually 1-seeded; leaves glabrous-oblong (4) **africana.**
Capsule 3–1-seeded; leaves hairy on both surfaces (5) **krebsiana.**
Capsule 3–1-seeded; leaves linear glabrous ... (6) **karooica.**

DISSECOCARPUS. Capsule 4-seeded, i.e. dorsal cell suppressed; petals blue.

Spathe oblique funnel-shaped ... ... (7) **eckloniana.**

Subgenus II. MONOON. Two anterior cells of ovary 1-ovulate; dorsal cell 1-ovulate or empty or suppressed; petals blue or white.

HETEROPYXIS. Capsule 3-seeded; dorsal cell indehiscent, or nearly so.

Capsule tough, dorsal cell tubercular scabrous ... (8) **Gerrardi.**
Capsule membranous, papery ... ... (9) **albescens.**

SPATHODITHYROS. Capsule 2-seeded.

Leaves linear ... ... (10) **Livingstoni.**

1. C. nudiflora (Linn. Sp. Pl. ed. 2, 61, not of Mant. 177); diffuse, nearly glabrous; stem 6–30 in. long, often rooting at the knots; leaves lanceolate; spathes scattered, peduncled, simply folded, ovate-lanceolate, middle nerve nearly straight; petals blue or nearly white, the anterior often much smaller, sometimes 0; capsule perfecting (nearly always)

5 seeds; surface of the 4 scattered seeds hexagonally reticulated. *Hook. f. Fl. Brit. Ind.* vi. 369. *C. agraria, Kunth, Enum.* iv. 38. *C. communis, Walter, Fl. Carol.* 68; *C. B. Clarke, Comm. et Cyrt. Beng. t.* 1 (*syn. C. africana excl.*).

VAR. β, **C. werneana** (Hassk. in Schweinf. Beitr. Fl. Æthiop. 206); leaves large, oblong-lanceolate, rounded at base, up to 3½ by 1¼ in.

EASTERN REGION: Natal; coast-land, *Dr. Sutherland!* Inanda, 1800 feet, *Wood*, 94! 1615! and without precise locality, *Gerrard*, 1838! Zululand, *Gerrard*, 1839! Var. β, Natal, *Sutherland! Sanderson!*

Distributed throughout Africa, and in nearly all warm countries. Var. β, in Africa only.

2. C. subulata (Roth, Nov. Pl. Sp. 23); erect, nearly glabrous; stem 6–12 in. long; leaves linear; spathes ⅓ in. long, scattered, nearly sessile, simply folded, ovate acute, falcate, mature pendent, middle nerve curved; lower cyme often wanting; flowers small; capsule perfecting 5 (or 3) seeds; surface of the scattered seeds transversely grooved, somewhat rugged. *Hook. f. Fl. Brit. Ind.* vi. 369.

KALAHARI REGION: Transvaal; Bosch Veld, *Rehmann*, 5346! Houtbosch, *Rehmann*, 5757! Hooge Veld, *Rehmann*, 6669!

Distributed throughout Africa, and in the Deccan of India.

3. C. benghalensis (Linn. Sp. Pl. ed. 2, 60); diffuse; stem 6–12 in. long, often rooting at the knots; leaves ovate, suddenly narrowed at base, the quasi-petioles often with bright fulvous-red hairs, sometimes nearly glabrous; spathes oblique funnel-shaped, sessile or peduncled, solitary or 2–3 close together; capsule perfecting (usually) 5 seeds; surface of seeds wrinkled or strongly reticulate. *Thunb. Fl. Cap. ed. Schultes*, 294; *C. B. Clarke, Comm. et Cyrt. Beng. t.* 4; *Hook. f. Fl. Brit. Ind.* vi. 370. *C. canescens, Vahl, Enum.* ii. 173; *Webb et Berth. Iles Canaries, Phyt.* iii. *pt.* ii. 358, *t.* 239.

SOUTH AFRICA: *Drège*, 8780! *Harvey!*
COAST REGION: Knysna Div., *Rehmann*, 504! Uitenhage Div., *Ecklon and Zeyher*, 594! *Harvey!* Port Elizabeth, *E.S.C.A. Herb.*, 208! Fort Beaufort Div., *Cooper*, 553!
CENTRAL REGION: Somerset Div.; Bosch Berg, *Burchell*, 3177! 3216!
EASTERN REGION: Tembuland; Bazeia, 2000 ft., *Baur*, 86! Natal; Inanda, *Wood*, 1237! and without precise locality, *Sanderson!* Zululand, *Gerrard*, 1840!

Found throughout Africa, and thence to Japan and the Moluccas.

This plant, both in Africa and India, often produces at base nearly leafless shoots bearing abnormal flowers; these burrow into the earth and there mature small abnormal capsules of 1–2 cells, 1–2 seeds.

4. C. africana (Linn. Sp. Pl. ed. 2, 60); diffuse; stem 6–20 in. long; leaves sessile, oblong, glabrous on both surfaces; spathes simply folded, peduncled, scattered, ovate acute, nearly glabrous, ½–1 in. long; corolla yellow; filaments ¼ in. long; capsule tough, perfecting 1 seed in the indehiscent dorsal cell 2–3 lines long; the 4 ovules of the two anterior cells enlarging, usually present, brown in the ripe capsule, but quite rudimentary. *Thunb. Fl. Cap. ed. Schultes*, 294; *Gaertn. Fruct.* i. 50, *t.* 15, *fig.* 1. *Hedwigia africana, Medicus in Röm. & Uster. Mag. St.* x. 124.

VAR. β, **lancispatha** (C. B. Clarke); larger than the typical *C. africana*; spathes longer, more acuminated, falcate, often 1½ in. long; capsule usually 1-seeded.

VAR. γ, **Barberæ** (C. B. Clarke); spathes ovate-oblong acute, not much acuminate nor much falcate, glabrous or with scattered several-celled hairs; capsule somewhat larger, 1-seeded, or more often 3-seeded, i.e. in each anterior cell the lower seed rudimentary, the upper 2–3 lines long, oblong, strongly reticulated. *C. karooica, var. β Barberæ, C. B. Clarke in DC. Monogr.* iii. 166.

SOUTH AFRICA: without locality, *Wallich! Pappe! Kitching!* Var. γ, *Zeyher*, 1728!

COAST REGION: Cape Div.; near Cape Town, *Burchell*, 75! Table Mt., *Harvey*, 860! *Wallich*, 315! Devils mountain, *Rehmann*, n. 938! Simon's Bay, *Wright!* Uitenhage Div., *Zeyher!* Algoa Bay, *Cooper*, 3322! Var. β, Alexandria Div.; Zuurberg Range, 2500–3500 ft., *Drège*, 8779!

CENTRAL REGION: Var. β, Graaff Reinet, 4300 ft., *Bolus*, 377! Hopetown Div., *Burchell*, 2685! Albert Div., *Cooper*, 586 partly! and 588 partly! Var. γ, Colesberg Div.; Carolus Poort, *Burchell*, 2753! Albert Div., *Cooper*, 586 partly! 1380!

KALAHARI REGION: Transvaal; Hooge Veld, *Rehmann*, 6860! (capsule perfecting 4 seeds). Var. β, Transvaal; Potchefstroom Dist., *Nelson*, 251! Barberton, 4000–5000 ft., *Galpin*, 1231! Var. γ, Vaal River, *Mrs. Barber!* Griqualand West, *Bowker*, 12! Orange Free State; Orange River, *Burke!* Wolve Kop, *Burke!* Transvaal; Streydpoort, *Rehmann*, 5392! (leaves linear).

EASTERN REGION: Natal, *Cooper*, 3324! Var. β, Transkei, 3000 ft., *Baur*, 85! Griqualand East; near Kokstad, 4300 ft., *Tyson*, 1799! Natal; Inanda, *Wood*, 829! Var. γ, Kaffraria, *Baur!* Drakensberg, *Cooper*, 3321! Natal; Inanda, *Wood*, 409A! Zaai Laager, 3800 ft., *Wood*, 3486! and without precise locality, *Sanderson! Gerrard*, 1439!

Distributed throughout Africa: Var. γ endemic.

Rehmann n. 6860 is, in other points, a typical *C. africana*; but each capsule (several good ones attached) perfects 2 short oblong reticulated seeds in each anterior cell, while the dorsal cell is very small and empty.

5. C. krebsiana (Kunth, Enum. iv. 40); leaves hispid with large, white, many-celled hairs on both surfaces; otherwise as *C. africana* (typical form). *C. africana, var. krebsiana, C. B. Clarke in DC. Monogr.* iii. 164.

VAR. β, **villosior** (C. B. Clarke); hairs similar to those in *C. krebsiana* type, but smaller more numerous, giving the plant a soft strigose, not hispidulous, appearance; leaves ovate, 9 by 5 lines. *C. barbata, var. β villosior, C. B. Clarke in DC. Monogr.* iii. 167, not of *Lam. C. africana, var. γ polyclada, C. B. Clarke in DC. Monogr.* iv. 165. There is also included here (*Cooper*, 965, *Gerrard*, 1842) a form in which the leaves are oblong, attaining 3 by ½ in.

SOUTH AFRICA: without locality, *Drège*, 3505!

COAST REGION: Var. β, Bathurst Div.; between Theopolis and Port Alfred, *Burchell*, 4038!

KALAHARI REGION: Griqualand West; between Knegts Fontein and Klip Fontein, *Burchell*, 2612! Transvaal; Pilgrim's Rest, *Greenstock!* Var. β, Orange Free State, *Cooper*, 965!

EASTERN REGION: Natal, *Gerrard*, 1440! Var. β, Natal, *Gerrard*, 1842! Sutherland!

Also in South-East Trop. Africa.

6. C. karooica (C. B. Clarke in DC. Monogr. iii. 166, excl. Var. β); leaves linear-oblong, nearly glabrous; corolla yellow, larger than in *C. africana*; filaments ½ in. long; capsule tough, usually perfecting 3 seeds, larger than the capsule of *C. africana* type; otherwise nearly as *C. africana*.

KALAHARI REGION: Griqualand West; Griqua Town, *Burchell*, 1929! between Griqua Town and Riet Fontein, *Burchell*, 1988! 1999! Orange Free State; Mud River Drift, *Rehmann*, 3577! Bechuanaland, *Holub*! Transvaal; Moiloa, *Holub*, 1496! 1497!
EASTERN REGION: Natal; Inanda, *Wood*, 409B!

7. C. **eckloniana** (Kunth, Enum. iv. 57); stem 12–18 in. long, diffuse; leaves narrow lanceolate with weak scattered hairs; spathes scattered, peduncled, obliquely funnel-shaped; corolla blue; capsule papery, square, dorsal cell very small empty or 0; seeds 4, globose, smooth.

SOUTH AFRICA: without locality, *Zeyher*!
KALAHARI REGION: Transvaal; Bosch Veld, *Rehmann*, 4997!
EASTERN REGION: Natal, *Sanderson*! Delagoa Bay, *Mrs. Monteiro*!

8. C. **Gerrardi** (C. B. Clarke in DC. Monogr. iii. 183); stem 12–18 in. long, diffuse; leaves narrow lanceolate, nearly glabrous; spathes 1–3 close together, oblique funnel-shaped, with short oblique points, hispid; petals blue or white; capsule with 3 globose seeds; dorsal cell tough, indehiscent, tubercular-scabrous. *L. angustifolia, Hassk. in Peters' Mossamb.* 528, *not of Mich.*

SOUTH AFRICA: without locality, *Gill*!
CENTRAL REGION: Albert Div., *Cooper*, 588!
EASTERN REGION: Natal, *Gerrard*, 1838! *Sanderson*!
Also in South-East Trop. Africa.

9. C. **albescens** (Hassk. in Schweinf. Beitr. Fl. Æthiop. 210); stem often 2 ft. long; leaves narrow lanceolate, up to 5 by $\frac{3}{4}$ in., nearly glabrous; spathes 1–3 close together, oblique funnel-shaped with acute or hooked point, hispid; petals blue or white; capsule membranous, with 3 ellipsoid, trigonous, smooth, puberulous, largish seeds. *Hook f. Fl. Brit. Ind.* vi. 373.

KALAHARI REGION: Transvaal; Barberton, *Galpin*, 599! 1188.
Distributed throughout Africa, and extends by Arabia and Beloochistan to Sind.

10. C. **Livingstoni** (C. B. Clarke in DC. Monogr. iii. 190, including Var. β); stem 1 ft. long, diffuse; leaves linear, nearly glabrous or more or less villous or shaggy; spathes 1–3, close together, oblique funnel-shaped, with short acute or hooked point, hispid; petals blue or white, "pale pinkish" (*Galpin*); capsule membranous, 2-celled, 2-seeded; seeds large, ellipsoid, smooth.

SOUTH AFRICA: without locality; *Zeyher*, 1729!
KALAHARI REGION: Bechuanaland; Hamapery, near Kuruman, *Burchell*, 2515! Transvaal, *McLea in Hb. Bolus*, 5796! Bosch Veld, *Rehmann*, 4866! Barberton, 2900 ft., *Galpin*, 808! Pilgrim's Rest, *Greenstock*! Mooi River, *Burke*, 339! Apies River, *Burke*!
EASTERN REGION: Natal; Inanda, *Wood*, 369! 409! and without precise locality, *Sanderson*!

Also in South-East Trop. Africa.

II. ANEILEMA, R. Br.

Panicle exserted, of many branches; bracts small, lowest not folded nor funnel-shaped. *Petals* 3, clawed, subequal, blue or whitish. *Stamens* 3–2 fertile, 3 sterile. *Ovary* subsymmetric, 3- or 2-celled; ovules 1–5 in each cell. *Seeds* 1–5 in each cell; hilum linear, vertical.

Annual succulent weeds; flowers evanescent.

DISTRIB: Species 55, Tropical or Warm-temperate zones, viz. 50 in the Old World, 5 in the New.

Subgenus I. TRICARPELLARIA. Capsule subequally 3-celled, 3-valved.

Capsule usually 6-seeded ... ... (1) **sinicum.**

Subgenus II. DICARPELLARIA. Capsule 2-celled, 2-valved (a very small third valve is rarely present).

Capsule truncate, 6- (or more) seeded ... ... (2) **æquinoctiale.**
Capsule small, rounded, 4–2-seeded ... ... (3) **dregeanum.**

1. A. sinicum (Lindl. in Bot. Reg. t. 659 char. emended); nearly glabrous, innovations more or less hairy; stem 12–18 in. long, with 2 or 3 cauline leaves; leaves linear, 4–5 by $\frac{1}{4}$ in., or the leaves at base and those on sterile tufts by the base much larger; panicle-branches subracemosely elongate, often scarred closely where the numerous infertile flowers have fallen; stamens 2 fertile; capsule 3-celled, scarcely $\frac{1}{4}$ in. long, pointed at tip; seeds usually 6 perfected, obscurely wrinkled. *Hook. f. Fl. Brit. Ind.* vi. 379. *A. secundum, Wight, Ic. Pl. Ind. Or. t.* 2075. *Commelina sinica, Roem. et Schult. Syst.* i. *Addit. Mant.* i. *Cl.* iii. 376.

EASTERN REGION: Natal; near Tugela River, *Wood*, 3854; near Nonoti River, 500 ft., *Wood!* Zululand, *Mrs. McKenzie!* and without precise locality, *Gerrard*, 1483!

Also distributed from the Niger and Angola to Canton and Malaya.

2. A. æquinoctiale (Kunth, Enum. iv. 72); stems 1–3 ft. long; leaves lanceolate, tapering at base, or ovate-lanceolate suddenly narrowed at base quasi-petioled, hairy beneath; panicle stout, pubescent; stamens 2 fertile, filaments not bearded; capsule 2-celled (or a third dorsal small cell added), at top truncate with two horns; seeds usually 3 in each cell, stony. *C. B. Clarke in DC. Monogr.* iii. 221, *all var. included. A. adhærens, Kunth, Enum.* iv. 72. *Commelina æquinoxialis, Beauv. Fl. Owar.* i. 65, *t.* 38. *Lamprodithyros æquinoctialis and L. adhærens, Hassk. in Schweinf. Beitr. Fl. Æthiop.* 211. *Amelina Wallichii, C. B. Clarke, Commel. Beng. t.* 26.

KALAHARI REGION: Transvaal, Houtbosch, *Rehmann*, 5759! Barberton, 2000-2900 ft., *Galpin*, 1186! *Thorncroft*, 265! *Wood*, 4507!

EASTERN REGION: Pondoland; St. John's River, *Drège*, 4466! Natal; Inanda, *Wood*, 540! Port Natal, *Grant!* near Durban, *Cooper*, 3323! *Molyneux!* and without precise locality, *Sutherland! Gerrard*, 1836!

Distributed throughout Tropical Africa and Arabia.

The type of *A. adhærens*, Kunth, is Drège, 4466, which has lanceolate leaves 4 by $\frac{3}{4}$ in., attenuate at base; and scarcely differs from Beauvois' tab. cited, the type of *A. æquinoctiale*. In Dr. Sutherland's examples, the leaves are ovate, 4 by $1\frac{1}{2}$ in., suddenly narrowed at base into a quasi-petiole. The third small cell of the capsule is present sometimes in both forms. The petals are stated in DC.

Monogr. Phanerog. iii. 221 to be "yellow," or "fide A. Rich. blue"; J. M. Wood has noted them "blue" in his 4507.

3. A. dregeanum (Kunth, Enum. iv. 73); stems 12–18 in. long; leaves elliptic-lanceolate, thinly hairy or nearly glabrous; panicle small, dense, primary axis erect; branches numerous, horizontal; petals small, blue or (when fresh) pink; stamens 3 fertile, 3–2 sterile; capsule $\frac{1}{4}$ in. long, 2-celled, quadrate-oblong, papery, perfecting 4 stony seeds. *Commelina comosa, Drège, Pfl. Dokumente,* 150, 174. *Lamprodithyros dregeanus, Hassk. in Peters' Mossamb.* 529.

VAR. β, **Galpini**; larger in all parts; leaves ovate, 4 by 2 in., suddenly narrowed at base into a quasi-petiole; panicle $2\frac{1}{2}$ by $1\frac{1}{2}$ in.; capsule quadrate-globose, perfecting 2 seeds; seeds nearly $\frac{1}{8}$ in. long.

KALAHARI REGION: Var. β, Transvaal; valley near Barberton, 2000 ft., *Galpin,* 1187!

EASTERN REGION: Pondoland; between Umtata River and St. Johns River, *Drège,* 4471! Natal; Inanda, *Wood,* 479 partly! Umzinyati, *Wood,* 1220! and without precise locality, *Mrs. Saunders! Gerrard,* 1837! Zululand, *Mrs. McKenzie!*

Also in South-East Trop. Africa.

III. COLEOTRYPE, C. B. Clarke.

Inflorescence axillary, piercing the base of the cylindric leaf-sheath. *Corolla-tube* linear, as long as the blue segments. *Stamens* 6, subequal, fertile, inserted towards the top of corolla-tube; filaments bearded with beaded hairs. *Ovary* sessile, free, ovate, 3-celled; ovules 2–1 in each cell. *Capsule* trigonous-ovoid, 3-celled, loculicidally 3-valved, hairy at top; seed 1 (rarely 2) in each cell.

DISTRIB. Species 1 in Natal, 3 or 4 in Madagascar.

1. C. natalensis (C. B. Clarke in DC. Monogr. iii. 239, t. 8); nearly glabrous, or innovations (sometimes densely) strigose; stems $1\text{–}2\frac{1}{2}$ ft. long, diffuse, branching; leaves 5 by 1 in., lanceolate; flowers densely clustered; sepals free, boat-shaped, $\frac{1}{3}$ in. long, tips linear strigose; corolla 1 in. long, blue; stamens nearly equal, the three posticous (and the dorsal cell of capsule) rather smaller; capsule $\frac{1}{4}$ in. long, pointed; seed 1 in each cell, oblong-ellipsoid.

EASTERN REGION: Natal; near Durban, *Wood,* 3396! *MacOwan and Bolus, Herb. Norm. Aust. Afr.,* 1005! *MacOwan, Herb. Aust. Afr.,* 1564! Inanda, *Wood,* 479! 964! and without precise locality, *Sanderson,* 438!

IV. CYANOTIS, D. Don.

Flowers in dense axillary and terminal clusters. *Petals* 3, free at insertion, then connate (more or less completely) into a cylindric tube, terminated by 3 round blue or pink-purple segments. *Stamens* 6, subequal, fertile, free or nearly free from the corolla; filaments bearded with beaded hairs. *Ovary* 3-celled, ovules 2 in each cell. *Capsule* 3-valved; seeds 2 perfected in each cell; embryostega (a circular disc) terminal, at the top of the upper seed in each cell, at the bottom of the lower, covering (as always in *Commelinaceæ*) the embryo.

DISTRIB. Species 30; warm regions of the Old World, from Africa to South China and North Australia.

As pointed out by R. Brown, the position of the embryo sharply differentiates this genus from all other Commelinaceæ. The flowers vary in colour in one species; and, where the young flowers are pink purple, they often fade (or dry) blue.

1. **C. nodiflora** (Kunth, Enum. iv. 106); nearly glabrous except the innovations, or villous, or shaggy; stems 6–48 in. long, diffuse, often branching from the root, where a sterile short stem is added; roots fibrous, sometimes bearing ovate-oblong tubers; leaves 2–5 in. long, oblong-linear, or those on the sterile short stem much larger, up to 9 by 1 in.; flower clusters dense, compound, of abbreviated scorpioid cymes interspersed with falcate folded bracts; flowers $\frac{1}{3}$ in. long; tube short; style linear, thickened just beneath the stigma; capsule $\frac{3}{4}$ lin. long; seeds 6, wrinkled. *Bot. Mag. t.* 5471. *C. speciosa, Hassk. Commel. Ind.* 108. *Tradescantia nodiflora, Lam. Encyc.* ii. 371. *T. speciosa, Linn. f. Suppl.* 192. *T. formosa, Willd. Sp. Pl.* ii. 20. *Commelina speciosa, Thunb. Fl. Cap. ed. Schultes,* 294.

SOUTH AFRICA: without locality, *Drège,* 8781 a, b, c!

COAST REGION: Riversdale Div; Vet River, *Gill!* Knysna Div.; near Keurbooms River, *Burchell,* 5180! Uitenhage Div.; between Uitenhage and Drostdy Farm, *Burchell,* 4460! and without precise locality, *Zeyher! Ecklon and Zeyher,* 610! Bathurst Div.; near Port Alfred, *Burchell,* 4008! Albany Div.; Woest Hill, near Grahamstown, 2500 ft., *Galpin,* 360! British Kaffraria, *Cooper,* 121! 388!

CENTRAL REGION: Karoo, *Bowie!* Somerset, *Bowker!*

KALAHARI REGION: Basutoland, *Cooper,* 3326! Bechuanaland; near Sirorume River, 3500 ft., *Holub!* Transvaal; near Barberton, 2000–5000 ft., *Galpin,* 547! 830! Pretoria, *Rehmann,* 4467! *Greenstock!* Orange Free State, *Molyneux!*

EASTERN REGION: Tembuland; Bazeia, 2500 ft., *Baur,* 389! Griqualand East; near Kokstad, 5000 ft., *Tyson,* 1894! *MacOwan and Bolus, Herb. Norm. Aust. Afr.,* 533! Natal; Inanda, *Wood,* 17! 51! 963! Port Natal, *Grant!* and without precise locality, *Sutherland!*

Also in South Trop. Africa, and Madagascar.

This species is sometimes covered nearly all over with white, almost silvery, soft hairs; sometimes the leaves and flower clusters are covered with much tawny or brown hair; sometimes (especially in the large forms) the plant is glabrous except a little hair on the inflorescence and about the mouth of the leaf-sheaths. The variability in size of the plant and leaves is also great; and being common in South Africa, collectors have preserved much of it supposing they had got several species. Several closely allied species occur in Trop. Africa, and in India; and in many of these the variability in size and in hairiness is great.

V. FLOSCOPA, Lour.

Racemes forming dense terminal panicles. *Petals* 3, free, obovate, sessile or scarcely clawed. *Stamens* 6–5, free, fertile, subequal, or 3–2 posticous rather smaller; filaments without beaded hairs. *Ovary* 2-celled, glabrous; 1 ovule in each cell. *Capsule* small, shortly stalked, compressed, membranous, 2-valved.

DISTRIB. Species 11, in warm regions of the world, 7 endemic in Africa.

1. **F. glomerata** (Hassk. Commel. Ind. 166); stems 1–3 ft. long, stout but weak, leafy; leaves 3 by ½ in., sessile, not narrowed (often eared) at base, nearly glabrous; inflorescence 1–2½ in. in diam., dense with flowers, very rufous-hairy; capsule 1 line in diam.; seeds ellipsoid, glaucous, smooth; embryostega central nearly opposite the hilum. *Fl. capensis, Hassk. Commel. Ind.* 166. *Tradescantia glomerata, Roem. et Schult. Syst.* vii. 1175. *Dithyrocarpus glomeratus and D. capensis, Kunth, Enum.* iv. 78.

EASTERN REGION: Natal; between Umtentu River and Umzimkulu River, *Drège*, 4472! Inanda, *Wood*, 907! and without precise locality, *Sutherland*, *Gerrard*, 359! *Cooper*, 3325!

Also in South Trop. Africa and Madagascar.

ORDER CXLI. FLAGELLARIEÆ.

(BY N. E. BROWN.)

Flowers hermaphrodite or diœcious, regular. *Perianth* 6-partite, persistent, the segments biseriate, small, subpetaloid or scarious, slightly unequal. *Stamens* 6, hypogynous, or shortly adnate to the base of the perianth-segments; filaments free; anthers basifixed, introrse, dehiscing by longitudinal slits. *Ovary* superior, 3-celled, with a solitary anatropous ovule in each cell; style short, or none; stigma 3-lobed and sessile, or of three linear, spreading branches. *Fruit* a 1–3-seeded berry. *Seeds* with a crustaceous testa, and a copious farinaceous albumen; embryo minute, lenticular, placed on the outside of the albumen, near the hilum.

Herbs of robust habit, with stout stems, bearing leaves up to the base of the inflorescence, sometimes climbing by means of tendrils at the tips of the leaves; leaves alternate, elongate, sheathing at the base, sometimes produced into a tendril at the apex, veins numerous, parallel; panicle terminal; flowers small, sessile or subsessile.

DISTRIB. A small Order of three genera, and about 8 species, confined to warm regions of the Old World. Only one species in South Africa.

I. FLAGELLARIA, Linn.

Flowers hermaphrodite. *Perianth* subpetaloid, the segments oblong, obtuse, the three outer slightly shorter than the three inner. *Stamens* 6, hypogynous, the filaments ultimately long and much exserted. *Ovary* with a short style and 3 linear stigmas. *Berry* one-seeded, rarely 2-seeded. *Seed* globose or ovoid with a crustaceous testa.

Stem sarmentose, climbing by means of the tendril-tipped leaves; panicle terminal; flowers small, spicate at the ends of the ultimate branchlets of the panicle.

DISTRIB. Three species, one African, one in Fiji, and the third widely dispersed in the tropics of the Old World, but apparently not occurring on the continent of Africa.

1. **F. guineensis** (Schumach. in Schumach. and Thonn. Beskr. Guin. Pl. 181); a tall climber, with a moderately stout herbaceous stem; leaves sheathing; the sheath open to the middle, with the thin membranous margins often closely overlapping but never connate to the top; the blade 4½–9 in. long, 6–13 lin. broad, elongate-lanceolate, gradually tapering from about the middle into a spiral tendril at the apex, abruptly contracted into a rounded base immediately above the sheath; panicle terminal, 3–4½ in. long, 2–4 in. broad, pyramidal, with ascending or spreading branches, the lower ones subtended by leafy bracts, the upper bractless; bracteoles minute, hyaline, ovate or suborbicular, obtuse or subacute; flowers subsessile along the rather slender, flexuose or zigzag ultimate branchlets of the panicle, usually with distinct internodes between them, whitish or pale yellowish; perianth campanulate; the segments 1–1½ lin. long, ⅔ line broad, the 3 outer ones shorter than the inner three, oblong obtuse, subpetaloid; stamens ultimately much exserted, with filaments 2–2½ lin. long, and linear-oblong anthers, sagittate at the base, 1¼–1⅓ lin. long; ovary narrow, trigonous; stigmas ⅔–¾ line long, linear, rather stout, exserted and recurving over the tips of the perianth-segments; berries globose, 2–3 lin. in diam., bright red. *F. indica, of authors, partly, not of Linnæus.*

EASTERN REGION: Pondoland, *Bachmann*, 253! Natal; Inanda, *Wood*, 824! Umbilo Waterfall, *Rehmann*, 8090! Umzimkulu River, and Natal Bay, *Drège*, 4484! and without precise locality, *Gerrard*, 403! *Peddie!*

Also in Tropical Africa.

This has been confused with *F. indica*, L., by almost all authors, but may be at once distinguished by the split leaf-sheaths, and the usually slender, zigzag axes, along which the flowers are spaced out with distinct internodes between them; whilst in *F. indica* the margins of the sheaths of the leaves are connate to the top, forming a closed tubular sheath, and the flowers are usually crowded together into very short subglobose spikes or glomerules on a rather thick, straight axis, without internodes, or with extremely short ones between them.

ORDER CXLII. JUNCACEÆ.

(By J. G. BAKER.)

Flowers regular, hermaphrodite in the Cape genera. *Perianth* inferior, cut down to the base into 6 subequal, biserial, glumaceous segments. *Stamens* usually 6, hypogynous or attached to the base of segments; filaments filiform or flattened; anthers dorsifixed or basifixed, dehiscing longitudinally. *Ovary* superior, 1–3-celled; ovules anatropous, usually many in a cell; style filiform, simple or trifurcate. *Fruit* capsular, splitting into 3 valves. *Seeds* ovoid, globose or angled by pressure, rarely compressed; testa thin or moderately thick, pale or black, the thin outer membrane sometimes produced beyond the nucleus; albumen copious, fleshy or hard; embryo small, placed near the hilum.

Perennial or annual herbs, rarely shrubs, with a woody caudex; stem often leafy only at the base; leaves terete or linear; flowers small, often clustered; bracts scariose, persistent.

DISTRIB. Cosmopolitan. Species 200.

I. **Juncus.**—Annual or perennial herbs, with basal glabrous leaves. Ovary more or less completely 3-celled, with many ovules in each cell.
II. **Luzula.**—Perennial herbs, with basal hairy leaves. Ovary 1-celled, with 3 nearly basal ovules.
III. **Prionium.**—Undershrub, with rigid serrated leaves in a dense rosette at the top of a woody caudex.

I. JUNCUS, Linn.

Perianth rigid, cut down to the base; segments subequal, ovate or lanceolate. *Stamens* 6, rarely 3, hypogynous or attached to the base of the perianth-segments; filaments filiform or flattened; anthers linear or oblong, basifixed. *Ovary* sessile, with 3 more or less intruded parietal placentas; style filiform, divided to the middle into 3 linear or filiform spreading stigmas. *Capsule* loculicidally 3-valved. *Seeds* very small, often tailed by the outer membrane of the testa being produced beyond the nucleus; albumen fleshy; embryo minute, placed near the hilum.

Perennial or annual herbs; leaves various, terete or flat, sometimes reduced to membranous sheaths, always glabrous; inflorescence terminal or apparently lateral, laxly panicled or congested into a head; perianth usually brown.

DISTRIB. Cosmopolitan. Species 170–180, according to Buchenau.

1. GENUINI. Leaves all reduced to sheaths; inflorescence lateral; seeds not tailed.

Stamens 3; pith continuous ... ... (1) **effusus.**
Stamens 6; pith interrupted ... ... (2) **glaucus.**

2. MARITIMI. Produced leaves few, resembling the stems; inflorescence lateral; seeds tailed.

Perianth-segments lanceolate, acute ... ... (3) **maritimus.**
Perianth-segments obtuse, inner emarginate ... (4) **acutus.**

3. ARTICULATI. Leaves septate; inflorescence terminal.

Produced leaf only one from the middle of the stem (5) **punctorius.**
Produced leaves more than one:
  Capsule not longer than the perianth:
    Style almost obsolete ... ... (6) **brevistilus.**
    Style produced ... ... (7) **oxycarpus.**
  Capsule longer than the perianth:
    Flowers many in a cluster ... ... (8) **exsertus.**
    Flowers few in a cluster ... ... (9) **rostratus.**

4. GRAMINIFOLII. Leaves not septate; inflorescence terminal.

Annuals:
  Style very short:
    Leaves subterete; clusters 2–5-flowered ... (10) **rupestris.**
    Leaves linear; clusters 8–15-flowered ... (11) **diaphanus.**
  Style as long as the ovary:
    Clusters usually solitary, 1–3-flowered:
      Perianth-segments subequal ... ... (12) **scabriusculus.**
      Inner perianth-segments longer than the outer:
        Perianth-segments whitish, with a brown keel ... ... (13) **parvulus.**
        Perianth-segments bright brown, with a pale edge ... ... (14) **polytrichos.**

Clusters 1–2, 2–6-flowered ... ... (15) **pictus.**
Clusters many, usually 1-flowered ... ... (16) **bufonius.**
Clusters 2–8; flowers 8–12 in a cluster:
Inner perianth-segments longer than the outer ... ... (17) **altus.**
Perianth-segments equal:
Perianth-segments acute ... ... (18) **cephalotes.**
Perianth-segments acuminate ... (19) **Sprengelii.**
Clusters many; flowers 3–12 in a cluster (20) **inæqualis.**
Perennial:
Leaves laterally compressed, resembling stems (21) **singularis.**
Leaves narrowly linear:
Clusters few, $\frac{1}{3}$–$\frac{1}{2}$ in. diam.:
Style very short .. .. (22) **dregeanus.**
Style not very short:
Clusters of flowers 1–3 ... ... (23) **sonderianus.**
Clusters of flowers 3–6 ... ... (24) **anonymus.**
Clusters many, smaller:
Stem subterete:
Rootstock short ... ... (25) **capensis.**
Rootstock long ... ... (26) **indescriptus.**
Stem angled ... ... (27) **acutangulus.**
Leaves broader ($\frac{1}{4}$–$\frac{1}{3}$ in. diam.) ... ... (28) **lomatophyllus.**

1. **J. effusus** (Linn. Sp. Plant. 326); perennial, densely tufted on a short creeping rhizome; leaves all reduced to green sheaths clasping tightly the base of the stem; stems green, terete, 2–3 ft. long, finely striated, with continuous pith; panicle dense, lateral, sessile, 1–2 in. diam., composed of many forked spikes; bracts small, ovate; perianth-segments rigid, lanceolate, green, $\frac{1}{12}$ in. long; stamens reduced to 3; capsule oblong, obtuse, cuspidate, dark brown, as long as the perianth; seeds not tailed. *Buchen. in Engl. Jahrb.* xii. 228. *J. communis var. effusus, E. Meyer, Monog. Junc.* 22; *Kunth, Enum.* iii. 320.

CENTRAL REGION: Somerset Div.; marshy places at the foot of the Bosch Berg, 2000 ft., *MacOwan*, 1964!
EASTERN REGION: Natal; Mooi River, in a marsh, *Wood*, 4062!
Cosmopolitan, concentrated in the North Temperate zone.

2. **J. glaucus** (Ehrh.) var. **acutissimus** (Buchen. in Abhand. Naturw. Ver. Brem. iv. 417, t. 5, "4"); perennial, densely tufted, with a short creeping rhizome; leaves all reduced to shining brown sheaths wrapped tightly round the lower part of the stem; stems terete, wiry, 2–3 ft. long, glaucous, with many fine grooves and interrupted pith; panicle lateral, very compound, composed of many dichotomously forked branches; bracts ovate, green, much shorter than the perianth; perianth-segments lanceolate, rigid, very acute, green, $\frac{1}{6}$ in. long; stamens 6; capsule oblong, dark brown, minutely cuspidate, nearly as long as the perianth; seeds not tailed. *Buchen. in Engl. Jahrb.* xii. 244.

COAST REGION: Queensland Div.; Shiloh, near the river, 3500 ft., *Baur*, 891! 1181!
CENTRAL REGION: Albert Div.; Stormberg Spruit, 5000 ft., *Drège*, 8796b! and Klein Buffels Vallei, near Gaatje, 4500–5000 ft., *Drège*, 8796c!

KALAHARI REGION: Orange Free State; near Winberg, *Buchanan*, 254!

The species occurs principally in the northern regions of the Old World. The variety is endemic at the Cape.

3. **J. maritimus** (Lam. Encyc. iii. 264); perennial, densely tufted on a short rhizome; leaves few, basal, terete, resembling the stems, dilated at the base into pale brown sheaths; stems terete, 2–3 ft. long, less robust and less pungent at the tip than in *J. acutus*; panicle very compound, produced from the side of the stem a short distance below the tip; flowers clustered; bracts ovate; perianth-segments lanceolate, acute, drab-green, $\frac{1}{12}$ in. long; stamens 6; capsule oblong, brown, as long as or a little longer than the perianth; seeds distinctly or indistinctly tailed. *Buchen. in Abhand. Naturw. Ver. Brem.* iv. 422, *t.* 5 ("4"); *Engl. Jahrb.* xii. 256. *J. Kraussii, Hochst. in Flora,* 1845, 342; *Buchen. in Abhand. Naturw. Ver. Brem.* iv. 418. *J. acutus, E. Meyer in Flora*, 1843, *Beig.* 51, 56. *J. spretus, Roem. et Schultes, Syst. Veg.* vii. ii. 1656, *in obs.*

COAST REGION: Cape Div.; marshy places near Cape Town, below 100 ft., *Bolus*, 4811! Salt River, *Burchell*, 677! Cape Flats, *Krauss.* Devils Mountain, *Ecklon*, 903! Camps Bay, *Burchell*, 336! between the Lions Head and Table Mountain, *Burchell*, 252/2! Stellenbosch Div.; Klapmut, *Rehmann*, 2266! Mossel Bay Div.; hills near the landing-place at Mossel Bay, *Burchell*, 6286! George Div.; Malgat River, *Krauss.* Uitenhage Div., *Ecklon and Zeyher*, 4, 647! Van Stadens River, *MacOwan*, 2085! Queenstown Div.; Shiloh, 3500 ft., *Baur*, 885!

CENTRAL REGION: Somerset Div.; between the Zuurberg Range and Klein Bruintjes Hoogte, 2000-2500 ft., *Drège!* Richmond Div.; vicinity of Stylkloof, 4000-5000 ft., *Drège!* Albert Div., *Cooper*, 1372!

KALAHARI REGION: Griqualand West; Lower Campbell, *Burchell*, 1812!

EASTERN REGION: Natal; near Durban, *Rehmann*, 8589! *Buchanan*, 126! 364! *Cooper*, 3330! *Gerrard*, 709!

Also in the North Temperate zone of both hemispheres.

4. **J. acutus** (Linn.) var. **Leopoldii** (Buchen. in Abhand. Naturw. Ver. Brem. iv. 421, t. 5, "4"); perennial, densely tufted on a short rhizome; leaves few, basal, resembling the stems; stems rigid, terete, 3–4 ft. long, very pungent at the tip; panicle ample, lax, very compound, produced from the side of the stem a short distance below the tip; flowers clustered; bracts ovate; segments of the perianth $\frac{1}{12}$ in. long, outer oblong, obtuse, inner conspicuously emarginate with a scariose edge; stamens 6; capsule subglobose, brown, obtuse, minutely apiculate, $\frac{1}{8}$–$\frac{1}{6}$ in. long and broad; seeds distinctly tailed, *Buchen. in Engl. Jahrb.* xii. 251. *J. Leopoldii, Parl. in Giorn. Bot. Ital.* 1846, ii. 1, 324. *J. macrocarpus, Nees in Linnæa*, xx. 243.

COAST REGION: Ceres Div.; Verkeerde Vlei, *Rehmann*, 2840! Worcester Div.; near Worcester, 800 ft., *Bolus*, 5271! *Rehmann*, 2550! Stellenbosch Div.; Somerset, *Zeyher*, 4308! Swellendam Div., *Mund.* Mossel Bay Div.; hills near the landing place, Mossel Bay, *Burchell*, 6286! Uitenhage Div.; near Zwartkops River, *Ecklon*, 783. Queenstown Div.; Shiloh, *Baur*, 885!

CENTRAL REGION: Somerset Div.; Biesjes Fontein, Loot's Kloof, 2300 ft., *MacOwan*, 1683! Graaff Reinet Div.; descent of the Voor Sneeuw Berg, *Burchell*, 2860!

WESTERN REGION: Little Namaqualand, *Zeyher!* near the mouth of the Orange River, *Ecklon and Zeyher*, 73!

Also Europe and North Africa.

5. **J. punctorius** (Linn. fil. Suppl. 208); perennial, tufted on a short rhizome; produced leaf one only from the middle of the stem, long, subulate, septate, the others reduced to sheaths; stem moderately stout, 2–4 ft. long; inflorescence ample, terminal, reaching ½ ft. long and broad; flowers many in a cluster; bracts ovate, shorter than the flowers; perianth-segments lanceolate, $\frac{1}{12}$ in. long; stamens 6, rather shorter than the perianth; capsule oblong, brown, acute, as long as the perianth; seeds not tailed. *Thunb. Prodr.* 66; *Roem. et Schultes, Syst. Veg.* vii. 191; *Kunth, Enum.* iii. 332; *Buchen. in Abhand. Naturw. Ver. Brem.* iv. 424, *t.* 8; *Engl. Jahrb.* xii. 277. *J. oxycarpus, Drège, Exsic. ex parte non E. Meyer. J. exaltatus, Desv. in Ann. Sc. Nat. sér.* 2, ii. 16. *J. Schimperi, Hochst. in Schimp. Pl. Abyss. No.* 56; *A. Rich. Fl. Abyss.* ii. 338. *J. acutiflorus, var. capensis, Spreng. Neue Entd.* ii. 107.

COAST REGION: Clanwilliam Div.; between Lange Vallei and Oliphants River, *Drège*, e! Wupperthal, *Drège*, f; Bosch Kloof, *Drège*, g. Cape Div.; about the ponds and at Salt River near Cape Town, *Burchell*, 674! between Lion Mountain and Table Mountain, *Burchell*, 253! *Ecklon*, 46, 47. Devils Mountain, *Ecklon*, 902! Paarl Div.; by the Berg River, *Drège*, aa. Tulbagh Div.; near Tulbagh waterfall, *Ecklon and Zeyher*, 5. Worcester Div.; mountains near Worcester, *Rehmann*, 2549! Stellenbosch Div.; Hottentots Holland, *Zeyher*, 104. Klapmut, *Rehmann*, 2265! Uitenhage Div., *Ecklon and Zeyher*, 648! *Drège*, h.

CENTRAL REGION: Aberdeen Div.; Camdeboo Mountain, *Drège*, c. Somerset Div.; banks of Little Fish River, 2800 ft., *MacOwan*, 1715!

KALAHARI REGION: Griqualand West; Griqua Town, *Burchell*, 1871! Transvaal; Wonderboom Poort, *Rehmann*, 4472! 4473!

EASTERN REGION: Pondoland; between Umtata River and St. Johns River, *Drège*, i.

6. **J. brevistilus** (Buchen. in Abhand. Naturw. Ver. Brem. iv. 433, t. 8); perennial, tufted; produced leaves more than one, short, subulate, septate; stem erect, compressed, smooth, moderately stout; inflorescence terminal; flowers 12–16 in a cluster; floral bracts ovate-lanceolate; perianth-segments lanceolate, $\frac{1}{6}$ in. long, inner broader; stamens 3, half as long as the perianth; style nearly obsolete; capsule oblong, not longer than the perianth; seeds not tailed. *Buchen. in Engl. Jahrb.* xii. 346.

SOUTH AFRICA: without locality. No specimen at Kew.

Described from a single imperfect specimen, probably gathered in Natal by Krauss.

7. **J. oxycarpus** (E. Meyer ex Kunth, Enum. iii. 336); perennial, densely tufted; produced leaves 3–5, short, ascending, subulate, separate; stem moderately stout, terete, 1½–2 ft. long; inflorescence terminal, much less ample and compound than in *J. punctorius*; flowers about 20 in a cluster; bracts ovate-acuminate, nearly as

long as the flowers; perianth-segments lanceolate, acute, $\frac{1}{6}$ in. long; stamens usually 3, much shorter than the perianth; style short, but distinctly produced; capsule oblong, dark brown, as long as the perianth; seeds not tailed. *Buchen. in Abhand. Naturw. Ver. Brem.* iv. 431, *t.* 8; *Engl. Jahrb.* xii. 336.

COAST REGION: Cape Div.; about the ponds and at Salt River, *Burchell*, 672! between Hout Bay and Wynberg, *Drège*, b. Paarl Div.; by the Berg River, *Drège*, a! Drakenstein Mountains, *Rehmann*, 2256! Worcester Div.; near Worcester, 800 ft., *Bolus*, 4217! *Rehmann*, 2551! Caledon Div.; Palmiet River, near Grabouw, 700 ft., *Bolus*, 4217! Riversdale Div.; Great Vals River, *Burchell*, 6551! Uitenhage Div.; by the Zwartkops River, *Ecklon and Zeyher*, 782.

EASTERN REGION: Natal; Umzinyate Falls, *Wood*, 1053! and without precise locality, *Buchanan*, 56! 80!

Also Angola, *Welwitsch*, 3008.

8. **J. exsertus** (Buchen. in Abhand. Naturw. Ver. Brem. iv. 435, t. 5, "4"); perennial, tufted on a short rhizome; produced leaves 3–4, short, ascending, subulate, septate; stems moderately stout, erect, 1–2 ft. long; inflorescence an ample terminal panicle; flowers 4–14 in a cluster; bracts ovate-cuspidate, shorter than the flowers; perianth-segments lanceolate, $\frac{1}{5}$ in. long; stamens 6, much shorter than the perianth; capsule oblong, dark brown, distinctly exserted, valves rigid; seeds not tailed. *Buchen. in Engl. Jahrb.* xii. 337.

COAST REGION: Worcester Div., *Zeyher*! Queenstown Div.; Shiloh, along the river, 3500 ft., *Baur*, 777! 852!

CENTRAL REGION: Somerset Div., *Bowker*! Graaff Reinet Div.; along the Sunday River, 2500 ft., *Bolus*, 188!

KALAHARI REGION: Griqualand West; Griqua Town, *Burchell*, 1906! Orange Free State; Bloemfontein, *Rehmann*, 3762! Transvaal; Hooge Veld, at Trigards Fontein, *Rehmann*, 6698! Pretoria; at Wonderboom Poort, *Rehmann*, 4471! Yster Spruit, Nelson, 320!

9. **J. rostratus** (Buchen. in Abhand. Naturw. Ver. Brem. iv. 437, t. 5, "4," ex parte); perennial, tufted on a short rhizome; produced leaves 4–8, short, ascending, subulate, septate; stem long, slender, spreading; inflorescence an ample lax terminal panicle; flowers 3–6 in a cluster; bracts ovate-cuspidate, small; perianth-segments lanceolate, $\frac{1}{8}$ in. long; stamens 6, much shorter than the perianth; capsule oblong, dark brown, much exserted, narrowed into a beak; seeds not tailed. *Buchen. in Engl. Jahrb.* xii. 338.

SOUTH AFRICA: without locality, *Drège*, 4465!

COAST REGION: Uitenhage Div.; by the Zwartkops River, *Ecklon and Zeyher.*

KALAHARI REGION: Transvaal; Houtbosch, *Rehmann*, 5741!

EASTERN REGION: Transkei; by the Bashee River, *Drège*, 4465. Natal, *Buchanan*, 67! 358!

10. **J. rupestris** (Kunth, Enum. iii. 344); annual, tufted; leaves all basal, linear-subulate, very slender, not septate, 1–2 in. long; stem very slender, deeply sulcate, naked, 2–4, rarely 6, in. long; clusters $\frac{1}{6}$ in. diam., 2–3, rarely 4–5, in a lax terminal panicle; flowers 2–5 in a cluster; bracts small, ovate; perianth $\frac{1}{10}$ in.

long; segments greenish-brown with a white edge, equal in length, all acute, outer lanceolate, inner ovate-lanceolate; stamens 6; style very short; capsule brown, oblong, as long as the perianth; seeds not tailed. *Buchen. in Abhand. Naturw. Ver. Brem.* iv. 441; *Engl. Jahrb.* xii. 460.

SOUTH AFRICA: without locality, *Harvey*, 377!

COAST REGION: Clanwilliam Div.; Ezels Bank, 3000–4000 ft., *Drège*, b; Gift Berg, *Drège*, c. Cape Div.; Simons Bay, *Wright*, 527!

WESTERN REGION: Little Namaqualand; Rood Berg, 2500–3500 ft., *Drège*, 2471a!

11. **J. diaphanus** (Buchen. in Abhand. Naturw. Ver. Brem. iv. 442, t. 7); annual, densely tufted; leaves all basal, linear, flat, acuminate, not septate, 3–4 in. long; stem slender, leafless, terete, reaching a foot long; clusters $\frac{1}{3}$ in. diam., 3–4 in a lax terminal panicle; flowers 8–15 in a cluster; bracts small, ovate; perianth $\frac{1}{6}$ in. long; segments equal in length, outer lanceolate, very acute, inner oblong, subobtuse; stamens 6, half as long as the perianth; style very short; capsule oblong, shorter than the perianth; seeds obovate, not tailed. *Buchen. in Engl. Jahrb.* xii. 460.

COAST REGION: Albany, *Bolus*, 188. No specimen in Kew Herbarium.

12. **J. scabriusculus** (Kunth, Enum. iii. 354); annual, not tufted; leaves all basal, not terete, very slender, not septate, 1–2 in. long; stem very slender, compressed, sulcate, leafless, 6–8 in. long, slightly scabrous upwards; cluster of flowers solitary, terminal, $\frac{1}{4}$–$\frac{1}{3}$ in. diam.; flowers 1–2, rarely 3, in a cluster; bracts ovate, pale, scariose; perianth $\frac{1}{4}$ in. long; segments stramineous, equal in length, lanceolate, outer acute, inner subobtuse; stamens 6, half as long as the perianth; style as long as the ovary; capsule oblong, shorter than the perianth; seeds obovate, not tailed. *Buchen. in Abhand. Naturw. Ver. Brem.* iv. 444, *t.* 6; *Engl. Jahrb.* xii. 457.

VAR. β, **subglandulosus** (Buchen. in Engl. Jahrb. xii. 458); more robust, with subglandular stem, clusters 3, and flowers 5–7 in a cluster. *J. subglandulosus, Steud. Glum.* ii. 303; *Buchen. in Abhand. Naturw. Ver. Brem.* iv. 459, *t.* 6.

COAST REGION: Cape Div.; foot of Table Mountain, *Ecklon*, 11 partly. Piquetburg Div.; Piquetburg Range near Groene Vallei, under 1000 ft., *Drège*, 8795! including Var. β.

According to Buchenau, Var. β was collected by Drège in Aliwal North Div., on the Witte Bergen at 5000–6000 ft., but this appears to be an error.

13. **J. parvulus** (E. Meyer et Buchen. in Abhand. Naturw. Ver. Brem. iv. 447, t. 6); annual, tufted; leaves all radical, erect, setaceous, not septate, $\frac{1}{4}$–$\frac{1}{2}$ in. long; stem very slender, leafless, an inch long; cluster of flowers solitary, terminal, $\frac{1}{16}$–$\frac{1}{12}$ in. diam.; flowers usually solitary; bracts ovate, pale, scariose; perianth $\frac{1}{12}$ in. long; segments whitish with a brown keel, outer oblong, acute, inner longer, obtuse; stamens 6, shorter than the perianth; style as long as the ovary; capsule ovoid-cuspidate, nearly as long as the perianth; seeds not tailed. *Buchen. in Engl. Jahrb.* xii. 458.

WESTERN REGION: Little Namaqualand; Modder Fontein Berg, 4000–5000 ft., *Drège*, 2472b!

14. J. polytrichos (E. Mey. et Buchen. in Abhand. Naturw. Ver. Brem. iv. 448, t. 6); annual, tufted; leaves all basal, erect, setaceous, not septate, $\frac{3}{4}$–1 in. long; stem very slender, sulcate, leafless, 2–4 in. long; cluster of flowers solitary, terminal, $\frac{1}{12}$–$\frac{1}{8}$ in. diam.; flowers 2–3, rarely 4, in a cluster; bract ovate, acute, pale, scariose; perianth $\frac{1}{8}$ in. long; segments bright brown with a pale margin, outer lanceolate, acute, inner longer, oblong, obtuse; stamens 6, shorter than the perianth; style long; capsule oblong, much shorter than the perianth; seeds not tailed. *Buchen. in Engl. Jahrb.* xii. 459.

WESTERN REGION: Little Namaqualand; hills near Lily Fontein, 4000–5000 ft., *Drège*, 2472aa!

15. J. pictus (Steud. Syn. Glum. ii. 305); annual, tufted; leaves all radical, setaceous, very slender, erect, not septate, 2–3 in. long; stem very slender, leafless, sulcate, slightly scabrous, 3–6, rarely 9–10, in. long; clusters 1–2, terminal, $\frac{1}{4}$–$\frac{1}{3}$ in. diam.; flowers 2–6 in a cluster; bracts ovate, pale; perianth $\frac{1}{6}$ in. long; segments stramineous, tipped with dark brown, outer lanceolate, acute, inner longer, oblong, obtuse; stamens 6, half as long as the perianth; style long; capsule oblong-trigonous, shortly cuspidate; seeds few, large, not tailed. *Buchen in Abhand. Naturw. Ver. Brem.* iv. 458, *t.* 6; *Engl. Jahrb.* xii. 457.

WESTERN REGION: Little Namaqualand; hills near Lily Fontein, *Drège*, 2472a!

16. J. bufonius (Linn. Sp. Plant. 328); annual, densely tufted; leaves slender, subulate, channelled down the face, the lower nearly as long as the stem, the upper shorter; stem slender, 3–12 in. long; flowers usually single, arranged in lax forked spikes; bracts ovate, scariose, minute; perianth-segments green, lanceolate; stamens much shorter than the perianth, sometimes 3 abortive; capsule oblong, brown, obtuse, minutely cuspidate, rather shorter than the perianth; seeds not tailed. *Kunth, Enum.* iii. 353; *Buchen. in Abhand. Naturw. Ver. Brem.* iv. 416; *Engl. Jahrb.* xii. 174. *J. dregeanus, Presl, Bot. Bemerk.* 117, *non Kunth. J. ranarius, Nees in Linnæa,* xx. 243, *non Song. et Perr. J. plebejus, Krauss in Flora* 1845, 342, *non R. Br.*

COAST REGION: Cape Div.; near Cape Town, *Bolus*, 4810! *Ecklon*, 26, 49, 84, 905! Cape Flats, *Krauss*, between Tygerberg and Simons Berg, below 500 ft., *Drège*, 8790; between Rondebosch and Wynberg, *Burchell*, 761! Simons Bay, *MacGillivray*, 425! Tulbagh Div.; Mitchells Pass, *Rehmann*, 2367! Paarl Div.; by the Berg River, under 500 ft., *Drège!* Stellenbosch Div.; Hottentots Holland, *Ecklon*, 20, 85, Somerset, *Ecklon*, 6.

CENTRAL REGION: Prince Albert Div.; by the Gamka River, 2500–3000 ft., *Drège*, Somerset Div.; between the Zuurberg Range and Klein Bruintjes Hoogte, *Drège*.

WESTERN REGION: Little Namaqualand; between Pedros Kloof and Lily Fontein, 3000–4000 ft., *Drège!* Rood Berg, 2500–3500 ft., *Drège*.

KALAHARI REGION: Griqualand West; Griqua Town, *Burchell*, 1858!

Cosmopolitan.

17. J. altus (Buchen. in Abhand. Naturw. Ver. Brem. iv. 457); annual; leaves all radical, linear, 4–5 in. long, not septate; stem leaves slender, sulcate, a foot long; clusters $\frac{1}{3}$–$\frac{1}{2}$ in. diam., 3–8 in a lax terminal panicle; flowers 8–10 in a cluster; bracts small, lanceolate; perianth $\frac{1}{5}$ in. long, outer segments lanceolate, mucronate, inner longer, oblong, obtuse; stamens 6, shorter than the perianth; style long; capsule oblong, shortly mucronate; seeds shortly apiculate. *Buchen. in Engl. Jahrb.* xii. 456.

COAST REGION: Swellendam Div.; between Puspas Vallei and Kochmans Kloof, 1000-4000 ft., *Ecklon and Zeyher*, 96. No specimen in Kew Herbarium.

18. J. cephalotes (Thunb. Prodr. 66, ex parte); annual, tufted; leaves all basal, linear, not septate, 2–3 in. long; stem terete, sulcate, leafless, 3–6 in. long; clusters 2–5, rarely 1, in a lax terminal panicle, $\frac{1}{4}$ in. diam.; flowers usually 8–12 in a cluster; perianth $\frac{1}{6}$ in. long; segments equal in length, brown, lanceolate, acute; stamens 6, much shorter than the perianth; style long; capsule oblong-mucronate, much shorter than the perianth; seeds not tailed. *Thunb. Fl. Cap. edit. Schult.* 337, *ex parte; Spreng. Neue Entd.* ii. 107; *Buchen. in Abhand. Naturw. Ver. Brem.* iv. 451, *t.* 7; *Engl. Jahrb.* xii. 454.

SOUTH AFRICA: without locality, *Harvey*, 353!
COAST REGION: Cape Div.; Table Mountain, *Ecklon*, 13, 901! Cape Flats, *Krauss*, Simons Bay, *Wright!*

19. J. Sprengelii (Nees in Linnæa, xx. 244); annual, tufted; leaves linear, all basal, not septate, 2–3 in. long; stem slender, leafless, 3–5 in. long; clusters $\frac{1}{8}$ in. diam., 2–5 in a lax terminal panicle; flowers 8–12 in a cluster; bracts ovate-lanceolate; perianth $\frac{1}{6}$ in. long; segments equal, pale brown, lanceolate-acuminate; stamens 6, half as long as the perianth; style long; capsule oblong-cuspidate, much shorter than the perianth; seeds not tailed. *Buchen. in Abhand. Naturw. Ver. Brem.* iv. 449, *t.* 10; *Engl. Jahrb.* xii. 453.

COAST REGION: Cape Div.; Camps Bay, *Ecklon and Zeyher*. Tulbagh Div.; near Tulbagh Waterfall, *Ecklon and Zeyher*, 11!

20. J. inæqualis (Buchen. in Abhand. Naturw. Ver. Brem. iv. 455, t. 7); annual, tufted; leaves all basal, linear, grass-like, not septate, 3–6 in. long, $\frac{1}{8}$ in. broad low down; stem slender, leafless, 6–8 in. long; clusters $\frac{1}{4}$ in. diam., many in a lax terminal panicle; flowers 8–12 in a cluster; bracts ovate-lanceolate, scariose; perianth $\frac{1}{6}$ in. long, outer segments lanceolate, much shorter than the oblong, obtuse inner ones; stamens 6, much shorter than the perianth; style long; capsule oblong-mucronate, shorter than the perianth; seeds not tailed. *Buchen. in Engl. Jahrb.* xii. 455. *J. isolepoides, Nees in Linnæa,* xx. 244, *ex parte.*

COAST REGION: Cape Div.; near Camps Bay, *Ecklon*, 12, 24. Stellenbosch Div.; Hottentots Holland, *Zeyher*, 46, *Ecklon*, 14. Swellendam Div.; by the Buffeljagts River, 1000–2000 ft., *Zeyher*, 4319!

21. J. singularis (Steud. Syn. Glum. ii. 302); perennial, tufted; leaves all basal, rigid, shorter than the stem, laterally compressed, almost ancipitous, not septate; stem leafless, angular, moderately stout, 1–1½ ft. long; panicle lax, ample, terminal; flowers 6–10 in a cluster; bracts minute; perianth ⅙ in. long, outer segments lanceolate, acute, inner oblong-lanceolate, obtuse; stamens 6, half as long as the perianth; style long; capsule brown, oblong, mucronate, shorter than the perianth; seeds not tailed. *Buchen. in Abhand. Naturw. Ver. Brem.* iv. 438, *t.* 9, *ex parte*; *Engl. Jahrb.* xii. 408.

COAST REGION: Uitenhage Div.; Van Stadens Berg and Bethelsdorp, *Drège*, 1604b, *partly*. No specimen in Kew Herbarium.

22. J. dregeanus (Kunth, Enum. iii. 344); perennial, tufted; leaves all basal, rigid, linear, much shorter than the stem, not septate; stem moderately stout, naked, terete, 1–1½ ft. long; panicle terminal, usually composed of 5–6 clusters; flowers many in a cluster; bracts ovate, minute; perianth ⅛ in. long, outer segments lanceolate, mucronate, inner oblong, obtuse; stamens usually 3, half as long as the perianth; style very short; stigma capitate; capsule oblong, brown, shorter than the perianth; seeds not tailed. *Buchen. in Abhand. Naturw. Ver. Brem.* iv. 462, *t.* 9; *Engl. Jahrb.* xii. 436. *J. cephalotes, Thunb. Prodr.* 66, *ex parte. J. submonocephalus, Steud. Syn. Glum.* ii. 303.

SOUTH AFRICA: without locality, *Drège*, 1604 i, 4387, 4447.

COAST REGION: Humansdorp Div.; near the Kromme River, below 1000 ft., *Drège*, 1604 f. Uitenhage Div.; between Van Stadens Berg and Bethelsdorp, below 1000 ft., *Drège*, 1604b. By the Zwartkops River, *Ecklon and Zeyher*, 13, 101, 779, and 899, partly. Bathurst Div.; by the Fish River, below 1000 ft., *Drège*, 1604 c. Assegai Bosch, near Sidbury, *Ecklon and Zeyher*, 10.

KALAHARI REGION: Transvaal; Houtbosch, *Rehmann*, 5742!

23. J. sonderianus (Buchen. in Abhand. Naturw. Ver. Brem. iv. 476, t. 10); perennial, densely tufted; leaves all radical, short, linear, not septate; stem slender, naked, 3–12 in. long; clusters of flowers 1–3, terminal, ⅓ in. diam.; flowers many in a cluster; perianth ⅙ in. long, outer segments lanceolate, acuminate, inner oblong, obtuse; stamens 6, half as long as the perianth; style and stigmas as long; capsule oblong, shorter than the perianth; seeds not tailed. *Buchen. in Engl. Jahrb.* xii. 440. *J. capensis var. angustifolius, E. Meyer, Syn. Junc.* 49, *ex parte. J. capensis var. capitatus, Nees in Linnæa*, xx. 244.

COAST REGION: near the landing-place at Mossel Bay, *Burchell*, 6237! 6247! Port Elizabeth Div.; Cape Recife, *Burchell*, 4386! *Ecklon and Zeyher*, 9, 780; near Port Elizabeth, *Drège*, e!

24. J. anonymus (Steud. Syn. Glum. ii. 304); perennial, tufted; leaves all basal, narrowly linear, not septate, much shorter than the stem; stem slender, naked, 1–2 ft. long; clusters of flowers 3–6 in a terminal panicle, ⅓–½ in. diam.; flowers 20–35 in a cluster; bracts lanceolate; perianth ⅙ in. long, outer segments lanceolate, mucronate, inner rather shorter, oblong, obtuse; stamens

6, half as long as the perianth; style not very short; capsule oblong, mucronate, shorter than the perianth; seeds not tailed. *Buchen. in Abhand. Naturw. Ver. Brem.* iv. 478; *Engl. Jahrb.* xii. 441.

COAST REGION: Worcester Div.; Dutoits Kloof, 3000-4000 ft., *Drège*, 1604a!

25. **J. capensis** (Thunb. Prodr. 66); perennial, densely tufted; rootstock short; leaves all radical, linear or linear-subulate, not septate, as long as, or shorter than, the stem; stem slender, naked, deeply sulcate, $\frac{1}{2}$–$1\frac{1}{2}$ ft. long; clusters many or few, in a lax terminal panicle, $\frac{1}{6}$ in. diam.; flowers up to 10 in a cluster; bracts ovate-mucronate, shorter than the flowers; perianth $\frac{1}{8}$ in. long, outer segments lanceolate, very acute, inner oblong, obtuse; stamens 6, shorter than the perianth; capsule oblong-trigonous, shorter than the perianth; seeds comparatively large, not tailed. *Thunb. Fl. Cap. edit. Schult.* 337; *Willd. Sp. Plant.* ii. 210; *E. Meyer, Syn. Junc.* 48; *Kunth, Enum.* iii. 342; *Buchen. in Abhand. Naturw. Ver. Brem.* iv. 482, *t.* 11; *Engl. Jahrb.* xii. 443 (*with several varieties*).

COAST REGION: Cape Div.; Table Mountain, *Ecklon*, 35, 897, 898! 900, *Drège! MacGillivray*, 423! *Milne*, 229! Cape Flats, *Ecklon*, 899! Devils Mountain, *Ecklon*, 35! Vicinity of Cape Town, *Burchell*, 8! 283! 471! Camps Bay, *Burchell*, 852! Paarl Div.; flats between Paarl and French Hoek, *Drège!* Berg River, *Drège!* Worcester Div.; Bains Kloof, *Rehmann*, 2306! 2307! George Div., *Zeyher*, 4317. Knysna Div.; in the forest, 500 ft., *Bolus*, 2497! Uitenhage Div.; near the mouth of the Zwartkops River, *Drège*, 1604d! Bathurst Div.; Assegai Bosch, near Sidbury, *Burchell*, 4175! Albany Div.; Howisons Poort, 1800 ft., *MacOwan*, 2019, 2020!

CENTRAL REGION: Somerset Div.; summit of Bosch Berg, 4800 ft., *MacOwan*, 1744!

26. **J. indescriptus** (Steud. Syn. Glum. ii. 304); perennial, tufted; rootstock long; leaves all radical, linear-subulate, shorter than the stem, not septate; stem slender, compressed, naked, 9–12 in. long; clusters many, in a lax terminal panicle, $\frac{1}{6}$ in. diam.; flowers usually 9–12 in a cluster; bracts ovate-mucronate, rather shorter than the flowers; perianth $\frac{1}{8}$ in. long, outer segments lanceolate, mucronate, inner oblong, obtuse; stamens 6, shorter than the perianth; style long; capsule ovoid-trigonous, shorter than the perianth; seeds large, not tailed. *Buchen. in Abhand. Naturw. Ver. Brem.* iv. 479; *Engl. Jahrb.* xii. 442.

COAST REGION: Paarl Div.; by the Berg River near Paarl, *Drège*, 1604h!

27. **J. acutangulus** (Buchen. in Abhand. Naturw. Ver. Brem. iv. 480); perennial, densely tufted; leaves all radical, narrowly linear, firm, falcate, not septate, 4–8 in. long; stem leafless, erect, triquetrous, a foot or more long; clusters 10–20 in a terminal panicle; flowers about 12 in a cluster; bracts ovate-lanceolate, mucronate; perianth $\frac{1}{8}$ in. long, outer segments lanceolate, shortly mucronate, inner oblong, obtuse; stamens 6, rather shorter than the perianth;

capsule oblong-triquetrous, shorter than the perianth; seeds not tailed. *Buchen. in Engl. Jahrb.* xii. 442.

COAST REGION: Stellenbosch Div.; Somerset West, *Zeyher*, 4318. Cape Flats, *Ecklon*, 100. No specimen in Kew Herbarium.

28. **J. lomatophyllus** (Spreng. Neue Entd. ii. 108); perennial, tufted and stoloniferous; leaves many, in a basal tuft, linear, flat, ½–1 ft. long, ¼–½ in. broad, not septate; stem moderately stout, naked, deeply sulcate, 1–2, rarely 3, ft. long; inflorescence terminal, laxly panicled, sometimes 6–8 in. long; flowers 6–12 in a cluster; bracts ovate-cuspidate, scariose; perianth-segments ⅛ in. long, outer lanceolate, acute, inner oblong-lanceolate; stamens 6, rather shorter than the perianth; capsule brown, oblong, shorter than the perianth; seeds not tailed. *Buchen. in Abhand. Naturw. Ver. Brem.* iv. 466, *t.* 10, *ex parte*; *Engl. Jahrb.* xii. 429. *J. cymosus, Lam. Encyc.* iii. 267, *ex parte*. *J. cephalotes, Thunb. Prodr.* 66, *ex parte*. *J. capensis var. latifolius, E. Meyer, Syn. Junc.* 48; *Kunth, Enum.* iii. 342.

COAST REGION: Cape Div.; Table Mountain, *Ecklon*, 25, 26, 50! 896, *Milne*, 225! *MacGillivray*, 420! 421! Paarl Div.; by the Berg River, *Drège!* Drakenstein Mountains, near Bains Kloof, 1500 ft., *Bolus*, 4080! Worcester Div.; Dutoits Kloof, 2000-3000 ft., *Drège!* Caledon Div.; Grabouw, near Palmiet River, 700 ft., *Bolus*, 4219! Uitenhage Div.; near Uitenhage, *Burchell*, 4271! by the Sunday River, *Gill!*

EASTERN REGION: Tembuland; Bazeia, 2000 ft., *Baur*, 543! Griqualand East; by streams around Clydesdale, 2500 ft., *Tyson*, 2866! Natal; Umgeni River, *Rehmann*, 8591! Umlaas River, *Krauss*, 416! Inanda, *Wood*, 221! Durban, *Rehmann*, 8590! and without precise locality, *Gerrard*, 494! *Buchanan*, 359!

II. LUZULA, DC.

Perianth rigid, cut down to the base; segments ovate, subequal. *Stamens* 6, shorter than the perianth, hypogynous, or attached to the base of the segments; filaments filiform; anthers oblong or linear, basifixed. *Ovary* sessile, 1-celled; ovules 3, arising from a short basal placenta; style filiform, trifurcate. *Capsule* 3-valved. *Seeds* 3, or fewer, erect, subglobose, or angled by pressure on the inner side; testa minutely striated, sometimes prolonged beyond the nucleus; albumen fleshy or farinaceous; embryo minute.

Densely-tufted perennial herbs; leaves grass-like, mostly radical, ciliated with soft hairs; inflorescence laxly or densely panicled, terminal; perianth brown, yellow, or white.

DISTRIB. Cosmopolitan. Species 50, according to Buchenau.

1. **L. africana** (Drège in Steud. Syn. Glum. ii. 294); perennial, densely tufted; leaves linear, flat, moderately firm, inconspicuously ciliated, the lower 4–8 in. long; stem slender, erect, 1–1½ in. long, bearing several reduced leaves; inflorescence congested into a small globose or ovoid terminal head; primary bracts linear-convolute; perianth ⅙ in. long; segments ovate-acuminate, brown, with a greenish edge, outer rather longer than the inner; capsule sub-

globose; seeds $\frac{1}{16}$ in. long, rounded on the back, flattened on the inner side. *Buchen. in Abhand. Naturw. Ver. Brem.* iv. 414, *t.* 5 ("4"), *ex parte*; *Engl. Jahrb.* xii. 150.

COAST REGION: Stockenstrom Div.; Katberg, in grassy places and swamps, *Drège*, 3963!

CENTRAL REGION: Basutoland; Mont aux Sources, 9500 ft., *Flanagan*, 2008!

III. PRIONIUM, E. Meyer.

Perianth rigid, cut down to the base; segments ovate, subequal. *Stamens* 6, as long as the perianth-segments; filaments filiform; anthers oblong, basifixed. *Ovary* sessile, globose, 3 celled; ovules axile, usually 2 in a cell; stigmas 3, sessile, spreading. *Capsule* rigid, loculicidally 3-valved. *Seeds* usually 1 in each cell, ovoid-oblong, not appendiculate; testa appressed; albumen fleshy; embryo straight, sometimes nearly as long as the albumen.

DISTRIB. Endemic. Monotypic.

1. P. Palmita (E. Meyer in Linnæa, vii. 131); leaves aggregated in a dense rosette at the top of a simple or forked stem, which is as thick as a man's arm, and reaches a length of 5–6 feet, linear, rigid, glabrous, 3–4 ft. long, an inch broad at the base, tapering gradually into a long point, serrated upwards on the edge and keel beneath; inflorescence a dense, large, terminal panicle on a long peduncle; branches ascending, subtended at the base by large, scariose, clasping, lanceolate bracts; branchlets spicate; perianth brown, $\frac{1}{8}$ in. long; bracts ovate, scariose, persistent, as long as the perianth. *Kunth, Enum.* iii. 315; *Hook. in Lond. Journ. Bot.* ix. (1857), 173; *Hook. fil. in Bot. Mag. t.* 5722. *P. serratum, Drège in Flora*, 1843, *Beig.* 10; *Buchen. in Abhand. Naturw. Ver. Brem.* iv. 408, *t.* 5 ("4"), *ex parte; Engl. Jahrb.* xii. 72, *fig.* 8. *Juncus serratus, Linn. fil. Suppl.* 208; *Thunb. Prodr.* 66. *Acorus Palmita, Lichten. Reise,* ii. 256.

COAST REGION: Clanwilliam Div.; Gift Berg, 1500-2500 ft., *Drège.* Cape Div.; Camps Bay, *Bergius, Ecklon and Zeyher*, 1; Table Mountain, *Ecklon*, 904. Paarl Div.; by the Berg River, *Drège!* Worcester Div.; Dutoits Kloof, 3000-4000 ft., *Drège.* Caledon Div.; Palmiet River, *Krauss.* Riversdale Div.; Great Vals River, *Burchell*, 6528! George Div.; between Malgat River and Great Brak River, *Burchell*, 6144! Uitenhage Div.; Zwartkops River, *Ecklon and Zeyher*, 786, Van Stadens River, *Drège*, Uitenhage, *Burchell*, 4442!

ORDER CXLIII. PALMÆ.

(BY C. H. WRIGHT.)

Flowers diœcious. *Calyx* inferior, 3-lobed. *Petals* 3, usually valvate in the male flowers, imbricate in the female. *Stamens* usually 6, hypogynous or perigynous. *Ovary* superior, with 3 connate or free carpels; stigma terminal or lateral; ovule solitary. *Fruit* drupaceous or baccate, often 1-celled by the abortion of two carpels. *Seed* oblong to spherical; spermoderm often adhering to the endocarp; albumen copious, cartilaginous, horny or oily, homogeneous or ruminate; embryo small, peripheral.

Trees with erect, simple, more rarely branched, stems, or woody climbers; leaves alternate, usually collected into a terminal crown, pinnate or flabellate; petiole sheathing at the base and often breaking up into a fibrous network on the decomposition of the leaves; inflorescence axillary; spathes one or more, usually woody; spadix branched, rarely simple.

DISTRIB. Genera about 140. Species about 1200, widely spread through the tropics and subtropics of both hemispheres.

I. **Phœnix.**—Leaves pinnate. Carpels distinct. Embryo dorsal.
II. **Hyphæne.**—Leaves flabellate. Carpels united. Embryo apical.

I. PHŒNIX, Linn.

Flowers diœcious. *Male flower* oblong. *Calyx* cupular, 3-toothed. *Petals* 3, slightly connate at the base, valvate. *Stamens* 6; filaments connate at the base; anthers linear-oblong, erect, dorsifixed. Rudimentary *ovary* small or none. *Female flower* globose. *Calyx* as in the male. *Petals* 3, rotundate, concave, widely imbricate. *Staminodia* 6, connate. *Carpels* 3, distinct; stigma sessile; ovule erect. *Fruit* oblong, 1-seeded; stigma terminal. *Seed* deeply grooved on the ventral side; embryo minute, dorsal; albumen cartilaginous.

Unarmed trees; leaves pinnate; pinnæ lanceolate, induplicate; spathe solitary; spadix intrafoliaceous, branched; flowers small, yellow.

DISTRIB. Species about 12, dispersed through tropical and subtropical Asia and Africa.

1. **P. reclinata** (Jacq. Fragm. i. 27, t. 24); stem erect, covered with the scars of the old leaves; leaves 3–5 ft. long, reclinate at the apex; petiole expanded at the base into a sheath with fibrous margins, flat above, slightly convex beneath; leaflets 30–40 on each side of the axis, equidistant or 2–4 together, lanceolate, acuminate, 1 ft. long, 12 lin. wide, lower ones subspinescent; male flower: calyx cupular; petals 3, oblong, acute; stamens 6; female spadix 2 ft. long; peduncle 1 ft. long, complanate, glabrous; branches 6–8 in. long; petals widely orbicular; drupe oblong, 7–8 lin. long, 4–5 lin. wide in the middle, tawny or reddish; seed oval, 5 lin. long, deeply sulcate ventrally; albumen cartilaginous, whitish; embryo slightly below the middle on the dorsal side. *Willd. Sp. Plant.* iv. 731; *Mart. Palm.* iii. 272, *t.* 164; *Kunth, Enum.* iii. 256; *Becc. Malesia,* iii. 349. *P. spinosa, Schum. & Thonn. Beskr. Guin. Pl.* 437; *Kunth, Enum.* iii. 256; *Kirk in Journ. Linn. Soc.* ix. 234. *P. leonensis, Lodd. ex Kunth, Enum.* iii. 256.

COAST REGION: Albany Div.; near the Karega and other rivers, *Bowie!* Algoa Bay, ex *Martius.* Fish River, ex *Martius.*
EASTERN REGION: Natal, *Garden!* Port Natal, *Krauss,* 82! Kaffraria, *Ecklon and Zeyher.*

Extending throughout Tropical Africa.

II. HYPHÆNE, Gærtn.

Flowers diœcious. *Male flower: Calyx* 3-lobed, imbricate. *Petals* widely ovate, concave, connate into a short stalk at the base, glumaceous, imbricate. *Stamens* 6; filaments short, subulate; anthers linear, 2-lobed at the base, basifixed. Rudimentary *ovary* none. *Female flower* shortly pedicelled, larger than the male. *Sepals* ovate orbicular, obtuse, imbricate. *Petals* slightly smaller than the sepals, widely ovate, obtuse, imbricate. *Staminodia* 6, connate into a membranous cup. *Ovary* subglobose, obscurely 3-lobed, 3-celled; stigmas 3, sessile; ovule laterally affixed near the base. *Fruit* sessile or stalked, 1-celled; stigma basilar; pericarp grumose with shining epidermis; endocarp woody. *Seed* adnate to the endocarp, which is intruded at the base, erect; raphe with reticulate branches; albumen bony, homogeneous; embryo apical.

Unarmed or slightly spiny medium or large sized trees; leaves flabellate; petiole ending in a short ligule; spadices with cylindrical spathes; branches alternate with subfastigiate branchlets; bracts semicircular, very densely imbricate; bracteoles membranous; flowers immersed in the branchlets, the male in pairs, the female solitary.

DISTRIB. Species about 9, dispersed through tropical Africa, Arabia, and the Mascarene Islands.

1. H. crinita (Gærtn. Fruct. ii. 13, t. 82, f. 4); stem simple, 8–20 ft. high; leaves flabellate with fibres between the laciniæ, covered on both surfaces with very fugacious, white tomentum; margin and upper side of nerves scabrous; petiole sheathing at the base, deeply channelled above; margin aculeolate; fruit obovate, depressed, shortly stalked, smooth; putamen subglobose. *Mart. Palm.* iii. 227; *Wendl. in Bot. Zeit.* 1881, 92; *Kirk in Journ. Linn. Soc.* ix. 234. *H. natalensis, G. Kunze in Linnæa,* xx. 15; *Gard. Chron.* 1890, viii. 381. *H. petersiana, Klotsch ex Mart. Palm.* iii. 227.

EASTERN REGION: Natal, *Gueinzius, Drège.*

Also found in the central part of Tropical Africa and in Madagascar.

The widely spread tropical African *H. thebaica,* Mart., differs in its branched stem and ovate fruit and putamen.

ORDER CXLIV. TYPHACEÆ.

(By N. E. BROWN.)

Flowers unisexual, monœcious, in dense bracteate heads or spikes. *Male flowers* with a perianth of 3–6 scales, or without a perianth, but irregularly intermingled with slender, narrowly clavate filaments or spathulate or cuneate scales, which are often variously toothed or lobed. *Stamens* with free or connate filaments and basifixed, linear, oblong, or cuneate, 2-celled anthers, opening by longitudinal slits, the connective sometimes produced beyond the cells at the apex; no rudiment of an ovary. *Female flowers* with a perianth of scales or of fine hairs, sometimes accompanied by slender spathulate or clavate bracteoles. *Ovary* superior, 1-celled, sessile or stalked; style simple, terminal, persistent; stigma unilateral, elongate; ovule solitary, pendulous from the apex of the cell, anatropous. *Fruit* sessile or stalked; the pericarp thin and membranous, or thick with a spongy outer layer, and hard, woody inner layer. *Seed* albuminous; testa thin; embryo cylindric, axile.

Perennial aquatic or marsh herbs, with creeping rhizomes; leaves alternate, linear or strap-shaped, sheathing at the base; veins parallel; flowers small or minute, sessile, bracteolate or ebracteolate, densely crowded into globose heads or cylindric spikes, along simple or branched axes, with or without leafy bracts at their base; the male inflorescence terminal.

A small order of two genera and about 25 species, very widely dispersed.

I. TYPHA, Linn.

Flowers densely crowded into simple, cylindric spikes, the male spike terminal and separated from the female spike or contiguous to it. *Male flowers* irregularly intermingled with variously shaped scales, or slender, clavate, curved filaments. *Stamens* with their filaments variously connate; anthers linear, basifixed, 2-celled, the connective produced beyond the cells; pollen simple or compound. *Female flowers* ebracteolate or mingled with slender clavate or spathulate bracteoles, and often with abortive clavate female flowers (*carpodia*) mixed with them. *Perianth* composed of several very fine simple or clavate hairs. *Ovary* superior, stalked, at least after fertilization, narrow, 1-celled, with a solitary pendulous ovule; style elongate, slender, erect; stigma linear or lanceolate. *Fruit* minute, stalked, ellipsoidal or subcylindric, with a thin membranous pericarp. *Seed* subcylindric or narrowly ellipsoidal, albuminous; testa thin; embryo axile.

Aquatic or marsh herbs, with creeping rhizomes and erect stems; leaves alternate, linear or strap-shaped, parallel-veined; flowering-stem erect, simple, terminated by the dense, cylindric, superposed, unisexual flower-spikes; bracts none, or linear and deciduous or caducous.

Species described about 18, but probably some are only varietal forms, widely distributed.

Female flowers bracteolate; stigmas linear ... ... (1) **australis.**
Female flowers usually ebracteolate; stigmas spathulate-lanceolate ... ... (2) **capensis.**

1. **T. australis** (Schum. and Thonn. Beskr. Guin. Pl. 401); plant attaining 5–6 ft. in height, glabrous; blade of leaf 1½–5 ft. long, 2–5 lin. broad, linear or strap-shaped, obtusely pointed, convex on the back and not keeled at its junction with the sheath; male and female spikes subequal, or the female longest, 4½–10 in. long, contiguous or separated by a short interval; stamens mixed with reddish-brown, spathulate scales or bracteoles, that are variously lobed and toothed in the dilated portion; pollen simple; female flowers bracteolate; bracteoles lanceolate-spathulate or fusiform-clavate, acute or acuminate, longer than the hairs and nearly as

long as the stigma, brownish; stigma linear, longer than the hairs, mottled with brown; hairs filiform. *Schum. and Thonn. in Danske Vidensk. Selsk. Afhandl.* 1829, 175; *Kronfeld in Verhandl. Zool.-Bot. Gesselsch. Wien*, 1889, 156, *t.* 5, *f.* 4. *T. angustifolia, var. australis*, *Rohrbach in Verhandl. Bot. Ver. Prov. Brandenb.* xi. 83.

SOUTH AFRICA: without locality, *Krebs*, *Masson*, *Bergius*.
COAST REGION: Uitenhage Div.; between Van Stadens Berg and Bethelsdorp, under 1000 ft., *Drège*. 8811!
CENTRAL REGION: Prince Albert Div.; between Drie Koppen and Blood River, *Drège*, 8810.

Also in Tropical and North Africa.

This differs from *T. angustifolia* in the form of the bracteoles, and the finer hairs of the female perianth.

2. T. capensis (Rohrb. in Verhandl. Bot. Ver. Prov. Brandenb. xi. 96); plant attaining 6 ft. in height, glabrous; blade of leaf 1½–4 ft. long, 2–7 lin. broad, linear or strap-shaped, obtusely pointed, convex on the back, not keeled at its junction with the sheath; male and female spikes subequal, or the female a little the longer, 3¾–7 in. long, contiguous or separated by a short interval; stamens mixed with brownish bracts or scales, that are linear-spathulate or cuneate-spathulate, entire and acute, or variously toothed; pollen simple; female flowers usually ebracteolate, or with a few narrow spathulate-lanceolate, colourless bracteoles mingled with them; stigmas spathulate-lanceolate, longer than the filiform hairs. *Kronfeld in Verhandl. Zoolog.-Bot. Gesselsch. Wien*, 1889, 180. *T. latifolia*, *Krauss in Flora*, 1845, 343, *not of Linn.*

SOUTH AFRICA: without locality, *Harvey*, 558! *Ecklon and Zeyher*, 913, 1222.
COAST REGION: Queenstown Div.; Shiloh, 3500 ft., *Baur*, 893!
EASTERN REGION: Natal; banks of rivers, *Krauss*. Inanda, *Wood*, 1378!

Also in Tropical Africa and Madagascar.

ORDER CXLV. AROIDEÆ.

(By N. E. BROWN.)

Flowers unisexual or hermaphrodite, with or without a perianth, sessile on a spadix enclosed within, adnate to, or subtended by a green or coloured spathe. *Spadix* monœcious or rarely unisexual, with or without a terminal barren appendix, and with or without neuter organs on various parts of it, or entirely covered with hermaphrodite flowers. *Perianth*, when present, of 4–9 free or connate segments. *Stamens* 4–6, rarely more or fewer (when the male flowers have no perianth, the stamens are crowded together so that the number belonging to each flower is often indeterminable), free or connate; filaments none, or broad and flat, or rarely filiform; anthers opening by terminal pores, or by short or long longitudinal slits; pollen often emitted in sausage-like strings. *Ovary* sometimes surrounded by staminodia, superior, or rarely inferior, with or without a style, and with an entire or lobed stigma, one-many-celled, with axile, parietal, basal or apical placentation; ovules 1-many in a cell, orthotropous, campylotropous or anatropous. *Fruit* a 1-many-seeded berry. *Seeds* albuminous or exalbuminous.

Herbs or shrubs, with a thick fleshy or tuberous rootstock, or with erect, creeping, or climbing stems, simple or branched; leaves alternate, radical or cauline, usually with sheathing petioles, net- or parallel-veined; spathe open to the base or the lower part convolute or tubular, wholly deciduous or persistent, or only the tubular part persistent.

DISTRIB. Genera about 100. Species about 1000, chiefly concentrated in the Tropics of both hemispheres, very few in the temperate regions.

Tribe 1. PISTIEÆ. *Female part of spadix* adnate to the spathe. *Flowers* unisexual.

I. **Pistia.**—A floating aquatic herb. *Leaves* in a small flattened rosette, cuneate-oblong.

Tribe 2. STYLOCHITONEÆ. *Spadix* free, without an appendix. *Flowers* unisexual. *Perianth* annular or cup-shaped.

II. **Stylochiton.**—Tuberous-rooted herbs. *Leaves* hastate, sagittate, or lanceolate. *Spathe* tubular below.

Tribe 3. PHILODENDREÆ. *Spadix* free, without an appendix. *Flowers* unisexual, without a perianth. *Ovaries* surrounded with staminodes or naked.

III. **Richardia.**—Tuberous-rooted herbs. *Leaves* hastate, sagittate, or lanceolate. *Spathe* convolute below.

I. PISTIA, Linn.

Spathe small, tubular below; limb open, oblique, constricted on each side at its base. *Spadix* shorter than the spathe, monœcious, with the female part adnate to the back of the spathe, and the male part free, stipitate, having two neuter organs at its base, the upper cup-shaped or frill-like, the lower suborbicular or subreniform in outline with the sides bent inwards or downwards. *Flowers* unisexual. *Perianth* none. *Female inflorescence* of a single ovary, apparently arising from the back of the spathe, very oblique, 1-celled; style obliquely erect from the top of the ovary; stigma capitate; ovules numerous, basal, orthotropous. *Male inflorescence* of a stipitate whorl of 3–8 flowers, each flower composed of 2 connate, 2-celled anthers, opening by short slits. *Fruit* ellipsoidal, with a thin pericarp, many-seeded. *Seeds* cylindric-oblong, truncate at each end, depressed at the apex and with an operculum closing the micropyle, rugulose, albuminous; testa thick, with a brown membranous outer and inner skin, and a thick, white cellular layer between them; embryo minute, apical.

A floating, stoloniferous, stemless herb, with a tuft of fibrous roots; leaves in a rosette, with flabellate veins, raised beneath; spathes axillary, with very short peduncles.

DISTRIB. A single species, variable in the form of its leaves, found throughout the tropical and subtropical regions of the globe, in still, fresh water.

1. **P. Stratiotes** (Linn. Sp. Pl. 963); leaves sessile in a rosette, 1–3 in. long, 1–2 in. broad, cuneate-obovate, notched or entire at the

broadly rounded or subtruncate apex, more or less pubescent on both surfaces, tomentose with long matted hairs at the base; spathe $3\frac{1}{2}$–5 lin. long, more or less hairy or villose outside, glabrous within, pale green; limb ovate, subobtuse, with a broad, white, ciliolate margin; ovary pale green, pubescent; neuter organs frill-like, pale green; anthers dirty-white. *Bot. Mag. t.* 4564; *Flore des Serres,* vi. 289, *t.* 625; *Engl. in DC. Monog. Phan.* ii. 634. *P. africana, Presl, Epimel. Bot.* 240; *Klotzsch in Abhandl. Akad. Wissenschaft. Berlin,* 1853, 355, *reprint* 27. *P. natalensis, Klotzsch in Abhandl. Akad. Wissenschaft. Berlin,* 1853, 354, *reprint* 26.

EASTERN REGION: Natal; Umgeni River, *Drège!* Umlaas River, *Krauss!* Widely distributed in all the warm regions of the earth.

II. STYLOCHITON, Leprieur.

Spathe tubular below; limb expanded or hooded, at length entirely deciduous. *Spadix* monœcious, free, usually with a naked space between the male and female parts, or the base of the male part more or less interrupted, without neuter organs, or appendix. *Female flowers* free or connate in a single cycle, or in 2–5 spirals, crowded, laterally compressed or angular from mutual pressure, the uppermost often imperfectly hermaphrodite. *Perianth* gamophyllous, cupular or suburceolate, truncate, often with a thickened margin. *Staminodes* none. *Ovary* superior or inferior, 1-celled with basal placentation or two parietal placentas, or 2–4-celled in the lower part, with axile placentation; ovules 2 to several in each cell, anatropous; style exserted; stigma discoid-capitate. *Male flowers* numerous, in a cylindric spike, crowded or lax. *Perianth* as in the female flowers, but less deep and not contracted at the mouth, usually laterally compressed. *Stamens* 4, inserted at the base of a rudimentary ovary, exserted; filaments filiform or clavate; anthers basifixed, with oblong or elliptic, parallel or divergent cells, opening by longitudinal slits. *Fruit* not seen, described as a berry containing 2 to several ovoid, slightly compressed seeds, with a thin, black, striated testa, copious, fleshy albumen, and an elongated, axile embryo.

Perennial herbs with the habit of an Arum; rhizome stout, fleshy, nodose or ringed; leaves all radical, petiolate, cordate, sagittate, hastate, or rarely entire at the base, contemporary with the flowers, or appearing after them; peduncle terminal or axillary, solitary, short.

DISTRIB. Species about 10, all natives of Tropical Africa except the two here described.

Leaves $2\frac{1}{2}$–$6\frac{1}{2}$ in. broad; spathe 3–5 in. long, cream-coloured within ... ... (1) **natalense.**
Leaves 5–$10\frac{1}{2}$ in. broad; spathe 5–8 in. long, dark puce within ... ... (2) **maximus.**

1. **S. natalense** (Schott, Aroid. i. 10, t. 14); leaves 3–6, quite glabrous; petiole 3–9 in. long, vaginate for $\frac{1}{4}$–$\frac{1}{3}$ its length, rounded

on the back, channelled down the face, green with faint darker green markings on the upper part, greyish with fuscous markings on the vaginate part; blade 3–6 in. long, 2½–6½ in. broad, varying from sagittate to hastate, obtuse or acute, with a subulate apiculus 1½–3 lin. long; basal lobes oblong, very obtusely rounded at the apex, more or less divergent; flowers contemporary with the leaves; peduncle 2½–3 in. long, 1½–2 lin. thick, terete, glabrous, greenish with fuscous markings; spathe with a tube 1–1¾ in. long, inflated at the base, green outside, whitish-green within; limb erect, oblong-lanceolate, acute or acuminate, 2–3 in. long, 8–12 lin. broad, cream-coloured within, green on the back; spadix sessile, 1¼–1½ in. long; base of the male part more or less interrupted but scarcely separated from the female flowers by a distinct naked space; female part 4–5 lin. long, of 4–5 crowded spirals of flowers, angular from mutual pressure; perianth ¾–1 line deep, cup-shaped, contracted at the mouth, truncate and thickened at the margin, cream-coloured; ovary oblong-ovoid, 2- rarely 3–4-celled with axile placentation, green; style ¾ line long, whitish-green; male perianth about ¼ line deep, laterally compressed, cream-coloured; stamens with filiform filaments and elliptic-oblong anthers, creamy-yellow. *Schott, Gen. Aroid.* 68; *Prod. Aroid.* 345; *Engl. in DC. Monog. Phan.* ii. 523. *Gueinzia natalensis, Sond. ex Schott, Aroid.* i. 10.

EASTERN REGION: Natal; Attercliffe, 800 ft., *Sanderson*, 309! Inanda, *Wood*, 271! and without precise locality, *Gerrard*, 750! *Gueinzius!* Delagoa Bay, cultivated specimen, *Mrs. Monteiro!*

2. **S. maximus** (Engl. Jahrb. xv. 466); leaves about 6, quite glabrous; petiole 6–12 in. or more (?) long, vaginate for ⅓–½ its length; blade 4–12 in. long, 5–10½ in. broad, deeply cordate or cordate-sagittate, acute or obtuse, apiculate; basal lobes elliptic, elliptic-ovate, or elliptic-oblong, about half as long as the front lobe, very obtusely rounded at the apex, with an obtuse sinus 2–3 in. deep between them; peduncle 2–3 in. long; spathe with a tube 2–3 in. long, inflated at the base, cylindric above; limb oblong-lanceolate, acute or acuminate, 3–5 in. long, ¾–1⅓ in. broad, green outside, dark puce within; spadix sessile, 2¼–3 in. long; basal portion of the male part more or less interrupted, but scarcely separated from the female part by a distinct naked space; female part 5–9 lin. long, of 4–5 crowded spirals of flowers, angular from mutual pressure; perianth 1 line deep, cup-shaped, contracted at the mouth, truncate and thickened at the margin; ovary subglobose, 2-celled with axile placentation; style 1 line long; male perianth ⅓ line deep, laterally compressed; filaments of stamens 1 line long, filiform, exserted.

EASTERN REGION: Delagoa Bay, under the shade of trees, *Forbes! Monteiro*, 2!

Mr. Monteiro on his label states that the flowers "have a powerful musky odour," whilst Mrs. Monteiro informed me that they are "deliciously and strongly scented like violets."

III. RICHARDIA, Kunth.

Spathe with a convolute, narrowly funnel-shaped tube, broadest at the mouth; limb oblique, open, suberect or recurving, terminated by a subulate point, persistent. *Spadix* monœcious, free, sessile or stipitate, much shorter than the spathe, everywhere densely covered with unisexual flowers; the basal part female; the upper part male; appendix none; staminodes sometimes mixed with the ovaries, but no other barren organs. *Perianth* none. *Ovaries* in several spirals, subglobose, angular from mutual pressure, 2–3-celled; placentas axile; ovules 2–4 in each cell, anatropous; style short or almost none; stigma discoid. *Stamens* free, crowded together without order; anthers sessile, oblong, laterally compressed, truncate at the apex, opening by terminal pores and emitting the pollen in sausage-like strings. *Fruit* a berry, 1–3-celled; cells 1–2-seeded. *Seed* subglobose or ovoid; testa rather thin; albumen copious; embryo axile.

Herbs with the habit of Arum; rootstock a thick fleshy rhizome; leaves all radical, with long petioles and hastate, sagittate, cordate or lanceolate blades, contemporary with the flowers; peduncle as long as or longer than the leaves; spathe large, showy, white, yellow or rosy, persisting and changing to green as the fruit develops.

DISTRIB. Species 7 or 8, of which one or two are Tropical African.

Leaves narrow lanceolate, acute at the base ... ... (1) **Rehmanni.**
Leaves hastate, sagittate or cordate at the base:
Leaf with the elongated-deltoid or oblong part above the basal lobes at least twice as long as broad; petiole smooth, without bristles; spathe blotched at the base:
Leaf without spots; spathe yellow ... ... (2) **angustiloba.**
Leaf with white spots; spathe white ... ... (3) **albomaculata.**
Leaf with the broadly ovate or deltoid part above the basal lobes usually much less than twice as long as broad,
Petiole with soft bristles below; spathe greenish-yellow, blotched at the base:
Leaf spotted with white ... ... (4) **melanoleuca.**
Leaf without spots ... ... (5) **hastata.**
Petiole smooth, without bristles; spathe entirely white ... ... (6) **africana.**

1. **R. Rehmanni** (N. E. Br. ex Harrow in Gard. Chron. 1888, iv. 570, as *R. Lehmanni* by error); petiole smooth, without bristles; blade of the leaf 7½–15 in. long, 1–2 in. broad, lanceolate, acuminate, subulate at the apex, cuneate-acute at the base, of an uniform green, or marked with short, linear, semitransparent white spots; spathe 3–4½ in. long; limb oblique, more or less recurved, light rosy-purple, darker but not blotched at the base within, or white or greenish-white to the base within, with rosy-tinted margins; spadix not half the length of the spathe, stipitate, cylindric; ovaries with a very short, stout style; anthers yellow. *Watson in Gard. Chron.* 1892, xii. 124; 1893, xiv. 770, *and* 1894 xvi. 364; *Krelage in*

Gartenfl. 1894, 12 *and* 15, *fig.* 7 ; *Bot. Mag. t.* 7436. *R. Lehmanni* (*by error*), *Krelage in Gard. Chron.* 1893, xiv. 564, *fig.* 94. *Zantedeschia Rehmanni, Engl. Jahrb.* iv. 63.

EASTERN REGION : Natal; stony hills, Weeneu County, 4000–5000 ft., *Wood,* 5204! dry hills near New Castle, *Rehmann,* 80! and cultivated specimens!

2. **R. angustiloba** (Schott in Journ. Bot. 1865, 35); petiole smooth, without bristles; blade of leaf 7½–17 in. long, 4–11 in. broad across the basal lobes, elongate-deltoid or elongate-oblong, acute or acuminate, hastate, with short or long and spreading basal lobes and a very open sinus, the part above the basal lobes elongate-deltoid or elongate-oblong, acute or acuminate, twice as long as broad or more, green, unspotted; spathe 4–4½ in. long; limb oblique, subhorizontal, clear deep gamboge-yellow, with a dark purple-brown blotch at the base inside, paler outside; spadix shortly stipitate, scarcely half as long as the spathe, cylindric; ovaries with a subsessile stigma, pale greenish-white; staminodia none; anthers yellow. *Engl. in DC. Monog. Phan.* ii. 329. *R. macrocarpa, Watson in Gard. Chron.* 1892, xii. 124. *R. Pentlandii, Whyte ex Watson in Gard. Chron.* 1894, xv. 590; *Bot. Mag. t.* 7397. *Calla Pentlandii, Whyte ex Watson in Gard. Chron.* 1892, xii. 124. *Zantedeschia angustiloba, Engl., and Z. macrocarpa, Engl. Jahrb.* iv. 64.

KALAHARI REGION : Transvaal; Trigards Fontein, *Rehmann,* 82, 83; Mapoch District, west of Lydenburg, cultivated specimen!

Also in Tropical Africa.

I have only seen a drawing of Rehmann's specimen of *R. macrocarpa* (Engler, Araceæ No. 27), and although the colour of the flowering spathe is unknown, yet so far as the drawing and description go, it appears to me in every way identical with *R. angustiloba,* Schott, and *R. Pentlandii,* Whyte. The Mapoch locality is given on the authority of Mr. E. E. Galpin, formerly of Barberton, Transvaal, who presented living plants of this species to Kew.

3. **R. albomaculata** (Hook. in Bot. Mag. t. 5140); petiole smooth, without bristles; blade of the leaf 6–18 in. long, 3–9 in. broad across the basal lobes, hastate or occasionally sagittate, with a wide open sinus, the part above the basal lobes very elongate-deltoid or elongate-oblong, acute or acuminate, more than twice as long as broad, green, with numerous elongated, semitransparent, white spots; spathe 2¼–4½ in. long; limb oblique, suberect, acuminate, milk-white, with a dark purple-brown blotch at the base inside; spadix shortly stipitate, scarcely half as long as the spathe, cylindric; ovaries with a sessile or subsessile stigma, pale greenish; staminodia none, or a few around the uppermost ovaries; anthers yellow. *Schott, Prod. Aroid.* 325; *Lem. in Ill. Hort.* 1860, *t.* 255; *Gartenfl.* 1865, *t.* 462; *Fl. des Serres,* xiii. 97, *t.* 1343, *and* xxi. 165, *t.* 2258; *Engl. in DC. Monog. Phan.* ii. 327. *Zantedeschia albomaculata, Baill. in Bull. Soc. Linn. Paris,* i. 254; *Engl. Jahrb.* iv. 64.

CENTRAL REGION : Aliwal North Div.; Witte Bergen, 5000–6000 ft., *Drège,* 3572!

EASTERN REGION: Tembuland; Bazeia, 2000–2500 ft., *Baur*, 433! Griqualand East; stony slopes around Kokstad, 4500 ft., *Tyson*, 1590!

4. **R. melanoleuca** (Hook. f. in Bot. Mag. t. 5765); petiole with soft bristles on the lower part; blade of the leaf 5–8 in. long, 2¼–6 in. broad across the basal lobes, deltoid or ovate-deltoid, acute, with a subulate point, hastate or sagittate at the base, with an open sinus, the part above the basal lobes usually much less than twice as long as broad, green, with numerous semitransparent white spots; spathe 2–3 in. long, obliquely subtruncate at the mouth, light yellow or greenish-yellow, with a dark purple-brown blotch at the base inside; spadix shortly stipitate, cylindric; ovaries with scarcely any style, pale greenish; stigma subsessile; staminodia none; anthers yellow. *Gard. Chron.* 1869, 1182; *Engl. in DC. Monog. Phan.* ii. 328. *Zantedeschia melanoleuca, Engl. Jahrb.* iv. 64.

EASTERN REGION: Natal; Inanda, *Wood*, 431! and without precise locality, *Buchanan! Gerrard*, 1525! *Sanderson*, 209!!

5. **R. hastata** (Hook. f. in Bot. Mag. t. 5176); petiole with soft bristle-like hairs on the lower part, which wither and often almost disappear in the dried state; blade of the leaf 7–10 in. long, 5–7½ in. broad across the basal lobes, hastate, acute, tipped with a subulate point, the length of the part above the basal lobes usually much less than twice its greatest breadth, green, unspotted; spathe 3–4 in. long.; limb oblique, nearly in a line with the tube, light greenish-yellow, with a dark purple-brown blotch at the base within; spadix sessile or shortly stipitate, about half as long as the spathe, cylindric; ovaries with scarcely any style; stigma subsessile; staminodia none; anthers yellow. *Schott, Prod. Aroid.* 325; *Engl. in DC. Monog. Phan.* ii. 328; *Garden, Dec.* 1889. *Calla oculata, Lindl. in Gard. Chron.* 1859, 788. *Zantedeschia hastata, Engl. Jahrb.* iv. 64.

KALAHARI REGION: Basutoland, *Cooper*, 3327! Transvaal; Lomati Valley, near Barberton, in swamps, 4000 ft., *Galpin*, 1360!
EASTERN REGION: Natal, *Sanderson! Wood*, 2!

6. **R. africana** (Kunth in Mém. Mus. Hist. Nat. Paris, iv. 433, t. 20); petiole smooth, without bristles; blade of the leaf 6–18 in. long, 4–10 in. broad, cordate or hastate, obtuse or acute, tipped with a subulate point, the length of the part above the basal lobes much less than twice its breadth, green, unspotted; spathe 4–10 in. long; limb oblique, recurving from the tube, milk-white, without any blotch at the base within; spadix sessile, about half as long as the spathe, or less, cylindric; ovaries narrowed into a distinct style ½–¾ line long, pale greenish-white; staminodia and anthers bright yellow. *Kunth, Enum. Pl.* iii. 58; *Schott, Synop. Aroid.* 131; *Gen. Aroid. t.* 62; *and Prod. Aroid.* 324; *Engl. in DC. Monog. Phan.* ii. 327. *Calla æthiopica, Linn. Sp. Pl. ed.* i. 968; *Bot. Mag. t.* 832; *Jacq. Fragm.* 29, *t.* 32, *fig.* 3 (*fruit*). *Colocasia æthiopica, Spreng. in*

Link Handb. i. 267. *Zantedeschia æthiopica, Spreng. Syst. Veg.* iii. 765; *Engl. Jahrb.* iv. 64.

COAST REGION: Cape Div.; common around Cape Town, *Pappe!* Tyger Berg, below 1000 ft., *Drège!* Riversdale Div.; Zoetemelks River, *Gill!*
EASTERN REGION: Natal; Inanda, *Wood*, 286!

A curious form of this species with a branched spadix is figured in the *Gardeners' Chronicle* 1895, xvii. 135, figs. 19, 20. A dwarf form about a foot high was introduced into European gardens in 1894 or 1895, where it is known under the name of "Little Gem."

ORDER CXLVI. LEMNACEÆ.

(By N. E. BROWN.)

Flowers very minute, seated in a cavity at the margin or in the upper surface of the frond, consisting of 1–2 stamens accompanied by a sessile ovary, either naked or enclosed in a membranous spathe (or perianth?), which ruptures irregularly as the stamens mature. *Stamens* exserted from the frond; filaments filiform; anthers 1–2-celled; cells subglobose, opening by transverse slits or by a slit across the top. *Ovary* narrowed into a style or with a subsessile stigma, 1-celled; placenta basal; ovulė solitary or several in an ovary, anatropous, semianatropous, or orthotropous. *Fruit* 1–several-seeded, indehiscent or opening transversely. *Seeds* minute, albuminous; embryo straight, axile.

Small or minute, gregarious, floating plants, consisting of suborbicular, elliptic, obovate, oblong or linear fronds of cellular tissue, with or without rudimentary vessels, usually 2 or more fronds connected together as one plant, flat on both sides or more or less convex beneath, sometimes as thick as broad, developing young fronds (which remain attached to the parent frond for some time) from a cavity or cavities placed near or at the basal end of the frond, rootless or producing one or more roots from the middle of the under surface.

DISTRIB. All warm and temperate countries, in still water. Genera 2; species 21.

I. **Lemna.**—*Frond* with one or more roots. *Flowers* seated in a cavity at the margin of the frond.
II. **Wolffia.**—*Frond* rootless. *Flowers* seated in a cavity in the upper surface of the frond.

I. LEMNA, Linn.

Flowers very minute, very rarely seen in some of the species, seated in a cavity at the margin of the frond, enclosed in a minute membranous spathe. *Stamens* 1–2; anthers 2-celled; cells opening by transverse slits.

Fronds small or minute, with one or more roots on their under surface, floating on still waters, thin and flat, or flat above and more or less convex beneath, orbicular elliptic, oblong, obovate or spathulate in outline, entire, or in one species minutely denticulate on the margin; roots tipped with a distinct sheath-like root-cap.

DISTRIB. All warm and temperate countries. Species 9.

Besides the two species described below, there may be a third one in South Africa, as there is a specimen of *L. oligorrhiza*, Kurz, in the Kew Herbarium, named by Hegelmaier, which is labelled "*Lemna gibba ?* Uitenhage, Ecklon and Zeyher, 512," but according to a note written by Hooker upon the label, and another written by Hegelmaier, it is doubtful if the label rightly belongs to the specimen. *L. oligorrhiza*, Kurz, resembles *L. minor*, it is of about the same size, but rather more oblong, and differs in having 2–4 roots to each frond, instead of a single one as in *L. minor*.

Frond thin, nearly flat on both sides ... ... (1) **minor.**
Frond thick, very convex and somewhat spongy beneath ... (2) **gibba.**

1. **L. minor** (Linn. Sp. Pl. ed. 1, 970); fronds $\frac{3}{4}$–$2\frac{1}{2}$ lin. long, $\frac{1}{2}$–$1\frac{1}{2}$ lin. broad, elliptic, elliptic-oblong, or somewhat obovate, rather thin, nearly flat on both sides, bright green; young fronds sessile; root solitary from the under surface of each frond; ovary flask-shaped, with a distinct style; ovule solitary, semianatropous. *Hegelmaier, Monog. Lemn.* 142, *t.* 9-10. *L. ovata, A. Braun, ex Krauss in Flora,* 1845, 344.

South Africa: without locality, *Ecklon*.
Coast Region: Uitenhage Div.; Zwartkops River, in stagnant places, *Krauss*, 1600.
Eastern Region: Natal; Umgeni River, *Drège!* Umlaas River, *Krauss!* Port Natal, *Gueinzius*.
Widely dispersed in all parts of the globe.

2. **L. gibba** (Linn. Sp. Pl. ed. 1, 970); fronds $1\frac{1}{2}$–3 lin. long, and nearly as broad, subrotund or broadly obovate, very thick, flat and bright green above, very convex, greenish-white and somewhat spongy beneath; young fronds sessile; roots solitary from the under surface of each frond; flowers rare. *Hegelmaier, Monog. Lemn.* 145, *t.* 11-13. *Telmatophace gibba, Schleid. in Linnæa,* xiii. 391.

South Africa: without locality, *Ecklon, Zeyher, Bergius*.
Coast Region: Uitenhage Div.; near the mouth of the Zwartkops River, *Drège*, 8812!
Eastern Region: Port Natal, *Gueinzius*.
Widely distributed within the range of the Order.

II. WOLFFIA, Hook.

Flowers rarely seen, microscopic, seated in a cavity in the upper surface of the frond, without a spathe. *Stamen* 1; anther 1-celled, opening by a slit across the top and the valves becoming reflexed.

Fronds small or minute, entirely without roots, but in two species a root-like hair, quite destitute of a root-cap, is developed from the under side, floating on still waters, thin and flat, or as thick as broad, subglobular, hemispherical, ellipsoidal, elliptic, oblong, or linear, entire, or rarely minutely toothed at the margin or end of the frond.

Distrib. The same as for *Lemna*. Species 12.

Frond minute, $\frac{1}{4}$–$\frac{1}{3}$ lin. long, $\frac{1}{4}$ lin. broad, nearly as thick as broad, elliptic ... ... (1) **Michelii.**
Frond 2–$2\frac{1}{4}$ lin. long, $\frac{1}{4}$ lin. broad, linear ... ... (2) **denticulata.**

1. **W. Michelii** (Schleid. Beitr. Bot. 233); frond minute, $\frac{1}{4}$–$\frac{1}{3}$ lin. long, about $\frac{1}{4}$ lin. broad, elliptic in outline, nearly as thick as broad,

rather dark green; flowers very rare. *W. arrhiza, Wimm. Fl. Schles. ed.* 3, 140; *Journ. Bot.* 1865, 113, *t.* 29; *Hegelmaier, Monog. Lemn.* 124. *Lemna arrhiza, Linn. Mant. alt.* 294.

KALAHARI REGION: Transvaal; Hooge Veld at Standarton, *Rehmann,* 6777!

Widely distributed in Europe, Asia, and Africa.

2. W. denticulata (Hegelmaier, Monog. Lemn. 133, t. 3, f. 16-23); frond 2–2½ lin. long, ¼ lin. broad, linear, minutely toothed at the end; young fronds sessile; flowers unknown.

SOUTH AFRICA: *Krauss.*

I have not seen this species. According to Hegelmaier it was collected by Krauss with *Lemna minor.*

ORDER CXLVII. NAIADACEÆ.

(By ARTHUR BENNETT.)

Flowers usually green (often coloured in Aponogeton), 1–2-sexual. *Perianth* 0, or tubular, or of 3–4, inferior, valvate segments. *Stamens* hypogynous; anthers 1–2-celled. *Ovary* of 1–4, 1-ovuled carpels; style long or short; stigma of many forms. *Fruit* of 1-seeded utricules, achenes, or drupelets. *Seed* exalbuminous; embryo curved, or straight, large at the radicular end.

Aquatic or marsh herbs; habit various; rootstock usually creeping; stems elongate (except Triglochin), simple, or branched; leaves erect, submerged or floating, sheathing at the base; stipules 0, or contained in the sheath.

DISTRIB. All climates.

* *Stigmas discoid or decurrent on the ovary.*

Perianth of sepals or scales. Flowers bisexual.

I. **Triglochin.**—Marsh herbs. Sepals 6, green.
II. **Aponogeton.**—Aquatics. Sepals 1–3, white or coloured.
III. **Potamogeton.**—Aquatics. Sepals 4, herbaceous, green.

Perianth 0. Flowers uni- or bi-sexual.

IV. **Ruppia.**—Aquatics. Stamens 2. Carpels stipitate.
V. **Zannichellia.**—Aquatics. Stamen 1. Carpels usually sessile.

** *Stigmas subulate or capillary.*

VI. **Zostera.**—Aquatics. Flowers sessile on a flat linear spadix.
VII. **Naia**.—Aquatics. Flowers axillary.

I. TRIGLOCHIN, Linn.

Perianth-segments 3 or 6, herbaceous, deciduous. *Stamens* 6, at the base of the perianth-segments; anthers didymous. *Carpels* 3–6, 1-celled, 1-ovuled, 3 often imperfect; styles short, sometimes connate; stigmas sessile or subsessile, plumose. *Fruit* of 3 or 6 free or connate achenes or follicles; tips recurved. *Seed* erect; embryo straight.

Marsh herbs with rush-like flat or terete leaves; flowers small, bisexual, spicate or racemose, 2-bracteate.

DISTRIB. Temperate regions.

Fruit linear; leaves canaliculate ... ... (1) **bulbosum.**
Fruit linear; leaves not canaliculate ... ... (2) **laxiflorum.**
Fruit subrotund ... ... (3) **striatum.**

1. **T. bulbosum** (Linn. Mant. alt. 226); stem simple, tuberous at the base, with interlaced fibres, forming a brown mass at the base; leaves half-cylindrical, finely striated, canaliculate, generally shorter than the stem; raceme elongating after flowering, 6–12-flowered; fruits erect, ascending (not appressed), larger at the base than the apex, of 3 carpels. *Bot. Mag. t.* 1445; *Thunb. Prod.* 67; *Fl. Cap. ed.* 2, ii. 347; *Micheli in DC. Monog.* iii. 99; *Buchenau in Engl. Jahrb.* ii. 510. *T. Barrelieri, Lois, Fl. Gall. ed.* 2, i. 264.

SOUTH AFRICA: without locality, *Zeyher*, 1735! *Oldenburg! Krauss*, 1398! 1399! 1400!

COAST REGION: Cape Div.; Cape Flats, *Bolus*, 2850! *Elliot*, 1175! Simons Bay, *Wright!* Paarl Div., Flats between Paarl and French Hoek, below 500 ft., *Drège!* Uitenhage Div., *Ecklon and Zeyher*, 620!

EASTERN REGION: Griqualand East; Zuurberg Range, 5000 ft., *Tyson*, 1866! Natal; Inanda, *Wood*, 997!

2. **T. laxiflorum** (Gussone, Ind. Sem. Hort. Boccadit. 1825); stem ascending, tuberous at the base; leaves linear, flat (not canaliculate); flowers distant; peduncles short, patent in fruit; carpels 3, triangular, tapering upwards. *Fl. Sic. Prod.* i. 451; *Micheli in DC. Monog.* iii. 101; *Buchenau in Engl. Jahrb.* ii. 510.

EASTERN REGION: Natal; Durban, *Rehmann*, 8581! *Wood*, 925!

Differs from *T. bulbosum*, Linn., in the ascending (not erect) stem, leaves nearly flat, flowers distant, and fruit pedicels shorter.

3. **T. striatum** (Ruiz & Pavon, Fl. Peruv. iii. 72); rootstock small, stoloniferous; leaves variable, narrowly linear to subulate, striate, generally shorter than the scape; stem (scape) without leaves, 3–12 in. tall; flowers shortly pedicelled, often many; carpels 3, orbicular, with 6 permanent ribs, about 1–1½ lin. long. *Micheli in DC. Monog.* iii. 101; *Buchenau in Engl. Jahrb.* ii. 510. *T. maritimum, Thunb. (not of Linn.) Fl. Cap. ed.* 2, ii. 346; *Durand and Schinz, Conspect. Fl. Afr.* 491.

COAST REGION: Cape Div.; about the ponds near Cape Town and at Salt River, *Burchell*, 691! Stinkwater near Cape Town, *Rehmann*, 1194! Cape Flats, *Rehmann*, 1832! Salt River, *Schlechter*, 253! Uitenhage Div., *Zeyher*, 620! 914! Zwartkops River, below 100 ft., *Drège!* Van Stadens River, *MacOwan*, 2051! Port Elizabeth, *E.S.C.A. Herb.*, 176! Bathurst Div.; near the source of Kasuga River, *Burchell*, 3906!

EASTERN REGION: Natal; Durban, *Rehmann*, 8580! *Gerrard*, 738!

Differs from the other two species in the carpels, which are short, and conspicuously ribbed, or angled.

II. APONOGETON, Thunb.

Perianth of 1–3 equal or unequal segments, or 0. *Stamens* 5–6 or more, with subulate filaments, persistent. *Carpels* 3–6, with 2 or more ovules; stigma on a short style, decurrent or persistent. *Fruit* of 3 or more hardened follicles. *Seeds* erect; embryo straight.

Submerged plants with tuberous rootstocks, in the early state very like Potamogeton; leaves either floating, or erect, variable in shape, from linear to oblong (or subulate in one species?); scape with simple or bifid sessile spikes of unilateral or distichous flowers; the young flower enclosed in a deciduous sheath.

DISTRIB. Asia, Africa, and Australia.

Leaves with a distinct lamina:
 Bracts longer than the flowers.
  Leaves 1–3 in. long, 1½–7 lin. broad ... ... (1) **angustifolium.**
  Leaves 4–8 in. long, ¾–2¼ in. broad ... ... (2) **distachyon.**
  Leaves 5–9 lin. long, 1–2½ lin. broad ... ... (3) **gracilis.**
 Bracts equalling or shorter than the flowers.
  Leaves oblong, subcordate or rounded at the base.
   Leaves 4¾–8 in. long; spikes 3–4 in. long ... (4) **Holubii.**
   Leaves 2¼–5 in. long; spikes ¾–2 in. long ... (5) **kraussianum.**
  Leaves lanceolate, tapering at the base, 2¾–5½ in. long ... ... (6) **natalense.**
  Leaves oblong or oblong-lanceolate, usually rounded at the base, 1¼–1½ in. long ... ... (7) **Rehmanni.**
Leaves not or only slightly dilated into a lamina:
 Leaves linear or subulate ... ... (8) **spathaceum.**
 Leaves filiform or capillary ... ... (9) **crinifolium.**

1. **A. angustifolium** (Ait. Hort. Kew. i. 495); leaves linear-lanceolate, submerged, opaque, variable in length and breadth; nervules indistinct; spike bifid, few-flowered; flowers pale red; bracts white; sepals linear-oblong; stamens 6; follicles smooth, attenuated into a beak $\frac{1}{6}$ the length of the ovary. *Bot. Mag. t.* 1268; *Engl. Jahrb.* viii. 272; *Durand and Schinz, Conspect. Fl. Afr.* v. 492.

SOUTH AFRICA: without locality, *Oldenburg!*
COAST REGION: Cape Div.; Riet Valley on Cape Flats, *Krauss!* Green Point near Cape Town, *Harvey!* Paarl Div.; Berg River, near Paarl, *Drège!*

2. **A. distachyon** (Linn. f. Supp. 215); stem base tuberous; leaves lanceolate-ovate, longly-petiolate; floating leaves obovate, ovate or variable (much like *Potamogeton natans*); spike bifid, dense-flowered, each 1–2 in. long; peduncle stout, thickening upwards; flowers white; sepals lanceolate-ovate; stamens 6–12. *Thunb. Nov. Gen.* 74; *Andr. Bot. Rep.* v. *t.* 290; *Bot. Mag. t.* 1293; *Engl. Jahrb.* viii. 272; *Durand and Schinz, Conspect. Fl. Afr.* v. 492.

SOUTH AFRICA: without locality, *Thunberg! Masson! Ecklon!*
COAST REGION: Cape Div.; ponds near Cape Town, *Burchell*, 670! Wynberg, *Drège!* Cape Flats, *Krauss*, 1605! *Rehmann*, 1872! Paarl Div.; streams near Paarl, *Thunberg*, Berg River, *Drège!* George Div.; near George, *Rogers!* Humansdorp Div.; Kromme River, *Burchell*, 4883! Uitenhage Div., *Ecklon and Zeyher*, 107!

3. **A. gracilis** (Schinz in Durand and Schinz, Conspect. Fl. Afr. v. 492, name only); leaves floating, linear-oval or elliptical, acute or obtuse at the base, 5–10 lin. long, 1–2½ lin. wide; petiole 4–6 in. long, filiform; spike 2–4 lin. long; flowers small; carpels 3; seeds oblong, ecostate.

KALAHARI REGION: Transvaal; Houtbosch, *Rehmann*, 5761!

This species resembles *A. Rehmanni*, Oliv., but is much more slender, and has 3 instead of 8–9 carpels. I am indebted to Dr. Hans Schinz for the diagnosis of this species.

4. **A. Holubii** (Oliver in Hook. Ic. Pl. t. 1470); leaves oblong-elliptical, 4½–6 in. × 2–3 in., apex shortly semiobtuse, base rounded, cordate, with elongated petioles, 7–9-veined, veining obscure; peduncles thick, tapering upwards; spike bifid, 2½–3½ in. long, dense-flowered, hermaphrodite; bracts 2, oblong or ovate-oblong, obtuse; stamens about 6; follicles oblong terete with a beak ¾ as long as the ovary. *Durand and Schinz, Conspect. Fl. Afr.* v. 492.

SOUTH AFRICA: without locality, *Mrs. O. Bowker!* Herb. Brit. Mus.

Also in Tropical Bechuanaland.

5. **A. kraussianum** (Hochst. ex Krauss in Flora 1845, 343); leaves longly-petioled, floating, oblong elliptical, 7–9-nerved; medium nerve 3-cleft, acute, base subcordate; peduncles long, tapering upwards; spike bifid; arms short, ¾–1½ in., dense-flowered; flowers white; stamens generally 6; anthers yellowish; ovaries 3–5 (rarely 6); follicles globose, with a thickened short beak. *Krauss, Beitr. Fl. Cap. und Natal.* 172. *A. leptostachyum, E. Meyer ex Baker in Trans. Linn. Soc.* xxix. 158 (*name only*); *Engl. Jahrb.* viii. 270. *A. desertorum, Zeyher herb.! ex Steud. Nomen. ed.* 2, i. 114.

COAST REGION: Uitenhage Div.; stagnant pools at the foot of Winterhoek Mountains, *Krauss*, 1604! and without precise locality, *Zeyher*, 912! *Bolus*, 1874!

CENTRAL REGION: Somerset Div., *Bowker!* Richmond Div.; between Richmond and Brak Vallei River, 3000–4000 ft., *Drège!*

6. **A. natalense** (Oliver in Hook. Ic. Pl. t. 1471a); leaves elongated, linear-oval, acute; lamina often narrow; spike bifid, many-flowered; male flowers 3–2, bracteate, linear-oblong, obtuse, unequal, or subequal; stamens 6–7; anthers broadly subrotund; follicles often 3. *Durand and Schinz, Conspect. Fl. Afr.* v. 493.

EASTERN REGION: Natal; York, *McKen*, 10! between Kar Kloof and Umgeni River, *Rehmann*, 7429! Liddesdale, *Wood*, 4250! and without precise locality, *Sutherland!*

7. **A. Rehmanni** (Oliver in Hook. Ic. Pl. t. 1471b); leaves oblong-oval, subacute, 1–1⅔ in. long, 3–5 lin. broad, subapiculate; spike bifid, ⅔ in. long, dense-flowered; flowers diœcious; bracts ovate-oblong, subacute; follicles 8–9, oblong. *Durand and Schinz, Conspect. Fl. Afr.* v. 493.

KALAHARI REGION: Transvaal; Bosch Veld, between Kleinsmit and Kameel Poort, *Rehmann*, 4835!

8. **A. spathaceum** (Hook. f. in Bot. Mag. sub t. 6399); leaves very narrow, or elongate-subulate, obtuse, or dilated into a linear-lanceolate lamina; spike bifid; arms short, dense-flowered; floral bracts 2, obovate-oblong, obtuse, 3-nerved, pale lilac; stamens 6–8; follicles 3–8. *E. Meyer in Zwei Pflanzengeogr. Documente* 137, *name only; Engl. Jahrb.* viii. 272; *Durand and Schinz, Conspect. Fl. Afr.* v. 493.

VAR. **junceum** (Hook. f. in Bot. Mag. t. 6399); tubers hemispherical, about the size of a hazel nut; stem subacute, obtusely trigonous, or semiterete with rounded angles; leaves erect, flexuose, 6–10 in. long, elongate-subulate; scapes usually shorter than the leaves, cylindric; spadix forked, with arms 1–2 in. long; flowers curved on the spadix, on some plants all female, on others hermaphrodite, rarely all male; bracts 2, imbricate, ovate or oblong, obtuse, obscurely 3-nerved; stamens 6–8; carpels 3–8; ovules about 4 in each carpel; fruit trigonous, tumid. *A. spathaceum, var., E. Meyer in Linnæa*, xx. 215. *A. junceum, Herb. Zeyher, teste Hooker f.*

COAST REGION: Var. β, Uitenhage Div., *Zeyher*, 1734! British Kaffraria, *Cooper*, 465!

CENTRAL REGION: Albert Div., 4500 ft., *Drège!* Colesberg Div.; "on the table mountain near the Horse's Grave," *Burchell*, 2688! Var. β, Somerset Div.; between the Zuurberg Range and Klein Bruintjes Hoogte, 2000–2500 ft., *Drège!* marshy places at the foot of the Bosch Berg, *MacOwan*, 2204!

KALAHARI REGION: Orange Free State; Great Vet River, *Burke!* Var. β, Basutoland, *Bowker!* Transvaal; Hooge Veld at Trigards Fontein, *Rehmann*, 6692! and without precise locality, *Sanderson!*

EASTERN REGION: Griqualand East; swamps near Kokstad, *Wood*, 4206! Var. β, Tembuland; swamps near Bazeia, 2200 ft., *Baur*, 579! Natal; Biggarsberg and Umzinyate River, 3000–4000 ft., *Sutherland!* Umhlanga, 600 ft., *Wood*, 1611!

9. **A. crinifolium** (Lehm ex Steud. Nomen. ed. 2, i. 114, name only); "spike bifid; bracts obovate, diaphanous; nerves bifid; leaves mostly capillary; flowers 6-androus." *Dr. Pappe in Herb. Brit. Mus.*

COAST REGION: Cape Flats in inundated places, *Pappe!* No specimen at Kew.

The leaves in this are extremely slender for the genus, and the material is too poor for a better description.

III. POTAMOGETON, Linn.

Perianth-segments 4, green, valvate. *Anthers* 4, sessile, didymous. *Carpels* 4 (many abortive), sessile, with one cell, and one ovule; ovules campylotropous. *Stigma* persistent, subsessile or decurrent, altering much in position in ripening; drupelets small, variable in shape and contour, coriaceous. *Seed* subreniform; radicle large.

Submerged water plants with creeping rootstocks, terete or compressed, often floating (rarely aerial) leaves; leaves alternate or opposite, entire, or toothed; stipules intrafoliaceous, free or adnate to the petiole; flowers small, in spikes; peduncles at the base in a membranous spathe, ebracteate.

Species 60–70, cosmopolitan.

*With floating coriaceous leaves:
- Fruit 3-keeled; stems moderately stout; submerged leaves few or none.
  - Leaves plicate at the base; stipules acute ... (1) **natans.**
  - Leaves not plicate at the base; stipules obtuse.
    - Leaves tapering at both ends ... ... (2) **fluitans.**
    - Leaves obtuse at the base.
      - Fruit dorsally tapering into the style (3) **americanus.**
      - Fruit dorsally subtruncate at the top (4) **Richardii.**
- Fruit 3-keeled; stems filiform; submerged leaves abundant ... ... (5) **javanicus.**
- Fruit with 1 keel; stems moderately stout; submerged leaves abundant ... ... (6) **alpinum.**

**Without floating leaves:
Leaves 2–12 lin. broad.
Leaves crisped, serrulate; fruit long-beaked ... (7) **crispum.**
Leaves flat, entire; fruit short-beaked ... ... (8) **lucens.**
Leaves less than 2 lin. broad.
Leaves linear, grass-like, 5-veined; stem flattened (9) **Friesii.**
Leaves linear, 3-veined; stem filiform ... ... (10) **pusillum.**
Leaves filiform, with adnate stipules at the base as a sheath ... ... (11) **pectinatum.**

1. **P. natans** (Linn. Sp. Pl. ed. 1, 126); stem stout, very rarely branched; submerged leaves 0 or reduced to blades; floating leaves long-petioled, from subrotund to lanceolate, cordate and plicate at the base; stipules long and acute; peduncles stout; spikes dense-flowered; drupelets large, convex on the ventral margin, semicircular on the dorsal, without any raised bosses. *Durand and Schinz, Conspect. Fl. Afr.* v. 494.

KALAHARI REGION: Transvaal; Yuckschyt River, sources of the Limpopo, *Nelson*, 514!
EASTERN REGION: Natal; in a pool at Inyangwine, *Wood*, 3020!

Found in all temperate regions; rare in the tropics.

2. **P. fluitans** (Roth, Tent. Fl. Germ. i. 72); stem stout; submerged leaves translucent, linear-lanceolate; floating leaves ovate, or ovate-lanceolate, long-stalked, tapering at either end, not plicate; stipules large, blunt; peduncles stout; spike dense-flowered; no ripe fruit seen on either African or European specimens. *P. natans, var. fluitans, Cham. Adnot. Fl. Berol.* 4.

COAST REGION: Worcester Div.; Braud Vlei, *Rehmann*, 2415! Uniondale Div.; Lange Kloof, *Krauss*, 1233!
KALAHARI REGION: Transvaal; Hooge Veld at Bronkers Spruit, *Rehmann*, 6578!

This is probably a hybrid between *P. natans* and *P. lucens*.
Widely distributed in Europe, Asia, North and Tropical Africa.

3. **P. americanus** (Cham. in Linnæa, ii. 226) var. **Thunbergii** (A. Benn.); stem stout, rarely branched; the submerged leaves translucent, elongate or linear-lanceolate; floating leaves variable, oblong-lanceolate, elliptical or ovate-lanceolate, not plicate at the base; stipules blunt, large; peduncles stout, slightly swollen upwards; spikes dense-flowered; fruit obovate, dorsally 3-carinate, unequal, central one acute. *P. Thunbergii, Cham. in Linnæa*, ii. 221. *P. natans, var. capensis, Durand and Schinz, Conspect. Fl. Afr.* v. 494. *P. natans, Thunb. Prod.* 32. *P. capensis, Scheele in Herb. Buchenau!*

COAST REGION: Cape Div.; Zeekoe River, *Thunberg!* Hartebeest Kraal, Brak River, *Mund and Maire!* Paarl Div.; Berg River, *Drège!* Worcester Div.; Verkeerde Vlei, *Thunberg!* Uitenhage Div., *Zeyher!*
EASTERN REGION: Port Natal, *Gueinzius!*

Durand and Schinz have mistaken the meaning of Chamisso in heading the sections as at p. 221 "β, capensis," which was merely intended for a geographical indication, not a botanical name.

The typical form of *P. americanus* is found in Europe, Asia (rare), North Africa, and North America.

4. **P. Richardii** (Solms in Schweinf. Beitr. Fl. Æthiop. 194 and 292); stem stout; submerged leaves few, finer in texture than the last; floating leaves ovate or oblong-lanceolate, not plicate at the base; stipules blunt, large; peduncles stout, slightly swollen upwards; spikes dense-flowered; fruit large, $2\frac{1}{2}$ lin. long; style long, at first bent back, when ripe nearly central; dorsal ridges 3, the two lateral with blunt, wavy, tooth-like margins, slightly convex on the ventral margin, and indications of small bosses, but none of Schimper's specimens are quite ripe. *Durand and Schinz, Conspect. Fl. Afr.* v. 496. *P. natans, A. Rich. Tent. Fl. Abyss.* ii. 354, *non L.*

KALAHARI REGION: Transvaal; Hooge Veld at Standarton, *Rehmann*, 6801!

Also in Abyssinia. The fruit is described from Abyssinian specimens collected by Schimper.

5. **P. javanicus** (Hassk. in Verh. Natuurk. Ver. Nederl. Ind. i. 26); stem branched, filiform; submerged leaves sessile, narrowly linear, acute; floating leaves lanceolate, long petioled; lower stipules slightly connate, the upper ones free, acuminate; peduncles slender; spikes varying from 3–8 lin. long; fruit compressed, oblique-obovate; style straight with the ventral margin; dorsal margin 3-carinate, and strongly repand denticulate, ventral nearly straight, with a projection in the centre tapering to either end, variable as to the teeth-like projections on various parts of the fruit, which are strongest in the Australian forms. *P. tenuicaulis, F. Mueller, Frag. Fl. Austr.* i. 90, 244. *P. parvifolia, Buchenau in Bremen, Abhandl.* vii. 32. *P. huillensis, Welw. in Herb. Kew.; Durand and Schinz, Conspect. Fl. Afr.* v. 495.

KALAHARI REGION: Transvaal; Nylstroom River, *Nelson*, 290!

EASTERN REGION: Natal; Umtshati River, 3000–4000 ft., *Wood*, 4300! near York, *Rehmann*, 9062!

Also in Tropical Africa, Australia, Malay Archipelago, India, North China, and Japan.

6. **P. alpinum** (Balbis, Mém. Acad. Tur. vii. 329); plant drying a dull reddish colour; stem moderately stout, unbranched, or very rarely branched; submerged leaves very thin and translucent, sessile, strap-shaped, elliptical, attenuated at the base and apex; floating leaves (not always present) obovate, elliptical-ovate or oblanceolate, subcoriaceous; stipules large, not winged; peduncles slender; spike many-flowered, dense; fruit yellowish-red, oval-ovoid, acuminate at the apex, semicircular and sharply keeled on the dorsal face, convex on the ventral. *Misc. Bot.* 13. *P. rufescens, Schrad. in Cham. Adnot. Fl. Berol.* 5.

SOUTH AFRICA: Brak River, Berlin Herbarium!

Also in Europe, Asia, and America.

7. **P. crispum** (Linn. Sp. Pl. ed. 1, 126); stem compressed, slender, branched; leaves all similar, sessile, semiamplexicaul, strap-shaped oblong, or oblong, variously undulate, and strongly serrate at the

apex, less so on the margins; stipules small, subobtuse, lower soon decaying; peduncles between the forks of the stem, rather stout, and tapering towards the apex; spike large, lax-flowered; fruit large, acuminate and compressed, obliquely-ovoid; beak very long and curved. *Durand and Schinz, Conspect. Fl. Afr.* v. 493.

KALAHARI REGION: Griqualand West; Modder River, *Barber*, 4! Vaal River, *Nelson*, 70!

EASTERN REGION: Natal; Inanda, *Wood*, 154! Little Umhlaaga, *Sanderson*, 2018! Umgeni Valley, *Krauss*.

Widely dispersed in various parts of the earth.

8. **P. lucens** (Linn. Sp. Pl. ed. 1, 126) var. **fluitans** (Coss and Germ. Fl. Env. Paris, ed. 1, 571); stem stout, branched; leaves all submerged (?), the lower ones sometimes 10 in. long, lanceolate, or lanceolate-linear, acuminate, the lowest sessile, the pedicels gradually lengthen until they become 1–1½ in. long; upper leaves lanceolate, acute, slightly firmer in texture than the lower ones; stipules long, beaked, or winged; peduncles 2–6 in. long, stout, tapering upwards; spikes 1¼–2 in. long, dense-flowered; fruits large, slightly acuminate, slightly convex on the ventral margin, nearly semicircular on the dorsal, slightly keeled. *P. longifolium, Gay in Poir. Encyc. Suppl.* iv. 535. *P. macrophyllus, Wolfg. in Roem. et Schultes, Syst. Mant.* iii. 358.

COAST REGION: Uitenhage Div., *Ecklon and Zeyher*, 640! *Zeyher!*

KALAHARI REGION: Transvaal; Schoon Spruit, near Klerksdorp, *Nelson*, 224!

EASTERN REGION: Natal; Mid Illovo, *Wood*, 1889! Umhlongwe, *Wood*, 3015! Delagoa Bay; Mathibis Kom, 500 ft., *MacOwan and Bolus, Herb. Norm. Aust. Afr.*, 1393!

P. lucens is widely distributed in Europe, Asia, Africa, and America, rare in Australia.

The var. *fluitans* also occurs in Tropical Africa and Europe, and rarely in Asia, N. America, and Australia.

The South African specimens probably belong to a new sub-species of *P. lucens*, Linn., but for the present it is best to retain them under the name above given, to which variety they bear the greatest affinity of any named form.

9. **P. Friesii** (Ruprecht, Beitr. Pfl. Russland. iv. 43); stem slender, compressed, slightly branched; leaves all similar, linear, narrowed towards the base, acuminate cuspidate, 5- very rarely 3- or 7-veined, generally with fascicles of smaller leaves in the axils; stipules acute or acuminate, scarious; peduncles slightly thickened upwards; spikes few-flowered, generally interrupted; fruit small, slightly compressed, convex and bluntly 3-keeled on the dorsal margin, less so on the ventral, similar to but larger than that of *P. pusillus*. *P. pusillus, var. major, Fries, Nov. Fl. Succ.* 48 (*non Mert. et Koch*). *P. Oederi, Meyer, Fl. Hann. Excurs.* 536.

COAST REGION: Cape Div.; Zout River, Kochmanns Kloof, *Bergius!* King Williamstown Div.; Buffalo River, near King Williamstown, *Leighton!* Queenstown Div.; Shiloh, 3500 ft., *Baur*, 919!

EASTERN REGION: Natal; streamlet between Umzinto and Ifafa, *Wood*, 3055! Umlazi River, *Drège*, 4458!

Also in Europe, Asia (very rare), and N. America.
Very rarely this produces spathulate upper leaves.

10. P. pusillum (Linn. Sp. Plant. ed. i. 127); stem slender, subterete, branched; leaves all similar, sessile, semiamplexicaul, narrowly linear, acuminate, acute, or subacute, 1–3-nerved, and mostly without fascicles of leaves in the axils; stipules small, acute or subacute; peduncles slender, variable in length; spike few-flowered, $\frac{1}{4}$–$\frac{1}{2}$ in. long, not interrupted (or very rarely so); fruit small, semioval, obovate, very little compressed, bluntly 3-keeled on the dorsal face, and slightly convex on the ventral, generally without teeth. *Durand and Schinz, Conspect. Fl. Afr.* 496.

COAST REGION: Cape Div.; ponds near Cape Town, *Burchell*, 669! sand flats between Tyger Berg and Blue Berg, *Drège*, 1206! Swellendam Div.; lower part of Zonder Einde River, *Zeyher*, 1733!
KALAHARI REGION: Transvaal; Apies Poort, near Pretoria, *Rehmann*, 4024!
EASTERN REGION: Natal; Umzinyate River, *Wood*, 94! Umhlasine River, *Wood*, 962!

Also in Tropical and North Africa, Europe, Asia, and America.
As with the last, this very rarely produces spathulate upper leaves, in which state it has been called *P. panormitanus* by Bivona.

11. P. pectinatum (Linn. Sp. Plant. ed. 1, 127); stem cylindrical, or subcompressed, repeatedly branched; leaves all similar, the upper sometimes setaceous, 1-nerved, lower long linear with 3–5 nerves; stipules long, united to the sheathing petiole of the leaf, free at the apex; peduncles variable, short or long, equal; spike often interrupted; upper flowers generally approximate; fruit large, olive or green, obliquely obovate, slightly compressed, 3-keeled on the dorsal margin, central one generally prominent (but variable), slightly convex, or nearly straight on the ventral margin. *Durand and Schinz, Conspect. Fl. Afr.* v. 495. *P. marinum, L.? ex Krauss in Flora*, 1845, 344.

COAST REGION: Bredasdorp Div.; Zoetendals Vlei, *Krauss!* Uitenhage Div.; Zwartkops River, *Zeyher*, 4328! *Ecklon and Zeyher*, 644!
KALAHARI REGION: Transvaal; Schoon Spruit, near Klerksdorp, *Nelson*, 225!

Widely dispersed in all parts of the earth.

IV. RUPPIA, Linn.

Perianth none. *Stamens* 2; anthers 2-celled. *Carpels* 4–1-ovuled; stigma sessile. *Fruit* of 4 long-stalked, ovoid, or oblique-ovoid achenes on a common peduncle. *Seeds* uncinate; radicle large.

Submerged slender brackish-water plants; leaves elongate, filiform; sheath stipuliform; flowers small, usually 2 together (2–6) within the leaf-sheath, on a short peduncle which mostly elongates, and becomes straight or spirally twisted.

DISTRIB. Temperate and tropical regions. Species 1 or more? or many subspecies?

1. **R. spiralis** (Dum. Fl. Belg. 164); sheaths inflated; peduncles spirally coiled; drupes nearly straight. *R. maritima var. spiralis, Aschers. in Aschers. and Schweinf. Ill. Flor. Égypte*, 144; *Durand and Schinz, Conspect. Fl. Afr.* 497.

COAST REGION: Cape Div.; shallow pools in the vicinity of Green Point, near Cape Town, *Harvey!*

Also in Egypt and along the coasts of Europe.

V. ZANNICHELLIA, Linn.

Male flowers: *Anther* 2–4-celled, with a slender filament. *Female flower* in a membranous perianth. *Carpels* variable; style long or short; stigma peltate, crenate. *Fruit* of 4, rarely less or more, long, incurved achenes. *Seed* pendulous, cotyledonary end folded on itself.

Slender submerged fresh, brackish, or salt water plants; leaves mostly opposite, linear or capillary, with stipular sheaths; flowers small, in pairs, or solitary, in a membranous perianth.

DISTRIB. Temperate and tropical regions. Species 4–5?

1. **Z. palustris** (Linn. Sp. Plant. ed. 1, 969); fruits subsessile, or sessile; beak about half as long as the rest of the fruit. *Z. stylaris, Presl, Bot. Bemerk.* 112.

SOUTH AFRICA: without locality, *Harvey*, 481!

WESTERN REGION: Little Namaqualand; near the mouth of the Orange River, *Drège*, 8801!

KALAHARI REGION: Griqualand West; Hay Div., Ongeluk, *Burchell*, 2648! Orange Free State; Kanon Fontein, *Rehmann*, 3558!

Sub-sp. **pedicellata** (Asch. Fl. Branden. 669); fruits with the common peduncle and pedicels elongated. *Durand and Schinz, Conspect. Fl. Afr.* v. 498. *Z. pedicellata, Fr. Novil. Mant.* i. 18; *Syme, Engl. Bot.* ix. 57.

EASTERN REGION: Natal, according to Durand and Schinz. No specimen at Kew.

VI. ZOSTERA, Linn.

Male flowers of single sessile anthers. *Female flower* of a solitary 1-ovuled carpel; style persistent; stigmas 2. *Fruit* membranous. *Embryo* grooved.

Submerged marine plants, with dark ribbon-like leaves; rootstocks matted, creeping; stem compressed; leaves sheathing, long, linear; stipules combining with the sheathing leaf-base; flowers in series of anthers and carpels on one surface of a linear spadix, enclosed in a leaf-like spathe.

DISTRIB. Coasts of Europe, Asia, and America.

Spadix without clasping bands; nuts striate ... ... (1) **marina.**
Spadix with clasping bands; nuts smooth ... ... (2) **nana.**

1. **Z. marina** (Linn. Sp. Plant. ed. 1, 968) var. **angustifolia** (Hornem. Fl. Dan. t. 1501); leaves $\frac{1}{8}$–$\frac{1}{4}$ in. broad, 1–3-nerved.

SOUTH AFRICA: without locality or collector's name, in Kew Herbarium!

The type has leaves $\frac{1}{4}$–$\frac{1}{2}$ in. broad and 3–7-nerved, and is found on the coasts of North Africa, Europe, America, and Asia?

2. Z. nana (Roth, Enum. Pl. Germ. 8); leaves narrowly linear, 1–3-nerved; peduncle of spathe filiform; spadix strap-shaped oblong, with small inflexed bands; fruit shorter than in *Z. marina*, and nearly smooth. *Z. minor, Nolte ex Reichb. Ic. Fl. Germ.* vii. 2. *Z. pumila, Le Gall, Congr. Sc. Fr.* xvi. i. 144. *Z. marina, L. var. minor, Mert. ex Krauss, in Flora,* 1845, 344.

EASTERN REGION: Natal Bay, *Krauss,* 70!

Also found on the coasts of Europe, N. Africa, and in the Caspian and Black Seas.

VII. NAIAS, Linn.

Male flower solitary. *Anther* 1–4-celled in 2 sheaths or tubes. *Female flower: Perianth* 0, or membranous and adherent. *Carpel* 1; stigmas 2–4, slender; ovule erect. *Achene* oblong with an erect seed, having a straight embryo.

Submerged plants with branched, filiform stems, smooth or muricate; leaves linear, entire, or toothed, alternate, opposite, or whorled; flowers diœcious or monœcious, axillary and small.

DISTRIB. Temperate and tropical regions.

1. N. interrupta (R. Schum. in Engl. Pfl. Ost-Afr. C. 94); stem slender; internodes elongated; leaves short; basal auricle truncate, entire; blade sparingly serrated with minute teeth; female flower with 2–3 stigmas; fruit fusiform, dull yellow.

KALAHARI REGION: Transvaal, *McLea in Herb. Bolus,* 6283!

Plant with much the habit of *N. tenuifolia,* R. Br.

ORDER CXLVIII. ERIOCAULEÆ.

(BY N. E. BROWN.)

Flowers very small, regular or irregular, unisexual, bracteate or ebracteate, densely crowded in globose, hemispherical, oblong or campanulate heads. *Calyx* of 2–3 free or variously connate, scarious or submembranous sepals, often ciliate and more or less hairy on the back near the apex, very rarely wanting. *Corolla* usually separated from the calyx by a distinct stipes; petals 2–3, free or variously connate, equal or unequal, sometimes rudimentary or absent, membranous and hyaline, or moderately thick and opaque, with or without a gland on their inner face, often ciliate or hairy. *Stamens* equal in number to the petals and inserted on them at or above their base, or 4 or 6 (or by abortion fewer) in two series, the one alternating with the petals, the other opposite to them; filaments filiform or slightly flattened, free; anthers small, dorsifixed, ovate, oblong or subquadrate, 1–2-celled, opening longitudinally. *Staminodes* in the female flowers rare, when present minute. *Pistil* in the male flowers very rudimentary, reduced to 2–3 minute glands. *Ovary* of the female flowers superior, 2–3-celled; style terminal, divided above into 3 simple or bifid, filiform branches, with or without 3 other branches or appendages alternating with them or arising from the style below them; ovules solitary in each cell, pendulous, orthotropous. *Fruit* a 2–3-celled capsule; cells opening longitudinally at the back. *Seed* solitary in each cell, pendulous, ellipsoidal or subglobose; testa thin, striate, reticulate or minutely tuberculate; albumen firm; embryo minute, lenticular, seated at the apex of the albumen opposite the hilum.

Perennial or annual herbs, growing in water, swamps, bogs, or on dry ground, stemless or with simple or branched leafy stems; leaves linear or subulate, alternate, arranged in a dense or lax rosette or scattered along the stem; scapes one to many, each with a tubular sheath at the base, one- or rarely several-headed; heads monœcious or rarely with the sexes in separate heads; outer or involucral bracts in 2–several series, imbricate, shorter than, equalling, or longer than the flowers, scarious or rigid; flowering bracts solitary under each flower, variable in form, often ciliate or hairy at the apex, rarely absent; receptacle flat, convex, subglobose, or elongated, glabrous or villose; flowers usually very numerous, very small or minute, pedicellate or sessile, the females in the lower or outer part of the head and the males in the central part, or both sexes irregularly intermingled, rarely separated in distinct heads.

DISTRIB. Genera 6. Species about 350. Found in all warm climates, most numerous in Tropical America, very few in the temperate regions.

* *Stamens double the number of the petals, 4 or 6, or fewer by abortion, in two alternating series. Style-branches 3, without alternating appendages.*

I. **Eriocaulon.**—*Petals* free, sometimes rudimentary, or rarely absent.

II. **Mesanthemum.**—*Petals* connate into a cylindric tube, but with free claws in the female flowers.

* *Stamens in one series, equal in number to the petals and opposite to them. Style-branches 6, 3 of them stigmatose, simple or bifid, and 3 others alternating with them or arising from the style lower down, non-stigmatic, and usually shorter and stouter.*

III. **Pæpalanthus.**—*Petals* of the male flowers connate into a minute funnel-shaped tube, of the female flowers free or connate into a tube at their middle or upper part, with free claws.

I. ERIOCAULON, Linn.

Sepals 2–3, free, or in the male flowers variously connate, concave or boat-shaped, rounded or keeled on the back; apex bearded, ciliate, or glabrous; margins sometimes ciliate. *Petals* 2–3, free, sometimes rudimentary or absent in the male, rarely in the female flowers, usually with a gland and often hairy on their inner face; apex often ciliate or bearded. *Stamens* in two series, double the number of the petals, or by abortion fewer; anthers 2-celled. *Staminodes* in the female flowers none. *Ovary* 2–3-celled; style-branches 3, simple, filiform, without alternating appendages.

Marsh or aquatic herbs, usually stemless, often cæspitose, with the leaves all radical, or in a few species with a simple or branched leafy stem; scapes one-headed, longer than the leaves; heads globose, hemispherical or oblong; flowering bracts oblong or obovate, flattish or concave. The other characters as for the Order.

DISTRIB. That of the Order. Species about 150.

Flower-heads 1–1¼ lin. diam., dark fuscous; bracts, sepals, and petals glabrous; leaves subulate ... (1) **abyssinicum.**
Flower-heads 2 lin. diam., dark fuscous; bracts and petals slightly ciliate; sepals glabrous; leaves linear-ensiform ... ... (2) **transvaalicum.**
Flower-heads 3 lin. or more in diam.:
Leaves tapering to a very fine acute point:
Leaves ¾–1½ in. long; flowering bracts spathulate-obovate, dark fuscous; receptacle villous ... ... (3) **Bauri.**
Leaves 1½–3 in. long; flowering bracts oblong-obovate, pale brownish with an olive-green spot on each side above the middle; receptacle villous ... ... (4) **sonderianum.**
Leaves not tapering to a very fine point, often obtuse:
Leaves 3–10 in. long, obtuse; sepals of the female flowers with an entire or toothed keel, of the male flowers cuneate-obovate, very obtuse, not keeled; receptacle more or less villous ... ... (5) **Dregei.**
Leaves ½–2½ in. long, subobtuse; sepals rounded on the back, gibbous in the female flowers, not keeled; receptacle with very minute points upon it ... ... (6) **africanum.**
Leaves 10–13 in. long, subacute; sepals of both sexes similar, obovate, obtuse, apiculate and denticulate at the apex, not keeled ... (7) **Woodii.**

1. **E. abyssinicum** (Hochst. in Flora, 1845, 341); leaves not very numerous, 4–9 lin. long, ¼–⅓ lin. thick, subulate, acute, glabrous; scapes several to a plant, ¾–4 in. long, filiform, obtusely quadrangular or trigonous, glabrous; basal sheath ¼–1 in. long, with an oblique mouth and shortly bifid apex, glabrous; flower-heads about 1–1¼ lin. in diam., at first subglobose, becoming oblong, monœcious, the sexes intermingled; involucral bracts ⅔ lin. long, ¼ lin. broad, oblong, acute, thin, pallid, glabrous; flowering bracts ½ lin. long, ⅙ lin. broad, lanceolate, acute, concave, glabrous, blackish or dark fuscous; receptacle conical or oblong, glabrous; female flowers subsessile; sepals 3, free, ½ lin. long, lanceolate, acute, boat-shaped, rounded on the back, fuscous, or dark olive-green, glabrous; stipes between the sepals and petals exceedingly short; petals 3, free, ⅓–½ lin. long, linear, obtuse or acute, glabrous, not ciliate, without glands, fuscous; ovary trigonous, glabrous; style with 3 filiform branches; male flowers subsessile; sepals 3, all free, or two or all of them more or less connate, ½ lin. long, linear or linear-oblong, acute, fuscous, quite glabrous, entirely wanting in some of the central flowers; stipes between the sepals and petals rather more than half as long as the sepals; petals rudimentary; stamens 4–6, anthers black. *Steud. Synops. Glum.* ii. 273; *Körnicke in Linnæa*, xxvii. 612.

CENTRAL REGION: Somerset Div.; summit of Bosch Berg, *MacOwan*, 2104!

EASTERN REGION: Natal; Griffins Hill near Eastcourt, *Rehmann*, 7297! and without precise locality, *Gerrard*, 491!

Also in Tropical Africa.

2. **E. transvaalicum** (N. E. Brown); leaves $2\frac{1}{2}$–$3\frac{1}{2}$ in. long, 2–$2\frac{1}{2}$ lin. broad, or larger? linear-ensiform, subacute, flat; scapes several to a plant, $4\frac{1}{2}$–6 in. long, $\frac{1}{2}$ lin. thick, terete, striate, glabrous; basal sheath about 2 in. long, very oblique at the mouth; apex obtuse, glabrous; flower heads globose, 2–$2\frac{1}{2}$ lin. diam., monœcious; involucral bracts $\frac{3}{4}$–$1\frac{1}{4}$ lin. long, about $\frac{1}{2}$ lin. broad, oblong-obovate, obtuse, glabrous, pale olive; flowering bracts 1–$1\frac{1}{4}$ lin. long, $\frac{1}{2}$ lin. broad, obovate, acute, glabrous, sparingly ciliate with minute white hairs, fuscous or dark olive-brown; receptacle globose, villous; female flowers with glabrous pedicels about $\frac{1}{6}$ lin. long; sepals $\frac{2}{3}$–$\frac{3}{4}$ lin. long, lanceolate, acute, boat-shaped, acutely keeled on the back, quite glabrous, dark olive-brown, or fuscous; stipes between the sepals and petals exceedingly short; petals 3, free, $\frac{2}{3}$–$\frac{3}{4}$ lin. long; $\frac{1}{6}$ lin. broad, linear-lanceolate, acute or obtuse, ciliate with a few very short hairs, and with a few longer hairs on their inner face, pale, fuscous, or olive-green, all without glands, or one with a minute gland near the tip; ovary trigonous, glabrous; style with 3 filiform branches; male flowers with glabrous pedicels $\frac{1}{4}$–$\frac{1}{3}$ lin. long; sepals 3, all combined into a funnel-shaped tube split open on one side, and shortly trifid at the apex, or one of them free, cuneate-oblong, obtuse, quite glabrous, pale, fuscous, or olive-green; stipes between the sepals and petals about $\frac{2}{3}$ as long as the former; petals rudimentary; stamens 6; anthers fuscous, or dark olive-green.

KALAHARI REGION: Transvaal; on the Bosch Veld at Buchenhouts Kloof Spruit, *Rehmann*, 4787!

3. **E. Bauri** (N. E. Brown); leaves $\frac{3}{4}$–$1\frac{1}{3}$ in. long, 1–$1\frac{1}{2}$ lin. broad at the base, subulate, tapering to a fine point, flat on the face, very convex on the back; scape solitary, $5\frac{1}{2}$–14 in. long, $\frac{1}{2}$ lin. thick, 5-angled, striate in the dried state; basal sheath 1–$1\frac{1}{2}$ in. long, 2-lobed at the top; flower-heads subglobose, 3–$3\frac{1}{2}$ lin. diam., monœcious; lower flowers female, upper male; involucral bracts small, $\frac{1}{2}$–1 lin. long, $\frac{1}{2}$–$\frac{2}{3}$ lin. broad, obovate, obtuse, concave, glabrous, pallid; flowering bracts 1–$1\frac{1}{4}$ lin. long, the outer ones $\frac{3}{4}$–1 lin. broad, the inner narrower, spathulate-obovate, obtuse or acute, concave, scarcely or very obtusely keeled on the back, the apical part blackish-green, or dark olive-green, with a tuft of white hairs; receptacle villous, with fine hairs, half as long as the flowers; female flowers subsessile, or very shortly pedicellate; sepals 3, free, about $1\frac{1}{4}$ lin. long, boat-shaped, acute, keeled, dark olive-green, with white hairs at the apex; margins ciliate with rather long hairs at the base; keel subacute; stipes between the sepals and petals exceedingly short, $\frac{1}{5}$–$\frac{1}{4}$ the length of the ovary; petals 3, free, 1–$1\frac{1}{4}$ lin. long, less than $\frac{1}{4}$ lin. broad, cuneate-linear, acute, ciliate at the apex, white, with a linear black gland a little below the apex; ovary trigonous, glabrous; style

with 3 filiform branches; male flowers with a short, stout pedicel; sepals 3, free or united into a 3-lobed tube, split open on one side, $\frac{3}{4}$–1 lin. long, obovate, obtuse, or the lobes rounded, dark olive-green, with a tuft of white hairs at the apex; stipes between the sepals and petals very long, $\frac{1}{2}$–$\frac{3}{4}$ as long as the sepals; petals unequal, $\frac{1}{3}$–$\frac{2}{3}$ lin. long, oblong-lanceolate, or narrowly obovate, obtuse, ciliate at the apex, and sometimes hairy on the face, white, with a linear black gland at the middle or a little above it; stamens 6; anthers black.

EASTERN REGION: Tembuland; Bazeia Mountain, 4000 ft., *Baur*, 622!

4. E. **sonderianum** (Körnicke in Linnæa, xxvii. 669); leaves 1$\frac{1}{2}$–3 in. long, 2–4 lin. broad at the base, whence they gradually taper to a very fine acute point, concave down the face; scapes 1–2 to a plant, 6–10 in. long, $\frac{1}{2}$–$\frac{3}{4}$ lin. thick, striate in the dried state; basal sheath 1$\frac{1}{4}$–2$\frac{1}{2}$ in. long, 2-lobed at the apex; flower-heads globose, 3–4 lin. in diam., monœcious; involucral bracts about 1 lin. long, $\frac{3}{4}$ lin. broad, oblong-obovate, glabrous, whitish or pallid; flowering bracts $\frac{3}{4}$–1 lin. long, $\frac{1}{2}$ lin. broad, oblong-obovate, subacute, concave, obtusely keeled on the back, pale brownish, with a fuscous or olive-green spot on each side, a little above the middle; apex densely bearded with white hairs; receptacle globose, villous; female flowers shortly pedicellate; sepals 3, free, 1 lin. long, rhomboidal-lanceolate, boat-shaped, acute, acutely keeled (not wing-keeled) on the back, pale olive-green, with the apex and keel whitish; apex ciliate, and more or less bearded with white hairs; petals 3, free, about 1 lin. long, $\frac{1}{4}$ lin. broad, cuneate-oblong or cuneate-linear, obtuse, ciliate, and a little hairy on the face at the apex, white, with a linear black gland a little below the apex; ovary trigonous, glabrous; style with 3 filiform branches; male flowers shortly pedicellate; sepals $\frac{2}{3}$–$\frac{3}{4}$ lin. long, $\frac{1}{3}$–$\frac{1}{2}$ lin. broad, cuneate-obovate, concave, not at all keeled on the back, obtuse or subacute, densely bearded with white hairs at the apex, whitish, with an olive-green spot on each side, a little below the apex; petals 3, free, unequal, the dorsal or larger one $\frac{2}{3}$–$\frac{3}{4}$ lin. long, $\frac{1}{4}$–$\frac{1}{2}$ lin. broad, cuneate-obovate, obtuse or subacute, white, rather thick, ciliate, and bearded with white hairs on the inner face, and with a linear black gland a little above the middle; stamens 6; anthers black.

KALAHARI REGION: Transvaal; Magalies Berg, northern slopes, 6000–7000 ft., *Zeyher*, 1731! *Burke*!

5. E. **Dregei** (Hochst. in Flora, 1845, 341); leaves 3–10 in. long, 1$\frac{1}{2}$–4 lin. broad, linear, tapering to an obtuse point, glabrous; scapes 1–4 to a plant, 9–18 in. long, $\frac{1}{2}$–1 lin. thick, sulcate-striate, glabrous; basal sheath 3–6$\frac{1}{2}$ in. long, obtusely bifid at the apex, glabrous; flower-heads globose, monœcious and 4–6 lin. diam., or male and about 3 lin. diam.; involucral bracts 1$\frac{1}{2}$–2 lin. long, $\frac{3}{4}$–1 lin. broad, oblong, obtuse, glabrous, fuscous or pale brown; flowering bracts 1$\frac{1}{4}$–1$\frac{3}{4}$ lin. long, $\frac{1}{3}$–$\frac{1}{2}$ lin. broad, linear-oblong, obtuse

or acute, concave, slightly and obtusely keeled on the back, pale fuscous or olivaceous, darker and densely bearded with white hairs at the apex; receptacle convex or globose, flat in the male heads, more or less villous; female flowers with short glabrous pedicels; sepals $1\frac{1}{2}$ lin. long, $\frac{1}{2}$–$\frac{2}{3}$ lin. broad, and $\frac{1}{2}$–$\frac{3}{4}$ lin. deep from front to back, lanceolate, boat-shaped as viewed from the front, falcate-lanceolate or semicordate as viewed from the side, acute, with a broad, acute, entire or toothed keel, dark fuscous or dark olive-green, ciliate with long hairs on the margins, with a few white hairs at the apex and sometimes some on the back; stipes between the sepals and petals about $\frac{1}{4}$ lin. long; petals $1\frac{1}{4}$–$1\frac{1}{2}$ lin. long, $\frac{1}{5}$–$\frac{1}{4}$ lin. broad, oblanceolate, obtuse, rather thick, white, ciliate and hairy on the inner face at the apex, and with a linear black gland a little below it; ovary trigonous, glabrous; style with 3 filiform branches; male flowers with glabrous pedicels $\frac{1}{4}$–$\frac{2}{3}$ lin. long; sepals 3, all free or variously connate, $\frac{3}{4}$–1 lin. long, $\frac{1}{3}$–$\frac{1}{2}$ lin. broad, cuneate-oblong or cuneate-obovate, obtuse, concave, thin, not at all keeled on the back, dark olive-green, densely bearded with white hairs at the apex; stipes between the sepals and petals $\frac{1}{6}$–$\frac{1}{2}$ lin. long; petals 3, free, unequal, in some of the flowers well developed and the dorsal or larger one $\frac{2}{3}$–$1\frac{1}{3}$ lin. long, $\frac{1}{3}$–$\frac{1}{2}$ lin. broad, cuneate-oblong or narrowly obovate, obtuse, thick, white, glandular, ciliate and hairy as in the female flowers, in others the petals are very small or rudimentary, with a dense fringe of white hairs at their apex; stamens 6; anthers black. *Steud. Synops. Glum.* ii. 272; *Körnicke in Linnæa,* xxvii. 672.

SOUTH AFRICA: without locality, *Drège,* 4101!
EASTERN REGION: Natal; in a swamp at Clairmont, *Wood,* 1427! among rocks in the bed of the Umbilo River, 1000 ft., *Sanderson,* 904!

This seems to be a somewhat variable species. In Wood's plant the leaves are much longer, and the sepals of the female flowers are distinctly semicordate at the base, and have a much broader, more distinctly toothed keel than those of Drège's specimen. But I believe these differences are due to luxuriance of growth, and are not of specific value, as the plants agree in all other points.

6. **E. africanum** (Hochst. in Flora, 1845, 340); leaves $\frac{1}{2}$–$2\frac{1}{2}$ in. long, $\frac{1}{2}$–$1\frac{1}{2}$ lin. broad, linear-subulate, subacute or somewhat obtuse at the apex, concave down the face; scapes 1–4 to a plant, $2\frac{1}{2}$–8 in. long, $\frac{3}{4}$–1 lin. thick, terete, striate in the dried state; basal sheath $\frac{3}{4}$–2 in. long, unequally 3–5-lobed at the apex; flower-heads subglobose, 3–4 lin. in diam., monœcious, lower flowers female, inner male; involucral bracts $\frac{3}{4}$–$1\frac{1}{4}$ lin. long, $\frac{1}{2}$–$1\frac{1}{4}$ lin. broad, elliptic or orbicular, obtuse, pallid, glabrous; flowering bracts 1–$1\frac{1}{3}$ lin. long, $\frac{2}{3}$–$\frac{3}{4}$ lin. broad, obovate, obtuse, or the inner ones acute, concave, glabrous, blackish; receptacle cushion-like, with very minute papilliform points upon it; female flowers shortly pedicellate; sepals 3, free, about $\frac{3}{4}$ lin. long, boat-shaped, rounded on the back and somewhat gibbous, acute or subacute and incurved at the apex,

blackish-green, ciliate and clothed on the back with white hairs; stipes between the calyx and corolla very short; petals 3, free, rather more than 1 lin. long, $\frac{1}{4}$–$\frac{1}{3}$ lin. broad, cuneate-obovate, obtuse, white, with an oblong or subquadrate black gland near the apex and ciliate with white hairs; ovary trigonous, glabrous; style with 3 filiform branches; male flowers with glabrous pedicels $\frac{1}{3}$–$\frac{1}{2}$ lin. long; sepals 3, about $\frac{3}{4}$ lin. long, $\frac{1}{2}$ lin. broad, cuneate-oblong, obtuse or subtruncate, very concave, rounded on the back, blackish-green, with white hairs at the tips; petals $\frac{1}{2}$–$\frac{3}{4}$ lin. long, $\frac{1}{3}$ lin. broad, obovate, obtuse, ciliate and hairy on the inner face at the apex, white, with an oblong black gland a little below it; stamens 6; anthers white. *Steud. Synops. Glum.* ii. 272; *Körnicke in Linnæa*, xxvii. 649.

EASTERN REGION: Griqualand East; banks of the Umzimkulu River, near Clydesdale and near Handcocks Drift, 2500 ft., *Tyson*, 1299! 2551! *MacOwan and Bolus, Herb. Norm. Aust. Afr.*, 1203! Natal; Umgeni River, near Pietermaritzburg, *Krauss*, 375!

7. **E. Woodii** (N. E. Brown); leaves 10–13 in. long, $1\frac{1}{2}$–$2\frac{1}{2}$ lin. broad, linear, tapering to an acute or subacute point, flat and flaccid, glabrous; scapes 1–5 to a plant, 1–$1\frac{1}{4}$ ft. long, about 1 lin. thick, terete, striate in the dried state, glabrous; basal sheath about 4 in. long, 2–4-fid at the apex, glabrous; heads about $3\frac{1}{2}$ lin. diam., monœcious, hemispherical; involucral bracts about $1\frac{1}{2}$ lin. long, $\frac{3}{4}$–1 lin. broad, elliptic, obtuse, whitish-brown, glabrous; flowering bracts $1\frac{1}{2}$ lin. long, $\frac{2}{3}$ lin. broad, obovate, acute, thin, flattish with the apex incurved, dark fuscous or blackish with a few, short, white hairs on the back and margins; female flowers shortly pedicellate; sepals $\frac{1}{2}$–$\frac{2}{3}$ lin. long, rather more than $\frac{1}{3}$ lin. broad, obovate, obtuse or subtruncate, apiculate, more or less irregularly denticulate, flattish or slightly concave, not keeled, very thin and flaccid, dark fuscous with white hairs on the back and margins; stipes between the sepals and petals very short; petals 3, free, equal or unequal, $\frac{1}{2}$–$\frac{2}{3}$ lin. long, linear-spathulate or narrowly obovate, white, densely fringed with white hairs, and with an elliptic black gland at their apex; ovary trigonous, glabrous; style very deeply divided into 3 filiform branches; male flowers with short glabrous pedicels; sepals 3, free or slightly connate at the base, like those of the female flowers, but rather less denticulate at the apex, and more densely ciliate and hairy on the back; stipes between the sepals and petals very short, or half as long as the sepals; petals like those of the female flowers but much smaller; stamens 6, or by abortion fewer; anthers white.

EASTERN REGION: Natal; in a swamp near Murchison, *Wood*, 3053!

This appears to be an aquatic species. In the Kew specimens, the flower heads of which are somewhat malformed, the receptacle is flat and appears to be glabrous. The flowers may not have attained their normal development, but the remarkably flaccid, denticulate sepals well distinguish this from all other South African species.

II. MESANTHEMUM, Körnicke.

Sepals 3, free, or shortly connate at the base in the male flowers, hyaline. *Petals* united into a cylindric tube, entire in the male flowers, divided at the base into 3 short claws in the female flowers, subtruncate or shortly 3-lobed at the apex, with 3 glands on the inside near the top of the tube. *Stamens* 6, in two alternating series, included within the tube. *Ovary* 3-celled ; style 3-branched; branches simple and filiform, or more or less enlarged and ramified near their apex, without alternating appendages.

Marsh herbs, stemless, or with a more or less elongating and rooting stem; leaves broadly linear, in a dense tuft; scapes one-headed, longer than the leaves; heads monœcious, with a campanulate involucre, flat-topped or, perhaps, at length convex or hemispherical; flowering bracts with a capillary stalk and a clavate head. Other characters as for the Order.

DISTRIB. Species 3, natives of Tropical Africa and Madagascar, one extending within the limits of the South African Flora.

1. **M. radicans** (Körnicke in Linnæa, xxvii. 573); leaves 12–16 in. long, $\frac{1}{3}$–$\frac{1}{2}$ in. broad, linear, tapering to an obtuse point, softly pubescent on both sides; scapes 1–2$\frac{1}{3}$ ft. long, scarcely 1 lin. thick, terete, striate, pubescent; basal sheaths 8–9 in. long, softly pubescent; apex obliquely produced into a point about 1$\frac{1}{2}$ in. long; heads about 5 lin. in diam., flat-topped, monœcious, the outer flowers female, the inner male; involucre campanulate, its scales about 2 lin. long and broad, broadly ovate, obtuse, rigid, pale whitish-brown, more or less pubescent, the inner ones a little longer and more pointed than the rest, radiating; receptacle flat, covered with long, dark grey hairs; flowering bracts 2 lin. long with a very fine capillary stalk and a linear-clavate head covered with short hairs, white; female flowers pedicellate; pedicels $\frac{1}{3}$–$\frac{1}{2}$ lin. long, with very long, fine grey hairs on the basal part; sepals $\frac{2}{3}$ lin. long, scarcely $\frac{1}{3}$ lin. broad, oblong, truncate, very concave, hyaline, white; no stipes between the sepals and petals; petals 1$\frac{3}{4}$ lin. long, connate into a cylindric tube with short, free claws, hairy on the outside just above the claws, ciliate at the slightly trifid apex, whitish with 3 linear brownish glands near the top of the tube within; ovary acutely triquetrous, glabrous; style long with 3 filiform branches, or the branches more or less enlarged and ramified near their apex; male flowers with pedicels 1 lin. long, hairy like those of the female flowers; sepals 3, free, 1$\frac{1}{3}$ lin. long, $\frac{1}{3}$ lin. broad, longer than the corolla, oblong, obtuse or subtruncate, and shortly ciliate at the apex, hyaline, whitish; stipes between the calyx and corolla about $\frac{1}{5}$ lin. long; corolla $\frac{3}{4}$ lin. long, tubular, subtruncate or obscurely 3-fid, ciliate at the apex, glabrous on the outside, white with 3 minute linear glands inserted on the middle of the tube within; stamens 6; anthers white. *Eriocaulon radicans, Benth. in Hook. Niger Fl.* 547; *Steud. Synops. Glum.* ii. 273. *E. giganteum, Afzel. ex Körnicke in Linnæa,* xxvii. 573. *E. guineense, Steud. Synops. Glum.* ii. 273.

WESTERN REGION: South of the Tropic, *Curror!*

III. PÆPALANTHUS, Mart.

Sepals 2–3, free, or rarely more or less connate. *Petals* 2–3; those of the female flowers free, or connate at the middle or upper part, with free claws; those of the male flowers connate into a campanulate or funnel-shaped tube, subtruncate or shortly (rarely deeply) bifid or trifid at the apex. *Stamens* 2 or 3, of the same number as the petals and opposite to them, free or adhering to the corolla-tube. *Ovary* 2–3-celled; style divided above into 2–3, simple or bifid, filiform stigmatose branches, with 2–3 other branches or appendages alternating with them, or arising from the style below them.

Marsh, bog, aquatic, or terrestrial perennial or annual herbs, stemless, or with simple or branched leafy stems; leaves linear or subulate. Scapes usually one-headed, rarely several-headed, in a few species shorter than the leaves; flowering-bracts oblong or obovate, flattish or concave. Other characters as for the Order.

DISTRIB. Species between 200 and 300, of which 3–4 are Tropical and South African, the rest Tropical American.

1. **P. Wahlbergii** (Körnicke in Fl. Bras. iii. pt. i. 459); leaves 3–9 lin. long, $\frac{1}{2}$ lin. thick at the base, subulate, very acute, flat on the face, convex on the back, glabrous, in dense tufted rosettes $\frac{1}{2}$–1 in. diam.; scapes several to a plant, $2\frac{1}{2}$–7 in. long, filiform, trigonous, often 3-striate in the dried state, glabrous; basal sheath 4–9 lin. long, oblique, with a short acute point at the apex, glabrous; flower-heads 2–$2\frac{1}{2}$ lin. in diam., hemispherical, monœcious, the sexes intermingled; outer involucral bracts about $\frac{2}{3}$ lin. long, the inner 1–$1\frac{1}{4}$ lin. long, $\frac{1}{3}$–$\frac{1}{2}$ lin. broad, oblong or oblong-lanceolate, obtuse or subacute, pale brown or fuscous; floral bracts none; receptacle convex, villous; flowers of both sexes with pedicels $\frac{1}{3}$ lin. long with long hairs on their basal part; sepals of both sexes subsimilar, free, $\frac{1}{2}$ lin. long, enlarging in fruit, lanceolate, boat-shaped, acute, keeled, glabrous, ciliate on the margins below, pale brown or fuscous; stipes between the sepals and corolla very short in the female flower, and about half as long as the sepals in the male; petals of the female flower 3, free at the base, connate above into a short tube, crenulate or shortly 3-lobed at the apex, rather more than $\frac{1}{2}$ lin. long, hairy outside, white, without glands; ovary triquetrous, glabrous; style with 6 filiform branches, three of them ending in globose knobs; corolla of male flower a minute funnel-shaped tube, irregular at the mouth; stamens 2–3, anthers white.

KALAHARI REGION: Transvaal; Magalies Berg, southern slopes, *Zeyher*, 1730! *Burke*, 74! in a bog near the Nylstroom River, *Nelson*, 294!

ORDER CXLIX. RESTIACEÆ.

(By M. T. MASTERS.)

Flowers diœcious, unisexual. *Perianth* superior, glumaceous, hexamerous, in two rows; segments all similar, or the outer three different from the inner, outer lateral segments generally conduplicate, more or less keeled, intermediate anterior segment flat, inner segments usually free, sometimes concrescent at the base. *Stamens* 3, opposite to the inner perianth-segments; filaments very slender; anthers 1-celled, linear-oblong, generally apiculate, dehiscing longitudinally and introrsely. *Pistillode* absent or with 2–3 rudimentary styles. *Female flower* generally like the male except in the 1-flowered species; staminodes 2–3, opposite the inner perianth-segments or absent. *Ovary* free, 3-celled, or 1–2-celled by abortion, in the 2-celled species the anterior cell is abortive, in the 1-celled the anterior and one of the lateral cells become obsolete; styles 1–2–3 free at the base, or more or less united, linear, with stigmatic plumose surface on the inner side. *Ovule* solitary, pendulous from the upper angle of each cell. *Fruit* capsular, 2–3-celled, or by abortion 1-celled, dehiscing by a longitudinal cleft in each cell, or nutlike, indehiscent, 1-celled. *Seed* solitary, pendulous; testa hard or membranous, smooth, or marked with ridges and tubercles; perisperm abundant, fleshy or floury. *Embryo* antitropous, lenticular, in a cavity opposite the hilum; micropyle and radicle inferior.

Perennial, rarely annual, rush- or sedge-like herbs, with tufted or creeping rootstock densely covered with coriaceous sheaths increasing in size upwards; stems erect, or prostrate, solid or fistular with, or rarely without, leaf-sheaths, simple or variously branched; leaf-sheaths convolute, free at the margins, sometimes prolonged at the apex into a linear, straight or curved leaf; inflorescence simple, or generally cymose, much or little branched, alike in both sexes, or different in the male and female; spikelets solitary or generally numerous, each with a sheath-like spathe at the base, and similar spathellæ at the base of the branches in the panicled inflorescence; bracts overlapping or ultimately spreading, all fertile, or some barren, generally longer than the flower. The flowers are wind-fertilized.

DISTRIB. An order for the most part confined to the Coast Region of South Africa, to the corresponding south-western portion of Australia, and in New Zealand and Tasmania. One species has been found in Cochin China and one in Chili. Two or three of the genera occur in Australia, and also at the Cape, but no species is native of both countries. Two species have been found in Natal, and specimens without flowers have been received from Sir H. Johnston from Mlanji, in S. E. tropical Africa, where they were obtained at a height of 7000 ft.

The diœcious character of these plants, and the fact that the appearance of the individuals of one sex often differs considerably from that of the other, render the discrimination of the species difficult, and may have led in some cases to errors in matching the sexes. In many instances, moreover, only one sex is known, and the material is insufficient for an adequate description. The artificial keys to the genera, and especially to the species, must therefore be considered as more or less approximate indications only, and by no means to be relied on without concurrent reference to the more detailed descriptions. The descriptions of the leaf-sheaths are taken from those in the middle of the stem or branches, not from the base or apex, unless otherwise stated. In the same way the bracts from the middle of the spikelet are generally the subject of description. In the tables the spathes, spikelets, and bracts are, where not otherwise mentioned, described from the male plants.

I am indebted to the courtesy of Dr. W. Brehmer, of Lubeck, for the privilege of examining the entire set of Drège's Restiaceæ, which were lent to Kew for the purpose of making this work more complete.

Tribe 1. *RESTIOIDEÆ.* *Ovary* 3-, 2-, or by abortion 1-celled. *Fruit* dehiscent.

I. **Restio.**—*Inflorescence* spicate or panicled. *Male flowers* compressed; outer lateral perianth-segments conduplicate, carinate. *Leaf-sheath* persistent.
II. **Askidiosperma.**—*Inflorescence* densely spicate; bracts setaceo-acuminate; perianth-segments lanceolate, acuminate. *Leaf-sheaths* deciduous.
III. **Dovea.**—*Male spikelets* very numerous, panicled; female spikelets fewer. *Male flowers* 3-sided; inner perianth-segments longest; ovary 3-celled. *Leaf-sheaths* deciduous.

Tribe 2. *WILLDENOVIEÆ.* *Ovary* always 1-celled. *Fruit* indehiscent.
IV. **Elegia.**—*Male* and *female spikelets* numerous, mostly closely panicled; spathes usually deciduous. *Flowers* 3-sided; inner perianth-segments longest. *Fruit* 3-sided. *Stigmas* 3. *Leaf-sheaths* usually deciduous.
V. **Lamprocaulos.**—*Spikelets* numerous, panicled. *Flowers* compressed; inner perianth-segments longest. *Stigmas* 2. *Fruit* compressed, oblong. *Sheaths* and spathes persistent, golden-brown.
VI. **Leptocarpus.**—*Male* and *female spikelets* panicled. *Flowers* compressed; outer perianth-segments longest; style short. *Stigmas* 3. *Fruits* 3-cornered. *Leaf-sheaths* persistent.
VII. **Thamnochortus.**—*Male spikelets* numerous, panicled. *Female spikelets* fewer, spicately cymose, many-flowered. *Flowers* compressed; outer lateral perianth-segments deeply winged. *Fruit* compressed, sessile.
VIII. **Hypolæna.**—*Male spikelets* panicled. *Female* solitary, or spicately cymose, 1-flowered. *Flowers* on a slender stalk, rarely sessile; perianth-segments subequal. *Fruit* sessile.
IX. **Hypodiscus.**—*Male and female spikelets* similar, solitary, or spicately cymose. *Female flowers* solitary, on a thick, fleshy stalk. *Ovary* sessile, or stalked, 1-celled. *Fruit* smooth or tubercled, generally raised on a fleshy-lobed disc or stalk.
X. **Camnamois.** *Spikelets* dissimilar. *Male* numerous, paniculate-cymose. *Female flowers* few, solitary or spicate. *Styles* 2. *Fruit* oblong, compressed, sessile.
XI. **Willdenovia.**—*Male and female spikelets* very dissimilar. *Males* loosely paniculate, cymose; bracts loose. *Female* solitary, or two. *Female flowers* solitary. *Fruit* raised on a fleshy stalk.
XII. **Ceratocarpum.**—*Male and female inflorescence* dissimilar. *Male spikelets* numerous, thyrsoid. *Female spikelets* 2–3, spicate, each 1-flowered. *Male perianth-segments* linear, acute. *Female* hyaline, subequal. *Styles* 2. *Fruit* sessile.

DOUBTFUL GENUS.

XIII. **Anthochortus,** Nees—Male plant only known.

I. RESTIO, Linn.

Flowers diœcious, 1-bracteate. *Perianth* usually of 6, sometimes of 4, segments, in two rows, rarely none; segments generally unequal, outer lateral ones conduplicate; rigid, more or less carinate; anterior intermediate segment generally flat, membranous; inner segments usually shorter, equal, flat, membranous. *Filaments* 3, opposite the inner segments. *Pistillode* absent, or with 3 rudimentary styles. *Female flower: Perianth* as in the male. *Staminodes* 3, or none. *Ovary* 2- or 3-celled; styles 2 or 3, united at the base, or free. *Ovules* solitary, pendulous. *Fruit* capsular, 2- or 3-celled, or by abortion 1-celled. *Seed* 1 in each cell.

Perennial herbs throwing up stems from a tufted or creeping stock, covered at the base with sheathing leaves; stems terete, compressed, or four-sided, with remote sheathing leaf-sheaths, usually more or less mucronate, sometimes prolonged into a linear, blunt leaf; male and female inflorescence similar, or dissimilar, spicate, spikes solitary, or spikelets numerous, in spicate or panicled cymes, each spikelet with a sheath or spathe at the base; bracts in many rows, imbricating, or ultimately patent, all, or only some, fertile.

DISTRIB. South-Western Africa, and approximately the same number in South-Western Australia; the species being in all cases different.

Section 1. DICARPIA. Styles two, concrescent, rarely free at the base; staminodes and pistillodes absent from the female and male flowers respectively.

* Female spikelets usually many-flowered:
† Leaf-sheaths loosely convolute, open or spreading at the tips:
Male spikelets usually in dense heads or clusters; female spikelets few or clustered:
Bracts obtuse, not awned ... ... (1) **pauciflorus.**
Bracts awned:
Bracts oblong, lanceolate, acuminate; awn spreading ... ... (2) **cuspidatus.**
Bracts and spathes oblong, subulate-mucronate ... ... (3) **fraternus.**
Spathe acuminate, aristate; bracts oblong-acuminate, aristate; awn short, straight ... ... (4) **ocreatus.**
Male spikelets 3–9, in loose spikes or panicles:
Spathes short, acuminate-aristate; mucro scarcely, if at all, projecting beyond the bract... ... ... (5) **ferruginosus.**
Spathes aristate; bracts shortly mucronate; spikelets nearly half an inch long, awned ... ... (6) **elatus.**
Spikelets very short, straight; spathes acuminate; bracts mucronate ... (7) **gaudichaudianus.**
Male spikelets 1–2, rarely more:
Branches curved, loose, open; spikelets oblong, acute ... ... (8) **arcuatus.**
Branches straight, verticillate; male spikelets cylindrical, acute ... ... (9) **subverticillatus.**
†† Leaf-sheaths, except at the ramifications, closely convolute, appressed at the tips:
Woolly scales projecting beyond the sheaths:
Male spikelets solitary or twin ... ... (10) **laniger.**
Male spikelets numerous:
Male spikelets subcapitate, elongate, straight ... ... (11) **distractus.**
Male spikelets panicled, curved, short; spathes aristate ... ... (12) **venustulus.**
Woolly scales not projecting beyond the sheaths:
Male spikelets 2–5:
Stem slender, but not filiform:
Spathe and bracts of male flower with recurved awns... (13) **squarrosus.**
Spathe of female spikelet mucronate; bracts acuminate... (14) **schœnoides.**
Stem filiform:
Pedicel shorter than spikelet... (15) **oblongus.**
Pedicel longer than the spikelet; awn of sheath very long (16) **crinalis.**
Male spikelets numerous, small, densely clustered; female spikelets 1–2; bracts long awned ... ... (17) **setiger.**
Male spikelets numerous, loosely spicate, or panicled; female spikelets few:

Spikelets unlike; male linear; female oblong or ovoid:
Male spikelets in linear cymes (see also *R. setiger*):
Male spikelets straight:
Spathes and bracts aristate ... ... (18) **vilis.**
Male spikelets curved:
Bracts subulate, mucronate ... ... (19) **intermedius.**
Spathes aristate; bracts mucronate... (20) **Sieberi.**
Male and female spikelets alike, cylindrical:
Bracts obsoletely mucronate; female spikes 2–3 .. ... (21) **triflorus.**
Bracts subulate-mucronate; female spike solitary ... (22) **Helenæ.**
Male spikelets in spreading or compact cymes:
Inner perianth-segments shorter than the outer:
Spikelets crowded; spathes aristate; bracts subulate-mucronate ... ... (23) **pycnostachyus.**
Spikelets loosely panicled; spathes aristate; bracts subulate ... ... (24) **subfalcatus.**
Inner perianth-segments as long as the outer:
Spathes and bracts mucronate (25) **Neesii.**
Male and female spikelets numerous, in loose panicled or spicate cymes:
Spathes half the length of the spikelet:
Spikelets straight, narrow, acute; spathes aristate; bracts mucronate ... ... (26) **Ecklonii.**
Spikelets short, oblong, curved; spathes and bracts shortly mucronate ... .(27) **Wallichii.**
Spikelets large; spathes and bracts with long, recurved awns ... ... (28) **hystrix.**
Spathes nearly as long as the spikelet; awns not recurved (29) **macer.**

** Female spikelets usually few-flowered (see also *R. gaudichaudianus* and *R. intermedius*):
Perianth of female flower abortive ... ... (30) **rottbœlloides.**
Perianth of six segments:
Male and female spikelets numerous:
Apex of leaf-sheath prolonged into a leaf ... ... (31) **graminifolius.**
Apex of leaf-sheath not leafy:
Styles 2, free to the base; stems filiform:
Spathes and bracts membranous, lanceolate, decurrent... (32) **leptostachyus.**
Spathes and bracts oblong, mucronate-aristate, sub-coriaceous ... ... (33) **depauperatus.**

Styles 2, united at the base:
Branches curved; female flowers 2 or more ... ... (34) **curviramis.**
Branches straight; female flowers solitary ... ... (35) **monanthos.**
Male spikelets numerous; female spike solitary; stem capillary ... ... (36) **capillaris.**
Male and female spikelets few:
Perianth of six segments:
Branches slender, much curved ... (37) **cincinnatus.**
Branches few, slender, straight, or nearly so:
Spathes and bracts mucronate (38) **Ludwigii.**
Branches numerous, fasciculate, or subverticillate:
Perianth-segments obtuse ... (39) **leptoclados.**
Perianth-segments acute ... (40) **Eleocharis.**
Perianth of four segments:
Branches capillary ... ... (41) **tenuissimus.**

Section 2. TRICARPIA.—Styles 3, free to the base; staminodes strap-shaped; pistillode with three rudimentary styles.

* Male and female spikelets numerous, in spicate or panicled cymes:
Stems four-sided (see also *R. quadratus*) ... (42) **tetragonus.**
Stems terete or compressed:
Sheaths with a well-marked leafy mucro ... (43) **quinquefarius.**
Sheaths mucronate, not leafy:
Spathes as long as the spikelets ... ... (44) **dispar.**
Spathes shorter than the spikelets:
Female flowers with a disc ... ... (47) **præacutus.**
Female flowers with no disc:
Spikelets several, pear-shaped, obtuse; bracts obtuse, white-edged ... (46) **bifurcus.**
Spikelets several, ovoid-lanceolate or cylindric-oblong:
Bracts obtuse, apiculate, with a dark brown border ... ... (45) **furcatus.**
Bracts acute:
Inner perianth-segments concrescent:
Spathe half as long as the spikelet, distinctly mucronate; bracts oblong-lanceolate, scarcely mucronate ... ... (48) **triticeus.**
Inner perianth-segments free:
Spathes acuminate-aristate, nearly as long as the spikelet; bracts oblong-mucronate ... (49) **pannosus.**
Spathes mucronate; bracts acuminate ... ... (50) **multiflorus.**
Spikelets few, ovoid-oblong; stem filiform ... ... (51) **filiformis.**
Spikelets numerous, flattened, in dense panicles ... ... (52) **festuciformis.**
Spikelets widely separate, minute, on long stalks; stems filiform ... (53) **subtilis.**
Male and female spikelets ovate-oblong in linear cymes ... ... (54) **subulatus.**

**Male and female spikelets few in number, alike in form:
Stems slender, not filiform; mature spikelets at least ½ in. long:
Spikelets globose:
Bracts obovate-orbicular ... ... (55) **strobilifer.**
Bracts with a long acumen ... (56) **pachystachyus.**
Spikelets oblong, cylindric; bracts lanceolate ... ... (57) **purpurascens.**
Spikelets compressed, oblong; bracts subulate-mucronate ... ... (58) **bifidus.**
Stems filiform, very slender; mature spikelets small, ¼–½ in. long:
Stems erect; branches straight, or nearly so:
Spathe lanceolate, as long as the oblong, obtuse spikelet; bracts muticous ... ... (59) **miser.**
Spathe and bracts acuminate; keel of perianth-segments glabrous ... (60) **sonderianus.**
Spathe, bracts, and perianth-segments obtuse, scarcely mucronulate; keel glabrous ... ... (61) **pedicellatus.**
Stems trailing; branches curved, entangled:
Perianth-segments linear-oblong, acute ... ... (62) **implexus.**
Perianth-segments broadly oblong (63) **perplexus.**
Stems creeping; branches erect, straight (64) **Harveii.**
***Male spikelets numerous, in panicled cymes; female spikelets solitary or few:
Stems and branches terete:
Spikelets flattened:
Bracts recurved ... ... (65) **bifarius.**
Bracts lanceolate, flat; keel glabrous ... ... (66) **bigeminus.**
Spikelets cylindric, not compressed:
Branches few:
Bracts oblong, acuminate ... (67) **egregius.**
Bracts rounded ... ... (68) **obtusissimus.**
Branches numerous, ascending:
Spikelets ⅜ in. long ... ... (69) **sarocladus.**
Spikelets ¼ in. long ... ... (70) **Rhodocoma.**
Stems or branches compressed; spikelets and bracts flattened:
Sheaths subulate-mucronate:
Male spikelets 6–9 ... ... (71) **callistachyus.**
Male spikelets 1–4 ... ... (72) **spinulosus.**
Sheaths with a leafy mucro ... ... (73) **compressus.**
Stems or branches 4-sided; male and female spikelets panicled ... ... (74) **quadratus.**
Stems filiform; bracts nearly or quite muticous; spathe aristate, equalling the spikelet ... ... (75) **debilis.**

IMPERFECTLY KNOWN SPECIES.

Stems slender, erect, much-branched; sheaths close; perianth-segments narrow; ovary 3-styled... ... (76) **Scopula.**
Stems very slender; sheaths loose, black; spikelets and bracts linear; female not known ... ... (77) **ambiguus.**

Stems branched; branches flabellate; sheaths close; male spikelets numerous, panicled; bracts oblong-ovate; female unknown ... ... (78) **sejunctus.**
Stems branched; leaf-sheaths closely convolute, prolonged into a thick leafy process; female spikelets in terminal, linear cymes; ovary by abortion 2-lobed ... ... (79) **protractus.**
Stems branching, slender; sheaths close; male spikelets panicled, oblong; bracts oblong... ... ... (80) **patens.**

1. **R. pauciflorus** (Poir. Encycl. vi. 168); stems cylindric, slender, puberulous, branched; branches slender, subflexuose, leafy; sheaths oval, membranous, acuminate; male spikelets clustered, sessile or stalked, ovate, obtuse; spathe short; bracts ovate obtuse; anthers oblong obtuse. *Kunth, Enum.* iii. 412; *Steud. Synops.* ii. 254; *Mast. in Journ. Linn. Soc.* viii. 225; *DC. Monog. Phan.* i. 235.

SOUTH AFRICA: without locality, *Poiret.*

A species of which I have seen no specimen.

2. **R. cuspidatus** (Thunb. in Hoffm. Phytog. Blätter, i. 8); stems tufted, erect, somewhat tubercled; fertile 1–2 ft. high, terete, simple or rarely branched; sterile shorter, repeatedly branched; ultimate branchlets filiform, curved, pointed at the apex; leaf-sheaths ¾ in. long, loosely convolute, coriaceous above, dividing into two hyaline lobes with a central awn; leaf-sheaths on the sterile branchlets ending in a leafy, curved mucro; male spikelets 2–5, clustered at the ends of the branches, each about ½ in. long, cylindric-lanceolate or oblong-acute, straight or curved, with a short sheath-like spathe at the base; bracts oblong, acuminate, aristate; awn spreading; flowers shorter than the bracts, ovate-lanceolate; perianth-segments oblong-lanceolate: female spikelets 2–3, approximate, ½–¾ in. long, oblong-lanceolate, afterwards ovoid; bracts exceeding the ovate-lanceolate, pedicellate flowers; ovary oblong, flattened, 2-celled; style single dividing into two stigmatic branches; capsule 2-celled, dehiscing at the margins; seeds oblong-obtuse, trigonous; testa white, thickly spotted with purple. *Thunb. in Weber and Mohr, Archiv.* i. 26; *Fl. Cap. ed. Schultes*, 87; *Thunb. Herb. partly, ex Mast. in Journ. Linn. Soc.* xiv. 418; *Mast. in DC. Monog. Phan.* i. 233. *Restio squarrosus, Spreng. Syst. Veg.* i. 186, *not of Poiret; Nees in Linnæa*, v. 643; *Steudel in Flora*, 1829, i. 133. *R. lucæanus, Kunth, Enum.* iii. 385, *male fl.* *R. Sprengelii, Mast. in Journ. Linn. Soc.* viii. 224.

SOUTH AFRICA: without locality, *Bergius!* *Harvey*, 388 b, ♂! *Sieber*, ♂! *Drège*, 44 partly!

COAST REGION: Cape Div.; Cape Flats, *Burchell*, 39/1, ♂! 209, ♂! *Bolus*, 4453, ♂! Lion Mountain, *Burchell*. 120, ♂! Table Mountain, *Milne*, 223, ♂! *Bolus*, 2885, ♂! *Ecklon*, 848, ♂! *MacOwan, Herb. Aust. Afr.* 1565, ♂! Camps Bay, *Burchell*, 854! Devils Mountain, *Bolus*, 4463, ♀! Paarl Div.; between Paarl and Lady Grey Railway Bridge, *Drège*, 48 partly! Ceres Div.; mountains near Ceres, 1800 ft., *Bolus*, 548?, ♂! Worcester Div., *Zeyher*, ♀! Caledon Div.;

between Genadendal and Donker Hoek, *Burchell*, 7926 partly, ♀! between Donker Hoek and Houw Hoek, *Burchell*, 8017! near Grietjes Gat, between Lowry's Pass and Palmiet River, 2000–4000 ft., *Zeyher!* Swellendam Div.; dry hills, by the Breede River, *Burchell*, 7474! Humansdorp Div.; Kromme River Heights, *Bolus*, 2499, ♀!

3. **R. fraternus** (Kunth, Enum. iii. 386); stems 2 ft. high, erect, branching, puncticulate; leaf-sheaths ½ in. long, coriaceous; apex membranous, mucronate; male spikelets clustered, sessile or slightly pedunculate, each about ¼ in. long, ovoid; spathe sheath-like; bracts ovate, acuminate, awned, with a deep white margin; perianth-segments oblong, acute; inner flatter, smaller; anthers oblong, apiculate; female spikelets solitary, linear-oblong. *Mast. in DC. Monog.* i. 235.

COAST REGION: Stellenbosch Div.; mountains near Stellenbosch, 3000–4000 ft., *Drège*, 45! Worcester Div.; Dutoits Kloof, 1000–2000 ft., *Drège*, 1623, ♂! 47, ♀!

4. **R. ocreatus** (Kunth, Enum. iii. 385); stems terete, solid, 2–3 ft. high, slightly wrinkled and furrowed, finely puncticulate; leaf-sheaths ¾ in. long, coriaceous, loosely convolute, acuminate, aristate, spreading near the apex; male spikelets numerous, crowded in dense spike-like cymes, each about 2 in. long; spathe sheath-like, longer than the oblong, acute, straight or curved spikelet; bracts shining brown, oblong, acuminate, point recurved; flowers oblong-lanceolate, shorter than the bracts; perianth-segments oblong, acute; lateral conduplicate, villous; inner nearly equal, oblong, hyaline; anthers ovate, acute; female spikelets one or two at the end of the branch, sessile or stalked, ¾ in. long, cylindric-oblong or subglobose; spathe oblong, not much longer than the spikelet; ovary ovoid, compressed, with 2 (rarely 3) styles; capsule as long as the persistent perianth-segments; seeds ovoid trigonous; testa grey, with white tubercles. *Mast. in Journ. Linn. Soc.* viii. 225; *DC. Monog. Phan.* i. 234. *Calopsis peronata, Kunth, Enum.* iii. 426, *male plant. R. Lehmanni, Nees in Linnæa,* xx. 241, *name only.*

SOUTH AFRICA: without locality, *Drège*, 42, ♂!

CENTRAL REGION: Calvinia Div.; Bokkeveld Mountains at Uien Vallei, 2000–2500 ft., *Zeyher*, 1207, ♂ and ♀! *Drège*, 1623? ♂ and ♀! *Drège*, 2505, ♀!

5. **R. ferruginosus** (Link ex Kunth, Enum. iii. 393); stems 2–3 ft. high, much branched; branches fasciculate, slender, puncticulate; leaf-sheaths about ½ in. long, ovate, acute, striate, membranous at the margins, spreading at the apex, smaller sheaths with a two-lobed hyaline apex and a foliaceous reflexed mucro; spikelets 1–5 or more in a spicate or paniculate cyme, each about ¼ in. long, curved, cylindric-lanceolate; spathe sheath-like, mucronate, much shorter than the spikelet; bracts oblong, shortly mucronate, coriaceous, brown; margins brown, membranous; perianth-segments linear-oblong; lateral conduplicate, villous-keeled; inner thinner and more

slender; female spikelets 1–3, each about ¼ in. long, oblong obpyramidal; spathe bract-like, shorter than the spikelet; bracts broadly ovate, acute; perianth-segments as in the male; ovary ovoid; capsule oblique, subglobose, by abortion 1-celled, as long as the persistent perianth-segments; seed compressed, pitted. *Steud. Synops.* ii. 252; *Mast. in Journ. Linn. Soc.* viii. 226. *R. ameles, Steud. Synops.* ii. 252. *R. dichotomus, β, Thunb. Herb. ex Mast. in Journ. Linn. Soc.* xiv. 420.

COAST REGION: Worcester Div.; Dutoits Kloof, 1000–4000 ft., *Drège*, 1619b! Caledon Div.; mountains of Baviaans Kloof, near Genadendal, *Burchell*, 7632!

6. **R. elatus** (Mast. in Journ. Linn. Soc. viii. 226); stems 3–4 ft. high, erect, terete, branched, puncticulate; leaf-sheaths about 1 in. long, coriaceous, ovate, acuminate, aristate; margins hyaline; smaller sheaths on the branching stems foliaceous-mucronate; male spikelets 6–8 in a spicate cyme, each about ½ in. long, cylindric-lanceolate, arcuate, erect or spreading; spathe bract-like, much shorter than the spikelet; bracts oblong, obtuse, subulate-mucronate; margins membranous; perianth-segments oblong; lateral conduplicate, villous-keeled; inner one-half shorter, membranous; anthers ovate, apiculate; female spikelets 4–6, approximate, sessile or pedicellate, each nearly ½ in. long, ovoid; bracts ovate-mucronate; perianth-segments as in the male, but broader; ovary oblong; style thick, terete, dividing into two stigmatic branches. *Mast. in DC. Monog. Phan.* i. 246.

COAST REGION: Paarl Div.; Paarl Mountains, 1000–2000 ft., *Drège*, 93, ♂ and ♀! Tulbagh Div.; New Kloof, near Tulbagh, *Burchell*, 1023, ♀! Worcester Div.; arid mountains near Hex River, *Bolus*, 4235, ♂! 4234, ♀!

7. **R. gaudichaudianus** (Kunth, Enum. iii. 387); stems 2–3 ft. high, branching from the middle; branches fasciculate, purple-spotted; leaf sheaths about ¾ in. long, closely convolute, coriaceous, subulate acuminate, membranous at the margins; male spikelets 3–9 or more, in linear, spicate cymes, each about ¼ in. long, cylindric, acute, straight or curved, with a short, oblong, acuminate spathe; bracts ovate-oblong, shortly mucronate, coriaceous, deep brown, margins slightly membranous; perianth-segments oblong, obtuse; outer lateral conduplicate, villous-keeled; anthers linear-oblong, mucronulate; female spikelets 2–3, sessile or pedicellate, each 1 line long; 1–3-flowered, ovoid, ultimately clavate; perianth-segments oblong, acute; lateral villous-keeled; capsule obliquely ovate, 1-celled by abortion; seed oblong, obtuse; testa grey, pitted and purple-spotted. *Steud. Synops.* ii. 250; *Mast. in Journ. Linn. Soc.* viii. 237; *DC. Monog. Phan.* i. 246.

VAR. β, **microstachys** (Mast. in DC. Monog. Phan., i. 247); stems more branching; sterile ones leafy; male spikelets very short. *Restio microstachys, Nees MSS. in herbaria.*

COAST REGION: Paarl Div.; Paarl Mountains, *Drège*, 1619a, ♂! Worcester Div.; Drakenstein Mountains near Bains Kloof, 1600–2000 ft., *Bolus*, 4107, ♀!

4096! Var. *β*, Tulbagh Div.; near Tulbagh Waterfall, *Zeyher*, ♂! Worcester Div.; Dutoits Kloof, 1000–2000 ft., *Drège*, 52, ♂, 58, ♂ and ♀!

A variable species differing much in stature and degree of branching; the size of the spikelets often varies considerably in the same specimen.

8. **R. arcuatus** (Mast. in DC. Monog. Phan. i. 247); stems 2–3 ft. high, branching; branchlets slender, ascending, curved, somewhat compressed, purple-spotted; sheaths $\frac{1}{4}$–$\frac{1}{2}$ in. long, loosely convolute, coriaceous, subulate-mucronate, hyaline at the apex; male spikelets about $\frac{1}{8}$ in. long, ovate-oblong, ultimately cuneate; spathe flat, oblong, mucronate, half the length of the spikelet; bracts oblong-ovate, coriaceous, hyaline at the margins, shortly mucronate; perianth-segments oblong, obtuse; lateral conduplicate, glabrous; inner shorter, hyaline.

COAST REGION: Swellendam Div.; on a mountain peak near Swellendam, *Burchell*, 7306, ♂!

Perhaps referable to *Hypolæna*.

9. **R. subverticillatus** (Mast. in Journ. Soc. Linn. viii. 227); stems erect, 3–4 ft. high, stout, terete, minutely purple-spotted, much branched; branches verticillate, dense, slender, filiform, spreading; sheaths about 1 in. long, coriaceous, striate above, membranous, spreading, acuminate; smaller sheaths foliaceous from beneath a two-lobed hyaline apex; male spikelets 1–6 in a simple spike-like cyme, each nearly $\frac{1}{2}$ in. long, erect or spreading, cylindric-lanceolate, straight or curved; spathe sheath-like, subulate, mucronate, much shorter than the spikelet; bracts oblong, acute, mucronate, brown; margins membranous, lacerate; perianth-segments broadly oblong; lateral villous-keeled; female plant like the male, but less branched; spikelets 1–3, sessile or pedunculate, $\frac{1}{4}$ in. long, elliptic, ultimately clavate; perianth-segments broader than in the male; ovary rounded, compressed; capsule obliquely ovate, 1-celled, 1-seeded. *Mast. in DC. Monog. Phan.* i. 248. *Ischyrolepis subverticillata*, *Steud. Synops.* ii. 249, *female plant*. *Restio ferruginosus*, *Kunth*, *Enum.* iii. 393, *as to the female plant, not of Link. R. casuariniformis, Nees in herbaria. Willdenovia teres of European gardens, not of Thunberg.*

SOUTH AFRICA: without locality, *Thom*, ♀!
COAST REGION: Worcester Div.; Drakenstein Mountains near Bains Kloof, 1600–2000 ft., *Bolus*, 4084, ♂! Dutoits Kloof, 2000–3000 ft., *Drège*, 1620, ♂ and ♀! Mountains above Worcester, *Rehmann*, 2572! Stellenbosch Div.; Hottentots Holland Mountains, 1000–3000 ft., *Zeyher!* Caledon Div.; Nieuw Kloof, Houw Hoek Mountains, *Burchell*, 8112, ♂! Genadendal, 2000–3000 ft., *Drège*, 1, ♂ and ♀!

Variable in stature and habit.

10. **R. laniger** (Kunth, Enum. iii. 386); stems cæspitose, 18–24 in. high, terete, slightly branched, yellowish; leaf-sheaths about $\frac{1}{2}$ in.

long, closely convolute, enclosing a woolly sheath, coriaceous; margins membranous; apex mucronate; male spikelets 2 or more, approximate, each about ½ in. long, erect or spreading, cylindric, acute, straight or slightly curved, twice the length of the obtuse, sheath-like spathe; bracts ovate-oblong, mucronate; perianth-segments oblong; lateral villous-keeled; anthers ovate, apiculate; female spikelets 1–3, like the males, ultimately ovoid; bracts oblong-ovate, coriaceous; margins membranous; apex acuminate-aristate; flowers oblong, ultimately longer than the bracts; perianth-segments oblong-lanceolate; lateral conduplicate, villous-keeled; inner flat or wrapping round the style; ovary ovoid; style bifid; capsule 1-celled by abortion, oblique; seed grey, pitted. *Steud. Synops.* ii. 250; *Mast. in Journ. Linn. Soc.* viii. 230; *DC. Monog. Phan.* i. 238.

COAST REGION: Clanwilliam Div.; Gift Berg, 1500–2500 ft., *Drège*, 51, ♂!
CENTRAL REGION: Prince Albert Div.; near Vrolykheid on the Great Zwart Bergen, 5000 ft., *Drège*, 36, ♀!

11. **R. distractus** (Mast.); stem 18–24 in. high, erect, terete, obsoletely tubercled, fastigiately branched; branches slender, spreading; leaf-sheaths about ½ in. long, closely convolute, lanceolate, aristate, coriaceous, except at the membranous edges, in many cases showing a woolly scale within; male spikelets 3–4 lin. long, cylindric-lanceolate, aggregate in terminal heads; spathe sheath-like, nearly as long as the spikelet; bracts oblong, acuminate, rather longer than the flower; perianth-segments linear-oblong; outer lateral conduplicate, villous-keeled; inner shorter; anthers linear, apiculate; female spikelet solitary or twin, at first cylindric-elliptic, ultimately ovoid; perianth nearly as in the male; inner segments partly convolute; styles 2.

CENTRAL REGION: Graaff Reinet Div.; among rocks at the top of Mount Gnadouw near Graaff Reinet, 6700 ft., *Bolus*, 2633, ♂ and ♀!

12. **R. venustulus** (Kunth, Enum. iii. 388); stems 3 ft. high, erect, terete, branched; branches ascending, puncticulate; leaf-sheaths about ½ in. long, coriaceous, ferruginous, prolonged into a long slender awn, and enclosing a villous scale; male spikelets numerous, in a linear cyme, each spikelet about ¼ in. long, cylindric, acute, with a short, aristate spathe at the base; bracts ovate, mucronate; flowers stipitate, ovoid; perianth-segments oblong, acute; lateral villous-keeled; inner smaller, flat; anthers linear-oblong, apiculate; female spikelets solitary or two, ¼ in. long, ovoid, with a short, bract-like spathe at the base; bracts oblong, coriaceous; flowers raised on short, villous pedicels; perianth-segments lanceolate; lateral villous-keeled; inner slightly shorter, slightly involute at the margins; ovary flattened, rounded, surmounted by a bifid style. *Steud. Synops.* ii. 250; *Mast. in Journ. Linn. Soc.* viii. 231; *DC. Monog. Phan.* i. 240. *R. scoparius*, *Kunth*, *Enum.*

iii. 390; *Steud. Synops.* ii. 251; *Mast. in Journ. Linn. Soc.* viii. 230; *DC. Monog. Phan.* i. 240.

SOUTH AFRICA: without locality, *Drège*, 73, ♀!
COAST REGION: Clanwilliam Div.; Ceder Bergen, near Ezels Bank, 4000–5000 ft., *Drège*, 2496, ♀! Worcester Div.; between the Bokke Veld and Hex River, 3000 ft., *Drège*, 9616, ♂! Dutoits Kloof, 3000–4000 ft., *Drège*, 1608 partly, ♀! Constable, 3000-3500 ft., *Drège*, 9450, ♂ and ♀!

13. R. squarrosus (Poir. in Lam. Encycl. vi. 174); stems 2 ft. high, slender, simple or sparingly branched, terete, coarsely tubercled; sheaths about ½ in. long, tightly convolute, terminating in a subulate, appressed mucro; male spikelets 2–5, clustered in a one-sided cyme, each about ¼ in. long, oblong, erect or spreading, supported at the base by a short, bract-like spathe; bracts ovate-lanceolate, prolonged at the apex into a long recurved awn; flowers oblong-lanceolate, shorter than the bracts; outer perianth-segments oblong-lanceolate; lateral conduplicate, villous-keeled; inner subequal, flatter; anthers linear-oblong, apiculate; ovary ovoid-globose; style bifid; capsule 2-celled. *Mast. in Journ. Linn. Soc.* viii. 228; *DC. Monog. Phan.* 236. *Restio echinatus, Kunth, Enum.* iii. 384; *Steud. Synop.* ii. 250.

COAST REGION: Worcester Div.; Dutoits Kloof, 3000–4000 ft., *Drège*, 49 partly, ♂!

14. R. schœnoides (Kunth, Enum. iii. 391); stems decumbent erect, 18 in. high, slender, simple or slightly branched; leaf-sheaths ½ in. long, closely convolute, elliptic, acute, coriaceous beneath, membranous and lacerate near the apex; male spikelets 3–4, each about ½ in. long, oblong-ovate, acute, with a short sheath-like spathe at the base; bracts oblong, acuminate, coriaceous, membranous at the edges; flowers curved; perianth-segments oblong, acute; lateral villous-keeled; inner smaller, hyaline; female spikelet solitary, ½ in. long, ovate-oblong, with a spathe of equal length at the base; bracts lanceolate, acuminate; margins membranous, lacerate; flowers lanceolate; perianth-segments lanceolate; lateral villous-keeled; inner thinner, flat or involute at the margins; ovary rounded; styles 2. *Steud. Synops.* ii. 251; *Mast. in Journ. Linn. Soc.* viii. 228; *DC. Monog. Phan.* i. 237.

CENTRAL REGION: Aliwal North Div.; Witte Bergen, 5000–6000 ft., *Drège*, 50, ♀!

15. R. oblongus (Mast. in Journ. Linn. Soc. viii. 229); stems 1 ft. high, very slender, decumbent, compressed, purple, with white spots, branching from near the base; leaf-sheaths ½–¾ in. long, coriaceous, mucronate-aristate; male spikelets 2–3, clustered, sessile, or pedicellate, each about ¼ in. long, oblong, shorter than the lanceolate, spreading spathe; bracts lanceolate, longer than the curved flowers; outer perianth-segments oblong, acute; lateral conduplicate, keeled; keel glabrous; inner similar, thinner, involute at the margins; anthers oblong-ovate. *Mast. in DC. Monog. Phan.* i. 239.

SOUTH AFRICA: without locality, *Drège*, 65, ♂!

This may be the male of some species of *Hypolæna*.

16. **R. crinalis** (Mast. in Journ. Linn. Soc. viii. 229); stems erect from a decumbent base, filiform, terete, coarsely tubercled, much branched from the base; leaf-sheaths coriaceous, closely convolute, about 1 in. long, with a hair-like awn of the same length; male spikelets clustered, sessile or pedicellate, each about ¼ in. long, oblong-ovate, or oblong, provided at the base with an open, sheath-like spathe, terminating in a very long filiform awn; bracts lanceolate, submembranous, slightly exceeding the flowers; perianth-segments lanceolate; outer lateral keeled; keel glabrous; inner shorter; anthers oblong. *Mast. in DC. Monog. Phan.* i. 239.

SOUTH AFRICA: without locality, *Drège*, 11, ♂!

Perhaps the male of a species of *Hypolæna*.

17. **R. setiger** (Kunth, Enum. iii. 385, female pl.); stem 2 ft. high, erect, simple or sparingly branched, puncticulate, white-spotted; sheaths about 1 in. long, elliptic, striate, tubercled, coriaceous beneath, hyaline above, and ending in a long awn; male spikelets clustered, each about ¼ in. long, cylindric-lanceolate; spathe open, sheath-like, aristate, as long as the spikelet; bracts oblong, hyaline above, ending in a long, spreading awn; flowers oblong-lanceolate; outer perianth-segments oblong, acuminate; lateral conduplicate, villous-keeled; inner oblong, flat, as long as the outer; anthers oblong; female spikelets solitary or two, each about 1 in. long, cylindric-lanceolate; spathe sheath-like, open, shorter than the spikelet; bracts oblong, hyaline at the margins, ending in a long, spreading awn; flowers oblong-lanceolate; outer perianth-segments oblong-lanceolate; lateral conduplicate, villous-keeled; inner flat or slightly involute at the margins, sometimes provided in the centre with a tubercle; ovary turbinate or compressed; style simple or dividing into 2 stigmatic branches; capsule 2-celled; seed tubercled. *Steud. Synops.* ii. 250; *Mast. in Journ. Linn. Soc.* viii. 228; *DC. Monog. Phan.* i. 236. *R. fuirenoides, Kunth, Enum.* iii. 386; *Steud. Synop.* ii. 250, *male plant.*

COAST REGION: Clanwilliam Div.; near Honig Valley on the Koude Berg, 3000–4000 ft., *Drège*, 2504, ♂! 2503, ♀!

18. **R. vilis** (Kunth, Enum. iii. 389); stems 2 ft. high, erect, simple or branched towards the base; branches erect, filiform, olivaceous, with disc-like white spots; leaf-sheaths ¾ in. long, closely convolute, coriaceous, ferruginous, deeply hyaline near the top, slenderly-aristate; male spikelets numerous, loosely aggregate into a linear, spicate cyme 1–2 in. long, each ¼ in. long, cylindric, acute, with a short, acuminate spathe at the base; bracts oblong, acute

acuminate; perianth-segments rather rigid, oblong, obtuse; lateral conduplicate, villous-keeled; female spikelets 2–3 sessile or pedunculate, each ¼ in. long, ovate-oblong; bracts acuminate; ovary oblong; style bifid. *Steud. Synops.* ii. 251; *Mast. in Journ. Linn. Soc.* viii. 231; *DC. Monog. Phan.* i. 241.

SOUTH AFRICA: without locality, *Drège*, 87!
WESTERN REGION: Little Namaqualand; Kamies Bergen, 3000–4000 ft., *Drège*, 2476, ♂ and ♀!

19. **R. intermedius** (Kunth, Enum. iii. 388); stems erect, 2–3 ft. high, slender, branched above the middle, terete, olive-coloured, faintly tubercled; leaf-sheaths about ½ in. long, closely convolute, coriaceous, subulate, mucronate; male spikelets 5–9 in linear, distichous spicate cymes, 2 in. long; each spikelet ¼–½ in. long, cylindric, acute, straight or curved, with a bract-like spathe at the base; bracts ovate, mucronate; flowers ovate-lanceolate, compressed; perianth-segments rigid, oblong-lanceolate; outer lateral conduplicate, villous-keeled; inner smaller; anthers oblong, apiculate; female spikelets in an erect, spike-like cyme, about 1 in. long; each spikelet oblong, becoming pyriform; ovary obliquely ovoid; style simple; stigma bifid; capsule 1-celled by abortion. *Steud. Synops.* ii. 251; *Mast. in Journ. Linn. Soc.* viii. 231; *DC. Monog. Phan.* i. 242.

SOUTH AFRICA: without locality, *Drège*, 81, ♀! 2470, ♂!
COAST REGION: Clanwilliam Div.; between Jackhals River and Klip Fontein, *Drège*, 2475, ♂ and ♀! Worcester Div.; mountains above Worcester, *Rehmann*, 2562, ♂!

20. **R. Sieberi** (Kunth, Enum. iii. 387); stems 12–18 in. high, terete, branched, olive-coloured, puncticulate; leaf-sheaths closely convolute, coriaceous, scarious at the margins, mucronate-aristate; male spikelets 1–5, each ¾ in. long, erect or spreading, straight or slightly curved, cylindric, acute, with a sheath-like, long-awned spathe at the base; bracts ovate-oblong, coriaceous, mucronate; flowers oblong; perianth-segments oblong, obtuse; lateral villous-keeled; intermediate flattish, villous in the middle above; inner smaller, flat; female spikes 1–2, ovate, acute, about ½ in. long; bracts oblong-ovate, subulate, mucronate; ovary ovoid; styles 2; capsule obliquely ovoid, 1-celled by abortion; seed large, grey, tuberculate. *Steud. Synops.* ii. 250; *Mast. in Journ. Linn. Soc.* viii. 235; *DC. Monog. Phan.* i. 244.

SOUTH AFRICA: without locality, *Sieber*, 228, ♂! *Drège*, 80, ♀! 9617, ♂!
COAST REGION: Cape Div.; Table Mountain, *Burchell*, 536, ♂! Worcester Div.; Drakenstein Mountains, near Bains Kloof, 1600–2000 ft., *Bolus*, 4090, ♀! 4090β, ♀! 4104, ♀! Dutoits Kloof, 1000–2500 ft., *Drège*, 1621, ♂! 9615, ♀!
CENTRAL REGION: Graaff Reinet Div.; summit of Koudeveld Berg, Sneeuw Berg Range, 5500 ft., *Bolus*, 2584, ♀!

21. **R. triflorus** (Rottboell, Descr. et Icon. 3, t. 2, fig. 2); stems cæspitose; sterile much branched, diffuse, leaf-bearing; fertile

1 ft. long, less branched, erect, terete, slender, rather coarsely tubercled; sheaths about ½ in. long, coriaceous, striate above, membranous, mucronate-acuminate; sheaths of the sterile branches shorter; apex hyaline, 2-lobed; mucro foliaceous, recurved; male spikelets 2–6 in linear, spicate cymes each about ½ in. long, cylindric-lanceolate; spathe bract-like; bracts ovate-oblong, obtuse, coriaceous, membranous above, shortly mucronate or muticous; flowers oblong-lanceolate; outer perianth-segments lanceolate, pointed; keel villous; inner subequal, smaller; anthers linear-oblong, apiculate; female spikelets 2–3, each about ¼ in. long, oblong, acute, many-flowered; bracts and perianth as in the male; ovary compressed, orbicular, 2-lobed; style terete, dividing into two branches. *Thunb. Diss.* 16; *Fl. Cap. ed. Schult.* i. 86; *Spreng. Syst. Veg.* i. 185; *Kunth, Enum.* iii. 391, *female pl.*; *Mast. in Journ. Linn. Soc.* viii. 234; *DC. Monog. Phan.* 249. *R. dichotomus, Thunb. var. β, Nees in Linnæa,* v. 641. *R. Kunthii, Steud. Synops.* ii. 251.

SOUTH AFRICA: without locality, *Sieber*, 115, ♀! *Thunberg! Masson!*
COAST REGION: Cape Div.; Devils Mountain, *Ecklon*, 845, ♀! Camps Bay, *Burchell*, 853, ♂! Paarl Div.; Paarl Mountains, 1000–2000 ft., *Drège*, 69, ♂ and ♀! 1617, ♂ and ♀! Caledon Div.; between Genadendal and Donker Hoek, *Burchell*, 7926, ♂! between Donker Hoek and Houw Hoek Mountains, *Burchell*, 8017, ♂! Riversdale Div.; hills near Zoetemelks River, *Burchell*, 6800, ♂! Alexandria Div.; Addo, *Drège*, 78!

[The plant collected on Devils Mountain by *Ecklon* (No. 845), and that collected at Addo by *Drège* (No. 78), are in every respect identical, and there is probably some error as to the locality given by Drège.—N. E. Brown.]

22. **R. Helenæ** (Mast. in Journ. Linn. Soc. viii. 233); stems 12–18 in. high, erect from a decumbent base, terete, branched above the base; branches compressed, olivaceous, with white tubercles, ultimate branches slender, flexuose, ascending; sheaths elliptic, cuspidate, coriaceous; margins membranous; smaller sheaths 2-lobed, membranous at the foliaceo-mucronate apex; male spikelets numerous, in linear cymes, each about ½ in. long, cylindric, acute; spathe flat, shorter than the spikelet; bracts oblong, mucronate, acuminate, coriaceous, brown; margins membranous; flowers oblong, obtuse; perianth-segments oblong, obtuse; lateral conduplicate, villous-keeled; inner smaller, hyaline; female spikelets solitary, ¾ in. long, cylindric-oblong, many-flowered; perianth-segments more acute than in the male; ovary rounded; styles 2. *Mast. in DC. Monog. Phan.* i. 249.

COAST REGION: Worcester Div.; Dutoits Kloof, 1000–2000 ft., *Drège*, 1616, ♂! 1618, ♀!

23. **R. pycnostachyus** (Mast. in Journ. Linn. Soc. viii. 232); stems cæspitose, erect, terete; sterile much branched, leaf-bearing; fertile long, faintly puncticulate; leaf-sheaths ½ in. long and upwards, elliptic, coriaceous, membranous at the apex, aristate; male spikelets numerous, in a dense, broad, paniculate

cyme, each spikelet $\frac{1}{4}$ in. long, cylindric-lanceolate; spathe ovate, acuminate, membranous at the edges, half the length of the spikelet; bracts ovate, acute, mucronulate; perianth-segments oblong, acute; lateral conduplicate, villous-keeled; inner shorter, membranous; anthers mucronulate. *Mast. in DC. Monog. Phan.* i. 243.

COAST REGION: Worcester Div.; Dutoits Kloof, 3000–4000 ft., *Drège*, 79, ♂ (not 75 as previously quoted)!

24. R. subfalcatus (Nees ex Mast. in Journ. Linn. Soc. viii. 231); sterile stems compactly branched, leaf-bearing; leaves linear, arcuate; fertile stems longer, less branched, puncticulate; leaf-sheaths nearly 1 in. long, closely convolute, elliptic, coriaceous, finely striated, membranous near the top, aristate; male spikelets very numerous arranged in a somewhat spreading, panicled cyme 2 in. long, each spikelet nearly $\frac{1}{2}$ in. long, cylindric-lanceolate, arcuate; spathe open, oblong-lanceolate, aristate, shorter than the spikelet; bracts oblong, acute, coriaceous, membranous at the top, subulate, mucronate; outer perianth-segments oblong, acute; lateral conduplicate, with a villous keel; inner 3 shorter, hyaline, flat; anthers oblong, mucronulate; female spikelets 3–4, ovoid, each about $\frac{1}{4}$ in. long, with an open, ovate-acute, aristate spathe; perianth-segments as in the male; ovary subglobose, compressed, 2-celled; style terete, dividing into two stigmatic branches. *Mast. in DC. Monog. Phan.* i. 243.

COAST REGION: Clanwilliam Div.; Oliphants River, *Zeyher*, ♂ and ♀! between Grasberg River and Watervals River, 2500–3000 ft., *Drège*, 74, ♂! and 2500, ♂, partly (mixed with *Leptocarpus peronatus*)! Worcester Div.; Dutoits Kloof, 3000–4000 ft., *Drège*, 75, ♂! Drakenstein Mountains, near Bains Kloof, 1600–2000 ft., *Bolus*, 4083, ♂!

25. R. Neesii (Mast. in Journ. Linn. Soc. viii. 232); stems 1$\frac{1}{2}$–2 ft. long, erect, branching in the centre, puncticulate; leaf-sheaths about $\frac{3}{4}$ in. long, oblong, coriaceous, with white spots, membranous at the apex, aristate; awn appressed; male spikelets numerous, in a dense, panicled cyme; each spikelet about 1 line long, with an open, sheath-like spathe at the base; bracts ovate, subulate-mucronate, coriaceous, dark brown, membranous at the margins; perianth-segments oblong, obtuse; lateral conduplicate, villous-keeled; inner flat, slightly longer than the outer; anthers linear-oblong, apiculate; female spikelets 3–5 in a broad, little-branched cyme, each about a line long, ovoid, eventually clavate; inner perianth-segments as long as the outer; ovary oblong, compressed; style terete; capsule obliquely ovate, 1-celled by abortion. *Mast. in DC. Monog. Phan.* i. 244.

COAST REGION: Tulbagh Div.; near Tulbagh Waterfall, *Ecklon and Zeyher*, ♀! Winter Hoek Mountain, *Ecklon and Zeyher*, ♂!

26. R. Ecklonii (Mast. in Journ. Linn. Soc. viii. 236); stems cæspitose, 1–2 ft. long, slender, terete, branched from the middle, olivaceous, tuberculate; leaf-sheaths ½–¾ in. long, closely convolute, subulate-mucronate, coriaceous; apex membranous; male spikelets 6–8 in linear erect cymes, each spikelet about ½ in. long, cylindric, acute; spathe oblong, aristate, half the length of the spikelet; bracts oblong, mucronate, dark brown, coriaceous; margin slightly membranous; flowers oblong-lanceolate; outer perianth-segments oblong-lanceolate; lateral conduplicate, villous-keeled; inner hyaline; anthers linear-apiculate. *Mast. in DC. Monog. Phan.* i. 250.

SOUTH AFRICA: without locality, *Ecklon*, 85, ♂!
COAST REGION: Clanwilliam Div.; Ceder Bergen, *Drège*, 2497, ♀ (not ♂ as previously quoted)! Cape Div.; vicinity of Cape Town, *Burchell*, 456, ♂! Caledon Div.; between Bot River and the Zwart Berg, *Ecklon and Zeyher*, ♂!

27. **R. Wallichii** (Mast. in Journ. Linn. Soc. viii. 234); stems cæspitose, 18–24 in. high, erect, slender, branched below the middle; branches ascending, faintly wrinkled; leaf-sheaths about ½ in. long, closely convolute, subulate, mucronate, coriaceous, striate; margins slightly membranous, lacerate; male spikelets very numerous in linear panicled cymes, each spikelet about ¼ in. long; spathe and bract subulate, mucronate; female spikelets in linear cymes, each about ½ in. long, cylindric, acute, erect or slightly spreading; spathe oblong, mucronate, membranous on the upper margins, half the length of the spikelet; bracts oblong, subulate; apex obtuse, membranous; flowers lanceolate; perianth-segments lanceolate; lateral conduplicate, villous-keeled; inner flatter, shorter, more membranous; ovary ovoid; style simple or dividing into two stigmatic branches. *Mast. in DC. Monog. Phan.* i. 251.

SOUTH AFRICA: without locality, *Wallich*, ♀!
COAST REGION: Cape Div.; Cape Flats, *Buchanan*, 157, ♂!

28. **R. hystrix** (Mast. in Journ. Linn. Soc. x. 276); stems 3–4 ft. high, erect, terete, slightly branched above the middle, yellowish, faintly wrinkled and puncticulate; leaf-sheaths ¾ in. long, closely convolute, aristate, coriaceous, upper portion membranous, lacerate; male spikelets numerous, in an erect much-branched panicled cyme; spathe oblong, coriaceous, half the length of the spikelet, brown, with a long awn; spikelet nearly ½ in. long, slender, cylindric-lanceolate; bracts like the spathe; flowers lanceolate; perianth-segments oblong, acute; lateral conduplicate, villous-keeled; inner thinner, shorter; anthers linear, apiculate; female spikelets numerous, in slightly branched, panicled cymes, each ¾–1 in. long, oblong, protected at the base by an open spathe; bracts and perianth-segments as in the male, but larger; capsule obliquely ovate, compressed, 1-celled by abortion, surrounded by the persistent perianth; style short; stigmas 2. *Mast. in DC. Monog. Phan.* i. 252.

COAST REGION: Riversdale Div.; lower part of the Lange Bergen, near Kampsche Berg, *Burchell*, 6991, ♂ and ♀!

29. R. macer (Kunth, Enum. iii. 390); stems cæspitose, 12–18 in. high, slender, erect, branching from below the middle; branches ascending, terete, yellowish, with white tubercles; sterile branches shorter, more branched; sheaths about ¼ in. long, closely convolute, mucronate, coriaceous, membranous and lacerate above; mucro sometimes leafy; male spikelets 4–9 in linear erect cymes; spathe oblong, mucronate, membranous at the margins, half as long as the spikelet; spikelet about ½ in. long, cylindric-lanceolate; bracts oblong, acute, mucronato-acuminate, coriaceous, membranous at the margins; flowers oblong; perianth-segments oblong, obtuse; lateral conduplicate, villous-keeled; inner similar, flattish, longer than the outer segments; anthers linear, apiculate; female spikelets like the males; inner perianth-segments shorter than the outer; ovary rounded, 2-celled; style bifid; capsule ovoid, bilocular. *Steud. Synops.* ii. 251; *Mast. in Journ. Linn. Soc.* viii. 235; *DC. Monog. Phan.* i. 251. *R. divaricatus*, *Mast. in Journ. Linn. Soc.* viii. 236.

COAST REGION: Clanwilliam Div.; near Groene River and Watervals River, 2500–3000 ft., *Drège*, 2487, ♂! Tulbagh Div.; near Tulbagh Waterfall, *Ecklon and Zeyher*, ♂! Worcester Div.; Dutoits Kloof, 3000–4000 ft., *Drège*, 1631! Swellendam Div.; right bank of the Zonder Einde River, *Burchell*, 7520, ♀!

WESTERN REGION: Little Namaqualand; Kamies Bergen, *Drège*, 2490!

30. R. rottboellioides (Kunth, Enum. iii. 394); stems tufted, 6–8 in. high, terete, sparingly branched, tubercled; leaf-sheaths about ½ in. long, closely convolute, subulato-mucronate, coriaceous, membranous at the apex; spikelets 4–5, in long, spike-like cymes, each about ¼ in. long, linear-oblong; spathe sheath-like, half as long as the spikelet; bracts oblong, deeply membranous, mucronate; oblong-lanceolate, membranous; uppermost one decurrent at the margin; male flowers oblong-lanceolate; perianth-segments oblong-lanceolate; lateral unequal, not keeled; inner smaller, thinner; anthers linear, apiculate; female flower solitary, destitute of perianth; ovary ovoid, wrapped round by the bract; styles 2. *Steud. Synops.* ii. 252; *Mast. in Journ. Linn. Soc.* viii. 239; *DC. Monog. Phan.* i. 261.

WESTERN DIV.; Little Namaqualand; Kamies Bergen, near Lily Fontein, 4000–5000 ft., *Drège*, 2494, ♂ and ♀!

31. R. graminifolius (Kunth, Enum. iii. 407); stems 12–18 in. high, filiform, much branched; branches capillary, slightly 2-winged, purple-spotted; leaf-sheaths ⅓–½ in. long, compressed, striated, purple-spotted, prolonged into a linear-oblong leaf; male spikelets 3–4, loose, pedicellate; pedicels flattened, shorter than the spikelet; spikelets about ⅛ in. long, cylindric-lanceolate; spathe shorter than the spikelet; bracts oblong-ovate, acute, mucronate, papery, ferruginous; flowers ovate-lanceolate, shortly stalked; outer perianth-segments submembranous, oblong, mucronulate; lateral conduplicate, with a glabrous keel; inner shorter and thinner than the outer; anthers oblong, apiculate. *Steud. Synops.* ii. 254; *Mast. in Journ. Linn. Soc.* viii. 239; *DC. Monog. Phan.* i. 265.

COAST REGION: Paarl Div.; French Hoek Mountains, 1000–2000 ft., *Drège*, 2021 partly, ♂!

A cellular cushion or pad occurs in the axils of the pedicels, and the leaf is also remarkable in the genus for its development.

32. R. leptostachyus (Kunth, Enum. iii. 407); stems 18 in. high, decumbent, trailing, very slender, capillary, sparingly branched; branches terete, white-tubercled; leaf-sheaths closely convolute, $\frac{1}{8}$ in. long, coriaceous, striate, subulate, mucronate; mucro appressed; apex two-lobed, hyaline; male spikelets 2–5, in linear, distichous, flexuose cymes; spathe lanceolate, papery, aristate; spikelets $\frac{1}{4}$ in. long, flattened, oblong or wedge-shaped; bracts papery, ferruginous, linear-oblong, mucronulate; perianth-segments linear-oblong, acute; lateral villous-keeled; inner shorter; anthers linear, apiculate; female spikelets like the male, 2-flowered; bracts decurrent at the margins; ovary shortly stalked, subclavate, 2-celled; styles 2; capsule obliquely elliptic, 1-celled by abortion, shorter than the persistent perianth. *Steud. Synops.* ii. 254; *Mast. in Journ. Linn. Soc.* viii. 237; *DC. Monog. Phan.* i. 262.

SOUTH AFRICA: without locality, *Drège*, 12, ♂ and ♀!

33. R. depauperatus (Kunth, Enum. iii. 405); stems filiform, much branched; branches somewhat compressed, puncticulate, purple-spotted; ultimate branchlets curved; leaf-sheaths about $\frac{1}{3}$ in. long, closely convolute, coriaceous, striate, subulate-mucronate; apex membranous; male spikelets 4–6, in linear cymes, each spikelet about $\frac{1}{6}$ in. long, erect, oblong, wedge-shaped; bracts ovate-oblong, mucronate; outer perianth-segments oblong-ovate, acute; lateral keeled; inner subequal, shorter than the outer; anthers linear-oblong, apiculate; female spikelets 1–2-flowered; perianth as in the male; staminodes 3, liguliform; ovary 2-celled; styles distinct; capsule somewhat compressed, roundish, surmounted by the remains of the styles. *Steud. Synops.* ii. 253; *Mast. in DC. Monog. Phan.* i. 262.

COAST REGION: Paarl Div.; French Hoek Mountains, 1000–2000 ft., *Drège*, 2021 partly, ♂! and ♀. Worcester Div.; Dutoits Kloof, 3000–4000 ft., *Drège*, ', ♂! 10, ♂! and ♀!

34. R. curviramis (Kunth, Enum. iii 395); stems 6–12 in. high, cæspitose, slender, branching; branches filiform, flexuose, ascending, puncticulate; leaf-sheaths about $\frac{1}{3}$ in. long, closely convolute, coriaceous, subulate, mucronate; apex 2-lobed, membranous; male spikelets numerous, distant, in linear cymes, each spikelet about $\frac{1}{4}$ in. long, cylindric-lanceolate, sessile or pedicellate, sometimes crowded; spathe sheath-like, open, half the length of the spikelet; bracts oblong, coriaceous, subulate, aristate; apex hyaline; male flowers lanceolate; perianth-segments oblong-lanceolate; lateral conduplicate, acute, villous-keeled; anthers linear, apiculate; female spikelets 2–4 (rarely 6–8), in terminal spikes, sessile or pedicellate, $\frac{1}{8}$ in. long, cylindric-lanceolate, ultimately ovoid; bracts ovate-lanceolate, acuminate, coriaceous; margins membranous; perianth-segments acute; ovary compressed, 2-celled; style short, dividing into two long stigmatic branches. *Steud. Synops.* ii. 252; *Mast. in Journ. Linn. Soc.* viii. 241; *DC. Monog. Phan.* i. 263.

SOUTH AFRICA: without locality, *Drège*, 54 ♂!

COAST REGION: Clanwilliam Div.; Blue Berg, 4000–5000 ft., *Drège*, 2498, ♂! Worcester Div.; Dutoits Kloof, 3000–4000 ft., *Drège*, 1626, ♂! 57, ♀! mountains above Worcester, *Rehmann*, 2567, ♂! and without precise locality, *Zeyher!* Caledon Div.; Nieuw Kloof, Houw Hoek Mountains, *Burchell*, 8064, ♀!

35. R. monanthos (Mast. in Journ. Linn. Soc. viii. 238); stems 18 in. high, slender, terete, sparingly branched; branches puncticulate; sheaths closely convolute, coriaceous, mucronato-subulate; apex thinner, ferruginous; female spikelets 6–8 in a linear cyme, subglobose, 1-flowered; bracts ovate-subulate, mucronate, coriaceous; apex membranous; outer perianth-segments rigid, oblong, acute; lateral conduplicate, villous-keeled; inner smaller, ovate, obtuse, ferruginous, slightly involute; ovary ovate-oblong, compressed 2-celled; style bifid; capsule 2-celled, longer than the persistent perianth. *Mast. in DC. Monog. Phan.* i. 264.

WESTERN REGION: Little Namaqualand; between Buffels River and Pedros Kloof, 2000–3000 ft., *Drège*, 2486, ♀!

36. R. capillaris (Kunth, Enum. iii. 405); stems 18 in. high, capillary, terete, purple-spotted, branched; sheaths tightly convolute, $\frac{1}{4}$ in. long, membranous, ferruginous, mucronato-aristate; awn appressed; apex hyaline; male spikelets numerous, in linear cymes, each about $\frac{1}{8}$ in. long, oblong-ovate, 2–4-flowered, with an open bract-like spathe at the base; bracts oblong, mucronato-aristate, hyaline at the apex; flowers ovate-lanceolate; perianth-segments rigid, linear-oblong; lateral conduplicate, villous-keeled; inner smaller; anthers linear-oblong; female spikelets solitary, 1-flowered; bracts lanceolate, mucronate; perianth as in the male, but segments more acute; capsule obliquely ovate-oblong, 1-celled by abortion, crowned by the rudiments of the style. *Steud. Synops.* ii. 253; *Mast. in DC. Monog. Phan.* i. 264.

COAST REGION: Cape Div.; Table Mountain, *Schlechter*, 348, ♂! Worcester Div.; Dutoits Kloof, 3000–4000 ft., *Drège*, 339, ♂ and ♀! Drakenstein Mountains, 1600–2000 ft., *Bolus*, 4095!

37. R. cincinnatus (Mast. in Journ. Linn. Soc. viii. 240); stem 12–18 in. high, terete, much branched from the centre upwards; branchlets very slender, curled, wrinkled and studded with white tubercles; leaf-sheaths nearly $\frac{1}{2}$ in. long, closely tubular, with two membranous lobes beneath the mucronate apex; male spikes solitary or two, each about $\frac{1}{4}$ in. long, cylindric-lanceolate; spathe oblong, obtuse, acuminate, half the length of the spike; bracts similar to the spathe, with one appressed mucro; flowers oblong; perianth-segments rigid, oblong; outer lateral villous-keeled; inner thinner, shorter; female spikes 1–2, about $\frac{1}{4}$ in. long, oblong-lanceolate, straight or slightly curved, clavate when in fruit; flower solitary; bracts and perianth segments as in the male; intermediate outer segment as well as the 3 inner somewhat involute at the margins; ovary globose; capsule obliquely ovate, subcompressed, 1-celled by abortion, surmounted by the remains of the style. *Mast. in DC. Monog. Phan.* ii. 267. *Restio vimineus β, Thunberg Herb. ex Mast. in Journ. Linn. Soc.* xiv. 420.

COAST REGION: Cape Div.; mountains near Simons Town, 1000–2000 ft., *Ecklon and Zeyher!* Ceres Div.; at the foot of mountains around Ceres, 1800 ft., *Bolus*, 5489, ♂!

38. R. Ludwigii (Steud. Synopsis, ii. 254); stems 8–12 in. high, very slender, branching in the middle; branchlets filiform, covered with white tubercles; leaf-sheaths closely convolute, $\frac{1}{4}$–$\frac{1}{2}$ in. long, with two membranous lobes beneath the subulate-mucronate apex; male spikes 1–2, $\frac{1}{8}$ in. long, erect, ovate, 1–2-flowered; spathe loose, ovate-subulate, mucronate, nearly as long as the spike; bracts loosely imbricate, oblong, obtuse, mucronate, tubercled; outer perianth-segments oblong, obtuse; outer lateral villous-keeled; female spikes 1–2, erect, ultimately clavate; flowers 1–2; outer perianth-segments oblong-lanceolate; inner three shorter, obtuse, hyaline, involute at the margins; ovary roundish, compressed; style dilated at the base; capsule oblique, 1-celled by abortion. *Mast. in Journ. Linn. Soc.* viii. 239; *DC. Monog. Phan.* i. 268. *Restio nutans, Steud. in Flora,* 1829, 134, *not of Thunberg.*

SOUTH AFRICA: without locality, *Ludwig*, in Oxford Herbarium! *Ecklon*, 83! 85!

COAST REGION: Cape Div., mountains near Simons Town, *Zeyher*, ♂ and ♀! Tulbagh Div.; Seven Fonteins, 900 ft., *MacOwan, Herb. Aust.-Afr.*, 1672, ♂! Tulbagh, 600 ft., *Schlechter*, 7517, ♂! 7518, ♀! Caledon Div.; mountains near Grietjes Gat, between Lowrys Pass and Palmiet River, 2000-4000 ft., *Ecklon and Zeyher!*

39. R. leptoclados (Mast. in Journ. Linn. Soc. viii. 241 and x. 279); stems 2 ft. high, erect, terete, olivaceous, purple-spotted, branching towards the middle; branches numerous, verticillate, filiform, ascending or spreading; leaf-sheaths about 1 in. long, closely convolute, coriaceous, slenderly aristate, membranous at the apex; male spikelet solitary, oblong, compressed, $\frac{1}{4}$ in. long; spathe sheath-like, open, half the length of the spikelet; bracts oblong, ferruginous, membranous above; lower subulato-mucronate; outer perianth-segments linear oblong; lateral conduplicate, villous-keeled; inner 3 oblong, hyaline; anthers apiculate; female spikelet solitary,

$\frac{1}{8}$ in. long, ovoid, pyriform, 1-flowered; bracts coriaceous, ovate-oblong, subulato-mucronate, membranous above; outer perianth-segments equalling the bract, linear oblong; lateral conduplicate, villous-carinate; inner rather shorter, thinner, involute at the margins; staminodes 3; ovary compressed, ovate, oblong, 2-celled; style simple or dividing into three branches; capsule 1-celled by abortion. *Mast. in DC. Monog. Phan.* i. 265.

SOUTH AFRICA: without locality, *Drège*, 3, ♀!
COAST REGION: George Div.; near the Touw River, *Burchell*, 5732, ♂! Knysna Div.; between Groene Vallei, and Zwart Vallei, *Burchell*, 5672, ♂! on sand hills at the west end of Groene Vallei, *Burchell*, 5652, ♂! near Goukamma River, *Burchell*, 5589, ♂! between Plettenberg Bay and Melville, *Burchell*, 5353, ♀!

40. R. Eleocharis (Nees ex Mast. in Journ. Linn. Soc. viii. 238); stems 18–24 in. high, erect, terete, slender, dichotomously branched; branches filiform, ascending; sheaths $\frac{1}{2}$ in. long, elliptic, coriaceous, mucronulate; margins membranous; male spikelet solitary, elliptic, acute, $\frac{1}{4}$–$\frac{1}{2}$ in. long; spathe bract-like, half the length of the spikelet; bracts oblong, ferruginous, coriaceous, mucronate; mucro often twisted; flowers ovate-lanceolate; outer perianth-segments rigid, oblong; lateral conduplicate, villous-carinate; inner subequal, oblong, ferruginous; anthers linear; female spikelets solitary, ovate, erect, $\frac{1}{8}$ in. long; ovary oblong, compressed, 2-celled; style short, dividing into two stigmatic branches. *Mast. in DC. Monog. Phan.* i. 266.

COAST REGION: Clanwilliam Div.; Ceder Bergen at Ezels Bank, 3000–4000 ft., *Drège*, 2489! Cape Div.; Sand-dunes near Cape Town, *Bolus*, 4433, ♂! *Schlechter*, 556, ♀! Zeekoe Vallei, *Zeyher*, 78! 984, ♂ and ♀! Cape Flats, *Ecklon*, 1! 560! *Rehmann*, 1802!

41. R. tenuissimus (Kunth, Enum. iii. 394); stems 12–14 in. high, capillary, much branched, white-spotted; ultimate branchlets curved; leaf-sheaths closely convolute, striate, coriaceous, membranous at the edges, subulato-mucronate at the apex; male spikes 1–2, about $\frac{1}{8}$ in. long, lanceolate, 1-flowered; spathe sheath-like; bract solitary; perianth-segments 4, lanceolate, hyaline; lateral villous-carinate or glabrescent; stamens 2; female spikes $\frac{1}{8}$ in. long, solitary or 2–3, linear, 1-flowered; bracts and perianth-segments as in the male; ovary ovoid, 2-celled; capsule obliquely ovate, compressed, 1-celled by abortion; seed oblong; testa covered with white tubercles. *Steud Synops.* ii. 252; *Mast. in Journ. Linn. Soc.* viii. 241; *DC. Monog. Phan.* i. 268.

SOUTH AFRICA: without locality, *Thom*, 1027!
COAST REGION: Tulbagh Div.; Waterfall near Tulbagh, *Drège*, 1970, ♂ and ♀! Tulbagh, 600 ft., *Schlechter*, 7517, ♂! 7518, ♀! Worcester Div.; Dutoits Kloof, *Drège*, 1625, ♂!

42. R. tetragonus (Thunb. Diss. Rest. 17); stems 2–3 ft. high and upwards, leafless or leafy; leafless stems four-sided, sparingly branched, wrinkled, and finely tubercled; leaf-sheaths about 1 in.

long, closely convolute, pale greenish-yellow, coriaceous, striated, with an appressed mucro at the apex; leafy stems shorter, much branched; ultimate branchlets filiform, curved; leaf-sheaths about ¼ in. long, with two deep hyaline lobes beneath the terminal, foliaceous mucro; male spikelets numerous, in linear panicled cymes, each cyme 4–8 in. long; spikelets less than 1 lin. long, some sessile; others stalked; peduncles flattened; spathe bract-like, half the length of the spikelet; bracts light brown, oblong lanceolate, membranous at the margin, shortly mucronate at the apex, and with a more or less prominent midrib; outer perianth-segments rigid, oblong, acute; inner shorter, hyaline; anthers purplish; rudiment of pistil minute, 3-lobed; female plant and inflorescence as in the male; spikelets many-flowered; staminodes 3; ovary 3-sided, dark brown, white at the angles; styles 3, distinct; capsule 2- or even 1-celled by abortion. *Thunb. Prod.* 15; *Fl. Cap. ed. Schult.* 87; *Thunb. Herb. ex Mast. in Journ. Linn. Soc.* xiv. 418; *Willd. Sp. Pl.* iv. 725; *Nees in Linnæa,* v. 642; *Kunth, Enum.* iii. 401; *Steud. Synops.* ii. 253; *Mast. in Journ. Linn. Soc.* viii. 245.

SOUTH AFRICA: without locality, *Sieber,* 118, ♀! *Thunberg! Masson! Drège,* 364 b!

COAST REGION: Cape Div.; Constantia, *Zeyher,* ♀! False Bay, *Robertson!* between Rondebosch and Hout Bay, *Drège,* 364 a, ♂! Devils Mountain, 500 ft., *Bolus,* 4460, ♂! 4461, ♀! 4462, ♀! Table Mountain, *MacGillivray,* 435! *Krauss;* vicinity of Cape Town, *Burchell,* 461! 499! Cape Flats near Rondebosch, *Burchell,* 186! Knysna Div.; on hills near Melville, *Burchell,* 5463, ♀!

43. **R. quinquefarius** (Nees in Linnæa, v. 639); stems 1–3 ft. high, cæspitose, slightly branched, olive-coloured, beset with rather large white tubercles; leaf-sheaths 1–2 in. long, closely convolute, prolonged at the apex into a long, linear, obtuse leaf; male spikelets 1–3 or more, sessile or pedicellate at the apex of the branches, erect, oblong, obtuse, ½–¾ in. long, many-flowered; spathe sheath-like, cuspidate, shorter than the spikelet; bracts quinquefarious, oblong, mucronate, striate; lower empty; flowers sessile, ovoid oblong; outer perianth-segments rigid, linear oblong; lateral conduplicate, villous-carinate; median flat, marked with a prominent midrib; inner shorter, thinner; anthers oblong, mucronulate; pistillode with 3 rudimentary styles; female plant, inflorescence and perianth as in the male; staminodes 3; ovary roundish, deep chestnut-coloured; capsule 2-celled by abortion. *Steud. Synops.* ii. 253; *Mast. in Journ. Linn. Soc.* viii. 242; *DC. Monog. Phan.* i. 278. *Restio xyridioides, Kunth, Enum.* iii. 397.

COAST REGION: Clanwilliam Div.; between Ezels Bank and Dwars River, 3000–4000 ft., *Drège,* 35, ♂! Cape Flats, *Zeyher,* ♀! *Burchell,* 188, ♂ and ♀! *Bolus,* 2888, ♀! 4732, ♂! 4760, ♀! *Ecklon,* 564! 564 b! 565! 566! Simons Town, *Wright!* Wynberg, *Ecklon,* 91!

44. **R. dispar** (Mast. in Journ. Linn. Soc. viii. 246); a tall-growing species with terete stems, and subcompressed branches of a deep olive colour, with white spots; leaf-sheaths about ¾ in. long, closely

convolute, coriaceous, lacerate-membranous at the apex; male spikelets numerous, in erect, linear, flexuose cymes, each spikelet about ½ in. long, oblong, acute, with an oblong, acuminate, rather open spathe at the base; bracts closely overlapping, oblong, obtuse, convex, subcarinate; outer perianth-segments unequal, oblong, obtuse; lateral villous-carinate; intermediate thinner, flat; inner subequal, shorter than the outer; anthers apiculate, purplish; pistillode minute; female spikelets many-flowered; perianth as in the male; ovary oblong, truncate, purplish, surmounted by 3 styles and encircled at the base by 3 staminodes; capsule 1-celled by abortion, obliquely ovoid; seed 3-angled, grey, purple-spotted. *Mast. in DC. Monog. Phan.* i. 274.

COAST REGION: Caledon Div.; mountain ridges between Babylons Tower and Caledon, 1000–2000 ft., *Ecklon and Zeyher!* Not in Kew Herbarium.

45. **R. furcatus** (Nees ex Mast. in Journ. Linn. Soc. viii. 242); tall-growing; stems terete, slightly branched, dark olive-coloured, with whitish spots; lower leaf-sheaths ¾–1 in. long, closely convolute, coriaceous, membranous towards the tip; membranous portion lacerate, deciduous; apex originally with a short, appressed, ultimately deciduous mucro; male spikelets 8–10, arranged in pairs on an erect, distichous, flexuose, linear cyme; spikelets about ¼ in. long, ovoid, acute; spathe oblong, obtuse, cuspidate, one-third shorter than the spikelet; bracts ovate, subulate, mucronate, coriaceous, ferruginous; margins and upper portion dull brown pitted; flowers ovate oblong; perianth-segments rigid, oblong, acute; lateral navicular, villous-carinate; central marked with a prominent midrib; inner 3 shorter, thinner; pistillodium with 3 styles; female inflorescence like the male; spikelets oblong, sessile or pedunculate; bracts oblong, mucronate, pitted near the apex, striate; lower generally sterile; perianth-segments oblong, acute, rigid; outer villous; median with a central rib, often villous near the top; inner 3 flat, smaller; staminodes 3; ovary 3-sided, with 3 styles; capsule 1–2-celled by abortion. *Mast. in DC. Monog. Phan.* i. 275. *R. bifurcus, Mast. in Journ. Linn. Soc.* viii. 247, *as to the female plant.*

COAST REGION: Cape Div.; mountains near Simons Town, *Zeyher,* ♂! Cape Flats, *Bolus,* 4452, ♀!

46. **R. bifurcus** (Nees ex Mast. in Journ. Linn. Soc. viii. 247 partly); stems 2–3 ft. high, terete, slightly branched, olivaceous, with circular white spots; leaf-sheaths about 1 in. long, closely convolute, coriaceous, brown, striate, membranous beneath the mucronate apex; male spikelets numerous, in linear, distichous, panicled cymes, each spikelet nearly ½ in. long, sessile or stalked, oblong, obtuse, ultimately subclavate, many-flowered; spathe much shorter than the spikelet; bracts closely overlapping, oblong or roundish, obtuse, coriaceous, pitted near the top and with a narrow, whitish, deciduous margin beneath the short mucro; nerves whitish; outer perianth-

segments oblong, obtuse; lateral navicular, villous-carinate; intermediate flattish, slightly villous at the back; inner 3 shorter, oblong, hyaline, free or more or less united and thickened at the base; pistillodium 3-lobed; female spikelets ovoid or obovoid, many-flowered, arranged in linear cymes; bracts as in the male, but with the white margin less obvious; perianth-segments oblong, obtuse; lateral villous-carinate; staminodes 3; ovary oblong, 3-styled. *Mast. in DC. Monog. Phan.* i. 275.

SOUTH AFRICA: without locality, *Thom*, 632, ♀!
COAST REGION: Cape Div.; Cape Flats, *Zeyher*, 1011, ♂! Caledon Div.; Nieuw Kloof, Houw Hoek Mountains, *Burchell*, 8119, ♂!

47. **R. præacutus** (Mast.); stems 2–3 ft. high, terete, dark olive, with white spots, sparingly branched; leaf-sheaths about 1 in. long, closely convolute, coriaceous, striate; apex membranous, deciduous; male spikelets 5–7, in an erect, panicled cyme, each spikelet linear-ovoid, acute, subtended by a bract-like spathe, $\frac{1}{3}$ shorter than the spikelet; bracts loosely imbricate, coriaceous, pitted near the apex; flowers pedicellate; outer perianth-segments rigid, lanceolate, acute; lateral villous-carinate; intermediate flat, cartilaginous, thickened at the apex and near the base; inner slightly shorter than the outer, lanceolate, thickened at the base; female inflorescence as in the male; spikelets erect, ovoid, acute, scarcely $\frac{1}{2}$ in. long; flowers on a short, villous pedicel; perianth-segments very acute; staminodes 3; disc cup-shaped; ovary oblong; styles 3; capsule 2-celled by abortion.

COAST REGION: Clanwilliam Div.; Gift Berg, *Drège*, 88, ♀! Worcester Div.; Dutoits Kloof, *Drège*, 46, ♀! 1608 partly, ♂ and ♀! according to *Drège, Docum.* p. 82, but the latter was collected in Caledon Div. at Genadendal, according to Drège's own label at Lubeck.

Confounded by myself and others with *R. furcatus*, from which it may be distinguished by its linear-oblong spikelets, its very acute perianth-segments, and by the presence of a disc.

48. **R. triticeus** (Rottboell, Progr. 11); stems tufted, erect, 18–24 in. high, slightly branched, olive-coloured, coarsely tubercled and white-spotted; leaf-bearing branches shorter, much subdivided; ultimate branches filiform, spreading; leaf-sheaths about $\frac{1}{2}$ in. long, closely convolute, coriaceous, with two deep hyaline, ultimately deciduous lobes beneath the subulate-mucronate apex; in the smaller sheaths the mucro is prolonged into a small acicular leaf; male spikelets varying in number from 2–9, in erect, spicate cymes, each spikelet about $\frac{1}{2}$ in. long, cylindric-lanceolate, with a short sheathing spathe at the base; bracts oblong-lanceolate, coriaceous, ferruginous, scarcely mucronate; flowers oblong, shortly stalked; perianth-segments oblong, obtuse; outer lateral narrow, boat-shaped, villous-carinate; central flat; inner 3 shorter, thinner, more or less concrescent at the base; anthers chestnut-coloured, with white

tubercles; pistillode with 3 rudimentary styles; female inflorescence and perianth as in the male; staminodes 3; ovary ovoid, 3-styled; capsule oblique, 2- or even 1-celled by abortion; seeds trigonous, studded with large white tubercles. *Rottboell, Descr. et Ic.* 7, *t.* 3, *f.* 1; *Thunberg, Diss.* 17; *Fl. Cap., ed. Schult.* 87; *Willd. Sp. Plant.* iv. 726; *Sprengel, Syst. Veg.* 1, 185; *Nees in Linnæa,* v. 640, *partly*; *Mast. in Journ. Linn. Soc.* viii. 243; *DC. Monog. Phan.* i. 277. *R. dichotomus, Thunberg Herb.! ex Mast. in Journ. Linn. Soc.* xiv. 420; *Thunb. Diss.* 314. *R. triticeus, β gracilis, Nees ex Drege in Linnæa,* xx. 241. *Calopsis triticea, Kunth, Enum.* iii. 424; *Steud. Synops.* ii. 257. *R. glumaceus, Klotzsch in several herbaria.*

SOUTH AFRICA: without locality, *Thom,* 1026, ♂! *Drège,* 85! 9451, ♂! *Bergius,* ♂! *Sieber,* 112, ♀!

COAST REGION: Cape Div.; False Bay, *Robertson!* Devils Mountain, 5000 ft., *Bolus,* 4464, ♀! *Rehmann,* 929! Table Mountain, *Drège,* 200, ♂! hills and flats near Cape Town, *Schlechter,* 802! *Zeyher,* 1742, ♂! Caledon Div.; Donker Hoek Mountain, *Burchell,* 8005, ♀!

Very variable in stature and in number of spikelets.

49. R. pannosus (Mast. in Journ. Linn. Soc. viii. 244); stems 2–3 ft. high, erect, slightly branched, coarsely tubercled, olivaceous; leaf-sheaths about 1 in. long, closely convolute, coriaceous, dividing below the long, acuminate or aristate apex into two deep, deltoid-lanceolate, hyaline lobes; male spikelets numerous, closely arranged in erect, linear, panicled cymes; each spikelet $\frac{1}{4}$–$\frac{1}{2}$ in. long, erect, many-flowered, linear-oblong; spathe sheath-like, almost as long as the spikelet; bracts closely overlapping, oblong, coriaceous, slightly membranous near the shortly mucronate apex; flowers shortly stalked; outer perianth-segments rather rigid, brownish; lateral conduplicate, villous-carinate; inner shorter and thinner, more or less concrescent at the base; anthers linear, ferruginous; pistillode minute; female plant, inflorescence and perianth nearly as in the male; ovary oblong-ovoid, compressed, 3-, or by abortion 2–1-celled; staminodes 3. *Mast. in DC. Monog. Phan.* i. 278. *R. triticeus, Thunb. Herb.* ♂ *ex Mast. in Journ. Linn. Soc.* xiv. 417, 418.

SOUTH AFRICA: without locality, *Thom,* 913, ♂!

COAST REGION: Cape Div.; Cape Flats, *Ecklon,* 50, ♀! *Ecklon and Zeyher! Thunberg,* ♂! False Bay, *Robertson!*

50. R. multiflorus (Spreng. Syst. Veg. i. 187); stems 2–3 ft. high, terete, branched; branches flabellate; ultimate branchlets filiform, olivaceous, wrinkled, white-spotted; leaf-sheaths $\frac{3}{4}$–1 in. long, closely convolute, coriaceous below, deeply hyaline above, lacerate; apex prolonged into a slender filiform mucro; male spikelets 4–8 in linear, panicled cymes; each spikelet about $\frac{1}{4}$ in. long, ovate-oblong; spathe sheath-like, open, half the length of the spikelet; bracts coriaceous, ovate, subulate-mucronate; outer perianth-segments oblong, acute; lateral conduplicate, villous-carinate;

intermediate outer and inner smaller, flat; female plant like the male; staminodes filiform; ovary trigonous, roundish, purplish; capsule 2-celled by abortion. *Nees in Linnæa*, v. 646; *Kunth, Enum.* iii. 412; *Steud. Synops.* ii. 254; *Mast. in Journ. Linn. Soc.* viii. 244; xxi. 575; *DC. Monog. Phan.* i. 279. *R. triticeus, β, foliosus, Nees in Linnæa*, v. 640. *R. triticeus var. destructus, Nees in various herbaria.*

COAST REGION: Cape Div.; Devils Mountain; *Drège*, 90, ♀! *Bolus*, 4469, ♀! *Ecklon and Zeyher!* Table Mountain, *Ecklon! Bolus*, 4442! Lion Mountain, *Mund!*

51. **R. filiformis** (Poir. Encycl. vi. 173); stems tufted, 1–2 ft. high, terete, slender, simple or slightly branched, olivaceous, white-spotted; leaf-sheaths $\frac{1}{4}$–$\frac{1}{2}$ in. long, closely convolute, coriaceous below, thinner above; apex subulate-mucronate; male spikelets 1–3, sessile or slightly stalked, oblong-ovate, acute, $\frac{1}{4}$–$\frac{1}{2}$ in. long; spathe bract-like, shorter than the spikelet; bracts ovate-oblong, coriaceous, pitted, thinner or white-membranous at the margins; apex subulate, mucronate; flowers oblong, stalked; outer perianth-segments rigid; lateral conduplicate, villous-carinate; inner shorter, thinner; anthers oblong; pistillodium 3-styled; female spikelets 3–4-flowered; perianth as in the male, sometimes more rigid; staminodes liguliform; ovary roundish, subtrigonous; styles 3; capsule 2-celled. *R. filiformis, Poir. in the Paris herbarium. R. garnotianus, Kunth, Enum.* iii. 392; *Steud. Synops.* ii. 251; *Mast. in Journ. Linn. Soc.* viii. 254; *DC. Monog. Phan.* i. 280. *R. bifidus, Nees in Linnæa*, v. 636; *Kunth, Enum.* iii. 409; *Steud. Synops.* ii. 250; *Mast. in Journ. Linn. Soc.* viii. 253, *not of Thunberg. R. triflorus, Thunb. Herb. ex Mast. in Journ. Linn. Soc.* xiv. 417. *Craspedolepis Verreauxii, Steud. Synops.* ii. 264. *C. fimbriata, Mast. in DC. Monog. Phan.* i. 280.

VAR. β, **oligostachyus** (Mast.); stem flexuose, 1–2-spiked; sheaths prolonged into a long, leafy mucro; female spikelets larger than in the type; perianth-segments more acute. *R. bifidus, var. β, Nees in Linnæa*, v. 637. *R. bifidus var. gracilis, Nees in herb. R. oligostachyus, Kunth, Enum.* iii. 399. *R. garnotianus, Kunth, var. oligostachyus, Mast. in DC. Monog. Phan.* i. 281.

VAR. γ, **monostachyus** (Mast.); stem slender; spikelets few, minute. *R. garnotianus, Kunth, var. monostachyus, Steud. ex Mast. in DC. Monog. Phan.* i. 281.

SOUTH AFRICA: without locality, *Bergius! Boivin! Drège*, 59!

COAST REGION: Clanwilliam Div.; Ceder Bergen, at Ezels Bank, 3000–4000 ft., *Drège*, 2473! Cape Div.; Table Mountain, *Drège*, 197, ♀! *Ecklon*, 77! between Cape Town and Table Mountain, *Burchell*, 925! near Cape Town, *Bolus*, 2887! Tulbagh Div.; New Kloof, near Tulbagh, 1000 ft., *MacOwan, Herb. Aust.-Afr.*, 1668, ♂! Ceres Div.; mountains near Ceres, 2400 ft., *Bolus*, 5292! Worcester Div.; Dutoits Kloof, 1000–4000 ft., *Drège*, 84 partly! 1628a! Caledon Div.; mountains of Baviaans Kloof, near Genadendal, *Burchell*, 7647! mountain near Grietjes Gat, between Lowrys Pass and Palmiet River, 1000–2000ft., *Ecklon and Zeyher!* Zwartberg, near Caledon, *Zeyher*, 4343! Alexandria Div.; Zwartwater Poort, *Burchell*, 3409! Var. β, Worcester Div.; Dutoits Kloof, 3000–4000 ft., *Drège*, 1624, ♀! Uniondale Div.; between Welgelegen and

Onzer, 1500–2000 ft., *Drège*, 37 partly, ♀ ! Var. γ, Cape Div.; Table Mountain, *Ecklon*, 840 ! 846 ! *Milne*, 218 ! *MacGillivray*, 436 !

Very variable in stature and in the number of the spikelets.

52. **R. festucæformis** (Nees ex Mast. in Journ. Linn. Soc. viii. 248); stems tufted, 12–18 in. high, erect, slender, straw-coloured, wrinkled; leaf-sheaths 1 in. long and upwards, closely convolute, coriaceous, striate, brownish, thinner at the margins, prolonged into a long, curved, filiform awn; male and female spikelets similar, clustered in an elongated, panicled cyme, each about ½ in. long, erect, cylindric-oblong, flattened from back to front, ultimately oblanceolate, shining yellow; spathe lanceolate, acuminate, membranous at the edges, shorter than the spikelet; bracts similar and with a central midrib; lower sterile; upper fertile, decurrent at the base; flowers stipitate, curved laterally; outer perianth-segments oblong, obtuse; lateral conduplicate, only slightly keeled, glabrescent; inner narrower; anthers linear, apiculate; staminodes 3, ovary ovoid-oblong or compressed; styles 3; fruit not seen. *R. festucæformis, Nees in herb. Sonder ! Mast. in DC. Monog. Phan.* i. 281. *R. ischæmoides, Nees MSS. in various herbaria.*

COAST REGION: Caledon Div.; Houw Hoek, 2600 ft., *Schlechter*, 7788, ♂ ! Zwartberg, near Caledon, *Ecklon and Zeyher*, ♂ and ♀! mountain near Grietjes Gat, between Lowrys Pass and Palmiet River, *Ecklon and Zeyher !*

53. **R. subtilis** (Nees ex Mast. in Journ. Linn. Soc. viii. 251); stems 18 in. high, very slender, filiform, subcompressed, sparingly branched, wrinkled, purple-spotted; leaf-sheaths about ½ in. long, closely convolute, ferruginous, paler and thinner above, prolonged into a short appressed mucro; male and female spikelets similar, numerous, loosely arranged in long, loose. linear cymes, sessile, or on long, slender stalks, each spikelet about 1 line long or less, oblong-ovoid; spathe open, mucronate, half the length of the spikelet; bracts imbricate, ovate, acute, coriaceous, thinner at the margins, shortly mucronate; flowers 2–3, ovoid, compressed; perianth-segments oblong, acute; lateral conduplicate, scarcely keeled, glabrous; inner flat, shorter; pistillode 3-styled; capsule 3-lobed, 3-celled; seeds grey. *Mast. in DC. Monog. Phan.* i. 282.

COAST REGION: Caledon Div., mountain near Grietjes Gat, between Lowrys Pass and Palmiet River, *Ecklon and Zeyher*, ♂ and ♀.

54. **R. subulatus** (Mast. in Journ. Linn. Soc. viii. 248); stems decumbent, 8–10 in. high, much branched; leaf-sheaths closely convolute, brown, coriaceous, membranous above; apex prolonged into a long, subulate mucro; fertile stems 18 in. high, compressed, olivaceous, pitted, white-spotted; male and female spikelets similar, 6–8 arranged in linear, spicate cymes, each ovate oblong, 3–4 lin. long; bracts loosely imbricate, oblong, coriaceous, mucronate; outer perianth-segments rigid, oblong, acute; lateral conduplicate, villous-carinate; inner smaller, hyaline; pistillode minute; stami-

nodes 3; capsule ovoid, compressed, 2-celled, coriaceous, ferruginous. *Mast. in DC. Monog. Phan.* i. 281.

COAST REGION: Caledon Div.; Zwartberg near Caledon, 1000–2000 ft., *Ecklon and Zeyher*! Not in Kew Herbarium.

55. **R. strobilifer** (Kunth, Enum. iii. 398); stems cæspitose, 2–3 ft. high, erect, terete, sparingly branched, dark olive, faintly speckled; leaf-sheaths ¾ in. long, closely convolute, coriaceous, dark brown, striate, membranous above, mucronate; membranous portion and mucro deciduous; male and female spikelets similar, solitary or twin, each about ½ in. long, ovoid or subglobose; spathe obovate, mucronate, much shorter than the spikelet; bracts broadly ovate, obtuse, subcoriaceous, striate, membranous and mucronate at the apex; male flowers oblong, obtuse, curved; perianth-segments oblong, acute; outer lateral rigid, conduplicate, villous-carinate; intermediate flattish, with a ferruginous midrib; inner 3 smaller, thinner, flat; pistillode 3-styled; female flowers stalked; stalk short, with a warty swelling at the apex [? always]; perianth as in the male; staminodes liguliform; capsule bilocular, suborbicular, compressed; seed oblong, obtuse, trigonous, blackish. *Steud. Synops.* ii. 252; *Mast. in DC. Monog. Phan.* i. 282.

COAST REGION: Clanwilliam Div.; Ceder Bergen, at Ezels Bank, 3000–4000 ft.; *Drège*, 2474, ♂ and ♀! Stellenbosch Div.; Sneeuw Kop, *Wallich*! Ceres Div.; at the foot of mountains around Ceres, 1800 ft., *Bolus*, 5484!

56. **R. pachystachyus** (Kunth, Enum. iii. 399); stems cæspitose, 2–3 ft. high, erect, terete, slightly branched, olivaceous, with white tubercles; leaf-sheaths closely convolute, coriaceous, ½–¾ in. long, nervose-striate, white-spotted; margins membranous; apex acuminate, prolonged; spikelets 1–3, sessile or stalked, about ½ in. long, cylindric, acute, becoming ovoid-acute; pedicel as long as the spikelet; spathe half the length of the spikelet, sheath-like, with a long, spotted acumen; bracts oblong, acuminate, coriaceous, nervose, membranous at the margins, sharply acuminate; flowers shortly pedicellate, ovoid. convex; outer perianth-segments rigid, oblong, acute; lateral conduplicate, villous-carinate; inner shorter, thinner; ovary roundish, 2-celled by abortion; styles 3. *Steud. Synops.* ii. 252; *Mast. in DC. Monog. Phan.* i. 283.

COAST REGION: Clanwilliam Div.; Wupperthal, 1500–2000 ft., *Drège*, 43, ♀! Worcester Div.; Drakenstein Mountains, near Bains Kloof, 1600–2000 ft., *Bolus*, 4093, ♀!

57. **R. purpurascens** (Nees ex Mast. in Journ. Linn. Soc. viii. 249); stems erect, 2–3 ft. high, terete, sparingly branched; branches ascending, olivaceous, densely covered with flattened, disc-like, white tubercles; leaf-sheaths 1½ in. long, closely convolute, coriaceous, membranous above, prolonged at the apex into a long, leafy point; spikelets 2–5, in linear, erect cymes, sessile or pedicellate, about 1 in. long; spathe as long as the spikelet, with a long, leafy, curved mucro; bracts oblong, acute; perianth-segments

oblong-lanceolate; lateral conduplicate, villous-carinate; intermediate flat, as long as, or even slightly longer than the lateral ones; inner 3 shorter, hyaline, united at the base; pistillodium 3-styled; female plant like the male; ovary subtrigonous, 2–3-celled; capsule 2-celled. *Mast. in DC. Monog. Phan.* i. 283.

SOUTH AFRICA: without locality, *Lind*, ♂!

COAST REGION: Stellenbosch Div.; Hottentots Holland, *Zeyher*, ♀! at the foot of Helderberg Mountain near Somerset West, 400 ft., *Bolus*, 5491! Caledon Div.; Nieuw Kloof, Houw Hoek Mountains, *Burchell*, 8068 partly!

58. R. bifidus (Thunb. in Hoffm. Phytog. Blätter, i. 7); stems cæspitose, erect, 12–18 in. high and upwards, slender, terete, simple or sparingly branched; branches erect, olivaceous, coarsely tubercled; leaf-sheaths ½–¾ in. long, tightly convolute, coriaceous, striate, membranous above, prolonged at the apex into a slender, foliaceous mucro; male spikelets 1–3, ½ in. long and upwards, erect, oblong-obtuse, ultimately flattened; spathe oblong, mucronate, shorter than the spikelet; bracts loosely imbricate, spreading, subcoriaceous, oblong, cuspidate; flowers oblong, obtuse; perianth-segments narrow, blunt, in the female plant somewhat broader than in the male; outer lateral conduplicate, villous-carinate; inner 3 hyaline; staminodes 3; ovary rounded, subtrigonous; styles 3. *Thunb. in Weber and Mohr, Archiv.* i. 25; *Fl. Cap. ed. Schultes*, 87; *Mast. in Journ. Linn. Soc.* xiv. 415; *DC. Monog. Phan.* i. 284. *R. vaginatus*, *Willd. Sp Pl.* iv. 719; *Spreng. Syst. Veg.* i. 184; *Kunth, Enum.* iii. 408; *Hochstetter in Flora*, 1845, 337; *Mast. in Journ. Linn. Soc.* viii. 250, *not of Thunb.* *R. pseudo-leptocarpus*, *Kunth, Enum.* iii. 399; *Steud. Synops.* ii. 252.

SOUTH AFRICA: without locality, *Drège*, 44 partly! *Sieber*, 221! *Burchell*, 5812! *Ludwig!*

COAST REGION: Cape Div.; Table Mountain, 3000 ft., *Ecklon! Burchell*, 541! 573! *Drège*, 28! *Bolus*, 4439, ♀! 4440, ♂! *Rehmann*, 627! Caledon Div.; Nieuw Kloof, Houw Hoek Mountains, *Burchell*, 8070, ♀! mountain near Grietjes Gat, between Lowrys Pass and Palmiet River, *Ecklon and Zeyher!*

59. R. miser (Kunth, Enum. iii. 392); stems 12–18 in. high, cæspitose, erect, filiform, olivaceous, with white tubercles; leaf-sheaths about ½ in. long, closely convolute, coriaceous, striate, membranous above, filaceo-mucronate; male spikelets 3–5 approximate, 2–3 lin. long, erect, oblong, obtuse; spathe ovate-lanceolate, as long as the spikelet; bracts oblong, obtuse, muticous; perianth-segments oblong, obtuse; lateral conduplicate, villous-carinate; inner hyaline, united at the base; pistillodium minute; female plant like the male; ovary trigonous, surrounded by 3 staminodes. *Steud. Synops.* ii. 252; *Mast. in Journ. Linn. Soc.* viii. 251; *DC. Monog. Phan.* i. 285.

COAST REGION: Worcester Div.; Dutoits Kloof, *Drège*, 1627, ♂ and ♀!

60. R. sonderianus (Mast. in Journ. Linn. Soc. viii. 252); stems tufted, erect, very slender, simple, terete, puncticulate, purple-spotted;

leaf-sheaths tightly convolute, coriaceous, membranous above, subulate-mucronate; spikelets 1–4, sessile or shortly stalked; each about ¼ in. long, ovate-oblong; spathe sheath-like, shorter than the spikelet; bracts ultimately spreading, oblong deltoid, coriaceous, striate, mucronate-aristate; outer perianth-segments rather rigid, oblong, rather obtuse; lateral conduplicate, carinate; keel glabrous; inner shorter, hyaline, ferruginous at the base; anthers linear; pistillodium 3-styled. *Mast. in DC. Monog. Phan.* i. 285.

COAST REGION: Clanwilliam Div.; Groen River and Watervals River, 2500–3000 ft., *Drège*, 92, ♂! Ceder Bergen at Ezels Bank, 3000–4000 ft., *Drège*, 1629 partly! 2483, ♂! Worcester Div.; Dutoits Kloof, 3000–4000 ft., *Drège*, 82 partly!

61. R. pedicellatus (Mast. in Journ. Linn. Soc. viii. 252); stems cæspitose, 12–14 in. high, erect, very slender, unbranched, covered with very minute yellow tubercles; leaf-sheaths ½ in. long, closely convolute, coriaceous, striate, pitted, subulato-mucronate; mucro sometimes prolonged into a long, appressed leaf; spikelets 1–2, about ¼ in. long, erect, cylindric, acute, ultimately oblong; spathe mucronate, shorter than the spikelet; bracts ultimately loosely imbricate, oblong, acute, castaneous, mucronulate; perianth-segments oblong, obtuse; lateral conduplicate, with a glabrous keel; inner hyaline; anthers linear; pistillode minute; female flower with liguliform staminodes; ovary trigonous, 2-celled by abortion; capsule roundish, 2-celled. *Mast. in DC. Monog. Phan.* i. 286.

COAST REGION: Worcester Div.; Dutoits Kloof, 3000–4000 ft., *Drège*, 82 partly, ♀! 91! 1629 partly, ♂! 1631b! Ceres Div.; at the foot of mountains around Ceres, 1800 ft., *Bolus*, 5483!

62. R. implexus (Mast.); stems 12–18 in. high, capillary, terete, purple-spotted; branches curving, intertwining; leaf-sheaths ⅛–¼ in. long, closely convolute, coriaceous, membranous above; apex subulate, mucronate; spikelets solitary, ⅛ in. long, oblong; spathe mucronate, half the length of the spikelet; bracts papery, ovate-oblong, acute, scarcely mucronate; outer perianth-segments linear-oblong, acute, cartilaginous, as long as or longer than the bracts; lateral conduplicate, only slightly villous, scarcely carinate; inner 3 shorter, oblong-lanceolate; ovary compressed, 2-celled styles. *R. perplexus, var. gracilis, Mast. in DC. Monog. Phan.* i. 287.

COAST REGION: Swellendam Div.; at the foot of the Lange Bergen, near Swellendam, *Burchell*, 7430, ♀! not 7420 as quoted in *DC. Monog. Phan.* i. 287, that number being an Umbellifer.

63. R. perplexus (Kunth, Enum. iii. 406); stems 12–18 in. high, filiform, terete, olivaceous, purple-spotted, much branched; branches capillary, arcuate, entangled; leaf-sheaths ¼–½ in. long, closely convolute, coriaceous, membranous near the top, mucronate; spikelets 1–3, remote, ovate-oblong, sessile or pedunculate, ¼ in. long; peduncles longer than the spikelets; spathe shorter than the spikelet; bracts loosely imbricate, broadly ovate-acute, sharply mucronate, coriaceous, hyaline at the margins; perianth-segments rigid; outer lateral conduplicate, villous, carinate; median flat, with the upper part of the midrib villous; inner shorter, hyaline; anthers linear-oblong, muticous; pistillode minute; female plant like the male; staminodes 3; ovary rounded, 2-celled by abortion; styles 2–3. *Steud. Synops.* ii. 253; *Mast. in DC. Monog. Phan.* i. 286. *R. vimineus, Thunb. Herb. ex Mast. in Journ. Linn. Soc.* xiv. 420.

SOUTH AFRICA: without locality, *Drège*, 8! *Mund!*

COAST REGION: Cape Div.; Table Mountain, *Drège*, 339a, ♂! *Cooper*, 3380! *Bolus*, 2886!

64. R. Harveii (Mast. in Journ. Linn. Soc. viii. 253, t. 15); rhizome creeping; stems slender, erect, 2–6 in. high, branching above the middle; branches curved, spreading, olivaceous, studded with small, whitish tubercles; leaf-sheaths $\frac{1}{4}$–$\frac{1}{2}$ in. long, closely convolute, coriaceous, membranous at the upper margins, subulato-mucronate; spikes solitary, oblong-cuneate, 2-flowered, about 1 line long; spathe oblong, acute, shortly mucronate; bracts spreading, oblong, shortly mucronate; outer perianth-segments subequal, obtuse, ciliolate at the margins; lateral conduplicate, villous-carinate; inner similar, thinner, with the midrib villosulous; staminodes 3; ovary oblong, subcompressed, 2-celled; stigmas 3. *Mast. in DC. Monog. Phan.* i. 287.

COAST REGION: near Cape Town, *Harvey*, ♀! in Dublin Herbarium.

65. R. bifarius (Mast. in Journ. Linn. Soc. x. 278); stems 3–4 ft. high, erect, terete, slightly branched; branches ascending, dark olive, finely tubercled; leaf-sheaths $\frac{1}{2}$ in. long, coriaceous, striate, thinner above, subulato-mucronate; male spikelets numerous, in a diffuse panicled cyme; peduncles flexuose, flattened, as long as the spikelet; spathe oblong, acuminate, short; spikelets oblong, flattened, about 1 in. long; bracts distichous, oblong, mucronate, spreading and recurved at the tip; flowers shortly stalked, oblong, obtuse, curved; perianth-segments oblong, acute; lateral conduplicate, villous-carinate; intermediate flat; inner shorter, thinner; anthers linear-oblong; pistillode none; female spikelets 1–3, terminal, erect, oblong-ovate, 1–2 in. long; spathe half the length of the spikelet; bracts oblong, mucronate, purplish-brown; perianth as in the male; ovary 3-celled; styles 3; capsule 2-lobed, 2-celled by abortion, scarcely shorter than the persistent perianth. *Mast. in DC. Monog. Phan.* i. 288.

COAST REGION: Caledon Div.; Nieuw Kloof, Houw Hoek Mountains, *Burchell*, 8068 partly, ♂ and ♀!

Burchell also collected *R. purpurascens* under the same number.

66. R. bigeminus (Nees ex Mast. in Journ. Linn. Soc. viii. 246); stems 2 ft. high, tufted, terete, slender, olivaceous, minutely punticulate, branching towards the middle; branches compressed, ascending;

leaf-sheaths about ½ in. long, closely convolute, coriaceous, membranous above, with an appressed foliaceous mucro; male spikelets 1–5, terminal, sessile or shortly stalked, ¾ in. long, elliptic, acute, somewhat compressed; spathe very short; bracts oblong, acute, scarcely mucronulate, striate; perianth-segments acutely pointed; outer lateral conduplicate, with a glabrescent keel; inner hyaline, shorter; anthers linear, apiculate; female spikelets 1–5, at the apex of the branches, in erect, racemose cymes, each about ½ in. long, elliptic, pedicellate, with a short spathe; bracts oblong-lanceolate, coriaceous, greyish-brown, shining, 7-nerved, mucronulate; perianth-segments somewhat rigid, oblong-lanceolate; lateral conduplicate, carinate; keel glabrescent; inner segments shorter, unequal, posterior one wider; staminodes 3; ovary roundish, truncate, ferruginous, 2-celled, with 3 stigmas; capsule roundish, 1–2-celled by abortion; seed oblong, ferruginous; testa studded with rather large tubercles. *Mast. in DC. Monog. Phan.* i. 289; *Journ. Linn. Soc.* xxi. 576. *R. micans, Nees MSS. in various herbaria.*

COAST REGION: Cape Div.; Cape Flats, *Zeyher*, 1123, ♀! *Bolus*, 4448, ♀! *Rehmann*, 1803, ♂!

67. **R. egregius** (Hochstetter in Flora, 1845, 337 adnot.); stems 2–3 ft. high, erect, terete, olivaceous, finely tubercled, branching in the middle; branches ascending; leaf-sheaths about 2 in. long, closely convolute, membranous above, subulato-mucronate; small spikelets numerous in terminal spicate or panicled cymes; peduncles as long as the spikelets, flattened, each nearly ½ in. long, ovoid-oblong; spathe oblong, mucronate; bracts oblong, shortly mucronate, coriaceous, chestnut-brown; flowers oblong; outer perianth-segments rigid, oblong, obtuse; lateral conduplicate, with a villous or glabrescent keel; inner shorter, flat, thickened, and ferruginous at the base; anthers oblong; pistillode 3-styled; female spike solitary, terminal, 1¼–1½ in. long, oblong, acute; spathe short; bracts in many rows, ovate-oblong, coriaceous, membranous above, prolonged at the apex into a long awn; flowers stalked, oblong-lanceolate; outer perianth-segments rigid, oblong-lanceolate, acute; lateral conduplicate, with a glabrous or glabrescent keel; inner shorter, thinner, posterior one widest; capsule obliquely ovate, coriaceous, ferruginous, 2-celled by abortion, shorter than the persistent perianth. *Steud. Synops.* ii. 253; *Mast. in Journ. Linn. Soc.* viii. 245; *DC. Monog. Phan.* i. 288.

VAR. β, **nutans** (Mast. in Journ. Linn. Soc. viii. 245); male spikelets smaller, in diffuse panicled cymes; peduncles deflexed; spikelets nodding. *Mast. in DC. Monog. Phan.* i. 289.

SOUTH AFRICA: without locality, *Thom.* 632! 906! Var. β, *Zeyher!*
COAST REGION: Stellenbosch Div.; Hottentots Holland Mountains, *Ecklon!* Caledon Div.; mountains near Caledon, *Zeyher!* Nieuw Kloof, Houw Hoek Mountains, *Burchell*, 8071! Var. β, Cape Div.; Table Bay, *Robertson!* False Bay, *Robertson*, ♂ and ♀!

68. R. obtusissimus (Steud. Synops. ii. 252); stems 2 ft. high, erect, terete, unbranched, olivaceous, with minute white spots; leaf-sheaths about 1 in. long, tightly convolute, coriaceous, with a thinner margin, subulate, mucronate; male spikelets 1–2–3 in linear cymes, each spikelet $\frac{1}{8}$ in. long, cylindric-oblong, obtuse; spathe half the length of the spikelet, subulate, mucronate; bracts slightly imbricate, suborbicular, cartilaginous, nervoso-striate, mucronulate; perianth-segments oblong, obtuse; lateral conduplicate, scarcely keeled, glabrous; intermediate anterior oblong, obtuse, with 2 ferruginous nerves, as long as or longer than the lateral segments; inner segments similar, hyaline, obtuse, shorter than the outer segments; anthers ovato-mucronate. *Mast. in DC. Monog. Phan.* i. 296. *R. digitatus, Nees in Linnæa,* v. 638; *Kunth, Enum.* iii. 410; *Mast. in Journ. Linn. Soc.* viii 228, *not of Thunb.*

SOUTH AFRICA: without locality, *Drège*, 22! *Ecklon and Zeyher*, 595!

COAST REGION: Cape Div.; Table Mountain, 1000 ft., *MacOwan, Herb. Aust.-Afr.*, 1673! Stellenbosch Div.; Hottentots Holland, *Zeyher!* *Ecklon*, 18! Lowrys Pass, *Burchell*, 8196! *Schlechter*, 7221! Caledon Div.; Nieuw Kloof, Houw Hoek Mountains, *Burchell*, 8072!

This may prove to be a species of *Hypolæna.*

69. R. sarocladus (Mast. in DC. Monog. Phan. i. 291); stems cæspitose, erect, 18 in. high, terete, much branched; branches erecto-patent, purple-mottled; leaf-sheaths about $\frac{3}{4}$ in. long, slightly convolute, coriaceous, striate, with a long leafy acumen between two membranous lobes; female spikes solitary, $\frac{3}{4}$–1 in. long, cylindric, lanceolate, finally compressed, wedge-shaped; spathe lanceolate, coriaceous, striate, rather shorter than the spikelet; bracts ultimately spreading, oblong-lanceolate, acuminate, decurrent at the base; perianth-segments oblong, rather obtuse; lateral conduplicate, villous-carinate; inner smaller, thinner, involute at the margins; ovary roundish, with 3 styles; capsule orbicular, compressed, 2-celled by abortion, surmounted by the remains of the styles.

COAST REGION: Cape Div.; Table Mountain, *Burchell*, 572, ♀! 605, ♀!

70. R. Rhodocoma (Mast. in Journ. Linn. Soc. x. 275); stems erect, stout, terete, 3–4 ft. high, olivaceous, puncticulate, subverticillately and repeatedly branching; ultimate branchlets filiform, rather compressed, erecto-patent; lower sterile; upper spike-bearing; leaf-sheaths 1–1$\frac{1}{2}$ in. long, slightly convolute, elliptic, acuminate, coriaceous, striate, purpurascent; upper margins deeply hyaline; apex prolonged into a long, ultimately deciduous awn; smaller sheaths more or less foliaceous; male spikelets numerous, in linear cymes, rarely solitary or two, usually forming a large, erect or spreading inflorescence 6–8 in. long, each spikelet about $\frac{1}{4}$ in. long, many-flowered; spathe oblong, coriaceous, membranous near the top; apex acuminate, aristate; bracts oblong, coriaceous, shining brown, hyaline above; apex subulate, mucronate; perianth-segments oblong, obtuse, rigid; lateral conduplicate, with a glabrous keel;

anthers linear-oblong; female spikelets and bracts as in the male; flower solitary, quasi-terminal, longer than the bract; staminodes strap-shaped; ovary rounded, 3-celled; styles 3; capsule coriaceous, 3-lobed, 3-celled; seeds large, trigonous. *Mast. in DC. Monog. Phan.* i. 294. *Rhodocoma, Nees in Lindl. Nat. Syst. ed.* 2, 450; *Kunth, Enum.* iii. 480. *R. capense, Nees ex Steud. Synops.* ii. 248. *R. Equisetum, Nees ex Mast. in Journ. Linn. Soc.* x. 275.

COAST REGION: Uitenhage Div.; Zwartkops River, *Ecklon*, 788, ♂ and ♀! Alexandria Div.; Zuurberg Range, *Drège*, 2016, ♂ and ♀!

CENTRAL REGION: Somerset Div.; Commadagga, *Burchell*, 3322, ♂! Alexandria Div.; on the rocks of Zwartwater Poort, *Burchell*, 3373! Albany Div.; near Riebeek, *Burchell*, 3509!

71. R. callistachyus (Kunth, Enum. iii. 400); stems 2–6 ft. high, erect, terete, with very numerous, ascending, slender branches, olivaceous, puncticulate; leaf-sheaths $\frac{1}{2}$–$\frac{3}{4}$ in. long, closely convolute, coriaceous, nervoso-striate, thinner at the margins; apex subulate, mucronate; mucro appressed; male spikelets 6–9 or more in linear spike-like cymes, each 1–2 lin. long; spathe sheath-like, acuminate, nearly equal to the spikelet; bracts ultimately spreading, oblong, mucronate, subcoriaceous, brown; perianth-segments membranous, oblong, acute; lateral conduplicate, villous-carinate; inner smaller; anthers linear, apiculate; pistillodium minute; female spikelets 1–3, about $\frac{3}{4}$ in. long, oblong, flattened; spathe oblong, cuspidate, half the length of the spikelet; bracts ultimately spreading, oblong, acuminate, rather flat, thin, wavy at the edges, chestnut-brown; perianth-segments as in the male, but broader; staminodes 3; ovary trigonous, dark chestnut, 2-celled by abortion; stigmas 3, recurved. *Steud. Synops.* ii. 253, *female. R. polystachyus, Kunth, Enum.* iii. 402; *Steud. Synops.* ii. 253, *male. R. concolor, Steud. Synops.* ii. 251; *Mast. in DC. Monog. Phan.* i. 241. *R. fastigiatus, Nees ex Mast. in Journ. Linn. Soc.* viii. 250, *not of R. Br. R. Mastersi, Muell. Fragm.* viii. 68; *Mast. in DC. Monog. Phan.* i. 292.

SOUTH AFRICA: without locality, *Drège*, 31, ♀! 32, ♂! *Burchell*, 5812, ♀!

COAST REGION: Tulbagh Div.; Winter Hoek, *Ecklon!* Stellenbosch Div.; Hottentots Holland, below 1000 ft., *Zeyher!* Caledon Div.; mountain slopes in the neighbourhood of Steenbrazen River, 1000 ft., *MacOwan, Herb. Aust.-Afr.*, 1669, ♂ and ♀! Humansdorp Div.; mountains near Kromme River, 1000–2000 ft., *Drège*, 30! north side of Kromme River, *Burchell*, 4843, ♀! Uitenhage Div., *Ecklon and Zeyher*, 539, ♀!

72. R. spinulosus (Kunth, Enum. iii. 402); stem 2–3 ft. high, erect, terete, dull olive, striate, purple-mottled, branching in the middle; branches virgate, ascending, compressed; leaf-sheaths 1 in. long, closely convolute, coriaceous, thinner at the margins, nervoso-striate; apex subulato-mucronate; male spikelets 1–4 in erect terminal spicate cymes, each about $\frac{1}{4}$ in. long, oblong, ultimately cuneate, compressed; spathe as long as the spikelet, ultimately deciduous; bracts loosely imbricate, ovate, oblong, coriaceous, mucronate; perianth-segments oblong, obtuse; lateral conduplicate, villous-carinate; intermediate flattish, with a prominent mid-nerve; anthers ovate; pistillode minute. *Steud. Synops.* ii. 253; *Mast. in DC. Monog. Phan.* i. 292.

COAST REGION: Uniondale Div.; between Welgelegen and Onzer, 1500–2000 ft., *Drège*, 37, ♂!

73. R. compressus (Rottboell, Descr. et Ic. 6, t. 2, fig. 4); stems 2–3 ft. high, solid, erect, compressed, olivaceous, minutely tubercled, branching about the centre; leaf-sheaths 1–1½ in. long, closely convolute, puncticulate, coriaceous, thinner towards the apex, mucronate; mucro often prolonged into a long, blunt leaf; male spikelets 5–12 in linear distichous cymes, each about ½ in. long; spathe oblong-lanceolate, shortly pointed, about the length of the oblong spikelet; bracts ultimately spreading at the tips, oblong-lanceolate, mucronulate, shining brown; flowers shortly stalked, ovoid; outer perianth-segments rigid, lanceolate; lateral conduplicate, villous-carinate; anthers linear-oblong, muticous; pistillode minute; female flowers similar; ovary subtrigonous, obtuse. *Willd. Sp. Pl.* iv. ii. 725; *Spreng. Syst.* i. 185, *excl. syn.*; *Kunth, Enum.* iii. 403; *Nees in Linnæa*, v. 642; *Steud. Synops.* ii. 253; *Mast. in Journ. Linn. Soc.* viii. 249; *DC. Monog. Phan.* i. 290.

VAR. β, **major** (Mast. in Journ. Linn. Soc. viii. 249); stems more robust; leaves more developed; spikelets nearly twice as large as in the type. *Mast. in DC. Monog. Phan.* i. 291.

SOUTH AFRICA: without locality, Var. β, *Sieber*, 224, ♂ and ♀!

COAST REGION: Malmesbury Div.; Groene Kloof, below 500 ft., *Drège*, 34 partly, ♂! Swellendam Div.; between Sparrbosch and Tradouw, 2000–3000 ft., *Drège*, 33, ♂! on a mountain-peak near Swellendam, *Burchell*, 7388, ♂! Riversdale Div.; on the Lange Bergen, near Riversdale, 1500 ft., *Schlechter*, 1914! George Div.; on the Post Berg, near George, *Burchell*, 5900, ♂! Var. β. Cape Div.; Table Mountain, 2200–3000 ft., *Bolus*, 4437, ♂ and ♀! 4370, ♂! *Burchell*, 568, ♀! *Ecklon*, 92, ♂! Cape Flats, *Buchanan*, 161, ♂! near Cape Town, *Drège*, 48 partly, ♂ and ♀! Caledon Div.; mountains near Grietjes Gat., 2000–4000 ft., *Zeyher*, 4350, ♂!

74. R. quadratus (Mast. in Journ. Linn. Soc. x. 277); stems erect, 4–5 ft. high, stout, 4-sided, olive-coloured, puncticulate, much and repeatedly branched; ultimate branches spreading, filiform; fertile branches longer than the sterile; leaf-sheaths 1–2 in. long, closely convolute, lanceolate, mucronate, coriaceous, thinner at the upper margins, nervoso-striate; smaller sheaths white-spotted, with 2 deep, hyaline lobes above; apex prolonged into a curved foliaceous mucro; male spikelets numerous in loose, terminal, slightly branched panicles, each ovato-oblong or turbinate, about 2 lin. long; spathe nearly as long as the spikelet; bracts oblong, acute, subcoriaceous, thinner at the margins, shortly mucronulate; perianth-segments oblong; lateral conduplicate, villous-carinate; keel glabrous (glabrescent?); inner similar, smaller, flat or slightly incurved at the margins; anthers oblong, apiculate; female spikelets in panicled cymes, each about 2 lin. long, oblong-ovate or turbinate, 1–2-flowered; bracts and perianth as in the male; staminodes strap-shaped, minute; ovary 3-lobed, 3-celled; capsule 3-lobed, or by abortion 2–1-celled, scarcely shorter than the persistent perianth. *Mast. in DC. Monog. Phan.* i. 293. *R. tetragonus, β exaltatus, Nees in herb.! Schœnus capensis, Burmann, Fl. Cap. Prod.* 3. *Equisetum, Breyn. Exot. Pl. Cent.* 176, *t.* 91.

SOUTH AFRICA: without locality, *Pappe*, 96, ♀! *Drège*, 364c! *Niven*, ♀!
COAST REGION: Cape Div.; Table Mountain, *Ecklon!* vicinity of Cape Town, *Burchell*, 408, ♂!

Variable in stature.

75. **R. debilis** (Nees in Linnæa, v. 641); stems cæspitose; branches ascending, very slender, minutely puncticulate; leaf-sheaths $\frac{1}{4}$–$\frac{1}{2}$ in. long, closely convolute, coriaceous, striate, membranous and 2-lobed above; apex subulate, mucronate; male spikelets numerous, in an erect, linear cyme, each about 1 lin. long, oblong, erect; spathe aristate, nearly as long as the spikelet; bracts oblong, obtuse, membranous at the apex, muticous or mucronulate; perianth-segments linear-oblong; lateral conduplicate, with a villous keel; anthers oblong, apiculate; pistillode minute; female spikelets solitary or few, ovoid, 2 lin. long; bracts ovate-oblong, mucronulate; perianth-segments oblong, obtuse; lateral conduplicate, villous-carinate; inner oblong, spathulate, wider than the outer; ovary oblong, obtuse; styles 3. *Kunth, Enum.* iii. 412; *Steud. Synops.* ii. 254; *Mast. in Journ. Linn. Soc.* viii. 250; *DC. Monog. Phan.* i. 290.

COAST REGION: Caledon Div.?; "near Knoblauch" (=Knoflooks Kraal?), *Zeyher!*

Imperfectly known species.

76. **R. Scopula** (Mast.); stems much branched; branches very slender, ascending, terete, greyish-olive, rather coarsely tubercled; leaf-sheaths $\frac{3}{4}$ in. long, tightly convolute, coriaceous, nervoso-striate; upper edges brown, membranous; apex tapering into a lanceolate, acuminate or mucronate process; spikelets $\frac{1}{8}$ in. long; linear-oblong, solitary, terminal, or 5–7 in linear cymes; spathe sheath-like, half the length of the spikelet; bracts exceeding the flowers, oblong-lanceolate, with a thin white margin; perianth compressed; segments cartilaginous, linear-oblong; outer lateral conduplicate; inner shorter; ovary 3-styled.

COAST REGION: Cape Div.; eastern slopes of Table Mountain, 2600 ft., *Bolus*, 4095 β, ♀!

77. **R. ambiguus** (Mast.); stems slender, terete, sparingly branched; branches spreading, terete, greyish, rugulose, of the thickness of a sparrow quill; sheaths $\frac{1}{2}$ in. long, rather loosely convolute, dark brown, coriaceous, pitted, slightly membranous towards the acute, shortly mucronate apex; male spikelets solitary or in linear spicate cymes, each spikelet about $\frac{3}{4}$ in. long, linear-cylindric, acute, supported at the base by an oblong, bract-like spathe half the length of the spikelet; bracts linear-oblong, acute, convolute at the margins,

blackish-brown, exceeding the shortly stipitate flowers; perianth-segments linear-oblong; two outer lateral conduplicate, glabrous; three inner equal, flattish, membranous, shorter than the outer; anthers blackish, apiculate; pistillode minute, 3-styled.

SOUTH AFRICA: without locality, *Zeyher!*

This plant, of which the female is not known, is very different from *R. distichus.*

78. R.? sejunctus (Mast.); stems erect, 2–3 ft. high, terete, flabellately branched; branches slender, ascending, olivaceous, somewhat coarsely tubercled; leaf-sheaths about ½ in. long, closely convolute, coriaceous, spotted, hyaline near the apex, and ending in a subulate mucro; male spikelets numerous, loosely arranged in linear, erect cymes, each spikelet about ¼ in. long, elliptic, ultimately cuneate; spathe sheath-like, half the length of the spikelet; bracts coriaceous, oblong-ovate, nearly as long as the oblong, somewhat triquetrous flowers; perianth 1½ lin. long, cartilaginous; outer lateral segments oblong, acute, villous-carinate; intermediate flattish, glabrous; inner 3 membranous, subequal, shorter than the outer; anthers linear ovoid; pistillode very minute.

EASTERN REGION: Natal, *Bolton!*

The specimen is in the herbarium of the British Museum, and a fragment at Kew. No other species of *Restio* is known at present to grow in Natal.

79. R. protractus (Mast.); stems 2–3 ft. high, erect, terete, much branched; branches filiform, fastigiate, ascending, rugulose, white-spotted; leaf-sheaths 1 in. long, closely convolute, coriaceous, white-spotted, protracted into a subclavate, leafy apex; sheaths on the branches smaller, deeply two-lobed; lobes hyaline; apex prolonged into a linear, obtuse leaf; female spikelets rather remote, in terminal, linear cymes, each about ¼ in. long, elliptic; spathe open, sheath-like, mucronate, half the length of the spikelet; bracts ovate-oblong, mucronate, longer than the flower; outer perianth-segments ovate-oblong, acute, conduplicate, villous; inner 3 shorter, more membranous; ovary roundish, 2-lobed by abortion, dehiscing at the angles.

COAST REGION: Cape Div.; in a valley near Table Mountain, 2800 ft., *Bolus,* 4443, ♀!

80. R. patens (Mast.); stems slender, fastigiately branched; branches very slender, terete, purplish, somewhat coarsely tubercled; leaf-sheaths about ¼ in. long, closely convolute, coriaceous, subulate, acuminate; margins membranous near the apex; male spikelets about ¼ in. long, 3–5 aggregate at the ends of the branches; spathes open, sheath-like, half the length of the spikelet; bracts oblong, acuminate, brownish, coriaceous, exceeding the flowers; outer perianth-segments linear-oblong, acute; 2 outer conduplicate, slightly villous-carinate; 3 inner equal, shorter than the outer, hyaline; anthers apiculate; pistillode very obscure.

COAST REGION: At the summit of Winterhoek Mountain, near Tulbagh, 6500 ft., *Bolus,* 7495, ♂!

81. R. lucens (Poir. Encycl. Meth. vi. 169); stems branching, subcompressed, glabrous, very smooth, jointed; sheaths cylindric, elongated, terminated by a setaceous filament, or sometimes lacerated at their summit; panicle terminal, large, diffuse, shining; branchlets filiform, somewhat lateral, almost verticillate; the common peduncle with a narrow, lanceolate, acute, membranous, bluish-grey sheath at its base; pedicels setaceous, thickened towards their summit, very smooth, flexible; spikelets about ½ in. long, narrow, lanceolate, acute, shining; bracts imbricate, ovate, membranous, with whitish or scarious margins. *Kunth, Enum.* iii. 414; *Mast. in DC. Monog. Phan.* i. 296.

VAR. minor (Mast. in DC. Monog. Phan. i. 296); panicles and spikelets smaller.

SOUTH AFRICA, in Lamarck's herbarium at Paris. I have not seen this plant.

82. R. glomeratus (Thunb. Diss. Rest. 18). A monstrous form of some species of *Willdenovia*. *Thunb. in Usteri, Delect.* i. 52; *Fl. Cap. ed. Schult.* 88; *Kunth, Enum.* iii. 414; *Mast. in Journ. Linn. Soc.* xiv. 418.

83. R. micans (Nees in Linnæa, v. 649). Under this name Nees describes some imperfect specimens which I am unable to identify. *Thamnochortus micans, Kunth, Enum.* iii. 441.

COAST REGION: Cape Div.; Cape Flats near Wynberg, collector not mentioned.

84. R. simplex (Forst. ex Thunb. Diss. Rest. 16). This has been wrongly stated by Kunth (Enum. iii. 414) to be a South African species, but according to Thunberg it is a native of New Zealand (see Thunb. in Usteri, Delect. i. 51, and 57 in note). On p. 51, under *R. triflorus*, the authority of *R. simplex* is credited to Forster by Thunberg.

85. R. squamosus (Thunb. in Hoffm. Phytog. Blätter, i. 8). This is not a Restiaceous plant, but perhaps a species of *Thesium*, and is too imperfect to be determined. *Thunb. in Weber and Mohr, Archiv.* i. 26; *Fl. Cap. ed. Schult.* 87; *Kunth, Enum.* iii. 414; *Mast. in Journ. Linn. Soc.* xiv. 418.

II. ASKIDIOSPERMA, Steud.

Outer perianth-segments subequal, oblong-lanceolate; lateral keeled and winged; inner purplish, flattish, longer than the outer. *Staminodes* narrow, strap-shaped. *Capsule* 2-lobed, 2-celled, dehiscing longitudinally. *Seed* pendulous.

Male unknown.

DISTRIB. Endemic.

1. A. capitatum (Steud. Synops. ii. 257); stem simple, erect, terete, slender, dull olive-coloured, wrinkled; leaf-sheaths about ½ in.

long, loosely convolute, deciduous, blueish-green, membranous and lacerate at the apex; inflorescence a solitary terminal spike, about 1 in. long; bracts oblong, with very long tails, coriaceous, with hyaline, deeply laciniate margins. *Mast. in Journ. Linn. Soc.* x. 247; *DC. Monog. Phan.* i. 304, *t.* 1, *fig.* 32–36.

COAST REGION: Clanwilliam Div.; Ceder Bergen, at Ezels Bank, 3000–4000 ft., *Drège*, 2510, ♀!

III. DOVEA, Kunth.

Flowers more or less 3-sided, longer than the subtending bract. *Perianth-segments* 6, in 2 rows; outer lateral conduplicate, carinate; inner longer, oblong-acute. *Stamens* 3. *Pistillode* with rudiments of 3 styles, or absent. *Female perianth* similar. *Staminodes* narrow, strap-shaped, or none. *Ovary* 3-lobed, 3-celled. *Stigmas* 3, sessile, revolute, plumose on the inner surface. *Fruit* hard, capsular, 3-lobed, 3-celled, dehiscing at the angles. *Seed* solitary, pendulous from the inner angle of each cell. *Testa* membranous, with wavy, membranous ridges.

Perennial herbs; stems springing from a rootstock covered with sheaths; upper sheaths deciduous; male and female inflorescence mostly alike, spicately or paniculately cymose, female less branched, often with large, flat, subpersistent spathes; spathes of male inflorescence caducous, convex, not greatly exceeding the primary branches of the inflorescence; spikelets 1–2-flowered.

DISTRIB. Endemic.

Male inflorescence elongate, linear, paniculate, 3–6 in. long ... ... (1) **tectorum.**
Male inflorescence dense, short, spicate, 1–2 in. ... (2) **cylindrostachya.**
Male inflorescence rather loose, linear paniculate, 1–3 in. long ... ... (3) **hookeriana.**
Male inflorescence elongate, loose, linear, paniculate, 2–3 in. long:
Male inflorescence loose ... ... (4) **recta.**
Male inflorescence dense:
  Bracts and perianth-segments ovate, acute ... (5) **aggregata.**
  Bracts and perianth-segments roundish ... ... (6) **macrocarpa.**
  Bracts and perianth-segments acute:
    Staminodes 3; fruit not constricted ... ... (7) **ebracteata.**
    Staminodes 0; fruit transversely constricted above the middle ... ... (8) **Bolusi.**
  Bracts papery, linear-lanceolate; perianth-segments aristate ... ... (9) **paniculata.**
  Bracts subcoriaceous, rounded; perianth-segments oblong, obtuse ... ... (10) **racemosa.**
  Bracts oblong, acuminate; perianth-segments acuminate; plant very robust ... ... (11) **mucronata.**
  Bracts and perianth-segments ovate, obtuse; plant very robust ... ... (12) **thyrsoidea.**

1. D. tectorum (Mast. in Journ. Linn. Soc. x. 249); stems 2–3 ft. long, cæspitose, erect, simple, terete, of the thickness of a crow-quill, puncticulate; leaf-sheaths closely convolute, deciduous; spikelets black, very numerous, in dense, erect, terminal, linear, much-branched panicles, 6–8 in. long; spathes deciduous; spikelets about 1 lin. long, oblong, ovoid; bracts ovate, mucronate, scarcely so long as the flowers; perianth-segments rigid, oblong-obtuse, the inner series longer than the outer; anthers apiculate; female flowers similar to the males, with 3 liguliform staminodes; ovary triangular; capsule 3-lobed, 3-celled. *Mast. in DC. Monog. Phan.* i. 306. *Restio tectorum, Linn. Suppl.* 425, *and in his herbarium! R. tectorum, β and γ, Thunb. Herb. ex Mast. in Journ. Linn. Soc.* xiv. 417. *Restio nudus, vars. α and β, Nees in Linnæa,* v. 651 *excl. Syn. Elegia racemosa, Pers. Synops.* ii. 607; *Kunth, Enum.* iii. 463.

COAST REGION: Clanwilliam Div.; Jakhals River, below 500 ft., *Drège*, 2506! Cape Div.; Cape Flats, *Thunberg*, ♀! *Burchell*, 428! 824, ♂! *Ecklon*, 559, *Burke*! *MacOwan*, Herb. Austr.-Afr., 1874! Caledon Div.; Nieuw Kloof, Houw Hoek Mountains, *Burchell*, 8127!

EASTERN REGION: Natal, *Rehmann*, 8596! locality probably an error.

2. **D. cylindrostachya** (Mast.); stems cæspitose, 15–18 in. long, unbranched, of the thickness of a crow-quill, erect, terete, dark olive, obsoletely puncticulate, clothed at the base with numerous crowded, overlapping, deep chestnut-coloured, shining scales, the lowest smallest, oblong-obtuse, mucronulate, gradually increasing in size upwards; leaf-sheaths numerous, remote, deciduous; male and female inflorescence in dense, terminal, spicate cymes, 1–1½ in. long; spathe oblong, shorter than the inflorescence; bracts ovate-mucronate; flowers 2 lin. long, trigonous; outer perianth-segments linear-oblong, glabrous, coriaceous; inner brown, twice the length of the outer, paler; anthers linear-oblong, acute; rudiment of the 3-styled pistil minute; female inflorescence rather longer, dense; spathes not conspicuous; bracts shorter than the flowers; outer perianth-segments subequal, oblong-ovate, acute, dark brown, shorter than the oblong, acute, paler-coloured inner segments; ovary 3-sided, 3-styled, surrounded by 3 staminodes.

SOUTH AFRICA: without locality, *Drège*, 9605, ♀! *Thom*, 908, ♀!

3. **D. hookeriana** (Mast. in Journ. Linn. Soc. x. 249); stems tufted, 1½–2 ft. long, rigid, erect, simple filiform, terete, solid, slightly wrinkled, puncticulate; leaf-sheaths closely convolute, deciduous, elliptic, coriaceous, membranous above, apex mucronato-aristate; male spikelets numerous, one lin. long, in erect, linear, slightly branched, panicled cymes, 1½–3 in. long; spathes deciduous; bracts suborbicular, acuminate, shorter than the oblong, trigonous flowers, 2 lin. long; perianth-segments rigid, ferruginous; outer 3 equal, boat-shaped, oblong-acute; inner similar, but longer; anthers apiculate; pistillode with 3 rudimentary styles; female plant like the male; spikelets roundish, the size of a small pea, in linear, spiciform cymes; spathes open, sheath-like; bracts ovate-mucronate, the lower ones sterile; staminodes 3; capsule 3-lobed, 3-celled, dehiscing

at the angles, surmounted by the remains of 3 styles. *Mast. in DC. Monog. Phan.* i. 306. *Restio tectorum, a, Thunb. Herb. ex Mast. in Journ. Linn. Soc.* xiv. 417.

SOUTH AFRICA: without locality, *Thom*, 901 and 1621, ♂! 632, ♀!
COAST REGION: Caledon Div.; Nieuw Kloof, Houw Hoek Mountains, *Burchell*, 8062, ♀!

4. **D. recta** (Mast.); rhizome creeping, giving off tufts of erect, wiry, terete, very slender stems, 10–12 in. high; basal sheaths persistent, oblong, obtuse, mucronulate, chestnut-brown, closely convolute; upper sheaths deciduous, subulate, mucronate; male inflorescence a loose, linear, spiciform panicle; spathes short, deciduous; flowers sessile, or shortly pedicellate, loosely approximate; spathe oblong-mucronulate, rather shorter than the spikelet; bracts oblong, acute, scarcely as long as the perianth; outer segments linear-oblong, acute, glabrous, shorter than the inner; anthers linear.

COAST REGION: Caledon Div.; between Donker Hoek and Houw Hoek Mountains, *Burchell*, 8007, ♂!

5. **D. aggregata** (Mast.); stems erect, cylindrical, as thick as a goose-quill, simple, 2–3 ft. long, faintly puncticulate; leaf-sheaths deciduous; female inflorescence a spicate panicle, 2–3 in. long; spikelets closely aggregate; bracts and spathellæ about the same length as the flowers, ovate-oblong, acute, shortly mucronate; perianth-segments cartilaginous, oblong, acute; outer lateral acute, conduplicate, villous-carinate; staminodes none; capsule 3-lobed, 3-celled, dehiscing at the angles, with a single pendulous seed from the apex of each cell; testa wrinkled; male plant not known.

COAST REGION: Cape Div.; near Cape Town, *Zeyher!* Caledon Div.; Genadendal, 3000–4000 ft., *Drège*, 1622, ♀!

6. **D. macrocarpa** (Kunth, Enum. iii. 458); stems 2–3 ft. long; terete or rather compressed, slightly branched, greenish puncticulate; sheaths 1–2 in. long, oblong, mucronate, deciduous; male spikelets numerous, blackish, in terminal, oblong, erect, spicate cymes each 3–4 in. long; spikelets 1–2 lin. long, roundish; spathes deciduous; bracts ovate, coriaceous, membranous at the margins; male flowers somewhat 3-sided, twice the length of the bracts; outer lateral perianth-segments boat-shaped, oblong-obtuse, coriaceous, with membranous margins; inner three equal, concave, twice the length of the outer; female spikelets 2–5, each about $\frac{1}{4}$–$\frac{1}{2}$ in. long.; bracts ovate, acute, coriaceous, nervoso-striate; perianth-segments rigid, oblong-acute; inner longer; capsule longer than the persistent perianth, 3-lobed, 3-celled; seed solitary, large, with a white minutely ribbed testa. *Steud. Synops.* ii. 248; *Mast. in Journ. Linn. Soc.* x. 250; *DC. Monog. Phan.* i. 307. *R. nudus, var. pauciflorus, Nees in Linnæa*, v. 651.

COAST REGION: Clanwilliam Div.; Ceder Bergen, *Zeyher*, 1740, ♂ and ♀!

Piquetberg Div.; near Piquiniers Kloof, *Dickson in Herb. Bolus*, 4243, ♀! sand hills between Pretor's and Piquiniers Kloof, 1000-1500 ft., *Drège*, 2523, ♀! Piquetberg, 1500-3000 ft., *Drège*, 2507, ♂!

7. **D. ebracteata** (Kunth, Enum. iii. 458); stems cæspitose, 12–18 in. long, covered at the base with deep brown scales, erect, terete, unbranched; sheaths alternately deciduous, $1\frac{1}{2}$ in. long, loosely convolute, deep brown, coriaceous, thinner towards the obtuse apex, aristulate; male spikelets numerous in terminal, erect, spicate cymes about 1 in. long; bracts ovate-subulate, mucronate, coriaceous, shorter than the flowers, which are about 1 line long, trigonous; outer perianth-segments rigid, oblong, acute, boat-shaped; inner oblong, acute, longer than the outer; pistillode minute; female plant as in the male; perianth-segments cartilaginous, cymbiform, ovate, acute, mucronulate; inner flat; capsule 3-lobed, longer than the persistent perianth. *Steud. Synops.* ii. 248; *Mast. in Journ. Linn. Soc.* x. 250; *DC. Monog. Phan.* i. 308. ? *D. microcarpa, Kunth, Enum.* iii. 459; *Steud. Synops.* vi. 248; *Mast. in Journ. Linn. Soc.* x. 248.

SOUTH AFRICA: without locality, *Sieber*, 232, ♀! *Drège*, 9602, ♂!

COAST REGION: Cape Div.; Table Mountain, *Milne*, 236, ♂! *MacGillivray*, 438, ♂! *Burchell*, 524, ♀! 526, ♀! 567, ♂! *Bolus*, 4445, ♂ and ♀! *Drège*, 72, ♂! Worcester Div.; Dutoits Kloof, *Drège*, 72 ?, ♂! 124, ♂!

8. **D. Bolusi** (Mast. in Journ. Linn. Soc. xxi. 576); stems 18 in. long, cæspitose, slender, erect, terete, unbranched, covered at the base with brownish, loosely convolute, aristulate sheaths; upper leaf-sheaths deciduous; male and female inflorescence compact, linear-cymose, about 2 in. long; spathes about $\frac{1}{2}$ in. long; bracts deep brown, ovate, mucronate, scarcely shorter than the flower; flowers oblong, 1 line long and upwards; outer perianth-segments navicular, acute; inner similar but longer; pistillode minute, 3-styled; female flower like the male; staminodes none; ovary oblong, 3-gonous, nearly as long as the inner perianth-segments, 3-lobed; lobes rounded, at first constricted in the middle, fleshy, yellow; styles 3.

COAST REGION: Cape Div.; Muizenberg, 1600 ft., *Bolus*, 3909, ♀! 3910, ♂!

This differs from *D. ebracteata* in the smaller proportions of all its parts.

9. **D. paniculata** (Mast. in Journ. Linn. Soc. xxi. 577); stems cæspitose, 2–3 ft. long, slender, erect, simple or sparingly branched, terete, wrinkled, puncticulate; basal sheaths coriaceous, chestnut-coloured, subulato-mucronate; upper sheaths deciduous; male spikelets very numerous, in loose, linear-oblong, much branched cymes, 2–3 in. long, with numerous flat, lanceolate, pale brown spathes; spathellæ linear-lanceolate; flowers 2 lin. long, stipitate, trigonous; outer perianth-segments oblong acuminate, of parchment-like consistence; inner similar but longer; female inflorescence about 2 in. long, compact, oblong, slightly branched, with few spikelets; spathes and spathellæ (bracts) as in the male; outer perianth-segments horny,

in shape as in the male; staminodes none; ovary oblong, obtuse, 3-lobed, 3-celled; cells 1-ovuled; style thickened at the base; fruit trigonous, dehiscent.

COAST REGION: Tulbagh Div.; on the Witsen Berg, near Tulbagh, *Burchell*, 8719, ♂! Worcester Div.; Dutoits Kloof, 3000–4000 ft., *Drège*, 125, ♀! 1643, ♂ and ♀! 1651, ♂! Drakenstein mts. near Bains Kloof, 1600–2000 ft., *Bolus*, 4082, ♂! 4098, ♂! 4081, ♀! 4097, ♀! Ceres Div.; mountains near Ceres, 2400 ft., *Bolus*, 5293, ♂! Caledon Div.; between Villiersdorp and French Hoek, *Bolus*, 7483, ♀!

The female specimen collected by Drège under No. 125 has hitherto been associated with Sieber's specimen No. 232 under *D. ebracteata*, Kth., but it is certainly not the same as Sieber's plant, and properly belongs to the present species. The two are easily distinguished by the colour of the annular persistent bases of the stem-sheaths, which are black in *D. ebracteata*, whilst in *D. paniculata* they are brown, and often scarcely darker than the stem.

10. **D. racemosa** (Mast. in Journ. Linn. Soc. xxi. 578); stems cæspitose, 2–3 ft. long, erect, terete, or slightly compressed, solid, unbranched, wrinkled, grey; sheaths mucronulate, deciduous; male spikelets numerous, 1-flowered, in long, erect, panicled cymes 3–6 in. long; spathes numerous, 1–2 in. long, flat, oblong, acute, coriaceous, minutely tuberculate, with membranous margins, shining on the inner surface; bracts oblong-obtuse, mucronulate, shorter than the oblong-obtuse flowers; outer perianth-segments oblong, obtuse, membranous; inner longer and narrower; female inflorescence (according to Lamarck's figure) like the male; perianth-segments acute; ovary 3-lobed, 3-celled. *Restio racemosus, Lam. Encycl. Meth.* vi. 177; *Lam. Ill.* iii. 400, *t.* 804, *fig.* 4.

SOUTH AFRICA: without locality, *Thom*, 904!

COAST REGION: Cape Div.; Table Mountain, *Burchell*, 574, ♂! Worcester Div.; Dutoits Kloof, 3000–4000 ft., *Drège*, 1647, ♂! Caledon Div.; Niew Kloof, Houw Hoek Mountains, *Burchell*, 8067, ♂! Riversdale Div.; lower part of the Lange Bergen near Kampsche Berg, above the waterfall at "Valley Rivers Poort," *Burchell*, 6999, ♂! Uitenhage Div.; rocky clefts near the top Witteklip Mountain, *MacOwan*, 2150, ♂!

11. **D. mucronata** (Mast. in Journ. Linn. Soc. x. 251); stems erect from a creeping rootstock, 5–7 ft. long, terete, of the thickness of the little finger, cinnamon-brown; upper leaf-sheaths 4 in. long, loosely convolute, spreading, apex acuminate or mucronate, shining on the inner surface; male spikelets very numerous, arranged in dense, much branched, panicled cymes, 2–3 in. long, with numerous broad, open, sheath-like spathes, each spikelet about 1 line long, 1-flowered; bracts oblong, acuminate, pale brown; flowers somewhat 3-sided; outer perianth-segments coriaceous, ferruginous, broadly ovate, acuminate; inner coriaceous, oblong-lanceolate, longer than the outer; female inflorescence like the male; spikelets 1-flowered; bracts ovate, acuminate; flowers somewhat 3-sided, longer than the bracts; outer perianth-segments rigid, equal, ovate, mucronate, carinate, concave, glabrous; inner similar, flatter, longer than the outer; capsule coriaceous, 3-lobed, 3-celled,

surmounted by the remains of the cells, dehiscent at the angles; seed solitary, pendulous from the inner angle of each cell of the capsule; testa membranous, with prominent wavy ridges. *Mast. in DC. Monog. Phan.* i. 308, *t.* 2, *figs.* 1–6, *t.* 5, *fig.* 6. *Restio mucronatus, Nees in Linnæa,* v. 660. *Elegia mucronata, Reich. ex Nees in Linnæa,* v. 660; *Kunth, Enum.* iii. 475; *Steud. Synops.* ii. 262. *Elegia panicoides, Kunth, Enum.* iii. 470; *Steud. Synops.* ii. 262.

SOUTH AFRICA: without locality, *Masson! Zeyher!*

COAST REGION: Cape Div.; Table Mountain, 3600 ft., *Bolus,* 2878, ♂! 4468, ♀! *Burchell,* 563, ♂ and ♀! *MacOwan,* 2514, ♂! Worcester Div.; Dutoits Kloof, 3000–4000 ft., *Drège,* 34. ♂! Riversdale Div.; Lange Bergen near Riversdale, 1800 ft., *Schlechter,* 1912, ♂ and ♀! and near Kampsche Berg, *Burchell,* 7026, ♂ and ♀!

12. **D. thyrsoidea** (Mast. in Journ. Linn. Soc. x. 251); stems stout, simple erect, 3–4 ft. long, terete or rather compressed, grey, impresso-puncticulate; sheaths loosely convolute, somewhat persistent, elliptic, coriaceous, brown, puncticulate, studded with minute pits; male inflorescence densely cymose, 2–3 in. long; pedicels 2, unequal, in the axil of a deciduous spathe; bracts ovate-concave, as long as, or longer than the flowers; outer perianth-segments oblong-acute, boat-shaped, rigid; inner oblong-obtuse, longer than the outer; female plant as in the male; inflorescence erect linear-oblong, 5–6 in. long, with numerous large, convolute, sheath-like spathes; perianth as in the male; ovary 3-lobed, 3-celled, surmounted by 3 revolute stigmas; capsule 3-lobed, 3-celled, dehiscing at the angles and surrounded by the persistent, accrescent perianth; seed oblong, testa in wavy folds. *Mast. in DC. Monog. Phan.* i. 309.

COAST REGION: George Div.; summit of Post Berg, near George, *Burchell,* 5895, ♂ and ♀!

IV. ELEGIA, Linn.

Male and female inflorescence similar, paniculate-cymose with numerous deciduous or persistent spathes. *Flowers* numerous, minute, 3-sided, arcuate or compressed; outer perianth-segments in the male subequal, much shorter than the inner; in the female, the outer and inner segments are nearly equal. *Stamens* 3; anthers 1-celled. *Pistillode* when present minute, 3-styled. *Female flower: Ovary* 1-celled, 1-seeded; styles 2–3, linear, distinct. *Fruit* triangular, indehiscent.

Rush-like herbs with creeping or tufted rootstock; leaf-sheaths generally deciduous and leaving a black annular scar.

DISTRIB. Endemic.

Branches verticillate:
  Inflorescence 6–8 in. long ... ... (1) **verticillaris.**
  Inflorescence 1–3 in. long ... ... (2) **equisetacea.**

Branches, if present, not verticillate:
- Stem compressed (at least when dry):
  - Leaf-sheaths sub-persistent:
    - Bracts rounded-mucronate; perianth-segments oblong ... ... (3) **coleura.**
    - Bracts aristate; perianth-segments acute (4) **glauca.**
  - Leaf-sheaths caducous:
    - Inflorescence 1–2 in. long, ovoid; bracts acuminate; ovary deeply 3-lobed ... (5) **thyrsifera.**
    - Inflorescence 2–3 in. long, linear; somewhat loosely panicled:
      - Bracts and perianth-segments acuminate ... ... (6) **acuminata.**
      - Bracts and perianth-segments acute (7) **membranacea.**
      - Spathes ovate-oblong, acute; bracts and perianth-segments obtuse ... (8) **propinqua.**
    - Inflorescence linear, 1–6 in. long, loosely panicled, with persistent ovate-lanceolate spathes ... ... (9) **juncea.**
    - Inflorescence 1–4 in. long, linear, panicled, very dense; spathes subcaducous:
      - Bracts aristate, entire... ... ... (10) **cuspidata.**
      - Bracts aristate; margins lacerate ... (11) **asperiflora.**
- Stem terete, fistular (see also *E. thyrsifera*) ... (12) **fistulosa.**
- Stem filiform, wiry, solid, except in *E. parviflora*:
  - Inflorescence rarely exceeding ½ in. in length:
    - Spathes subpersistent, conspicuous:
      - Rootstock creeping; ovary ribbed ... (13) **squamosa.**
      - Rootstock cæspitose; ovary smooth ... (14) **vaginulata.**
    - Spathes deciduous or inconspicuous:
      - Bracts mucronate; male flower curved; perianth-segments acute ... ... (15) **deusta.**
      - Bracts obtuse; male flower straight; perianth-segments obtuse ... ... (16) **filacea.**
  - Inflorescence (male) 1–3 in. long:
    - Spathes conspicuous:
      - Sheaths persistent, flat ... ... (17) **stipularis.**
      - Sheaths persistent, convolute ... (18) **obtusiflora**
      - Sheaths deciduous:
        - Stem solid; perianth-segments obtuse ... ... (19) **spathacea.**
        - Stem fistular; perianth-segments acute ... ... (20) **parviflora.**
        - Stem solid intermediate; perianth-segments spathulate ... (21) **rigida.**
    - Spathes deciduous or inconspicuous:
      - Sheaths persistent; inflorescence lax, linear ... ... (22) **Verreauxii.**
      - Sheaths deciduous:
        - Inflorescence dense ... ... (23) **nuda.**
        - Inflorescence interrupted ... (24) **elongata.**

1. E. verticillaris (Kunth, Enum. iii. 471); stems tall, solid, white-spotted, giving off numerous thread-like, erect branchlets in whorls at the nodes; leaf-sheaths 2 in. long, coriaceous, ultimately deciduous, smaller sheaths on the branchlets about ¼ in. long, loosely convolute or open, coriaceous, with two hyaline lobes at the apex, shortly mucronate; male and female inflorescence similar, 6–8 in. long, much branched, linear-paniculate, cymose, many-flowered, interspersed with numerous, deciduous, sheath-like spathes; bracts shorter than the flowers, broadly ovate, acute, hyaline at the tips; perianth-segments oblong, obtuse; outer navicular; inner much longer than the outer; anthers oblong, apiculate; perianth-segments of the female flower oblong, obtuse; inner scarcely longer than the outer; fruit 3-sided, with 3 styles, indehiscent, 1-seeded, surrounded at the base by the persistent perianth. *Mast. in Journ. Linn. Soc.* x. 244; *DC. Monog. Phan.* i. 351. *Restio verticillaris, Linn. f. Suppl.* 425; *Steud. Synops.* ii. 262; *Linn. Herb.! Mast. in Journ. Linn. Soc.* xxi. 590; *Thunberg, Diss.* 19, *fig.* 7; *Usteri, Delect.* i. 53, *t.* 2, *fig.* 7; *Fl. Cap. ed. Schult.* 88; *Thunb. Herb.! Mast. in Journ. Linn. Soc.* xiv. 418, 419; *Nees in Linnæa*, v. 661.

SOUTH AFRICA: without locality, *Masson! Mund and Maire*, ♀! *Harvey! Thunberg*, ♀ and ♂!
COAST REGION: Clanwilliam Div.; Oliphants River, *Zeyher*, ♂ and ♀! Tulbagh Div.; margins of streams below Winter Hoek near Tulbagh, *Thunberg!* Worcester Div.; Bains Kloof, 1000 ft., *Bolus*, 4101, ♀! Dutoits Kloof, 3000–4000 ft., *Drège*, 1609, ♂ and ♀! mountains above Worcester, *Rehmann*, 2568, ♀! Caledon Div.; banks of the Bot River, near Genadendal, *Krauss*, ♀; Swellendam Div.; mountain ridges along the lower part of Zonder Einde River, 500–2000 ft., *Zeyher*, 4344, ♂ and ♀! Riversdale Div.; lower part of the Lange Bergen, about the waterfall at "Valley Rivers Poort," near Kampsche Berg, *Burchell*, 6992, ♂ and ♀! Mossel Bay Div.; Attaquas Kloof, *Thunberg!* Humansdorp Div.; north side of Kromme River, *Burchell*, 4845, ♀!

2. **E. equisetacea** (Mast. in Journ. Linn. Soc. xxi. 583); stems 2–3 ft. long, terete, olivaceous, white-spotted, simple or verticillately branched at the nodes; branches again verticillately ramulose; leaf-sheaths 1¼ in., spreading; male spikelets numerous, linear-oblong, in loosely paniculate cymes; cymes 4–5 in. long; spathes 1¼ in. long, flat, ovate, lanceolate; bracts shorter than the flowers, ovate, acuminate; perianth ⅛ in. long; outer segments linear-oblong, shorter than the inner; anthers apiculate; female inflorescence 3–4 in. long, linear-oblong, spicate-cymose; flowers three-sided; perianth-segments nearly equal, linear-oblong; fruit ovoid, somewhat 3-cornered, as long as the persistent perianth, 1-celled, 1-seeded. *Elegia propinqua, Kunth, var. equisetacea, Mast. in Journ. Linn. Soc.* x. 242; *DC. Monog. Phan.* i. 357.

SOUTH AFRICA: without locality, *Drège*, 103, ♂!
COAST REGION: Cape Div.; Table Mountain and Devils Mountain, under 1000 ft., *Drège*, 367, ♂! Caledon Div.; near Palmiet River, *Drège*, 116! Swellendam Div.; mountains near Swellendam, *Ecklon and Zeyher!* at the foot of the Lange Bergen, near Swellendam, *Burchell*, 7429, ♀! Uitenhage Div.; on the Van Stadens Berg nearest Galgebosch, *Burchell*, 4715, ♂!

3. **E. coleura** (Nees ex Mast. in DC. Monog. Phan. i. 358); rootstock creeping; stems erect, simple, flattened; leaf-sheaths loosely convolute, subpersistent, 1–1½ in. long, coriaceous, membranous

at the apex, acute, spreading; male inflorescence $1\frac{1}{2}$–$2\frac{1}{2}$ in. long, oblong, densely panicled cymose; spathes flat, coriaceous, pale brown, ovate-lanceolate, aristulate; bracts ovate, mucronate, scarcely shorter than the 3-sided flowers; perianth-segments oblong, acute, slightly rough; outer half the length of the inner; female inflorescence oblong, 1–$1\frac{1}{2}$ in. long, spicately or subpaniculately cymose; flowers ovoid, 1 line long; perianth-segments subequal, ovate, mucronate, asperulate; ovary ovoid, trigonous; styles (2 ?) 3. *Mast. in Journ. Linn. Soc.* xxi. 586.

SOUTH AFRICA: without locality, *Drège*, 9599, as to the male specimen! COAST REGION: Clanwilliam Div.; Ceder Bergen, on sand flats near Ezels Bank, 3000 ft., *Drège*, 2519! Cape Div.; Cape Flats, *Ecklon and Zeyher!* near Rondebosch, *Burchell*, 168! Table Mountain, *Burchell*, 571, ♂! Paarl Div.; bank of the Berg River in Klein Drakenstein Mountains, below 500 ft., *Drège*, 1648, as to the male specimen! Tulbagh Div.; Tulbagh, *Zeyher!* Worcester Div.; mountains above Worcester, *Rehmann*, 2570, ♂! 2571! Caledon Div.; Genadendal, 3000–4000 ft., *Drège*, 1638, ♂!

4. **E. glauca** (Mast. in Journ. Linn. Soc. xxi. 579); stems cæspitose, 2–3 ft. long, erect, slightly branched, compressed, solid, glaucous; leaf-sheaths persistent, $1\frac{1}{2}$ in. long, loosely convolute, subcoriaceous, membranous at the margins, mucronato-aristate; male inflorescence about 1 in. long, oblong, compact, spicately cymose, with oblong, flat, papery spathes, the smaller ones with very long spreading awns; spikelets many-flowered; flowers rather more than 1 line long, trigonous, arcuate; perianth-segments all navicular, acuminate; outer half the length of the inner; female inflorescence 1–$1\frac{1}{2}$ in. long, oblong, with open lanceolate spathes of the same length; perianth-segments nearly equal, oblong, mucronulate; inner rather obtuse; fruit pyriform, three-sided, shining, black, surmounted by 3 styles.

COAST REGION: Worcester Div.; mountains above Worcester, *Rehmann*, 2564, ♀! Stellenbosch Div.; Hottentots Holland Mountains, near Grietjes Gat, *Bolus*, 4221, ♀! Riversdale Div.; summit of Kampsche Berg, *Burchell*, 7105, ♂!

5. **E. thyrsifera** (Pers. Syn. ii. 607, excl. syn.); stems erect, slightly compressed, simple, fistular, wrinkled; leaf-sheaths $1\frac{1}{2}$ in. long, loosely convolute, chestnut-coloured, shining on the inner surface; male inflorescence $2\frac{1}{2}$ in. long, oblong, densely paniculate-cymose; spathes oblong, flattened, about 1 in. long; flowers ovoid, 1 line long; bract suborbicular, navicular, shorter than the flower; outer perianth-segments oblong, obtuse, glabrous; inner similar, but twice the length; female inflorescence $1\frac{1}{2}$–2 in. long, compact, ovoid, paniculate-cymose; spathes oblong, ovate, flat, coriaceous, purple-spotted; spathellæ much shorter, with long tapering points; flowers nearly $\frac{1}{4}$ in. long, slightly compressed, straight; perianth-segments subequal, subcoriaceous, lanceolate; inner thinner, scarcely shorter; fruit oblong, obtuse, 3-winged, surrounded by the persistent perianth. *Mast. in Journ. Linn. Soc.* xxi. 585. *Restio*

thyrsifer, *Rottböll*, *Prog.* 11; *Descript. et Icon. Rar.* 8, *t.* 3, *fig.* 4!

SOUTH AFRICA: without locality, *Masson*, ♂ in British Museum! *Koenig*, ♀ in Rottböll's Herbarium!

6. **E. acuminata** (Mast. in Journ. Linn. Soc. xxi. 580); stems cæspitose, 2–6 ft. high, simple, erect, compressed, solid; sheaths deciduous; male inflorescence 2–3 in. long, densely paniculate-cymose, oblong, erect, many-flowered, with numerous flat, elliptic, subcoriaceous, pale brown spathes; spathellæ with very long acumens; spikelets oblong, many-flowered; bracts oblong, acuminate; perianth arcuate, compressed; outer segments oblong, mucronulate, subequal, half the length of the inner, oblong, acute segments; female inflorescence as in the male; perianth-segments all equal, cartilaginous, chestnut-coloured, oblong, mucronate, 1 line and upwards in length; ovary obovoid, acutely 3-winged; stigmas 3.

SOUTH AFRICA: without locality, *Thom*, ♀! *Drège*, 104, ♂!

COAST REGION: Cape Div.; Table Mountain above Kirstenbosch, 2400 ft., *Bolus*, 4640, ♂ and ♀! Swellendam Div.; at the foot of the Lange Bergen near Swellendam, *Burchell*, 7435, ♂! Riversdale Div.; lower part of the Lange Bergen, about the waterfall at "Valley Rivers Poort," near Kampsche Berg, *Burchell*, 6993, ♂ and ♀!

7. **E. membranacea** (Kunth, Enum. iii. 474); stems cæspitose, erect, 2–3 ft. high, terete, solid, thinly puncticulate, simple or scarcely branched; sheaths deciduous; male inflorescence linear-oblong, erect, 2–3 in. long, loosely paniculate cymose, with numerous, oblong-lanceolate, flat, chartaceous spathes; spikelets many-flowered; perianth straight; outer segments ovate, acute, half the length of the similar inner; female inflorescence wedge-shaped; perianth-segments equal, acute, mucronulate; fruit pyriform, 3-sided, shining black, smooth. *Mast. in Journ. Linn. Soc.* xxi. 581. *Restio membranaceus*, *Nees in Linnæa*, v. 657.

COAST REGION: Cape Div.; Table Mountain, *Ecklon*, 836 partly, ♂! *Zeyher*, ♂, *Burchell*, 564, ♂ and ♀! Cape Town, *Harvey*, 200, ♂!

8. **E. propinqua** (Kunth, Enum. iii. 473); rootstock cæspitose; stems 2–3 ft. high, ascending, simple, solid, lightly rugose sulcate, white-spotted; sheaths less than 1 in. long, loosely convolute, oblong, coriaceous, mucronulate, ultimately deciduous; male inflorescence 2 in. long, oblong, densely paniculate-cymose, with elliptic acute sheath-like shining brown spathes; flowers ovoid, stipitate, scarcely $\frac{1}{8}$ in. long, exceeding the ovate-obtuse bract; outer perianth-segments oblong, obtuse; inner similar in shape, but twice the length; anthers apiculate; female inflorescence oblong, in spike-like cymes, about 1 in. long, spathes oblong-lanceolate, flat, less than 1 in. long, bracts shorter than the oblong compressed flowers; perianth-segments equal, about $\frac{1}{8}$ in. long, oblong, equalling the ovary; fruit three-cornered, shining, black, puncticulate, surmounted by 3 styles. *Steud. Synops.* ii. 261; *Mast. in Journ. Linn. Soc.* x. 242; xxi. 582;

DC. Monog. Phan. i. 356. *E. Kraussii, Hochst. in Flora,* 1845, 340. *Restio propinquus, Nees in Linnæa,* v. 653. *R. erectus, Nees in Linnæa,* v. 655, *excluding synonym, not of Thunb.*

VAR. **minor** (Mast.); smaller in all parts.

SOUTH AFRICA: without locality, *Bergius! Drège,* 113! *Thom,* 1023!
COAST REGION: Cape Div.; Camps Bay, *Burchell,* 353, ♀! Table Mountain, *Ecklon, MacGillivray,* 432, ♀! *Drège,* 89! Lion Mountain, *Ecklon.* Worcester Div.; Drakenstein mountains near Bains Kloof, 1600–2000 ft., *Bolus,* 4087, ♀! Dutoits Kloof, 3000–4000 ft., *Drège,* 112! 120 partly! Stellenbosch Div.; Hottentots Holland, *Zeyher!* Caledon Div.; Nieuw Kloof, Houw Hoek Mountains, *Burchell,* 8066! Humansdorp Div.; Kromme River, below 1000 ft., *Drège,* 115! Uitenhage Div.; sides of Winter Hoek Mountains, *Krauss.* Port Elizabeth Div.; Witte Klip near Port Elizabeth, *MacOwan,* 2152! 2153! *Var. minor,* Cape Div.; Flats near Rondebosch, *Burchell,* 822, ♂!

9. **E. juncea** (Linn. Mant. alt. 297, and in herb.!); stems cæspitose, 3–4 ft. high, solid, erect, simple or slightly branched, terete or rather compressed, puncticulate, faintly gyrose-sulcate; sheaths deciduous; male inflorescence 7–8 in. long, linear-oblong, compact, many-flowered, paniculate cymose, with numerous oblong or oblong-lanceolate, flat spathes, brown externally, shining grey within; spathellæ oblong-lanceolate, acuminate; bracts oblong, obtuse, scarcely shorter than the rounded flowers; outer perianth-segments oblong, obtuse, chestnut-brown, shorter than the inner; anthers linear-oblong, apiculate; female inflorescence linear-oblong; bracts oblong-obtuse, scarcely shorter than the flowers; perianth-segments linear-oblong, subequal; fruit linear-clavate, somewhat 3-sided, as long as the perianth. *Mast. in DC. Monog. Phan.* i. 357; *Journ. Linn. Soc.* x. 241; xxi. 583. *Elegia juncea, Kunth, Masters and others, all in part and with doubtful synonymy. Restio Elegia, Linn. Syst. Veg. ed.* 13, 738. *Restio thyrsifer, Lam. Encycl. Meth.* vi. 177; *Ill.* iii. 399, *t.* 804, *fig.* 3, *not of Rottböll.*

SOUTH AFRICA: without locality; *Sieber,* 229, ♀! *Harvey,* 389!
COAST REGION: Cape Div.; Devils Mountain, 500 ft., *Bolus,* 4456, ♂! 4457, ♀! Table Mountain, 1800 ft., *Bolus,* 2884! Cape Town, *Rehmann,* 1455, ♂! Tulbagh Div.; mountains in the vicinity of New Kloof, near Tulbagh, 1000 ft., *MacOwan, Herb. Aust.-Afr.,* 1676! Caledon Div.; Nieuw Kloof, Houw Hoek Mountains, *Burchell,* 8132, ♀! Babylons Tower, 3000–4000 ft., *Zeyher,* 4340, ♀!

10. **E. cuspidata** (Mast. in Journ. Linn. Soc. x. 240); stems about 3 ft. high, simple, compressed, olive-coloured, puncticulate; leaf-sheaths 2–3 in. long, loosely convolute, elliptic, mucronate, ultimately deciduous; male spikelets very numerous, densely compacted in terminal spike-like or panicled cymes, 1½–3 in. long, with numerous, open, sheath-like, ultimately deciduous spathes; bracts broadly ovate, coriaceous, brown, ending in long spreading awns; flowers three-sided; outer perianth-segments equal, rigid, oblong, acute, boat-shaped, 1-nerved, glabrous; inner oblong-lanceolate, longer than the outer; anthers oblong, apiculate; female inflorescence generally less branched than the male; bracts and perianth-segments as in the male; fruit oblong, 3-sided, shining

black, crowned with the vestiges of the styles. *Mast. in DC. Monog. Phan.* i. 354.

COAST REGION: Cape Div.; False Bay, *Robertson!* Table Mountain, *Milne*, 233! Simon's Bay, *MacGillivray*, 437, ♀! *Wright*, 484, ♂!

11. E. asperiflora (Kunth, Enum. iii. 474); stems 2–3 ft. high, cæspitose, covered at the base with scales, erect, compressed, simple or sparingly branched; branches ascending, purple-spotted; leaf-sheaths 1–4 in. long, loosely convolute, subcoriaceous, fuscous, striolate, aristate; male inflorescence terminal, 1½ in. long, densely paniculately cymose, pluristachyate, with numerous flat sheath-like spathes; spikelets nearly ¼ in. long, roundish, stalked, many-flowered; bracts ovate-oblong, hyaline and lacerate at the margins, tapering at the apex into a long acumen; flowers 3-sided, shortly stalked; outer perianth-segments equal, oblong, boat-shaped, pale brown, ciliolate, median nerve serrulate, apex prolonged into a long awn; inner longer than the outer, oblong-acute; anthers linear-oblong, apiculate; female inflorescence similar to the male but smaller and less branched; bracts oblong, with a long acumen; perianth-segments boat-shaped, oblong-lanceolate, ciliolate; inner scarcely exceeding the outer; fruit 3-cornered, turbinate, surmounted by traces of three styles. *Steud. Synops.* ii. 262; *Mast. in Journ. Linn. Soc.* x. 244; *DC. Monog. Phan.* i. 355, *t.* 3, *fig.* 29–32, *and t.* 5, *fig.* 14. *E. dregeana, Kunth, Enum.* iii. 469. *Restio asperiflorus, Nees in Linnæa*, v. 656.

SOUTH AFRICA: without locality, *Drège*, 106! 107! 9599, as to the female specimen!

COAST REGION: Clanwilliam Div.; Wupperthal, *Drège*, 1648, as to the female specimen! Cape Div.; Cape Flats, *Ecklon.* Tulbagh Div.; mountains of New Kloof, 1000–2000 ft., *Drège*, 1641, ♀! near Tulbagh Waterfall, *Ecklon and Zeyher*, ♀! Worcester Div.; mountains above Worcester, *Rehmann*, 2563! Dutoits Kloof, 2000–3000 ft., *Drège*, 1640, ♀! Caledon Div.; Nieuw Kloof, Houw Hoek Mountains, *Burchell*, 8057, ♀! 8140, ♂ and ♀! Caledon Div.; Palmiet River Valley near Grabouw, 700 ft., *Bolus*, 4223, ♂! Riversdale Div.; between Vet River and Krombeks River, *Burchell*, 7179 ♂ and ♀! Uniondale Div.; between Welgelegen and Onzer, 1500–2000 ft., *Drège*, 102, ♂! Bathurst Div.; near Port Alfred, *Burchell*, 3996 partly, ♂!

WESTERN REGION: Little Namaqualand; Ezels Fontein, 3500–4000 ft., *Drège*, 2518!

12. E. fistulosa (Kunth, Enum. iii. 467); stems 3 ft. high, erect, terete, simple, puncticulate, fistular; sheaths 1¼ in. long, loosely convolute, subcoriaceous, deciduous; male inflorescence 2–6 in. long, oblong, compact, much-branched, with numerous, open, lanceolate spathes, brownish externally, shining within, aristate; bracts ovate, mucronate, carinate, glabrous, shorter than the three-sided flowers; outer perianth-segments oblong, obtuse, carinate; inner similar, but twice the length; female inflorescence 1–3 in. long, wedge-shaped, with numerous flat oblong-lanceolate spathes; bracts ovate-acute, carinate, shorter than the three-sided flowers; outer perianth-segments oblong, obtuse, glabrous; inner oblong or subspatulate, about the length of the outer; ovary three-sided pyriform; stigmas 3. *Steud. Synops.* ii. 261; *Mast. in. Journ. Linn. Soc.* x. 243; *DC. Monog. Phan.* i. 356.

SOUTH AFRICA: without locality, *Ecklon and Zeyher!*

COAST REGION: Cape Div.; near Wynberg, 100 ft., *Bolus*, 3903, ♀! Paarl Div.; Paarl Mountains, 1000–2000 ft., *Drège*, 117! Cape Flats, *Ecklon*, 853!

13. **E. squamosa** (Mast. in Journ. Linn. Soc. x. 244); rootstock creeping; stems slender, erect, 4–6 in. high, terete, solid, unbranched; leaf-sheaths deciduous, loose, shortly-mucronate; male flowers subtrigonous, in an oblong, erect, panicled cyme, $\frac{1}{4}$ in. long; spathes flat, ovate-oblong, acute, mucronate, as long as the spikelet, ultimately deciduous; bracts broadly-ovate, acute, coriaceous, membranous at the apex, subulate-mucronate; outer perianth-segments ovate-acute, coriaceous, puncticulate; inner oblong, acute, twice the length of the outer; female flowers in compact panicles, intermixed with flat spathes; perianth-segments all subequal, oblong, mucronulate, 1-nerved; ovary stipitulate, pear-shaped, coriaceous, 1- (or 2-celled ?); styles 2–3; fruit indehiscent, oblong, coriaceous, 1- (2-?) celled; cells 1-seeded. *Mast. in DC. Monog. Phan.* i. 359.

SOUTH AFRICA: without locality, *Pappe*, 105, ♂!

14. **E. vaginulata** (Mast. in Journ. Linn. Soc. xxi. 586); stems cæspitose, erect from a decumbent base, 9 in. long, slender, terete, solid, wrinkled; leaf-sheaths less than $\frac{1}{2}$ in. long, flat, ovate, oblong, acute, shining on the inner surface, ultimately deciduous; male and female inflorescence similar, of very short spike-like cymes; spathes sheath-like, open, as long as, or rather longer than the spike; bracts ovate, mucronulate, shorter than the flowers; male flowers 1 line long, trigonous, sessile; outer perianth-segments oblong, acute, navicular; inner longer and broader; female flowers rather larger than the male; perianth-segments subequal, oblong, acute; ovary barrel-shaped, smooth, surmounted by 3 styles, 1-celled, 1-seeded.

COAST REGION: Ceres Div.; mountains near Ceres, 1800 ft., *Bolus*, 5480, ♀! Riversdale Div.; at the foot of the Lange Bergen, near Kampsche Berg, *Burchell*, 7141, ♂!

15. **E. deusta** (Kunth, Enum. iii. 460); stems cæspitose, 1–2 ft. long, erect, unbranched, slender, rigid, compressed, solid; leaf-sheaths persistent, $1\frac{1}{2}$–2 in. long, closely convolute, coriaceous, filiform or aristate; male spikelets in dense, terminal, paniculate cymes, $\frac{3}{4}$–1 in. long; spathes flat, lanceolate, deciduous; bracts oblong, coriaceous, blackish chestnut, shining, subulate, mucronate, shorter than the arcuate trigonous flowers; perianth-segments rigid, ferruginous, boat-shaped or oblong, acute, apiculate; inner exceeding the outer; anthers apiculate; pistillode minute, 3-styled; female inflorescence as in the male; flowers arcuate, oblong, two or three times longer than the ovate, oblong, mucronate bracts; outer perianth-segments ovate-oblong, acute, subcarinate, glabrous; inner oblong,

twice the length of the outer; staminodes 3, liguliform; ovary oblong, trigonous, purple, 1-celled, with 3 stigmas. *Mast. in Journ. Linn. Soc.* x. 239; *DC. Monog. Phan.* i. 352. *Chondropetalum deustum, Rottb. Descr. et Ic.* 10, *t.* 3, *fig.* 2; *Linn. herb.! Mast. in Journ. Linn. Soc.* xxi. 590. *Restio Chondropetalum, Nees in Linnæa*, v. 652.

SOUTH AFRICA: without locality, *Sieber*, 226! *Zeyher! Bergius!*
COAST REGION: Cape Div.; Table Mountain, 2800–3000 ft., *Burchell*, 565! *Bolus*, 4444! *Drège*, 123! *Ecklon*, 839!

16. **E. filacea** (Mast. in Journ. Linn. Soc. xxi. 589); stems cæspitose, erect, 14–16 in. high, very slender, wiry, little branched; leaf-sheaths long, persisting, tightly convolute, about ½ in. long, coriaceous, thinner towards the lanceolate acuminate apex; inflorescence about ¾ in. long, linear oblong, densely paniculately cymose; spathe sheath-like, flat, oblong, acuminate; flowers very numerous, minute; perianth-segments oblong, obtuse; inner slightly longer and thinner; perianth of female flower less than 1 line in length; segments all subequal, cymbiform, 1-nerved; fruit equal to the persistent perianth, three-sided, 1-celled, 1-seeded.

SOUTH AFRICA: without locality, *Drège*, 110, ♂! Stellenbosch Div.; Lowrys Pass, 1000–2000 ft., *Drège*, 126! Caledon Div.; Nieuw Kloof, Houw Hoek Mountains, *Burchell*, 8121, ♀!

17. **E. stipularis** (Mast. in Journ. Linn. Soc. xxi. 587); stems tufted, erect, 12–18 in. and upwards high, slender, terete, branching towards the middle; leaf-sheaths about 1 in. long, flat, elliptic, acute, persistent; male inflorescence elongate, spike-like, erect, densely and compactly branched, and bearing numerous, flat, oblong spathes; perianth-segments oblong, obtuse, cymbiform; inner similar but longer; female inflorescence similar to the male; perianth-segments acute; staminodes small; fruit triangular.

SOUTH AFRICA: without locality, *Masson!*
COAST REGION: Clanwilliam Div.; Oliphants River, *Gill!* Onder Bokke Veld, *Drège*, 2517 (distributed also as 118)! Caledon Div.; mountains near Grietjes Gat, *Ecklon and Zeyher!* Mossel Bay Div.; hills near the landing-place at Mossel Bay, *Burchell*, 6290!

Often confounded with *E. parviflora*, but distinguishable by the flat, open, persistent sheaths and other characteristics.

18. **E. obtusiflora** (Mast.); stems cæspitose, erect, 15–18 in. high, of the thickness of a crow quill, terete or slightly flattened, greyish-green; leaf-sheaths ¾ in. long, pers stent, loosely convolute, spreading and sometimes flat at the apex, membranous, pale brown, thinner at the acute apex, shining on the inner surface; male and female inflorescences similar, in linear, rather dense, cymose panicles, with numerous, flat, sheath-like, lanceolate spathes; spikelets minute, arranged in clustered panicles; spathellæ membranous, obtuse, shorter than the spikelet; bracts rounded, scarcely shorter than the flower; outer perianth-segments somewhat ovate, shorter than the

oblong inner segments; anthers apiculate; female inflorescence less dense than the male; perianth similar.

COAST REGION: Tulbagh Div.; near Tulbagh, *Zeyher*, ♂ and ♀!

19. E. spathacea (Mast. in Journ. Linn. Soc. xxi. 588); stems cæspitose, slender, erect, 2 ft. high, wiry, terete, solid, unbranched; leaf-sheaths deciduous; male inflorescence 1–2 in. long, oblong, panicled, many-flowered, with numerous flat, oblong-acute spathes; flowers minute, triquetrous; perianth-segments oblong, obtuse; outer shortest; female inflorescence like the male; flowers 1 line long; perianth-segments subequal, oblong, obtuse, chestnut-brown; fruit clavate, three-sided, 1-celled, 1-seeded, surrounded by the persistent perianth.

COAST REGION: Tulbagh Div.; on the Witsen Berg near Tulbagh, *Burchell*, 8647, ♀! Caledon Div.; on Donker Hoek Mountain, *Burchell*, 7953, ♀! tops of the mountains of Baviaans Kloof, near Genadendal, *Burchell*, 7710, ♂!

20. E. parviflora (Kunth, Enum. iii. 467); stems cæspitose, erect, 12–18 in. high, very slender, wiry, fistular; leaf-sheaths deciduous; male and female inflorescence similar, varying in size, consisting of a compact, linear, spicate or paniculate cyme; spathes open, coriaceous, oblong-acute, shorter than the inflorescence; spikelets many-flowered; bracts ovate, acute, coriaceous, membranous at the apex, muticous or scarcely mucronulate; perianth-segments rigid; outer boat-shaped, obtuse; inner longer than the outer; flowers of female plant rather larger, and inner segments about the same length as outer; fruit pyriform, triangular, puncticulate, as long as the persistent perianth, crowned by the remains of 3 styles. *Steud. Synops.* ii. 262; *Mast. in Journ. Linn. Soc.* x. 244; *DC. Monog. Phan.* i. 353. *Restio parviflorus, Thunb. Diss.* 13; *Fl. Cap. ed. Schult.* i. 85; *Willd. Sp. Pl.* iv. 722; *Nees in Linnæa*, v. 655.

SOUTH AFRICA: without locality, *Zeyher*, 1739!
COAST REGION: Cape Div.; sand dunes near Cape Town, *Bolus*, 4432! between Wynberg and Devils Mountain, *Drège*, 150! Paarl Div.; Little Drakenstein Mountains, below 1000 ft., *Drège*, 1646! Tulbagh Div.; near Tulbagh Waterfall, *Ecklon and Zeyher!* mountains in the vicinity of New Kloof, 1000 ft., *MacOwan, Herb. Austr. Afr.*, 1675! Worcester Div.; Dutoits Kloof, *Drège*, 120! 1639! 1649! Drakenstein Mountains, 3000–4000 ft., *Drège*, 121! Caledon Div.; Donker Hoek and Ezelsjagt Mountains, 1000–2000 ft., *Drège*, 118!

21. E. rigida (Mast. in Journ. Linn. Soc. xxi. 587); stems cæspitose, 14–18 in. high, very slender, terete, scarcely branched; leaf-sheaths caducous; female inflorescence about 1 in. long, linear-oblong, with numerous convolute, oblong spathes; perianth about 1 line long; segments subequal; lateral conduplicate, acuminate; intermediate spathulate; ovary trigonous, shining, 1-celled, as long as the persistent perianth; styles 3, recurved.

COAST REGION: Tulbagh Div.; New Kloof, near Tulbagh, *Zeyher!* Worcester

Div.; Drakenstein Mountains, near Bains Kloof, 1600–2000 ft., *Bolus*, 4100, ♀! mountains above Worcester, *Rehmann*, 2565!

22. **E. Verreauxii** (Mast. in Journ. Soc. Linn. xxi. 589); rootstock horizontally creeping; stems 14–18 in. high, erect, very slender, rigid, unbranched; leaf-sheaths subpersistent, $\frac{1}{2}$ in. long, loosely convolute, mucronate, spreading at the apex; inflorescence linear, $\frac{1}{2}$–1 in. long, many-flowered; spathes aristate; flowers minute, loosely arranged, blackish; perianth-segments subequal, chestnut-brown, oblong, obtuse, equalling the 3-sided blackish fruit.

SOUTH AFRICA: without locality, *Verreaux!* in the British Museum.
COAST REGION: Caledon Div.; Nieuw Kloof, Houw Hoek Mountains, *Burchell*, 8114!

23. **E. nuda** (Kunth, Enum. iii. 462); rootstock creeping; stems 1–2 ft. high, erect, filiform, compressed, solid; leaf-sheaths caducous; male and female inflorescence in oblong, spicate cymes, each about 1 in. long; spathes oblong, mucronate, loosely convolute, blackish, deciduous; bracts in both sexes broadly ovate, mucronate, coriaceous, blackish-brown, shining, shorter than the flowers; perianth in both sexes similar; outer segments ovate-pointed, subcarinate; inner oblong-ovate, acute, twice the length of the outer; ovary trigonous; styles 3; fruit 3-gonous, blackish-purple, 1-celled, 1-seeded, crowned by the remains of the styles. *Steud. Synops.* ii. 261; *Mast. in Journ. Linn. Soc.* x. 239; *DC. Monog. Phan.* i. 353, *partly. Chondropetalum nudum, Rottb. Descr. et Ic.* 11, *t.* 3, *fig.* 3. *Restio nudus, Nees in Linnæa,* v. 651, *partly.*

SOUTH AFRICA: without locality, *Sieber*, 111, ♀! *Drège*, 128, partly, ♀!
COAST REGION: Cape Div.; Cape Flats, *Burke*, ♀!

24. **E. elongata** (Mast.); stems cæspitose, $2\frac{1}{2}$ ft. high, terete, very slender, unbranched, clothed at the base with dark shining scales, greenish-grey, obsoletely puncticulate; leaf-sheaths numerous, remote, deciduous; male inflorescence 1–3 in. long, linear, paniculate, interrupted; spathes inconspicuous; spathellæ short, acuminate; bracts blackish, shorter than the flower, ovate, mucronate; outer perianth-segments oblong, glabrous, half the length of the inner oblong segments; anthers apiculate; female inflorescence rather shorter; perianth as in the male; ovary 3-sided, surrounded by 3 minute staminodes.

COAST REGION: Cape Div.; between Wynberg and Devils Mountain, *Drège*, 147a, ♂! Paarl Div.; between Paarl and Lady Grey Bridge, below 1000 ft., *Drège*, 9454, ♂ and ♀!

V. LAMPROCAULOS, Mast.

Male and female inflorescence somewhat similar, provided with large sheath-like, ultimately deciduous, spathes. *Male flowers* in panicled cymes. *Bracts* small, linear. *Outer perianth-segments* linear; inner segments larger and broader than the outer. *Anthers* 1-celled. *Female spikelets* 2-flowered. *Female flowers* in spicate cymes, curved. *Bracts* linear. *Outer perianth-segments* oblong, obtuse; inner longer, oblong spathulate. *Fruit* surrounded at the base by the persistent perianth, oblong, truncate, compressed, 1-celled, with one seed pendulous from the top; styles 2, thick at the base and proceeding from the angles of the fruit at the top.

Rush-like perennials with coarsely-tubercled, brilliantly-coloured epidermis and persistent leaf-sheaths.

DISTRIB. Endemic.

Stem golden-bronze coloured; spathes 2 in. long ... ... (1) **grandis.**
Stem rich brown; spathes $1\frac{1}{2}$ in. long ... ... (2) **Neesii.**

1. L. grandis (Mast. in DC. Monog. Phan. i. 349, t. 3, fig. 22–28, and t. 5, fig. 13); stems 3 ft. high, erect, terete or compressed, fistular, slightly-branched, stems and branches with a golden sheen; leaf-sheaths 2 in. long, loosely convolute, shining, golden-coloured outside, subulate-mucronate; male flowers very numerous, in terminal, much-branched panicles intermingled with large, convolute, mucronate, ultimately deciduous, spathes; bracts small, linear; perianth slightly longer than the bract; outer segments equal, linear; inner membranous, spathulate, longer than the outer; anthers oblong; female spikelets 5–6 in linear oblong cymes, each spikelet 2-flowered, with a sheath-like spathe at the base; flowers about 2 lin. long, slightly stalked, curved; outer perianth-segments equal, oblong-obtuse; inner longer, oblong-spathulate; fruit oblong, truncate, compressed, 1-celled, 1-seeded, surrounded by the persistent perianth; style branches 2; seed pendulous. *Restio grandis, Sprengel fide Nees in Linnæa,* v. 660. *Elegia grandis, Kunth, Enum.* iii. 475; *Steud. Synops.* ii. 262; *Mast. in Journ. Linn. Soc.* x. 245.

COAST REGION: Worcester Div.; Dutoits Kloof, 3000–4000 ft., *Drège*, 1650, ♂! Caledon Div.; Genadendal, *Zeyher! Ecklon*, 34! Houw Hoek Mountains, 1000–3000 ft., *Zeyher*, 4337, ♂ and ♀!

2. L. Neesii (Mast. in DC. Monog. Phan. i. 350); stems 18 in.–2 ft. high, erect from a decumbent base, fistular, branching from near the base, tubercled, golden-brown; sheaths 1–2 in. long, loosely convolute, minutely scabrid, rich golden-brown, with a long subulate mucro; male flowers numerous in a much-branched cyme, intermixed with large, deciduous, sheath-like spathes, each pedicellate about $\frac{1}{8}$ in. long, protected by a small linear bract; outer perianth-segments linear-subulate; inner oblong subspathulate, longer than the outer ones; anthers oblong; female spikelets 2-flowered, in a linear cyme, each covered by a large sheath-like spathe; flowers curved, with a small linear bract; outer perianth-segments rigid, oblong, concave; inner wider; lateral conduplicate; fruit indehiscent, 1-celled, 1-seeded, oblong, compressed, truncate; styles 2, proceeding from the angles of the fruit; seed pendulous. *Elegia Neesii, Mast. in. Journ. Linn. Soc.* x. 246. *Restio grandis, var. β, Nees in Linnæa,* v. 661.

COAST REGION: Cape Div.; Table Mountain, *Burchell*, 569, ♂! *Ecklon!* *Bolus*, 4441, ♂! Cape Flats, *Ecklon*, 560b, ♀!

VI. LEPTOCARPUS, R. Brown.

Male and female inflorescence similar. *Spikelets* 1- or many-flowered. *Perianth-segments* 6; outer lateral conduplicate, usually villous on the keel, rarely glabrous. *Stamens* 3. *Pistillodium* minute, 3-styled or absent. *Female flower: Staminodes* none or minute. *Ovary* more or less triangular, 1-celled, 1-ovuled; styles 3, free at the base or slightly united. *Fruit* 3-sided, coriaceous, indehiscent.

Rush-like, much or sparingly branched plants, the basal portion thickly covered with brown scales; upper sheaths mucronate or aristate, sometimes foliaceous on the sterile branches.

DISTRIB. Several species inhabit Tropical and South-west Australia. Isolated species occur in Tasmania, New Zealand, Chile, and Cochin China.

Female spikelets many-flowered:
  Male and female inflorescence much-branched; plant leafy ... ... (1) **paniculatus.**
  Male and female inflorescence sparingly branched; plant rarely leafy:
    Spathes much shorter than the spikelets:
      Spikelets 7–9, squarrose ... ... (2) **neglectus.**
      Spikelets 1–5 ... ... (3) **Burchellii.**
    Spathes about the same length as the spikelets:
      Leaf-sheaths rounded, aristate ... ... (4) **incurvatus.**
      Leaf-sheaths straight, acute at the apex, subulate-mucronate ... ... (5) **oxylepis.**
      Leaf-sheaths recurved, spreading, obtuse, mucronate ... ... (6) **peronatus.**
Female spikelets 1-flowered:
  Spikelets obtuse ... ... (7) **modestus.**

1. L. paniculatus (Mast. in Journ. Linn. Soc. x. 221); stems 2–3 ft. high, erect, much-branched; branches compressed, olive-coloured, purple-spotted, ultimate branches leaf-bearing; leaf-sheaths 1 in. long, tightly convolute, striate, smaller ones deeply bi-lobed; lobes hyaline; apex mucronate, leafy; male and female spikelets very numerous, in much-branched, spreading, panicled cymes, each spikelet about ¼ in. long, obovate-oblong; spathe bract-like; bracts oblong, obtuse, hyaline near the acuminate apex; flowers ovate, rather shorter than the bracts; perianth-segments oblong, obtuse; lateral conduplicate; inner smaller, flattish; anthers oblong, apiculate; pistillodium small; female spikelets like the male; staminodes 3; ovary oblong, triquetrous; styles 3; fruit 3-angled. *Mast. in DC. Monog. Phan.* i. 330. *Restio paniculatus, Rottboell, Descr. et Ic.* 4, *t.* 2, *fig.* 3; *Linn. Herb. ex Mast. in Journ. Linn. Soc.* xxi. 590.

Restio fruticosus, Thunb. Diss. 16, *n.* 14; *Kunth, Enum.* iii. 413. *Restio ramiflorus, Nees in Linnæa,* v. 644. *Calopsis paniculata, Desv. in Ann. Sc. Nat.* xiii. (1828) 44, *t.* 3, *fig.* 2; *Kunth, Enum.* iii. 421; *Steud. Synops.* ii. 257.

SOUTH AFRICA: without locality, *Thunberg,* ♂! *Burchell,* 5813! *Mund! Thom! Drège,* 5! 167b!

COAST REGION: Cape Div.; near Cape Town, *Burchell,* 405! Table Mountain, *Scott-Elliot,* 140! *Ecklon,* 840! Cape Flats, *Ecklon,* 568! between Wynberg and Devils Mountain, *Drège,* 167! Paarl Div.; Paarl Mountains, 1000–2000 ft., *Drège,* 167a! 167b! Caledon Div.; banks of rivers near Houw Hoek, 950 ft., *MacOwan, Herb. Aust. Afr.,* 1667, ♂! Uitenhage Div.; Zwartkops River, *Zeyher,* 4346! *Ecklon and Zeyher,* 540! 922!

EASTERN REGION: Tembuland; Bazeia Mountain, 4000 ft., *Baur,* 506! Natal; rocky valley near Bevaan Falls, *Wood,* 3195!

Varies in stature and in the form of the bracts. The form with very small spikelets occurring in Tembuland may turn out to be the type of a distinct species.

2. **L. neglectus** (Mast. in Journ. Linn. Soc. x. 225); stems 2–3 ft. high, erect, terete, slightly branched, olive-coloured, covered with small white tubercles; leaf-sheaths $\frac{1}{2}$–$\frac{3}{4}$ in. long, tightly convolute, strongly subulate-mucronate; male and female inflorescence pluristachyate, paniculate, cymose; spikelets $\frac{1}{2}$–$\frac{3}{4}$ in. long, sessile or stalked, squarrose; spathe bract-like, shorter than the spikelet; bracts coriaceous, oblong, mucronate; flowers arcuate, ovate, stipitulate; perianth-segments oblong, acute; lateral conduplicate, villous on the keel; inner 3 smaller, hyaline; anthers linear, apiculate; ovary 3-styled, roundish, surrounded by 3 staminodes. *Mast. in DC. Monog. Phan.* i. 334. *Calopsis neglecta, Hochstetter in Flora,* 1845, 338; *Steud. Synops.* ii. 258.

SOUTH AFRICA: without locality, *Drège,* 68, ♀!

3. **L. Burchellii** (Mast. in Journ. Linn. Soc. x. 222); stems 3 ft. high, erect, branching; sterile stems leafy; branches olive-coloured, coarsely tubercled; leaf-sheaths 1 in. long, loosely convolute, membranous at the edges, subulate-mucronate; smaller sheaths $\frac{1}{4}$ in. long, foliaceous-mucronate; male spikelets 1–5, aggregate in a spike-like cyme, each $\frac{1}{2}$ in. long, oblong, acute, many-flowered, protected at the base by a short bract-like, aristate spathe; bracts at length loosely overlapping, oblong-lanceolate, coriaceous, mucronate; outer perianth-segments oblong, purple-spotted; lateral conduplicate, villous on the keel; inner similar, smaller, flattish; anthers linear-oblong, apiculate; female spikelets 4–5, in terminal cymes, each many-flowered, oblong-lanceolate, about $\frac{1}{4}$ in. long, protected by an aristate spathe; perianth-segments oblong-lanceolate; lateral conduplicate, not villous; ovary trigonous-oblong; stigmas 3. *Mast. in DC. Monog. Phan.* i. 331.

COAST REGION: Riversdale Div.; at the foot of the Lange Bergen near Kampsche Berg, *Burchell,* 7146, ♂! between Vet River and Krombeks River, *Burchell,* 7185, ♂ and ♀! George Div.; Wolf Drift, Malgat River, *Burchell,* 6101, ♂!

4. **L. incurvatus** (Mast. in Journ. Linn. Soc. x. 223); stems decumbent, 2–3 ft. long, sparingly branched; branches ascending, spotted; sterile branches with leafy sheaths; leaf-sheaths about ½ in. long, loosely convolute, recurved at the filaceous-aristate apex; smaller sheaths foliaceous-mucronate; male spikelets 4–8, in linear spike-like cymes, each spikelet ¼ in. and upwards long, oblong-ovoid, with a sheath-like open spathe at the base; bracts ultimately loosely spreading, oblong, acute, coriaceous, subulate-mucronate beneath the membranous apex; perianth-segments oblong, obtuse; lateral conduplicate, villous on the keel; anthers linear-oblong; female spikelets 2–4, sessile at the apex of the branch, similar in form to the male spikelets, many-flowered; perianth-segments obovate-spathulate; ovary stipitulate; fruit oblong, obtuse, three-sided. *Mast. in DC. Monog. Phan.* i. 332. *Restio incurvatus, Thunb. Fl. Cap.* i. 88. *Restio vimineus, Rottb. Descr. et Ic.* 4, *t.* 2, *fig.* 1. *Calopsis incurvata, Kunth. Enum.* iii. 427; *Steud. Synops.* ii. 258, ♂. *Calopsis festucacea, Kunth, Enum.* iii. 425; *Steud. Synops.* ii. 258, ♀.

COAST REGION: Clanwilliam Div.; plains between Jakhals River and Klip Fontein, below 500 ft., *Drège,* 2481, ♀! Cape Div.; False Bay, *Robertson!* Cape Flats near Rondebosch, *Burchell,* 832, ♀! Devils Mountain, *Ecklon,* 847, ♂! Pampoens Kraal, *Ecklon!* Constantia mountains, 2500 ft., *Schlechter,* 541! Stellenbosch Div.; Lowrys Pass, 900 ft., *Schlechter,* 7298! Paarl Div.; by the Berg River, *Drège,* 61! Tulbagh Div.; near Tulbagh Waterfall, 1200 ft., *Bolus,* 7498, ♂!

WESTERN REGION: Little Namaqualand; Kamies Bergen, 3000–4000 ft., *Drège,* 62! 63a!

5. **L. oxylepis** (Mast. in Journ. Linn. Soc. x. 223); stems 2–3 ft. long, erect from a decumbent base, sparingly branched, glabrescent or tomentose; leaf-sheaths about ½ in. long, loosely convolute, membranous at the apex, mucronate-aristate; male spikelets approximate at the ends of the branches, each cylindric-lanceolate, nearly ½ in. long, protected at the base by a spathe of equal length; bracts elliptic, acute, shining, shortly mucronate; perianth-segments oblong, acute, conduplicate, villous on the keel; anthers ovate, acute, apiculate; female spikelets 3–6 in linear, spike-like cymes, each about ½ in. long, many-flowered, oblong, protected by an open spathe of equal length; bracts oblong, acute; outer perianth-segments oblong, obtuse, purple-spotted; lateral carinate, villous; inner obovate-spathulate, ultimately longer than the outer; fruit trigonous, clavate. *Mast. in DC. Monog. Phan.* i. 332. *Calopsis oxylepis, Kunth, Enum.* iii. 427; *Steud. Synops.* ii. 258.

COAST REGION: Clanwilliam Div.; near Groene River and Watervals River, 2500–3000 ft., *Drège,* 38! Ceder Bergen, on sandy flats near Ezels Bank, 3000 ft., *Drège,* 39! 2501! between Grasberg River and Watervals River, 2500–3000 ft., *Drège,* 2500 partly!

6. **L. peronatus** (Mast. in Journ. Linn. Soc. x. 224); stems 2 ft. high, terete, slightly branched; branches ascending; leaf-sheaths 1 in. long, loosely convolute or recurved, mucronate-aristate; male spikelets numerous, in linear little-branched cymes; each spikelet

$\frac{1}{4}$–$\frac{1}{2}$ in. long, cylindric-lanceolate, sometimes curved, protected by a sheath-like spathe of equal length; bracts oblong elliptic, shining, shortly mucronate; outer perianth-segments oblong; lateral conduplicate, villous on the keel; inner 3 smaller, flat, united at the base; anthers oblong, obtuse, apiculate; female spikelets numerous, in erect, linear, much-branched cymes; spikelets $\frac{1}{2}$ in. long, cylindric-lanceolate, many-flowered; bracts elliptic, mucronate; perianth-segments oblong, obtuse; lateral conduplicate, villous on the keel; inner shorter, oblong, obtuse, flat or slightly involute; ovary oblong, obtuse, somewhat 3-sided, surmounted by a yellow epigynous disc; stigmas 3. *Mast. in DC. Monog. Phan.* i. 333. *Calopsis peronata, Kunth, Enum* iii. 426; *Steud. Synops.* ii. 258, ♀.

VAR. **hirtellus** (Mast. in Journ. Linn. Soc. x. 224); stems and branches pubescent. *Mast. in DC. Monog. Phan.* i. 334. *Calopsis hirtella, Kunth, Enum.* iii. 426; *Steudel, Synops.* ii. 258.

COAST REGION: Clanwilliam Div.; between Grasberg River and Watervals River, 2500–3000 ft., *Drège*, 2499, ♀! both the glabrous type and the pubescent variety are distributed under this number. Var. *β*, from the same locality, *Drège*, 2500 partly!

7. **L. modestus** (Mast. in Journ. Linn. Soc. x. 225); stems 1–1½ ft. high, terete, purple-spotted, slightly branched; branches erect or spreading, filiform; leaf-sheaths $\frac{3}{4}$ in. long, elliptic, coriaceous, membranous at the apex, prolonged into a long, deciduous awn; smaller sheaths with two hyaline lobes and a leafy awn; male spikelets 5–7, in spicate cymes, each about $\frac{1}{4}$ in. long, oblong, obtuse, slightly exceeding the bract-like spathe; bracts closely overlapping, oblong-lanceolate, coriaceous, with a prominent midrib; flowers oblong, stipitulate; perianth-segments linear-oblong, obtuse; lateral conduplicate, with a glabrous keel; inner rather shorter, thinner; anthers linear-oblong; female spikelets 1-flowered; staminodes minute; fruit 3-cornered, coriaceous, 1-celled, with 3 stigmas. *Mast. in DC. Monog. Phan.* i. 337. *Thamnochortus modestus, Kunth, Enum.* iii. 434. *Restio paleaceus, Nees partly, in various herbaria ex Mast. in Journ. Linn. Soc.* x. 225. *R. fruticosus, Thunberg herb. ex Mast. in Journ Linn. Soc.* xiv. 417.

SOUTH AFRICA: without locality, *Drège*, 7! *Ecklon and Zeyher*, 82!
COAST REGION: Cape Div.; Table Mountain or Devils Mountain below 1000 ft., *Drège*, 138b! Albany Div.; Grahamstown, *Ecklon!* between Donker Hoek and Houw Hoek Mountains, *Burchell*, 8016!
CENTRAL REGION: Albany Div.; near Riebeek, *Burchell*, 3470!

VII. THAMNOCHORTUS, Bergius.

Perianth-segments 6 in two rows; outer lateral generally longer, navicular, more or less keeled or winged. *Stamens* 3; filaments filiform; anthers oblong-linear, 1-celled, dehiscing lengthwise. *Female perianth* as in the male, but the lateral segments more deeply winged. *Staminodes* none, or only 1-celled, 1-seeded;

styles 1–3. *Fruit* indehiscent, enclosed within the persistent perianth, compressed, 1-celled, 1-seeded, surmounted by the remains of the styles.

Herbaceous perennials, with a creeping or contracted stock; stems erect, covered at the base with sheathing scales; leaf-sheaths closely convolute, sometimes prolonged into a linear leaf; male and female inflorescence alike or dissimilar, male usually with many spikelets in a panicled cyme; female spikelets 1–2 or cymose.

DISTRIB. Endemic.

Style 1:
- Male inflorescence panicled, much-branched, elongate; spikelets numerous; female inflorescence a long spicate cyme; spikelets numerous, oblong:
  - Male spikelets erect or spreading:
    - Male spikelets linear, oblong, acute; bracts mucronate ... ... (1) **spicigerus.**
    - Male spikelets ovoid-oblong; bracts oblong, nearly muticous ... ... (2) **giganteus.**
    - Male spikelets small, ovate-oblong; bracts oblong, mucronate, or acuminate ... ... (3) **scirpiformis.**
  - Male spikelets pendulous:
    - Spikelets ½ in. long, oblong, acute ... ... (4) **fruticosus.**
    - Spikelets 1 in. long, silvery, linear-clavate ... (5) **argenteus.**
- Male inflorescence loosely panicled, short, occupying the upper part only of the stem:
  - Male spikelets ½ in. long and upwards:
    - Bracts of the male flower very acute, gradually tapering ... ... (6) **elongatus.**
    - Bracts of the male flower ovoid-oblong, abruptly acuminate ... ... (7) **dichotomus.**
  - Male spikelets less than ½ in. long:
    - Male spikelets loose, pendulous:
      - Bracts oblong, mucronate ... ... (8) **erectus.**
      - Bracts lanceolate; inner perianth-segments (male) longer than the outer ... (9) **platypteris.**
    - Male spikelets crowded, spreading:
      - Bracts oblong, acute, submuticous ... (10) **Burchellii.**

Styles 2–3:
- Male and female inflorescence dissimilar; male spikes numerous; female spikes few; leaf-sheaths obtuse at the apex, with a prolonged mucro:
  - Male spikelets about ½ in. long, flattened, pear-shaped:
    - Stems wrinkled, spotted ... ... (11) **umbellatus.**
    - Stems not wrinkled, very minutely spotted (12) **imbricatus.**
  - Male spikelets suborbicular, less than ½ in. long:
    - Stems prominently tubercled ... ... (13) **cernuus.**
- Male and female inflorescence similar:
  - Spikelets at least ½ in. long; bracts lanceolate ... (14) **distichus.**
  - Spikelets about ¼ in. long; bracts oblong ... (15) **caricinus.**

IMPERFECTLY KNOWN SPECIES.

- Inflorescence panicled:
  - Bracts lanceolate ... ... (16) **gracilis.**
  - Bracts ovate-oblong ... ... (17) **floribundus.**
  - Bracts oblong-lanceolate, acuminate; spikelets cylindric ... ... (18) **occultus.**
- Inflorescence in terminal elongated spikes; stems striated ... ... (19) **striatus.**

1. **T. spicigerus** (R. Br. Prod. 244, in note); stems 3–4 ft. high, erect, terete, faintly puncticulate; leaf-sheaths $1\frac{1}{2}$–2 in. long, tightly convolute, coriaceous, acuminate, striate, ultimately lacerate at the free end; male spikelets very numerous, in loose, much-branched, erect, panicled cymes 6–10 in. long; branches spreading; each spikelet about $\frac{1}{2}$ in. long, at first linear-cylindric, ultimately oblong; spathe lanceolate, bract-like, much shorter than the spikelet; bracts in 6 rows, ultimately spreading, coriaceous, ferruginous, lanceolate, scarcely mucronate, about the length of the oblong stipitulate flower; perianth-segments rigid; outer lateral boat-shaped, the rest rather flat, oblong, acute; the inner shortest; anthers linear-oblong, apiculate; female plant like the male; spikelets less numerous, arranged in a linear, erect, slightly branched cyme, each spikelet about $\frac{3}{4}$ in. long, many-flowered; spathe and bracts lanceolate, sub-acuminate, coriaceous, chestnut-brown; flowers suborbicular, arcuate, flattened; lateral perianth-segments glabrous, deeply keeled, the rest flattish; ovary obovoid; style 1. *Kunth, Enum.* iii. 440; *Steud. Synops.* ii. 260; *Mast. in DC. Monog. Phan.* i. 314. *Restio spicigerus, Thunb. Diss.* 11, *fig.* 56; *in Usteri, Delect.* i. 46, *t.* 2, *fig.* 5–6; *Flora Cap. ed. Schultes,* 84; *Thunb. herb. ex Mast. in Journ. Linn. Soc.* xiv. 419, 420.

SOUTH AFRICA: without locality, *Thunberg,* ♂ and ♀! *Masson,* ♂!
COAST REGION: Malmesbury Div.; near Groene Kloof, 300 ft., *Bolus,* 4236, ♀! Cape Div.; Cape Flats, *Ecklon,* 567! *Ecklon and Zeyher,* ♂ and ♀! *Bolus,* 4446, ♀! 4450, ♀! Salt River, *Drège,* 227, ♂ and ♀!

2. **T. giganteus** (Kunth, Enum. iii. 435); stems 3–4 ft. high and upwards, stout, terete, slightly puncticulate, giving off numerous tufts of slender branches at the nodes; lower sheaths about 2 in. long, tightly convolute except at the tips, lanceolate, coriaceous, membranous and torn at the tips; upper sheaths similar, but smaller; sheaths on the branchlets with two hyaline lobes at the tip, and prolonged into a curved, linear leaf; male spikelets numerous, in loose, branching, panicled cymes nearly a foot long; branches spreading or pendulous; spikelet $\frac{1}{4}$–$\frac{1}{2}$ in. long, oblong, spindle-shaped, somewhat flattened, sessile or pedicellate; spathe coriaceous, half the length of the spikelet, oblong, aristate; bracts oblong, coriaceous, puncticulate, shortly mucronate; outer perianth-segments oblong, acute; lateral boat-shaped, glabrous; pistillode 1-styled; female plant not recognized. *Steud. Synops.* ii. 259; *Mast. in DC. Monog. Phan.* i. 315.

COAST REGION: Riversdale Div.; lower part of the Lange Bergen, about the waterfall at "Valley Rivers Poort," near Kampsche Berg, *Burchell,* 6994! George Div.; between Lange Vallei and Touw River, *Burchell,* 5711/2! Humansdorp Div.; Kromme River, below 1000 ft., *Drège,* 2!

3. **T. scirpiformis** (Mast. in Journ. Linn. Soc. x. 228); stems erect, terete, slender, olivaceous, puncticulate, wrinkled; leaf-sheaths 1–$1\frac{1}{2}$ in. long, tightly convolute, acuminate, coriaceous, striate, thinner, and lacerate, membranous at the margins; male spikelets

numerous, in terminal, panicled cymes; spikelets many-flowered, about $\frac{1}{4}$ in. long; bracts oblong, coriaceous, dull brown, subulate-mucronate; upper acuminate; outer perianth-segments oblong, acute; outer lateral conduplicate, keeled, glabrous; other segments linear-oblong, acute, flattish; female plant like the male; spikelets 3–4, subglobose, the size of a large pea, arranged in a linear, spicate cyme; bracts oblong-lanceolate; outer lateral perianth-segments deeply-winged; ovary compressed, suborbicular. *Mast. in DC. Monog. Phan.* i. 318. *Restio scirpiformis, Nees in various herbaria.*

COAST REGION: Sand-dunes near Cape Town, *Ecklon and Zeyher*, ♂ and ♀!
No specimen in Kew Herbarium.

4. T. fruticosus (Bergius, Fl. Cap. 353, t. 5, f. 8); rootstock creeping, densely clad with deep chestnut-brown sheaths; sterile stems 6–8 in. long, much-branched; branchlets fascicled, filiform, purple-spotted; fertile stems $1\frac{1}{2}$–2 ft. high, erect, terete, covered with velvety pubescence, rarely glabrous, unbranched, or sometimes branching, as in the sterile stem; leaf-sheaths about 1–$1\frac{1}{2}$ in. long, closely convolute, coriaceous, lanceolate, acute; smaller sheaths with two hyaline, membranous lobes near the apex, and with numerous cilia projecting from the inner surface; apex prolonged into a linear leaf; male spikelets numerous, in erect, loosely branched, panicled cymes 8–9 in. long; branches spreading or deflexed; spathe bract-like, lanceolate, acuminate; spikelets nearly $\frac{1}{2}$ in. long, oblong, acute, flattened, sessile, or on a slender pedicel as long as itself; bracts lanceolate, acuminate, subcoriaceous, thinner at the edge; perianth-segments linear-oblong, acute; lateral slightly winged; wing glabrous; anthers apiculate; female plant as in the male; inflorescence less branched; spikelets 5–9 in erect, linear, spicate cymes about $2\frac{1}{2}$ in. long, sessile or pedunculate, erect or appressed, each about $\frac{1}{2}$–$\frac{3}{4}$ in. long; spathe lanceolate, longer than the spikelet; bracts lanceolate, coriaceous, membranous at the margins; perianth-segments rigid, linear-oblong, acute; outer lateral with a narrow, glabrous wing; ovary ovoid, with a single style. *Mast. in Journ. Linn. Soc.* x. 229; *DC. Monog. Phan.* i. 316. *Restio dichotomus, Linn. Syst. Nat. ed.* 12, ii. 735, *not of Rottboell! R. scariosus, Thunb. Diss.* 15? ; *in Usteri, Delect.* i. 49? ; *Fl. Cap. ed. Schult.* 86? *Thamnochortus scariosus, R. Brown, Prod.* 244, *in note*; *Kunth, Enum.* iii. 430; *Steud. Synops.* ii. 259. *R. eriophorus, Reichb. in herb. Sieber. R. Thamnochortus, Thunb. herb. partly ex Mast. in Journ. Linn. Soc.* xiv. 420.

VAR. β, **glaber** (Mast. in Journ. Linn. Soc. x. 229); stems glabrous; female spikelets larger than in the type; outer perianth-segments narrowly winged.

SOUTH AFRICA: without locality, *Sieber*, 230, ♂! *Thom*, 916, ♂ and ♀! *Masson! Thunberg*, ♂ and ♀!
COAST REGION: Cape Flats, *Drège*, 362! *Burchell*, 180, ♂! near Wynberg, *Drège*, 130, ♂! *Wallich*, ♂! Table Mountain near Constantia, *Zeyher!* at the foot of Devils Mountain, 200 ft., *Bolus*, 4722, ♀! Caledon Div.; Nieuw Kloof,

Houw Hoek Mountains, *Burchell*, 8131! Var. β, Knysna Div.; near Melville, *Burchell*, 5462 (not 5642 as formerly quoted), ♂! and 5548, ♂ and ♀! Port Elizabeth Div.; along the Baakens River near Port Elizabeth, *Burchell*, 4363, ♀! Port Elizabeth, *E.S.C.A. Herb.* 160, ♂! 497, ♀! Albany Div.; near Brookhuizens Poort, 1800 ft., *MacOwan*, 633, ♀!

5. **T. argenteus** (Kunth, Enum. iii. 432); stems tufted, 2–3 ft. high, erect, terete, covered with soft, velvety tomentum, with tufts of slender branchlets at the nodes; leaf-sheaths 1½–2 in. long, closely convolute, coriaceous, acute, deeply membranous and lacerate above; smaller sheaths much smaller, hyaline and lacerate above; apex prolonged into a small curved leaf; male spikelets numerous, arranged in loosely branched, linear, panicled cymes 9–10 in. long; branches remote, spreading or deflexed, each spikelet about ¾–1 in. long, linear club-shaped, sessile, or on a slender stalk of the same length; spathe lanceolate, shorter than the spikelets; bracts numerous, ultimately somewhat spreading, oblong-lanceolate, chartaceous, white; perianth-segments oblong, somewhat rigid; outer lateral cymbiform, apiculate, deeply winged, glabrous, other segments smaller, flattish; anthers apiculate; female plant like the male; spikelets in less branched cymes, erect, each spikelet 1½ in. long; flowers orbicular, flattened, arcuate, stipitate; outer perianth-segments obtuse, apiculate; keel deeply winged, glabrous; fruit compressed, orbicular, surmounted by a single style. *Steud. Synops.* ii. 259; *Mast. in DC. Monog. Phan.* i. 317. *Restio argenteus*, *Thunb. Diss.* 14. *R. scariosus*, *Thunb. herb. ex Mast. in Journ. Linn. Soc.* xiv. 420.

South Africa: without locality, *Masson!*

Coast Region: Cape Div.; hills below Table Mountain, *Thunberg*, ♂ and ♀! Caledon Div.; Genadendal, 3000–4000 ft., *Drège*, 1611a, ♀! Swellendam Div.; between Sparrbosch and Tradouw, 1000–2000 ft., *Drège*, 132a! Riversdale Div.; lower part of the Lange Bergen, about the waterfall at "Valley Rivers Poort," *Burchell*, 6978, ♂ and ♀! Humansdorp Div.; Kromme River, under 1000 ft., *Drège*, 1611b, ♂! Uitenhage Div.; Van Stadens Berg, 1500 ft, *Bolus*, 1509, ♂! *Ecklon and Zeyher*, 291, ♀! Port Elizabeth, *E.S.C.A. Herb.*, 479, ♂!

Central Region: Prince Albert Div.; Great Zwart Bergen, near Klaarstroom, 3000–4000 ft., *Drège*, 132b, ♂! and 9607, ♂!

6. **T. elongatus** (Mast. in Journ. Linn. Soc. x. 226 and xiv. 416); stems tufted, 2–3 ft. high, slender, erect, terete, white spotted; sterile stems much branched; ultimate branches filiform, curved, leafy; leaf-sheaths tightly convolute, 1–1½ in. long, coriaceous, dull brown, striate, membranous at the margins; apex gradually acuminate, ultimately lacerate; male spikelets numerous, in somewhat rounded panicled cymes, sessile or peduncled, erect or sometimes drooping; spikelets ½–¾ in. long, oblong, flattened; spathe lanceolate, half the length of the spikelet; bracts ultimately spreading, deltoid-lanceolate, acuminate, coriaceous, chestnut-brown, membranous at the margins; flowers linear-oblong; perianth-segments linear-oblong, acute; lateral conduplicate, winged-carinate, glabrous; female plant like the male; spikelets 3–7, in erect spike-like cymes 2–4 in. long,

each spikelet oblong, $\frac{1}{2}$–$\frac{3}{4}$ in. long, spathe sheath-like, lacerate, as long as the spikelet; bracts ultimately loose at the tips, deltoid-lanceolate, acuminate, shining, brown, membranous at the margins; flowers orbicular, compressed, arcuate; outer lateral perianth-segments deeply keeled; ovary ovoid, globose. *Mast. in DC. Monog. Phan.* i. 320. *Restio elongatus, Thunb. in Hoffm. Phytogr. Blätt.* i. 7; *in Weber and Mohr, Archiv.* i. 25; *Fl. Cap.* i. *ed Schultes*, 83 ?; *Thunb. herb. ex Mast. in Journ. Linn. Soc.* xiv. 416. *R. simplex, Steudel in Flora*, 1829, i. 134, *not of Forster. R. nutans, Thunberg herb.* ♂! *not of Fl. Cap. ed. Schultes*, 84; *Mast. in Journ. Linn. Soc.* xiv. 416. *Thamnochortus dichotomus, Kunth Enum.* iii. 433, *partly*; *Mast. in Journ. Soc. Linn. Lond.* x. 229, *partly. Restio paleaceus, Nees partly, in various herbaria.*

SOUTH AFRICA: without locality, *Thunberg*, ♀! *Bergius!*
COAST REGION: Cape Div.; Table Mountain, *Drège*, 73, ♂! 98, ♀! *Bolus*, 2883, ♂ and ♀! *Rehmann*, 628, ♀! *Burchell*, 525, ♂! Devils Mountain, *Rehmann*, 917! 918! Worcester Div.; Drakenstein Bergen, near Bains Kloof, 1600 ft., *Bolus*, 4102, ♀! Caledon Div.; mountain near Genadendal, *Burchell*, 8629! Mossel Bay Div.; hills near the landing-place at Mossel Bay, *Burchell*, 6284!

Very variable and often confounded with *T. dichotomus*, from which it differs in the larger spikelets, and in the spathes and bracts being gradually and sharply, not abruptly, acuminate. Under the same numbers the two species are frequently confounded in herbaria.

7. **T. dichotomus** (R. Br. Prodr. 244, in note); rootstock creeping, covered with dark brown scales; stems 1–3 ft. high, erect, slender, terete, olivaceous, with white disc-like markings; fertile stems unbranched or only slightly forked; sterile stems shorter, much branched, leafy; leaf-sheaths tightly convolute, 1–1$\frac{1}{2}$ in. long, coriaceous, dull brown, striate, membranous and lacerate at the tips; smaller sheaths villous and prolonged into a small leaf; male spikelets numerous, in terminal, much-branched, panicled cymes; main branches remote, spreading; spikelets $\frac{1}{2}$ in. long, sessile or pedunculate, at first cylindric, acute, subsequently oblong, many-flowered; spathe bract-like, half the length of the spikelet; bracts ultimately loosely spreading at the tips, oblong, abruptly acuminate, pale, ferruginous, hyaline at the margins; outer lateral perianth-segments conduplicate, apiculate, scarcely keeled, the others oblong flattish; female plant like the male; female spikelets 7–12, loosely arranged in a terminal, linear, little-branched cyme each $\frac{1}{4}$–$\frac{1}{2}$ in. long, oblong subglobose; spathe lanceolate acute; bracts lanceolate, abruptly acute; perianth suborbicular, compressed, arcuate; outer lateral segments deeply keeled, glabrous; the remainder shorter, oblong, flattish; fruit subglobose. *Kunth, Enum.* iii. 433, *partly; Steud. Synops.* ii. 259; *Mast. in Journ. Linn. Soc.* x. 229; *DC. Monog. Phan.* i. 318. *Restio dichotomus, Rottboell, Descr. et Ic.* 2, *t.* 1, *fig.* 1, *not of Linnæus. R. Thamnochortus, Thunb. Diss.* 15; *in Usteri, Delect.* i. 50; *Flor. Cap. ed. Schultes*, 86; *Thunb. herb. partly ex Mast. in Journ. Linn. Soc.* xiv. 420. *R. paleaceus, Nees partly, in various herbaria.*

Thamnochortus consanguineus, Kunth, Enum. iii. 432. *T. ecklonianus,* ♀, *Kunth, Enum.* iii. 430. *T. bromoides, Kunth, Enum.* iii. 432.

SOUTH AFRICA: without locality, *Sieber,* 114, ♀! 116, ♂! 223! *Drège,* 136! 191!

COAST REGION: Clanwilliam Div.; between Grasberg River and Watervals River, 2500–3000 ft., *Drège,* 135, ♂! Cape Div.; Devils Mountain, 1800 ft., *Bolus,* 4758! Table Mountain or Devils Mountain, *Drège,* 138a, ♀! Table Mountain, *Drège,* 49 partly! *Ecklon,* 838, ♀! sandy places near Cape Town, *Bolus,* 2882, ♂! 4431, ♂! 4436, ♂! mountains near Constantia, 3000 ft., *Schlechter,* 501, ♂! Cape Flats, *Burke,* ♂! between Cape Town and Table Mountain, *Burchell,* 924, ♀! Camps Bay, *Zeyher,* 1741, ♂ and ♀! *Burchell,* 333, ♂! Tulbagh Div.; mountains in the vicinity of New Kloof, 1000 ft., *MacOwan, Herb. Aust. Afr.,* 1678, ♂! Caledon Div.; Hottentots Holland, *Ecklon and Zeyher!* Worcester Div.; Drakenstein Mountains, near Bains Kloof, 1600–2000 ft., *Bolus,* 4105, ♂! 4091, ♀! Dutoits Kloof, 3000–4000 ft., *Drège,* 133, ♂ and ♀! 1614, ♂! Mossel Bay Div.; Attaquas Kloof, *Gill,* ♂! Caledon Div.; Nieuw Kloof, Houw Hoek Mountains, *Burchell,* 8126, ♂ and ♀! *MacOwan,* 3046, ♀! mountains of Baviaans Kloof, near Genadendal, *Burchell,* 7595, ♂ and ♀!

Very variable and sometimes difficult to distingush from *T. elongatus,* with which it is frequently associated by collectors and in herbaria. The bracts of the latter plant are more straight-sided, deltoid, and not curved at the margins.

8. **T. erectus** (Mast. in Journ. Linn. Soc. xiv. 419); stems 2–3 ft. high, slender, yellowish, puncticulate, white-spotted; sheaths about 1 in. long, closely convolute, coriaceous, dull brown, striate, thinner at the sharply acuminate apex; male spikelets numerous, in terminal, open, much-branched, panicled cymes; main branches slender, spreading or deflexed; each spikelet about $\frac{1}{4}$ in. long, oblong, acute, somewhat compressed, sessile or pedicellate; pedicel slender, as long as the spikelet; spathe bract-like, shorter than the spikelets; bracts oblong, acute, shining brown, hyaline near the mucronate apex; male flowers oblong, stipitulate, almost as long as the bracts; perianth-segments linear-oblong; lateral conduplicate, scarcely keeled, glabrous; female plant like the male; spikelets in erect, linear cymes; spathes lanceolate; spikelets sessile or stalked, each $\frac{1}{4}$–$\frac{1}{2}$ in. long, oblong, acute, ultimately suborbicular; bracts as in the male; perianth-segments oblong; lateral erect, deeply winged; inner slightly convolute; ovary ovoid. *Mast. in DC. Monog. Phan.* i. 321. *Restio erectus, Thunb. Diss.* 14; *in Usteri, Delect.* i. 48; *Fl. Cap. ed. Schult.* 85; *Thunb. herb. ex Mast. in Journ. Linn. Soc.* xiv. 419. *R. paleaceus, Nees partly in various herbaria.*

SOUTH AFRICA: without locality, *Thunberg,* ♂! *Grey,* ♂!

COAST REGION: Cape Div.; Cape Flats, *Rehmann,* 1813! 1814! Tulbagh Div.; near Tulbagh, *Ecklon and Zeyher,* ♂ and ♀!

9. **T. platypteris** (Kunth, Enum. iii. 429, ♀); stems 2–3 ft. high, simple, terete, of the thickness of a crow's quill, unbranched, white-spotted; leaf-sheaths $1\frac{1}{2}$–2 in. long, closely convolute, lanceolate, thinner at the edges; male spikes numerous, in short, terminal, erect or nodding panicles; spathes as long as or longer than the spikelets, lanceolate, membranous, chestnut-brown; spikelets $\frac{1}{4}$ in.

long, sessile or on slender stalks of the same length, oblong, compressed; bracts lanceolate, acuminate; flowers obovate, slightly shorter than the bracts; perianth-segments oblong, acute; lateral slightly winged; inner flattish, linear-oblong, longer than the outer segments; female plant as in the male; spikelets 2–5, aggregate at the ends of the stems, subglobose, or somewhat flattened, about ½ in. long; spathes lanceolate, acuminate, longer than the spikelet; bracts lanceolate, acuminate; flowers stipitulate, flattened, as long as the bracts; lateral perianth-segments very deeply winged, with glabrous keel; inner shorter; fruit oblique, ovoid, 1-celled, 1-seeded; style persistent. *Steud. Synops.* ii. 258; *Mast. in DC. Monog. Phan.* i. 322.

COAST REGION: Clanwilliam Div.: Oliphants River, *Zeyher!* between Grasberg River and Watervals River, 2500–3000 ft., *Drège*, 2502a, ♂! 2512, ♀! Gift Berg, 1500–2500 ft., *Drège*, 2502b, ♂! 139, ♀!

10. **T. Burchellii** (Mast. in DC. Monog. Phan. i. 322, t. 2, fig. 16-27; t. 5, fig. 8, a-b); rootstock creeping; stems 2–3 ft. long, erect, terete, of the thickness of a crow's quill, olivaceous, puncticulate; leaf-sheaths about 1 in. long, tightly convolute, coriaceous, upper edge membranous; apex mucronulate-lacerate; male spikelets numerous, spreading, in terminal, short, much-branched, panicled cymes, about 2 in. long; spathes lanceolate, acuminate, half the length of the spikelet; spikelets about 4 lin. long, sessile or pedicellate, oblong, compressed, arcuate; bracts convex, coriaceous, oblong, acute, scarcely mucronate; flowers stipitulate, nearly as long as the bracts; perianth-segments linear-oblong; lateral conduplicate, scarcely keeled, glabrous; female plant as in the male; spikelets many, in terminal subpaniculate cymes, each about ½ in. long, oblong, subglobose, many-flowered; bracts lanceolate, ultimately spreading; outer lateral perianth-segments deeply winged; inner three involute, shorter than the outer; ovary ovoid.

SOUTH AFRICA: without locality, *Drège*, 129, ♂ and ♀!
COAST REGION: Malmesbury Div.; near Groene Kloof, 300 ft, *Bolus*, 4237, ♂! Cape Div.; Cape Flats, *Bolus*, 4449, ♂! Mossel Bay Div.; between Great Brak River and Little Brak River, *Burchell*, 6169, ♂ and ♀!

11. **T. umbellatus** (Kunth, Enum. iii. 440); stems cæspitose, procurrent, 2–3 ft. long, slender, erect, unbranched, terete, papillose; leaf-sheaths 1 in. long, coriaceous, fuscous, nervoso-striate, membranous at the ovate, acute apex, mucronulate; male spikelets pendulous, pedicellate, oblong or obovate, flattened, scarcely ½ in. long; bracts oblong-lanceolate, scarcely mucronulate, much exceeding the arcuate, oblong flowers; perianth-segments oblong, acute; lateral conduplicate, with a narrowly winged keel; female plant as in the male; spikelets 1–2, terminal, erect, cylindric or oblong, many-flowered, rather less than 1 in. long; spathe acute; bracts squarrose, oblong-lanceolate, submuticous, coriaceous, nervoso striate, longer than the flowers; outer perianth-segments arcuate, oblong, acute, conduplicate, winged-carinate; inner oblong-lanceolate; ovary turbinate, compressed, coria-

ceous, 1-celled; styles 3. *Steud. Synops.* ii. 260; *Mast. in Journ. Linn. Soc.* x. 232; *DC. Monog. Phan.* i. 324. *Restio umbellatus, Thunb. Diss.* ii. *fig.* 3; *Usteri, Delect.* i. 45, *t.* 2, *fig.* 3; *Fl. Cap. ed. Schult.* 84, *but not of his herbarium. R. distachyos, Rottb. Descr. et Ic.* 8, *t.* 3, *fig.* 5, *as to the female. Leptocarpus distachyos, R. Br. Prod.* 250. *Staberoha distachya, Kunth, Enum.* iii. 444; *Steud. Synops.* ii. 257.

SOUTH AFRICA: without locality, *Drège*, 23, ♂! 25, ♂!
COAST REGION: Malmesbury Div.; near Groene Kloof, 300 ft., *Bolus*, 4239, ♂! Cape Div.; Table Mountain, 2800 ft., *Bolus*, 4455, ♀! *Ecklon*, 841, ♂! Cape Flats, near Rondebosch, *Burchell*, 196! Caledon Div.; between Genadendal and Donker Hoek, *Burchell*, 7931, ♀! Mossel Bay Div.; between Mossel Bay and Zout River, *Burchell*, 6319, ♂ and ♀!

12. T. imbricatus (Mast. in Journ. Linn. Soc. x. 231); stems cæspitose, procurrent, 2–3 ft. high, slender, unbranched, wrinkled, brownish; leaf-sheaths 1½ in. long, closely convolute, coriaceous, thinner at the margins; apex abruptly tapering, sometimes aristate; sterile stems shorter, branching; male spikelets 2–7, at the apex of the branches, approximate, pedunculate; peduncles erect, or generally nodding, as long as the spikelet; spikelets ¾ in. long, suborbicular, compressed; spathe bract-like, shorter than the spikelet; bracts oblong, acute, ultimately spreading; perianth-segments hyaline; the outer lateral acute, slightly keeled, glabrous; inner 3 smaller; female plant as in the male; spikelets 1 or 2, sessile, or stalked, erect, 1¼ in. long, oblong, obtuse, many-flowered; spathe sheath-like, open, half the length of the spikelet; bracts at first close, afterwards spreading at the tips, ovate-oblong, acute, coriaceous, thinner along the upper margin, undulate, ultimately deciduous; outer perianth-segments subequal; lateral navicular, conduplicate, deeply winged; wing sometimes undulate and erose at the margin; ovary pyriform, with 2 styles; fruit oblong, obtuse, surmounted by the remains of the styles. *Mast. in DC. Monog. Phan.* i. 323. *Restio imbricatus, Thunb. Diss.* 9, *fig.* 1; *Usteri, Delect.* i. 43, *t.* 2, *fig.* 1; *Fl. Cap. ed. Schult.* 83; *Thunb. herb. ex Mast. in Journ. Linn. Soc.* xiv. 421. *R. umbellatus*, ♂, *β, in Thunb. herb. ex Mast. in Journ. Linn. Soc.* xiv. 415. *R. spicigerus, β, culmo monostachyo, Nees in Linnæa,* v. 647. *Leptocarpus imbricatus, R. Brown, Prod.* 250. *Staberoha imbricata, Kunth, Enum.* iii. 442, ♀. *Thamnochortus æmulus, Kunth, Enum.* iii. 439, ♂.

VAR. *β*, **stenopterus** (Mast. in Journ. Linn. Soc. x. 231, and xiv. 416); outer lateral perianth-segments with a narrow wing. *Staberoha stenoptera, Kunth, Enum.* iii. 443. *Restio tetrasepalus, Steud. Synops.* ii. 251. *Restio vaginatus, Thunb. Diss.* 10; *in Usteri, Delect.* i. 44; *Fl. Cap. ed. Schult.* 83; *Thunberg herb. ex Mast. in Journ. Linn. Soc.* xiv. 416.

SOUTH AFRICA: without locality; Var. *β*, *Thunberg*, ♀! *Drège*, 29, ♀! 83, ♀!
COAST REGION: Clanwilliam Div.; Ceder Bergen, *Drège*, 24, ♂! 2511, ♀! Malmesbury Div.; near Groene Kloof, 300 ft., *Bolus*, 4238, ♀! Mamre, 100–200 ft., *Baur*, 1180, ♂! Cape Div.; between Wynberg and Constantia, *Burchell*, 789, ♂! Cape Flats, *Zeyher*! *Bolus*, 4430, ♀! 4451, ♀! *Burke*, ♂! Table

Mountain, *Milne*, 227, ♀! Simons Bay, *MacGillivray*, 433, ♀! Stellenbosch Div. ?; top of Sneeuw Kop, *Wallich!* Worcester Div.; Breede River Valley, near Darling Bridge, 800 ft., *Bolus*, 2880, ♂ and ♀! Ceres Div.; Flats near Ceres, 1500 ft., *Bolus*, 7458, ♀! Stellenbosch Div.; Hottentots Holland, *Zeyher*, ♀! Var. β, Worcester Div.; Dutoits Kloof, 3000–4000 ft., *Drège*, 27, ♀! 1636, ♀! 1652, ♂! Sweet Valley (? Zoete Melks Vallei in Caledon Div.), *Wallich!*

13. T. cernuus (Kunth, Enum. iii. 439, ♂); stems erect, terete, slender, 18 in. long, coarsely tubercled, with flat, regular, disc-like prominences; sterile stems when present shorter, sparingly branched; branches filiform, curved; cauline sheaths 1–1½ in. long, tightly convolute, thinner at the apex, gradually tapering, obsoletely mucronate; leafy sheaths smaller, with two hyaline lobes near the apex; male spikelets in loose, terminal, panicled cymes; peduncles nodding; spikelets ¼–½ in. long, cylindric-oblong, ultimately clavate; spathe oblong, acute, mucronate, as long as the spikelet; bracts at first closely overlapping, ultimately loosely imbricate, oblong-obovate, coriaceous, ferruginous, shortly mucronate; flowers arcuate, shorter than the bracts; perianth-segments linear-oblong; lateral conduplicate, with a narrow keel; female plant like the male; spikelets solitary or twin, many-flowered, cuneate-oblong, ¾ in. long; bracts loosely imbricate, oblong-lanceolate, longer than the orbicular flowers; perianth-segments membranous, with purple lines; outer lateral deeply winged-carinate, with lacerate keel; intermediate oblong, obtuse, membranous, lacerate and fringed; inner 3 oblong, obtuse, lacerate; fruit oblong, obtuse, coriaceous, 1-celled, 1-seeded, indehiscent; styles 3. *Mast. in Journ. Linn. Soc.* x. 232; *DC. Monog. Phan.* i. 325, *t.* 2, *fig.* 28–37; *t.* 5, *fig.* 8, *c–d. Restio cernuus, Thunb. Diss.* 10, *fig.* 2; *Usteri, Delect.* i. 45, *t.* 2, *fig.* 2; *and Thunb. herb. ex Mast. in Journ. Linn. Soc.* xiv. 415. *R. spicigerus, Lam. Ill. t.* 804, *fig.* 2. *R. umbellatus, Thunb. herb. partly ex Mast. in Journ. Linn. Soc.* xiv. 415, *not of Thunb. Diss.*

COAST REGION: Cape Div.; Cape Flats, *Drège*, 368, ♂! near Cape Town, *Bolus*, 2881, ♂! Camps Bay, *Burchell*, 340, ♂ and ♀! Table Mountain, *MacGillivray*, 434, ♂! hills below Table Mountain, *Thunberg!* near Simons Town, *Milne*, 217, ♂! Worcester Div.; Drakenstein Mountains, near Bains Kloof, 1600–2000 ft., *Bolus*, 4103, ♂! Riversdale Div.; lower part of the Lange Bergen, near Kampsche Berg, *Burchell*, 6963, ♂ and ♀!

This and the two preceding species belong to the genus *Staberoha* of Kunth and others. They are very difficult of discrimination; the points of distinction lie in the markings of the stem, the form of the leaf-sheaths and bracts, and the size of the spikelets. Had they not been already separated by Kunth, it might have been better to have considered them as forms of one variable species.

14. T. distichus (Mast. in Journ. Linn. Soc. x. 233); stems 1–2 ft. high, slender, terete, simple or slightly branched, olivaceous, white-tubercled; leaf-sheaths ¾ in. long, closely convolute, coriaceous, mucronate, aristate, deeply membranous near the apex; sterile stems shorter, more branched, leafy; male spikelets 2–5 at the apex of the stem, each about ½ in. long, oblong, compressed; spathe oblong, acuminate, aristate, as long as the spikelet; bracts ultimately loosely spreading, lanceolate, convex, coriaceous, longer than the flowers; outer perianth-segments oblong, obtuse; lateral boat-shaped, minutely villous; inner 3 oblong-lanceolate, united at the base; pistillode minute, 3-styled; female plant as in the male; spikelets about ½ in. long, cylindric, acute, ultimately compressed, oblong, many-flowered; spathe lanceolate, acuminate, shorter than the spikelet; bracts ultimately loosely imbricate, lanceolate, coriaceous; margins membranous, decurrent; flowers oblong, trigonous; perianth-segments equal, rather rigid, obtuse; outer keeled; inner shorter, flatter, united at the base; staminodes none; ovary cuneate, trigonous, 2-celled; stigmas 3. *Mast. in DC. Monog. Phan.* i. 326. *Restio distichus, Rottb. Progr.* 11; *Descr. et Ic. Gram.* 6, *n.* 6, *t.* 2, *fig.* 5; *Willd. Sp. Plant.* iv. 2. 725; *Sprengel, Syst.* i. 185; *Nees in Linnæa,* v. 637; *Kunth, Enum.* iii. 409; *Steud. Synops.* ii. 254; *Mast. in Journ. Linn. Soc.* viii. 243 *and* x. 233. *Restio punctulatus, Nees ex Mast. in Journ. Linn. Soc.* viii. 242. *R. vimineus, Linn. herb.! Mast. in Journ. Linn. Soc.* xxi. 590.

COAST REGION: Cape Div.; Table Mountain, *Drège*, 64, ♂! Cape Flats, *Bolus*, 4434, ♂! Pampoens Kraal, *Zeyher*, ♀! Tulbagh Div.; Witsen Berg and Skurfde Berg, near Tulbagh, *Zeyher*, 35! 1737, ♂! Riversdale Div.; lower part of the Lange Bergen, near Kampsche Berg, *Burchell*, 7040, ♂!

15. **T. caricinus** (Mast. in DC. Monog. Phan. i. 327); stems erect, terete, simple, olive-coloured, puncticulate; leaf-sheaths closely convolute, about 1 in. long, coriaceous, striate, tapering, ultimately lacerate at the tip; male spikelets numerous, in terminal panicled cymes; spikelets stalked, ¼ in. long, oblong, obtuse, many-flowered; bracts overlapping, oblong, coriaceous, thinner and lacerate at the margins, about the length of the somewhat three-sided flowers; perianth-segments oblong, muticous; outer lateral narrowly winged, glabrous; the remaining segments flattish; anthers apiculate; female unknown.

SOUTH AFRICA: without locality, *Masson! Thom*, 900, ♂!
COAST REGION: Cape Div.; Simons Bay, *Wight!*

16. **T. gracilis** (Mast. in DC. Monog. Phan. i. 327); stems 2–3 ft. high, slender, erect, terete; leaf-sheaths ¾ in. long, closely convolute, coriaceous, lacerate at the tip; female spikelets in an erect, sparingly branched cyme, 2–3 in. long; each spikelet 2 lines long, erect, many-flowered, oblong, obtuse; spathe lanceolate, acuminate, as long as the spikelet; bracts coriaceous, shining brown, lanceolate, acuminate; upper margin white, membranous; flowers orbicular, compressed; outer perianth-segments oblong, obtuse; lateral boat-shaped, deeply keeled; keel glabrous; staminodes none; ovary ovoid-globose; style solitary.

COAST REGION: Caledon Div.; Mountains of Baviaans Kloof, near Genadendal, *Burchell*, 7894 partly, ♀!

17. **T. floribundus** (Kunth, Enum. iii. 435); stems simple, erect, terete, minutely puncticulate; spikelets oblong, 2–3 lin. long; bracts loosely imbricate, shining; perianth-segments ferruginous; two outer lateral shorter than the others, mucronate, with a shallow, glabrous keel; anthers 1-celled. *Steud. Synops.* ii. 259; *Mast. in Journ. Linn. Soc.* x. 231, *and in DC. Monog. Phan.* i. 328.

COAST REGION: *Reynaud.* No specimen at Kew.

18. ? **T. occultus** (Mast.); stem erect, 2 ft. high, terete, olivaceous, with whitish, flat tubercles, sparingly branched; leaf-sheaths ¾ in. long, coriaceous, closely convolute, membranous above, prolonged into a long acumen; inflorescence a terminal, cymose panicle; branches spreading or pendulous, flattened; male spikelets about ½ in. long, oblong, acute, with an open sheath-like spathe at the base; bracts oblong-lanceolate, coriaceous, compressed, chestnut-brown, exceeding the flowers; outer perianth-segments linear-oblong; two lateral conduplicate, glabrous; intermediate flat, with a midrib; inner 3 membranous, rather shorter than the outer series; anthers apiculate; rudiment of pistil minute, with traces of three styles.

COAST REGION: Worcester Div.; Dutoits Kloof, 2000–3000 ft., *Drège*, 1612, ♂!

19. **T. striatus** (Hochst. in Flora, 1845, 339); stems 3–4 ft. high, erect, striate; leaf-sheaths 1½ in. long, tightly convolute, acuminate, mucronate, striated; female spikelets numerous, in long spike-like cymes, each nearly half an inch long, obovate; bracts lanceolate, mucronate; outer perianth-segments acute; lateral navicular, with a deep keel; inner 3 flat, smaller, all deep brown; ovary compressed, roundish; style simple. *Steud. Synops.* ii. 260; *Mast. in Journ. Linn. Soc.* x. 231, *and in DC. Monog. Phan.* i. 328. *T. elongatus, Hochst. in Flora,* 1845, 339, *not of Mast.*

COAST REGION: Cape Div.; Cape Flats, *Krauss.* No specimen at Kew.

Allied to *T. spicigerus* ?

VIII. HYPOLÆNA, R. Br.

Male inflorescence 1-pluristachyate. *Spikelets* 1–many-flowered, subtended by a spathe. *Perianth-segments* 6, in two rows; outer larger, thicker; lateral conduplicate. *Stamens* 3; anthers 1-celled. *Female spikes* solitary, or few, 1-flowered. *Perianth* stipitate, 6-partite, biseriate; segments appressed to the sessile, 1-celled ovary. *Styles* 2. *Fruit* ovoid or trigonous, 1-celled, indehiscent, sometimes with an epigynous disc. *Seed* solitary, pendulous from the apex of the single cavity.

Perennials with erect, branching stems, and close leaf-sheaths.

DISTRIB. South-west Australia.

Stems narrowly winged ... ... (1) **anceps.**
Stems not winged:
  Male spikelets in spicate or panicled cymes:
    Male spikelets placed edgewise to the axis:
      Spikelets flattened:
        Spikelets 4–4½ lin. long, somewhat crowded; stems wiry, erect ... (2) **impolita.**
        Spikelets 2–3 lin. long, remote:
          Stems very slender, erect, sparingly branched; spikelets broad ... ... (3) **browniana.**
          Stems filiform, spreading, diffusely branched; spikelets narrow ... ... (4) **diffusa.**
      Spikelets cylindric:
        Spikelets clustered, subcapitate:
          Bracts acute, mucronulate ... (5) **eckloniana.**
          Bracts oblong, obtuse ... ... (6) **incerta.**
        Spikelets loosely panicled:
          Bracts aristate ... ... (7) **laxiflora.**
          Bracts acuminate ... ... (8) **aspera.**
        Spikelets in linear, spicate cymes:
          Spathes and bracts acuminate, aristate... ... ... (9) **tenuis.**
          Spathes and bracts shortly mucronate ... ... (10) **filiformis.**
    Male spikelets with one surface turned towards the axis:
      Spikelets remote; bracts coriaceous ... (11) **virgata.**
      Spikelets rather crowded; bracts scarious at the tips... ... ... (12) **Burchellii.**
  Male spikelets solitary at the ends of the branches... (13) **gracilis.**

1. **H. anceps** (Mast. in Journ. Linn. Soc. x. 267); stems filiform, compressed, narrowly winged, slightly and loosely branched; leaf-sheaths closely convolute, subulate, mucronate; mucro subfoliaceous; male spikelets 2–3, in terminal spikes; spathe mucronate; bracts loose; perianth-segments oblong-lanceolate, 1-nerved; inner 3 shorter, hyaline; anthers linear oblong, apiculate. *Mast. in DC. Monog. Phan.* i. 373.

COAST REGION: Riversdale Div.; towards the summit of Kampsche Berg, *Burchell*, 7080 partly, ♂! George Div.; on the Post Berg, near George, *Burchell*, 5896 partly, ♂!

2. **H. impolita** (Mast. in Journ. Linn. Soc. x. 264); stems tufted, about 18 in. high, sparingly branched, roughly tubercled; leaf-sheaths ¾ in. long, loosely convolute, roughly tubercled, with a deep hyaline margin; male spikes in terminal, panicled, distichous cymes, compressed, wedge-shaped, with a lanceolate spathe at the base; bracts loosely arranged, oblong-lanceolate; lateral perianth-segments opposite the rachilla, oblong, conduplicate, keeled; remaining segments flat, hyaline; anthers linear-oblong, apiculate; female inflorescence less branched; spikes larger than in the male; flowers curved; lateral perianth-segments deeply winged; wings striate; inner segments linear-oblong, membranous; ovary linear-oblong, 3-sided; style short; stigmas 2. *Mast. in DC. Monog. Phan.* i. 370. *Restio*

impolitus, Kunth, Enum. iii. 404 ; *Mast. in Journ. Linn. Soc.* viii. 249. *R. triticeus, Thunb. herb. male specimen! Mast. in Journ. Linn. Soc.* xiv. 418.

SOUTH AFRICA: without locality, *Thunberg*, ♂ ! *Bachmann*, ♀ !

COAST REGION: Clanwilliam Div.; between Lange Vallei and Heeren Logement, below 500 ft., *Drège*, 67, ♂ ! Tulbagh Div.; mountains in the vicinity of New Kloof, 1000 ft., *MacOwan, Herb. Aust. Afr.*, 1677! Caledon Div.; Houw Hoek Mountains, *Zeyher*, 4349 !

No female specimen at Kew.

3. **H. browniana** (Mast.); stems erect, dichotomously branching; branches erect, terete, of the thickness of a crow-quill, yellowish with minute white spots; leaf-sheaths about 4–5 lin. long, tightly convolute, coriaceous, brown, subulate-mucronate; male spikelets 4–8, in erect distichous spikes, each spikelet many-flowered, roundish, somewhat compressed, 3–4 lin. in. diam.; spathe nearly as long as the spikelet, similar to the sheaths; bracts coriaceous, oblong, obtuse, or the lower one shortly mucronate-subulate, as long as the oblong compressed flowers; outer perianth-segments unequal, coriaceous, conduplicate, with a hairy keel, hairs reddish-brown; intermediate shortest, flattish, with a central rib; inner equal, oblong, chartaceous; stamens 3; pistillode minute.

COAST REGION: Caledon Div.; on stony mountains around Houw Hoek, 1500 ft., *MacOwan, Herb. Aust. Afr.*, 1727, ♂ !

Although the female plant has not been detected, yet it seems probable, as Mr. N. E. Brown has suggested, that the above represents a new species of *Hypolæna* allied to *H. impolita*, Mast.

4. **H. diffusa** (Mast.); stems filiform, erect, 6–8 in. high, terete, coarsely tubercled, divaricately branched; branches numerous, very slender, wiry; leaf-sheaths $\frac{1}{4}$–$\frac{1}{2}$ in. long, closely convolute, coriaceous, lanceolate, acuminate, deeply hyaline above; inflorescence long, linear, flexuose; male spikelets in linear cymes, remote, sessile or pedicellate, $\frac{1}{8}$ in. long, 1–2-flowered: spathes oblong, acuminate, subcoriaceous, membranous at the apex; bracts oblong, acute, exceeding the flower; perianth-segments linear-oblong, obtuse; outer subcoriaceous; inner subequal, membranous; stamens 3. *Hypolæna aspera, Mast. in Journ. Linn. Soc.* x. 264 *partly.*

COAST REGION: Caledon Div.; Nieuw Kloof, Houw Hoek Mountains, *Burchell*, 8065, ♂ !

5. **H. eckloniana** (Mast. in Journ. Linn. Soc. x. 263, t. 7, fig. A); stems slender, erect, branched; branches ascending, spreading; leaf-sheaths $\frac{1}{2}$ in. long, tightly convolute, subulate, mucronate; male spikelets in dense terminal clusters, erect or spreading, cylindric, oblong, each placed edgewise to the rhachis, nearly $\frac{1}{2}$ in. long; spathe acuminate, open, shorter than the spike; bracts convex, ovate-oblong, obtuse, loosely packed; perianth-segments equal, linear-oblong, hyaline; anthers apiculate; female spikes 1–3 at the ends of the branches, cylindric, oblong, ultimately pear-shaped,

nearly ½ in. long; perianth-segments oblong, spathulate, obtuse, striolate, equal and appressed to the ovary; fruit cylindric-oblong, coriaceous; style simple, short; stigmas 2, linear. *Mast. in DC. Monog. Phan.* i. 369. *Restio digitatus, Thunb. herb.! Mast. in Journ. Linn. Soc.* xiv. 417, 418. *Restio paniculatus, Linn. herb.! Mast. in Journ. Linn. Soc.* xxi. 590.

SOUTH AFRICA: without locality, *Thom*, 632 partly, ♂ and ♀!

COAST REGION: Stellenbosch Div.; Hottentots Holland, *Thunberg*, ♂! Caledon Div.; mountain near Grietjes Gat, *Ecklon and Zeyher!* Donker Hoek Mountain, *Burchell*, 7989, ♂! Houw Hoek Mountains, 1000–3000 ft., *Zeyher*, 4348, ♂!

6. H. incerta (Mast.); stems 2–3 ft. high, erect, terete, sparingly branched; branches fasciculate, ascending, olive-coloured, slightly rugose, white-spotted; leaf-sheaths about 3–4 in. long, tightly convolute, coriaceous, striate; apex membranous, rather acute, with a subulate mucro; male spikes 5–7, aggregate at the ends of the branches, each about ½ in. long, oblong, ovate, acute, provided at the base with a small bract-like spathe; bracts convex, boat-shaped; upper nearly muticous; lower acuminate, longer than the flower; perianth compressed; outer segments linear-oblong, cartilaginous; inner membranous, shorter; anthers linear, apiculate.

SOUTH AFRICA: without locality, *Thom*, 1031, ♂!

COAST REGION: Stellenbosch Div.; Lowrys Pass, *Burchell*, 8267, ♂! Caledon Div.; mountains of Baviaans Kloof, near Genadendal, *Burchell*, 7844, ♂!

7. H. laxiflora (Nees in Linnæa, v. 663); stems filiform, capillary, slightly compressed, branched, purple-spotted; leaf-sheaths ¼ in. long, somewhat loosely convolute, subulato-mucronate; male spikes numerous, arranged at the ends of the branches in loose paniculate cymes, each about ⅛ in. long, 1–2-flowered, subtended by a spathe as long as itself; outer perianth-segments subcoriaceous, oblong, scarcely keeled; inner hyaline, flattish; anthers apiculate; female spike solitary, terminal, 1-flowered, ¼ in. long; bracts loosely imbricate, oblong, with long setaceous points; perianth-segments equal, oblong, 1-nerved; inner smaller, hyaline; ovary oblong, obtuse; styles 2, free, sometimes dilated at the base into a yellow stylopod. *Kunth, Enum.* iii. 451; *Steud. Synops.* ii. 265; *Mast. in Journ. Linn. Soc.* x. 263, *and in DC. Monog. Phan.* i. 369.

COAST REGION: Cape Div.; Table Mountain, in fissures of rocks towards the top, *Ecklon*, 843, ♂ and ♀! Caledon Div.! mountain near Grietjes Gat, between Lowrys Pass and Palmiet River, *Ecklon and Zeyher*, ♂!

8. H. aspera (Mast. in Journ. Linn. Soc. x. 264 partly); stems tufted, filiform, erect, much branched, rigid, coarsely tubercled; leaf-sheaths about ½ in. long, closely convolute, acuminate, deeply hyaline at the tips; male spikes numerous, in much-branched, terminal, panicled cymes; each linear-oblong, with a loose spathe at the base; bracts lanceolate, acuminate, hyaline at the tips; perianth-segments unequal, linear-oblong; outer lateral conduplicate, carinate, opposite the axis of the spike; remaining segments smaller, thinner, subequal; anthers linear-oblong, apiculate. *Mast. in DC. Monog. Phan.* i. 371.

SOUTH AFRICA: without locality, *Thom*, 632a, ♂!

COAST REGION: Stellenbosch Div.; Hottentots Holland Mountains at Lowrys Pass, *Ecklon and Zeyher*, ♂! Caledon Div.; Nieuw Kloof, Houw Hoek Mountains, *Burchell*, 8069, ♂!

Variable in habit and stature.

9. **H. tenuis** (Mast. in Journ. Linn. Soc. x. 265); culms tufted, erect, filiform, 18 in. long, striate or subangular; leaf-sheath ½ in. long, tightly convolute, setaceo-aristate, longitudinally sulcate; male spikelets in long terminal spikes; each with a long, open, setaceo-aristate spathe; flowers facing the axis of the spikelet; perianth-segments linear, oblong, subequal; anthers apiculate; female spikes solitary; perianth-segments oblong-obovate; ovary trigonous; styles 2, deciduous. *Mast. in DC. Monog. Phan.* i. 372.

COAST REGION: Swellendam Div.; summit of a mountain peak near Swellendam, *Burchell*, 7360, ♀! Riversdale Div.; lower part of the Lange Bergen, near Kampsche Berg, *Burchell*, 7028, ♂! on the Kampsche Berg, towards the summit, *Burchell*, 7080 partly, ♂! George Div.; on the Post Berg near George, *Burchell*, 5896 partly!

10. **H. filiformis** (Mast. in Journ. Linn. Soc. x. 267); stems erect, slender, sparingly branched, purplish, white-spotted; leaf-sheath closely convolute, about ½ in. long, setaceo-aristate; male spikelets numerous, in erect, terminal, elongated cymes; each spikelet about ¼ in. long, oblong, with a long aristate spathe at the base; bracts loosely packed, oblong, convex, brown; outer perianth-segments oblong, ferruginous; lateral conduplicate, villous; inner hyaline; anthers apiculate. *Mast. in DC. Monog. Phan.* i. 372.

SOUTH AFRICA: without locality, *Thom*, ♂!

COAST REGION: Caledon Div.; Houw Hoek Mountains, *Zeyher*, 4349 partly, ♂!

11. **H. ? virgata** (Mast. in Journ. Linn. Soc. x. 268); culms erect, slender, moderately branched, olive-coloured, thinly tubercled and white-spotted; leaf-sheaths about ½ in. high, tightly convolute, striate, with a long mucro; male spikelets 8–10, in terminal, spike-like cymes; each spikelet oblong, about ¼ in. long; spathe elliptic, mucronate; bracts oblong, obtuse, aristate; flowers compressed, stipitate; outer perianth-segments rigid; lateral unequal, conduplicate, keeled; inner 3 hyaline; anthers apiculate. *Mast. in DC. Monog. Phan.* i. 374.

COAST REGION: Caledon Div.; mountains of Baviaans Kloof, near Genadendal, *Burchell*, 7817, ♂! Genadendal, *Drège*, 1613, ♂!

12. **H. ? Burchellii** (Mast. in Journ. Linn. Soc. x. 268); stems 2–3 ft. high, tufted, erect, terete, olive-coloured, white-spotted

branching in the middle; branches ascending, virgate; leaf-sheaths $\frac{1}{4}$ in. long, tightly convolute, hyaline at the tip; male spikelets numerous, in terminal elongated cymes; each spikelet about $\frac{1}{8}$ in. long, with a subtending spathe; bracts loosely imbricate, oblong-lanceolate, filaceo-aristate, deeply hyaline; outer perianth-segments oblong-lanceolate, cartilaginous; lateral conduplicate; inner smaller, thinner, oblong. *Mast. in DC. Monog. Phan.* i. 374.

COAST REGION: Caledon Div.; Nieuw Kloof, Houw Hoek Mountains, *Burchell*, 8116, ♂! mountains of Baviaans Kloof, near Genadendal, *Burchell*, 7632 partly, ♂! 7894 partly, ♂!

13. **H. gracilis** (Mast. in Journ. Linn. Soc. x. 266); stems tufted, erect, very slender, much branched, purplish, with white spots; leaf-sheaths $\frac{1}{4}$–$\frac{1}{2}$ in. long, closely convolute, aristate; male spikelets solitary at the ends of the branches, each $\frac{1}{4}$ in. long, oblong-ovate, acute, subtended by an open mucronate spathe of the same length as the spikelets; bracts loose, oblong, acuminate; outer perianth-segments oblong; lateral keeled, villous; inner hyaline, broader than the outer; anthers apiculate. *Mast. in DC. Monog. Phan.* i. 375.

COAST REGION: Cape Div.; mountains near Cape Town, *Zeyher*, 4347, ♂! near Simons Town, *Zeyher*, 1006, ♂! *Wright*, 509! Muizen Berg, 200–600 ft., *Bolus*, 4466, ♂! *MacOwan, Herb. Aust. Afr.*, 1670!

IX. HYPODISCUS, Nees.

Male inflorescence solitary or paniculate-cymose. *Perianth-segments* 6, in two rows; outer lateral conduplicate. *Female inflorescence* of 1 or few spikelets in linear, spicate cymes. *Flower* solitary, sessile or stipitate. *Perianth-segments* small, hyaline, subequal, sometimes wanting. *Ovary* on a fleshy stalk, 1-celled, smooth, lobed or tubercled; style short, thick, sometimes distended at the base into an epigynous disc, dividing into two linear stigmas, feathery on the inner surface. *Fruit* bony, indehiscent, 1-celled, with a single pendulous ovule, surrounded at the base by the persistent perianth, and surmounted by the disc.

Rush-like herbs with creeping rootstocks and tufted stems, covered at the base with overlapping leaf-sheaths.

DISTRIB. Endemic.

Stems compressed:
  Male spikelet solitary, oblong; spathe short, aristate (1) **Willdenovia.**
  Male spikelets clustered, subglobose; spathe long... (2) **nitidus.**
Stems terete:
  Upper leaf-sheaths wanting ... ... (3) **oliverianus.**
  Upper leaf-sheaths present:
    Female perianth wanting; male spikes panicled, silvery ... ... (4) **argenteus.**
    Female perianth accrescent; male and female spikelets solitary, oblong, or aggregated ... (5) **aristatus.**

Female perianth minute:
Stems fistular:
Male perianth-segments awned; male inflorescence panicled ... ... (6) **Neesii.**
Male perianth-segments not awned ... ... (7) **binatus.**
Stems solid:
Stems striated ... ... (8) **striatus.**
Stems smooth, not striated:
Ovary surmounted by long, filiform appendages ... ... (9) **synchroolepis.**
Ovary surmounted by a cup-shaped, toothed disc ... ... (10) **alboaristatus.**
Ovary surmounted by a few, short teeth ... (11) **rugosus.**

1. **H. Willdenovia** (Mast. in Journ. Linn. Soc. x. 259); stems tufted, 8–12 in. high, erect, slender, flattened, sulcato-striate, simple; leaf-sheaths tightly convolute, subulate, mucronate, about 1 in. long; male and female spikes of like shape, $\frac{1}{2}$–$\frac{3}{4}$ in. long, solitary, terminal, oblong, compressed, provided at the base with an open, aristate spathe, somewhat shorter than itself; bracts closely overlapping, elliptic, coriaceous, maculate beneath an obtuse apex, mucronate-aristate; flowers oblong-lanceolate; outer perianth-segments ferruginous, mucronate; lateral conduplicate, glabrous; inner irregular, shorter than the outer, hyaline, combined into a tube at the base; anthers linear-oblong, apiculate; female flower solitary; perianth-segments hyaline, oblong, free; ovary oblong, obtuse, shortly stalked, surmounted by a fleshy disc; styles 2. *Mast. in DC. Monog. Phan.* i. 389, *t.* 4, *figs.* 7–15, *and t.* 5, *fig.* 17. *Willdenovia striata, Spreng. Syst. Veg.* i. 188. *Lepidanthus Willdenovia, Nees in Linnæa,* v. 665, *as to the male plant. Restio sulcatus, Kunth, Enum.* iii. 404; *Steud. Synops.* ii. 253. *R. cuspidatus, Thunb. herb. as to female specimen! Mast. in Journ. Linn. Soc.* xiv. 418.

COAST REGION: Cape Div.; Cape Flats, *Burchell,* 8545, ♂! *Zeyher,* ♀! *Ecklon and Zeyher,* ♂ and ♀! at the foot of Muizen Berg, near Fish Hoek, 200 ft., *Bolus,* 4465, ♂! Stellenbosch Div.; Hottentots Holland, *Zeyher!* Riversdale Div.; near Zoetemelks River, *Burchell,* 6662, ♂! Knysna Div.; near the west end of Groene Vallei, *Burchell,* 5646, ♂ and ♀!

2. **H. nitidus** (Mast. in Journ. Soc. Linn. x. 259); stems 12–18 in. high, erect, simple, compressed, brownish, rugulose; leaf-sheath $1\frac{1}{2}$ in. long, tightly convolute, setaceo-mucronate; male spikelets numerous, with close terminal panicles, protected by a large open sheath-like spathe, each obovoid; bracts loosely imbricate, with a long acumen; outer perianth-segments subequal, ferruginous, oblong, acute; lateral conduplicate; inner shorter, ovate, obtuse, apiculate, hyaline; anthers linear, apiculate. *Mast. in DC. Monog. Phan.* i. 383.

COAST REGION: Tulbagh Div.; New Kloof near Tulbagh, *Ecklon,* ♂! at the foot of mountains in the vicinity of New Kloof, near Tulbagh, 850 ft., *MacOwan, Herb. Aust. Afr.* 1680! summit of Winterhoek Mountain, *Ecklon!*

3. **H. oliverianus** (Mast. in Journ. Linn. Soc. x. 254); stems tufted, slender, erect, unbranched, 2–3 ft. high, with leaf-sheaths at the very base only, purple-spotted; male spikes 2–3, approximate at the ends of the stems, rarely solitary, each oblong or top-shaped, about ¼ in. diam., protected at the base by an open shining yellow spathe, as long as the spike; bracts tightly imbricate, cartilaginous, chestnut-brown, ending in a long, white, spreading awn; flowers arcuate, compressed; perianth-segments unequal, lanceolate; anthers apiculate; female spikes 2–3 at the end of the branches, 1-flowered, cylindric-lanceolate, with an open, yellowish spathe at the base; bracts oblong, chestnut-brown, with a long, straight, or twisted awn; perianth minute, hyaline; segments subequal, ovate; ovary on a fleshy stalk, oblong, surmounted by a yellow, fleshy, lobed disc; style short; stigmas 2, linear. *Mast. in DC. Monog. Phan.* i. 381.

COAST REGION: Stellenbosch Div.; Hottentots Holland Mountains, near Grietjes Gat, *Bolus*, 4222, ♂! Caledon Div.; Nieuw Kloof, Houw Hock Mountains, *Burchell*, 8118, ♂ and ♀!

4. **H. argenteus** (Mast. in Journ. Linn. Soc. x. 261); stems tufted, erect, 2–3 ft. high, sparingly branched, terete, faintly striate; leaf sheaths 1 in. high, tightly convolute, acuminate; male spikes very numerous, collected in terminal, close, much-branched panicles, surrounded by an open sheath-like spathe, each roundish, about ¼ in. diam.; bracts oblong, acuminate, membranous at the tips, concealing the stipitate flowers; outer perianth-segments oblong-lanceolate, acuminate; inner scarcely shorter, membranous; anthers oblong, apiculate; female spikes 1–3, terminal wedge-shaped, about ¾ in. long, with an open sheath-like spathe at the base; bracts oblong-lanceolate, acuminate, purplish at the tips, nervoso-striate; flower solitary; perianth absent; fruit oblong, obtuse, bony, blackish, placed on a fleshy, lobulate stalk. *Mast. in DC. Monog. Phan.* i. 383. *Leucoplœus argenteus, Nees in Lindl. Nat. Syst. ed.* 2, 450. *Restio argenteus, Thunberg, Diss.* 14; *in Usteri, Delect.* i. 49; *Thunb. Herb.! Mast. in Journ. Linn. Soc.* xiv. 417.

SOUTH AFRICA: without locality, *Drège*, 99, ♂! 9612, ♀! *Thunberg*, ♂!
COAST REGION: Caledon Div.; Genadendal, 2000–3000 ft., *Drège*, 1642, ♂! mountains of Baviaans Kloof, near Genadendal, *Burchell*, 7901, ♂! Nieuw Kloof, Houw Hoek Mountains, *Burchell*, 8117, ♂! 8129, ♀! Palmiet River, *Ecklon*, 972, ♂! Riversdale Div.; on the Kampsche Berg, *Burchell*, 7058, ♀ and ♂! Ceres Div.; mountain slopes near Ceres, 2400 ft., *Bolus*, 5294, ♀!
WESTERN DIV.: Little Namaqualand; rocky hills by the Hartebeest River, *Zeyher*, 4338, ♀!

5. **H. aristatus** (Nees in Lindl. Nat. Syst. ed. 2, 450); stems tufted, erect, cylindric, olive-coloured, white-spotted, sparingly branched or quite simple; leaf-sheaths 1¼ in. long, tightly convolute, except at the apex, subulato-mucronate; male spikelets subglobose or oblong, solitary or more commonly clustered at the end of the branches; bracts with long spreading awns; flowers compressed, arcuate; outer perianth-segments oblong or obovate-lanceolate;

lateral conduplicate, villous; anthers apiculate; female spikes solitary or few, approximate at the ends of the branches, ovate-oblong, about 1 in. long, subtended by a sheath-like spathe; bracts ovate-lanceolate, yellow-margined, with long terminal awns; flower solitary, terminal; perianth on a fleshy stalk; segments subequal, membranous, oblong-lanceolate or obovate, pointed; style short; stigmas 2, linear; fruit oblong, obtuse, smooth, bony, purplish, surrounded at the base by the persistent perianth; seed solitary, pendulous. *Mast. in Journ. Linn. Soc.* x. 252, *and in DC. Monog. Phan.* i. 380. *Restio aristatus, Thunb. Diss.* 10, *No.* 3, *fig.* 4, *male; in Usteri, Delect.* i. 44, *t.* 2, *fig.* 4. *Fl. Cap. ed. Schult.* 83; *Thunb. Herb.! male plant, Mast. in Journ. Linn. Soc.* xiv. 415; *Nees in Linnæa,* v. 636; *Kunth, Enum.* iii. 383 *male plant, excl. syn.; Steud. Synops.* ii. 249, *male plant excl. syn.*

VAR. bicolor (Mast. in Journ. Linn. Soc. x. 253); awns and margins of the leaf-sheaths, spathes and bracts golden-yellow. *Mast. in DC. Monog. Phan.* i. 381.

SOUTH AFRICA: without locality, *Thunberg,* ♂! *Drège,* 21, ♂! *Sieber,* 113, ♂!
COAST REGION: Piquetberg Div.; Piquet Berg, 1500–3000 ft., *Drège,* 2513, ♂! Cape Div.; Camps Bay, *Burchell,* 339, ♀! Tulbagh Div.; New Kloof, near Tulbagh, 1000 ft., *MacOwan, Herb. Aust.-Afr.,* 1680, ♂ and ♀! 1681, ♂! Worcester Div.; Drakenstein Mountains near Bains Kloof, 1600–2000 ft., *Bolus,* 4106, ♂! Dutoits Kloof, *Drège,* 20, ♂! 199, ♂! 1656, ♀! mountains above Worcester, *Rehmann,* 2555! Stellenbosch Div.! Hottentots Holland Mountains, 1000–2000 ft., *Zeyher,* 4332, ♂! Caledon Div.; mountains near Grietjes Gat, between Lowrys Pass and Palmiet River, 2000–4000 ft., *Zeyher,* ♂ and ♀! Houw Hoek Mountains, 1000–3000 ft., *Zeyher,* 4332, ♀! mountains of Baviaans Kloof, near Genadendal, *Burchell,* 7594, ♂ and ♀! Donker Hoek Mountain, *Burchell,* 7963, ♂! Swellendam Div.; on a mountain peak near Swellendam, *Burchell,* 7373, ♀! Riversdale Div.; on the Kampsche Berg, *Burchell,* 7057, ♀! George Div.; on mountains near George, *Drège,* 3941, ♂ and ♀! lower part of the Post Berg, near George, *Burchell,* 6027, ♂ and ♀! Var. bicolor; Clanwilliam Div.; Oliphants River, *Zeyher,* ♂ and ♀! Pretoris Kloof, between Piquiniers Kloof and Oliphants River, *Drège,* 2509, ♂ and ♀! Worcester Div.; Dutoits Kloof, 3000–4000 ft., *Drège,* 140, ♀!

6. **H. Neesii** (Mast. in Journ. Linn. Soc. x. 260); stems tufted, erect, slender, cylindric, fistular, sulcato-striate; leaf-sheaths tightly convolute, about 1½ in. long, membranous at the tip and prolonged into a long awn; male spikes less than ½ in. long, oblong, aggregate into a long, linear, spicate cyme about 2 in. long; bracts with a long whitish acumen; perianth-segments subequal, lanceolate, acuminate; anthers aristulate; female spikes 2–3, terminal, protected by a long sheath-like spathe; bracts imbricate, acuminate, aristate, rigid; flower solitary; perianth-segments subequal, minute, oblong, acute; ovary oblong, tubercled near the apex; stigmas 2, linear. *Mast. in DC. Monog. Phan.* i. 384.

COAST REGION: Clanwilliam Div.; near Brak Fontein, *Zeyher!* Pretoris Kloof, between Piquiniers Kloof and Oliphants River, 1000–1500 ft., *Drège,* 2492, ♂! 2493, ♀!

7. **H. binatus** (Mast. in Journ. Linn. Soc. x. 258); stems tufted, erect, 2 ft. high, slender, simple, fistular, olive-coloured, purple

spotted; leaf-sheaths about 1 in. long, tightly convolute; apex yellowish, subulato-mucronate; male spikelets 4–6, in an erect, linear, panicled cyme, each protected at the base by an oblong, sulcato-striate spathe shorter than itself; bracts ovate, acute, chestnut-brown, thinner at the apex; perianth rigid, ferruginous; segments oblong-lanceolate, unequal; lateral conduplicate, glabrous; filaments flattened; anthers apiculate; female spike cylindric-lanceolate, $\frac{1}{4}$ in. long; perianth-segments hyaline, subequal, ovate; ovary oblong, shortly stalked, surmounted by a yellow, warty disc; style short, chestnut-brown, stigmas long. *Mast. in DC. Monog. Phan.* i. 388. *Dovea binata, Steud. Synops.* ii. 248, *female only.*

COAST REGION: Piquetberg Div.; Piquet Berg, 2000 ft., *Drège*, 2477, ♂! 2478, ♀! Tulbagh Div.; mountains in the vicinity of New Kloof, near Tulbagh, 1000 ft., *MacOwan, Herb. Aust. Afr.*, 1679! Caledon Div.; on Donker Hoek Mountain, *Burchell*, 7958, ♂! Houw Hoek, *MacOwan!*

8. **H. striatus** (Mast. in Journ. Linn. Soc. x. 258); rootstock creeping, covered with brown scales; stems 18–24 in. high, erect, simple or sparingly branched, slender, terete, sulcato-striate; leaf-sheaths tightly convolute, aristate; male spikelets 3–6, subsessile at the apex of the branches, protected by open, sheath-like, deciduous spathes, spreading, cylindric, acute, many-flowered, rather less than $\frac{1}{2}$ in. long; bracts tightly imbricate, oblong, acute, coriaceous, chestnut-brown; outer perianth-segments linear-oblong, apiculate; inner broadly oblong-obovate; anthers apiculate; female spikes solitary or twin, erect, about $\frac{1}{2}$ in. long, cylindric-lanceolate, straight or curved; bracts oblong-acute, cartilaginous, brown; outer perianth-segments oblong, acute; inner shorter, broader; staminodes 3, separated from the perianth-segments by a short internode; ovary oblong, coriaceous, surmounted by an epigynous, horny, yellow disc; styles 2, deciduous; fruit stalked, oblong cylindric, tubercled at the apex, yellowish, 1-celled, surrounded by the persistent perianth-segments. *Mast. in DC. Monog. Phan.* i. 385. *Boeckhia striata, Kunth, Enum.* iii. 449.

SOUTH AFRICA: without locality, *Drège*, 2484 bb, ♂! 9613, ♂!

COAST REGION: Clanwilliam Div.; Pretoris Kloof, between Piquiniers Kloof and Oliphants River, *Drège*, 2484, ♀! Worcester Div.; near Touws River Railway Station, 3000 ft., *Bolus*, 7456, ♂! Uniondale Div.; Wagenbooms River in Lange Kloof, *Drège*, 2022, ♂! Humansdorp Div.; Kromme River, below 1000 ft., *Drège*, 2484b, ♀! Uitenhage Div.; mountains near Elands River, 1000–4000 ft., *Zeyher!*

CENTRAL REGION: Prince Albert Div.; Great Zwart Bergen near Klaarstroom, 3000–4000 ft., *Drège*, 9452, ♀!

WESTERN REGION: Little Namaqualand; Kamies Bergen at Ezels Fontein, 3500–5000 ft., *Drège*, 2479, ♂! 2480, ♀! 2491, ♂!

9. **H. synchroolepis** (Mast. in Journ. Linn. Soc. x. 256, t. 7, fig. C); stems tufted, erect, 2–3 ft. high, slender, rigid, unbranched; leaf-sheaths 1 in. long, loosely convolute at the apex, yellow at the margins, aristate; male spikelets 6–8, arranged in pairs in an erect, linear cyme; spikelet many flowered, ovate-oblong, about $\frac{1}{4}$ in. long,

protected at the base by a short sheath-like spathe; bracts chestnut-brown, ovate-oblong, with long terminal white awns; perianth-segments unequal, oblong-lanceolate; lateral conduplicate, glabrous; inner shorter; female spikelet cylindric-lanceolate, $\frac{1}{4}$ in. long; perianth minute; segments subequal, hyaline, ovate, acute; ovary oblong, obtuse, exceeding the segments, surmounted by numerous, erect, fleshy linear lobes; style dilated at the base, dividing into two linear stigmas; fruit oblong, hard, indehiscent, 1-seeded, surrounded at the base by a persistent perianth. *Mast. in DC. Monog. Phan.* i. 387.

COAST REGION: Humansdorp Div.; near Kromme River, *Drège*, 17, ♀! Uitenhage Div.; Van Stadens Berg, *Burchell*, 4705, ♂! *Ecklon and Zeyher*, 817, ♂ and ♀! *Zeyher*, 4334, ♂ and ♀!

10. H. alboaristatus (Mast. in Journ. Linn. Soc. x. 257); stems tufted, erect, 2 ft. high, slender, unbranched, purple spotted; leaf-sheaths about 1 in. long, tightly convolute, aristulate; male spikelets 2–3, approximate, oblong or subglobose; bracts oblong, acuminate, whitish; outer perianth-segments oblong acute; anthers linear, apiculate; female spikelets 2–5, approximate, oblong; ovary oblong, stipitate, surmounted by a lobed disc; fruit oblong, surrounded at the base by the persistent perianth. *Mast. in DC. Monog. Phan.* i. 382. *Restio alboaristatus, Nees in Linnæa,* v. 635; *Kunth, Enum.* iii. 407; *Steud. Synops.* ii. 249, *as to the male plant. Restio aristatus, Thunb. herb.! female plant, ex Mast. in Journ. Linn. Soc.* xiv. 415. ? *Hypodiscus duplicatus, Hochst. in Flora,* 1845, 338. *Boeckhia lævigata, Kunth, Enum.* iii. 450, *female plant.*

SOUTH AFRICA: without locality, *Thom*, 1060, ♂! *Drège*, 19! *Masson! Thunberg*, ♂! *Zeyher*, 972!

COAST REGION: Cape Div.; Table Mountain, *Drège*, 199a, ♂! *Schlechter*, 725; Devils Mountain, 1000 ft., *Bolus*, 4447, ♂ and ♀! Cape Flats, *Zeyher!* Worcester Div.; Dutoits Kloof, 2000–3000 ft., *Drège*, 15! 1653! Caledon Div.; mountains near Hemel en Aarde, 500–2000 ft., *Zeyher*, 4333! Uitenhage Div.; Van Stadens Berg, *Ecklon and Zeyher*, 816, ♂, 818, ♀! Mountains of Baviaans Kloof, near Genadendal, *Burchell*, 7655, ♀! Genadendal, 2000-3000 ft., *Drège*, 16! 9608!

11. H. rugosus (Mast. in. Journ. Linn. Soc. x. 255); stems cæspitose, erect, 8–12 in. long, simple, olivaceous, white spotted; leaf-sheaths about 1 in. long, tightly convolute, mucronato-aristate; female spikelets 2–3, aggregated at the ends of the stems, each about $\frac{1}{2}$ in. long, oblong-turbinate, protected at the base by a lanceolate, aristate spathe; bracts oblong-lanceolate; flower solitary, terminal; perianth sessile or stipitate; perianth-segments equal, oblong-lanceolate; ovary shortly stalked, oblong, obtuse, tubercled; style short; stigmas two; fruit ovate-oblong, obtuse, tubercled; surrounded at the base by the persistent perianth and surmounted at the apex of the deciduous style by six small teeth. *Mast. in DC. Monog. Phan.* i. 386.

SOUTH AFRICA: without locality, *Ecklon and Zeyher*, 85, ♀!
COAST REGION: Swellendam Div.; Voormans Bosch, *Zeyher*, 4336, ♀!
Male plant unknown.

X. CANNOMOIS, Beauvois.

Male spikelets numerous, paniculate cymose, with numerous open deciduous spathes. *Perianth-segments* 6, in two rows; outer larger; lateral navicular; filaments free; anthers linear-oblong. *Pistillode* none. *Female spikes* 1–3 at the apex of the branches, surrounded by one or more permanent spathes. *Perianth-segments* oblong, appressed to the fruit. *Staminodes* none. *Ovary* oblong, obtuse, 1-celled; styles 2, free, deciduous. *Fruit* stipitate, oblong, obtuse, more or less compressed, coriaceous or woody, indehiscent, 1-celled, 1-seeded. *Seed* pendulous from the apex of the funicle.

Stems rush-like, tufted or erect from a creeping rootstock; sheaths closely convolute.

DISTRIB. Endemic.

In my previous writings this genus is erroneously spelled *Cannamois*.

Spathes of male inflorescence deciduous.
  Plant robust, much branched; male spikelets oblong, $\frac{1}{4}$–$\frac{1}{2}$ in. long ... ... (1) **virgata.**
  Plant simple or but slightly branched; male spikelets $\frac{1}{8}$ in. long.
    Bracts roundish, usually acute or slightly acuminate ... ... (2) **scirpoides.**
    Bracts all acuminate; spikelets roundish, $\frac{1}{8}$ in. long ... ... (3) **simplex.**
Spathes of male inflorescence persistent; bracts with long whitish points ... ... (4) **congesta.**

1. C. cephalotes (Beauv. in Ann. Sc. Nat. xiii. (1828), 43, t. 3, fig. 1 ♀); stems erect from a creeping rootstock, tall, fistular, much branched, olive-coloured, covered with minute whitish spots; leaf-sheaths 1 in. long, tightly convolute, mucronate, yellowish at the margins; male inflorescence terminal, paniculate, spikelets about $\frac{1}{4}$ in. long, pedicellate, ovoid-oblong; bracts broadly ovate, acute, chestnut-brown, white-edged; perianth-segments unequal, oblong, obtuse; outer shorter; anthers oblong; female spikes solitary, or 2–3 aggregate, 1 in. or more long, cylindric, oblong, acute; bracts oblong-ovate, acuminate; perfect flowers 3–5; perianth tubular at the base, dividing deeply into 6 equal, suborbicular segments, 3 outer, 3 inner; ovary oblong, obtuse, compressed, surmounted by a fleshy disc; styles 2 very long; fruit oblong-ovate, plano-convex. *Steud. Synops.* ii. 263; *C. virgata Hochst. in Flora,* 1845, 340; *Steud. Synops.* ii. 263; *Mast. in Journ. Linn. Soc.* x. 234, *and in DC. Monog. Phan.* i. 361. ♂ *Restio virgatus, Rottb. Descr. et Ic.* 5, *t.* 1, *fig.* 2 ♂ (1773); *Thunb. Diss.* 20; *in Usteri, Delect.* i. 54; *Fl. Cap. ed. Schult.* 89; *Thunb. Herb.! Mast. in Journ. Linn. Soc.* xiv. 419. *R. Scopa, Thunb. Diss.* 20; *in Usteri, Delect.* i. 54; *Fl. Cap. ed. Schult.* 88; *Thunb. Herb.! Mast. in Journ. Linn. Soc.* xiv. 417. *Restio elegans, Poir.*

Encyl. vi. 171. *Elegia paniculata, Pers. Synops.* ii. 607 (*fid. synonym*). *Thamnochortus robustus, Kunth, Enum.* iii. 436 *Steud. Synops.* ii. 263 *in note. T. virgatus, Kunth, l.c.*, ♂ *and* ♀ *Mesanthus macrocarpus, Nees in Lindl. Nat. Syst. Bot. ed.* 2, 451 1836 ; *Kunth, Enum.* iii. 485 ; *Steud. Synops.* ii. 264. *Willdenovia compressa, Thunb. in Vet. Akad. Handl. Stockh.* xi. 1790, *t.* 2, *fig.* 3 ; *Fl. Cap. ed. Schult.* 82 ; *and of Thunberg herb. ! Mast. in Journ. Linn. Soc.* xiv. 421.

SOUTH AFRICA : without locality, *Thunberg*, ♂ and ♀ ! *Burchell*, 5810, ♂ ! 5811, ♂ ! *Masson !*

COAST REGION : Clanwilliam Div. ; Gift Berg, 1500–2000 ft., *Drège*, 139 partly, ♂ ! Onder Bokke Veld, *Drège*, 1606 ?, ♂ and ♀ ! Tulbagh Div. ; on the Witsen Berg, *Burchell*, 8697, ♀ ! Ceres Div. ; mountains around Ceres, 2100 ft., *Bolus*, 5481, ♀ ! Worcester Div. ; Drakenstein Mountains, near Bains Kloof, 1600–2000 ft. *Bolus*, 4088, ♀ ! Dutoits Kloof, *Drège*, 1605, ♀ ! 1606, ♂ and ♀ ! 1607, monstrosity ! Caledon Div. ; by the Zonder Einde River, near Appels Kraal, *Zeyher*, 1738, ♂ ! Swellendam Div. ; Voormans Bosch, *Zeyher*, 4345 ! Riversdale Div. ; between Vet River and Krombeks River, *Burchell*, 7163, ♂ ! near the foot of the Lange Bergen, near Kampsche Berg, *Burchell*, 7139, ♂ and ♀ ! Port Elizabeth Div. ; Witteklip Mountain, near Port Elizabeth, *MacOwan*, 2151, monstrosity ! Alexandria Div. ; Zuurberg Range, *Drège*, 2024, monstrosity !

Variable in the size of the parts and in the shape of the male spikelets.

2. C. scirpoides (Hochst, in Flora, 1845, 340) ; stems tufted, erect, fistular, moderately branched, somewhat coarsely rugulose ; leaf-sheaths about 1 in. long, closely convolute, mucronate ; male inflorescence panicled ; spikelets numerous, crowded, each about $\frac{1}{4}$ in. long, ovoid, acute ; bracts lanceolate, markedly acuminate, chestnut-brown, shining ; perianth-segments ovate, apiculate, membranous ; female spike solitary, about $1\frac{1}{2}$ in. long, ovoid, cylindric, surrounded at the base by an open spathe nearly as long as itself ; bracts concave, acuminato-aristate, coriaceous ; terminal bract sterile ; flowers 2–3, one only maturing ; perianth-segments oblong, obtuse, membranous ; fruit oblong, obtuse, compressed, stipitate ; seed solitary, pendulous. *Mast. in Journ. Linn. Soc.* x. 236, *and in DC. Monog. Phan.* i. 362. *Thamnochortus scirpoides, Kunth. Enum.* iii. 438. *Restio parviflorus, Thunb. Diss.* 13 ; *in Usteri, Delect.* i. 48 ; *Fl. Cap. ed. Schult.* 85 ; *and Thunb. herb. ! male plant, Mast. in Journ. Linn. Soc.* xiv. 416.

SOUTH AFRICA : without locality, *Thunberg !*

COAST REGION : Clanwilliam Div. ; Brak Fontein, *Zeyher*, ♂ and ♀ ! Worcester Div. ; mountains above Worcester, *Rehmann*, 2674 ♂ ! Caledon Div. ; Houw Hoek Mountains, 1000–3000 ft., *Zeyher*, 4337, ♂ and ♀ ! mountains of Baviaans Kloof, near Genadendal, *Burchell*, 7874, ♂ ! Uniondale Div. ; Welgelegen, in Lange Kloof, 2000 ft., *Drège*, 2023 !

CENTRAL REGION : Prince Albert Div. ; Great Zwart Bergen, near Vrolykheid, 4000–5000 ft., *Drège*, 100, ♂ !

3. C. simplex (Kunth. Enum. iii. 448) ; stems erect from a creeping rhizome, fistular, slender, scarcely branched, faintly striate, obscurely rugulose ; leaf-sheaths about $1\frac{1}{4}$ in. long, closely convolute, mucronate ; male spikelets numerous, terminal, in close, much branched panic les, each about $\frac{1}{8}$ in. long ; spathe open, lanceolate acuminate, nearly as long as the panicle; bracts ovate-oblong, acuminate, chestnut-brown, white-margined; perianth-segments oblong, obtuse; female spikes few, sessile near the apex of the stem, $\frac{1}{2}$–$\frac{3}{4}$ in. long; spathes oblong, acute, as long as the spikelet; bracts ovate oblong, acute, shining brown; flowers solitary or two or three in a spike; perianth-segments equal, hyaline shorter than the oblong-ovoid, blackish fruit. *Mast. in Journ. Linn. Soc.* x. 237, *and in DC. Monog. Phan.* i. 363, *t.* 3, *figs.* 33–41, *and t.* 5, *fig.* 15. ♂ *Restio acuminatus, Thunb. Diss.* 13; *in Usteri, Delect.* i. 47; *Fl. Cap. ed. Schult.* 84; *Thunb. herb.! Cucullifera dura, Nees in Lindl. Nat. Syst. ed.* 2, 451, *Thamnochortus strictus, Kunth, Enum.* iii. 438; *Steud. Synops.* ii. 259, ♂!

SOUTH AFRICA: without locality, *Thunberg*, ♂!
COAST REGION: Clanwilliam Div.; between Grasberg River and Watervals River, 2500–3000 ft., *Drège*, 101, ♂! 2514, ♀! Malmesbury Div.; near Groene Kloof, 300 ft., *Bolus*, 4240, ♂! Tulbagh Div.; Tulbagh, *Zeyher* ♂! Ceres Div.; mountains around Ceres, 1800 ft., *Bolus*, 5479 ♀!

4. C. congesta (Mast.); stems erect, 2–3 ft. high, of the thickness of a crow-quill or rather thicker, terete, unbranched, olivaceous, nearly smooth, obscurely puncticulate; leaf-sheaths about 1 in. long, tightly convolute, coriaceous; apex lanceolate, shortly mucronate; male flowers numerous, in dense, compact, paniculate heads at the end of the branches, intermixed with long, sheathing, acuminate, coriaceous spathes; bracts oblong, acuminate, longer than the oblong perianth; lateral segments oblong, conduplicate; inner shorter; anthers apiculate; female inflorescence similar to the male; female flowers 3–4 in the spikelet; bracts oblong, acuminate, aristate; perianth-segments obsolete; ovary cylindric, surmounted by a smooth dome-like top from whence proceed the two styles. *Mesanthus Ricinus, Nees MSS. in various Herbaria.*

COAST REGION: Clanwilliam Div.; Ceder Bergen, near Ezels Bank, 4000–5000 ft., *Drège*, 2508, ♂ and ♀! Tulbagh Div.; stony places in the vicinity of New Kloof, 1000 ft., *MacOwan, Herb. Aust.-Afr.*, 1682! Caledon Div.; Donker Hoek Mountain, *Burchell*, 7960 ♀! Zwartberg, near Caledon, *Ecklon and Zeyher*, ♂ and ♀!

XI. WILLDENOVIA, Thunb.

Male inflorescence much branched many-flowered; *female inflorescence* in spicate cymes; *spikelets* 1-flowered. *Perianth-segments* 6. *Stamens* 3; *anthers* 1-celled. *Pistillode* none. *Female flowers* supported on a short, fleshy, often lobed, stalk. *Perianth-segments* 6, equal, persistent. *Staminodes* none. *Ovary* 1-celled, generally surmounted by an epigynous disc. *Styles* 2, deciduous. *Fruit* cylindric, bony, 1-celled, indehiscent. *Seed* solitary, pendulous.

Rootstock creeping or tufted, covered with sheatl like scales and producing erect, more less branched stems.

DISTRIB. Endemic.

Stems slender:
  Leaf-sheaths twisted-aristate, deeply membranous; female perianth-segments very short ... ... (1) **humilis.**
  Leaf-sheaths acuminate, aristate, not twisted; female perianth-segments nearly as long as the fruit ... (2) **cuspidata.**
Stems robust, generally much branched:
  Stems sulcato-striate:
    Leaf-sheaths loosely convolute, acuminate, aristate; female perianth-segments ovate, shorter than the fruit ... ... (3) **striata.**
    Leaf-sheaths closely convolute, acuminate; perianth-segments nearly as long as the fruit ... ... (4) **sulcata.**
  Stems smooth, not sulcato-striate:
    Leaf-sheaths deeply membranous, with long awns:
      Spathe elliptic-aristate, deeply membranous at the tip ... ... (5) **arescens.**
      Spathe lanceolate, gradually acuminate, aristate, not membranous at the tip ... ... (6) **lucæana**
      Spathe oblong-ovate, abruptly acuminate ... (7) **teres.**
      Spathe ovate-lanceolate, fimbriate at the edges (8) **fimbriata.**
Imperfectly known ... ... (9) **brevis.**

1. **W. humilis** (Mast. in Journ. Linn. Soc. x. 272); stems erect from a decumbent base, about 1 ft. high, slightly branched, terete, fistular, slender, purple-spotted; leaf-sheaths 1½ in. long, tightly convolute, edges membranous, apex aristate; male spikelets numerous, in an erect, simple, linear cyme; spathes deciduous; flowers loose; bracts oblong-lanceolate, membranous, ferruginous; perianth-segments subequal, ciliiform, twisted; anthers linear-oblong; female spikelets 1–3 in a linear, erect cyme, provided with a sheath-like spathe; bracts oblong-lanceolate, mucronate, membranous; female flower solitary sessile; perianth-segments 4?–6, ovate, hyaline, shorter than the fruit; ovary on a short, fleshy stalk and with an epigynous, 4-lobed disc; ovule pendulous. *Mast. in DC. Monog. Phan.* i. 396.

COAST REGION: Cape Div.; Cape Flats, *Ecklon*, 867!; *Ecklon and Zeyher! Viellard*, ♂ and ♀!

2. **W. cuspidata** (Mast. in Journ. Linn. Soc. x. 271); stems erect, about 2 ft. high, terete slender with spreading branches; leaf-sheaths tightly convolute, membranous above, acuminate; female spikelets solitary or 2, each nearly ½ in. long, turbinate; with an open, oblong, acuminate spathe about as long as the spikelet; bracts numerous, oblong, coriaceous, with a long acumen; flower solitary on a thick, six-lobed disc; perianth-segments subequal, oblong-lanceolate, membranous, scarcely as long as the fruit; fruit stipitate, oblong, obtuse, rugose, punctate; styles deciduous. *Mast. in DC. Monog. Phan.* i. 396.

COAST REGION: Clanwilliam Div.; near Groen River and Watervals River, 1500–2000 ft., *Drège*, 2516, ♀!

3. **W. striata** (Thunb. in Vet. Akad. Handl. Stockh. xi. 1790, 57, t. 2, fig. 1); root-stock creeping; stems a yard high, terete, sulcato-striate, branched; leaf-sheaths ¾ in. long, loosely convolute, coriaceous, deeply membranous, aristate, membranous portion and awn deciduous;

inflorescence 1 in. long, densely paniculately cymose; cymes erect, with numerous open deciduous sheath-like spathes; bracts linear-lanceolate, membranous; perianth-segments membranous, linear-lanceolate, subequal; anthers linear-oblong; female spikes 1–3, ½–¾ in. long; bracts loose, oblong, acuminate, coriaceous, membranous at the margins; perianth-segments equal, membranous, oblong, obtuse, sometimes mucronate; ovary with an epigynous disc; fruit on a very short stalk, ovoid, horny, blackish-purple, pitted, marked above with the remains of the disc and surrounded at the base by the persistent perianth. *Thunb. herb.! in Journ. Linn. Soc.* xiv. 421; *Kunth, Enum.* iii. 453; *Steud. Synops.* ii. 262; *Mast. in Journ. Linn. Soc.* x. 270, *and in DC. Monog. Phan.* i. 394. *Nematanthus Eckloni, Nees in Linnæa,* v. 662! *Willdenovia neglecta, Steud. Synops.* ii. 263.

SOUTH AFRICA: without locality, *Thunberg*, ♀! *Drège*, 960 ♀!
COAST REGION: Clanwilliam Div.; Oliphants River, *Zeyher*, 1736, ♂ and ♀! Pitquetberg Div.; between Twenty-four Rivers and Piquiniers Kloof, below 1000 ft.. *Drège*, 2520, ♂! Malmesbury Div.? Lange Fontein, *Zeyher*, monstrosity! Cape Div.; between Wynberg and Constantia, *Burchell*, 787, ♂! Cape Flats, *Bolus*, 4467, monstrosity! Muizenberg, near Fish Hoek, *Bolus*, 4459, ♂! 4458, ♀! Paarl Div.; by the Berg River, *Drège*, 1645, ♂! Worcester Div.; between the Bokke Veld and Hex River, *Drège*, 9610, ♂!

4. **W. sulcata** (Mast. in Journ. Linn. Soc. x. 270); stems erect, 2–3 ft. high, terete, fistular, slightly sulcate, sparingly branched; leaf-sheaths about 1 in. long, tightly convolute, lanceolate, acuminate; male spikelets in a terminal, paniculate cyme, each with an open deciduous sheath-like spathe at the base; bracts linear-lanceolate, longer than the stalked flower; perianth-segments 6, equal, linear; anthers linear, apiculate; female spikelets 1–3 in linear erect cymes, ¼–½ in. long, cylindric-oblong, ultimately turbinate, with a deciduous spathe; bracts numerous, oblong obtuse, mucronate; flower solitary on a short, fleshy, unlobed stalk; perianth-segments oblong, obtuse, membranous, nearly as long as the fruit; ovary oblong, stipitate, puncticulate; styles 2, deciduous; fruit blackish, retuse, shorter than the bracts. *Mast. in DC. Monog. Phan.* i. 395.

COAST REGION: Cape Div.; Cape Flats, *Ecklon*, 930, ♂ and ♀! *Bolus*, 4435, ♂! Camp Ground, near Cape Town, *Bolus*, 7223, ♀! Worcester Div.; mountains above Worcester, *Rehmann*, 2556, ♀!

5. **W. arescens** (Kunth, Enum. iii. 454); stems erect, terete, moderately branched; leaf-sheaths ½ in. long, loosely convolute, coriaceous, upper portion deeply membranous, aristate, ultimately deciduous; male inflorescence, about 2 in. long, terminal, oblong, with numerous sheath-like, open, oblong, acuminate spathes; bracts linear-lanceolate, membranous; perianth-segments linear, membranous; anthers oblong, apiculate, purplish at the back; female spikes solitary or geminate, terminal; spathe oblong, coriaceous, shining brown, upper portion membranous, with a long twisted awn; bracts lanceolate, acute; perianth raised on a fleshy, 6-lobed stalk; segments oblong, emarginate, appressed to the ovary;

ovary oblong, obtuse, smooth; stigmas 2, deciduous; fruit oblong, horny, tubercled. *Steud. Synops.* ii. 262; *Mast. in Journ. Linn. Soc.* x. 271, *and in DC. Monog. Phan.* i. 393.

COAST REGION: Clanwilliam Div.; Ceder Bergen, 2000–2500 ft., *Drège*, 2522, ♂ and ♀! and 2521, monstrosity!

6. W. lucæana (Kunth, Enum. iii. 455); stems a yard or more high, terete, fistular, olive-coloured, white spotted, slightly branched; leaf-sheaths 1–1½ in. long, tightly convolute; margins membranous; apex aristate; male inflorescence about 2 in. long, loosely panicled cymose, provided with oblong-lanceolate, acuminate spathes; bracts linear-lanceolate, membranous, ferruginous; flowers stipitate; perianth-segments linear-lanceolate, equal, longer than the stamens; anthers oblong, apiculate; female spikelets 2–3, sessile, each provided with an open spathe; bracts shining ferruginous, oblong, subulate-acuminate; female flower solitary, stipitate; stalk fleshy, 6-lobed; perianth-segments membranous, equal, oblong-lanceolate, 1-nerved, mucronate; ovary cylindrical, with an epigynous, subglobose, horny disc; stigmas 2; fruit oblong, obtuse, bony, purple, puncticulate near the top; seed oblong, obtuse. *Steud. Synops.* ii. 262; *Mast. in Journ. Linn. Soc.* x. 271, *t.* 8, *fig. D, and in DC. Monog. Phan.* i. 392, *t.* 4, *figs.* 21–29, *and t.* 5, *fig.* 19. *Spirostytis Ecklonii, Nees in various herbaria.*

COAST REGION: Clanwilliam Div.; Pretoris Kloof, between Piquiniers Kloof and Oliphants River, 1000–1500 ft., *Drège*, 2515, ♀! 2515a, ♀! Tulbagh Div.; near Tulbagh, *Zeyher!* Worcester Div.; Drakenstein Mountains, near Bains Kloof, 1600–2000 ft., *Bolus*, 4085, monstrosity! 4094, ♀! Caledon Div.; Donker Hoek Mountain, *Burchell*, 7961, ♂ and ♀! between Donker Hoek and Houw Hoek Mountains, *Burchell*, 8010, monstrosity!

7. W. teres (Thunb. in Vet. Akad. Handl. Stockh. 1790, 28, t. 2, fig. 2); stems 3–4 ft. high, erect, terete, fistular, olive-coloured, covered with fine white spots, sparingly branched; upper leaf-sheaths 1 in. long, loosely convolute, membranous at the apex and prolonged into a short awn; male inflorescence 2–3 in. long, paniculately cymose, with large, open, coriaceous, ferruginous, ovate-oblong, acuminate spathes; spikelets 1-flowered; bracts linear-lanceolate, acuminate; perianth-segments linear, equal; anthers linear-oblong; female spikes 1-3, sessile at the apex of the branches, erect, ¾ in. long, with open spathes; bracts oblong, coriaceous, shining, chestnut-brown, mucronate; flower stipitate; stalk fleshy, sulcate, 6-lobed; perianth-segments obovate, retuse, membranous, purple-spotted, with the median nerve prominent; styles 2; fruit oblong, obtuse, blackish, puncticulate. *Thunb. Fl. Cap. ed. Schult.* 82, *and Thunb. herb.! Mast. in Journ. Linn. Soc.* xiv. 421; *Kunth, Enum.* iii. 452; *Steud. Synops.* ii. 262; *Mast. in Journ. Linn. Soc.* x. 269, *and in DC. Monog. Phan.* i. 392. *Restio dichotomus, Gaertn. Fruct.* ii. 12, *t.* 82, *fig.* 3, *excl. Syn.*

SOUTH AFRICA: without locality, *Thunberg*, ♀! *Veillard!*

COAST REGION: Clanwilliam Div.; between Jakhals River and Klip Fontein, below 500 ft., *Drège*, 2482, ♀! Onder Bokke Velde, *Drège!*

8. **W. fimbriata** (Kunth, Enum. iii. 455); stems erect, 2 ft. high, terete, unbranched, yellowish, puncticulate; leaf-sheaths 1½ in. long, tightly convolute, coriaceous, brown, fimbriate and ultimately lacerate at the margins, lanceolate, acuminate at the apex; male and female inflorescences similar, about 3 in. long, oblong, with 3–5 large, ovate-lanceolate, shining, deep brown spathes; male inflorescence in linear cymes; spikelets numerous; bracts thin, membranous, pale brown, linear-lanceolate, nearly ½ in. long; flowers half the length of the bract; perianth-segments all linear-oblong, with a conspicuous midrib; inner 3 rather thinner, scarcely shorter; anthers linear, apiculate; female spikelets 3–5–7, aggregated; bracts acute-rigid; perianth-segments membranous, oblong, obtuse or retuse, 1-nerved, appressed to the ovary; fruit oblong, horny, smooth or tubercled, raised on a fleshy, lobed stalk; ovule pendulous, *Steud. Synops.* ii. 263; *Mast. in Journ. Linn. Soc.* x. 271, *and in DC. Monog. Phan.* i. 394.

COAST REGION: Worcester Div.; Dutoits Kloof, 3000–4000 ft., *Drège*, 6611, ♂! 1635b, ♀! Caledon Div.; Genadendal, 1500–2000 ft., *Drège*, 1635a, ♀!

9. **W. brevis** (Nees ex Mast. in Journ. Linn. Soc. x. 269 in note); stems 1–1½ ft. high, decumbent, terete, branching; leaf-sheaths about 1 in. long, loosely convolute, aristate, deeply membranous at the edges, membranous portion ultimately deciduous; male spike solitary (?), erect, oblong, about ½ in. long, with an open coriaceous spathe; bracts oblong lanceolate, acute, membranous; perianth-segments linear, hyaline; anthers linear, apiculate. *Mast. in DC. Monog. Phan.* i. 397.

COAST REGION: Cape Div.; Mosselbanks River, *Zeyher!*

Perhaps a form of *W. teres.*

Excluded Species.

W. compressa, Thunb. in Vet. Akad. Handl. Stockh. 1790, t. 2, fig. 3 = *Cannomois cephalotes*, Beauv.

XII. CERATOCARYUM, Nees.

Male and female inflorescence different; male densely paniculate-cymose, many-spiked, with large leathery spathes. *Bracts* white, lanceolate. *Perianth-segments* in two rows. *Anthers* linear-oblong. *Female inflorescence* spicate. *Spikelets* 1-flowered. *Perianth-segments* appressed to the fruit. *Fruit* sessile, bony, subglobose, 1-celled, 1-seeded; styles 2.

Perennial rush-like plants, with stout ascending stems.

DISTRIB. Endemic.

Stems solid ... ... (1) **argenteum.**
Stems fistular ... ... (2) **fistulosum.**

1. **C. argenteum** (Kunth, Enum. iii. 483 male plant); tall, little-branched perennial; stems erect, cylindric, purplish, minutely rugulose, solid; leaf-sheaths 2 in. long, closely convolute; male

inflorescence terminal, 2–2½ in. long, densely paniculate-cymose, oblong, erect, nearly covered by the coriaceous, deep-brown, oblong, acute spathes; bracts lanceolate, acuminate, twisted, 1-nerved, membranous, whitish; flower arcuate, stipitate; perianth-segments lanceolate, acuminate; inner shorter; anthers linear-oblong; female inflorescence terminal, linear-oblong, cymose, 2–4 in. long, pluristachyate, with 3–4 coriaceous, striate, shining, purplish spathes; spikelets 1–flowered, oblong-pyramidal; bracts loosely imbricate, lanceolate, membranous, yellowish, 1-nerved; flower solitary, shorter than the bracts; perianth minute, of two rows; segments all suborbicular, acute; ovary globose; fruit bony, subglobose, smooth, indehiscent, 1-celled, 1-seeded; styles 2. *Steud. Synops.* ii. 264; *Mast. in Journ. Linn. Soc.* x. 273, *and in DC. Monog. Phan.* i. 390. *Restio (Elegia) argenteus, herb. Willd. fide Nees in Linnæa*, v. 656, *not of Thunberg. Ceratocaryum speciosum, Nees in various herbaria.*

SOUTH AFRICA: without locality, *Drège*, 105!
COAST REGION: Caledon Div.; Nieuw Kloof, Houw Hoek Mountains, *Burchell*, 8074, ♂! mountains of Baviaans Kloof, near Genadendal, *Burchell*, 7656, ♂! 7905, ♂! mountains near Grietjes Gat, between Lowrys Pass and Palmiet River, *Ecklon and Zeyher*, ♂ and ♀!

2. C. fistulosum (Mast. in Journ. Linn. Soc. x. 274, t. 8, fig. E); stems a yard or two high, fistular, little branched; leaf-sheaths 1–2 in. long, tightly convolute, membranous and lacerate at the tips; male infloresence terminal, pluristachyate, cymose, with large, open, purplish, ultimately deciduous spathes; spikelets ¼ in. long; bracts oblong, acuminate, membranous, whitish; perianth-segments similar to the bracts; inner shortest; female spikelets 4–6 in linear, erect, spike-like cymes, 2–3 in. long, each pear-shaped, erect, pluribracteate, 1-flowered; bracts lanceolate, membranous, 1-nerved; perianth-segments subequal, oblong-ovate, cartilaginous with membranous margins; fruit globose, bony, tubercled, 1-celled, surrounded at the base by the persistent perianth. *Mast. in DC. Monog. Phan.* i. 391, *t.* 4, *figs.* 16–20, *and t.* 5, *fig* 18.

COAST REGION: Riversdale Div.; on the Kampsche Berg, *Burchell*, 7095, ♂ and ♀!

XIII. ANTHOCHORTUS, Nees.

Male spikelets in terminal cymes 1–3 ft. long. *Perianth* of male flower stipitate; segments oblong; outer lateral conduplicate, acutely keeled; others smaller. *Filaments* free; anthers linear, *Pistillodium* none.

A genus adopted by Kunth and others, but very imperfectly known and probably referable to *Hypolæna*. Bentham suggests that it may be a *Willdenovia*, perhaps *W. striata*, Thunb.

1. A. Ecklonii (Nees in Lindl. Syst. ed. 2, 451); stems filiform, slender, curved, four-angled, branched; leaf-sheath about ½ in. long, coriaceous, elliptic, acute, membranous; male spikelets in elongated cymes; spathes oblong-lanceolate, acuminate; flowers stipitate; perianth-segments linear-oblong; lateral conduplicate, keeled; inner 3 subequal; anthers linear, apiculate. *Kunth, Enum.* iii. 486; *Mast. in Journ. Linn. Soc.* x. 274, *and in DC. Monog. Phan.* i. 398.

COAST REGION: Swellendam Div.; mountains near Swellendam, *Ecklon and Zeyher!*

Probably the male plant of some species of *Hypolæna.*

ORDER CL. CYPERACEÆ.

(By C. B. CLARKE.)

Flowers glumaceous, 2- or 1-sexual. *Perianth* hypogynous, of 6 or fewer small scales or bristles, not petaloid, frequently irregular or imperfect, often 0. *Stamens* 3–1, free, all anterior, or in a few species 6 or 8. *Ovary* superior, ovoid, 1-celled; ovule 1, basal; style 1, linear (base often abruptly thickened); branches 3 or 2, in *Tetraria* often 4, linear or very rarely (in *Rynchospora*) very short or obsolete (i.e. style appearing simple). *Fruit* a nut. *Seed* obovoid or ovoid; testa thin; embryo minute, obpyramidal, at the base of the usually floury albumen.

Herbs; stems solid; leaves narrow, grass-like, usually very tough and inedible; sheaths often cylindric entire; flowers many, few or 1 in spikelets (the axis of which bearing the glumes being the rhachilla of the spikelet) which are 1- or 2-sexual, with empty glumes either at bottom or top, or both; spikelets 1 or many or very numerous, solitary, or in clusters or heads (such compound inflorescences usually designated as *spikes*); spikelets or spikes arranged in corymbs or in apparent umbels or in various ways.

DISTRIB. Species about 3300, extending throughout the world, growing especially in damp places.

Sub-Order I. SCIRPO-SCHOENEÆ. *Fertile flowers* all with perfect stamens, except in *Scirpus spathaceus* and *Tetraria crinifolia.* [See also Sub-Order II. *Mapanieæ*, in which Bentham regards the spikelet as possibly a 2-sexual flower.]

Tribe 1. *CYPEREÆ.* *Empty glumes* at the base of the spikelet 2 or 1; fertile glumes many, few or 1 to the spikelet, 2-ranked. [Rhachilla of spikelet itself rarely twisted; after the glumes and nuts are fallen, the notches on the rhachilla can be seen to be exactly 2-ranked.] *Hypogynous bristles* 0.

I. **Kyllinga.**—*Style* 2-branched. *Nut* compressed laterally. *Spikelet* bearing 1 or 2 (rarely more) nuts.
II. **Pycreus.**—*Style* 2-branched. *Nut* compressed laterally. *Spikelet* bearing several or many nuts.
III. **Juncellus.**—*Style* 2-branched. *Nut* compressed dorsally.
IV. **Cyperus.**—*Style* 3-branched. *Rhachilla* of spikelet persistent.
V. **Mariscus.**—*Style* 3-branched. *Rhachilla* of spikelet caducous.

Tribe 2. *SCIRPEÆ.* *Empty glumes* at the base of the spikelet 2–0; *fertile glumes* arranged spirally, many, often very numerous.

VI. **Eleocharis.**—*Stem* leafless, with one terminal spikelet.
VII. **Fimbristylis.**—*Hypogynous bristles* 0. *Style-base* constricted above the nut, persistent, or deciduous without leaving a button.
VIII. **Bulbostylis.**—*Style* deciduous, leaving a button on the nut; otherwise as *Fimbristylis.*
IX. **Scirpus.**—*Style* passing gradually and continuously into the nut.
X. **Eriophorum.**—*Hypogynous bristles* flat, divided nearly to the base, appearing very numerous; otherwise as *Scirpus.*
XI. **Ficinia.**—*Ovary* on a more distinct obpyramidal gynophore; otherwise as *Scirpus.*
XII. **Fuirena.**—*Spikelets* conspicuously hairy; otherwise as *Scirpus.*
XIII. **Lipocarpha.**—*Hypogynous scales* 2, hyaline, standing fore and aft within the glume; otherwise as *Scirpus.*
XIV. **Ascolepis.**—*Hypogynous scale* 1, posticous, within the glume and longer than it, thickened upwards, almost enveloping the nut; otherwise as *Scirpus.*

Tribe 3. *SCHOENEÆ. Empty glumes* at the base of the spikelet usually more than 2; fertile glumes few, very often 1.

* *Style 2-fid.*

XV. **Rynchospora.**—*Branches of style* 2, long linear, or exceedingly short (style then nearly entire).

** *Style 3-fid. Fertile glumes, and empty glumes below them, more or less 2-ranked.*

XVI. **Carpha.**—Lowest *flower* perfecting a nut. Hypogynous *bristles* 6, long, simple.
XVII. **Ecklonea.**—*Hypogynous bristles* feathered at the base, 3-fid at top; otherwise as *Carpha.*
XVIII. **Schoenus.**—Lowest *flower* perfecting a nut. *Rhachilla* above perfect flower lengthened.
XIX. **Epischoenus.**—Lowest *flower* (or 2 lowest flowers) not perfecting a nut. *Rhachilla* above the perfect flower lengthened.
XX. **Costularia.**—Lowest *flower* male. *Rhachilla* not elongated above the fertile flower.
XXI. **Tetraria.**—Lowest *flower* with a pistil that does not perfect a nut; otherwise as *Costularia.* [In *T. cuspidata,* the lowest flower perfects a nut.]

*** *Style 3-fid. Glumes spirally placed. Lowest flower perfecting a nut.*

XXII. **Macrochnetium.**—*Hypogynous bristles* long. *Stamens* 6.
XXIII. **Cladium.**—*Hypogynous bristles* 0 (in the Cape sp.). *Stamens* 3–2.

Sub-Order II. MAPANIEÆ. *Spikelet* of one terminal female flower without a perianth. Lower *glumes,* some with 1 stamen, some empty.

XXIV. **Chrysithrix.**—*Style* long linear; branches 3. *Glumes* next outside the pistil empty.

Sub-Order III. CARICEÆ. *Flowers* all 1-sexual. *Spikelets* 1-sexual or 2-sexual; if bisexual then consisting of 1 basal female flower, and 1 or many upper male flowers.

XXV. **Scleria.**—*Spikelet* solitary or clustered. *Hypogynous bristles* 0. *Nut* bony, shining, on a gynophore.
XXVI. **Eriospora.**—*Spikelets* crowded in small Scirpus-like spikes. *Hypogynous bristles* capillary. *Nut* small, not bony.
XXVII. **Schoenoxiphium.**—*Nut* enclosed by a bottle-like bract, deeply split down, often containing a rhachilla or a male spike. No complete utricles.
XXVIII. **Carex.**—*Nut* completely enclosed in a utricle. [In several androgynous species the utricle is split down in such spikelets as have the upper male portion or its rhachilla fully developed.]

I. KYLLINGA, Rottb. partly.

Spikes ovoid or cylindric, dense, with many small compressed spikelets. *Spikelets* of 4–7 distichous glumes, 1–4-flowered; lowest (or 2 lowest) flower 2-sexual, perfecting a nut; upper flowers sterile, male, or uppermost glume empty; 2 lowest glumes smaller, empty (bracts); rhachilla (wingless) in fruit falling off by disarticulation from a cushion below the lowest fertile flower. *Stamens* 1–3, anterior; anthers narrow, oblong, not crested. *Style* slender; branches 2, linear, in a plane passing through the rhachilla. *Nut* oblong or ellipsoid, compressed laterally, smooth.

Glabrous (in *K. Lehmanni* and *K. alba*, var. β, the top of the culm is minutely scabrous-pubescent); leaves green, long (except in *K. pungens*), all close to the base of the stem; inflorescence a head of 1–3 terminal, absolutely sessile, spikes; supported by 3–6 leaf-like bracts (character of genus, as in genera following, narrowed to Cape species).

DISTRIB. Species 43, in all hot and temperate regions, except Europe. A specially African genus.

Subgenus I. THRYOCEPHALUM. Keel of fertile glume winged in fruit.

Head globose, white ... ... (1) **alba.**

Subgenus II. EU-KYLLINGA. Keel of fertile glume not winged. Spikelets perfecting 1 (rarely 2) nut.

Rhizome horizontal. Head of 1 (rarely more) spike. Bracts rarely exceeding 3.

Spike of a few spikelets ... ... (2) **pauciflora.**
Spike dense, often somewhat golden ... ... (3) **erecta.**

Rhizome, horizontal or oblique, thick. Bracts often 4 or more.

Bracts 6–5, very long; middle spike cylindric ... (4) **elatior.**
Bracts 4–3, long; spike ovoid ... ... (5) **melanosperma.**

Rhizome hardly any.

Middle spike cylindric, straw-coloured ... ... (6) **cylindrica.**

Subgenus III. PSEUDO-PYCREUS. Keel of fertile glume not winged. Spikelets often perfecting 2 (or more) nuts.

Middle spike cylindric, chestnut-brown ... ... (7) **pulchella.**
Spike globose, somewhat golden-brown ... ... (8) **tetragona.**
Spike large, globose, greenish-white ... ... (9) **Lehmanni.**
Spike globose, straw-coloured ... ... (10) **Buchanani.**

1. **K. alba** (Nees in Linnæa, x. 140); glabrous (but see var. β); stems 16–18 in. long, bases thickened by sheaths close together on a thick, very short, woody rhizome; leaves more than half the length of the stem, $\frac{1}{8}$ in. broad; bracts 2–5 (usually 3), spreading, lowest 2–4 in. long, similar to the leaves; spike 1, $\frac{1}{2}$ in. long, globose or short ellipsoid, white (but see var. β); spikelet $\frac{1}{5}$ in. long, ovoid, compressed, 2-flowered (lower flower perfecting a nut); fertile glume ovate, acute, boat-shaped; keel with a white wing of loose cellular tissue, usually ciliate with many-celled hairs (this wing or crest hardly discernible in the young flowers); style 2-fid; nut half as long as the glume, oblong-ellipsoid, compressed, dark chestnut-brown. *Boeck. in Linnæa,* xxxv. 430. *K. cristata, Kunth, Enum.* ii. 136. *Kyllinga sp., Burch. Travels,* i. 538, *in note.*

VAR. β, **alata** (C. B. Clarke in Durand and Schinz, Conspect. Fl. Afr. v. 526);

stem minutely scabrous pubescent at the top; spike dull golden-white or greenish-white. *K. alata*, *Nees in Linnæa*, ix. 286; *Boeck. in Linnæa*, xxxv. 430. *K. aurea*, *Krauss in Flora*, 1845, 757.

SOUTH AFRICA: without locality, *Zeyher*, 1760! 1764!

COAST REGION: Queenstown Div.; between Table Mountain and Wildschuts Berg, 4000 ft., *Drège*, 3930! Shiloh, 3500 ft., *Drège!* Komgha Div.; Kei River, *Eklon! Flanagan*, 927 partly! Var. β, Bathurst Div.; near the mouth of the Kasuga River, 50 ft., *McOwan*, 727! Uitenhage Div.; between Van Stadens Berg and Galgebosch, *Burchell*, 4681! and without precise locality, *Ecklon and Zeyher*, 450! *Zeyher*, 4367!

WESTERN REGION: Little Namaqualand, *Schinz*. 363!

KALAHARI REGION: Griqualand West; between Witte Water and Riet Fontein, *Burchell*, 1997! Orange Free State: Caledon River, *Burke*, 202! Basutoland, *Cooper*, 919! Transvaal; Crocodile River, *Burke!* Houtbosh, *Rehmann*, 5633! 5636! Magalies Berg, *Zeyher*, 1761! and without precise locality, *McLea in Herb. Bolus*, 6020! 6021! Var. β: Griqualand, *Rehmann*, 3397! Orange Free State, *Rehmann*, 3673! Caledon River, *Burke*, 301!

EASTERN REGION: Natal; Durban, *Kuntze*, 233! Var. β, Natal; margins of woods around Durban Bay, *Krauss*, 292! near Durban, *Wood*, 4014! without precise locality, *Gerrard*, 485! *Rehmann*, 8595!

Also in South Tropical Africa.

2. **K. pauciflora** (Ridley in Trans. Linn. Soc. ser. 2, Bot. ii. 147, t. 23, figs. 1–4); glabrous; stems 18 in. long, slender, triquetrous; bases slender, close together on a woody rather slender rhizome; leaves half the length of the stem (or only 2 in.), hardly $\frac{1}{8}$ in. broad; bracts 3, spreading, lowest 1–2 in. long, similar to the leaves; spike 1, ovoid, of 6–8 spikelets, fuscous or somewhat golden; spikelets $\frac{1}{5}$ in. long, narrow-lanceolate, perfecting 1 nut; fertile glume lanceolate, acute; keel wingless, green, smooth; style 2-fid.

EASTERN REGION: Natal; *Buchanan*, 329! 330!

Also in Angola. This species is hardly separable from the world-wide *K. brevifolia*, Rottb.

3. **K. erecta** (Schumach. Beskr. Guin. Pl. 42); glabrous; rhizome horizontal, herbarium pieces often 3–4 in. long, varying from $\frac{1}{10}$–$\frac{1}{5}$ in. in thickness, clothed with bright brown, horny, ovate, striate scales; stems 4–18 in. long, slender; bases hardly thickened, close together or $\frac{1}{4}$–$\frac{1}{2}$ in. apart on rhizome; leaves often $\frac{1}{2}$ the length of the stem, $\frac{1}{8}$ in. broad; bracts usually 3, spreading or suberect, lowest 2–5 in. long, similar to the leaves; spike 1, ovoid, dense, golden-brown or fuscous; spikelets $\frac{1}{8}$–$\frac{1}{6}$ in. long, ellipsoid, perfecting 1 nut; fertile glume elliptic, keel wingless, green, minutely excurrent into a mucro, smooth or with 1 or 2 microscopic bristles; style 2-fid; nut oblong, black. *K. aurata*, *Nees in Linnæa*, x. 139, *excl. Nees' reference to Linnæa*, vii. 512, ? *in Linnæa*, ix. 286; *Boeck. in Linnæa*, xxxv. 422, *excl. var.* γ. *K. consanguinea*, *Kunth*, *Enum.* ii. 135.

SOUTH AFRICA: without locality, *Zeyher*, 1763!

COAST REGION: Uitenhage Div.; *Ecklon*, 882! *Zeyher!* Albany Div.; Grahamstown, *Ecklon!* Glenfilling, 1000 ft., *Drège*, 4387! Bothas Hill, 2200 ft. *MacOwan*, 1351! Komgha Div.; near Komgha, *Flanagan*, 917! King Williamstown Div.; Toise River, 3500 ft., *Kuntze*, 228! Queenstown Div.; Table Mountain, 4000–5000 ft., *Drège*, 3931!

KALAHARI REGION: Orange Free State; Caledon River, *Burke*, 425! Transvaal; *Rehmann*, 4475! Bosch Veld, *Rehmann*, 5350! Hooge Veld, *Rehmann*, 6612! Pretoria, 5000 ft., *Kuntze*, 261!

EASTERN REGION: Transkei, 1000–2000 ft., *Drège!* Natal; Inanda, *Wood*, 1076! near Durban, *Wood*, 3157! and without precise locality, *Gerrard*, 706! 707! *Buchanan*, 99! *Wood*, 459! Delagoa Bay, *Junod*, 240!

Common throughout Tropical Africa, and the Mascarene Isles.

4. **K. elatior** (Kunth. Enum. ii. 135); glabrous; rhizome obliquely descending, thick but hardly woody; scales herbaceous; stems 1–2 ft. long, thick, triquetrous, somewhat remote on the rhizome; leaves $\frac{1}{6}$–$\frac{1}{4}$ in. broad, short, i.e. uppermost usually 1–3 in. long, but sometimes elongate, $\frac{1}{2}$ as long as the stem; bracts 6–5, $\frac{1}{6}$–$\frac{1}{4}$ in. broad, green, lax, spreading, lowest 4–8 in. long; spikes 3–1, middle one cylindric, $\frac{1}{2}$–$\frac{3}{4}$ in. long, dirty-white; spikelet $\frac{1}{8}$ in. long, elliptic, 2- (rarely 3-) flowered, rarely perfecting more than lowest nut; keel of fertile glume wingless, minutely excurrent into a green mucro, smooth or very nearly so; style 2-fid; nut ellipsoid, brown, less than $\frac{1}{2}$ the length of the glume. *Krauss in Flora*, 1845, 757; *Boeck. in Linnæa*, xxxv. 422.

COAST REGION: King Williamstown Div.; Perie Forest, *Kuntze*, 270! Komgha Div.; near the mouth of the Kei River, 300 ft., *Flanagan*, 1789!

KALAHARI REGION: Orange Free State, *Buchanan*, 83!

EASTERN REGION: Pondoland; between Umtata River and St. John's River, 1000–2000 ft., *Drège*, 4384! Natal; banks of the Umlaas River, *Krauss*, 32! and without precise locality, *Buchanan*, 98a! 324! Zululand; near Amatikulu River, *Wood*, 3993!

Also in South-east Tropical Africa.

5. **K. melanosperma** (Nees in Wight, Contrib. 91); stems 18 in. long, slender, hardly triquetrous; bracts 4–3, $\frac{1}{6}$ in. broad; head solitary, ovoid; keel of fertile glume often scabrous; nut finally black [otherwise as *K. elatior*, Kunth]. *Boeck. in Linnæa*, xxxv. 419; *C. B. Clarke in Hook. f. Fl. Brit. Ind.* vi. 588.

COAST REGION: Albany Div.; Assegai Bosch, near Sidbury, *Burchell*, 4173! Brookhuizens Poort, near Grahamstown, 2000 ft., *MacOwan*, 673!

CENTRAL REGION: Graaff Reinet Div.; Sunday River, *Bolus!*

KALAHARI REGION: Transvaal, *Fehr!*

EASTERN REGION: Tembuland; Bazeia, 2000 ft., *Baur*, 46! Griqualand East; near Clydesdale, 2500 ft., *Tyson*, 2864! 2865! Natal; Durban Flat, *Wood*, 4100! Newcastle, *Buchanan*, 188! Howick, 1000 ft., *Junod*, 223! and without precise locality, *Rehmann*, 8596! *Buchanan*, 325! 326! 327!

Also in East Tropical Africa, Mascarene Islands, Indian Peninsula, and Malaya.

6. **K. cylindrica** (Nees in Wight, Contrib. 91); glabrous; stems 6–8 in. long, slender, hardly thickened at the base, tufted on a very short rhizome; leaves $\frac{1}{2}$ as long as the stem, $\frac{1}{6}$ in. broad; bracts 3, spreading, lowest $1\frac{1}{2}$ in. long, similar to the leaves; spikes 3–1; middle one $\frac{1}{2}$ by $\frac{1}{4}$ in. cylindric, dense, straw-coloured; spikelets $\frac{1}{8}$ in. long, perfecting 1 nut; fertile glume ovate, hardly acute, keel wingless, smooth; style 2-fid; nut broad, ellipsoid, $\frac{2}{3}$ as long as the glume, yellow-brown when ripe (finally nearly black). *Boeck. in Linnæa*, xxxv. 415; *C. B. Clarke in Hook. f. Fl. Brit. Ind.* vi. 588.

KALAHARI REGION: Transvaal, *Rehmann*, 5634!

EASTERN REGION: Pondoland, *Bachmann*, 102! Natal; Inanda, *Wood*, 1421! and without precise locality, *Buchanan*, 98! 323!

Scattered throughout Tropical Africa, the Mascarene Islands, India, and the warmer part of Australia; but *K. odorata*, Vahl, which is abundant in Tropical and Sub-Tropical America, is hardly specifically separable.

7. **K. pulchella** (Kunth, Enum. ii. 137); glabrous; stolons short, very slender; stem 6–15 in. long, slender, subsolitary, hardly thickened at the base; leaves often as long as the stem, $\frac{1}{8}$ in. broad; bracts 3–4, spreading, lowest often 4 in. long, similar to the leaves; spikes 3–1, middle one cylindric, $\frac{3}{4}$ by $\frac{1}{3}$ in. (but small examples occur with one ovoid spike $\frac{1}{4}$ in. long), dense; spikelets $\frac{1}{6}$–$\frac{1}{5}$ in., 5–7-glumed, perfecting 3–1 nuts; glumes boat-shaped, chestnut-brown; keel green, wingless, smooth, scarcely excurrent; style 2-fid; nut $\frac{1}{2}$ as long as the glume, ellipsoid, dark-brown. *Hochst. in Flora*, 1844, 102; *Boeck. in Linnæa*, xxxv. 405. *K. atrosanguinea*, *Steud. in Flora*, 1842, 598.

SOUTH AFRICA: without locality, *Zeyher*, 1755!

COAST REGION: Komgha Div.; among rocks near Komgha, 2000 ft., *Flanagan*, 1261!

CENTRAL REGION: Albert Div.; 4500 ft., *Drège*, 7384! Somerset Div.; Bosch Berg, 2500 ft., *MacOwan*, 1351!

KALAHARI REGION: Orange Free State; Riet River, *Burke*, 433! Transvaal; Hooge Veld, at Trigards Fontein, *Rehmann*, 6670! Houtbosh, *Rehmann*, 5620!

Also in Abyssinia.

8. **K. tetragona** (Nees in Linnæa, vii. 512); small, glabrous or the top of the stem minutely scabrous-pubescent; rhizome woody, erect, short, often divided at the top, covered with scales torn into brown fibres; stem solitary, 3–7 in. long, slender, thickened at the base by leaf-sheaths; leaves $\frac{1}{2}$ as long as the stem, $\frac{1}{10}$ in. broad; bracts 3, spreading, lowest 1–2 in. long, similar to the leaves; spike 1, ovoid, $\frac{1}{4}$–$\frac{1}{3}$ in. long, golden-brown; spikelets $\frac{1}{8}$–$\frac{1}{6}$ in. long, usually perfecting 2 nuts; glumes golden-brown; keel wingless, smooth or scabrous, green upwards, minutely excurrent; style 2-fid; nut $\frac{1}{2}$ as long as the glume, ellipsoid (not seen ripe). *K. aurata*, *Kunth, Enum.* ii. 137, *partly*. *K. inaurata*, *Boeck. in Linnæa*, xxxv. 406.

SOUTH AFRICA: without locality, *Zeyher*!

COAST REGION: Queenstown Div.; between Table Mountain and the Zwart Kei River, *Drège*, 3931!

9 **K. Lehmanni** (Nees in Linnæa, x. 139); glabrous except the minutely scabrous pubescent top of the stem; rhizome woody, short, suberect; scales torn into fibres; stem 5–11 in. long, medium stout, thickened at the base by leaf-sheaths; leaves often $\frac{2}{3}$ the length of the stem, $\frac{1}{8}$ in. broad; bracts 3, spreading, lowest 1–3 in. long, similar to the leaves; spike 1, globose, $\frac{1}{3}$–$\frac{1}{2}$ in. in diam., greenish-white; spikelets $\frac{1}{6}$–$\frac{1}{4}$ in. long, usually perfecting two nuts; glumes elliptic-lanceolate, acute; keel wingless, green, smooth or sparingly bristly, excurrent as a mucro; style 2-fid; nut only seen young. *Boeck. in Linnæa*, xxxv. 408. *K. ciliata*, *Kunth, Enum.* ii. 136.

SOUTH AFRICA: without locality, *Ecklon and Zeyher*, 28!
COAST REGION: Albany Div.; *Zeyher*, 4384! Komgha Div.; near the mouth of the Kei River, *Flanagan*, 927 partly!
EASTERN REGION: Tembuland; Bazeia, *Baur*, 274! Port Natal, 500 ft., *Drège*, 4386!

10. K. Buchanani (C. B. Clarke); glabrous; rhizome hardly any; stems 6–15 in. long, tufted, hardly thickened at the base; leaves often $\frac{2}{3}$ as long as the stem, $\frac{1}{8}$–$\frac{1}{6}$ in. broad; bracts 3, spreading, lowest 4 in. long, similar to the leaves; spike 1, globose or subovoid, $\frac{1}{3}$ in. long, straw-coloured; spikelets $\frac{1}{5}$ in. long, elliptic-oblong, often perfecting 2 nuts; glumes 1-coloured, hardly acute, with compressed glandular dots; keel wingless, not at all green, smooth; style 2-fid; nut $\frac{1}{2}$ as long as the glume, yellow-brown.

EASTERN REGION: Natal; Umzula, *Mudd!*
Also in the Shire Highlands, British Central Africa.
The keel of the fertile glume is wingless; yet this species may prove only an extreme state of *K. alba*.

II. PYCREUS, Beauv.

Spikelet of many distichous glumes; 2 lowest empty, 5 at least (usually many) succeeding glumes 2-sexual and perfecting nuts, uppermost male or sterile; rhachilla persistent, not disarticulating below the lowest fertile glume, not winged. *Stamens* 1–3, anterior; anthers narrow-oblong, not crested. *Style* slender; branches 2, linear, in a plane passing through rhachilla. *Nut* oblong or ellipsoid, compressed laterally, smooth.

Glabrous; leaves green, long, all close to the base of the stem; but in *P. Mundtii* the stem is covered nearly half its length by leaf-sheaths; inflorescence (as in *Cyperus*) a corymbose-panicle, shortened into a false umbel or head.

DISTRIB. Species 64; in all warm and temperate regions.

ZONATÆ. Superficial cells of the nut longitudinally oblong; nut often appearing zonate by reason of the narrow ends of the cells running into an undulating or broken horizontal line.

Spikelets $1\frac{1}{2}$ lin. broad, yellow or dirty purple (1) **flavescens.**
Spikelets 2 lin. broad, chestnut ... ... (2) **rehmannianus.**
Spikelets 3 lin. broad, chestnut ... ... (3) **macranthus.**

PUNCTICULATÆ. Superficial cells of the nut square; nut often appearing regularly dotted by reason of the light reflected from the convex surface of each cell.

Stem clothed, often for $\frac{1}{2}$ its length, by leaf-sheaths ... ... (4) **Mundtii.**
Spikelets narrow, red-tinged or pale, never chestnut-brown; stolons 0:
Spikelets linear, glumes closely imbricate ... ... (5) **polystachyus.**
Glumes more distant, rather larger, brighter coloured .. .. (6) **ferrugineus.**
Spikelets chestnut-brown:
Rhizome oblique descending; spikes umbelled; spikelets 2 lin. broad ... ... (7) **umbrosus.**
As preceding, but glumes acute ... ... (8) **cakfortensis.**

Stoloniferous; spikelets 1 lin. broad; glumes close ... ... (9) **betschuanus.**
As last, but stolons 0; glumes more distant; keel green ... ... (10) **elegantulus.**
Leaves stout, inrolled; spikelets 1½ lin. broad, hard, shining ... ... (11) **Cooperi.**
Spikes umbelled; spikelets yellow:
Spikelets 2 lin. broad, elliptic-lanceolate, straw-coloured ... ... (12) **angulatus.**
Spikelets 1⅓ lin. broad, linear, parallel-sided, golden ... ... (13) **chrysanthus.**

1. **P. flavescens** (Reichb. Fl. Germ. Excurs. 72); glabrous, annual; stems 1–12 in. long; leaves ⅔ as long as the stem, $\frac{1}{16}$–⅛ in. broad; spikes loosely umbelled, or stem with 1 spike, sometimes with only 3 spikelets; bracts 3, lowest 2–6 in. long, similar to the leaves; spikes of 3–12 closely spicate spikelets, ebracteate; spikelets commonly ½ by $\frac{1}{12}$ in. (in form *abyssinica* ¾ by ⅛ in.) compressed, parallel-sided, straw-coloured more or less red-tinged, 8–36-nutted; glumes close-placed, ovate, boat-shaped, inflated, obtuse; rhachilla persistent in fruit, glumes falling off regularly from its base; style-branches 2; nut obovoid, compressed, shining-black, hardly ½ as long as the glume, transversely white-muriculate by reason of the sub-persistent ends of the small longitudinally-oblong superficial cells. *C. B. Clarke in Hook f. Fl. Brit. Ind.* vi. 589. *Cyperus flavescens, Linn. Sp. Plant. ed.* 2, 68, *not Linn. Herb.*; *Boeck in Linnæa*, xxxv. 438. *C. abyssinicus, Steud. Syn. Pl. Glum.* ii. 4; *Boeck. in Linnæa*, xxxv. 440.

KALAHARI REGION: Orange Free State; (forma *abyssinica*) *Buchanan*, 110! Transvaal, Houtbosh, *Rehmann*, 5640!
EASTERN REGION: Natal; Biggarsberg, 5000 ft., *Kuntze*, 222! and without precise locality, *Gerrard*, 489! (forma *abyssinica*) *Buchanan*, 309!
Extends from England and Denmark to Cabul and the Cape, and widely scattered in America. The large bright form *abyssinica* (admitted as a species by Boeckeler) extends from South Europe to South Africa.

2. **P. rehmannianus** (C. B. Clarke in Durand and Schinz, Conspect. Fl. Afr. v. 542, excluding several numbers cited); glabrous; roots fibrous, tough; stems tufted, 8–11 in. long; leaves 7 by ⅙ in.; umbel ¾–2 in. in diam., contracted, of 7–22 spikelets; bracts 3, lowest 2½ in. long, similar to the leaves; spikes of 3–7 loosely spicate spikelets, ebracteate; spikelets ⅓ by ⅛ in., much compressed, linear-lanceolate, chestnut-brown, 12-nutted; glumes ovate, boat-shaped, obtuse, 1-nerved; style-branches 2; nut ellipsoid, compressed, shining black, obscurely transverse-undulate by reason of the large longitudinally-oblong superficial cells. *Cyperus rehmannianus, Boeck. ex C. B. Clarke in Durand and Schinz, Conspect. Fl. Afr.* v. 542.

KALAHARI REGION: Transvaal; Houtbosh, *Rehmann*, 5651!
EASTERN REGION: Natal; Howick, *Schlechter*, 6789!

3. **P. macranthus** (C. B. Clarke in Durand and Schinz, Conspect. Fl. Afr. v. 538); glabrous; roots fibrous, tough; stems tufted, 1–2 ft. long; leaves 12 by ⅛–⅙ in. (rarely ¼ in.) broad; umbel 3–1 in.

in diam., contracted often into 1 head of 6–30 spikelets; bracts 3, lowest 3–8 in. long, similar to the leaves; spikes of 3–8 somewhat loosely spicate spikelets, ebracteate; spikelets often up to 1 by $\frac{1}{4}$ in., compressed, linear-oblong, chestnut-brown, 28-nutted; glumes ovate, boat-shaped, obtuse, inflated; style-branches 2; nut small, obovoid, turgidly biconvex, black, scarcely $\frac{1}{4}$ as long as the glume, obscurely transversely lineolate by reason of the small longitudinally-oblong superficial cells. *Cyperus macranthus, Boeck. in Linnæa,* xxxv. 462. *C. lanceus var. β macrostachya, Kunth, Enum.* ii. 8.

SOUTH AFRICA: without locality, *Zeyher,* 1745!

COAST REGION: Komgha Div.; near Komgha, 2000 ft., *Flanagan,* 930! 1261! *Drège,* 4394! King Williamstown Div.; Toise River Station, *Kuntze,* 224!

CENTRAL REGION: Somerset Div.; Bosch Berg, 4500 ft., *MacOwan,* 1362! 1365!

KALAHARI REGION: Transvaal, Houtbosh, *Rehmann,* 5653! Magalies Berg, *Burke!* Johannesburg, *Barber.*

EASTERN REGION: Tembuland; Bazeia, 2000 ft., *Baur,* 307! Natal; Mohlamba Range, 5000–6000 ft., *Sutherland!* Inanda, *Wood,* 302!

Also in South Tropical Africa.

4. **P. Mundtii** (Nees in Linnæa, ix. 283; x. 131); glabrous; rhizome obliquely descending, 2–6 in. long; stem solitary, 8–24 in. long, covered for often half its length by leaf-sheaths; leaves 3–8 in. by $\frac{1}{6}$–$\frac{1}{4}$ in.; umbel simple or compound or reduced to 1 head; bracts 3–4, lowest 2–4 in. long, similar to the leaves; rays few or many; spikes of 3–7 somewhat loosely-spicate spikelets, ebracteate; spikelets $\frac{1}{2}$ by $\frac{1}{8}$ in., compressed, linear, brown or chestnut-red or pale brown, 20-nutted; glumes ovate, boat-shaped, obtuse, somewhat inflated; style-branches 2; nut obovoid, compressed, biconvex, brown, scarcely half the length of the glume, closely and regularly marked by raised dots. *Cyperus Mundtii, Kunth, Enum.* ii. 17; *Boeck. in Linnæa,* xxxv. 448. *C. distichophyllus, Steud. in Flora,* 1842, 582; *Boeck. in Linnæa,* xxxv. 488. *C. turfosus, Krauss in Flora,* 1845, 754. *C. Eragrostis, A. Rich. Tent. Fl. Abyss.* ii. 475; *Willk. et Lange, Fl. Hisp.* i. 138; *Kunth, Enum.* ii. 7, *not of Vahl.*

SOUTH AFRICA: without locality, *Ecklon and Zeyher,* 2! *Harvey,* 383!

COAST REGION: Cape Town, *Rehmann,* 1779! Port Elizabeth, *E.S.C.A. Herb.,* 235!

KALAHARI REGION: Transvaal; Marico Dist., Matebe River, *Holub,* 1897–1990!

EASTERN REGION: Natal; Umlaas River, *Krauss,* 415! Camperdown, *Rehmann,* 7736! Umbilo River, near Pinetown. 800 ft., *Wood,* 4016! Delagoa Bay, *Forbes! Scott!*

Common throughout Africa and the Mascarene Islands, also in Spain.

5. **P. polystachyus** (Beauv. Fl. Owar, ii. 48, t. 86, fig. 2); glabrous, annual; stems tufted, 8–24 in. long; leaves all close to the base of the stem, 10 by $\frac{1}{5}$ in.; spikes dense, of very many spikelets (but see var. β) in compound (few or 1) heads; bracts 3–5, lowest 2–6 in. long, similar to the leaves; spikelets densely spicate, $\frac{1}{2}$–$\frac{3}{4}$ by

$\frac{1}{12}$ in., linear, compressed, yellow more or less brown-red-tinged, 20-nutted; glumes ovate boat-shaped, obtuse, closely imbricated even in fruit; style-branches 2; nut oblong-ellipsoid, compressed, black, half the length of the glume, closely and regularly marked by raised dots. *Nees in Linnæa*, x. 130; *C. B. Clarke in Hook. f. Fl. Brit. Ind.* vi. 592. *Cyperus polystachyus, R. Br. Prod.* 214; *Thunb. Prod.* 18; *Fl. Cap. ed. Schult.* 102; *Nees in Linnæa*, vii. 515; *Boeck. in Linnæa*, xxxv. 477; *not of Rottb. C. odoratus, Linn. Sp. Plant.* 46, *ed.* 2, 68, *and Linn. Herb. mainly.*

VAR. β, **laxiflora** (Benth. Fl. Austral. vii. 261); umbel open; spikelets solitary, i.e. loosely spicate, spreading at right angles. *C. B. Clarke in Hook. f. Fl. Brit. Ind.* vi. 592.

SOUTH AFRICA: without locality, *Sieber*, 143! *Drège*, 4404! 4405! *Bergius*, 166! *Harvey*, 70! 371! *Ecklon and Zeyher*, 84!

COAST REGION: Clanwilliam Div.; Markus Kraal, near the Oliphants River, *Drège!* Cape Div.; near Cape Town, *Thunberg*, *Drège*, 9! *Bolus*, 3309! *Burchell*, 683! 831! Camps Bay, *Burchell*, 851! False Bay, *Robertson!* Tulbagh Div.; near Tulbagh, *Burchell*, 1045/1! and 8643/2! Worcester Div.; *Rehmann*, 2416! Caledon Div.; Zoetemelks Valley, *Burchell*, 7567! Uitenhage Div.; by the Zwartkops River, *Burchell*, 4433! *Zeyher*, 4358! Albany Div.; Grahamstown, 2000 ft., *MacOwan*, 569! Komgha Div.; Kei River, *Flanagan*, 959!

EASTERN REGION: Natal; Igogondwane, *Sutherland!* Durban Flats, *Buchanan*, 21! 37! 112! 310! Umlaas River, *Krauss*, 213! Var. β, Natal; Durban, *Rehmann*, 8606! 8608! Pondoland, *Drège*, 4403!

Abundant in all warm and temperate regions, except Europe.

There occurs every gradation from the dense tassel-like spikes of *P. polystachyus* type to the extreme form of the var. β, *laxiflora*.

6. P. ferrugineus (C. B. Clarke in Hook. f. Fl. Brit. Ind. vi. 593); spikelets rather larger, up to $\frac{3}{4}$ by $\frac{1}{8}$ in., more brightly coloured, often somewhat lanceolate (not linear) at the base; glumes rather more distant, their tips less closely imbricated in fruit; nut rather larger (otherwise as *P. polystachyus*, Beauv. var. β ***laxiflora***, Benth.). *Cyperus ferrugineus, Poir in Lam. Encyc.* vii. 261. *C. micans, Kunth, Enum.* ii. 12; *Krauss in Flora*, 1845, 754. *C. polystachyus, vars. ferrugineus and macrostachyus, Boeck. in Linnæa*, xxxv. 479. *C. polystachyus, vars. ferrugineus, micans, and filicina, C. B. Clarke in Journ. Linn. Soc.* xxi. 54, 55.

SOUTH AFRICA: without locality, *Drège*, 4419!

COAST REGION: Komgha Div.; Kei River, *Flanagan*, 961!

EASTERN REGION: Caffraria, *Schultz!* Pondoland, *Drège*, 4419! Natal; marshes near Umlaas River, *Krauss*, 189! Coast land, *Sutherland!* Howick, *Schlechter*, 6790! and without precise locality, *Gueinzius*, 12! *Buchanan*, 103! 111! 129! 319!

Scattered throughout Africa and South-east Asia: in Tropical America and extending north to Canada, i.e. the American *Cyperus Nuttallii*, C. Spreng., seems to me identical with the South African *Pycreus ferrugineus*.

7. P. umbrosus (Nees in Linnæa, x. 130); glabrous; stolons long, stout, clothed by distant scales $1\frac{1}{4}$ in. long; stem solitary, 4–20 in. long, triquetrous at the top; leaves all near the base of the stem, often 12 by $\frac{1}{4}$ in.; umbel 3–4 in. in diam., or contracted to 1 compound head; bracts 3–4, lowest 3–8 in. long, similar to the leaves; spikelets 3–9 together, spicate, ebracteate, $\frac{1}{2}$ by $\frac{1}{6}$ in., much compressed, 18-nutted, brown, or chestnut-brown; glumes ovate boat-shaped, obtuse; style-branches 2; nut small, obovoid, $\frac{1}{4}$–$\frac{1}{3}$ the length of the glume, dark-brown, superficial cells round-hexagonal. *Cyperus lanceus, Thunb. Prod.* 18; *Fl. Cap. ed. Schult.* 101; *Nees in Linnæa*, vii. 517; x. 135; *Steud in Flora*, 1829, 153. *C. atratus, Krauss in Flora*, 1845, 753. *C. umbraticola, Kunth, Enum.* ii. 13; *Boeck. in Linnæa*, xxxv. 443. *C. permutatus, Boeck. in Linnæa*, xxxv. 477.

SOUTH AFRICA: without locality, *Krebs!* *Bergius*, 172! *Harvey*, 81! 351! 375! 382! *Ecklon and Zeyher*, 1! 76!

COAST REGION: Clanwilliam Div., *Zeyher!* Cape Div.; Table Mountain, *Ecklon*, 885! *Milne*, 224! *MacGillivray*, 417! Cape Flats, *Thunberg*, *Burchell*, 55! 676! *Krauss!* Simons Bay, *Wright!* Paarl Div.; Berg River, *Drège*, 1598! Worcester Div.; Brand Vley, *Thom*, 1063! Caledon Div.; Grabouw, near Palmiet River, 700 ft., *Bolus*, 4231! Riversdale Div.; between Little Vet River and Kampsche Berg, *Burchell*, 6862! Knysna Div.; hills near Melville, *Burchell*, 5453! Uitenhage Div.; Van Stadens Hoogte, *MacOwan*, 2083! *Zeyher*, 4357! and without precise locality, *Ecklon and Zeyher*, 634! 715! King Williamstown Div.; Toise River Station, 3500 ft., *Kuntze*, 277!

KALAHARI REGION: Transvaal; Pretoria, *Rehmann*, 4772! 4773! Hooge Veld, at Donkers Hoek, *Rehmann*, 6151! 6553!

EASTERN REGION: Griqualand East, near Matatiele, 5000 ft., *Tyson*, 1614! Pondoland; Umtsikaba River, *Drège!* Natal; Umlaas River, *Krauss*, 205! Durban Flat, *Wood*, 4099! and without precise locality. *Buchanan*, 97! 308! *Drège*, 4400!

Also in Tropical Africa and Madagascar.

8. P. oakfortensis (C. B. Clarke); stems 1–2 ft. long, with leaves and inflorescence as in the narrow-leaved forms of *P. umbrosus;* spikelets rigid, shining black; nut shining black; glumes lanceolate, acute (otherwise as *P. umbrosus*, Nees). *Cyperus oakportensis, Boeck. ex C. B. Clarke in Durand and Schinz, Conspect. Fl. Afr.* v. 543, *by error.*

KALAHARI REGION: Transvaal; Lomati Valley, 4000 ft., *Galpin*, 1364.

EASTERN REGION: Natal; Oakford, *Rehmann!* and without precise locality, *Buchanan*, 97a! 99! 100!

Entered as a form of *P. umbrosus*, Nees, in *Durand and Schinz, Conspect. Fl. Afr.* v. 543.

9. P. betschuanus (C. B. Clarke in Durand and Schinz, Conspect. Fl. Afr., v. 535); glabrous; stolons horizontal, slender, clothed by very pale brown scales $\frac{1}{2}$ in. long; stems 6–14 in. long, slender; leaves all near the base of the stem, $\frac{2}{3}$ the length of the stem, $\frac{1}{10}$ in. broad; umbel $\frac{1}{2}$–$2\frac{1}{4}$ in. broad, simple, rays 3–5; bracts 3, lowest 2–4 in. long, similar to the leaves; spikes of 4–8 shortly-spicate spikelets, ebracteate; spikelets $\frac{1}{2}$ by $\frac{1}{12}$ in., much compressed, hard, shining, 22–nutted, dark chestnut-brown; glumes close-packed, ovate, boat-shaped, obtuse; keel of fertile glume, paler yellowish, not green; style-branches 2; nut obovoid, scarcely $\frac{1}{2}$ the length of the glume, outermost cells subquadrate. *Cyperus betschuanus, Boeck. in Engl. Jahrb.* xi. 406. *C. globosus var. nilagirica, C. B. Clarke in Journ. Linn. Soc.* xxi. 49.

KALAHARI REGION: Griqualand West; Hay Div., between the Asbestos Mountains and Witte Water, *Burchell*, 2081! Transvaal; Houtbosh, *Rehmann!* Bechuanaland, 4000 ft., *Marloth*, 1027!

Some dark chestnut-brown Indian mountain forms of *P. globosus, Reichb.*, resemble the present plant very closely, but never show its slender stolons.

10. **P. elegantulus** (C. B. Clarke in Durand and Schinz, Conspect. Fl. Afr. v. 536); glabrous; stems tufted on a very short rhizome, 8–24 in. long, slender or medium; leaves all near the base of the stem, $\frac{2}{3}$ its length, $\frac{1}{6}$ in. broad where the stem is medium, usually much narrower, sometimes filiform; umbel 2–3 in. in diam., dense or lax, or with few or only 1 spike; bracts 3, lowest 2–8 in. long, similar to the leaves; spikes of 4–8 closely-spiked spikelets, ebracteate; spikelets $\frac{1}{4}$–$\frac{1}{3}$ by $\frac{1}{12}$–$\frac{1}{10}$ in., much compressed, 6–10-nutted; glumes ovate, boat-shaped, obtuse, black chestnut-brown with prominent green or yellowish keel, in fruit scarcely imbricate; style-branches 2; nut ellipsoid, $\frac{2}{3}$ the length of the glume, brown, closely and regularly marked by raised dots. *Cyperus atronitens, Hochst. in Flora*, 1841, *Band* 1, *Intell.* 20; *Boeck. in Linnæa*, xxxv. 456. *C. elegantulus, Steud. in Flora*, 1842, 583.

EASTERN REGION: Natal; Durban Flat, *Buchanan*, 50! and without precise locality, *Buchanan*, 101! 307!

Also in Tropical Africa, and from Mexico to Peru.

11. **P. Cooperi** (C. B. Clarke in Durand and Schinz, Conspect. Fl. Afr. v. 535); glabrous, stout; roots fibrous, very thick (probably grew in water); stems densely packed, thickened at the base by large leaf-sheaths, 8–20 in. long.; leaves longer than the stem, trigonous-terete, $\frac{1}{12}$ in. in diam., all very close to the base of the stem; umbel $\frac{1}{2}$–$1\frac{1}{2}$ in. in diam., condensed nearly into 1 compound spike; bracts 3, lowest 3–6 in. long, similar to the leaves; spikelets $\frac{1}{3}$–$\frac{1}{2}$ by $\frac{1}{12}$–$\frac{1}{10}$ in., rigid, moderately compressed, shining chestnut-black; glumes ovate, boat-shaped, obtuse, closely imbricate; style-branches 2; nut narrowly-ellipsoid, compressed, half as long as the glume, brown; superficial cells quadrate. *Cyperus nilagiricus, Boeck. ex C. B. Clarke in Durand and Schinz, Conspect. Fl. Afr.* v. 535. *C. lanceus Thunb.? Drège, Pflanzengeogr. Documente* 177.

COAST REGION: Stockenstrom Div.; Kat Berg, 4000–5000 ft., *Drège!*
KALAHARI REGION: Orange Free State, *Cooper*, 912! Transvaal; Houtbosh, *Rehmann*, 5652!

12. **P. angulatus** (Nees in Linnæa, ix. 283); glabrous, somewhat robust; rhizome creeping; stems 8–24 in. long; leaves all near the base of the stem, $\frac{2}{3}$ its length, $\frac{1}{8}$–$\frac{1}{4}$ in. broad; umbel 2–4 in. in diam., simple, or frequently of 1 spike; bracts 3, lowest 3–10 in. long, similar to the leaves; spikelets 4–10 together, spicate, yellow or brownish yellow, $\frac{1}{2}$ by $\frac{1}{6}$–$\frac{1}{4}$ in., much compressed, 8–20-nutted; glumes ovate, boat-shaped, subobtuse or with very short triangular points, one-coloured, inflated; style-branches 2; nut small, obovoid, black, $\frac{1}{4}$–$\frac{1}{3}$ the length of the glume, superficial cells square. *C. B. Clarke in Hook. f. Fl. Brit. Ind.* vi. 593. *C. bromoides, Link, Jahrb.* iii. 85; *Boeck. in Linnæa*, xxxv. 463. *C. angulatus, Nees*

in Wight, Contrib. 73; *Boeck. in Linnæa,* xxxv. 465. *C. pseudo-bromoides, Boeck. in Linnæa,* xxxv. 464.

COAST REGION: Komgha Div.; Kei River *Flanagan,* 958! Queenstown Div.; Klipplaat River, 3500 ft., *Drège,* 3958!
KALAHARI REGION: Transvaal; Pretoria, *Rehmann,* 4326!

EASTERN REGION: Tembuland; Bazeia, 2000 ft., *Baur,* 371! Pondoland; between Umtata River and St. Johns River, 1000–2000 feet., *Drège,* 4397!

Widely distributed in all tropical and warm-temperate lands.

13. P. chrysanthus (C. B. Clarke in Durand and Schinz, Conspect. Fl. Afr. v. 534); glabrous, somewhat stout; stems 15–24 in. long; leaves all near the base of the stem, $\frac{2}{3}$ its length, $\frac{1}{6}$–$\frac{1}{5}$ in. broad; umbel 3–4$\frac{1}{2}$ in. broad, simple; bracts 3–4, lowest 3–8 in., similar to the leaves; spikelets 6–16 together, loosely spicate, ebracteate, up to $\frac{3}{4}$ by $\frac{1}{10}$ in., linear with exactly parallel sides, 12–22-nutted, much compressed; glumes ovate, boat-shaped, golden yellow, one-coloured; keel microscopically excurrent and often scabrous; style-branches 2; nut oblong-ellipsoid, $\frac{1}{2}$ the length of the glume, black; superficial cells square. *C. chrysanthus, Boeck. in Linnæa,* xxxv. 476. *C. lanceus var. γ? mucronata, Kunth, Enum.* ii 8.

SOUTH AFRICA: without locality, *Harvey,* 80!
KALAHARI REGION: Transvaal; *Rehmann,* 5145!
EASTERN REGION: Natal, *Grant!* Pondoland; Umtsikaba River, below 500 ft., *Drège,* 4409!

III JUNCELLUS, C. B. Clarke.

Spikelet of many distichous glumes; 2 lowest empty, 5 at least (usually many) succeeding glumes 2-sexual perfecting nuts, uppermost male or sterile; rhachilla persistent, not disarticulating below the lowest fertile glume, scarcely winged. *Stamens* 1–3, anterior. *Style* slender; branches 2, linear, in a plane at right angles to the median plane of the spikelet. *Nut* oblong or ellipsoid, triangular, the anterior angle flattened, smooth.

Differs from *Cyperus* only in having the style 2-fid, not 3-fid.

Species 13, scattered through nearly all warm and temperate regions.

1. J. lævigatus (C. B. Clarke in Hook. f. Fl. Brit. Ind. vi. 596); glabrous; rhizome creeping; stems 6–24 in. long, somewhat fleshy, roundish; leaves usually short, often hardly any, sometimes exceeding the stem, $\frac{1}{12}$–$\frac{1}{8}$ in. broad, upper part terete-trigonous; spikelets 1–30, in one apparently lateral head; bracts 2, lower 1–3 in. long, suberect, similar to the leaves; spikelets $\frac{1}{3}$ by $\frac{1}{8}$ in., straw-coloured, often more or less purple-tinged, 12–24-flowered, compressed but thick; glumes very close-packed, broad-elliptic, obtuse, rounded on the back, falling seriatim from the base of the persistent rhachilla; stamens 3, anthers oblong, with a short lanceolate red crest; undivided part of style about as long as the nut; branches 2 as long as the nut; nut plano-convex, $\frac{1}{2}$–$\frac{2}{3}$ the length of the glume, obovoid, obtuse, smooth, brown, plane posticous face pressed against the rhachilla, anterior face convex or somewhat ridged. *Cyperus lævigatus, Linn. Mant.* 179; *Rottb. Descr. et Ic.* 19, *t.* 16, *fig.* 1; *Thunb. Prod.* 18, *and Fl. Cap. ed. Schult.* 102; *Boeck. in Linnæa*, xxxv. 486; *C. B. Clarke in Journ. Linn. Soc.* xxi. 77, *t.* 3, *fig.* 20, 21, *and t.* 4, *fig.* 33. *C. mucronatus, Rottb. Descr. et Ic.* 19, *t.* 8, *fig.* 4. *Pycreus lævigatus, Nees in Linnæa*, x. 130.

SOUTH AFRICA: without locality, *Ecklon and Zeyher!*
COAST REGION: Malmesbury Div.? Mooresbury, *Bachmann*, 759! Cape Div.; Cape Town, *Thunberg, Harvey*, 172! 188! *Rehmann*, 1786! Nieuwernoolen, *Bergius*, 159! Port Elizabeth, *Drège*, 4382! *E.S.C.A. Herb.*, 273!
CENTRAL REGION: Graaff Reinet Div.; by the Sunday River, 2500 ft., *Bolus*, 715!
EASTERN REGION: Natal; near Durban, *Wood*, 1365! *Rehmann*, 8597! *Kuntze*, 225a! 225b! near Umgeni, *Wood*, 4007! and without precise locality, *Gerrard*, 488!

Found in all warm and temperate regions. The Var. *junciformis*, C. B. Clarke (*C. junciformis*, Desfont. Fl. Atlant. i. 42, t. 7, fig. 1), with few, hard, black-chestnut spikes is also spread over most of the world, but no examples have been yet seen from South Africa.

IV. CYPERUS, Linn.

Spikelet of many (rarely 6–5) distichous glumes; 2 lowest empty, 3 at least (usually many) succeeding glumes 2-sexual perfecting nuts, uppermost male or sterile; rhachilla persistent, not disarticulating below the lowest fertile glume; fertile glumes and nuts falling seriatim beginning with the lowest. *Stamens* 3–1 anterior. *Style* slender, sometimes short; branches 3, long linear, or in *C. semitrifidus* (and in some non-Cape species) short, weak or obsolete. *Nut* triangular, or convexo-plane by reason that the anticous angle is more or less flattened, from narrow-oblong to obovoid in longitudinal section, smooth; superficial cells nearly square (except in one Australian species).

Usually glabrous; leaves all near the base of the stem; inflorescence a terminal corymb depressed into an apparent umbel, or reduced to a single head or spikelet.

DISTRIB. Species 300, in all warm and temperate regions.

According to Nees (in Linnæa, x. 134), *Cyperus Haspan*, Linn., was collected by Ecklon at Uitenhage; but the species has not been found in Ecklon's plants, nor has it been communicated by anyone from South Africa.

Subgenus I. **Pycnostachys.**—Inflorescence umbellate throughout or capitate; spikelets digitate, clustered or solitary, not spicate; rhachilla hardly winged.
Sect. 1. Inflorescence of 1 head (a simple umbel once seen in *C. compactus*, Lam.):
Small annuals; head of 1–3 spikelets:
Spikelets 1 line broad ... ... (1) **tenellus.**
Spikelet 2 lines broad ... ... (2) **micromegas.**
Tufted annuals; head of 4–15 spikelets:
Very small; spikelet ½ line broad ... (3) **leucoloma.**
Stems thickened at the base, not annual; spikelets brown, red or chestnut-brown:
Glumes with long recurved points ... (4) **meyerianus.**

Glumes large, bright-coloured, with white points ... ... (5) **Teneriffæ.**
Glumes chestnut-brown or dark red; nut ½ the length of the glume ... (6) **rupestris.**
As last, but leaves setaceous, cylindric, obtuse ... ... (7) **amnicola.**
As *C. rupestris*, but spikelets narrower; nut very small ... ... (8) **parvinux.**
Stout plants with woody rhizome:
Glumes straw-coloured, yellow or yellow-brown, densely packed ... ... (9) **compactus.**
Glumes straw-coloured, broad, a little inflated, less dense ... ... (10) **margaritaceus.**

Sect. 2. Spikes simply umbelled (1-headed stems often occur); spikelets digitate or clustered, rarely solitary; leaf-bearing:
Umbel contracted; rays very short:
Spikelets dull grey-red; nut obovoid ... (11) **bellus.**
Spikelets dull; glumes muticous; nut oblong ... ... (12) **fuscescens.**
Umbel open:
Spikelets pale-reddish; style long, with very short branches ... ... (13) **semitrifidus.**
Spikelets black-chestnut; style-branches long ... ... (14) **tenax.**

Sect. 3. Umbel compound or apparently simple, of fascicled, often densely agglomerated, spikelets; leaf-bearing:
Spikelets very small; glumes minute, obtuse ... ... (15) **difformis.**
Spikelets much compressed; glumes pointed ... ... (16) **hæmatocephalus.**

Sect. 4. Umbel usually compound, of apparently simple digitate spikes; plants rather stouter, leaf-bearing:
Stems triquetrous at the top; spikelets oblong, rather thick ... ... (17) **pulcher.**
Stem trigonous at the top; spikelets linear, compressed ... ... (18) **sphærospermus.**

Sect. 5. Leafless, somewhat large plants:
Bracts 2–4, scarcely overtopping the umbel:
Stem triquetrous at the top ... ... (19) **denudatus.**
Stem obtusely trigonous or nearly terete at the top ... ... (20) **marginatus.**
Bracts numerous, exceeding the umbel, usually 8 or more:
Stem round at the top ... ... (21) **textilis.**
Stem with 6 angles at the top ... ... (22) **sexangularis.**
Rays of umbel equal, usually 40 or more, exceeding the bracts:
Stem triquetrous at the top ... ... (23) **isocladus.**

Sect. 6. Solitary spikelets frequent in the compound umbel; leaves long, flat; bracts long, usually more than 6:
Leaves broad, grass-like, 3-nerved, flat ... (24) **albostriatus.**
Leaves and bracts narrower, more rigid, obscurely, nerved:
Glumes minutely hairy on the edges ... (25) **prasinus.**
Glumes glabrous on the edges ... ... (26) **leptocladus.**
Spikelets 8–16 together, spicate rather than digitate ... ... (27) **subchoristachys.**

Subgenus II. **Choristachys.**—Spikelets spicate, not digitate, nor solitary, nor densely clustered.

Sect. 7. Rhachilla of spikelets hardly winged (narrowly winged in *C. distans*):
Perennials, with stolons:
Stem stout; glumes imbricate ... (28) **latifolius.**
Umbel large, open; spikelets slender; glumes distant ... ... (29) **distans.**
Annuals:
Medium-sized; umbel usually simple ... (30) **compressus.**
Small; glumes with a long recurved point (31) **aristatus.**

Sect. 8. Rhachilla of spikelets winged; plant stoloniferous:
Stolons very slender, producing a bulb or tuber at their ends:
Slender; umbel contracted; stolons bearing bulbs ... ... (32) **usitatus.**
Umbel open; stolons bearing tubers ... (33) **esculentus.**
Stolons stout, hardening into tough, woody rhizomes:
Leaves and bracts short (occasionally long):
Spikelets narrow; stem obscurely transverse-marked ... ... (34) **corymbosus.**
Spikelets broader, hard, shining ... (35) **natalensis.**
Leaves and bracts long:
Stem erect, thickened at the base; rhizomes tuberiferous ... ... (36) **rotundus.**
Stem decumbent at the base; rhizome irregularly thickened ... (37) **longus.**

Sect. 9. Rhachilla of spikelets winged; stolons 0; plant tall; umbel compound, large; bracts overtopping the umbel:
Bracts to secondary umbel 0–¼ in. long, setaceous ... ... (38) **fastigiatus.**
Bracts to secondary umbels 1–4 in. long (39) **immensus.**

Sect. 10. Rhachilla of spikelets winged; rhizome creeping; plant tall; bracts short:
Umbel of 50–100 subequal rays ... ... (40) **madagascariensis.**

1. C. tenellus (Linn. f. Suppl. 103); glabrous, slender, annual: stems 1–7 in. long, setaceous; leaves few, ½ as long as the stem, setaceous, weak; head of 1–3 (rarely 4) digitate spikelets; bracts 1–2, setaceous, short, lower continuing the stem, rarely attaining ½–1 in.; spikelets compressed, exactly parallel-sided, with very regularly-placed glumes, commonly ¼ by $\frac{1}{12}$ in. with 10 nuts, but very variable in length, sometimes ⅝ in. long with 20 nuts, sometimes with only 3–4 nuts; rhachilla persistent, stout for so small a plant; glumes ovate, boat-shaped, obtuse, strongly 7–9-ribbed, falling off seriatim from the base of the spikelet, white, green, reddish or brown; stamens 2–1; anthers small, linear-oblong, muticous; nut ½–⅔ as long as the glume, triangular, ovoid, with acutely pyramidal apex, black; style linear, shorter than the nut; branches 3, linear, long. *Nees in Linnæa*, vii. 513, x. 132; *Steud. in Flora*, 1829, 152; *Benth. Fl. Austral.* vii. 265. *C. lateralis*, *Linn. f. Suppl.* 102. *C. pygmæus*, *Lam. Ill* i. 143. *C. minimus*, *Thunb. Prod.* 18; *Fl.*

Cap. ed. Schult. 99; *Boeck. in Linnæa,* xxxv. 523, *not of Linn. C. nudiusculus, Nees in Linnæa,* x. 132. *Pycreus lateralis, Nees in Linnæa,* ix. 283.

VAR. β, **gracilis** (Nees in Linnæa, x. 132); stem 1 in. with one or two ovate 3-flowered spikelets.

SOUTH AFRICA: without locality, *Sieber,* 99! *Bergius,* 164! *Harvey,* 271! 355! *R. Brown! Ecklon and Zeyher,* 6! 7! 8! 85! Var. β, *Masson! Harvey,* 341!

COAST REGION: Cape Div.; Table Mountain, *Ecklon,* 891! Constantia, *Zeyher!* Simons Bay, *Wright!* Paarl Div.; Paarl Mountains, below 1000 ft., *Drège,* 7422! Worcester Div., *Ecklon!* Riversdale Div., near Zoetemelks River, *Burchell,* 6703! Knysna Div.; Vlught, 500 ft., *Bolus,* 2690! Uitenhage Div.; Van Stadens Hoogte, *MacOwan,* 2175! Var. β, Cape Div.; Hout Bay, 700 ft., *Schlechter in Herb. Bolus,* 7199! Simons Bay, *Wright,* 519! Uitenhage Div.; *Ecklon and Zeyher!*

CENTRAL REGION: Somerset Div.; summit of Bosch Berg, 4500 ft., *MacOwan,* 2103!

The var. *gracilis* was *C. minimus,* Roth. MS. according to Nees; but in the old Cape collections (as in Drège's collection named by Kunth) *Cyp. tenellus,* Linn. f., was distributed mixed with the small form of *Scirpus antarcticus,* Linn., to which last the synonym *Cyperus minutus,* Roth. (and other depending on this), belongs.

Also in temperate Australia and New Zealand.

2. C. micromegas (Nees in Linnæa, x. 131); glabrous, slender, annual; stem 3–4 in. long, setaceous; leaves very few, $\frac{1}{2}$ as long as the stem, setaceous; head of 1–2 sessile spikelets; lower bract setaceous, $\frac{1}{4}$ in. or less; spikelets much compressed, exactly parallel-sided, with very regular closely placed glumes, 2 lin. broad, $\frac{1}{2}$ in. long with 36 nuts at most, usually shorter, with fewer nuts; glumes ovate, boat-shaped, obtuse, strongly 7–9-ribbed, falling off seriatim from the base of the persistent rhachilla; nut $\frac{1}{4}$–$\frac{1}{3}$ as long as glume, triangular, ovoid, with depressed pyramidal top, shining black; superficial cells scarious, white, punctured; style branches 3, linear, long. *Boeck. in Linnæa,* xxxv. 524. *C. tenellus, Linn. f. var., F. Muell. Fragm. Phyt. Austral.* viii. 261.

SOUTH AFRICA: without locality, *Ecklon and Zeyher,* 9!

COAST REGION: Clanwilliam, *Zeyher!* Cape Div.; Table Mountain, fide *Nees.*

This is structurally the same as *C. tenellus,* Linn. f., with the spikelets twice as broad; the nut is differently shaped, and very much smaller in proportion to its glume.

3. C. leucoloma (Nees in Linnæa, ix. 284, x. 133); glabrous, very small, annual; stem $\frac{1}{2}$ in. long, setaceous; leaf 1, as long as the stem, setaceous; head of 4–8 digitate spikelets; bracts 2, lower $\frac{1}{2}$ in. long, setaceous; spikelets compressed, $\frac{1}{6}$ by $\frac{1}{20}$ in., 6–8-flowered; glumes closely imbricated, boat-shaped, ovate, minutely mucronate, 7-ribbed, brown, with conspicuous scarious white margins; stamens 3; nut $\frac{1}{2}$ as long as the glume, triangular, pyramidal at both ends, smooth, black; style short; branches 3, linear, long. *Boeck. in Linnæa*, xxxv. 506.

SOUTH AFRICA: without locality, *Bergius*, 174! No specimen at Kew.

4. **C. meyerianus** (Kunth, Enum. ii. 49); glabrous, weak; stems 2–3 in. long, tufted on the apex of a shortly-divided, slender rhizome, thickened at the base by numerous pale, rust-coloured sheaths; leaves nearly as long as the stem, $\frac{1}{20}$ in. broad, flaccid; bracts 2, lower up to 1 in. long, similar to the leaves; spikelets 7–1, sessile in one head, $\frac{1}{4}$–$\frac{1}{3}$ in. long and broad, much compressed, 12-flowered; glumes close-packed, boat-shaped, ovate, strongly 7-nerved, rose-red, surmounted by a subterminal, lanceolate-linear, recurved mucro longer than the glume; stamens 3–2; nut $\frac{1}{2}$ as long as the glume without the mucro, triquetrous, ovoid, acute, subacuminate at the top, dark-brown, almost rough from the reticulated white, superficial cells; style shorter than nut; branches 3, long linear. *Boeck. in Linnæa*, xxxv. 508; *Steud. Syn. Pl. Glum.* ii. 29 (*meyenianus, by typog. error in index*).

SOUTH AFRICA: without locality, *Drège*, 7421!

5. **C. Teneriffæ** (Poir. in Lam. Encyc. vii. 245); glabrous, often reddish; stolons 0; stems 3–12 in. long, medium-sized, trigonous at the apex, smooth; bases thickened by leaf-sheaths, clustered on a very short rhizome; leaves $\frac{2}{3}$ the length of the stem, $\frac{1}{8}$–$\frac{1}{6}$ in. broad, usually flaccid; sheaths inflated, scarious-red striate, occasionally rigid, very little inflated; bracts 2, lowest 2–3 in. long, similar to the leaves; spikelets 3–20 in a head, large, much compressed, 10–36-flowered, up to 1 by $\frac{1}{4}$ in.; glumes close-packed, boat-shaped, ovate, rose-red or pale-red, 9–15-ribbed; keel excurrent in a strong, erect, acute-pyramidal point; stamens 3; anthers linear, not crested; nut $\frac{1}{4}$–$\frac{1}{3}$ as long as the glume, obovoid, black, reticulated by white, superficial cells; style as long as the nut; branches 3, linear, hardly so long as the style. *C. B. Clarke in Hook. f. Fl. Brit. Ind.* vi. 601. *C. rubicundus, Kunth, Enum.* ii. 49; *Webb and Berth. Iles Canaries, Phyt.* iii. *pt.* iii. 361, *t.* 240 (*nut depicted much too acute at the top*); *Drège, Pflanzengeogr. Documente* 147; *Boeck. in Linnæa*, xxxv. 507, *not of Vahl.*

SOUTH AFRICA: without locality, *Bolus*, 5812! *Ecklon and Zeyher*, 17!
COAST REGION: Albany Div.; *Ecklon.* East London Div.; cultivated specimen! Komgha Div.; among rocks near Komgha, 2000 ft., *Flanagan*, 1013! Uitenhage Div., 300 ft., *Schlechter*, 2485!
KALAHARI REGION: Transvaal, Bosch Veld at Klippan, *Rehmann*, 5349!
EASTERN REGION: Transkei, *Drège*, 5349. Natal, *Drège*, 4395!

Extends from Teneriffe throughout Africa and Madagascar to the Madras Peninsula.

This species has been greatly confused, being still marked in most European herbaria, and by South African collectors, *C. rubicundus*, Vahl. From Vahl's description (as also from his West Indian locality) it is sufficiently clear that his *C. rubicundus* was not *C. Teneriffæ*; and from the type specimen in the Copenhagen herbarium, Vahl's *C. rubicundus* was *C. brunneus*, Swartz.

6. C. rupestris (Kunth, Enum. ii. 52); glabrous; stolons 0; stems 2–6 in. long, triquetrous at the top, smooth, united at the base, with hardly any rhizome, thickened at the base by brown leaf-sheaths, of which the lower are resolved into fibrils; leaves often as long as the stem, hardly $\frac{1}{16}$ in. broad; apex cylindric, setaceous, many-striate, microscopically rough-tuberculate (in Kunth's example of *C. rupestris* exactly as described by him in *C. cognatus*); bracts 2–3, lowest up to 3 in. long, similar to the leaves; spikelets 2–9 in one head, compressed, shining, black-chestnut, or chestnut-red, $\frac{1}{2}$ by $\frac{1}{6}$–$\frac{1}{5}$ in., 8–16-flowered; glumes overlapping, boat-shaped, ovate, somewhat inflated, 5-ribbed; keel subexcurrent into a microscopic point; stamens 3; anthers linear, not crested; nut $\frac{1}{2}$ the length of the glume, trigonous, obovoid, dusky black; style much shorter than the nut; branches 3, linear, long. *Boeck. in Linnæa,* xxxv. 510. *C. cognatus, Kunth, Enum.* ii. 52; *Boeck. in Linnæa,* xxxv. 511.

COAST REGION: Cape Div.; Table Mountain, 2500 ft., *Schlechter*, 421! Stockenstrom Div.; Kat Berg, 4000–5000 ft., *Drège!* Komgha Div.; between Zandplaat and Komgha, 2000–3000 ft., *Drège,*, 4393. Cathcart Div.; between Windvogel Mountain and Zwart Kei River, 3000–4000 ft., *Drège*, 7395!

CENTRAL REGION: Graaff Reinet Div.; summit of Tandjes Berg, near Graaff Reinet, 4700 ft., *Bolus*, 760! Somerset Div.; summit of Bosch Berg, 5000 ft., *MacOwan*, 1993!

EASTERN REGION: Natal; Inanda, *Wood*, 344!

The type specimens of Kunth's *C. cognatus* and *C. rupestris* in the Kew herbarium appear identical.

7. C. amnicola (Kunth, Enum. ii. 52); leaves more fleshy, obtuse at tip; spikelets a trifle narrower; glumes hardly mucronate; otherwise as *C. rupestris*, Kunth. *Boeck. in Linnæa,* xxxv. 509.

EASTERN REGION: Pondoland; between St. Johns River and Umtsikaba River, 1000–2000 ft., *Drège*, 4392! 7394! *Bachmann*, 88!

Also occurs in Angola, according to Ridley.

8. C. parvinux (C. B. Clarke); glabrous; stem 1–8 in. long, united at the base on a very short rhizome, somewhat thickened by leaf-sheaths; leaves often as long as the stem, setaceous; margins of sheaths prominently white-scarious; bracts 2, lower up to 2 in. long, similar to the leaves; spikelets 5–14 in 1 head, up to $\frac{5}{8}$ by $\frac{1}{10}$ in., 8–36-flowered, compressed, bright chestnut-red or brownish-red; glumes imbricate, boat-shaped, ovate, with 5 fine pale striations; keel excurrent into a minute acute-pyramidal mucro; stamens 2 or 3; anthers oblong; connective shortly produced; nut very small, scarcely $\frac{1}{3}$ the length of the glume, plane-convex, obovoid obtuse, smooth, dark-brown, glistening by reflection from the inflated surfaces of the superficial cells; style as long as the nut; branches 3, long, linear.

CENTRAL REGION: Albert Div.; Molteno, 5000–6000 ft., *Kuntze*, 264!

KALAHARI REGION: Transvaal; Houtbosch, *Rehmann*, 5648! Bosch Veld at Elands River, *Rehmann*, 4998

9. C. compactus (Lam. Ill. i. 144, not of Retz.); glabrous, stout; rhizome short, oblique, often $\frac{1}{4}$ in. in diam., clothed by black, hard, shining sheaths, breaking up into strong fibres; stems 4–18 in. long; apex trigonous smooth; bases thickened, becoming woody, close-packed; leaves often as long as the stem, sometimes quite short, $\frac{1}{12}$–$\frac{1}{6}$ in. broad, rigid, tough, convolute and subcylindric upwards; bracts 3, dilated at the base, lowest up to 2–6 in. long, similar to the leaves; spikelets 5–20, in the head (in Gerrard, 459 there are 4 rays up to 2 in. long), stramineous, or dirty-white (see var. β), large; compressed, 8–18-flowered; glumes very close-packed, $\frac{1}{4}$ in. long, boat-shaped, ovate, obtuse, conspicuously 11–17-nerved; margins often incurved in dry ripe spikelets; stamens 3; anthers linear, not crested; nut medium-sized, but scarcely $\frac{1}{3}$ the length of the glume, triquetrous, obovoid, dull-black, reticulated by the white superficial layer of cells; style shorter than the nut; branches 3, linear, long, but not much exserted. *C. obtusiflorus, Vahl, Enum.* ii. 308; *Boeck. in Linnæa,* xxxv. 528. *C. sphærocephalus var. β, leucocephalus, Kunth, Enum.* ii. 45; *Krauss in Flora,* 1845, 754.

VAR. β, **flavissimus** (C. B. Clarke in Durand and Schinz, Conspect. Fl. Afr. v. 552); spikelets fine golden-yellow, or yellowish, or yellow-brown or full brown. *C. obtusiflorus, var. β, flavissimus, Boeck. in Linnæa,* xxxv. 529. *C. flavissimus, Schrad. Anal. Fl. Cap.* 5, *t.* 2, *fig.* 2; *Nees in Linnæa,* x. 131. *C. sphærocephalus, Vahl, Enum.* ii. 310; *Krauss in Flora,* 1845, 754; *Oliver in Trans. Linn. Soc.* xxix. 164, *t.* 108. *A.*

SOUTH AFRICA: without locality, *Zeyher,* 1749! *Burke!* Var. β, *Menzies!*

COAST REGION: Cathcart Div.; between Windvogel Mountain and Zwart Kei River, 3000–4000 ft., *Drège!* Var. β, Clanwilliam Div., *Zeyher!* Uitenhage Div.! *Ecklon and Zeyher,* 651! Alexandria Div.; Zuurberg Range, *Drège,* 2036; Bathurst Div.; between Blue Krantz and Kaffir Drift, *Burchell,* 3691! near Theopolis, *Burchell,* 4146! Albany Div.; near Grahamstown, 2000 ft., *Cooper,* 3334! *MacOwan,* 1205! *Atherston,* 52! Fort Beaufort Div., *Cooper,* 459! *Drège.!* Komgha Div.; Kei River, *Flanagan,* 996; Cathcart Div.; near Cathcart, 4500 ft., *Kuntze!*

KALAHARI REGION: Orange Free State, *Buchanan,* 132! Transvaal; Makapans Berg, at Streydpoort, *Rehmann,* 5389! Pretoria, at Wonderboom Poort, *Rehmann,* 4476! Maquasi Spruit, *Nelson,* 7*! var. β, Orange Free State, *Buchanan,* 131! *Cooper,* 918! Transvaal; Macalisberg, *Burke!* Pretoria, at Wonderboom Poort, *Rehmann,* 4477! Wit Waters Rand, *Nelson,* 63*! banks of the Mooi River, *Zeyher,* 1744! and without precise locality, *MacLea in Herb. Bolus,* 5816!

EASTERN REGION: Tembuland; between Bashee River and Morley 1000–2000 ft., *Drège,* 3955; Indwe, 3500–4000 ft., *Baur,* 1170! Pondoland, *Drège,* 4388! *Bachmann,* 78! 79! 80; Natal; Coast-land, *Sutherland!* Durban Flat, *Buchanan,* 306! Inanda, *Wood,* 28! around Durban Bay, *Krauss,* 324! and without precise locality, *Gueinzius! Gerrard,* 459! *Buchanan,* 101! Delagoa Bay, *Forbes! Junod,* 395! Var. β, Transkei; between Kei River and Gekau, 1000–2000 ft., *Drège.* Tembuland; Bazeia, 2000 ft., *Baur,* 280! Pondoland, *Bachmann,* 85! Griqualand East; near Kokstad, 4300 ft., *MacOwan and Bolus, Herb. Norm. Aust.-Afr.,* 1236! Natal; summit of Houtbosch Rand Mountain between Umgeni River and Mooi River, *Krauss,* 16! Durban Flat, *Buchanan,* 47! Howick, 1000 ft., *Junod,* 310. Klip River Div., 3500 ft., *Sutherland!* and without precise locality, *Gerrard,* 484! *Buchanan,* 54! 88!

Distributed throughout Tropical Africa and the Mascarene Islands.

This abundant African species varies greatly in size as well as in colour; medium dimensions are given above; in Buchanan, 131, the spikelets are $\frac{1}{3}$ in. broad,

shining brown; in Krauss, 324 (and in many others), the spikelets are $\frac{3}{16}$ in. broad and dirty-white.

10. **C. margaritaceus** (Vahl, Enum. ii. 307); glabrous; stems 4–20 in. long; rather slender, thickened at the base, clustered on a very short woody rhizome, clothed by black hard scales; leaves $\frac{1}{2}$ as long as the stem, $\frac{1}{12}$ in. broad, usually narrower than in *C. compactus*; bracts 3, lowest up to 2–3 in. long, similar to the leaves; spikelets 3–7 in 1 head, $\frac{3}{4}$ by $\frac{1}{4}$ in., compressed but much thicker than in *C. compactus*, shining straw-coloured, 10–20-flowered; glumes boat-shaped, ovate, obtuse, strongly 11–15-nerved, less closely packed but more inflated than in *C. compactus*; margins not incurved; nut large, $\frac{1}{3}$–$\frac{2}{5}$ the length of the glume, ellipsoid, acute at the top, strongly triquetrous with concave faces, dull black. *Boeck. in Linnæa*, xxxv. 529.

SOUTH AFRICA: without locality, *Zeyher*, 175! *Marloth*, 1022!

WESTERN REGION: South of the Tropic, *Curror!*

KALAHARI REGION: Griqualand West; between Griqua Town and Witte Water, *Burchell*, 1992! between Knegts Fontein and Klip Fontein, *Burchell*, 2613! Hebron, by the Vaal River, *Nelson*, 183! Orange River, *Curzon!* Transvaal; Bosch Veld at Menaars Farm, *Rehmann*, 4855! Magalies Berg, *Burke!* South African Gold Fields, Bains! and without precise locality, *McLea in Herb. Bolus*, 6014! Bechuanaland; Kosi Fontein, *Burchell*, 2575!

Distributed throughout Tropical Africa.

This species is tolerably uniform in South Africa, but in Tropical Africa it varies greatly in the width of the leaves and in the size of the spikelets. Some examples have needle-like stem and leaves, and one very small spikelet, others have leaves $\frac{1}{5}$ in. wide, others have spikelets 1 by $\frac{3}{8}$ in. The distribution will require emendation, if the species be subdivided.

11. **C. bellus** (Kunth, Enum. ii. 52); glabrous, annual; stems tufted, 2–7 in. long, slender, trigonous; leaves half as long as the stem, subsetaceous; head of 3–30 spikelets; rays 0–$\frac{1}{3}$ in. long; bracts 2–3, setaceous, slightly dilated at the base, lowest up to 2 in. long, suberect; spikelets fascicled, $\frac{1}{2}$ by $\frac{1}{12}$ in., 26-flowered, or much smaller as in the examples described by Kunth, compressed, with parallel sides, dusky red; glumes boat-shaped, in fruit scarcely overlapping, ovate, obtuse, scarcely mucronate, strongly 5–7-ribbed; rhachilla not winged; stamens 2 or 1; nut small, $\frac{1}{3}$–$\frac{2}{5}$ the length of the glume, trigonous, obovoid, subobtuse, smooth, dull ashy black; style shorter than the nut; branches 3, linear, shortly exserted from the glume. *Boeck. in Linnæa*, xxxv. 509.

SOUTH AFRICA: without locality, *Drège*, 7396!

KALAHARI REGION: Orange Free State; Great Vet River, *Burke! Zeyher*, 1748!

12. **C. fuscescens** (Link, Jahrb. iii. 83); glabrous, annual; stems tufted, 4–7 in. long; leaves often $\frac{2}{3}$ the length of the stem, $\frac{1}{10}$ in. broad; head of 5–15 spikelets; rays 0–$\frac{1}{4}$ in. long; bracts 2–3, similar to the leaves, lowest $1\frac{1}{2}$ in. long, suberect; spikelets fascicled, $\frac{1}{3}$ by $\frac{1}{16}$ in., 12-flowered, brown; nut oblong, $\frac{2}{3}$ the length of the glume, trigonous, pyramidal at each end, black reticulated with white; style short; branches 3, linear, much exserted from the glume. *Kunth, Enum.* ii. 51; *Boeck. in Linnæa*, xxxv. 527.

EASTERN REGION: Kaffraria; without collector's name in the British Museum.

Closely allied to *C. bellus*, but the nut is much larger.

13. C. semitrifidus (Schrad. Anal. Fl. Cap. 6); glabrous; rhizome woody, decumbent; stems 3–10 in. long, trigonous at the top, bulbous, woody at base, enclosed by fibrillose remains of leaf-sheaths; leaves often $\frac{2}{3}$ the length of the stem, $\frac{1}{8}$ in. broad; rays of umbel 3–0, 0–$1\frac{1}{2}$ in. long; bracts 3, similar to the leaves, lowest up to 3 in. long, suberect; spikelets 3–10 together, fascicled, up to $\frac{3}{4}$ by $\frac{1}{5}$ in., usually smaller, compressed, narrow-lanceolate, 20-flowered, reddish to brown; glumes boat-shaped, close-packed, ovate, pointed, 9–13-ribbed; rhachilla not winged; stamens 3; anthers linear, not crested; nut $\frac{2}{3}$ the length of the glume, trigonous, narrowly obovoid, subobtuse, black reticulated with white; undivided part of style very long, much exserted from the glume; branches 3–2, much shorter than the style, usually very short indeed, sometimes 0. *Kunth, Enum.* ii. 107; *Boeck. in Linnæa*, xxxv. 513. *C. crinitus, Spreng. in Flora*, 1829, *Band* i. *Beil.* 5 (*not C. cruentus, Rottb.*). *C. herbivagus, Kunth, Enum.* ii. 53. *C. herbicagus, Drège, Pflanzengeogr. Documente* 132, 177. *C. solidus, Kunth, Enum.* ii. 76, *partly* (*at least as to syn. cited*). *C. usitatus, Nees in Linnæa* vii. 516 (*not of Burchell*).

SOUTH AFRICA: without locality, *Ecklon and Zeyher*, 3! 127! *Bowie*, 5!

COAST REGION: Uitenhage Div.; by the Zwartkops River, *Ecklon! Zeyher*, 4364! Alexandria Div.; Addo, 1000–2000 ft., *Drège*, 2035! Komgha Div.; near Komgha, 1800 ft., *Flanagan*, 1016!

CENTRAL REGION: Somerset Div.; Bothas Berg, 2200 ft., *MacOwan*, 1505! Wodehouse Div., *Ecklon and Zeyher!*

KALAHARI REGION: Transvaal, *Rehmann*, 4998! 5648!

14. C. tenax (Boeck. in Linnæa, xxxv. 504); glabrous, medium-sized, tough; rhizome woody, with fibrillose remains of leaf-sheaths; no stolons seen; stems tufted, 6–16 in. long, slender but wiry; leaves often $\frac{2}{3}$ the length of the stem, $\frac{1}{6}$ in. broad, much enrolled when dry, tough; umbel of 5–10 rays, 2–3 in. in diam., more or less compound, or frequently contracted, with rays $\frac{1}{2}$ in. long; bracts 6–3, similar to the leaves, usually much overtopping the umbel; spikelets 3–16, digitate in each spike, much compressed, shining chestnut or nearly black, $\frac{1}{3}$ by $\frac{1}{12}$ in., 8–20-flowered; glumes boat-shaped, not close-packed, ovate, short-pointed, 3-nerved; rhachilla not winged; stamens 3; anthers linear-oblong, not crested; nut $\frac{1}{2}$ the length of the glume, trigonous, ellipsoid, pyramidal at either end, dull black; style hardly as long as the nut; branches 3, linear, long. *C. B. Clarke in Durand and Schinz, Conspect. Fl. Afr.* v. 578.

KALAHARI REGION: Transvaal; Pretoria, at Wonderboom Poort, *Rehmann*, 4481!

EASTERN REGION: Natal; *Guienzius*, 13! *Gerrard*, 699!

Occurs throughout Tropical Africa.

15. C. difformis (Linn. Amoen. Acad. iv. 302); glabrous, annual, medium-sized, but rather weak; stems tufted, 4–20 in. long, triquetrous at the top; leaves often $\frac{2}{3}$ the length of the stem, $\frac{1}{6}$ in. broad, flaccid; rays of umbel 3–8, unequal, up to 1–2 in. long, or often very short, each terminated by a dense compound spike of usually 20–60 spikelets; bracts 3–4, similar to the leaves, lowest 2–10 in. long; spikelets $\frac{1}{5}$ by $\frac{1}{30}$ in., 10–30-flowered, dark red; glumes minute, round, obtuse, concave, not imbricated in fruit; stamen 1, rarely 2; anther small, oblong, obtuse; nut equally trigonous, subglobose, as long as the glume or nearly so, pale brown; style much shorter than the nut; branches 3, linear, not elongate. *Sp. Plant, ed.* 2, 67; *Rottb. Descr. et Ic.* 24, *t.* 9, *fig.* 2; *Nees in Linnæa*, x. 138; *Kunth, Enum.* ii. 38; *Boeck. in Linnæa*, xxxv. 586, *and in Flora*, 1879, 550; *C. B. Clarke in Hook. f. Fl. Brit. Ind.* vi. 599.

COAST REGION: Albany Div.; near Grahamstown, 2000 ft., *MacOwan*, 1347! Komgha Div.; *Flanagan*, 1262!

CENTRAL REGION: Somerset Div.; mountain near Commadagga, *Burchell*, 3350!

KALAHARI REGION: Transvaal; Bosch Veld, *Rehmann*, 5151!

EASTERN REGION: Transkei; banks of Bashee River, below 1000 ft., *Drège*, 3948! Pondoland; between Umtata River and St. Johns River, 1000-2000 ft., *Drège*, 4428! 4429! Natal, *Gerrard*, 702! *Wood*, 4953! *Kuntze*, 223!

A rice-field weed in all warm and warm-temperate regions of the Old World, also in Mexico.

16. **C. hæmatocephalus** (C. B. Clarke in Durand and Schinz, Conspect. Fl. Afr. v. 564, name only); glabrous, medium-sized; stems 1–2 ft. long, triquetrous at the top, with lateral shoots at the base; leaves often $\frac{2}{3}$ the length of the stem, $\frac{1}{5}$ in. broad; umbel 1–5 in. in diam., compound; spikes often aggregated near the end of the rays; bracts 3–4, similar to the leaves, lowest attaining 4–8 in. in length; spikelets 20–40, densely and globosely fascicled, chestnut-red, $\frac{1}{4}$ by $\frac{1}{12}$ in., strongly compressed, 8–12-flowered, hard, shining; glumes boat-shaped, 3–1-nerved, mucronate, regularly and closely imbricate; stamens 3; nut $\frac{1}{2}$ the length of the glume, trigonous, ellipsoid, narrowed at each end; style-branches 3, long.

KALAHARI REGION: Orange Free State, *Buchanan*, 124! Transvaal; Hooge Veld, *Rehmann*, 6779! 6783!

EASTERN REGION: Natal; Biggars Berg, 5000 ft., *Kuntze*, 241! and without precise locality, *Schlechter*, 6356! 6767!

17. **C. pulcher** (Thunb. Prod. 18, not of D. Don); glabrous, stout; rhizome (seen) short; stems approximate, 1–2$\frac{1}{2}$ ft. long, acutely triquetrous at the top; leaves often $\frac{2}{3}$ the length of the stem, $\frac{1}{4}$–$\frac{1}{3}$ in. broad; umbel compound or decompound, 2–5 in. in diam.; bracts 3–6, similar to the leaves, lowest overtopping the umbel; spikelets 3–5, digitate, rose-red, $\frac{1}{5}$ by $\frac{1}{12}$ in., moderately compressed, 8–12-flowered; glumes boat-shaped, ovate, shortly acute, nearly nerveless; stamens 3; anthers linear, not crested; nut nearly $\frac{1}{2}$ the length of the glume, triquetrous, pyramidal at either end; style not $\frac{1}{2}$ the length of the nut; branches 3, linear, exserted a little from the glume. *Fl. Cap. edit. Schult.* 100; *Nees in Linnæa*, x. 133,

excluding syn.; Kunth, Enum. ii. 114; *Boeck. in Linnæa,* xxxv. 583. *C. ingratus, Kunth, Enum.* ii. 31. *C. dregeanus, Kunth, Enum.* ii. 31.

SOUTH AFRICA: without locality, *Thunberg, Ecklon and Zeyher,* 10!

COAST REGION: Albany Div.; near Grahamstown, 1000–2200 ft., *Drège! MacOwan,* 652! Glenfilling, 1000 ft., *Drège!* King Williamstown Div.; near Toise River Station, 3600 ft., *Kuntze,* 273! 274! Komgha Div., *Flanagan,* 901! Queenstown Div.; Shiloh, 3500 ft., *Baur,* 884! 899!

CENTRAL REGION: Somerset Div.; near Somerset East, 3000 ft., *MacOwan,* 652! *Atherston,* 176!

KALAHARI REGION: Orange Free State, *Buchanan,* 79! Transvaal; Pretoria, at Wonderboom Poort, *Rehmann,* 4482! and without precise locality! *Rehmann,* 4833!

EASTERN REGION: Transkei; Bashee River, 1000 ft., *Drège!* Tembuland; Bazeia, 2500 ft., *Baur,* 361! between Bashee River and Morley, *Drège!*

18. C. sphærospermus (Schrad. Anal. Fl. Cap. 8); glabrous, medium-sized; rhizome horizontal, thick, woody, often several in. long, with ovate scales; stems 4–20 in. long, slender and obtusely trigonous at the top, smooth; leaves $\frac{2}{3}$ the length of the stem, $\frac{1}{8}$–$\frac{1}{6}$ in. broad; umbel compound or decompound, 1–6 in. in diam.; bracts 3–5, similar to the leaves, lowest overtopping the umbel; spikelets 3–7, digitate, brown, brown-red or dusky straw-colour, up to $\frac{1}{2}$ by $\frac{1}{6}$–$\frac{1}{8}$ in., much compressed, 16–30-flowered; glumes boat-shaped, elliptic, obscurely 3-nerved, varying greatly in the closeness of packing; keel excurrent in a very short obtuse mucro; stamens 3; anthers linear, not crested; nut $\frac{1}{4}$–$\frac{1}{3}$ the length of the glume, globose, triquetrous, pale-brown (in this species as in some of its neighbours great numbers of the nuts are infertile); style not $\frac{1}{2}$ the length of the nut; branches 3, linear, exserted from the glume. *Nees in Linnæa,* viii. 78, x. 133; *Kunth, Enum.* ii. 106; *Krauss in Flora,* 1845, 754. *C. corymbosus, Steud. in Flora,* 1829, 153, *not of Rottb. C. flavissimus, Steud. in Flora,* 1829, 152, *not of Schrader. C. denudatus, Boeck. in Linnæa,* xxxv. 576 *partim; C. B. Clarke in Journ. Linn. Soc.* xxi. 124. *C. tristis, d, Drège, Pflanzengeogr. Documente* 177. *C. arrhizus, Boeck. ex C. B. Clarke in Durand and Schinz, Conspect. Fl. Afr.* 577.

VAR. β, **triqueter** (Boeck. ex C. B. Clarke in Durand and Schinz, Conspect. Fl. Afr. v. 578); stems very rigid, acutely trigonous; leaves half as long as the stem, upper 3 in. long, trigonous, solid; bracts as long as the umbels; spikelets (not ripe) small, slender.

SOUTH AFRICA: without locality, *Drège,* 1597! 2029! 2466! *Harvey,* 384! *Ecklon and Zeyher,* 21! 128! 130!

COAST REGION: Clanwilliam Div., *Zeyher!* Cape Div.; Table Mountain, *Ecklon,* 890! Zeekoe Vallei, *Ecklon,* 82! Cape Flats, *Schlechter,* 231! Paarl Div.; Berg River in Klein Drakenstein Mountains, below 500 ft. (mixed with *C. denudatus*), *Drège!* Swellendam Div.; Zonder Einde River, *Burchell,* 7514! Humansdorp Div.; Diep River, *Bolus,* 2691! Uitenhage Div.; by the Zwartkops River, *Burchell,* 4432! *Gill!* Bathurst Div.; near Theopolis, *Burchell,* 4095! Albany Div.; Glenfilling, *Drège,* 3960! Komgha Div.; near the mouth of the Kei River, *Flanagan,* 1785!

KALAHARI REGION: Bechuanaland; Eastern Bamanguato Territory, *Holub!* Transvaal; Houtbosh, *Rehmann,* 5645! Bosch Veld, between Kameels Poort and Elands River, *Rehmann,* 4800! Var. β, Transvaal, *Fehr.*

EASTERN REGION: Natal; Umlaas River, *Krauss*, 215! and without precise locality, *Buchanan*, 105! 123! *Gerrard*, 700! *Drège*, 4436! Delagoa Bay, *Speke*, 2! *Menyhart*, *Junod*, 87! *Kuntze*, 220!

Also in Tropical Africa and Madagascar.

This species hardly differs from *C. denudatus*, but by the evolution of leaves; Var. β is intermediate between the two species, and might be interpreted to indicate that *C. sphærospermus* is only the leafy state of *C. denudatus*—as Kunth treated it in his naming of Drège's collection.

19. C. denudatus (Linn. f. Suppl. 102); glabrous, stout; rhizome horizontal, thick, woody, often several in. long, with ovate scales; stems 2 ft. long, acutely triquetrous at the top; leaves none, uppermost sheath produced on one side 1–3 in., greenish, acute; umbel compound or decompound, 1–6 in. in diam.; bracts 2, lower erect not exceeding the umbel, bayonet-tipped; spikelets (glumes, stamens, style, nut) as in *C. sphærospermus*.—*Steud. in Flora*, 1829, 152; *Schrad. Anal. Fl. Cap.* 8; *Nees in Linnæa*, vii. 514, x. 134; *Kunth. Enum.* ii. 36; *Boeck. in Linnæa*, xxxv. 576 *partim. C. tristis* α, *Kunth, Enum.* ii. 35. *C. spretus, Steud. Syn. Pl. Glum.* ii. 21. *C. amphibolus, Steud. Syn. Pl. Glum.* ii. 22. *C. lanceus, Schultes in Roem. et Schultes, Syst.* ii. *Mant.* 106, *not of Thunb.*

SOUTH AFRICA: without locality, *Wallich!* *Sieber*, 109! *Harvey*, 361! *Ecklon and Zeyher*, 14! 15!

COAST REGION: Clanwilliam Div.; Gift Berg, 1500–2500 ft., *Drège!* between Piquiniers Kloof and Oliphants River, 1000–1500 ft., *Drège*, Cape Div.; Table Mountain, *Ecklon*, 888! *Burke*, 329! Paarl Div.; by the Berg River, *Drège!* Worcester Div.; mountains above Worcester, *Rehmann*, 2578! Caledon Div.; Grabouw, near Palmiet River, 700 ft., *Bolus*, 4232! Genadendal and Zonder Einde River, below 1000 ft., *Drège*. Riversdale Div.; Great Vals River, *Burchell*, 6535! Uitenhage Div.; near the mouth of the Zwartkops River, *Drège*. Albany Div.; near Grahamstown, 1000–2000 ft., *Drège*, *MacOwan*, 1363! Glenfilling, *Drège!* King Williamstown Div.; 2000 ft., *Kuntze*, 269! Komgha Div.: *Flanagan*, 976!

KALAHARI REGION: Orange Free State, *Buchanan*, 108! Transvaal; Houtbosch, *Rehmann*, 5643!

EASTERN REGION: Tembuland; between Bashee River and Morley, *Drège*, 4408! Natal; Inanda, 1200 ft., *Wood*, 1398! and without precise locality, *Buchanan*, 313! 321!

Also in South Tropical Africa and Madagascar.

Spikelets very variable in length, usually $\frac{1}{4}$–$\frac{1}{3}$ in., sometimes $\frac{3}{4}$–1 in.; not varying much in breadth.

20. C. marginatus (Thunb. Prod. 18); glabrous, stout; rhizome horizontal, thick, woody, often several in. long, with ovate scales; stems 1–2$\frac{1}{2}$ ft. long, terete at the top; leaves none; uppermost sheath produced $\frac{1}{4}$–1 in., greenish; umbel compound or simple or congested into 1 head, 1–5 in. in diam.; bracts 2, lower suberect, much shorter than the umbel; spikelets 3–7, digitate (in Baur, 886, nearly all solitary), chestnut-brown, glistening, up to $\frac{3}{4}$ by $\frac{1}{8}$ in. (usually less than $\frac{1}{2}$ in.), 6–38-flowered, strongly compressed, hard; glumes boat-shaped, elliptic, closely imbricate, scarcely nerved; keel excurrent in a short point; margins often glistening, hyaline; stamens 3; anthers linear, not crested; nut $\frac{2}{5}$ the length of the glume, ellipsoid, trigonous, brown; style as long as the nut; branches 3, about as long as the style. *Fl. Cap. ed. Schult.* 100; *Nees in Linnæa*, x. 134; *Boeck. in Linnæa*, xxxv. 571. *C. blandus*, *Kunth, Enum.* ii. 36; *Boeck. in Linnæa*, xxxv, 570. *C. fonticola*, *Kunth, Enum.* ii. 36. *C. prolifer*, *Nees in Linnæa*, vii. 494, *at least in part.* *C. prionodes*, *Steud. Syn. Pl. Glum.* ii. 22.

SOUTH AFRICA: without locality, *Thunberg*, *Zeyher*, 1747! *Ecklon and Zeyher*, 16! 135!

COAST REGION: Clanwilliam Div.; Bosch Kloof, 1000 ft., *Drège.* Worcester Div.; mountains near Hex River, 2500 ft., *Bolus*, 5273! Port Elizabeth Div.; *E.S.C.A. Herb.* 237! Komgha Div.; by the Kei River, 500 ft., *Drège.* Cathcart Div.; Blesbok Flats, 3000–4000 ft., *Drège.* Queenstown Div.; Shiloh, 3500 ft., *Baur*, 886! Klaarsmits River, 3500 ft., *Baur*, 983!

CENTRAL REGION: Calvinia Div.; Lospers Flats. 3000–4000 ft., *Zeyher*, 1753! Beaufort West Div.; Nieuw Veld Mountains, 3000–5000 ft., near Beaufort West, *Drège!* Somerset Div.; near the Little Fish River, 2300–2800 ft., *MacOwan*, 1916! 1955! 2019! 2019b! Richmond Div., vicinity of Styl Kloof, near Richmond, 4000–5000 ft., *Drège!* Colesberg Div., *Shaw*, 20! Albert Div.; near the Orange River, *Cooper*, 1373! Burghersdorp, *Flanagan*, 1667! Aliwal North Div.; bank of the Orange River, near Aliwal North (Buffel Vallei), *Drège*, 4407! Leeuwen Spruit, between Kraai River and the Witte Bergen, 4500 ft., *Drège*, 7392!

WESTERN REGION: Little Namaqualand; between Verleptpram and the mouth of the Orange River, 1000 ft., *Drège!*

KALAHARI REGION: Griqualand West; Lower Campbell, *Burchell*, 1814! 1815! Kimberley, *Marloth*, 874! Orange Free State; near Kaffir Fontein, *MacLea in Herb. Bolus*, 5815! near Winburg, *Buchanan*, 252! Caledon River, *Burke!* and without precise locality, *Buchanan*, 125! Basutoland, *Zeyher*, 8! Bechuanaland; near the source of Kuruman River, *Burchell*, 2454! Transvaal; Hooge Veld, *Rehmann*, 6635! Vaal River, *Nelson*, 198!

EASTERN REGION: Natal; Colenso, *Kuntze*, 239! and without precise locality, *Drège*, 4407.

Also in Angola, and at Walfisch Bay.

21. C. textilis (Thunb. Prod. 18); glabrous, tall, leafless; rhizome horizontal, woody, $\frac{1}{6}$ in. diam., with ovate, acute, black striations, $\frac{3}{4}$ in. long; stem up to 2–3 ft. long; top terete or round trigonous, smooth not striate; uppermost sheath produced scarcely an in., shortly acute; bracts 8 or 10, subequal, overtopping the umbel, lanceolate (not caudate) at the tip; umbel compound; rays 8 or 10; bracteoles to the secondary umbels or corymbs inconspicuous, not $\frac{1}{4}$ in. long; spikelets 2–7, digitate, rusty pale red, compressed, up to $\frac{1}{3}$ by $\frac{1}{10}$ in., 6-16-flowered; glumes imbricate, boat-shaped, subacute, irregularly ribbed; stamens 3; anthers linear, hardly crested; nut $\frac{1}{2}$ the length of the glume, triquetrous, ellipsoid or obovoid, pyramidal at each end, brown; style much shorter than the nut; branches 3, linear, shortly exserted from glume. *Fl. Cap. ed. Schult.* 100; *Nees in Linnæa*, vii. 513; x. 135; *Kunth, Enum.* ii. 32; *Boeck. in Linnæa*, xxxv. 567. *C. Burchellii*, *Schrad. Anal. Fl. Cap.* 11; *Nees in Linnæa*, x. 135; *Kunth, Enum.* ii. 33. *C. Smithii*, *Schrad. Anal. Fl. Cap.* 12 *in Obs.*; *Nees in Linnæa*, x. 135; *Kunth, Enum.* ii. 115. *C. asperifolius*, *Desfont. Cat. Hort. Paris ed.* 3, 387; *Kunth, Enum.* ii. 114. *C. punctorius*, *Schrader ex Boeck. in Linnæa*, xxxv. 567.

SOUTH AFRICA: without locality, *Thunberg, Thom*, 890! *Bergius*, 171! *Ecklon and Zeyher*, 12! 129!

COAST REGION: Clanwilliam Div.; Ebenezer, *Drège*. Cape Div.; Cape Flats, *Rehmann*, 1775! *Burke! Bolus*, 4860! Salt River; *Burchell*, 507! 687! Tygerberg, below 1000 ft., *Drège!* Wynberg, *Schlechter*, 508! Paarl Div.; Berg River, near Paarl, *Drège*, 7389! Worcester Div.; near Mord Kuil, on the Doorn River, below 1000 ft.; *Drège*. Riversdale Div.; Kaffirkuils River, *Gill!* Uitenhage Div.; *Ecklon and Zeyher*, 170! Bathurst Div.; near Theopolis, *Burchell*, 4055!

This is a colonist near Bordeaux, and has been collected in Algiers by Munby. Cultivated in Mauritius, according to Bojer.

22. **C. sexangularis** (Nees in Linnæa, ix. 284, x. 135); stem at the top, and nearly its whole length, 6-angled, i.e. triquetrous with a line or rib down the centre of each face; otherwise as *C. textilis, Thunb.—Kunth, Enum.* ii. 32; *Krauss in Flora*, 1845, 754; *Boeck. in Linnæa*, xxxv. 568. *C. webbianus, Steud. Syn. Pl. Glum.* ii. 20.

SOUTH AFRICA: without locality, *Zeyher*, 1746! *Ecklon and Zeyher*, 11!

COAST REGION: King Williamstown Div.; Yellowwood River, 1000–2000 ft., *Drège*! Bedford, *MacOwan*, 1668!

CENTRAL REGION: Somerset Div.; Fish River, *Burke!*

WESTERN REGION: Little Namaqualand, *Scully!*

KALAHARI REGION: Transvaal; Pretoria, *Rehmann*, 4775!

EASTERN REGION: Transkei; Bashee River, below 1000 ft., *Drège*, 4443! 4444! Natal; Umlaas River, *Krauss*, 46! near Pieter Maritzburg; *Krauss*, 357! Sydenham, 300 ft., *Wood*, 4029! and without precise locality, *Buchanan*, 126!

Also in Tropical Africa.

In Wood, 4029, from Natal, the stem is trigonous, irregularly or obscurely 6-angled, and resembles much that of the nearly-allied *C. flabelliformis*, Rottb.; but the spikelets in Wood, 4029, are rather those of *C. sexangularis*.

23. **C. isocladus** (Kunth, Enum. ii. 37); glabrous; rhizome creeping, $\frac{1}{8}$–$\frac{1}{6}$ in. in diam., with ovate brown scales $\frac{1}{2}$ in. long; stems 1–2$\frac{1}{2}$ ft. long., triquetrous at the top, often minutely scabrous; leaves 0; uppermost sheath produced on one side an in., hardly acute, not green; bracts very much shorter than the umbel; umbel simple; rays 50–100, 1–4 in. long, subequal, slender; spikelets 1–5, digitate, up to $\frac{1}{2}$ by $\frac{1}{12}$ in., compressed, brown or dirty straw-coloured, 6–12-flowered; glumes ovate, truncate, 3-nerved; keel hardly excurrent as a point; stamens 3; anthers linear-oblong, not crested; nut minute, scarcely $\frac{1}{4}$ the length of the glume, trigonous, obovoid, white or pale-brown; style slender about as long as the nut; branches 3, linear, shortly exserted from the glume. *C. æqualis, Krauss in Flora*, 1845, 754. *C. æqualis, var. β, Boeck.! in Peters, Mossamb.* 538, *and in Linnæa*, xxxv. 578. *C. jocladus, Drège and E. Meyer, Pflanzengeogr. Documente* 177. *C. iocladus, Drège and E. Meyer, l.c.* 151, 157. *C. esculentus, Drège l.c.* 177.

EASTERN REGION: Pondoland; between Umtentu River and St. Johns River, *Drège*, 4430! 4431! 4432! *Bachmann*, 94! 95! 96! 97! Natal; Coast-land, *Sutherland!* ponds near Umlaas River, *Krauss*, 165! Clairmont, 100 ft.; *MacOwan and Bolus, Herb. Norm. Aust. Afr.*, 1032! Durban, *Rehmann*, 8623! Umhlanga Valley, *Wood*, 1331! and without precise locality; *Buchanan*, 106! 360!

C. prolifer, Lam. (*C. æqualis*, Vahl), is a frequent plant in South-east Tropical Africa and the Mascarene Islands, and has the stem trigonous or nearly terete at the top. From this *C. isocladus* only differs in its triquetrous stem. Boeckler may be judicious in uniting the two; but several of the preceding species are kept distinct on the same single character, viz. stem triquetrous or stem roundish.

24. C. albostriatus (Schrad. Anal. Fl. Cap.7); glabrous; stolons clothed by dark brown, elliptic scales, hardening into woody rhizomes $\frac{1}{8}$–$\frac{1}{6}$ in. in diam.; stems 8–20 in. long, rather slender but tough, trigonous at the top; leaves numerous, all near the base of the stem, as long as the stem or only $\frac{1}{2}$ that length, $\frac{1}{3}$–$\frac{3}{4}$ in. broad, grass-like, with 2 lateral, white, more or less prominent nerves; tip lanceolate, not caudate; umbel compound; rays often 8–24, slender, 1–4 in. long, several usually nearly equal; bracts about 8, overtopping the umbel, several of the lower nearly equal, similar to the leaves; spikelets often all solitary, pedicelled, but in some examples many 2–4-digitate occur, $\frac{1}{3}$ by $\frac{1}{16}$–$\frac{1}{12}$ in., compressed, pale-brown, 8–24-flowered; glumes ovate, acute or obtuse, not keeled, many-nerved; margins scabrous; stamens 3; anthers linear, not crested; nut $\frac{3}{4}$ the length of the glume, trigonous, ellipsoid, brown; style scarcely $\frac{1}{3}$ the length of the nut; branches 3, linear, rather short. *Kunth, Enum.* ii. 34; *Krauss in Flora*, 1845, 754; *Boeck. in Linnæa*, xxxv. 581. *C. Mariscus, Nees in Linnæa*, vii. 515, x. 134; *Kunth, Enum.* ii. 106. *C. pulcher, Spreng. in Flora*, 1829, *Band* i. *Beil.* 5; *not of Thunb.*

SOUTH AFRICA: without locality, *Zeyher! Drège*, 3960 partly! *Ecklon and Zeyher*, 13!

COAST REGION: Port Elizabeth, *E.S.C.A. Herb.*, 189! Alexandria Div.; Enon, below 1000 ft., *Drège*, 2028! Albany Div.; Glenfilling, *Drège*, 3961! Komgha Div., *Flanagan*, 986! King Williamstown Div.; near Toise River Station, 3500 ft., *Kuntze*, 275! Bathurst Div.; Port Alfred, *Hutton*, 60!

CENTRAL REGION: Somerset Div.; Bosch Berg, *Burchell*, 3133! *MacOwan*, 1690!

KALAHARI REGION: Transvaal; Houtbosch, *Rehmann*, 5659!

EASTERN REGION: Tembuland, Bazeia, 2500–3000 ft., *Baur*, 578! Pondoland, *Bachmann*, 85! Griqualand East; Umzimkulo Div., at Emyembi, 5000 ft. *Tyson*, 2546! Natal; Coast-land, *Sutherland!* Clairmont, *Kuntze*, 260! Nottingham, 4000–5000 ft., *Buchanan*, 129! near Durban, *Wood in Natal Governm. Herb.*, 749! *Buchanan*, 48! Umbilo River, *Rehmann*, 8093! Howick, *Junod*, 307! Inanda, *Wood*, 1088! and without precise locality, *Buchanan*, 361! *Krauss* 10! 230!

25. C. prasinus (Kunth, Enum. ii 31); leaves 20 by $\frac{1}{5}$ in., acuminate, tough, many-nerved; bracts narrow, acuminate similar to the leaves; otherwise as *C. albostriatus, Schrad.—Boeck. in Linnæa*, xxxv. 584.

COAST REGION: East London, *Kuntze*, 271!

KALAHARI REGION: Transvaal; Houtbosch, *Rehmann*, 5658!

EASTERN REGION: Pondoland, *Drège*, 4440!

Boeckler distinguishes this plant from *C. albostriatus* by its being monandrous; I find three stamens in the lower, and two in the upper flowers. Moreover, the nut, and the remarkable scabrous margins to the glumes, are the same in both species. Indeed *C. prasinus* has been reduced to *C. albostriatus* in herb. Kew, though the broad thin leaves of the latter differ at first sight widely from the narrow tough leaves of *C. prasinus*.

26. **C. leptocladus** (Kunth, Enum. ii. 32); glabrous; rhizome woody; stems 12–20 in. long, rather slender, trigonous at the top; leaves often nearly as long as the stem, $\frac{1}{5}$ in broad, more or less 3-nerved; umbel lax, compound, 3–8 in. in diam., irregular, sometimes proliferous; bracts about 8, somewhat overtopping the umbel, similar to the leaves; spikelets mostly solitary, pedicelled, $\frac{1}{4}$ by $\frac{1}{12}$ in. compressed, 8–14-flowered, greenish, ultimately dirty straw-coloured; glumes boat-shaped, obscurely 3-nerved, green on the back, keel subexcurrent in a mucro, margins broadly scarious not scabrous on the edge; stamens 3; anthers linear-oblong, hardly crested; nut $\frac{3}{4}$ the length of the glume, ellipsoid, trigonous, brown; style $\frac{1}{3}$ the length of the nut, branches 3 shortish. *Boeck. in Linnœa*, xxxv. 581 *partly*.

EASTERN REGION: Natal; Coast-land, 0–1000 ft., *Sutherland!* Durban, *Kuntze*, 227! and without precise locality, *Drège*, 4441! *Buchanan*, 318!

In Durand and Schinz, Conspect. Fl. Afr. v. 566, I have united my *C. Balfourii* (Journ. Linn. Soc. xx. 289) and some other Mascarene examples with *C. leptocladus*. The type example of *C. Balfourii* does not exactly match *C. leptocladus*, having narrower spikelets, and the other Mascarene "*C. leptocladus*" differs much more.

27. **C. subchoristachys** (C. B. Clarke); glabrous; roots fibrous; stems 12–20 in. long tufted, slender, trigonous at the top; leaves very long, often overtopping the stem, $\frac{1}{8}$–$\frac{1}{6}$ in. broad; bracts 3–4, patent, similar to the leaves, lowest up to 6–8 in. long; umbel in appearance simple, rays 3–4 up to $1\frac{1}{2}$ in. long; spikelets 8–20, clustered without bracts at the head of each ray, forming a short spike or compound close head; spikelets $\frac{1}{3}$ by $\frac{1}{16}$ in. 8–14-flowered; glumes ovate-oblong, obtuse, distant, not imbricate in fruit, green, 3–5 nerved on the back, sides ferruginous 2-nerved, margins broad scarious; rhachilla with narrow, oblong, scarious wings; stamens 3; style 3-fid; nut $\frac{2}{3}$ the length of the glume, obovoid, triquetrous, black.

COAST REGION: Cape Town, *Spielhaus*! in herb. Luebeck. No specimen at Kew.

This plant is closely allied to the ubiquitous *C. diffusus*, Vahl. It has much narrower leaves and bracts, the spikelets are *not* in few-flowered digitate clusters and the glumes more obtuse.

28. **C. latifolius** (Poir. in Lam. Encyc. vii. 268); glabrous, large; stolons long, $\frac{1}{8}$ in. in diam., clothed with sheathing, lanceolate scales, often more than 1 in. long; stems often 3 ft. long, triquetrous at the top, $\frac{1}{5}$ in. in diam. (slenderer in var. β); leaves 18 by $\frac{3}{4}$ in. thick, flat in the typical form (but see var. β); umbel 6–12 in. in diam., somewhat dense, compound, with numerous spikes; bracts 3–4, overtopping the umbel, similar to the leaves; spikelets 4–12 to a spike, lower often $\frac{1}{12}$ in. apart on the sub-flexuose, glabrous rhachis, $\frac{3}{4}$ by

$\frac{1}{10}$ in., compressed, pale or reddish, 8–16-flowered; rhachilla not winged; glumes loosely overlapping, hardly keeled, ovate-oblong, very obtuse, obscurely nerved, margins and tip scarious; stamens 3; anthers linear, not crested; nut $\frac{1}{2}$ the length of the glume, trigonous, narrow-obovoid, brown; style $\frac{3}{4}$ the length of the nut, branches 3 linear, moderately exserted from the glume. *Kunth, Enum.* ii. 75; *Krauss in Flora*, 1845, 754; *Boeck. in Linnæa*, xxxv. 602; *C. B. Clarke in Journ. Linn. Soc.* xxi. 152; *not of Decaisne. C. scoparius, Poir. in Lam. Encyc.* vii. 253; *Kunth, Enum.* ii. 75; *Willd. MS. not of Decaisne (see Presl. in Oken, Isis* xxi. 271).

VAR. β, **angustifolia** (Krauss in Flora, 1845, 754); slenderer in all parts, leaves 15 by $\frac{1}{4}$ in. *C. retusus, Nees in Linnæa,* ix. 285; *Kunth, Enum.* ii. 115. *C. pilosus, var. mutica, Boeck. ex Clarke in Durand and Schinz Conspect. Fl. Afr.* v. 566.

KALAHARI REGION: Orange Free State; between Harrismith and Leribe, *Buchanan*, 215! Transvaal; Nylstroom River, Nelson 2*! Var. β, Transvaal; Houtbosch, *Rehmann*, 5660!

EASTERN REGION: Transkei; Bashee River, *Drège*, 4442! Griqualand East; near Clydesdale, 2500 ft., *Tyson*, 2867! Natal; Umlaas River, *Krauss*, 79! Durban Flat, *Buchanan*, 5! *Wood*, 5808! and without precise locality, *Buchanan*, 314! Delagoa Bay, *Scott!* Var. β, Natal; swamps near Umlaas River, *Krauss*, 211! and without precise locality, *Buchanan*, 110!

Also in South Tropical Africa and the Mascarene Islands. Var. β, in Madagascar.

29. **C. distans** (Linn. f. Suppl. 103); glabrous, large or medium-sized; stolons $\frac{1}{20}$ in. in diam., clothed by black-brown acute scales $\frac{1}{3}$ in. long, but as in other stoloniferous species by no means always produced by the plant, rarely present in herbaria; stems 12–25 in. long, trigonous at the top, smooth, somewhat slender; leaves nearly as long as the stem, $\frac{1}{5}$–$\frac{1}{3}$ in. broad; umbel usually 4–8 in. in diam. compound (small examples with condensed umbel occur); bracts 3–6, overtopping the umbel, similar to the leaves; spikelets loosely spicate, compressed, pale or reddish, often $\frac{3}{4}$ by $\frac{1}{20}$–$\frac{1}{12}$ in., very variable in breadth, and up to $1\frac{3}{4}$ in. in Rehmann, 7752, 10–20-flowered; rhachilla narrowly winged, the linear-lanceolate, scarious-white wings finally dehiscing from the base (exactly as in *C. rotundus, fastigiatus, &c.*); glumes distant, hardly boat-shaped, elliptic, obtuse, 3-nerved, green on the back; stamens 3; anthers linear-oblong, not crested; nut $\frac{1}{2}$–$\frac{2}{3}$ the length of the glume, trigonous, oblong or somewhat obovoid, brown; style much shorter than the nut, branches 3 linear, a little exserted from the glume. *Beauv. Fl. Owar.* i. 35, *t.* 20; *Kunth, Enum.* ii. 93; *Boeck. in Linnæa*, xxxv. 612; *C. B. Clarke in Journ. Linn. Soc.* xxi. 144, *and in Hook. f. Fl. Brit. Ind.* vi. 607. *C. elatus, Presl. in Oken, Isis*, xxi. 271; *Boeck. in Flora*, 1879, 551; *not of Linn.*

COAST REGION: Komgha Div.; *Flanagan*, 1264!

KALAHARI REGION: Transvaal; Cave Mountains, *Nelson*, 95! Houtbosch, *Rehmann*, 5656!

EASTERN REGION: Tembuland; between Morley and the Umtata River, *Drège*, 4434! 4435! Natal; Clairmont, *Kuntze*, 231! Durban, *Buchanan*, 19!

Camperdown, *Rehmann.* 7752! and without precise locality, *Buchanan,* 116! 117!

Found in all tropical and warm-temperate regions.

30. C. compressus (Linn. Sp. Plant. ed. ii. 68); annual, glabrous, green; stems 4–16 in. long, trigonous at the top, smooth; leaves about as long as the stem, $\frac{1}{8}$–$\frac{1}{5}$ in. broad, grass-like; umbel simple, 3–10 in. in diam., or sometimes of 1 head; bracts 3–4, overtopping the umbel, similar to the leaves; spikelets 3–10, closely spicate, much compressed, $\frac{1}{2}$–$\frac{3}{4}$ by $\frac{1}{8}$–$\frac{1}{6}$ in., 4–30-flowered, green or green with red patches, ultimately straw-coloured; rhachilla not winged; glumes closely imbricated, boat-shaped, ovate, acute, conspicuously many-nerved; stamens 3; anthers linear-oblong, not crested; nut $\frac{1}{3}$ the length of the glume, broadly obovoid, triquetrous with concave faces; style shorter than the nut, branches 3 linear, shortly exserted from the glume. *Kunth, Enum.* ii. 23; *Boeck. in Linnæa,* xxxv. 517; *C. B. Clarke in Journ. Linn. Soc.* xxi. 97, *and in Hook. f. Fl. Brit. Ind.* vi. 605; *not of Jacquin. C. caffer, G. Bertol. in Rendiconto Ist. Bologna,* 1853–4, 33, *and in Mem. Acad. Sci. Bologna,* v. 464, *t.* 23.

EASTERN REGION: Kaffraria, *Schultz!* Natal; Coast-land, *Sutherland!* Umlazi River, below 200 ft., *Drège,* 4396! and without precise locality, *Gerrard,* 704! *Kuntze,* 228! Delagoa Bay, *Kuntze,* 218!

Found in all tropical and warm-temperate regions.

31. C. aristatus (Rottb. Descr. et. Ic. 23, t. 6, fig. 1); annual, glabrous; stems tufted, 1–6 in. long, trigonous at the top, smooth; leaves about as long as the stem, $\frac{1}{12}$–$\frac{1}{10}$ in. broad; umbel simple or reduced to 1 head, up to 5 in. in diam.; bracts 3–5, overtopping the umbel, similar to the leaves; spikelets 5–40, closely spicate, compressed, green or brown, $\frac{1}{2}$ by $\frac{1}{6}$ in., 6–24-flowered; rhachilla not winged; glumes boat-shaped, ovate-lanceolate, 7–9-ribbed over nearly their entire width, keel green produced as a recurved bristle; stamen 1; anther oblong, not crested; nut $\frac{1}{3}$ as long as the glume (bristle included), trigonous, narrow-obovoid, dull brown; style shorter than the nut, branches 3 linear, shortly exserted from the glume. *Kunth, Enum.* ii. 23; *Steud. in Flora,* 1842, 585; *Boeck. in Linnæa,* xxxv. 500; *C. B. Clarke in Journ. Linn. Soc.* xxi. 91, *and in Hook. f. Fl. Brit. Ind.* vi. 606. *C. squarrosus, Linn. Amoen. Acad.* iv. 303; *Sp. Plant. ed.* ii. 66 *partly. Scirpus intricatus, Linn. Mant.* 182; *Thunb. Prod.* 18; *Fl. Cap. ed. Schult.* 98.

SOUTH AFRICA: without locality, *Thunberg, Fleck,* 817!

KALAHARI REGION: Transvaal; Hooge Veld, between Porter and Trigards Fontein, *Rehmann,* 6647! and without precise locality, *McLea in Herb. Bolus,* 6016!

Found in all tropical and warm-temperate regions.

32. C. usitatus (Burchell, Trav. S. Afr. i. 417 in note); glabrous; stolons $\frac{1}{20}$ in. in diam., arising from the stem-base, elongate, clothed by pale linear scales $\frac{1}{2}$ in. long, producing bulbs at their extremities, then withering; bulbs globose or ovoid, $\frac{1}{2}$–$\frac{3}{4}$ in. long, dark-brown, clothed by numerous imbricate scales, outer striate, inner shining smooth, in germination producing a simple shoot from the vertex (1–3 by $\frac{1}{16}$ in. in the dried examples), which grows erect to near the surface of the soil where it divides and forms the new stem-base; stem 4–10 in. long, slender, trigonous at the top, smooth; leaves overtopping the stems, narrow with whip-like ends but much widened (usually $\frac{1}{5}$ in. wide) close to the base; umbel $\frac{3}{4}$–$1\frac{1}{2}$ in. in diam., with 6–30 spikelets, sub-corymbose, very imperfectly umbelled, lowest ray often $\frac{1}{8}$–$\frac{1}{6}$ in. below the next; bracts 3–2, longer than the umbel, similar to the leaves; spikelets $\frac{1}{3}$–$\frac{1}{2}$ by $\frac{1}{8}$ in., 8–12-flowered, compressed, shining dark-red; rhachilla with hyaline wings; glumes hardly keeled, ovate, subacute, strongly 9-ribbed; stamens 3; anthers linear, not crested; nut (only seen young, perhaps seldom perfecting fruit) small, trigonous, obovoid; style shorter than the nut, branches 3 linear, long, much exserted from the glume. *Nees in Linnæa*, vii. 516, ix. 285, *excluding syn.*, x. 136; *Kunth, Enum.* ii. 107; *Boeck. in Linnæa*, xxxv. 511; *C. B. Clarke in Journ. Linn. Soc.* xxi. 176, *excluding syn. C. semitrifidus. C. solidus, Kunth, Enum.* ii. 76 *partly. C. bulbifex, E. Meyer MS.*; *Drège in Linnæa*, xx. 245, *and Pflanzengeogr. Documente* 177, (*bulbifer*) 63.

COAST REGION: Paarl Div.; Paarl Mountains, 2000–3000 ft., *Drège*, Uitenhage Div.; *Ecklon and Zeyher!* Bathurst Div.; Port Alfred, *Hutton*, 42.

CENTRAL REGION: Prince Albert Div.; between the Dwyka River and Zwartbulletje, *Drège!* Somerset Div.; banks of the Little Fish River, *MacOwan*, 2035! Somerset East, *Atherston*, 57! Hopetown Div.; between "Bare Station and Gnu Halt," *Burchell*, 2684! Albert Div.; Burghersdorp, *Cooper*, 1376! 3366! Aliwal North Div.; bank of the Orange River, near Aliwal North, 4300 ft., *Drège!* Colesberg Div.; near Colesberg, *Shaw*, 6!

WESTERN REGION: Little Namaqualand, *Drège*, 4391c!

KALAHARI REGION: Griqualand West; between Griqua Town and the Asbestos Mountains, *Burchell*, 1990! 2082! Orange Free State; Caledon River, *Zeyher*, 1743! *Burke*, 302! Draai Fontein, *Rehmann*. 3084! Bloem Fontein, *Kuntze*, 245! Transvaal; Bamboes Spruit, *Nelson*, 97! and without precise locality, *McLea in Herb. Bolus*, 6013!

33. C. esculentus (Linn. Sp. Plant. ed. ii. 67); glabrous; stolons $\frac{1}{16}$ in. in diam., clothed by pale ferruginous lanceolate scales $\frac{1}{3}$ in. long, terminated often by zonate, ellipsoidal, woody tubers 1 in. long, not hardening ultimately into tough woody rhizomes; stems 8–16 in. long, erect at the base, rather slender at the top, trigonous, smooth; leaves often as long as the stem, $\frac{1}{5}$ in. broad; umbel 2–6 in. in diam., usually once compound; bracts 3–4, overtopping the umbel, similar to the leaves; spikelets spicate, $\frac{1}{3}$–$\frac{1}{2}$ by $\frac{1}{10}$ in., compressed, pale, often somewhat golden, 8–14-flowered; rhachilla scarious-winged; glumes ovate, lightly keeled, subobtuse, conspicuously (in dried plants) 7–9-ribbed; stamens 3; anthers linear, not crested; nut scarcely half the length of the glume, trigonous, obovoid; style much shorter than the nut, branches 3 linear, long. *Kunth, Enum.* ii. 61; *Boeck. in Linnæa*, xxxvi. 287; *C. B. Clarke in Journ. Linn. Soc.* xxi. 178, *and in Hook. f. Fl. Brit. Ind.* vi. 616. *C. Tenorii*,

Presl, Fl. Sicul. xliii.; *Krauss in Flora*, 1845, 754. *C. retusus, Nees ex Krauss in Flora*, 1845, 754. *C. Buchanani, Boeck. Cyp. Novæ*, i. 1888, 4.

KALAHARI REGION: Orange Free State; between Harrismith and Leribe, *Buchanan*, 212! Transvaal; Pretoria, *Rehmann*, 4776! Houtbosch, *Rehmann*, 5654!

EASTERN REGION: Griqualand East; marshes near Clydesdale, 2500 ft., *Tyson*, 2593! Kaffir Kraal, *Wood*, 1581! near Umlaas River, *Krauss*, 97! Colenso, *Rehmann*, 7147! and without precise locality, *Buchanan*, 83! 316!

In all tropical and warm-temperate regions, except Malaya, Australia, and Oceania.

34. **C. corymbosus** (Rottb. Descr. et Ic. 42, t. 7, fig. 4); glabrous; stolons stout, becoming a thick woody rhizome; stems 2–3 ft. long, stout, trigonous at the top, very obscurely transverse-septate; leaves 0, or uppermost short; rays of umbel up to 3–6 in. long; bracts 3–4, usually 1½–3 in. long, suberect, keeled; spikelets numerous, loosely or closely spicate, ⅓–1 by $\frac{1}{16}$ in., compressed, 10–18-flowered, straw-coloured with red marks; rhachilla scarious-winged; glumes ovate, subobtuse, hardly keeled; stamens 3; anthers linear, not crested; nut scarcely ½ the length of the glume, trigonous, obovoid; style shorter than the nut, branches 3 linear, longish. *C. B. Clarke in Journ. Linn. Soc.* xx. 292, xxi. 158, *and in Hook. f. Fl. Brit. Ind.* vi. 612. *C. articulatus, Kunth, Enum.* ii. 53, *partly, i.e. the Drège examples with the septations of the stems obsolete.*

EASTERN REGION: Natal; near the mouth of the Umzimkulu River, *Drège*, 4406!

Throughout the tropical and warm-temperate regions of both hemispheres.

35. **C. natalensis** (Hochst. ex. Krauss in Flora, 1845, 755); glabrous; rhizome long, ⅛–⅙ in. thick, clothed by ovate, acute, pale ferruginous scales 1 in. long; stems solitary, 2 ft. long, trigonous, smooth; leaves usually hardly any, sometimes 2–6 in. long, in Krauss' type specimen up to 26 in. (Krauss defines the species as leafless); umbel condensed more or less, or rays sometimes up to 7 in. long; bracts 3, shorter than the umbel, but 8 in. long in one example; spikelets numerous, up to 1 by $\frac{1}{10}$ in., hardly compressed, 14-flowered, hard rigid; rhachilla scarious-winged; glumes ovate, obtuse, not keeled, 9–11-nerved on back, shining, pale or brownish; nut scarcely ½ the length of the glume; style as long as the nut, branches 3 linear. *Boeck. in Linnæa*, xxxvi. 343.

SOUTH AFRICA: without locality, *Harvey*, 71!

KALAHARI REGION: Orange Free State, *Buchanan*, 109!

EASTERN REGION: Natal; marshy places, Umlaas River, *Krauss*, 207! Durban Flat, *Wood*, 4102! *Buchanan*, 111! 127! Clairmont, *Kuntze*, 229! and without precise locality, *Buchanan*, 311! *Gerrard*, 493!

Also occurs at the mouth of the Zambesi River.

Leafless examples of this species are easily recognized; long-leaved examples are exceedingly like *C. rotundus*, var. *platystachys*, differing chiefly by their harder, more shining, less compressed spikelets.

36. C. rotundus (Linn. Sp. Plant. ed. ii. 67); glabrous; rhizome woody, creeping, with tuberous thickenings; stem 8–24 in. long, erect at the base, trigonous at the top, smooth; leaves $\frac{2}{3}$ the length of the stem, $\frac{1}{6}$–$\frac{1}{4}$ in. broad; umbel 1–8 in. diam., compound simple or of one spike; bracts 3–4, overtopping the umbel; spikelets spicate, 1 by $\frac{1}{12}$–$\frac{1}{10}$ in. (in the large Cape form), much compressed, 12–24-flowered, chestnut-red; rhachilla scarious-winged; glumes boat-shaped, ovate, pointed, 1–3 nerves forming the keel; stamens 3; anthers linear, not crested; nut $\frac{1}{2}$ the length of the glume, trigonous, obovoid, dark-brown; style shorter than the nut, branches 3 linear, longish. *Kunth, Enum.* ii. 58; *Boeck. in Linnæa,* xxxvi. 283; *C. B. Clarke in Journ. Linn. Soc.* xxi. 167, *and in Hook. f. Fl. Brit. Ind.* vi. 614.

VAR. β, **centiflora** (C. B. Clarke in Journ. Linn. Soc. xxi. 171); spikelets greatly elongated, 2½ in. long, with 68 flowers in *Buchanan,* 312. *Hook. f. Fl. Brit. Ind.* vi. 615.

VAR. γ, **platystachys** (C. B. Clarke in Durand and Schinz, Conspect. Fl. Afr. v. 575); spikelets large, clustered, suberect. *C. tuberosus, Boeck. in Linnæa,* xxxvi. 285, *in great part, i.e. the African examples.*

CENTRAL REGION: Albert Div., *Cooper,* 1365!

WESTERN REGION: Little Namaqualand; near the mouth of the Orange River, *Drège,* 2468!

EASTERN REGION: Natal, *Kuntze,* 297! Var. β, Natal, *Buchanan,* 312! Delagoa Bay, *Junod,* 211! Var. γ, Kaffraria, *Schultz.* Natal, *Gerrard,* 705! *Drège,* 4420!

Widely distributed in all tropical and warm-temperate regions—a rice-field pest.

In Durand and Schinz, Conspect. Fl. Afr. v. 575, I followed Mr. N. E. Brown, and sorted Buchanan, 109, 311, as *C. rotundus* var. *platystachys*; I now follow Mr. Baker and sort these as *C. natalensis.*

37. C. longus (Linn. Sp. Plant. ed. ii. 67); stem decumbent at the base; rhizome stout, woody, irregular, but hardly producing tubers; umbel compound, straggling; otherwise nearly as the large forms of *C. rotundus.*—*Kunth, Enum.* ii. 60; *Boeck. in Linnæa,* xxxvi. 279, *excluding var.* β, ε, ζ, η; *C. B. Clarke in Journ. Linn. Soc.* xxi. 163, *excluding var.* β, γ, *and in Hook. f. Fl. Brit. Ind.* vi. 614. (*Type form not known from South Africa.*)

VAR. β, **tenuiflorus** (Boeck. in Linnæa, xxxvi. 281); umbel less straggling, with less unequal rays; spikes denser; spikelets neater, more brightly coloured, more brown or chestnut. *C. tenuiflorus, Rottb. Descr. et Ic.* 30, *t.* 14, *fig.* 1. *C. badius, Desfont. Fl. Atlant.* i. 45, *t.* 7, *fig.* 2. *C. emarginatus, Schrad. Anal. Fl. Cap.* 5; *Kunth, Enum.* ii. 60. *C. lateriflorus, Steud. in Flora,* 1829, 152, *and Syn. Pl. Glum.* ii. 33 *in obs.*; *Nees in Linnæa,* vii. 517; x. 136. *C. amœnus, Kunth, Enum.* ii. 58. *C. longus var. badius, Cambess. in Mém. Mus. d'Hist. Nat.* xiv. 323; *Boeck. in Linnæa,* xxxvi. 280; *C. B. Clarke in Journ. Linn. Soc.* xxi. 165. *C. longus var. elongata, C. B. Clarke in Journ. Linn. Soc.* xxi. 166 *partly* (= *C. elongatus in various authors and herbaria, not of Sieber*). *C. rotundus, Drège, Zwei Pflanzengeogr. Documente* 177 *partly.*

SOUTH AFRICA: *Drège,* 2469! *Bergius,* 170! *Ecklon and Zeyher,* 19! 64!

COAST REGION: Clanwilliam Div.; Wupperthal, *Drège!* Cape Div.; Cape Flats, *Bolus,* 3926! 3926b! *Rehmann,* 1777! 1778! Camps Bay, *Burchell,* 385! about the ponds near Cape Town, *Burchell,* 675! Botany Bay; near Cape Town, *Kuntze,* 254! Paarl Div.; by the Berg River, below 500 ft., *Drège.*

Tulbagh Div., *Drège*, 2467! Worcester Div.; mountains above Worcester, *Rehmann*, 2576! and without precise locality, *Ecklon!* Zwellendam Div., *Zeyher.* Queenstown Div.; Shiloh, *Baur*, 1182! Klaas Smits River, *Baur*, 86!

CENTRAL REGION: Prince Albert Div.; near Weltevrede, *Drège.* Richmond Div.; between Richmond and Brak Vallei River, 3000–4000 ft., *Drège!* Albert Div.; near Braam Berg, *Cooper*, 1365! Aliwal North Div., 4500 ft., *Drège*, 7386!

WESTERN REGION: Little Namaqualand; near the mouth of the Orange River, below 600 ft., *Drège!*

KALAHARI REGION: Hopetown Div.; by the Orange River, *Burchell*, 2651! Griqualand West; Griqua Town, *Burchell*, 1937! Hebron, by the Vaal River, *Nelson*, 86*! Transvaal; Pretoria, at Wonderboom Poort, *Rehmann*, 4483! and without precise locality, *McLea in Herb. Bolus*, 5813!

Typical *C. longus* is a native of Europe, North Africa and West Asia. The form *amœnus* is endemic in South Africa.

All the South African "*longus*" belongs to that form of var. *tenuiflorus* called a sp. (*amœnus*) by Kunth, which is very uniform, and easily distinguished from *C. rotundus*, Linn., by the much narrower neater spikelets with tightly imbricated harder glumes. *C. longus* is generally separated from *C. rotundus* by the stem decumbent at the base, passing into an equally thick, scarcely tuberous rhizome; but some examples of *C. amœnus* have the stem nearly erect at the base with a woody bulb narrowed suddenly into the rhizome; these Kunth named *rotundus.*

38. **C. fastigiatus** (Rottb. Descr. et Ic. 32, t. 7, fig. 2); glabrous; stolons 0; stem 2–3 ft. long, rather stout, triangular at the top, smooth; leaves often as long as the stem, $\frac{1}{4}$–$\frac{1}{3}$ in. broad; umbel 6–10 in. in diam.; bracts 3–4, similar to the leaves, lowest often 12–18 in. long; umbellules (secondary umbels) nearly or quite bractless; spikes elongate, often 1$\frac{1}{2}$ in. long, of 36 spikelets, nodding, more or less corymbosely compound at the base; spikelets (even in fruit) suberect, $\frac{1}{3}$ by $\frac{1}{16}$–$\frac{1}{12}$ in., 8–20-flowered, compressed, brown; wings of rhachilla yellow, lanceolate, acute; soon disarticulate, caducous; glumes closely imbricate, ovate, acute, 5–7-nerved, hardly keeled; stamens 3; anthers linear-oblong, connective somewhat produced in a short triangle; nut hardly $\frac{2}{5}$ the length of the glume, triangular, oblong-ellipsoid, pyramidal-topped, pale brown; style as long as the nut, branches 3 linear, longish. *Boeck. in Linnæa*, xxxvi. 311; *cf. R. Br. Prod.* 217 *in Obs. C. fastigiosus*, *Linn. Syst. Nat. ed. Gmelin*, vii. 134. *C. flabellaris*, *Nees in Linnæa*, vii. 519, x. 137; *Kunth, Enum.* ii. 69. *C. exaltatus*, *Nees in Wight, Contrib.* 84 *partly, not of Retz. C. egregius*, *Kunth, Enum.* ii. 69; *Krauss in Flora*, 1845, 754. *C. venustus*, *Kunth, Enum.* ii. 68, *as to the Cape plant. C. semiangulatus*, *Boeck. in Linnæa*, xxxviii. 367. *Papyrus venustus*, *Nees in Linnæa*, x. 138 *partly, i.e. plant only.*

SOUTH AFRICA: without locality, *Drège*, 1600! 7386! *Zeyher*, 1754! 4377! *Ecklon and Zeyher*, 22! *Wallich!*

COAST REGION: Clanwilliam Div.; Wupperthal, *Drège*, 7387! Cape Div.; near Cape Town, below 100 ft., *Bolus*, 4815! Paarl Div.; Berg River, below 500 ft., *Drège!* Stellenbosch Div.; between Stellenbosch and Somerset West, below 1000 ft., *Drège!* Riversdale Div.; Great Vals River, *Burchell*, 6533! Uitenhage Div., *Ecklon and Zeyher*, 437! Albany Div.; near Grahamstown,

1000–2000 ft., *Drège!* Bothas Hill, 2000 ft., *MacOwan*, 1270! King Williamstown Div.; near Toise River Station, 3700 ft., *Kuntze*, 281! Queenstown Div.; Shiloh, 3500 ft., *Baur*, 888! 898!

CENTRAL REGION: Aliwal North Div.; Kraai River, 4500 ft., *Drège!*

KALAHARI REGION: Griqualand West; by the Vaal River, *Burchell*, 1773! Orange Free State; near Winburg, *Buchanan*, 241! Transvaal; Schoon Spruit, near Klerksdorp, *Nelson*, 57! Hooge Veld, between Porter and Trigards Fontein, *Rehmann*, 6664!

EASTERN REGION; Transkei; between Gekau and the Bashee River, 1000–2000 ft., *Drège.* Natal; Umlaas River, *Krauss*, 65! 161! Nottingham, *Buchanan*, 135! Durban Flat, *Drège*, 4412! *Wood*, 4017! 4088! and without precise locality, *Buchanan*, 91! 317!

39. C. immensus (C. B. Clarke in Journ. Linn. Soc. xx. 294); glabrous, huge; stem 3–6 ft. long, triangular at the top, $\frac{1}{4}$–$\frac{1}{2}$ in. in diam., smooth; leaves 2–4 ft. by $\frac{1}{2}$–1 in., very stout; rays of umbel often a foot long; bracts 4–8, similar to the leaves, lowest often 24 by 1 in.; secondary umbels with many rays and bracts 1–4 in. long; spikes 1$\frac{1}{2}$ by $\frac{3}{4}$ in., cylindric, dense, of 30–70 spikelets, often more or less compound at the base; spikelets $\frac{1}{3}$ by $\frac{1}{10}$ in., compressed, yellow or straw-coloured, 10–20-flowered; wings of rhachilla linear-lanceolate, or linear, yellow, soon separating, caducous; glumes closely and somewhat rigidly imbricate, boat-shaped, pointed; stamens 3; anthers small, oblong, not crested; nut about $\frac{1}{3}$ the length of the glume, trigonous, ellipsoid, pyramidal at the ends, brown; style as long as the nut, branches 3 linear. *C. alopecuroides, var. a dives, Boeck. in Linnæa*, xxxvi. 321.

KALAHARI REGION: Orange Free State, *Buchanan*, 103! Transvaal; Cave Mountains, Great Spelonke, *Nelson*, 69*!

EASTERN REGION: Natal; swamp near Sydenham, *Wood*, 4093! 5807! and without precise locality, *Drège*, 4446! *Buchanan*, 335! Delagoa Bay, *Forbes! Kuntze*, 299!

Also in Tropical Africa and Madagascar.

This differs from *C. exaltatus*, Retz. (*C. alopecuroides*, Boeck.), not only by its great size and large spikelets, but by the deciduous yellow wings of the rhachilla; which character (as the structure and habit generally) bring it next the Indian *C. digitatus*, Roxb.

40. C. madagascariensis (Roem. and Schultes, Syst. ii. 876); glabrous; rhizome woody, far creeping; stems 5–12 ft. long, stout, trigonous at the top, smooth; leaves 0 (except on sterile shoots); umbel of 50–135 rays each 3–10 in. long; bracts 1–3 in. long, narrow-triangular, brown; spikelets (in closely corymbose spikes) $\frac{1}{3}$ by $\frac{1}{16}$ in., 8–16-flowered; bracts to secondary umbels not overtopping the inflorescence, often nearly obsolete; wings of rhachilla lanceolate, yellow, quickly caducous; glumes rather distant, pointed, hardly keeled; nut $\frac{1}{2}$ the length of the glume, obtusely trigonous, ellipsoid, dusky black. *Kunth, Enum.* ii. 64. *C. Papyrus, forma, Vahl, Enum.* ii. 366 *in Obs.; Boeck. in Linnæa*, xxxvi. 304. *Papyrus madagascariensis, Willd. in Abhandl. Akad. Berlin*, 1812, 72.

EASTERN REGION: Delagoa Bay, *Forbes!*

Also in the Mascarene Islands.

The description of the species here given is taken shortly from Madagascar examples. The fragments of the young plant collected by Forbes are either *C. madagascariensis* or *C. Papyrus*, Linn., which hardly differs but by the long bracts to the secondary umbels.

V. MARISCUS, Gaertn.

Spikelet of few, or many distichous glumes; 2 lowest empty, 1 few or many succeeding glumes 2-sexual perfecting nuts; uppermost male or sterile; rhachilla disarticulating below the lowest fertile glume from a cushion, falling off in one piece. *Stamens* 3–2, anterior. *Style* with 3 linear branches. *Nut* triangular (or plano-convex from the flattening of the anticous angle), narrowly oblong, elliptic, or oboval in longitudinal section, smooth; superficial cells nearly square.

Usually glabrous; leaves all near the base of the stem; inflorescence of all the forms occurring in *Cyperus*.

DISTRIB. Species 170, in all tropical and warm-temperate regions; no species extends to Europe; in the New World a few reach Canada and a few Patagonia.

Sect. I. **Bulbocaules.** Medium-sized. No stolons. Stems thickened at the base by the conspicuous scarious or brown leaf-sheaths. Umbel simple, small, or often of 1 head. Spikelets mostly ripening 1 or few nuts (in *M. vestitus* 4–8 nuts). See also *M. rehmannianus* in Sect. III.

Spikelets perfecting 1 nut (rarely as in *M. albo-marginatus var.* 2 nuts):
  Inflorescence of 1 head of 1–3 sessile or nearly sessile spikes; spikelets ovoid; nut ellipsoid ... (1) **capensis.**
  Dense cylindric spikes sessile or peduncled, in a simple umbel; spikelets lanceolate; nut narrow-oblong ... ... (2) **albomarginatus.**
  Spikes short-cylindric, peduncled, in a simple umbel; spikelets obtuse; nut ellipsoid... ... (3) **Marlothii.**
Spikelets usually perfecting 2 or more nuts:
  Inflorescence in one dense head, straw-coloured ... (4) **dregeanus.**
  Spikes peduncled shortly in a simple umbel, reddish ... ... (5) **vestitus.**

Sect. II. **Eu-Marisci.** Medium-sized. No stolons. Basal leaf-sheaths not conspicuously inflated. Umbel simple or nearly so. Spikelets ripening 1 or few nuts (in *M. luzuliformis* 4–8 nuts), green or yellowish, not red.

Leaves and bracts long; spikelets maturing 1–2 (rarely 3–4) nuts. These are all one species, *M. umbellatus* (or *M. flavus*), Vahl, in the opinion of some:
  Spikelets linear-lanceolate; rays as long as the spikes ... ... (6) **sieberianus.**
  Spikelets oblong; spikes large, long, dense, cylindric ... ... (7) **nossibeensis.**
  Spikelets linear-lanceolate, small; spikes shorter than the rays ... ... (8) **umbellatus.**
  Spikelets oblong, 2–1-flowered, 1-nutted; rays often long... ... ... (9) **radiatus.**
  Spikelets oblong, 1-flowered; rays hardly any .. (10) **macer.**
  Spikelets oblong, 2–3-nutted ... ... (11) **macrocarpus.**
Leaves and bracts long; spikelets maturing 3–5 nuts (12) **luzuliformis.**
Leaves and bracts short, narrow; very slender plant; spikelets minute ... ... (13) **deciduus.**

Sect. III. Thunbergiani. Robust biennials or perennials; tall (except *M. tabularis*, var. *humilis*). Spikelets very numerous, brown-red or much marked with red. Leaves and bracts long.

Spikelets linear, many-flowered; glumes reddish, with green or pale back:
  Leaves not transversely veined ... ... (14) **congestus.**
Spikelets 3-4-nutted, oblong, turgid, red-brown; umbel compound:
  Leaves transversely veined ... ... (15) **umbilensis.**
Spikelets 2-10-nutted, lanceolate, compressed, red-brown:
  Leaves and bracts transversely veined (somewhat spongy):
    Spikes clustered on the primary rays of the umbel:
      Stem obtusely trigonous at the top ... (16) **riparius.**
      Stem acutely triquetrous at the top; glumes mucronulate ... ... (17) **elatior.**
      Leaves (dry) inrolled; spikelets with 10 nuts ... ... (18) **involutus.**
    Umbel more distinctly compound:
      Spikelets densely spicate, 4-nutted ... (19) **Grantii.**
      Spikelets loosely spicate, 4-nutted ... (20) **Owani.**
      Spikelets 10-nutted ... ... (21) **Gueinzii.**
      Spikelets 10-nutted; umbel 4½ ft. in diam. ... ... (22) **elephantinus.**
  Leaves and bracts not transversely veined (of close tissue):
    Leaves (when dry) flat, midrib strong ... (23) **tabularis.**
    Leaves (when dry) rolled up, no distinct midrib ... ... (24) **durus.**
Spikelets terete, 3-4-nutted; upper glumes long acuminate, more or less recurved:
  Leaves narrow, flaccid; spikelets very flexuose ... (25) **rehmannianus.**
  Leaves ¼ in. broad, flat; spikelets nearly straight (26) **Cooperi.**

1. M. capensis (Schrad. Anal. Fl. Cap. 13); glabrous; stolons 0; stem 4-14 in. long, medium or slender, trigonous at the top, smooth; leaves often as long as the stem, $\frac{1}{8}$-$\frac{1}{6}$ in. broad, weak; basal sheaths inflated, brown; stem-base appearing much thickened, oblong or ovoid; inflorescence subcapitate, straw-coloured, of 1-4 short cylindric spikes, sessile or a ray rarely up to $\frac{1}{3}$ in. in length; bracts 3-5, similar to the leaves, lowest 3-5 in. long; spikelets numerous, densely packed in each spike, $\frac{1}{8}$-$\frac{1}{5}$ in. long, ovoid, of 4 glumes, dehiscing from a cushion below the lowest (3rd) fertile glume; fertile glume ovoid, hardly keeled, subacute, strongly 13-ribbed throughout its breadth; rhachilla above the fertile glume winged, usually topped by a rudimentary sterile glume (4th), the whole remarkably simulating a single glume, but there are no ribs on its sides; less often the upper (4th) glume is more developed, bearing a male flower; very occasionally the 4th glume bears a bisexual flower (not perfecting a nut), but is then much smaller than the 3rd; stamens 3-2; anthers oblong, not crested; nut $\frac{3}{4}$ the length of the glume (or more), obtusely trigonous, broadly ellipsoid, black; style $\frac{1}{2}$ the length of the nut; branches 3, linear, long, much exserted from the glume. *Nees* *in Linnæa,* vii. 520, x. 139 ; *Kunth, Enum.* ii. 122. *M. uitenhagensis, Steud. Syn. Pl. Glum.* ii. 317 ; *cf. Boeck. in Flora,* 1859, 66. *Kyllinga capensis, Steud. in Flora,* 1829, 153. *Cyperus capensis, Boeck. in Linnæa,* xxxvi. 378. *C. Marlothii, Boeck. in Eng. Jahrb.* xi. 407 *partly.*

SOUTH AFRICA: without locality, *Zeyher,* 1765! *Ludwig,* 266! *Harvey! Ecklon and Zeyher,* 26!

COAST REGION: Uitenhage Div., *Ecklon!*

CENTRAL REGION: Somerset Div.; Bosch Berg, 2500 ft.! *MacOwan,* 2036! Albert Div., *Cooper,* 3331!

KALAHARI REGION: Orange Free State; Caledon River, *Burke,* 303! Bechuanaland; stony plains near Kuruman, 4000 ft., *Marloth,* 1108! Transvaal, *McLea in Herb. Bolus,* 6022!

EASTERN REGION: Delagoa Bay, *Junod,* 236! 233 partly!

2. M. albomarginatus (C. B. Clarke in Durand and Schinz, Conspect. Fl. Afr. v. 584); rays of umbel 3–5, the lowest usually $\frac{1}{4}$–$1\frac{1}{2}$ in. long (but sometimes 0), each bearing a dense cylindric spike of spikelets; spikelets lanceolate; nut $\frac{2}{3}$ the length of the glume, linear oblong; otherwise as *M. capensis,* Schrad.

VAR. β, **binucifera** (C. B. Clarke); spikelets often perfecting 2 nuts; rhachilla nearly as long as the glume.

COAST REGION: Uitenhage Div.; between Van Stadens River and Galgebosch, *Burchell,* 4680!

KALAHARI REGION: Griqualand West; between Griqua Town and Moses Fontein, *Burchell,* 2034!

EASTERN REGION: var. β, Pondoland; St. Johns River, 4500 ft., *Schlechter,* 6433!

Both forms also occur in Nyasaland.

3. M. Marlothii (C. B. Clarke in Durand and Schinz, Conspect. Fl. Afr. v. 590); glabrous; stolons 0; stem 6–9 in. long, rather slender, trigonous at the top, smooth; leaves as long as the stem, $\frac{1}{10}$–$\frac{1}{8}$ in. broad; rays of umbel 2–5, 0–$1\frac{1}{2}$ in. long, rather stiff; bracts 3–5, similar to the leaves, suberect, lowest 3–5 in. long; spikes $\frac{1}{2}$ by $\frac{1}{4}$ in., very dense, greenish; spikelets (even when nearly ripe) suberect, 1-nutted, $\frac{1}{8}$–$\frac{1}{6}$ in. long, oblong-ellipsoid, very similar to those of *M. capensis,* but greener. *Cyperus Marlothi, Boeck. in Eng. Jahrb.* xi. 407 *partly.*

VAR. β, **globospica** (C. B. Clarke); rays of umbel slender; spikes globose, $\frac{1}{3}$ in. long; spikelets divaricate (not suberect as in *M. Marlothii* type).

SOUTH AFRICA: without locality, *Alexander.*

COAST REGION: var, β, Uniondale Div.; rocky hill near Groot River, *Burchell,* 5020!

KALAHARI REGION Transvaal; Pretoria, *Rehmann,* 4035! 4728!

The var. β may be treated as a distinct species; but no differences have been found other than the slight ones given above.

4. M. dregeanus (Kunth, Enum. ii. 120); glabrous; stolons 0; stem 6–14 in. long, slender, trigonous at the top, smooth, bearing 1 dense compound head of spikelets; leaves often as long as the stem, $\frac{1}{10}$–$\frac{1}{6}$ in. broad, weak, grass-like; sheaths of basal leaves inflated, scarious or coloured, forming an oblong thickening to the stem-base; inflorescence ovoid, $\frac{1}{2}$–$\frac{3}{4}$ in. long, of 1 or 2–5 confluent spikes; bracts 3–4, similar to the leaves, lowest 3–7 in. long; spike of very numerous, densely crowded spikelets, straw-coloured or fuscous; spikelets $\frac{1}{6}$ in. by $\frac{1}{12}$ in., not compressed, somewhat angular, 2–6-(usually 3–4) nutted; rhachilla with broad scarious persistent wings, disarticulating below the lowest fertile flower, usually falling with the glumes and nuts; glumes ovate, hardly acute, not keeled, strongly 13–15-ribbed over nearly their whole breadth; stamens 3–2; anthers oblong, not crested; nut about $\frac{1}{2}$ the length of the glume, trigonous, oblong-obovoid, dusky black; style shorter than the nut; branches 3, linear, longish. *C. B. Clarke in Hook f. Fl. Brit. Ind.* vi. 620. *M. Kraussii, Hochst. in Flora,* 1845, 756. *M. kyllingiæformis, Boeck. in Flora,* 1859, 443, 496. *Cyperus dubius, Rottl. in Neue Schr. Gesell. Nat. Freunde Berlin,* iv. 1803, 193; *Boeck. in Linnæa,* xxxvi. 336 *partly*; *C. B. Clarke in Journ. Linn. Soc* xxi. 197, *not of Rottb. C. kyllingæoides, Vahl, Enum.* ii. 312; *Kunth, Enum.* ii. 94. *Schœnus coloratus, var. β Linn. Sp. Plant. ed.* ii. 64. *S. niveus, Linn. Syst. Veget. ed.* xiii. 81.

VAR. β, **Buchanani** (C. B. Clarke in Durand and Schinz, Conspect. Fl. Afr. v. 587); spikelets elongated up to $\frac{1}{3}$ in., the flowers not being more numerous but more distant on the rhachilla.

KALAHARI REGION: Transvaal; Houtbosch, *Rehmann,* 5637!

EASTERN REGION: Transkei; Bashee River, 1000 ft., *Drège!* Tembuland; between the Bashee River and Morley, 1000–2000 ft., *Drège!* Pondoland, *Bachmann,* 99! 100! 101! Natal; Umlaas River, *Krauss,* 6! Durban Flat, *Buchanan,* 31! Clairmont, *Kuntze,* 232! 259! and without precise locality, *Buchanan,* 82! *Gerrard,* 460! *Grant!* Delagoa Bay, *Junod,* 151! 233 partly!

VAR. β, Natal; Durban Flat, *Buchanan!*

Also in Tropical Africa, the Mascarene Islands, and Tropical India to Borneo.

The var. β resembles much *M. macrocarpus* below; but has the spikelets less green and much more densely crowded; it looks very different from the ordinary form of *M. dregeanus,* but a similar state occurs in Madras.

5. M. vestitus (C. B. Clarke in Durand and Schinz, Conspect. Fl. Afr. v. 595); glabrous, stolons 0; stem 6–14 in. long, rather slender, trigonous at the top, smooth, much thickened at the base by the inflated coloured leaf-sheaths; leaves nearly as long as the stem, $\frac{1}{8}$–$\frac{1}{6}$ in. broad, weak; umbel simple; rays 2–7, $\frac{1}{8}$–$1\frac{1}{2}$ in. long; bracts 3–5, similar to the leaves, lowest 3–5 in. long; spikelets closely spicate, 7–20 in a spike, $\frac{1}{3}$ by $\frac{1}{12}$ in., 6–10-flowered, scarcely compressed; glumes ovate, strongly 13-nerved, reddish, with an excurrent green mucro; stamens 3–2; anthers linear, not crested; nut nearly half as long as the glume, trigonous, ellipsoid, black; style shorter than the nut, branches 3 linear, longish. *Cyperus vestitus, Hochst. ex Krauss in Flora,* 1845, 755. *C. usitatus, Boeck. in Linnæa,* xxxv. 511 *partly; C. B. Clarke in Journ. Linn. Soc.* xxi. 176, *not of Burchell.*

EASTERN REGION: Natal; margin of woods around Durban Bay, *Krauss,* 287!

Durban Flat, *Wood*, 4087 ! Umbilo River, *Rehmann*, 8443 ! and without precise locality, *Gerrard*, 701 !

Also in South-east Tropical Africa.

6. **M. sieberianus** (Nees in Linnæa, ix. 286); glabrous; rhizome very short, of woody nodules or often hardly any; stem 1–2 ft. long, not stout, trigonous at the top, smooth, not conspicuously thickened at the base by inflated leaf-sheaths; leaves nearly as long as the stem, $\frac{1}{8}$–$\frac{1}{6}$ in. broad, green; umbel simple; rays 5–12, up to 1–4 in. long, or fewer or very short; bracts 5–10, similar to the leaves, longer than the umbel; spikes solitary, $\frac{1}{2}$–1 by $\frac{1}{3}$ in., cylindric, rarely much shorter than the ray; spikelets very numerous, spreading in fruit at right angles or somewhat deflexed, nearly terete, not red, linear-lanceolate, bearing 1 or 2 nuts; rhachilla separating below the lowest fertile glume; glume 1 (lowest) small, triangular, with a longish seta (like a bract); glume 2 shorter than 1, subquadrate; glume 3 twice as long as 2, nut-bearing, ovate, obtuse, strongly 7–41-nerved; stamens 3–2; anthers linear-oblong, not crested; nut $\frac{2}{3}$–$\frac{3}{4}$ the length of the glume, trigonous, linear-oblong, finally nearly black; style $\frac{1}{2}$ as long as the nut, branches 3 linear, longish. *C. B. Clarke in Hook. f. Fl. Brit. Ind.* vi. 622. *M. umbellatus, Vahl, Enum.* ii. 376 *partly; Kunth, Enum.* ii. 118 *almost wholly. Scirpus cyperoides, Linn. Mant.* 181. *Cyperus cylindrostachys, Boeck. in Linnæa,* xxxvi. 383.

VAR. β, **evolutior** (C. B. Clarke in Hook. f. Fl. Brit. Ind. vi. 622); spikelets ripening 2–4 nuts, linear.

SOUTH AFRICA: without locality, *Harvey*, 85!
COAST REGION: Komgha Div.; *Flanagan*, 1016!
EASTERN REGION: Tembuland; between the Bashee River and Morley, 1000–2000 ft., *Drège!* Pondoland, *Bachmann*, 91! Natal; Coast-land, *Sutherland!* Inanda, *Rehmann*, 82 bis! Howick, 1000 ft., *Junod*, 215! and without precise locality, *Buchanan*, 154! 333! *Schlechter*, 6349! Var. β, Tembuland; hillsides, Bazeia, 2000 ft., *Baur*, 445! Griqualand East; near Clydesdale, 2500 ft., *Tyson*, 2592!

Found in all tropical and warm countries; not in Europe, rare in America.

This species is abundant throughout the Old World and Oceania; and Mr. Bentham would consider the 5 species following here enumerated as mere varieties of it. Kunth regarded *M. sieberianus* and *M. umbellatus* as two forms of one species; consequently, as the Drège collections at Kew have been issued with Kunth's name, usually without numbers or locality, it is impossible in the present case to say where Drege's example of *M. sieberianus*, Nees type, came from.

7. **M. nossibeensis** (Steud. Syn. Pl. Glum. ii. 63); resembling *M. sieberianus*, but larger in all its parts; stem stouter; rays numerous, up to 4 in. long; spikes $1\frac{1}{4}$ by $\frac{1}{3}$ in., very dense; spikelets oblong; nut ellipsoid, much thicker than in *M. sieberianus*, Nees.

EASTERN REGION: Natal, *Drège*, 4425! No South African specimen at Kew.

Also in Tropical Africa and Madagascar.

8. **M. umbellatus** (Vahl, Enum. ii. 376 partly); resembling *M. sieberianus*, but spikelets very small, numerous; spikes usually $\frac{1}{4}$–$\frac{1}{2}$ the length of the rays, with the innumerable spikelets densely stellately spreading, so that the lower are much deflexed. *Kyllinga umbellata, Rottb. Descr. et Ic.* 15, *t.* 4, *fig.* 2, *excluding some syn.; Linn. f. Suppl.* 105. *Cyperus ovularis, Boeck. in Linnæa,* xxxvi. 376, *form a partly. C. umbellatus, C. B. Clarke in Journ. Linn. Soc.* xxi. 200, *form a partly; Boeck. in Engl. Jahrb.* v. 91.

EASTERN REGION: Natal; around Durban Bay, *Drège*, 4449!

Abundant in Tropical Africa and the Mascarene Islands!

The form figured by Rottboell l.c. is abundant in Tropical Africa, but has not yet been seen from India. Rottboell assigns his plant to India, but his characteristic figure may have been taken from an African plant. Whether the typical original "*umbellata*" be esteemed a species or a var., I entirely dissent from Boeckeler's plan of uniting it with the American *M. ovularis*, which I regard as well distinct from any Old World *Mariscus*.

9. **M. radiatus** (Hochst. ex Krauss in Flora, 1845, 757); resembling *M. sieberianus*, but spikelets oblong-obovoid, conspicuously trigonous. *Cyperus Krausii, Boeck. in Linnæa,* xxxvi. 379.

SOUTH AFRICA: without locality, *Harvey*, 88! *Bowie*, 7!
EASTERN REGION: Natal; Umlaas River, *Krauss*, 35! and without precise locality, *Buchanan!* Delagoa Bay, *Junod*, 239! *Kuntze*, 216!

In the type example of this species, Krauss 35, the rays are short, but they are sometimes long and occasionally very long. The spikelets are 2–1-flowered, but generally mature 1 nut; they are greenish when young, white when ripe.

10. **M. macer** (Kunth, Enum. ii. 121); resembling *M. sieberianus*, but spikelets oblong-obovoid, conspicuously trigonous, always 1-flowered, the 4th glume (i.e. that above the 3rd nut-bearing glume) rudimentary, not easily to be distinguished from the glume-like rhachilla. *C. cylindrostachys, Boeck. in Linnæa,* xxxvi. 383 *partly.*

KALAHARI REGION: Orange Free State; *Buchanan*, 153 partly!
EASTERN REGION: Natal, *Buchanan*, 108 partly!

Also in the mountains of Tropical Africa at 3000–5000 ft.

Spikes long, cylindric, nearly sessile in all the examples; but this species does not really differ from *M. radiatus* except by the suppression of the upper flower. There is a "representative" species of the *M. sieberianus* group abundant in Tropical America, viz. *M. flavus*, Vahl, admitted distinct by all authors, including Bentham. I can find no character except the lowest glume I., which is longer (like a setaceous bract) in the American group. *M. macer* appears to me extraordinarily like *M. flavus*, Vahl.

11. **M. macrocarpus** (Kunth, Enum. ii. 120); resembling *M. sieberianus*, but spikelets maturing 2–3 broad-oblong nuts. *C. macrocarpus, Boeck. in Linnæa,* xxxvi. 380. *C. flavus, Ridley in Trans. Linn. Soc. ser.* 2, ii. 144.

EASTERN REGION: Natal; between the Umtentu River and Umsamculo River, *Drège*, 4421! *Buchanan*, 153 partly!

Frequent in South Tropical Africa.

This form (as the two preceding) differs from *M. sieberianus* by the much broader nuts, giving the ripe spikelets a conspicuously triangular section. *M. macrocarpus* is exceedingly like *M. flavus*, Vahl, var. *humilis*, Benth. (which var. has 2–3 nuts to the spikelet), differing only by the less bract-like basal glume; it is less surprising that Ridley unites them than that Bentham separates them.

12. **M. luzuliformis** (C. B. Clarke in Durand and Schinz, Conspect. Fl. Afr. v. 589); glabrous; rhizome 0; stem 10 in. long, trigonous, smooth at the top; leaves overtopping the stem, $\frac{1}{6}$ in. broad; bracts 6–7, similar to the leaves, several of the lower 6 in. long; umbel $1\frac{1}{4}$ in. diam., of numerous spikelets, simple; rays many, hardly $\frac{1}{4}$ in. long; spikes of 8–12 spikelets spicately arranged, diverging at right angles; spikelets $\frac{1}{2}$ by $\frac{1}{12}$ in., maturing 3–5 nuts, nearly terete, greenish-white; rhachilla winged, deciduous below the lowest fertile glume; glumes rounded on the back, conspicuously 9–13-striate; stamens 3; anthers linear-oblong, not crested; nut $\frac{1}{2}$ as long as the glume, trigonous, broad-oblong, black. *Cyperus indecorus, Kunth, Enum.* ii. 85. *C. luzuliformis, Boeck. in Linnæa,* xxxvi. 356.

SOUTH AFRICA: without locality, *Harvey*, 89!

EASTERN REGION: Transkei, Bashee River, *Drège*, 4424!

The long spikelets with 5–7 flowers (3–5 nuts) would appear to distinguish this from the other *Eu-Marisci*, but the plant is so like *M. radiatus* type (Krauss 35), except as to number of flowers, that its distinctness may be doubted.

13. **M. deciduus** (C. B. Clarke); glabrous, slender; rhizome (seen) very short, horizontal, woody, slender; stem 12–18 in. long, very slender, smooth; leaves about $\frac{1}{3}$ the length of the stem, hardly $\frac{1}{8}$ in. broad; umbel of 3–5, slender, unequal rays, each terminated by very small condensed corymb of spikelets; bracts 3, similar to the leaves, lowest shorter than the umbel; spikelets $\frac{1}{10}$ in. long, very green, 2–4-flowered, early caducous above the 2 empty basal glumes; glumes broad-oblong, truncate, plane-convex on the back, not nervose; rhachilla wingless; stamens 3; anthers linear-oblong, slightly apiculate; style as long as the young ovary, branches 3, linear, long. *Cyperus deciduus, Boeck. in Flora,* 1879, 547; *C. B. Clarke in Durand and Schinz, Conspect. Fl. Afr.* v. 555.

KALAHARI REGION: Transvaal; Houtbosch, *Rehmann*, 5642!

Also in Dammara-land.

As the style has 3 linear branches, Boeckeler must be far wrong as to the affinity of this plant in putting it with *Pycreus*. But the wingless rhachilla, non-striate glumes, small leaves and bracts, and whole habit, indicate that it is not in its right place here among the *Eu-Marisci*.

14. **M. congestus** (C. B. Clarke); glabrous, medium- to large-sized; rhizome short, horizontal, woody (no stolons seen except in var. β); stem 9–24 in. long, triquetrous at the top, smooth, throwing short lateral shoots at the base; leaves often as long as the stem, $\frac{1}{6}$–$\frac{1}{4}$ in. broad, tough, smooth, without transverse lineolations; umbel usually open-simple (cf. varr. β, γ); rays 2–7, up to 4–6 in. long; bracts 3–6, overtopping the umbel, similar to the leaves; spike (at the

end of a ray) of many spikelets, usually compound, more or less corymbose, dense, with very small or no bract; spikelets spicate, $\frac{2}{3}$ by $\frac{1}{10}$ in., linear, compressed, maturing often 10–16 nuts, scarcely deflexed even in fruit; rhachis somewhat flexuose, finally (but very late) deciduous above the two lowest empty glumes from a small cushion, wings oblong, hyaline, persistent; glumes rather distant, elliptic-oblong, obtuse, scarcely keeled, 7–9-striate on the back, green or pale, sides red; stamens 3; anthers linear-oblong, not crested; nut $\frac{2}{5}$ the length of the glume, oblong-ellipsoid, obtuse, trigonous, nearly black; style slender, rather shorter than the nut, branches 3 as long as the style. *Cyperus congestus, Vahl, Enum.* ii. 358; *Schrad. Anal. Fl. Cap.* 9; *Nees in Linnæa,* vii. 518; x. 136 (*excl. var β*); *Kunth, Enum.* ii. 87; *Boeck. in Linnæa,* xxxvi. 347; *C. B. Clarke in Journ. Linn. Soc.* xxi. 182; *not of Poiret. C. badius, Steud. in Flora,* 1829, 153; *not of Desfont.*

VAR. β, **glandulifera** (C. B. Clarke); larger, brighter; umbel often compound; primary rays up to 6 in. long; spikelets up to 1 in. long, 20-nutted, bright brown-red with yellowish keel; a yellow gland beneath the two lowest empty glumes of each spikelet. *Cyperus congestus, var. glanduliferus, C. B. Clarke in Durand and Schinz, Conspect, Fl. Afr.* v. 553.

VAR. γ, **brevis** (C. B. Clarke); stem robust, short or long, bulbous, woody at the base, with stolon 4 in. long; umbel nearly condensed into a compound head; spikelets very numerous, large, yellow-brown. *Cyperus congestus, var. brevis, C. B. Clarke in Durand and Schinz, Conspect. Fl. Afr.* v. 554. *C. congestus, var. a, Nees in Linnæa,* x. 137. *C. brevis, Boeck. in Linnæa,* xxxvi. 341.

SOUTH AFRICA: without locality, *Drège*, 4422! *Ecklon and Zeyher*, 24! *Bergius*, 168! *Krebs*, 13! *Harvey*, 368! *Thom*, 1028! Toise River, 3500 ft., *Kuntze*, 272! Var. γ, *Ecklon and Zeyher*, 134!

COAST REGION: Cape Div.; Cape Town, *Rehmann*, 1776! Table Mountain, *Ecklon*, 886! Paarl Div.; between Paarl and Lady Grey Bridge, *Drège.* Worcester Div.; mountains above Worcester, *Rehmann*, 2577! Khomga Div.; Kei River, *Flanagan*, 957! Cathcart Div.; near Cathcart, 4500 ft., *Kuntze*, 265! Queenstown Div.; Shiloh, 3500 ft., *Baur*, 845! 850! Var. β, Albany Div.; Grahamstown, 2000 ft., *MacOwan*, 496! Var. γ, Uitenhage Div.; Zwartkops River, *Ecklon and Zeyher*, 504! *Ecklon! Zeyher*, 16!

CENTRAL REGION: Graaff Reinet, *MacOwan*, 1019! Albert Div.; *Cooper*, 656!

KALAHARI REGION: Griqualand West, Ongeluk, *Burchell*, 2646! Orange Free State, *Buchanan*, 84! Transvaal; Pretoria, *Kuntze*, 244! *Rehmann*, 4037! 4774! Houtbosch, *Rehmann*, 5655! and without precise locality, *McLea in Herb. Bolus*, 5814!

EASTERN REGION: Pondoland; between the Umtentu River and Umzimkulu River, *Drège*, *Bachmann*, 90! Natal; Coast-land, 0–5000 ft., *Sutherland!* Mohlamba Range, 5000–6000 ft., *Sutherland!* Mooi River Station, *Kuntze*, 240! Howick, 1000 ft., *Junod*, 267! and without precise locality, *Buchanan*, 336! *Schlechter*, 6351! Var. β, Natal, *Buchanan*, 315!

Also in St. Helena, Mediterranean Region, and Australia.

This species, though closely allied to the twelve following, is, as to technical diagnosis, poorly separated from the genus *Cyperus*; the rhachilla of the spikelet below the lowest fertile flower only disarticulating very late. The geographical distribution is remarkable: the plant, identical in habit, is frequent in Australia; and the examples from Diana's Peak (St. Helena) are probably indigenous. The species has been for nearly a century cultivated in Europe. There are so many examples sent from the Mediterranean Region, especially its eastern end, Turkey and Asia Minor, that it would appear indigenous there; but examples (marked as wild) from Potsdam, Geneva Lake, Coimbra, may probably be escapes.

15. **M. umbilensis** (C. B. Clarke ex W. Watson in Gard. Chron. 1891, x. 190); glabrous, robust; stem 2–3 ft. long, triquetrous at the top, often scabrous; leaves ⅔ the length of the stem, ¼–⅓ in. broad, transversely lineolate, somewhat spongy; midrib often scabrous beneath; umbel large, compound, 5–9 in. in diam.; bracts 4–8, up to 1–2 ft. long, similar to the leaves; spikes ¼–1 in. long, cylindric, of many spikelets, sometimes very dense; spikelets oblong, as thick as broad, perfecting usually 3–4 nuts (sometimes 1–2 only); rhachilla disarticulating below the lowest fertile flower; wings broad, scarious; stamens 3; nut ½–¾ the length of the glume, oblong or ellipsoid, trigonous, black; style much shorter than nut, branches 3 linear, long. *M. Bolusi, C. B. Clarke in Durand and Schinz, Conspect. Fl. Afr.* v. 585. *Cyperus umbilensis, Boeck. ex Watson in Gard. Chron.* 1891, x. 190. *C. Bolusi, Boeck. ex C. B. Clarke in Durand and Schinz, Conspect. Fl. Afr.* v. 585.

South Africa: without locality, *Harvey*, 90!

Central Region: Somerset Div.; Bosch Berg, 3000 ft., *MacOwan*, 496b!

Kalahari Region: Orange Free State, *Buchanan*, 141!

Eastern Region: Pondoland, *Bachmann*, 107! Natal, *Buchanan*, 334! *Kuntze*, 298!

Also in Hereroland.

16. **M. riparius** (Schrad in Goett. Gel. Anz. iii. 1821, 2067); glabrous, robust; stem 1½–3 ft. long, ⅙ in. in diam., obtusely trigonous, smooth; leaves often as long as the stem by ⅓ in. broad, nearly flat when dry, thick, somewhat spongy, transversely lineolate, often scabrous on the margins near the tip and on the midrib beneath; rays of umbel 6–14, up to 3–5 in. long; bracts 6–8, up to 20 in. long, similar to the leaves; spikes very dense, 1–3, confluent at the apex of each ray in the normal form (but see var. β, γ, δ), often 1 by ½ in.; spikelets ⅕ by 1/16 in., maturing 3–1 nuts, dull red-brown, suberect even in fruit, not much compressed; rhachilla disarticulating below the lowest fertile glume, wings elliptic; glumes elliptic-oblong, obtuse, hardly keeled, 9-striate; stamens 3; nut ⅔ as long as the glume, trigonous, linear-oblong, black; style short, branches 3 linear. *M. Thunbergii, Schrad. Anal. Fl. Cap.* 13; *Nees in Linnæa*, viii. 79. *Cyperus Thunbergii, Vahl, Enum.* ii. 371; *Nees in Linnæa*, vii. 520 *in Obs.*; x. 137, *excl. several syn.*; *Kunth, Enum.* ii. 76; *Boeck. in Linnæa*, xxxvi. 330.

Var. β, **robustior** (C. B. Clarke in Durand and Schinz, Conspect. Fl. Afr. v. 592); larger in all its parts; spikes often 1½ by ½ in., very dense.

Var. γ, **trisumbellatus** (C. B. Clarke in Durand and Schinz, Conspect. Fl. Afr. v. 592); very large; primary rays of umbel up to 8 in. long; secondary rays 2 in. long, conspicuously deflexed (even when young); tertiary rays up to ⅕ in. long. *Cyperus alopecuroides, Thunb. Prod.* 18, *and Fl. Cap. ed. Schult.* 101, *not of Rottb.*

Var. δ, **Gillii** (C. B. Clarke in Durand and Schinz, Conspect. Fl. Afr. v. 592); stem and umbel rays very thick; umbel contracted; the innumerable spikelets and spikes densely confluent; young spikelets nearly ⅓ in. long, 5–8-flowered.

South Africa: Var. β, *Drège! Bergius*, 165! *Thom*, 909! *Ecklon and Zeyher*, 25! Var. γ, *Thunberg.*

Coast Region: Malmesbury Div.; Laauws Kloof, near Groene Kloof, 1000 ft., *Drège!* Cape Div.; near Cape Town, *Burchell*, 282! Paarl Div.; Berg River,

Drège! Tulbagh Div.; near Tulbagh, *Burchell*, 1027! Uitenhage Div.; Zwartkops River, *Pappe!* Var. β, Cape Div.; Table Mountain, *Ecklon*, 105! *Milne*, 215! Simons Bay, *MacGillivray*, 409! Uitenhage Div.; *Ecklon and Zeyher*, 312! Bathurst Div.; between Theopolis and Port Alfred, *Burchell*, 4053! Albany Div.; *Williamson*. Var. γ, Clanwilliam Div.; Oliphants River, *Zeyher!* Mossel Bay Div.; Great Vals River, *Burchell*, 6530! Bathurst Div.; near Port Alfred, *Burchell*, 3992! Var. δ, Swellendam Div.; Buffeljagts River, *Gill!*

17. **M. elatior** (C. B. Clarke in Durand and Schinz, Conspect. Fl. Afr. v. 587); stem triquetrous at the top; spikes large, dense, compound; spikelets spreading, lower deflexed; glumes distant, microscopically mucronate; otherwise as *M. riparius*. *Cyperus elatior, Boeck. in Linnæa*, xxxvi. 327. *C. solidus, α elatior, Kunth, Enum.* ii. 76. *C. congestus, var. γ, Nees in Linnæa*, x. 137, *but not C. multiceps, Link (according to Boeck.)*.

SOUTH AFRICA: without locality, *R. Brown!*

COAST REGION: Paarl Div.; Paarl Mountains, 2000–3000 ft., *Drège*, 4410!

EASTERN REGION: Natal; Durban Flat, *Buchanan*, 44! and without precise locality, *Buchanan*, 69! 140!

18. **M. involutus** (C. B. Clarke in Durand and Schinz, Conspect. Fl. Afr. v. 589); glabrous; rhizome (seen) 2 in. long, horizontal, thick; stem 18–30 in. long, obtusely trigonous at the top; leaves $\frac{3}{4}$ the length of the stem, $\frac{1}{6}$–$\frac{1}{4}$ in. broad, transversely lineolate, in dried examples involute into a tube, many-striate, without keel; bracts 4–6, up to 10 in. long, similar to the leaves; rays of umbel 7–9, up to 3–4 in. long, but very unequal, with a condensed corymb of close spikes at the apex of each; spikelets $\frac{1}{2}$ by $\frac{1}{10}$ in., compressed, hard, dull chestnut-red, maturing about 10 nuts; glumes elliptic, obtuse, many-striate; keel obscure, paler, hardly excurrent; nut $\frac{3}{5}$ the length of the glume, trigonous, narrow-oblong; style not $\frac{1}{3}$ the length of the nut, branches 3 linear, shortly exserted from the glume.

COAST REGION: Uitenhage Div.; Van Stadens River, *Ecklon and Zeyher*, 174! *MacOwan*, 2084! King Williamstown Div.; Toise River Station, 3500 ft., *Kuntze*, 276!

The much longer spikelets, often maturing 10 nuts, separate this from the preceding species; the inflorescence, though much looser, does not in essence differ.

19. **M. Grantii** (C. B. Clarke in Durand and Schinz, Conspect. Fl. Afr. v. 588); spikes $\frac{1}{3}$ in. in diam., ovoid, dense, ferruginous brown, peduncled in a compound umbel; spikelets spreading, lower deflexed; otherwise as *M. riparius, Schrad. var. γ trisumbellata.*

EASTERN REGION: Natal, *Grant! Cooper*, 3332!

Flowered in Kew Gardens, May, 1890.

This plant might be arranged as a var. of *M. riparius*, but the spreading or deflexed spikelets give it a different appearance.

20. **M. Owani** (C. B. Clarke in Durand and Schinz, Conspect. Fl. Afr. v. 590); umbel once-compound or imperfectly twice-compound; spikes loose; spikelets spreading or deflexed; otherwise as *M. riparius* and *M. Grantii*. *Cyperus ligularis, Thunb. Prod.* 18; *Fl.*

Cap. ed. Schult. 100; *not of Linn. C. dactyliformis, Boeck. in Linnæa,* xxxvi. 329 *partly. C. tabularis, var.* β *major, Nees in Linnæa,* x. 137, *according to Boeck. C. Owani, Boeck. in hb. Schinz ex Clarke in Durand and Schinz, Conspect. Fl. Afr.* v. 590, *scarcely in Flora,* 1878, 29.

SOUTH AFRICA: without locality, *Drège,* 4411! *Ecklon and Zeyher,* 131! *Thunberg.*

COAST REGION: Albany Div.; near the Kowie River, *MacOwan,* 669 partly! Komgha Div.; near Komgha, *Flanagan,* 921!

EASTERN REGION: Natal; near Sydenham, 300 ft., *Wood,* 4004!

Also in Madagascar.

The loose spreading spikelets look very unlike the heads of *M. riparius* with crowded suberect spikelets; but here, again, I can find no difference in glumes, rhachilla, nut, style, or any trustworthy character. *C. Owani, Boeck. in Flora,* 1878, 29, as to MacOwan, n. 496b. in herb. Kew, I sort it with *M. umbilensis.*

21. **M. Gueinzii** (C. B. Clarke in Durand and Schinz, Conspect. Fl. Afr. v. 588); large; umbel compound, rays 8–12 up to 8 in. long; spikes dense, mostly short-peduncled; spikelets $\frac{1}{2}$ in. long, maturing often 10 nuts; otherwise as *M. tabularis,* Schrad. *Cyperus dactyliformis, Boeck. in Linnæa,* xxxvi. 329 *partly.*

COAST REGION: Uitenhage Div.; by the Zwartkops River, *Burchell,* 4431! *Zeyher,* 15! Komgha Div.; *Flanagan,* 977!

EASTERN REGION: Natal; near Durban, *Wood,* 4098! *Buchanan,* 130! and without precise locality, *Gueinzius!*

The umbel of this species much resembles large umbels of *M. congestus*; but *M. Gueinzii* (and the preceding species) have spongy leaves with transverse reticulations—remote from those of *M. congestus.*

22. **M. elephantinus** (C. B. Clarke); glabrous; stem $\frac{1}{4}$ in. in diam. at the top, obtusely trigonous, smooth; umbel $4\frac{1}{2}$ ft. in diam.; bracts 8–10, exceeding the umbel, $\frac{3}{4}$ in. broad, thick, spongy, transversely lineolate, scabrous-edged; ultimate umbels 2-4 in. long, with bracts up to 2 in. long; spikelets innumerable, rather loosely spicate, $\frac{1}{3}$ by $\frac{1}{12}$ in., 6–10-nutted, pale brown; rhachilla disarticulating below the lowest fertile glume, wings conspicuous, persistent, yellowish; glumes ovate, obtuse, striate, obscurely keeled; nut $\frac{1}{2}$ the length of the glume, oblong-ellipsoid, trigonous, black; style shorter than the nut, branches 3 linear, longish. *Cyperus elephantinus, C. B. Clarke in Durand and Schinz, Conspect. Fl. Afr.* v. 559.

EASTERN REGION: Natal, *Buchanan,* 113! 320!

I arranged this plant in *Cyperus* in Durand and Schinz, looking at the huge umbel with bracts to the secondary umbels; but overlooking the disarticulating spikelets. The spikelets (with the nuts and styles) I find now similar to those of *Mariscus* (Sect. *Thunbergiani*).

23. **M. tabularis** (C. B. Clarke in Durand and Schinz, Conspect. Fl. Afr. v. 594); glabrous, robust, stoloniferous; stems 8–24 in. long; leaves $\frac{2}{3}$ the length of the stems, $\frac{1}{4}$ in. broad, flat when dry, tough, without transverse reticulations, closely striate lengthwise;

umbel simple or nearly so, 2–5 in. in diam.; bracts 4–5, up to 10 in. long, similar to the leaves; spikes $\frac{1}{2}$ in. long, globose or ovoid, dense, dark-red; spikelets $\frac{1}{3}$ by $\frac{1}{12}$ in., somewhat compressed, maturing 4–7 nuts; rhachilla disarticulating below the lowest fertile glume; glumes ovate, very obtuse, 13–15-striate; nut $\frac{1}{2}$ the length of the glume, oblong-obovoid, trigonous; style shorter than the nut, branches 3 linear, longish. *Cyperus tabularis, Schrad. Anal. Fl. Cap.* 10; *Nees in Linnæa*, x. 137, *excluding var. β*; *Kunth, Enum.* ii. 107; *Boeck. in Linnæa*, xxxvi. 324. *C. solidus, Drège, Pflanzengeogr. Documente* 177 (*at least c*). *C. Ecklonii, Boeck. in Linnæa*, xxxvi. 325.

VAR. β, **Zeyheri** (C. B. Clarke in Durand and Schinz, Conspect. Fl. Afr. v. 594); slenderer; rays of umbel nearly 4 in. long; spikes more elongated, less dense. *M. humilis, Nees ex C. B. Clarke, l. c.*

VAR. γ, **humilis** (C. B. Clarke in Durand and Schinz, Conspect. Fl. Afr. v. 594); stems $1\frac{1}{2}$–$2\frac{1}{2}$ in. long, stout; leaves far overtopping the stem. *M. humilis, Zeyher ex Kunth, Enum.* ii. 107. *Cyperus tabularis, var. β, Kunth, Enum.* ii. 107. *C. solidus, var. β, humilior, Kunth, Enum.* ii. 76. *C. Ecklonii, Boeck. in Linnæa*, xxxvi. 325 *partly*.

SOUTH AFRICA: without locality, *Drège*, 4401! 4402! *MacOwan*, 669 partly! *Ecklon and Zeyher*, 23! 76! Var. β, *Zeyher!*

COAST REGION: Cape Div.; Table Mountain, *Hesse*. Uitenhage Div.; Zwartkops River, *Ecklon*, 18! *Zeyher!* Bathurst Div.; Port Alfred, *Hutton!* Albany Div.; near the Bushmans River, below 1000 ft., *Drège!* King Williamstown Div.; near Toise River Station, 3600 ft., *Kuntze*, 280! Cathcart Div.; near Cathcart, 4500 ft., *Kuntze*, 263! Var. γ, Albany Div.; on the Flats near Grahamstown, 2000 ft., *Drège*, 3952!

CENTRAL REGION: Somerset Div.; mountains above the Spring of Commadagga, *Burchell*, 3352!

24. M. durus (C. B. Clarke in Durand and Schinz, Conspect. Fl. Afr. v. 587); glabrous, stoloniferous; stems $1\frac{1}{2}$–$2\frac{1}{2}$ ft. long, obtusely trigonous at the top, smooth; leaves $\frac{3}{4}$ the length of the stem, hard, semicircular-backed, concave-faced (even close to the base), scarcely $\frac{1}{6}$ in. broad; umbel simple, 1–3 in. in diam., of 3–8 very unequal rays; bracts 3–5, up to 6–10 in. long, similar to the leaves; spikes very dense, compound, $\frac{1}{2}$–$\frac{3}{4}$ in. in diam., dark-red; spikelets maturing 4–5 nuts, similar to those of *M. tabularis*. *Cyperus durus, Kunth, Enum.* ii. 76; *Boeck. in Linnæa*, xxxvi. 326.

SOUTH AFRICA: without locality, *Harvey! Zeyher*, 14!

COAST REGION: Knysna Div.; Ruigte Vallei, below 500 ft., *Drège*, 3951! Uitenhage Div., *Zeyher!* Albany Div.; between the source of the Kasuga River and Sidbury, *Burchell*, 4169!

Also in Angola.

25. M. rehmannianus (C. B. Clarke in Durand and Schinz, Conspect. Fl. Afr. v. 591); glabrous; stems 6–18 in. long, trigonous at the top, smooth, thickened at the base by coloured strongly purple-striate leaf-sheaths; leaves nearly as long as the stem, $\frac{1}{8}$ in. broad, weak; rays of umbel 6, up to $1\frac{1}{2}$ in. long, each bearing 1 ebracteate spike; bracts 3–5, rather longer than the umbel, similar to (but rather stouter than) the leaves; spikes $\frac{2}{3}$ in. in diam., dense; spikelets $\frac{1}{3}$ in. long, nearly terete, 3–4-nutted, whitish, with red marks;

rhachilla flexuose, disarticulating below the lowest fertile flower, wings elliptic, persistent, finally holding the nut; fertile glumes elliptic, acuminate, round-backed, 13-striate; margins scarious, decurrent on the rhachilla-wings; upper glumes infertile, distant, with long, acuminate, recurved points; stamens 3; nut ½ the length of the glume, trigonous, oblong-ellipsoid, slightly curved; style ½ as long as the nut, branches 3 linear, longish. *Cylindrolepis, Boeck. in Bot. Centralblatt*, xxxix. 73.

KALAHARI REGION: Transvaal; lagoon near the Pinaars River, *Nelson*, 13*! Pretoria, *Rehmann*, 4479! and without precise locality, *Holub!*

Boeckeler found the margins of each glume united into a cylinder embracing the nut, and established his genus *Cylindrolepis* on this character. I find the margins of the glumes decurrent on the wings of the rhachilla, which are closely wrapped round and hold fast the ripe nut, as is the case very commonly in *Mariscus*, where the scattering of the nuts is provided for by the disarticulation of the rhachilla itself.

26. M. Cooperi (C. B. Clarke in Durand and Schinz, Conspect. Fl. Afr. v. 586); stem 32 in. long, rather stout; leaves long, ¼ in. wide, flat, not flaccid; rays of umbel up to 5 in. long, with 3–1 sessile spikes at the end of each ray, supported by linear bracts up to 1½ in. long; rhachilla of spikelet straight; fertile glumes slightly acuminate; upper sterile glumes acuminate into recurved points; otherwise as *M. rehmannianus*.

EASTERN REGION: Natal, *Cooper*, 3333!

The structure of the spikelets, with the remarkable recurved points of the upper infertile glumes, is so exactly that of the spikelets in *M. rehmannianus*, that I feel doubtful whether this solitary piece of Cooper's (3333), which is without the base of stem so characteristic in *M. rehmannianus*, may not be merely the low-level, luxuriant, over-ripe state of that species.

VI. ELEOCHARIS, R. Br.

Spikelet of many spirally imbricate glumes; lowest 1 or 2 empty, shorter than the spikelet, many succeeding glumes 2-sexual perfecting nuts, uppermost male or sterile. *Stamens* 3 or 2, anterior. *Style* glabrous; base much thickened, distinguishable from apex of nut, but persistent; branches 3–2. *Nut* triangular, or flat (dorsally compressed).

Glabrous; stem with one terminal spikelet (not rarely proliferous at the base); leaves 0; uppermost sheath truncate or very shortly produced on one side; hypogynous scabrous bristles usually (in the African species always) present, 7–3 in number, representing the sepals, petals and posticous stamens.

DISTRIB. Species 115, scattered over nearly the whole world; prevalent (80 species) in America.

Sect. I. LIMNOCHLOA (Nees). *Stem* stout. *Spikelet* hardly wider than the stem. *Glumes* subrigid:

Stem triquetrous at the top. Spikelet pale ... ... (1) **fistulosa.**

Sect. II. ELEOGENUS (Nees). *Spikelet* wider than the stem. *Glumes* membranous. *Style* 2-fid:

Stem terete at the top. Spikelet brown or chestnut ... (2) **palustris.**

Sect. III. EU-ELEOCHARIS. *Spikelet* wider than the stem. *Glumes* membranous. *Style* 3-fid :

Stem terete at the top. Spikelet brown ... ... (3) limosa.

1. E. fistulosa (Link, Jahrb. iii. 78 Obs.); stoloniferous; stem 1–3 ft. long, acutely triquetrous at the top; spikelet nearly 1 in. by $\frac{1}{8}$–$\frac{1}{5}$ in., hardly broader than the stem, pale-coloured; glumes obovate, obtuse, without keel, with numerous fine lines on the back; bristles 6 (i.e. 3 sepals, 3 petals), as long as the nut, linear, retrorse-scabrous, rusty-red; style 3-fid, or sometimes according to authors 2-fid; nut $\frac{1}{2}$ as long as the glume, obovoid, trigonous with the anticous angle flattened, longitudinally striate cancellate (i.e. the external cells are arranged in 24–30 vertical series); style-base scarcely $\frac{1}{3}$ the length of the nut, ovoid. *C. B. Clarke in Hook. f. Fl. Brit. Ind.* vi. 626. *Scirpus fistulosus, Poiret in Lam. Encyc.* vi. 749. *Heleocharis fistulosa, Boeck. in Linnæa,* xxxvi. 472.

COAST REGION: Komgha Div.; Kei River, *Flanagan*, 982! No South African specimen at Kew.

Also in Tropical Africa, Asia and Australia.

Boeckeler unites with this species *E. mutata*, R. Br., abundant in Tropical America; and which certainly differs very slightly. If this view be taken the distribution area is widened.

2. E. palustris (R. Br. Prod. 224 in note); stems 4–20 in. long, on a creeping rhizome, usually distant, terete; spikelet $\frac{1}{3}$–$\frac{2}{3}$ by $\frac{1}{6}$–$\frac{1}{5}$ in., subcylindric, dense; glumes obtuse, without keel, brown or chestnut, with scarious margins; bristles 6 or often fewer, as long as the nut; style-branches 2, longish; nut small, obovoid, dorsally compressed, smooth, yellow or brown; style-base ovoid or conic, $\frac{1}{3}$–$\frac{1}{2}$ the length of the nut. *Nees in Linnæa,* vii. 509; *Kunth. Enum.* ii. 147; *C. B. Clarke in Journ. Bot.* 1887, 267, *and in Hook. f. Fl. Brit. Ind.* vi. 628. *E. limosa a (not b, c) Drège, Pflanzengeogr. Documente* 181. *E. dregeana, Steud. Syn. Pl. Glum.* ii. 78. *Scirpus palustris, Linn. Sp. Plant. ed.* 2, 70 *partly. Heleocharis palustris, Lindl. Syn. Brit. Fl.* 280; *Boeck. in Linnæa,* xxxvi. 466 *partly, i.e. excluding all examples with 3-fid style. Limnochloa capensis, Nees in Linnæa,* x. 185; *not of Krauss.*

SOUTH AFRICA: without locality, *Drège* 4376! *Verreaux!*

COAST REGION: Komgha Div.; between Zandplaat and Komgha, 2000–3000 ft., *Drège.*

CENTRAL REGION: Somerset Div.; Bosch Berg, 3000 ft., *MacOwan*, 1741! *Burchell*, 3046! Richmond Div.; vicinity of Styl Kloof, near Richmond, 4000–5000 ft., *Drège!* Aliwal North Div., 5000 ft., *Flanagan*, 1661!

KALAHARI REGION: Griqualand West; Ongeluk, *Burchell*, 2649!

Distributed throughout the World except Australia; rare in tropical regions.

3. E. limosa (Schultes in Roem. et Schultes, Syst. ii. Mant. 87); stems 1–2 ft. long, clustered on a short woody rhizome, terete; spikelet $\frac{1}{2}$–1 by $\frac{1}{8}$–$\frac{1}{5}$ in., cylindric or somewhat long-conic, dense; glumes obtuse, without keel, brown with scarious margins; bristles 6–5 (i.e. 3 sepals, 3–2 petals), linear, retrorsely scabrous, brown, as long as the nut with style-base; style-branches 3, longish; nut small, obovoid, trigonous, brown, smooth; style-base oblong-obovoid, ½ the length of the nut. *Kunth, Enum.* ii. 148; *Drège b, c (not a) in Linnæa*, xx. 246. *E. sororia, Kunth, Enum.* ii. 148. *E. multicaulis, Drège, Pflanzengeogr. Documente* 47. *E. capensis, Nees ex Boeck. in Linnæa*, xxxvi. 467. *Scirpus limosus, Schrad. Anal. Fl. Cap.* 29, *t.* 2, *fig.* 1. *Limnochloa limosa, Nees in Linnæa*, x. 185. *L. capensis, Krauss in Flora* 1845, 759, *not of Nees. Limnocharis limosa, Kunth, Enum.* ii. 148, *in citat. Heleocharis palustris, Boeck. in Linnæa*, xxxvi. 466 *partly (i.e. the African examples with 3-fid style), not of R. Br.*

SOUTH AFRICA: without locality, *Drège*, 3940! 7406! *Verreaux! Pappe! Boivin! Gueinzius*, 211a! *Ecklon and Zeyher*, 31! 32!

COAST REGION: Malmesbury Div., *Bachmann*, 2187! Cape Div.; Cape Town, *Hesse.* Swellendam Div.; mountains along the lower part of the Zonder Einde River, *Zeyher*, 4423b! Riversdale Div.; by the Zoetemelks River, *Burchell*, 6607; Uitenhage Div.; Zwartkops River, below 500 ft., *Zeyher*, 4423! *Drège, Ecklon and Zeyher*, 438! Alexandria Div.; Enon, below 500 ft., *Drège.* Bathurst Div.; near Port Alfred, *Burchell*, 3802! Komgha Div., *Flanagan*, 903! Cathcart Div.; Blesbok Flats, 3000–4000 ft., *Drège.*

CENTRAL REGION: Richmond Div.; Winter Veld, *Drège*, 826!

EASTERN REGION: Natal; Umlaas River, *Krauss*, 80! Durban Flat, *Buchanan*, 128!

Also in Madagascar.

Very closely allied to the Abyssinian *E. marginulata*, Steud. It varies much in stoutness of stem and of spikelet.

VII. FIMBRISTYLIS, Vahl.

Spikelet of many glumes spirally imbricate, or (in *F. monostachya*) sub-distichous, lowest 1 or 2 empty, many or several succeeding glumes 2-sexual perfecting nuts, uppermost male or sterile. Hypogynous *bristles* 0. *Stamens* 3–2, anterior. *Style* often more or less hairy, deciduous with its bulbous or little enlarged base; branches 3 or 2. *Nut* triangular, or flat, dorsally compressed.

Stolons 0; leaves all near the base of the stem; inflorescence a terminal umbel, simple or compound, or a single terminal spikelet.

DISTRIB. Species 130, distributed over the tropical and warm-temperate parts of the World, especially abundant in the tropics of S.E. Asia and of Australia.

This genus differs from *Scirpus* (Sect. *Isolepis*, R. Br.) only by the style being articulated at the base, the transverse line separating it from the apex of the nut being visible even in the young stage.

Sect. I. DICHELOSTYLIS, Benth.—Lowest fertile *glumes* spirally imbricate. *Umbel* compound or simple. *Style* 2-fid.

Nut smooth. Long hairs pendent from the style-base ... ... (1) **squarrosa.**
Nut ribbed. Spikelets with many angles ... ... (2) **dichotoma.**
Nut ribbed. Spikelets very smoothly rounded ... (3) **diphylla.**
Nut smooth. Glumes puberulous on the back ... (4) **ferruginea.**

Sect. II. TRICHELOSTYLIS (Genus), Lestib.—Lowest fertile *glumes* spirally imbricate. *Umbel* compound or simple. *Style* 3-fid.

Whole plant hairy ... ... (5) **exilis.**
Glabrous. Spikelets pedicelled, solitary ... ... (6) **complanata.**
Glabrous. Spikelets in dense heads ... ... (7) **obtusifolia.**

Sect. III. ABILDGAARDIA (Genus), Vahl.—Lowest fertile *glumes* distichous, or very nearly so. *Stem* bearing 1 (rarely 2–3) spikelet.

Spikelet strongly compressed, yellowish ... ... (8) **monostachya.**

1. F. squarrosa (Vahl, Enum. ii. 289); annual, puberulous or nearly glabrous; stem 2–8 in. long; leaves $\frac{1}{2}$–$\frac{3}{4}$ the length of the stem, $\frac{1}{16}$ in. broad; umbel usually compound; bracts 3–4, often as long as the umbel, similar to the leaves; spikelets all pedicelled, $\frac{1}{5}$ by $\frac{1}{12}$ in., dense; glumes elliptic, keeled, dusky, the 3 nerves coalescing into a curved excurrent mucro; stamens usually 2; style small, hairy; branches 2; nut $\frac{1}{3}$ as long as the glume, obovoid, biconvex, pale straw-colour, smooth; hairs 10–18, simple, white, $\frac{1}{2}$–$\frac{2}{3}$ the length of the nut, hanging from the style-base closely round the nut, falling off with the deciduous style. *Kunth, Enum.* ii. 224; *Boeck. in Linnæa*, xxxvii. 10; *C. B. Clarke in Hook. f. Fl. Brit. Ind.* vi. 635. *F. Ecklonii, Nees in Linnæa*, ix. 290, x. 145; *Kunth, Enum.* ii. 226. *Scirpus squarrosus, Poir. Encyc. Suppl.* v. 100, *not of Linn.*

COAST REGION: Clanwilliam Div.; Doorn River, below 1000 ft., *Drège!* Ebenezer, near the Oliphants River, below 100 ft., *Drège!*
WESTERN REGION: Clanwilliam Div.; near the Oliphants River, below 1000 ft., *Zeyher*, 1779 partly! *Ecklon and Zeyher!*

In all tropical and warm-temperate regions.

2. F. dichotoma (Vahl, Enum. ii. 287); annual, puberulous or minutely pubescent; stem 2–8 in. long; leaves often as long as the stem, $\frac{1}{12}$ in. broad; umbel usually compound (or decompound); bracts 3–4, often as long as the umbel, similar to the leaves; spikelets all pedicelled, $\frac{1}{5}$ by $\frac{1}{12}$ in., dense; glumes boat-shaped, ovate, acute, 3–1-nerved; stamens 2–1 (or sometimes 3 according to Boeckeler); style nearly always hairy, branches 2, shortish; nut $\frac{1}{3}$ the length of the glume, obovoid, biconvex, pale or scarcely brown, with 5–9 ribs on each face due to the cells being vertically superposed. *Kunth, Enum.* ii. 225; *Boeck. in Linnæa*, xxxvii. 12 *partly; C. B. Clarke in Hook. f. Fl. Brit. Ind.* vi. 635. *Scirpus dichotomus, Linn. Sp. Plant. ed.* 2, 74; *Rottb. Descr. et Ic.* 57, *t.* 13, *fig.* 1 (*cf. Hoppe in Flora*, 1825, 177–181).

WESTERN REGION: Clanwilliam Div.; near the Oliphants River, below 1000 ft., *Zeyher*, 1779 partly!

Ranges from Portugal and West Africa to Hong Kong and New South Wales.

This very common weed is difficult to distinguish from *F. diphylla*. Boeckeler's varieties of *F. dichotoma* and his American examples are here referred to *F. diphylla*. In *F. dichotoma*, as here understood, the glumes are distinctly keeled, so that the spikelets have numerous short longitudinal angles.

3. F. diphylla (Vahl, Enum. ii. 289); glabrous or more or less pubescent; stem commonly 12–20 in. long; umbel compound, though reduced sometimes to a few spikelets; glumes ovate, glabrous, concave, without distinct keel; style 2-fid; nut with 5–10 longitudinal ribs on each face; otherwise as *F. dichotoma.*—*C. B. Clarke in Hook. f. Fl. Brit. Ind.* vi. 636. *F. communis, Kunth, Enum.* ii. 234. *F. dregeana, Kunth, Enum.* ii. 232. *F. polymorpha, Boeck. in Linnæa,* xxxvii. 14. *Scirpus diphyllus, Retz. Obs.* v. 15.

SOUTH AFRICA: without locality, *Harvey,* 65!

KALAHARI REGION: Orange Free State, *Buchanan,* 147! Transvaal; Houtbosch, *Rehmann,* 5622! 5623! 7714! and without precise locality, *Kuntze,* 243!

EASTERN REGION: Tembuland; Umtata River, below 1000 ft., *Drège!* Pondoland; between the Umtentu River and Umzimkulu River, *Drège!* and without precise locality, *Bachmann,* 113! 125! Griqualand East, near Clydesdale, 2500 ft., *Tyson,* 1191! Natal, *Drège,* 4371! 4373! *Buchanan,* 81! 118! 119! 338! 339! 340! 341! *Gerrard,* 487! *Wood,* 6012!

All tropical and warm-temperate regions.

The most abundant and widely-distributed of sedges, having (when most narrowly diagnosed) 140 names. The Cape examples are mostly tall, with broader leaves than in *F. dichotoma; F. dregeana,* Kunth, is 3–4 in. only, but has much stouter, rounder spikelets than *F. dichotoma.*

4. **F. ferruginea** (Vahl, Enum. ii. 291); stems closely cæspitose on a very short, or obsolete, rhizome, 8–30 in. long; leaves short, 4–8 in. long, or hardly any, hairy or glabrate, $\frac{1}{10}$ in. broad or usually less; umbel compound or simple, usually of 5–15 pedicelled spikelets, sometimes of 1 spikelet; bracts 3–2, usually short, occasionally as long as the umbel; spikelets $\frac{1}{3}$ by $\frac{1}{8}$–$\frac{1}{6}$ in., ovoid, rusty-brown; glumes round-backed, ovate, nearly always minutely pubescent in the upper half of the back; stamens 3–2; style longer than the nut, hairy, branches 2, rather short; nut scarcely $\frac{1}{2}$ the length of the glume, biconvex, obovoid, smooth, pale brown, not, or very obscurely, longitudinally ribbed. *Kunth, Enum.* ii. 236; *Boeck. in Linnæa,* xxxvii. 16; *C. B. Clarke in Hook. f. Fl. Brit. Ind.* vi. 638. *F. sieberiana, Kunth, Enum.* ii. 237. *Scirpus ferrugineus, Linn. Sp. Plant. ed.* 2, 74.

SOUTH AFRICA: without locality, *Harvey!*

COAST REGION: Bathurst Div.; Fish River, below 1000 ft., *Drège!* Komgha Div.; Kei River, *Flanagan,* 971!

KALAHARI REGION: Orange Free State, *Buchanan,* 145! Transvaal; Bosch Veld, *Rehmann,* 4793!

EASTERN REGION: Pondoland; near the mouth of the Umtsikaba River, *Drège!* and without precise locality, *Bachmann,* 112! Natal; Durban Flat, *Buchanan,* 2! 25! 55! Coast-land, *Sutherland!* Delagoa Bay, *Kuntze,* 221!

All tropical and warm-temperate regions.

This abundant *Fimbristylis* is usually correctly named in herbaria from the minute white or greyish hairs on the top of the backs of the glumes; but the glumes are occasionally absolutely glabrous. Harvey's 7 stems have only 1 spikelet on each.

5. **F. exilis** (Roem. et Schultes, Syst. Veg. ii. 98); annual, hairy; stems tufted, 4–16 in. long, rather slender, hairy or rarely nearly

glabrous; leaves $\frac{1}{3}$–$\frac{1}{2}$ the length of the stem, $\frac{1}{12}$ in. broad, hairy or rarely glabrate; sheaths always hairy; umbel of 3–14 pedicelled spikelets; bracts 4–3, about as long as the umbel; spikelets $\frac{1}{4}$–$\frac{1}{3}$ by $\frac{1}{8}$ in., 10–20-flowered; glumes boat-shaped, ovate, acute, chestnut-brown, puberulous; keel green, excurrent in a point; stamens 3–2; style long, slender, glabrous, branches 3, long; nut rather large, triquetrous, obovoid, pale, its subconcave faces often obscurely marked with transverse wavy lines, yellowish but finally ashy-brown. *F. hispidula, Kunth, Enum.* ii. 227; *Krauss in Flora*, 1845, 757; *Boeck. in Linnæa*, xxxvii. 27. *Scirpus hispidulus, Vahl, Enum.* ii. 276. *Cyperus hirtus, Thunb. in Hoffm. Phytogr. Blaett.* i. 6, *and Fl. Cap. edit. Schult.* 100; *Kunth, Enum.* ii. 106. *Chætospora distachya, Nees in Linnæa*, x. 192?

SOUTH AFRICA: without locality, *Drège*, 4372!

KALAHARI REGION: Kalahari, *Schinz*, 371! Transvaal; near Klerks-dorp, 4000 ft., *McLea in Herb. Bolus*, 6018!

EASTERN REGION: Transkei! banks of the Bashee River, below 1000 ft., *Drège!* Natal; near the Umlaas River, *Krauss*, 185! Sydenham, near Durban, *Wood*, 1956! and without precise locality, *Buchanan*, 341! *Kuntze*, 23! Zululand, *Gerrard*, 490!

Common throughout Africa and Madagascar, rare in Tropical America.

This species is the connecting-link between *Fimbristylis* and *Bulbostylis*; its hairy stem is common in *Bulbostylis*, very rare in *Fimbristylis*; the spikelets, with hairy glumes, resemble those of *Bulbostylis*. It differs from *Bulbostylis* in that the style is somewhat persistent; and, when it does fall, it takes the whole style-base, and does not leave behind the dark-coloured button so conspicuous on the nut of *Bulbostylis*.

6. F. complanata (Link, Hort. Berol. 1827, i. 292); glabrous; stems tufted on a very short, or obsolete, rhizome, 6–24 in. long, often much flattened at the top; leaves often 4–8 by $\frac{1}{5}$ in., obtuse; umbel with 6–180 pedicelled spikelets; bracts 2, shorter than the umbel, often suberect, like the leaves, with obtuse tip; spikelets $\frac{1}{4}$ by $\frac{1}{12}$ in., dense, chestnut-red or brownish-red; glumes boat-shaped, ovate, acute; stamens often 3; style longer than the nut, linear, glabrous, branches 3, linear; nut very small, narrow-obovoid, round trigonous, straw-coloured, slenderly longitudinally ribbed because the transversely oblong cells are in vertical series. *Kunth, Enum.* ii. 228; *C. B. Clarke in Hook. f. Fl. Brit. Ind.* vi. 646. *F. autumnalis, var. α, Boeck. in Linnæa*, xxxvii. 38. *Scirpus complanatus, Retz. Obs.* v. 14. *Trichelostylis complanata, Nees in Linnæa*, x. 146.

VAR. β, **kraussiana** (C. B. Clarke in Hook. f. Fl. Brit. Ind. vi. 646); stems slenderer, often scarcely at all flattened at the apex; umbel more compact, of 10–30 spikelets. *F. kraussiana, Hochst. ex Krauss in Flora*, 1845, 757.

VAR. γ, **consanguinea** (C. B. Clarke in Durand and Schinz, Conspect. Fl. Afr. v. 602); spikelets often in clusters of 2–5 on the ultimate rays of the umbel. *F. consanguinea, Kunth, Enum.* ii. 228; *Boeck. in Linnæa*, xxxvii. 36.

SOUTH AFRICA: without locality, *Drège*, 6448! Var. γ, *Drège*, 4414! 4418! 7404! *Ecklon and Zeyher*, 62!

COAST REGION: Komgha Div., 2000 ft., *Flanagan*, 960! Var. γ, Uitenhage Div.; Van Stadens Hoogte, *MacOwan*, 1354! 1354b! Alexandria Div.; between Hoffmanns Kloof and Drie Fontein, 2000–3000 ft., *Drège!* Albany Div.; Howisons Poort, 1800 ft., *MacOwan*, 1352!

KALAHARI REGION: Var. β, Orange Free State, *Buchanan*, 142! Transvaal; Pretoria, *Rehmann*, 4771! Magalies Berg, *E.S.C.A. Herb.*, 309! and without precise locality, *Rehmann*, 5619!

EASTERN REGION: Pondoland, *Bachmann*, 108! Natal; Umzinyati, 1200 ft., *Wood*, 1399! Durban Botanic Gardens, *Wood*, 3151! Durban Flats, *Wood*, 4003! Nottingham, *Buchanan*, 132! Var. β, Natal; marsh near the Umlaas River, *Krauss*, 180! Nottingham, *Buchanan*, 104! and without precise locality, *Gerrard*, 696! *Buchanan*, 343! *Rehmann*, 8603! 8604! Var. γ, Tembuland; between the Bashee River and Morley, 1000–2000 ft., *Drège*. Pondoland; near the mouth of the Umtsikaba River, *Drège*. Natal, *Gueinzius*, 215!

Found throughout the tropical and warm-temperature regions of the World.

The character "solitary" or "clustered" spikelets has been used as of "sectional" value in *Fimbristylis*; and Kunth therefore put forward *F. consanguinea* as a new species. But the character is not of specific value; nor is so the character taken from the flattening of the top of stem. Boeckeler, perhaps rightly, unites this whole series of plants under *F. autumnalis*, Roem. et Schult. (i.e. *Scirpus autumnalis*, Linn.). Where the species, or varieties, are so close together, it is not possible to state the distribution, except as matter of opinion.

7. **F. obtusifolia** (Kunth, Enum. ii. 240); glabrous, rigid; rhizome short, thick, woody; stems 4–18 in. long; leaves many, usually about ½ the length of the stem, $\frac{1}{10}$–$\frac{1}{8}$ in. broad, hard, obtuse; umbel 1–3 in. in diam., simple or compound, or condensed into one head; bracts 3–2, lowest ½–2½ in. long, shorter or longer than the umbel, rigid, similar to the leaves; spikelets usually clustered, ⅛ in. long, ellipsoid, pale, dense; glumes ovate, obtuse; style nearly as long as the nut, glabrous, branches 3 (at least in the lower flowers), as long as the style; nut hardly ½ the length of the glume, obovoid, round-trigonous, dark-coloured, not longitudinally ribbed. *F. glomerata, Boeck. in Linnæa*, xxxvii. 47 *partly. Isolepis obtusifolia, Beauv. Fl. d'Owar*. ii. 38, *t*. 81, *fig*. 1. *Trichelostylis obtusifolia, Nees in Linnæa*, ix. 290.

EASTERN REGION: Pondoland; between the Umtentu River and Umzimkulu River, below 500 ft., *Drège*, 4415! Natal, *Rehmann*, 8602! Delagoa Bay, *Kuntze*, 215!

Also in Tropical Africa, the Mascarene Islands, and America; near salt or brackish water.

8. **F. monostachya** (Hassk. Pl. Jav. Rar. 61); glabrous; stems clustered on a very short woody rhizome, 3–11 in. long, slender; leaves often ⅔ the length of the stem, $\frac{1}{15}$–$\frac{1}{10}$ in. broad; spikelet 1 (rarely 2 or 3), ½ by ⅕ in. and few-flowered, or in fruit 1 in. long and many-flowered; bract suberect, rarely much overtopping the spikelet; spikelet greenish-white or yellowish, in the young state much flattened (resembling *Cyperus*) with the fertile glumes apparently distichous; in ripe spikelets, though the lowest glumes may be nearly 2-ranked, the scars of the glumes are perfectly spiral on the rhachilla; glumes boat-shaped, ovate, without nerves except those forming keel, minutely mucronate; stamens 3; anthers oblong-linear, muticous; nut rather large, less than ½ the length of the glume, triquetrous, obovoid, almost stalked, straw-coloured or scarcely brown, without striations, but often minutely tubercled; style longer than the nut, hairy, slightly thickened at the base, com-

pletely deciduous with its base, branches 3 linear, shortish. *C. B. Clarke in Hook. f. Fl. Brit. Ind.* vi. 649. *Cyperus monostachyos, Linn. Mant.* 180; *Rottb. Descr. et Ic.* 18, *t.* 13, *fig.* 3. *Abildgaardia monostachya, Vahl, Enum.* ii. 296; *Kunth, Enum.* ii. 247; *Boeck. in Linnæa,* xxxvii. 53.

SOUTH AFRICA: without locality, *Drège*, 4377! *Zeyher*, 1752!
COAST REGION: Albany Div.; near Grahamstown, *MacOwan*, 188! Komgha Div.; *Flanagan*, 914! Queenstown Div.; Shiloh, 3000–4000 ft., *Drège*.
KALAHARI REGION: Transvaal; Magalies Berg, *Burke!* and without precise locality, *Rehmann*, 4327!
EASTERN REGION: Transkei; between Gekau and the Bashee River, 1000–2000 ft., *Drège*. Tembuland; between the Bashee River and Morley, 1000–2000 ft., *Drège!* Bazeia, 2000 ft., *Baur*, 64! 412! Natal; Umgeni River, *Rehmann*, 8593! Colenso, 3400 ft., *Kuntze*, 238!

Widely distributed in the tropical and warm-temperate regions of the globe.

VIII. BULBOSTYLIS, Kunth.

Spikelet of many spirally imbricate glumes, lowest 1 or 2 empty, many or several succeeding glumes 2-sexual perfecting nuts, uppermost male or sterile. Hypogynous *bristles* 0. *Stamens* 3–2, anterior. *Style* glabrous, linear, deciduous with its slightly enlarged base, leaving a dark-coloured button on the nut; branches 3 or 2, linear. *Nut* triangular or flat (dorsally compressed), obovoid, very obtuse or truncate, pale till quite ripe, the dark-coloured button finally falling off.

Stolons 0; leaves all near the base of the stem, very narrow, often hairy, nearly always ciliate-hairy in the mouth of the sheaths; inflorescence a terminal umbel, simple or compound, or a single terminal spikelet; flower-glumes nearly always pubescent.

DISTRIB. Species 130; in all tropical and warm-temperate regions, especially abundant in Africa and America, one species reaching north to Canada. [See also *Fimbristylis exilis.*]

Stem with 1 (or 1–2) spikelets:
  Style-branches 2; plants nearly glabrous:
    Nut smooth ... ... (1) **humilis.**
    Nut longitudinally striate ... ... (2) **striatella.**
  Style-branches 3; plant hairy ... ... (3) **breviculmis.**
Stem with 1 dense head of several spikelets; style-branches 3:
  Stem minutely and densely hairy at the top; spikelets small; leaf-sheaths fine woolly; head 1, small (4) **filamentosa.**
  Stem glabrous; leaf-sheaths hardly ciliate even in their throats:
    Leaves longish; flower-glumes very obtuse; anthers crested ... ... (5) **schœnoides.**
    Leaves longish; flower-glumes acuminate ... (6) **scleropus.**
    Leaves hardly any; nut minute ... ... (7) **parvinux.**
Stem with 1, (or 1–4) umbellate, heads of spikelets; leaf-sheaths with long white hairs, often conspicuously hairy in the throat:
  Head 1, with long white hairs; nut smooth, not zonate ... ... (8) **collina.**

Head 1, without long white hairs; nut not zonate ... (9) **cardiocarpa.**
Head 1, with few long white hairs; nut transversely lineolate ... ... (10) **Burkei.**
Heads 3–1, simply umbellate; flower-glumes white-margined ... ... (11) **Zeyheri.**
Heads 3–1, simply umbellate, flower-glumes not white-margined ... ... (12) **cinnamomea.**
Spikelets many, solitary; umbel often compound:
Spikelets 1–2 (sometimes 3) together; umbel often compound ... ... (13) **Kirkii.**
Spikelets nearly all solitary:
Annual, with a little long white hair ... ... (14) **capillaris.**
Rhizome woody, with long white hair in the leaf-sheaths ... ... (15) **Burchellii.**

1. **B. humilis** (Kunth, Enum. ii. 207, cf. 205); glabrate or scarcely hairy, annual, tufted; stem ½–4 in. long, slender, glabrous, carrying 1 (rarely 1–2) spikelets; leaves often overtopping the stems, setaceous, glabrous, or obscurely ciliate at the base, sometimes microscopically scabrous-hairy; spikelets ¼ by ½ in., of 3–6 fertile flowers, pale, with a browner lanceolate tip of male glumes; basal spikelets above ground but concealed among the tufted stems are often added; bracts shorter or considerably longer than the spikelet, similar to the leaves; glumes elliptic, with a green 3-nerved keel excurrent in a curved mucro; stamens 3–2; nut ⅗ the length of the glume, subsessile, obovoid, dorsally compressed, pale; external layer of cells, nearly square, obscure, small, so that the nut appears smooth, microscopically reticulated, without longitudinal ribs; style as long as the nut, linear, smooth, with 2 linear branches, falling off early, leaving a minute discoloured button or small tubercle on the pale nut. *Fimbristylis arenaria, Nees in Linnæa,* x. 146; *Kunth, Enum.* ii. 244. *Isolepis humilis, Steud. Syn. Pl. Glum.* ii. 100. *I. arenaria, Drège in Linnæa,* xx. 247. *Scirpus arenarius, Boeck. in Linnæa,* xxxvi. 741.

SOUTH AFRICA: without locality, *Drège*, 3946! 4374! *Verreaux! Boivin! Harvey!*

COAST REGION: Cape Div.; Simons Bay, *Wright*, 514! Ceres Div.; near Ceres, *Rehmann*, 2366! Uitenhage Div.; Zwartkops River, *Zeyer*, 4385! Port Elizabeth, *Herb. Schinz*, 158! *E.S.C.A. Herb.*, 105!

CENTRAL REGION: Somerset Div.; Bosch Berg, *MacOwan*, 1348! 1258! Aliwal North Div.; Witte Bergen, 5000–6000 ft., *Drège!*

KALAHARI REGION: Orange Free State, *Rehmann*; 3623!

EASTERN REGION: Transkei; Bashee River, below 1000 ft., *Drège*. Natal Inanda, *Wood*, 1587!

2. **B. striatella** (C. B. Clarke in Durand and Schinz, Conspect. Fl. Afr. v. 616); nut with 18–20 longitudinal ribs on each convex face, somewhat trabeculate by reason of the transversely oblong cells vertically superposed in one series between each two ribs; otherwise as *B. humilis*. *Isolepis humillima, Hochst. ex C. B. Clarke in Durand and Schinz, Conspect. Fl. Afr.* v. 616.

SOUTH AFRICA: without locality, *Schlechter*, 381!

COAST REGION: Komgha Div.; rocks near Komgha, 2000 ft., *Flanagan*, 1266!

KALAHARI REGION: Orange Free State; rocks near Harrismith, 5000 ft., *Wood*, mixed with 4762!
EASTERN REGION: Natal, *Buchanan*, 86! *Cooper*, 3361!

Also in Abyssinia.

This species is usually taller (up to 3–6 in. high) and more perfectly glabrous than *B. humilis*; but it has spikelets added similarly among the bases of the stems, and is so exceedingly like *B. humilis* that I do not think the two distinguishable except by the nut. These two species in their 2-fid style and glabrescent habit differ from the whole genus *Bulbostylis*.

3. **B. breviculmis** (Kunth, Enum. ii. 207, cf. 205); hairy, tufted, annual; stem 1–2½ in. long, slender, hairy, carrying 1 spikelet; basal spikelets sometimes added as in the two preceding species; leaves as long as the stem, setaceous; spikelets $\frac{1}{4}$–$\frac{1}{3}$ by $\frac{1}{10}$ in., about 10-flowered; bracts about as long as the spikelet; glumes ovate-acuminate, rusty-brown, green 3-nerved keel excurrent in a long mucro; style hardly as long as the nut, linear, glabrous, branches 3 linear, longish; nut $\frac{2}{5}$ the length of the glume (mucro included), subequally trigonous, obovoid, yellow-brown, not striate, slightly verrucose; tubercle or button (left on nut by the deciduous style) hardly darker in colour than the nut. *Isolepis breviculmis, Steud. Syn. Pl. Glum.* ii. 100. *Scirpus breviculmis, Boeck. in Linnæa*, xxxvi. 742.

CENTRAL REGION: Aliwal North Div.; between Aliwal North (Buffel Vallei) and the Kraai River, 4500–5000 ft., *Drège*, 3947! Herb. Berlin and Delessert.

No specimen at Kew.

4. **B. filamentosa** (Kunth, Enum. ii. 210, cf. 205); stem densely tufted, 1 ft. long, very slender, minutely densely hairy at the top, 1-headed; leaves nearly as long as the stem, setaceous, hairy; sheaths with long, white, thin hairs; head $\frac{1}{4}$ to $\frac{1}{3}$ in. in diam., dense with many small spikelets, chestnut-brown; bracts $\frac{1}{4}$ to $\frac{1}{3}$ in. long, setaceous; spikelets $\frac{1}{6}$ by $\frac{1}{12}$ in., 6–10-flowered; glumes boat-shaped, ovate, minutely hairy, without long white hair; keel 3–1-nerved, paler, excurrent in a mucro; stamens 3; anthers linear, dark-red, with a small linear crest; nut $\frac{2}{5}$ the length of the glume, broadly obovoid, white, finally pale brown, obscurely transversely wavy lineolate (owing to the small longitudinal external cells); style as long as the nut, linear, glabrous, branches 3, rather longer than the style; style-base ovoid-trigonous, persistent after the style drops, hardly darker than the nut, rather large for the genus. *B. Rehmanni, C. B. Clarke in Durand and Schinz, Conspect. Fl. Afr.* v. 615. *Scirpus filamentosus, Vahl, Enum.* ii. 262; *Boeck. in Linnæa*, xxxvi. 747, *excluding American syn. and plants. Isolepis filamentosa, Roem. et Schult. Syst. Veg.* ii. 113.

KALAHARI REGION: Transvaal; Houtbosch, *Rehmann*, 5611! 5617!

Also found in West Tropical Africa from Sierra Leone to Angola.

In Rehmann's examples the nut is narrower than in the Niger typical examples, and without transverse wavy lineolations, the external cells being nearly quadrate.

5. **B. schœnoides** (Kunth, Enum. ii. 208, cf. 205); nearly glabrous, except the flower glumes; rhizome woody; stems approximate, 4–10 in.

long, slender, glabrous at the top, 1-headed; leaves $\frac{1}{3}$–$\frac{2}{3}$ the length of the stem, setaceous, channelled, minutely scabrous on the edges, without long white hairs; lower leaf-sheaths weak, pale rusty-brown, nearly glabrous, spikelets 3–5, $\frac{1}{4}$ by $\frac{1}{8}$ in., very obtuse, chestnut-brown; bracts 2–1, hardly longer than the head, not green; glumes ovate, obtuse, almost truncate, narrowly white margined at the almost fimbriate tip; keel hardly reaching the tip; stamens 3; anthers linear-oblong, rather large, not crested; style linear, shorter than the young ovary, branches 3 linear, longish; style-base bulbous, exactly of young *Bulbostylis;* nut unknown. *C. B. Clarke in Durand and Schinz, Conspect. Fl. Afr.* v. 616 *partly. Isolepis schœnoides, Steud. Syn. Pl. Glum.* ii. 100. *Scirpus schœnoides, Boeck. in Linnæa,* xxxvi. 745.

COAST REGION: King Williamstown Div.; between the Yellowwood River and Zandplaat, 1000–2000 ft., *Drège.* Stockenstrom Div.; Kat Berg, 4000–5000 ft., *Drège,* 3936!

CENTRAL REGION: Aliwal North Div.; Witte Bergen, 7000–8000 ft., *Drège,* 4378!

Sir J. D. Hooker first separated the Tropical African *Schœnus ? erraticus* with hesitation, and then united it specifically; it has not the very obtuse glumes of *B. schœnoides.*

6. B. scleropus (C. B. Clarke); nearly glabrous, except the flower-glumes; stems tufted; enclosed by $\frac{1}{5}$ in. broad, harsh, chestnut-black, striate leaf-sheaths, 10–16 in. long, glabrous at the top, 1-headed; leaves often $\frac{2}{3}$ the length of the stem, narrowly linear, channelled, minutely scabrous on the edges, without long white hairs; spikelets 6–10, crowded in 1 head, $\frac{2}{5}$ by $\frac{1}{6}$ in., cylindric-lanceolate, chestnut-black; bracts 2–1, shorter than the head; glumes elliptic-oblong, hardly acute; stamens 3; anthers large, linear-oblong, not crested; nut obovoid, trigonous, pale, finally brown, obscurely transversely lineolate; style linear, longer than nut, deciduous, leaving a dark-coloured button on the nut, branches 3 linear, as long as the style. *B. schœnoides, C. B. Clarke in Durand and Schinz, Conspect. Fl. Afr.* v. 616 *partly.*

KALAHARI REGION: Transvaal; Apies River, *Burke!*

EASTERN REGION: Tembuland, Bazeia, 2000 ft., *Baur,* 285!

7. B. parvinux (C. B. Clarke); nearly glabrous, except the flower-glumes; stem 20 in. long, slender, 1-headed, glabrous at the top, almost leafless, the glabrous, pale-brown, weak leaf-sheaths produced on one side about 1 in., not green; head 1 in. in diam., dense, pale-brown, the numerous spikelets spreading on all sides, concealing the small bracts; spikelets nearly $\frac{1}{2}$ by $\frac{1}{8}$ in., many-flowered; glumes ovate, subacute, minutely pubescent; keel hardly excurrent; stamens 3; anthers large, linear-oblong, not crested; nut minute, hardly $\frac{1}{4}$ the length of the glume, obovoid, round-trigonous, white, ultimately yellow-brown, smooth, not, or most obscurely, lineolate; style linear, smooth, twice as long as the nut, deciduous, leaving the brown style-base on the nut, branches 3 linear, shorter than the style.

EASTERN REGION: Delagoa Bay, *Kuntze,* 217!

The spikelets greatly resemble those of *B. cinnamomea* (below), which differs in having ciliate-hairy leaf-sheaths, and very much larger nuts.

8. **B. collina** (Kunth, Enum. ii. 208, cf. 205); rhizome short, woody; stems closely tufted, 4–10 in. long, glabrous at the top, 1-headed; leaves numerous, short, $\frac{1}{4}$–$\frac{1}{2}$ the length of the stem. setaceous, hairy; leaf-sheaths with long white hair in their throats; head $\frac{1}{2}$ in. in diam., of 5–12 spikelets, more or less hairy with long white hair; bracts scarcely longer than the head; spikelets $\frac{1}{3}$ by $\frac{1}{6}$ in., oblong-cylindric, chestnut-brown; glumes elliptic-oblong, minutely pubescent, and often also with long white hairs; keel green scarcely excurrent; stamens 3; anthers linear-oblong, not cristate; nut $\frac{2}{5}$ the length of the glume, rather broadly obovoid trigonous, pale, ultimately brown, smooth, not transversely wavy lineolate (outermost cells being subquadrate); style linear, about as long as the nut, glabrous, deciduous leaving a small dark-coloured tubercle on the nut, branches 3 linear, about as long as the style. *C. B. Clarke in Durand and Schinz, Conspect. Fl. Afr.* v. 613 *partly. Ficinia contexta, Nees in Linnæa,* ix. 292. *Trichelostylis contexta, Nees in Linnæa,* x. 146. *Fimbristylis contexta, Kunth, Enum.* ii. 245. *Isolepis collina, Steud. Syn. Pl. Glum.* ii. 101. *Scirpus collinus, Boeck. in Linnæa,* xxxvi. 746.

COAST REGION: Uitenhage Div., *Zeyher!* Alexandria Div.; Addo, 1000–2000 ft., *Drège,* 2037! Albany Div.; Blue Krantz, *Burchell,* 3629! Grahamstown, 2000 ft., *MacOwan,* 1972! near the Bushman River, below 1000 ft., *Drège;* Bathurst Div.; between Blue Krantz and Kaffir Drift Military Post, *Burchell,* 3686!

KALAHARI REGION: Transvaal; hills above the Apies River, *Rehmann,* 4329!

EASTERN REGION: Swaziland; Havelock Concession, 4000 ft., *Saltmarshe in Herb. Galpin,* 1021b!

Also in South Tropical Africa?

There are numerous very closely allied species. *B. collina* is here described from Drège's specimens in Herb. Kew, distributed as "*Isolepis collina,* Kth. *a*"; and the examples cited are only those in Herb. Kew which exactly match them. If the species be taken in a wider sense as in Durand and Schinz, Conspect. Fl. Afr. v. 613, its distribution will be wider.

9. **B. cardiocarpa** (C. B. Clarke in Durand and Schinz, Conspect. Fl. Afr. v. 612); leaf-sheaths with long white hair in their throats, but much less than in *B. collina;* heads and flower-glumes without long white hair; spikelets rather narrowly lanceolate, acute; nut narrower than that of *B. collina;* otherwise as *B. collina. Fimbristylis cardiocarpa, Ridley in Trans. Linn. Soc. ser.* 2, ii. 154; *not of F. Mueller.*

EASTERN REGION: Tembuland; Cofimvaba, 2500 ft., *Baur,* 478! 479!

Also in South Tropical Africa.

This may be made a var. of *B. collina;* but the two plants do not match; in *B. collina* the heads are subovoid; in *B. cardiocarpa* the lanceolate separate tips of the spikelets stick out almost stellately from the heads.

10. **B. Burkei** (C. B. Clarke); heads less dense than those of *B. collina,* with hardly any long white hair; nut oblong, pear-shaped, somewhat narrowed at the top, yellow-brown, transversely lineolate,

almost zonate, by reason that the outermost cells are rather large and longitudinally oblong; otherwise as *B. collina*, Kunth.

SOUTH AFRICA: without locality, *Zeyher*, 1769!
KALAHARI REGION: Orange Free State; Caledon River, *Burke*, 332!

From Kunth's description of the nut of *B. collina*, it seems likely that he had before him the nut of *B. Burkei* as well. The nut of *B. Burkei* has a different structure from that of *B. collina*, but is extremely like the nut of *B. Zeyheri*, of which species *B. Burkei* may possibly be a form.

11. **B. Zeyheri** (C. B. Clarke in Durand and Schinz, Conspect. Fl. Afr. v. 616); stems 4–10 in. long, tufted on a very short woody rhizome; leaves ½ as long as the stem, setaceous; leaf-sheaths weak, with much long white hair in their throats; heads 1–4, each of 2–7 spikelets; rays ½–1½ in. long glabrous; spikelets chestnut-brown, with little, or no, long white hair; glumes ovate, acute, minutely pubescent on the sides, white-margined; keel of 3–1 green nerves; style-branches 3; nut oblong, pear-shaped, somewhat narrowed at the top, yellow-brown, transversely lineolate, almost zonate by reason that the outermost cells are rather large and longitudinally oblong. *Scirpus Zeyheri, Boeck. in Linnæa*, xxxvi. 752. *S. macrolepis, Boeck. ex C. B. Clarke in Durand and Schinz, Conspect. Fl. Afr.* v. 616 *partly.*

KALAHARI REGION: Transvaal; Magalies Berg, *Zeyher*, 1768! *Burke!* hills above the Apies River, *Rehmann*, 4325! Wonderboom Poort, *Rehmann*, 4478!
EASTERN REGION: Natal; Durban Flat, *Wood*, 4008!

Also in South East Tropical Africa.

This species is perhaps not distinct from *B. Burkei*, which has an exactly similar nut. One-headed examples of *B. Zeyheri* have more long white hairs and margined flower-glumes.

12. **B. cinnamomea** (C. B. Clarke in Durand and Schinz, Conspect. Fl. Afr. v. 612); stems 1–2 ft. long, densely tufted on a very short woody rhizome; leaves ⅓ as long as the stem, setaceous, more or less hairy; leaf-sheaths weak, with some long white hair in the throat; heads 1–3, each of 2–9 spikelets; rays ½–1½ in. long, glabrous; spikelets ferruginous-brown, with few, or no, long white hairs; glumes ovate, acute, minutely pubescent, not white-margined; keel green or yellowish; style-branches 3; nut obovoid, almost truncate, trigonous, yellow-brown, transversely lineolate by reason that the outermost cells are longitudinally oblong; style deciduous, leaving an ovoid black-purple tubercle on the nut. *Scirpus cinnamomeus, Boeck. in Engl. Jahrb.* v. 505. *S. transvaalensis, Boeck. ex C. B. Clarke in Durand and Schinz, Conspect. Fl. Afr.* v. 613. *S. macrolepis, Boeck. ex C. B. Clarke in Durand and Schinz, Conspect. Fl. Afr.* v. 616, *as to Rehmann*, 8620.

KALAHARI REGION: Transvaal; Houtbosch, *Rehmann*, 5618! 5621!
EASTERN REGION: Natal; Inanda, *Wood*, 1352! *Rehmann*, 8620! and without precise locality, *Gerrard*, 697! 698! *Buchanan*, 337!

Also in Nyassaland and Madagascar.

13. **B. Kirkii** (C. B. Clarke in Durand and Schinz, Conspect. Fl.

Afr. v. 614); umbels somewhat lax, often compound with secondary rays ½ in. long; spikelets solitary or paired; otherwise as *B. cinnamomea.*

EASTERN REGION: Natal; Inanda, *Wood*, 1576!

14. **B. capillaris** (Kunth, Enum. ii. 211, cf. 205); annual; stems tufted, 3–15 in. long, setaceous, glabrous; leaves ⅔ the length of the stem, setaceous, nearly glabrous; leaf-sheaths more or less long, white-hairy in their throat; umbel simple or compound, or reduced to one spikelet, 1–8 in. in diam.; bracts shorter than the umbel; spikelets all or mostly solitary, scarcely ⅙ in. long, chestnut-brown; glumes ovate, acute, most minutely pubescent, or glabrous; keel usually green, subexcurrent; stamens often 2; anthers not aristate; nut less than ½ the length of the glume, obovoid, obtuse, pale-brown, scarcely transversely undulate; style slender, glabrous, longer than the nut, deciduous, leaving a black-red tubercle on the nut, branches 3, longer than the style. *C. B. Clarke in Hook. f. Fl. Brit. Ind.* vi. 652. *Scirpus capillaris, Linn. Sp. Plant. ed.* 2, 73; *Boeck. in Linnæa,* xxxvi. 759. *Isolepis capillaris, Roem. and Schultes, Syst. Veg.* ii. 118; *Mant.* ii. 68, 533.

EASTERN REGION: Pondoland, 4500–7200 ft., *Sutherland!*

Abundant in the warm regions of both hemispheres.

15. **B. Burchellii** (C. B. Clarke in Durand and Schinz, Conspect. Fl. Afr. v. 612); stems tufted closely on a woody rhizome, very slender, glabrous; leaf-sheaths conspicuously white-hairy in the throat; spikelets when in flower about ⅕ in. long, hardly different from those of *B. capillaris,* but in fruit (in examples from different localities) elongating to ⅓ in. (sometimes nearly ½ in.) long; otherwise as *B. capillaris. Fimbristylis Burchellii, Ficalho and Hiern in Trans. Linn. Soc. ser.* 2, ii. 28, *t.* 6 *B, fig.* 7–15. *F. hispidula, Boeck. in Linnæa,* xxxvii. 27 *partly.*

KALAHARI REGION: Griqualand West; Klip Fontein, *Burchell,* 2151; Bechuanaland; Kosi Fontein, *Burchell,* 2589! 2598! Transvaal; Pretoria Div., hills above the Apies River, *Rehmann,* 4330! Potchefstroom; by the Mooi River, *Nelson,* 57!

Also in South Tropical Africa.

IX. SCIRPUS, Linn.

Spikelet of many, rarely few, spirally imbricate glumes, lowest 1 or 2 empty, many or several succeeding glumes 2-sexual perfecting nuts (cf. *Sc. dioicus*), uppermost male or sterile. Hypogynous *bristles* 0 or 2–6, undivided. *Stamens* 3–2, anterior. *Style* glabrous, linear, passing gradually into the nut (except in *Sc. Hystrix and Sc. Isolepis* where it is very short and deciduous); branches 3 or 2, linear.

Ovary sessile, or often much narrowed at the base, pyriform, but not on a minute obpyramidal gynophore. *Nut* triangular or flat (dorsally compressed).

Leaves all near the base of the stem (in the Cape species) except the *Fluitantes*, linear, glabrous; flower-glumes perfectly glabrous. *Ficinia* only differs from the *Scirpi* without hypogynous bristles in having a most minute obpyriform gynophore. The distinction is artificial, nor do the best Cyperologists see it alike.

DISTRIB. Species 130, very generally distributed in all climates and countries.

Subgenus I. FLUITANTES. Stems with leaf-bearing nodes or leafless. Spikelets solitary. Hypogynous bristles none. Style and its branches elongate. Fertile flowers usually 2-sexual.

Style-branches 2:
  Spikelets lanceolate:
    Peduncles subsolitary:
      Nut small, obovoid ... ... (1) **fluitans.**
      Nut larger, ellipsoid ... ... (2) **capillifolius.**
    Peduncles clustered ... ... (4) **Ludwigii.**
  Spikelets globose ... ... (3) **globiceps.**
Style-branches 3 ... ... (5) **tenuissimus.**

Subgenus II. ISOLEPIS (Genus, R. Br.). Stem with leaves at the base only or leafless. Spikelets solitary or in one head, or heads umbellate. Hypogynous bristles always entirely wanting. Style and its branches elongate. Fertile flowers usually 2-sexual, except in 32 *S. diœcus.* (Species of *Eu-Scirpus* which have no hypogynous bristles cannot be diagnosed from this Subgenus.)

Group 1. Style 2-fid. Spikelet solitary. Nut smooth (except 6 *S. sororius*).
  Spikelet linear-oblong ... ... (8) **leptostachyus.**
  Spikelet elliptic-lanceolate:
    Glumes with a long narrow tip ... ... (9) **Burchellii.**
    Glumes subobtuse:
      Leaves produced ... ... (6) **sororius.**
      Leaves hardly any ... ... (7) **verrucosulus.**

Group 2. Style 3-fid. Spikelets in one head or solitary. Nut trabeculate.
  Leaves usually present:
    Small annual; spikelets $\frac{1}{10}$ in. long ... ... (10) **setaceus.**
    Stouter with a slender, at length woody, rhizome; spikelets $\frac{1}{4}$ in. long ... ... (11) **diabolicus.**
  Leaves hardly any:
    Stems very slender; spikelets not exceeding $\frac{1}{8}$ in. long ... ... (12) **macer.**
    Stems much stouter; spikelets $\frac{1}{4}$ in. long ... (13) **costatus.**

Group 3. Style 3-fid. Spikelets in one head, or solitary, occasionally proliferous. Nut neither longitudinally ribbed nor trabeculate.
  Small plants; stem with 1–3 (rarely 4) small spikelets:
    Spikelet 1, or in *S. venustulus* rarely 2:
      Leaves hardly any:
        Spikelet ovoid or broadly oblong ... (14) **venustulus.**
        Spikelet only $\frac{1}{16}$ in. broad ... ... (15) **bulbiferus.**
      Leaves $\frac{2}{3}$ the length of the stem ... ... (16) **tenuis.**

Spikelets 3–1 (rarely 4):
  Leaves evolute:
    Spikelets ellipsoid to lanceolate:
      Glumes hardly mucronate:
        Stems green, setaceous ... (17) **cernuus.**
        Stems black when dry, capillary ... ... (18) **subprolifer.**
        Stems black when dry, thick, flaccid ... ... (19) **rivularis.**
      Glumes mucronate ... ... (20) **karroicus.**
    Spikelets globose ... ... (23) **pinguiculus.**
  Leaves hardly any:
    Nut smooth or punctate ... ... (21) **Neesii.**
    Nut tuberculate ... ... (22) **trachyspermus.**
Small plants; stem with a head of 3–10 (rarely 1–2) small spikelets.
  Glumes herbaceous, not shining:
    Leaves $\frac{2}{3}$ the length of the stem ... ... (25) **incomtulus.**
    Leaves 0–$\frac{1}{3}$ in. long:
      Root fibrous; stem capillary, black... (24) **griquensium.**
      Rhizome very slender; stem pale ... (26) **expallescens.**
  Glumes hard, shining, with prominent curved red linear marks:
    Glumes muticous ... ... (27) **antarcticus.**
    Glumes mucronate ... ... (28) **dregeanus.**
Plants stouter than in the two preceding groups, with larger spikelets:
  Leaves present:
    Leaves long, linear, very flaccid ... ... (29) **flaccifolius.**
    Leaves 1–3 in. long, rigid ... ... (30) **membranaceus.**
  Leaves none:
    Bract overtopping the head:
      Spikelets ellipsoid ... ... (31) **nodosus.**
      Spikelets linear-oblong ... ... (32) **diœcus.**
    Bract much shorter than the head ... (33) **prolifer.**

Group 4. Style 3-fid. Heads of spikelets umbellate (often solitary in *S. Holoschœnus*).
  Leaves short; bracts 3–1, very short ... ... (34) **Holoschœnus.**
  Leaves long; bracts 4–8, long ... ... (35) **Burkei.**

Subgenus III. EU-SCIRPUS. Stem with leaves at the base only or leafless. Spikelets solitary or panicled, or in one head, or heads umbellate. Hypogynous bristles often present. Style and its branches elongate. Fertile flowers usually 2-sexual.

Hypogynous bristles frequently wanting or rudimentary.
  Head solitary, lateral:
    Spikelets few:
      Nut transversely wavy ... ... (36) **supinus.**
      Nut obscurely wavy ... ... (37) **quinquefarius.**
    Spikelets many ... ... (38) **articulatus.**
  Heads several, corymbose, or one compound:
    Glume acuminate into a very long mucro ... (39) **varius.**
    Glume acute or minutely mucronate ... ... (40) **corymbosus.**
Hypogynous bristles $\frac{1}{2}$ as long as the nut, nearly always present:
  Leaves longer than the stem, linear-setaceous:
    Head 1, small ... ... (41) **ficinioides.**
    Heads several, usually agglomerated ... ... (42) **falsus.**

Leaves hardly any:
Stem triquetrous:
Glumes notched; style-branches 2 ... (43) **triqueter.**
Glumes entire, acute; style-branches 3 ... (44) **paludicola.**
Stem terete or trigonous:
Hypogynous bristles setaceous ... ... (45) **lacustris.**
Hypogynous bristles feathery ... ... (46) **littoralis.**
Leaves longer than the stem, flat $\frac{1}{5}$ in. broad ... (47) **maritimus.**

Subgenus IV. MICRANTHÆ. Stem with basal leaves only or leafless. Spikelets solitary, or small and capitate. Hypogynous bristles none. Style hardly any, branches short linear. Fertile flower usually 2-sexual.

Spikelet solitary; glumes very obtuse ... ... (48) **Isolepis.**
Spikelets capitate; glumes very long-awned ... (49) **Hystrix.**

Subgenus V. PSEUDOSCHŒNUS. Leafless. Spikelets in an elongated branched panicle. Hypogynous bristles 5. Style and its branches elongate. Flowers polygamo-diœcious ... ... (50) **spathaceus.**

1. **S. fluitans** (Linn. Sp. Plant. ed. 2, 71); glabrous; stem weak, branched, floating or creeping in wet places, often 6–12 in. long, with numerous leaf-bearing nodes nearly throughout its entire length; leaves 1–4 by $\frac{1}{20}$ in., sheathing at the base; peduncles axillary, 1–5 in. long, scattered, rarely several together, each carrying 1 spikelet; bract 0, i.e. lowest glume (often empty) like the flower-bearing glumes and early deciduous; spikelet $\frac{1}{4}$ by $\frac{1}{12}$ in., cylindric-lanceolate, many-flowered (the frequent state in the Cape examples, but some Cape examples are short, few-flowered, green, as the common European form); glumes ovate, concave, hardly keeled, hardly acute, green or more or less purple-marked; hypogynous bristles 0; stamens usually 2; anthers oblong-linear, not crested; style linear, nearly as long as the nut, branches 2, nearly as long as the style; nut very small, scarcely $\frac{2}{5}$ the length of the glume, obovoid, biconvex, dorsally compressed, smooth, straw-coloured, finally yellow-brown, outermost cells minute, subquadrate, withering scarious and peeling off, so that the nut under a common lens is smooth, hardly at all longitudinally striate. *Sowerby, Engl. Bot. t.* 216; *Boeck. in Linnæa*, xxxvi. 485, *excluding var.* γ *and Isolepis Ludwigii* β; *C. B. Clarke in Hook. f. Fl. Brit. Ind.* vi. 653. *Isolepis fluitans*, *R. Br. Prod.* 221; *Kunth, Enum.* ii. 188. *I. fascicularis*, *Kunth, Enum.* ii. 188. *Eleogiton fascicularis*, *Nees in Linnæa*, x. 165.

COAST REGION: Tulbagh Div.; Mitchells Pass, *Rehmann*, 2372! Worcester Div.; Drakenstein Mountains, 3000–4000 ft., *Drège*, 7409! King Williamstown Div.; Perie Forest, *Kuntze*, 268!

CENTRAL REGION: Somerset Div.; wet places on the summit of Bosch Berg, *MacOwan*, 1885!

KALAHARI REGION: Transvaal, Hooge Veld, *Rehmann*, 6551!

EASTERN REGION: Natal; Van Reenans Pass, 5800 ft., *Kuntze*, 234! and without precise locality, *Buchanan*, 345! 346!

Widely dispersed in Europe, Africa, South East Asia, and Australia.

This species, in its narrowly-defined form, is doubtless frequent in South Africa; many special localities are not given, from the difficulty of distinguishing young specimens from the 3 succeeding species.

2. S. capillifolius (Parl. Fl. Ital. ii. 83); as *S. fluitans*, but the nut much larger, ellipsoid, very little narrowed at the base, appearing under a common lens to have 45 longitudinal striolations on each face. *S. fluitans, var. γ robustus, Boeck. in Linnæa,* xxxvi. 486. *Eleogiton striatus, Nees in Linnæa,* ix. 291; x. 165. *Isolepis striata, Kunth, Enum.* ii. 189. *I. robustula, Steud. Syn. Pl. Glum.* ii. 90.

COAST REGION: Cape Div.; Cape Town, *Harvey*, 185! Table Mountain, near Wynberg, 2200 ft., *Bolus*, 4731! Worcester Div.; Dutoits Kloof, 1000–2000 ft., *Drège!*

All the examples of this (subspecies of *S. fluitans*?) have exactly similar nuts, which differ considerably from the nut of *S. fluitans*. It must be observed, however, that the exterior cells in the latter are really exactly similar to and placed in longitudinal rows like those of *S. capillifolius*; there is no real structural difference.

3. S. globiceps (C. B. Clarke in Durand and Schinz, Conspect. Fl. Afr. v. 622); as *S. fluitans*, but the spikelets are $\frac{1}{6}$ in. broad, subglobose.

COAST REGION: Swellendam Div.; mountains along the lower part of the Zonder Einde River, 500–2000 ft., *Zeyher*, 4390! 4390b!

The nuts are rather longer and browner than those of *S. fluitans*; the spikelets are very red. This plant is considered by Boissier to be *Eleogiton rubicundus*, Nees (in Linnæa, x. 164), i.e. *Isolepis rubicunda*, Kunth (Enum. ii. 188); but this latter is described as having a "subulate-mucronate bract as long as the spikelet," which Zeyher, 4390b, has not; and Boeckeler refers the *Eleogiton rubicundus* to *Scirpus Ludwigii*.

4. S. Ludwigii (Boeck. in Linnæa, xxxvi. 486); as *S. fluitans* but leaves and peduncles in clusters at the more or less proliferous nodes. *Fimbristylis Ludwigii, Steud. in Flora,* 1829, i. 139; *Nees in Linnæa,* vii. 491. *Trichelostylis Ludwigii, Nees in Linnæa,* ix. 290; x. 146. *Isolepis Ludwigii, Kunth, Enum.* ii. 189.

VAR. β, tenuior (Kunth, Enum. ii. 189); slender, less rigid, peduncles in less numerous clusters. *S. lenticularis, Boeck. in Linnæa,* xxxvi. 483 *partly. S. fluitans forma, Boeck. in Linnæa,* xxxvi. 486. *Isolepis tenuior, Steud. Syn. Pl. Glum.* ii. 91.

SOUTH AFRICA: Without locality, *R. Brown*; *Harvey*, 363! 374! *Bolus*, 6019!

COAST REGION: Cape Div.; Claremont, *Schinz*, 29! Table Mountain, *Rehmann*, 651! Cape Flats, near Rondebosch, below 100 ft., *Burchell*, 736! *Bolus*, 4898! and at Doorn Hoogte, *Zeyher*, 1778! Simons Bay, *Wright!* Paarl Div.; French Hoek Kloof, 1000–2000 ft., *Drège*, 1593! Stellenbosch Div., *Zeyher!* Caledon Div.; between Houw Hoek Mountains and the Palmiet River, *Burchell*, 8163! Swellendam Div.; near Swellendam, 1000 ft., *Kuntze*, 251! Knysna Div.; *Rehmann*, 373! George Div.; Wolf Drift, Malgat River, *Burchell*, 6122! Albany Div.; near Grahamstown, 2200 ft., *MacOwan*, 1350! Var. β, Alexandria Div.; Enon, below 500 ft., *Drège!* Uitenhage Div.; *Ecklon and Zeyher*, 521!

KALAHARI REGION: Var. β, Orange Free State; Caledon River, *Burke!*

Spikelets usually many-flowered, lanceolate-conic, in colour varying from pale-straw to brown-red. The type-form, with rigid peduncles 8–20 in a cluster, appears well-marked, but Kunth's var. *β tenuior* ends where *S. fluitans* begins.

5. **S. tenuissimus** (Boeck. in Linnæa, xxxvi. 481); stems with the habit of *S. fluitans*, but very slender, some herbarium pieces more than 2 ft. long; leaves 1–3 in. long, capillary; peduncles 1 in. long, capillary, often 2–5 from one node, each bearing one spikelet; spikelet $\frac{1}{10}$ in. long, ellipsoid, 6–10-flowered; lowest glume usually empty, very similar to the flower-glumes, the midrib sometimes excurrent in a green point; flower-glumes ovate, hardly acute, obscurely 1–5-ribbed; hypogynous bristles 0; nut more than $\frac{1}{2}$ the length of the glume, ellipsoid, trigonous; style $\frac{1}{2}$ the length of the nut; style-branches 3, about as long as the nut. *S. confervoides, Poir. in Lam. Encyc.* vi. 755 *partly; Kunth, Enum.* ii. 173 *partly* (*i.e. diagnosis, not description*). *Isolepis tenuissima, Nees in Linnæa,* ix. 291 *partly; Kunth, Enum.* ii. 190. *I.? aquatilis, Kunth, Enum.* ii. 215.

COAST REGION: Paarl Div.; Berg River in Klein Drakenstein Mountains, below 500 ft., *Drège,* 1589! Swellendam Div.; in the Buffeljagts River at Zuurbraak, *Burchell,* 7264! Uitenhage Div.; water-pools near the Zwartkops River, *Ecklon and Zeyher,* 643!

Also in Madagascar.

There is such confusion involved in the name of *S. confervoides* for this species that it is better avoided. It was originally described by Poiret from Madagascar; he describes the two long barren glumes enclosing the spikelet (which will not at all do for *S. tenuissimus*); Kunth, following, describes the spikelets as 1-flowered. It seems probable that Poiret and Kunth partly described from *S. submersus,* Sauv., an aquatic species which occurs in Madagascar.

6. **S. sororius** (C. B. Clarke in Durand and Schinz, Conspect. Fl. Afr. v. 630); small, glabrous, tufted, annual; stems 2–4 in. long, setaceous, each with 1 or 2 sessile spikelets; leaves 1–2 in. long, setaceous, or often hardly any; spikelets $\frac{1}{10}$–$\frac{1}{8}$ in. long, ovoid, many-flowered; bracts sometimes 1 in. long, more often less than $\frac{1}{8}$ in.; glumes ovate, obtuse, pale, more or less brown-purple marked; hypogynous bristles 0; style shorter than the nut, linear, branches 2 long-linear; nut minute, less than $\frac{1}{2}$ the length of the glume, obovoid, flattened, irregularly trabeculate by reason of the large outer cells being transversely oblong but not arranged in regular vertical rows. *S. setaceus, Boeck. in Linnæa,* xxxvi. 500 *partly. S. minutissimus, Boeck. ex C. B. Clarke in Durand and Schinz, Conspect. Fl. Afr.* v. 630. *Isolepis sororia, Kunth, Enum.* ii. 192. *I. natans, Nees in Linnæa,* ix. 291 *partly.*

COAST REGION: Riversdale Div.; Zoetemelks River, *Burchell,* 6707!

This species is (as many other small Cape *Isolepis*) generally like *S. setaceus,* Linn., and *S. cernuus,* Vahl. The nut is much contracted at the base, where it is surrounded by a wide circular yellow disc with which it falls.

7. **S. verrucosulus** (Steud. in Flora, 1829, i. 145); small, glabrous, tufted, annual; stems 2–4 in. long, each with 1 spikelet; leaves 0, i.e. uppermost sheath carrying sometimes a green point $\frac{1}{8}$ in. long; spikelet $\frac{1}{10}$–$\frac{1}{5}$ in. long, ovoid, sometimes produced in fruit, many-flowered, pale; bract shorter than the spikelet; hypogynous bristles 0; style linear, nearly as long as the nut, branches 2 as long as the style; nut nearly $\frac{1}{2}$ the length of the glume, obovoid flattened; outermost cells minute, quadrate, becoming white, scarious and breaking up, thus appearing as numerous rows of points on the black nut. *S. setaceus, var. γ ecklonianus, Boeck. in Linnæa*, xxxvi. 503. *Eleogiton verrucosula, Dietr. Sp. Pl.* ii. 98. *Isolepis eckloniana, Schrad. Anal. Fl. Cap.* 15, *t.* 1, *fig.* 3; *Nees in Linnæa*, viii. 79; x. 149; *Kunth, Enum.* ii. 192. *I. verruculosa, Nees in Linnæa*, vii. 495. *I. ptycholeptos, Steud. Syn. Pl. Glum.* ii. 93.

SOUTH AFRICA: without locality, *Drège!* *Bergius!*

COAST REGION: Cape Div.; near Cape Town, 60–100 ft., *Bolus*, 4900! 7919! marshes on Table Mountain, *Ecklon*, 875! Worcester Div.; *Zeyher!* Port Elizabeth, *E.S.C.A. Herb.*, 108!

Also in Sénegal?

This plant is like the common *S. cernuus*, Vahl., but differs in having a 2-fid style and flattened nut. The "Sénegal" example was seen in Steudel's own herbarium.

8. **S. leptostachyus** (Boeck. in Linnæa, xxxvi. 485); glabrous; rhizome thread-like; stem 8 in. long, capillary, with 1 spikelet; leaf 2 in. long, very narrow; spikelet $\frac{5}{12}$ by $\frac{1}{16}$ in., linear-cylindric, erect, dense, dirty straw-colour; bract hardly $\frac{1}{16}$ in. long; glumes round-ovate; hypogynous bristles 0; style linear, shorter than the nut, branches 2 linear, longer than the style; nut small, less than $\frac{1}{2}$ the length of the glume, obovoid, flattened, smooth, pale. *Isolepis leptostachya, Kunth, Enum.* ii. 190.

COAST REGION: Paarl Div.; Klein Drakenstein Mountains, by the Berg River, below 500 ft., *Drège*.

No specimen at Kew.

9. **S. Burchellii** (C. B. Clarke in Durand and Schinz, Conspect. Fl. Afr. v. 618); small, glabrous, tufted, annual; stems 1–2 in. long, each with 1 spikelet; leaves $\frac{1}{4}$–$\frac{1}{2}$ in. long, setaceous; spikelet $\frac{1}{5}$ by $\frac{1}{12}$ in., cylindric, dense, pale; bract about as long as the spikelet, but deciduous; glumes ovate, boat-shaped, 7-nerved, suddenly narrowed into the triangular-linear 3-nerved prolongation; hypogynous bristles 0; style linear, shorter than the nut, branches 2 linear, longer than the style; nut scarcely $\frac{1}{3}$ the length of the glume, obovoid, flattened, pale-brown, with minute rows of tubercles arising from the lax outermost cells.

COAST REGION: Riversdale Div.; near the Zoetemelks River, "in a walk to the White-clay Pit," *Burchell*, 6705!

The nut is like that of *S. trachyspermus*, Nees, but not trigonous; the prolonged glumes are quite different.

10. S. setaceus (Linn. Sp. Plant. ed. 2, 73, partly); small, glabrous, tufted, annual; stems 1–7 in. long, slender, each bearing 1–3 spikelets in a small head; leaves 1–3 in. long, setaceous, usually present, though the uppermost sheath in the Cape examples is often leafless or very nearly so; spikelets 1–3, sessile, $\frac{1}{10}$ in. long, ovoid, many-flowered, pale, more or less marked with chestnut-red; bract usually $\frac{1}{4}$–$\frac{1}{2}$ in. long, suberect, resembling the leaves, persistent; flower-glumes ovate, hardly acute, somewhat keeled, falling off successively from the base of the spikelet; hypogynous bristles 0; stamens usually 2; anthers linear-oblong, scarcely apiculate; style nearly as long as the nut, linear, glabrous, branches 3 as long as the style; nut $\frac{2}{3}$ the length of the glume, sessile, trigonous, ellipsoid, pyramidal at the top and bottom, yellow-brown, appearing under a low magnifier to have 12–18 longitudinal ribs and to be somewhat 12–18-gonal, by reason that the outermost cells of the nut are transverse-oblong, and superimposed in regular vertical series. *Rottb. Descr. et Ic.* 47 (*var. β, γ*), *t.* 15, *figs.* 5, 6; *Sowerby, Engl. Bot. t.* 1693; *Boeck. in Linnæa*, xxxvi. 500, *excl. var. β and γ*; *Benth. Fl. Austral.* vii. 327; *C. B. Clarke in Hook. f. Fl. Brit. Ind.* vi. 654. *Isolepis setacea, R. Br. Prod.* 222; *Kunth, Enum.* ii. 193. *I. plebeia var. Schrad. Anal. Fl. Cap.* 18, *t.* 1, *fig.* 6; *Nees in Linnæa*, viii. 82, *t.* 3, *fig.* 2, *a*; x. 153 *partly*. *I. expallescens, b, Drège, Zwei Pflanzengeogr. Documente* 55, 195.

COAST REGION: Uitenhage Div., *Zeyher!*
CENTRAL REGION: Graaff Reinet Div.; Cave Mountain, near Graaff Reinet, 3800 ft., *Bolus*, 712! Sneeuw Berg Range, 4000–5000 ft., *Drège!*
KALAHARI REGION: Transvaal; Hooge Veld, *Rehmann*, 6648! Houtbosch Berg, 7000 ft., *Schlechter*, 4703!

Also in Europe, West Asia, North and Tropical Africa, Indo China, and a var. in Australia.

11. S. diabolicus (Steud. in Flora, 1829, i. 147); glabrous; rhizome slender, becoming woody; stems 6–15 in. long, each bearing a head of 3–9 spikelets; leaves often as long as the stem, $\frac{1}{12}$ in. broad, inrolled in the dried examples; lower bract $\frac{1}{2}$–$1\frac{1}{2}$ in. long, suberect so that the head appears lateral; spikelets $\frac{1}{4}$ in. long or more, ellipsoid, dense, chestnut-brown; glumes broad-ovate, shortly acuminate, with radiating pale red nerves (exceedingly like those of *Ficinia*); nut as of *S. setaceus*, but a little larger. *Boeck. in Linnæa*, xxxvi. 505. *Isolepis diabolica, Schrad. Anal. Fl. Cap.* 21. *I. antarctica, Nees in Linnæa*, vii. 505; viii. 83; x. 162. *Ficinia antarctica, Nees in Linnæa*, ix. 292; *Kunth, Enum.* ii. 255. *F. schinziana, Boeck. in Abhandl. Ver. Brandenb.* xxix. 47.

SOUTH AFRICA: without locality, *Zeyher! Verreaux! Pappe*, 94!
COAST REGION: Cape Div.; marshy places at the foot of Devils Mountain, *Ecklon*, 871! Table Mountain, *Schinz!*
CENTRAL REGION: Somerset Div.; Bosch Berg, 2500 ft., *MacOwan*, 1743!

12. S. macer (Boeck. in Engl. Jahrb. v. 503); glabrous; rhizome very slender; stems 6–15 in. long, slender, each bearing a head of 1–12 spikelets; leaves 0; spikelets scarcely more than $\frac{1}{8}$ in. long, often less; nut as of *S. setaceus.*

KALAHARI REGION: Orange Free State, *Buchanan*, 87! 88! Transvaal; Houtbosch, *Rehmann*, 5615!

EASTERN REGION: Natal; near Durban, *Wood*, 1963! Nottingham, *Buchanan!* and without precise locality, *Buchanan*, 90!

Also in Madagascar.

Closely allied to the two preceding species.

13. **S. costatus** (Boeck. in Linnæa, xxxvi. 511); glabrous; stems 8–24 in. long, rather stout, each bearing a head of 3–12 spikelets; leaves 0; spikelets usually about $\frac{1}{4}$ in. long, much stouter than those of *S. macer;* nut as of *S. setaceus*, Linn. *C. B. Clarke in Durand and Schinz, Conspect. Fl. Afr.* v. 620.

COAST REGION: Queenstown Div.; Shiloh, *Baur!*

CENTRAL REGION: Somerset Div.; Bosch Berg, 4000 ft., *MacOwan*, 1919a! 1919b! 1919c!

EASTERN REGION: Tembuland; Isikoba River, 2000 ft., *Baur*, 725!

Also in Tropical Africa, Madagascar, and Tasmania.

This species is very closely allied to the three preceding.

14. **S. venustulus** (Boeck. in Linnæa, xxxvi. 479); glabrous; rhizome oblique or horizontal; stems 2–10 in. long, with 1 (or very rarely 2) spikelet, quadrangular; leaves hardly any; uppermost sheath truncate, sometimes carrying a green, linear point $\frac{1}{4}$ in. long; spikelet $\frac{1}{6}$–$\frac{1}{5}$ in. long, ovoid or broadly oblong, 6–12-flowered; lowest glume empty, with a green point, not overtopping the spikelet, persistent; flower glumes ovate, red with paler lines on the sides curving outwards; hypogynous bristles 0; stamens 3; filaments when dry very rugose; anthers rather large, linear-oblong, yellow, terminated by a white mucro; style linear, rather shorter than the nut, branches 3 linear, longer than the style; nut obovoid, trigonous, smooth, finally nearly black, much narrowed below, but with a yellow disc of loose tissue at the base, not on an obpyramidal gynophore. *Isolepis bicolor, Nees in Linnæa*, ix. 291; x. 148. *I. venustula, Kunth, Enum.* ii. 192.

SOUTH AFRICA: without locality, *Drège! Pappe*, 79! *Wagenau! Wallich!*

COAST REGION: Malmesbury Div.; Hopefield, *Bachmann*, 1605! 1606! Cape Div.; near Cape Town, *R. Brown! Bolus*, 4813! Constantia, *Bergius*, 155! Uitenhage Div., *Ecklon!*

This has altogether the habit of a *Ficinia*, and the dark-red glumes with paler lines curving outwards at the top of the sides are as those of *Ficinia.* It is put in *Scirpus* Sect. *Isolepis only* because the nut has no distinct gynophore; the dilated ring of yellow lax tissue at its base falls off with the nut, as the gynophore in *Ficinia* often does.

15. **S. bulbiferus** (Boeck. in Linnæa, xxxvi. 482); stems 2–5 in. long, capillary; roots capillary; leaf up to $\frac{1}{2}$ in. long, capillary; spikelet hardly $\frac{1}{15}$ in wide, $\frac{1}{8}$–$\frac{1}{6}$ in. long; nut ellipsoid, narrowed conically at the top; otherwise as *S. venustulus*, Boeck.

COAST REGION: Cape Div.; Sand-dunes near Cape Town, *Zeyher*, 1777 partly!

Zeyher, n. 1777, contains also *S. venustulus* and *S. Savii* mixed. In the tangled capillary roots of *S. bulbiferus* are included (both in Kew and Berlin Herbaria) bulbils $\frac{1}{8}$ in. long, ellipsoid, shining chestnut; these in the Kew specimen of 1777 are not attached to the *Scirpus*. The bulbils in the Berlin example of 1777 look the same, and do not appear to be *Cyperaceous*. *S. bulbiferus*, therefore, if distinct from *C. venustulus*, must stand on its slenderness, slender spikelets, and narrow nut.

16. S. tenuis (Spreng. in Flora, 1829, i. Beil. 12); glabrous, slender, annual; stems tufted, 2–4 in. long, capillary, with 1 head; leaves $\frac{2}{3}$ the length of the stem, capillary; spikelet $\frac{1}{4}$ by $\frac{1}{15}$ in., narrow-cylindric, 10–15-flowered; bracts 0, i.e. lowest glume, though usually empty, hardly different from the others; flower-glumes broad-elliptic, obtuse, not keeled, broadly green on the back, marked with red towards the edges; margins narrowly scarious; stamens often 2; style linear, shorter than the nut, branches 3 as long as the style; nut sessile, oblong, slightly wider at the top, trigonous, black, but covered with one layer of marcescent white glistening subpapillose cells in regular lines. *Boeck. in Linnæa*, xxxvi. 478, *not of Willd. Isolepis tenuis, Schrad. Anal. Fl. Cap.* 15; *Nees in Linnæa*, viii. 79; x. 148; *Kunth, Enum.* ii. 190. *I. pusilla, Kunth, Enum.* ii. 190. *I. atro-purpurea, Nees in Linnæa*, vii. 495; *not of Roem. et Schult.*

SOUTH AFRICA: without locality, *Zeyher*, 507! *Harvey*, 367!
COAST REGION: Worcester Div.; Drakenstein Mountains, 3000–4000 ft., *Drège*, 7413!

17. S. cernuus (Vahl, Enum. ii. 245); glabrous, small, tufted, annual; stems 2–7 in. long, slender, with 3–1 (rarely 4) spikelets; leaves setaceous, $\frac{1}{4}$–3 in. long, usually much shorter than the stem; spikelets $\frac{1}{10}$–$\frac{1}{5}$ in. long, ovoid, 6–20-flowered, pale or chestnut-red, sometimes elongated, or with a cylindric sterile top; lowest glume often bract-like, suberect, longer than the spikelet which then appears lateral; glumes ovate, hardly keeled or acute; hypogynous bristles 0; stamens 3–2–1; anthers not crested; style linear, glabrous, hardly so long as the nut, branches 3 linear, longer than the style; nut nearly $\frac{1}{2}$ the length of the glume, obovoid, obtuse, trigonous, smooth, black; outermost cells small, subquadrate, in very numerous regular rows, becoming on the ripe nut inflated, white, and scarious (so that the nut is described sub-tubercular white), finally breaking up in their centres (when the nut is described as porose). *S. Savii, Sebast. et Mauri, Prod. Fl. Rom.* 22; *Sowerby, Engl. Bot. Suppl. t.* 2782. *S. setaceus, Linn. herb. propr. et Mant.* 321 (*not Sp. Plant. ed.* 2, 73); *Rottb. Descr. et Ic.* 47, *t.* 15, *fig.* 4; *Boeck. in Linnæa*, xxxvi. 500, *var. α, β partly, not γ. Isolepis riparia, R. Br. Prod.* 222; *cf. Schrad. Anal. Fl. Cap.* 16 *in note. I. leptalea, Schultes in Roem. et Schult. Syst. Veg.* ii. *Mant.* 62; *Schrad. Anal. Fl. Cap.* 16 *in note, t.* 1. *fig.* 2. *I. pygmæa, Kunth, Enum.* ii. 191. *I. chlorostachya, Nees in Linnæa*, ix. 291; x. 149. *I. microcarpa, Nees in Linnæa*, x. 150; *Kunth, Enum.* ii. 214. *I. rupestris, Kunth, Enum.* ii. 193.

VAR. β, **subtilis** (C. B. Clarke in Durand and Schinz, Conspect. Fl. Afr. v. 619); stems and leaves capillary; spikelet often very small. *S. tenuissimus var. β capillaris, Boeck. in Linnæa*, xxxvi. 481. *Isolepis subtilis, Kunth, Enum.* ii. 191 *partly*. *I. perpusilla, Nees in Linnæa*, ix. 291; *Zeyher in Linnæa*, xx. 247.

SOUTH AFRICA: without locality, *Ecklon*, 892! *Wallich!* Var. β, *Bergius*, 162!

COAST REGION: Cape Div.; Lion Mountain, *Mund*, 24! Cape Flats, *Zeyher*, 1776! near Cape Town, *Kuntze*, 255! *Rehmann*, 1452! 1799! Robertson Div.; Cogmans Kloof, *Kuntze*, 293! Mossel Bay Div.; hills near the landing place at Mossel Bay, *Burchell*, 6287! Port Elizabeth Div.; near Port Elizabeth, below 100 ft., *Drège!* Albany Div.; damp places at the foot of Bothas Hill, 2000 ft., *MacOwan*, 1269! Fort Beaufort Div.; Kat River, *Baur!* King Williamstown Div.; Yellowwood River, 1000–2000 ft., *Drège*, 4375! Cathcart Div.; Blesbok Flats, 3000–4000 ft., *Drège!* Queenstown Div.; *Shiloh, Baur!* Var. β, Paarl Div.; between Paarl and Lady Grey Bridge, below 1000 ft., *Drège!* Swellendam Div.; *Zeyher!* Uitenhage Div.; in the Zwartkops River, *Ecklon and Zeyher*, 643! Komgha Div.; Gonubie River, 2000 ft., *Schlechter*, 6142!

CENTRAL REGION: Somerset Div.; summit of Bosch Berg, 4500 ft., *McOwan*, 2118! Var. β, Aliwal North; Witte Berg, 5000–6000 ft., *Drège*, 7411!

WESTERN REGION: Little Namaqualand, *Drège*, 7417!

KALAHARI REGION: Transvaal, *Rehmann!*

EASTERN REGION: Tembuland; Bazeia, 2000 ft., *Baur*, 757! Natal, *Krauss!* Var. β, Natal; Inanda, 2000 ft, *Wood*, 1575!

Common in all warm and temperate climes, except in South-east Asia.

18. S. subprolifer (Boeck. in Linnæa, xxxvi. 492); stems 4–7 in. long, capillary, straight, black in drying, with 3–1 sessile spikelets; leaves short; spikelets small; otherwise as *S. cernuus*, Vahl, and its var. *subtilis*.

SOUTH AFRICA: without locality, *Harvey!*

EASTERN REGION: Natal, *Gerrard*, 492!

This is hardly proliferous; as in some other allied species the spikelets elongate in fruit, and from the 1 or 2 riper spikelets in the head the lower glumes fall, giving the semblance of a head of 1 sessile and 1 or 2-stalked spikelets. *S. subprolifer* may be esteemed a form of *S. cernuus*, Vahl; but in *S. cernuus* the weak examples (as also its var. *subtilis*) have usually but one spikelet, while the stems are curved, and green when dry. *S. subprolifer* is very like in colour and inflorescence to *S. rivularis*, in which, however, the stems are thicker than in strong *S. cernuus*.

19. S. rivularis (Boeck. in Linnæa, xxxvi. 504); rootstock weak, often decumbent, rooting; stems 2–12 in. long, usually thicker than in *S. cernuus*, weak, flaccid, usually drying black; nut minutely apiculate; otherwise as *S. cernuus*. *S. natans, Thunb. Prod.* 17; *Fl. Cap. ed. Schult.* 95?; *Spreng. in Flora*, 1829, i. *Beil.* 12, *not of Griseb.* *S. setaceus var., Spreng. in Flora*, 1829, i. 145 *partly*. *Isolepis rivularis, Schrad. Anal. Fl. Cap.* 19, *t.* 1, *fig.* 5; *Kunth, Enum.* ii. 194. *I. palustris, Schrad. Anal. Fl. Cap.* 17, *t.* 1, *fig.* 7; *Nees in Linnæa*, viii. 80. *I. pallida, Nees in Linnæa*, viii. 81, *t.* 3, *fig*, 2, *c*; x. 153; *Kunth, Enum.* ii. 195. *I. natans, Dietr. Sp. Pl.* ii. 106?; *Nees in Linnæa*, vii. 497, *excl. var.* β; viii. 80; x. 151, *excl. syn. R. Br.*

SOUTH AFRICA: without locality, *Thunberg; Zeyher*, 345! *Wallich! Petit Thouars!*

COAST REGION: Cape Div.; Table Mountain, *Ecklon*, 878! Alexandria Div.; between Hoffmanns Kloof and Drie Fontein, 1000–2000 ft., *Drège*, 7410! Albany Div.; Grahamstown, *Schœnland*, 584!

CENTRAL REGION: Somerset Div.; damp places on the summit of Bosch Berg, 4800 ft., *MacOwan*, 1742!

WESTERN REGION: Little Namaqualand; Kamies Bergen, near Ezels Fontein and at Rood Berg, 3500–4000 ft., *Drège*, 2463!

EASTERN REGION: Natal; Inanda, *Wood*, 227! Umgeni River, *Rehmann*, 7453!

This appears to creep on mud or on the margin of water. The whole plant, and the spikelets, are blacker than *S. cernuus*.

20. **S. karroicus** (C. B. Clarke in Durand and Schinz, Conspect. Fl. Afr. v. 624); glabrous, small, tufted, annual; stems 3 in. long, setaceous, with 1–3 spikelets; leaves hardly $\frac{1}{2}$ in. long, setaceous; spikelets $\frac{1}{8}$–$\frac{1}{6}$ in. long, ovoid, 6–12-flowered, straw-coloured; lowest glume often as long as the spikelet, bract-like, sometimes short; flower glumes ovate, boat-shaped, the keel excurrent in a short, often recurved point; nut less than half the length of the glume, trigonous, ellipsoid, pyramidal at both ends, sessile, black, smooth (unless highly magnified); outermost cells longitudinally oblong so that the nut (highly magnified) has faint transverse wavy lines.

CENTRAL REGION: Worcester Div.; near Constable, "*A.R.*" *in Herb. Bolus*, 5811! Karroo; Witte Berg, Babians Krantz, *Rehmann*, 2882!

This resembles *S. cernuus*, Vahl; the curved points of the glume sticking out from the spikelets on all sides give it a different appearance; the structure of the nut is quite diverse from that of *S. incomtulus*.

21. **S. Neesii** (Boeck. in Linnæa, xxxvi. 504); stem 8 in. long, slender, nearly leafless, with 2–1 spikelets; spikelet $\frac{1}{6}$ in. long, ovoid, 10-flowered, straw-coloured; bract nearly as long as the spikelet; flower-glumes ovate, obtuse; hypogynous bristles 0; style 3-fid; nut nearly as of *S. cernuus*, Vahl (but not seen ripe). *Isolepis compressa, Nees in Linnæa*, x. 157; *Kunth, Enum.* ii. 215.

SOUTH AFRICA: without locality, *Drège! Ecklon and Zeyher*, 91!

No specimen in the Kew Herbarium.

This differs from *S. cernuus*, Vahl, by want of leaves; while the plant is too strong for *S. venustulus*, nor has it the Ficinia-like habit and glumes of *S. venustulus*.

22. **S. trachyspermus** (C. B. Clarke in Durand and Schinz, Conspect. Fl. Afr. v. 633); small, glabrous, annual; stem 1–3 in. long, setaceous, with 1–3 spikelets; leaf hardly $\frac{1}{2}$ in. long, setaceous, spikelet $\frac{1}{5}$ in. long, oblong-ellipsoid, many-flowered; bract shorter than the spikelet; glumes ovate, striate, submucronate; style 3-fid; nut $\frac{1}{2}$ the length of the glume, narrowly obovoid, trigonous, yellow-brown, minutely tuberculated (i.e. outermost cells rather larger than those of *S. cernuus*). *Isolepis trachysperma, Nees in Linnæa*, x. 152; *Kunth, Enum.* ii. 194. *Scirpus setaceus, Boeck. in Linnæa*, xxxvi. 501, *so far as regards syn. Isolepis trachycarpa, Boeck. (error for trachysperma).*

SOUTH AFRICA: without locality, *Drège! Burmann! Ecklon and Zeyher*, 48! No specimen in the Kew Herbarium.

This may be made a var. of *S. cernuus*, Vahl.

23. S. pinguiculus (C. B. Clarke in Durand and Schinz, Conspect. Fl. Afr. v. 629); rhizome creeping, branching, very slender, but becoming almost woody; stems 1–3 in. long, flattened, broader (as is also the bract) than in *S. cernuus;* spikelets ovoid, broader than those of *S. cernuus*, brightly coloured brown-red and green; otherwise as *S. cernuus*, Vahl.

CENTRAL REGION: Fraserburg Div.; by the Dwaal River, *Burchell*, 1475!
KALAHARI REGION: Bechuanaland; Moshowa River, near Takun, *Burchell*, 2253!

24. S. griquensium (C. B. Clarke in Durand and Schinz, Conspect. Fl. Afr. v. 623); glabrous, slender, tufted, annual; stems 5–8 in. long, capillary, straight, black when dried, carrying 1–10 spikelets, sessile in one head; leaves hardly any; uppermost leaf-sheath somewhat inflated, produced lanceolate on one side; spikelets $\frac{1}{10}$ by $\frac{1}{16}$ in., in ripe fruit elongated, $\frac{1}{5}$ by $\frac{1}{16}$ in., of numerous minute flowers; bracts 2–1, lowest not so long as the fruiting spikelet; glumes minute, ovate, obtuse, green on the back, more or less red on the sides; hypogynous bristles 0; stamens often 2; anthers linear-oblong, minutely apiculate; nut $\frac{1}{2}$ the length of the glume, sessile, broad-oblong, trigonous, black, outermost cells nearly quadrate, so that the nut looks "punctate" in regular rows; style linear, shorter than the nut, branches 3 linear, longer than the style.

EASTERN REGION: Griqualand East; streams near Clydesdale, *Tyson*, 2861!

25. S. incomtulus (Boeck. in Linnæa, xxxvi. 693); glabrous, slender, tufted, annual; stems 1–3 in. long, carrying 2–7 spikelets sessile in one head; leaves often $\frac{2}{3}$ the length of the stem, linear, flaccid; uppermost sheath sometimes nearly leafless; lower sheaths red; spikelets $\frac{1}{6}$ by $\frac{1}{10}$ in., many-flowered; lowest bract linear, leaf-like, often $\frac{1}{2}$–$\frac{1}{3}$ in. long; glumes boat-shaped, ovate, obtuse, with a short excurrent, sometimes recurved, mucro; nut minute, less than $\frac{1}{2}$ the length of the glume, broadly obovoid, obtuse, smooth, outermost cells subquadrate inconspicuous; style-branches 3. *Isolepis incomtula, Nees in Linnæa*, ix. 291; x. 154; *Kunth, Enum.* ii. 196. *I. Echinidium, Nees in Linnæa*, ix. 291; x. 155; *Kunth, Enum.* ii. 196. *I. exilis, Nees in Linnæa*, ix. 291; x. 147; *Kunth, Enum.* ii. 214. *I. kunthiana, Steud. Syn. Pl. Glum.* ii. 95. *I. bergiana, e, Drège, Pflanzengeogr. Documente* 101, 195.

SOUTH AFRICA: without locality, *Drège! R. Brown! Bergius*, 151! *Lalande*, 362! *Reynaud! Ecklon and Zeyher*, 47! 55!
COAST REGION: Clanwilliam Div.; Ceder Bergen, *Shaw!* Karree Bergen, 1500 ft., *Schlechter*, 8210! Malmesbury Div.; between Groene Kloof and Saldanha Bay, below 500 ft. *Drège!* Cape Div.; near Cape Town, *Harvey!* between Claremont and False Bay, *Schlechter*, 432! Devils Mountain, *Rehmann*, 912! Paarl Div.; Klein Drakenstein Mountains and Dal Josephat, below 1000 ft., *Drège!*

26. S. expallescens (Boeck. in Linnæa, xxxvi. 509); glabrous, rather slender; rhizome very slender, pale, descending; stems approximated, 3–9 in. long, carrying 3–7 spikelets sessile in a head; leaves weak, mostly about 1/8 in. long, but some 2 in., narrowly linear; spikelets 1/4 by 1/8 in., many-flowered, pale; bracts 2, lower leaf-like usually overtopping the inflorescence; glumes ovate, obtuse; hypogynous bristles 0; style-branches 3; nut 1/3 the length of the glume, ellipsoid, trigonous, smooth. *Isolepis expallescens, Kunth, Enum.* ii. 196.

CENTRAL REGION: Victoria West Div.; Nieuw Veld, between the Brak River and Uitvlugt, 3000–4000, ft., *Drège*, 7415! Graaff Reinet Div.; Sneeuw Berg Range, 4000–5000 ft., *Drège.*

27. S. antarcticus (Linn. Mant. 181, not of Thunb.); glabrous, tufted, annual; stem 5–10 in. long in the type form of Linnæus (which is *I. seslerioides*, Kunth), more commonly 2–6 in. (*S. cartilagineus*, Poir., or *S. bergianus*, Spreng.), bearing a head of 1–13 (commonly 3–8) spikelets; leaves often 1/2 the length of the stem, narrowly linear; lowest bract 1/4–1 in. long, leaf-like, usually longer than the inflorescence; spikelets 1/4 by 1/8 in., angular, not terete; glumes often 3-ranked or the lower nearly 2-ranked, rigid, keeled, ovate, tip triangular or scarcely pointed, back curved, each side with 4–6 curved lines, yellow, with a larger or smaller chestnut-red patch near the top; hypogynous bristles 0; stamens 3, anthers linear-oblong, scarcely apiculate; nut 1/2 the length of the glume, ellipsoid, trigonous, pyramidal at the top, minutely tubercular, somewhat transversely undulate or smooth, golden or becoming black; outermost cells short, longitudinal-oblong, more or less inflated; style linear, nearly as long as the nut, branches 3 longer than the style. *S. bergianus, Spreng. Syst.* i. 212; *Steud. in Flora*, 1829, i. 149; *Boeck. in Linnæa*, xxxvi. 693. *S. setaceus, Thunb. Prod.* 17; *Fl. Cap. ed. Schult.* 95. *S. setaceus var. Steud. in Flora*, 1829, i. 145 *partly*. *S. ficiniæformis, Boeck. ex C. B. Clarke in Durand and Schinz, Conspect. Fl. Afr.* v. 617. *Isolepis bergiana, Schultes in Roem. et Schult. Syst. Veg.* ii. *Mant.* 532; *Nees in Linnæa*, vii. 500; viii. 82; x. 291; *Kunth, Enum.* ii. 194; *Boeck. in Flora*, 1860, 33. *I. chrysocarpa, Nees in Linnæa*, vii. 499, viii. 80, *in Obs. t.* 3, *fig.* 2 *b*; x. 152. *I. plebeia, Schrad. Anal. Fl. Cap.* 18, *t.* 1, *fig.* 1 (*fig.* 6 *excl.*); *Nees in Linnæa*, viii. 80 *partly*; x. 153. *I. setacea, Nees in Linnæa*, x. 151. *I. phæocarpa, Nees in Linnæa*, x. 153. *I. sphærocarpa, Kunth, Enum.* ii. 195. *I. seslerioides, Kunth, Enum.* ii. 195. *Cyperus minutus, Roth, Nov. Pl. Sp.* 32; *Nees in Linnæa*, vii. 513. *C. rothianus, Roem. et Schult. Syst. Veg.* ii. *Mant.* 95; *Kunth, Enum.* ii. 25; *cf. Boeck. in Flora*, 1860, 33. *Ficinia antarctica, Drège in Linnæa*, xx. 249.

SOUTH AFRICA: without locality, *R. Brown! Drège*, 1599! 7418! *Sieber*, 98! *Bergius*, 36! 37! *Harvey*, 344! 348! 373! 376! *MacGillivray*, 426! *Ecklon and Zeyher*, 49! 50! *Rehmann*, 912! 1798! *Bolus*, 5272! *Mund and Maire!*

COAST REGION: Clanwilliam Div.; Ebenezer, below 100 ft., *Drège!* Cape Div., near Cape Town, *Thunberg, Harvey*, 190! *Pappe!* Table Mountain, *Ecklon*, 880! gardens and roadsides, *Ecklon*, 879a! Wynberg Sand Flats, *Bolus*

2862! Cape Flats, near Rondebosch, *Burchell*, 183! Simons Bay, *Wright! Milne*, 228! Constantia Mountain, *Schlechter*, 545; Swellendam Div.; near Swellendam, *Zeyher*, 4411! Riversdale Div.; near the Zoetemelks River, *Burchell*, 6724! between the Little Vet River and Kampsche Berg, *Burchell*, 6899! George Div.; Montagu Pass, *Rehmann*, 72! Humansdorp Div.; Kromme River Heights, *Bolus*, 1339! Port Elizabeth Div., *E.S.C.A. Herb.*, 196! Albany Div.; near Howisons Poort, 1800 ft., *MacOwan*, 1353! Cathcart Div.; Blesbok Flats, 3000–4000 ft., *Drège!*

Also in St. Helena, Extra-Tropical Australia and New Zealand.

28. **S. dregeanus** (C. B. Clarke); stems 1–1½ in. long; glumes obtuse; keel excurrent in a mucro, often subrecurved; nut ovoid, triangular, pyramidal at each end, very smooth, shining black; exterior cells long-oblong longitudinal; otherwise as the very small examples of *S. antarcticus*, Linn. *Isolepis bergiana*, *Drège, Pflanzengeogr. Documente* 120.

COAST REGION: Caledon Div.! Genadendal, and by the Zonder Einde River, below 1000 ft., *Drège!*

29. **S. flaccifolius** (Steud. Syn. Pl. Glum ii. 83 char. emended); glabrous; stems 6–20 in. long, clustered, weak, thickened; oblong at the base by the leaf-sheaths, with one head of 4–1 spikelets; leaves often as long as the stem, linear, very weak, flaccid; bracts 1–3, up to ⅙ in. long, ovate, membranous; spikelets up to ½ by ⅙ in., but often much shorter, dense with many flowers, terete, pale brown; glumes ovate, obtuse, thin, keel stout, green, excurrent as a bristle; hypogynous bristles 0; stamens 3; nut ½ the length of the glume, broad-oblong, trigonous, pyramidal at the top, brown, smooth; style ½ the length of the nut, branches 3 long, linear. *S. digitatus*, *Boeck. in Linnæa*, xxxvi. 691. *Isolepis digitata*, *Schrad. Anal. Fl. Cap.* 20; *Kunth, Enum.* ii. 202. *I. dubia*, *Kunth, Enum.* ii. 216. *I.? dissoluta*, *Kunth, Enum.* ii. 219. *Schœnus aggregatus*, *Spreng. in Flora*, 1829, i. *Beil.* 12, *not of Thunb.* *Eleogiton digitatus*, *Nees in Linnæa*, ix. 291; x. 163. *E. dissolutus*, *Nees in Linnæa*, ix. 291. *E. rubicundus var. β*, *Nees in Linnæa*, x. 164, *fide Boeck.* *E. longifolius*, *Nees in Linnæa*, ix. 291; x. 164. *Eleocharis flexifolia*, *Reichenb. ex C. B. Clarke in Durand and Schinz, Conspect. Fl. Afr.* v. 622.

SOUTH AFRICA: without locality, *Zeyher*, 338! *Ecklon and Zeyher*, 57.

COAST REGION: Worcester Div; Dutoits Kloof, 2000–3000 ft., *Drège!* Drakenstein Mountains, 1000–2000 ft., *Drège!* and without precise locality, *Zeyher!* Caledon Div.; submersed in the Palmiet River, Houw Hoek Mountains, *Burchell*, 8077! Riversdale Div.; at the waterfall in "Valley Rivers Poort," near Kampsche Berg, *Burchell*, 7053!

Steudel and Nees placed their "*flaccifolius*" in Sect. *Eleogiton*, to which they assign a 2-fid style. As their descriptions are not only insufficient, but radically misleading, some will prefer to call this plant *S. digitatus*, Boeck.

30. **S. membranaceus** (Thunb. Prod. 17); glabrous, except the minutely ciliate margins of glumes; rhizome ¼ in. thick, horizontal, woody; stems 1–2 ft. long, approximate on the rhizome,

rather stout, each with one head of 1–7 spikelets; leaf on the uppermost sheath 1–3 in. long, rigid, narrow, almost trigonous in cross-section (leaves 5–8 in. long occur at the base of the barren stems); bracts 2 (1–3), $\frac{1}{2}$–2 in. long, similar to the uppermost leaf; spikelets up to 1 by $\frac{1}{3}$ in., always large, ellipsoid-cylindric, obtuse, terete, dense, brown; glumes orbicular, flat on the back, with a scarious somewhat fimbriate margin; hypogynous bristles 0 (in Drège 3943 two large lateral boat-shaped scarious "glumellæ" $\frac{1}{5}$ in. long enclose the flower); stamens 3, anthers linear-oblong, not crested; nut scarcely $\frac{1}{3}$ the length of the glume, obovoid, plane-convex, brown, smooth; style hardly any, branches 3 long linear. *Fl. Cap. ed. Schult.* 98; *Boeck. in Linnæa,* xxxvi. 695; *Benth. in Benth. et Hook. f. Gen. Pl.* iii. 1051. *Isolepis membranacea, Nees in Linnæa,* ix. 291; x. 159. *Ficinia membranacea, Kunth, Enum.* ii. 252 (*exclud. Scirpus trispicatus, Thunb.*). *Hellmuthia restioides, Steud. Syn. Pl. Glum.* ii. 90.

SOUTH AFRICA: without locality, *Thunberg, Brossard!* *Ecklon,* 1903! 1904! 1905! *Ecklon and Zeyher,* 66!

COAST REGION: Malmesbury Div.; sand dunes near Groene Kloof, 300 ft., *Bolus,* 4357! *Drège,* 3943! Cape Div.; Cape Flats at Doorn Hoogte, *Zeyher,* 1775! damp sandy places between Retreat and Muizenberg, 30 ft., *Bolus,* 7187! Mossel Bay Div.; sand dunes in Fish Bay, below 500 ft., *Drège,* 3943! Knysna Div.; near the west end of Groene Vallei, *Burchell,* 5645!

This is so remarkable a plant that there can be no question but that Drège, 3943, is merely a form of it where the "glumellæ" are present by atavism. Bentham in Gen. Pl. iii. 1051, suggests that Drège, 3943, indicates a connection between *Scirpus* and *Hypolytrum.* I regard the squamellæ here present as analogous to the broad scale-like hypogynous setæ sometimes developed in *S. littoralis,* Schrad. In *S. membranaceus* the anterior angle of the nut is greatly flattened down; so that weak flowers may be expected to show a 2-fid style: the species seems to me a *Eu-Scirpus,* near *S. littoralis,* Schrad., rather than an *Isolepis.*

31. S. nodosus (Rottb. Descr. et Ic. 52, t. 8, fig. 3); glabrous; rhizome $\frac{1}{4}$ in. in diam., horizontal, woody; stems 1–3 ft. long, approximate, terete, each with 1 globose dense brown head of numerous spikelets; leaves 0; lowest bract 1–2 in. long, erect, as though a continuation of the stem with the head lateral; spikelets often 50–80 in the head, $\frac{1}{3}$ by $\frac{1}{10}$ in., many-flowered; glumes broad-ovate, not mucronate; hypogynous bristles 0; stamens 3, anthers hardly crested; nut less than $\frac{1}{2}$ the length of the glume, obovoid, trigonous, brown, smooth; style linear, hardly so long as nut, branches 3 linear, long. *Boeck. in Linnæa,* xxxvi. 718. *Fimbristylis textilis, Beatson, S. Helena* 309. *Isolepis nodosa, R. Br. Prod.* 221; *Nees in Flora,* 1828, 296; *Linnæa,* x. 159; *Kunth, Enum.* ii. 199. *Holoschœnus nodosus, Dietr. Sp. Pl.* ii. 165.

SOUTH AFRICA: without locality, *Bergius,* 50! *Verreaux!*

COAST REGION: Cape Div.; Camps Bay, *Burchell,* 850! marshy places on Cape Flats, near the Zwart River, *Bolus,* 3308! Botany Bay, near Cape Town, *Kuntze,* 253! Uitenhage Div.; near the mouth of the Van Stadens River, *MacOwan,* 2086! near the mouth of the Zwartkops River, *Ecklon and Zeyher,* 503! *Zeyher,* 4394! *Drège!*

KALAHARI REGION: Orange Free State, *Rehmann*, 3557!

Also in the islands of St. Helena, and St. Paul, Temperate Australia, New Zealand, and Temperate S. America.

32. **S. diœcus** (Boeck. in Linnæa, xxxvi. 719); head straw-coloured or pale brown; male perfect, or female with small sterile stamens; glumes oblong-obovoid, very obtuse; style-branches 3, exserted, with numerous several-celled linear-oblong papillæ; nut brown, enclosed in a white net (i.e. the more-persistent margins of the outermost layer of round hexagonal cells); otherwise as *S. nodosus*, Rottb. *Isolepis diœca, Kunth, Enum.* ii. 199, *not of Steud.*

VAR. β, **macrocephala** (Boeck. in Linnæa, xxxvi. 720); stem nearly $\frac{1}{8}$ in. in diam.; head 2 in. in diam. with 90 spikelets; spikelets 1 by $\frac{1}{16}$ in., curved, pendent on all sides;—perhaps an accidental monstrous state, but the head abounds in ripe nuts.

SOUTH AFRICA: without locality, *Drège*, 717! *Zeyher*, 4421!
COAST REGION: Clanwilliam Div., *Zeyher!*
CENTRAL REGION: Murraysburg Div.; Englishmans Kraal, near Murraysburg, 4000 ft., *Tyson!* Graaf Reinet Div.; banks of the Sunday River, 2500 ft., *Bolus*, 713! Aliwal North Div.; between Aliwal North and the Kraai River, 4500 ft., *Drège;* Colesberg Div., *Shaw*, 19!
WESTERN REGION: Little Namaqualand! near the mouth of the Orange River, below 600 ft., *Drège!* Var. β, Great Namaqualand, *Wandres*, 4!
KALAHARI REGION: Griqualand West; Herbert Division at Lower Campbell, *Burchell*, 1814! Hay Division at Griqua Town, *Burchell*, 1874! Orange Free State; Kanon Fontein, *Rehmann*, 3557! and without precise locality, *Buchanan*, 91!

Also in Tropical Africa.

This species is distinct from *S. nodosus*, Rottb., but closely allied to it. The plant collected by Zeyher at "Clanwilliam" and distributed by Reichenb. f. as "*Holoschœnus nodosus*, A. Dietr." is *Scirpus diœcus*, Boeck.

33. **S. prolifer** (Rottb. Descr. et Ic. 55, t. 17, fig. 2); glabrous; rhizome hardly any, or short oblique descending, not thick; stems 8–30 in. long, leafless, each bearing one (very often proliferous) head, varying much in thickness, being sometimes slender, sometimes as stout as that of *S. nodosus;* one smaller proliferous head is present in nearly every collection; but two or more additional heads are rare; bract about $\frac{1}{3}$ in. long, oblong, obtuse; spikelets 5–20 in a head, up to $\frac{1}{2}$ by $\frac{1}{12}$ in., cylindric, but of irregular lengths in nearly every head, brown, many-flowered; glumes ovate, obtuse; hypogynous bristles 0; stamens 3; anthers oblong-linear, not crested; nut less than half the length of the glume, broad, ovoid, triquetrous, brown, smooth; style-branches 3, long. *Boeck. in Linnæa*, xxxvi. 692. *Cyperus punctatus, Lam. Ill.* i. 144. *C. prolifer, Thunb. Prod.* 18 *partly; Fl. Cap. ed. Schult.* 99; *Steud. in Flora*, 1829, 150 *chiefly*. *Isolepis prolifer, R. Br. Prod.* 223; *Nees in Flora*, 1828, 294; *Linnæa*, vii. 493; x. 158; *Kunth, Enum.* ii. 201.

SOUTH AFRICA: without locality, *Roxburgh! Thom! Miller! Chamisso*, 219! *Wallich!*
COAST REGION: Cape Div.; near Cape Town, *Thunberg*, *Rehmann*, 1791!

Cape Flats, near Rondebosch, *Burchell*, 176! Table Mountain, *Ecklon*, 892! 893! 895! *Milne*, 235! *MacGillivray*, 419! *Drège*; Botany Bay, *Kuntze*, 256! Wynberg. below 500 ft., *Drège*, 142; Paarl Div.; by the Berg River, near Paarl, below 500 ft., *Drège!* Paarl Mountains, 1000–2000 ft., *Drège*; Alexandria Div.; between Hoffmanns Kloof and Drie Fontein, 1000–2000 ft., *Drège*; Uitenhage Div.; in the bed of the Vans Stadens River, *MacOwan*, 2052! Albany Div.; Assegai Bosch, near Sidbury, *Burchell*, 4174! near Grahamstown, 2000 ft., *MacOwan*, 1865! *Schoenland!*

KALAHARI REGION: Orange Free State, *Buchanan*, 85!

EASTERN REGION: Natal; Inanda, *Wood*, 223; 1049! 1608! marshy flats between Umlaas River and Durban Bay, *Krauss*, 190! and without precise locality, *Gerrard*, 490! *Buchanan*, 322! 331! 363!

Also found in St. Helena, Extra-tropical Australia and New Zealand.

34. **S. Holoschœnus** (Linn. Sp. Plant. ed. 2, 72) var. **Thunbergii** (C. B. Clarke in Durand and Schinz, Conspect. Fl. Afr. v. 623); glabrous; rhizome woody, horizontal; stems approximate, 6–30 in. long, round or trigonous; leaves short or (in the Cape form) hardly any; heads few in a simple umbel, or many in a compound umbel, or only 1, globose of many spikelets; bracts 3–1, short, but sometimes over-topping the inflorescence; spikelets $\frac{1}{8}$–$\frac{1}{5}$ in. long ellipsoid, many-flowered; glumes ovate, broadly white-scarious-margined at the top, glabrous; hypogynous bristles 0; stamens 3; anthers linear-oblong, crested; nut $\frac{1}{2}$ the length of the glume, ovoid, trigonous, pyramidal at the top, smooth, leaden-black; style shorter than the nut, branches 3 linear, longer than the style. *S. Thunbergii, Boeck. in Linnæa,* xxxvi. 719. *Cyperus prolifer, Thunb. Prod.* 18 *partly, fide Boeck. C. glomeratus, Thunb. fide Vahl, Enum.* ii. 299 *and Spreng. Syst.* i. 215, *but the name has not been found. Isolepis Thunbergii, Schrad. Anal. Fl. Cap.* 22; *Nees in Linnæa,* viii. 85; x. 160. *I. thunbergiana, Nees in Linnæa,* vii. 508; *Kunth, Enum.* ii. 200. *Holoschœnus Thunbergii, Dietr. Sp. Pl.* ii. 164.

SOUTH AFRICA: without locality, *Zeyher! Verreaux! Harvey*, 378!

COAST REGION: Malmesbury Div.; *Bachmann*, 1608! Cape Div.; wet places near Cape Town, *Burchell*, 685! *Bolus*, 3310! *Rehmann*, 1790; Stellenbosch Div., *Zeyher!* Uitenhage Div.; wet sandy places, Van Stadens Hoogte, *MacOwan*, 2132! Port Elizabeth Div.; along the coast, *E.S.C.A. Herb.*, 173! Bathurst Div.! near Theopolis, *Burchell*, 4070! Komghu Div.; marshy places near the mouth of the Kei River, *Flanagan*, 1779!

35. **S. Burkei** (C. B. Clarke in Durand and Schinz, Conspect. Fl. Afr. v. 618); glabrous; stems $1\frac{1}{2}$–$2\frac{1}{2}$ ft. long, slender, wiry, smooth, obscurely 3–5-gonous at the top; leaves all near the base of the stem, 12–20 by scarcely $\frac{1}{8}$ in.; heads 30–60 brown, in a compound umbel; bracts 4–8, long, over-topping the umbel, similar to the leaves; spikelets 3–20 in a globose head, $\frac{1}{6}$ by $\frac{1}{12}$ in., dense; glumes ovate, acute; hypogynous bristles 0; stamens 2; filaments persistent, somewhat elongate; anthers linear-oblong, crested; style hardly any, branches 3 linear, long; nut small, ellipsoid, unequally trigonous, apiculate, ultimately nearly black, smooth. *S. schinzianus, Boeck. ex C. B. Clarke in Durand and Schinz, Conspect. Fl. Afr.* v. 618.

SOUTH AFRICA: without locality, *Zeyher*, 1766! 1767!

KALAHARI REGION: Orange Free State; Caledon River, *Burke*, 231! and without precise locality *Buchanan*, 95! Basutoland; Leribe, *Buchanan*, 150! Transvaal; Magalies Berg, *Burke!* Johannesburg, *E.S.C.A. Herb.*, 366! and without precise locality, *Fehr! Rehmann*, 4694! 6889!

36. S. supinus (Linn. Sp. Plant. ed. 2, 73) var. **leucosperma** (C. B. Clarke); glabrous, annual; stems tufted, 1–6 in. long, terete, with one lateral head of 3–1 spikelets; leaves 1–2 in. long, linear; bracts 1–2, lower erect 1–4 in. long, as though a continuation of the stem; spikelets $\frac{1}{6}$ by $\frac{1}{12}$ in., 6–9-flowered, straw-coloured; glumes boat-shaped, ovate, subacute with a green keel; hypogynous bristles 0 or rudimentary; stamens 3; anthers linear-oblong, hardly crested; nut nearly $\frac{1}{2}$ the length of the glume, broadly obovoid, trigonous, with transverse wavy lines, pale, ultimately black; style linear, nearly as long as the nut, branches 3 longer than the style. *S. Thunbergii, A. Spreng. Tent. Suppl. Syst.* 4. *S. leucanthus, Boeck. in Abhandl. Ver. Brandenb.* xxix. 46. *Isolepis supina, R. Br. var. leucosperma, Nees ex Drège, Zwei Pflanzengeogr. Documente* 195.

COAST REGION: Clanwilliam Div.; Ebenezer, by the Oliphants River, below 100 ft., *Drège*, 7414!

WESTERN REGION: Great Namaqualand; Fish River, *Schinz*, 379!

37. S. quinquefarius (Hamilt. ex Boeck in Linnæa, xxxvi. 701); glabrous, annual; stems 4–10 in. long, cæspitose, terete, continuous or pseudoseptate when dry, with one lateral head of 1–9 spikelets; leaves hardly any; bracts 2, lower erect 2–15 in. long as though a continuation of the stem; spikelets $\frac{1}{2}$ by $\frac{1}{5}$ in., straw- or cinnamon-coloured; glumes broad-ovate, hardly acute; hypogynous bristles 0 (or uncertain rudiments); nut $\frac{1}{2}$ the length of the glume, broadly obovoid, trigonous much dorsally compressed, black, slightly transversely wavy owing to the small longitudinally-oblong outermost cells; style 3-fid. *C. B. Clarke in Hook. f. Fl. Brit. Ind.* vi. 657. *S. rehmannianus, Boeck. ex C. B. Clarke in Durand and Schinz, Conspect. Fl. Afr.* v. 629. *Isolepis lupulina, Nees in Wight, Contrib.* 107; *Kunth, Enum.* ii. 197.

KALAHARI REGION: Transvaal; Boshveld, towards M'Cabes Vley, *Rehmann*, 5147!

Also in Tropical Africa, West Asia, and India.

38. S. articulatus (Linn. Sp. Plant. ed. 2, 70); glabrous, annual; stems tufted, 4–12 in. long, terete, usually pseudoseptate when dry, with one lateral dense head of 4–20 pale-brown spikelets; leaves hardly any; bracts 2, lower 6–12 in. long, as though a continuation of the stem; spikelets $\frac{1}{3}$ by $\frac{1}{6}$ in., very obtuse (in the Cape form); glumes nearly hemispheric, neither pointed nor keeled; hypogynous bristles 0; stamens 3; anthers oblong-linear, not crested; nut less than $\frac{1}{2}$ the length of the glume, ovoid, subequally triquetrous, nearly black, strongly transversely wrinkled; style nearly as long as the nut, branches 3. *Boeck. in Linnæa*, xxxvi. 702; *C. B. Clarke in Hook. f. Fl. Brit. Ind.* vi. 656. *Isolepis articulata, Nees in Wight, Contrib.* 108; *Kunth, Enum.* ii. 198. *I. senegalensis, Steud. Syn. Pl. Glum.* ii. 96.

KALAHARI REGION: Transvaal; Boshveld, towards M'Cabes Vley, *Rehmann*, 5150!

Also in Tropical Africa, the Mascarene Islands, India, Malaya, and North Australia.

The Cape specimen is the form *senegalensis*, Steud., in which the glumes and spikelets are very obtuse; in the abundant Indian "type" form the glumes are somewhat acute.

39. S. varius (C. B. Clarke in Durand and Schinz, Conspect. Fl. Afr. v. 634); glabrous; rhizome becoming woody; stem 20 in. long, trigonous at the top, smooth, with one compound dense head (a small short-peduncled head rarely added); leaves nearly as long as the stem, $\frac{1}{10}$ in broad; bracts 3 or more, lowest 2–5 in. long, similar to the leaves, erect or divaricate; head $\frac{1}{3}$–$\frac{2}{3}$ in. in diam.; spikelets very numerous, hardly $\frac{1}{6}$ in. long; glumes elliptic, acuminate, 5-nerved, brown, keel green produced into a long arista; hypogynous glumes rudimentary or 0; stamens 3; anthers oblong, small; nut $\frac{1}{3}$ the length of the glume (including its arista), ovoid, triquetrous, narrowed at either end, brown, nearly smooth; style linear, rather shorter than the nut, branches 3 linear, longer than the style.

KALAHARI REGION: Basutoland; Leribe, *Buchanan*, 225! Transvaal; Pretoria, at Apies Poort, *Rehmann*, 4036! Houtbosch, *Rehmann*, 5635!

This is remarkable by the heads being densely echinate from the long points of the glumes. It does not appear allied to any other *Scirpus*, except *S. mexicanus*, Britton.

40. S. corymbosus (Roth, Nov. Pl. Sp. 28); glabrous; rhizome short; stems 2–3 ft. long, thick, terete, leafless, not transversely septate; umbel apparently lateral, 1–6 in. in diam.; bract apparently a continuation of stem, 1–6 in. long; spikelets clustered, $\frac{1}{3}$ by $\frac{1}{8}$ in., pale brown; glumes boat-shaped, ovate, acute, entire; hypogynous bristles 0; stamens 3; anthers linear-oblong, crested; nut less than $\frac{1}{2}$ the length of the glume, ovoid, trigonous, pyramidal-topped, black, smooth, obscurely transversely wavy-lined. *Boeck. in Linnæa*, xxxvi. 706; *C. B. Clarke in Hook. f. Fl. Brit. Ind.* vi. 657.

KALAHARI REGION: Transvaal; Matebe River, *Holub!*

Also in Tropical Africa, the Mascarene Islands, and India.

41. S. ficinioides (Kunth, Enum. ii. 172); glabrous; roots fibrous fide Kunth; stems 6–10 in. long, slender, tough; leaves all basal, longer than the stem, linear-setaceous; spikelets 3–9, in one close apparently lateral head, $\frac{1}{5}$ by $\frac{1}{10}$ in., black-red; lower bract as though a continuation of stem, 2–4 in. long; glumes boat-shaped, ovate, obtuse, black-red, with yellowish keel, and pale-red longitudinal nerves; hypogynous bristles 6 or 5, much longer than the young ovary, slender, white, retrorsely scabrous; stamens 3; anthers linear-oblong apiculate; style linear, longer than the young ovary, branches 3 linear, longer than the style. *S. Kunthii*, *Boeck. in Linnæa*, xxxvi. 508.

COAST REGION: Stockenstrom Div.; Kat Berg, 4500 ft., *Drège*, 3937! 3938!

This plant has all the appearance of a *Ficinia*; Drège's examples are so very young that nothing can be said about the gynophore or the nut. But as authors do not admit any species with hypogynous bristles into *Ficinia*; both Kunth and Boeckeler keep this in *Scirpus*.

42. **S. falsus** (C. B. Clarke); glabrous; rhizome woody; stems 8 in. long, tufted, trigonous at the top, smooth, with one compound lateral head; leaves longer than the stem, terete, hardly $\frac{1}{20}$ in. broad immediately above the stout sheath; bracts 3–4, lowest suberect, 3 in. long, dilated at the base similar to the leaves; head $\frac{1}{3}$ in. in diam. and upwards, made up of several heads of small spikelets, lowest sometimes $\frac{1}{16}$ in. distant; glumes boat-shaped, ovate, acute, bright-red, with paler longitudinal striations; hypogynous bristles 6, longer than the nut, antrorsely scabrous, white; stamens 3; anthers linear-oblong, crested; young nut ovoid, triquetrous, pyramidal at both ends, nearly sessile; style short, branches 3 (in upper weak flowers sometimes 2) linear long.

KALAHARI REGION: Basutoland; Mont aux Sources, 9500 ft., *Flanagan*, 2010! 2011!

43. **S. triqueter** (Linn. Mant. 29); glabrous, stoloniferous; stem 10–30 in. long, nearly leafless, triquetrous at the top, with 1 lateral head, or a simple umbel of heads; lowest bract as though a continuation of the stem, triquetrous, 1–3 in. long, acute; spikelets 3–10, up to $\frac{1}{2}$ by $\frac{1}{5}$ in., cylindric, obtuse, brown; glumes ovate, scarious at the tip, notched, with a minute bristle in the notch; hypogynous bristles 3–4 (or 6), often overtopping the nut, dark brown, obscurely scabrous; stamens 3; anthers linear-oblong, not crested; nut more than $\frac{1}{2}$ the length of the glume, obovoid, much dorsally compressed, pale, finally brown, smooth; style shorter than the nut, branches 2 long, sparingly papillose, slightly fusiform. *Sowerby, Eng. Bot. t.* 1694; *Kunth, Enum.* ii. 163; *C. B. Clarke in Hook. f. Fl. Brit. Ind.* vi. 658. *S. Pollichii, Gren. et Godr. Fl. France*, iii. 374; *Boeck. in Linnæa*, xxxvi. 711.

CENTRAL REGION: Albert Div.; near hot springs, *Cooper*, 1371!
KALAHARI REGION: Griqualand West; Bloems Fontein, *Burchell*, 2641!

Also in Europe and Asia.

The South African examples have only a single lateral head; *S. mucronatus*, Linn., differs by its 3-fid style and pointed, less-brown spikelets.

44. **S. paludicola** (Kunth, Enum. ii. 163); glabrous; rhizome woody, not thick; stems tufted, 1–2 ft. long, terete, slender, not transversely septate; leaves hardly any, sometimes 1 in. long; lowest bract as though a continuation of the stem, 1–6 in. long; inflorescence lateral, of a single head or a contracted umbel of several heads; spikelets 2–10 together, $\frac{1}{4}$ by $\frac{1}{12}$ in., chestnut-brown; glumes boat-shaped, ovate, acute, entire; hypogynous bristles 5–6, as long as the nut in the typical *S. paludicola*, fewer, slender, imperfect, hardly scabrid, or rarely 0 in the form *decipiens* (sp.) Nees; nut

hardly ½ the length of the glume ovoid, triquetrous, pyramidal at each end, smooth or nearly so, shining black when ripe; style linear, rather shorter than the nut, branches 3 linear, as long as the style. *Boeck. in Linnæa,* xxxvi. 703. *Scirpus pulchellus, Boeck. in Linnæa,* xxxvi. 698. *Isolepis paludicola, Kunth, Enum.* ii. 198. *I. decipiens, Nees in Linnæa,* ix. 291 ; x. 157 ; *Kunth, Enum.* ii. 198. *Ficinia pulchella, Kunth, Enum.* ii. 261 ; *C. B. Clarke in Durand and Schinz, Conspect. Fl. Afr.* v. 641. *Schœnoplectus paludicola, Palla in Eng. Jahrb.* x. 299.

SOUTH AFRICA: without locality, *Drège,* 3959! 4426! *MacOwan,* 1743! *Rehmann,* 3621! *Boivin! Ecklon and Zeyher,* 59!

COAST REGION: Swellendam Div., *Zeyher,* 4386! Alexandria Div.; between Enon and Bushmans River, below 500 ft., *Drège!* Bathurst Div.; near Fish River, below 1500 ft., *Drège!* King Williamstown Div.; near Toise River Station, 3700 ft., *Kuntze,* 278! near King Williamstown, *Kuntze,* 277! Komgha Div., 2000 ft., *Flanagan,* 1259! Cathcart Div.; Blesbok Flats, 3000–4000 ft., *Drège;* Queenstown Div.; between Table Mountain and Wildschuts Berg, 4000 ft., *Drège!* Shiloh, 3500 ft., *Baur,* 778! Bathurst, 1000 ft., *Drège,* 4427!

CENTRAL REGION: Beaufort West Div.; between Beaufort West and Rhinoster Kop, 2500–3000 ft., *Drège!* Somerset Div.; Bosch Berg, 2000 ft., *MacOwan,* 1964b! Albert Div., 4500 ft., *Drège!*

KALAHARI REGION: Orange Free State, *Buchanan,* 163!

EASTERN REGION: Griqualand East; wet places around Clydesdale, 2500 ft., *Tyson,* 2600! Natal, *Buchanan,* 357!

Also in Madagascar.

45. S. lacustris (Linn. Sp. Plant. ed. 2, 72); glabrous except the glumes; rhizome horizontal; stems 2–6 ft. long, terete or somewhat trigonous at the top, usually leafless; lowest bract as though a continuation of the stem, 1–3 in. long; inflorescence lateral, umbel simple or compound; spikelets clustered or solitary, ⅓ by ⅛–⅙ in., brown; glumes ovate, concave, keeled, top scarious, emarginate, minutely hairy, with often a minute point in the notch; hypogynous bristles 6–5, as long as the nut, retrorsely scabrous, often reduced, irregular, rarely 0; stamens 3; anthers linear-oblong, hardly crested; nut more than ½ the length of the glume, obovoid, dorsally compressed, smooth, pale, finally brown; style nearly as long as the nut, branches 3 or 2 as long as the style. *Sowerby, Engl. Bot. t.* 666; *Kunth, Enum.* ii. 164; *Boeck. in Linnæa,* xxxvi. 712; *C. B. Clarke in Hook. f. Fl. Brit. Ind.* vi. 658. *S. Tabernæmontani, Gmel. Fl. Badens.* i. 101; *Kunth, Enum.* ii. 164.

KALAHARI REGION: Griqualand West; Hay Division, at Bloems Fontein, *Burchell,* 2642! Orange Free State, *Buchanan,* 94! 275!

Found in nearly all warm and temperate regions, except South America.

46. S. littoralis (Schrad. Fl. Germ. i. 142, t. 5, fig. 7); glabrous except the glumes; rhizome horizontal; stems 1½–4 ft. long, trigonous at the top; leaves short or 0; lowest bract as though a continuation of the stem, 1–3 in. long; umbel lateral, usually compound; spikelets solitary, ½ by ⅛–⅙ in., pale brown; glumes ovate, tip notched thinly ciliate with often a small mucro in the notch;

hypogynous bristles 4–7, as long as the nut, pale brown, wide or almost petaloid, feathery by reason of many multicellular hairs; stamens 3; anthers linear-oblong, crested; nut ½ as long as the glume, obovoid, much dorsally compressed, smooth, pale or finally brown; style linear, nearly as long as the nut, branches 2 longer than the style. *Kunth, Enum.* ii. 166; *C. B. Clarke in Hook. f. Fl. Brit. Ind.* vi. 659. *S. subulatus, Vahl, Enum.* ii. 268; *Boeck. in Linnæa*, xxxvi. 715. *S. triqueter, Gren. et Godr. Fl. France*, iii. 373; *Boeck. in Linnæa*, xxxvi 716; *not of Linn.* *S. Pterolepis, Kunth, Enum.* ii. 166; *Pterolepis scirpoides, Schrad. Anal. Fl. Cap.* 30. *Malacochæte Pterolepis, Nees in Linnæa*, ix. 292; x. 184.

SOUTH AFRICA: without locality, *Drège*, 2451! 7401! *Verreaux! Bolus! Zeyher*, 1771!
COAST REGION: Cape Div.; Salt River, *Burchell*, 694! Uitenhage Div.; Zwartkops River, below 100 ft., *Drège! Harvey*, 13!
WESTERN REGION: Little Namaqualand; near the mouth of the Orange River, below 500 ft., *Drège*, 2465!
EASTERN REGION: Natal; salt marshes at the mouth of the Umlaas River, *Krauss*, 33! Umgeni River, in a swamp, 50 ft., *Wood*, 4009!

Throughout the warmer parts of the Old World, from the Atlantic to New Caledonia.

47. S. maritimus (Linn. Sp. Plant. ed. 2, 74); glabrous except the glumes; rhizome woody, creeping; stem 1–3 ft. long, trigonous; leaves all basal, often exceeding the stem, ⅕ in. broad, flat, green; umbel simple or somewhat compound, or reduced to 3–1 spikelets in a head; bracts 3–4, up to 4–8 by ⅕ in., similar to the leaves; spikelets attaining 1 by ⅓ in., often much smaller, dark-brown; glumes up to ¼ in. long, ovate, notched at the top, with a minute mucro in the notch, more or less pubescent at the top, sometimes (especially the outer ones) minutely brown-hairy all over; hypogynous bristles 6–3, hardly so long as the nut, retrorsely scabrous, sometimes reduced, very rarely 0; stamens 3; anthers linear-oblong, apiculate; nut scarcely ⅓ the length of the glume, obovoid, dorsally compressed or nearly biconvex, smooth, brown; style linear, as long as the nut, branches 3 or 2 linear, longer than the style. *Thunb. Prod.* 17; *Fl. Cap. ed Schult.* 97; *Sowerby, Engl. Bot. t.* 542; *Nees in Linnæa*, vii. 509; x. 184; *Kunth, Enum.* ii. 167; *Boeck. in Linnæa*, xxxvi. 722; *C. B. Clarke in Hook. f. Fl. Brit. Ind.* vi. 658. *S. capensis, Burm. Prod. Fl. Cap.* 3.

SOUTH AFRICA: without locality, *Drège*, 2452! *Zeyher! Harvey! Verreaux! Krauss!*
COAST REGION: Clanwilliam Div.; by the Oliphants River, near Ebenezer, below 100 ft. *Drège*, 7402! Cape Div.; at Salt River, *Burchell*, 514! near Cape Town, below 100 ft., *Thunberg*, *Bolus*, 4814! Cape Flats, *MacOwan and Bolus, Herb. Norm. Aust.-Afr.*, 998! near Nord Hoek, *Milne*, 251! Uitenhage Div.; near the mouth of the Zwartkops River, *Drège!* Kowie River, *MacOwan*, 709! Komgha Div.; Kei River, *Flanagan*, 983
CENTRAL REGION: Calvinia Div.; Bokkeveld Mountains, at Uien Vallei, 2000-2500 ft., *Drège!*

Found in all temperate and warm regions.

48. S. Isolepis (Boeck. in Linnæa, xxxvi. 498); small, glabrous, tufted, annual; stem 1–5 in. long, setaceous, carrying one apparently lateral spikelet; leaves few, near the base of the stem, ½–1 in. long, linear; lower bract as though a continuation of the stem, ¼–½ in. long; spikelet horizontal, ⅙–⅕ in. long, ellipsoid, obtuse, dense, hard; glumes ovate, obtuse; hypogynous bristles 0, but a small oblong lateral hyaline scale is sometimes present within the glume; stamen 1; anther very small, oblong; nut ½ the length of the glume, oblong-obovoid, biconvex, smooth, reticulate, finally black; style hardly any, branches 2 linear, much shorter than the nut. *C. B. Clarke in Hook. f. Fl. Brit. Ind.* vi. 663. *Isolepis minima, Schrad. Anal. Fl. Cap.* 17, *t.* 1, *fig.* 4; *Nees in Linnæa,* vii. 498 *in Obs.;* x. 151. *Hemicarpha Isolepis, Nees in Edinb. New Phil. Journ.* xvii. 263; *Kunth, Enum.* ii. 268; *Benth. in Benth. et Hook. f. Gen. Pl.* iii. 1053.

COAST REGION: Cape Div.; Table Mountain, *Hesse* (fide Schrader).

Also in Tropical Africa and India.

49. S. Hystrix (Thunb. Prod. 17); small, glabrous, tufted, annual; stem 1–5 in. long trigonous, carrying 1 dense head; leaves near the base of the stem, often as long as the stem, up to $\frac{1}{16}$–$\frac{1}{12}$ in. broad; bracts 3–6, often as long as the stem, similar to the leaves; spikelets ¼ in. long, ellipsoid, very dense, hispid from the long awns; glumes lanceolate, 5-striate, red-brown, acuminated into stout green curved awns longer than themselves; hypogynous bristles 0; stamens 1–2; anthers very small, oblong; nut narrowly obovoid, trigonous, smooth, reticulated, brown or black; style hardly any, branches 3 linear, hardly so long as the minute nut. *Fl. Cap. ed. Schult.* 96; *Boeck. in Linnæa,* xxxvi. 735. *S. natalensis, Boeck. ex C. B. Clarke in Durand and Schinz, Conspect. Fl. Afr.* v. 623. *Isolepis Hystrix, Schrad. Anal. Fl. Cap.* 23; *Nees in Linnæa,* vii. 496; viii. 85; x. 156; *Kunth, Enum.* ii. 204. *I. dregeana, Kunth, Enum* ii. 204, *excluding var. β.*

SOUTH AFRICA: without locality, *Zeyher! Drège,* 1601! *Harvey,* 126! 358!

COAST REGION: Cape Div.; inundated places near Cape Town, *Bolus,* 4529! *Rehmann,* 1797! near streams below Table Mountain, *Thunberg;* Cape Flats, *Zeyher,* 4371! Rondebosch, below 1000 ft., *Drège;* Paarl. Div.; by the Berg River, below 500 ft., *Drège!* Worcester Div., *Zeyher!*

WESTERN REGION: Little Namaqualand; Bushmans Karoo, 3000–4000 ft., *Drège;* between Pedros Kloof and Lily Fontein, 3000–4000 ft., *Drège!*

EASTERN REGION: Natal; Griffins Hill, near Eastcourt, *Rehmann,* 7305! 7315! Biggars Berg, 4000–5000 ft., *Sutherland! Kuntze,* 242!

Also found in Angola.

50. S. spathaceus (Hochst. in Flora, 1845, 759); glabrous, polygamo-diœcious; stem 1–2½ ft. long, stout, terete, without nodes; uppermost sheath slightly inflated, obtusely triangular on one side; panicle 6 by 1½ in., rusty-red, with numerous branches springing from stiff pale red bracts ½–1 in. long; branches flexuose, dividing, resembling very much the genera *Schœnus* and *Cladium;* spikelets 12–24-flowered, small, ¼ by $\frac{1}{10}$–⅛ in., sometimes up to ½ by $\frac{1}{10}$–⅛ in., cylindric, brown; glumes boat-shaped, elliptic, subacute, rusty-brown; hypogynous bristles 5, as long as the nut, retrorsely scabrous, brown; stamens 3; anthers linear-oblong, crested (in female flowers the anthers are sometimes reduced, linear, and empty, with the crest fully developed); nut $\frac{1}{2}$ the length of the glume, narrowly ellipsoid, trigonous, pyramidal at the top, smooth, brown-black; style linear, as long as the nut, branches 3 longer than the style. *S. inanis, Steud. Syn. Pl. Glum.* ii. 86; *Boeck. in Linnæa*, xxxvi. 714. *Schœnus inanis, Thunb. Prod.* 16; *Fl. Cap. ed. Schult.* 94; *Kunth, Enum.* ii. 338. *S. tegetalis, Burchell, Travels*, i. 260 *in note* (*name only*).

SOUTH AFRICA: without locality, *Thunberg*, *Ecklon and Zeyher*, 152!
COAST REGION: Clanwilliam Div.; near the Groene River and Watervals River, 2500–3000 ft., *Drège!* Queenstown Div.; by the Klipplaat River, near Shiloh, 3500 ft., *Drège!* Zwart Kei River, *Baur*, 97!
CENTRAL REGION: Beaufort West Div.; on the Nieuw Veld, between Rhinoster Kop and Ganze Fontein, 3500–4500 ft., *Drège*, 689! Somerset Div.; banks of streams at the foot of the Bosch Berg, 2300 ft., *MacOwan*, 1716! Graaf Reinet Div.; near Graaf Reinet, *Bolus*, 266! Sutherland Div.; between Kuilen Berg and the Great Riet River, *Burchell*, 1346! Great Riet River Kloof, *Burchell*, 1387!
WESTERN REGION: Little Namaqualand; between Pedros Kloof and Lily Fontein, 3000–4000 ft., *Drège!*
KALAHARI REGION: Griqualand West; Kimberley, *Flanagan*, 1649!

This species has the aspect of *Cladium* or one of the large *Schœni*, and the nut and setæ suit the genus *Schœnus* well enough. But it is placed in *Scirpus* now by all authorities, because the flowers are numerous in the spikelet. The tribe of *Schœneæ* differs mainly from *Scirpeæ* proper by having few flowers to the spikelet.

Imperfectly known species.

S. pilosus (Thunb. Prod. 18); stem flattened; leaves radical, convolute, ciliate on the margins and keel; head ovate; bracts about 4, up to 2–3 in. long, reflexed; glumes lanceolate, ciliate, yellow with green edges. *Fl. Cap. ed. Schult.* 98; *Kunth, Enum.* ii. 174, *not of Retz.*

SOUTH AFRICA: without locality, *Thunberg!*

No author seems to have been able to divine what this could have been.

X. ERIOPHORUM, Linn.

Hypogynous *bristles* 6, strap-shaped, cut down to the base into several narrow segments, elongate in fruit, rendering the spikelet comose; otherwise as *Scirpus*.

DISTRIB. Species 10; 8 North Temperate or Arctic, 2 in India and China.

1. E. angustifolium (Roth, Neue Beitr. 94); glabrous, stoloniferous; stem 8–24 in. long, smooth, usually with nodes throughout its length; leaves often $\frac{1}{2}$ the length of the stem, $\frac{1}{8}$–$\frac{1}{4}$ in. broad; rays of umbel 1–7 (rarely reduced to 1 spikelet), 0–3 in. long, quite smooth; lowest bract a little overtopping the inflorescence; spikelet $\frac{1}{4}$–$\frac{2}{3}$ in. long, ellipsoid, of very many flowers; glumes obtuse, scarious-black; style long, branches 3; hypogynous bristles in fruit long, exserted, discoloured; nut small, oblong. *Sowerby, Engl. Bot. t.* 564; *Kunth, Enum.* ii. 178; *Boeck. in Linnæa*, xxxvii. 95. *E. polystachyon, Linn. Sp. Plant.* 76 *partly; Host, Gram.* i. 29, *t.* 37.

KALAHARI REGION: Transvaal, *McLea in Herb. Bolus*, 6024!

Spread through the North Temperate and Arctic Regions of both Hemispheres.

XI. FICINIA, Schrad.

Spikelet of many (rarely few) spirally imbricate glumes (or in *Hemichlæna* the lower glumes almost distichous), lowest 1 or 2 empty, many or several succeeding glumes 2-sexual perfecting nuts (cf. *F. ixioides*), uppermost male or sterile. Hypogynous *bristles* 0. *Stamens* 3–2, anterior. *Style* glabrous, linear, passing gradually into the nut, branches 3 or (in *F. lateralis*) 2 linear (cf. *F. radiata*). *Ovary* on a minute obpyramidal gynophore. *Nut* triangular, or (in *F. lateralis*) dorsally compressed.

This genus differs from *Scirpus* Sect. *Isolepis* only by the small obpyramidal gynophore. It is difficult to draw the line between this gynophore and the expanded foot of the nut of many species of *Scirpus*: thus *Ficinia pulchella* of Kunth is *Scirpus pulchellus*, Boeck. This small gynophore is present or not in various species of *Fimbristylis*, and varies greatly in development in one species of *Fimbristylis*; so that it is difficult to attribute much real importance to it.

DISTRIB. Species 58; of which 57 are confined to South Africa (except 2 also on Kilimanjaro), and 1 is endemic in the mountains of Abyssinia.

Subgenus I. SICKMANNIA (Genus, Nees). Leaves and nodes all near the base of the stem; spikelets 2-sexual; glumes spiral; style long, undivided or microscopically notched.

Stem short, stout, with 1 head; bracts many ... (1) **radiata.**

Subgenus II. Leaves and nodes all near the base of the stem; spikelets with the central nut-bearing, 1–3 lateral male; glumes spiral; style-branches 3, long.

Leaves and bracts scarious-edged ... ... (2) **ixioides.**

Subgenus III. EU-FICINIA. Leaves and nodes all near the base of the stem; spikelets 2-sexual; glumes spiral even at the base of the spikelet; style-branches 3, long (or 2 in *F. lateralis*).

* Stem with 1 spikelet:
 Stem stout; spikelet large, often 1 in. long ... (3) **scariosa.**
 Stems very slender; spikelet $\frac{1}{4}$ in. long ... ... (4) **Zeyheri.**
  Spikelet $\frac{1}{6}$ in. long; uppermost sheath leafless ... ... (5) **pusilla.**
  Spikelet $\frac{1}{6}$ in. long; leaf on uppermost sheath long, flat near the base ... ... (6) **micrantha.**

** Stem with several spikelets in a head:
 Style 2-fid (leafless) ... ... (7) **lateralis.**
 Style 3-fid:
  † Leafless; rhizome horizontal (none in *F. rugulosa*):
   Glumes rounded-obtuse; stems round ... (8) **repens.**
   Glumes rounded-obtuse; stems very angular ... ... (9) **sylvatica.**
   Glumes obtuse; stems setaceous ... (10) **rugulosa.**
   Glumes ovate, acuminate; bracts longer than the head ... ... (11) **leiocarpa.**

Glumes ovate, acuminate; bracts shorter than the head ... ... (12) **punctata.**

†† Leaves usually present; spikelets 1–5 (occasionally more in *F. albicans*), digitate:

Flower-glume very obtuse (midrib hardly excurrent in *F. albicans*):

Spikelets very small; scarcely $\frac{1}{8}$ in. long:

Spikelets chestnut-brown; leaf-sheaths white margined ... ... (13) **filiformis.**

Spikelets chestnut-brown; leaf-sheaths pale brown ... ... (14) **bergiana.**

Spikelets greenish-white; glumes pointed ... ... (18) **albicans.**

Spikelets rather larger, $\frac{1}{8}$–$\frac{1}{4}$ in. long:

Stolons present ... ... (15) **stolonifera.**

Stolons none seen; stems 4–8 in. long ... ... (16) **tristachya.**

Stolons none seen; stems 12–16 in. long ... ... (17) **involuta.**

Flower-glume oblong, narrowed at the top, or acute, or mucronate:

Upper sheaths leafless; basal leaves flat, $\frac{1}{8}$ in. broad ... ... (20) **quinquangularis.**

Upper sheath with a long leaf:

Glumes triangular, acute, hardly mucronate ... ... (19) **subacuta.**

Glumes with a strong green excurrent mucro ... ... (21) **tribracteata.**

Glumes obtuse, with a most minute red mucro ... ... (22) **acuminata.**

††† Leaves usually present; spikelets 5–10 or more, at least in some of the heads. Glumes mucronate or acute:

Uppermost leaf $\frac{1}{3}$–$\frac{2}{3}$ in. long ... ... (23) **elongata.**

Uppermost leaf elongate:

Stems capillary:

Glumes not white-edged ... ... (24) **kunthiana.**

Glumes white-edged ... ... (25) **MacOwani.**

Stems slender, rigid:

Nut smooth ... ... (26) **cinnamomea.**

Nut transversely wavy ... ... (27) **tenuifolia.**

†††† Leaves present; head compound; bracts often more than 2, somewhat dilated at the base:

Flower-glumes fimbriate:

Leaf-sheath with white edges much torn or woolly ... ... (28) **paradoxa.**

Leaves flat, $\frac{1}{20}$ in. broad; leaf-sheaths less torn ... ... (29) **laciniata.**

Flower-glumes not fimbriate:

Leaves flat, thick, $\frac{1}{10}$–$\frac{1}{6}$ in. broad, truncate:

Leaves with white scarious edges ... (30) **truncata.**

Leaves without white scarious edges (31) **præmorsa.**

Leaves very narrow:

Nut smooth:

Stem and leaves long, rather slender ... ... (32) **ecklonea.**

Stem and leaves stout ... ... (33) **lævis.**

Nut transversely lineolate:

Heads apparently simple, dense (35) **acrostachys.**

Heads sub-compound:
Leaves very short... ... (34) **brevifolia.**
Leaves elongate ... ... (37) **gracilis.**
Nut strongly transversely rugose, truncate ... ... (38) **latinux.**
Nut unknown; leaves flat, narrow; head very compound, dense ... (36) **dasystachys.**

††††† Leaves present; head compound; bracts usually numerous, much dilated, enclosing the head at the base; spikelets numerous:
Stout rigid plants:
Bracts hardly scarious-edged; leaf-sheaths scarcely scarious-edged ... (40) **lithosperma.**
Bracts broadly scarious-edged; leaf-sheaths scarious-edged ... (41) **lucida.**
Bracts hardly scarious-edged; leaf-sheaths with broad scarious edges ... (42) **pinguior.**
Small woody plant 2–3 in. high ... ... (43) **pygmæa.**
Rather slender plants 6–15 in. high:
Head dense, chestnut-brown ... ... (39) **setiformis.**
Head longer; glumes with thin brown recurved tips ... ... (44) **bracteata.**
Head green; spikelets subspicate ... (45) **fastigiata.**

*** Stem with several, more or less distinct, spikelets [in 45 *fastigiata*, the lowest spikelet is often $\frac{1}{8}$ in. distant]:
Spikelets $\frac{1}{6}$ by $\frac{1}{16}$ in., chestnut-brown:
Bracts not white scarious at the base ... (46) **fascicularis.**
Bracts white scarious at the base ... ... (49) **bulbosa.**
Spikelets $\frac{1}{4}$–$\frac{1}{3}$ in.:
Spikelets few in a cluster:
Bracts not white scarious at the base ... (47) **anceps.**
Bracts white scarious at the base ... (48) **monticola.**
Spikelets mostly solitary in long secund spikes (50) **secunda.**
Spikelets $\frac{2}{5}$ by $\frac{1}{10}$ in., pale ... ... (51) **compasbergensis.**

Subgenus IV. ACROLEPIS (Genus, Schrader). Stems elongate, branching, with many nodes and leaves through their whole length; spikelets very small; glumes spiral; style-branches 3, long.

Peduncles with 1 spikelet; leaf-sheaths nearly entire (52) **trichodes.**
Peduncles with 1 spikelet; leaf-sheaths scarious fimbriate ... ... (53) **ferruginea.**
Peduncles with 1–3–5 spikelets ... ... (54) **ramosissima.**

Subgenus V. HEMICHLÆNA (Genus, Schrader). Stem branched or not; spikelets 2-sexual, elongate; lower flower-glumes distichous, somewhat remote; style-branches 3, long.

Stems tufted, simple, nodes few close above the base; spikelets digitate, $\frac{1}{2}$ in. long ... ... (55) **angustifolia.**
Stems much branched, with nodes often far above the base:
Leaves $\frac{1}{10}$ in. broad; spikelets $\frac{1}{8}$–$\frac{1}{6}$ in. broad ... (56) **longifolia.**
Leaves capillary; spikelets linear ... ... (57) **capillifolia.**

1. F. radiata (Kunth, Enum. ii. 260); glabrous; rhizome $\frac{1}{5}$ in. in diam., horizontal, woody; stems 2–6 in. long, thick, smooth, grooved, about $\frac{1}{2}$ in. apart on the rhizome, each with 1 head; leaves all near the base of the stem, usually much overtopping the stem, $\frac{1}{8}$–$\frac{1}{5}$ in. broad at the base, deeply channelled, very tough, thick edges scabrous cutting, narrowed and triquetrous at the top; head $\frac{1}{2}$–$\frac{3}{4}$ in.

in diam., compound, dense, pale brown; bracts 5–22, radiate, unequal, up to 1–3 in. long, resembling abbreviated leaves; spikelets $\frac{1}{4}$–$\frac{1}{3}$ in. long, 6–16-flowered, usually male at the top (some wholly male spikelets occur); glumes elliptic-lanceolate, acuminate, tip blunt, keel weak, otherwise nerveless; stamens 3; anthers linear-oblong, not crested; style as long as the glume, linear-cylindric, entire or microscopically notched at the tip; nut $\frac{1}{3}$ the length of the glume, obovoid, round trigonous, very smooth, shining black; gynophore small, obconic. *Boeck. in Linnæa,* xxxvii. 81, *exclud. syn. Melancranis radiata; cf. Schrad. Anal. Fl. Cap.* 51 *in Obs. Schœnus radiatus, Linn. f. Suppl.* 101. *Scirpus radiatus, Thunb. Prod.* 18; *Fl. Cap. ed. Schult.* 98. *Isolepis radiata, Roem. et Schult. Syst.* ii. 113. *Sickmannia radiata, Nees in Linnæa,* ix. 292; x. 183.

SOUTH AFRICA: without locality, *Sieber,* 149! 263! *Pappe,* 104! *Bergius,* 41! *Stanger!*

COAST REGION: Piquetberg Div.; on the Piquet Berg, 1500–3000 ft., *Drège,* 1658c! Cape Div.; mountain tops between Nord Hoek and False Bay, *Thunberg,* Cape Flats, *Zeyher,* 1772! *Burchell,* 195! Table Mountain, *Milne,* 230! Simons Bay, *Harvey,* 308! *MacGillivray,* 408! between Tyger Berg and Simons Berg, *Drège,* 1658a! Paarl Div.; Paarl Mountains, 1000–2000 ft., *Drège;* Stellenbosch Div.; Lowrys Pass, 1600 ft., *MacOwan and Bolus, Herb. Norm. Aust.-Afr.,* 1394! Caledon Div.; Nieuw Kloof, Houw Hoek Mountains, *Burchell,* 8153! Grabouw, near the Palmiet River, 800 ft., *Bolus,* 4228!

2. **F. ixioides** (Nees in Linnæa, ix. 292); glabrous; stems tufted, 3–10 in. long, with 1 head; leaves all near the base of the stem, 2–6 by $\frac{1}{16}$ in., rather rigid; sheaths scarious-edged; bracts 3, suberect, lowest 1–1$\frac{1}{2}$ in. long, the scarious-edged dilated base nearly as long as the head; head of 2–4 subsessile spikelets $\frac{1}{3}$ in. long; central spikelet producing nuts in its basal portion; lateral spikelets male; glumes elliptic-lanceolate, membranous, aristate, chestnut, scarious-edged; stamens 3; anthers linear, minutely crested; nut small, obovoid, plane-convex, smooth, on an obconic gynophore; style long, linear, branches 3 long, linear. *Linnæa,* x. 180; *Kunth, Enum.* ii. 264; *Boeck. in Linnæa,* xxxvii. 57. *F. scariosa?, Drège, Pflanzengeogr. Documente* 82, 83, 185, *not of Nees.*

SOUTH AFRICA: without locality, *R. Brown! Drège,* 564! *Ecklon and Zeyher,* 65!

COAST REGION: Worcester Div.; Drakenstein Mountains and Dutoits Kloof, 3000-4000 ft., *Drège,* 7382!

3. **F. scariosa** (Nees in Linnæa, ix. 292); glabrous; rhizome short, horizontal, stout, woody; stems close together, 4–20 in. long, robust, with one spikelet; leaves often $\frac{1}{2}$ as long as the stem, subcylindric, hardly $\frac{1}{16}$ in. in diam., sheaths scarious-edged; lower bract erect, shorter than the spikelet, linear; spikelet $\frac{3}{4}$–1$\frac{1}{2}$ by $\frac{1}{5}$–$\frac{1}{3}$ in.; glumes very many, $\frac{1}{3}$ in. long, elliptic, obtuse, convex, scarious-edged; stamens 3; anthers long-linear, hardly apiculate; gynophore obconic; nut $\frac{1}{3}$ the length of the glume, obovoid, plane-convex, conic-tipped, smooth, black; style longer than the nut, slender, branches 3 very long, linear. *Linnæa,* x. 182, *excluding syn. of Thunb.*; *Kunth,*

Enum. ii. 251; *Boeck. in Linnæa*, xxxvii. 57. *F. conifera, Nees in Linnæa*, x. 181; *Kunth, Enum.* ii. 251. *Scirpus trigynus, Linn. Mant.* 180, *excluding habitat. S. bulbosus, Rottb. Descr. et Ic.* 46, *t.* 16, *fig.* 2. *Schænus scariosus, Vahl, Enum.* ii. 210, *not of Thunb.; Steud. in Flora*, 1829, 134. *S. deustus, Berg, Fl. Cap.* 10, *fide Vahl. Isolepis ? scariosa. Nees in Linnæa*, vii. 501; viii. 85. *Eleogiton scariosa, Dietr. Sp. Pl.* ii. 98.

SOUTH AFRICA: without locality, *Drège*, 290! *Sieber*, 8! 103! *Pappe*, 117! *Ecklon and Zeyher*, 139! 140! *Bergius!*
COAST REGION: Piquetberg Div.; on the Piquet Berg, 1000–2000 ft., *Drège!* Cape Div.; mountains near Cape Town, *Zeyher*, 1773! Devils Mountain, 300 ft., *Bolus*, 4073! *Burchell*, 8491! Cape Flats, *Pappe! Burke!* near Cape Town, *Schlechter*, 407! Table Mountain, *Ecklon*, 101! *Bergius*, 183! Caledon Div.; Houw Hoek Mountains, 1800 ft., *MacOwan, Herb. Aust.-Afr.*, 1726! George Div.; between Zwart Vallei and the West end of Lange Vallei, *Burchell*, 5695!
WESTERN REGION: Little Namaqualand; between Uitkomst and Geelbeks Kraal, 2000–3000 ft., *Drège.*

4. **F. Zeyheri** (Boeck. in Linnæa, xxxvii. 58); glabrous; rhizome very short, horizontal, woody; stems densely tufted, 8–11 in. long, very straight, slender, wiry, with 1 spikelet; leaves all near the base of the stem, $\frac{1}{2}$ the length of the stem, wiry, filiform; sheaths rich-brown, lacerate, uppermost carrying a leaf; bracts 2, lower erect, slightly overtopping the spikelet, linear; spikelet $\frac{1}{4}$ by $\frac{1}{6}$ in., ovoid, few-flowered, brown; glumes ovate-oblong, subobtuse, convex on the back; stamens 3; anthers linear, not crested; gynophore obconic; nut scarcely $\frac{2}{5}$ the length of the glume, obovoid, plane-convex, smooth, brown; style a little shorter than the nut, slender, branches 3 long, linear.

SOUTH AFRICA: without locality, *R. Brown!*
COAST REGION: Caledon Div.; on the Zwart Berg, near Caledon, 1000–2000 ft., *Zeyher*, 75! 4379!

5. **F. pusilla** (C. B. Clarke in Durand and Schinz, Conspect. Fl. Afr. v. 641); glabrous; roots fibrous; stems 2 in. long, setaceous, with 1 spikelet; leaves all near the base of the stem, often $\frac{1}{2}$ the length of the stem, setaceous; uppermost sheath leafless, its mouth slightly dilated, notched; bracts 2, lower erect, $\frac{1}{3}$ in. long, linear; spikelet $\frac{1}{8}$–$\frac{1}{6}$ in. long, ellipsoid, 6-flowered; glumes boat-shaped, ovate, obtuse, many-striate, chestnut or chestnut-red; gynophore obconic; style linear, branches 3 linear, very long. *F. bergiana, var. β, capillaris, C. B. Clarke in Durand and Schinz, Conspect. Fl. Afr.* v. 636. *F. capillaris, Nees ex C. B. Clarke l. c.* 636.

COAST REGION: Cape Div.; near Cape Town, *Rehmann*, 1182!
WESTERN REGION: Little Namaqualand; Harde Veld, 2000–3000 ft., *Zeyher*, 4380!

No specimen in the Kew Herbarium.

Zeyher's 4380 is quoted by Drège in Linnæa, xx. 251, as an *Acrolepis.*

The species is not allied to *F. Zeyheri*; it looks like a starved example of *F. setiformis*, Schrader; but the leafless top sheath is quite unlike. The flowers are very young.

6. **F. micrantha** (C. B. Clarke); glabrous; rhizome thin, woody; stems tufted, 2–4 in. long, wiry, with 1 spikelet; leaves ½ the length of the stem, in the lower part flat, 1/24 in. wide, in the upper part thick, terete, rigid; sheaths stout, red, uppermost bearing a leaf; bracts 2, suberect, lower as though a continuation of the stem, ⅓–½ in. long; spikelet ⅛ by 1/12 in., 6-flowered, white, hard; glumes boat-shaped, ovate, very thick, stiff, the sides nearly occupied by the close thick white ribs, the thick keel excurrent in a short thick obtuse point; nut nearly as long as the glume, long-ellipsoid, trigonous, glistening with series of points (i.e. the outermost layer of quadrangular cells); gynophore conspicuous, narrowly ob-pyramidal, ending in 3 lanceolate teeth opposite the 3 angles of the nut-base. *F. albicans, C. B. Clarke in Durand and Schinz, Conspect. Fl. Afr.* v. 635 *partly*.

COAST REGION: Cape Div.; Table Mountain, *Rehmann*, 652!

From its white colour this has a superficial resemblance to *F. albicans*, Nees; but the flat-based leaves and very large nut distinguish it.

7. **F. aphylla** (Nees in Endl. Prodr. Fl. Norf. 23); glabrous; rhizome sometimes short, sometimes 8 in. long, erect, slender, woody, black, clothed by scales, dividing, carrying tufts of stems where its divisions approach the surface of the soil; stems 2–4 in. long, tufted, rigid, with 1 head; sheaths all near the base of the stem, leafless, often with a lanceolate brown process on one side; lower bract ¼–1¼ in., as though a continuation of the stem, terete, stiff; head ¼ in. in diam., of 3–8 ellipsoid spikelets; glumes ovate, obtuse, many-striate, brown, closely incurved; stamens 3; anthers linear-oblong, apiculate; nut ⅖ the length of the glume, obovoid, biconvex, smooth, black; gynophore obconic, microscopically 3-toothed at the top; style shorter than the nut, branches 2 (or rarely 3 fide *Nees*), linear longer than the style. *Endl. Atakta*, 12, *t.* 12. *F. lateralis, Kunth, Enum.* ii. 254; *Boeck. in Linnæa*, xxxvii. 63. *Schœnus lateralis, Vahl, Enum.* ii. 211. *S. filiformis, Roem. et Schultes, Syst. Veg.* ii. *Mant.* 528. *Schœnidium laterale, Nees in Linnæa*, ix. 292; x. 166. *Scirpus truncatus, Sieber ex Nees in Endl. Prodr. Fl. Norf.* 23. *Isolepis arcuata, Ecklon ex Kunth, Enum.* ii. 254.

SOUTH AFRICA: without locality, *Sieber*, 109! *Bergius! Chamisso*, 218! *Pappe*, 89!

COAST REGION: Cape Div.; near Cape Town, *Harvey*, 187! boggy places above Simons Bay, *Milne*, 219! Cape Sandhills, *MacGillivray*, 410! sand-flats between Tyger Berg and Blue Berg, below 500 ft., *Drège*, 214! Paarl Div.; Paarl Mountains, 1000–2000 ft., *Drège;* Worcester Div.; Hex River Kloof, 1000–2000 ft., *Drège;* Uitenhage Div.; near the mouth of Van Stadens River, *MacOwan*, 2124! Port Elizabeth Div.; sand hills along the coast, *E.S.C.A. Herb.*, 171! 193! Komgha Div.; grassy slopes near the mouth of the Kei River, *Flanagan*, 1781!

8. **F. repens** (Kunth, Enum. ii. 255); glabrous; rhizome horizontal, ⅙ in diam. (2½ in. fragments seen), clothed by hard dark-brown nearly-quadrate scales; stems solitary (or in separate clusters) on the rhizome, 4–16 in. long, medium-sized, not setaceous, nearly

round, many-striate; sheaths basal, brown, produced ones lanceolate, brown on one side, not linear; spikelets 3–8, sessile, $\frac{1}{6}$ by $\frac{1}{12}$ in., ellipsoid, obtuse, many-flowered, brown; bract hardly any; glumes ovate, rounded-obtuse, concave, 17-striate, margins incurved; nut $\frac{2}{5}$ the length of the glume, obovoid, unequally trigonous, smooth, chestnut-black; style scarcely $\frac{1}{2}$ the length of the nut, branches 3 linear, slightly exserted; gynophore obconic, exceedingly short, microscopically 3-toothed. *Boeck. in Linnæa,* xxxvii. 64. *Isolepis repens, Nees in Linnæa,* ix. 291; x. 158.

SOUTH AFRICA: without locality, *Drège,* 7405!
COAST REGION: Uitenhage Div.; by the Zwartkops River, 50–500 ft., *Zeyher,* 4392! *Ecklon and Zeyher,* 660!

9. **F. sylvatica** (Kunth, Enum. ii. 254); glabrous; rhizome (seen) very short, much divided; stems 10–22 in. long, medium-sized, not setaceous, strongly 4–5-angular (in Bolus, 2693, slenderer, much less angular); leaves reduced to sheaths; sheaths basal, dark-red, produced on one side, lanceolate, rarely a linear red bristle $\frac{1}{20}$–$\frac{1}{6}$ in. long is added; spikelets 2–5, sessile, $\frac{1}{6}$–$\frac{1}{3}$ by $\frac{1}{10}$ in., cylindric, obtuse, many-flowered, pale brown; bract very short; glumes, nut, style and gynophore nearly as in *F. repens,* Kunth. *Boeck. in Linnæa,* xxxvii. 65. *F. membranacea, var. β, Kunth, Enum.* ii. 252. *Scirpus trispicatus, Linn. f. Suppl.* 103; *Thunb. Prod.* 17; *Fl. Cap. ed. Schultes,* 96. *S. Sparmanni, Lam. Ill.* i. 140.

SOUTH AFRICA: without locality, *Thunberg, Zeyher,* 4387!
COAST REGION: Swellendam Div.; near Swellendam, 1000 ft., *Kuntze,* 246! George Div.; near George, below 1000 ft., *Drège!* Knysna Div.; in damp situations, *Bolus,* 2693! Uitenhage Div.; Van Stadens Berg, 1000 ft., *Drège,* 3949! *Zeyher!*

10. **F. rugulosa** (C. B. Clarke); glabrous; rhizome 0; stems 8 in. long, densely tufted, setaceous, strongly 5–6-angular, minutely rough granular, with 1–3 spikelets; sheaths near the base, long, narrow, scarious-brown, closely wrapped round the stem; lower bract as though a continuation of the stem, $\frac{1}{4}$ in. long; spikelets $\frac{1}{8}$ in. long, ellipsoid, obtuse, dark-brown; glumes elliptic, hardly keeled, concave with incurved margins; style 3-fid.

COAST REGION: Knysna Div.; on sand-hills near the west end of Groene Vallei, *Burchell,* 5666!

This plant has the general appearance of *F. filiformis* and its numerous neighbours, which are often nearly leafless; but the smoothly ellipsoid spikelets with the edges of the glumes incurved do not suit the *F. filiformis* group. The rugulose filiform stem is peculiar to this species, and the strongly-marked *F. quinquangularis,* Boeck.

11. **F. leiocarpa** (Nees in Linnæa, x. 171); rhizome not seen; stems 16–24 in. long, slender; sheaths quite leafless; head of 3–8 spikelets, globose; bracts 2, spreading, lower 1–2 in. long; glumes ovate, acuminate, mucronate; otherwise as *F. sylvatica,* Kunth. *Boeck. in Linnæa,* xxxvii. 77. *F. semibracteata, Nees in Linnæa,* ix. 292.

SOUTH AFRICA: without locality, *Drège*, 75!

COAST REGION: Uitenhage Div.; *Ecklon!* Algoa Bay, *Forbes!*

No specimen in the Kew Herbarium.

The nut is more orbicular than that of *F. sylvatica*, but not more smooth. The gynophore is still smaller, reduced to a minute tubercle.

12. F. punctata (Hochst. in Flora, 1845, 758); head of 10–12 spikelets; bract $\frac{1}{4}$–$\frac{1}{3}$ in.; otherwise as *F. leiocarpa*, but a little larger in all its parts.

COAST REGION: Uitenhage Div.; at the sources of streams in the Winterhoek Mountains, *Krauss in Herb. Boissier!*

No specimen in the Kew Herbarium.

The spikelets are a good deal larger than those of *F. leiocarpa*, the nut larger, the gynophore larger and 3-toothed; it may be only a large state of *F. leiocarpa*, but the bract is very much shorter and different in character.

13. F. filiformis (Schrad. Anal. Fl. Cap. 46); glabrous; stolons and rhizome 0; stems 2–8 in. long, densely tufted, setaceous, with 4–5 elevated strong angles (between these apparently dotted), carrying 1 head of 1–3 (rarely more) sessile spikelets; leaf-sheaths scarious, white or discoloured, broad, rolling up and wearing away (not lacerate-fimbriate), uppermost leafless; leaves as though a prolongation of the midrib of the sheath, sometimes $\frac{2}{3}$ the length of the stem, terete, very similar to the stem but slenderer; bracts not more than 2, lower as though a continuation of the stem $\frac{1}{5}$–1 in. long; spikelets $\frac{1}{8}$–$\frac{1}{6}$ by $\frac{1}{12}$–$\frac{1}{8}$ in., 4–12 flowered, chestnut or brown; glumes boat-shaped, with no definite keel, ovate, obtuse or truncate, chestnut-red with 13 paler curved striations, central nerve not reaching the tip of the glume, margins and tip of glume scarious, not incurved; stamens 3; anthers linear-oblong, yellow, crested; style much shorter than the nut, linear, deciduous, branches 3 linear, longish; gynophore distinct, obpyramidal, margin 3-lobed; nut $\frac{2}{5}$ the length of the glume, round-trigonous, obovoid-truncate, about as broad as long, smooth, finally a leaden-black green-tinged. *Nees in Linnæa*, x. 173, *excluding var. contorta; Kunth, Enum.* ii. 253 *partly; Boeck. in Linnæa*, xxxvii. 59. *Schœnus filiformis, Lam. Ill.* i. 135. *S. oliganthos, Kunth, Enum.* ii. 253, *in syn. Cyperus crinitus, Poir. in Lam. Encyc.* vi. 752. *Scirpus marginatus, Thunb. Prod.* 17; *Fl. Cap. ed. Schultes*, 96. *S. crinitus, Roem. et Schultes, Syst. Veg.* ii. 125. *S. obliganthus, Steud. in Flora*, 1829, i. 146. *S. oliganthus, Steud. Nomencl.* i. 634 *in syn. and* ii. 540. *Isolepis oliganthes, Nees in Linnæa*, vii. 503. *I. marginata, Dietr. Sp. Pl.* ii. 110. *I. pumila, Steud. Syn. Pl. Glum.* ii. 90. *Bæothryon crinitum, Dietr. Sp. Pl.* ii. 94.

SOUTH AFRICA: without locality, *Zeyher! Thunberg, Bergius! Pappe*, 90! 109! *MacGillivray*, 427! *Drège!*

COAST REGION: Cape Div.; Table Mountain, *Ecklon*, 873! *Milne*, 237! *Mudd!* Simons Bay, *Wright*, 509! *MacGillivray*, 411! Devils Mountain, *Ecklon*, 874! Tulbagh Div.; mountains in the vicinity of New Kloof, 1000 ft., *MacOwan, Herb. Aust.-Afr.*, 1689! Worcester Div.; Dutoits Kloof, 2000–3000 ft., *Drège!* Caledon Div.; Houw Hoek Mountains, 2000 ft., *Schlechter*, 7425!

CENTRAL REGION: Worcester Div.; near Touws River Station, *Bolus*, 7459!

Also grows on Kilimanjaro at an elevation of 10,000 ft.

Several of the succeeding *Ficinias* differ very little from this species; and the synonyms cited may belong partly to them.

14. **F. bergiana** (Kunth, Enum. ii. 254); stems 10–15 in. long, capillary (i.e. slenderer than those of *F. filiformis*), very numerous; leaf-sheaths scarious, pale-brown, uppermost wrapped round the stem; nut $\frac{1}{2}$–$\frac{2}{3}$ the length of the glume, ellipsoid, nearly twice as long as broad; margin of gynophore nearly entire; otherwise as *F. filiformis*, Schrad. *F. tristachya, Boeck. in Linnæa*, xxxvii. 60 *partly.*

South Africa: without locality, *Sieber*, 101! *Harvey*, 347!
Coast Region: Cape Div.; near Cape Town, *Burchell*, 433! Simons Bay, *Wright!* Muizen Berg, 1000 ft., *Kuntze*, 252! Worcester Div.; Dutoits Kloof, 2000–3000 ft., *Drège!*

The nut of this species much resembles that of *F. tristachya* to which species Boeckeler reduces it; but *F. tristachya* has spikelets much larger than those of *F. filiformis* and *F. bergiana.*

15. **F. stolonifera** (Boeck. in Linnæa, xxxvii. 60); stolons long, covered by striate pale-brown lanceolate scales; stems and leaves rather stouter than those of *F. filiformis;* spikelets up to $\frac{1}{4}$–$\frac{1}{3}$ by $\frac{1}{8}$ in.; margin of gynophore nearly entire; otherwise as *F. filiformis*, Schrad. *F. filiformis, var. contorta, Nees in Linnæa*, x. 173.

South Africa: without locality, *Harvey! Zeyher!*
Coast Region: Cape Div.; Simons Bay, *Wright*, 501! Riversdale Div.; between Little Vet River and Kampsche Berg, *Burchell*, 6896!
Central Region: Somerset Div.; on the Bosch Berg, 4000 ft., *MacOwan*, 1974!
Eastern Region: Natal; Van Reenens Pass, 5600 ft., *Kuntze*, 235!

The examples of Burchell, MacOwan and Wright show the characteristic stolons. It is possible that many of these *Ficinias* flower before they throw stolons, and it is possible that, in some cases where the plants had stolons, the collector did not collect them. It is not certain but that species, the specimens of which in our herbaria nowhere present stolons, may produce them sometimes.

16. **F. tristachya** (Nees in Linnæa, ix. 292); leaf-sheaths fimbriated; spikelets $\frac{1}{5}$ by $\frac{1}{8}$ in.; nut ellipsoid, longer than broad; gynophore margin nearly entire; otherwise as *F. filiformis*, Schrad. *Linnæa*, x. 175; *Kunth, Enum.* ii. 252 *partly; Boeck. in Linnæa*, xxxvii. 60. *F. involuta, Drège in Linnæa*, xx. 248. *Scirpus tristachyos, Rottb. Descr. et Ic.* 48, *t.* 13, *fig.* 4; *Linn. f. Suppl.* 103; *Thunb. Prod.* 17; *Fl. Cap. ed. Schultes*, 97. *Isolepis tristachya, Roem. et Schultes, Syst. Veg.* ii. 110, *not of Mant.* 64; *Nees in Linnæa*, vii. 502.

South Africa: without locality, *Sieber*, 97! *Pappe*, 106!
Coast Region: Cape Div.; Mountains near Cape Town, 2000 ft., *Thunberg, Zeyher*, 1774! Table Mountain, *Ecklon*, 877! Cape Flats, at Doorn Hoogte, *Zeyher*, 1772! Worcester Div.; Drakenstein Mountains, 1600 ft., *Bolus*, 4074! Riversdale Div.; near the Zoetemelks River, *Burchell*, 6706 partly!

This differs from *F. filiformis* and *F. bergiana* in its larger spikelets; it differs from *F. stolonifera* in that none of the examples in the Herbarium have stolons, and that the nut is considerably longer than broad.

17. F. involuta (Nees in Linnæa, ix. 292); no stolons seen; stems 11–14 in. long; leaf-sheaths pale brown, not fimbriated nor with white edges, the uppermost bearing a leaf; spikelets $\frac{1}{4}$ by $\frac{1}{8}$–$\frac{1}{6}$ in. (rather larger and brighter than in *F. tristachya*, Nees); nut obovoid ellipsoid, a little longer than broad; otherwise as *F. tristachya*, Nees. *Linnæa*, x. 174; *Kunth, Enum.* ii. 253; *Boeck. in Linnæa*, xxxvii. 61.

COAST REGION: Worcester Div.; Hex River Mountains, *Drège*, 576! Drakenstein Mountains, 1500 ft., *Bolus*, 4075! 4078! Uitenhage Div.; near Van Stadens Berg, *MacOwan!* 2176.

This is a much taller plant than *F. tristachya*, with brighter, rather larger spikelets. Zeyher, 1774, which Drège took to be *F. involuta*, matches identically the examples in Linnæus' Herbarium marked by Linnæus "*tristachya.*" But, it may be doubted whether the two can be kept distinct.

18. F. albicans (Nees in Linnæa, x. 175); stolons frequently present, covered by rigid lanceolate nearly white scales; stem 2–7 in. long, carrying 1–5 or sometimes 5–10 spikelets; spikelets $\frac{1}{5}$ by $\frac{1}{10}$ in., when young very white, when in flower green with more or less chestnut on sides; glumes ovate, thick in texture from the numerous close ribs, the thick green-white keel subexcurrent as an obtuse mucro; otherwise as *F. tristachya*, Nees. *Kunth, Enum.* ii. 253. *F. tristachya, var. β, Boeck. in Linnæa*, xxxvii. 61. *Scirpus Pseudoschœnus, Steud. in Flora*, 1829, 147. *Isolepis Pseudoschœnus, Dietr. Sp. Pl.* ii. 103. *I. tristachya var., Nees in Linnæa*, vii. 502.

SOUTH AFRICA: without locality, *Harvey*, 357! *Zeyher!*
COAST REGION: Cape Div.; Cape Flats, near Rondebosch, *Burchell*, 156! Devils Peak, 1300 ft., *Kuntze*, 258! Cape Div.; Table Mountain, *Sieber*, 104; Paarl Div.; Paarl Mountains, 2000–3000 ft., *Drège*, 1585! 1586! Caledon Div.; near Caledon, 1300 ft., *Kuntze*, 248! Swellendam Div.; near Swellendam, 950 ft., *Kuntze*, 249! Riversdale Div.; near the Zoetemelks River, *Burchell*, 6706! between Great Vals River and Zoetemelks River, *Burchell*, 6577! Uitenhage Div., *Ecklon and Zeyher*, 667!

19. F. subacuta (C. B. Clarke); glabrous; stolons none seen; stems densely tufted, 1–3 in. long, curved, rather rigid, each carrying 1 head of 2–4 spikelets; leaf-sheaths dusky brown, lacerate; leaves 1–2 in. long, setaceous; bracts 2, lower $\frac{1}{4}$–$\frac{3}{4}$ in. long, spreading, not appearing as a continuation of the stem; spikelets scarcely $\frac{1}{6}$ in. long, rather smaller than those of *F. tristachya*; nut-bearing glumes ovate-triangular, chestnut-red, keel in the upper part green, extending to the point of the glume; nut $\frac{2}{5}$ the length of the glume, as broad as long, round-trigonous, very obtuse, green-yellow; gynophore conspicuous, margin nearly entire; style 3-fid.

COAST REGION: Mossel Bay Div.; between the landing-place at Mossel Bay and Cape St. Blaize, *Burchell*, 6271!

This is very near *F. tristachya*, Nees, but has much more acute glumes.

20. F. quinquangularis (Boeck. in Linnæa, xxxvii. 63); glabrous, no stolons nor rhizome present; stems densely tufted,

4–11 in. long, 4–5-angular, densely rough by upward-pointing minute papillæ, carrying 1 head of 4–1 chestnut-red spikelets; upper sheaths lanceolate, scarious-brown, leafless, not fimbriate, some of the lower sheaths bearing leaves, 4–5 by $\frac{1}{10}$–$\frac{1}{8}$ in., flat, striate; bracts 2, lower $\frac{1}{4}$–1 in. long, as though a continuation of the stem; spikelets $\frac{1}{6}$–$\frac{1}{4}$ in. long, few-flowered; nut-bearing glumes elliptic-oblong, produced but obtuse-tipped, much longer than in the 7 preceding species; anthers large, linear-oblong, crest very short, white; style 3-fid.

COAST REGION: George Div.; on the Post Berg, near George, *Burchell*, 5932! on mountain flats near George, *Drège*, 3933!

A strongly-marked species, but only young examples have been seen.

21. F. tribracteata (Boeck. in Linnæa, xxxvii. 62); glabrous; no rhizome nor stolon present; stems densely tufted, 4–12 in. long, each carrying 1 head of 2–5 dull brown-red spikelets; leaf-sheaths nearly all with leaves, reddish-brown; leaves often as long as the stems, setaceous; bracts 3, or oftener 2, lower $\frac{1}{2}$–2 in. long, as though a continuation of the stem; spikelets $\frac{1}{5}$–$\frac{1}{8}$ in. long, 4–6-flowered; nut-bearing glume boat-shaped, the strong green keel excurrent as a lanceolate-linear, sometimes rough and curved, mucro; anthers linear-oblong, crest short linear; gynophore very short, margin subentire; nut obovoid-ellipsoid, round-trigonous, very obscurely transversely wavy-lineolate; style very short, branches 3 long, linear. *F. tristachya, Kunth, Enum.* ii. 252 *partly; Drège, Pflanzengeogr. Documente* 116. *F. filiformis, Drège l.c.* 88, 185.

SOUTH AFRICA: without locality, *Pappe*, 103!
COAST REGION: Cape Div.; Table Mountain, 1000–2000 ft., *Drège!* Caledon Div.; Genadendal, 3000–4000 ft., *Drège!*

22. F. acuminata (Nees in Linnæa, ix. 292); stems 7–20 in. long; leaves often $\frac{1}{2}$ as long as the stem, but leaf of the uppermost sheath often short; nut-bearing glumes obtuse, with a most minute chestnut or red excurrent mucro; nut smooth or obscurely and minutely reticulate; otherwise as *F. tribracteata*, Boeck. *Kunth, Enum.* ii. 256 *partly; Boeck. in Linnæa*, xxxvii. 66. *F. ignorata, Boeck. Cyp. Novæ*, ii. 18. *Scirpus acuminatus, Steud. in Flora*, 1829, 146. *Isolepis acuminata, Nees in Linnæa*, vii. 501; x. 162.

SOUTH AFRICA: without locality, *Pappe*, 111!
COAST REGION: Cape Div.; Table Mountain, 2000–3000 ft., *Ecklon*, 858! *Bolus*, 2863! 4768!

This plant resembles *F. involuta*, Nees, and might be a form of it with the glumes microscopically mucronate. Boeckeler says the spikelets are sometimes 5–10 in a head; but these examples in Herb. Kew are probably referable to the next (*tenuifolia*) series.

23. F. elongata (Boeck. in Linnæa, xxxvii. 65); glabrous; stolons clothed with rigid short-lanceolate scales, hardening into tough rhizomes $\frac{1}{10}$ in. in diam.; stems 1–2 ft. long, very slender, each carrying one globose head of 10–30 spikelets in the type plant (probably sometimes fewer); leaf-sheaths narrow, brown-red, not fimbriated, lower leafless, uppermost usually with a narrow linear green leaf $\frac{1}{3}$–$\frac{2}{3}$ in. long (but sometimes leafless); bracts 2, lower $\frac{1}{4}$–$\frac{3}{4}$ in. long; spikelets $\frac{1}{6}$ in. long, dusky brown-reddish; glumes ovate, obtuse, keel most minutely excurrent or running to the microscopic tip; anthers hardly crested; nut $\frac{2}{5}$ as long as the glume, obovoid, trigonous, smooth; style shorter than the nut, branches 3 linear, long; gynophore exceedingly short. *F. lateralis a, Drège, Pflanzengeogr. Documente* 85, 185.

COAST REGION: Paarl Div.; Paarl Mountains, 1000–2000 ft., *Drège!* Worcester Div.; Dutoits Kloof, 2000–4000 ft., *Drège!*

This differs from *F. acuminata,* Nees, in the leaves, and in the numerous spikelets to the head. But some of Drège's *F. acuminata* seems to belong to the present species, though the spikelets are only 5 to the head.

24. F. kunthiana (Boeck. in Linnæa, xxxvii. 67); stem 6–9 in. long; leaves long, setaceous, that on the uppermost sheath usually $\frac{2}{3}$ the length of the stem (sometimes overtopping it); lowermost bract 1–2 in. long; spikelets 5–10 in the head; nut ellipsoid, fully $\frac{1}{2}$ the length of the glume, finally black, minutely white reticulated (by the outermost layer of withered cells); otherwise much as *F. elongata,* Boeck., or *F. acuminata,* Nees. *F. acuminata, Kunth, Enum.* ii. 256 *partly.*

EASTERN REGION: "Between Cape Colony and Natal," *Drège,* 3939! 4390!

The above locality is taken from the label on Drège's specimens at Berlin.

25. F. MacOwani (C. B. Clarke in Durand and Schinz, Conspect. Fl. Afr. v. 640); stems 1–2 ft. long, very slender; leaves $\frac{1}{3}$–$\frac{1}{2}$ the length of the stems, setaceous; spikelets 10–30 in a head; glumes ovate, acute, conspicuously white-edged; otherwise much as *F. elongata,* Boeck., and *F. kunthiana,* Boeck.

COAST REGION: Swellendam Div.; mountains near Swellendam, 1000–1400 ft., *MacOwan and Bolus, Herb. Norm. Aust.-Afr.,* 1399! *Kuntze,* 247!

This does not match any one of the three preceding species, but the four might be united. If this however be done, and the diagnosis widened to include *F. acuminata* (the forms with few spikelets to the head), the difficulty would then arise of separating several species (such as *F. involuta,* Nees) in the preceding group.

26. F. cinnamomea (C. B. Clarke in Durand and Schinz, Conspect. Fl. Afr. v. 637); glabrous; rhizome very short; stems 4–16 in. long, tufted, very slender, each with 1 head of 5–10 spikelets; uppermost sheath leafless, lower pale brown with long setaceous leaves; bracts 2, lower 1–3 in. long, suberect; spikelets $\frac{1}{4}$ in. long, pale brown; glumes ovate; keel excurrent; anthers linear-oblong; crest small, lanceolate; nut $\frac{3}{5}$ the length of the glume, obovoid, trigonous, smooth, leaden-black minutely-white reticulated by the outermost layer of withered cells; style shorter than the nut, branches 3 linear, long; gynophore rather prominent, obpyramidal, margin entire.

CENTRAL REGION: Aliwal North Div.; Witte Bergen, *Cooper*, 635!
EASTERN REGION: Natal; Inanda, *Wood*, 1939! and without precise locality, *Buchanan*, 71!

This differs from the very similar preceding species by the distinct gynophore.

27. F. tenuifolia (Kunth, Enum. ii. 257); glabrous; rhizome woody, $\frac{1}{10}$ in. in diam., obliquely descending; stems clustered, 10–16 in. long, slender, each with one head; leaves $\frac{1}{3}$–$\frac{2}{3}$ the length of the stem, setaceous; bracts 2, lower 1–3 in. long, dilated to $\frac{1}{8}$ in. broad at the base, oblique, not appearing to be a continuation of the stem; spikelets $\frac{1}{5}$ in. long, 6–10 packed very closely in the head, fine brown; glumes ovate, keel distinctly excurrent; nut $\frac{1}{2}$ the length of the glume, narrowly obovoid, trigonous, obtuse, lead-coloured, with rather obscure transverse wavy lines (the outermost cells being short-oblong longitudinally); style rather shorter than the nut, branches 3 long, linear; gynophore very short (a mere rim dilated at the base of the nut). *Boeck. in Linnæa*, xxxvii. 67. *F. filamentosa, Nees in Linnæa*, ix. 292. *F. Ludwigii, Boeck. in Flora*, 1882, 15. *F. gracilis, Schrader ex Kunth, Enum.* ii. 256. *Isolepis filamentosa, Nees in Linnæa*, vii. 503; viii. 84; x. 162, *not of Roem. et Schultes. I. fibrosa, Nees in Linnæa*, ix. 292. *I. ficinioides, Steud. Syn. Pl. Glum.* ii. 98. *Scirpus elatus, Boeck. in Linnæa*, xxxvi. 694.

SOUTH AFRICA: without locality, *Ecklon and Zeyher*, 72! *Sieber*, 207!
COAST REGION: Cape Div.; Table Mountain, 1100 ft., *Bolus*, 2864! Worcester Div.; mountains near Bains Kloof, 1500 ft., *Bolus*, 5309!

In this species, Kunth says, "gynophore 0," and Boeckeler says it is "obsolete." It is a ring of looser pale tissue at the very foot of the nut, which I can in no wise distinguish from what I see in numerous species of *Scirpus*. The allied species, *F. gracilis*, Schrader, has a similar "gynophore." *F. tenuifolia*, Kunth, is much more closely allied to *F. gracilis*, Schrader, and to *F. acrostachys*, than to the preceding species, 23–25.

28. F. paradoxa (Nees in Linnæa, x. 178); glabrous; rhizome very short; stems 6–16 in. long, each with one dense often sub-compound head; leaf-sheaths with prominent scarious margins usually much torn, often forming a white wool (but sometimes very little torn as in Burchell, 6246), uppermost bearing a long leaf; leaves $\frac{1}{2}$ as long as the stems, setaceous, terete below, 5–7-angled above; bracts often more than two, lowest $\frac{1}{4}$–$1\frac{1}{4}$ in. similar to the leaf, pointed, base dilated, with scarious fimbriated edge; spikelets often 20–30 in the head (sometimes on the same rhizome only 3 or 4) $\frac{1}{6}$–$\frac{1}{5}$ in. long, 6–8-flowered; glumes boat-shaped, ovate, obtuse, chestnut-red, nerves 6–9 on each side of the midrib, excurrent from the scarious edge as linear-lanceolate teeth, ferruginous or glistening white; stamens 3; anthers linear-oblong; crest large, lanceolate, white, scabrous; nut $\frac{1}{2}$ the length of the glume, obovoid, trigonous, smooth, lead-black, white-reticulated by the outermost layer of withered cells; style shorter than the nut, branches 3 linear, longish; gynophore somewhat long, obpyramidal, margin entire. *Kunth, Enum.* ii. 258; *Boeck. in Linnæa*, xxxvii. 72. *F. laciniata, var.*

paradoxa, Nees in Linnæa, ix. 292. *Isolepis paradoxa, Schrad. Anal. Fl. Cap.* 22. *I. eckloniana, var. β, Nees in Linnæa*, vii. 506, *excluding syn. Scirpus laciniatus, Thunb.*

VAR. β, **minor** (C. B. Clarke in Durand and Schinz, Conspect. Fl. Afr. v. 641); heads much smaller, hardly 1/5 in. in diam.; bracts dilated at the base, without scarious fimbriated edges. *F. argyropus, var. β minor, Boeck. in Linnæa*, xxxvii. 71. *F. fimbriata, E. Meyer ex C. B. Clarke in Durand and Schinz, Conspect. Fl. Afr.* v. 641.

VAR. γ, **argyropus** (C. B. Clarke in Durand and Schinz, Conspect. Fl. Afr. v. 641); glumes white fimbriate-ciliate on the edge; bracts not fimbriate-edged at the base. *F. argyropus, Nees in Linnæa*, x. 177; *Kunth, Enum.* ii. 258; *Boeck. in Linnæa*, xxxvii. 70.

SOUTH AFRICA: without locality, *Sieber*, 6! 153! *Bergius*, 184! *Pappe*, 92! *Drège*, 2454! Var. β, *Drège*, 436! Var. γ, *Drège*!

COAST REGION: Cape Div.; Cape Flats, near Wynberg, *Drège*! at Doorn Hoogte, *Zeyher*, 4419! and between Cape Town and Simons Bay, *Burchell*, 8519! Table Mountain, *Schlechter*, 584! near Tygerberg, *Burchell*, 973! Mossel Bay Div.; sand-hills near the landing-place at Mossel Bay, *Burchell*, 6246!

The variety *argyropus* is not in the Kew Herbarium.

29. F. laciniata (Nees in Linnæa, ix. 292); leaves flat, 1/20 in. broad; sheaths scarious, brown or white, more or less torn; lowest bract flattened, not pointed at tip, scarious, little fimbriated at the base; otherwise as *F. paradoxa*, Nees. *Linnæa*, x. 177; *Kunth, Enum.* ii. 259; *Boeck. in Linnæa*, xxxvii. 73. *Scirpus laciniatus, Thunb. Prod.* 17; *Fl. Cap. ed. Schultes*, 97.

SOUTH AFRICA: without locality, *Thunberg*, *Drège*! *Wallich*!

COAST REGION: Cape Div.; near Cape Town, *Zeyher*! Uitenhage Div., *Zeyher*, 2367! 2368! Port Elizabeth Div., *Herb. Schinz*, 152! 153! *E.S.C.A. Herb.* 471!

EASTERN REGION: Pondoland, *Bachmann*, 106; Natal; Clairmont Flat, 150 ft., *Wood*, 3998! and without precise locality, *Gerrard*, 458!

30. F. truncata (Schrad. Anal. Fl. Cap. 43, t. 2, fig. 3); glabrous; stolons rigid, densely clothed by lanceolate striate hard brown scales, hardening into wiry rhizomes; stems 4–12 in. long, rather rigid, each carrying one dense subglobose head; leaves 2 by 1/8 in., parallel-sided, flat, truncate, thick, with scarious white margins which wear away, on the upper face 6–10-ribbed, densely covered by prominent stomata; bracts 3–2, lowest 1/4–1/3 in. shorter than the head, dilated at the base, with a linear obtuse termination; head 1/3–1/2 in. in diam., of 6–15 spikelets; glumes ovate, obtuse, chestnut, scarious-edged; anthers linear-oblong, not crested; nut hardly 1/2 the length of the glume, obovoid, trigonous, smooth; style much shorter than nut, branches 3 linear, long; gynophore rather long, narrow obpyramidal, margin very lightly 3-toothed. *Nees in Linnæa*, x. 167; *Kunth, Enum.* ii. 260; *Boeck. in Linnæa*, xxxvii. 80. *Scirpus truncatus, Thunb. Prod.* 17; *Fl. Cap. ed. Schult.* 97. *Isolepis truncata, Nees in Linnæa*, vii. 508.

COAST REGION: Riversdale Div.; Hocge Kraal, near the Zoetemelks River, below 500 ft., *Drège*! Mossel Bay Div.; between the landing-place at Mossel Bay and Cape St. Blaize, *Burchell*, 6262! Uitenhage Div.; between the Coega and Sunday Rivers, below 1000 ft., *Drège!* near the mouths of the Coega and Zwartkops Rivers, *Zeyher*, 4418! near Van Stadens River, *Thunberg.*

31. **F. præmorsa** (Nees in Linnæa, x. 167); leaves without scarious margins, 5–6-ribbed on the upper face, without prominent stomata; otherwise as *F. truncata*, Schrad. *Kunth, Enum.* ii. 270; *Boeck. in Linnæa*, xxxvii. 80. *F. truncata, var. præmorsa, C. B. Clarke in Durand and Schinz, Conspect. Fl. Afr.* v. 645.

COAST REGION: Bredasdorp Div.; Elim, 300 ft., *Schlechter*, 7706! Uitenhage Div., *Zeyher! Ecklon! Drège!*

32. **F. ecklonea** (Nees in Linnæa, viii. 91); glabrous; rhizome hardly any seen; stems 12–20 in. long, tufted, rather stouter than in the species preceding, each with one head; uppermost leaf-sheath scarious, torn, usually bearing a long leaf; leaves 2–10 in. long, setaceous; bracts 3 or more, spreading, lowest 1–3 in. long, similar to the leaves, sometimes much dilated at the base, with wide scarious edges sometimes very little dilated; head $\frac{1}{2}$–$\frac{2}{3}$ in. diam. (smaller heads occur), subglobose of 12–24 spikelets; spikelets $\frac{1}{3}$ by $\frac{1}{8}$ in., cylindric, obtuse, dull brown; glumes ovate, obtuse, the white margin exceedingly narrow, microscopically toothed, appressed; anthers apiculate, i.e. hardly crested; nut $\frac{2}{5}$ the length of the glume, obovoid, trigonous, smooth; style shorter than the nut, branches 3 linear, long; gynophore obconic, margin entire. *Nees in Linnæa*, x. 178; *Boeck. in Linnæa*, xxxvii. 74. *F. Steudelii, Kunth, Enum.* ii. 255. *Scirpus eckloneus, Steud. in Flora*, 1829, 148. *Isolepis Steudelii, Schrad. Anal. Fl. Cap.* 20. *I. eckloniana, Nees in Linnæa*, vii. 506, *var. a. I. Ecklonii, Dietr. Sp. Pl.* ii. 116.

SOUTH AFRICA: without locality, *Bergius*, 195! *R. Brown! Krauss! Harvey*, 567!

COAST REGION: Cape Div.; Simons Bay, *Wright!* Table Mountain, *Ecklon*, 869! Stellenbosch Div.; Lowrys Pass, 700 ft., *Kuntze*, 250! Caledon Div.; Houw Hoek Mountains, 1000-3000 ft., *Zeyher*, 4420! Uitenhage Div., *Ecklon and Zeyher*, 177!

33. **F. lævis** (Nees in Linnæa, ix. 292); glabrous, except the back of the glumes; rhizome (in Burchell 427) long, $\frac{1}{8}$ in. diam., straight, clothed by distant lanceolate dull-brown scales, $\frac{3}{4}$ in. long; stems 12–16 in. long, clustered, stout, each with 1 head; sheaths entire, uppermost with a leaf 2–6 in. long, $\frac{1}{12}$ in. wide at the base, trigonous, solid at the top, rigid; some of the lower leaf-sheaths produce much longer and broader leaves; bracts 2–3, spreading, lowest $\frac{1}{2}$–$1\frac{1}{2}$ in. long, similar to the leaves, dilated at the base; head $\frac{1}{3}$–$\frac{1}{2}$ in. in diam., dense, brown; glumes ovate, apiculate or mucronate, glabrous in other examples, but in Kunth's *F. dregeana* hispid on the back in the upper part; nut $\frac{1}{2}$ the length of the glume, ellipsoid trigonous, pyramidal at either end, smooth, dull intense black; style-branches 3. *Nees in Linnæa*, x. 170; *Boeck. in Linnæa*, xxxvii. 75. *F. dregeana, Kunth, Enum.* ii. 259. *F. præusta, Nees in Linnæa*, ix. 292.

COAST REGION: Cape Div.; near Cape Town, *Burchell*, 427! Cape Flats, at

Doorn Hoogte, *Zeyher*, 4400! Table Mountain, *Fleck!* Riversdale Div.; Hooge Kraal, near the Zoetemelks River, below 500 ft., *Drège!* Uitenhage Div.; between the Coega and Sunday Rivers, below 1000 ft., *Drège!*

CENTRAL REGION: Calvinia Div.; Bokke Veld Mountains at Uien Vallei, 2000–2500 ft., *Drège*, 2458!

The localities (except Drège's) refer to the typical *F. lævis*, Nees, which has the flower-glumes (as throughout the genus) glabrous. This is closely allied to *F. ecklonea* (with which it agrees as to points not mentioned in the above description), but has stouter stems and leaves, and less cylindric spikelets.

The plant of Drège, which Kunth made a species, *F. dregeana*, differs from *F. lævis*, Nees, by having the flower-glumes hispid on the upper part of the back. Future gatherings of it may cause it to be admitted as a distinct species, though Boeckeler has united it with *F. lævis.*

34. F. brevifolia (Nees in Linnæa, ix. 292); glabrous; rhizome hardly any seen; stems 1–2 ft. long, tufted, somewhat stout, each with 1 head; sheaths scarious, torn, uppermost with a short setaceous leaf (up to 2 in. long), lower sheaths often with much longer setaceous leaves; bracts 3–4 or more, spreading, lowest 1–4 in. long, setaceous, usually much dilated orbicular and scarious at the base; head frequently $\frac{1}{2}$ by $\frac{1}{3}$ in., conical, evidently compound, dense, with numerous dark-brown spikelets, but there are numerous much smaller heads, some only $\frac{1}{5}$ in. long with 3 spikelets; glumes ovate, acute, apiculate; anthers linear-oblong, with linear apiculation; nut half the length of the glume, narrowly ellipsoid, almost lanceolate, trigonous, top acutely pyramidal, black, with minute transverse and interrupted lines (owing to the outermost cells being short-oblong placed longitudinally); style much shorter than the nut, branches 3 linear, long; gynophore obpyramidal, yellow-brown, margin often with 3 obtuse triangular teeth. *Kunth, Enum.* ii. 260; *Boeck. in Linnæa*, xxxvii. 76. *F. composita, Nees in Linnæa*, ix. 292; x. 172; *Kunth, Enum.* ii. 263. *Melancranis radiata, Vahl, Enum.* ii. 239; *cf. Schrad. Anal. Fl. Cap.* 51 *in Obs. Hypolepis composita, Nees in Linnæa*, vii. 525 *partly.*

VAR. β, **atrocastanea** (C. B. Clarke in Durand and Schinz, Conspect. Fl. Afr. v. 636); spikelets and heads shining black-chestnut. *F. striata α, Drège, Pflanzengeogr. Documente* 82, 185.

COAST REGION: Cape Div.; Table Mountain, 1000–3500 ft., *Drège*, 1588! *Bolus*, 4761! *MacGillivray*, 415! *Schinz!* Devils Mountain, 1200-1400 ft., *Bolus*, 3850! *MacOwan and Bolus, Herb. Norm.* 1398! Simons Bay, *Milne*, 234; *MacGillivray*, 416! *Wright!* Paarl Div.; Paarl Mountains, 1000–2000 ft., *Drège!* Worcester Div.; Dutoits Kloof, 2000–4000 ft., *Drège!* Var. β, Cape Div.; mountains near Cape Town, below 1000 ft., *Drège!*

This species varies greatly in size; it is usually stout with stout setaceous leaves and bracts; but the collections made by Drège in Worcester Division, at 2000–4000 ft., have very slender stems, small heads of 3–5 spikelets, and filiform bracts. One of Drège's examples has a node, with 3 leaves, 23 in. from the base of the stem—an abnormality.

35. F. acrostachys (C. B. Clarke in Durand and Schinz, Conspect. Fl. Afr. v. 635); glabrous; stem exceeding 8 in. in length, very slender; lowest bract placed more than $\frac{1}{3}$ in. below the head, 3 in. long, filiform; two bracts close under the head, spreading, lower 1 in.

filiform, dilated at the base; head $\frac{1}{3}$ in. broad, depressed ovoid, dense, chestnut; glumes ovate, mucronate; nut $\frac{2}{5}$ the length of the glume, obovoid-ellipsoid, trigonous, black, minutely transversely lineolate; style-branches 3 linear; gynophore small, obpyramidal, margin entire. *F. composita, Kunth, Enum.* ii. 263 *partly. Hypolepis composita, Nees in Linnæa,* vii. 525 *partly. Isolepis acrostachys, Schrad. Anal. Fl. Cap.* 23 *in note. Scirpus acrostachys, Steud. in Flora,* 1829, 148.

COAST REGION: Cape Div.; Table Mountain, *Ecklon*, 870!

Steudel, who originally described this species, had only the upper part of the stem, which is all Ecklon's specimens show. The plant, however, does not match any other species; it is nearest, perhaps, to *F. tenuifolia*, Kunth, but has more numerous spikelets and a more compound head.

36. **F. dasystachys** (C. B. Clarke); glabrous; stems 16 in. long, tufted on a short woody rhizome, each with one head; uppermost sheath not torn, truncate, with a leaf as long as the stem $\frac{1}{20}$ in. wide, ligulate curved, long, not setaceous; bracts 3 or more, spreading, lowest $1\frac{1}{2}$ in. long, similar to the leaves, very little dilated at the base; head $\frac{1}{2}$ in. broad, manifestly compound, dense, with numerous spikelets, pale brown; glumes ovate, apiculate; anthers short for the genus, with small linear-lanceolate white scabrous crest; nut (not ripe) $\frac{2}{5}$ the length of the glume, narrowly obovoid, trigonous; style shorter than the nut, branches 3 linear, long; gynophore small, obpyramidal, margins entire.

COAST REGION: Komgha Div.; pastures near Komgha, 2000 ft., *Flanagan*, 922!

The long leaves of this plant are unlike those of any neighbouring species; they are exceedingly narrow, flat on the face, rounded on the back, and 3–4-nerved to the tip, where they terminate suddenly (as do the bracts) in a white callosity.

37. **F. gracilis** (Schrad. Anal. Fl. Cap. 44); glabrous; rhizome 3 by $\frac{1}{10}$ in., clothed by rigid, lanceolate, striate, dull brown scales $\frac{1}{4}$ in. long (usually 0 in herbarium examples); stems 6–16 in. long, tufted, each with 1 head; leaf on the uppermost sheath long, often $\frac{2}{3}$ the length of the stem, $\frac{1}{20}$ in. broad, channelled; broader leaves from the lower sheaths are sometimes present; bracts often 3 or more, spreading, lowest $\frac{1}{2}$–$1\frac{1}{2}$ in. similar to the leaves, not much dilated at the base; head $\frac{1}{3}$–$1\frac{1}{2}$ in. in diam., manifestly compound of 6–20 spikelets, greenish with chestnut spots when young, brownish-red when old; spikelets $\frac{1}{4}$ in. long; glumes ovate, obtuse, or obtuse with a hardly excurrent mucro, or subacute; nut $\frac{2}{5}$ the length of the glume, obovoid, trigonous, ashy-black, minutely transversely interrupted lineolate; style shorter than the nut, branches 3 linear, long; gynophore small obpyramidal, margin minutely 3-lobed. *Nees in Linnæa,* x. 172; *Kunth, Enum.* ii. 256; *Boeck. in Linnæa,* xxxvii. 68. *F. Poiretii, Kunth, Enum.* ii. 255; *Boeck. in Linnæa,* xxxvii. 75. *Scirpus gracilis, Poir. in Lam. Encyc.* vi. 763, *not in Encyc. Suppl.* v. 99, 102; *Spreng. in Flora,* 1829, i., *Beil.* 11. *Isolepis gracilis, Nees in Linnæa,* vii. 493; x. 161, *not in Wight, Contrib.*

109. *I. lineata, Nees in Linnæa,* x. 160 ; *Kunth, Enum.* ii. 215. *I. Poiretii, Steud. Syn. Pl. Glum.* 98.

VAR. β, **commutata** (C. B. Clarke in Durand and Schinz, Conspect. Fl. Afr. v. 639); head looser, subspicate; lowest bract $\frac{1}{12}$ in. distant. *F. commutata, Kunth, Enum.* ii. 256; *Boeck. in Linnæa,* xxxvii. 71. *Isolepis commutata, Nees in Linnæa,* x. 161. *Ficinia gracilis, Nees in Linnæa,* viii. 91. *Trichelostylis gracilis, Nees in Linnæa,* vii. 492.

SOUTH AFRICA: without locality, *Petit Thouars,* 17! *Harvey,* 360!

COAST REGION: Riversdale Div.; between Great Vals River and Zoetemelks River, *Burchell,* 6576! Mossel Bay Div.; Attaquas Kloof, *Gill!* Humansdorp Div.; grassy places in Lange Kloof, near Kromme River Heights, *Bolus,* 2692! Uitenhage Div.; near Geelhoutboom, below 1000 ft., *Drège,* 7419! Albany Div.; tops of hills near Grahamstown, 2000 ft., *MacOwan,* 186! 1973! Var. β, Stellenbosch Div., *Ecklon,* 1894!

CENTRAL REGION: Somerset Div.; Bosch Berg, 4000 ft., *MacOwan,* 1970!

EASTERN REGION: Pondoland, *Bachmann,* 115! Natal, *Buchanan,* 89! 332!

Also found on Kilimanjaro in Tropical Africa.

38. F. latinux (C. B. Clarke); glabrous; stems 5–6 in. long, rigid; sheaths pale brown, fimbriate; uppermost leaf 2 in. long, linear, rigid; head scarcely $\frac{1}{3}$ in. in diam., dense, with 10–12 spikelets, compound (a second peduncled head added on one stem); bracts 5, up to 1 in. long, linear, rigid, spreading, a little dilated at the base; glumes ovate-lanceolate, sub-obtuse, much striated, entire; style 3-fid, scarcely dilated at the base; gynophore small; nut broad-obovoid, trigonous, truncate at the top, black, transversely rugose.

COAST REGION: Uitenhage, *Ecklon and Zeyher!* in Lubeck Herbarium.

No specimen at Kew.

On account of the transversely rugose nut, this was associated in herb. Lubeck with *F. tenuifolia,* Kunth; from which it differs in the 5 rigid bracts (implying a compound head) and the rigid thicker leaves. The truncate, almost emarginate nut is unlike most *Ficinias,* which have a nut with a depressed-pyramidal apex.

39. F. setiformis (Schrad. Anal. Fl. Cap. 45); glabrous; stolons clothed by pale-brown striate lanceolate scales $\frac{1}{2}$ in. long, hardening into tough rhizomes $\frac{1}{10}$ in. in diam., up to 4–8 in. long in herbarium examples; stems 4–12 in. long, tufted, slender but rigid, not setaceous, each carrying one head; sheaths subentire, uppermost truncate, with a well-developed leaf; leaves $\frac{1}{3}$–$\frac{2}{3}$ the length of the stem, $\frac{1}{20}$ in. or less broad, with margins more or less inrolled; head $\frac{1}{4}$–$\frac{1}{3}$ in. in diam., ovoid, compound, with 3–5 bracts when fairly developed (but small heads less than $\frac{1}{8}$ in. in breadth with few flowers occur); chestnut, more or less green-marked; lowest bract $\frac{1}{2}$–$2\frac{1}{2}$ in. long, similar to the leaves, much dilated and subquadrate at the base; margins chestnut, striate, entire, not scarious; flower-glumes ovate, acute, mucronate, microscopically toothed at the top; upper glumes of the spikelet male or empty, often obtuse, more or less scarious; stamens 3; anthers linear, with small linear-lanceolate white crests; nut $\frac{2}{5}$ the length of the glume, obovoid, trigonous, smooth, black, microscopically reticulate; style about $\frac{1}{2}$ the length of the nut, branches 3 linear, long; gynophore obconic, dark-brown, margin with 3 obtuse lobes in well-developed examples. *F. atrata, Nees in Linnæa*, ix. 292; x. 167. *F. pallida, Nees in Linnæa*, ix. 292? *F. pallens, Nees in Linnæa*, x. 169. *F. striata, Kunth, Enum.* ii. 257; *Boeck. in Linnæa*, xxxvii. 69. *Schœnus striatus, Thunb. Prod.* 16; *Fl. Cap. ed Schult.* 91; *Schrad. Anal. Fl. Cap.* 25 *in note. S. indicus, Lam. Ill.* i. 135; *Encyc.* i. 740, *excluding the figures cited. S. coronatus, Steud. in Flora*, 1829, 136. *S. atratus, Schrad. Anal. Fl. Cap.* 24, *t.* 4, *fig.* 1. *S. pallens, Schrad. Anal. Fl. Cap.* 25; *Nees in Linnæa*, viii. 85. *Chætospora striata, Dietr. Sp. Pl.* ii. 27. *Hypolepis atrata, Nees in Linnæa*, vii. 525 *in note*; viii. 85, 93.

VAR. β, **Capitellum** (C. B. Clarke in Durand and Schinz, Conspect. Fl. Afr. v. 643); bracts 2, rigid, much dilated at the base, hardened, enclosing tightly the small head. *F. Capitellum, Nees in Linnæa*, ix. 292; x. 168. *Schœnus Capitellum, Thunb. Prod.* 16; *Fl. Cap. ed Schult.* 91; *Steud. in Flora*, 1829, 137. *S. subserratus, Steud. in Flora*, 1829, 126. *Chætospora Capitellum, Dietr. Sp. Pl.* ii. 27. *Hypolepis Capitellum, Nees in Linnæa*, vii. 522; viii. 93.

SOUTH AFRICA: without locality, *Sieber*, 100! 104! *Pappe*, 91! 98! 100! *Bergius*, 57! 77! 190! *Boivin*, 486! *Rehmann*, 1180! 1791! Var. β, *Thunberg*, *Bergius*! *R. Brown*!

COAST REGION: Malmesbury Div.; Hopefield, *Bachmann*! Cape Div.; near Cape Town, *Thunberg*, *Burchell*, 460! *Harvey*, 189! *Bolus*, 4738! Cape Flats, near Rondebosch, *Burchell*, 187! and near Wynberg, *Drège*! Cape Sand-dunes, *Zeyher*! Table Mountain, *Ecklon*, 857! mountains near Cape Town, below 1000 ft., *Drège*! Sea Point, *Bolus*, 4862! Swellendam Div.; mountains by the lower part of the Zonder Einde River, 500–2000 ft., *Zeyher*, 4396! Var. β, Cape Div.; near Cape Town, *MacGillivray*, 428! *Bolus*, 4016! *Rehmann*, 1796! Table Mountain, *Ecklon*, 856! *Burchell*, 597! *Rehmann*, 651! marshy ground above Simons Town *Milne*, 221.

In one head of the collection marked by Drège, "*F. radiata*, Kunth, *a*," the lowest bract is ½ in. distant.

40. F. lithosperma (Boeck. in Linnæa, xxxvii. 72); glabrous; rhizome (seen) hardly 1 in. long, very stout; stems 5–11 in. long, stout, rigid, each with 1 dense compound head; leaf-sheaths pale-brown, torn, uppermost bearing a leaf; leaves $\frac{1}{3}$–$\frac{2}{3}$ the length of the stem, very rigid, $\frac{1}{16}$ in. wide at the base, channelled, but nearly solid, triquetrous for their whole length; bracts 3–5 (head evidently compound), green or yellowish, lowest up to 2½ in. long, dilated at the base with green margins, above the base similar to the leaves, very rigid; head $\frac{3}{5}$ in. in diam.; glumes more than ¼ in. long, oblong-lanceolate, upper male or empty, scarious; stamens 3; anthers long, linear, with short linear-lanceolate white crests; nut $\frac{2}{5}$ the length of the glume (i.e. large for the genus), obovoid, obtuse, smooth, black, microscopically marked transversely with undulating lines; gynophore short, obpyramidal. *F. Bolusii, Boeck. in Engl. Jahrb.* v. 506.

VAR. β, **compacta** (C. B. Clarke); head ½ in. broad, very tightly packed, chestnut-coloured (as are the shining bracts); leaf-sheaths not conspicuous. *F. striata? a, Drège, Pflanzengeogr. Documente* 82, 185.

SOUTH AFRICA: without locality, *Sieber*, 102! *Pappe*, 93!

COAST REGION: Cape Div.; Muizen Berg, 1800 ft., *Bolus*, 4233! *Schlechter*, 596! Caledon Div.; Nieuw Kloof, Houw Hoek Mountains, *Burchell*, 8152! *Schlechter*, 7427! Var. β, Worcester Div.; Dutoits Kloof 3000–4000 ft., *Drège*, 2459!

Bolus, 4233 (the type of *F. Bolusii*, Boeck.), is identical with Burchell, 8152 (the type of *F. lithosperma*, Boeck.). The plant collected by Drège in Worcester, and also referred to *F. lithosperma* by Boeck., differs a good deal in appearance from all the others, but Mr. Bolus says it is only a "form" of the type.

41. F. lucida (C. B. Clarke in Durand and Schinz, Conspect. Fl. Afr. v. 640); bracts brown with scarious margin; nut-bearing glumes more than $\frac{1}{4}$ in. long, scarious, except the keel; half-ripe nut narrow-obovoid; gynophore narrow, as long as the half-ripe nut; otherwise much as slender examples of *F. lithosperma.*

COAST REGION: Clanwilliam Div.; Ceder Bergen, 2900 ft., *Shaw in Herb. Bolus*, 6023!

42. F. pinguior (C. B. Clarke in Durand and Schinz, Conspect. Fl. Afr. v. 641); as though a stouter form of *F. lithosperma;* leaf-sheaths with very large white-scarious edges; bracts dilated at the base with white-scarious edges; spikelets leaden brown; nut-bearing glume obtuse; nut large, obovoid, trigonous, smooth; otherwise nearly as *F. lithosperma.*

SOUTH AFRICA: without locality, *Harvey*, 348!

COAST REGION: Cape Div.; Muizen Berg, 800–1200 ft., *MacOwan and Bolus, Herb. Norm. Aust.-Afr.*, 1396! *Schlechter*, 592! False Bay, *Robertson! Lind!*

43. F. pygmæa (Boeck. in Linnæa, xxxvii. 82); glabrous; rhizome horizontal, 1–2$\frac{1}{2}$ in. long in the dried examples, $\frac{1}{6}$ in. thick, densely clothed by ovate-lanceolate, pale-brown scales, $\frac{1}{3}$ in. long; stems $\frac{1}{3}$–1 in. long, standing solitary, $\frac{1}{2}$–1 in. apart, each with 1 head; leaves tufted, 1$\frac{1}{2}$–2$\frac{1}{2}$ by $\frac{1}{20}$ in., channelled near the base, solid upwards, trigonous; bracts 3–2, lower $\frac{1}{2}$–1 in. long, similar to the leaves, dilated at the base; head $\frac{1}{2}$ in. in diam., with many spikelets, pale-brown; nut-bearing glume $\frac{1}{4}$ in. long, lanceolate elongate, acute; stamens 3; anthers linear, with short lanceolate scabrous white crest; style slender, long, branches 3 linear; pistil very young, carpophore of the genus minute, distinct.

COAST REGION: Clanwilliam Div.; Lange Vallei, near Rhinoster Fontein, 1000–1500 ft., *Drège*, 2464!

44. F. bracteata (Boeck. in Linnæa, xxxvii. 83); glabrous; rhizome seen 0–$\frac{3}{4}$ in. long; stems 6–20 in. long, tufted, slender, each with 1 head; leaves $\frac{1}{3}$–$\frac{2}{3}$ the length of the stem, $\frac{1}{20}$ in. broad or less at the base; lower 2–3 bracts $\frac{1}{2}$–2 in. long, similar to the leaves, dilated, sheathing at the base, passing by degrees into numerous spirally imbricate upper bracts $\frac{1}{4}$ to $\frac{1}{3}$ in. long, scarious, brown, with triangular usually recurved tips; head (fairly developed) $\frac{1}{3}$ in. in diam., but many examples occur with heads $\frac{1}{6}$ in. broad and upper bracts fewer and obscure without dissection; spikelets 1 (very rarely 2) in the bracts, very small, weak, 4–7-flowered; nut-bearing glumes $\frac{1}{8}$ in. long, lanceolate, acuminate, pale-brown, very thin, hardly striate, resembling the upper male or empty glumes in other species; nut $\frac{2}{5}$ as long as the glume, obovoid, trigonous, smooth, black; style not longer than the nut, branches 3 long, linear; gynophore short, obconic, teeth 3, often rather long. *Schœnus scariosus, Thunb. Prod.* 16; *Fl. Cap. ed. Schult.* 91. *S. filiformis, Willd. Sp. Pl.* i. 260, *excluding syn. of Lam. Melancranis scariosa, Vahl, Enum.* ii. 239; *Schrad. Anal. Fl. Cap.* 49, *t.* 2, *fig.* 4; *Nees in Linnæa,* x. 140; *Kunth, Enum.* ii. 264; *Steud. in Flora,* 1829, 153. *M. nigrescens, Schrad. Anal. Fl. Cap.* 50; *Nees in Linnæa,* x. 141. *M. gracilis, Nees in Linnæa,* ix. 287; *Kunth, Enum.* ii. 265. *M. rigidula, Nees in Linnæa,* x. 141; *Kunth, Enum.* ii. 265. *Scirpus scariosus, Roem. et Schultes, Syst.* ii. 61; *Mant.* 40, *not of Thunb. Hypolepis scariosa, Nees in Linnæa,* vii. 521. *H. nigrescens, Nees in Linnæa,* vii. 522.— *Pluk. Alm. t.* 416, *fig.* 1.

SOUTH AFRICA: without locality, *Thunberg, Sieber,* 105! *Bergius! R. Brown! Boivin,* 490! *Hope! Menzies! Harvey,* 386! *Mund and Maire!*

COAST REGION: Clanwilliam Div.; Lange Vallei, below 1000 ft., *Drège!* Cape Div.; near Cape Town, *Rehmann,* 1179! Devils Mountain, *Burchell,* 8494! *Rehmann,* 909! Table Mountain, 2000–3000 ft., *Pappe! Bolus,* 4741! *Rehmann,* 908! *MacGillivray,* 413! *Drège! MacOwan and Bolus, Herb. Norm. Aust.-Afr.,* 1395! *Ecklon,* 866! Cape Flats, between Cape Town and Simons Bay, *Burchell,* 8544! Worcester Div.; mountains above Worcester, *Rehmann,* 2579! 2672! Caledon Div., *Zeyher!* Bredasdorp Div.; Elim, 400 ft., *Schlechter,* 7682! Knysna Div.; near the Keurbooms River, *Burchell,* 5138! at the Ford of the Knysna River, *Burchell,* 5521! Uitenhage Div., *Ecklon and Zeyher,* 948!

CENTRAL REGION: Somerset Div.; at Commadagga, *Burchell,* 3293! Bosch Berg, 4000 ft., *MacOwan,* 1971!

WESTERN REGION: Little Namaqualand! Harde Veld, 2000–3000 ft., *Zeyher,* 4378! near Mieren Kasteel, below 1000 ft., *Drège.* Modder Fontein Mountains, 4000–5000 ft., *Drège!* Kamies Bergen, 3000–4000 ft., *Drège!*

It is not easy to understand how any confusion arose between this and *F. scariosa,* Nees, as there is no similarity in structure, unless it be said that a bract of the present plant resembles a flower-glume of *F. scariosa.*

45. **F. fastigiata** (Nees in Linnæa, x. 170); glabrous; rhizome hardly any seen (fide Boeckeler is slender creeping); stems 8–12 in. long, tufted, slender, each with 1 head; leaves 4–12 in. long, $\frac{1}{20}$ in. broad at the base, that on the uppermost sheath shorter; bracts $\frac{1}{2}$–$1\frac{1}{2}$ in. long, similar to the leaves, dilated and sheathing at the base; head $\frac{1}{3}$ by $\frac{1}{4}$ in., grey-green, lower spikelet in its bract often $\frac{1}{16}$–$\frac{1}{8}$ in. distant; nut-bearing glume ovate, mucronate; anthers linear, crest minute; nut $\frac{1}{2}$ the length of the glume or more, (large for the genus), trigonous, smooth, brown; style shorter than the nut, branches 3 long, linear; gynophore narrowly obpyramidal, rather long, margin toothed. *Kunth, Enum.* ii. 258; *Boeck. in Linnæa,* xxxvii. 78. *F. picta, Nees in Linnæa,* ix. 292. *Schœnus filiformis, Thunb. Prod.* 16; *Fl. Cap. ed. Schult.* 91, *not of Lam. Scirpus fastigiatus, Thunb. Prod.* 18; *Fl. Cap. ed. Schult.* 99.

SOUTH AFRICA: without locality, *Thunberg, Pappe,* 101! *Drège! Ecklon and Zeyher,* 560!

COAST REGION: Cape Div.; near Wynberg, *Drège,* 164!

46. F. fascicularis (Nees in Linnæa, x. 178); glabrous; rhizome 1–2 in. or more long, $\frac{1}{8}$ in. diam., woody, oblique; stems 1–2 ft. long, approximate, slender, each bearing a small panicle, frequently reduced to a spike or almost a head;, leaves 2 in. long, setaceous; bracts setaceous, very little dilated at the base, lowest 1–2 in. long; panicle, the largest seen, $\frac{1}{2}$ by $\frac{1}{3}$ in.; spikelets $\frac{1}{6}$ by $\frac{1}{16}$ in., uniform dull brown; glumes ovate, keel excurrent in a mucro; stamens 3; anthers linear, with small linear white crest; nut hardly $\frac{2}{5}$ the length of the glume, obovoid, trigonous, smooth; style not longer than the nut, branches 3 linear, longish; gynophore very small. *Kunth, Enum.* ii. 263; *Boeck. in Linnæa,* xxxvii. 78.

SOUTH AFRICA: without locality, *R. Brown! Drège!*
COAST REGION: Knysna Div.; near Melville, *Burchell,* 5473! Albany Div.; damp places in woods near Grahamstown, 2000 ft., *MacOwan,* 1240!
CENTRAL REGION: Somerset Div.; Bosch Berg, 4000 ft.! *MacOwan,* 1240! 1975!

47. F. anceps (Nees in Linnæa, x. 179); glabrous; rhizome short (as seen), woody; stems 4–8 in. long, tufted, angular; leaves longer than the stems, $\frac{1}{16}$ in. wide, edges much incurved, minutely scabrous; sheaths brown, torn; inflorescence spicate, with the lowest bract $\frac{1}{4}$ in. distant, or very often of one head only; bracts 2 or 3, similar to the leaves, at the base lanceolate-dilated, hardly scarious, lowest up to 2–3 in. long; spikelets $\frac{1}{4}$ in. long, dull red; glumes ovate, acuminate acute; nut $\frac{2}{5}$ the length of the glume, obovoid, trigonous, smooth; style-branches 3 linear, longish; gynophore narrowly obpyramidal, margin slightly 3-lobed. *Kunth, Enum.* ii. 263; *Boeck. in Linnæa,* xxxvii. 79.

SOUTH AFRICA: without locality, *Drège! Zeyher!*
COAST REGION: Cape Div.; Muizen Berg, in fissures of rocks, 1300 ft., *MacOwan and Bolus, Herb. Norm. Aust.-Afr.,* 1397! Simons Town, *Ecklon!*

48. F. monticola (Kunth, Enum. ii. 261); glabrous; rhizome (seen) short, woody; stems 6–14 in. long, tufted; leaves usually $\frac{2}{3}$ the length of the stem, $\frac{1}{16}$ in. broad, edges inrolled, sheaths white, scarious, torn; inflorescence a narrow panicle or spike, up to $2\frac{1}{4}$ by $\frac{1}{3}$ in.; lowest bract often $\frac{1}{2}$ in. distant, overtopping the inflorescence, similar to the leaves, much dilated at the base, chestnut, with white scarious edge (often very conspicuous); glumes elliptic-lanceolate, acute; nut nearly $\frac{1}{3}$ the length of the glume, globose-obovoid, trigonous, smooth; style about as long as the nut, branches 3 linear; gynophore narrow, rather long, margin subentire. *Boeck. in Linnæa,* xxxvii. 84. *F. montana, Drège, Pflanzengeogr. Documente* 117, 185.

COAST REGION: Swellendam Div.; between Sparrbosch and Tradouw, 3000–4000 ft., *Drège,* 7403! near Swellendam, below 4000 ft., *Zeyher,* 4433!

49. F. bulbosa (Nees in Linnæa, viii. 91); glabrous; stolons elongate, clothed by lanceolate pale scales $\frac{3}{4}$ in. long, hardening into woody rhizomes $\frac{1}{16}$–$\frac{1}{12}$ in. in diam.; stems 4–16 in. long, tufted; leaves $\frac{2}{3}$ the length of the stem, setaceous; sheaths with white scarious torn margins, sometimes brown and much less conspicuous; spikes $\frac{1}{4}$ in. diam., globose, chestnut-brown, sessile, $\frac{1}{4}$–$\frac{3}{4}$ in. apart, simply spicate; bracts 1–4 in. long, similar to the leaves, overtopping the inflorescence, dilated at the base, white, scarious, very conspicuous, or sometimes hardly dilated at all; spikelets $\frac{1}{8}$ in. ovoid, 4–10 in each spike; glumes ovate, nerve excurrent; nut hardly $\frac{2}{5}$ the length of the glume, obovoid, trigonous, smooth; style shorter than the nut, branches 3 linear, long; gynophore small. *Nees in Linnæa*, x. 176; *Kunth, Enum.* ii. 261; *Boeck. in Linnæa*, xxxvii. 85. *Schœnus bulbosus, Linn. Mant.* 178. *S. spicatus, Thunb. Prod.* 16; *Fl. Cap. ed. Schult.* 94; *Schrad. Anal. Fl. Cap.* 26. *Scirpus capensis, Rottb. Descr. et Ic.* 53, *t.* 16, *fig.* 3; *Steud. in Flora,* 1829, 147. *S. vaginatus, Thunb. Prod.* 17; *Fl. Cap. ed. Schult.* 96. *S. bicapitatus, Poir. in Lam. Encyc.* vi. 761. *S. biceps, Roem. et Schultes, Syst. Veg.* ii. 134. *Isolepis bulbosa, Nees in Linnæa*, vii. 507, *excluding syn. Schœnus bulbosus, Thunb.*

SOUTH AFRICA: without locality, *Thunberg, Sieber,* 110! 163! *R. Brown! Bergius,* 181! *Petit Thouars,* 18! *Boivin,* 485! *Harvey,* 344! 360!

COAST REGION: Cape Div.; near Cape Town, *Bolus,* 4864! *Rehmann,* 1796! Camps Bay, *Burchell,* 326! Lions Rump, *Pappe!* Constantia, *Zeyher!* Table Mountain, *Ecklon,* 112! *Drège! Schlechter,* 347! near Rondebosch, *Bolus,* 4493! Worcester Div.; Hex River Mountains, 1000–2000 ft., *Drège,* 285! 562! slopes of the Drakenstein Mountains, near Bains Kloof, 1600 ft., *Bolus,* 4076! Swellendam Div.; between the Breede and Zonder Einde Rivers, *Burchell,* 7492! Knysna Div.; sand-hills near the west end of Groene Vallei, *Burchell,* 5661! Uitenhage Div., *Ecklon and Zeyher,* 666! Port Elizabeth Div.; along the coast, *E.S.C.A. Herb.,* 172!

50. F. secunda (Kunth, Enum. ii. 262); glabrous; stolons long (seen up to 6 in. long), clothed by pale-brown striate lanceolate scales $\frac{2}{3}$ in. long, hardening into wiry rhizomes $\frac{1}{10}$ in. in diam.; stems 4–16 in. long, tufted; leaves often as long as the stem, hardly $\frac{1}{16}$ in. broad, channelled, solid and triangular in the upper half; leaf-sheaths white or brown, often much torn, sometimes less conspicuous; inflorescence a simple spike 2–5 in. long, in which the spikelets, mostly solitary, stand $\frac{1}{8}$–$\frac{1}{4}$ apart subspirally, sessile; bracts similar to the leaves, usually overtopping the spike; spikelets $\frac{1}{5}$ in. long, ellipsoid, hard, shining chestnut or chestnut-red; glumes ovate, keeled, tip triangular, hardly mucronate; nut $\frac{2}{5}$ the length of the glume, obovoid-ellipsoid, trigonous, smooth, black; style much shorter than the nut, branches 3 linear, long; gynophore obconic, margin subentire. *Boeck. in Linnæa*, xxxvii. 86. *Schœnus bulbosus, Thunb. Prod.* 16; *Fl. Cap. ed Schult.* 94. *S. bulbosus β, Lam. Ill.* i. 136. *S. dispar, Spreng. Neue Entd.* iii. 8. *S. secundus, Vahl, Enum.* ii. 215; *Nees in Linnæa*, vii. 526. *Chætospora dispar, Dietr. Sp. Pl.* ii. 32. *Pleurachne secunda, Schrad. Anal. Fl. Cap.* 47, *t.* 4, *fig.* 3; *Nees in Linnæa*, viii. 91; x. 190. *P. Sieberi, Schrad. Anal. Fl. Cap.* 48; *Nees in Linnæa*, x. 190.

SOUTH AFRICA: without locality, *Thunberg, Sieber,* 109! 115! *Sonnerat! Pappe,* 97!

COAST REGION: Clanwilliam Div.; between Lange Vallei and Heeren Loge.

ment, below 500 ft., *Drège!* Cape Div.; near Cape Town, *Bolus*, 4863! Cape Flats, *Zeyher*, 1781! Knysna Div.; near the west end of Groene Vallei, *Burchell*, 5642! Uitenhage Div., *Harvey*, 126! 127!

51. F. compasbergensis (Drége, Pflanzengeogr. Documente 185); glabrous; rhizome short, woody; stems 12–20 in. long, tufted, slender; leaves 4–8 in. long, $\frac{1}{12}$ in. broad, channelled near the base, solid upwards; sheaths brown fimbriate; panicle narrow-oblong (nearly reduced to a compound spike), $2\frac{1}{2}$ by $\frac{3}{4}$ in., lowest bract often $\frac{3}{4}$ in. distant; bracts often 4–6 in. long, similar to the leaves, lower lanceolate at the base, upper often much dilated at the base; spikelets up to $\frac{1}{3}$ in. long, cylindric, often very bright (white with a few chestnut spots), sometimes dull straw-coloured; glumes rather loosely packed, oblong-acute, keel excurrent in a mucro; stamens 3; anthers linear, minutely white-crested; nut $\frac{1}{3}$ the length of the glume, obovoid, trigonous, smooth; style shorter than the nut, branches 3 linear, long; gynophore small, narrow, margin subentire. *F. comparbergensis, Steud. Syn. Pl. Glum.* ii. 124.

CENTRAL REGION: Graaff Reinet Div.; Compass Berg, 6000–7000 ft., *Drège*, 2034! Sneeuw Berg Range, 4500–8000 ft., *Bolus*, 704! Colesberg Div., *Shaw*, 23!

52. F. trichodes (Benth. in Gen. Pl. iii. 1053); glabrous; stems 6–18 in. long, slender, branching from the base to the top; upper leaves 2 in. long, capillary; lower stem leaves sometimes $\frac{1}{16}$ in. broad; peduncles 1–2 in. long, axillary, with 1 spikelet; lower bract $\frac{1}{2}$–1 in. long, similar to the leaves, dilated at the base; spikelet scarcely $\frac{1}{6}$ in. long, 3–5-flowered, pale brown; glumes ovate, long accuminate; nut $\frac{1}{3}$ the length of the glume, narrowly obovoid, trigonous, smooth; style rather shorter than the nut, branches 3 linear, long; gynophore narrowly obconic, margin subentire. *Acrolepis trichodes, Schrad. Anal. Fl. Cap.* 42, *t.* 2, *fig.* 5; *Nees in Linnæa,* viii. 90; x. 129; *Kunth, Enum.* ii. 331; *Boeck. in Linnæa,* xxxvii. 90. *Hypophialium capillifolium, Nees in Linnæa,* viii. 90. *Hemichlæna capillifolia, Nees in Linnæa,* vii. 530, *excluding syn. H. trichodes, Nees in Linnæa,* viii. 90.

SOUTH AFRICA: without locality, *Sieber*, 95! *Ecklon and Zeyher*, 116! *Zeyher*, 2426, 2427.

COAST REGION: Cape Div.; near Cape Town, *Burchell*, 411! Table Mountain, *Hesse* (ex *Schrader*). Caledon Div.; Houw Hoek Mountains, 2500 ft., *Schlechter*, 7551!

This and the two following closely allied species differ greatly from *Ficinia* in habit; but the spikelets are altogether as of *Eu-Ficinia*. See Benth. in Benth. et Hook. f. Gen. Pl. iii. 1053.

53. F. ferruginea (C. B. Clarke in Durand and Schinz, Conspect. Fl. Afr. v. 638); leaf-sheaths scarious-edged, torn, conspicuous, so that the stems appear woolly; bracts to the spikelet more prominent, ferruginous; otherwise as *F. trichodes. Acrolepis ferruginea, Boeck. in Linnæa,* xxxvii. 91. *A. trichodes, var. β ferruginea, Nees in Linnæa,* x. 129.

SOUTH AFRICA: without locality, *Zeyher*, 2425! *Ecklon and Zeyher*, 116 β! No specimen in Kew Herbarium.

54. F. ramosissima (Kunth, Enum. ii. 262); rather stouter than *F. trichodes* in all parts; lower stem leaves $\frac{1}{12}$ in., or even $\frac{1}{8}$ in. broad; peduncles with 1–3 spikelets, or even 5–6 in a minute panicle; otherwise as *F. trichodes*. *Acrolepis ramosissima, Boeck. in Linnæa,* xxxvii. 89.

COAST REGION: Cape Div.; near Cape Town, *Burchell,* 419! Caledon Div.; Niew Kloof, Houw Hoek Mountains, *Burchell,* 8135! Knysna Div.; sand-hills near the west end of Groene Vallei, *Burchell,* 5667! Uniondale Div.; mountains by the Klip River, near Keurbooms River, 2000–3000 ft., *Drège!* Uitenhage Div., *Harvey,* 123! Port Elizabeth Div.; on sand-hills by the sea-shore at Port Elizabeth, *Burchell,* 4296! Albany Div., *Williamson!*

CENTRAL REGION: Aberdeen Div.; Camdeboo Mountain, 4000–5000 ft., *Drège.* Somerset Div.; upper part of Bruintjes Hoogte, *Burchell,* 3083! Bruintjes Hoogte and Bosch Berg, 4500 ft., *MacOwan,* 1901!

55. F. angustifolia (C. B. Clarke in Durand and Schinz, Conspect. Fl. Afr. v. 635); glabrous; stolons short, clothed by pale-brown lanceolate striate scales $\frac{1}{2}$ in. long, hardening into a short rhizome; stems 4–16 in. long, slender, approximate on the rhizome, not divided above the rhizome, each with 1 head; leaves usually $\frac{3}{4}$ the length of the stem, often $\frac{1}{10}$ in. broad (not narrower than in the next species); nodes all close to the base of the stem; leaf-sheaths not torn (stouter than in *Eu-Ficinia*), but some examples show few leaves and those not $\frac{1}{3}$ the length of the stem; bracts 2–3, similar to the leaves, lowest up to 2 in. long, dilated (or very little dilated) at the base; spikelets $\frac{1}{2}$ by $\frac{1}{8}$–$\frac{1}{6}$ in., 1–7 clustered subdigitately, compressed, 8–14-flowered, chestnut or leaden-grey; flower-glumes distant, nearly distichous, uppermost (as indeed in *Cyperus*) going off into a spire; glumes ovate, obtuse, scarious-edged; stamens 3; anthers linear; crest very small, lanceolate, scabrous, white; nut about $\frac{1}{3}$ the length of the glume, ellipsoid, trigonous, smooth, dull-black, outermost cells very minute, subquadrate; style much shorter than the nut, filiform, branches 3 linear, long; gynophore obconic, small margin with 3 oblong lobes. *Hemichlæna angustifolia, Schrad. in Goett. Gel. Anz.* iii. 2066; *Anal. Fl. Cap.* 41; *Nees in Linnæa,* vii. 531; x. 130; *Kunth, Enum.* ii. 330 *partly; Boeck. in Linnæa,* xxxvii. 87 *partly. H. fascicularis, Steud. Syn. Pl. Glum.* ii. 2. *Fimbristylis vexata, Steud. Syn. Pl. Glum.* ii. 318; *cf. Boeck. in Flora,* 1860, 178. *Isolepis podocarpa, Boeck. in Flora,* 1860, 179.

SOUTH AFRICA: without locality, *Bergius,* 179! *Harvey! Ecklon and Zeyher,* 114!

COAST REGION: Cape Div.; Table Mountain, *MacGillivray,* 414! *Rehmann,* 699! 714! Paarl Div.; Paarl Mountains, 1000–2000 ft., *Drège,* 7407! Worcester Div., *Zeyher!* Tulbagh Div.; Mosterts Berg near Mitchells Pass, 2000 ft., *Bolus,* 5310!

This species, with the two following closely allied ones, forms a very natura group; but the flowers are so exactly those of *Eu-Ficinia* that it is not convenient to treat the group as a genus, on habit only, after Bentham has thrown the *Acrolepis* group into *Ficinia.*

56. F. longifolia (C. B. Clarke in Durand and Schinz, Conspect. Fl. Afr. v. 640); stems branched, leafy 1–6 in. above their bases, carrying 1–2 spikelets; spikelets rather larger; glumes more brightly white-edged; otherwise nearly as *F. angustifolia. Hemichlæna longifolia, Nees in Linnæa,* ix. 283; x. 129; *Kunth, Enum.* ii. 330; *Boeck. in Linnæa,* xxxvii. 88. *H. angustifolia, Kunth, Enum.* ii. 330 *partly; Boeck. in Linnæa,* xxxvii. 87 *partly. Schœnus caricoides, Steud. in Flora,* 1829, 137. *Eleogiton caricoides, Dietr. Sp. Pl.* ii. 99.

SOUTH AFRICA: without locality, *Zeyher,* 68!
COAST REGION: Cape Div.; Table Mountain, *Burchell,* 638! *Ecklon,* 864! *Milne,* 216! *MacGillivray,* 412! *Bolus,* 4736.

57. **F. capillifolia** (C. B. Clarke in Durand and Schinz, Conspect. Fl. Afr. v. 637); stem slender, long (up to 30 in. in herbarium specimens), branched and leafy nearly throughout its length; leaves very narrow, upper capillary; spikelets rather slenderer; otherwise as *F. longifolia. Hemichlæna capillifolia, Schrad. in Goett. Gel. Anz.* iii. 2066; *Anal. Fl. Cap.* 40, *t.* 3, *fig.* 1; *Nees in Linnæa,* viii. 90; x. 130 (*not in Linnæa,* vii. 530); *Kunth, Enum.* ii. 330; *Boeck. in Linnæa,* xxxvii. 88.

SOUTH AFRICA: without locality, *Mund and Maire! Ecklon.*
COAST REGION: Cape Div.; Table Mountain, *Hesse* (ex *Schrader*). George Div.; on the lower part of the Post Berg, near George, *Burchell,* 6021! Outeniqua Mountains, *Rehmann,* 69!

Imperfectly known species.

58. **F. nuda** (Boeck. in Linnæa, xxxviii. 383); stolons horizontal, as thick as a pigeon's quill; stems 8–7 in. long, filiform, tufted; leaves (1–2 upper) 2–1 in. long, solid, rigid, rusty-brown; spikelets 4–1, clustered, $\frac{1}{10}$ in. long, 4-flowered; bracts 0; glumes clustered, linear-lanceolate; style 3-fid; perigynium cylindric, margin shortly 3-lobed; nut $\frac{1}{3}$ the length of the glume, narrowly obovoid, obtusely triangular, rusty straw-colour, with raised points.

SOUTH AFRICA: without locality, *Petersen* (ex *Boeckeler*).

59. **F. tenuis** (C. B. Clarke in Durand and Schinz, Conspect. Fl. Afr. v. 645); stems filiform, glabrous; leaves filiform, overtopping the stems; heads terminal, solitary, oblong; bracts keeled-convex, aristate, lower distichous, all with 4-flowered spikes; glumes 8, distichous, many-nerved, keeled, mucronate, 3 lowest empty; ovary trigonous, on an entire disc; style-branches 3, very long. *Melancranis tenuis, Ecklon ex Kunth, Enum.* ii. 265; *but Kunth adds "scarcely a Melancranis."*

SOUTH AFRICA: without locality, collector not indicated.

XII. **FUIRENA**, Rottb.

Spikelet of many imbricate glumes, lowest 2 empty, many succeeding glumes 2-sexual, perfecting nuts, aristate, hairy, uppermost male or sterile. Hypogynous *bristles* 3, outer (sepals) small linear or 0,

3 inner (petals) obovate, often irregular or small or in Sect. *Pseudo-Scirpeæ* setaceous or 0. *Stamens* 3 anterior. *Style* glabrous, linear, not dilated at the base; branches 3, long. *Nut* sessile, obovoid, trigonous, in the typical species enclosed by the 3 persistent petals.

Rhizome (in the Cape species) woody, horizontal; stems with nodes and leaves throughout their length [except in *F. enodis*]; inflorescence paniculate, not umbellate; but in numerous species the panicle is short, reduced to 1 or 2 subterminal spikes.—The very hairy aristate flower-glumes and stems leafy upwards make this genus easy to distinguish even when the characteristic "petals" fail or are like the bristles of *Scirpus*.

DISTRIB. Species 25; in all warm countries, very alike in general appearance.

Sect. 1. PSEUDO-SCIRPEÆ.—No obovate or hastate interior bristles (petals).

Petals 0, or minute linear:
- Panicle short; nut opaque, smooth, white ... (1) **pubescens.**
- Panicle often longer; nut white, transversely granular ... ... (2) **pachyrrhiza.**

Hypogynous bristles 4–6, setaceous, simple, scabrous [as of many *Scirpus*], usually as long as the greenish-brown nut ... ... (3) **chlorocarpa.**

Sect. 2. EU-FUIRENA.—Petals with an obovate lamina (sometimes narrow) but not mere linear bristles (in *F. cærulescens* very rarely wanting altogether).

Nut subtrabeculate, by reason of the short, oblong, transverse, outermost cells:
- Petals as long as the nut, obtriangular, broad ... (4) **microlepis.**
- Petals ½ the length of the nut, nearly square, clawed ... ... (5) **gracilis.**
- Petals as long as the nut, aristate, elliptic or narrow obovate ... ... (6) **cærulescens.**
- Glabrous except the back of the glumes; leaves all basal ... ... (7) **enodis.**

Nut smooth, or obscurely and most minutely reticulate:
- Nut stalked, linear-beaked, yellow-brown; leaves nearly glabrous ... ... (8) **glabra.**
- Nut subsessile, obovoid, white; beak small, conic ... (9) **Ecklonii.**
- Nut yellow-brown; very hairy; heads dense comose (10) **hirta.**

1. **F. pubescens** (Kunth, Enum. ii. 182); more or less hairy (as are all the succeeding species); stems 4–16 in. long, triquetrous; leaves 2–6 by $\frac{1}{10}$–$\frac{1}{5}$ in. flat; inflorescence of 4–20 spikelets, dull (or somewhat glaucous) green-brown, usually in 2 or 3 loose subterminal spikes, but a long-peduncled axillary spike is sometimes added; spikelets $\frac{1}{2}$ by $\frac{1}{6}$ in., dense, prominently hairy and hispid by the soft excurrent points of the glumes; glumes frequently appearing spirally 4–5-ranked; anthers linear-oblong, hardly crested; hypogynous bristles 0, or minute, rudimentary, irregular; nut hardly $\frac{2}{5}$ the length of the glume, triquetrous, opaque white, very smooth, with a short microscopically scabrid cone at the top; outermost cells very minute, rectangular, so that the nut bears small nearly round reticulations, which in this species are exceedingly obscure in the ripe nut; style about $\frac{1}{2}$ the length of the nut, branches 3 linear, much longer than the style. *Boeck. in Linnæa*, xxxvii. 104 *partly; C. B. Clarke in Hook. f. Fl. Brit. Ind.* vi. 665. *Scirpus pubescens, Lam.*

Ill. i. 139; *Desfont. Fl. Atlant.* i. 52, *t.* 10. *S. ciliaris, Pers. Syn.* i. 69 *partly. Carex pubescens, Poir. Voy. en Barb.* ii. 254, 317. *C. Poireti, Gmel. Syst. Nat.* i. 140. *Isolepis pubescens, Roem. et Schultes, Syst.* ii. 118, *not of Mant.* 67.

SOUTH AFRICA: without locality, *Burke!*
COAST REGION: Albany Div.; Grahamstown, 2000 ft., *MacOwan*, 1271! Cathcart Div.; near Cathcart, 4500 ft., *Kuntze*, 284!
CENTRAL REGION: Somerset Div.; on a mountain above the spring of Commadagga, *Burchell*, 3351! Bosch Berg, 4000 ft., *MacOwan*, 1356!
KALAHARI REGION: Transvaal; Pretoria, *Rehmann*, 4770! Matebe Valley, *Holub!* Lomati Valley, near Barberton, 4000 ft., *Galpin*, 1363!
EASTERN REGION: Tembuland; Bazeia, 2000 ft., *Baur*, 286! Griqualand East; banks of streams near Kokstad, 5000 ft., *Tyson*, 1587!

Extends from Portugal and Corsica throughout Africa to the Punjab.

There are many young examples from South Africa, besides those cited here, referred to *F. pubescens* in herb. Kew, from the inflorescence and general aspect, probably rightly. In the account here given of Cape species of *Fuirena*, examples so young that nothing can be made out about the nut or petals are in general not cited.

2. **F. pachyrrhiza** (Ridley in Trans. Linn. Soc. ser. 2, ii. 161); rather stouter; panicle often 4–8 in. long, with several axillary long-peduncled heads; spikelets a very little larger; nut small, white, minutely transversely granular; otherwise as in *F. pubescens*, Kunth. *F. macrostachya, Boeck. in Eng. Jahrb.* v. 507.

KALAHARI REGION: Transvaal; Houtbosch, *Rehmann*, 5625! Houtbosch Berg, *Nelson*, 21*!
EASTERN REGION: Delagoa Bay, *Junod*, 368!

Widely distributed in Tropical Africa.

Wood 1194, Natal (very young), may also belong to *F. pachyrrhiza*.

3. **F. chlorocarpa** (Ridley in Trans. Linn. Soc. ser. 2, ii. 159); rather slenderer; spikelets narrower, cylindric; glumes very little pointed (for genus *Fuirena*), the points not at all recurved; hypogynous bristles 4–6, about as long as the nut, exactly linear-setaceous retrorsely hispid, altogether as of *Scirpus*; nut dull-brown of a greenish tinge; otherwise as the two preceding species.

KALAHARI REGION: Transvaal; Houtbosch, *Rehmann*, 5614!
EASTERN REGION: Pondoland, *Bachmann*, 118! Natal, *Buchanan*, 342! 362!

Also in Tropical Africa and Madagascar.

4. **F. microlepis** (Kunth, Enum. ii. 182 partly); spikelets $\frac{1}{4}$ in. long, in rather dense subgloboso heads; petals broadly obtriangular, brown, 3-nerved, as long as the nut, nearly sessile, hardly microscopically apiculate; nut reticulated by short-oblong transverse cells, tip minute; otherwise as the 3 preceding species. *F. glabra, Ecklon ex Krauss in Flora*, 1845, 757, *not of Kunth. F. pubescens, Boeck. in Linnæa*, xxxvii. 104 *partly.*

EASTERN REGION: Natal; Flats near Durban, *Drège*, 4339! Umlaas, *Krauss*, 59! 191! near Durban, *Wood*, 213! Delagoa Bay, *Junod*, 37! 232!

5. **F. gracilis** (Kunth, Enum. ii. 181); petals minute, about $\frac{1}{3}$ the length of the nut, clawed, nearly square, hairy, nerveless; nut with a conic scabrous tip; otherwise as the preceding species. *Boeck. in Linnæa,* xxxvii. 100. *F. microlepis, Kunth, Enum.* ii. 182 *partly. F. Eckloniii, Nees in Linnæa,* x. 143 *partly.*

SOUTH AFRICA: without locality, *Zeyher*, 1757!
COAST REGION: Queenstown Div.; Shiloh, 3500 ft., *Baur*, 892!
EASTERN REGION: Tembuland; St. Marks Division, on the banks of the Isikoba River, 2000 ft., *Baur*, 724! Transkei; between Gekau and the Bashee River, 1000–2000 ft., *Drège*, 4341! 4338!

Kunth referred Drège 4338, 4339, to his *F. microlepis*, while he named Drège 4341, *F. gracilis*. At all events 4341 is identical with 4338 (they were collected at the same spot, see Drège Pflanzengeogr. Documente 146) and is entitled to the name *gracilis*. I have therefore (in Durand and Schinz, Conspect. Fl. Afr. v. 647) applied the name *microlepis* to Drège 4339 to avoid introducing a new name. It may be doubted whether there is more than one species here altogether; indeed Boeckeler has taken Drège 4338 and 4339 as only *F. pubescens*—a view quite defensible, but which will cut the genus *Fuirena* down to 4 or 5 species.

6. **F. cœrulescens** (Steud. in Flora, 1829, 153); petals as long as the nut, clawed, elliptic, pinnatifid, brown, with a linear scabrous point overtopping the nut, often irregular, unequal or reduced; nut reticulated by short-oblong transverse cells; otherwise as *F. pubescens*, Kunth. *Syn. Pl. Glum.* ii. 125; *Schrad. Anal. Fl. Cap.* 51; *Nees in Linnæa,* vii. 510; x. 142; *Kunth, Enum.* ii. 180; *Boeck. in Linnæa,* xxxvii. 102 *excluding the Indian examples.*

VAR. β, **Buchanani** (C. B. Clarke); stout; panicle 4 in. long, with numerous heads.

SOUTH AFRICA: without locality, *Ecklon and Zeyher*, 36b! 36c!
COAST REGION: Cape Div.; near Cape Town, *Harvey*, 177! Cape Flats, *Ecklon*, 868!
EASTERN REGION: Natal; between Umtentu and Umzimkulu Rivers, *Drège*, 4340! Inanda, *Wood*, 1620! and without precise locality, *Bolton, Rehmann*, 8443! 8600! Delagoa Bay, *Junod*, 86! Var. β; Natal, *Buchanan*, 120!

The petals in *Fuirena* are probably to be regarded as reduced organs in process of disappearing; the present plentiful species shows that they must not be relied on too absolutely for the discrimination of species. The 3 petals on one nut of *F. cœrulescens* sometimes differ entirely in size and form; in some examples they nearly or quite disappear. Var. β in its nut and long panicle resembles *F. pachyrrhiza*, but then it has fairly developed petals.

7. **F. enodis** (C. B. Clarke in Durand and Schinz, Conspect. Fl. Afr. v. 646); glabrous, except the backs of the flower-glumes; rhizome slender, woody, long-creeping; stems numerous, in a series on the rhizome, 3–7 in. long, leafless except near the base, each with a lateral head of 1–3 spikelets; lowest bract $\frac{1}{4}$–$\frac{1}{3}$ in. long, as though a continuation of the stem; spikelets about $\frac{1}{3}$ in. long (with the glumes), as of *Fuirena;* sepals 3, linear, setaceous, retrorsely hispid; petals 3, as long as the nut, clawed, round-elliptic, 3-nerved, brown, hairy on the margin, with a linear hispid tip; nut dirty straw-colour,

reticulated by minute short-oblong transverse cells; tip conic scabrous.

KALAHARI REGION: Griqualand West; Hay Division, at Griqua Town, *Burchell*, 1865!

One of the few well-marked species in the Genus: its glabrousness and inflorescence would suggest *Scirpus* as the genus.

8. **F. glabra** (Kunth, Enum. ii. 182, not of Ecklon); stems and leaves more than usually glabrous; petals elliptic, brown, 3-nerved, the linear points overtopping the nut, claw hairy; nut small, stalked, globose, acutely 3-angled, very smooth, yellow-brown, beak long, subulate hispid; otherwise as *F. pubescens* or *F. cærulescens*, Steud. *Boeck. in Linnæa*, xxxvii. 101. *Cyperus? hirsutus, Berg. Descr. Cap.* 11; *Kunth, Enum.* ii. 114. — *Juncus, Buxbaum, Cent.* iii. 29, *t.* 53.

EASTERN REGION: Tembuland: bank of the Umtata River, below 1000 ft., *Drège*, 4343! Griqualand East; Ibisi River, *Wood*, 3144!

9. **F. Ecklonii** (Nees in Linnæa, ix. 288); leaves with some hair on them, at least underneath; petals nearly as long as the nut, elliptic, toothed or almost lobate, claw hairy; nut nearly sessile, obovoid, smooth, white, beak small, conic; otherwise nearly as the preceding species. *Linnæa*, x. 143 *partly*. *F. mollicula, including var. β, Kunth, Enum.* ii. 182. *F. cærulescens, Boeck. in Linnæa*, xxxvii. 102 *partly*. *F. glabra, Ecklon ex Drège, Pflanzengeogr. Documente* 186.

COAST REGION: Knysna Div.; Koratra, below 1000 ft., *Drège!* Uitenhage Div.; between Van Stadens Berg and Bethelsdorp, below 1000 ft., *Drège!* and without precise locality, *Zeyher!* Alexandria Div.; Zuur Berg Range, 2000–3000 ft., *Drège*, 2039!

CENTRAL REGION: Aberdeen Div.; Camdeboo Mountain, 4000–5000 ft., *Drège!*

EASTERN REGION: Pondoland or Natal; between the Umtentu and Umzimkulu Rivers, *Drège!*

The description of *F. ecklonianα*, Boeck. (in Linnæa, xxxvii. 109), does not agree with that of *F. Ecklonii*, Nees, here given. A sheet in the Berlin Herbarium contained formerly a plant of *F. hirta*, Vahl, on which was pinned the original description of *F. Ecklonii* in Nees' own hand—some drawings by Kunth being also superimposed.

10. **F. hirta** (Vahl, Enum. ii. 387); more hairy than the preceding species; the long excurrent midrib of the glumes long-pilose, so that the dense spikes are comose; petals as long as the nut, ovate, hairy; nut smooth, yellow-brown; otherwise as *F. pubescens*, *F. cærulescens*, and most of the preceding species. *Schrad. Anal. Fl. Cap.* 52; *Steud. in Flora*, 1829, 153; *Nees in Linnæa*, vii. 510; x. 142; *Kunth, Enum.* ii. 181; *Boeck. in Linnæa*, xxxvii. 108. *F. cephalotes, Schrad. in Goett. Gel. Anz.* iii. 2071. *F. erioloma, Nees in Linnæa*, ix. 288; x. 142; *Kunth, Enum.* ii. 181. *F. intermedia, Kunth, Enum.* ii. 181; *Boeck. in Linnæa*, xxxvii. 101.

Scirpus hottentottus, *Linn. Mant.* 182; *Thunb. Prod.* 18; *Fl. Cap. ed. Schult.* 98.

South Africa: without locality, *Thunberg*, *Sieber*, 96! *Drège*, 4342! *Wallich!* *Thom*, 907! *Bergius*, 177! *Harvey*, 340! *Ecklon and Zeyher*, 136! 137!

Coast Region: Clanwilliam Div.; Jan Dissels Valley, below 1000 ft., *Drège!* Cape Div.; flats near Cape Town, *Burchell*, 57! *Harvey*, 175! Table Mountain, *Milne*, 232! *MacGillivray*, 418! *Burke*, 314! Devils Mountain, *Ecklon*, 34! 881! 882! Paarl Div.; Paarl Mountains, *Drège*, 7383! Worcester Div., *Zeyher!* Caledon Div.; Grabouw, near the Palmiet River, 700 ft., *Bolus*, 4227! Mossel Bay Div.; sand-hills near the landing-place at Mossel Bay, *Burchell*, 6249! Knysna Div.; Vlugt, *Bolus*, 2500! Koratra, below 1000 ft., *Drège!* Uitenhage Div.; near Uitenhage, *Burchell*, 4230! *Ecklon and Zeyher*, 168! Albany Div.; mountains near Grahamstown, 2000 ft., *MacOwan* 1356!

XIII. LIPOCARPHA, R. Br.

Spikelet of many spirally imbricate glumes, lowest 2 empty, many succeeding glumes 2-sexual perfecting nuts, uppermost male or sterile. *Bracteoles* (?) 2, very thin, scarious, ovate, parallel with the glume, alternate, anterior lower, enclosing the upper, which encloses the nut. Hypogynous *bristles* 0, unless the bracteoles represent them. *Stamens* 3–1; anthers small, oblong. *Style* linear, not dilated at the base, deciduous; branches 3–2 small.

Stems with leaves at the base only, each bearing a dense head of 1–5 (rarely more) spikelets; glumes very densely packed. This genus altogether resembles the *Micranthæ* Sect. of *Scirpus*, and is confounded much therewith. This is probably the true affinity, as in *Scirpus Isolepis*, Boeck., we have small scales sometimes present intermediate between the bristles of *Scirpus* and the bracteoles of *Lipocarpha*. The view of Bentham (in Gen. Pl. iii. 1054) that *Lipocarpha* is allied to the *Mapanieæ*, or that the fore-and-aft bracteoles of *Lipocarpha* can be in any way compared with the two lowest male glumes in a head of *Mapania* is untenable.

Distrib. Species 14; in the warmer portions of both Hemispheres.

Spikelets cylindric; glumes obtuse, tips closely appressed ... (1) **argentea.**
Spikelets squarrose; glumes with recurved excurrent points (2) **pulcherrima.**

1. L. argentea (R. Br. in Tuckey, Congo, Append. 459); glabrous, annual; stems 6–24 in. long, tufted, nearly round, each with 1 head; leaves $\frac{1}{3}$–$\frac{2}{3}$ the length of the stem, $\frac{1}{10}$ by $\frac{1}{5}$ in. broad, almost without a midrib; bracts 3–4, lowest 1–4 in. long, spreading; spikelets 1–5 in the head, sometimes (as in Buchanan, 114) 10 in the head, usually $\frac{1}{4}$ by $\frac{1}{6}$ in. (but they lengthen in fruit sometimes up to $\frac{3}{4}$ in. long), dense, with numerous flower-glumes, white or straw-coloured; rhachilla persistent; glumes narrowly obovate-truncate, rather thick, deciduous seriatim from the base of the spikelet leaving the rhachilla prominently covered by lozenge-shaped scars; lower bracteole $\frac{2}{3}$ the length of the glume, hyaline, obtuse, 5-nerved, central nerve sometimes green (so thin that if the nut is placed in water the bracteoles enclosing it may be easily overlooked under a microscope); upper bracteole similar but narrower; stamens usually 2; nut sessile, shorter than the bracteoles, obovoid-ellipsoid, trigonous, smooth, dark brown, reticulate; style linear, branches 3 linear, rather short. *Kunth, Enum.* ii. 266; *Boeck. in Linnæa,* xxxvii. 114 (*excluding the American examples*); *Goebel in Ann. Jard. Buitenz.* vii. 131, *t.* 14, *fig.* 18, *and t.* 15, *figs.* 19, 20; *C. B. Clarke in Hook. f. Fl. Brit. Ind.* vi. 667. *Hypælyptum argenteum, Vahl, Enum.* ii. 283.

KALAHARI REGION: Transvaal; Macalis Berg, *Zeyher,* 1759! *Burke,* 76!
EASTERN REGION: Natal; Umlaas River, *Krauss,* 13! Inanda, *Wood,* 513! and without precise locality, *Buchanan,* 114! 348! *Rehmann,* 8172!

Also in Tropical Africa, the Mascarene Isles, South East Asia, Malaya, and Queensland. A very common plant.

2. L. pulcherrima (Ridley in Trans. Linn. Soc. ser. 2, ii. 162); stems 3–8 in. long, slender; leaves slender; heads of 1–5 spikelets; spikelets $\frac{1}{6}$ by $\frac{1}{12}$ in., squarrose; glumes obovate, dark chestnut, with a long, linear, yellow-green, recurved mucro; bracteoles hardly longer than the nut; style hardly any, branches 3 minute, linear; otherwise resembling very small examples of the preceding species. *L. tenera and L. atropurpurea, Boeck. Cyp. Novæ,* i. 21.

KALAHARI REGION: Transvaal; Pretoria, at Koedus Poort, *Rehmann,* 4639! and without precise locality, *McLea in Herb. Bolus,* 6025!
EASTERN REGION: Natal, *Rehmann,* 7364!
Scattered throughout South Tropical Africa.

XIV. ASCOLEPIS, Steud.

Spikelet of many spirally imbricate glumes, lowest 2 empty, many succeeding glumes 2-sexual perfecting nuts, uppermost male or sterile. *Scale* anterior, parallel with the glume, longer than it, thickened, in the Cape species utricular enclosing the flower, perhaps representing two lateral partially connate bracteoles. Hypogynous *bristles* 0, unless the scale represents them. *Stamens* 3–2, anterior to the glume; anthers linear-oblong, not crested. *Style*-branches 2 or 3 (sometimes in one spikelet). *Nut* small, oblong or obovoid, unequally trigonous or nearly flat, smooth.

Stems with leaves at the base only, each bearing a head of 1–5 spikelets. It is difficult to say positively what the scale is, as there is nothing much like it in the whole Order.

DISTRIB. Species 8; in Africa and the Mascarene Islands; one of which is abundant through nearly the whole of South America.

1. A. capensis (Ridley in Trans. Linn. Soc. ser. 2, ii. 164); glabrous; rhizome short, weak; stems 8–16 in. long, tufted, striated, hardly trigonous, each with 1 head of 1–3 spikelets; leaves as long as the stem, or only half as long, $\frac{1}{8}$ in. broad, enrolled in the dry state; bracts 3, spreading, similar to the leaves, with a broad short, dilated base, lowest 1–4 in. long; heads $\frac{1}{3}$–$\frac{2}{3}$ in. in diam., white or white straw-colour; spikelets with the tips of the numerous scales spreading on all sides, completely obscuring the narrow oblong small glumes; scale utricular, flattened, with a slit on the posticous face through which the style peeps, and a large, solid, linear-conic, obtuse, white beak; style-branches 2; nut $\frac{1}{2}$–$\frac{2}{3}$ the length of the scale (without its beak), black. *Platylepis capensis, Kunth, Enum.* ii. 269; *Boeck. in Linnæa,* xxxvii. 119. *P. dioica, Steud. Syn. Pl. Glum.* ii. 131.

SOUTH AFRICA: without locality, *Zeyher,* 1762!
COAST REGION: Stockenstrom Div.; Kat Berg, 4000–5000 ft., *Drège!*
KALAHARI REGION: Orange Free State; Nelsons Kop, *Cooper,* 911! Transvaal; Pretoria, at Koedus Poort, *Rehmann,* 4638! Magalies Berg, *Burke!* Hooge Veld, by the Komati River, *Nelson,* 12!
EASTERN REGION: Tembuland; damp places at the foot of Bazeia Mountain, 2300 ft., *Baur,* 133! Pondoland; near the mouth of the Umtentu River, *Drège!* Natal; near Durban, *Sutherland!* Noods Berg, *Wood,* 126! Inanda, *Wood,* 307! and without precise locality, *Buchanan,* 89! 305! *Gerrard,* 486!

XV. RYNCHOSPORA, Vahl.

Spikelet of spirally imbricate glumes, 3–4 lowest empty, smaller, 1–2 succeeding glumes 2-sexual perfecting nuts, uppermost thinner, sterile or empty. Hypogynous *bristles* 6, as long as the nut, linear, scabrous (or in extra-South-African species 0 or various). *Stamens* 3–2 (in the South African species), anterior. *Style* long, linear, with 2 long linear branches or nearly entire; base dilated, persistent. *Nut* obovoid, flattened parallel to the glume (in the South African species).

Stem with leaves and nodes above the base in the South African species.

DISTRIB. Species 183 [i.e. including *Psilocarya,* but excluding *Dichromena* and *Pleurostachys*]; few comparatively in the Old World; very numerous in America, from New England to the Argentine provinces.

Series 1. HAPLOSTYLEÆ, Benth. Style nearly undivided; the two linear branches less than $\frac{1}{4}$ the length of the undivided part of the style.

Polycephalæ.—Globular heads of spikelets usually more than 1.
Heads usually 3–5, often subumbelled ... ... (1) cyperoides.
Calyptrostylis.—Spikelets paniculate (clustered or solitary).
Clusters of spikelets many; nut with beak $\frac{1}{4}$ by $\frac{1}{16}$–$\frac{1}{12}$ in. (2) aurea.
Clusters of spikelets much fewer; nut with beak $\frac{1}{3}$ by $\frac{1}{8}$ in. (3) spectabilis.

Series 2. DIPLOSTYLEÆ, Benth. Style-branches 2, linear, as long or longer than the undivided part of the style.

Hypogynous bristles 6, scabrous, with teeth pointed upwards ... ... (4) glauca.

1. R. cyperoides (Mart. in Denkschr. Akad. Wissen. Muench. vi. 149 sub *Rhynchosporâ*); glabrous or nearly so; rhizome horizontal, short, arising from basal offsets (very short stolons); stem 1–2 ft. long, trigonous at the top; leaves often overtopping the stem, $\frac{1}{8}$–$\frac{1}{5}$ in. broad, uppermost node and leaf often 6–12 in. above the base; inflorescence when fully developed an elongate panicle, sometimes with 20–40 heads, but often (by the suppression of the axillary peduncles) reduced to an umbel of 3 or 4 globose heads (as in all the Cape examples); heads $\frac{1}{3}$–$\frac{1}{2}$ in. diam., dense, straw-coloured; spikelets perfecting 1 or 2 nuts; hypogynous bristles 6, nearly as long as the nut, brown, with teeth pointed upwards; nut $\frac{1}{3}$ as long as the glume, obovoid, compressed, chestnut-coloured; beak linear, longer than the nut, white, scabrid; style very long, linear, exserted, entire, or microscopically 2-fid at the tip. *R. polycephala, Kunth, Enum.* ii. 291; *Boeck. in Linnæa,* xxxvii. 552. *R. triceps, Roem. et Schultes, Syst. Veg.* ii. *Mant.* 50; *Hochst. in Flora,* 1845, 759, *not of Boeck. Schœnus fragiferus, Rudge, Guiana* 15, *t.* 17. *Mariscus piluliferus, G. Bertol. in Rendiconti Ist. Bologna,* 1853–4, 33; *in Mem. Accad. Sci. Bologna,* v. (1854) 466, *t.* 24; *cf. Boeck. in Flora,* 1861, 336. *Cephaloschœnus oligocephalus, Hochst. in Flora,* 1845, 759; —*Pluk. Mant.* 97, *t.* 417, *fig.* 3.

EASTERN REGION: Pondoland; between St. Johns River and Umtsikaba River, 1000 ft., *Drège*; Natal; ponds near Umlaas River, *Krauss,* 206! Durban, *Rehmann,* 8607!

Rare in Tropical Africa and the Mascarene Islands, abundant in America from Florida and Mexico to the Argentine provinces.

2. **R. aurea** (Vahl, Enum. ii. 229); glabrous or nearly so; stem 3–6 ft. long; leaves elongate, $\frac{1}{4}$–$\frac{1}{3}$ in. broad, robust; panicle oblong, often 1 ft. long, the axillary corymbs compound, subumbellate with numerous clusters of lanceolate, brown or yellowish spikelets; hypogynous bristles 5–6, as long as the nut, brown with teeth pointed upwards; nut (including beak) $\frac{1}{4}$ by $\frac{1}{16}$–$\frac{1}{12}$ in., obovoid, flattened, brown, smooth (sometimes with a notch on each side); beak very narrow-triangular, longer than the nut; style about $\frac{1}{3}$ in. long, very narrow, entire or nearly so. *Beauv. Fl. d'Owar.* ii. 39, *t.* 81, *fig.* 2; *Kunth, Enum.* ii. 293; *Boeck. in Linnæa,* xxxvii. 626; *C. B. Clarke in Hook f. Fl. Brit. Ind.* vi. 670. *Scirpus corymbosus, Linn. Amoen. Acad.* iv. 303; *Sp. Plant. ed.* 2, 76. *Schœnus surinamensis, Rottb. Descr. et Ic.* 68, *t.* 21, *fig.* 1.

EASTERN REGION: Natal, *Herb. Schinz!* No South African specimen at Kew.

Widely distributed in the Tropical and Sub-tropical regions of both Hemispheres, an abundant plant; but not plentiful in Africa.

3. **R. spectabilis** (Hochst. in Flora, 1845, 760); panicle long, axillary; spikes condensed nearly into heads; spikelets larger and browner than in *R. aurea*; nut (including beak) $\frac{1}{3}$ by $\frac{1}{8}$ in.; otherwise as *R. aurea,* Vahl. *R. macrocarpa, Boeck. in Linnæa,* xxxvii. 629. *Calyptrostylis Rudgei, Hochst. in Flora,* 1845, 760. *C. macrocarpa, Nees ex Boeck. in Linnæa,* xxxvii. 630.

EASTERN REGION: Natal; ponds near Umlaas River, *Krauss,* 210! marsh near Sydenham, 300 ft., *Wood,* 4001!

This may be only a form of the variable and ubiquitous *R. aurea.* The type specimens collected by Krauss have the nut unusually large, and the inflorescence collected almost into distant heads. As to Wood 4001, matched with Krauss 210 in Herb. Kew, it is too young to show the size of the nut, and the inflorescence is more like that of *R. aurea.*

4. **R. glauca** (Vahl, Enum. ii. 233); glabrous; rhizome very short, horizontal, from a basal offset; stem 1–2$\frac{1}{2}$ ft. long, rather

slender, subtrigonous, node with leaf often 6–12 in. above the base of the stem; leaves often 8–16 by $\frac{1}{8}$ in.; inflorescence a compound oblong panicle (sometimes reduced to a compound quasi-terminal head); spikelets numerous, $\frac{1}{8}$–$\frac{1}{6}$ in. long, subsolitary, ovoid, maturing 1–2 nuts; hypogynous bristles 6, about as long as the nut, brown, with up-pointed teeth; stamens usually 2; nut $\frac{1}{2}$–$\frac{3}{4}$ the length of the glume, broad-ellipsoid flattened, chestnut-brown, obscurely transversely undulate; beak a depressed cone hardly $\frac{1}{3}$ the length of the nut; style-branches 2, long, linear. *Kunth, Enum.* ii. 297; *Boeck. in Linnæa,* xxxvii. 585; *C. B. Clarke in Hook. f. Fl. Brit. Ind.* vi. 671. *R. laxa, R. Br. Prod.* 230; *Kunth, Enum.* ii. 298; *Nees in Flora,* 1828, 291, *and in Linnæa,* viii. 94. *R. ferruginea, Roem. et Schultes, Syst. Veg.* ii. 85; *Nees in Linnæa,* vii. 529; *Steud. in Flora,* 1829, 138.

COAST REGION: Cape Div.; Table Mountain, *Ecklon,* 867!

KALAHARI REGION: Transvaal; Houtbosch, *Rehmann,* 5740! Lomati Valley, near Barberton, 4000 ft., *Galpin,* 1365!

EASTERN REGION: Natal; Inanda, *Wood,* 1596! and without precise locality, *Buchanan,* 148! 356!

Also in Algeria and the Mascarene Islands, Australia and Polynesia; very abundant in South East Asia and the warm parts of America.

XVI. CARPHA, R. Br. partly.

Spikelet of 5–7 glumes, 3 lowest empty, 1–2 succeeding subdistichous (not distant) bisexual nut-bearing, upper male or empty; axis of spikelet short, not flexuose. Hypogynous *bristles* 6, as long as the nut, sometimes irregular. *Stamens* 3, anterior. *Style* long, linear, with 3 long branches. *Nut* sessile, ellipsoid, trigonous, pyramidal at either end, acuminate more or less hispid at the top.

Heads of spikelets in oblong panicles or few; stems with nodes and leaves frequently some distance above the base.

DISTRIB. Species 9; confined to the Southern Hemisphere, viz. South Africa, the Mascarene Islands, Australia, and Temperate South America.

Heads of spikelets numerous, in a large oblong panicle; leaves often $\frac{1}{4}$ in. broad or more ... ... (1) **glomerata.**
Heads of spikelets few, in a short panicle; leaves less than $\frac{1}{4}$ in. broad:
  Bracts to heads not conspicuous, lanceolate, narrow at the base ... ... (2) **capitellata.**
  Bracts to heads ovate or subcordate at the base ... (3) **bracteosa.**

1. **C. glomerata** (Nees in Linnæa, vii. 529); robust, glabrous, except the sparsely hispid flower-glumes; rhizome not seen in the herbarium, probably as in the next species arising from a long-creeping stolon; stem 3–6 ft. long, trigonous, with nodes and leaves (or bracts) throughout its length, nearly smooth; leaves 4–20 by $\frac{1}{6}$–$\frac{1}{2}$ in., tough, nearly smooth, often transversely lineolate; inflorescence often 10 by 3 in. of 30 heads in a compound panicle; primary bracts like the leaves but scarcely overtopping the panicle,

secondary bracts lanceolate; heads $\frac{1}{3}$–$\frac{1}{2}$ in. in diam., ovoid, of 16–30 spikelets, chaffy, straw-coloured or brown; spikelets $\frac{1}{4}$ in. long, broadly lanceolate; glumes all close together, nut-bearing glumes boat-shaped, elliptic, lanceolate-tipped, 1-nerved; hypogynous bristles 6 (3 "sepals," 3 "petals"), linear, rigid, scabrous with up-pointed teeth, straw-coloured, not rarely irregular (several short or almost rudimentary); nut $\frac{1}{12}$ in. long, ellipsoid, subacutely trigonous, dark-brown; style-base when young pyramidal, hispid, as the fruit ripens the style-base becomes confluent with the nut as a pointed pyramidal tip, scabrous on the angles (or nearly smooth), the style breaking off just above; outer cells of the nut quadrangular, rather large, prominent in regular series. *Boeck. in Linnæa*, xxxviii. 265. *Schœnus glomeratus, Thunb. Prod.* 17; *Fl. Cap. ed. Schultes* 94. *S. dactyloides. Vahl, Enum.* ii. 224. *Chætospora dactyloides, Dietr. Sp. Pl.* ii. 32. *Asterochæte glomerata, Nees in Linnæa*, ix. 300; x. 194, *exclud. var.* β; *Kunth, Enum.* ii. 311.

SOUTH AFRICA: without locality, *Roxburgh!* *Villette!* *Wallich*, 216! *Ecklon and Zeyher*, 108.

COAST REGION: Cape Div.; near Cape Town, *Thunberg*, *Rehmann*, 1176! between Wynberg and Constantia, *Burchell*, 819! Paarl Div.; Paarl Mountains, 1000–2000 ft., *Drège!* Worcester Div.; near Worcester, *Pappe!* Caledon Div.; by the Palmiet River, near Grabouw, 700 ft., *Bolus*, 4229! Riversdale Div.; by the Great Vals River, *Burchell*, 6529! Uitenhage Div.; by the Van Stadens River, *MacOwan*, 651! Albany Div.; near Grahamstown, *Schoenland*, 138! Howisons Poort, 2100 ft., *Schoenland*, 760! by the rivulet at Grahamstown, *Burchell*, 3563!

EASTERN REGION: Pondoland, *Bachmann*, 123!

2. C. capitellata (Boeck. in Linnæa, xxxviii. 266); much slenderer in all parts than *C. glomerata;* stem 12–18 in. long; leaves $\frac{1}{10}$–$\frac{1}{8}$ in. broad; panicle 2–6 in. long, of 3–8 heads; secondary bracts lanceolate, narrowed at the base, not prominent; otherwise as *C. glomerata*, Nees. *Asterochæte capitellata, Nees in Linnæa*, ix. 300; x. 194; *Kunth, Enum.* ii. 312. *A. glomerata, var.* β *minor, Nees in Linnæa*, x. 194. *A. tenuis, Kunth, Enum.* ii. 312. *A. angustifolia, Nees in Linnæa*, ix. 300.

SOUTH AFRICA: without locality, *Ecklon and Zeyher*, 146!

COAST REGION: Alexandria Div.; Zuur Berg Range, 2000–3000 ft., *Drège!* Riversdale Div., 900 ft., *Schlechter*, 1905! Komgha Div.; marshy places near Komgha, 2000 ft., *Flanagan*, 920!

3. C. bracteosa (C. B. Clarke in Durand and Schinz, Conspect. Fl. Afr. v. 656); bracts at the base elliptic or broad-ovate, sheathing the heads of spikelets; otherwise as *C. capitellata*.

COAST REGION: Worcester Div.; in the valley of the Breede River, near Bains Kloof, 800 ft., *Bolus*, 2867!

CENTRAL REGION: Somerset Div.; marshy places on the summit of Bosch Berg, 4500 ft., *MacOwan*, 1616! 2187!

Imperfectly known species.

Asterochæte Ludwigii (Hochst. in Flora, 1845, 759, in note); style-base hispid, persistent, forming a beak.

SOUTH AFRICA: without locality, *Ludwig* (ex *Hochst.*).

Hochstetter says that this species is close to *C. capitellata*, Boeck., and his full description agrees very well with *C. capitellata* except as to the style-base.

XVII. ECKLONEA, Steud.

Hypogynous *bristles* 3, feathered at the base, 3-fid at the tip; otherwise as *Carpha*.

DISTRIB. Species 1; endemic.

1. E. capensis (Steud. in Flora, 1829, 138); glabrous; rhizomes $\frac{1}{12}$–$\frac{1}{2}$ in. long, slender, white, often terminating in a minute globose bulbil $\frac{1}{10}$ in. in diam.; stems 2–6 in. long, with leaves or bracts throughout their length; leaves 2–4 by $\frac{1}{10}$–$\frac{1}{8}$ in.; panicle 2–3$\frac{1}{2}$ in. long, linear, of 1–4 distant heads; bracts like the leaves, slightly overtopping the inflorescence; spikelets 3–8 in a head, loosely clustered, $\frac{1}{5}$ in. long, elliptic, green-white; glumes usually 6 in the spikelet, lowest 3 empty, next bisexual producing a nut, upper male or sterile; flower-glume boat-shaped, elliptic, with many obscure nerves; hypogynous bristles 3 (petals, i.e. they are opposite the faces of the nut as the "petals" of *Fuirena*), much longer than the nut, feathered with cinnamon or pink hairs in the lower half, in the upper half yellowish rigid, ending in 3 points of which the middle is much the longest; nut obovoid, trigonous, yellow-brown, reticulated owing to the large subquadrate outermost cells; style linear, branches 3 linear; style-base small, conic, persistent (almost confluent with) the nut. *Flora*, 1830, 539; *Syn. Pl. Glum.* ii. 139; *Schrad. Anal. Fl. Cap.* 34; *Nees in Linnæa*, ix. 299; x. 143; *Kunth, Enum.* ii. 287; *Boeck. in Linnæa*, xxxviii. 229. *Uncinia spartea, Spreng. in Flora*, 1829, i. *Beil.* 13. *Trianoptiles capensis, Harvey, Gen. S. Afr. Pl. ed.* 2, 422; *Benth. in Hook. f. Ic. Pl.* xiv. 34, *t.* 1348. *Carpha capensis, Steud. Nomencl. ed.* 2, i. 300. *Carex dregeana, Herb. Drège*, *partly*.

SOUTH AFRICA: without locality, *R. Brown!* *Ludwig!* *Zeyher!* *Drège*, 8152!
COAST REGION: Cape Div.; Table Mountain, *Ecklon*, 853! 854! near Cape Town, *Harvey*, 196 partly!

The structure of the spikelet and flower in this plant is so exactly that of *Carpha* that it might be better to sink the genus in *Carpha*.

XVIII. SCHŒNUS, Linn. partly.

Spikelet of few (or not very many) glumes, lowest few (or several) empty; 1–3 succeeding bisexual, nut-bearing, distichous (or nearly so), distant on the axis of spikelet; upper glumes close together, male or empty; the part of the axis bearing the fertile glumes is elongated, flexuose, so that the lowest nut stands in an excavation or deep bend of it. Hypogynous *bristles* various or 0. *Stamens* 6–1. *Style* linear, continuous with the nut; style-base small, conic, or hardly any; branches 3, linear, or by accident 2. *Nut* sessile or stalked, triquetrous or the anterior angle rounded.

Species 59; very varied in habit—only 1 (the English *S. nigricans*) in South Africa. The genus is now limited by Bentham's character of the elongation of the nut-bearing portion of the axis of the spikelet; it comprises the two common European species, spread in various countries, and a large mass of Australian species, mostly endemic, but 2 or 3 extending to Malaya.

1. **S. nigricans** (Linn. Sp. Plant. ed. 2, 64); glabrous; stolons 0; stems stoutly tufted, 6–20 in. long, leafy only near the base, each with 1 head; basal sheaths often shining chestnut; leaves $\frac{1}{3}$–$\frac{2}{3}$ the length of the stem, $\frac{1}{16}$ in. broad, channelled at the base, solid trigonous upwards; bracts 2, lower 1–4 in. long, similar to the leaves, dilated and scarious at the base; head $\frac{1}{2}$ in. in diam., of 1–15 spikelets, dense, chestnut-brown or sometimes pale brown; spikelets $\frac{1}{3}$ in. long, 3–1-nutted; glumes minutely scabrous on the keel; hypogynous bristles 3–5, hardly $\frac{1}{2}$ the length of the nut, rigid, yellow-brown; stamens 3; nut $\frac{1}{3}$ the length of the glume, sessile, ellipsoid, trigonous, smooth, pale. *Sowerby, Engl. Bot. t.* 1121; *Thunb. Fl. Cap. ed. Schult.* 92; *C. B. Clarke in Hook. f. Fl. Brit. Ind.* vi. 673. *S. aggregatus, Thunb. Fl. Cap. ed. Schult.* 92. *S. hypomelas, Spreng. Neue Entd.* iii. 8; *Syst.* i. 190. *Chætospora nigricans, Kunth, Enum.* ii. 323; *Boeck. in Linnæa,* xxxviii. 290. *Elynanthus spathaceus, Nees in Linnæa,* x. 186; *Kunth, Enum.* ii. 309.

COAST REGION: Cape Div.; near Cape Town, *Thunberg, Ecklon! Rehmann,* 1784! 1785! Cape Flats, *Zeyher*, 1782! Mossel Bay Div.; hills near the landing-place at Mossel Bay, *Burchell,* 6285! Komgha Div.; near Komgha, *Flanagan,* 916!

KALAHARI REGION: Griqualand West; Lower Campbell, *Burchell,* 1816!

Widely distributed; Europe, North Africa, Abyssinia, West and Central Asia to the Punjab; United States, California, Surinam (possibly introduced in the New World).

Imperfectly known species.

The subjoined species of *Schœnus* from South Africa are perhaps most of them included among the species of *Tetraria* below; but, in the absence of the original specimens, I have not been able to identify them from the descriptions.

2. **S. aristatus** (Thunb. Prod. 16; Fl. Cap. ed. Schultes 92); culm terete, 2 ft. long, leafless; spikes terminal, several agglomerated; involucre of one (or two?) leaves, erect, twice the length of the spikes, at the base concave sheathing; glumes about 6, outer ovate concave, inner lanceolate, all entire glabrous, ferruginous. *Kunth, Enum.* ii. 336.

SOUTH AFRICA: without locality, *Thunberg.*

3. **S. lævis** (Thunb. Prod. 17; Fl. Cap. ed. Schultes 95); culm trigonous, 2 ft. long, leafy; leaves as long as the stem, sheathing, convolute, smooth, glabrous; heads 4 or 5 lateral, peduncled, clustered, ovate, scarcely leafy; spikelets very many, ovate; outermost glumes (or bracts) larger, longer-cuspidate, the rest of the glumes ovate, subemarginate, mucronate, entire.

SOUTH AFRICA: without locality, *Thunberg.*

4. **S. tristachyus** (Thunb. Prod. 16; Fl. Cap. ed. Schultes, 92); culm terete, 1 ft. long, filiform, erect, jointed; heads 3, terminal, ovate, glabrous, smooth, the size of a pea.

SOUTH AFRICA: without locality, *Thunberg.*

5. **S. Hystrix** (Vahl, Enum. ii. 226); stem terete, leafless, with nodes; leaf-sheaths 1 in. long, ending in a short yellow point; spike 1, terminal, obovate, dusky-black, size of a hazel-nut, imbricated; bract half the length of the spike, ovate-lanceolate; glumes ovate, rigid, smooth, white-margined, acuminate, squarrose, aristate; arista longer than the glumes very scabrid. No flowers among the lower glumes. *Bæothryon Hystrix, Dietr. Syn. Pl.* ii. 92.

SOUTH AFRICA: without locality or mention of the collector.

6. **S. spicatus** (Burm. Fl. Cap. Prod. 3); stem terete; leaves sheathing, setaceous; panicle of flowers narrow. *S. Burmanni, Vahl, Enum.* ii. 227; *Kunth, Enum.* ii. 337.

SOUTH AFRICA: without locality or mention of the collector.

7. **S. tener** (Spreng. Neue Entd. iii. 9); stems cæspitose, a span high, capillary; leaves capillary, erect, glabrous, shorter than the stem, sheaths loose scarious; spikelets in threes, lateral, clustered, purplish-red, 4-flowered, size of a lentil; glumes lanceolate, obtuse, striate, glabrous, lower sterile, upper very many fertile; anthers erect, subulate, long exserted; ovary densely supported by bristles. (Might be mistaken for *Sch. tristachyus*, Thunb., but that has a jointed stem.) *Chætospora tenera, Dietr. Syn. Pl.* ii. 32; *Kunth, Enum.* ii. 328.

COAST REGION: Stellenbosch; no mention of the collector.

XIX. EPISCHŒNUS, C. B. Clarke.

Spikelet of 7 subdistichous glumes; 3 lowest small empty, 4th and 5th male, 6th bisexual, perfecting a nut, 7th small empty; axis of spikelet above the insertion of the nut elongated, thickened, curved round the nut. *Nut* subglobose, obscurely trigonous, smooth, marble-white; otherwise as *Schœnus.*

DISTRIB. Species 1, endemic.

The two lowest flowers of the spikelet have 3 stamens each only, but in one spikelet examined a rudimentary pistil was found in the lowest flower but one. This plant, from the marble-white nut has been doubtfully referred to *Lepidosperma;* but the structure, especially the elongated rhachilla curved round the nut, is much nearer *Schœnus* in which genus Bentham included it.

1. **E. quadrangularis** (C. B. Clarke in Durand and Schinz, Conspect. Fl. Afr. v. 657); glabrous; rhizome seen very short, woody; stems tufted, 10–20 in. long, slender, conspicuously quadrangular at the top, with nodes only near the base, leaf-sheaths bright red; leaves $\frac{1}{3}$–1 in. long, flat at the base, red or pale-brown, scarious, occasionally green; bracts 2 lower suberect, 1–$1\frac{1}{2}$ in. long, linear, acute; inflorescence $\frac{1}{2}$–$1\frac{1}{2}$ in. long, of 4 spikelets in a linear raceme; spikelets 1–2 in each bract, $\frac{1}{3}$ by $\frac{1}{12}$ in., dull-brown; glumes elliptic-

lanceolate, minutely scabrous on the keel; hypogynous bristles 0; nut $\frac{1}{12}$ in. long, sessile; style deciduous, base linear, branches 3. *Schœnus quadrangularis*, *Boeck. in Linnæa*, xxxviii. 274; *cf. Benth. and Hook. f. Gen. Pl.* iii. 1063.

COAST REGION: Cape Div.; Table Mountain, *Burchell*, 557! Swellendam Div.; mountain peak near Swellendam, *Burchell*, 7324! Caledon Div.; Houw Hoek, 2500 ft., *Schlechter*, 7402! Riversdale Div.; on the Kampsche Berg, *Burchell*, 7076!

XX. COSTULARIA, C. B. Clarke.

Spikelet of several (or many) subdistichous glumes, lowest 3–4 (or more) empty; lowest flower male, next above it bisexual, nut-bearing (rarely two nut-bearing glumes), upper glumes male or empty; all the glumes near together; axis of spikelet above the insertion of the nut neither elongated nor thickened curved. Hypogynous *bristles* 6 (or 5), long, slender. *Stamens* 3. *Nut* oblong-obovoid, trigonous, crowned by the pyramidal persistent style-base. *Style* long, branches 3.

Rather stout plants; stems with nodes bearing leaves or bracts through their whole length; leaves tough; panicle oblong, compound, with many spikelets.—From their habit, the species have often been referred to *Cladium*.

DISTRIB. Species 6 or 7; in the Mascarene Islands, Natal and Australia.

Stems 3 ft. long; leaves 12–18 in. long ... ... (1) **natalensis.**
Stems 3–4 in. long; leaves 9 in. long ... ... (2) **brevicaulis.**

1. C. natalensis (C. B. Clarke in Durand and Schinz, Conspect. Fl. Afr. v. 658); glabrous (except edges of glumes); stolon horizontal, clothed by rigid lanceolate scales 1 in. long; stems tufted, 3 ft. or more long, round, tough, rather slender, striated; leaves 12–18 by $\frac{1}{4}$–$\frac{1}{3}$ in., very tough, scabrous-edged, upper leaves much smaller passing into the bracts; panicle 20 by $1\frac{1}{2}$ in., nearly continuous with short oblong dense axillary erect branches; bracts 1–4 in. long, not conspicuous; spikelets crowded, $\frac{1}{4}$ by $\frac{1}{12}$ in., chestnut-red, compressed, hard; glumes not spreading, boat-shaped, elliptic-lanceolate, hardly acute; margins microscopically white-hispid; hypogynous bristles exceeding the nut, weak, white, with scattered weak simple hairs in the upper part; stamens 3, anterior; anthers linear-oblong, rather large, apiculate; nut less than $\frac{1}{3}$ the length of the glume, brown, smooth; style-base $\frac{1}{2}$ the length of the nut, pyramidal, white hispid in the upper half.

KALAHARI REGION: Transvaal; Saddleback Mountain, near Barberton, 4500 ft., *Galpin*, 1316!

EASTERN REGION: Natal, *Buchanan*, 152! 354!

The Mascarene species are closely allied; *C. Baroni*, C. B. Clarke, hardly differs except by the laxer panicle with slender flexuose branches.

2. C. brevicaulis (C. B. Clarke); glabrous; stems woody, $\frac{1}{10}$ in. in diam., erect, sometimes divided close to the base; branches (or whole height) 3–4 in., erect; leaf-sheaths distichous, equitant, imbricate, exceedingly close together, firm, green or brownish; leaves 9 by $\frac{1}{8}$ in., flat, tough, unsymmetrically bent laterally, strongly striate, without a midrib; inflorescence of short axillary 1–3-flowered racemes, exserted 1 in. from the leaf-sheaths, flexuose; bracts of the raceme 1–3 in. long, similar to the leaves; pedicels about $\frac{1}{4}$ in. long, solitary in each bract; spikelet $\frac{1}{3}$ by $\frac{1}{8}$–$\frac{1}{6}$ in., turgid, green, hard, 8–10-glumed; glumes exactly distichous, closely rigidly imbricated, boat-shaped, ovate-triangular; lower 6 (or 7) glumes empty, 7th glume male, without any trace of pistil, 8th bisexual, perfecting a nut, 9th very small or 0; hypogynous bristles 3–5, $\frac{2}{3}$ the length of the nut; stamens 3; young ovary obscurely 6-ribbed; style-base large, bulbous; style long, linear, brown-red, branches 3 linear; nut large, $\frac{1}{10}$ in. long, ovoid, smooth, brown, crowned by a boss (the persistent style-base). *Tetraria brevicaulis, C. B. Clarke in Durand and Schinz, Conspect. Fl. Afr.* v. 659.

SOUTH AFRICA: without locality, *Carmichael!* in the British Museum.

The 3 specimens collected by *D. Carmichael* are very fine complete examples with fruit; it is nevertheless difficult to find a genus for them.

XXI. TETRARIA, Beauv.

Spikelet 2-flowered, of several (or many) subdistichous glumes, lowest 4–10 empty; lower flower not perfecting a nut, male (or the glume empty in *T. cuspidata*); upper flower bisexual, perfecting a nut (or the spikelets all 1-sexual in *T. crinifolia*); upper glume empty, often small or 0; all the glumes close together; axis of spikelet not elongated nor curved round the nut. Hypogynous *bristles* various or 0. *Stamens* 3, sometimes 4, more rarely 6 or 8; anthers linear-oblong, often with ears at the base of lax tissue barren of pollen; crest lanceolate-linear scabrous. *Style* continuous with the nut, dilated at the base, linear, branches 3 or 4, in a few species 6 or 8, long, linear. *Nut* obovoid, trigonous or subtetragonous; style-base scabrous, often pyramidal on the young pistil, in the ripe nut sometimes narrow-conic, sometimes confluent with the rounded apex of the nut.

DISTRIB. Species 32; endemic except *T. circinalis*, which is also found in Usambara.

Sect. 1. **Aulacorhynchus** (Genus, Nees in Linnæa, ix. 305).—Spikelets all 1-sexual, monœcious, female 1-flowered; glumes not aristate; stamens 3; style-branches 3; stem and leaves slender ... ... (1) **crinifolia.**

Sect. 2. **Hemischœnus.**—Fertile flower 2-sexual; lowest flower perfecting a nut; spikelets usually 1-flowered; glumes not aristate; stamens 3; style 3-fid; stem and leaves slender, the latter all close to the base of the stem ... ... (2) **cuspidata.**

Sect. 3. **Elynanthus** proper, of authors.—Fertile flower 2-sexual; lowest flower not producing a nut, next above it fertile; glumes not aristate, scarcely mucronate (except in 13, *aristata*); stamens 3 or 4; style-branches 3 or 4; slender or medium-slender plants with narrow leaves (but see 18, *robusta*):

Stems without any node between the lowest bract and the nearly basal leaves:
 Leaf-sheaths not manifestly fimbriate:
  Stems very slender:
   Leafless or nearly so ... ... (3) **dregeana.**
   Leaves numerous, long, setaceous (4) **Bolusi.**
  Stems medium-sized ($\frac{1}{16}$ in. in diam.):
   Bracts not dilated at the base:
    Flower-glumes hard, shining brown ... ... (5) **sylvatica.**
    Flower-glumes (young) scarious ... ... (6) **ligulata.**
   Bracts much dilated at the base (7) **picta.**
 Leaf-sheaths fimbriate or cancellate:
  Panicle lax ... ... (8) **pleosticha.**
  Panicle contracted ... ... (9) **wallichiana.**
Stems with one or more nodes between the lowest bract and the basal leaves:
 Leaf-sheaths fimbriate:
  Leaves hemi-cylindric channelled immediately above the sheath, $\frac{1}{30}$–$\frac{1}{25}$ in. broad:
   Glumes woolly on the edges ... (10) **fimbriolata.**
   Glumes glabrous ... ... (11) **fasciata.**
   Glumes minutely hairy on the edges:
    Glumes not aristate ... (12) **capillacea.**
    Glumes aristate ... ... (13) **aristata.**
  Leaves flat immediately above the sheath, often $\frac{1}{12}$ in. broad, rolled up to the very base when dry:
   Spikelets $\frac{1}{4}$ in. long:
    Stems 8–10 in. long ... (14) **Burmanni.**
    Stems 15–24 in. long ... (17) **flexuosa.**
   Spikelets $\frac{1}{3}$ in. long:
    Stems 12–16 in. long ... (15) **nigrovaginata.**
    Stems 3–10 in. long ... (16) **circinalis.**
 Leaf-sheaths not fimbriate ... ... (18) **robusta.**

Sect. 4. **Lepisia** (Genus, Presl, Symb. Bot. 9, t. 5).—Fertile flower 2-sexual; lowest flower not producing a nut, next above it fertile; lower empty glumes aristate; stamens 3 or 4; style-branches 3 or 4; large plants with flat leaves (except 27, *ustulata*); stems with nodes bearing either leaves or leaf-like bracts scattered throughout their length:
 Stem trigonous or triquetrous; basal leaf-sheaths entire, very firm:
  Hypogynous bristles present:
   Stem trigonous:
    Stem very stout ... ... (19) **thermalis.**
    Stem slender ... ... (20) **eximia.**
   Stem acutely triquetrous ... ... (21) **triangularis.**
  Hypogynous bristles none ... ... (22) **secans.**
 Stem terete; basal leaf-sheaths scarious fimbriate:
  Hypogynous bristles 6, minutely scabrid:
   Leaves $\frac{1}{6}$ in. broad:
    Spikelets many in a cluster ... (23) **Rottbœllii.**
    Spikelets 1–4 in a cluster ... (25) **involucrata.**
   Leaves $\frac{1}{8}$ in. broad ... ... (24) **rottbœllioides.**
  Hypogynous bristles short, feathery ... (26) **spiralis.**
  Hypogynous bristles none ... ... (27) **ustulata.**

Sect. 5. **Eu-Tetraria.**—Fertile flower 2-sexual; lower flower male with a rudimentary pistil, not producing a nut; glumes distichous, not aristate; stamens 6 or 8; style-branches 3 or 4; stout or medium-sized plants:

Stem medium-sized, without stem-leaves; spikelets brown ... ... (28) **compar.**
Stem stout, with a stem-leaf; spikelets pale ... (29) **Thuarii.**

Sect. 6. **Buekia.**—Fertile flower 2-sexual; lowest flower not producing a nut; glumes hardly distichous, minutely mucronate; stamens 3 or 6, anthers crested; style-branches 6 or 8; stem robust, with nodes between the basal leaves and the panicle:

Stem 3 ft. long, terete; stamens 3 ... ... (30) **punctoria.**
Stem 2 ft. long, triquetrous; stamens 6 ... (31) **MacOwani.**

1. **T. crinifolia** (C. B. Clarke in Durand and Schinz, Conspect. Fl. Afr. v. 660); glabrous; stems 1 ft. long, slender, terete, striate, with distant nodes carrying leaves or bracts; leaves usually as long as the stem, linear-setaceous; lower bracts similar; inflorescence of 3 or 4 very distant heads; terminal head (sometimes 2) of 10–20 male spikelets, subcompound; lower heads axillary, peduncle bearing 1–4 female spikelets; male spikelets $\frac{1}{5}$ in. long, lanceolate, dark-brown, usually 2-flowered; female spikelets similar, each with 1 flower without stamens; glumes elliptic, acute, not aristate; hypogynous bristles 0; nut $\frac{1}{10}$–$\frac{1}{8}$ in. long, sessile, obovoid, trigonous, pale. smooth; beak scarcely $\frac{1}{3}$ the length of the nut; style-branches 3, linear. *Aulacorhynchus crinifolius, Nees in Linnæa*, x. 199; *Kunth, Enum.* ii. 535; *Boeck. in Linnæa*, xxxix. 2.

SOUTH AFRICA: without locality, *Ecklon and Zeyher*, 95!
COAST REGION: Tulbagh Div.; near Tulbagh Waterfall, *Drège!* near Tulbagh, 2000 ft., *Schlechter*, 7469!

In habit this plant goes well with *T. cuspidata.*

2. **T. cuspidata** (C. B. Clarke in Durand and Schinz, Conspect. Fl. Afr. v. 660); glabrous; rhizome (seen) short, obliquely descending; stems 1–2 ft. long, nearly terete, slender; basal sheaths brown-red, not torn; leaves all near the base of the stem, often 6 in. long, linear-setaceous; panicle sometimes 3–4 by 1 in., compound, with numerous spikelets, but more often 1 by $\frac{1}{3}$ in. with few spikelets (reduced almost to a small oblong spike); bracts not dilated at the base, often shorter than the panicle, sometimes 4–5 in. long, setaceous, much overtopping the panicle; spikelets $\frac{1}{6}$–$\frac{1}{5}$ in. long, lanceolate, dark-brown, scarcely compressed, of 5–7 glumes, usually in small close clusters; lowest 3 (or 4) glumes empty, 4th containing a bisexual nut-bearing flower, 5th usually empty (sometimes male), small; fertile glume elliptic-lanceolate, not aristate, hard, rounded on the back, more or less scarious on the edges; hypogynous bristles 0; stamens 0; nut $\frac{3}{5}$ the length of the glume, sessile, obovoid, trigonous, yellowish; outer cells in regular rows, quadrate, so that the nut is cancellate; beak narrow-pyramidal, passing into the style, scabrous, especially when young; style-branches 3 linear. *Schœnus cuspidatus, Rottb. Descr.*

et Ic. 66, *t.* 18, *fig.* 3; *Thunb. Prod.* 16; *Fl. Cap. ed. Schult.* 92. *Chætospora cuspidata, Nees in Linnæa,* vii. 529. *Elynanthus cuspidatus, Nees in Linnæa,* ix. 298; x. 188; *Kunth, Enum.* ii. 309; *Boeck. in Linnæa,* xxxviii. 258. *E. gracilis, Nees in Linnæa,* ix. 298; x. 189. *E. microstachyus, Boeck. in Linnæa,* xxxviii. 261 *partly, i.e. as to herb. Drège a α. Scirpus cuspidatus, Roem. et Schultes, Syst. Veg.* ii. 67 *in citation (not of Rottb.).*

VAR. β **lorea** (C. B. Clarke in Durand and Schinz, Conspect. Fl. Afr. v. 660); lower empty glumes hardly shorter than the nut-bearing glume, very acute or mucronate, the lowest bract-like aristate. *Elynanthus loreus, Nees in Linnæa,* ix. 298; x. 188; *Kunth, Enum.* ii. 310; *Boeck. in Linnæa,* xxxviii. 257.

SOUTH AFRICA: without locality, *Thunberg, Sieber,* 6! 107! 108! *Pappe,* 95! 108! *Ecklon and Zeyher,* 97! 98! Var. β, *Drège,* 7378! *Harvey,* 390! *Ecklon and Zeyher,* 99! *MacGillivray,* 431!

COAST REGION: Clanwilliam Div.; Ceder Bergen at Ezels Bank, 3000–4000 ft., *Drège!* Cape Div.; Cape Flats, *Ecklon,* 865! near Cape Town, *Burchell,* 501! *Harvey,* 198 partly! Table Mountain, 3500 ft., *Bolus,* 4742! *Burchell,* 570! *Bergius! Rehmann,* 619! Devils Mountain, 500 ft., *Bolus,* 4861! Muizen Berg, 800 ft., *Bolus,* 7178! Paarl Div.; Simons Berg, near the Waterfall, 2000 ft., *Drège!* Worcester Div.; Dutoits Kloof, 1000–3000 ft., *Drège!* Stellenbosch Div.; Lowrys Pass, 1500 ft., *Schlechter,* 7284! Caledon Div.; Palmiet River, 1200 ft., *Schlechter,* 7325! Houw Hoek, 3000 ft., *Schlechter,* 7341! 7404! 7429! between Genadendal and Donker Hoek, *Burchell,* 7920! 7926! Bredasdorp Div.; Elim, 500 ft., *Schlechter,* 7656! George Div.; near George, *Burchell,* 6033/2! Port Elizabeth Div.; near Port Elizabeth, *E.S.C.A. Herb.,* 191! Bathurst Div.; near Port Alfred, *Burchell,* 3997! East London Div.; grassy valleys near East London, 300 ft., *Flanagan,* 1786! Var. β: Cape Div.; Table Mountain, *Milne!* Caledon Div.; Houw Hoek, 2000 ft., *Schlechter,* 7403!

CENTRAL REGION: Somerset Div.; Bosch Berg, 4800 ft., *MacOwan,* 1594! Graaff Reinet Div.; Koudeveld Berg, 5500 ft., *Bolus,* 2585! Albany Div.; near Riebeek, *Burchell,* 3501!

Besides the var. β *lorea,* there are here included under *T. cuspidata* several forms which Mr. Bentham thought probably distinct species. There is a form (Burchell, 3997, 6033/2) with stems and leaves much stouter than in the typical wiry slender *T. cuspidata.* There is another form with viscid spikes and the bracts dilated lanceolate at the base (Bolus, 4742); this was marked *T. sylvatica* by Boeckeler (which it certainly is not).

3. **T. dregeana** (C. B. Clarke in Durand and Schinz, Conspect. Fl. Afr. v. 661); glabrous; rhizome (seen) slender, woody; stems 2 ft. long, tufted, very slender, without any node between the lowest bract and the sub-basal leaf-sheaths; leaf-sheaths dark-red, entire (i.e. not torn), lanceolate, produced on one side, leafless or very nearly so; panicle $2\frac{1}{2}$ by $\frac{3}{8}$ in., very lax and slender, with few (6–10) spikelets; bracts 1–3 in. long, setaceous, hardly overtopping the panicle, not dilated at the base; spikelets pedicelled, $\frac{1}{4}$ in. long, lanceolate when young; lowest flower male, upper bisexual; hypogynous bristles 0; stamens 3; young pistil obovoid, crowned by the large hemispheric style-base; nut not seen. *Elynanthus dregeanus, Boeck. in Linnæa,* xxxviii. 262.

COAST REGION: Worcester Div.; Dutoits Kloof, *Drège,* 1632a! Riversdale Div.; on the lower part of the Lange Bergen, near Kampsche Berg, *Burchell,* 7137!

4. T. Bolusi (C. B. Clarke in Durand and Schinz, Conspect. Fl. Afr. v. 659); glabrous; rhizome (seen) slender, woody; stems 8–12 in. long, tufted, slender, without any node between the lowest bract and the sub-basal non-fimbriate leaf-sheaths; leaves $\frac{1}{2}$–$\frac{2}{3}$ the length of the stem, setaceous, hemi-cylindric, scabrous-edged; panicle 1$\frac{1}{2}$ by $\frac{1}{3}$ in., lax, very slender, with 6–15 spikelets; bracts usually overtopping the panicle, setaceous, not dilated at the base; spikelets subsolitary, $\frac{1}{8}$–$\frac{1}{6}$ in. long, ovate, acute, much flattened, chestnut-red; lowest flower male (with rudimentary pistil); upper bisexual perfecting a nut; glumes subacute, not mucronate; hypogynous bristles 0; stamens 3; nut very small, hardly $\frac{2}{5}$ the length of the glume, obovoid, trigonous, brown, obscurely coarsely reticulated (outermost cells rather large, round, quadrate), crowned by the large pyramidal hispid style-base; style linear, branches 3, long, linear.

SOUTH AFRICA: without locality, *R. Brown! Ecklon and Zeyher*, 112!

COAST REGION: Cape Div.; Devils Mount, *Rehmann*, 919! 921! Stellenbosch Div.; eastern slopes of Hottentots Holland Mountains, near Grabouw, 1500 ft., *Bolus*, 4230!

5. T. sylvatica (C. B. Clarke in Durand and Schinz, Conspect. Fl. Afr. v. 663); glabrous; rhizome $\frac{1}{8}$ in. in diam., horizontal; stems 10–16 in. long, tufted, medium-sized, $\frac{1}{16}$ in. in diam., terete, rigid, without any node between the lowest bract and the sub-basal, non-fimbriate, bright orange-red leaf-sheaths; leaves 4–8 by $\frac{1}{20}$–$\frac{1}{16}$ in., subterete, channelled, rigid, stiffly pointed; panicle 2 by $\frac{1}{3}$ in. with 9 spikelets (or smaller), nearly reduced to a raceme; bracts similar to the leaves, scarcely dilated at the base, lowest overtopping the panicle by $\frac{1}{2}$–4 in.; spikelets 0–$\frac{1}{8}$ in. pedicelled, $\frac{1}{4}$–$\frac{1}{3}$ by $\frac{1}{8}$ in., somewhat flattened, shining brown, very hard; empty lower glumes 6, distichous; lower flower-glume male, with a rudimentary pistil, upper bisexual, perfecting a nut; hypogynous bristles 0; stamens 3; anthers linear-oblong, with a white lanceolate scabrous crest, but hardly eared at the base; nut trigonous subglobose, together with its conoid beak $\frac{1}{5}$ in. long, brown-yellow, nearly smooth; style linear, branches 3 long linear. *Elynanthus sylvaticus, Nees in Linnæa*, ix. 298; x. 188; *Kunth, Enum.* ii. 310; *Boeck. in Linnæa*, xxxviii. 256. *E. auritus, Nees in Linnæa*, ix. 298; x. 187; *Kunth, Enum.* ii. 309; *E. compar, var. β, Boeck. in Linnæa*, xxxviii. 256.

SOUTH AFRICA: without locality, *Drège*, 3945! *Ecklon and Zeyher*, 96! 100!

COAST REGION: Cape Div.; near Cape Town, *Schlechter*, 406! Caledon Div.; Houw Hoek, 2500 ft., *Schlechter*, 7385! Bredasdorp Div.; Elim, 1600 ft., *Schlechter*, 7655!

6. T. ligulata (C. B. Clarke in Durand and Schinz, Conspect. Fl. Afr. v. 661); leaf-sheaths membranous-ligulate at the top; young flower-glumes scarious, except the keel; otherwise as *T. sylvatica*. *Elynanthus ligulatus, Boeck. in Linnæa*, xxxviii. 262.

SOUTH AFRICA: without locality.

The whole material is an example in Herb. Berlin without precise locality or name of collector, and very young. The plant might be a young state of *T.*

sylvatica, except that the bracts have come off in the drying; it is not supposed that even bad drying would bring off the tough bracts of *T. sylvatica*.

7. **T. picta** (C. B. Clarke in Durand and Schinz, Conspect. Fl. Afr. v. 662); bracts much dilated at the base; panicle contracted, about 1 in. long; spikelets (young) more than $\frac{1}{3}$ in. long, oblong-lanceolate, little compressed (the glumes only obscurely distichous); otherwise as *T. sylvatica*. *Elynanthus pictus, Boeck. in Linnæa*, xxxviii. 254.

COAST REGION: Tulbagh Div.; mountains near New Kloof, 2000 ft., *MacOwan, Herb. Aust.-Afr.*, 1683! Worcester Div., Drakenstein Mountains, near Bains Kloof, 1800 ft., *Bolus*, 4077! Dutoits Kloof, 2000-4000 ft., *Drège*, 1654a! 1654aa!

Boeckeler says the style is (? always) 2-branched; it is 3-branched at Kew; a 2-branched style must be a very rare abnormality in *Tetraria*. There are present in *T. picta*, 3 minute linear hypogynous bristles, but such are likely to occur occasionally in any species of the present section.

8. **T. pleosticha** (C. B. Clarke in Durand and Schinz, Conspect. Fl. Afr. v. 662); glabrous; rhizome $\frac{1}{8}$ in. in diam., horizontal; stems 12–16 in. long, tufted, medium-sized to slender, terete, without any node between the lowest bract and the sub-basal fimbriate brown sheaths; leaves 8–12 in. long, setaceous, hemi-cylindric, channelled; panicle 4–7 by 1 in., with often 20 spikelets or more; lowest peduncle exserted 1–2 in., suberect, from the sheathing close-cylindric black base of the leaf-like bract; spikelets $\frac{1}{3}$–$\frac{1}{2}$ by $\frac{1}{12}$ in., oblong-lanceolate, hardly compressed, dusky brown-black; glumes very obscurely distichous, lanceolate, acute, scarcely mucronate; hypogynous bristles 0; stamens 3; anthers (of the genus) with sterile ears at the base; style 3-fid; nut, including beak, $\frac{1}{6}$ in. long, trigonous, nearly smooth, yellow-white; beak narrow conic. *Chætospora circinalis c, Drège, Pflanzengeogr. Documente* 171.

COAST REGION: Worcester Div.; Ratel River, *Thom*! Caledon Div.; Houw Hoek, 2000 ft., *Schlechter*, 7396! 7411! Riversdale Div.; on the lower part of the Lange Bergen, near Kampsche Berg, *Burchell*, 6952! Humansdorp Div.; Kromme River, below 1000 ft., *Drège*!

9. **T. wallichiana** (C. B. Clarke in Durand and Schinz, Conspect. Fl. Afr. v. 664); glabrous; stems 12–18 in. long, tufted, medium-sized to slender, terete, without any node between the lowest bract and the sub-basal sheaths; leaf-sheaths fine brown, $\frac{1}{5}$ in. broad, striate, with many pale-brown fimbriations on the edges; leaves nearly a foot long, near the base $\frac{1}{24}$ in. broad, flat; panicle 2 by $\frac{1}{2}$ in., close; 2 lower bracts 1 by $\frac{1}{4}$–$\frac{1}{3}$ in. lanceolate, striate, chestnut-coloured, with pale-brown thin margins and linear weak tips $\frac{1}{4}$–$\frac{1}{2}$ in. long; spikelets $\frac{1}{4}$ in. long, clustered; lower flower male with imperfect pistil, upper bisexual, perfecting a nut; hypogynous bristles 0; stamens 3; anthers (of the genus) with sterile ears at the base; young pistil obovoid, hemispheric at the top, hispid (indicating a style-base confluent with pistil ?); style short, linear, branches 3 long linear.

COAST REGION: Cape Div.; near Cape Town, *Harvey*, 198! Stellenbosch Div.? Sneeuw Kop, *Wallich!*

Though the material is young, this plant must be a *Tetraria*, but it does not appear closely allied to any other species. The inflorescence and bracts recall *Ficinia monticola* and the neighbouring *Ficinias*.

10. T. fimbriolata (C. B. Clarke in Durand and Schinz, Conspect. Fl. Afr. v. 661); glabrous except the glumes; rhizome $\frac{1}{6}$ in. in diam., horizontal; stems 12–18 in. long, tufted, slender, terete, with 1–3 leaf-bearing nodes between the lowest bract and the basal leaves; basal leaf-sheaths pale-brown, fimbriate; basal leaves 4–8 in. long, setaceous; stem leaves similar, but shorter, from tight cylindric black sheaths; panicle 3–5 in. long, lax, of 10–16 spikelets; lowest bract like the stem leaves; peduncle exserted 1–3 in.; spikelets subsolitary, $\frac{1}{5}$ by $\frac{1}{8}$ in., much compressed; lower flower male with imperfect pistil; upper 2-sexual, perfecting a nut; glumes manifestly distichous, broadly ovate, obtuse, with minute white wool on the edge; hypogynous bristles 4, hardly $\frac{1}{3}$ the length of the nut, linear, minutely scabrid; stamens 3; anthers (of the genus) with sterile basal ears; nut (not ripe) small, obovoid; style-base pyramidal, confluent with the nut, hardly scabrid; style linear, short, branches 3 linear, very papillose. *Chætospora fimbriolata, Nees in Linnæa,* x. 191; *Kunth, Enum.* ii. 327; *Boeck. in Linnæa,* xxxviii. 300.

SOUTH AFRICA: without locality, *R. Brown! Ecklon!*
COAST REGION: Caledon Div.; Houw Hoek, 2500 ft., *Schlechter*, 7392! and without precise locality, *Zeyher*, 81! Swellendam Div.; mountains along the lower part of the Zonder Einde River, *Zeyher*, 4429!

11. T. fasciata (C. B. Clarke in Durand and Schinz, Conspect. Fl. Afr. v. 661); glabrous; rhizome $\frac{1}{6}$ in. in diam., horizontal; stems 12–18 in. long, tufted, slender, terete, with several leaf-bearing nodes between the lowest bract and the basal leaves; basal leaf-sheaths brown, much fimbriate; basal leaves 2–5 in. long, setaceous; stem leaves similar, but shorter, from tight cylindric black sheaths; panicle 4–8 in. long, slender, lax, with capillary branches; bracts like the stem-leaves, much shorter than the panicle; spikelets few, solitary, $\frac{1}{6}$ in. long, lanceolate; glumes pale, ovate-lanceolate, microscopically scabrous; style 3-fid. *Schœnus fasciatus, Rottb. Descr. et Ic.* 67, *t.* 16, *fig.* 5. *Elynanthus microstachyus, Boeck. in Linnæa,* xxxviii. 261 *mainly.*

SOUTH AFRICA: without locality, *R. Brown! Drège!*
COAST REGION: Cape Div.; False Bay, *Robertson!* Caledon Div.; Houw Hoek, 2500 ft., *Schlechter*, 7390! 7394!

All the examples seen are young; the above description is from the type plant in herb. Vahl, which agrees identically with Drège's *Elynanthus cuspidatus* bb in herb. Kew. Boeckeler founded his *E. microstachyus* on *E. cuspidatus* aa and bb in herb. Drège, but his description applies to the present plant only. Unfortunately, Drège, in his Documente, has omitted *E. cuspidatus* bb altogether, so that its precise locality cannot be ascertained.

12. T. capillacea (C. B. Clarke in Durand and Schinz, Conspect. Fl. Afr. v. 659); glabrous; rhizome $\frac{1}{6}$ in. in diam., horizontal; stems 12–18 in. long, tufted, slender, terete, with usually 1 leaf-bearing node between the lowest bract and the basal leaves; basal leaf-sheaths chestnut-brown and fimbriate; basal leaves 5–11 in. long, setaceous; stem leaf similar, shorter, from a tight cylindric black sheath; panicle 6 by $\frac{1}{3}$–$\frac{1}{2}$ in., dense with somewhat numerous spikelets clustered on very short erect rather rigid branches, lowest peduncle often long exserted; spikelets $\frac{1}{4}$ in. long, ellipsoid, not much flattened; glumes distichous, ovate, acute, under the microscope white powdery and minutely hairy on the edges; lower flower male with an imperfect pistil; upper 2-sexual, perfect; stamens 3; anthers (of the genus) with sterile basal ears; style 3-fid; young style-base depressed-pyramidal, confluent with the top of the ovary. *Schœnus capillaceus, Thunb. Prod.* 16; *Fl. Cap. ed. Schult.* 93. *Chætospora capillacea, Nees in Linnæa*, x. 192; *Kunth, Enum.* ii. 325; *Boeck. in Linnæa*, xxxviii. 305. *C. flexuosa, Drège in Linnæa*, xx. 251. *C. flexuosa, var. β gracilis, Boeck. in Linnæa*, xxxviii. 304.

SOUTH AFRICA: without locality, *Bergius!* *Thunberg.*
COAST REGION: Swellendan Div.; *Zeyher*, 80! Caledon Div.; Klein River Mountains, *Zeyher*, 4437! Houw Hoek, 2000 ft., *Schlechter*, 7420! George Div.; near the Touw River, *Burchell*, 5749!

T. flexuosa has the panicle and spikelets very like those of the present species, but has (instead of strictly setaceous leaves hardly $\frac{1}{80}$ in. in diam. at their base) very much stouter leaves plano-convex near their base.

13. T. aristata (C. B. Clarke in Durand and Schinz, Conspect. Fl. Afr. v. 659); young spikelets linear-lanceolate; flower-glumes conspicuously aristate by curved aristæ $\frac{1}{16}$ in. long; otherwise as *T. capillacea*. *Elynanthus aristatus, Boeck. in Linnæa*, xxxviii. 260. *Chætospora capillacea (status abnormis), Boeck. in Linnæa*, xxxviii. 306. *C. flexuosa, forma*, Nees in Linnæa*, x. 190.

SOUTH AFRICA: without locality, *Ecklon!* in herb. Nees, *Drège!* in herb. Delessert, *Ecklon and Zeyher*, 145!

No specimen in the Kew Herbarium.

The strongly aristate glumes are unlike those of the preceding closely-allied species. But the examples are all very young, and it may be an abnormal state of *T. capillacea* as Boeckeler finally thought. The out-growth of the glumes in the monstrous state of *T. flexuosa* suggests this explanation.

14. T. Burmanni (C. B. Clarke in Durand and Schinz, Conspect. Fl. Afr. v. 659); glabrous; rhizome hardly $\frac{1}{10}$ in. diam., horizontal, stems 8–10 in. long, tufted, slender, terete, with nodes (bearing leaves or bracts) scattered throughout their length; basal leaf-sheaths brown, more or less torn into fibres; basal leaves 3–8 in. long, $\frac{1}{16}$ in. broad at the base, flat but (when dry) rolled up; stem-leaves and bracts similar, but shorter, from tight cylindric chestnut-brown or pale sheaths; panicle elongate, narrow, often carried down nearly to the base of the stem; branches axillary, 1–4 in. long, carrying near the top 3 or 4 spikelets; spikelets $\frac{1}{4}$ by $\frac{1}{8}$ in., elliptic-lanceolate, hard, somewhat flattened, rusty-brown; lower empty glumes 6–5, small; flower-glumes hardly acute; lower flower male with an imperfect pistil, upper bisexual, producing a nut; hypogynous bristles 3–4, minute, setaceous (according to Schrader and so seen at Kew), but 6–3 capillary plumose ferruginous according to Boeckeler; stamens 3; anthers of the genus hardly eared at the base; style 3-fid. *Lepidosperma Burmanni, Spreng. Syst.* i. 194, *excluding syn.; Flora,* 1829, i. *Beil.* 8. *Chætospora Burmanni, Schrad. Anal. Fl. Cap.* 32, *t.* 3, *fig.* 4; *Nees in Linnæa,* x. 191; *Kunth, Enum.* ii. 324; *Boeck. in Linnæa,* xxxviii. 302. *C. circinalis* α, *Drège, Pflanzengeogr. Documente,* 171. *Schœnus Burmanni, Roem. et Schultes, Syst. Veg.* ii. *Mant.* 528. *S. aristatus, Steud. in Flora,* 1829, 135, *according to the number cited. Schœnopsis? Burmanni, Nees in Linnæa,* vii. 528; viii. 86.

SOUTH AFRICA: without locality, *Zeyher!* *Ecklon and Zeyher,* 113!

COAST REGION: Worcester Div.; Dutoits Kloof, 2000–3000 ft., *Drège,* 1634! Swellendam Div., 500–1600 ft., *Zeyher,* 1059!

15. T. nigrovaginata (C.B. Clarke in Durand and Schinz, Conspect. Fl. Afr. v. 662); glabrous; rhizome $\frac{1}{8}$ in. diam., horizontal; stems 12–16 in. long, tufted, slender (rather stouter than in the 4 preceding species), terete, with a leaf-bearing node between the panicle and basal leaves; basal sheaths brown-red, fimbriate; leaves 4–6 by $\frac{1}{16}$–$\frac{1}{12}$ in.; stem-leaves and bracts much shorter, from cylindric tight black or reddish sheaths; panicle 4–10 by $\frac{1}{2}$–1 in., with 12–18 brown spikelets; lowest peduncle exserted, 1–4 in.; spikelets nearly $\frac{1}{3}$ in. long, obtuse, ellipsoid, hardly at all flattened; lower empty glumes 6–8; flower-glumes acute; lower flower male with imperfect pistil, upper bisexual, perfecting a nut; hypogynous bristles 3, linear, scabrid, as long as the young nut; stamens 3; anthers (of the genus) obscurely eared at the base; style linear, branches 3. *Cyathocoma nigrovaginata, Nees in Linnæa,* x. 196; *Drège in Linnæa,* xx. 250; *Kunth, Enum.* ii. 323; *Boeck. in Linnæa,* xxxviii. 313; *Decalepis dregeana, Boeck. in Engl. Jahrb.* v. 509.

SOUTH AFRICA: without locality, *Drège,* 1615! *Ecklon and Zeyher,* 105!

COAST REGION: Worcester Div.; Brand Vley, *Rehmann,* 2421! Tulbagh Div.; mountains near New Kloof, 2000 ft. *MacOwan, Herb. Aust.-Afr.,* 1687! near Tulbagh, 2000 ft., *Schlechter,* 7463!

16. T. circinalis (C. B. Clarke in Durand and Schinz, Conspect. Fl. Afr. v. 659); glabrous; rhizome $\frac{1}{8}$ in. in diam., horizontal; stems 3–10 in. long, tufted, slightly angular in the panicled part, with nodes (bearing leaves or bracts) scattered throughout their length; basal sheath brown or reddish, fimbriate or cancellate; basal leaves usually as long as the stem, $\frac{1}{16}$–$\frac{1}{8}$ in. broad; panicle 2–6 by $\frac{1}{2}$–$1\frac{1}{2}$ in., with 15–60 spikelets, often rather dense; spikelets $\frac{1}{3}$ by $\frac{1}{8}$ in., nearly cylindric, brown or pale; lower empty glumes 6, upper longer, hardly acute, glabrous or microscopically hispid on the edges; lower flower male with an imperfect pistil, upper 2-sexual, perfecting a nut; hypogynous bristles 4–3 linear, scabrous, shorter than the nut; stamens 3; anthers (of the genus) with sterile ears at the base; nut (not seen ripe) small, obovoid, style-base depressed-hemispheric, confluent with the top of the pistil, minutely scabrous. *Schœnus circinalis, Schrad. in Roem. et Schultes, Syst. Veg.* ii. *Mant.* 43. *S. microstachys, Vahl, Enum.* ii. 220, *fide Kunth. Chætospora circinalis, Schrad. Anal. Fl. Cap.* 31, *t.* 3, *fig.* 2; *Nees in Linnæa*, x. 191; *Kunth, Enum.* ii. 324; *Boeck. in Linnæa*, xxxviii. 303 (?).

SOUTH AFRICA: without locality, *Bergius!*

COAST REGION: Cape Div.; Table Mountain, 2500 ft., *Bolus*, 4643! Muizen Berg, 1100 ft., *Bolus*, 4630! Camps Bay, *Burchell*. 337! Simons Bay, *Wright*, 525! Caledon Div.; on Donker Hoek Mountain, *Burchell*, 7981! Houw Hoek, 2000 ft., *Schlechter*, 7387! 7390! 7426! Lowrys Pass, 2800 ft., *Schlechter*, 4833! Riversdale Div.; hills near the Zoetemelks River, *Burchell*, 6801!

Also in Tropical Africa.

This plant, and its neighbours, are exceedingly like many Australian species of *Schœnus*, and are doubtless closely allied thereto, differing by the technical generic character of flower-glumes close together on the axis of the spikelet. Boeckeler describes his *Chætospora circinalis* as having a solitary stem and ciliate glumes, so that the specimen he described from could hardly have been the present species.

17. T. flexuosa (C. B. Clarke in Durand and Schinz, Conspect. Fl. Afr. v. 661); glabrous; rhizome $\frac{1}{6}$ in. in diam., horizontal; stems 15–24 in. long, tufted, stouter than those of the preceding species, terete, with leaf-bearing nodes between the panicle and the basal leaves; basal leaf-sheaths dark brown, more or less torn; leaves often 8 by $\frac{1}{10}$ in., tough, harsh; sheaths of the stem-leaves long, close, cylindric, black; panicle 4–6 by $\frac{3}{4}$–1 in., compound, with short suberect branches; spikelets clustered, dull-brown or black, $\frac{1}{4}$ in. long, not much flattened; empty glumes 6–7; flower-glumes longer, acute or submucronate, hardly microscopically puberulous; lower flower male, with an imperfect pistil, upper 2-sexual, perfecting a nut; hypogynous bristles (about) 3, very short, linear, scabrous; stamens 3; anthers (of the genus) shortly eared at the base; young pistil obovoid, crowned by the (then relatively) large hemispheric style-base; style linear, branches 3. *Schœnus flexuosus, Thunb. Prod.* 16; *Fl. Cap. ed. Schult.* 93. *Chætospora flexuosa, Schrad. Anal. Fl. Cap.* 33, *in Obs. t.* 3, *fig.* 3; *Nees in Linnæa*, viii. 87 *in Obs.*; x. 190; *Kunth, Enum.* ii. 325; *Boeck. in Linnæa*, xxxviii. 304. *C. capillacea, var. β major, Boeck. in Linnæa*, xxxviii. 306. *Schœnopsis flexuosa, Nees in Linnæa*, vii. 528.

VAR. β **abortiva** (C. B. Clarke in Durand and Schinz, Conspect. Fl. Afr. v. 661); spikelets elongate, $\frac{3}{8}$ in. long, lanceolate, in a monstrous state from the effect of a smut.

SOUTH AFRICA: without locality, *Sieber! Bergius! Verreaux!*

COAST REGION: Cape Div.; Table Mountain, *Thunberg, Burchell*, 566! *Krebs! Rehmann*, 615! Caledon Div.; Houw Hoek, 2000 ft., *Schlechter*, 7420! Var. β: Caledon Div.; Nieuw Kloof, Houw Hoek Mountains, *Burchell*, 8063! Riversdale Div.; lower part of the Lange Bergen, above the waterfall at Valley Rivers Poort, *Burchell*, 7000!

The examples of var. β, being destroyed by smut, are doubtfully determined, and may belong to *T. circinalis*. The spikelets would do for those of *T. circinalis*, but the tall stems 16–30 in. high are more like those of *T. flexuosa*.

18. **T. robusta** (C. B. Clarke in Durand and Schinz, Conspect. Fl. Afr. v. 662); glabrous; stems 3 ft. long, stout, trigonous, nearly smooth, with a leaf-bearing node far above the base; basal leaf-sheaths entire, firm; leaves 10–20 by $\frac{1}{4}$–$\frac{1}{3}$ in., stout, flat; panicle 12–18 by 1–1$\frac{1}{2}$ in., compound, dense with numerous upright branches; bracts like the stem-leaves, not overtopping the panicle; spikelets $\frac{1}{2}$ by $\frac{1}{8}$ in., slightly compressed, hard dull-brown; empty glumes 6–8, distichous; flower-glumes ovate, acute, submucronate; lower flower male with an imperfect pistil, upper 2-sexual, perfecting a nut; hypogynous bristles 3–4, small, linear, scabrous; stamens 3; anthers linear-oblong, crest lanceolate, scabrous; basal sterile ears hardly any; pistil obovoid; style-base confluent with the young nut; style linear, branches 3 linear. *Chætospora robusta, Kunth, Enum.* ii. 325; *Boeck. in Linnæa,* xxxviii. 306.

COAST REGION: Humansdorp Div.; mountains near the Kromme River, 1000–2000 ft., *Drège*, 7375!

This plant is quite different from the preceding and naturally belongs to the next series, but the glumes are not aristate.

19. **T. thermalis** (C. B. Clarke in Durand and Schinz, Conspect. Fl. Afr. v. 663); glabrous; rhizome stout, horizontal, woody; stem 3–6 ft. long, trigonous, stout, nearly smooth, with nodes scattered throughout its length; basal leaves often 3 ft. long, $\frac{1}{3}$ in. broad, flat, stout, scabrous on the margins; sheaths entire, firm; stem-leaves similar but smaller, from cylindric sheaths, passing into the similar bracts; panicle 12–24 by 1–2 in., of numerous axillary branches, somewhat nodding at the top; spikelets dusky-brown, clustered in the axils of secondary ovate-lanceolate bracts, $\frac{1}{2}$ in. long, elliptic-lanceolate; empty lower glumes about 6, $\frac{1}{3}$ in. long, subdistichous, long acuminate, acute, puberulous; lower flower male, with an imperfect pistil, upper 2-sexual, perfecting a nut; hypogynous bristles 6, up to $\frac{1}{5}$ in. long, subulate, white, scabrid (not dilated at the base nor feathered); stamens 3; anthers of the genus with small sterile basal ears; style-base when young ovoid, broader than the young pistil, scabrous, ultimately contracted into a small boss at the top of the large nut; nut (including the stalk) nearly $\frac{1}{8}$ in. long, ovoid, smooth, dusky-brown; style persistent, its linear undivided part nearly $\frac{1}{3}$ in. long, branches 3 or 4, linear, long. *Schœnus capensis, Linn. herb. propr. S. thermalis, Linn. Mant.* 179; *Thunb. Prod.* 17; *Fl. Cap. ed. Schult.* 95; *not of Rottb. Lepidosperma thermale, Schrad. Anal. Fl. Cap.* 38, *t.* 4, *fig.* 5; *Boeck. in Linnæa,* xxxviii. 337. *Sclerochætium thermale, Nees in Linnæa,* viii. 88; x. 198; *Kunth, Enum.* ii. 321 *in small part. S. giganteum, Steud. Syn. Pl. Glum.* ii. 159.

SOUTH AFRICA: without locality, *Ecklon and Zeyher*, 111! *Nelson!*

COAST REGION: Cape Div.; Table Mountain, *Thunberg, Rehmann*, 614! *Bolus*, 4717! Caledon Div.; near the Palmiet River and Houw Hoek, 1000–

2000 ft., *Drège*, 3965b! mountains of Baviaans Kloof, near Genadendal, *Burchell*, 7780! Klein River Mountains, 1000–2000 ft., *Zeyher*, 4439! Riversdale Div.; lower part of the Lange Bergen, near Kampsche Berg, *Burchell*, 6957!

20. T. eximia (C. B. Clarke in Durand and Schinz, Conspect. Fl. Afr. v. 661); leaves $\frac{1}{5}$–$\frac{1}{3}$ in. broad; hypogynous bristles about 4, small, narrow-ligulate, white-scabrous; style-base hemispheric, nearly smooth, confluent with the young pistil; otherwise as though a very slender *T. thermalis.*

SOUTH AFRICA: without locality, *Hooker*, 315!
COAST REGION: Cape Div.; False Bay, *Robertson!*

The stem and leaf are much slenderer than those of *T. thermalis*, the spikelets (and clusters of them) nearly as large.

21. T. triangularis (C. B. Clarke in Durand and Schinz, Conspect. Fl. Afr. v. 663); stem of medium thickness, acutely triangular; leaves $\frac{1}{5}$ in. broad; panicle 8 by 1 in., very thin, of about 12 subsolitary bright ferruginous-brown spikelets; otherwise as though a very slender *T. thermalis. Lepidosperma triangulare, Boeck. in Linnæa*, xxxviii. 336.

SOUTH AFRICA: without locality, *Ecklon and Zeyher* (ex Boeckeler).
COAST REGION: Cape Div.; Table Mountain, *Burchell*, 659!

The examples of this species are exceedingly young; it is clearly closely allied to *T. thermalis.*

22. T. secans (C. B. Clarke in Durand and Schinz, Conspect. Fl. Afr. v. 663); glabrous; stems 3–4 ft. (or more) long, trigonous, with nodes scattered throughout their length; basal sheaths, entire, firm; leaves 12–20 by $\frac{1}{2}$ in., but above the base soon narrowed, long acuminate caudate "cutting like a knife" (Burchell note); stem-leaves similar but shorter, passing into the similar shorter bracts; panicle 12–16 by 1–$\frac{1}{2}$ in., compound with very numerous clustered spikelets; spikelets $\frac{1}{3}$ in. long, ellipsoid, dusky-brown, lower empty glumes 6–5, aristate, scarious-edged, somewhat puberulous; lower flower male, with an imperfect pistil, upper 2-sexual, perfecting a nut; hypogynous bristles 0; stamens 3; anthers of the genus scarcely eared at the base; nut, including the short stalk, nearly $\frac{1}{4}$ in. long, ovoid-ellipsoid, smooth, bony; style-base when young large, pyramidal, hispid, but in the ripe nut nearly absorbed.

COAST REGION: George Div.; in and near the forest by the Touw River, *Burchell*, 5712! 5733!

23. T. Rottbœllii (C. B. Clarke in Durand and Schinz, Conspect. Fl. Afr. v. 662); glabrous; rhizome very short, woody; stems 3–4 ft. long, tufted, terete, with nodes (bearing leaves or bracts) scattered throughout their length; basal leaf-sheaths fimbriate; leaves 12 by $\frac{1}{6}$ in., tough, stem-leaves and bracts similar, shorter; panicle sometimes in appearance simple, 12 by 1 in., with a single peduncled compound head of spikelets from each axil, sometimes much more compound with 1–3 peduncles, up to 4 in. long, each bearing a spike of compound heads; compound heads of spikelets often 1 by $\frac{2}{3}$ in., grey-brown, with many aristate secondary bracts; spikelets $\frac{1}{8}$ in. long, ellipsoid, acute, often more or less white scarious at the top; empty glumes 6, subdistichous, aristate, scabrid-puberulous; lower flower male, with an imperfect pistil, upper 2-sexual, perfecting a nut; hypogynous bristles 6, regularly placed, equal, a little longer than the stalk of the nut, rigid, ligulate, straw-yellow, somewhat scabrous-edged; stamens 3; anthers (of the genus) with very small basal ears; nut, including the stalk, $\frac{1}{6}$ in. long, ovoid, smooth, hard, finally dusky-black; style-base ultimately a depressed boss; style linear long, branches 3 (Schrader says 2, which has not been verified even as an accident). *Schœnus thermalis, Willd. Sp. Pl.* i. 267 *partly. S. bromoides, Lam. Ill.* i. 137; *Encyc.* i. 740, *according to Schrader. Schœnus sp. n.* 83, *Rottb. Descr. et Ic.* 63, *t.* 18, *fig.* 2. *Lepidotosperma Rottbœllii, Schrad. in Roem. et Schultes, Syst. Veg.* ii. *Mant.* 474. *Lepidosperma Rottbœllii, Schrad. Anal. Fl. Cap.* 37, *t.* 4, *fig.* 4; *Boeck. in Linnæa*, xxxviii. 335. *L. thermale, Spreng. Syst.* i. 194. *Sclerochætium thermale, Nees in Linnæa*, vii. 512 *partly; Kunth, Enum.* ii. 321 *mostly. S. Rottbœllii, Nees in Linnæa*, viii. 88; x. 198. *S. Kœnigii, Nees ex Hochst. in Flora*, 1845, 763.

SOUTH AFRICA: without locality, *Harvey*, 283! *Thom*, 1022! *Chamisso*, 217! *Ecklon and Zeyher*, 103! 109! 110! *Nelson!*
COAST REGION: Cape Div.; near Cape Town, *Burchell*, 28! *Rehmann*, 1183; Constantia, *Wallich*, 141! Devils Mountain, *Rehmann*, 917! Camps Bay, *Burchell*, 356! Paarl Div.; Paarl Mountains, 1000–2000 ft., *Drège*, 7376! Worcester Div.; mountains above Worcester, *Rehmann*, 2558! Stellenbosch Div.; near Stellenbosch, 200 ft., *Bolus*, 2868!

The hypogynous bristles in this species and in *T. involucrata* resemble greatly those of the closely-allied Australian genus *Lepidosperma*.

24. T. rottbœllioides (C. B. Clarke in Durand and Schinz, Conspect. Fl. Afr. v. 663); leaves hardly attaining $\frac{1}{8}$ in. in breadth; compound spikes $2\frac{1}{2}$ by $\frac{2}{3}$ in.; young spikelets ferruginous; otherwise as *T. Rottbœllii. Sclerochætium angustifolium, Hochst. in Flora*, 1845, 762.

SOUTH AFRICA: without locality, *Thom! Bauer!*
COAST REGION: Cape Div.; Constantia, *Zeyher!* False Bay, *Robertson!* near Cape Town, *Rehmann*, 1183!

The examples seen are all very young, and the species may prove to be only a variety of *T. Rottbœllii.* The general appearance differs in the elongate-oblong compound heads of brighter-brown spikelets.

25. T. involucrata (C. B. Clarke in Durand and Schinz, Conspect. Fl. Afr. v. 661); panicle very compound; with often several peduncles from each leaf-sheath; spikelets smaller than those of *T. Rottbœllii* and in very much smaller clusters (only 1–4 together); otherwise much as *T. Rottbœllii. Schœnus involucratus, Rottb. Descr. et Ic.* 64, *t.* 19, *fig.* 1. *Lepidosperma involucratum, Spreng. Syst.* i. 194; *Schrad. Anal. Fl. Cap.* 35; *Boeck. in Linnæa*, xxxviii. 334. *Sclerochætium involucratum, Nees in Linnæa*, vii. 511; x. 197.

SOUTH AFRICA: without locality, *Zeyher*, 4438! *Sieber*, 231.

COAST REGION: Cape Div.; between Rondebosch and Hout Bay, *Drège!* Tulbagh Div.; on the Witsen Berg, *Burchell*, 8698! Worcester Div.; Dutoits Kloof, 1000–4000 ft., *Drège!* mountains above Worcester, *Rehmann*, 2593! 2595! Caledon Div.; mountains of Baviaans Kloof, near Genadendal, *Burchell*, 7755! Swellendam Div.; on the summit of a mountain peak near Swellendam, *Burchell*, 7364! Riversdale Div.; lower part of the Lange Bergen, above the waterfall at Valley Rivers Poort, near Kampsche Berg, *Burchell*, 7015!

26. T. spiralis (C. B. Clarke in Durand and Schinz, Conspect. Fl. Afr. v. 663); leaves very narrowly linear; hypogynous bristles short, feathered; otherwise as as *T. involucrata*. *Sclerochætium spirale, Hochst. in Flora*, 1845, 761. *Lepisia ustulata, Hochst. in Flora*, 1845, 761, *not of Presl.*

COAST REGION: Caledon Div.; mountain summits near Genadendal, *Krauss.*

I have not seen Krauss' plant, but the description is very full and good. With this agrees, perfectly so far as it goes, *Burchell* 5986, collected on the Post Berg near George, which is a complete but very young specimen; in this the leaves are $\frac{1}{10}$ in. wide close to the base, incurved, very rigid, the back convex without keel, but striate. Whether *Burchell's* 5986 is really the *Sclerochætium spirale* of Hochstetter or no, it indicates a species close to but distinct from *T. involucrata*, and undescribed, unless Hochstetter's description of *S. spirale* be its description.

27. T. ustulata (C. B. Clarke in Durand and Schinz, Conspect. Fl. Afr. v. 664); glabrous; horizontal rhizome slender but woody; stems often clustered, 15–30 in. long, slender, terete, striate, thick, globose and woody at the base, with nodes between the panicle and basal leaves; basal leaf-sheaths a rich brown, fimbriate and usually cancellate; basal leaves 12–16 in. long, setaceous, i.e. close to the base $\frac{1}{20}$ in. diam., nearly terete, channelled; stem-leaves and lower bracts similar but shorter, from black cylindric tight sheaths; panicle 3–7 by 1–1½ in., of about 2–8 fine-brown compound heads; spikelets rather loosely clustered in the heads, ½ in. long; lower flower male, with an imperfect pistil, upper 2-sexual, perfecting a nut; lower empty glumes 5–6, distichous, conspicuously aristate; hypogynous bristles 0, or most minute; stamens 3 or 4; style-branches 3 or 4; nut including its stalk nearly ¼ in. long, broad ellipsoid, stony, somewhat wrinkled or obscurely longitudinally ribbed; style-base (ultimately) small, pyramidal, scabrous, confluent with the nut. *Schœnus ustulatus, Linn. Mant.* 178; *Thunb. Prod.* 16; *Fl. Cap. ed. Schult.*, 93; *Nees in Linnæa*, vii. 527, *excluding syn. S. capillaceus Thunb. Schœnus n.* 82, *Rottb. Descr. et Ic.* 63, *t.* 18, *fig.* 1. *Lepisia ustulata, Presl, Symb. Bot.* 10, *t.* 5; *Kunth, Enum.* ii. 308. *Elynanthus ustulatus, Nees in Linnæa*, x. 189, 207; *Boeck. in Linnæa*, xxxviii. 260.

SOUTH AFRICA: without locality, *Sieber*, 117! *Thom*, 1025! *Grey*, 50! *Pappe*, 115!

COAST REGION: Cape Div.; near Cape Town, *Thunberg*, *Burchell*, 926! Camps Bay, *Zeyher*, 1780! *Burchell*, 365! Simons Bay, *Wright*, 523! Table Mountain, *Zeyher* 89! *Rehmann*, 616! 617! False Bay, *Robertson*! Tulbagh Div.; mountain tops near New Kloof, 2000 ft., *MacOwan, Herb. Aust.-Afr.*, 1684! Worcester Div.; Dutoits Kloof, *Drège*, 1610! Drakenstein Mountains near Bains Kloof, 1600 ft., *Bolus*, 4079! mountains above Worcester, *Rehmann*, 2554! Caledon Div.; Houw Hoek, 1500 ft., *Schlechter*, 7445!

28. T. compar (Lestib. Essai Cyp. 1819, 36); glabrous; rhizome short, horizontal; stems 8–30 in. long, tufted, medium-sized, terete, without nodes between the panicle and basal leaves; leaf-sheaths coloured, with scarious edges, not fimbriate; leaves 4–12 by $\frac{1}{20}$–$\frac{1}{12}$ in., rigid, concave-convex; panicle 1–2 by $\frac{1}{2}$ in., rigid, condensed, of 6–15 spikelets or sometimes reduced to 1–3 spikelets; bracts 1–8 in. long, resembling the leaves, not dilated nor sheathing at the base; spikelets $\frac{1}{3}$–$\frac{1}{2}$ by $\frac{1}{6}$ in., nearly sessile, erect, rigid, a fine brown, often viscid; empty lower glumes 4–5, ovate, hardly acute, distichous, rigid; lower flower male, with imperfect pistil (or sometimes 2 lower flowers male); upper 2-sexual, perfecting a nut; hypogynous bristles 6–5, $\frac{1}{2}$ the length of the nut, linear, slenderly scabrous-hairy; stamens 6 (or often 8), sometimes 3–4 only in the nut-bearing flower; anthers linear, nearly $\frac{1}{6}$ in. long, with long lanceolate crest, not eared at the base; nut $\frac{1}{12}$–$\frac{1}{8}$ in. long, ovoid, obscurely 3–4-angular, brown, continuous with the pale conic beak, smooth, very minutely reticulated; style linear, branches 3 or 4, long, linear. *Schœnus compar, Linn. Mant.* 177; *Thunb. Prod.* 16; *Fl. Cap. ed. Schult.* 93. *S. arenarius, Schrad. Anal. Fl. Cap.* 27, *t.* 4, *fig.* 2. *S. viscosus, Schrad. Anal. Fl. Cap.* 26. *Schœnus sp. n.* 85, *Rotto. Descr. et Ic.* 65, *t.* 18, *fig.* 4. *Rhynchospora nitida, Spreng. Neue Entd.* iii. 10; *Syst.* i. 195. *Elynanthus compar, Nees in Linnæa*, vii. 520; x. 187; *Kunth, Enum.* ii. 308; *Boeck. in Linnæa*, xxxviii. 255. *E. arenarius, Nees in Linnæa*, ix. 298. *E. viscosus, Nees in Linnæa*, x. 186.

SOUTH AFRICA: without locality, *Thunberg*, *Sieber*, 106! *Verreaux! Ecklon! Ecklon and Zeyher*, 101! 102!

COAST REGION: Cape Div.; Cape Flats, *Pappe!* Table Mountain, *Drège*, 198! *Burchell*, 8417! *Rehmann*, 618! 620! *Bolus*, 4642! False Bay, *Robertson*! Constantia, *Zeyher*, 85! Tulbagh Div.; mountain tops near New Kloof, 2000 ft., *McOwan, Herb. Aust.-Afr.*, 1685! Worcester Div.; Dutoits Kloof, 2000–4000 ft., *Drège!* Caledon Div.; Houw Hoek, 2000–2500 ft., *Schlechter*, 7383! 7413! Riversdale Div.; lower part of the Lange Bergen, about the waterfall at Valley Rivers Poort, near Kampsche Berg, *Burchell*, 6990!

29. T. Thuarii (Beauv. in Mém. Inst. Fr. 1812 [ed. 1816] ii. 54, 55); glabrous; stem 1½–3 ft. long, obscurely trigonous, with 1 or 2 cylindric sheaths between the panicle and basal leaves; stem-leaf 12–24 by $\frac{1}{3}$–$\frac{1}{2}$ in.; panicle 12 by 1–2 in., compound, with numerous erect branches and spikelets; bracts similar to the stem-leaf but shorter; spikelets approximate but hardly clustered, nearly $\frac{1}{3}$ by $\frac{1}{10}$ in., pale dusky-brown; glumes subdistichous, ovate-oblong, hardly acute; lower flower male, with imperfect pistil; upper 2-sexual, perfecting a nut; hypogynous bristles 6–5, rather longer than the nut, linear, scabrous; style-branches 3 or 4; nut obovoid, smooth; style-base confluent with it, conical, white, minutely hispid. *Cyathocoma Ecklonii, Nees in Linnæa*, ix. 300; x. 195; *Kunth, Enum.* ii. 323; *Boeck. in Linnæa*, xxxviii. 311. *Ideleria capensis, Kunth, Enum.* ii. 311.

VAR. β **gracilior** (C. B. Clarke in Durand and Schinz, Conspect. Fl. Afr. v. 663); slenderer; stems 12–20 in. long with no leaves except at the base; leaves very narrow, sometimes almost threadlike, terete, but rigid; panicle 4–6 in. long, with 10–40 spikelets; spikelets red-brown, more deeply coloured than in *T. Thuarii* proper.

SOUTH AFRICA: without locality, *Burke! Pappe!*

COAST REGION: Clanwilliam Div.; Ceder Bergen, on sand-flats near Ezels Bank, 3000 ft., *Drège*, 2448! Lamberts Hoek, near Clanwilliam, *Wallich!* Tulbagh Div.; damp places at the foot of the mountains around New Kloof, 850 ft., *MacOwan, Herb. Aust.-Afr.*, 1686! near Tulbagh, 600 ft., *Schlechter*, 7512! Worcester Div., *Ecklon and Zeyher*, 106!

EASTERN REGION: Var. β: Pondoland, *Bachmann*, 76! 77!

The var. β is not at Kew for comparison; but from my description, as also from the locality, it may probably be some very distinct species.

30. T. punctoria (C. B. Clarke in Durand and Schinz, Conspect. Fl. Afr. v. 662); glabrous; stem 3 ft. long, terete, striate, often with a node between the base and panicle; basal leaf-sheaths entire; "uppermost with a terete leaf often as long as the stem" (Boeckeler); panicle 3 by 1 in., compound, rather dense, of 15–30 dull-brown spikelets; lowest bract erect, 3–8 in. long, folded at the base $\frac{1}{4}$ in. broad, terete above; spikelets $\frac{1}{2}$ by $\frac{1}{10}$ in.; glumes ovate, acute; keel green, hardly excurrent; lower flower male, with imperfect pistil; upper 2-sexual, perfecting a nut; hypogynous bristles longish subulate, thinly scabrous; stamens 3; anthers linear, crested, hardly eared at the base; nut (including its stalk) nearly $\frac{1}{4}$ in. long, ellipsoid scarcely trigonous; style-base fused with the nut as a boss; style short, branches 6 (or 8) long. *Schœnus punctorius, Vahl, Enum.* ii. 217. *Chætospora punctoria, Dietr. Sp. Pl.* ii. 28 (*cf. Schrad. Anal. Fl. Cap.* 33 *in note*). *Buekia punctoria, Nees in Linnæa*, x. 197; *Kunth, Enum.* ii. 310; *Boeck. in Linnæa*, xxxviii. 310.

SOUTH AFRICA: without locality, *Ecklon and Zeyher*, 104! *Masson! Verreaux!*

COAST REGION: Cape Div., between Wynberg and Constantia, *Burchell*, 818! between Wynberg and Hout Bay, below 1000 ft., *Drège!* Table Mountain, *Zeyher*, 91! False Bay, *Robertson!* Stellenbosch Div.; Lowrys Pass, *Schlechter*, 4835! and without precise locality, *Zeyher!* Tulbagh Div.; at the foot of mountains near New Kloof, 900 ft., *MacOwan Herb. Aust.-Afr*, 1688! Caledon Div.; mountains of Baviaans Kloof, near Genadendal, *Burchell*, 7892!

31. T. MacOwani (C. B. Clarke in Durand and Schinz, Conspect. Fl. Afr. v. 661); glabrous or nearly so; rhizome woody; stems 2 ft. long, acutely triquetrous, with nodes between the basal leaves and panicle; basal leaves numerous, nearly as long as the stem, $\frac{1}{4}$ in. and more wide, but much rolled up when dry even at the base, long tapering; sheaths stout, not torn; stem-leaves similar to the lower bract but very much shorter; panicle 10 by 1 in., lax, thin, of 8–20 dull-brown spikelets; spikelets $\frac{1}{3}$ by $\frac{1}{8}$ in.; glumes ovate; keel green, hardly excurrent; lower flower male, with imperfect pistil; upper 2-sexual, perfect; hypogynous bristles very small, linear, smooth; stamens 6; anthers crested; nut (including its stalk) $\frac{1}{6}$ in. long, ellipsoid, obscurely 3- (or 4-) gonous; style short, linear, branches 6 (or 8) long. *Chætospora hexandra, Boeck. in Flora*, 1878, 37.

CENTRAL REGION: Somerset Div.; on the summit of Bosch Berg, 4800 ft., *MacOwan*, 1864!

XXII. MACROCHÆTIUM, Steud.

Spikelet of 4 or 5 spirally imbricate glumes, with one perfect flower, an upper male flower being sometimes added. Hypogynous *bristles* 6, long, linear. *Stamens* 6. *Style* with a dilated base forming a beak on the nut, long, linear, branches 3 long. *Nut* rather small, obovoid trigonous, sessile.

Species 1, endemic, with the habit of the large species of *Tetraria;* from which it differs in the few non-distichous glumes, and the lowest flower perfecting a nut. It differs from *Carpha* and *Costularia* by the 6 stamens.

1. **M. Dregei** (Steud. Syn. Pl. Glum. ii. 159); glabrous; rhizome short, horizontal, woody, thick; stem 1–3 ft. long, stout, terete, striate; basal leaves 12 by $\frac{1}{6}$ in., stout, striate, subdistichous, sheaths very coarse, entire, dilated, often more than 1 in. wide, strongly striate; panicle often 1 or 2 ft. long, 1–3 in. broad; branches from distant axils, lowest sometimes near the base of the stem; lowest bract like the leaves, but much shorter; spikelets almost clustered, chestnut-red, $\frac{1}{4}$ in. long or more, ellipsoid, of 4 (or 5) glumes; lowest 1 (or 2) glume small, empty, bractlike, usually aristate, next glume ovate-oblong, obtuse, containing a perfect nut-bearing flower, next glume similar to this, but bearing a male flower or empty; hypogynous bristles 6, linear, longer than the nut (including its beak), thinly scabrous-hairy; nut sessile, including the style-base scarcely $\frac{1}{12}$ in. long, brown, smooth, very obscurely transverse-lineolate; style-base broad-conic, hispid, half as long as the nut; style $\frac{1}{12}$ in. linear, smooth, branches 3 linear, about $\frac{1}{10}$ in. long. *Carpha hexandra, Nees in Linnæa,* ix. 300; x. 193. *Ideleria? Neesii, Kunth, Enum.* ii. 311. *Cyathocoma Neesii, Boeck. in Linnæa,* xxxviii. 312. *Elynanthus Kraussii, Krauss in Flora,* 1845, 761.

SOUTH AFRICA: without locality, *Burke! Drège,* 3944! *Ecklon and Zeyher,* 107! *Pappe! Verreaux!*

COAST REGION: Swellendam Div.; near Swellendam, 1000–4000 ft., *Zeyher,* 4434! Riversdale Div.; at the foot of the Lange Bergen, near Kampsche Berg, *Burchell,* 7142! Uitenhage Div.; Van Stadens Mountains, *MacOwan,* 2020!

XXIII. CLADIUM, P. Browne.

Spikelet of 4–11 spirally imbricate glumes, with 1–7 flowers, the lowest bisexual perfecting a nut. *Stamens* 3–2. *Style* linear, branches 3 long; base dilated, ultimately confluent with the nut. *Nut* medium-sized, ovoid, trigonous.

Species 44; mostly Insular or near the sea; one in South Africa, which is nearly Cosmopolitan.

1. **C. Mariscus** (R. Br. Prod. 236); glabrous, stoloniferous; stem 3–8 ft. long, stout, roundish, with nodes (bearing leaves or bracts) throughout its length; leaves 2–4 ft. by $\frac{1}{4}$–$\frac{1}{2}$ in., tough, scabrous (cutting the hand) on the margins and keel; panicle 1–2 ft. by 3–5 in., of numerous axillary compound (often dense) corymbs;

bracts like the leaves but shorter; spikelets numerous (sometimes 1200–1500), mostly in clusters of 2–8, $\frac{1}{8}$ in. long, ellipsoid, acute, brown, usually 2-flowered; glumes 6–7, very close together, 3–4 lowest empty, ovate, subacute; hypogynous bristles 0; stamens 2 (or according to authors sometimes 3); nut sessile, $\frac{1}{16}$ in. long, cavity extending upwards into the ovoid beak (style-base). *C. jamaicense, Crantz, Inst.* i. 362. *C. germanicum, Schrad. Fl. Germ.* i. 75, *t.* 5, *fig.* 7**. *C. Mariscus, Kunth, Enum.* ii. 303; *Boeck. in Linnæa,* xxxviii. 232. *Schœnus Mariscus, Linn. Sp. Plant. ed.* ii. 62; *Sowerby, Engl. Bot. t.* 950.

COAST REGION: Uitenhage Div.; Van Stadens River, below 200 ft., *Drège,* 3964! Port Elizabeth Div.; banks of the Krakakama River, 200 ft., *MacOwan*!
KALAHARI REGION: Griqualand West: Herbert Div.; at Lower Campbell, *Burchell,* 1811! Bechuanaland, *Marloth,* 1000! Transvaal, *Rehmann,* 4040! 4484!
EASTERN REGION: Natal; Umlaas River, *Krauss,* 162! Clairmont, 150 ft., *Wood,* 3846!

Widely dispersed in warm and temperate regions.

XXIV. CHRYSITHRIX, Linn.

Spikelets several, 1-flowered, 1-sexual, in a dense spike; central terminal flower a pistil without bracts, surrounded by numerous linear bracts, some empty, a few carrying 1 stamen each. Hypogynous *bristles* 0. *Anther* linear, with long linear crest. *Style* long, linear, not dilated at the base, branches 3 linear. *Nut* ovoid, subglobose, pale, longitudinally many-striate.

Stem without nodes, except close to the base; head 1, lateral, of a few spikes, almost enclosed by coloured broad bracts; lowest bract as though a continuation of the stem.

DISTRIB. Species 2, endemic.

This is the only genus of the Sub-Order *Mapanieæ* found in South Africa; and in this, as in all the *Mapanieæ,* Bentham (Gen. Pl. iii. 1057) appears to lean to the view that the spikelet above described is a single hermaphrodite flower. As the inner sterile bracts, 1 or 2 at least, which come between the staminiferous bracts and the pistil, are evidently homologous with the staminiferous bracts, Bentham's view is hardly possible; and it is negatived also by the study of the development of the spikes made by Goebel (Ann. Jard. Buitenz. vii. 128).

Leaves flat ... ... (1) **capensis.**
Leaves terete ... ... (2) **junciformis.**

1. C. capensis (Linn. Mant. 304); glabrous; rhizome, stout, woody, creeping, clothed by hard ovate striate brown-red scales $\frac{1}{2}$ in. long; stem 12–20 in. long, rather slender, tough, smooth, slenderly striated, much flattened under the head; leaves $\frac{2}{3}$ the length of the stem, $\frac{1}{12}$–$\frac{1}{4}$ in. broad except near the folded base, falcate, with no midrib except near the folded base; lowest bract 2–4 in. long, erect, similar to the leaves, from a folded base; head $\frac{1}{2}$–$\frac{2}{3}$ by $\frac{1}{4}$–$\frac{1}{3}$ in., enclosed by ovate obtuse shining chestnut or black-red bracts; inner bracts $\frac{2}{5}$ in. long; nut $\frac{1}{8}$–$\frac{1}{6}$ in. long. *Thunb. Fl. Cap. ed. Schult.* 431; *Nees in Linnæa,* vii. 537; x. 144; *Kunth, Enum.* ii. 365;

Boeck. in Linnæa, xxxvii. 139. *Wildenovia compressa, Steud. in Flora*, 1829, 134 (*cf. Nees in Linnæa*, vii. 619) *not of Thunb.*

VAR. β **subteres** (C. B. Clarke in Durand and Schinz, Conspect. Fl. Afr. v. 668); stem round-trigonous at the top, obscurely slenderly striated; leaves long, narrow. *C. junciformis, Drège in Linnæa*, xx. 251 *partly.*

SOUTH AFRICA: without locality, *Carmichael! Villette! Masson!*

COAST REGION: Cape Div.; Cape Flats in front of Table Mountain, *Thunberg*, Table Mountain, *Burchell*, 600! *Bergius! Ecklon*, 850! *Rehmann*, 676! Swellendam Div.; near Swellendam, 1000–4000 ft., *Zeyher*, 4424b! Riversdale Div.; on the Kampsche Berg, *Burchell*, 7062! George Div.; on the Post Berg, near George, *Burchell*, 5948! near George, 1000–2000 ft., *Drège*, 3942b! Var. β: Worcester Div.; Dutoits Kloof, 3000–4000 ft., *Drège*, 3942a!

2. C. junciformis (Nees in Linnæa, x. 144); stem throughout its whole length terete, strongly striated; leaves terete, resembling barren stems; otherwise as *C. capensis*, Linn. *Kunth, Enum.* ii. 365; *Boeck. in Linnæa*, xxxvii. 140.

COAST REGION: Riversdale Div.; summit of Kampsche Berg, *Burchell*, 7103!

XXV. SCLERIA, Berg.

Flowers all 1-sexual. *Spikelets* bisexual or unisexual; bisexual spikelets of 1 basal female flower, and 1 or more male flowers above; unisexual female spikelets like the bisexual but the upper male portion reduced to 1 or 2 empty glumes; unisexual male spikelets like the bisexual but without the basal female flower. *Glumes* concave, open, i.e. margins not connate. Hypogynous *bristles* 0. *Stamens* 3–1; anthers linear-oblong, often mucronate. *Style* linear, not dilated at the base, branches 3 linear. *Nut* bony, ovoid, on a gynophore; apex of gynophore often dilated into a disc, simple trigonous, or compound, or with evolute lobes.

Stems with nodes throughout their length, generally trigonous; leaves and bracts at the base sheathing the stem, narrow, often very scabrous at the edges, cutting.

DISTRIB. Species 160; common in Tropical, frequent in warm Temperate Regions, extending to Lake Ontario and New Zealand, but absent from Europe and the whole Mediterranean and Orient Region.

Subgenus I. HYPOPORUM. Spikelets many 2-sexual; margin of disc annular and minute, or hardly any. Plants slender or scarcely medium-sized, with slender woody horizontal rhizomes (in the Cape species), very narrow leaves and small spikelets.

Clusters of spikelets usually distant on a nearly undivided spike:

Spikelets nodding in fruit:

Basal leaves short ... ... (1) **hirtella.**

Basal leaves long ... ... (2) **catophylla.**

Spikelets always erect:

Stem slender at the base... ... ... (3) **meyeriana.**

Stem with a woody bulb at the base ... (4) **Buchanani.**

Clusters of spikelets sessile on a more or less compound spike or panicle :
Panicle slender with few branches :
Nut longer than broad ... ... (5) **dregeana.**
Nut broader than long ... ... (6) **Rehmanni.**
Panicle compound :
Panicle lax with capillary branches ... (7) **Woodii.**
Panicle stouter, more rigid ... ... (8) **holcoides.**

Subgenus II. EUSCLERIA. Spikelets all unisexual; margin of disc entire or with 3 rounded lobes. Plants stouter than in the Subgenus *Hypoporum*, leaves broader and tapering symmetrically to the point.
Nut large, ovoid, smooth, white, often black-tipped ... ... (9) **melanomphala.**
Nut medium-sized, ellipsoid, strongly reticulated ... ... (10) **natalensis.**

Subgenus III. SCHIZOLEPIS. Spikelets all unisexual; margin of disc with many lanceolate lobes. Plants stout, leaves broad, the margins præmorse at unequal distances from the tip ... ... (11) **angusta.**

1. **S. hirtella** (Swartz, Prod. 19); hairy, sometimes only the inflorescence; rhizome $\frac{1}{8}$–$\frac{1}{5}$ in. in diam., horizontal, woody; stems 8–20 in. long, slender, very slender at the base, with distant nearly leafless sheaths; leaves 6–10 by $\frac{1}{10}$ in., lower gradually shorter; spike 2–5 in. long; clusters of spikelets $\frac{1}{4}$–1 in. apart, drooping, at least when ripening; spikelets $\frac{1}{6}$–$\frac{1}{5}$ in. long, many bisexual; nut $\frac{1}{16}$–$\frac{1}{12}$ in., ovoid, round-trigonous, white, smooth, shining (but see var. β); gynophore obpyramidal, margin obscurely 3-lobed, confluent with the nut, disc scarcely discernible. *Fl. Ind. Occid.* i. 93; *Kunth, Enum.* ii. 353; *Boeck. in Linnæa*, xxxviii. 439 *part of a only*. *S. nutans, Kunth, Enum.* ii. 351. *S. cenchroides, Kunth, Enum.* ii. 352. *Hypoporum hirtellum, Nees in Linnæa*, ix. 303.

VAR. β **tuberculata** (Boeck. ex C. B. Clarke in Durand and Schinz, Conspect. Fl. Afr. v. 671); nut somewhat tubercled, and with wavy ribs, not shining.

COAST REGION : Komgha Div.; mouth of the Kei River, *Flanagan*, 2363!

KALAHARI REGION: Transvaal; Magalies Berg, *Zeyher*, 1758! Var. β: Transvaal; Magalies Berg, *Burke*, 62! *Zeyher*, 2334!

EASTERN REGION : Pondoland or Natal; between the Umtentu and Umzimkulu Rivers, below 500 ft., *Drège*!

The typical form is also found in Tropical Africa and America.

2. **S. catophylla** (C. B. Clarke in Durand and Schinz, Conspect. Fl. Afr. v. 670); basal leaves close together, elongate, with long hair and recurved margins in the dry specimens; otherwise as *S. hirtella*.

EASTERN REGION : Natal; Clairmont, *Wood*, 1423!

Also in Tropical Africa.

3. **S. meyeriana** (Kunth, Enum. ii. 354); clusters of spikelets erect even in fruit; otherwise as *S. hirtella*, Sw. *Boeck. in Linnæa*, xxxviii. 441.

KALAHARI REGION: Bechuanaland; near the source of the Kuruman River, *Burchell*, 2463!

EASTERN REGION: Pondoland, *Drège*, 4364!

Also in Angola.

Nearly glabrous, whereas *S. hirtella* is hairy; the spikelets are ferruginous with green marks, a colour not seen in *S. hirtella*.

4. **S. Buchanani** (Boeck. Cyp. Novæ i. 33, character emended); stems with a bulbous woody thickening at the base, $\frac{1}{4}$–$\frac{1}{3}$ in. in diam.; otherwise as *S. meyeriana.*

EASTERN REGION: Tembuland; Umnyolo, 3500 ft., *Baur*, 759! Pondoland, *Bachmann*, 117!

Also in Tropical Africa and Madagascar.

5. **S. dregeana** (Kunth, Enum. ii. 354); sparingly hairy; rhizome slender, creeping, woody, branching; stems 12–30 in. long, slender, at the base very slender with distant short leaves; leaves 12 by $\frac{1}{16}$–$\frac{1}{12}$ in.; spike 1–5 in. long, with 1 or 2 branches at the base or nearly unbranched; clusters of spikelets $\frac{1}{4}$–1 in. apart, erect, brown or nearly black; spikelets nearly $\frac{1}{4}$ in. long, many 2-sexual; nut $\frac{1}{16}$ in. long, ellipsoid (longer than broad) trigonous, white, shining, smooth or somewhat rough on the shoulders; gynophore obpyramidal, disc obscure. *Boeck. in Linnæa*, xxxviii. 443. *S. setulosa, Boeck. Cyp. Novæ*, i. 33.

COAST REGION: Komgha Div.; marshy places near Komgha, 2000 ft., *Flanagan*, 1260! Stockenstrom Div.; Kat Berg, 4000–5000 ft., *Drège*, 3943!

KALAHARI REGION: Bastuland, *Cooper*, 3365!

EASTERN REGION: Tembuland; Bazeia, 2000–2500 ft., *Baur*, 311! Pondoland, *Bachmann*, 121!

Also in South Tropical Africa.

The example collected by *Cooper*, 3365, has no ripe nut, therefore it is difficult to say if it may not be *S. meyeriana.*

6. **S. Rehmanni** (C. B. Clarke in Durand and Schinz, Conspect. Fl. Afr. v. 674); spike more branched; spikelets $\frac{1}{6}$–$\frac{1}{5}$ in. long, chestnut-brown; nut minute, depressed ovoid (broader than long), white, shining; otherwise as *S. dregeana.*

KALAHARI REGION; Transvaal; Houtbosch, *Rehmann*, 5626!

7. **S. Woodii** (C. B. Clarke in Durand and Schinz, Conspect. Fl. Afr. v. 675); panicle compound, lax, with capillary branches; spikelets in clusters, about $\frac{1}{5}$ in. long, pale brown or rusty brown; bracts and bractlets (lowest empty glumes) aristate; nut obovoid, rather longer than broad, white, smooth or obscurely reticulate; otherwise as *S. dregeana.*

COAST REGION: Komgha Div., *Flanagan*, 954!

KALAHARI REGION: Orange Free State; Drakens Berg Range, near Harrismith, *Buchanan*, 114! Transvaal; Oliphants River, 5000 ft., *Schlechter*, 3803!

EASTERN REGION: Natal; Biggars Berg, 4000 ft., *Wood*, 4757! (spikelets monstrous, grown out); Zululand; near Inyoni River, *Wood*, 3994!

Also in Angola and Madagascar.

8. **S. holcoides** (Kunth, Enum. ii. 354); nearly glabrous, except the leaf-sheaths; rhizome horizontal, woody; stems 18–30 in. long, somewhat stouter than in any of the 7 preceding species; slenderer at the base, i.e. basal leaf-sheaths distant, nearly leafless; leaves 12–15 by $\frac{1}{6}$–$\frac{1}{5}$ in.; panicle 3–4 by 1–2 in., compound, usually more rigid and dense than in any of the preceding species; spikelets clustered, $\frac{1}{4}$ in. long, rusty-brown, many bisexual; nut $\frac{1}{16}$–$\frac{1}{12}$ in. long, obovoid, trigonous, white or somewhat lead-coloured, smooth, but usually with a few tubercles on the shoulders; gynophore obpyramidal, disc hardly discernible. *Boeck. in Linnæa*, xxxviii. 445.

EASTERN REGION: Pondoland or Natal; between Umtentu River and Umzimkulu River, below 500 ft., *Drège*, 4381! Natal; Durban Flat, *Buchanan*, 3! and without precise locality, *Buchanan*, 349!

9. **S. melanomphala** (Kunth, Enum. ii. 345); nearly glabrous, except the inflorescence; rhizome creeping, $\frac{1}{10}$ in. diam., woody; stems 2–3 ft. long, triquetrous, sometimes very scabrous, sometimes nearly smooth; leaves up to 3 ft. by $\frac{1}{3}$ in., somewhat 3-nerved; edges scabrous, cutting the hand; partial panicles 1–3, 3 by $\frac{2}{3}$ in., rather dense, long-peduncled; spikelets all unisexual; female $\frac{1}{4}$ in. long; male rather longer, several flowered, chestnut-brown; nut $\frac{1}{6}$ in. long, ovoid, smooth, shining, white, often with a black tip (where the black style base is confluent); gynophore obpyramidal; disc truncate, i.e. a horizontal ring, or an exceedingly short cylindric tube. *Boeck. in Linnæa*, xxxviii. 476. *S. macrantha, Boeck. in Flora*, 1879, 572, *but not S. macrantha, Boeck. in Flora*, 1858, 647.

COAST REGION: Komgha Div.; Kei River, *Flanagan*, 988!
EASTERN REGION: Tembuland; between the Bashee River and Morley, 1000–2000 ft., *Drège*, 4369! Pondoland, *Bachmann*, 124! Natal; Coast-land, Izikane, below 1000 ft., *Sutherland!* Inanda, *Wood*, 1597! near the Umlaas River, *Krauss*, 42! and without precise locality, *Buchanan*, 351!

Also in Tropical Africa and the Mascarene Islands.

10. **S. natalensis** (C. B. Clarke in Durand and Schinz, Conspect. Fl. Afr. v. 673); medium-sized; nut $\frac{1}{12}$ in. long, fenestrate; disc (at the apex of the gynophore) of 3 broad, short, rounded, subreflexed white, earshaped lobes; otherwise as *S. melanomphala*.

EASTERN REGION: Natal, *Buchanan*, 352! *Gerrard*, 450! *Rehmann*, 8000! 8259!

Closely allied to *S. melanomphala*, of similar habit and ragged panicle; but not likely to be mistaken for it, as it is considerably smaller in all its parts.

11. **S. angusta** (Nees in Linnæa, ix. 303); nearly glabrous; rhizome $\frac{1}{8}$–$\frac{1}{6}$ in. in diam., horizontal, woody, clothed by chestnut-red ovate striate scales $\frac{1}{2}$ in. long; stem often 3 ft. high; sheaths of leaves and bracts triquetrous or almost 3-winged; leaves 1–2 ft. by $\frac{1}{3}$–$\frac{2}{3}$ in., 3-nerved, the marginal portions (outside the 2 strong lateral nerves) suddenly terminating unequally; panicle 1 ft. long, of 3–5 axillary compound corymbs; spikelets all 1-sexual, $\frac{1}{6}$ in. long, pale; nut $\frac{1}{16}$–$\frac{1}{12}$ in. long, ovoid, smooth, white or purpurascent; gynophore

obconic, large, black-purple, glandular with a short scarious white 3-lobed edge (outer disc of anthers); disc above this margin (inner disc of anthers) of 12–24 lanceolate, linear, suberect, thin lobes, reaching about half the height of the nut. *Kunth, Enum.* ii. 346; *Boeck. in Linnæa,* xxxviii. 527.

EASTERN REGION: Pondoland or Natal; between the Umtentu and Umzimkulu Rivers, below 500 ft., *Drège,* 4246! Natal; Ungoya Forest, in a swamp, *Wood,* 3863! Delagoa Bay, *Junod,* 356!

Also in Madagascar.

XXVI. ERIOSPORA, A. Rich.

Spikelets very small, 2–3-flowered, collected in close spikes resembling the spikelets of *Scirpus,* mostly bisexual, about 4-glumed; lowest flower female, upper 1–2 male. *Glumes* ovate, boat-shaped, obscurely distichous, lowest empty, next female. Hypogynous *bristles* numerous, linear. *Stamens* 3–1; anthers not crested. *Nut* ovoid, trigonous, passing gradually into the conical style-base; linear part of the style short, persistent, branches 3 long.

Perennials with linear leaves; stems with nodes bearing leaves or bracts; panicle oblong usually large and lax, smaller in the South African species, with slender peduncles to the heads.

DISTRIB. Species 7; scattered through Tropical Africa and the Mascarene Islands.

1. E. rehmanniana (C. B. Clarke in Durand and Schinz, Conspect. Fl. Afr. v. 676); nearly glabrous, rather slender; rhizome woody; stems 10 in. long, clustered, obscurely trigonous, smooth; leaves 8–10 in. long, setaceous, subtrigonous in section; mouths of the split sheaths bearded by many white hairs; panicle 2–3 in. long, lax, with 8–16 spikes; lower bract like the stem leaf, but only 1–2 in. long; spikes $\frac{1}{4}$ in. long, ellipsoid, dense with very many spirally-arranged spikelets, dull brown; spikelets $\frac{1}{10}$–$\frac{1}{8}$ in. long, ellipsoid-lanceolate; glumes scarious, the yellow keel excurrent in a mucro; hypogynous bristles more than half the length of the nut, rigid, rather stout, papillose; style 3-fid. *Trilepis rehmanniana, Boeck. ex C. B. Clarke in Durand and Schinz, Conspect. Fl. Afr.* v. 676.

KALAHARI REGION: Transvaal; Houtbosch, *Rehmann,* 5624! Pretoria, *Rehmann,* 4469!

This African genus *Eriospora* is united by Boeckeler with the American genus *Fintelmannia* of Benth. and Hook. f. Gen. Pl. iii. 1068. In *Eriospora* the spikelets are 2-sexual, in *Fintelmannia* all are 1-sexual.

XXVII. SCHŒNOXIPHIUM, Nees.

Spikelets, some bisexual on nearly every plant, unisexual (male or female) often added. Bisexual *spikelet* with one female flower at the base and several males on a stalk springing obliquely within the so-called glume to the female flower. *Glume* to female ovate;

margins free or only connate a very little way up from the base. Axis of spikelet below the male inflorescence, elongate, flattened, scabrous on the margin. *Style* 3-fid. Otherwise as *Carex*.

Carex only differs from this genus by having the female utricles completely closed: but in a few androgynous Carices bisexual spikelets with the deeply-split utricle occur; and there is no line between these species and *Schœnoxiphium dregeanum*.—Stem nodose; panicle long; branches axillary distant; spikelets very numerous; lowest bracts cylindric, long-sheathing.

DISTRIB. Species 4, endemic in South Africa.

Glume entire:
- Nut ovoid ... ... (1) **rufum.**
- Nut narrowly ellipsoid ... ... (2) **dregeanum.**
- Nut linear-oblong ... ... (3) **capense.**

Glume bifid at the tip ... ... (4) **sickmannianum.**

1. **S. rufum** (Nees in Linnæa, x. 201); glabrous, slightly scabrous; rhizome woody, divided at the apex; stems 10–30 in. long, rather stout, somewhat trigonous; leaves 8–16 by $\frac{1}{6}$–$\frac{1}{4}$ in., flat, tough; panicle 6–10 in. long; spikelets very numerous, approximate or clustered, brown, $\frac{1}{3}$ in. long, lanceolate, some bisexual mostly in the lower spikes, 3–7-flowered; female glume nearly $\frac{1}{8}$ in. long, ovate, strongly 13-nerved, enclosing the nut, but the edges free to the base; rhachilla below the lowest male flower nearly as long as the nut, flat, 2–3-nerved, margined, with two earshaped processes at the top; nut scarcely $\frac{1}{12}$ in. long, ovoid, trigonous, smooth, pyramidal at the base and tip; style not $\frac{1}{2}$ the length of the nut, linear, branches 3 linear, longish. *Kunth, Enum.* ii. 532; *Boeck. in Linnæa,* xli. 354 *partly. S. Thunbergii, Nees in Linnæa,* x. 201 *partly. S. Ludwigii, Hochst. in Flora,* 1845, 764. *S. dregeanum, Drège, Zwei Pflanzengeogr. Documente* 49, *not of Kunth. Carex capensis, Thunb. Prod.* 14; *Fl. Cap. ed. Schult.* 90; *Nees in Linnæa,* vii. 535, *not of Schkuhr.*

SOUTH AFRICA: without locality, *Thunberg, Ecklon and Zeyher,* 150!
COAST REGION: Port Elizabeth, *Ecklon!* Queenstown Div.; on the Storm Bergen, 5000–6000 ft., *Drège!*
CENTRAL REGION: Somerset Div.; Bosch Berg, 4000 ft., *MacOwan,* 1866! Philipstown Div., *Ecklon!*
KALAHARI REGION: Orange Free State, *Buchanan,* 133!
EASTERN REGION: Pondoland, *Bachmann,* 116! Natal; Durban Flat, *Buchanan,* 43! and without precise locality, *Buchanan,* 94!

2. **S. dregeanum** (Kunth, Enum. ii. 529); margins of the leaves (in dried specimens) much recurved; rhachilla to the male part of the bisexual spikelet not much widened or eared at the top; nut $\frac{1}{8}$ in. long, narrowly ellipsoid; otherwise as *S. rufum. S. rufum, Boeck. in Linnæa,* xli. 354 *partly. S. Burkei, C. B. Clarke in Journ. Linn. Soc.* xx. 386, *t.* 30, *fig.* 8 (*nut shown rather too narrow*).

SOUTH AFRICA: without locality, *Ecklon and Zeyher,* 149! *Zeyher,* 1783!
CENTRAL REGION: Queenstown Div.; on the Storm Bergen, 5000–6000 ft., *Drège,* 7399! Cradock Div.; near Cradock, *Burke,* 211!

This species is united by Boeckeler with the preceding (*S. rufum*), and is doubtfully specifically distinct. Some utricles, in Zeyher, 1873, and in other examples, are very little split.

3. **S. capense** (Nees in Linnæa, vii. 533, excluding synonyms); robust, 3 ft. high; leaves more than $\frac{1}{4}$ in. broad, flat; edges very scabrous, cutting the hand; nut nearly $\frac{1}{6}$ in. long, linear-oblong; otherwise nearly as *S. rufum*, Nees. *Nees in Linnæa*, x. 202 *partly; Boeck. in Linnæa*, xli. 353 *partly. S. meyerianum, Kunth, Enum.* ii. 530, *not of C. B. Clarke. Hemicarex meyeriana, Benth. in Journ. Linn. Soc.* xviii. 367.

South Africa: without locality, *Bergius! Mundt! Ecklon and Zeyher*, 126!
Coast Region: Cape Div.; Table Mountain, 2000–3000 ft., *Drège*, 329! Swellendam Div.; on mountains along the lower part of the Zouder Einde River, 500–2000 ft., *Zeyher*, 4442!

4. **S. sickmannianum** (Kunth, Enum. ii. 530); plant rather slenderer than *S. capense*, with narrower, less scabrid leaves; spikelets rather larger; female glume $\frac{1}{4}$ in. long, elliptic, bifid; otherwise as *S. capense. S. capense, Boeck. in Linnæa*, xli. 353 *partly. S. meyerianum, C. B. Clarke in Journ. Linn. Soc.* xx. 386, *not of Kunth. Schœnus lanceus, Thunb. Prod.* 17; *Fl. Cap. ed. Schult.* 95. *Hemicarex sickmanniana, Benth. in Journ. Linn. Soc.* xviii. 367.

South Africa: without locality, *Thunberg! Sieber*, 119!
Coast Region: Cape Div.; Table Mountain, 1200–1800 ft., *Ecklon*, 851! *Bolus*, 2861! 7938! *Rehmann*, 678! *Schlechter*, 582! Worcester Div.; Dutoits Kloof, 2000–3000 ft., *Drège!*

XXVIII. CAREX, Linn.

Spikelets unisexual or rarely bisexual, female 1-flowered, male (at least in appearance) many-flowered, arranged in compound spikes or (the male often) solitary. *Female flower* completely enclosed in a utricle (really the bract to a division of the axis). *Stamens* 3–1; anthers linear-oblong. *Style*-branches 3 or 2. *Nut* trigonous or flattened.

Perennial herbs; leaves grass-like. Bisexual spikelets occur only in Subgenus II. *Eu-Carex*, Sect. 1.

Distrib. Species 1400; scattered throughout the World, but especially in moist places; hence fewer in South Africa than elsewhere.

Subgenus I. Vignea. Style 2-fid.
- Spikes small, ovoid, numerous, many 2-sexual, clustered:
  - Inflorescence a small oblong spike; utricle dull-brown ... ... (1) **divisa.**
  - Inflorescence larger; utricle bright-chestnut ... ... (2) **glomerata.**
  - Spike subpaniculate paler; utricle yellow-brown ... ... (3) **vulpina.**
- Spikes cylindric, 4–8, solitary, peduncled, terminal wholly male ... ... (4) **Phacota.**

Subgenus II. Eu-Carex. Style 3-fid.

Sect. 1. Spikes female at the base, male at the top, very rarely any wholly male. [Several of these species were formerly referred to *Schœnoxiphium.*]

Medium-sized plants, with few (or no) nodes between the basal-leaves and the inflorescence; spikes about 6–20 on a stem.
Utricles oblong; nut oblong or linear:
Utricle oblong-ellipsoid; nut oblong (5) **bisexualis.**
Utricle oblong-lanceolate; nut nearly linear... ... ... (6) **Zeyheri.**
Utricles ovoid or ellipsoid; nut ovoid, triquetrous:
Utricle not beaked:
Terminal spikes clustered ... (7) **dregeana.**
Inflorescence very narrow, straggling ... ... (8) **Bolusi.**
Utricle very short-beaked ... ... (9) **spartea.**
Utricle with long linear beak... ... (10) **esenbeckiana.**
Stout plants with many nodes; spikes 50–150 on a stem:
Utricle hairy:
Partial panicles pyramidal, open ... (11) **spicato-paniculata.**
Partial panicles of dense oblong spikes (12) **condensata.**
Utricle glabrous (enclosing usually a rhachilla)... ... ... (13) **Buchanani.**

Sect. 2. Spikes 3–8 (in *C. petitiana* sometimes 8–12) on a stem, terminal wholly male, lower female often male at the tip. But, in *C. cognata* (and others) occasionally, in *C. petitiana* nearly always, the terminal male spike is female at the top.

Female spikes 4–7 in. long, slender, flexuose, pendulous ... ... (14) **petitiana.**
Female spikes less (usually much less) than 3 in. long:
Utricle narrowed conically upward, hardly beaked:
Rather slender; terminal male spike 1 (15) **burchelliana.**
Stout, 2–3 ft. high; terminal male spikes 2 ... ... (16) **acutiformis.**
Utricle more or less suddenly contracted into a beak:
Spikes sessile near together, one lowest sometimes distant, but then with peduncle very shortly exserted:
Female spikes ovoid or short-cylindric:
Beak of utricle short ... (17) **extensa.**
Beak of utricle long ... (18) **flava.**
Female spikes $1\frac{1}{2}$ in. long, narrowly cylindric... ... ... (19) **cognita.**
Spikes (at least the lower female) distant, lowest often long peduncled:
Utricle, even when very young, brown ... ... (20) **æthiopica.**
Utricle, at least till nearly ripe, green:
Female spikes thick, erect, peduncle scarcely exserted (21) **clavata.**
Female spikes nodding, peduncles much exserted ... (22) **drakensbergensis.**

1. C. divisa (Huds. Fl. Angl. 348); glabrous; rhizome 1–8 in. long, horizontal, woody; stems 6–18 in. long; leaves often $\frac{3}{4}$ the length of the stem, $\frac{1}{10}$–$\frac{1}{6}$ in. broad; inflorescence $\frac{1}{2}$–$1\frac{1}{2}$ by $\frac{1}{4}$–$\frac{1}{2}$ in., a compound spike, lowest bract $\frac{1}{4}$ in. long, or sometimes much overtopping the inflorescence; spikes ovoid, many bisexual, female at the base, male at the top; female glumes ovate, acuminate or short-mucronate, brown; style 2-fid; utricle ovate, acuminate, flattened, moderately nerved; beak short, bifid, scabrous-edged; nut flat, ovoid, truncate, much shorter than the utricle. *Kunth, Enum.* ii. 372; *Boott, Carex,* iv. 186; *Boeck. in Linnæa,* xxxix. 54; *C. B. Clarke in Hook. f. Fl. Brit. Ind.* vi. 701. *C. rivularis, Schkuhr, Riedgr.* i. 30, *t. Cc fig.* 87. *C. consanguinea, Kunth, Enum.* ii. 374. *Vignea divisa, Reich. Fl. Excurs.* 58.

WESTERN REGION: Little Namaqualand; between Pedros Kloof and Lily Fontein, 4000 ft., *Drège,* 2450!
Widely distributed.

In the South African form (*C. consanguinea,* Kunth) the angles of the stem are closely scabrous, and the anthers have a long linear white crest; in the typical European *C. divisa,* Hudson, the angles of the stem are remotely scabrous or smooth, and the anthers are muticous or scarcely apiculate.

2. C. glomerata (Thunb. Prod. 14); glabrous; rhizome 1–$1\frac{1}{2}$ in. long, horizontal, woody; stems 10–20 in. long, smooth or nearly so on the 3 angles at the top; leaves often $\frac{3}{4}$ the length of the stem, $\frac{1}{8}$–$\frac{1}{6}$ in. broad; the interrupted compound spike 1 by $\frac{1}{3}$–$\frac{1}{2}$ in. (in fruit); lowest bract $\frac{1}{4}$–$\frac{3}{4}$ in. long, setaceous, spreading, rarely overtopping the inflorescence; spikes ovoid, many bisexual, female at the base, male at the top; female glumes ovate, acuminate or short-mucronate, brown; style 2-fid; utricle ovate, acuminate, plane-convex, green then chestnut, moderately nerved on both faces; beak short, bifid, with a green scabrous margin at the base; nut flat, ovoid truncate, much shorter than the utricle. *Thunb. Fl. Cap. ed. Schult.* 90; *Nees in Linnæa,* vii. 534; x. 203; *Kunth, Enum.* ii. 384; *Boott, Carex,* ii. 81, *t.* 222; *Boeck. in Linnæa,* xxxix. 59 *partly (i.e. the African plant and synon.); not of Schkuhr nor of Host. C. vulpina, var. β, Wahl. in Act. Holm.* 1803, 144.

SOUTH AFRICA: without locality, *Verreaux!*
COAST REGION: Riversdale Div.; Great Vals River, *Burchell,* 6554! Albany Div.; near Grahamstown, *MacOwan!* Uitenhage Div.; Zwartkops River, *Ecklon and Zeyher,* 187!
KALAHARI REGION: Orange Free State, *Buchanan,* 97!

This species differs little from *C. divisa;* it is rather larger, with a larger inflorescence; the chief obvious difference is the larger, brighter-chestnut, green-margined utricle, which also separates *C. glomerata* from the closely allied *C. muricata,* Linn., which has yellow-brown utricles and the spike more interrupted. As to the synonymy given above, it refers at least partly to the following species.

3. C. vulpina (Linn. Sp. Plant. ed. i. 973); rather stouter than *C. glomerata;* stems scabrous at the top; leaves broader, usually $\frac{1}{4}$ in.

and more broad; inflorescence larger, paler, often 2 in. long or more with the lower branches of the panicle growing out, frequently $\frac{1}{2}$ in. long and more; utricle yellow, finally brown, not shining chestnut-brown; edges upward and base of beak strongly margined; otherwise as *C. glomerata*, Thunb. *Schkuhr, Riedgr.* i. 17 *and* ii. 12, *t. O fig.* 10; *Kunth, Enum.* ii. 383; *Boeck. in Linnæa*, xxxix. 89. *C. glomerata, Drège, Zwei Pflanzengeogr. Documente* 56, 58, *and probably of other authors in part.*

COAST REGION: Queenstown Div.; Shiloh, 3500 ft., *Baur*, 823! 1136!
CENTRAL REGION: Victoria West Div.; Nieuw Veld, between Brak River and Uitvlugt, 3000–4000 ft., *Drège!* Richmond Div.; vicinity of Styl Kloof, near Richmond, 4000–5000 ft., *Drège!*

Also in Europe, West Asia and North Africa.

4. **C. Phacota** (Spreng. Syst. iii. 826); glabrous; rhizome woody, creeping; stem 1–2$\frac{1}{2}$ ft. long; leaves often as long as the stem, $\frac{1}{6}$–$\frac{1}{4}$ in. wide; spikes 3–8, long-peduncled, 1–2$\frac{1}{2}$ in. long, linear-cylindric, dense, terminal wholly male, the others female, frequently male at the top; lowest bract overtopping the inflorescence or sometimes shorter than it; female glumes 3-nerved, with a rough excurrent lanceolate tip, greenish; style 2-fid, short; utricle flat, ovoid, beakless, triangular-pointed, nerveless, rough to its base with red glands. *Kunth, Enum.* ii. 420; *Boott, Carex*, i. 63, *t.* 168; *Boeck. in Linnæa*, xl. 435; *C. B. Clarke in Hook. f. Fl. Brit. Ind.* vi. 708.

KALAHARI REGION: Transvaal; Pretoria, *Rehmann*, 4039! Orange Free State; Nelsons Kop, *Cooper*, 909! 3335!
EASTERN REGION: Natal; in a marsh by the Mooi River, *Wood*, 4038! *MacOwan, Herb. Aust.-Afr.*, 1690!

Also in India, Malaya and Japan.

The South African examples differ from the Indian in having the spikes paler, larger (often $\frac{1}{4}$ in. wide), the utricles longer and more acuminate, and the female glumes considerably longer, often much exceeding the utricles; Boeckeler, however, has marked the Cape plant "*C. Phacota*, Spreng." simply, and the differences indicate a "geographic form" only.

5. **C. bisexualis** (C. B. Clarke); nearly glabrous; rhizome woody; stem 4–12 in. long, rather slender, trigonous, without any node between the basal-leaves and the inflorescence; leaves often nearly as long as the stem, $\frac{1}{8}$ in. broad; spike compound, 1–1$\frac{1}{2}$ by $\frac{1}{8}$ in., rather dense, of 10–40 spikelets; lowest bract 2–4 in. long, similar to the leaves, hardly sheathing at the base; spikelets mainly 1-sexual, a few 2-sexual (i.e. the rhachilla to the male upper part of the spikelet being within the utricle); 2-sexual spikelet $\frac{1}{4}$ in. long, greenish; utricle $\frac{1}{8}$–$\frac{1}{6}$ in. long, trigonous, oblong-ellipsoid subacute, smooth, many-nerved, greenish; nut nearly filling the utricle, narrowly ellipsoid. *Schœnoxiphium Thunbergii, Nees in Linnæa*, x. 201 *partly; Kunth, Enum.* ii. 531; *Boeck. in Linnæa*, xli. 355 *mainly; C. B. Clarke in Journ. Linn. Soc.* xx. 388, *t.* 30, *fig.* 9–11 (*nut drawn too broad*). *S. Ecklonii, Nees in Linnæa*, x. 200 *partly. Hemicarex Thunbergii, Benth. in Journ. Linn. Soc.* xviii. 367.

COAST REGION: Cape Div.; Lion Mountain, *Ecklon*, 853! *Pappe!* near Cape Town, *Harvey*, 196!

This is placed in *Carex* because the utricle is complete, and more especially because it is very near *C. dregeana.* The utricles in the type example are ripe and all unisexual; the drawing (in Journ. Linn. Soc. xx. t. 30) was taken from Pappe's example which has a few unripe bisexual spikelets: it consequently shows the utricle split on one side at the top.

6. **C. Zeyheri** (C. B. Clarke); leaves scarcely $\frac{1}{10}$ in. broad; inflorescence shortened, subovoid, of 15–30 spikelets; utricles $\frac{1}{4}$ in. long and upwards, narrow-oblong, acuminate; nut $\frac{1}{6}$ in. long, linear-oblong curved; otherwise as *C. bisexualis. Schænoxiphium Ecklonii, Nees in Linnæa,* x. 200 *partly; Kunth, Enum.* ii. 531; *C. B. Clarke in Journ. Linn. Soc.* xx. 388. *S. Thunbergii, Nees in Linnæa,* x. 201 *partly; Boeck. in Linnæa,* xli. 355 *partly. Hemicarex Ecklonii, Benth. in Journ. Linn. Soc.* xviii. 367.

COAST REGION: Swellendam Div.; on mountains along the lower part of the Zonder Einde River, 500–2000 ft., *Zeyher*, 4441! Uitenhage Div.; Grass Ridge, *Ecklon!*

7. **C. dregeana** (Kunth, Enum. ii. 511); glabrous; rhizome woody, short; stems clustered, 6–14 in. long, with 0 or 1 node between the basal-leaves and inflorescence; leaves $\frac{1}{2}$–$\frac{3}{4}$ the length of the stem by $\frac{1}{10}$ in. wide; inflorescence long, interrupted, with axillary peduncles 3–6 in. distant, or reduced to an oblong quasi-terminal dense compound spike 2 by $\frac{1}{2}$ in., of 10–50 spikelets; bracts usually overtopping the inflorescence; spikelets all unisexual; female spikes somewhat dense; utricles $\frac{1}{10}$ in. long, obovoid, trigonous, many-striate, smooth, ultimately yellow-brown, suddenly narrowed near the top into a short bifid cone, not beaked, not containing a rudiment in any one examined. *Drège, Zwei Pflanzengeogr. Documente* 136, 138; *in Linnæa,* xx. 252. *C. spartea, Boeck. in Linnæa,* xl. 370 *excl. syn.; Schkhur, Riedgr. t. Bb fig.* 86, *i only.*

VAR. **major** (C. B. Clarke); larger in all its parts; spikes and spikelets almost turgid; nut attaining $\frac{1}{10}$ by $\frac{1}{12}$ in.

COAST REGION: Tulbagh Div.; between Tulbagh and "the Drostdy," *Burchell*, 1041! Swellendam Div.; near Swellendam, 1000–4000 ft., *Zeyher*, 4440! Alexandria Div.; Zuur Berg Range, 2000–3000 ft., *Drège!* Albany Div.; Grahamstown, *Alexander!* Var. *β*: Bathurst Div.; between Port Alfred and Kaffir Drift, *Burchell*, 3853! Cathcart Div.; near Cathcart, 4500 ft., *Kuntze!*

CENTRAL REGION: Somerset Div.; Klein Bruintjes Hoogte, 3000–4000 ft., *Drège*, 2033!

EASTERN REGION: Tembuland; Umnyolo, near Bazeia, 3000 ft., *Baur*, 744! Natal; Inanda, *Wood*, 1353!

This species is here diagnosed by the beakless approximated utricles; several spikes are close together, forming an oblong or obpyramidal terminal head, while 1–3 distant axillary spikes are sometimes added. No bisexual spikelets have been found, but the analogy of the allied species shows this character to be of small value.

8. C. **Bolusi** (C. B. Clarke); stems up to 20 in. long, very slender; inflorescence 5–10 in. long, very slender, with remote, thin spikes; utricles somewhat distant, subsolitary; otherwise as *C. dregeana*.

COAST REGION: Bathurst Div.; Port Alfred, *Hutton!*
CENTAL REGION: Graaff Reinet Div.; Cave Mountain, 4300 ft., *Bolus*, 1974!
EASTERN REGION: Natal, *Buchanan*, 328! *Rehmann*, 8262!

This has the beakless utricle of *C. dregeana*, but not the terminal compound close inflorescence of that species.

9. **C. spartea** (Wahlenb. in Vet. Akad. Nya Handl. Stockholm, 1803, 149); stems 15–24 in. long, slender; inflorescence 5–10 in. long, lax; terminal spike pyramidal, of approximated spikelets; spikelets many unisexual; utricles frequently containing a rudimentary process, sometimes enclosing the rhachilla of a male spike; utricle ellipsoid, trigonous, acuminated into a very short conic beak; otherwise as in *C. drègeana*, Kunth. *Schkuhr, Riedgr.* ii. 40, *t. Bb fig.* 86 (*except perhaps i*); *Thunb. Fl. Cap. ed. Schult.* 90; *C. indica, Schkuhr, Riedgr.* i. 37, *not of Linn. C. Sprengelii, Boeck. in Linnæa*, xl. 371. *Uncinia spartea, Nees in Linnæa*, x. 205. *U. Sprengelii, Nees in Linnæa*, x. 205.

SOUTH AFRICA: without locality, *Thunberg.*
COAST REGION: Riversdale Div.; near the Zoetemelks River, in a walk to the White-clay Pit, *Burchell*, 6643!
KALAHARI REGION: Transvaal; Pretoria, *Rehmann*, 4041!

The Tropical African *C. schimperiana* hardly differs except by the absence (in the small quantity of it seen) of bisexual spikelets.—Hardly separable from the preceding species.

10. C. esenbeckiana (Boeck. in Linnæa, xl. 372); stems 15–24 in. long, slender; leaves $\frac{2}{3}$ the length of the stem, weak, $\frac{1}{8}$–$\frac{1}{5}$ in. broad, green; inflorescence 5–10 in. long, very lax, thin, with remote straggling axillary spikes; spikelets loose and irregular, bisexual and unisexual; utricles subsolitary, often (beak included) $\frac{1}{6}$–$\frac{1}{4}$ in. long; beak $\frac{1}{2}$ the length of the utricle, linear; otherwise as *C. spartea. C. spartea, Kunth, Enum.* ii. 511 *excl. syn.*; *Drège, Zwei Pflanzengeogr. Documente* 134. *Uncinia Lehmanni, Nees in Linnæa*, x. 206. *Schœnoxiphium? Lehmanni, Kunth, Enum.* ii. 528; *Steud. Syn. Pl. Glum.* ii. 245.

COAST REGION: Alexandria Div.; Enon, 1000–2000 ft., *Drège!* Albany Div.; woods near Grahamstown, 2000 ft., *MacOwan*, 1206! King Williams Town Div.; Perie Forest, 2000 ft., *Kuntze*, 287!
EASTERN REGION: Natal; Van Reenens Pass, 5800 ft., *Kuntze*, 290!

There are adjacent in one spike—(1) linear male whitish spikelets $\frac{1}{2}$ in. long, (2) utricles with the linear beak shortly bifid at the top, either solitary or with a male spikelet rising close to their base outside, (3) "bisexual" schœnoxiphioid spikelets in which the rhachilla of the $\frac{1}{2}$ in. white male spikelet is included in the utricle, which then is split down sometimes deeply on one side to allow the rhachilla to emerge.

11. C. spicato-paniculata (C. B. Clarke in Durand and Schinz, Conspect. Fl. Afr. v. 690); stout; stem 2–3 ft. long; leaves rather numerous, 18 by $\frac{1}{4}$–$\frac{1}{3}$ in., scabrous on the margins, and often on the keel beneath; inflorescence 1 ft. long, of 4–5 axillary, peduncled, pyramidal, compound panicles; panicle-branches densely pilose; spikes 50–150 to a stem, $\frac{1}{3}$ in. long, ovoid in fruit, subsolitary, ferruginous when young, in fruit with green utricles and ferruginous glumes, many bisexual, male at the top and female at the base with 4–8 utricles, style 3-fid; utricle, including beak, $\frac{1}{6}$ in. long, ellipsoid, trigonous, many-ribbed, hispid nearly to the base, straight; beak linear, half the length of the utricle, round-notched, with 2 linear teeth; nut filling the utricle except the beak.

KALAHARI REGION: Orange Free State, *Buchanan*, 98! on the Drakens Berg Range, *Cooper*, 1066! Transvaal; Houtbosch, *Rehmann*, 5627!
EASTERN REGION: Natal; Inanda, *Wood*, 1190! and without precise locality, *Buchanan*, 350! 355!

Also in Tropical Africa.

The examples of this species are exactly alike, and easily distinguished from any other South African *Carex*. But this section of *Carex* (*Indicæ* of authors) comprises about 20 species in India and 10 in Tropical Africa, abounding in individuals and varieties, and very closely allied. The nearest African species to *C. spicato-paniculata* is *C. ramosa*, Schkuhr, which has narrower, more elongate, curved, less hairy utricles. The nearest Indian species is *C. cruciata*, Wahl., which has more ovoid, more suddenly-narrowed, less hairy utricles.

12. **C. condensata** (Nees in Wight, Contrib. 123); axillary peduncled panicles oblong, dense, ferruginous; spikes usually with 1–4 utricles at the base and numerous males above; utricles ovoid, more or less hairy; otherwise as *C. spicato-paniculata*, C. B. Clarke. *Boott, Carex*, ii. 86, *tt.* 247, 248; *C. B. Clarke in Hook. f. Fl. Brit. Ind.* vi. 716.

KALAHARI REGION: Orange Free State, *Buchanan*, 150!
EASTERN REGION: Tembuland; in forests on Bazeia Mountain, 2000–2500 ft., *Baur*, 444! 1156! Natal, *Buchanan*, 149! 353!

Also in Northern India.

In this species nearly all of the fruiting spikes have a conspicuous lanceolate male top, while in *C. spicato-paniculata* the fruiting spikes appear to be entirely female, or with a very few ferruginous glumes at the top.

13. **C. Buchanani** (C. B. Clarke); panicle 13 by 2 in., interrupted, of 4–5 distant, rather loose, axillary, peduncled panicles; branches more or less hairy; spikelets $\frac{1}{4}$–$\frac{1}{3}$ in. long, elliptic, brown, male, female, or bisexual; utricle in the female spikelet oblong-ellipsoid, scarcely trigonous, much narrowed at the top, scarcely beaked, nearly nerveless, glabrous, scarcely split on one side (but in the bisexual spikelet more or less split); nut nearly $\frac{1}{8}$ in. long, broad-oblong, suddenly narrowed into a short-cylindric neck. *Schœnoxiphium Buchanani, C. B. Clarke in Durand and Schinz, Conspect. Fl. Afr.* v. 676.

EASTERN REGION: Natal, *Buchanan*, 134!

The material consists merely of the top of one stem. The utricle, when female, usually contains a flat rudimentary rhachilla. The plant is closely allied to *Schœnoxiphium rufum*. The genus is doubtful, but the nut renders it specifically distinct.

14. C. petitiana (A. Rich. Tent. Fl. Abyss. ii. 513); glabrous; stems stout, 2–5 ft. long; leaves 12–24 by $\frac{1}{2}$–$\frac{2}{3}$ in., closely striate longitudinally, without transverse lineolation; inflorescence 6–14 in. long, of 4–9 rusty-brown spikes; uppermost sometimes wholly male, frequently female at the top, lower female 4–7 by $\frac{1}{4}$–$\frac{1}{3}$ in. cylindric, pendulous peduncled; glumes lanceolate, acuminate or mucronate, brown, much overtopping the utricles; utricle $\frac{1}{10}$ in. long, ellipsoid, scarcely trigonous, stalked, slenderly few-ribbed, smooth, yellowish-brown, often with microscopic glandular dots; beak hardly $\frac{1}{6}$ the length of the utricle, linear, scarcely notched; style 3-fid; nut obovoid, trigonous, much shorter than the utricle. *Boott, Carex,* ii. 88, *t.* 259; *Boeck. in Linnæa,* xl. 411. *C. pendula, C. B. Clarke in Durand and Schinz, Conspect. Fl. Afr.* v. 688 *partly, i.e. as to all the South Africa plants cited.*

COAST REGION: King Williamstown Div.; Perie Forest, 1900 ft., *Kuntze!* British Kaffraria, *Cooper,* 288!
CENTRAL REGION: Somerset Div.; in woods on the sides of Bosch Berg, 4000 ft., *MacOwan,* 1608!
KALAHARI REGION: Orange Free State, *Cooper,* 3336!
EASTERN REGION: Tembuland; forests on Bazeia Mountain, 2500 ft., *Baur* 443! Natal, *Buchanan,* 68!

Also in Abyssinia.

This has altogether the habit of *C. pendula,* Hudson, and the South African plants have been so named hitherto. Boott l.c. doubts whether *C. petitiana,* A. Rich., is other than a form of *C. pendula,* Hudson. In the latter the terminal spike is (except accidentally) wholly male; the glumes to the utricles are much shorter and brighter-coloured; the utricle is less stalked; and the leaves are transversely lineolate. At all events, if *C. petitiana* be kept up as a species, as is done by Boott and Boeckeler, all the South African examples agree minutely with it—not with *C. pendula.*

15. C. burchelliana (Boeck. in Linnæa, xli. 234); glabrous; stems 10–18 in. long, nearly smooth; leaves 4–10 by $\frac{1}{10}$–$\frac{1}{6}$ in., rather tough, thinly scabrous on the margins and keel beneath; spikes 3–5, all within 2 inches of the top of the stem or (more frequently) the lowest 4–6 in. distant; terminal spike $\frac{1}{2}$–$1\frac{1}{4}$ in. long, oblong, brown, male or (in two cases) with a few utricles at the base; male glumes oblong, obtuse, with the green keel sometimes shortly excurrent; female spikes $\frac{1}{2}$–$\frac{3}{4}$ by $\frac{1}{4}$ in., suberect, peduncle subincluded; glumes oblong or elliptic, brown, with the 3-nerved green keel excurrent in a rough mucro, hardly so long as the utricle; style 3-fid; utricles $\frac{1}{10}$ in. long, ellipsoid, trigonous, narrowed upwards into a very short beak, 8–10-nerved, dotted with red glands, glabrous; beak and tip of the utricle minutely papillose-scabrous; nut obovoid, trigonous, on a slender stalk. *C. flavescens, Burchell, Trav.* i. 467, *name only.*

KALAHARI REGION: Griqualand West; Herbert Div., at Upper Campbell, *Burchell,* 1831! Hay Div., at Griqua Town, *Burchell,* 1911!

This plant resembles *C. diluta,* M. Bieb., which has been found to be identical with the European *C. punctata,* Gaud. (cf. C. B. Clarke in Hook. f. Fl. Brit. Ind. vi. 737). It differs, as Boeckeler states, by the slender stalk to the nut which cannot be found in *C. diluta.*

16. **C. acutiformis** (Ehrh. Beitr. iv. 43); glabrous; stems 3–2 ft. long, robust, scabrous at the top; leaves often $\frac{2}{3}$ the length of the stem, $\frac{1}{4}$–$\frac{1}{3}$ in. broad; spikes 3–7, distant, erect, male 2–1, $1\frac{1}{2}$ by $\frac{1}{4}$ in., chestnut, females $1\frac{1}{2}$–2 by $\frac{1}{4}$ in., lowest on a 1–3 in. peduncle, but very erect; glumes of female spikes oblong-lanceolate, caudate, brown or chestnut-brown, overtopping (in Cape examples) the utricles; style-branches 3, rather short; utricles ovoid, trigonous, 7–9-nerved, glabrous, minutely granular, especially upwards, triangular acuminate, hardly beaked; nut broadly obovoid-globose, scarcely $\frac{2}{3}$ the length of the utricle. *Boeck. in Linnæa,* xli. 289; *C. B. Clarke in Hook. f. Fl. Brit. Ind.* vi. 740. *C. paludosa, Gooden. in Trans. Linn. Soc.* ii. 202; *Kunth, Enum.* ii. 487.

CENTRAL REGION: Somerset Div.; on the summit of Bosch Berg, 4500 ft., *MacOwan,* 1963!

KALAHARI REGION: Transvaal; Marico District, on the banks of the Matebe River, *Holub,* 1558! 1559!

Widely distributed in the North Temperate Zone.

The South African plant differs a little from the European in having the female glumes much more aristate, the utricles nearly or quite beakless, the nut broader than long and very truncate. Holub's three examples have the terminal spike female at the top.

17. **C. extensa** (Gooden. in Trans. Linn. Soc. ii. 175, t. 21, fig. 7); glabrous; stems 6–18 in. long, nearly smooth; leaves usually as long as the stem, $\frac{1}{10}$–$\frac{1}{5}$ in. broad, thinly scabrous; spikes 3–4 within an inch of the top of the stem, $\frac{1}{2}$–$\frac{3}{4}$ in. long, terminal male, the others female; bracts long, often overtopping the inflorescence by 4–7 in.; female spikes $\frac{1}{2}$ by $\frac{1}{4}$ in. (but, in Buchanan 167, $1\frac{1}{3}$ by $\frac{1}{4}$ in., not crowded, lowest $1\frac{3}{4}$ in. distant); glumes oblong, often truncate, brown, with the 3-nerved brown keel often excurrent as a tough point hardly reaching the top of the utricle; male glumes nearly the same as the female; style 3-fid; utricle $\frac{1}{6}$ in. long, ellipsoid trigonous, 10–13-nerved, glabrous, narrowed into a short conic beak about $\frac{1}{4}$ the length of the utricle. *Schkuhr, Riedgr.* i. 74; ii. 56, *t. Xx.* 72, *and t.* V. 72; *Kunth, Enum.* ii. 447; *Boeck. in Linnæa,* xli. 288. *C. Ecklonii, Nees in Linnæa,* x. 203; *Kunze, Suppl. Riedgr.* 25, *t.* 5; *Kunth, Enum.* ii. 517.

SOUTH AFRICA: without locality, *Bergius! Verreaux! Harvey,* 349! *Pappe! Auerswald!*

COAST REGION: Port Elizabeth Div.; common along the coast, *E.S.C.A. Herb.,* 177!

EASTERN REGION: Natal; near Greytown, *Buchanan,* 167!

Widely distributed in Europe, North Africa, the Orient, and North and South America.

The South African material agrees with the largest European examples; having the keel of the glumes strongly excurrent, the beak rather long, and has both narrow and broader leaves. Boott agrees with Boeckeler that *C. Ecklonii* cannot be kept distinct from *C. extensa.*

18. **C. flava** (Linn. Sp. Plant. ed. ii. 1384); glabrous; stems 4–24 in. long; leaves $\frac{2}{3}$ the length of the stem, narrow; spikes 3–5, all (or

all but one) usually clustered, subterminal; terminal spike male, the rest female, or with a few males at the top, $\frac{1}{3}$–$\frac{2}{3}$ in. long, short-cylindric or globose, dense, yellow; glumes ovate-oblong, obtuse; utricles subglobose, inflated, glabrous, minutely granular, ribbed, suddenly narrowed into the beak; beak $\frac{1}{3}$–$\frac{2}{3}$ the length of the utricle, linear, curved, often deflexed. *Schkuhr, Riedgr.* i. 72; ii. 56, *t. H, fig.* 36; *Engl. Bot. t.* 1294; *Kunth, Enum.* ii. 446; *Boeck. in Linnæa,* xli. 272. *C. Œderi, Willd. in Act. Berol.* 1794, 44, *t.* 1, *fig.* 2; *Schkuhr, Riedgr.* i. 67, *t. F, fig.* 26.

KALAHARI REGION: Basutoland; Mont aux Sources, 9500 ft., *Flanagan,* 2013!

Widely distributed in Europe and the Orient to Kashmir, and in North America.

19. C. cognata (Kunth, Enum. ii. 502); stem 3 ft., top triquetrous scabrous; leaves 1–2$\frac{1}{2}$ ft. long, $\frac{1}{3}$–$\frac{1}{2}$ in. broad, midrib sharply projecting beneath, tissue lax transversely lineate; spikes about 7, pale, close together near the top of the stem, sessile, erect (lowest 1 in. distant), topmost wholly male (or occasionally female at the top); spikes in fruit 1$\frac{1}{2}$ by nearly $\frac{1}{3}$ in., dense; glumes from near middle of spike, lanceolate, green with white margins, the hispid arista long but hardly reaching the top of the utricle; utricles spreading, not deflexed, ovoid-subglobose, suddenly narrowed into a slender beak much shorter than the utricle, with about 12 strong nerves, tissue thin shining glabrous; beak smooth with 2 rather long spreading smooth linear teeth; nut smaller than the glume, trigonous, dull-brown, obovoid hardly longer than broad. *Boeck. in Linnæa,* xli. 299. *C. retrorsa, Nees in Linnæa,* x. 204, *not of Schweinitz. C. Pseudo-cyperus, var., Boott, Carex,* iv. 141.

COAST REGION: Swellendam and George, *Mund* fide *Nees.*
EASTERN REGION: Delagoa Bay; *Junod,* 414.

This differs from *C. Pseudo-cyperus,* Linn., by the broader utricles and nuts; it is very near *C. sphærogyne,* Baker (from Madagascar), which has narrower leaves.

20. C. æthiopica (Schkuhr, Riedgr. i. 107, t. Z, fig. 83); glabrous; stems 1$\frac{1}{2}$–4 ft. long; basal leaves 1–2$\frac{1}{2}$ ft. by $\frac{1}{4}$–$\frac{1}{2}$ in.; spikes 3–7, lowest 3–9 in. distant, top 1–2 male, others female; male spike 1–1$\frac{1}{2}$ by $\frac{1}{6}$ in., rusty-brown; keel of glumes excurrent; female spikes 1–2$\frac{1}{4}$ by $\frac{1}{4}$ in., greenish, finally pale brown; female glumes oblong-elliptic, 3-nerved; keel excurrent, overtopping the utricle; style 3-fid; utricles scarcely $\frac{1}{6}$ in. long, ellipsoid, trigonous, narrowed into a short smooth shortly bifid beak, glabrous, slenderly nerved, even when very young red-brown, in fruit with red glands; nut obovoid, trigonous, sessile, $\frac{2}{3}$ the length of the utricle exclusive of its beak. *Boott, Carex,* iii. 110, *t.* 341–343; *Boeck. in Linnæa,* xli. 285. *C. iridifolia, Kunth, Enum.* ii. 492.

VAR. β **latispica** (C. B. Clarke); spikes larger, female $\frac{1}{3}$ in. broad and upwards; utricles more than $\frac{1}{6}$ in. long, acutely triquetrous, narrowed into a longer beak.

SOUTH AFRICA: without locality, *Zeyher*, 743!

COAST REGION: Cape Div.; Devils Mountain, near the waterfall, 200 ft., *Bolus*, 3848! George Div.; in the forest near George, *Burchell*, 6048! Knysna Div.; Ruigte Valley, below 500 ft., *Drège*, 7398! Uitenhage Div.; in the forests on Van Stadens Berg, *Ecklon and Zeyher*, 684! Var. β: Bathurst Div.; near Kaffir Drift, *Burchell*, 3869! Albany Div.; in woods on Bothas Hill, 2200 ft., *MacOwan*, 1013!

The synonymy of this species, called *C. lævigata* by Wahl. and Kunth, is given by Boott (Carex, iii. 110), where it can also be seen that the type of Schkuhr's *C. æthiopica* was collected by Thunberg in South Africa, and not in the Island of Bourbon as originally stated by Schkuhr.

21. C. clavata (Thunb. Prod. 14); glabrous; stems 2–4 ft. long; stout, triquetrous at the top, rough; leaves 1–2 ft. long, $\frac{1}{4}$–$\frac{1}{3}$ (sometimes more than $\frac{1}{2}$) in. broad; spikes 4–6, lower distant, suberect; bracts very long; peduncles short, lowest sometimes 3 in. long, but not much exserted; terminal spike (sometimes the next also) male, bright-brown, sometimes with a few females at the top; female spikes 1–2$\frac{1}{2}$ by $\frac{1}{3}$ in., dense; female glumes ovate; arista reaching nearly to the top of the beak; utricles (including beak) $\frac{1}{5}$ in. long, ellipsoid, narrowed rather suddenly into the beak, striated, glabrous, when young greenish, in fruit sometimes deflexed; beak about $\frac{1}{4}$ the length of the utricle, deeply bifid into nearly linear teeth; nut $\frac{1}{10}$ in. long, obovoid, triquetrous, dark-brown, nearly filling the utricle. *Fl. Cap. ed. Schult.* 90; *Schkuhr, Riedgr.* ii. 55; *Kunze, Suppl. Riedgr.* 67, *t.* 17; *Nees in Linnæa*, vii. 535; x. 204; *Kunth, Enum.* ii. 495; *Boeck. in Linnæa*, xli. 298. *C. vesicaria, Thunb. Prod.* 14; *Fl. Cap. edit. Schult.* 90, *not of Linn.* *C. lutensis, Kunth, Enum.* ii. 487. *C. macrocystis, Boeck. Cyp. Novae*, i. 50 (?).

SOUTH AFRICA: without locality, *Thunberg*, *Pappe*, 77! *Harvey*, 346! 365! *Thom*, 894! *R. Brown!* *Drège*, 4367!

COAST REGION: Malmesbury Div.; Groene Kloof, below 500 ft., *Drège!* Cape Div.; between the Lions Head and Table Mountain, *Burchell*, 275! by the river leading from Table Mountain, *Milne*, 239! Table Mountain, 800 ft., *Bolus*, 7939! Simons Bay, *MacGillivray*, 424! *Wright!* Wynberg, *Wallich!* Paarl Div.; Paarl Mountains, 1000–2000 ft., *Drège*, 1563! Port Elizabeth Div.; common along the coast, *E.S.C.A. Herb.*, 174! Bathurst Div.; near Theopolis, *Burchell*, 4137! at the source of the Kasuga River, *Burchell*, 3904!

KALAHARI REGION: Orange Free State, *Buchanan*, 139!

No authentic example of *C. macrocystis*, Boeck., has been seen. As the utricle was $\frac{1}{4}$ in. long, it must either have been *C. clavata*, Thunb., or some entirely new large species, which latter alternative is improbable, as Boeckeler found his *C. ma crocystis* in the collection of Ecklon and Zeyher.

22. C. drakensbergensis (C. B. Clarke); stems 3 ft. long, stout, triquetrous at the top, rough; leaves 1–2 ft. by $\frac{1}{4}$–$\frac{1}{3}$ in., closely longitudinally striate, not (or most obscurely) transversely lineolate; spikes 5–7, lower distant 3–8 in.; bracts very long; lower peduncles often 3–4 in. long (and top female spike usually distinctly peduncled); terminal spike male, or with a very few females at the top; glumes $\frac{1}{3}$ in. long, narrow-lanceolate, pale-brown; female spikes 1–3 by $\frac{1}{4}$–$\frac{1}{3}$ in., dense, pale-brown in fruit; female glumes elliptic, bright-brown; arista reaching the top of the beak; style 3-fid;

utricle $\frac{1}{6}$ in. long (beak included), ellipsoid, rather suddenly narrowed, with 1–8 very strong ribs, glabrous; beak $\frac{1}{3}$ the length of the utricle or rather more, deeply bifid into almost linear teeth; nut nearly filling the utricle.

KALAHARI REGION: Orange Free State; on the Drakens Berg, near Harrismith, *Buchanan*, 112! 136! Transvaal; Mooi River, near Potchenstroom, *Nelson*, 72*!

EASTERN REGION: Griqualand East; river banks near Kokstad, *Haygarth in Herb. Wood*, 4201! Natal, *Buchanan*, 137!

ORDER CLI. GRAMINEÆ.

(BY O. STAPF.)

Partial inflorescences (*spikelets*) consisting of an axis (*rhachilla*) and, typically, of 3 or more alternate, distichous, more or less heteromorphous bracts, of which the 2 lowest (*glumes*) form an involucre to the spikelet and are empty, whilst the following (*valves*) bear in their axils subsessile flowers, subtended by a usually hyaline, 2-keeled or 2-nerved dorsal bract (*pale*); valves differing usually in structure and size from the glumes, and forming with the pale and the flower proper false flowers (*florets*), which are alike or different in structure and sex, and often more or less reduced toward the top of the spikelet. *Flowers* hermaphrodite, or unisexual (often with the rudiments of the other sex), consisting of 3 or, usually, of 2 (anterior) minute hyaline or fleshy, nerved or nerveless scales (*lodicules*), representing a perianth, and of stamens or of a pistil, or of both. *Stamens* usually 3, rarely 6, 4, 2 or 1, very rarely more, hypogynous; filaments very slender, nearly always free; anthers versatile, consisting of 2 parallel cells, opening longitudinally by a slit, rarely by a terminal pore. *Ovary* entire, 1-celled; styles 2, lateral, rarely 3 or 1, free or more or less united, sometimes very short; stigmas as many as the styles, with simple or branched stigmatic hairs, exserted from the sides or the top of the florets; ovule 1, anatropous, often more or less adnate to the posterior side of the carpel. *Fruit* with the thin pericarp adnate to the seed (*caryopsis, grain*; rarely a delicate dehiscing or rupturing utricle enclosing a free seed, still more rarely a nut or a berry), with 2 marks, an anterior indicating the position of the embryo, and a posterior (*hilum*) free within the valve and pale, or adhering to the latter, rarely to both, usually forming with them a *false fruit* which becomes free by the disarticulation of the rhachilla. *Seed* erect; albumen copious, starchy; embryo usually small on the anterior face at the base of and outside the albumen; cotyledon shield-like (*scutellum*), closely attached by its inner side to the albumen, having the plumule and the descending radicle in front, and sometimes also a small anterior appendage opposite it (*epiblast*).

Herbs, rarely suffruticose, or in *Bambuseæ* often tall shrubs or trees, annual or perennial by means of rhizomes, rarely of aerial woody stems. Stems nearly always (often repeatedly and profusely) branched at the base, very rarely simple, thus forming fascicles or tufts of erect, ascending, prostrate or creeping, simple or ramified branches, which in the annual species are all more or less alike, having usually much shortened basal and lengthened upper internodes, terminating with an inflorescence (*culms*) or, in the perennial species, consist of culms and short, leafy, usually biennial shoots (*innovation shoots*) which grow into culms only in the second season; innovation shoots either piercing the subtending sheath at the base and growing up outside it, often as runners or stolons (*extravaginal*), or inside the sheaths, which may or may not be thrown aside (*intravaginal*); culms jointed, internodes usually hollow, closed at the nodes, with or without an annular swelling above the nodes and within the sheaths (*culm nodes*); all the branches and their leaf-supported ramifications with a 2-keeled dorsal, usually hyaline, leaflet at the base. Leaves alternate, usually 2-ranked, rarely pseudo-opposite owing to the alternation of long and very short internodes, very often crowded in tufts or fan-shaped bunches at the base of the culms, or in some cases also of their upper branches; in the perfect form (foliage leaves or "*leaves*" simply) consisting of sheath, ligule and blade; sheaths with the margins free (*open sheaths*) or more or less connate (*closed sheaths*), clasping each other or the culm, finally often loosened or sometimes slipping from the culm and more or less spreading, of the same structure throughout, or with an annular succulent swelling at the base (*sheath nodes*), which becomes at length hardened and persistent, or partly shrinks, leaving a depressed, often dark-coloured annular mark; ligules placed transversely at the inside at the junction of the sheath and the blade, consisting of a membrane or of a fringe of hairs, rarely altogether absent; blades usually long and narrow, entire, parallel-nerved, rarely ovate, cordate or sagittate, usually more or less gradually passing into the sheath, rarely articulated with it or constricted at the base into a petiole, folded or convolute in the bud, and often folding or rolling up in the mature state as they become dry, usually much reduced or quite suppressed in the lowest leaves which, in the perennial species, act as bud-scales, sometimes also in the upper leaves. Inflorescence terminal, rarely terminal and lateral, built up of the variously arranged spikelets, panicled, racemose, capitate, simply or compoundly spicate, very rarely consisting of a single spikelet, nearly always ebracteate. Spikelets all alike or heteromorphous, differing in sex and (in correlation with the sex) more or less also in the general structure, bisexual with all the flowers ⚥, or with ⚥ and ♂, or ♀ and ♂ flowers in the same spikelet, or unisexual (monœcious or diœcious). Mature spikelets falling entire from the tips of the pedicels, or together with a part of the pedicel or of the rhachis, or breaking up above the glumes into as many false fruits as there are fruiting florets, rarely persistent and shedding the grains. In the first case the glumes, in the second the valves are often decurrent into a callous swelling or extension (*callus*) at the insertion on the pedicel or rhachilla respectively.

About 325 genera, comprising 3000 to 3500 species in all parts of the world.

The typical structure of the spikelets is sometimes more or less obscured by the reduction or suppression, or by peculiar modifications of certain parts, generally in obvious correlation to the loss of functions, or the assumption of functions other than usual. The morphological character of those parts may, however, usually be recognized from their position in the spikelet and from comparison with allied, less modified species. Reduction or suppression is frequent in the glumes, less so in the pales, and it is extremely rare in the valves, except the reduced florets. The lower valves are, in certain tribes, frequently without a flower; but they very often enclose a rudiment of a floral branchlet, in the shape of a perfect or reduced pale, thereby indicating their homology with typical valves. In this case, they often lose some of the characteristics of the fertile valves, and approach either to the glumes in their general structure, or assume some special structure differing from that of the glumes as well as from that of the fertile valves. The nervation of the valves is very constant in nearly all the genera, and often throughout the greater part of a tribe; but in order to see it clearly, it is always advisable to flatten the valve and to examine it by transmitted light in a drop of water. Where the sexual conditions of the florets are of importance, it should be kept in mind, that many grasses are very distinctly protandrous, i.e. that they shed their anthers some time before the stigmas expand. Such flowers have frequently been taken to be female, whilst they were actually hermaphrodite. To avoid this error, young spikelets should, if possible, be examined beside the fully developed ones, or the filaments which usually remain around the ovary should be sought for.

[In the sequence of genera Dr. Stapf has found it necessary to deviate from that adopted in Bentham and Hooker's *Genera Plantarum*, as well as in the *Flora Australiensis*, and in Sir Joseph Hooker's enumeration of Indian grasses recently published in the *Flora of British India*. In the series of colonial floras issued from Kew, it has been thought advisable, for the most part, to avoid as far as possible breaking new ground and to adhere closely to a somewhat conventional sequence of orders and genera. The object is to make the floras of the different areas easily comparable. In the present case, however, further research has shown some changes to be indispensable. To these Sir Joseph Hooker, who has kindly allowed me to consult him, now assents. In accordance with general practice, *Andropogoneæ* stands at the head of the order, while a different position is assigned to *Zoysiæ* and to *Oryzeæ*.—W. T. T. D.]

SERIES I. Mature *spikelets* falling entire from their pedicels or with them (rarely subpersistent on a flat, indistinctly and tardily disarticulating rhachis: *Stenotaphrum*), all alike or differing in sex and structure; perfect spikelets with 2 heteromorphous florets, the upper ⚥, the lower ♂ or barren; rhachilla not continued beyond the upper floret.

Spikelets falling entire, singly or in clusters, occur in the following genera belonging to the second series: *Holcus, Chætobromus, Polypogon, Perotis, Tragus, Desmostachya, Spartina, Lamarckia, Fingerhuthia, Urochlæna, Tetrachne, Entoplocamia.*

Tribe 1. *ANDROPOGONEÆ. Spikelets* usually in pairs, one sessile, the other pedicelled (see *Trachypogon*), rarely 3-nate or solitary on the axes of variously arranged, often spikelike, racemes. *Glumes* more or less rigid and firmer than the valves, and the lower always longer than the florets. *Valves* membranous, often hyaline, that of the upper floret generally awned or reduced to an awn.

* *Spikelets all alike. Racemes in compound panicles (except Pollinia); joints of the rhachis slender.*

Subtribe 1. SACCHAREÆ.

I. **Imperata.**—*Racemes* in a spiciform silky panicle; rhachis not fragile. *Spikelets* awnless.

II. **Saccharum.**—*Racemes* in large much branched silky panicles; rhachis fragile. *Spikelets* awnless.

III. **Erianthus.**—*Racemes* in large much branched silky panicles; rhachis fragile. *Spikelets* awned.

IV. **Pollinia.**—*Racemes* 2-nate, digitate or approximate on a short main axis.

** *The two spikelets of each pair differing in sex and structure (in the South African species, except in Ischæmum fasciculatum, where the difference is limited to the lower glume).*

Subtribe 2. ISCHÆMEÆ. Sessile *spikelets* not sunk in hollows. *Racemes* spiciform, solitary, 2-nate, digitate or approximate on a short main axis. Lower *floret* of the sessile spikelet always ♂.

V. **Ischæmum.** The only South African genus.

Subtribe 3. ROTTBŒLLIEÆ. Sessile *spikelets* sunk in hollows formed by the contiguous joint of the rhachis and the appressed or adnate pedicel of the pedicelled spikelet of the same pair.

VI. **Rottbœllia.**—All spikelets awnless.

VII. **Urelytrum.**—Pedicelled spikelets long awned.

Subtribe 4. **Euandropogoneæ.**—*Rhachis* as in *Sacchareæ* and *Ischæmeæ*. *Racemes* variously arranged. Lower *floret* of all spikelets empty ; upper floret usually awned.

VIII. **Trachypogon.**—*Rhachis* tough, indistinctly articulate. *Spikelets* all pedicelled; the fruiting deciduous.
IX. **Elionurus.**—*Rhachis* of the spikelike solitary raceme fragile. One *spikelet* of each pair sessile, the other pedicelled, both awnless.
X. **Andropogon.**—*Racemes* solitary, 2-nate, digitate or panicled; rhachis fragile. The sessile *spikelet* awned, the pedicelled if present awnless or awned, but not from the upper valve. No whorl of ♂ spikelets at the base of the raceme, or where an imperfect one occurs the racemes are always paired, and each pair is subtended by a spathe.
XI. **Anthistiria.**—*Racemes* short, crowded in spathaceous panicled fascicles each with a whorl of 4 or more ♂ or barren spikelets at the base, and supported by a proper spathe.

Tribe 2. *PANICEÆ*. *Spikelets* in usually continuous spikes, racemes or panicles. *Glumes* herbaceous or membranous, the lower smaller, very small or suppressed. Lower *valve* generally resembling (except *Anthephora*) the glumes in structure and nervation, the upper firmer, at length rigid, often chartaceous to crustaceous, awnless, very rarely mucronate.

* *Glumes and lower valve entire, awnless or caudate- or subulate-aristate. Fruiting valve subchartaceous to crustaceous.*

† Mature spikelets falling entire and singly from the tips of their pedicels.

XII. **Paspalum.**—*Spikelets* awnless. Lower *glume* 0. Lower *valve* quite empty; nerves 5 or fewer, side-nerves curved, usually submarginal.
XIII. **Digitaria.**—*Spikelets* small, awnless. Lower *glume* minute, rarely 0. Lower *valve* generally with a minute pale; nerves 5–7, close, straight, prominent.
XIV. **Panicum.**—*Spikelets* awnless, or the glumes and the lower valve caudate-mucronate or caudate-aristate. Lower *glume* distinct. Lower *valve* with a 2-nerved or a hyaline rudimentary pale or quite empty; nervation different from that in *Digitaria*.
XV. **Oplismenus.**—*Spikelets* fascicled or solitary on a simple axis, or on the branches of a panicle. *Glumes* and lower *valve*, or at least the lower glume subulate-aristate.
XVI. **Axonopus.**—*Spikelets* of *Panicum*. Upper *glume* fimbriate. Lower *valve* with a deeply cleft short pale and a ♂ flower.
XVII. **Setaria.**—*Spikelets* of *Panicum*, but subtended by bristles.

†† Mature spikelets not falling singly, or not from the tips of their pedicels.

XVIII. **Pennisetum.**—*Spikelets* surrounded singly or in small clusters by an involucre (formed by naked or plumose bristles) and falling with it.
XIX. **Stenotaphrum.**—*Spikelets* subpersistent on the flat indistinctly and tardily disarticulating rhachis of a false spike.
XX. **Anthephora.**—*Spikelets* in deciduous spicate clusters with a false involucre, formed from the lower very hard glume of each spikelet.

** *Glumes and lower valve 2-lobed or emarginate, with a fine awn or mucro from the sinus, rarely all muticous. Fruiting valve rigidly membranous.*

XXI. **Tricholæna.**—Upper *glume* and lower *valve* 5-nerved; nerves usually very faint, hidden by copious and long silky hairs, and anastomosing below the obtuse tips.
XXII. **Melinis.**—Upper *glume* 7-, lower *valve* 5-nerved; nerves straight, conspicuous, not anastomosing below the acute tips.

SERIES II. Mature *spikelets* breaking up, leaving the persistent or subpersistent glumes on the pedicel, or if falling entire, then not consisting of 2-heteromorphous florets as in Series I.

A. Blades never transversely veined (in the South African species, except sometimes in Arundo), nor articulated on the sheath. (XXIII.—XCVII.)

a. *Awns, if present, kneed and twisted below the knee, or straight in reduced forms.* (XXIII.—XLIX.)

i. Florets 2 or more. (XXIII.—XL.)

Tribe 3. *ARUNDINELLEÆ. Florets* 2, heteromorphous, the lower awnless ♂ or barren. *Rhachilla* not continued beyond the upper floret. Lower *valve* awnless, rather resembling the glumes; upper generally (always in the African species) awned, at length firm or hard; awn from the sinus between 2 sometimes minute or bristlelike lobes, rarely from the entire, obtuse tip, usually kneed and twisted below the knee.

* *Spikelets solitary on distinct pedicels.*

XXIII. **Arundinella.**—*Spikelets* small (2½ lin. to less than 1 lin.). *Valve* of the upper floret 2-setose, minutely 2-toothed or entire, awn sometimes reduced or absent. Flaps of the *pale* auricled.

XXIV. **Trichopteryx.**—*Spikelets* large (2 lin. to 1½ in.). *Valve* of the upper floret always distinctly 2-toothed or 2-lobed, awn always kneed. Flaps of the *pale* not auricled.

** *Spikelets in clusters of* 3.

XXV. **Tristachya.**—*Spikelets* as in *Trichopteryx.*

Tribe 4. *AVENEÆ. Florets* 2 to many, all alike (except the uppermost which often are gradually reduced) or more or less heteromorphous (slightly in *Holcus, Chætobromus,* and sometimes also in *Aira,* very distinctly in *Anthoxanthum*). *Glumes* generally hyaline or scarious and shining. *Valves* membranous or subherbaceous with hyaline shining margins, 5- or more nerved, rarely 3-nerved (with the side nerves delicate and not submarginal); awn, if present, from the back or from a sinus, or from between bristles, kneed and usually twisted below the knee. (See *Kœleria.*)

* *Valves awnless or awned; awn, if present, from the back of the valve.*

XXVI. **Prionanthium.**—*Florets* 2, equal. *Glumes* cartilaginous on the back; keel with obtuse, pectinate teeth. *Valves* awnless.

XXVII. **Achneria.**—*Florets* 2, equal. *Glumes* firmly membranous, shining; keel smooth or almost so. *Valves* awnless.

XXVIII. **Aira.**—*Florets* 2. *Rhachilla* obscurely continued beyond the upper floret. *Valves* awned (occasionally the lower awnless).

XXIX. **Holcus.**—*Spikelets* deciduous as a whole. *Florets* 2; the upper usually ♂, awned, the lower ⚥, awnless.

XXX. **Anthoxanthum.**—*Florets* 3, heteromorphous; the lower 2 male or barren, awned, the terminal hermaphrodite, awnless. *Rhachilla* not continued beyond the upper floret.

XXXI. **Kœleria.**—*Florets* 2 or more, the uppermost gradually reduced. *Rhachilla* glabrous or almost so. *Valves* 3–5-nerved, awnless or mucronate or minutely awned from, or from below the tip. *Panicle* spiciform.

XXXII. **Trisetum.**—*Florets* 2 or more, the uppermost gradually reduced. *Rhachilla* ciliate. *Valves* 3–5-nerved, awned from the back, tips often with 2 fine bristles. *Ovary* glabrous.

XXXIII. **Avenastrum.**—*Spikelets* erect or suberect. *Florets* 2 or more, the uppermost gradually reduced. *Glumes* usually 1–3-nerved. *Valves* herbaceous, 5–9-nerved, rather firm, awned from the back. *Ovary* top hairy. Perennial.

XXXIV. **Avena.**—*Spikelets* large, pendulous. *Florets* 2 or more, the uppermost gradually reduced. *Glumes* 7–9-nerved. *Valves* herbaceous, 5–9-nerved, firm, awned from the back. *Ovary* top hairy. Nearly always annual.

** *Valves awned, rarely mucronate; awn or mucro from the sinus of the more or less distinctly 2-lobed valve, lobes often bristlelike.*

XXXV. **Pentaschistis.**—*Florets* 2. Continuation of *rhachilla* minute or bristle-like. Lobes of the *valve* 2-, rarely 4-fid, both or all 4 divisions bristlelike; or only the inner, with the outer lobule minute and more or less adnate to it.

XXXVI. **Pentameris.**—*Florets* 2, as in *Pentaschistis*. *Grain* globose-ellipsoid; pericarp crustaceous; seed free or almost free.

XXXVII. **Danthonia.**—*Florets* 3 to many, the uppermost gradually reduced. Lobes of the *valves* larger, more or less triangular or lanceolate, with or without an awnlike bristle.

XXXVIII. **Chætobromus.**—*Spikelets* falling entire with a part of the disarticulating pedicel; lowest valve without, the rest with side bristles.

Tribe 5. *ARUNDINEÆ*. *Florets* 2 to many, enveloped by very long hairs, springing either from a long and slender callus or from the back of the valves. *Glumes* and *valves* membranous, often hyaline, awnless or minutely awned from the tips.

XXXIX. **Arundo.**—Lowest *floret* like the rest, hairy from the back of the valve.

XL. **Phragmites.**—*Florets* heteromorphous; the lowest ♂ or barren, the rest ⚥, hairy from the long slender callus, otherwise glabrous.

ii. Floret 1. (XLI.—XLIX.)

Tribe 6. *AGROSTEÆ*. *Floret* 1. *Rhachilla* not continued beyond the floret or only as a more or less distinct point or bristle. *Valve* membranous (in the South African species) or thinly herbaceous, not or hardly changed when mature, usually truncate, 5-, very rarely 3-nerved, all the nerves or the outer side nerves often slightly excurrent, parallel or at least not anastomosing; awn, if present, from the back, rarely from the truncate tip.

XLI. **Lagurus.**—*Glumes* covered with long soft hairs. *Panicle* spiciform, oblong or subglobose, plumose.

XLII. **Polypogon.**—*Glumes* awned, usually from a minute notch or sinus.

XLIII. **Agrostis.**—*Glumes* not awned. Callus glabrous or shortly bearded.

XLIV. **Calamagrostis.**—*Glumes* not awned. Callus with long fine hairs, several times longer than the valve.

Tribe 7. *STIPEÆ*. *Floret* 1. *Rhachilla* not continued beyond the floret. *Valve* hardened when mature, tightly enveloping the fruit; nerves joining or closely approaching at the tip; awn terminal, rarely 0.

XLV. **Aristida.**—*Awns* 3 from the entire tip, or one, simple below and 3-branched above, very rarely quite simple. *Ligule* a fringe of hairs.

XLVI. **Stipa.**—*Floret* cylindric or linear-oblong; awn solitary, simple, kneed and twisted below the knee. *Ligule* membranous.

XLVII. **Oryzopsis.**—*Floret* broader than in *Stipa*, awn fine, straight, very caducous, or 0. *Ligule* membranous.

Tribe 8. *ZOYSIEÆ*. Mature *spikelets* falling entire and singly, or in clusters. *Floret* 1. *Rhachilla* not continued beyond the floret. *Glumes* equal, or the lower much smaller or suppressed. *Valve* small, delicately membranous, 3–1-nerved. *Spikelets* in slender spiciform panicles or racemes.

XLVIII. **Perotis.**—*Spikelets* falling singly. *Glumes* equal, long awned.

XLIX. **Tragus.**—*Spikelets* falling in clusters of 2–4. Upper (outer) *glume* beset with hooked spines or short coarse bristles.

b. *Awns, if present, never kneed and twisted below the knee.* (L.—XCVII.)

i. Valves 3-nerved, rarely 1-nerved. (See also *Eleusine*). (L.—LXVI.)

Tribe 9. *SPOROBOLEÆ*. *Glumes* and *valves* very similar. *Floret* 1. *Rhachilla* very rarely continued beyond the floret. *Valve* membranous, acute or obtuse, not changed when ripe, 1- or more or less distinctly 3-nerved, awnless, usually olive-green or olive-grey; side-nerves, if present, delicate, evanescent above. *Seed* often free in the delicate pericarp.

L. **Sporobolus.**—The only genus.

Tribe 10. *ERAGROSTEÆ.* *Spikelets* variously panicled, sometimes spicate or subspicate. *Florets* usually numerous and far exserted from the glumes. *Glumes* and *valves* rather similar in general appearance. *Valves* membranous to chartaceous, very often olive-green or olive-grey, entire or 3-cleft, 3-nerved; nerves evanescent above or excurrent into bristles; side-nerves submarginal, glabrous or pubescent or finely ciliate below. *Pales* often persistent or subpersistent.

LI. **Pogonarthria.**—*Spikelets* densely crowded on the flattened rhachis of stiff secund false spikes, which are spirally arranged on a long axis. Tips of *rhachilla* joints fringed with minute hairs. *Valves* acuminate, entire.

LII. **Diplachne.**—*Spikelets* more distant or even remote on the slender rhachis of often flexuous false spikes. *Valves* usually 2-toothed or minutely notched.

LIII. **Eragrostis.**—*Spikelets* usually panicled, rarely in simple spikes. *Rhachilla* often persistent. *Valves* usually numerous and closely imbricate, broad, entire, awnless. *Pales* persistent or deciduous, but usually not falling together with the valves.

LIV. **Desmostachya.**—*Spikelets* crowded on and falling entire from the flattened axis of stiff spreading secund spikes, which are closely arranged on a long common axis.

Tribe 11. *CHLORIDEÆ.* *Spikelets* usually in 2-ranked, secund spikes or spike-like racemes, rarely distinctly pedicelled and panicled. *Florets* 1 to many. *Valves* generally membranous, truncate, emarginate or toothed, 3-nerved; nerves distant, subparallel, distinct, percurrent or excurrent and often ciliate all along, lateral submarginal (see *Spartina* with 1-nerved valves, and *Eleusine* which has sometimes additional side-nerves close to the middle nerve of the valve); awn, if present, straight, usually from a truncate or toothed tip.

* *Spikelets crowded in 2-ranked, secund spikes.*

† Floret 1, awnless. Rhachilla not continued beyond the floret or only as a minute point or a naked bristle.

LV. **Spartina.**—Mature *spikelets* falling entire from the rhachis. *Valve* 1-nerved.

LVI. **Cynodon.**—*Glumes* shorter than the glabrous, subchartaceous valve.

LVII. **Microchloa.**—*Glumes* longer than the ciliate, membranous valve.

†† Florets several, one ⚥, the others ♂ or barren.

LVIII. **Ctenium.**—Upper *glume* generally with a stiff bristle from the back. Lowest 2 florets barren or the second ♂, the third ⚥.

LIX. **Harpechloa.**—*Spike* solitary, terminal. *Spikelets* awnless. Lowest floret ⚥.

LX. **Chloris.**—*Spikes* 2 to many, digitate or fascicled. *Spikelets* awned or awnless. Lowest *floret* ⚥.

††† Several to many ⚥ florets in each spikelet. Seeds free within the delicate, dehiscing or rupturing pericarp.

LXI. **Eleusine.**—*Spikes* terminated by a spikelet.

LXII. **Dactyloctenium.**—Point of the rhachis of the *spikes* naked, projecting.

** *Spikelets panicled, pedicelled (though often very shortly) or subspicate, but not on the flattened rhachis of a 2-ranked, secund spike.*

LXIII. **Lophacme.**—*Spikelets* in long very slender approximate spikelike racemes. *Valves* pubescent, the upper 2–4 barren, forming a tuft of awns.

LXIV. **Leptocarydium.**—*Spikelets* subsessile in dense spiciform panicles. *Florets* distinctly (often long) exserted from the glumes. *Valves* long and finely awned; side-nerves not excurrent. *Blades* constricted and rounded at the base.

LXV. **Crossotropis.**—*Spikelets* subsessile in racemously arranged erect or spreading false spikes. *Florets* equalling or slightly exceeding the glumes or much shorter. *Valves* very shortly awned; side-nerves not excurrent. *Blades* not constricted at the base.

LXVI. **Triraphis.**—*Spikelets* usually distinctly pedicelled, often in compound and dense panicles. *Valves* finely awned; side-nerves excurrent into long or short bristles.

ii. Valves 5-many-nerved, very rarely (as in *Triphlebia*) 3-nerved. (LXVII.—XCVII.)

Tribe 12. *PAPPOPHOREÆ*. *Valves* broad, cleft into 3 to many sometimes subulate lobes with or without alternating fine straight awns from the sinuses, usually many-nerved.

LXVII. **Enneapogon.**—Fertile *floret* 1. *Valve* cleft into 9 subulate, awn-like lobes.

LXVIII. **Schmidtia.**—Fertile *florets* 3–5. *Valves* cleft into 4 hyaline lobes, alternating with five straight awns from the sinuses.

Tribe 13. *ORYZEÆ*. *Spikelets* all alike or more or less heteromorphous and unisexual. Fertile *floret* 1, awnless, very rarely caudate-aristate, terminal with 2 minute empty florets (valves) below it or solitary. *Glumes* very minute or confluent into an annular rim or suppressed. *Pales* 3–9-nerved. *Stamens* usually 6, rarely more or 3–1.

LXIX. **Potamophila.**—Fruiting *valve* membranous to subherbaceous, smooth, with 2 minute, nerveless empty valves below it.

LXX. **Leersia.**—Fruiting *valve* chartaceous, rigidly ciliate (in the African species) without empty valves below it.

Tribe 14. *PHALARIDEÆ*. *Spikelets* all alike. Fertile *floret* 1, awnless, terminal with 2 empty florets (valves) below it. *Glumes* distinct, often equalling or exceeding the terminal floret. *Pales* 2-nerved to nerveless. *Stamens* 6 or 3 (in the African species), or 4 or 2.

LXXI. **Ehrharta.**—Both empty *florets* or at least the upper larger than the fertile, exceeding or equalling the glumes.

LXXII. **Phalaris.**—Empty *florets* reduced to minute scales, enclosed together with the fertile floret by the glumes.

Tribe 15. *FESTUCEÆ*. *Glumes* more or less resembling the valves in general appearance. Fruiting *florets* 2 to many, very rarely 1, often much exserted from the glumes. *Valves* 5- or more nerved, very rarely 3- or 1-nerved, then with the side-nerves not submarginal nor with the spikelets in 2-ranked spikes; awns, if present, terminal or subterminal, never kneed.

Subtribe 1. MELICEÆ. *Glumes* more or less equalling the florets, membranous. Uppermost 2–3 *valves* small, empty, enclosing each other and forming a club or spindle-shaped body.

LXXIII. **Melica.**—The only South African genus.

Subtribe 2. DACTYLIDEÆ. *Spikelets* in panicled clusters or more or less spiciform panicles, or true spikes, very rarely in loose panicles (*Briza*). *Valves* membranous to papery or subchartaceous, 5–9-nerved, very rarely 3-nerved (*Triphlebia*), side-nerves conniving or joining below the tip; awns, if any, fine, short, terminal or subterminal.

* *Spikelets of 2 kinds, the fertile surrounded by the sterile, consisting solely of numerous bracts.*

LXXIV. **Lamarckia.**—Mature *spikelets* falling in fascicles, consisting of 1 fertile and several empty spikelets. Barren *bracts* obtuse, membranous.

LXXV. **Cynosurus.**—Fertile *spikelets* hidden by the persistent barren ones, breaking up, when mature. Barren *bracts* very narrow, rigid, acute or awned.

** *Spikelets all alike, sometimes the lowest in the inflorescence or on the branch more or less reduced and barren.*

† Lowest floret like the rest, ⚥.

LXXVI. **Fingerhuthia.**—*Spikelets* falling entire and singly from the pedicels of a compact spiciform panicle. *Glumes* 1-nerved, narrow.

LXXVII. **Schismus.**—*Spikelets* in contracted panicles, shining. *Florets* 6–7, small, only the upper ones exserted from the glumes. *Valves* 2-cleft, with or without a fine mucro from the sinus; margins hyaline, white.

LXXVIII. **Dactylis.**—*Spikelets* in dense secund clusters, arranged in contracted or open panicles. *Valves* 5-nerved, firmly herbaceous; tips mucronate or shortly awned.

LXXIX. **Triphlebia.**—*Spikelets* in dense cylindric softly hairy panicles. *Glumes* and *valves* equally hairy all over, acuminate. *Valves* 3-nerved.

LXXIX. **Lasiochloa.**—*Spikelets* generally hispid, in spiciform often lobed panicles. *Florets* not exserted from the glumes. *Valves* 5–9-nerved; hairs acute.

LXXX. **Urochlæna.**—*Panicles* spiciform or capitellate, falling as a whole together with the subtending sheath. *Glumes* and *valves* caudate-aristate.

LXXXI. **Brizopyrum.**—*Spikelets* in spiciform panicles or in 2-ranked spikes. *Florets* exserted from the glumes. *Valves* 5–9-nerved; hairs generally clavately tipped.

LXXXII. **Briza.**—*Spikelets* in lax panicles or racemes, generally nodding on long capillary pedicels. *Glumes* and *valves* more or less boat-shaped or saccate and scarious.

†† Florets heteromorphous; the 2 lowest barren, consisting of valves which resemble the glumes more than the fertile valves. Spikelets falling entire and singly from the flattened rhachis of short secund spikes.

LXXXIII. **Tetrachne.**—*Glumes* and barren valves 1-nerved. Fertile *valves* 5-nerved. *Styles* short.

LXXXIV. **Entoplocamia.**—*Glumes* 3–5-nerved. Barren *valves* 6–8-nerved; fertile valves with 9–11 very marked nerves. *Styles* very long.

Subtribe 3. POEÆ. *Spikelets* generally loosely panicled. *Florets* exserted from the glumes. *Valves* generally rather broad, ovate or oblong, often obtuse, membranous to herbaceous, often with broad and variegated hyaline margins or tips, awnless; side-nerves 2, sometimes 3 or 1 on each side, often faint.

LXXXVI. **Poa.**—*Glumes* and *valves* keeled. *Florets* often with a tuft of long curled wool at the base.

LXXXVII. **Atropis.**—*Glumes* and *valves* rounded on the back. *Valves* rather broad, obtuse, hyaline and variegated towards the tips, equally 5-nerved.

LXXXVIII. **Scleropoa.**—*Glumes* and *valves* rounded on the back, or slightly keeled. *Valves* rather narrow, acute or subacute, firm except at the very tip, 5-nerved; outer side-nerves much more distinct than the faint inner.

Subtribe 4. EUFESTUCEÆ. *Spikelets* generally loosely panicled or racemose or spuriously spicate. *Florets* exserted from the glumes. *Valves* herbaceous to chartaceous, usually acute or acuminate, often more or less mucronate or awned from the entire or 2-toothed tips, rarely from below the tips. (See also *Scleropoa.*)

* *Spikelets panicled.*

LXXXIX. **Festuca.**—Perennial. *Panicle* various. *Glumes* subequal (in the South African species), rarely conspicuously unequal. *Valves* lanceolate, rounded on the back or keeled towards the tip, 5- rarely 7-nerved; mucro or awn, if present, terminal, straight. *Styles* terminal or rarely subterminal on the glabrous or hairy ovary top.

XC. **Vulpia.**—Annual. *Panicle* contracted, spiciform or racemiform, usually secund. *Glumes* generally very unequal; the lower small or sometimes obsolete. *Valves* subulate-lanceolate, passing into a straight, often long, awn, rounded on the back, 5-nerved. *Ovary* glabrous; styles terminal.

XCI. **Bromus.**—Annual or perennial. *Panicle* various. *Glumes* more or less unequal. *Valves* lanceolate to broadly oblong, rounded on the back or keeled, 5–9-nerved; awn close to or somewhat distant from the, often 2-toothed, tip, rarely 0. *Styles* distinctly lateral on a hairy 2–3-lobed appendage of the ovary.

** *Spikelets subsessile in a simple raceme (false spike).*

XCII. **Brachypodium.**—*Valves* 7–9-nerved; awn terminal, rarely 0.

Tribe 16. *HORDEÆ.* *Spikelets* sessile, singly or in clusters on the notches of a simple spike, sometimes partially sunk in hollows of the same. *Florets* 1 or more.

Subtribe 1. LOLIEÆ. *Spikelets* solitary at the nodes of the spike, with their plane radial to the rhachis. Lower glume, if present, contiguous with the rhachis.

XCIII. **Lolium.**—*Florets* 3 to many. Lower *glume* generally suppressed, except in the terminal spikelet, upper rigidly herbaceous, many-nerved, longer or shorter than the florets. *Valves* herbaceous to chartaceous, 5–7-nerved.

XCIV. **Lepturus.**—*Florets* 1–2. Lower *glume* minute or 0, upper exceeding the florets, rigid, many-nerved. *Valves* membranous, 3–1-nerved, side-nerves submarginal.

XCV. **Oropetium.**—*Florets* 1–3. Lower *glume* minute or 0, upper exceeding the florets, nerves 3, confluent in a broad rigid midrib, sides membranous. *Valves* membranous, 3-nerved, side-nerves submarginal, callus hairy.

Subtribe 2. TRITICEÆ. *Spikelets* solitary at the nodes of the spike, with their plane tagential to the rhachis.

XCVI. **Agropyrum.**—The only South African genus.

Subtribe 3. ELYMEÆ. *Spikelets* 2 or more, collateral or fascicled at the nodes of the spike.

XCVII. **Hordeum.**—The only South African genus.

B. Blades transversely veined (in the South African species). (XCVIII.—C.)

Tribe 17. *PHAREÆ.* *Spikelets* heteromorphous, unisexual, monœcious, the ♂ small. *Floret* 1. One or both *glumes* of the ♂ often minute or obsolete. Fruiting *valve* coriaceous to cartilaginous. *Lodicules* 3. *Stamens* 3–6. *Blades* flat, broad, many-nerved, often petioled.

XCVIII. **Olyra.**—The only South African genus.

Tribe 18. *BAMBUSEÆ.* *Spikelets* all of one kind. *Florets* few to many, rarely 1. *Glumes* distinct or indistinctly differentiated, i.e. passing below into more or less numerous bracts and sometimes having like them flowering branchlets or spikelets in their axil, and at the same time resembling the valves. *Valves* subherbaceous to subcoriaceous, 5- to many-nerved, generally awnless. *Lodicules* usually 3. *Stamens* 3, 6 or more. *Styles* 2 or 3. Shrubs or trees. *Blades* flat, many-nerved, articulated on the sheath.

Subtribe 1. ARUNDINARIEÆ. *Stamens* 3. *Fruit* a true caryopsis.

XCIX. **Arundinaria.**—The only South African genus.

Subtribe 2. EUBAMBUSEÆ. *Stamens* 6. *Fruit* a true caryopsis.

C. **Bambusa.**—*Florets* many. *Pales* 2-keeled. *Filaments* free.

I. IMPERATA, Cyr.

Spikelets all alike, generally 2-nate (one subsessile, the other distinctly pedicelled) on the continuous branches (racemes) of a spiciform or narrowly thyrsiform panicle, disarticulating from the pedicels. *Florets* 2; the lower reduced to an empty valve (very rarely ♂?). *Glumes* subequal, membranous, 3–9-nerved, rarely nerveless, enveloped by very long silky hairs from the obscure callus and the lower portion of both glumes. Lower *valve* generally much smaller than the glumes, hyaline; upper valve still smaller, hyaline, awnless, rarely 0. *Pale* broad, hyaline, nerveless. *Lodicules* 0. *Stamens* 1–2. *Styles* connate below; stigmas linear, exserted from the top of the spikelet. *Grain* oblong; embryo half the length of the grain or more.

Perennial; basal leaves crowded and like those of the innovation shoots long; panicle silvery-silky.

DISTRIB. Species 5, mostly closely allied, in the warm regions of both hemispheres.

1. I. arundinacea (Cyr. Pl. Rar. Neap. fasc. ii. 26, t. 11, and in Usteri, Ann. Bot. xiii. 61); culms 1½–4 ft. long, erect, 3–4-noded, glabrous; sheaths rather loose, glabrous, the lowest at length usually breaking up into fibres; ligules membranous, short, hairy; blades linear, from a very narrow base, tapering to an acute point, the lowest ½–1 ft. by 1–2 lin., the upper very short, rigid, usually convolute, glabrous or bearded at the base, smooth below, margins scabrid, midrib stout; panicle spiciform, 2–8 in. long, cylindric, very dense; branches and branchlets very numerous, crowded, appressed; pedicels fine with clavate tips, ½–1½ lin. long., with long fine hairs below; spikelets about 2½ lin. long, pale or purplish, enveloped by hairs 5–6 lin. long; glumes ovate-oblong to ovate-lanceolate, obtuse or subobtuse with the tips ciliate, usually 5–7-nerved; lower valve oblong, obtuse, denticulate, ciliate, nerveless, upper ovate, acute, glabrous, nerveless; anthers 1½–2¼ lin. long; stigmas 1½–2½ lin. long, purple. *Host, Gram. Austr.* iv. *t.* 40; *Kunth, Enum.* i. 477; *Trin. Androp. in Mém. Ac. Pétersb. sér.* vi. ii. 330; *Steud. Syn. Pl. Glum.* i. 405; *Reichb. Ic. Fl. Germ.* i. *fig.* 1504; *Anders. in Öfvers. K. Vet. Akad. Stockh.* 1855, 159; *Hack. Androp. in DC. Monogr. Phan.* vi. 92; *Hook. f. Fl. Brit. Ind.* vii. 106; *I. cylindrica, Beauv. Agrost.* 165, *t.* v. *fig.* 1, *Explan. planch.* 5; *Durand & Schinz, Consp. Fl. Afr.* v. 693. *Lagurus cylindricus, Linn. Syst. ed.* x. ii. 878. *Saccharum cylindricum, Lamk. Encycl.* i. 594; *Illustr.* i. 155, *t.* 40, *fig.* 2. *Calamagrostis Lagurus, Koeler, Descr. Gram.* 112.

VAR. β **Thunbergii** (Hack. l.c. 94); sheaths usually glabrous at the nodes; ligules very short, truncate; blades usually flat, 2–4 lin. broad, otherwise as in the type. *I. Thunbergii, Beauv. l.c.* 165. *Roem. and Schult. Syst.* ii. 289; *Nees, Fl. Afr.-Austr.* 89; *Steud. Syn. Pl. Glum.* i. 405. *I. arundinacea, var. africana, Anders. l.c. I. cylindrica, var. Thunbergii, Durand & Schinz, l.c.; Hack. in Bull. Herb. Boiss.* iv. *App.* iii. 10. *Saccharum spicatum, Thunb. Prod. Cap.* 50, *non Linn. (except in Herb.).* *S. Thunbergii, Retz. Obs.* v. 17.

VAR. γ **Kœnigii** (Benth. Fl. Hongk. 419); culms slender; sheaths generally long bearded at the nodes; ligules very short, truncate; blades less rigid than in the type, or almost flaccid, subglaucous; panicle less dense, particularly when ripe; spikelets 1½–2 lin. long, hairs 5–7 lin. long. *Hack. l.c.* 94. *I. Kœnigii, Beauv. l.c.* 165; *Roem. & Schult. l.c.* 289; *Trin. l.c.* 331; *Nees l.c* 89; *Steud. l.c. I. arundinacea, var. indica, Anders. l.c.* 160. *I. cylindrica, var. Kœnigii, Durand & Schinz, l.c.* 694. *Saccharum Kœnigii, Retz, Obs.* v. 16.

COAST REGION: Var. β: Clanwilliam Div.; Olifants River, *Ecklon and Zeyher!* Cape Div.; Constantia, *Bergius!* Cape Peninsula, Muizenberg Vlei, *Dodd*, 2360! Humansdorp Div.; hills near the Kromme River, in bush, *Drège!* Albany Div., *Zeyher*, 873! British Kaffraria, *Cooper*, 295!

KALAHARI REGION: Var. β: Griqualand West; Upper Campbell, *Burchell*, 1833! Bechuanaland; Hamapery near Kuruman, *Burchell*, 2509! Transvaal; Pretoria, *Rehmann*, 4777! Matebe Valley, *Holub!*

EASTERN REGION: Var. β: Tembuland; Bazeia, *Baur*, 321! Natal; by streams near Pietermaritzburg, *Krauss*, 260! Umpumulo to the coast, *Buchanan*, 164! Var. γ: Pondoland; between Umtata River and St. Johns River, *Drège*, 4253! Natal; Umtente, *Sutherland!* near Ladysmith, *Gerrard*, 158! and without precise locality, *Gerrard*, 478! *Buchanan*, 113!

Var. β occurs also through tropical Africa. Var. γ in the coast region of East Africa, in the Comoros and throughout Southern and Eastern Asia to Polynesia and Australia. The type is limited to the Mediterranean Region as far as Turkestan and North-west India. These varieties and the type, although on the whole pretty distinct within their areas, often pass into each other, chiefly along the confines of their areas, or they seem to lose their distinctive characters under particular local conditions, when their separation becomes almost impossible.

II. SACCHARUM, Linn.

Spikelets all alike or at least similar, usually 2-nate (one sessile, the other pedicelled, rarely both pedicelled) on the articulate, usually fragile, rhachis of panicled spiciform racemes, the pedicelled falling from their pedicels, the sessile deciduous together with the contiguous joint of the rhachis and pedicel. *Florets* 2; lower reduced to an empty valve; upper ⚥ or sometimes ♀ in the pedicelled spikelets, which are then slightly smaller. *Glumes* equal, membranous to coriaceous; lower with inflexed margins and, in the sessile spikelet, with an even number of nerves; upper 1–3–5-nerved. *Valves* hyaline, muticous or mucronate, upper smaller or 0. *Pales* small, hyaline, nerveless, or 0. *Lodicules* 2, cuneate. *Stamens* 3. *Stigmas* laterally exserted. *Grain* oblong to subglobose; embryo half the length of the grain or more; hilum basal.

Perennial, often very tall; leaves various; panicle large, often silky and showy, much and densely branched, or contracted to spiciform; spikelets often surrounded by long silky hairs from the base.

DISTRIB. Species about 14, tropical or subtropical.

The only South African species belongs to the Subgenus *Eriochrysis* (Hack.), characterized by the spikelets being slightly heteromorphous, and the pedicelled smaller with a ♀ upper floret.

1. S. munroanum (Hack. Androp. in DC. Monogr. Phan. vi. 124); culms 2 ft. long, 3-noded; lower leaves crowded; sheaths rather lax, scantily hairy at the nodes, otherwise glabrous, the lowest shining as if lacquered; ligule short with hairs from the base behind, otherwise glabrous; blades narrow, linear, acute, flat, 4–5 in. long by 2 lin. (or by less than 1 lin. in the more or less convolute innovation leaves), rigid, puberulous on both sides, margins smooth; panicle oblong 4–5 in. long, dense, fulvous or ferruginous, shining; racemes ½ to almost 1 in. long, rather stout, joints and pedicels stout, ciliate; spikelets crowded, lanceolate, the sessile 2–2½ lin. long, enveloped by rigid hairs; lower glume chartaceous, entire, subobtuse, 2-keeled and 4-nerved between the keels, which are long ciliate above the middle; upper glume more acute, 1–3-nerved, ciliate along the margins and on the keel above the middle; lower valve 1½–1¾ lin. long, lanceolate; upper valve very small, ovate, nerveless, tips ciliate; pale 0; anthers 1 lin. long; grain obovoid-globose, ½ lin. long; pedicelled spikelets similar, but smaller (1½–1¾ lin. long), enveloping hairs more copious; anthers rudimentary, very minute. *Durand & Schinz, Consp. Fl. Afr.* v. 694. *Eriochrysis pallida, Munro in Harvey, Gen. S. Afr. Pl. ed.* ii. 440.

KALAHARI REGION: Transvaal; Magalies Berg, near springs, *Zeyher*, 1793! *Burke*, 75!

EASTERN REGION: Natal; Drakensberg Range, at Coldstream, *Rehmann*, 6876! Umpumulo, in marshes, 2000 ft., *Buchanan*, 212!

III. ERIANTHUS, Michx.

Spikelets all alike, 2-nate (one sessile, the other pedicelled, very rarely both unequally pedicelled) on the articulate and usually fragile rhachis of panicled racemes, the pedicelled falling from their pedicels, the sessile deciduous together with the contiguous joint of the rhachis and pedicel. *Florets* 2; lower reduced to an empty valve; upper always ⚥. *Glumes* equal, membranous to coriaceous; lower dorsally flattened, more or less distinctly 2-keeled, with inflexed margins and a variable number of intracarinal nerves, upper 1–3-nerved, keeled. *Valves* hyaline, lower muticous or mucronate, upper generally awned. *Pale* small; hyaline, nerveless. *Lodicules* 2, cuneate. *Stamens* 2–3. *Stigmas* usually laterally exserted. *Grain* oblong to linear-oblong; embryo half the length of the grain.

Perennial, erect; blades long, narrow, midrib very stout, ligules membranous; panicle much branched, often large, silky and showy; spikelets usually villous, and with an involucre of hairs from the base.

DISTRIB. Species about 20, mainly tropical, few temperate.

Blades not terete:
  Blades flat, 1–3 ft. by 4–6 lin., densely bearded at the base inside, and above the short ligule; rhachis and branches of panicle glabrous ... ... (1) **capensis.**
  Blades convolute or canaliculate, almost reduced to the midrib below, 1–2 ft. or more by 2 lin., glabrous or very scantily hairy at the base inside the elongate ligule; rhachis and branches of panicle pubescent at the nodes ... ... (2) **Sorghum.**
Blades subterete, solid ... ... (3) **junceus.**

1. E. capensis (Nees, Fl. Afr. Austr. 93, excl. var. β); culms up to 3 ft. and more long, stout, sheathed nearly all along, glabrous; sheaths glabrous or scantily and appressedly hirsute near the margins; ligules about 1 lin. long, rounded; blades broadly linear from a narrow base, tapering to a long acute point, 1–3 ft. by 4–6 lin., flat, firm, smooth, green, densely bearded at the base inside, otherwise glabrous; panicle 1–1½ ft. long, linear-oblong, usually sheathed at the base, much branched; branches and branchlets suberect, filiform, like the rhachis quite glabrous, the lowest 3–4, rarely to 6 in. long; racemes obscurely articulate, tough, joints equalling or exceeding the spikelets, bearded at the base; pedicels unequal, one very short, the other almost as long as the spikelets, both filiform; spikelets oblong to oblong-lanceolate, 2–2½ lin. long, pale or reddish-brown, loosely, hairy to ½ lin. long; glumes chartaceous, the lower minutely 2-toothed, usually with 2 intracarinal nerves, the upper broadly lanceolate, boat-shaped, subacute, 3-nerved, keeled; lower valve lanceolate, about as long as the lower glume, hyaline above, 1–3-nerved, ciliate, upper valve linear-oblong, acutely 2-bifid, 1–1¼ lin. long, 3-nerved, ciliate, awn 2–4 lin., column short; pale very small long ciliate. *Hack. Androp. in DC. Monogr. Phan.* vi. 149; *Durand & Schinz, Consp. Fl. Afr.* v. 695. *E. Ecklonii, Nees, l.c.; Krauss in Flora,* 1846, 117. *Saccharum capense and S. Ecklonii, Steud. Syn. Pl. Glum.* i. 408. *Miscanthus capensis, Anders. in Öfvers. K. Vet. Akad. Förh. Stockh.* 1855, 165.

VAR. β **villosa** (Stapf); tops of culms, sheaths and blades, and lower part of the rhachis villous to tomentose or the leaves glabrescent; spikelets loosely hairy almost all over.

SOUTH AFRICA: without locality, *Drège*, 4234!

COAST REGION: Humansdorp Div.; between Gamtoos River and Melk River, *Burchell*, 4800! and North side of the Kromme River, *Burchell*, 4868! Uitenhage Div.; banks of Zwartkops River, *Ecklon and Zeyher*, 621! Albany Div.; Howisons Poort, near Grahamstown, *MacOwan*, 1273! and Bothas Hill, *MacOwan*, 487! Komgha Div.; banks of Kei River, *Drège!*

KALAHARI REGION: Griqualand West; Hay Div., at Griquatown, *Burchell*, 1901! Herbert Div., at Lower Campbell, *Burchell*, 1810!

EASTERN REGION: Natal; among reeds by the Umlazi River, *Krauss*, 159! feeder of Tugela River, 1000 ft., *Buchanan*, 271! Var. β: Natal; Durban Flat, *Buchanan*, 45! Umpumulo, common near water, 2000–2500 ft., *Buchanan*, 270! Riet Vlei, 5000 ft., *Buchanan*, 272! Umsinga and base of Biggars Berg, *Buchanan*, 103! Ladysmith, *Gerrard!* and without precise locality, *Gerrard*, 689!

2. E. Sorghum (Nees, Fl. Afr. Austr. 92); culms to 4 ft. and more long, rather stout, sheathed nearly all along, pubescent or glabrescent below the nodes; sheaths appressedly hirsute near the margins, bearded at the nodes or ultimately often glabrous; ligules oblong, subacute, up to 3 lin. long; blades narrowly linear, the lower from a long, still narrower base, which is sometimes almost reduced to the very stout midrib, convolute, or canaliculate, rarely the upper flat, all tapering to a long setaceous point; 1-3 ft. or more by 2 lin., scantily hairy at the base inside, otherwise glabrous; panicle 1–1½ ft., somewhat lax and nodding or contracted and stiff; branches and branchlets flexuous or strict and erect, filiform, the lowest 3–4 in. long, primary branches like the rhachis, pubescent at the nodes; racemes tough, bearded at the nodes, joints equalling or exceeding the spikelets; pedicels unequal, one shorter to very short or obsolete, the other often exceeding the spikelet, both filiform; spikelets lanceolate to oblong, 2–3 lin. long, usually densely hairy; otherwise very like those of *E. capensis*, Nees. *Hack. Androp. in DC. Monogr. Phan.* vi. 148; *Durand & Schinz, Consp. Fl. Afr.* v. 695. *E. capensis, var. angustifolius, Nees, l.c.* 94. *Saccharum Sorghum, Steud. Syn. Pl. Glum.* i. 408.

SOUTH AFRICA: without locality, *Drège*, 3876!

COAST REGION: Tulbagh Div.; near New Kloof on the Klein Berg River, *Drège!* Queenstown Div.; Shiloh, in moist spots, 3500 ft., *Baur*, 890!

CENTRAL REGION: Graaff Reinet Div.; Sneeuw Berg Range, 4000–5000 ft., *Drège! Bolus*, 285! and on the banks of Sunday River, above Graaff Reinet, 2500 ft, *Bolus* 285!

KALAHARI REGION: Orange Free State; sanddrift Spruit, *Burke*, 229!

EASTERN REGION: Tembuland; Bazeia Mountain, *Baur*, 100!

3. **E. junceus** (Stapf); culms rather stout, over 3 ft. long, sheathed all along or nearly so, glabrous; sheaths glabrous, firm, with a broad scarious margin near the mouth, continuing along the base of the blade; ligule short, rounded; blades subterete, solid, tapering to a long setaceous point, 2–3 ft. by $\frac{1}{2}$–$\frac{3}{4}$ lin. in diam., grooved at the very base and villous in the groove, glaucous, smooth, glabrous; panicle linear, about 1 ft. long; branches and branchlets erect, filiform, like the rhachis quite glabrous, the lowest 2–3 in. long; racemes obscurely articulate, tough, joints exceeding the spikelets, bearded at the base; pedicels unequal, one about $\frac{1}{2}$ as long as the joint, the other exceeding it, both filiform; spikelets oblong to lanceolate, 2 lin. long, pale or purplish, loosely hairy to the middle, basal beard short; glumes chartaceous; awn 2–2$\frac{1}{2}$ lin. long, fine; otherwise very like those of the 2 preceding species.

KALAHARI REGION: Basutoland; Leribe, *Buchanan*, 228! Transvaal; Olifants River, *Nelson*, 73*! Jungle streamlet, *Nelson*, 77*!

IV. POLLINIA, Trin.

Spikelets all alike or nearly so, 2-nate (one sessile, the other pedicelled, very rarely both unequally pedicelled), on the articulate and usually fragile rhachis of 2-nate, digitate or fascicled, spikelike racemes, the pedicelled falling from their pedicels, the sessile deciduous together with the contiguous joint of the rhachis and pedicel. *Florets* 2; lower ♂ (sometimes without a valve), or reduced to an empty valve; upper always ⚥. *Glumes* equal, membranous to coriaceous; lower dorsally flattened or concave, more or less 2-keeled with inflexed margins; upper 1–3-nerved, keeled. *Valves* hyaline; lower muticous, sometimes suppressed; upper very short, 2-lobed or entire, generally awned. *Pale* small or 0. *Lodicules* 2, small, cuneate. *Stamens* 3, rarely fewer.

Stigmas linear, laterally exserted. *Grain* oblong; embryo almost half the length of the grain, or longer; hilum basal, punctiform.

Perennial or annual, of very different habit.

DISTRIB. Species about 32, in the tropical and subtropical regions of the Old World.

I have followed here, as in *Andropogoneæ* generally, Hackel's excellent monograph, in the definition of the genera, but in this and in other cases, they seem to me too heterogeneous, comprising as they do easily distinguishable and natural groups which are held together only by few ill-definable and often unreliable characters.

Subgenus I. EULALIA. Perennial; culms erect; blades not contracted at the base; glume dorsally flattened or slightly depressed ... ... (1) **villosa.**

Subgenus II. LEPTATHERUM. Annual; culms decumbent, rooting; blades lanceolate, conspicuously contracted at the base; lower glume grooved ... ... (2) **nuda.**

1. **P. villosa** (Spreng. Syst. i. 288, not Benth.); perennial; culms $1\frac{1}{2}$–3 ft. long, rather stout, 3–4-noded, appressedly hirsute or pubescent, at least below the nodes and at the top; sheaths usually exceeding the internodes except the uppermost, terete, tight, scantily hairy to villous or glabrescent, bearded at the nodes; ligules membranous, $\frac{1}{2}$–1 lin. long, glabrous; blades lanceolate-linear, gradually passing into the sheath, shortly acute, 4–7 in. by 3–6 lin., flat, suberect, firm, glabrous or more or less appressedly hairy; racemes (false spikes) 2–7 on a short, common axis, suberect, stout, 2–6 in. long, joints and pedicels linear, stout, with long silvery grey hairs mainly along the margins and round the base, about $1\frac{1}{2}$–$2\frac{1}{2}$ lin. long; spikelets crowded, ferruginous, lanceolate, $2\frac{1}{2}$–$3\frac{1}{2}$ lin. long; glumes coriaceous, the lower minutely truncate, dorsally depressed, margins broadly inflexed, keels long and densely ciliate; upper glume lanceolate, boat-shaped, shining, 3-nerved, keel ciliate above; lower valve empty, almost as long as the subtending glume, membranous below, hyaline above, 2-nerved, margins inflexed, ciliate; upper valve ovate-oblong, about $1\frac{1}{2}$ lin. long, acutely bifid to the middle, lobes ciliate, awn 7–10 lin. long, kneed above the middle; pale small, oblong, obtuse, ciliate; anthers $1\frac{1}{2}$ lin. long. *Hack. Androp. in DC. Monogr. Phan* vi. 157; *Durand & Schinz, Consp. Fl. Afr.* v. 696. *Andropogon villosum, Thunb. Prodr.* 20; *Fl. Cap. ed. Schult.* 108; *Kunth, Enum.* i. 499. *Eulalia villosa, Nees, Fl. Afr. Austr.* 91; *Steud. Syn. Pl. Glum.* 412.

COAST REGION: Clanwilliam Div.; on the Olifants River and near Brack Fontein, *Ecklon.* "Swellendam and George District," *Mund.* Uitenhage Div.; rocky slopes of Van Stadens Berg Range, 1000–1500 ft., *Drège!* between the Kromme River and the Krakakamma Mountains, *Ecklon.* Albany Div.; stony hills near Grahamstown, *MacOwan*, 698! Stockenstrom Div.; near Philippstown, *Ecklon.* Komgha Div.; banks of the Kei River, *Drège!*

KALAHARI REGION: Transvaal; West of Lyden Berg Dorp, *Nelson*, 30*!

EASTERN REGION: Tembuland; Bazeia, *Baur*, 325! Griqualand East; rocky hills near Kokstad, *Tyson*, 1474! Natal; Inanda, *Wood*, 1591! Umpumulo, *Buchanan*, 210! and without precise locality, *Buchanan*, 51!

Also in Madagascar, according to Hackel.

Very closely allied to *P. quadrinervis*, Hack of India, particularly to var. *Wightii*, Hook. f., and perhaps not specifically distinct from the latter.

2. P. nuda (Trin. Androp. in Mém. Ac. Pétersb. sér. vi. ii. 307); annual; culms very slender, decumbent, rooting, branched below, 2–5 ft. long, many-noded, grooved; sheaths as long as, or the upper shorter than, the internodes, terete, tight, finely hairy or glabrescent; ligules membranous, glabrous; blades spreading, lanceolate from a constricted base, acuminate, 1–2½ in. by 2–7 lin., thin, often flaccid, glabrous or with a few tubercle-based hairs; racemes (false spikes) 3–6, somewhat distant on a common axis, very slender, flexuous, 1–4 in. long, green, joints equalling or exceeding the spikelets, glabrous; pedicels like the joints but shorter; spikelets rather distant, linear-lanceolate, 2 lin. long, minutely bearded at the base; glumes membranous, the lower bicuspidate, dorsally concave, 2-keeled, keels scaberulous; upper glume acuminate, 1-nerved, obscurely keeled, ciliolate; lower valve empty, linear-lanceolate, 1½ lin. long, hyaline, glabrous, nerveless; upper valve linear, very narrow, 1 lin. long, hyaline, 1-nerved, awn terminal, finely capillary, flexuous, 6–9 lin. long; pale 0; anthers ½ lin. long; grain 1 lin. long. *Hack. Androp. in DC. Monogr. Phan.* vi. 178; *Durand & Schinz, Consp. Fl. Afr.* v. 696; *Hook. f. Fl. Brit. Ind.* vii. 117. *Leptatherum royleanum, Nees in Proc. Linn. Soc.* i. 93. *L. japonicum, Franch. & Savat. Enum. Pl. Jap.* ii. 190 *and* 609. *Psilopogon capensis, Hochst. in Flora*, 1846, 117. *Eulalia capensis, Hochst. l.c.; Steud. Syn. Pl. Glum.* i. 412.

COAST REGION: Knysna Div.; in woods on the banks of streamlets, *Krauss*, 92! and between Plettenberg Bay and Melville, *Burchell*, 5365!

EASTERN REGION: Natal; Riet Vlei, in bush, 6000 ft., *Buchanan*, 292!

Also from North India to China and Japan.

There is no difference between the South African and the Indian specimens, the latter exhibiting often the same characters which are supposed to distinguish the South African var. *capensis*, Hack.

V. ISCHÆMUM, Linn.

Spikelets of each pair alike, or differing only in sex, or distinctly heteromorphous, 2-nate (one sessile or subsessile, the other pedicelled), on the articulate fragile rhachis of solitary, 2-nate, digitate or fascicled, spike-like racemes, the pedicelled falling from their pedicels, the sessile deciduous together with the contiguous joint of the rhachis and pedicel. *Florets* 2; lower generally ♂; upper ⚥, or sometimes ♂ in the pedicelled spikelets. *Glumes* equal or subequal, coriaceous to chartaceous; lower dorsally flattened or subdepressed, more or less 2-keeled with inflexed margins; upper keeled at least above, sometimes awned. *Valves* rigidly membranous to hyaline; lower muticous, upper usually bifid and awned from the sinus, rarely mucronate or muticous. *Pales* more or less equalling their valves, hyaline. *Lodicules* 2;

cuneate. *Stamens* 3; sometimes smaller or barren in the ⚥ flower. *Stigmas* linear, laterally exserted. *Grain* various.

Generally perennial; blades convolute when young, at length flat, ligules generally membranous; racemes compressed, joints flattened or subconcave on the inner side and often stout; spikelets sessile, dorsally compressed, often rather broad, the pedicelled sometimes apparently laterally compressed with a median keel owing to the more or less complete suppression of one side.

DISTRIB. Species about 45, mainly in the tropics of the Old World.

Both S. African species belong to the subgenus *Eu-Ischæmum* (Hack.) characterized thus:—racemes 2-nate, digitate or fascicled; pedicelled spikelets generally with 2 florets, very rarely rudimentary; grain dorsally slightly compressed, not grooved.

Blades linear-lanceolate, 3–6-lin. broad; racemes stout; lower glume flat, 5-ribbed, keels winged near the 2-toothed reddish tip ... ... (1) **fasciculatum.**

Blades narrowly linear, 2 lin. broad; racemes slender, glaucous; lower glume dorsally depressed, not ribbed, acute, keels not winged... ... ... (2) **glaucostachyum.**

1. I. **fasciculatum** (Brongn. Voy. Coq. Bot. 73.) var. **arcuatum** (Hack. Androp. in DC. Monogr. Phan. vi. 235); perennial; culms ascending or suberect from a slender creeping rhizome, more or less branched and fascicled near the base or simple, 1–3 ft. long, rather slender, glabrous, 5–many-noded; sheaths exceeding or more or less equalling the internodes, subterete, glabrous, rarely hairy; ligules very short, truncate, ciliate; blades linear-lanceolate, setaceously acuminate, 3–8 in. by 3–6 lin., flat, quite glabrous or hairy on the back near the base, smooth, midrib white; racemes (false spikes) fascicled, 2–5, stout, strict or flexuous, 2–5 in. long, hairy, joints stout, triquetrous, $1\frac{1}{4}$–$1\frac{1}{2}$ lin. long, slightly curved, keel (outer angle) ciliate with yellowish hairs; subsessile spikelets $2\frac{1}{2}$–3 lin. long, reddish above, pedicel $\frac{1}{2}$ lin. long or shorter, bearded at the base; lower glume lanceolate-oblong, acutely and shortly 2-toothed, coriaceous below, flat and more or less distinctly 5-ribbed on the back, scaberulous, scantily hairy, keels winged near the tips, scabrid, intracarinal nerves 5–9; upper glume slightly longer than the lower, lanceolate, acuminate, mucronate, 3–5-nerved, keeled above; lower valve lanceolate, almost as long as the subtending glume, rigidly membranous, glabrous, 3-nerved, with an almost equal pale and a ♂ flower; upper valve oblong, deeply 2-fid, 2 lin. long, delicately membranous, 3-nerved, lobes very broad, hyaline, ciliate, awn 4–5 lin. long, twisted below; pale as long as or slightly longer than the valve, acute; anthers 1 lin. long; pedicelled spikelets $2\frac{1}{2}$ lin. long, reddish; pedicel 2 lin. long, more or less hairy on the back; lower glume lanceolate, boat-shaped, keeled, 7–9-nerved, loosely hairy, the rest as in the subsessile spikelets. *Durand & Schinz, Consp. Fl. Afr.* v. 697. *Spodiopogon arcuatus*, *Nees, Fl. Afr. Austr.* 97. *Andropogon arcuatus*, *Steud. Syn. Pl. Glum.* i. 374.

COAST REGION: Komgha Div.; by the Kei River, *Drège*.

EASTERN REGION: Natal; coast region, *Sutherland!* Durban Flat, *Buchanan*, 32! 109! in woods and marshy places along the coast between the

Umtentu River and the Umzimkulu River, *Drège!* Umpumulo, near streamlets, *Buchanan*, 209! and without precise locality, *Gerrard*, 679! *Harvey*, 49!

The type, which is a weaker and much more hairy plant with longer and more exserted awns, occurs in the Mascarene Islands.

2. **I. glaucostachyum** (Stapf); perennial; culms erect, slender, up to 3 ft. long, glabrous, terete, simple; sheaths terete, very tight, glabrous, smooth, the upper shorter than the internodes; ligules membranous 2½–3 lin. long, oblong, acute, ciliate; blades linear, tapering to a setaceous point, up to 6 in. by 2 lin., flat, erect, rigid, glaucous, glabrous, smooth, margins scabrid; racemes (false spikes) 3–4, fascicled, glaucous, 3–6 in. long, distant from the uppermost sheath, on slender glabrous peduncles 3–7 lin. long; joints and pedicels cuneate (the latter more slender) 1½–2 lin. long, hollow, convex on the back, glabrous, margins ciliate; sessile spikelets lanceolate 2½ lin. long, bearded at the obtuse base; lower glume herbaceous-chartaceous, acute to subacute, dorsally depressed, scantily hairy along the middle, keels scabrid, intracarinal nerves 2–4; upper glume subchartaceous, ovate-lanceolate, boat-shaped, acute, 5-sub-7-nerved, nerves anastomosing, keel ciliate; valves hyaline, equal, 2–2½ lin. long, the lower lanceolate, acute, nerveless, glabrous with an equal ciliate 2-nerved pale, and a ♂ flower, the upper oblong, deeply bifid, 3-nerved, lobes lanceolate, ciliolate, awn very slender, 3 lin. long, kneed at the middle, column smooth; pale oblong, 1½ lin. long; anthers 1½ lin. long; pedicelled spikelets about 1½ lin. long, reduced to the lower glume which is similar to that of the sessile spikelets.

KALAHARI REGION: Transvaal; Pinaars River, *Nelson*, 17*!

Very near to *I. brachyatherum*, Fenzl ex Hack., from Abyssinia and the Sudan, but differing in the slender, more hairy racemes, the much less inflated rhachis joints and the smaller spikelets.

VI. ROTTBŒLLIA, Linn. f.

Spikelets all alike, or more or less heteromorphous, 2-nate, rarely 3-nate (one, rarely 2, sessile, the other pedicelled and sometimes rudimentary, with the pedicel free and appressed or adnate to the contiguous joint, whence the spikelets appear to be in true spikes), on the articulate and usually fragile rhachis of solitary, digitate or panicled, spike-like racemes, the sessile usually deciduous with the contiguous joint of the rhachis. *Florets* 2; lower ♂ or barren; upper ⚥, or ♂ in the pedicelled spikelets, or the latter quite barren or rudimentary. *Glumes* equal or subequal; lower more or less coriaceous, at least along the 2 keels, dorsally flattened or subconvex, muticous, obtuse, rarely acuminate; upper chartaceous to membranous, muticous. *Valves* hyaline or the lower membranous, muticous. *Pales* almost equalling their valves, hyaline or obsolete. *Lodicules* 2, cuneate. *Stamens* 3. *Stigmas* generally laterally exserted. *Grain* broadly oblong; embryo ½ the length of, or almost as long as, the grain.

Perennial or annual, of various habit.

DISTRIB. Species about 35, in both hemispheres.

The only S. African species belongs to the subgenus *Hemarthria* (Hack.) characterized by solitary spikes, terminal on the culms or their branches, compressed, tardily or not at all disarticulating; joint tips truncate, not hollowed; spikelets all alike in form or nearly so; lower floret reduced to an empty valve.

1. **R. compressa** (Linn. f. Suppl. 114) var. **fasciculata** (Hack. Androp. in DC. Monogr. Phan. vi. 286), perennial; culms erect from a decumbent rooting base, branched, 1–5 ft. long, compressed, glabrous; sheaths shorter than the internodes, compressed, keeled, often ciliate at the mouth, otherwise glabrous; ligules membranous, very short, ciliate; blades linear, gradually tapering, acute, very variable in length and width, uppermost almost suppressed, somewhat rigid, glabrous; racemes often fascicled, rather stout, straight or curved, with the appearance of true spikes owing to the pedicels of the upper spikelets being adnate to the contiguous joint, more or less fragile; spikelets linear-oblong to oblong, 3–3½ lin. long, glabrous; glumes equal, lower coriaceous, constricted below the obtuse tips or in the pedicelled spikelets more acuminate, finely 7–9-nerved, upper broadly lanceolate, acuminate, membranous and 3-nerved in the sessile, subchartaceous and 5–7-nerved in the pedicelled spikelets; valves subequal, nerveless, 2–2½ lin. long; pale linear, 1–1¼ lin. long; anthers 1–1½ lin. long. *Durand & Schinz, Consp. Fl. Afr.* v. 699; *Hook. f. Fl. Brit. Ind.* vii. 153. *R. fasciculata, Lam. Illustr.* i. 204; *Desf. Fl. Atl.* i. 110, *t.* 36. *Lodicularia fasciculata, Link, Hort. Berol.* i. 6. *L. capensis, Nees, Fl. Afr. Austr.* 128. *Lepturus fasciculatus, Trin. Fund. Agrost.* 123. *Hemarthria fasciculata, Kunth, Rév. Gram.* i. 153; *Enum.* i. 465, *Suppl.* 375; *Steud. Syn. Pl. Glum.* i. 359; *Hack. in Mart. Fl. Bras.* II. iii. 314, *t.* 72, *fig.* 2. *H. capensis, Trin. Androp. in Mém. Acad. Pétersb. sér.* vi. ii. 248.

COAST REGION: Clanwilliam Div.; Ebenezar, on sandy hills, *Drège.* Cape Div.; between Constantia and Wynberg, *Burchell,* 815! by the Salt River and at the Ponds near Cape Town, *Burchell,* 686! Cape Peninsula, near Claremont and in Muizenberg Vlei, *Dodd,* 2350! 2380! Paarl Div.; Klein Drakenstein Mountains, by the Berg River, *Drège!* Tulbagh Div.; near Tulbagh Waterfall, *Verreaux!* Worcester Div.; by the Hex River, *Mund and Maire!* Stellenbosch Div.; Hottentotts Holland, *Ecklon.* Caledon Div.; between Genadendal, and Zonder Einde River, *Drège.* Swellendam Div.; by the Zonder Einde River, *Burchell,* 7510! Uitenhage Div.; by the Zwartkops River, *Drège!* Albany Div.; Glenfilling, *Drège.*

CENTRAL REGION: Graaff Reinet Div.; banks of Sunday River, near Graaff Reinet, 2500 ft., *Bolus,* 536!

EASTERN REGION: Tembuland; banks of Qumancu River, 2000 ft., *Baur,* 723! Natal; by the Umlazi River, *Krauss,* 14! between St. Johns River and the Great Waterfall, *Drège!* by streamlets near Umpumulo, *Buchanan,* 198! and without precise locality, *Buchanan,* 74! *Gerrard,* 677!

Throughout the warm countries of both hemispheres, in many localities evidently introduced. The type through India to South-west China.

VII. URELYTRUM, Hack.

Spikelets heteromorphous, 2-nate (one sessile, the other pedicelled with the pedicels appressed to the rachis) on the very fragile rhachis of solitary, terminal, subcylindric, spike-like racemes, the sessile deciduous with the contiguous joint of the rhachis and pedicel; joints obliquely truncate with appendaged tips. *Sessile spikelets: Florets* 2, lower ♂, upper ⚥. *Glumes* equal or subequal; lower coriaceous, dorsally flattened, 2-keeled, muticous; upper subchartaceous or membranous, boat-shaped, keeled. *Valves* hyaline, subequal, slightly shorter than the glumes, muticous, lower 2-, upper 3-sub-5-nerved. *Pales* almost equalling their valves, hyaline, 2-nerved. *Lodicules* 2, cuneate. *Stamens* 3. *Stigmas* linear, laterally exserted. *Pedicelled spikelets* with 2 ♂ florets and with the lower glume narrowed into a long subulate awn, otherwise as in the sessile spikelets, or more or less reduced, sometimes to a pair of rudimentary glumes, the lower of which is represented by a long awn, flattened and widened at the base.

Perennial; culms erect, few-noded, simple; blades narrow, linear, flat or convolute; racemes long, strict.

DISTRIB. Species 1 or 2 in South Africa.

1. **U. squarrosum** (Hack. Androp. in DC. Monogr. Phan. vi. 272); compactly tufted, culms 2–2½ ft. long, glabrous, few-noded; sheaths exceeding the internodes except the uppermost, tight, terete, glabrous, smooth, the lower widened, very firm, shining, persistent; ligules oblong, 2–3 lin. long, glabrous or with a few long hairs behind; blades very narrow, linear, tapering to a setaceous point, lowest up to 1 ft. by ¾–1¼ lin., flat convolute, rather rigid, glabrous, prominently nerved and finely scaberulous above; raceme (false spike) 5–6 in. long, 1½–2 lin. thick, almost glabrous to villous, joints 3–4 lin. long, silky at the base, scantily pubescent to villous above, appendage up to 2 lin. long, unequally lobed or toothed, ciliolate; sessile spikelets about 4 lin. long; lower glume oblong to lanceolate, obtuse to acute, glabrous and smooth or punctate, or pubescent to villous, keels acute at least above, distantly (sometimes obscurely) spinulous muricate at the middle, rigidly ciliate above, with 1–3 intracarinal nerves which are more or less prominent towards the tips; upper glume lanceolate, acute, firmly membranous, 3-nerved, side-nerves submarginal, keel rigidly ciliate above, margins softly ciliate; valves about 3 lin. long, softly ciliate, lower lanceolate, acute, upper ovate-lanceolate, obtuse to subacute, mucronulate; anthers 2–1½ lin. long; pedicelled spikelets varying in hairiness like the sessile, narrow, and usually smaller than them, with 2 ♂ florets, or (often in the same raceme) more or less reduced, sometimes to an awn representing the lower glume and a minute rudiment of the upper; awn squarrose, flattened below, scabrid, 1½–3½ in. long. *Durand & Schinz, Consp. Fl. Afr.* vi. 701.

Rottbœllia hordeoides (*name only*), *Munro in Harvey, Gen. S. Afr. Pl. ed.* ii. 442.

VAR. β., **robusta** (Stapf); stouter; blades over 1 ft. by 2–2½ lin., scabrid above; raceme very stout, 2½–3 lin. in diam.; spikelets on the whole broader, never quite glabrous and smooth, otherwise varying as in the type; pedicelled spikelets always with 2 ♂ florets. *Vossia sp. Munro in Harvey, l.c.*

KALAHARI REGION: Orange Free State; between Harrismith and Leribe, *Buchanan*, 213! Transvaal; Pretoria, *Rehmann*, 4729! East of the Maquasi Mountains, *Nelson*, 38*! and Mooi River, *Burke* ex *Munro*. Var. β: Bechuanaland; Pellat Plains, between Matlareen River and Takun, *Burchell*, 2200!

VIII. TRACHYPOGON, Nees.

Spikelets heteromorphous, 2-nate (one very shortly, the other long pedicelled) on the tough, slender, rhachis of solitary or digitate racemes. *Lower* (subsessile) *spikelet* persistent, dorsally compressed, awnless: *Florets* 2; lower reduced to an empty valve; upper ♂. *Glumes* equal; lower subcoriaceous, 2-keeled; upper chartaceous, grooved on either side of the fine keel. *Valves* hyaline, linear-lanceolate, subequal, slightly shorter than the glumes; lower acute, 2-nerved; upper 2-toothed or 2-fid, mucronulate, 3-nerved. *Pale* obsolete. *Lodicules* 2, cuneate, small. *Stamens* 3. *Ovary* rudimentary; styles filiform; stigmas 0. *Upper* (long pedicelled) *spikelets* deciduous, subcylindric, long-awned: *Florets* 2; lower reduced to an empty valve; upper ⚥. *Glumes* subequal, chartaceous; lower obscurely 2-keeled below the tip, margins involute, base decurrent on the pedicel, forming a densely bearded pointed callus. *Valves* linear; upper hyaline below, coriaceous above, passing into a stout, kneed awn. *Stigmas* laterally exserted. The rest as in the lower spikelets.

Perennial, cæspitose; blades long, narrow, more or less rigid; ligules membranous; racemes slender.

DISTRIB. Species 1, very polymorphous, throughout tropical and subtropical America, in South Africa as far as the Congo, and in Madagascar.

1. T. polymorphus (Hack. in Mart. Fl. Bras. II. iii. 263) var. **capensis** (Hack. Androp. in DC. Monogr. Phan. vi. 326); culms erect, slender, simple, 1½–3 ft. long, terete, glabrous, few-noded; leaves mainly crowded near the base; sheaths terete, tight, generally exceeding the internodes, bearded at the nodes, otherwise glabrous or scantily and fugaciously hairy; ligules firm, up to 2 lin. long; blades very narrow, linear, tapering to a long setaceous point, 4–10 in. by ¾–2 lin., usually convolute, glabrous or hirsute at the base, rarely hairy all over; racemes solitary, 3–7 in. long, strict or subflexuous, rhachis subterete, glabrous; subsessile (♂) spikelets oblong, 3–4 lin. long, hairy; lower glume obtuse or 2-toothed, keels crested, ciliolate, intracarinal nerves about 7, margins very narrowly inflexed; upper glume linear-oblong, subacute, subglabrous, 3-nerved, margins softly ciliate; anthers 1½–2 lin. long; pedicelled (⚥) spikelets, 3½–4½ lin. long, pedicel very slender, about 1½ lin. long, scantily long-hairy; lower glume oblong-linear, minutely truncate, ciliate near the membranous reddish tips, otherwise more or less hairy, nerves 8–11, transversely anastomosing, callus 1 lin. long; upper glume lanceolate-linear, obtuse, strongly 3-nerved, scantily hairy above; awn of upper valve 1½–2½ in. long, softly hairy with hairs decreasing upwards, or the bristle scabrid. *T. polymorphus var. truncatus, Hack. Androp. in DC. Monogr. Phan.* vi. 327. *T. capensis, Trin. Androp. in Mém. Acad. Pétersb. sér.* vi. ii. 257; *Nees, Fl. Afr. Austr.* 100. *T. truncatus, Anderss. in Öfvers. K. Vet. Akad. Förh. Stockh.* 1857, 49. *Stipa capensis and St. spicata, Thunb. Prod.* 19-20; *Fl. Cap. ed. Schult.* 106-7. *Heteropogon truncatus, Nees, Fl. Afr. Austr.* 102. *Andropogon spicatus and A. truncatus, Steud. Syn. Pl. Glum.* i. 368.

COAST REGION: Riversdale Div.; near the Zoetemelks River, *Burchell*, 6680! Uitenhage Div.; between Kromme River and Krakakamma Mountains, *Ecklon and Zeyher!* Port Elizabeth Div.; Port Elizabeth, *E.S.C.A. Herb.* 98! Alexandria Div.; Zuur Berg Range, 2000–3000 ft., *Drège!* Bathurst Div.; Kasuga River, *MacOwan*, 1014! Albany Div.; mountain sides near Grahamstown, 2200 ft., *MacOwan*, 1287!

KALAHARI REGION: Orange Free State, *Cooper*, 3389! Transvaal; Johannesburg, common on open Veld, *E.S.C.A. Herb.* 312! Houtbosch, *Rehmann*, 5670! 5671!

EASTERN REGION: Tembuland; Bazeia, *Baur*, 318, partly! Natal; Umpumulo, *Buchanan*, 153! Riet Vlei, 4000–5000 ft., *Buchanan*, 154! 155! Inanda, *Wood*, 1621! and without precise locality, *Buchanan*, 125! *Gerrard*, 476!

Also in Madagascar and on the Congo.

IX. ELIONURUS, Humb. and Bonpl.

Spikelets similar, usually awnless, but differing in sex, 2-nate (one sessile, the other pedicelled) on the articulate fragile rhachis of solitary spike-like racemes, the sessile deciduous with the contiguous joint of the rhachis and the pedicel. *Florets* 2; lower reduced to an empty valve; upper ⚥ in the sessile, ♂, rarely barren, in the pedicelled spikelet. *Glumes* equal; lower subcoriaceous to herbaceous, often 2-toothed or 2-fid, rarely awned, dorsally flattened, 2-keeled, usually with fine, filiform, transparent oil glands close to the ciliate or penicillate keels; upper membranous, lanceolate, acute, rarely awned. *Valves* hyaline, awnless. *Pale* obsolete or 0. *Lodicules* 2, cuneate. *Stamens* 3. *Stigmas* laterally exserted. *Grain* oblong, dorsally compressed; embryo about ½ the length of the grain.

Generally perennial, cæspitose, aromatic; blades flat or folded; ligules membranous, very short; racemes erect, joints strongly compressed, usually villous, tips oblique, not appendaged.

DISTRIB. Species about 15, in the tropics and the subtropical regions of both hemispheres.

1. **E. argenteus** (Nees, Fl. Afr. Austr. 95); densely cæspitose; innovation shoots intravaginal; culms slender, simple, rarely with a lateral flowering branch from the upper part, 1–2 ft. long, more or

less compressed below, glabrous, 2–3 noded; leaves mainly crowded at the base; lower sheaths compressed, villous at the very base, appressedly and usually fugaciously hirsute above, or glabrous except near the mouth; upper terete, tight, shorter than the internodes, the uppermost sometimes tumid with a minute blade; ligules truncate, ciliolate with long hairs from behind; blades very narrow, linear, acute, generally tightly convolute and filiform, $\frac{1}{3}$–1 ft. by about 1 lin., erect, rigid, sulcate, flexuous or curved, glabrous or hirsute at the base; raceme (false spike) 3–6 in. long, rather stout, strict or flexuous, whitish silky, joints rather stout or slender, to 2 lin. long, long and densely villous on the back; sessile spikelets lanceolate, acuminate, 4–5 lin. long; lower glume herbaceous to chartaceous, bi-cuspidate, villous to subglabrous on the back, intracarinal nerves about 6, evanescent below, keels acute, long and densely ciliate, in the lower part with tubercle-based tufts of hairs, and with oil-glands nearly all along, callus short, obtuse, hairy; upper glume acute, 3-nerved, pubescent along the fine prominent keel, margins softly ciliate; lower valve lanceolate, $2\frac{1}{2}$ lin. long, 2-nerved; upper oblong-lanceolate, 2 lin. long, 3-nerved; margins of both softly ciliate; pale very minute; anthers 2 lin. long; pedicelled spikelets $2\frac{1}{2}$–3 lin. long, pedicels up to 2 lin. long, like the joints; lower glume entire or subentire, one margin inflexed, the other spreading, intracarinal nerves 5–6; upper glume 3–5-nerved; otherwise as in the sessile spikelet, except the sex. *Hack. Androp. in DC. Monogr. Phan.* vi. 339; *Durand & Schinz, Consp. Fl. Afr.* v. 702. *Andropogon tenuifolius, Steud. Syn. Pl. Glum.* i. 365.

VAR. **thymiodora** (Stapf); culms very slender, to 1 ft. long; blades setaceous; racemes 1–$1\frac{1}{2}$ in. long; spikelets purplish, 3–$3\frac{1}{2}$ lin. long; lower glume shortly bifid or 2-toothed, and keels beset with more numerous tubercle-based tufts of hairs in the sessile spikelets; the rest as in the type, but all parts smaller and with fewer or fainter nerves. *E. thymiodorus, Nees, l.c.* 95; *Hack. l.c.* 340; *Durand & Schinz, l.c.* 703. *Andropogon thymiodorus, Steud. Syn. Pl. Glum.* i. 365.

SOUTH AFRICA: without locality, *Drège*, 2042! 4317! *Zeyher*, 1796!

COAST REGION: Uitenhage Div.; sandy hills near the Zwartkops River, *Zeyher*, 462! Alexandria Div.; Zuur Veld, *Gill!* Zuur Berg Range, 2000–3500 ft., *Drège!* Albany Div.; Grahamstown, grassy slopes, 2000 ft., *MacOwan*, 1316! Fort Beaufort Div.; Kat River, 2500–3000 ft., *Drège.* Stockenstrom Div.; Winter Berg, near Philippstown, *Drège.*

CENTRAL REGION: Albert Div., *Cooper*, 1367! Aliwal North Div.; Witte Bergen, moist, rocky places, 5000–6000 ft., *Drège!*

KALAHARI REGION: Orange Free State; Caledon River, *Burke*, 200! and Draai Fontein, *Rehmann*, 4491! Bechuanaland; Kosi Fontein, *Burchell*, 2581! Kuruman, *Burchell*, 2183! and between Kosi Fontein and Knegts Fontein, *Burchell*, 2607! Transvaal; Houtbosch, *Rehmann*, 5672! Wonderboom Poort, *Rehmann*, 4491! and Christiana on the Vaal River, *Nelson*, 65*!

EASTERN REGION: Tembuland; Bazeia, 2000 ft., *Baur*, 284! Natal; Umpumulo, 2500 ft., and Riet Vlei, 4000–5000 ft., *Buchanan*, 161! between Umtentu River and Umzimkulu River, *Drège*, and without precise locality, *Gerrard*, 768! *Buchanan*, 52! Zululand, *Buchanan*, 161a! Var. β: Uitenhage Div.; Van Stadens Berg, 1000–3000 ft., *Zeyher*, 4475! Albany Div.; mountain tops near Grahamstown, *MacOwan*, 671! Fort Beaufort Div.; Kat River Mountains above the wood zone, *Ecklon.*

X. ANDROPOGON, Linn.

Spikelets similar or heteromorphous, 2-nate (one sessile, the other pedicelled) on the fragile rhachis of solitary, 2-nate, digitate, fascicled or panicled, spike-like racemes, the sessile deciduous with the contiguous joint of the rhachis and the pedicel. *Florets* 2; lower reduced to an empty valve; upper ☿ in the sessile, ♂ or barren, or quite suppressed in the pedicelled spikelet. *Glumes* equal or subequal; lower coriaceous to subchartaceous, 2-keeled, muticous; upper usually less firm, with a median keel, muticous, rarely with a terminal bristle. *Valves* hyaline, the upper at least so at the base, and usually awned. *Pale* various or 0. *Lodicules* 2, cuneate. *Stamens* 3–1. *Stigmas* laterally exserted. *Grain* various.

Perennial or annual, of very varied habit.

DISTRIB. Over 200 species in the warm parts of the world.

The genus, as defined by Hackel, consists of very heterogeneous elements, and I have no doubt that several of the old genera of which it is made up, will have to be restored, as for instance, *Chrysopogon*, *Cymbopogon* and *Heteropogon*. However, I am not yet ready to define the genus *Andropogon* (sensu strictiore) in a satisfactory manner, and therefore prefer to follow Hackel for the present in this respect.

A. ISOZYGI. Sessile spikelets of all the pairs alike in sex and form:
- *Racemes solitary, terminal on the culms and their branches, often in spurious leafy or spathaceous panicles:
  - Subgenus I. SCHIZACHYRIUM. Joints of the rhachis stout, tips cupular with toothed margins; upper valve deeply bifid ... ... (1) **hirtiflorus.**
  - Subgenus II. HYPOGYNIUM. Joints of the rhachis filiform, tips truncate, upper valve entire, or very minutely 2-toothed ... ... (2) **ceresiæformis.**
- **Racemes 2-nate, digitate or panicled, terminal on the culms and their branches:
  - Subgenus III. ARTHROLOPHIS. Lateral racemes sessile:
    - Pedicelled spikelets suppressed; racemes silvery plumose ... ... (3) **eucomus.**
    - Pedicelled spikelets ♂ (in the South African species), very rarely barren:
      - Lower glume dorsally concave or grooved; keels not winged:
        - Lower glume dorsally concave, not grooved... (4) **appendiculatus.**
        - Lower glume with a narrow groove corresponding to a keel on the inner side:
          - Blades linear, 1–5 lin. broad; pedicelled spikelets 3½–6 lin. long:
            - Blades narrow at the base:
              - Culms simple; leaves mostly crowded near the base, blades very narrow (5) **schirensis.**
              - Culms branched from the middle and upper nodes; leaves scattered along the culm, blades 2–3½ lin. broad ... ... (6) **Schinzii.**
            - Blades rounded or subcordate and subamplexicaul at the base ... ... (7) **amplectens.**
          - Blades filiform; pedicelled spikelets 6–8 lin. long... ... ... (8) **filifolius.**

Lower glume dorsally flattened; keels broadly winged above ... ... (9) **distachyus.**

Subgenus IV. AMPHILOPHIS. Racemes all peduncled, joints of the rhachis many (in the South African species) to few, like the pedicels linear, compressed, with a translucent channel and thickened margins:

Racemes digitate or fascicled, lower exceeding the common rhachis:

Lower glume of sessile spikelets not pitted ... (10) **Ischæmum.**

Lower glume of sessile spikelets pitted ... ... (11) **pertusus.**

Racemes panicled, lower shorter than the common rhachis ... ... (12) **intermedius.**

Subgenus V. SORGHUM. Racemes all peduncled, joints of the rhachis 8–1, and like the pedicels opaque; racemes usually 3–8-, sometimes 2–1-articulate; sessile spikelets and grains dorsally compressed; lodicules ciliate (in the South African species) or glabrous:

Perennial; rhachis of raceme fragile ... (13) **halepensis.**

Annual; rhachis of raceme tough ... ... (14) **Sorghum.**

Subgenus VI. CHRYSOPOGON. Racemes 1-articulate (i.e. reduced to 1 sessile ⚥ and 2 pedicelled ♂ or barren spikelets), jointed on the bearded tips of the branches of the panicle ... (15) **monticola.**

B. HETEROZYGI. Sessile spikelets of the lowest 1 or more pairs, differing from those above in sex or form (if the racemes are 2-nate, this is the case at least in one of the racemes):

Subgenus VII. DICHANTHIUM. Racemes usually many, digitate or panicled; all the spikelets alike in form, but differing in sex ... ... (16) **annulatus.**

Subgenus VIII. HETEROPOGON. Racemes solitary, terminal on the culms and their branches; lower sessile spikelets very unlike the upper sessile which are cylindric ... ... (17) **contortus.**

Subgenus IX. CYMBOPOGON. Racemes 2-nate, terminal on the culms and their branches, both together subtended by spathiform sheaths, frequently in spurious leafy or spathaceous panicles; spikelets usually differing in sex and form:

Column of awn glabrous:

Culms densely tufted, with intravaginal innovation shoots; lower sheaths firm, fugaciously hairy to tomentose at the base, persistent; blades narrow at the base, $\frac{1}{2}$–$2\frac{1}{2}$ lin. broad:

Lower glume of ⚥ spikelets lanceolate, dorsally flattened or shallowly depressed; keels more or less winged above ... ... (18) **Nardus.**

Lower glume linear-lanceolate, concave below between the rounded keels which are scarcely winged above ... ... (19) **plurinodis.**

Culms fascicled, frequently with extravaginal innovation shoots, or annual; sheaths quite glabrous, subherbaceous; lower glume of ⚥ spikelets dorsally flattened with a narrow groove from the middle downwards corresponding to a keel inside ... ... (20) **Schœnanthus.**

Column of awn pubescent or hirsute:

Callus of ⚥ spikelets very short or obscure, obtuse:

†Proper spathes narrow, linear to lanceolate, 1–3 in. long:
Spikelets of the lowest pair of the peduncled raceme of different sex:
Racemes long, 5–many-jointed:
Spikelets silvery villous:
Common peduncles equalling or more often exceeding the long and very narrow spathes; culm slender; ligules short up to 1½ lin. long ... (21) **hirtus.**
Common peduncles shortly and laterally exserted from the linear-lanceolate spathes; culms robust; ligules up to 3½ lin. long ... (22) **auctus.**
Spikelets not silvery villous:
Spurious panicle decompound; lower sheaths quite glabrous below; culms stout, tall:
Common peduncles enclosed in the lanceolate spathes or shortly and laterally exserted; spikelets glabrous or with whitish to greyish hairs ... ... (23) **Schimperi.**
Common peduncles more or less, often long, exserted from the long and very narrow spathes; spikelets glabrous or pubescent with short rigid rufous hairs ... ... (24) **rufus.**
Spurious panicle narrow, remotely and scantily branched; culms rather slender; lower sheaths fugaciously tomentose at the base; common peduncles exserted, strongly curved or cirriform ... (25) **dregeanus.**
Racemes short, 2–5-jointed, shortly laterally or subterminally exserted, falling from the spreadingly bearded tips of the common peduncles ... ... (26) **dichrous.**
Spikelets of the lowest 2 pairs of the peduncled raceme alike in form and sex (28) **Buchanani.**
††Spathes broadly lanceolate to ovate ½–1 in. long, often vividly coloured; racemes very short, 1–3-jointed ... ... (27) **Cymbarius.**
Callus of ⚥ spikelets slender, acute, up to 1 lin. long, densely bearded:
No scarious bract-like appendage below the ⚥ spikelets:
Racemes short, 5–6 lin. long, each with 1 ⚥ spikelet ... ... (29) **filipendulus.**
Racemes 6–12 lin. long, both or at least one of each pair with 2–4 ⚥ spikelets:
Culms simple; racemes 6–8 lin. long, both or at least one of each pair with 2–3 ⚥ spikelets; pedicelled spikelets ♂ ... (30) **transvaalensis.**
Culms branched all along; racemes about 1 in. long, both of each pair with 3–4 ⚥ spikelets; anthers of pedicelled spikelets barren ... ... (31) **pleiarthron.**
⚥ spikelets subtended by a bract-like appendage to the tip of the joint or peduncle ... (32) **Ruprechtii.**

Subgenus 1. SCHIZACHYRIUM (Benth.). Racemes solitary; joints stout, tips cupular with toothed margins. Spikelets of each pair similar, the sessile ⚥, wedged in between the joint and pedicel. Lower glume 2-keeled, dorsally compressed; upper more or less lanceolate, keeled, muticous or with a short bristle. Lower valve hyaline, slightly shorter than the glumes, 2-nerved or nerveless; upper deeply bifid with a slender awn from the sinus. Pedicelled spikelets usually more or less reduced and barren, rarely ♂, muticous or with a bristle from the lower glume.

1. **A hirtiflorus** (Kunth, Rév. Gram. ii. 569, t. 198) var. **semiberbis** (Stapf); perennial, tufted; culms 1–3 ft. long, glabrous, more or less branched or almost simple; sheaths tight, glabrous or hairy, particularly near the mouth; ligules membranous, short, glabrous; blades linear, acute, 2–12 in. by 1–2 lin., folded, firm, smooth except the margins, glabrous or hairy near the base, glaucous or reddish, the uppermost very much reduced or suppressed; peduncles never more than 1 from each sheath, the lateral generally exserted; racemes slender, stiff, 2–5 in. long, joints up to 20, 3–3½ lin. long, with white silky hairs at the base, or more or less ciliate along the edges, pedicels slightly shorter, ciliate along one or both edges; sessile spikelets linear, 3½–4 lin. long; lower glume bicuspidate, coriaceous, subconvex, smooth and often shining, keels scaberulous, intracarinal nerves 2–5; upper chartaceous, 3-nerved, margins ciliate, keel narrowly crested above; lower valve lanceolate, 2-nerved, ciliate; upper bifid beyond the middle, 1-nerved, lobes ciliate, awn 5–8 lin. long, lower twisted portion scarcely exserted; pale very minute or 0; pedicelled spikelets barren, 1½–2½ lin. long, middle and carinal nerves strong, often with a fine awn from between the mucros; valves hyaline, 3–1-nerved or suppressed. *A. semiberbis, Kunth, Enum. Pl.* i. 489; *Hack. Andr. in DC. Monogr. Phan.* vi. 369; *Durand & Schinz, Consp. Fl. Afr.* v. 723. *A. Pseudograya, Steud. Synops. Pl. Glum.* i. 365; *Hack. Andr. l.c.* 370; *Durand & Schinz, l.c.* 720. *A. leptostachyus, Benth. in Hook. Niger Fl.* 571.

KALAHARI REGION: Bechuanaland; Kuruman, *Burchell*, 2187! Transvaal; Elands River, *Nelson*, 49*! Houtbosch, *Rehmann*, 5727! 3728!
EASTERN REGION: Natal; Umpumulo, *Buchanan*, 199!

Tropical regions of both hemispheres.
The typical form, which has the spikelets as well as the joints and pedicels hairy all over, does not occur in the Old World.

Subgenus 2. HYPOGYNIUM (Hack.). Racemes solitary, each generally supported by a spathe; joints slender, tips oblique. Spikelets of each pair very similar, the sessile ⚥, rarely ♀, dorsally compressed. Lower glume 2-keeled, upper lanceolate, keeled, muticous or shortly awned (in the South African species). Lower valve hyaline or suppressed; upper entire or bifid (in the South African species), hyaline below, passing into the awn above, rarely muticous. Pedicelled spikelets ♂ or barren, awnless.

2. **A. ceresiæformis** (Nees, Fl. Afr. Austr. 109); perennial, tufted, culms erect or ascending; very slender, 1–4 ft. long, glabrous, smooth, many-noded, simple below, branched above; branches solitary or 2–4-nate, often subpendulous, filiform, with bearded tips, collected into a spathaceous raceme or panicle; sheaths terete, tight,

glabrous or hairy to villous, shorter than the internodes; ligules very short, rounded; blades linear, tapering to an acute point, 2–6 in. by 1–2 lin., flat, erect, somewhat firm, glabrous or hairy, turning red, midrib white; spathes boat-shaped, acuminate, $\frac{3}{4}$–$1\frac{1}{4}$ in. long, membranous, reddish, glabrous, many-nerved; racemes about 7–8 lin. long, shortly peduncled, surrounded by the spathe except the pedicellate spikelets, joints 6–8, filiform, villous above, about $\frac{3}{4}$ lin. long, pedicels similar, finer, villous all along; spikelets oblong, yellowish-green or brown, hairy or glabrous and shining on the back, the sessile $1\frac{1}{2}$–2 lin. long; lower glume chartaceous, narrowly truncate, intracarinal nerves 2, evanescent below; upper glume membranous below, hairy above, 3-nerved, tips generally awned; lower valve lanceolate-oblong, faintly 2-nerved, margins softly ciliate; upper oblong-linear, bifid, almost $1\frac{1}{2}$ lin. long, 1-nerved, lobes very narrow, glabrous, awn fine, about 5–7 lin. long, kneed below the middle; pale very minute; anthers 1 lin. long; grain oblong, dorsally slightly compressed; pedicelled spikelets obtuse, about 2 lin. long; lower glume 7–9-nerved, nerves unequal; upper 5-nerved; lower valve 3-nerved, ciliate; upper very narrow, linear, nerveless, awnless; pale 0. *Steud. Syn. Pl. Glum.* i. 383; *Hack. in Hook. Icon. Pl. t.* 1870; *Hack. Androp. in DC. Monogr. Phan.* vi. 398.

SOUTH AFRICA: without locality, *Drège*, 4263!

KALAHARI REGION: Orange Free State, *Cooper*, 3373! Transvaal; on the summit of Houtbosch Berg, *Nelson*, 25*! Makapans Bergen at Streyd Poort, *Rehmann*, 5382! Johannesburg, *E.S.C.A. Herb.*, 308!

EASTERN REGION: Pondoland; between St. Johns River and Umtsikaba River, *Drège!* Griqualand East; Mount Pumugwan, near Clydesdale, 3500 ft., *Tyson*, 3088! Natal; coast region, *Sutherland!* common near Umpumulo, 2000 ft., *Buchanan*, 225! Riet Vlei, 6000 ft., *Buchanan*, 224! and without precise locality, *Plant*, 60! *Gerrard*, 676.

In the typical form extending to the Zambesi, and in more or less distinct varieties throughout tropical Africa.

Subgenus 3. ARTHROLOPHIS (Hack.). Racemes 2-nate(one sessile, the other peduncled), digitate or paniculate, then partly sessile, partly peduncled, from spathiform sheaths; joints numerous, rarely 4–6, ancipitous opaque (confer *A. distachyus*) tips often appendaged. Spikelets of each pair heteromorphous, the sessile generally all alike, ⚥, usually awned. Lower glume 2-keeled; upper lanceolate (in the South African species) or oblong, keeled. Lower valve hyaline, as long as or slightly shorter than the glumes, linear to oblong; upper hyaline, oblong or linear, generally bifid and awned (in the South African species) rarely muticous. Stamens 3 (in the South African species) to 1. Pedicelled spikelets ♂ or reduced and barren, or suppressed, very rarely ⚥.

3. A. eucomus (Nees, Fl. Afr. Austr. 104); perennial, densely tufted, slender, 1–3 ft. long, glabrous, 4–6- or more-noded, simple to or beyond the middle, then distantly branched; branches long, very slender, solitary, or 2–3-nate, simple or again branched; leaves mainly basal; sheaths shorter than the internodes, glabrous or hairy near the mouth, rarely lower down, the lower compressed, keeled, the upper distant, subtumid; ligules membranous, very short, truncate, ciliolate; blades linear, acute, folded, from 3 in. to 1 ft. by 1–$1\frac{1}{2}$ lin. (unfolded), glabrous or hairy, pale green; spathe linear, setaceously acuminate, 2 in. long, glabrous, usually exceeding the filiform glabrous peduncle; racemes 2–4-nate, 1–1½ in. long, very slender, flexuous, suberect or nodding, silvery plumose, joints filiform, shorter than the spikelets, silky with soft long (about 5 lin.) hairs; pedicels similar, exceeding the spikelets, barren; spikelets lanceolate-oblong, 1–1½ lin. long, often purplish; glumes membranous, acute; lower nerveless between the scabrid keels, callus minute, long-bearded; upper 1-nerved; lower valve nerveless, ciliate; upper lanceolate-oblong, lobes very fine, awn a fine bristle, 6–9 lin. long; pale very minute; anthers ¼ lin. long; grain over ½ lin. long. *Steud. Syn. Pl. Glum.* i. 390; *Oliv. in Trans. Linn. Soc.* xxix. 176; *Hack. Andr. in DC. Monogr. Phan.* vi. 421; *Durand & Schinz, Consp. Fl. Afr.* v. 711. *Eriopodium Kraussii, Hochst. ex Krauss in Flora,* 1846, 115.

COAST REGION: Clanwilliam Div.; Olifants River, *Zeyher!* Wupperthal, 1500–2000 ft., *Drège!* Cape Div.; near Capetown, *Harvey*, 168! Cape Peninsula, Vyges Kraal, *Dodd*, 2211! Paarl Div.; Klein Drakenstein Mountains, by the Berg River, below 500 ft., *Drège!* Tulbagh Div.; New Kloof, on wet rocks, 900 ft., *MacOwan*, 1691! Tulbagh Div.; near Ceres Road Station, *Schlechter*, 288! Port Elizabeth Div.; Port Elizabeth, *E.S.C.A. Herb.*, 277! Albany Div.; Sidbury, near Grahamstown, *Baur*, 1073!

KALAHARI REGION: Griqualand West; Hay Division at Griquatown, *Burchell*, 2102! Bechuanaland; Hamapery, near Kuruman, *Burchell*, 2523! Transvaal; Nile River, near Mantatus Kraal, *Nelson*, 87*! Houtbosch, *Rehmann*, 5675!

EASTERN REGION: Natal; abundant on the coast flats, common at 1000 ft., *Buchanan*, 211! Durban Flat, *Buchanan*, 26! near Durban, *Wood*, 1654! *M'Ken*, 126! between Umlazi River and Durban Bay, *Krauss*, 163! by the Umlazi River, *Drège!* and without precise locality, *Gerrard*, 683!

4. A. appendiculatus (Nees, Fl. Afr. Austr. 105); perennial, densely tufted; culms erect, 2–3 ft. long, glabrous, compressed or terete, 5–6-noded, simple below, remotely branched from the upper 2–3 nodes; branches solitary, long, strict, glabrous; leaves mostly crowded near the base; sheaths glabrous or bearded at the mouth, lower very firm, strongly compressed, keeled, persistent, upper shorter than the internodes, tight, uppermost subspathaceous; ligules membranous, very short, truncate, ciliolate; blades linear, acute, usually folded, lower 4–12 in. by 1–2½ lin. (unfolded), uppermost very short or obsolete, rigid, glabrous, or hairy at the base, acutely keeled, margins scabrid; peduncles generally long exserted, glabrous; racemes 2–8, fascicled, unequal, 1½–4 in. long, flexuous, slender, loose, joints 2–3 lin. long, thicker upwards, hairy on the convex back, or only along the margins, hairs white, particularly above, tips subcupular with irregularly toothed margins, pedicels similar, more slender, produced into a subulate appendage facing the upper glume; sessile spikelets lanceolate, 3–3½ lin. long, often purplish; glumes subcoriaceous to chartaceous, glabrous, lower acute or acuminate, dorsally concave, keels acute, scabrid, or rigidly ciliate, callus scantily bearded; upper glume boat-shaped, 1-nerved; lower valve oblong-lanceolate, faintly 2-nerved, softly ciliate; upper valve 2 lin. long, bifid to ½, 1-nerved, lobes very narrow, ciliate, awn 5–7 lin., kneed below the middle; pale ½ lin. long, ciliate; anthers 1 lin. long; pedicelled spikelets ♂, narrowly lanceolate, 3–3½ lin. long, purplish, glabrous; lower glume acuminate, with a median keel in the upper part; upper 1-nerved, resembling the entire, lanceolate, muticous, reversedly ciliate valves. *Steud. Syn. Pl. Glum.* i. 379; *Hack. Androp. in DC. Monogr. Phan.* vi. 436. *A. Ischæmum, Thunb. Prodr. Pl. Cap.* 20, *and Fl. Cap. ed. Schult.* 108, *not Linn., ex Nees.*

SOUTH AFRICA: without locality, *Zeyher*, 1801! *Drège*, 14!
COAST REGION: Clanwilliam Div.; Wupperthal, 1500-2000 ft., *Drège*. Tulbagh Div.; near Tulbagh Waterfall, *Ecklon!* Worcester Div.; Dutoits Kloof, 3000–4000 ft., *Drège!* Uitenhage Div.; *Ecklon and Zeyher*, 696! Van Stadens Berg Range, *Drège*. Port Elizabeth Div.! Upper Leadmine River, *Burchell*, 4622! Albany Div.; moist places near Grahamstown, *MacOwan*, 1314! Fort Beaufort Div.; Winter Berg, *Ecklon*. Queenstown Div.; Zwart Key River, *Ecklon*.
CENTRAL REGION: Albert Div.; near Gaatje, 5000 ft., *Drège!*
KALAHARI REGION: Orange Free State; Caledon River, *Burke!* Basutoland, *Cooper*, 923! Transvaal; Bothas Berg, Steel Poort, *Nelson*, 59*! Magalies Berg, in marshy places, *E.S.C.A. Herb.*, 305! Matjesgoed Spruit, *Nelson*, 60*!
EASTERN REGION: Tembuland; Bazeia, 2000 ft., *Baur*, 295! Natal; Durban Flat, *Buchanan*, 41! Riet Vlei, 6000–7000 ft., *Buchanan*, 192! Umpumulo, *Buchanan*, 191!

There is no South African specimen referable to this species or to *A. Ischæmum*, Linn., in Thunberg's herbarium.

5. **A. schirensis** (Hochst. ex Rich. Tent. Fl. Abyss. ii. 456) var. **angustifolia** (Stapf); perennial, tufted; culms erect, slender, 2–3 ft. long, glabrous, 3–4-noded, simple, rarely with an additional flowering branch from one of the uppermost nodes; leaves mainly crowded near the base; sheaths terete, tight, glabrous, lowest more or less persistent; ligules membranous, very short, truncate; blades narrow, linear, tapering to a fine, sometimes setaceous point, 4–7 in. by 1 rarely 2 lin., rather firm, glabrous, rarely hairy below, margins scabrid; racemes 2-nate, rarely solitary, rather slender, 2–3½ in. long, strict or subflexuous, joints and pedicels very similar, cuneate-linear, 2–3 lin. long, tips deeply hollowed and produced into a short irregularly toothed appendage, long ciliate along the margins; sessile spikelets strongly laterally compressed, wedged in between the pedicel and the joint, 2½–3 lin. long, callus slender, pungent, densely bearded, 1 lin. long, sunk in the hollow of the preceding joint; glumes subcoriaceous or chartaceous, glabrous, smooth, lower linear, subacute, keels narrow with a deep narrow groove (fold) between them, scaberulous above; upper boat-shaped, finely mucronate, 1-nerved; lower valve oblong-linear, obtuse, 2-nerved, ciliate; upper deeply bifid, 1-nerved, lobes lanceolate, acute, ciliolate, awn ¾–1¼ in. long, slender, kneed and pubescent below the middle; pale oblong, 1½ lin. long, nerveless or sub-2-nerved; anthers 1 lin. long; pedicelled spikelets ♂, dorsally compressed, lanceolate, 4–5 lin. long, often reddish or purple; glumes subherbaceous, acuminate, glabrous; lower acutely 2-keeled, keels scabrid or rigidly ciliate above, intracarinal nerves 8; upper linear-lanceolate, 3-nerved, ciliate; lower valve as in the sessile spikelets; upper linear, shortly 2-toothed, muticous; pale linear, 1–1½ lin. long.

SOUTH AFRICA: without locality, *Zeyher*, 1802! 292!

KALAHARI REGION: Orange Free State; Drakens Berg Range, near Harrismith, *Buchanan*, 120! near Thaba Unchu, *Burke*, 434! Basutoland; Leribe, *Buchanan*, 144!

EASTERN REGION: Natal; Umpumulo, 2000–2500 ft., *Buchanan*, 195! Riet Vlei, 4000-6000 ft., *Buchanan*, 193! 196! and without precise locality, *Buchanan*, 116 partly!

The species was described from specimens collected by Schimper in Abyssinia. I have not seen them; but Schweinfurth's and Barter's specimens from Central Africa and the Niger, which Hackel (Andr. in DC. Monogr. Phan. vi. 452) refers to it, represent a much taller and stouter plant, attaining a height of 10 ft. and having leaves 3–7 lin. broad and very long, and racemes up to 6 in. long, whilst the calli of the spikelets are generally shorter and more obtuse, and the anthers up to 2 lin. long. Another form, intermediate between this robust state and var. *angustifolius*, occurs in East Africa from the Zambesi to Usambara.

6. A. Schinzii (Hack. Androp. in DC. Monogr. Phan. vi. 458); perennial; culms erect, slender, 2–4 ft. long, glabrous, terete, branched from the middle and upper nodes; branches usually simple, leafy, long, erect, flowering; leaves scattered along the culm; sheaths exceeding the internodes except the uppermost, terete, tight, glabrous, auricled at the mouth; ligules membranous, reddish, truncate, laterally produced and adnate to the auricles of the sheath, ½–1 lin. long; blades linear from a slightly narrower base, long tapering to a fine point, 4–8 in. by 2–3½ lin., flat, rigid, glaucous, turning reddish, glabrous or hairy at the base, scaberulous; racemes 2-nate, 2–3 in. long, rather stout, strict or flexuous, rather long exserted from the uppermost sheath which bears a subsetaceous blade; joints and pedicels equal or subequal and similar, cuneate-linear, 2–2½ lin. long, stout, very long and rigidly ciliate along the margins, tips hollowed, scarious, 2-lobed and unequally toothed; sessile spikelets laterally compressed, wedged in between joint and pedicel, 3–3½ lin. long, glabrous, callus short, obtuse, shortly bearded, sunk in the hollow of the tip of the preceding joint; lower glume subcoriaceous to chartaceous, linear, minutely 2-toothed, keels narrow, acute, spinulously ciliate above, with a deep narrow groove (fold) between them, and with 4 nerves close to each keel; upper glume firmly membranous, boat-shaped, 3-nerved, with a terminal bristle about as long as the spikelet, keel scabrid, margins ciliate; valves equal, lower linear-oblong, obtuse, 2-nerved, ciliate; upper broadly oblong, deeply bifid, 3-sub-5-nerved, lobes lanceolate, ciliolate, awn slender, about 1 in. long, kneed just below the middle, scaberulous below the knee; pale linear, almost as long as the valve, nerveless, glabrous; anthers 1½ lin. long; pedicelled spikelets ♂, dorsally compressed, linear-lanceolate, acute, 3½ lin. long, purplish or reddish; lower glume herbaceous, minutely 2-toothed with an interposed bristle of about the same length, 2-keeled, keels scabrid, intracarinal nerves 7–9; upper glume lanceolate, 3-nerved, ciliolate, aristulate like the lower; valves oblong-linear, acute or mucronulate, 3-nerved; lower finely 2-keeled, keels rigidly ciliate; upper long and softly ciliate, muticous. *Durand & Schinz, Consp. Fl. Afr.* v. 721.

KALAHARI REGION: Griqualand West; Hay Div. at Klip Fontein, *Burchell*, 2164/2!

Also in Amboland, German South-West Africa.

7. **A. amplectens** (Nees, Fl. Afr. Austr. 104); perennial, densely tufted; culms erect, slender, 2–2½ ft. long, glabrous, 3–5-noded, simple or with 1–3 flowering branches from the upper nodes; leaves mostly crowded at the base; sheaths terete, tight, glabrous, lowest widened, subpersistent; ligules membranous, short, truncate or rounded; blades linear, tapering to a long setaceous point, broad, rounded at the base or subcordate and subamplexicaul, 4–8 in. by 3–5 lin., flat, or convolute above, rather firm, glabrous, margins scabrid; racemes 2-nate, rather slender, 2–3½ in. long, strict or subflexuous; joints and pedicels very similar, sublinear, 3 lin. long, shortly ciliate along the margins, tips hollowed, denticulate; sessile spikelets laterally compressed, wedged in between the pedicel and the joint, 3½–4 lin. long, glabrous, callus short, acute, bearded, sunk in the hollow of the preceding joint; glumes coriaceous; lower linear-lanceolate, acuminate, 2-toothed, keels rounded and broad below, acute and scabrid near the tips, with a distinct groove (fold) extending between them to the acumen and with 2 fine lateral furrows in the upper third; upper 1-nerved, mucronate, or aristulate; valves subequal; lower lanceolate, 3-nerved, softly ciliate; upper linear-lanceolate, deeply bifid, lobes lanceolate, ciliolate, awn stout, 1¼–2 in. long, kneed and pubescent below the middle; pale linear-oblong, 1 lin. long, nerveless, ciliate; anthers 2½ lin. long; pedicelled spikelets ♂, dorsally compressed or subterete, lanceolate, 5–6 lin. long; glumes subherbaceous, acuminate, often aristulate; lower acutely 2-keeled, keels scabrid or rigidly ciliate above, intracarinal nerves many (to 19), middle nerve stronger; upper linear-lanceolate, 3-sub-5-nerved, ciliate; valves linear 3-nerved; lower acuminate, 4½ lin. long, ciliate; upper 3½ lin. long, 2-toothed, muticous; pale linear, 2 lin. long, nerveless. *Steud. Syn. Pl. Glum.* i. 372; *Hack. Androp. in DC. Monogr. Phan.* vi. 453; *Durand & Schinz, Consp. Fl. Afr.* v. 705.

KALAHARI REGION: Basutoland, *Marloth*, 1114! Transvaal; Houtbosch, *Rehmann*, 5678! Makapans Bergen, at Streyd Poort, *Rehmann*, 5380! Dronk Fontein, *Nelson*, 76*! Apies River, *Nelson*, 41*!

EASTERN REGION: Tembuland; near Morley, 1000–2000 ft., *Drège!* Natal; Umpumulo, 2000 ft., *Buchanan*, 194! between Umtata River and St. Johns River, 1000–2000 ft., *Drège!* and without precise locality, *Gerrard*, 769! *Buchanan*, 116 partly!

8. **A. filifolius** (Steud. Syn. Pl. Glum. i. 374); perennial, densely tufted; culms erect, slender, 1–2½ ft. long, glabrous, terete, about 3-noded, simple; leaves mainly crowded at the base; sheaths shorter than the internodes, terete, tight, glabrous, lowest persistent, at length breaking up into fibres; ligules membranous, very short, truncate; blades very narrow, convolute, filiform, acute, lowest ⅔–1 ft. long, upper very short, firm, flexuous, glabrous or hairy at the base;

racemes 2-nate, 2½–3½ in. long, stout, strict, joints and pedicels subequal and similar, cuneate-linear, 2½–3½ lin. long, densely and shortly villous along the margins, tips hollowed, unequally toothed; sessile spikelets laterally compressed, wedged in between joint and pedicel, 3½ lin. long, callus slender, 1–1½ lin. long, acute, densely bearded; glumes subcoriaceous, glabrous, smooth; lower linear, obtuse, tips hyaline, keels narrow, rounded, smooth, almost contiguous, with a deep very narrow groove (fold) between them; upper boat-shaped, 1-sub-3-nerved, ciliate; valves subequal; lower oblong-linear, 2-nerved, ciliate; upper linear, deeply bifid, 3-nerved, lobes oblong, ciliolate, awn stout, 2–2¼ in. long, kneed at the middle, pubescent below; pale ovate, acute, 1½ lin. long, nerveless, glabrous; anthers 2¼ lin. long; pedicelled spikelets ♂, dorsally compressed or subterete, lanceolate, 6–8 lin. long, glabrous, often purplish; lower glume herbaceous, 2-toothed, one tooth often prolonged into a bristle (to 3 lin. long), acutely 2-keeled, intracarinal nerves many (to 20); upper finely acuminate, 3-nerved, ciliate; lower valve linear-oblong, acute, 4 lin. long, 3–2-nerved, ciliate; upper linear 1-nerved, glabrous, muticous; pale linear, 2 lin. long, nerveless; anthers 3½ lin. long. *Hack. Androp. in DC. Monogr. Phan.* vi. 453; *Durand & Schinz, Consp. Fl. Afr.* v. 712. *Heteropogon filifolius, Nees, Fl. Afr. Austr.* 102.

COAST REGION: Port Elizabeth; *E.S.C.A. Herb.*, 96! Alexandria Div.; Zuur Berg Range, 2000–3500 ft., *Drège!* Bathurst Div.; near Theopolis, *Burchell*, 4096! between Port Alfred and Kaffir Drift, *Burchell*, 3852!

KALAHARI REGION: Transvaal; Houtbosch, *Rehmann*, 5679!

EASTERN REGION: Natal; Riet Vlei, 4000–5000 ft., *Buchanan*, 197! near Newcastle, *Buchanan*, 195!

9. A. distachyus (Linn. Spec. Plant. 1046); perennial, tufted; culms erect, slender 1–2½ ft. long, glabrous, terete, about 3-noded, simple or branched below; sheaths terete, tight, glabrous or, particularly the lower, hairy, lowest reduced to villous scales; ligules membranous, about ½–¾ lin. long, ciliolate; blades linear, tapering to a long and sometimes very fine point, 2–8 in. by ¾–1½ lin., flat, subrigid to flaccid, more or less hairy to villous or subglabrous, margins rough; racemes 2-nate, very rarely 3–5-nate, distant from the uppermost sheath, 2–4 in. long, rather stout, strict or curved; joints cuneate, 2–2½ lin. long, stout, translucent along the middle, finely pubescent on the back, ciliate along the outer margin, tips denticulate, pedicels similar, slightly longer, firmer and less translucent, tips produced into a denticulate lobe facing the upper glume; sessile spikelets 5–5½ lin. long, pale green, tips often purplish, callus short, obtuse, bearded; lower glume subherbaceous, broadly lanceolate, dorsally flattened, glabrous or puberulous, rarely villous, keels broadly winged above, wings membranous, whitish or purplish, intracarinal nerves about 7–11, partly evanescent below; upper glume distinctly shorter than the lower, rigidly membranous, boat-shaped, 3-nerved, softly ciliate, tips minutely 2-toothed with an interposed bristle, about as long as the glume; valves equal, 3 lin. long; lower 2-nerved, ciliate; upper oblong, 2-fid to $\frac{2}{3}$, firmer below, 3-nerved, lobes lanceolate, glabrous, awn slender, about 1 in. long, kneed much below the middle, scaberulous below the knee; pale minute or obsolete; anthers $1\frac{1}{2}$ lin. long; grain oblong, 1 lin. long; pedicelled spikelets ♂ similar to the sessile, about 4 lin long, but narrower; lower glume less acuminate and less distinctly winged, with a short terminal bristle; upper thinly membranous; both valves delicately hyaline, upper shortly bifid, muticous. *Jacq. Ic. Pl. Rar.* iii. 630; *Host, Gram. Austr.* iii. 2; *Sibth. and Sm. Fl. Graec.* i. 53, *t.* 69. *Kunth, Enum.* i. 491; *Steud. Syn. Pl. Glum.* i. 372; *Hack. Androp. in DC. Monogr. Phan.* vi. 461; *Durand & Schinz, Consp. Fl. Afr.* v. 710. *Pollinia distachya, Spreng. Pug. Pl. Nov.* ii. 12; *Reichb. Ic. Fl. Germ.* i. 21, *fig.* 1501.

KALAHARI REGION: Orange Free State; on the Witte Bergen, near Harrismith, *Buchanan*, 263!

EASTERN REGION: Natal; Newcastle, *Buchanan*, 179! and without precise locality, *Buchanan*, 15!

Common in the Mediterranean countries, also in Abyssinia and on Cameroon Peak.

Subgenus 4. AMPHILOPHIS (Hack.). Racemes fascicled or panicled, peduncled; oints 1 to many, like the pedicels linear, flattened, chanelled, and translucent between the thickened, ciliate margins; tips truncate. Spikelets of each pair very similar; the sessile dorsally compressed, callus short. Lower glume 2-keeled; upper more or less lanceolate, keeled, muticous, generally 3-nerved. Lower valve hyaline; upper very narrowly linear, generally passing gradually into or almost wholly reduced to the awn; pale 0 (in the South African species) or minute. Pedicelled spikelets usually as long as the sessile, but narrower, ♂ or barren.

10. A. Ischæmum (Linn. Spec. Pl. 1047) var. **radicans** (Hack. Androp. in DC. Monogr. Phan. vi. 476); perennial; culms erect or ascending from a prostrate branched base, $1–2\frac{1}{2}$ ft. long, rather stout, glabrous, 3–6-noded, simple above, or with 1–2 flowering branches from the uppermost nodes; sheaths compressed, rather loose, bearded at the nodes and the mouths; ligules truncate, ciliolate; blades linear, gradually tapering to a fine point, $1\frac{1}{2}–3$ in. by $1\frac{1}{2}–2$ lin., flat, rigid, more or less hirsute, with tubercle-based hairs, margins cartilaginous; racemes 5–7 or more, fascicled, $1\frac{1}{2}–3$ in. long, pale or purplish, silky, shortly peduncled, joints and pedicels slightly exceeding half the spikelet, long ciliate; sessile spikelets 2 lin. long, oblong-lanceolate, rather acute, callus minutely hairy or subglabrous; lower glume chartaceous, dorsally flattened or obscurely channelled, 7–9-nerved, hairy below the middle; upper acute or subobtuse, keel rigidly ciliate above; lower valve oblong-lanceolate acute; upper almost reduced to a kneed awn, 5–9 lin. long; anthers $1–1\frac{1}{4}$ lin. long; pedicelled spikelets ♂, glabrous, awnless. *Durand & Schinz, Consp. Fl. Afr.* v. 715. *A. radicans, Lehm. in Ind. Sem. Hort. Hamb.* 1828; *Kunth, Enum.* i. 499; *Nees, Fl. Afr. Austr.* 106; *Steud. Syn. Pl. Glum.* i. 380.

CENTRAL REGION: Graaff Reinet Div.; near Graaff Reinet, on stony hills, 2600 ft., *Bolus*, 518; Somerset Div.; by the Little and Great Fish Rivers, 2000–3000 ft., *Drège! Ecklon.*

The typical form which has glabrous nodes and shorter, finer awns, ranges from Central and South Europe to Eastern Asia.

11. A. pertusus (Willd. Spec. Plant. iv. 922); perennial; culms erect or ascending from the branched often prostrate base, simple above, 1–2½ ft. long, glabrous, 5–6-noded; sheaths rather terete, except the compressed lower ones, usually bearded at the nodes; ligules truncate, up to ½ lin. long, often with hairs from behind; blades linear, tapering to a fine point, 2–6 in. by 1–2 lin., flat, glabrous or more or less hairy, scabrid; racemes digitate, rarely more distant on a common rhachis not longer than the lowest racemes, about 2 in. long; peduncles usually glabrous except their bearded axils, short, joints and pedicels equalling ½ of the spikelet or slightly longer, ciliate, upper cilia as long as the joint; sessile spikelets about 2 lin. long, lanceolate-oblong, pale, callus bearded; lower glume subchartaceous, minutely truncate, villous near the base, faintly 5–9-nerved, pitted on the back, keels spinulously ciliate above; upper acute, sparsely ciliate or glabrous, 3-nerved; lower valve oblong, obtuse; upper almost reduced to a kneed awn, 6–10 lin. long; pedicelled spikelets usually ♂, not or shallowly pitted, purplish. *Beauv. Agrost.* 131, *t.* 23, *fig.* 2; *Kunth, Enum.* i. 498; *Steud. Syn. Pl. Glum.* i. 364; *Hack. Androp. in DC. Monogr. Phan.* vi. 479; *Durand & Schinz, Consp. Fl. Afr.* v. 718; *Hook. f. Fl. Brit. Ind.* vii. 173.

VAR. β **capensis** (Hack. l.c. 482); culms up to 4 ft. long; blades 4–12 in. by 2–3½ lin.; racemes few to 20; lower glume of sessile spikelets rather firm, glabrous or very scantily hairy, pale, shining, 2–2¼ lin. long; pedicelled spikelets usually 2–3-pitted. *Durand & Schinz, Consp. Fl. Afr.* v. 718. *A. pertusus, Nees, Fl. Afr. Austr.* 107.

COAST REGION: Var. β, Komgha Div.; banks of the Kei River, *Drège!*

EASTERN REGION: Natal; without precise locality, *Gerrard*, 692! Var. β, Natal; at the foot of Table Mountain, near Pietermaritzburg, *Krauss*, 29! Umpumulo, *Buchanan*, 190! 295 partly! and without precise locality, *Buchanan*, 201!

The species of which Hackel distinguishes 8 varieties, for the greatest part very slightly differing, ranges all over Africa and tropical Asia to North Australia.

12. A. intermedius (R. Br. Prodr. 202) var. **punctatus** (Hack. Andr. in DC. Monogr. Phan. vi. 487); perennial; culms erect or ascending, usually simple, to 5 ft. long, rather stout; sheaths glabrous except the sometimes appressedly hairy nodes, terete, tight, longer than the internodes or the upper shorter; ligules very short, truncate, often with hairs from behind; blades linear, tapering to a long fine point, 4–30 in. long by 1½–4 lin., rather rigid, glabrous or sometimes hispid at the base, usually smooth except the margins; panicle compound, oblong, dense, 4–7 in. long, usually purplish, common rhachis 2–5 in. long, branches like the short peduncles glabrous except for the bearded axils; racemes slender, often flexuous, joints and pedicels about half the length of the spikelets or slightly longer, ciliate, cilia usually shorter than the joint; sessile spikelets linear-oblong, 1½–1¾ lin. long, purplish, callus minutely bearded;

lower glume rather thin, slightly truncate, pitted, glabrous or more or less hairy below the middle, intracarinal nerves faint 4–7, keels spinulously ciliate above, rarely smooth; upper glume acute or subacute, 3-nerved, scantily ciliate or glabrous; lower valve oblong, subobtuse, nerveless, glabrous; upper almost reduced to a very slender kneed awn 5–7 lin. long; pedicelled spikelets ♂ or barren, usually not pitted, sometimes reduced to the glumes. *Durand & Schinz, Consp. Fl. Afr.* v. 715; *Hook. f. Fl. Brit. Ind.* vii. 176. *A. punctatus, Trin. Spec. Gram. Ic. t.* 328; *Steud. Syn. Pl. Glum.* i. 391, *not Roxb. A. perfossus, Nees ex Steud. l.c.*

KALAHARI REGION: Transvaal; Pretoria Div. at Kamel Poort, *Nelson*, 9! Lydenburg Div.; by the Komati River, *Nelson*, 46!

EASTERN REGION: Transkei: banks of the Bashee River, *Drège!* Griqualand East; Clydesdale, 2500 ft., *Tyson*, 3104! Natal; banks of the Umlazi River, *Drège!* Umpumulo, on hills, *Buchanan*, 295 partly! and without precise locality, *Gueinzius!*

Also in tropical Africa and Asia as far as China; the typical form which differs in the less compound panicle and in slightly larger, not pitted, spikelets extends throughout Australia to India and China.

Subgenus 5. SORGHUM (Hack.). Racemes panicled, all peduncled; rhachis fragile or tough in the cultivated varieties; joints filiform, truncate, margins ciliate. The spikelets of each pair similar or heteromorphous; all the sessile ☿, dorsally compressed; lower glume usually coriaceous and shining, 2-keeled near the tip, upper muticous; lower valve hyaline, as long as or slightly shorter than the glumes; upper oblong or linear, 2-fid or 2-toothed, awned from the sinus, rarely muticous. Lodicules usually ciliate. Stamens generally 3. Grain oblong, dorsally compressed, or subglobose (in the cultivated varieties). Pedicelled spikelets ♂ or barren, often much reduced or quite suppressed.

13. A. halepensis (Brot. Fl. Lusit. i. 89) var. **effusus** (Stapf in Hook. f. Fl. Brit. Ind. vii. 183); perennial; culms erect, usually very tall, up to 10–16 ft. long, stout, simple or scantily branched; sheaths glabrous, except the minutely silky nodes, strongly striate; ligules membranous, short, ciliate, hairy inside; blades linear-lanceolate or linear from an often rounded base, long tapering to a fine point, 1–2 ft. long, $\frac{3}{4}$–$2\frac{1}{2}$ in. broad, flat, glabrous, or with a silky line on the back at the union with the sheath, margins serrulate, midrib stout; panicle decompound, very large, up to $1\frac{1}{2}$ ft. long, effuse, nodding, lower branches up to 1 ft. long, often undivided to the middle; rhachis and branches or at least the ultimate branchlets scabrid and minutely bearded at the nodes; racemes $\frac{1}{2}$–1 in. long, linear; joints 3–7, more than half as long as the sessile spikelets, more or less ciliate, pedicels very similar; sessile spikelets ovate-lanceolate, 3–$4\frac{1}{2}$ lin. long, 1–$1\frac{1}{2}$ lin. broad, pale, ultimately sometimes darker or even black, shining; lower glume more or less hairy, at least on the sides, 7–13-nerved, callus shortly bearded; upper lanceolate, acuminate, shining, 5–7-nerved; upper valve broadly oblong or ovate, 2-lobed, half as long as the glumes, ciliate, 1-nerved, awn 4–6 lin. long, rarely longer, kneed, sometimes reduced to a bristle or suppressed; pale linear-oblong, slightly shorter than the valve; anthers $1\frac{1}{4}$–$1\frac{1}{2}$ lin. long; grain obovate or obovate-oblong, $\frac{1}{3}$ shorter than the glumes;

pedicelled spikelets almost as long as the sessile, but narrower, ♂ or barren; lower glume herbaceous, glabrous, 5–9-nerved, keels aculeolate or scabrid; upper similar, 3–5-nerved; valves, when present, hyaline, ciliate, 2–1-nerved. *A. Sorghum, subsp. halepensis, var. effusus, Hack. Androp. in DC. Monogr. Phan.* vi. 503; *Durand & Schinz, Consp. Fl. Afr.* v. 724. *A. arundinaceus, Willd. Spec.* iv. 906. *Rhaphis arundinaceus, Desv. Opusc.* 69. *Sorghum halepense, Nees, Fl. Afr. Austr.* 88.

COAST REGION: Knysna Div.; hills near Melville, *Burchell*, 5465! Komgha Div., banks of the Kei River, below 500 ft., *Drège*.
WESTERN REGION: Little Namaqualand; in thickets by the Orange River, near Verleptpram, below 500 ft., *Drège!*
EASTERN REGION: Tembuland; by the Umtata River, *Drège*, Pondoland; between the Umtata River and St. Johns River, 1000–2000 ft., *Drège*, by the St. Johns River, in coffee plantations, *Drège*. Natal; near Durban, *Drège*, banks of Tugela River, *Buchanan*, 296! by the Umlazi River, *Krauss*, 184! Umhlanga, *Wood*, 1332! and without precise locality, *Gerrard*, 690!

Throughout the tropics, but particularly in Africa. The form from which the species was originally described is apparently only a weaker state with smaller spikelets.

14. **A. Sorghum** (Brot. Fl. Lus. i. 88); annual; leaves as in *A. halepensis*, but the ligule often glabrous or glabrescent; panicle very variable, from effuse to compact; rhachis of racemes tough, joints and pedicels half as long as the sessile spikelets or very often much shorter, more or less ciliate; sessile spikelets very variable in shape and size, on the whole broader than in *A. halepensis*, $2–3\frac{1}{2}$ lin. by $1\frac{1}{4}–2\frac{1}{2}$ lin., pale, reddish, brown or at length black, usually shining; lower glume coriaceous, or more or less herbaceous, particularly towards the tips, rarely quite thin except at the base, often prominently nerved in the herbaceous part, hairy, rarely quite glabrous, otherwise as in *A. halepensis*, awn $2\frac{1}{2}–7\frac{1}{2}$ lin. long, more or less reduced or 0; grain obovate to globose; pedicelled spikelets ♂ or more frequently barren and more or less reduced. *Kunth, Enum.* i. 501; *Steud. Syn. Pl. Glum.* i. 393; *Körnicke und Werner, Handb. Getreidebau.* i. 294-315; *Hook. f. Fl. Brit. Ind.* vii. 183; *K. Schum. in Engl. Pfl. Ost-Afr. B.* 34. *A. Sorghum, subsp. sativus, Hack. Androp. in. DC. Monogr. Phan.* vi. 505. *Holcus Sorghum, Linn. Spec. Plant.* 1047; *Gærtn. Fruct.* ii. 2, *t.* 80; *Lamk. Illustr. t.* 838.

Cultivated in numberless forms in the tropical and subtropical regions, particularly in the Old World, and in the warmer parts of the temperate zones of both hemispheres. A great many varieties have been described; of these 7 are said to occur in South Africa. As I have, however, seen only a few specimens from that region, I must rely mainly on Hackel's, Körnicke's, and K. Schumann's descriptions which, unfortunately, are not always reconcilable. The following disposition is therefore only a provisional attempt, principally intended to give the student an idea in which directions the variation takes place chiefly, and to enable him to collect judiciously further material of this most important native cereal.

* Glumes exceeding and enveloping the grain (*Obtectæ*, K. Schum.).

VAR. α, **saccharatus** (Körn. in Körn. und Wern. Handb. d. Getreidebaues 310); panicle effuse, branches drooping, the lower undivided for 2–3 in.; spikelets elliptic, over 2½ lin. long, straw-coloured to brownish, awned (or awnless?). *A. saccharatus, Kunth, Enum.* i. 502, *not Roxb. Holcus saccharatus, Linn. Spec. Plant.* i. 1047. *H. Caffrorum, Thunb. Prod. Pl. Cap.* 20; *Fl. Cap. ed. Schult.* 109 (*from the description*). *H. Dochna, Forsk. Fl. Ægypt. Arab.* 174. *Sorghum saccharatum, Pers. Syn.* i. 101. *Host, Gram. Austr.* iv. 3, *t.* 4. *Nees, Fl. Afr. Austr.* (*excl. var. β*) 86. *S. Arduini, Jacq. Ecl. Gram. t.* 18.

VAR. β, **rufescens** (Hack. Androp. in DC. Monogr. Phan. vi. 511); panicle dense, rhachis glabrous, lower branches short, undivided for ½–1 in.; spikelets elliptic, 3 lin. long, reddish below, yellowish above; lower glume quite coriaceous, or the very tips herbaceous, hairy only along the margins and below the tips; grain pale reddish. *Sorghum saccharatum β rubens, Nees, Fl. Afr. Austr.* 87 (*excl. Syn.*).

VAR. γ, **Usorum** (Körn. l.c. 312); panicle dense, rhachis and branches pubescent, villous at the nodes, branchlets often ciliate along the angles, lower branches short, divided almost from the base; spikelets broadly elliptic, over 2½ lin. long, hairy, yellowish; grain yellowish or reddish. *Hack. l.c.* 512. *Sorghum Usorum, Nees, Fl. Afr. Austr.* 87. *Andropogon Usorum, Steud. Syn. Pl. Glum.* i. 392.

** Glumes shorter than the grain, adpressed to it (*Seminudæ*, K. Schum.).

VAR. δ, **Cafer** (Körn. l.c. 307); panicle effuse, rhachis short, branches up to 6 in. long, nodding, undivided to or beyond the middle; spikelets hairy; grains reddish, much longer than the glumes. *Holcus Cafer, P. Arduino, Saggi Padov.* i. 119, *t.* i. *f.* i. *Panicum Caffrorum, Retz. Obs.* ii. 7 (?).

VAR. ε, **Schenkii** (Körn. in Baumann, Usambara, 319); panicle contracted; branches glabrous except the scantily pubescent nodes; glumes yellowish, glabrous or subglabrous; grain yellowish-reddish, 2½ lin. long.

VAR. ζ, **Neesii** (Körn. l.c. 315); panicle compact, erect or recurved, rhachis and ramifications minutely hairy to villous, lower branches undivided for ½–1 in., short; spikelets suborbicular, very obtuse, 2 lin. long, minutely villous, awnless; lower glume quite coriaceous, dark-brown or black when mature; grain subglobose, white. *Sorghum bicolor, Nees, Fl. Afr. Austr.* 86, *not Willd.*

VAR. η, **melanospermus** (Hack. l.c. 518); like the preceding, but spikelets straw-coloured, scantily hairy, grain black.

The cereal of the Kaffirs, generally known as the Kaffir corn; occasionally grown in gardens in the western parts of the Cape Colony. I have seen only a few South African specimens, and as some of them are rather young or otherwise imperfect, the following determinations will have to be revised along with those by other authors, as soon as more complete material is at hand.

Var. *saccharatus* is very probably represented by a young specimen in the Kew herbarium distributed by *Drège* under the name "*Sorghum bicolor*," and without locality. It is said to be the commonest form grown in tropical Africa, and is also cultivated in tropical Arabia, India, Southern Europe and the United States of North America. Var. *Usorum* is indicated by Nees from Dutoits Kloof, in Worcester Div., and from the country of the Koosa Kaffirs ("tUs" Kaffirs of Nees, whence the name, approximately between the Great Fish River and Key River), and a flowering specimen collected by *Wood* (3990) near the Umvoti River, Natal, belongs very likely to it. Var. *Schenkii* was collected by *Schenk* on the Bosch Veldt, north of the Magalies Mountains, in the Transvaal, whilst var. *Neesii* is recorded from the Koosa Country and from Wilgebosch Spruit, in the Orange Free State (?). To the latter variety are evidently referable some specimens gathered by *Burchell* (5278!) at "Kaatje's Kraal," in Knysna Div., and at Takun in Bechuanaland (2214!), as well as a specimen in Thunberg's herbarium, marked *Sorghum Caffrorum*. The Knysna specimens represent, however, 2 forms, one having almost black mature glumes, whilst, in the other, their colour is the same as in the flowering state. The Takun specimen is in flower.

All of Burchell's specimens agree otherwise exactly. Their panicles are erect, oblong, the rhachis and the branches are villous at the nodes and more or less hairy above them. The spikelets answer the description of those of var. *Neesii;* they are, however, not always very obtuse, but often apiculate. Var. *Schenkii* as well as var. *Neesii* is said to be grown also in various parts of tropical Africa.

Subgenus 6. CHRYSOPOGON (Hack.). Racemes panicled, peduncled, nearly always reduced to 1 sessile ⚥ and 2 pedicelled ♂ or barren spikelets; tips of peduncles obconic, obliquely truncate, bearded; pedicels filiform, not grooved. Sessile spikelets usually laterally compressed, awned. Lower glume coriaceous or chartaceous, complicate or involute; upper generally keeled and awned. Lower valve hyaline; upper linear minutely 2-toothed or entire, hyaline below, passing into the awn above. Lodicules glabrous. Grain linear, laterally compressed. Pedicelled spikelets dorsally compressed, awnless or aristulate from the glumes.

15. A. monticola (Schult. Mant. iii. 665) var. **Trinii** (Hook. f. Fl. Brit. Ind. vii. 193); perennial; culms in dense tufts, simple or branched, 1–3 ft. high, usually very slender; lower sheaths compressed, crowded, more or less flabellate, upper keeled, shorter than the internodes; ligules almost reduced to a fringe of hairs; blades linear, acute, 3–12 in. by 1½–4 lin., folded or flat, rigid, glaucous, glabrous or minutely hairy, or ciliate towards the base; panicle ovate or oblong, erect, 2–4½ in. long, rhachis smooth or scabrid, branches few or many in each whorl, capillary, flexuous, spreading, at length erect; sessile spikelets linear or linear-oblong, 3–3½ lin. long, pale, callus bearded, hairs whitish to reddish; lower glume chartaceous, complicate, linear, glabrous except the keel, which is scabrid or ciliate below the tip, 4-nerved; upper obtuse, entire or submarginate, aristulate, keel scantily ciliate above; lower valve ciliate; upper shorter, awn 7–15 lin. long; pedicelled spikelets lanceolate, about as long as the sessile, glabrous; lower glume 5–7-nerved; upper 3-nerved and like the valves hyaline and ciliate; pedicels about ⅓ the length of the spikelets, subclavate, fulvous or reddish ciliate along the margins. *A. Trinii, Steud. Syn. Pl. Glum.* i. 395; *Hack. Androp. in DC. Monogr. Phan.* vi. 558. *Chrysopogon serrulatus, Trin. Androp. in Mém. Acad. Pétersb. Sér.* vi. ii. (1833) 318; *Sp. Gram. Ic. t.* 331.

KALAHARI REGION: Bechuanaland; Kuruman, *Marloth*, 1064; Hopetown Div.; by the Orange River, near "Saltpan Station," *Burchell*, 2655!

Also in India; the typical form which differs in the upper glume being covered for more or less of its length with long, usually rigid rufous hairs, is only known from Southern India.

Marloth's plant was described as var. γ *simplicior* of *A. Trinii* by Hackel, l.c. 559; but I cannot see how it differs from Burchell's which Hackel refers to typical *A. Trinii*.

Subgenus 7. DICHANTHIUM (Hack.). Racemes digitate, rarely solitary or panicled; joints many, linear, opaque or translucent along the middle line, ciliate, obliquely truncate; pedicels very similar. Lowest 1–4 sessile spikelets like the pedicelled in form and sex, ♂ or barren, the others ⚥, rarely all ⚥ (sometimes in *A. annulatus*). ⚥ spikelets dorsally compressed, callus short, rarely long. Lower glume usually very obtuse, rarely minutely truncate, 2-keeled; upper keeled, awnless. Lower valve hyaline, mostly nerveless; upper very narrow, gradually passing into the awn. Pale 0 (in the African species), rarely present, very small. Pedicelled spikelets similar to the ⚥, awnless.

16. A. annulatus (Forsk. Fl. Ægypt. Arab. 173); perennial, cæspitose; stem erect or ascending, simple, rarely with 1–2 flowering branches from the upper nodes, 3 ft. long; sheaths terete or obscurely keeled above, tight, glabrous except the bearded nodes, margins sometimes ciliate above; ligules 1–2 lin. long, obtuse, glabrous; blades linear, tapering to a long fine point, 2–10 in. by 1½–3 lin., rigid, glaucous, scabrid, glabrous or with tubercle-based hairs; racemes 5–7, rarely fewer, or solitary, or many (15), slender 1½–3½ in. long, rather flaccid, pale or purplish, peduncles short, glabrous, joints and pedicels about half the length of the spikelets, ciliate; all sessile spikelets ⚥, or the lowest one like the pedicelled and ♂; ⚥ spikelets oblong, 1½–2½ lin. long, callus minute; lower glume very obtuse, concave, usually more or less hairy all over the back, with tubercle-based hairs, intracarinal nerves 5–9, evanescent below the tip, keels spinulously ciliate; upper lanceolate; lower valve as long as the upper glume, linear-oblong, obtuse, nerveless, glabrous; upper stipitiform, awn slender, scabrid, 8–12 lin. long; pedicelled spikelets hairy all over, flatter than the ⚥; lower glume usually long, ciliate all along the margins, 7–11-nerved; valves more or less reduced, or the upper or both suppressed. *Délile, Fl. Égypte, t. 7, fig. 2; Hack. Androp. in DC. Monogr. Phan.* vi. 570; *Hook. f. Fl. Brit. Ind.* vii. 196. *A. Bladhii, Retz. Obs.* ii. 27; *Kunth, Enum.* i. 498; *Trin. Spec. Gram. t.* 325. *A. comosus, Link, Hort. Berol.* i. 239, *not Spreng. A. garipensis, Steud. Syn. Pl. Glum.* i. 379. *Lipeocercis annulata, Nees, Fl. Afr. Austr.* 98.

WESTERN REGION: Little Namaqualand; Verleptpram on the Orange River, *Drège!*

CENTRAL REGION: Calvinia Div.; Hantam Mountains, *Meyer.*

KALAHARI REGION: Griqualand West; Herbert Div., at Blaauwbosch Drift on the Vaal River, *Burchell*, 1748!

Throughout the tropics of the Old World.

Subgenus 8. HETEROPOGON (Endl.). Racemes solitary, subcylindric, dense; rhachis fragile or tough in the lower part; joints numerous, opaque, the lower glabrous, the upper ciliate, obliquely truncate. Lowest 1–15 sessile spikelets like the pedicelled, ♂ or barren, awnless, very different from the rest which are ♀ or ⚥, subcylindric and awned. Lower glume involute; upper oblong, more or less keeled, awnless. Lower valve oblong to linear, hyaline, nerveless; upper stipitiform, hyaline at the base, gradually passing into the awn. Pale minute or 0. Grain linear-oblong to oblong. Pedicelled spikelets usually imbricate and the upper concealing the fruiting spikelets.

17. A. contortus (Linn. Spec. Plant. 1045); perennial; culms erect, stout, simple, or branched from the upper nodes, often sheathed all along, 1–3 ft. long; sheaths compressed, keeled, glabrous; ligules short, truncate, ciliolate; blades linear, acute or tapering to a long, fine point, 3–8 in. by 1½–3 lin., flat or channelled, rigid, scabrid, especially above, sparingly hairy at the base or all over; racemes 1½–3 in. long, straight or curved, rhachis tough between the lowest 2–6 pairs, the spikelets of which are ♂ and like the pedicelled of the upper pairs, joints very short, unequally bearded; sessile spikelets of the heteromorphous pairs ♀ 3–3½ lin. long, callus 1½ lin. long, rufous-bearded; lower glume linear-oblong, truncate, shortly hairy, obscurely nerved; upper membranous, 3-nerved; awn stout, 3½–7 in. long, rufous-hairy below; ♂ spikelets obliquely lanceolate, 4–5 lin. long, dull green; lower glume herbaceous, subobtuse, glabrous, or hirsute with tubercle-based hairs, many-nerved, keels unequally winged; upper 3-nerved; lower valve minute, upper suppressed; anthers 1½ lin. long. *Lamk. Illustr. t.* 840; *All. Fl. Pedem.* ii. 260, *t.* 91, *fig.* 4; *Kunth, Enum.* i. 486; *Steud. Syn. Pl. Glum.* i. 367; *Baker, Fl. Maurit.* 444; *Hack. Androp. in DC. Monogr. Phan.* vi. 585, *and in Bull. Herb. Boiss.* iv. *App.* iii. 11; *Hook. f. Fl. Brit. Ind.* vii. 199. *Heteropogon hirtus, Pers. Syn.* ii. 533. *H. glaber, Pers. l.c.*; *Beauv. Agrost.* 134, *t.* 23, *fig.* 8. *H. contortus, Roem. and Schult. Syst.* ii. 836; *Nees, Fl. Afr. Austr.* 101. *H. Allionii, Roem. and Schult. l.c.* 835; *Reichenb. Icon. Fl. Germ.* i. *t.* 53, *fig.* 1496-7.

SOUTH AFRICA: without precise locality, *Drège*, 2041!

COAST REGION: Swellendam Div., *Mund.* Riversdale Div.; Great Vals River, *Burchell*, 6536! between Zoetemelks River and Little Vet River, *Burchell*, 6839! George Div.; Outeniqua Mountains, near Roodemur, 2000–3000 ft., *Drège!* Knysna Div.; Pletten Berg Bay, *Drége* (ex *Nees*). Uitenhage Div., in fields near the Zwartkops River, *Ecklon and Zeyher*, 185! Port Elizabeth Div.; Port Elizabeth, *E.S.C.A. Herb.*, 120! Alexandria Div.; North of the Zuur Berg Range, 2500–3000 ft., *Drège.* Fort Beaufort Div.; near Fort Beaufort, *Ecklon and Zeyher!* Stockenstrom Div.; Winter Berg, near Philippston, *Ecklon.*

CENTRAL REGION: Prince Albert Div.; Great Zwarte Bergen, near Klaarstroom, 3000–4000 ft., *Drège.* Aberdeen Div.; Camdeboo Mountain, 3000–4000 ft., *Drège.* Somerset Div.; Klein Bruintjes Hoogte, 3000–4000 ft., *Drège!* Graaff Reinet Div.; on stony hills near Graaff Reinet, 2600 ft., *Bolus*, 460! Albert Div.; without precise locality, *Cooper*, 1366!

KALAHARI REGION: Griqualand West; Hay Div., Asbestos Mountains, *Burchell*, 2046! Orange Free State; near Winburg, *Buchanan*, 233! Caledon River, *Burke!* Transvaal; Pretoria, at Wonderboom Poort, *Rehmann*, 4494!

EASTERN REGION: Tembuland; Bazeia, *Baur*, 2000–2500 ft., 318! Natal; Port Natal, *Herb. Harvey!* Umsinga and base of the Biggars Berg, *Buchanan*, 91! Delagoa Bay, *Forbes!*

Mediterranean regions and tropics generally.

Subgenus 9. CYMBOPOGON (Hack.). Racemes 2-nate often divaricate or deflexed, both sessile or shortly peduncled, or one sessile and the other peduncled, terminating a common peduncle, which is more or less sheathed by a proper spathe; joints few or many, dorsally convex, linear or clavate, tips obliquely truncate or cupular with toothed margins, pedicels very similar. Lowest 1 or several sessile spikelets of one or of both racemes like the pedicelled in form and sex, ♂ or barren, the others ⚥ or ♀. ⚥ spikelets oblong to lanceolate, dorsally compressed or subcylindric; lower glume dorsally flat, excavate or pitted, awnless, 2-keeled; upper boat-shaped, keeled, awnless, rarely awned. Valves hyaline, lower 1–3-nerved or nerveless; upper smaller, narrow, 2-toothed or 2-lobed, rarely subentire, nearly always awned from the cleft. Pale minute or 0. Pedicelled spikelets awnless or aristulate from the lower glume.

Flowering branches from the axis of spathiform, bladeless or almost bladeless sheaths rarely of perfect leaves, arranged in spurious panicles.

18. A. Nardus (Linn. Sp. Plant. 1046) var. **marginatus** (Hack. Androp. in DC. Monogr. Phan. vi. 606); perennial, densely tufted, innovation shoots intravaginal; culms slender, erect, 1–2½ ft. long, glabrous, smooth, simple and 1–2-noded below the spurious panicle, longest internode 6–8 in., rarely up to 1 ft. long; leaves mainly crowded near the base, sheaths tight, glabrous except the lowermost, which are fugaciously hairy to tomentose at the very base and 2–4 in. long, very firm and persistent, the upper 1–2 much shorter than the internodes; ligules very firm, rounded, up to 1 lin. long; blades narrow, linear, very long, tapering to a fine point, 4 in. to almost 1 ft. by 1–2½ lin., flat, rigid, glabrous, margins scabrid; panicles more or less compound, spathaceous, usually narrow, 4–8 in. long; racemes ½–¾ in. long, finally deflexed or horizontally spreading, subtended by lanceolate, boat-shaped, many-nerved, scarious, often reddish spathes ¾–1¼ in. long, on slender common peduncles, which are 4–8 in. long; joints and pedicels linear, slender, about 1 lin. long, densely hairy along the margins, tips cupular, irregularly toothed; spikelets of the lowest pair of the sessile raceme ♂ like the pedicelled, the sessile of all the other pairs ⚥; ⚥ spikelets lanceolate, 2½–3 lin. long, reddish above; lower glume sub-chartaceous, minutely 2-toothed or subacute; dorsally flattened or slightly depressed, glabrous, keels widened above into narrow or broad, often serrulate, scarious wings, intracarinal nerves 2–4, unequal, evanescent below, callus short, minutely bearded; upper glume lanceolate, acute or mucronate, 1-nerved, glabrous, keel narrowly winged; valves hyaline, the lower lanceolate-linear, 1½–2 lin. long, ciliate; upper linear, very narrow, bifid, 1¼–1¾ lin. long, 1-nerved, lobes fine, ciliate, awn slender, 6–7 lin. long, kneed at or below the middle, glabrous below; pale 0; anthers 1 lin. long; grain 1¼ lin. long; pedicelled spikelets ♂, lanceolate, acute, 2½ lin. long; lower glume many-nerved, upper 3-nerved, keels of both scabrid; lower valve linear, obtuse, nerveless or 2-nerved at the base, 1¼ lin. long, upper 0; pale 0. *Hack. in Bull. Herb. Boiss.* iv. *App.* iii. 11; *Durand & Schinz, Consp. Fl. Afr.* v. 718. *A. Schœnanthus, Thunb. Prod.* 20; *Fl. Cap. ed. Schult.* 108, *not Linn. A. marginatus, Steud. in Flora,* 1829, 472. *A. pseudo-hirtus, Steud. l.c.* 471. *A. Iwarancusa, Nees, Fl. Afr. Austr.* 117; *Steud. Syn. Pl. Glum.* i. 388 (*Iwarankusa*), *not Blane. Trachypogon Schœnanthus, Nees in Linnæa,* vii. 281.

VAR. β, **prolixus** (Stapf); culms 3–4 ft. long, slender, 3-noded, internodes very long, the longest frequently over 1–1½ ft. long; lowest sheaths ½–1 ft. long; ligules 3–5 lin. long, rarely less, sometimes up to 8 lin. long; blades 1–2 ft. by 2–2½ lin., often very rigid; spurious panicles, narrow, usually more compound and more contracted than in the preceding variety, 3–6 in. long, with a more or less distant flowering branch below it, joints and pedicels usually less hairy.

VAR. γ, **validus** (Stapf); culms 4–7 ft. long, stout, 4–5-noded, internodes often 1 ft. long or longer; lower sheaths ¾–1 ft. long; ligules 2–3 lin. long; blades up to 2 ft. by 2½–4½ lin.; panicle narrow, much more compound than in the first variety, densely contracted, ½–1 ft. long, with a distant flowering branch below it; proper spathes 7–9 lin. long; spikelets 2¼–2½ lin. long.

The African specimens of *A. Nardus* have a peculiar facies which distinguishes them in most cases from the Indian, which include the type. It is mainly due to the less compound structure of the panicle and the, on the whole, slightly larger spikelets. The lower glumes of the ⚥ spikelets of the Indian varieties have also generally 1–2 shallow pits on the back, a peculiarity which I have never observed in the African material. The 3 varieties described above are very different, if extreme forms are compared; they seem, however, so closely linked by intermediate states, that I cannot separate them specifically.

SOUTH AFRICA: without precise locality, Var. β, *Drège*, 4354! Var. γ, *Zeyher*, 1800!

COAST REGION: Var. α: Cape Div.; Table Mountain, *Milne*, 240! Devils Mountain, *Ecklon*, 919! near Cape Town, *Burchell*, 470! Cape Flats, near Rondebosch, *Burchell*, 200! Simons Bay, *Wright! MacGillivray*, 402! Orange Kloof, *Dodd*, 2219! 2959! top of Kloof between Millers Point and Smithwinkel, *Dodd*, 2974! Low hills, South of Kommetje, *Dodd*, 1571! Ladies Mile, *Dodd*, 2071! Tokay Plantation, *Dodd*, 377! Worcester Div.; Dutoits Kloof, 2000–3000 ft., *Drège!* Caledon Div.; Nieuw Kloof, Houw Hoek Mountain, *Burchell*, 8054! Riversdale Div.; near the Zoetemelks River, *Burchell*, 6692/1! 6790/2! near Riversdale, on the Lange Bergen Range, *Schlechter*, 1867! Var. β: Mossel Bay Div.; mouth of the Gauritz River, *Mund and Maire!* Port Elizabeth Div.; Port Elizabeth, *E.S.C.A. Herb.* 118! Bathurst Div.; between Port Alfred and Kaffir Drift, *Burchell*, 3845! Albany Div., moist places on mountains near Grahamstown, 2200 ft., *MacOwan*, 116! and without precise locality, *Atherstone*, 51! Fort Beaufort Div.; Nieuw Veld, near Bok Poort, 3500–4500 ft., *Drège*, 700! Cathcart Div.; Windvogels Berg, 4000 ft., *Baur*, 1114!

CENTRAL REGION: Var. α: Aliwal North Div.; between Krai River and Witte Bergen, *Zeyher!* Var. β: Aliwal North Div. (?); by the Orange River, 5000–6000 ft., *Ecklon and Zeyher!*

KALAHARI REGION: Var. β: Orange Free State, *Cooper*, 1065!

EASTERN REGION: Var. β: Natal; Riet Vlei, 5000 ft., *Buchanan*, 230! Umpumulo, *Buchanan* 232! Biggars Berg, *Rehmann*, 7108! Var. γ: Natal; near Durban, *Williamson*, 52! in damp places throughout the colony, *Krauss*, 87! Inanda, at Umtamtootu, *Wood*, 1622! Pietermaritzburg, *Rehmann*, 7600!

Var. δ occurs also in Tropical Transvaal.

19. A. plurinodis (Stapf); perennial, compactly tufted; innovation shoots intravaginal; culms slender, erect, 1–3 ft. long, glabrous, smooth, simple below, 4–6- (rarely 3-) noded, longest internode 6–10 in. long, rarely longer; sheaths tight, shorter than the internodes except the lowermost, which are fugaciously hairy to tomentose at the very base, 1½–3 in. long, firm and persistent, the upper usually much shorter; ligules firm, very short, up to 1 lin. long, rounded; blades very narrow, linear, filiform or setaceous in the upper part, 4–8 in. by ½–1½ lin., flat or partly folded, rather rigid below, flexuous above, glabrous, glaucous, margins scabrid; panicles narrow, rather lax, usually scantily branched, 3–6 in. long; racemes ⅔–1 in. long, at length horizontally spreading or deflexed on slender peduncles, which are 3–5 lin. long and subtended by lanceolate, acuminate, many-nerved, scarious, reddish spathes, ¾–1¼ in. long; joints and pedicels linear, slender, 1–1½ lin. long, densely hairy along the margins, tips cupular, irregularly toothed; spikelets of the lowest pair of the sessile raceme alike, ♂, the sessile of all the other pairs ⚥, the pedicelled ♂; ⚥ spikelets very narrow, lanceolate, acute, 3 lin. long; lower glume chartaceous, acute or 2-toothed, more or less concave between the keels which are rounded and smooth near the base, obscurely winged and scabrid towards the reddish or purplish tips, intracarinal nerves 2–5, the middle one or all evanescent below; upper glume subchartaceous, lanceolate, acute, 1-nerved, glabrous; valves hyaline; lower linear, 2–2½ lin. long, ciliate; upper slightly shorter, linear-oblong to linear, deeply bifid, lobes lanceolate to subulate, awn slender, 5–8 lin. long, kneed at or below the middle; pale 0; anthers $1\frac{1}{3}$ lin. long; pedicelled spikelets dorsally depressed, lanceolate, acuminate, $3-3\frac{1}{4}$ lin. long, purplish; lower glume many-nerved, keels scabrid; upper 3–5-nerved, with or without a median keel; lower valve linear-lanceolate, $2\frac{1}{2}$ lin. long, 2-nerved, ciliate; upper and pale 0.

COAST REGION: Fort Beaufort Div.; Kat River, *Baur!* Queenstown Div.; Imvani River, *Baur*, 78!

CENTRAL REGION: Graaff Reinet Div.; near Graaff Reinet, 4000 ft., *Bolus*, 243! Albert Div., *Cooper*, 1363! 3369!

KALAHARI REGION: Griqualand West; Herbert Div.; between Spruigslang and the Vaal River, *Burchell*, 1709/2! Hay Div.; near Klip Fontein, *Burchell*, 2166! and at Griqua Town, *Burchell*, 1862! Orange Free State; between Harrismith and Leribe, *Buchanan*, 207! Bechuanaland; between Kosi Fontein and Knegts Fontein, *Burchell*, 2601! and at Hampery, near Kuruman, *Burchell*, 2520! Transvaal, near Lydenburg, *Atherstone!*

EASTERN REGION: Natal; near Colenso, on the Tugela River, *Rehmann*, 7158! Biggars Berg, *Rehmann*, 7112! Mooi River, *Wood*, 4318!

Not unlike meagre specimens of *A. Nardus*, var. *marginatus*, but distinguished by the more numerous nodes, very narrow leaves, scantier panicles, narrower, usually distinctly concave lower glumes to the ⚥ spikelets and larger pedicelled spikelets.

20. A. Schœnanthus (Linn. Sp. Plant. 1046) var. **versicolor** (Hack. Andr. in DC. Monogr. Phan. vi. 610); perennial, frequently with extravaginal innovation shoots beside some intravaginal ones, or annual (or at least flowering the first year); culms fascicled, erect or shortly ascending, stout or rather so, 2–4 ft. long, simple and 4–7-noded below the spurious panicle; sheaths tight, quite glabrous, subherbaceous or the lowermost firmer and sometimes subpersistent, 2–3 in. long; upper shorter than the internodes; ligules very short, rounded; blades linear to linear-lanceolate from a broader rounded or subcordate base, tapering to a long setaceous point, $\frac{1}{2}$–1 ft. by 3–6 lin. (in the South African specimens), flat, rather flaccid, dull-green or subglaucous, glabrous, smooth or the margins scabrid; panicles more or less compound, narrow, often interrupted, 4–6 in. long, rarely longer; racemes 4–9 lin. long, 2-nate, at length reflexed, or horizontally spreading on slender peduncles, which are 2–4 in. long, and subtended by lanceolate boat-shaped many-nerved scarious reddish spathes $\frac{1}{2}$ to 1 in. long; joints and pedicels linear, slender or the lowest somewhat stout, $1-1\frac{1}{4}$ lin. long, densely hairy, mainly along the margins, tips minutely cupular, toothed; spikelets of the lowest pair of the sessile raceme alike, ♂, the sessile of all the other pairs ⚥, the pedicelled ♂; ⚥ spikelets lanceolate, $2-2\frac{1}{4}$ lin. long, often variegated; lower glume subacute or obscurely 2-toothed, dorsally flattened with a narrow groove from the middle downward corresponding to a keel on the inner side, keels narrowly or obscurely winged and scabrid above, evanescent below; upper glume lanceolate, acute, 1-sub-3-nerved; valves hyaline, lower oblong, $1\frac{3}{4}$ lin. long, 2-nerved, ciliate; upper scarcely shorter, very narrow, linear, deeply bifid, lobes very fine, awn 5–7 lin. long, slender, bent just below the middle, glabrous below the bend; pale 0; anthers 1 lin. long;

pedicelled spikelets oblong, subacute, 2 lin. long; lower glume 9–11-nerved, upper 3-nerved; lower valve oblong, almost equalling the glumes, 2-nerved, upper and pale 0; anthers 1½ lin. long. *Durand & Schinz, Consp. Fl. Afr.* v. 722. *A. connatus, Hochst. ex A. Rich. Tent. Fl. Abyss.* ii. 464. *A. nardoides β minor, Nees, Fl. Afr. Austr.* 116. *A. excavatus, Hochst. in Flora,* 1846, 116. *A. pruinosus and A. versicolor, Steud. Syn. Pl. Gram.* i. 388. *A. foliatus, Steud. l.c.* 389. *A. Schœnanthus, Baker, Fl. Maurit.* 446. *Gymnanthelia connata, Aschers. and Schweinf. in Schweinf. Beitr Fl. Aeth.* 310.

SOUTH AFRICA: without precise locality, *Drège*, 4358! *Zeyher*, 1798!

KALAHARI REGION: Griqualand West; Hay Div., at Klip Fontein, *Burchell*, 2165! Orange Free State; Caledon River, *Burke*, 199! near Winburg, *Buchanan*, 235! Bloemfontein, *Rehmann*, 3729! Bechuanaland; between Kosi Fontein and Knegts Fontein, *Burchell*, 2602! Transvaal; Johannesburg, *E.S.C.A. Herb.* 311! Apies River, *Nelson*, 70*! Houtbosch, *Rehmann*, 5682!

EASTERN REGION: Griqualand East; rocky slopes, near Kokstad, *Tyson*, 1412! near Clydesdale, 2500 ft.; *Tyson*, 3121! Natal; very common on the plains between Boshmanns Rand and the Drakensberg Range, *Krauss*, 26! Umpumulo, 2400 ft., *Buchanan*, 231!

Also in the Mascarene Islands and in India.

The specimens from Bloemfontein and Johannesburg have very narrow leaves, and meagre inflorescences, like those of var. *cæsius* (Hackel); but I consider them only as slight variations due to a particularly dry habitat. Hackel indicates also the typical form, which is a more robust plant having longer and much broader blades and larger more compound panicles, from "Caffraria." I have not seen it from extra-tropical South Africa, unless a specimen, collected by Buchanan, 134 in part, on the Natal Coast is referable to it. A specimen, collected by Fleck at Rehoboth, Great Namaqualand is referred by Hackel (in Bull. Herb. Boiss. iv. App. iii. 12) to *A.* "*Schœnanthus*, L. sens. ampl." Thunberg enumerates *A. Schœnanthus* in Prod. Pl. Cap. 20 and in Fl. Cap. ed. i. 407 and ed. Schult. 108. Nees refers it to his *A. Iwarancusa* which is *A. Nardus* var. *marginatus*. There is, however, no such plant in Thunberg's Cape collection so far as I have seen it; but the localities given by Thunberg in his Flora Capensis, namely, Lange Kloof and Kromme River, render it very probable that Nees was right.

21. A. hirtus (Linn. Sp. Plant. 1046); perennial, tufted; innovation shoots mostly intravaginal; culms erect, rather slender, 1–3 ft. long, glabrous, simple or more or less branched and 3–6-noded below the panicle, longest internode generally less than ½ ft. long; sheaths tight, glabrous, the lowest crowded, compressed, firm, persistent, the upper terete, shorter than the internodes; ligules oblong, obtuse, up to 1½ lin. long; blades narrow, linear, tapering to a long fine point, 4 in. to more than 1 ft. by ¾–1½ lin., glabrous, finely scaberulous, more or less glaucous, turning reddish, midrib white; panicle spathaceous, lax, contracted, sometimes reduced to a few simple branches, ½ to more than 1 ft. long; racemes ¾–1½ lin. long, silvery, villous, on very slender, usually nodding, pubescent or villous, rarely glabrous peduncles, which are 1–3 lin. long, and more or less exserted from the long very narrow finely acuminate glabrous or scantily hairy reddish spathes; joints filiform, obliquely truncate, up to 2 lin. long, densely ciliate, pedicels very similar, often produced (in the South African species) into a linear or subulate appendage facing the upper glume; spikelets of the lowest pair of the sessile raceme alike, ♂, the sessile of all the other pairs ⚥, the pedicelled ♂; ⚥ spikelets linear-oblong, 2–3 lin. long, pale or purplish; lower glume chartaceous, minutely truncate, dorsally flattened, villous, keels obscure except close to the membranous reddish tips, intracarinal nerves about 5, evanescent below; upper glume lanceolate-oblong, obtuse, mucronate, 3-nerved, margins ciliate, keel hirsute above; lower valve hyaline, linear-oblong, obtuse, faintly 2-nerved, softly ciliate; upper very narrow, linear, bifid, about $1\frac{1}{2}$ lin. long, ciliate, base margins and lobes hyaline, awn stout, pubescent, $1-1\frac{1}{2}$ in. long, kneed below the middle; pale very minute or 0; anthers $\frac{3}{4}-1$ lin. long; pedicelled spikelets lanceolate, $2\frac{1}{2}-3$ lin. long; lower glume herbaceous, 7–11-nerved, mucronate or shortly awned; upper lanceolate-oblong, acute, 3-nerved, softly ciliate; lower valve as in the ⚥ spikelets, but 1–3-nerved; upper linear, ciliate, nerveless, usually much shorter than the lower or obsolete. *Thunb. Prodr.* 20; *Fl. Cap. ed. Schult.* 108; *Host, Gram. Austr.* iv. *t.* 1; *Kunth, Enum.* ii. 492; *Reichb. Ic. Fl. Germ.* i. *fig.* 1498; *Nees, Fl. Afr. Austr.* 110; *A. Rich. Tent. Fl. Abyss.* ii. 459; *Steud. Syn. Pl. Glum.* i. 384; *Hack. Androp. in DC. Monogr. Phan.* vi. 618, *and in Bull. Herb. Boiss.* iv. *App.* iii. 11; *Durand & Schinz, Consp. Fl. Afr.* v. 714. *A. pubescens, Vis. in Flora,* 1829, i. *Erg. Bl.* 3; *Fl. Dalm. t.* ii. *fig.* 2; *Nees, Fl. Afr. Austr.* 111. *Trachypogon hirtus, Nees, Agrost. Bras.* 346. *Heteropogon hirtus and H. pubescens, Anderss. in Schweinf. Beitr. Fl. Aeth.* 310.

VAR. β, **podotrichus** (Hack l.c. 620 in part); culms 3–4 ft. long, rather slender; longest internode over $\frac{1}{2}$ ft. long; blades 8 in. to almost 2 ft. by 2 lin.; ligules $\frac{1}{2}-1\frac{1}{2}$ lin. long; panicle $1-1\frac{3}{4}$ ft. long, the lower branches very distant, common peduncles always with long tubercle-based hairs near the curvature. *A. podotrichus, Hochst. ex Steud. Syn. Pl. Glum.* i. 384. *Hyparrhenia podotricha, Anderss. in Schweinf. Beitr. Fl. Aeth.* 310.

SOUTH AFRICA: without precise locality; Var. β, *Drège,* 4345!

COAST REGION: Cape Div.; mountains near Cape Town, *Thunberg! Ecklon,* 87! *Burchell,* 7! at the foot of Table Mountain and Devils Mountain, below 1000 ft., *Drège!* Lions Head, *Dodd,* 2347! Claremont Flats, *Schlechter,* 327! cornfield below Klein Constantia, *Dodd,* 2070! Simons Bay, *Wright! MacGillivray,* 390! *Milne,* 238! Worcester Div.; Brand Vlei, *Rehmann,* 2419! Stellenbosch Div.; Lowrys Pass, 500 ft., *Schlechter,* 7821! Riversdale Div.; near Riversdale, 6000 ft., *Schlechter,* 1869! Uitenhage Div.; in the rocky bed of Zwartkops River, *Ecklon and Zeyher,* 485! Albany Div.; mountain sides near Grahamstown, *MacOwan,* 494! King Williamstown Div.; between Yellowwood River and Zandplaat, 1000–2000 ft., *Drège!*

CENTRAL REGION: Graaff Reinet Div.; stony hill-sides, near Graaff Reinet, 2700 ft., *Bolus,* 462! Cradock Div.; without precise locality, *Cooper,* 3376! Aliwal North Div.; by the Orange River, 5000–6000 ft., *Zeyher!*

KALAHARI REGION: Griqualand West; Hay Div., at Klipfontein, *Burchell,* 2152! Doorn River, *Burchell,* 2136/1! and at Griquatown, *Burchell,* 1887! Orange Free State; Bloemfontein, *Rehmann,* 3735! Bechuanaland; between Takun and Molito on the Moshowa River, *Burchell,* 2316! Transvaal; Bosch Veld, between Elands River and Klippan, *Rehmann,* 5110!

EASTERN REGION: Natal; without precise locality, *Gerrard,* 670! 674! Var. β: Natal, Umpumulo, 2000–2500 ft., *Buchanan!* Riet Vlei, 6000 ft., *Buchanan!* Biggars Bergen, *Rehmann,* 7120! 7121! Fark Kop, *Rehmann,* 7668! and without precise locality, *Buchanan,* 62! 121!

The typical form very common throughout the Mediterranean region; var. β through East Africa to Abyssinia.

Peduncles having long tubercle-based hairs occur also occasionally in the typical form in the Cape as well as in the Mediterranean countries; but they do not differ otherwise from it, whence I exclude them from the var. *podotrichus*, confining this to the tall, long-leafed form which represents Hochstetter's *A. podotrichus*.

22. **A. auctus** (Stapf); perennial; culms erect, robust, up to 2 lin. thick, 3–5 ft. long, glabrous, simple or more or less branched and 3–5-noded below the panicle, longest internode over ½ ft. long; sheaths tight, glabrous or hairy, the lowest crowded, compressed, subpersistent, the upper terete, shorter than the internodes; ligules oblong, up to 3½ lin. long; blades linear from a narrow, stoutly ribbed base, tapering to a long fine point, up to 1½ ft. by 1–2½ lin. glabrous or scantily hairy, glaucous; panicle lax, narrow, up to 1½ ft. long; racemes ¾–1½ lin. long, very villous, on slender shortly and laterally exserted peduncles, having numerous long tubercle-based hairs near the curvature; spathes, joints, pedicels and spikelets as in *A. hirtus*, except for the pedicelled spikelets having a muticous lower glume and subequal valves.

COAST REGION: Uitenhage Div.; *Ecklon and Zeyher*, 626! Queenstown Div.; near Shiloh, 3500 ft., *Baur*, 897!

KALAHARI REGION: Orange Free State; Thaba Unchu, *Burke*, 427! *Zeyher*, 1799!

EASTERN REGION: Griqualand East; grassy slopes near Clydesdale, 2500 ft., *Tyson*, 3105! Natal; Riet Vlei, 6000 ft., *Buchanan*, 222!

23. **A. Schimperi** (Hochst. ex A. Rich. Tent. Fl. Abyss. ii. 466); perennial (always?); culms erect, sometimes geniculate, stout, up to 5 ft. long, glabrous, up to 7-noded; sheaths quite glabrous or the uppermost more or less hairy, those of the innovation shoots compressed, the others terete or slightly keeled in the upper part, the lowest whitish, almost scarious; ligules obtuse, scarious up to 2 lin. long; blades linear, tapering to a long fine point, ½–1½ ft. by 3–7 lin., subrigid to almost flaccid, flat, glabrous, rarely scantily hirsute near the base, scabrid, at least upwards, margins scabrid to subspinulous; panicle large, oblong to ovate, lax, decompound, up to 1½ ft. long; spathes lanceolate, acuminate, narrow or broad, 1–1½ in. long, scarious, reddish; common peduncles filiform, glabrous except the curved tips, which are beset with long yellowish tubercle-based hairs and enclosed in the spathe or shortly and usually laterally exserted; racemes ½–1 in. long, dense, scarcely spreading, very shortly and unequally peduncled; joints 5 to many, filiform, obliquely truncate, up to 1 lin. long, shortly whitish ciliate, pedicels very similar; spikelets of the lowest pair of the sessile raceme alike, ♂, the sessile of all the other pairs ⚥, the pedicelled ♂; ⚥ spikelets linear-oblong, 2–2½ lin. long, pale; lower glume subchartaceous to almost membranous, minutely truncate, dorsally flattened, glabrous or more or less villous, intracarinal nerves 5, some at least evanescent below, keels spinulous ciliate above, callus bearded; upper membranous, obtuse, 3-nerved, nerves scabrid near the tip; lower valve oblong, obtuse, faintly 2–3-nerved, ciliate; upper shortly 2-fid, 1-nerved, lobes

oblong, obtuse, subglabrous, awn about 10–15 lin. (rarely 2–2½ in.) long, pubescent and kneed below the middle; pale 0; anthers 1 lin. long; pedicelled spikelets lanceolate, usually pale, 2½–4 lin. long, glabrous or hairy; lower glume subherbaceous, acutely acuminate, 9–11-nerved, often with a terminal bristle (up to 2½ lin. long), keels spinulous ciliate; upper cuspidate, 3-nerved, ciliate; lower valve linear-oblong, obtuse, equalling the glumes, 1–3-nerved; upper narrow, linear-cuneate, 1–3 lin. long, 1-nerved; anthers 1½–2 lin. long. *Steud. Syn. Pl. Glum.* i. 384; *Hack. Androp. in DC. Monogr. Phan.* vi. 623 (*in part*); *Durand & Schinz, Consp. Fl. Afr.* v. 721. (*in part*). *A. formosus, Klotzsch ex Hack. l.c. A. giganteus, Hort. ex Hack. l.c. Hyparrhenia Schimperi, Anderss. in Schweinf. Beitr. Fl. Aeth.* 300.

EASTERN REGION: Natal; without precise locality, *Gerrard*, 691!

A rather variable plant, extending throughout East Africa to Abyssinia. Gerrard's specimen represents a robust state with hairy spikelets like those of Schimper's no. 897 (coll. 1853) and 1052 (coll. 1863-8), and with long-aristulate ♂ spikelets as in Schimper's no. 1052 (coll. 1838).

24. A. rufus (Kunth, Enum. i. 492); perennial; culms erect, stout, 3–8 ft. long, glabrous, 5–7-noded below the panicle; sheaths quite glabrous (in the South African species) or hairy along the upper margins, terete, shorter than the internodes (except the lowest); ligules rounded, 1–2½ lin. long, sometimes with hairs from behind; blades linear from a narrow base, long tapering to a fine point, 1–2½ ft. by 2–6 lin., flat, rigid, erect, glabrous, more or less scabrid, margins very rough, midrib stout, white above; panicle usually large and much branched, lax, oblong, contracted or rather open, 1–1½, sometimes to 2 ft. long; spathes very narrow, linear-lanceolate to linear, finely acuminate, 1½–2 in. long, glabrous, scarious, reddish, erect or spreading; common peduncles finely filiform, pubescent, sometimes with spreading yellowish tubercle-based hairs towards the tips, flexuous or pendulous, more or less (often long) exserted; racemes slender, ¾–1 in. long, not spreading, often drooping; joints 6–10, filiform, obliquely truncate, up to 1 lin. long, shortly and densely ciliate, lower hairs white, upper reddish; pedicels very similar; all the sessile spikelets ⚥, or sometimes the lowest of the sessile raceme like the pedicelled ♂ or the latter sometimes barren; ⚥ spikelets linear-oblong, 2–2½ lin. long, yellowish or reddish; lower glume subchartaceous, minutely truncate, dorsally flattened, glabrous or pubescent with short rather rigid fulvous or ferruginous hairs or subvillous, intracarinal nerves about 5, evanescent below, callus shortly bearded, keels rigidly ciliate above; upper glume membranous, obtuse, 3-nerved; lower valve oblong, obtuse, ciliate, faintly 2-nerved; upper shortly bifid, 1-nerved, lobes oblong, ciliate, awn ¾–1 in. long, bent at the middle, pubescent below the bend; anthers 1 lin. long; pedicelled spikelets lanceolate-linear, 2¼–2½ lin. long, reddish or purplish, glabrous or pubescent to villous; lower glume subherbaceous, acute, 7-nerved, keels rigidly ciliolate; upper acute 3-nerved, ciliate; lower valve oblong, subacute, 1½ lin. long, 1-sub-3-nerved; upper narrow, linear, 1-nerved ciliate or both valves more or less reduced or the upper quite suppressed. *Hack. Androp. in DC. Monogr. Phan.* vi. 621; *Durand & Schinz, Consp. Fl. Afr.* v. 721. *A. fulvicomus, Hochst. ex A. Rich. Tent. Fl. Abyss.* ii. 463. *A. altissimus, Hochst. in Flora,* 1841, 277. *A. hirtus, Baker, Fl. Maurit.* 446. *Hyparrhenia fulvicoma, Anderss. in Schweinf. Beitr. Fl. Aeth.* 310.

EASTERN REGION: Zululand; moist places near Inyezaan River, *Wood*, 3927!

Throughout tropical Africa and the Mascarenes; also in Brazil.

25. **A. dregeanus** (Nees, Fl. Afr. Austr. 112); perennial, compactly tufted; culms erect, rather slender, 3–4 ft. long, glabrous, shining, about 5-noded below the panicle; sheaths of innovation shoots compressed, fugaciously tomentose at the base, sheaths of the culms terete, quite glabrous or the uppermost scantily hairy, smooth; ligules obtuse, up to 2½ lin. long, hairy behind; blades linear from a narrow base, tapering to a fine point, up to 1 ft. by 2–3 lin. long, erect, rather rigid, flat, glabrous, or the uppermost scantily hairy, subglaucous, turning reddish, scabrid, margins spinulous; panicle narrow, 1–1½ ft. long, remotely branched, contracted, spathes narrow, lanceolate, acute, about 1½ in. long, scarious, reddish; common peduncles finely filiform, long exserted, glabrous except the strongly curved or curled tips, which are densely beset with long spreading, yellowish, tubercle-based hairs; racemes ½–1 in. long, lax, scarcely spreading; joints 5–10, filiform, obliquely truncate, up to 1½ lin. long, shortly ciliate; pedicels very similar; spikelets of the lowest pair of the sessile raceme alike, ♂, the sessile of all the other pairs ⚥, the pedicelled ♂ or sometimes with stamens and an apparently perfect pistil; sessile ⚥ spikelets linear-oblong, 2 lin. long, purplish, pale below; lower glume subchartaceous above, membranous near the base, minutely truncate, dorsally flattened, glabrous or minutely and scantily pubescent or villous, intracarinal nerves about 5, evanescent below the tip, keels rigidly ciliate above, callus shortly bearded; upper glume membranous, 3-nerved, nerves scabrid above; lower valve oblong, obtuse, faintly 2-nerved, ciliate; upper shortly bifid, 1-nerved, lobes oblong, obtuse, ciliate, awn about 6–8 lin. long, very slender, pubescent and kneed below the middle; pale ovate, obtuse, ½ lin. long; anthers ¾ lin. long; pedicellate spikelets oblong-linear, 2½ lin. long, dull purplish, glabrous or hairy; lower glume subherbaceous, acute, 9–11-nerved; upper very acute, 3-nerved, long ciliate; lower valve linear-oblong, obtuse, equalling the glumes, 3-nerved; upper ¼–⅓ shorter, spathulate-linear from an extremely fine base, sub-1-nerved; anthers more than 1 lin. long. *A. Schimperi, Hack. Androp. in DC. Monogr. Phan.* vi. 623, *and Durand & Schinz, Consp. Fl. Afr.* v. 721 (*in part*).

EASTERN REGION: Transkei; between the Kei River and Gekau (Gcua or Geuu River), 1000–2000 ft., *Drège!* Natal; between Umtata River and St. Johns River, *Drège*, between Umzimkulu River and Umkomanzi River, *Drège*, and Griffins Hill at Eastcourt, *Rehmann*, 7310!

A. Schimperi, to which Hackel referred *A. dregeanus*, is a more robust plant, having much larger, and more compound panicles, shorter, less curved common peduncles, which are usually not exserted at all, and more conspicuously nerved and setulously ciliate lower glumes. The plant distributed by Schimper from Abyssinia as *A. dregeanus* is also distinct from Nees' *A. dregeanus*.

26. A. dichroos (Steud. Syn. Pl. Glum. i. 389); perennial; culms rather slender, erect, over 2 ft. long, glabrous, terete, 4–5-noded and simple below the panicle; sheaths terete or slightly keeled in the upper part, tight, glabrous, shorter than the internodes, except the subpersistent lowest ones; ligules membranous, rotundate-ovate, ciliolate, up to 1 lin. long; blades linear from a rounded base, long tapering to a fine point, over $\frac{1}{2}$ ft. by $1\frac{1}{2}$–$3\frac{1}{2}$ lin., glabrous or scantily hairy at the base, scabrid below and along the margins, turning reddish; panicle oblong, about 1 ft. long, lax; spathes very narrow, lanceolate, acuminate, $1\frac{1}{2}$–2 in. long, obliquely erect or divaricate, reddish, glabrous or spreadingly hairy, particularly below; common peduncles filiform, more or less curved at the upper end or at length strict, shortly exserted from near the tip of the spathe or shortly exceeding it, with long spreading tubercle-based hairs below the tip; racemes 4–10 lin. long, at length horizontally spreading or deflexed on very short hirsute peduncles, falling together from the tip of the common peduncle, joints 2–5, filiform, obliquely truncate, $1\frac{1}{4}$ lin. long, shortly ciliate, pedicels very similar; spikelets of the lowest pair of the subsessile raceme alike, barren, the sessile of all the other pairs ⚥, the pedicelled barren; ⚥ spikelets oblong-lanceolate, 2 lin. long, pale; lower glume subchartaceous above, membranous towards the base, truncate, or minutely 2-toothed, dorsally flattened, scantily pubescent to almost villous, with whitish or above with reddish hairs, intracarinal nerves 5–7, evanescent below, keels rigidly ciliate above, callus shortly bearded; upper glume rigidly membranous, obtuse 3-nerved, ciliate above; lower valve oblong, truncate, faintly 2-nerved, ciliate; upper shortly 2-toothed, glabrous, awn slender, $\frac{3}{4}$–1 in. long, kneed and pubescent below the middle; pale 0; anthers $\frac{2}{3}$–$\frac{3}{4}$ lin. long; pedicelled spikelets linear-lanceolate, up to $2\frac{3}{4}$ lin. long, rufously hirsute; lower glume acuminate, sometimes mucronate, 9–11-nerved, upper 3-nerved, ciliate above; valves 0 or the lower broadly oblong, up to $\frac{3}{4}$ lin. long, ciliate and the upper reduced to a microscopic ciliate scale. *Hack. Andr. in DC. Monogr. Phan.* vi. 622; *Durand & Schinz, Consp. Fl. Afr.* v. 710. *A. bicolor, Nees, Fl. Afr. Austr.* 113, *not Roxb.*

EASTERN REGION: Natal; between St. Johns River and Umtsikaba River, 1000–2000 ft., *Drège*, 4357! Umpumulo, 2000 ft., *Buchanan!* Biggars Bergen, *Rehmann*, 7116! and without precise locality, *Buchanan*, 303!

27. A. cymbarius (Linn. Mant. ii. 303, not Hack.); perennial (?); culms erect, usually from a very slender and sometimes ascending base, 3–6 ft. or more long, often rooting from the lower nodes, terete, glabrous, 6–10-noded, simple or branched; sheaths glabrous, rarely hirsute with tubercle-based hairs or villous at the nodes, terete or keeled in the upper part, the lowest whitish, withering or thrown off, the middle and upper usually shorter than the internodes; ligules membranous, rounded or truncate, very short; blades linear from a narrow base, long tapering to a fine point, ½ to over 1½ ft. by 3–6 lin., rarely narrower, rigid to subflaccid, flat, glabrous or subhirsute at the base from tubercle-based hairs, scaberulous or smooth except the scabrid or spinulous margins, glaucous; panicle leafy, almost overtopped by its blades, linear-oblong, decompound, usually dense, ½–2 ft. long; spathes obliquely ovate, acute or acuminate, 5–9 lin. long, scarious, often brilliantly red or purple, glabrous; common peduncles filiform, enclosed, spreadingly hirsute above; racemes usually half exserted from the middle of the spathe and at a right angle to it, 2½–4½ lin. long, dense, subcontiguous, subsessile with a tuft of long rigid yellowish hairs at the base of the lower, joints 1–3, filiform, truncate, ½ lin. long, ciliate, pedicels very similar, the terminal up to 1 lin. long; spikelets of the lowest pair of the lower raceme alike, ♂ or barren, the sessile of all the other pairs ⚥, the pedicelled ♂ or barren; ⚥ spikelets oblong, 1½–2 lin. long, pale; lower glume thinly chartaceous, minutely truncate, dorsally flattened, subglabrous or pubescent, intracarinal nerves 3–5, mostly evanescent below the tip, keels spinulously ciliate above, callus very short, bearded; upper glume membranous, obtuse, 3-nerved, nerves scabrid or ciliate above; lower valve linear-oblong, truncate, obscurely 2-nerved or nerveless, subglabrous; upper minutely 2-toothed, almost reduced to a reddish fine geniculate awn, 5–10 lin. long; pale 0; stamens up to 1 lin. long; pedicelled spikelets oblong-lanceolate to oblong, 2–3 lin. long, pale to deep reddish, glabrous or pubescent; lower glume subherbaceous, acutely acuminate, 7–9-nerved, keels spinulously ciliate; upper acute, 3-nerved, ciliate; lower valve oblong, obtuse, up to 1½ lin. long, faintly 1-sub-3-nerved; upper linear to cuneate-linear, 1 lin. long or reduced, or 0; anthers to 1 lin. long, if present. *Spreng. in Mém. Acad. Pétersb.* ii. (1810), 305, *t.* 9. *Cymbopogon elegans, Spreng. Pug. Pl. Nov.* ii. 14; *Roem. & Schult. Syst.* ii. 833. *Anthesteria cymbaria, Trin. Androp. in Mém. Acad. Pétersb. sér.* vi. ii. (1833), 323, *not Roxb. Anthistiria latifolia, Anderss. in Peters, Reise Mossamb.* ii. 562.

VAR. **lepidus** (Stapf); blades 2–3½ lin. broad; panicle much less compound, lax or very lax, narrow; spathes broadly lanceolate, acuminate, 9–12 lin. long, awn 8–10 lin. long. *A. lepidus, Nees, Fl. Afr. Aust.* 113. *A. lepidus var. typicus (in part), and var. umbrosus and intonsus, Hack. Androp. in DC. Monogr. Phan.* vi. 625; *Durand & Schinz, Consp. Fl. Afr.* v. 716. *A. intonsus, Nees, l.c.* 114. *A. umbrosus, Hochst. ex A. Rich. Tent. Fl. Abyss.* ii. 467. *Hyparrhenia umbrosa, Anderss. in Schweinf. Beitr. Fl. Aeth.* 300.

SOUTH AFRICA: without precise locality; Var. β, *Drège*, 4360! 4263!

EASTERN REGION: Natal; Umpumulo, *Buchanan*, 228! Drakens Berg Range, at Laings Nek, *Rehmann*, 6939! Inanda, *Wood*, 1304! and without precise locality, *Cooper*, 3346! Var. β: Pondoland; between St. Johns River and Umtsikaba River, 1000–2000 ft., *Drège!* Natal; between Umzimkulu River and Umkomanzi River, below 500 ft., *Drège!* between Umtentu River and Umzimkulu River, *Drège!* and Umpumulo, *Buchanan*, 229! Zululand; Entumeni, *Wood*, 3991!

The typical form is common throughout East Africa as far as Abyssinia, and in the Mascarene Islands,—the var. *lepidus* also occurs in Usambara and Abyssinia, and is perhaps only a starved state with narrow leaves and spathes.

Koenig's specimen from which Linnæus described the species was most probably collected like other plants in Koenig's herbarium in the Mascarene Islands. There is also a specimen in Rottler's herbarium without any indication of locality. The plant is not known from India. *A. cymbarius*, Hack. l.c. 629, is quite a different species.

28. A. Buchananii (Stapf), perennial (?); culms erect, slender, over 2 ft. long, glabrous; sheaths (of the upper leaves, which alone are known) terete, strongly striate, glabrous; ligules membranous, up to 1½ lin. long, glabrous; blades linear, acute, up to 8 in. by 1 lin., flat, rather rigid, glabrous or hairy at the very base, smooth or scaberulous; flowering branches remote, up to 1 ft. long, erect, strict, scantily branched from the axils of perfect leaves; spathes very narrow, linear-lanceolate, acuminate, 2–2½ in. long, glabrous, or very scantily hairy; common peduncles filiform, quite enclosed in the spathe, strict, finely puberulous, 4–5 lin. long; racemes contiguous slender, 6–12 lin. long, peduncled; peduncles connate below whence one raceme subsessile, glabrous, up to 2 lin. long; joints 4–6, filiform, lower short, glabrous or subglabrous, obliquely truncate, the upper like the pedicels, much finer and densely whitish ciliate; lowest 2 sessile spikelets (or lowest 1 in the subsessile raceme) ♂, similar to the barren pedicelled spikelets, all the other sessile spikelets ⚥; ⚥ spikelets oblong-linear, 2–2½ lin. long, callus acute, ½–¾ lin. long, beard dense, white; lower glume subchartaceous, truncate, densely hairy, particularly in the upper part, hairs rigid, reddish, up to 1 lin. long, intracarinal nerves about 5–7, evanescent below; upper membranous, truncate, 3-nerved, reversely ciliate; lower valve linear-oblong, 1½–1¾ lin. long, ciliate, faintly 2-nerved; upper linear, shortly bifid, awn about 1½ in. long, bent above the middle, rufous-hairy below the bend, scabrid above; pale 0; anthers 1 lin. long; ♂ sessile spikelets linear-lanceolate, subacute, 3 lin. long, rufous-hairy; lower glume subherbaceous, intracarinal nerves about 5, percurrent, keels spinulous ciliate above; upper acute, 3-nerved, reversely ciliate; lower valve oblong, obtuse, 2½ lin. long, faintly 3-nerved, reversely ciliate; upper ovate-oblong, about 1½ lin. long, 1-nerved, ciliate; anthers 1 lin. long or smaller and barren; barren pedicelled spikelets similar to the ♂, but narrower and smaller; upper valve minute or 0; anthers 0 or rudimentary.

EASTERN REGION: Natal; Umpumulo, 2000–2500 ft., *Buchanan!*

29. A. filipendulus (Hochst. in Flora, 1846, 115); perennial (always?); culms erect, slender, 2–10 ft. long, glabrous, more or less branched; branches erect, intravaginal, leafy; sheaths terete, subcarinate above, tight or ultimately slipping from the culm, glabrous, or the lower sparingly hairy; ligules membranous, truncate, about 1 lin. long; blades linear, tapering to an acute, often very fine point,

1–2 ft. by ¾–3 lin., flat or subconvolute, glabrous, rarely sparingly hairy, smooth or scaberulous, margins rough; panicle narrow, ½–1¼ ft. long, contracted; branches erect, strict, filiform, from lanceolate to linear long and setaceously acuminate sheaths; spathes very narrow, linear, 2 in. long, glabrous or finely hairy along the margins; common peduncles capillary, enclosed in the spathe except the flexuous or pendulous or finally strict upper part, which is pubescent and very long but very scantily bearded; racemes subcontiguous, slender, unequally peduncled (the longer peduncle 2–5 lin. long, glabrous or hairy), 5–6 lin. long; joints filiform, up to ¾ lin. long, glabrous or subglabrous, pedicels somewhat longer, the upper ciliate; sessile spikelets 2 in the lower, 3 in the upper raceme, in both only the upper one ⚥, the others ♂ like the pedicelled spikelets; ⚥ spikelets oblong-linear, 2–2½ lin. long, pale, tips reddish, callus fine, acute, up to 1 lin. long, densely bearded; lower glume subchartaceous, truncate, glabrous or more or less hairy, intracarinal nerves about 5, prominent all along or only upwards, not pitted; upper membranous, truncate, 3-nerved, glabrous; lower valve linear-oblong, 1½ lin. long, nerveless or almost so, reversely ciliate; upper very finely cuneate linear, obscurely 2-toothed, passing into a stout awn, 1½–2½ in. long, rufous-hispidulous below; pale 0; anthers 1½ lin. long; ♂ spikelets lanceolate to linear, subacuminate, 2½–3½ lin. long, dull purplish, glabrous; lower glume 9–11-nerved, muticous in the lower, aristulate in the upper pairs; upper glume acute, 3-nerved, reversely ciliate; valves oblong, obtuse, lower 1-nerved; upper nerveless, up to 1½ lin. long. *A. filipendulus var. calvescens, Hack. Androp. in DC. Monogr. Phan.* vi. 635; *Durand & Schinz, Consp. Fl. Afr.* v. 712. *A. filipendulinus, Steud. Syn. Pl. Glum.* i. 389.

EASTERN REGION: Natal; throughout the colony, *Krauss*, 28! near Durban Bay, *Krauss*, 164 partly! Umpumulo, common, 2000–2500 ft., *Buchanan*, 223!

Also in tropical Africa and Ceylon, and in a somewhat different form in Australia.

Hochstetter's var. *pilosus* (l.c.) is based on Krauss, 164; but what I have seen under this number, is exactly like 28. Hackel describes, however, *A. filipendulus* var. *pilosus* as having 2 ⚥ spikelets in the long-peduncled raceme; I have not seen any such specimen from Africa.

30. A. transvaalensis (Stapf); perennial, tufted; culms erect, subgeniculate, simple, slender, over 3 ft. long, terete, glabrous, 5-noded below the panicle; sheaths tight, glabrous, the lower firm, short, persistent, keeled above, fugaciously and adpressedly hairy, the upper much shorter than the internodes; ligules membranous, rounded, up to 1 lin. long; blades narrow, linear, tapering to a long setaceous point, up to 1 ft. by 1½ lin., flat or convolute above, rigid, glabrous, scabrid in the upper part; panicle consisting of about 12 erect 2–3-nate simple, long filiform branches from long narrow spathiform sheaths bearing filiform or setaceous blades; spathes finely linear, acute, 2–2½ in. long, glabrous, reddish; common peduncles filiform, exserted near the tip of the spathe, the exserted part flexuous,

$\frac{1}{4}$–1 in. long, pubescent and bearded with long tubercle-based hairs, dark purple; racemes contiguous, 6–8 lin. long, one sessile, the other on a fugaciously hairy purple peduncle, joints filiform, obliquely truncate, 1$\frac{1}{2}$ lin. long, dark purple, densely ciliate with rigid white hairs; pedicels very similar, usually produced into a fine subulate membranous appendage, facing the upper glume; sessile spikelets 2–4 in each raceme, ⚥ with the exception of the lowest, which is ♂ like the pedicelled; ⚥ spikelets linear-oblong, 2$\frac{1}{2}$ lin. long, purple, hairy, callus acute, bearded, up to $\frac{1}{2}$ lin. long; lower glume subchartaceous, truncate, dorsally flattened, sometimes shallowly pitted, intracarinal nerves 5, prominent above almost throughout or evanescent below, hairs scattered all over or mainly near upper margins; upper glume obtuse, 3-nerved, hairy above; lower valve almost equalling the glumes, linear, obtuse, sub-2-nerved, scantily ciliate; upper shortly 2-lobed, ciliate, 3-nerved near the base, awn about 1 in. long, slender, pubescent and kneed below the middle; pale 0; ♂ spikelets narrowly lanceolate, up to 3–3$\frac{1}{2}$ lin. long, muticous, hairy or lowest glabrous, 7–9-nerved, keels scantily and rigidly ciliate above; upper acute, hairy or glabrous, long ciliate; valves almost equalling the glumes, ciliate, lower 3-, upper 1-nerved.

KALAHARI REGION: Transvaal; Johannesburg, common, *E.S.C.A. Herb.*, 301!

Intermediate between *A. dregeanus* and *A. filipendulus.*

31. A. pleiarthron (Stapf); perennial; culms erect, somewhat stout, 2–2$\frac{1}{2}$ ft. long, glabrous, 4–5-noded, with intravaginal erect leafy flowering branches throughout; sheaths terete, loose, thin, the lowest withering away, glabrous; ligules rounded, membranous, about 1 lin. long; blades (upper only known) linear, tapering almost from the rounded base, $\frac{1}{2}$ ft. by 2 lin., erect, subrigid, glabrous, scabrid in the upper part; panicle very narrow, stiff, over 1 ft. long; branches strict, erect from very long and narrow spathiform sheaths; spathes finely linear, acute, up to 2$\frac{1}{2}$ in. long, glabrous, reddish; common peduncle finely filiform, strict, exserted $\frac{1}{2}$–1$\frac{1}{2}$ in. from the tip of the spathe, pubescent with long fine white tubercle-based hairs in the upper part; racemes slender, contiguous, usually held together by the intertwining awns, about 1 in. long, one subsessile, the other on a more or less hairy peduncle about 2$\frac{1}{2}$ lin. long, both with 4–5 sessile spikelets, of which the lowest is like the pedicelled spikelets ♂, the others ⚥; joints filiform, up to 2 lin. long, obliquely truncate, densely ciliate with short yellowish hairs, pedicels very similar; ⚥ spikelets linear-oblong, 2$\frac{1}{2}$ lin. long, pale fulvous hairy, callus slender, acute, bearded, up to 1 lin. long; lower glume subchartaceous except the membranous base, truncate, intracarinal nerves 5–7, faintly prominent above, evanescent below the middle, not pitted, hairs up to more than 1 lin. long; upper glume obtuse, 3-nerved, hairy above; lower valve equalling the glumes, linear, obtuse, nerveless, ciliate; upper linear, 1$\frac{3}{4}$ lin. long, 2-toothed, subglabrous, awn stout, about 1$\frac{1}{2}$ in. long, column rufous-pubescent, shorter than the bristle; pale minute oblong, glabrous; anthers 1 lin. long; pedicelled spikelets linear-lanceolate, subacute, muticous, 2½–3 lin. long, hairy; lower glume with 5–9 intracarinal nerves, keels very scantily and minutely spinulous near the tip; upper acute, ciliate; lower valve ciliate, up to 3 lin. long, and 3-nerved in the lowest pair, much shorter, broadly oblong, 1-nerved or nerveless in the upper ones; upper valve cuneate linear, up to 2½ lin. long, 1-nerved in the lowest pair, suppressed in the upper ones; anthers rudimentary in the pedicelled, up to 1¼ lin. long in the sessile spikelet of the lowest pair.

KALAHARI REGION: Transvaal; Krans Kop or Modimulle, near the Nylstroom River, *Nelson*, 15*!

32. A. Ruprechti (Hack. Androp. in DC. Monogr. Phan. vi. 645); perennial, densely tufted; culms erect, stout, up to 6 ft. long, glabrous, terete, simple or with a leafy branch from one of the uppermost internodes; sheaths glabrous, pale, striate, tight or somewhat loose, the lowest and those of the innovation shoots keeled, subpersistent, the upper keeled in the upper part, otherwise terete; ligules membranous, obtuse, up to 1½ lin. long; blades narrow, linear from a narrow base, tapering to a fine point, ½–1 ft. by 1½–2½ lin., flat or revolute along the margins, rigid, glabrous, smooth or scabrid, midrib rather stout; panicle linear to oblong, erect, stiff, up to more than 1 ft. long; branches filiform, erect, glabrous, from long narrow glabrous or hairy spathiform sheaths; spathes very narrow, linear-lanceolate, acuminate, 1½–2 in. long, subscarious to almost herbaceous, pale-reddish or greenish, usually glabrous; common peduncles filiform, strict, glabrous, enclosed; racemes about 10 lin. long, laterally exserted from the spathe, one subsessile with 1 sessile ♂ spikelet at the base and 1 sessile ⚥ above, the other peduncled with 1 sessile ⚥ and 2 pedicelled ♂ spikelets; peduncle of the upper and joint of the subsessile raceme filiform, glabrous, 2 and 1 lin. long respectively, with a hyaline lanceolate 1–2-toothed bract-like appendage about 3 lin. long; lowest pedicel very short, glabrous, upper up to 2 lin. long, very densely and shortly ciliate; ⚥ spikelets subcylindric, over 3 lin. long, pale, callus very slender, pungent, 1½ lin. long, densely bearded; glumes subcoriaceous, glabrous except the hispidulous tips, the lower oblong, 2-cuspidate (points up to 1½ lin. long), finely channelled, inconspicuously 8–10-nerved, margins implicate near the tips, otherwise involute; upper glume lanceolate-oblong, sub-5-nerved; lower valve linear, obtuse, 2½ lin. long, 2-nerved, scantily ciliate; upper slightly shorter, stipitiform, obscurely 2-toothed, hyaline and 3-nerved at the base, awn 2–3 in. long, bent near the middle, rufous-pubescent below the bend; pale 0; anthers 1–1¼ lin. long; ♂ spikelets narrow, linear-lanceolate, about 5 lin. long, glabrous; lower glume subherbaceous, acute or in the upper spikelets narrowed into a scabrid bristle (up to 2 lin. long), 7–11-nerved, keels spinulously ciliate; upper glume 3-nerved, ciliate; valves subequal, 3-nerved, ciliate; anthers 2 lin. long. *A. anthesterioides, Rupr. ex Galeotti in Bull. Ac. Brux.* ix. 245, *not Hochst. A. macrolepis, Hack. l.c.* 646, *and Durand & Schinz, Consp. Fl. Afr.* v. 717, *in part. Hyparrhenia Ruprechti, Fourn. Pl. Mex.* ii. 67.

KALAHARI REGION: Transvaal; Dorum (Doorn?) River, Cave Mountains, *Nelson*, 10*! Pretoria, at Wonderboom Poort, *Rehmann*, 4488!

Also in tropical Africa, Madagascar and Mexico.

Hackel (Bull. Herb. Boiss. iv. App. iii. 11) indicates *A. macrolepis* from the Transvaal. *A. macrolepis*, as originally described by Hackel, is, however, an annual with longer spikelets and appendages, known to me only from Jurland, Central Africa; Barter's plant referred by Hackel to *A. macrolepis* is *A. Ruprechti*, and so are all other specimens of this little group which I have seen from tropical Africa.

XI. ANTHISTIRIA, Linn. f.

Spikelets heteromorphous, 2-nate, clustered on the articulate fragile rhachis of short solitary racemes, subtended by proper spathes and crowded in panicled fascicles; lowest 2 pairs of each raceme closely approximate, ♂ or barren, awnless, sessile or subsessile, usually persistent, forming a spurious tetramerous whorl enveloping the upper 1–3 pairs, each of which consists of a sessile ⚥ and a pedicelled ♂ spikelet, the latter very much resembling those of the involucre. *Florets* 2, the lower reduced to an empty valve, the upper ⚥ in the sessile upper spikelets, ♂ in the involucral and pedicelled spikelets or these more or less reduced and barren. *⚥ spikelets: Glumes* equal or subequal, lower usually coriaceous, at length hardened and often dark-brown to almost black, upper obtusely keeled, coriaceous and channelled along the keel, margins membranous. Lower *valve* hyaline, upper very narrow, passing from a hyaline base into a usually stout awn, very rarely linear and awnless. *Pale* obsolete or 0. *Lodicules* 2, cuneate, glabrous. *Stamens* 3. *Stigmas* laterally or subterminally exserted. *Grain* linear-obovate, biconvex, with 2 grooves on the anterior side. *Involucral* and *pedicelled spikelets: Glumes* equal or subequal, lower herbaceous, dorsally flattened, 2-keeled, many-nerved, upper membranous, lanceolate, acute, 3-nerved, margins ciliate. Lower *valve* hyaline 1-nerved or like the upper suppressed.

Usually perennial, tall, stout; blades long, narrow; panicle occupying one-half or more of the culm; spathes cymbiform, keeled, many-nerved.

Species about 10, in the tropical and subtropical regions of the Old World.

Perennial; involucral spikelets 4–6 lin. long; their lower glume glabrous or beset in the upper part with short tubercle-based hairs ... ... (1) **imberbis.**
Annual; involucral spikelets 3–3½ lin. long; their lower glume beset along the lateral flexures with rigid bristles springing from coarse tubercles ... ... (2) **ciliata.**

1. **A. imberbis** (Retz, Obs. iii. 11); perennial, densely tufted; innovation shoots intravaginal; culms erect or geniculate and

ascending, 1–3 ft. long, rather slender, glabrous, simple or branched; sheaths compressed, keeled, firm, glabrous or the lower sometimes hairy and bearded at the nodes, exceeding the internodes, except the upper; ligules membranous, very short, obtuse or truncate, ciliolate; blades linear, long tapering to a fine point, 2–8 in. by 1–2½ lin., very rarely broader, flat, rather rigid, glabrous or hairy below, panicle narrow, often nodding, 3–6 in. long, scantily branched, branches solitary or the upper 2–3-nate, filiform, glabrous, bearing capituliform, usually dense, fascicles of racemes; spathes lanceolate, obtusely acuminate, about 1 in. long, rarely longer, usually bearded at the base, glabrous or hairy, greenish or reddish; racemes 6–9 lin. long; involucral spikelets whorled sessile, persistent, lanceolate, acute or acuminate, 3½–5 lin. long, rarely longer, glabrous or with scattered short tubercle-based hairs, ♂; their lower glume not or very obscurely winged on one side only; lower valve generally present; pedicelled spikelets linear-lanceolate, glabrous, on short glabrous or subglabrous pedicels; ☿ spikelet 1, not exserted from the involucre, linear-oblong, subcylindric, 2½–3½ lin. long, callus pungent, up to 1½ lin. long, glabrous in front, otherwise densely bearded with shining reddish or purplish hairs; lower glume obtuse or emarginate, smooth and shining except the rigidly pubescent or scabrid tip, obscurely 7–9-nerved, upper glabrous; lower valve glabrous, somewhat shorter than the glumes; awn of upper valve 2½ in. long; anthers 1 lin. long; grain about 2 lin. long. *Thunb. Fl. Cap.* i. 402, *ed. Schult.* 107; *Kunth, Enum.* i. 481; *Steud. Syn. Pl. Glum.* i. 401; *Hook. f. Fl. Brit. Ind.* vii. 211. *A. ciliata, Retz, Obs.* iii. 11; *Anderss. in Peters, Reise Mossamb. Bot.* 562; *Oliv. in Trans. Linn. Soc.* xxix. 176, *not Linn. f. A. ciliata, var. β, Nees in Linnæa,* vii. 285. *A. ciliata, vars. hispida and ciliata, Nees, Fl. Afr. Austr.* 121. *A. arguens, Nees, l.c.* 124, *not Willd. A. vulgaris, Hack. in Engl. & Prantl. Nat. Pflanzenf.* ii. *Abt.* ii. 29. *Stipa arguens, Thunb. Prodr.* 20. *Themeda Forskalii, vars. vulgaris, imberbis and major, Hack. Androp. in DC. Monogr. Phan.* vi. 660–662; *Durand & Schinz, Consp. Fl. Afr.* v. 731.

VAR. β, **mollicoma** (Stapf); leaves, spathes and involucral spikelets more or less densely villous with long soft often tubercle-based hairs, otherwise like the type. *A. ciliata γ mollicoma, Nees, Fl. Afr. Austr.* 121. *Themeda Forskalii, var. mollissima, Hack. l.c.* 661; *Durand & Schinz, l.c.* 730.

VAR. γ, **argentea** (Stapf); leaves silvery with soft closely appressed hairs; spathes and involucral spikelets glabrous or scantily hairy. *A. argentea, Nees, l.c.* 124. *Themeda Forskalii, var. argentea, Hack. l.c.* 661; *Durand & Schinz, l.c.* 730.

VAR δ, **Burchellii** (Stapf); panicle laxer than in the type; spathes rather longer (often up to 2 in. or more); fascicles of 2–3 racemes only; involucral spikelets 4–6 lin. long, glabrous or subglabrous. *Themeda Forskalii, var. Burchellii, Hack. l.c.* 661; *Durand & Schinz, l.c.* 730.

SOUTH AFRICA: without precise locality, *Thunberg!* *Zeyher*, 4487! *Ecklon*, 922! *Bergius! Mund & Maire! Harvey*, 284!

COAST REGION: Clanwilliam Div.; Honig Vallei and on the Konde Berg, near Wupperthal, 3000–4000 ft., *Drège!* Cape Div.; between Cape Town and Table Mountain, *Burchell*, 62! Lions Rump *Burchell*, 127! Devils Peak, below

Blockhouse, *Dodd*, 2749! Lion Mountain, over Sea Point, *Dodd*, 1609! Kasteels Poort, *Dodd*, 1365! Tokay Plantation, *Dodd*, 1964! Simons Bay, *Wright!* Worcester Div.; Dutoits Kloof, 2000–3000 ft., *Drège!* Riversdale Div.; Duivenhoeks River, *Gill!* George Div.; Outeniqua Mountains, near Roodemur, 2000–3000 ft., *Drège;* Port Elizabeth Div.; Port Elizabeth, *E.S.C.A. Herb.* 100! Alexandria Div.; Sam Tees Vlakte, near Enon, *Drège!* near Ado, 1000–2000 ft., *Drège*, on the rocks of Zwartwater Poort, *Burchell*, 3365! Albany Div.; near Grahamstown, *MacOwan*, 46! *Drège*, between Grahamstown and Bothas Hill, 1000–2000 ft., *Drège;* Beaufort West Div.; Nieuw Veld Mountains, near Beaufort West, *Drège;* Queenstown Div.; Table Mountain, 5000–6000 ft., *Drège!* British Kaffraria, *Cooper*, 1799! Var. β: Riversdale Div.; near the Zoetemelks River, *Burchell*, 6692/3! between the Zoetemelks River and Great Vals River (intermediate between the type and var. β), *Burchell*, 6565! George Div.; near George, *Alexander!* Outeniqua Mountains at Montague Pass, *Rehmann*, 81! 82! Lower Albany; Glenfilling, below 1000 ft., *Drège.* Var. γ: Uitenhage Div.; between Coega River and Sunday River, *Drège!*

CENTRAL REGION: Albert Div.; *Cooper*, 1782! Aberdeen Div.; Camdeboo Mountain, 3000–5000 ft., *Drège!* Graaff Reinet Div.; near Graaff Reinet, 3000–4000 ft., *Drège*, at 2700 ft., *Bolus*, 553! Sneeuw Berg Range, 4000–5000 ft., *Drège!* Aliwal North Div.; Witte Bergen, 7000–8000, *Drège.* Var. δ: Aliwal North Div. (?); by the Orange River, 5000–6000 ft., *Zeyher!* Var. ε: Colesberg Div.; near Colesberg, *Shaw*, 5!

KALAHARI REGION: Var. β: Griqualand West; fields to the north of Kimberley, *Nelson*, 90*! Transvaal; Wonderfontein, *Nelson*, 1*! Umhlenga's Kraal, Cave Mountains, *Nelson*, 56*! Var. ε: Griqualand West; Hay Div., between Kloof Village and Witte Water, *Burchell*, 2095! near Griquatown, *Burchell*, 1844! and at Doorn River, *Burchell*, 2136/2! Orange Free State; by the Orange River, *Burke*, 197! by the Caledon River (intermediate between var. β and var. ε), *Burke*, 423! and without precise locality, *Cooper*, 915! *Hutton!* Transvaal; Bosch Veld, between Elands River and Klippan, *Rehmann*, 5108! 5353! near Lydenburg, *Atherstone!* by the Nylstroom River at Moord Drift, *Nelson*, 40*!

EASTERN REGION: Tembuland; Bazeia, 2000–2500 ft., *Baur*, 369! 275! Pondoland; between Umtata River and St. Johns River, *Drège:* Natal; bank of the Tugela River (a form with very broad leaves), *Buchanan*, 227! between Umzimkulu River and Umkomanzi River, *Drège!* Var. β: Natal; Inanda, *Rehmann*, 8244! Var. ε: Umpumulo, 2000 ft. (intermediate between var β and var. ε), *Buchanan*, 226!

Durand & Schinz, l.c. 731, indicate also "*Themeda Forskalii*, var. *glauca*, Hack.," from Bechuanaland (*Marloth*, 995) and from the Cape Colony (*Marloth*, 807). I have not seen those specimens, but I suspect, they represent merely an unusually glaucous state of *A. imberbis.*

The species ranges throughout the tropics of the Old World. In the western parts of the Cape Colony, it covers vast tracts of land at low or high altitudes.

2. **A. ciliata** (Linn. f. Suppl. 113, not Retz); annual; culms suberect or geniculately ascending and rooting from the lower nodes, very slender, terete, 1–3 ft. long, glabrous, simple or branched; sheaths glabrous, shorter than the internodes or the upper with scattered tubercle-based hairs towards the mouth; ligules membranous, rounded, up to 1 lin. long, glabrous; blades linear, acute, 6–12 in. by 2–3 lin., flat, rather flaccid, glabrous or hairy; panicle suberect, occupying $\frac{1}{2}$–$\frac{2}{3}$ of the culm, usually rather dense, compound; lower branches solitary or 2–3-nate, filiform, glabrous, undivided often to the middle, then bearing, at equal distances, usually shortly peduncled dense clusters of racemes; spathes linear to subulate from a broad lanceolate base, $\frac{2}{3}$–1 in. long, glabrous or with scattered

tubercle-based bristles; racemes 4–5½ lin. long, erect; involucral spikelets whorled, sessile, persistent, lanceolate to linear-lanceolate, acute or acuminate, 2½–3½ lin. long, reddish, barren or imperfectly ♂; lower glume beset, chiefly along the asymmetrically winged keels, with stiff bristles which spring from large tubercles, otherwise glabrous; lower valve generally present; pedicelled spikelets narrow, symmetrical, not winged, glabrous; ⚥ spikelet 1, not exserted from the involucre, narrow, linear-lanceolate, 2–2½ lin. long, callus very short, subobtuse, bearded with short reddish hairs; lower glume obtuse, brown, shining, glabrous, or puberulous, particularly upwards, obscurely 6–7-nerved, upper glabrous; lower valve shorter than the glumes, often 2-toothed, awn of upper valve 1–1½ in. long, slender; anthers ½ lin. long. *Gærtn. Fruct.* ii. 465, *t.* 175; *Lam. Illustr. t.* 841, *fig.* 1; *Beauv. Agrost. t.* 23, *fig.* 7; *Kunth, Enum.* i. 481; *Steud. Syn. Pl. Glum.* i. 401; *Baker, Fl. Maurit.* 448; *Hook. f. Fl. Brit. Ind.* vii. 213. *A. scandens, Roxb. Fl. Ind. ed. Carey,* i. 248; *Nees, Fl. Afr. Austr.* 125. *A. semiberbis, Nees, l.c. Themeda ciliata, Hack. Androp. in DC. Monogr. Phan.* vi. 664.

COAST REGION: Cape Town, in gardens (evidently introduced), *Drège,* 1990!
A native of India, introduced elsewhere.

XII. PASPALUM, Linn.

Spikelets orbicular to oblong, obtuse, rarely acute or acuminate falling entire from the very short or obscure pedicels, secund and usually 2-ranked on the flattened or triquetrous rhachis of false spikes, plano-convex; lower floret barren, reduced to the valve, rarely with a rudimentary pale, upper floret ⚥. *Lower glume* 0 (occasionally present and small in *P. Digitaria*), upper membranous, as long as the valves, rarely shorter or obsolete, usually with 1–2 submarginal side-nerves on each side, with or without a middle-nerve, rarely nerveless. *Valves* equal or subequal, lower resembling the upper glume, usually 3–5-, rarely 7-nerved, with the side-nerves curved, close, mostly submarginal and distant from the middle-nerve, when present, upper chartaceous to subcoriaceous, faintly 5–7-nerved. *Pale* subequal to and of the same texture as the valve, 2-nerved. *Lodicules* 2, cuneate. *Stamens* 3. *Styles* distinct, slender; stigmas laterally exserted near the tip of the floret. *Grain* tightly enclosed by the slightly hardened valve and pale, dorsally subcompressed; hilum basal, punctiform; embryo less than half the length of the grain.

Perennial or annual, of various habit; false spikes solitary, 2-nate, digitate or panicled; pedicels usually very short and adnate to the dilated rhachis; spikelets imbricate or contiguous, rarely distant, solitary or sometimes 2-nate.

Species over 150, with very few exceptions, natives of the New World.

The genus has been divided into 3–7 distinct sections by different authors. The South African species belong to the section *Eu-Paspalum* which is characterized by having the lower valve, which is usually very flat, turned away from the rhachis.

Culms usually erect, fascicled on a short rhizome; spikelets broadly-elliptic to orbicular, obtuse or subobtuse ... (1) **scrobiculatum.**
Culms ascending from a creeping base; spikelets acute or acutely acuminate:
Blades 1–3 lin. broad, flat; lateral false spikes sessile or subsessile; spikelets ovate-oblong, acute ... (2) **Digitaria.**
Blades ½–1½ lin. broad, involute to subulate; false spikes peduncled, articulate on the top of the culm; spikelets oblong, acute to acuminate ... ... (3) **distichum.**

1. P. scrobiculatum (Linn. Mant. 29); perennial; culms fascicled on a very short præmorse rhizome, innovation shoots few, usually intravaginal; culms erect, rarely ascending from a prostrate, rooting base, 1 to several feet long, usually sheathed throughout, glabrous; leaves glabrous or more or less softly hairy; sheaths lax, rather thin, the lower often purplish; ligules membranous, short; blades linear to linear-lanceolate, acute to acuminate, 4–8 in. (rarely longer in the African specimens) by 2–4 lin., soft, flat or with involute margins; false spikes 2–3 (in the South African specimens) or more, rhachis herbaceous, 1–1½ lin. broad, margins rigidly ciliate; pedicels very short, almost wholly adnate; spikelets 2-, rarely 3–4-ranked, imbricate, broadly elliptic to orbicular, obtuse, 1–1½ lin. long, glabrous; lower glume 0, upper convex, 3–7-nerved; lower valve flat, often obscurely pitted or wrinkled near the margin, 5–7-nerved, submarginal nerves 2 on each side, very close; upper valve subcoriaceous, brown, shining; flaps of pale auricled; anthers over ½ lin. long. *Fluegge, Gram. Monogr.* 86; *Trin. Pan. Gen.* 56, *and in Mém. Acad. Petersb. sér.* vi. iii. 144; *Spec. Gram. Ic. t.* 143; *Kunth, Enum.* i. 53; *Steud. Syn. Pl. Glum.* i. 21; *Baker, Fl. Maurit.* 432; *Durand & Schinz, Consp. Fl. Afr.* v. 738; *Hook. f. Fl. Brit. Ind.* vii. 10. *P. Commersonii, Lamk. Illustr.* i. 175, *t.* 43, *fig.* 1. *P. Kora, Willd. Sp. Pl.* i. 332. *P. dissectum, Nees, Fl. Afr. Austr.* 15, *not Agrost. Bras. nor Linn. P. mauritanicum, Nees ex Steud. Syn. Pl. Glum.* i. 26 (?).

COAST REGION: Uitenhage Div.; by the Zwartkops River, below 100 ft., *Drège!* Bathurst Div.; by the Fish River, near Trompetters Drift, below 500 ft., *Drège.* Komgha Div.; near the Kei River, *Drège!*
EASTERN REGION: Natal; in swamps by the Umlazi River, *Krauss*, 147! in grassy plains between Durban Bay and Umlazi River, *Krauss*, 204! common near Umpumulo, 2000 ft., *Buchanan*, 184! and without precise locality, *Gerrard*, 587!

Common throughout the tropics of the Old World.

2. P. Digitaria (Poir. Encycl. Suppl. iv. 316); perennial; culms ascending from a creeping, often rooting and branching long base, many-noded, sheathed throughout; leaves numerous, distichous, imbricate; sheaths compressed, keeled, striate, thin, pale, glabrous except the finely ciliate upper margins, rather loose; ligules membranous, short, truncate; blades linear, acute 2–6 in. by 1–3 lin., spreading, flat, rarely involute, glabrous or scantily hairy towards the base; false spikes usually 2 (rarely 3 or 4), the lateral sessile or subsessile, villous at the base, rhachis herbaceous, ½–1 lin. broad, margins scabrid; pedicels very short, stout, scabrid; spikelets solitary, rarely 2-nate, ovate-oblong, acute, 1–1½ lin. long, subimbricate, adpressed to the rhachis, green; lower glume 0 or more or less developed, very small to ¼ the size of the lower valve, often oblique, membranous, rounded to lanceolate; upper glume convex, usually pubescent, 5- to sub-7-nerved; lower valve usually 3-nerved, rarely with 1–2 additional side-nerves, middle-nerve percurrent; pale minute or 0; upper valve subcoriaceous, 5-nerved, smooth; pale not auricled; anthers ¾ lin. long. *Gray, Man. Fl. Un. Stat. ed.* v. 645. *P. michauxianum, Kunth, Rév. Gram.* i. 25. *P. Elliottii, Wats. in Gray, Man. Fl. Un. Stat. ed.* vi. 629. *P. paspaloides, Scribn. in Mem. Torr. Bot. Club.* v. 29. *Digitaria paspalodes, Mich. Fl. Bor. Amer.* i. 46; *Duby, Bot. Gall.* i. 501.

COAST REGION: Cape Div.; roadside between High Constantia and Nek, *Dod*, 2477! near Claremont Station, *Dod*, 2381!

Of American origin, introduced into Southern Europe, India, Australia and New Zealand, but very rare.

3. P. distichum (Linn. Amoen. Acad. v. 391); perennial; culms ascending from a creeping, rooting, often very long and branched, base, many-noded, sheathed throughout; leaves numerous, distichous, imbricate below, sheaths thin, pale, glabrous except the often bearded mouth, the lower at length loose; ligules very short, truncate, with fine hairs from behind; blades linear, acute, 2–4 in. by ½–1½ lin., spreading, involute, rarely flat, glabrous; false spikes 2-nate, both peduncled and articulate on the top of the culm, rarely 3-nate, often spreading, rhachis herbaceous, ½ lin. broad, margins scabrid; spikelets subsessile, solitary, oblong, acute to acuminate, 1–2 lin. long, dorsally flattened, imbricate and adpressed to the rhachis, glabrous, pale; lower glume 0; upper slightly convex, 5- or 4-nerved (middle-nerve suppressed), side-nerves close, submarginal; lower valve very like the glume, middle-nerve always percurrent, side-nerves 2–3 on each side, submarginal; upper valve distinctly shorter, subcoriaceous, 5-nerved, smooth, pale; pale obscurely auricled; anthers ¾ lin. long; grain ½ lin. long. *Burm. Fl. Ind.* 23; *Gærtn. Fruct.* ii. 2, *t.* 80; *Kunth, Enum.* i. 52; *Steud. Syn. Pl. Glum.* i. 29; *Baker, Fl. Maurit.* 431; *Durand & Schinz, Consp. Fl. Afr.* v. 737; *Hook. f. Fl. Brit. Ind.* vii. 12. *P. littorale, R. Br. Prodr.* 188; *Trin. Diss.* ii. 95; *Spec. Gram. Ic. t.* 112; *Kunth, l.c.* 51. *P. longiflorum, Beauv. Fl. Owar.* ii. 46, *t.* 85, *fig.* 2, *not Baker nor Durand & Schinz. P. vaginatum, Sw. Prodr. Veg. Ind. Occ.* 21, *not Beauv.*; *Trin. Diss.* ii. 94; *Spec. Gram. Ic. t.* 120; *Kunth, l.c.* 52; *Steud. l.c.* 20. *P. squamatum, Steud. l.c.* 21. *Digitaria paspaloides var. longipes, Lge. in Willk. & Lge. Prodr. Fl. Hisp.* i. 45.

VAR. β, **nanum** (Doell in Mart. Fl. Bras. II. ii. 75); 2–3 in. high; blades subulate, ¾–1¾ in. long; racemes 5–7 lin. long on very slender peduncles; spikelets 1 lin. long.

COAST REGION: Var. β: Cape Div.; shore at Camps Bay, *Dod*, 2343

EASTERN REGION: Natal; coast marshes, *Buchanan*, 84! and without precise locality, *Gerrard*, 590!

The typical form is widely distributed throughout the tropics and introduced into North-West Spain; var. *nanum*, a very marked dwarf variety, is otherwise known only from extra-tropical South America.

XIII. DIGITARIA, Rich.

Spikelets lanceolate to elliptic-oblong, flat in front, convex on the back, falling entire from the pedicels, usually 2–3-nate and pedicelled on the triquetrous or dilated rhachis of very slender and usually spiciform racemes or false spikes; lower floret barren reduced to the valve and a very minute rudimentary pale, upper floret ⚥. *Glumes* very dissimilar, the lower usually hyaline, minute or sometimes quite suppressed, the upper membranous, half as long to as long as the ⚥ floret, rarely exceeding it or less than half its length, 3-, rarely 1- or 5-nerved or nerveless. *Valves* equal or subequal, lower resembling the upper glume, 7–9-nerved, rarely 5-, very rarely 3-nerved, nerves close, parallel, straight, prominent; upper valve chartaceous to subchartaceous, 3- very rarely sub-5-nerved. *Pale* of upper floret subequal to the valve, 2-nerved, of the same texture. *Lodicules* 2, minute, broadly cuneate. *Stamens* 3. *Styles* distinct; stigmas laterally exserted near the tip of the floret. *Grain* tightly enclosed by the slightly hardened rigid valve and pale, oblong, slightly dorsally compressed; hilum basal, punctiform; embryo less than ½ the size of the grain.

Perennial or annual; blades linear to linear-lanceolate, usually flat, often flaccid; ligules membranous; racemes sessile, digitate or more or less distant on a common axis, or solitary, or rarely peduncled, and panicled, simple or sometimes compound near the base; rhachis triquetrous, angles often herbaceously winged, particularly the lateral whence the rhachis becomes flattened; pedicels unequal, the longer from half as long as to longer than the spikelet; spikelets closely adpressed and imbricate, rarely lax or very lax, usually silky, though often apparently glabrous in consequence of the extremely fine hairs being very tightly adpressed and arranged in lines between the nerves, rarely really glabrous.

Species over 50, mostly in the warm parts of the Old World; 1 widely spread as a weed all over the globe except the arctic region.

Section 1. EU-DIGITARIA (Stapf). Racemes simple or nearly so, not setose, on a short or long common axis, rarely solitary; spikelets ⅔–1, rarely to 1½ lin. long, with extremely fine adpressed or sometimes spreading silky hairs, rarely glabrous; upper glume usually 5- or 3-nerved; lower valve 7-, rarely 5- or 9-nerved.

Racemes solitary, dense; pedicels and spikelets densely villous ... ... (1) **monodactyla.**

Racemes 2 to many, digitate, subdigitate or more distant on a long common axis:

* Upper glume and lower valve more or less silky, sometimes apparently glabrous:

Racemes very slender, 2–3-nate, often so closely contiguous as to imitate a solitary raceme, often compound at the base ... ... (2) **argyrograpta.**

Racemes never contiguous, except in a very young state, quite simple:
- Perennial:
  - Culms rather stout, $1\frac{1}{2}$-3 ft. long; lowest sheaths tomentose at the very base; blades $\frac{1}{2}$–1 ft. by 1–2 lin. ... ... (3) **ériantha.**
  - Culms very slender, $\frac{1}{3}$–1 ft. long; sheaths glabrous; blades setaceous, up to 2–6 in. long ... ... (4) **setifolia.**
- Annual:
  - Spikelets $\frac{3}{4}$–$1\frac{1}{2}$ lin. long; pedicels angular, scabrid:
    - Pedicels shortly hairy near the tips; spikelets obtuse, silky with clavate hairs in very slender long racemes ... (5) **ternata.**
    - Pedicels glabrous; hairs of spikelets not clavate:
      - Upper glume long acuminate, conspicuously exceeding the upper valve ... ... (6) **debilis.**
      - Upper glume shorter than the upper valve:
        - Spikelets $\frac{3}{4}$–1 lin. long, linear-oblong; lower glume obsolete or 0; lower valve very closely nerved (7) **horizontalis.**
        - Spikelets 1–$1\frac{1}{4}$ lin. long, oblong; lower glume usually distinct; the inner lateral nerves of the lower valve rather distant from the middle-nerve ... ... (8) **sanguinalis.**
  - Spikelets $\frac{2}{3}$–$\frac{3}{4}$ lin. long, pedicels terete, smooth, tips discoid; racemes extremely slender ... ... (10) **tenuiflora.**

** Spikelets quite glabrous; upper glume 1-sub-3-nerved; lower valve 7-9-nerved ... (9) **diversinervis.**

Section 2. SETARIOPSIS (Stapf). Racemes simple, several to many on a long common rhachis, setose; spikelets surrounded by stiff hairs from the tip of the pedicel; lower glume 0, upper minute or 0; lower valve 3-nerved or almost nerveless ... ... (11) **diagonalis.**

Section 3. TRICHACHNE (Stapf). Racemes usually rather lax, compound near the base, and whorled on a long common axis, rarely subdigitate (*D. tricholænoides*) or truly panicled (*D. flaccida*); spikelets $1\frac{1}{2}$–$2\frac{1}{2}$ lin. long, densely silky, villous with long often purplish hairs conniving into a brush-like point.
- Racemes sessile or subsessile, subdigitate; spikelets 2–$2\frac{1}{2}$ lin. long ... ... (12) **tricholænoides.**
- Racemes peduncled in a scantily branched panicle; spikelets $1\frac{1}{2}$–2 lin. long ... ... (13) **flaccida.**

1. D. monodactyla (Stapf); perennial, compactly tufted, innovation shoots intravaginal; culms very slender, erect, 1–$1\frac{1}{2}$ ft. long, glabrous, 1–2-noded above the base, simple; leaves crowded at the base; sheaths very tight, glabrous or scantily hairy, the lower firm, persistent, at length sometimes breaking up into fibres; ligules very short, truncate; blades setaceously convolute, filiform, 2–4 in. long, rather rigid or flexuous, glabrous or scantily hairy, smooth; raceme spikelike, solitary, terminal on the long exserted culm, very slender, 2½–5 in. long; rhachis wavy, very narrow, ciliate, midrib obtuse, stout; pedicels 2-nate, densely ciliate, unequal, the longer up to ¾ lin. long; spikelets subimbricate, adpressed to the rhachis, the superposed ones distant almost by their own length, oblong-lanceolate, up to 1½ lin. long, with silky and adpressed hairs; lower glume 0, upper linear-oblong, ¾ to almost 1 lin. long, long and densely ciliate, 3–2-nerved; lower valve oblong-lanceolate, subacuminate, up to 1½ lin. long, 5–7-nerved, densely and long ciliate, particularly along the marginal nerves; upper subchartaceous, ovate-lanceolate, subacuminate, 1 lin. long, smooth, brown; flaps of pale very broad below, overlapping; anthers ½ lin. long. *Panicum monodactylum, Nees, Fl. Afr. Austr.* 21; *Steud. Syn. Pl. Glum.* i. 56; *Durand & Schinz, Consp. Fl. Afr.* v. 755.

COAST REGION: Alexandria Div.; grassy hills near Ado, *Ecklon.*

CENTRAL REGION: Aliwal North Div.; at the foot of the Witte Bergen, between Leeuwen Spruit and Riet Vallei, 5000 ft., *Drège!*

KALAHARI REGION: Orange Free State; Drakens Berg Range, near Harrismith, *Buchanan*, 121! Winburg, *Buchanan*, 234! Transvaal; Apies River, *Nelson*, 102*!

2. **D. argyrograpta** (Stapf); perennial; culms erect or ascending from a short præmorse rhizome, very slender, 1–1½ ft. long, glabrous, much branched near or above the base from usually swollen nodes, or simple and 1–3-noded near the base; leaves mostly crowded at the base of the branches or of the culms; sheaths glabrous or scantily hairy, or the lowest fugaciously tomentose at the very base and subpersistent or withering; ligules very short, truncate; blades very narrow, linear, often tapering to a very long fine flexuous point, 1–2½ in. by ½ lin., usually more or less folded, flaccid or rather firm, flexuous, curled or even twisted, glabrous or scantily hairy or minutely puberulous above; racemes 2–3-nate, often contiguous, very rarely solitary, erect, slender, 2–4 in. long, dense or lax, often subcompound near the base; rhachis very slender, more or less wavy, triquetrous, angles not winged, smooth or scantily scabrid; pedicels 2-nate, finely filiform, angular, scabrid, glabrous, unequal, the longer 1½ to 2 lin. long, adpressed or spreading; spikelets oblong, acute, 1–1¼ lin. long, greenish, silky and woolly; lower glume membranous, minute, ovate, or rotundate, nerveless, upper oblong-lanceolate, 1–1¼ lin. long, 3-nerved, with lines of very fine adpressed or spreading acute silky hairs between the nerves and along the margins; lower valve oblong, acute, very closely and prominently 7–9-nerved, more or less silky along the margins and between 1 or 2 pairs of the side-nerves, hairs short or long as in the upper glume; upper valve subchartaceous, greenish, oblong, acute, slightly shorter than the lower; anthers 1 lin. long. *Panicum argyrograptum, Nees, Fl. Afr. Austr.* 27; *Steud. Syn. Pl. Glum.* i. 40. *P. commutatum var. argyrograptum, Hack. in Durand & Schinz, Consp. Fl. Afr.* v. 743.

COAST REGION: Uitenhage Div.; in the bed of the Zwartkops River, *Ecklon.* King Williamstown Div.; Amatola Mountains, *Buchanan,* 8!

KALAHARI REGION: Orange Free State; near Kommissie Drift on the Caledon River, 4000 ft., *Zeyher,* 1790! *Burke,* 208! Jackals Fontein, *Zeyher,* 1791! *Burke,* 24! between Kimberley and Bloemfontein, *Buchanan,* 278!

Nees describes the lower valve as 5-nerved; but this is probably due to an oversight, some of the side-nerves being often covered by the silky tomentum.

3. **D. eriantha** (Steud. in Flora, 1829, 468); perennial, densely tufted; rhizome short, præmorse; culms erect or geniculate, rather stout, 1½–3 ft. long, glabrous, 2–4-noded, usually simple, sometimes scantily branched below; sheaths striate, rather loose, the upper shorter than the internodes, the lowest more or less hirsute-tomentose at the base, otherwise glabrous or scantily hirsute; ligules hyaline between the more or less produced margins of the sheath-mouth, 1½–4 lin. long; blades linear, tapering to a fine point, ½–1 ft. long by 1–2 lin., or the lowest sometimes much shorter, flat or with the margins revolute, rigid, rarely flaccid, glaucous, glabrous or scantily hirsute at the base, margins scabrid above; racemes 4 to many, subdigitate, slender, erect or suberect, 4–8 in. long, strict or flexuous; rhachis very slender, triquetrous, angles scarcely winged, scabrid, internodes up to 1 lin. long, rarely longer; pedicels 2-nate, filiform, scabrid, unequal, the longer 1 lin., rarely 1½ lin. long; spikelets oblong- or ovate-lanceolate, 1¼–1½ lin long, subadpressed, pale green, often silvery; lower glume membranous, ovate, acute or obtuse, up to ¼ lin. long, nerveless; upper lanceolate, acute, about 1 lin. long, 3-nerved, with 4 lines of very fine, usually adpressed, long silky hairs; valves equal, lower oblong- or ovate-lanceolate, acute, 7–9-nerved, inner 4 side-nerves like the middle-nerve very prominent, marginal nerves faint, lines of usually adpressed dense long silky hairs between the inner pair of side-nerves and along the margins; upper subchartaceous, oblong, shortly acuminate, 3-nerved, dull-green; anthers ¾ lin. long. *Panicum commutatum, Nees in Linnæa,* vii. 274; *Fl. Afr. Austr.* 25 (*excl. most synonyms*), *not Nees in Wight & Arn. Glum. Ind. Or. ined. No.* 3; *Durand & Schinz, Consp. Fl. Afr.* v. 743 (*in part*).

VAR. β, **stolonifera** (Stapf); throwing out long or short rooting runners, culms very slender; blades 1–2½ in. by 1–1½ lin.; racemes very slender; spikelets smaller than in the type, 1–1¼ lin. long.

SOUTH AFRICA: without precise locality, *Drège,* 4207! *Zeyher,* 1792!

COAST REGION: Uitenhage Div.; by the Zwartkops River, *Drège!* between Coega River and Sunday River, *Ecklon,* and without precise locality, *Zeyher!* Alexandria Div.; grassy hills near Ado, *Ecklon.* Bathurst Div.; Port Alfred, *Hutton!* Albany Div.; grassy places near Grahamstown, 2000 ft., *MacOwan,* 1290! King Williamstown Div.; Yellow Wood River, 1000–2000 ft., *Drège!* Queenstown Div.; between Kat Berg and Klipplaats River, *Drège!*

CENTRAL REGION: Prince Albert Div.; Great Zwarte Bergen, on stony hills, 2000–3000 ft., *Drège.* Somerset Div.; Blyde River, *Burchell,* 2982! Graaff Reinet Div.; on stony hills near Graaff Reinet, 2600 ft., *Bolus,* 684! *Ecklon.* Aliwal North Div.; between Aliwal North and Kraai River, *Drège.*

KALAHARI REGION: Griqualand West; Hay Div., between Griquatown and Witte Watter, *Burchell,* 1974! Prieska Div.; between Modder Fontein and

Keikams Poort, *Burchell*, 1612/8! Orange Free State; without precise locality, *Buchanan*, 269! Transvaal; Pienaars River, *Nelson*, 33*! and without precise locality, *Holub*! Var. β: Transvaal; Houtbosch, *Rehmann*, 5712! 5712/b!

EASTERN REGION: Tembuland; Bazeia, 2000 ft., *Baur*! Natal; Port Natal, *Plant*, 62! between Umzimkulu and Umkomanzi River, *Drège*, coast region near Umpetezane, *Sutherland*! Nottingham, *Buchanan*, 133! very common near Umpumulo, *Buchanan*, 208! and without precise locality, *Gerrard*, 467! *Buchanan*, 19!

Also in tropical Transvaal, and as far as the Zambesi; in a slightly differing variety in the Mediterranean region from the Canaries to North-west India. A very distinct species, which has, curiously enough, been confounded by Nees with *Panicum ciliare*, Roxb., a variety of *Digitaria sanguinalis*, of which specimens were distributed by Wight and Arnott with Nees's determination as *Panicum commutatum*. Nees enumerates, in Fl. Afr. Austr. 26, *P. filiforme*, *Thunb.* Prodr. Cap. 19; Fl. Cap. ed. Schult. 103), as a synonym of his *P. commutatum*; but there is no plant named *P. filiforme* in Thunberg's herbarium, nor a specimen of *D. eriantha*. Moreover, as Thunberg refers to Linnæus as author, who had an American plant in view, when describing his *P. filiforme*, I think that the plant was included in Thunberg's Prodromus and Flora Capensis by error.

4. **D. setifolia** (Stapf); perennial, densely cæspitose; culms erect, very slender, ⅓–1 ft. long, glabrous; leaves glabrous except the scantily hairy mouth of the sheath or more or less hairy, all crowded at the base except the uppermost which is reduced to a very long almost bladeless sheath; basal sheaths firm, persistent, at length breaking up into fibres; ligules very short, truncate; blades filiform, setaceously convolute, up to 2–6 in. long, rather firm, flexuous, smooth; racemes 2–3, slender, erect, 1–2 in. long; rhachis very slender, angular, smooth, internodes 1½–2 lin. long; pedicels 2-nate, one very short, the other up to 1¼ lin. long, angular, scantily scabrid above; spikelets oblong, obtuse, 1½ lin. long, greenish; lower glume minute, rotundate, delicately hyaline, upper oblong, subacute, 1¼ lin. long, 3-nerved, with lines of adpressed or slightly spreading rufous capitellate hairs between the nerves and along the margins, upper marginal hairs exceeding the glume; lower valve 7–9-nerved, nerves prominent, except the submarginal, hairs similar to those of the upper glume, in 4 dorsal and 2 dense marginal lines, upper valve subchartaceous, oblong, subacuminate, chestnut-brown except the hyaline whitish margins; anthers ½ lin. long.

COAST REGION: Albany Div.; on mountain slopes near Grahamstown, 2000 ft., *MacOwan*, 1300!

EASTERN REGION: Tembuland: moist spots near Bazeia, 2000 ft., *Baur*, 287!

5. **D. ternata** (Stapf); annual; culms fascicled, erect from a geniculate base, slender, simple, ½–2 ft. long, glabrous or more or less beset with fine long spreading hairs below the racemes, about 2-noded; sheaths glabrous or scantily fimbriate at the mouth; ligules short, truncate; blades linear-lanceolate to linear, acute, 2–9 in. by 3 lin., flat, flaccid, glabrous or sparingly hairy near the base; racemes 2–7, subdigitate, erect or spreading, very slender, strict, 2–12 in. long; rhachis very narrow, linear, up to ½ lin. broad, margins scabrid,

internodes about $1\frac{1}{2}$ lin. long; pedicels 2–3-nate unequal, up to more than 1 lin. long, shortly hairy towards the scarcely thickened tips; spikelets adpressed to the rhachis, oblong-elliptic, obtuse or subobtuse, about 1 lin. long, pale; lower glume 0; upper very delicate, narrower and shorter than the upper valve, 3-nerved, sides densely villous with clavate hairs; valves subequal, lower prominently 5-nerved, very densely and adpressedly silky villous with clavate hairs, upper chartaceous, ovate-oblong, subacute, dark-brown to black, except the whitish hyaline margins; anthers $\frac{1}{2}$ lin. long. *Panicum ternatum, Hochst. in Flora,* 1841, i. *Intell.* 19; *Steud. Syn. Pl. Glum.* i. 40; *Engl. Hochgebirgsfl. Trop. Afr.* 118; *Durand & Schinz, Consp. Fl. Afr.* v. 766. *P. phæocarpum, var. gracile, Nees, Fl. Afr. Austr.* 23. *Cynodon ternatum, A. Rich. Tent. Fl. Abyss.* ii. 405. *Paspalum ternatum, Hook. f. Fl. Brit. Ind.* vii. 17.

COAST REGION: Queenstown Div.; Shiloh, *Baur*, 780!

EASTERN REGION: Transkei; near the Gekau (Gcua or Geuu) River, below 1000 ft., *Drège!* Natal; near Umpumulo, 2000–2500 ft., *Buchanan*, 203! 205! and without precise locality, *Buchanan*, 77!

Also in Abyssinia and India.

6. D. debilis (Willd. Enum. Hort. Berol. 1890, 91); annual; culms ascending from a geniculate base, scantily branched below, 1–2 ft. long, glabrous, 5- or more noded, upper node by far the longest; leaves glabrous or hairy; sheaths rather thin, striate; ligules rounded, $1\frac{1}{2}$–2 lin. long; blades linear from a subcordate base, tapering to a fine point, 3–5 in. by $1\frac{1}{2}$–$2\frac{1}{2}$ lin., flat, flaccid, margins scabrid; racemes 3–10, subdigitate or on a scabrid angular common axis (1–2 in. long), singly or the lower subverticillate, erect or spreading, very slender, strict, 4–8 in. long; rhachis filiform, angular, very scabrid, internodes up to 2 lin. long; pedicels 2-nate, one very short, the other up to 1 lin. long, fine, angular, scabrid; spikelets lanceolate, about $1\frac{1}{2}$ lin. long; lower glume very minute, rounded hyaline, upper linear-lanceolate, long acuminate, exceeding the lower valve by $\frac{1}{3}$, strongly 7-nerved, finely and adpressedly silky between the outer nerves and along the margins; lower valve oblong, shortly acuminate, rather over 1 lin. long, strongly 7-nerved, finely and adpressedly silky between the outer nerves and along the margins, upper subchartaceous, slightly shorter than the lower, pale; anthers $\frac{3}{8}$ lin. long; grain $\frac{1}{2}$ lin. long. *Coss. & Durieu, Expl. Scient. Algér.* ii. 33. *D. decipiens, Fig. & De Notaris, in Mem. Acc. Torin. sér.* ii. xiv. (1853) 359, *t.* 24. *Panicum debile, Desf. Fl. Atlant.* i. 59; *Steud. Syn. Pl. Glum.* i. 41; *Durand & Schinz, Consp. Fl. Afr.* v. 746. *P. filiforme, Poir. Voy. en Barb.* ii. 93, *not Linn. P. reimarioides, Anderss. in Peters, Reise Mossamb. Bot.* 547. *Paspalum debile, Poir. Encycl.* v. 34 (*excl. syn.*); *Fluegge, Gram. Monogr.* 136; *Kunth, Enum.* i. 45.

EASTERN REGION: Natal; Umpumulo, *Buchanan*, 202! and without precise locality, *Gerrard*, 693!

Also in tropical Africa, Madagascar, and the Mediterranean countries, from Algeria and South Italy to Portugal.

7. D. horizontalis (Willd. Enum. Pl. Hort. Berol. 1809, 92); annual; culms ascending from a geniculate or more or less prostrate, sometimes rooting, branched base, slender, 1 to several ft. long, glabrous, few to many-noded, upper internode by far the longest; leaves glabrous or hairy; sheaths thin, the lower withering; ligules very short; blades linear-lanceolate to linear, acute or gradually tapering from below the middle, 2–5 in. by 2–4 lin. (in the South African specimens), flat, flaccid, margins scabrid; racemes 4- to many, subdigitate or on an angular common axis of variable length (1 to several in. long), singly or the lower subverticillate, erect or spreading, slender, strict, 1½–6 in. long, usually villous or bearded at the base, often with a few scattered stiff long very fine hairs from the common axis; rhachis very narrow, usually wavy, triquetrous, lateral angles more or less winged, scabrid, internodes about 1½ lin. long (in the South African specimens) or longer; pedicels 2-nate, one very short, the other up to ¾ lin. long, triquetrous, scabrid; spikelets linear-oblong, acute, ¾–1 lin. long, greenish; lower glume extremely minute or suppressed, upper lanceolate-oblong, equalling ½–⅘ of the upper valve, 3-nerved with lines of extremely fine adpressed silky hairs between the nerves and along the margins or with the marginal hairs spreading; valves equal or subequal, lower prominently and closely 7-nerved, with lines of extremely fine adpressed silky hairs between the outer nerves and along the margins or with the marginal hairs spreading; upper valve subchartaceous, acute, pale-greenish. *D. setosa, Hamilt. Prod. Veget. Ind. Occ.* 6. *D. umbrosa, Link, Hort. Bot. Berol.* i. 227. *D. undulata, Schrank ex Nees, Fl. Afr. Austr.* 24. *Panicum horizontale, G. F. W. Meyer, Prim. Fl. Esseq.* 54; *Kunth, Enum.* i. 81; *Nees, Fl. Afr. Austr.* 24; *Hook. Niger Fl.* 560; *Steud. Syn. Pl. Glum.* i. 39. *P. Hamiltonii, Kunth, Enum.* i. 84. *P. Zeyheri, Nees, Fl. Afr. Austr.* 25; *Steud. l.c.* 40; *Durand & Schinz, Consp. Fl. Afr.* v. 767. *P. fenestratum, Hochst. ex A. Rich. Tent. Fl. Abyss.* ii. 361; *Steud. Syn. Pl. Glum.* i. 40; *Engl. Hochgebirgsfl. Trop. Afr.* 117; *Durand & Schinz, l.c.* 749. *P. porranthum, Steud. l.c.* 42. *P. sanguinale, var. distans, Doell in Mart. Flor. Bras.* ii. ii. 134. *P. sanguinale, vars. cognatum and horizontale, Hack in Bull. Herb. Boiss.* ii. *App.* ii. 18; *Durand & Schinz, Consp. Fl. Afr.* v. 763. *P. sanguinale, var. fenestratum, Schweinf. in Bull. Herb. Boiss.* ii. *App.* ii. 18. *P. sanguinale, var. filiforme, Durand & Schinz, l.c.*

EASTERN REGION: Transkei; banks of the Bashee River, below 1000 ft., *Drège!* Tembuland; Bazeia, 2000 ft., *Baur*, 177! Natal; near Durban, *Drège! Williamson*, 27! Umpumulo, *Buchanan*, 202! 204!

Throughout tropical Africa, in the Mascarene Islands and tropical America.

8. D. sanguinalis (Scop. Fl. Carn. ed. II. i. 52); annual; culms ascending from a geniculate or prostrate, often rooting, branched base, 1 to several ft. long, glabrous, few- to many-noded, upper node by far the longest; leaves glabrous or hairy; sheaths thin, herbaceous, loose, sometimes bearded at the nodes; ligules truncate, up to more than ½ lin. long; blades linear-lanceolate to linear, acute, 1–5 in. by 2–4 lin., flat, flaccid, margins scabrid; racemes few to many, subdigitate, solitary or 2–3-nate on a short angular scaberulous common axis, erect or spreading, rather stout for the genus, usually strict, 1–6 in. long, often finely villous at the base; rhachis triquetrous, lateral angles winged, scabrid, internodes over 1 lin. long, pedicels 2-nate, one very short, the other up to ¾ lin. long, triquetrous, scabrid; spikelets oblong, acute, 1–1¼ lin. long, greenish or purplish; lower glume ovate, acute, about ⅙ lin. long; upper ovate-lanceolate, acute, equalling ½ or less of the upper valve, 3-nerved, with lines of extremely fine adpressed silky hairs between the nerves and along the margins, lower valve oblong acute, 7-nerved, the inner lateral nerves somewhat distant from the middle-nerve, very prominent, the outermost submarginal, faint lines of very fine adpressed silky hairs along the margins and often also between the lateral nerves; upper valve subchartaceous, oblong, subacuminate or acute, greenish or purplish, slightly shorter than the lower valve; anthers up to ½ lin. long. *Reichb. Icon. Fl. Germ.* i. *fig.* 1407; *Johnson & Sowerby, Brit. Grass. t.* 143. *D. ægyptiaca, Willd. Enum. Hort. Berol.* 1809, 93. *Panicum sanguinale, Linn. Spec Plant.* 57; *Schreb. Beschr. Graes. t.* 16; *Fl. Dan. t.* 388; *Host, Gram. Austr.* ii. *t.* 17; *Sowerby, Engl. Bot. t.* 849; *Trin. Spec. Gram. Ic. t.* 93; *Knapp, Gram. t.* 12; *Kunth, Enum.* i. 82. *Trin. Pan. Gen.* 112, *and in Mém. Acad. Pétersb. sér.* vi. iii. 200 (*var.* 1 *a.*); *Nees, Fl. Afr. Austr.* 28; *Steud. Syn. Pl. Glum.* i. 39; *Durand & Schinz, Consp. Fl. Afr.* v. 761. *P. sanguinale, var. ægyptiacum, Durand & Schinz, l.c. P. filiforme, Jacq. Obs.* iii. 18, *t.* 70. *P. ægyptiacum, Retz. Obs.* iii. 8; *Kunth, Enum.* i. 83; *Nees, l.c.*; *Steud. l.c.* 39. *Paspalum sanguinale, Lamk. Illustr.* i. 176; *Hook. f. Brit. Ind.* vii. 13 (*in part*). *Paspalum ægyptiacum, Poir. Encycl. Suppl.* iv. 314. *Syntherisma vulgare, Schrad. Fl. Germ.* 161.

SOUTH AFRICA: without precise locality, *Drège*, 32!

COAST REGION: Clanwilliam Div.; Wupperthal, *Wurmb.* Cape Div.; in gardens near Capetown, *Ecklon*, 963! Table Mountain, *Drège.* "Swellendam and George District," *Mund.* Knysna Div.; Knysna Forest, *E.S.C.A. Herb.*, 425! Uitenhage Div.; in cultivated grounds by the Zwartkops River, *Ecklon.*

CENTRAL REGION: Graaff Reinet Div.; near Graaff Reinet, in cultivated ground, *Bolus*, 684*!

EASTERN REGION: Natal; near Durban, *Williamson*, 10!

KALAHARI REGION: Griqualand West; Hay Div., at Griqua Town, *Burchell*, 1959/1!

Durand & Schinz (l.c. 762) indicate also *Panicum sanguinale*, var. *ciliare* (Doell), from the Cape Colony and Bechuanaland (*Marloth*, 997). I have not seen South African specimens of this variety, which is distinguished by having spreading hairs along the margins of the spikelet.

9. **D. diversinervis** (Stapf); perennial (?); culms ascending from a prostrate, rooting, loosely branched base, very slender, weak, up to 1½ ft. long, glabrous, 4–6-noded; uppermost internode by far the

longest, long exserted; leaves glabrous or very sparingly hairy; sheaths thin, the lower withering; ligules very short; blades linear-lanceolate, acute, 1–3 in. by 1½–2½ lin., flat, flaccid, scaberulous in the upper part, margins scabrid; racemes subdigitate, slender, erect or spreading, 1¼–2 in. long, minutely villous at the insertion; rhachis very narrow and wavy, triquetrous, angles scabrid, internodes up to 2 lin. long; pedicels 2-nate, one very short, the other up to 1 lin. long, triquetrous, scaberulous; spikelets oblong, acute, 1½ lin. long, greenish, glabrous; glumes thin, lower broadly ovate, up to ⅜ lin. long, 1-nerved or nerveless, upper ovate, acute, 1 lin. long, 1–sub-3-nerved; valves equal, lower 7–9-nerved, upper subchartaceous, acuminate, pale to dark-brown. *Panicum diversinerve, Nees, Fl. Afr. Austr.* 23; *Steud. Syn. Pl. Glum.* i. 40; *Durand & Schinz, Consp. Fl. Afr. Austr.* v. 748.

EASTERN REGION: Natal; in woods near Durban, *Drège!* *Plant,* 56! *McKen,* 10! Durban Flat, *Buchanan!* and without precise locality, *Gerrard,* 695! 589!

10. D. tenuiflora (Beauv. Agrost. 51); annual or subperennial; culms fascicled, very slender, prostrate and rooting, or ascending or suberect, ½–1½ ft. long, glabrous, many-noded, often branched; sheaths rather lax, or the lower more or less hairy, sometimes bearded at the nodes; ligules very short, membranous; blades lanceolate to linear, acute, ½–4 in. long, spreading, flat, glabrous, rarely hairy; racemes usually 2–3, rarely more, very slender, 1–3 in. long, erect or spreading; rhachis finely linear, margins scabrid; pedicels 2- rarely 3-nate, unequal, tips discoid; spikelets adpressed to the rhachis, the superposed distant by rather less than their own length, elliptic-oblong, acute or subacute, ⅔–¾ lin. long, pale or purplish; lower glume 0; upper 5–3-nerved, very finely and adpressedly silky with wrinkled hairs; lower valve similar and subequal to the upper glume, flat, 7-nerved; upper valve subchartaceous, ovate-oblong, slightly shorter than the lower; flaps of pale overlapping below; anthers ¼ lin. long. *D. Pseudo-Durva, Schlechtend. in Linnæa,* xxvi. 458; *Miq. Fl. Ind. Bat.* iii. 439. *D. linearis, Schult. f. Mant.* ii. 264, *not Roem. & Schult. Paspalum longiflorum, Retz. Obs.* iv. 15; *Baker, Fl. Maurit.* 431; *Durand & Schinz, Consp. Fl. Afr.* v. 737; *Hook. f. Fl. Brit. Ind.* vii. 17. *P. brevifolium, Fluegge, Gram. Monogr.* 150; *Kunth, Enum.* i. 48. *P. tenuiflorum, R. Br. Prodr.* 193. *P. parvulum, Trin. Pan. Gen.* 117, *and in Mém. Acad. Pétersb. sér.* vi. iii. (1835) 305; *Steud. Syn. Pl. Glum.* i. 41. *P. Pseudo-Durva, Nees, Fl. Afr. Austr.* 21; *Steud. l.c.* 41.

EASTERN REGION: Natal; near Durban, *Drège,* Durban Flat, *Buchanan,* 35! near Umpumulo, 1000 ft., *Buchanan,* 204a! and without precise locality *Gerrard,* 588!

Through the tropics of the Old World.

Nees distinguishes 3 varieties of his *Panicum Pseudo-Durva* of which *α majus* and *β minus* are said to be Indian and to have a very short retuse lower glume, whilst *γ gracillimum,* based on Drège's Natal specimens, are without one. I have not been able to find a lower glume in the Indian or the African specimens.

11. D. diagonalis (Stapf); perennial; culms tufted on a short præmorse rhizome, erect, rather firm, simple, up to more than 3 ft. long, glabrous or hairy below the racemes, 1–3-noded, the uppermost internode by far the longest; leaves mainly crowded near the base; lower sheaths rather firm, strongly striate, adpressedly hairy to tomentose at the base, persistent, at length breaking up into fibres, upper thinner, hairy or glabrous except the bearded nodes, uppermost very long, rather loose; ligules up to 1½ lin. long, obtuse or truncate; blades linear, tapering to a fine point, ½–1 ft. by 2–4 lin., flat or with involute margins, rigid or almost flaccid, scabrid, glabrous or softly hairy; racemes 5- to many, solitary or fascicled on a scabrid or hirsute angular common rhachis of variable length (the whole inflorescence ½–1 ft. long), slender, strict or flexuous, erect or more or less spreading, 2–5 in. long, villous at the base; rhachis subtriquetrous, wavy, very narrow, angular, angles rigidly ciliate, internodes up to 1½ lin. long; spikelets in fascicles of 3–6, oblong, obtuse to subacute, 1–1¼ lin. long, pedicels unequal, up to 1½ lin. long, setulose, uppermost hairs equalling the spikelets; lower glume 0; upper rotundate-ovate, obtuse, ⅓–½ lin. long, hyaline, nerveless or 1-nerved; valves equal, lower membranous, whitish, oblong, glabrous, 3-nerved; upper chartaceous like the pale, brown to almost black, somewhat shining; anthers ½ lin. long. *Panicum diagonale, Nees, Fl. Afr. Austr.* 23; *Steud. Syn. Pl. Glum.* i. 40; *Engl. Hochgebirgsfl. Trop. Afr.* 117; *Durand & Schinz, Consp. Fl. Afr.* v. 746. *P. uniglume, Hochst. in Flora*, 1841, i. *Intell.* 19 (*name only*); *A. Rich. Tent. Fl. Abyss.* ii. 370. *P. densiglume, Steud. l.c.* 41 (*by error*).

COAST REGION: Albany Div.; near Grahamstown, 1800–2000 ft., *MacOwan!*
EASTERN REGION: Pondoland; between Umtata River and St. Johns River, 1000–2000 ft., *Drège*, 4312! Griqualand East; grassy slopes near Kokstad, 5000 ft., *Tyson*, 1411! Natal; near Durban, *Drège*, very common near Umpumulo, *Buchanan*, 206! hill tops near Riet Vlei, 6000 ft., *Buchanan*, 207! and without precise locality, *Buchanan*, 72!

Through Central and East Africa to the Nile region and Abyssinia.

12. D. tricholænoides (Stapf); perennial; rhizome short, oblique, densely covered by the persistent imbricate bases of the old sheaths; culms erect, simple, 1–1½ ft. long, glabrous; leaves about 4 at the base, one from the only suprabasal node, reduced to a long almost bladeless sheath; sheaths hairy, the uppermost narrow, silky bearded at the node, the lower widened; ligules short, truncate; blades linear to linear-lanceolate, tapering to an acute point, 2–3½ in. by 2–3 lin., flat, firm, glaucous, glabrous or hairy; racemes 3–6, subdigitate, erect, strict or slightly nodding, densely silky, often purplish, 2½–3 in. long, compound below; rhachis triquetrous, narrowly winged, scabrid, internodes 1½–3 lin. long; spikelets 2–5-nate or in the lowest fascicles with the central 2–3 on a very short branchlet, oblong, acute, 2–2½ lin. long (exclusive of the hairs), densely silky, unequally pedicelled, longer pedicels up to 3 lin. long, filiform, flexuous, glabrous, tips subdiscoid; lower glume delicately hyaline, whitish, truncate or rotundate, up to ½ lin. long, upper oblong, acute to obtuse, 1½–1¾ lin. long, 3-nerved, densely villous with long soft acute hairs, upper margins broadly and very delicately hyaline; lower valve oblong, acuminate, 7-nerved, glabrous along the middle, otherwise densely villous with soft, very long hairs (up to 2 lin.); upper chartaceous, oblong, acutely acuminate, 2 lin. long; anthers over 1 lin. long.

COAST REGION: King Williamstown Div.; Amatola Mountains, *Buchanan*, 11!

EASTERN REGION: Tembuland; Tabase, near Bazeia, 2500 ft. *Baur*, 317! Natal; Umsinga and base of the Biggars Berg, *Buchanan*, 87!

The dense, often purplish, silky hairs of the spikelets give this grass the appearance of a *Tricholæna;* the structure of the spikelets is, however, as in *Digitaria*, section *Trichachne*. The American *D. leucophæa*, Stapf (=*Panicum leucophæum*, Sw.), is easily distinguished by the long axis of the inflorescence. It was indicated from the Cape Colony by Kunth and others who were misled by Thunberg's erroneous identification of a specimen of *Tricholæna* with *Andropogon insularis*, L., which is synonymous with *D. leucophæa*.

13. D. flaccida (Stapf); culms very slender, over 1 ft. long; uppermost sheath long, narrow, with a rudimentary blade, glabrous; racemes simple or compound below, silky villous, 3–7 lin long, on the slender branches of a narrow, flaccid, scantily branched panicle, 2–3 in. long; rhachis finely filiform, triquetrous, smooth, internodes ¾–1½ lin. long; spikelets 2-nate or the lowest on 3–4-spiculate short branchlets, oblong, subacute, 2–1½ lin. long, unequally pedicelled, pedicels finely filiform, smooth and glabrous or with very few rigid hairs near the discoid tips; lower glume delicately hyaline, ovate, obtuse, up to ¼ lin. long, nerveless, glabrous or scantily hairy; upper lanceolate-oblong, subacute, 1 lin. long, 3-nerved, densely and long villous, margins delicately hyaline, rather broad; lower valve oblong, subacute, 1½ to almost 2 lin. long, 7-nerved, glabrous along the middle-nerve, densely villous on the sides, particularly along the upper edge of the inflexed margins, hairs acute, somewhat rigid, often purplish, exceeding the valve; upper valve subchartaceous, oblong, acuminate, up to 1½ lin. long.

COAST REGION: King Williamstown Div.; Amatola Mountains, *Buchanan*, 9!

EASTERN REGION: Natal; Umsinga and base of the Biggars Berg, *Buchanan*, 38!

This species is remarkable for having 2 short, nerved, cuneate lodicules in the lower floret although the pale itself is extremely reduced and scarcely exceeds them. I do not know any other similar case in the genus.

XIV. PANICUM, Linn.

Spikelets variously shaped, usually more or less ovate or oblong, falling entire from the pedicels, loosely panicled, solitary, 2-nate or fascicled on the rhachis of the spike-like branches (false spikes) of a panicle. Lower *floret* usually ♂, or reduced to the valve and a pale; upper floret ⚥. *Glumes* similar or dissimilar, membranous to herbaceous, lower generally much smaller, thinner, sometimes minute, upper equalling the upper floret or almost so, sometimes cuspidate or caudate-aristate, 5–13-nerved. Lower *valve* equalling the upper or longer, more or less resembling the upper glume; upper valve coriaceous to crustaceous, rarely chartaceous, 5–7-nerved. *Pales* usually subequal to their valves, flat, 2-nerved; hyaline, sometimes more or less reduced in the lower, of the same substance as the valve in the upper floret. *Lodicules* 2, broadly cuneate. *Stamens* 3. *Styles* distinct; stigmas laterally exserted below the tip of the floret. *Grain* tightly enclosed by the hardened valve and pale, oblong or ellipsoid; hilum basal, punctiform, or orbicular; embryo equalling about $\frac{1}{2}$ of the grain.

Perennial or annual of various habit; ligules usually reduced to a ciliate rim or a fringe of hairs, rarely a distinct membrane, or 0; panicle various, more or less lax, contracted or effuse, or consisting of usually secund, 2–4-, more rarely 1- or many-ranked spike-like branches, rarely converted into dense, cylindric false spikes.

Species 200 to 250, mainly in the warm regions of the world.

I am inclined to restrict the genus *Panicum*, so far as it is represented in Africa, to the sections *Brachiaria*, *Echinochloa*, *Eu-Panicum* of Hooker and Bentham's Genera Plantarum, and my section *Vilfoideæ*. Even thus defined, *Panicum* is still larger than any other genus of *Gramineæ*, except *Andropogon* (in the sense of Hackel), and comprises grasses of very different aspect. The reasons for separating *Digitaria* and *Tricholæna* and treating them as distinct genera, will appear from a comparison of the diagnoses of these genera. *Ptychophyllum* has been sunk in *Setaria* on account of the presence of bristle-like branchlets below the spikelets, of exactly the same nature as in *Setaria*. The only character by which the species of the section *Ptychophyllum* and the open-panicled species of *Setaria* (e.g. *S. macrostachys*, H. B. K., and *S. forbesiana*, Hook f.) can be distinguished, is the remarkable fan-like folding up of the leaves in the former; but this is equally foreign to the genus *Panicum*, and would, if relied upon, rather justify the generic separation of *Ptychophyllum* than its union with *Panicum*. The affinities of nearly all the larger natural groups of the tribe *Paniceæ*, whether called genera, subgenera or sections, are very close and intricate, and if we insist on absolute characters for the definition of genera, little would be left of this tribe outside a huge and very heterogeneous genus *Panicum*.

Section 1. BRACHIARIA (Trin.). Spikelets in racemosely arranged secund or subsecund simple or almost simple usually 2-, rarely 1- or 4-ranked false spikes; glumes and lower valve rarely cuspidate, never awned.

Spikelets usually 1-ranked, $2\frac{1}{2}$ lin. long, turgid, very obtuse, glabrous; fruiting valve finely pitted ... (1) **brizanthum.**
Spikelets 2-, rarely 4-ranked, less than $2\frac{1}{2}$ lin. long:
  Spikelets turgid, silky-villous, upper hairs gathered in tufts equalling or exceeding the spikelets:
    Spikelets constricted at the base into a short glabrous black stalk, 2 lin. long ... ... (2) **nigropedatum.**
    Spikelets not constricted at the base, $1\frac{1}{4}$–$1\frac{1}{2}$ lin. long with pinkish or purplish tufts of hairs... (3) **serratum.**
  Spikelets not turgid, hairy (but not silky-villous nor with subterminal tufts of hairs or only with very minute ones in *P. Marlothii*), or quite glabrous:
    Spikelets 1 lin. long, finely pubescent:
      False spikes $\frac{1}{2}$–1 in. long, 2-ranked, more or less adpressed to a much longer common axis ... ... (4) **Isachne.**

False spikes less than ½ in. long, 4-ranked on a very short common axis; spikelets with very minute subterminal tufts of hairs ... (5) **Marlothii.**
Spikelets more than 1 lin. long, not finely pubescent:
Fruiting valve transversely wrinkled:
Spikelets with 1–3 bristles from the centre of the lower glume; lower valve rigidly ciliate along one or both margins ... (6) **trichopus.**
Spikelets glabrous:
Lower glume turned away from the rhachis; fruiting valve mucronate. (7) **helopus.**
Lower glume facing the rhachis; fruiting valve not mucronate ... ... (8) **arrectum.**
Fruiting valve smooth; false spikes very short in small interrupted spiciform panicles; spikelets loosely white hairy; glumes thin; tips of blades recurved, subspinous... ... ... (9) **glomeratum.**

Section 2. ECHINOCHLOA (Beauv.). Spikelets in racemosely arranged secund or subsecund simple or compound 4- or more (rarely 2-) ranked false spikes; upper glume and lower valve cuspidate (often obscurely in *P. Holubii*) or caudate-aristate with 5–7 strong, more or less hispid nerves.

Ligule a fringe of hairs:
False spikes erect, usually adpressed to the slender axis of a linear, interrupted spike-like panicle; glumes and lower valve finely woolly-ciliate in the upper part ... ... (10) **Holubii.**
False spikes obliquely erect or spreading; glumes and lower valve not woolly-ciliate:
Panicle secund; false spikes solitary, often nodding; spikelets 2–3 lin. long, rarely less; lower valve caudate-acuminate, usually more or less awned ... ... (11) **stagninum.**
Panicle erect, rarely secund, linear-oblong; false spikes very numerous, partly fascicled or crowded; spikelets 1½–2 lin. long, shortly cuspidate ... ... (12) **pyramidale.**
Ligule 0:
Culms stout, like the sheaths terete; upper blades often long decurrent, like the others firm, rarely flaccid; false spikes compound in linear-oblong dense panicles; fruiting valve ovate-oblong to oblong ... ... (13) **Crus-pavonis.**
Culms and sheaths compressed below; blades slightly or not at all decurrent, flaccid; false spikes simple or subsimple; fruiting valve elliptic-ovate; an annual weed ... ... (14) **Crus-galli.**

Section 3. EU-PANICUM (Bentham). Spikelets distinctly pedicelled and more or less loosely or effusely panicled (see also *P. curvatum* in the following section).

* Culms very slender, ascending from a usually trailing or rambling base, many-noded, often more or less divaricately branched; blades lanceolate to lanceolate-linear, rarely linear, 1½–3 in. long, more or less spreading:
Lower valve 7–9-nerved; fruiting valve smooth:

Lower branches of panicle ½–2½ in. long; pedicels ½–1 lin. long:
Spikelets obtuse, glabrous:
Leaves glabrous (except at the ciliolate margins and mouth of the sheath and the very obscurely pubescent nodes); lower glume equalling ⅓ of the spikelet or less; anthers ⅜ lin. long (15) **filiculme.**
Leaves finely hairy; lower glume equalling ½ of the spikelet; anthers ⅜ lin. long ... ... (16) **hymeniochilum.**
Spikelets acute to acuminate, often pubescent; glumes subequal ... ... (17) **æquinerve.**
Lower branches of panicle 3–6 in. long; pedicels ½ to more than 1 in. long ... ... (18) **perlaxum.**
Lower valve 5-nerved:
Panicle scanty; spikelets 1½–2 lin. long; fruiting valve wrinkled:
Blades contracted into a short petiole; longer pedicels up to 1½ lin. long; spikelets oblong ... ... (19) **chusqueoides.**
Blades not petioled; longer pedicels up to 4 lin. long; spikelets lanceolate to oblong-lanceolate ... ... (20) **obumbratum.**
Panicle large, effuse, very delicately branched; spikelets 1 lin. long; fruiting valve finely pitted .. .. (21) **laticomum.**
** Culms usually robust, often very tall; blades linear to linear-lanceolate, ⅓–2 ft. by 2–8 lin.:
Spikelets obtuse or subobtuse:
Spikelets 2–3 lin. long, turgid:
Tips of glumes and lower valve callous, laterally compressed; lower valve 5-nerved ... ... (22) **zizanioides.**
Tips of glumes and lower valve not callous; lower valve 7-nerved ... (23) **deustum.**
Spikelets 1–1½ lin. long, not turgid:
Perennial; lower valve 5-nerved, upper finely wrinkled ... ... (24) **maximum.**
Annual; lower valve 7–9-nerved, upper quite smooth ... ... (25) **lævifolium.**
Spikelets distinctly acute or acuminate:
Pedicels very short with discoid tips; spikelets glaucous; upper glume and lower valve faintly 5-nerved ... ... (26) **meyerianum.**
Pedicels gradually thickened upwards, the longer 2–6 lin. long; spikelets not glaucous; upper glume and lower valve distinctly 7–13-(rarely 5-)nerved:
Spikelets oblong to oblong-lanceolate, acuminate, 1–1¼ lin. long:
Panicle contracted, or when open pyramidal or ovate with contracted branchlets; longer pedicels up to 2½ lin. long; upper glume and lower valve 7–9-nerved ... (27) **proliferum.**
Panicle obversely pyramidal or obovate, ultimately divaricate, very lax; longer pedicels up to more than 6 lin. long; upper glume and lower valve 5–7-nerved ... (28) **capillare.**

Spikelets ellipsoid to oblong, acute 1¼–1⅔ lin. long, in usually dense, contracted, nodding panicles; upper glume 11–13-nerved, lower valve 9-nerved ... (29) **miliare.**

*** Perennial, compactly cæspitose, rarely long stoloniferous and creeping or ascending from a branched, decumbent base; culms 1–2 ft. high, rather or very slender, often wiry; blades convolute, or when flat usually firm and rarely over 5 in. long; spikelets 1–1½ lin. long:

Glumes and lower valve entire:

Culms many-noded, from a long creeping rhizome; stoloniferous; lower leaves distichously imbricate; blades 1½–3 lin. broad, usually rigid and involute ... ... (30) **repens.**

Culms 4- to many-noded, from an often decumbent, branched base, not stoloniferous; leaves more distant; blades 2½–4½ lin. broad, flat, often rather flaccid ... ... (31) **coloratum.**

Culms 1–3-(rarely 4-)noded, simple, densely cæspitose; blades very narrow, usually convolute:

Lower glume ⅓–⅔ the size of the spikelet; upper glume 7-nerved:

Spikelets not oblique, nor gaping except when flowering; lower valve 9-nerved ... ... (32) **minus**

Spikelets somewhat oblique, gaping; lower valve 5-nerved ... ... (33) **dregeanum.**

Glumes and lower valve equal, very similar, 5-nerved ... ... (34) **natalense.**

Glumes and lower valve 3–5-toothed ... ... (35) **Ecklonii.**

Section 4. VILFOIDEÆ (Stapf). Spikelets in slender dense cylindric (rarely lax and open, *P. curvatum*) panicles, often more or less curved; upper glume and lower valve usually strongly 9-ribbed; fruiting floret conspicuously shorter than the spikelet.

Panicle cylindric; axis stout:

Spikelets oblong, olive-green with dark tips, 1½ lin. long; blades flaccid ... ... (36) **interruptum.**

Spikelets ovate, slightly curved, purplish, scarcely 1 lin. long; blades rigid ... ... (37) **typhurum.**

Panicles contracted or open; axis very slender; spikelets strongly curved ... ... (38) **curvatum.**

1. **P. brizanthum** (Hochst. in Flora, 1841, Intell. i. 19 (name only), and ex A. Rich. Tent. Fl. Abyss. ii. 363); perennial; culms erect or geniculately ascending from a short rhizome, stout in tall specimens, 1½–6 ft. long, glabrous or hirsute with tubercle-based hairs, 4–6- or sometimes many-noded, upper internodes exserted; leaves glabrous or hirsute with fine tubercle-based hairs; sheaths rather tight, terete, striate; ligule a narrow, fimbriate rim; blades linear to sublanceolate, acute to acuminate, 2–15 in. by 3–8 lin., usually flat, firm, light green, margins cartilaginous, spinulous; false spikes solitary, or more usually 2–8 on a slender triquetrous scabrid or rarely almost smooth, and often hairy, common axis, distant, often curved and more or less spreading, rather stout, 2–6 in. long; rhachis slender, linear, dorsally keeled, scabrid and more or less fringed with

tubercle-based hairs, villous at the base, internodes 1½–2½ lin. long; pedicels solitary, alternate from near the edges, very short, stout; spikelets usually 1-ranked, contiguous, ellipsoid, obtuse, turgid, about 2½ lin. long, pallid with purple tips, glabrous; lower glume membranous, very broad, clasping, obtuse, 1–1¼ lin. long, 7–11-nerved, often purple; upper firmly membranous, oblong, very concave, 2–2½ lin. long, 7–9-nerved, nerves anastomosing above; lower floret ♂; valve similar to the upper glume, 5-nerved; pale broad, elliptic, obtuse, 2-keeled; anthers 1½ lin. long; ⚥ floret equalling the upper glume, oblong; valve crustaceous with an obscure, callous tip, very finely pitted, 5-nerved, whitish; flaps of pale very broad above the base. *Steud. Syn. Pl. Glum* i. 63; *Oliv. in Trans. Linn. Soc.* xxix. 170, *t.* 112, *fig.* 1; *Engl. Hochgebirgsfl. Trop. Afr.* 120; *Durand & Schinz, Consp. Fl. Afr.* v. 742.

EASTERN REGION: Natal; near Umpumulo and Mapumulo, *Buchanan*, 183! Inanda, *Wood*, 1579! 1601! Fenton Tacy, *Wood*, 1577!

Also in tropical Africa.

1a. P. mesocomum (Nees, Fl. Afr. Austr. 34); perennial; culms ascending, repeatedly geniculate, over 1½ ft. long, terete, branched at least from the lower nodes and sulcate on the side facing the branch, glaucous, waxy below the nodes, glabrous; leaves glabrous; sheaths rather loose, striate, often slipping from the stem and rolling inwards, shorter than the internodes, those at the base of the branches reduced to tomentose scales; ligule a fringe of minute hairs; blades linear, subacute, ½–2 in. by 1 lin., involute, rigid, glaucous, glabrous or minutely puberulous upwards; panicle erect, contracted, linear, 3–5 in. long; axis slender, flexuous, glabrous; branches 5–6, erect, ½–1½ in. long, simple or often compound at, and bearing spikelets from, the base, filiform, flexuous, glabrous, lower as long as, upper longer than the internodes; pedicels finely filiform, glabrous or minutely puberulous, curved, the lateral up to ¾ lin. long, tips discoid; spikelets ovoid, acuminate, 3 lin. long, contracted at the base into a very short pallid stalk, not compressed, whitish with long silky hairs; lower glume ovate-oblong to lanceolate, acute or subobtuse, 3 lin. long, subhyaline, 3–5-nerved, pubescent below, ciliolate; upper glume oblong-lanceolate, rostrate-acuminate, 3 lin. long, membranous, 5-nerved, with a dense row of hairs attached across the middle, which reach to the tip of the glume, pubescent below, glabrous above, beak scaberulous; lower floret ♂; valve similar and subequal to the upper glume, depressed and subhyaline along the middle, the fringe broken up into 2 tufts; pale oblong, subacuminate, 2-keeled, keels scaberulous; anthers 1 lin. long; ⚥ floret oblong, subacute or obtuse; valve crustaceous; straw-coloured smooth, shining, 1½ lin. long, 5-nerved. *Steud. Syn. Pl. Glum.* i. 88; *Durand & Schinz, Conspectus Fl. Afr.* v. 754.

WESTERN REGION: Little Namaqualand; by the Orange River, near Verlaptpraam, *Drège*, 2537!

This very peculiar species was considered to be a *Tricholæna* at the time the clavis was drawn up. Further examination shows it to be a *Panicum*, nearest allied to *P. nigropedatum*, Munro.

2. P. nigropedatum (Munro ex Hiern in Trans. Linn. Soc, ser. 2, ii. 29); perennial, rhizome short, oblique, densely beset with fascicles of culms and intravaginal innovation shoots; culms erect, 1-2½ ft. long, rather slender, glabrous or finely pubescent, about 3-noded, simple, internodes exserted, except the lowest; sheaths tight, terete, rather firm, the outer and lowermost densely tomentose, the following gradually less hairy or spreadingly hirsute, the uppermost finely pubescent or subglabrous; ligule a fringe of short rigid hairs; blades linear from a narrow base, long tapering to a fine point, 4-10 in. by 1½-3 lin., erect, subrigid to flaccid, softly pubescent, sometimes also hirsute, light or yellowish green; false spikes 3-8, secund or subsecund on a slender triquetrous puberulous common axis, erect or spreading, subequal or decreasing upwards, slightly longer than the internodes, stout, 1-½ in. long; rhachis slender, almost straight, convex and striate on the back, puberulous or minutely hirsute, rigidly ciliate along the margins, tomentose at the base; pedicels solitary, very short, stout, bristly, tips slightly thickened; spikelets 2-ranked, secund or subsecund, ellipsoid, cuspidate, turgid, constricted at the base into a short glabrous black stalk, 2 lin. long, pallid, silky-villous, contiguous or imbricate; lower glume facing the rhachis, subhyaline, broadly ovate, acute or acuminate, 1 lin. long, 3-5-nerved, shortly hairy; upper glume membranous, elliptic, rostrate-acuminate, 2 lin. long, very concave, faintly 5-nerved, laxly hairy to the base of the glabrous tip, uppermost hairs longest, equalling the tip; lower floret ♂; valve equal and similar to the upper glume, hyaline and subglabrous along the depressed middle-line, with a tuft of long hairs on each side below the tip; pale elliptic, subacuminate, equalling the valve; anthers 1 lin. long; ☿ floret elliptic, cuspidate or mucronulate, nearly 1½ lin. long; valve subcoriaceous, 5-nerved, whitish, finely pitted. *P. melanostylum, Hack. in Engl. Bot. Jahrb.* xi. 398; *Durand & Schinz, Consp. Fl. Afr.* v. 756.

KALAHARI REGION: Griqualand West, Hay Div.; between Knegts Fontein and Klip Fontein, *Burchell*, 2610! Herbert Div.; St. Clair, Douglas, *Orpen*, 255! Bechuanaland; on Ga Mhana (Kamhanin) Mountain, near Kuruman, 4000 ft., *Marloth*, 1091! Kosi Fontein, *Burchell*, 2577! on rocks in Chue Vley, *Burchell*, 2391! Transvaal; on the Bosh Veld at Klippan, *Rehmann*, 5366!

Extending into tropical Africa as far as the upper Zambesi region.

3. P. serratum (Spreng. Syst. i. 309); perennial; rhizome short, thick, tomentose, with equally tomentose innovation buds; culms erect or geniculately ascending, simple or often much branched above the base, slender, ½-2 ft. long, terete, pubescent above; leaves often crowded, distichously imbricate towards the base; sheaths tight, terete, the lowest tomentose, sometimes glabrescent, the following gradually less hairy, glabrescent or glabrous, except the pubescent or villous margins and nodes; ligule a fringe of short rigid hairs; blades linear to linear-lanceolate with a callous point, 1–4 in. by 1½–2½ lin., rigid, flat or involute, glabrous or hairy, particularly near the base, margins cartilaginous, wavy, spinulous; false spikes 4–10 secund on a filiform angular glabrous or puberulous common axis, erect or spreading, usually longer than the internodes, gradually decreasing upwards, dense, 12–3 lin. long, rhachis filiform, angular, very wavy, rigidly pubescent to subhirsute, pedicels solitary or the lowest 2-nate, filiform with discoid tips, curved, the lower very short, the upper up to 1 lin. long, hairy; spikelets usually 2-ranked, contiguous or subcontiguous, obovoid-ellipsoid, obtuse, turgid, 1¼–1½ lin. long, pallid, overtopped by a purplish tuft of silky hairs; lower glume facing the rhachis, subhyaline, broadly ovate to rotundate-ovate, obtuse, ¾–1 lin. long, often purplish, 3-nerved, hairy; upper glume membranous, broadly-elliptic, acute or subcuspidate, 1¼–1½ lin. long, very concave, faintly 5-nerved, minutely villous or glabrescent in the centre, with a dense transverse fringe of purplish hairs below the glabrous tip; lower floret ♂; valve equal and similar to the upper glume, but more distinctly cuspidate, hyaline and glabrous along the middle nerve, transverse fringe broken up into 2 broad tufts; pale broad, almost equalling the valve; anthers over 1 lin. long; ⚥ floret elliptic, cuspidate or mucronulate 1–1¼ lin. long; valve elliptic, coriaceous, pallid, glabrous or ciliate near the tip, finely pitted. *Kunth, Rev. Gram.* i. 215, *t.* 19; *Enum.* i. 98; *Trin. Pan. Gen.* 146, *and in Mém. Acad. Pétersb. sér.* vi. iii. 233; *Nees, Fl. Afr. Austr.* 31; *Steud. Syn. Pl. Glum.* i. 57; *Durand & Schinz, Consp. Fl. Afr.* v. 765 (*includ. var. holosericeum, Hack.*). *P. scopuliferum, Trin. Sp. Gram. t.* 165; *Kunth, Enum.* i. 99; *Durand & Schinz, l.c.* 764. *P. holosericeum, Nees, l.c.* 30, *not R. Br. Holcus serratus, Thunb. Prodr.* 20; *Fl. Cap. ed. Schult.* 110. *Sorghum serratum, Roem. & Schult. Syst.* ii. 839.

SOUTH AFRICA: without precise locality, *Thunberg!* *Drège*, 2048!

COAST REGION: Swellendam Div.; Puspus Valley and Keurboom River, &c., near Swellendam, *Ecklon*. Riversdale Div.; near the Zoetemelks River, *Burchell*, 6665! Mossel Bay Div.; between Little Brak River and Hartenbosch, *Burchell*, 6206! Humansdorp Div.; on Kromme River Heights, *Bolus*, 2696! Uitenhage Div.; hills by the Zwartkops River, *Ecklon*. Alexandria Div.; Zuurberg Range, 2500–3000 ft., *Drège!* Bathurst Div.; Port Alfred, *Hutton*, 43/a! Albany Div.; mountains near Grahamstown, 2000 ft., *MacOwan*, 213! 1307! Coldspring near Grahamstown, *Flanagan*, 758! Fort Beaufort Div.; Kat River Poort, 2000 ft., *Drège*. Stockenstrom Div.; Philipton, *Ecklon & Zeyher!* Cathcart Div.; between Kat Berg and Klipplaat River, 3000–4000 ft., *Drège!* Queenstown Div.; Finchams Nek, near Queenstown, 4000 ft., *Galpin*, 2380!

CENTRAL REGION: Somerset Div.; Klein Bruintjes Hoogte, 3000–4000 ft., and between Little Fish River and Great Fish River, 2000–3000 ft., *Drège*. Albert Div.; without precise locality, *Cooper*, 3344!

KALAHARI REGION: Orange Free State; without precise locality, *Cooper*, 910! 3343! Basutoland; without precise locality, *Cooper*, 921! Bechuanaland; Kuruman, *Burchell*, 2186! and between Hamapery and Kosi Fontein, *Burchell*, 2543! Transvaal; Wonderfontein, *Nelson*, 11*! Matebe Valley, *Holub!* near Lydenburg, *Wilms*, 1664!

EASTERN REGION: Tembuland: Bazeia, *Baur*, 319! Natal; Umpumulo,

2400 ft., *Buchanan*, 213! Umsinga and base of the Bigars Berg, *Buchanan*, 105! and Riet Vlei, 4000–5000 ft., *Buchanan*, 214!

Also in Nyasaland, and a very similar variety (*P. serratum* var. *gossypinum*, Hack.) in Abyssinia.

4. P. Isachne (Roth ex Roem. & Schult. ii. 458); annual; culms geniculately ascending from a sometimes decumbent and rooting base, very slender, 1–2 ft. long, terete or angular, glabrous, 3- to many-noded, much branched below, simple above; sheaths rather tight, terete or subterete, strongly striate, softly hairy or glabrous except at the minutely villous nodes, the upper shorter than the internodes; ligule a dense fringe of stiff hairs; blades linear-lanceolate from a rounded base, acute, 1–2½ in. by 1–2½ lin., flat or involute, rather rigid, softly hairy or quite glabrous, margins very scabrid; false spikes 2–10, secund or subsecund, on and usually adpressed to a filiform scabrid axis 1–3 in. long, as long as or longer than the internodes, very slender, ½–1 in. long; rhachis filiform, wavy, triquetrous, scabrid; pedicels solitary, very short, stout, scabrid or bristly, tips thickened with hyaline margins; spikelets 2-ranked, secund, contiguous, oblong, subacute or obtuse, about 1 lin. long, pallid with purplish tips, softly pubescent; lower glume facing the rhachis, minute, nerveless, rarely ½ lin. long and 1-nerved; upper membranous, oblong, 5-nerved, minutely hairy; lower floret usually barren, sometimes ♂; valve like the upper glume, but narrower, 5–3-nerved; pale very obtuse, subequal to the valve in ♂, shorter in barren florets; ⚥ floret elliptic-oblong, rounded at both ends, ¾–⅘ lin. long; valve subcoriaceous, faintly 5-nerved, shining, whitish, finely granulate; anthers ½ lin. long; grain oblong-ellipsoid, ½ lin. long; hilum orbicular, ⅓ the length of the grain. *Roth, Nov. Pl. Spec.* 54; *Schult. Mant.* ii. 252; *Kunth, Enum.* i. 97; *Steud. Syn. Pl. Glum.* i. 57; *Hook. f. Fl. Brit. Ind.* vii. 28. *P. eruciforme, Sibth. & Sm., Fl. Græc.* i. *t.* 59; *Fl. Græc. Prodr.* i. 40; *Kunth, l.c.* 78; *Baker, Fl. Maurit.* 434; *Aschers. & Schweinf. Illustr. Fl. Egypte*, 159; *Engl. Hochgebirgsfl. Trop. Afr.* 120; *Durand & Schinz, Consp. Fl. Afr.* v. 748. *P. caucasicum, Trin. Sp. Gram. Ic. t.* 262; *Panic. Gen.* 149, *and in Mém. Acad. Pétersb. sér.* vi. iii. 237. *P. Wightii, Nees, Fl. Afr. Austr.* 29; *A. Rich. Tent. Fl. Abyss.* ii. 364; *Steud. l.c.* 58. *P. pubinode, Hochst. ex A. Rich. l.c.* 363; *Steud. l.c.* 57. *Echinochloa eruciformis, Reichb. Ic. Fl. Germ.* i. *t.* 29, *fig.* 1413.

KALAHARI REGION: Basutoland; Leribe, *Buchanan*, 145! 230! Bechuanaland; Batlapin Country, *Holub!*

EASTERN REGION: Transkei; near the Gekau (Gcua or Geuu) River, below 1000 ft., *Drège!* Natal; Riet Vlei, 4000–6000 ft., *Buchanan*, 215! 216!

Also in the Mediterranean countries, India and Abyssinia.

5. P. Marlothii (Hack. in Engl. Bot. Jahrb. xi. 398); perennial; culms geniculately ascending from a long decumbent rooting and much branched base, very slender, ½ ft. long, angular, often compressed below; sheaths rather lax, subcompressed, shorter than the internodes, striate, glabrous or with scanty tubercle-based hairs, slightly bearded at the nodes, ciliate along the outer margin, minutely villous at the nodes; ligule a dense fringe of stiff hairs; blades linear-lanceolate from a rounded base, acute, 7–15 lin. by 1–2 lin., firm, green, puberulous above, glabrous below except for some tubercle-based hairs along the midrib, margins cartilaginous, distantly fimbriate with long rigid tubercle-based bristles; false spikes 2–4, crowded on a very short angular scabrid common axis, secund or subsecund, oblong, 3–5 lin. long, 4-ranked; pedicels 2-nate, unequal, the longer up to ½ lin. long, scabrid and with 1–3 stiff bristles; spikelets imbricate, elliptic, obtuse or subapiculate, 1 lin. long, pallid, minutely hairy; lower glume facing the rhachis in the subsessile spikelets forming the inner 2 ranks, turned away in the others, broadly ovate, acute, ½ lin. long, subhyaline, 1-nerved; upper glume membranous, oblong, whitish with 5 green nerves, scantily puberulous and with 2 little transverse tufts of hairs ¼ below the tip; lower floret ♂; valve like the upper glume, rather narrower; pale subequal; anthers ⅖ lin. long; ⚥ floret broadly elliptic, rounded at both ends, ¾ lin. long; valve finely pitted, faintly 5-nerved, whitish; anthers ½ lin long. *Durand & Schinz, Consp. Fl. Afr.* v. 753.

KALAHARI REGION: Griqualand West; on hills near Barkley West, *Marloth*, 1147! and in Bechuanaland, on stony plains near Maneering, *Marloth*, 1147 (ex *Hackel*).

Nearly allied to *P. Isachne.*

6. **P. trichopus** (Hochst. in Flora, 1844, 254); perennial (?); culms ascending from a decumbent geniculate base, 1–2 ft. long, slender, glabrous or pubescent, 4- to many-noded, simple or scantily branched above the base; leaves finely hirsute with tubercle-based hairs, rarely glabrescent; sheaths rather lax, bearded at the nodes, ciliate along the outer margin; ligules membranous, very short, fimbriate; blades linear to linear-lanceolate from a broad rounded or subcordate base, acute to acuminate, 2–4 in. by 2–6 lin., flat, flaccid, margins often wavy, scabrid or fimbriate; false spikes 3–12, rather distant on a compressed or angular pubescent, sometimes bristly, common axis, suberect or spreading, secund, 2–4-ranked, 1–2½ in. long; rhachis linear, ½–¾ lin. broad, flat on the back, with a very prominent midrib on the face, scabrid or spinulously ciliate and often with long fine tubercle-based bristles as well, villous at the base; pedicels solitary, or 2-nate and then unequal, very short, stout, scabrid, pubescent, with 1–3 bristles near the thickened tips; spikelets contiguous or imbricate, elliptic-ovate, acutely acuminate, 2 lin. long, pale; lower glume turned away from the rhachis, elliptic, obtuse, equalling ¾–⅘ of the spikelet, subhyaline, 3-nerved, with 1–3 bristles from a central tubercle; upper glume membranous, elliptic-ovate, acutely acuminate, 2 lin. long, 5-nerved, scantily pubescent, sometimes with a few rigid cilia on one side; lower floret ♂; valve like the upper glume, 3–5-nerved, rigidly and adpressedly ciliate on one

or on both sides; pale equal, acuminate; anthers $\frac{1}{2}$–$\frac{3}{4}$ lin. long; ☿ floret broadly elliptic, rounded, 1–1$\frac{1}{4}$ lin. long, whitish; valve with a scabrid mucro up to $\frac{1}{3}$ lin. long, 5-nerved, finely and transversely rugose; anthers $\frac{3}{4}$ lin. long. *Steud. Syn. Pl. Glum.* i. 100; *K. Schum. in Engl. Pfl. Ost-Afr. C.* 101. *P. trichopodon, A. Rich. Tent. Fl. Abyss.* ii. 369; *Durand & Schinz, Consp. Fl. Afr.* v. 766. *P. dorsisetum, Hack. ex Durand & Schinz, l.c.* 748 (*name only*).

KALAHARI REGION: Transvaal; Zoutpans Berg, Mateassa's Kraal, *Nelson*, 18*! Bosh Veld, between Elands River and Klippan, *Rehmann*, 5121!
EASTERN REGION: Natal; Umpumulo, 1000 ft., *Buchanan*, 218!

Also in tropical Africa.

7. **P. Helopus** (Trin. in Spreng. Neu. Entdeck. ii. 84) var. **glabrescens** (K. Schum. in Engl. Pfl. Ost-Afr. C. 101); annual; culms suberect or ascending from a geniculate, sometimes decumbent and rooting, base, 1–2 ft. long, glabrous, striate, 4- to many-noded, branched or simple except at the base; leaves more or less finely hirsute with tubercle-based hairs, rarely glabrescent or glabrous; sheaths rather lax, pubescent to villous at the nodes, ciliate above; ligules very short, membranous, fimbriate; blades linear-lanceolate to lanceolate from a broad rounded or cordate base, acute to acuminate, 2–4 in. by 3–8 lin., flat, flaccid, margins often wavy, scabrid or fimbriate; false spikes 3–12, rather distant on a compressed or angular, scabrid and often pubescent common axis, suberect or spreading, secund, 2-ranked; rhachis linear, $\frac{1}{4}$–$\frac{1}{2}$ lin. broad, flat on the back, with a very prominent midrib on the face, scabrid or spinulously ciliate, and often with scattered fine long tubercle-based bristles as well; pedicels usually solitary, very short, stout, scabrid, pubescent, with or without 2–3 bristles near the thickened tips; spikelets contiguous or subimbricate, ovate to ovate-oblong, acute, 2–2$\frac{1}{2}$ lin. long, pallid, glabrous; lower glume turned away from the rhachis, broadly ovate, clasping, obtuse, equalling $\frac{1}{4}$–$\frac{1}{3}$ of the spikelet, hyaline, 3–5-nerved; upper glume membranous, ovate to ovate-oblong, subacute to subacuminate, prominently 7–9-nerved; lower floret ♂ or barren; valve ovate, acuminate, 5–7-nerved; pale equal, subacute; ☿ floret elliptic, rounded at both ends, 1$\frac{1}{2}$ lin. long, pallid; valve with a scabrid mucro, up to $\frac{1}{2}$ lin. long, transversely wrinkled, 5-nerved; anthers scarcely $\frac{1}{2}$ lin. long; grain elliptic, dorsally compressed; embyro $\frac{3}{4}$ the length of the grain. *P. Helopus, Durand & Schinz, Consp. Fl. Afr.* v. 750 (*in part*). *P. hochstetterianum, A. Rich. Tent. Fl. Abyss.* ii. 369. *P. controversum, Steud. Syn Pl. Glum.* i. 60. *P. javanicum, Baker, Fl. Maurit.* 434 (*in part*); *Hook. f. Fl. Br. Ind.* vii. 35 (*in part*), *not Poir.*

COAST REGION: Queenstown Div.; plains near Queenstown, 3500 ft., *Galpin*, 2351!

KALAHARI REGION: Orange Free State; between Harrismith and Leribe, *Buchanan*, 211!

EASTERN REGION: Natal; without precise locality, *Buchanan*, 78!

Also in the Mascarene Islands, East Africa, Abyssinia, and India; the typical form with densely pubescent spikelets has not been found on the African continent.

8. P. arrectum (Hack. ex Durand and Schinz, Consp. Fl. Afr. v. 741); perennial, quite glabrous; culms ascending from a prostrate, rooting base, 1½–2 ft. long, compressed below, terete in the upper part, glabrous, many-noded, scantily branched; sheaths somewhat loose, striate, smooth, the lower withering; ligule a dense fringe of hairs; blades linear, acute, 2½–4 in. by 1½–2 lin., usually more or less convolute, rigid, green, smooth except the scabrid margins; false spikes 2–4, distant on a triquetrous smooth axis up to 3 in. long, secund, 2-ranked, 1–1½ in. long; rhachis linear, ⅓–½ lin. broad, flat on the back, with a very prominent wavy midrib in front, smooth; pedicels solitary, very short, stout, tips subdiscoid, with 1–2 spreading hairs; spikelets contiguous, oblong, subacute, 1½–1¾ lin. long, green or tinged with purple; lower glume facing the rhachis, thinly membranous, elliptic to rotundate-ovate, obtuse, ½–¾ lin. long, sub-7-nerved; upper glume membranous, oblong, conspicuously 7–9-nerved, nerves green; lower floret ♂; valve like the upper glume, but narrower, 7-nerved; pale equal, obtuse; anthers over 1 lin. long; ⚥ floret broadly elliptic, obtuse, 1¼–1½ lin. long; valve 7-nerved, transversely wrinkled. *P. subquadriparum, Nees, Fl. Afr. Austr.* 29, *not Trin.*

COAST REGION: Uitenhage Div.; without precise locality, *Zeyher.* Albany Div.; between Assegay Bosh and Botram, 1000–2000 ft., *Drège!* Komgha Div.; near the Kei River, below 1000 ft., *Drège!*

EASTERN REGION: Natal; without precise locality, *Gerrard,* 686!

9. P. glomeratum (Hack. in Verh. Ver. Prov. Brandenb. xxx. 141); annual; culms erect or ascending, very slender, ⅓–1 ft. long, compressed, pubescent; leaves softly pubescent to villous all over; sheaths lax, thin, whitish; ligule a fringe of short hairs; blades linear-lanceolate, acuminate, up to 2½ in. by 3 lin., flat, subglaucous, margins wavy, tips spinous, recurved; false spikes oblong, 2–5 lin. long, very dense, pubescent with mostly tubercle-based hairs, crowded on an angular pubescent very slender axis, into a linear or linear-oblong panicle, 2–4 in. long and interrupted below; pedicels very short, tips thickened with hyaline margins; spikelets densely clustered, oblong-elliptic, acute, almost 1½ lin. long, pale or variegated with purple, hairy; glumes loosely beset with fine, long, white hairs, lower broadly ovate, acute, ⅜–½ lin. long, hyaline, 1-nerved; upper thinly membranous, oblong, acute or subacuminate, almost 1½ lin. long, faintly 5-nerved; lower floret ♂; valve very similar to the upper glume; pale equal, obtuse; anthers ¾ lin. long; ⚥ floret elliptic-oblong, subobtuse, 1 lin. long, whitish; valve 5-nerved, crustaceous, smooth; anthers ¾ lin. long. *Hack. in Engl. Bot. Jahrb.* xi. 398; *Durand & Schinz, Consp. Fl. Afr.* v. 750.

WESTERN REGION: Great Namaqualand, Gubub, near Aus, *Schinz,* 640!

Also in Mossamedes, near the Coroca River.

10. P. Holubii (Stapf); perennial; culms erect or geniculate from a slender rhizome, 2–3 ft. long, terete, glabrous, 4–5-noded, sheathed all along or the upper internodes exserted; leaves glabrous; sheaths rather loose, terete, striate, bases of the lowest persistent, breaking up into fibres; ligule a fringe of fine hairs; blades linear, scarcely narrowed or constricted at the base, long-tapering to a fine point, 4–10 in. by 2–3 lin., more or less glaucous, smooth except in the upper part, margins finely cartilaginous, smooth or scabrid; false spikes adpressed to the slender compressed scabrid axis of a linear usually interrupted erect panicle from 4–6 in. long, compact, 4-ranked, $\frac{1}{3}$–$\frac{3}{4}$ in. long; rhachis slender, scabrid, slightly wavy, villous at the base: pedicels 2–3-nate, extremely short, tips discoid; spikelets oblong, subacute or cuspidate, $1\frac{1}{4}$–$1\frac{1}{2}$ lin. long, pallid or variegated with purple; glumes very thin; lower broadly ovate, cuspidate, about $\frac{1}{2}$ the length of the spikelet, 5-nerved, finely woolly-ciliate on the sides; upper oblong, very concave, tips short, slightly compressed, 5- or (at the tips) 7-nerved, nerves scantily or minutely hispid, margins very shortly woolly; lower floret ♂; valve very like the upper glume; pale subequal to the valve, keels scabrid; anthers $\frac{1}{2}$ lin. long; ⚥ floret equalling the ♂, elliptic-oblong, shortly acuminate or cuspidate, light straw-coloured, back strongly convex; valve subcoriaceous, 5-nerved; anthers slightly over $\frac{1}{2}$ lin. long.

COAST REGION: King Williamstown Div.; Amatola Mountains, *Buchanan*, 1! KALAHARI REGION: Bechuanaland; Kuruman, *Burchell*, 2128/7! Transvaal; Bosh Veld at Klippan, *Rehmann*, 5369; and without precise locality, *Holub!*

Also on the upper Zambesi (*Holub*).

Burchell's specimen was grown in England from seeds collected by him near Kuruman. It is, in all parts, larger, the leaves attaining more than 1 ft. in length, and the false spikes almost to 2 in. The spikelets are distinctly cuspidate, and the anthers up to 1 lin. long.

11. P. stagninum (Koenig, Naturf. xxiii. 201); annual or perennial; culms erect from a geniculate or prostrate base, terete or subterete, up to 6 ft. high, in tall specimens to more than 3 lin. thick below, often rooting from the lower nodes, sheathed all along or some of the internodes at length exserted, often branched in the lower part; sheaths finely striate, smooth, terete or subcarinate above, quite glabrous, rarely pubescent at the lowest nodes; ligule a fringe of rather long stiff hairs, or sometimes 0 in the uppermost leaves; blades linear from a scarcely narrowed usually not decurrent base, long-tapering to a fine point, $\frac{1}{2}$ to more than $1\frac{1}{2}$ ft. by $2\frac{1}{2}$–7 lin., flat, rigid or flaccid, glabrous, light green or glaucous, smooth above, scabrid below, particularly in the upper part, margins cartilaginous, scabrid to spinulous; panicle erect or nodding, 4–10 in. long, secund; axis slender, more or less flexuous, convex or flat on the back, usually hispidulous with scattered bristles, rarely glabrous except on the scabrid angles; branches few to many, distant or rather crowded, alternate, suberect or nodding, 1–2 in. long, forming often stout dense 2–4-ranked simple secund sessile false spikes; rhachis like the axis, but more slender; pedicels 4–2-nate, extremely short, tips discoid; spikelets crowded, ovate-oblong to lanceolate-ovate, 2–3 lin. long, rarely less, pallid, hispid; glumes thin, lower ovate, acuminate or mucronate, about ½ the length of the spikelet, 3- to sub-5-nerved, upper oblong, equalling the spikelet, concave, caudate-acuminate or produced into a short, scabrid, compressed awn, 5-nerved or 7-nerved at the tips, pubescent between the hispidulous nerves; lower floret ♂ or sometimes barren; valve similar to the upper glume, but flat or depressed on the back, subhyaline except the herbaceous sides, awn 2–12 lin. long; pale oblong, keels scabrid; anthers, when present, 1 lin. long; ⚥ floret oblong to lanceolate-oblong, mucronate-acuminate, 1½–2½ lin. long excluding the scabrid mucro, straw-coloured, smooth, shining; valve 5-nerved. *Retz. Obs.* v. 17; *Roxburgh, Fl. Ind.* i. 298; *Nees, Agrost. Bras.* 261; *Trin. Pan. Gen.* 128, *and in Mém. Acad. Pétersb. sér.* vi. iii. 216; *Steud. Syn. Pl. Glum.* i. 47. *P. Galli, Thunb. Prod.* 18; *Fl. Cap. ed.* i. 389; *ed Schult.* 103. *P. Crus-galli, var. stagninum, Hack. ex Durand & Schinz, Consp. Fl. Afr.* v. 745. *P. scabrum, Lam. Illustr.* i. 171, *and Encycl.* iv. 744; *Nees, l.c.*; *Steud. l.c.*; *Durand & Schinz, l.c.* 764. *P. pictum, Nees, Fl. Afr. Austr.* 59, *not Agrost. Bras., nor Koen. Echinochloa stagnina, Beauv. Agrost.* 161. *E. scabra, Roem. & Schult. Syst.* ii. 479. *Orthopogon stagninus, Spreng. Syst.* i. 307. *Oplismenus stagninus, Kunth, Rév. Gram.* i. 44; *Enum.* i. 144 (*in part*). *O. scaber, Kunth, Rév. Gram.* i. 44; *Enum.* i. 145.

South Africa: without precise locality, *Thunberg!*

Coast Region: Alexandria Div.; near Enon, below 500 ft., *Drège!* Komgha Div.; sandy banks of Kei River, *Flanagan*, 950!

Eastern Region: Transkei Div.; near Gekau (Gcua or Geuu) River, *Drège*, 4238! Delagoa Bay, *Forbes!*

Through tropical Africa, from the Senegal to Abyssinia, in Madagascar and India.

The few South African specimens, which I have seen, represent a state with somewhat smaller spikelets than is usual in this species. Nees' description of *P. pictum* in Fl. Afr. Austr. 59, is copied from Agrost. Bras. 260, and fits very well the Indian plant so named by Koenig, but not Drège's South African specimens, referred by Nees to *P. pictum*. However, *P. pictum*, Koenig, is scarcely specifically distinct from *P. stagninum*, being rather a variety of it distinguished by its broader, turgid spikelets. It seems to be much more common in India than the type.

12. P. pyramidale (Lam. Illustr. i. 171, excl. var. β); perennial; culms erect from a geniculate or prostrate base, or floating, terete, up to 15 ft. high, in tall specimens very robust, as thick below as the middle finger or thicker, often with whorls of long roots from the submerged nodes, sheathed all along or some of the internodes at length exserted, many-noded; sheaths striate, smooth, glabrous, rarely hispid, terete; ligule a fringe of hairs; blades linear from a rather broad and rounded, or from a slightly attenuate and decurrent base, long tapering to a fine point, 1–2 ft. by 3–12 lin., flat, firm, glabrous, often more or less glaucous, smooth above, scabrid below in the upper part, margins cartilaginous, spinulous or scabrid, or smooth below, midrib usually broad, whitish; panicle erect, rarely nodding, usually linear-oblong, dense, ½–1 ft. long, facing all sides or sometimes subsecund; axis stout, 3–4-angular, sulcate, hispidulous or glabrous and smooth except the scabrid angles, usually with a fringe of hairs at the nodes; branches very many, some solitary, others 2-nate or fascicled, the lowest distant, the others rather close, suberect, strict or flexuous, rarely nodding, 1–3 in. long, forming moderately dense simple or subsimple spikes; rhachis slender, triquetrous, hispidulous, or glabrous; pedicels fascicled, very short, tips discoid; spikelets ovoid, cuspidate, 1½–2 lin. long, greenish or variegated with purple; glumes herbaceous-membranous; lower broadly ovate, clasping at the base, acute, about ½ the length of the spikelet, 5-nerved, margins scabrid or ciliate; upper glume ovate to ovate-oblong, shortly acuminate, very concave, scarcely shorter than the spikelet, 5–7-nerved, minutely and rigidly pubescent or subglabrous between the scabrid or spinulous nerves; lower floret ♂; valve similar to the upper glume, flat on the back; pale oblong, subacuminate, keels scabrid; anthers ½–1 lin. long; ⚥ floret usually elliptic, rarely oblong, cuspidate, 1½–2 lin. long, straw-coloured, smooth; valve coriaceous, 5-nerved. *Kunth, Rév. Gram.* i. 223, *t.* 23; *Enum.* i. 93 (*excl. var. β*); *Trin. Panic. Gen.* 157, *and in Mém. Acad. Pétersb. sér.* vi. iii. 245; *Steud. Syn. Pl. Glum.* i. 62; *Durand & Schinz, Consp. Fl. Afr.* v. 760. *P. frumentaceum, v. cuspidatum, Nees in E. Meyer, Zwei Pflanzengeogr. Docum.* 207.

VAR. β, **hebetatum** (Stapf); all parts weaker; blades 2–3 lin. broad, flaccid; ligule replaced by a pubescent line or a fringe of short, fine hairs; panicle interrupted, slender; branches up to 1 in. long; spikelets longer cuspidate than in the type, 1½ lin. long; lower floret barren.

COAST REGION: Var. β: Uitenhage Div.; without precise locality, *Ecklon & Zeyher!*

EASTERN REGION: Natal; valley of the Umgeni River, *Drège*, 4242!

The typical form is common throughout tropical Africa, sometimes covering large areas in and near stagnant water.

13. P. Crus-pavonis (Nees, Agrost. Bras. 259) var. **rostratum** (Stapf); perennial; culms erect, stout, terete, up to 5 ft. long, and to 3 lin. thick below, glabrous, smooth, about 5-noded, sheathed all along or the internodes at length more or less exserted; leaves quite glabrous; sheaths terete, striate, smooth; ligules 0, junction of sheath and blade quite glabrous inside or scantily and very minutely pubescent; blades lanceolate-linear from a slightly narrowed base, which is usually long decurrent in the upper leaves, tapering to a very fine point, 3–10 in. by 5–9 lin., flat, rather firm, smooth above, scaberulous below, at least in the upper part, margins cartilaginous, scabrid to spinulous, midrib broad, white; panicle erect, linear-oblong, 4–10 in. long, dense; axis rather stout, triquetrous, very scabrid; branches solitary or 2-nate, distant below, close above, the lower 1½–4½ in.

long, forming sessile, stout, very dense, simple or compound, subsecund false spikes; rhachis usually beset with tubercle-based bristles; pedicels 2-nate or fascicled on very short branchlets, very short up to $\frac{1}{2}$ lin. long, scabrid, tips obscurely discoid; spikelets in compact clusters, elliptic-oblong, caudate-acuminate, $1\frac{1}{4}$–$1\frac{1}{2}$ lin. long, light green or tinged with purple; lower glume very broadly ovate, acute to subacuminate, clasping at the base, $\frac{1}{2}$–$\frac{2}{3}$ lin. long, 3–5-nerved, scaberulous; upper glume herbaceous-membranous, broadly oblong, very concave, cuspidate-acuminate, equalling the spikelet, 5-nerved, rigidly pubescent between the scabrid or spinulous nerves; lower floret barren; valve similar to the upper glume, but flat or depressed on the back, tips rostrate, laterally compressed, up to $\frac{1}{2}$ lin. long; pale oblong, keels scaberulous above; ⚥ floret ovate-oblong, subacuminate or cuspidate, up to $1\frac{1}{2}$ lin. long, greenish-yellow, smooth; valve crustaceous, 5-nerved; anthers $\frac{1}{2}$ lin. long; grain obovate-oblong, very broad, $\frac{3}{4}$ lin. long. *P. Crus-pavonis, Nees, Fl. Afr. Austr.* 59, *and Durand & Schinz, Consp. Fl. Afr.* v. 745 (*as to the African plant*).

EASTERN REGION: Natal; on sand-flats near the mouth of the Umsimkulu River, *Drège!* Durban Flats, *Buchanan*, 4! Umhlali, *Wood*, 3992! and without precise locality, *Gerrard*, 496! *Buchanan*, 269!

Also in marshes in Northern Nyasaland.

The typical form, which has less compactly crowded and slightly larger spikelets with longer awns, is common throughout tropical America. Nees' description, in Fl. Afr. Austr., is drawn up from it, and does not agree with Drège's specimen. Minute arrested anthers and lodicules occur sometimes in the lower floret.

14. **P. Crus-galli** (Linn. Sp. Plant. 56); annual; culms geniculately ascending, compressed below, 1–3 ft. high, glabrous, smooth, 3–5-noded, sheathed all along or the internodes at length more or less exserted, often branched below; sheaths striate, smooth, the lower often strongly compressed, whitish, glabrous except the lowest which are pubescent at the very base; ligules 0, junction of blade and sheath glabrous inside; blades linear from a scarcely narrowed base, tapering to an acute point, 3–8 in. by 3–6 lin., flat, subflaccid, glabrous, more or less glaucous, smooth above, scaberulous below, particularly towards the tip, margins finely cartilaginous, scabrid to almost smooth, midrib narrow; panicle erect, strict, or flexuous, 3–8 in. long; axis triquetrous, 3–5-angled, scabrid; branches few to about 15, solitary or 2-nate, suberect or spreading, distant except the uppermost, the lower 1–$2\frac{1}{2}$ in. long, forming rather stout, dense, simple or subsimple, subsecund, sessile false spikes; rhachis triquetrous, scabrid, coarsely bristly, particularly near the nodes; pedicels fascicled or 2-nate, very short, up to $\frac{1}{2}$ lin. long, scabrid, bristly at the base, tips obscurely discoid; spikelets crowded, ovoid-ellipsoid, cuspidate, $1\frac{1}{4}$–$1\frac{1}{2}$ lin. long, greenish or tinged with purple; lower glume membranous, very broadly ovate, clasping at the base, obtuse to subcuspidate, $\frac{1}{2}$ lin. long, 5-nerved, scaberulous; upper

glume herbaceous-membranous, very broadly ovate-oblong, concave, acute or cuspidate, 1½ lin. long, 5- or (near the tips) 7-nerved, rigidly pubescent between the scabrid and spinulous nerves; lower floret barren; valve similar to the upper glume, but flat or depressed on the back; cuspidate or produced into a scabrid, often long, awn, 7-nerved throughout or only towards the tips; pale elliptic, shorter by ¼ than the valve, keels scaberulous above; ⚥ floret elliptic-ovate, cuspidate, over 1 lin. long, whitish or yellowish, smooth; valve subcoriaceous, 5-nerved; anthers oblong, scarcely ⅜ lin. long; grain broadly elliptic, ¾ lin. long. *Fl. Dan. t.* 1564; *Host, Gram. Austr.* ii. 15, *t.* 19; *Engl. Bot. t.* 876; *Knapp, Gram. Britt.* xi.; *Trin. Sp. Gram. Ic. t.* 161, 162; *Nees. Fl. Afr. Austr.* 58; *Steud. Syn. Pl. Glum.* i. 47; *Durand & Schinz, Consp. Fl. Afr.* v. 744 (*excl. var.*). *P. hispidulum, Nees, Fl. Afr. Austr.* 57, *and Durand & Schinz, Consp. Fl. Afr.* v. 750 (*the African plant, excl. syn.*). *Echinochloa Crus-galli, Beauv. Agrost.* 161; *Reichb. Ic. Fl. Germ.* i. *t.* 29, *fig.* 1411, 1412. *Oplismenus Crus-galli, Dumort. Agrost. Belg.* 138; *Kunth, Rév. Gram.* i. 44; *Enum.* i. 143 (*excl. syn. P. zonale*).

COAST REGION: Cape Div.; Lion Mountain, *Drège!* near Rondebosch, *Ecklon.* Near Claremont, *Dod,* 2423! Uitenhage Div.; near the Zwartkops River, *Ecklon & Zeyher!* Knysna Div.; Knysna Forests, *E.S.C.A. Herb.* 308! Queenstown Div.; moist spots near Shiloh, 3500 ft., *Baur,* 889!

CENTRAL REGION: Richmond Div.; Styl Kloof, in the Winter Veld, near Richmond, *Drège!* Somerset Div.; near the Blyde River, *Burchell,* 2968!

EASTERN REGION: Natal; near Durban, *Williamson,* 13!

The typical form of *P. Crus-galli,* as described above, occurs as a weed throughout the temperate zones of both hemispheres, and more rarely in the tropics. All the South African specimens belong to the muticous state, except that collected by Baur near Shiloh. In Burchell's specimen (2968), the junction-line of blade and sheath is perfectly glabrous as usual, with the exception of one leaf where it is pubescent. I have not met with another case of this kind in the very numerous specimens of *P. Crus-galli* I have examined.

15. **P. filiculme** (Hack. in Bull. Herb. Boiss. iii. 377); culms decumbent, very slender and weak, up to 1½ ft. long or longer, quite glabrous, many-noded, branched from most of the nodes, internodes exserted; leaves almost perfectly glabrous; sheaths thin, rather tight, at length dry, loosened and thrown aside, strongly striate, outer margin and mouth as well as the nodes minutely hairy (under the microscope); ligule a minute ciliolate rim; blades sublanceolate-linear from a rounded base, tapering to an acute point from below the middle, 1–2 in. by 1½–2 lin., flat, thin, somewhat rigid, margins scabrid; panicles scanty, 1–2 in. long, consisting of few 6–2-spiculate short branches, up to ¾ in. long; axis filiform, and like the more or less capillary flexuous branches and pedicels obscurely scaberulous, lateral pedicels short; spikelets oblong, obtuse, 1¼ lin. long, green, glabrous, finely (but distinctly) nerved; lower glume ovate-lanceolate from a broad base, 1-nerved, equalling about ⅓ of the spikelet or shorter to very short and then hyaline and nerveless; upper glume thinly herbaceous, oblong, obtuse, almost 1¼ lin. long, 7–9-nerved,

lower floret ♂ (always ?); valve like the upper glume, $1\frac{1}{4}$ lin. long; pale slightly shorter, keels almost quite smooth, evanescent below the hyaline tip; ☿ floret oblong, obtuse, 1 lin. long, shining, smooth (or finely granulate under a high power), greyish; valve thinly chartaceous, 5-nerved; anthers $\frac{3}{4}$ lin. long.

EASTERN REGION: Natal, between Pinetown and Umbilo River, *Rehmann*, 8049!

Very closely allied to the following species.

16. **P. hymeniochilum** (Nees, Fl. Afr. Austr. 46); culms decumbent, very slender and weak, $1–1\frac{1}{2}$ ft. long, more or less finely hairy or glabrescent, many-noded, branched from some or most of the nodes, internodes exserted; leaves finely hairy; sheaths thin, rather tight, at length often loose and thrown aside, strongly striate, tubercled between the nerves; ligule an obscure ciliolate rim; blades lanceolate to linear-lanceolate, tapering almost from the broad clasping base to a fine point, 1–3 in. by $1\frac{1}{2}$–3 lin., flat, flaccid, margins scabrid, sometimes callously serrulate and often with a few bristles towards the base; panicles very scanty, flaccid, $1–1\frac{1}{2}$ in. long, consisting of few 5–2-spiculate short branches, up to $\frac{1}{2}$ in. long; axis, branches and pedicels filiform, angular, finely hairy or glabrescent, subscaberulous, lateral pedicels very short; spikelets oblong, subobtuse, $1\frac{1}{4}$ lin. long, greenish, glabrous, finely but prominently nerved; lower glume lanceolate to subulate from a broader base, 1-nerved, hyaline, equalling $\frac{1}{2}$ of the spikelet; upper glume thinly herbaceous, oblong, subobtuse, 1 lin. long, 7–9-nerved; lower floret ♂; valve like the upper glume, $1\frac{1}{4}$ lin. long, 7–9-nerved; pale slightly shorter, keels scabrid above, evanescent below the hyaline tip; ☿ floret oblong, subobtuse, 1 lin. long, smooth, whitish; valve thinly chartaceous, 5-nerved; anthers $\frac{3}{8}$ lin. long. *Steud. Syn. Pl. Glum.* i. 83; *Durand & Schinz, Consp. Fl. Afr.* v. 751.

VAR. β, **glandulosum** (Nees, l.c. 47); more robust; blades up to 3 in. by 3 lin., more densely hairy to villous; panicle oblong to ovate, up to 3 in. long, much more divided; axis, branches and branchlets with scattered gland-tipped hairs, branches suberect or spreading, up to more than 1 in. long; pedicels longer; spikelets more scattered. *P. lætum, Durand & Schinz, Consp. Fl. Afr.* v. 752 (*in part*), *not Kunth. P. lætum* (?) *var. β Nees, l.c.* 45.

×EASTERN REGION: Natal; between the Umzimkulu River and the Umkomanzi River, *Drège*, 4247! Var. β: Natal: near the Umlazi River, below 200 ft., *Drège*, 4292! on the flats near Durban, *Drège*, 4248! coastland, *Sutherland*!

17. **P. æquinerve** (Nees, Fl. Afr. Austr. 40, in part); perennial; culm suberect or ascending, very slender, $1–1\frac{1}{2}$ ft. long, laxly branched, many-noded, glabrous, internodes usually exserted; leaves glabrous or hairy; sheaths thin, tight, striate; ligule an obscure minutely ciliate rim; blades linear to lanceolate-linear from a suddenly contracted, often subauricled, base, tapering to a fine point, 1–2 in. by $1\frac{1}{2}$–2 lin., flat, thin, smooth, margins scabrid; panicle scanty, lax, erect, 2–4 in. long; axis filiform, smooth; branches very few, 2-nate or solitary, distant, at length spreading, the longer $1\frac{1}{2}–3\frac{1}{2}$ in. long, simple or almost so, 5–2-spiculate, finely filiform to capillary,

flexuous, smooth; pedicels ½–1 lin. long, tips subcupular; spikelets oblong, acute to subacuminate, about 1½ lin. long, greenish, glabrous or pubescent, prominently nerved; lower glume oblong, subobtuse, very slightly shorter than the upper, 5-nerved, margins and tip hyaline; upper glume similar, broader, subacuminate, 7-nerved; lower floret barren, valve equal to and like the upper glume, but 5-nerved; pale ½ the length of the valve; ⚥ floret oblong, acute or subacuminate, 1–1¼ lin. long, yellowish; valve subcoriaceous, 5-nerved, smooth; anthers over ½ lin. long. *Steud. Syn. Pl. Glum.* i. 79; *Durand & Schinz, Consp. Fl. Afr.* v. 740 (*both in part*).

KALAHARI REGION: Transvaal; Houtbosch, *Rehmann*, 5731!
EASTERN REGION: Natal; Pondoland; between St. Johns River and Umtsikaba River, in swampy places, below 1000 ft., *Drège!* and near Umpumulo, 2000–2500 ft., *Buchanan*, 260!

18. P. perlaxum (Stapf); perennial; rhizome slender; culms suberect or ascending, very slender, 1–2 ft. long, more or less branched, 5- to many-noded, glabrous, internodes mostly exserted; leaves glabrous or hairy; sheaths thin, striate; ligule an obscure, minutely ciliate rim; blades linear to linear-lanceolate from a suddenly contracted, subauriculate base, tapering to an acute point, 2 in. by 1½–3 lin., flat, thin, margins scabrid; panicle very lax and often very scanty, erect, 5–8 in. long; axis filiform, smooth; branches 3–5, usually solitary, remote, at length spreading, the longer 3–6 in. long, simple, 4–1-spiculate, or remotely and sparingly divided, with long 2–3-spiculate branchlets, finely filiform to capillary, flexuous, smooth; lateral pedicels ½–1 in. long, tips cupular; spikelets oblong, acute to subacuminate, 1½–2 lin. long, greenish, glabrous or pubescent, prominently nerved; glumes thinly herbaceous, equal, 1½–2 lin. long, the lower narrow, oblong, acute or subacute, 7–5-nerved; upper ovate-oblong, acute, very concave 7- to sub-9-nerved; lower floret barren; valve similar to the upper glume, but slightly shorter and 5-nerved or 7-nerved near the tip, subhyaline along the middle; pale ½ the length of the valve or more; ⚥ floret oblong, acute or subacuminate, equalling the lower or a very little shorter, yellowish; valve coriaceous, smooth, 5-nerved; anthers ¾ lin. long. *P. æquinerve, Nees, Fl. Afr. Austr.* 40 (*in part*).

COAST REGION: Albany Div.; Coldspring near Grahamstown, *Flanagan*, 766!
EASTERN REGION: Tembuland; near Morley, 1000–2000 ft., *Drège!* Natal; coastland, *Sutherland!*

19. P. chusqueoides (Hack. in Bull. Herb. Boiss. iii. 377); perennial; culms divaricately branched, rambling, very slender, 2 ft. long, glabrous, many-noded, internodes exserted; sheaths very tight, terete, glabrous or ciliate along the margins, sometimes scantily dotted with tubercles; ligules very short, truncate, obscurely ciliate; blades spreading, linear-lanceolate from a broad rounded and suddenly contracted subpetiolate base, 2–3 in. by 2–4 lin., flat, thin, very finely nerved, glabrous, smooth, margins scabrid; panicle up to 5 in. long, consisting of 3–5 distant, suberect or spreading, glabrous, filiform angular branches, which are up to 2 in. long and bear spikelets from 1–3 lin. above the base; pedicels solitary or 2-nate, unequal, one very short, the other up to $1\frac{1}{2}$ lin. long, angular, smooth; spikelets oblong, acute, $1\frac{1}{2}$–$1\frac{3}{4}$ lin. long, greenish, glabrous; lower glume hyaline above, herbaceous below, rotundate-ovate, very obtuse, equalling $\frac{1}{3}$–$\frac{1}{2}$ of the spikelet, 3-nerved; upper glume thinly herbaceous, oblong, subapiculate, $1\frac{1}{2}$–$1\frac{3}{4}$ lin. long, 7- to sub-9-nerved; lower floret barren; valve equal and very similar to the upper glume, but 5-nerved; pale subequal; ⚥ floret oblong, subapiculate, slightly shorter than the lower; valve coriaceous, transversely wrinkled, 5-nerved, light green or yellowish; anthers $\frac{3}{4}$ lin. long.

EASTERN REGION: Natal; near Durban, *Rehmann*, 8648! *Williamson*, 11!

20. P. obumbratum (Stapf); perennial (?); culms very slender and weak, ascending from a prostrate base, more than 1 ft. long, many-noded, branched from all the nodes except the uppermost, internodes exserted; sheaths tight, thin, sparingly beset with long very fine spreading hairs, particularly along the margins; ligules very short, minutely ciliolate; blades linear-lanceolate from a contracted base, tapering to a fine point, 1–2 in. by $1\frac{1}{2}$–3 lin., flat, very thin, with some long fine hairs near the base, margins serrulate-scabrid; panicle much reduced, very narrow, consisting of a few distant, 4–1-spiculate branches, 1–2 in. long; axis very slender, filiform, like the branches scaberulous (at least above) and sparingly beset with long spreading hairs; pedicels 1–4 lin. long, tips slightly thickened; spikelets lanceolate or lanceolate-oblong, acute, 2 lin. long, glabrous, greenish; lower glume very thin, broadly ovate, subobtuse, clasping and equalling $\frac{1}{3}$ of the spikelet, 3-nerved; upper glume thinly herbaceous, oblong, acute, 2 lin. long, 7-nerved; lower floret ♂; valve equal and very similar to the lower glume, 5-nerved; pale subequal; anthers $\frac{1}{2}$–$\frac{2}{3}$ lin. long; ⚥ floret oblong, acute, yellowish, $1\frac{1}{2}$ lin. long; valve coriaceous, very finely transversely wrinkled, 5-nerved.

COAST REGION: Alexandria Div.; Zuurberg Range, in shady places by streamlets, *Drège!*

The description is drawn up from specimens in Drège's herbarium at Lübeck. They belong to the species which is described by Nees as "*Panici species, procumbenti affinis*" in a note to *P. diffusum*, Sw. (Fl. Afr. Austr. 32), and are identical with some others in the same herbarium, collected by Ecklon & Zeyher, but without indication of the locality. These latter are possibly (at least partly) those which Nees refers, l.c., to *Panicum diffusum*, Sw., a West Indian plant, of which I have not seen any examples from South Africa. Nees says they were collected in Uitenhage Div., by the Zwartkops River and in the primeval forest of the Krakakama Mountains.

21. P. laticomum (Nees, Fl. Afr. Austr. 43); perennial, culms ascending, divided above the base into somewhat spreading long leafy flowering branches, slender, $1\frac{1}{2}$–2 ft. long, minutely hairy or glabrous, many-noded, internodes exserted; sheaths tight, thin, striate, hairy, often with tubercle-based hairs or glabrescent except at the nodes and near the junction with the blade; ligule a minutely ciliolate rim; blades spreading, lanceolate from a rounded base, acutely acuminate, 1½–3 in. by ¼–½ in., flat, very thin, sparingly and finely hairy, margins scabrid; panicle erect, very lax, delicately and divaricately branched, about ½ ft. long; axis filiform, terete and smooth below, angular and finely scabrid above; branches in fascicles of 4–2 or solitary, unequal, at length spreading, finely filiform to capillary, very laxly divided, often from 1–2 in. above the base; branchlets and pedicels extremely fine, scaberulous, lateral pedicels 1–4 lin. long, tips scarcely thickened; spikelets oblong, acute at both ends, slightly more than 1 lin. long, glabrous, green; glumes very thinly herbaceous, lower broadly ovate, subobtuse, ½ lin. long, 3-nerved; upper glume somewhat remote from the lower, oblong, acute, almost 1 lin. long, 5-nerved, middle nerve scaberulous above; lower floret barren; valve like the upper glume, but slightly longer; pale ½ the length of the valve; ⚥ floret oblong, obliquely apiculate or acute, equalling or slightly exceeding the upper glume; valve subcoriaceous, whitish, faintly 5-nerved; anthers ½ lin. long; grain obovoid, ½ lin. long, white. *Steud. Syn. Pl. Glum.* i. 84; *Durand & Schinz, Consp. Fl. Afr.* v. 752; *K. Schum. in Engl. Pfl. Ost-Afr. B.* 81.

EASTERN REGION: Natal; shady woods near Durban, *Drège*, 4289! Coast-land, *Sutherland!* and without precise locality, *Gerrard*, 89!

The fruiting florets often separate from the remainder of the spikelet, which remains for a while attached to the pedicel, but falls at length as a whole. The habit is not unlike that of *Isachne albens*, Trin.

22. P. zizanioides (H. B. K. Nov. Gen. et Spec. i. 100); perennial; culms rather slender, firm, ascending from an often long prostrate and rooting base, up to 8 ft. long and to 2–4 ft. high, glabrous, many-noded, more or less branched; sheaths rather tight or the lower loose and deciduous, striate, glabrous or finely hairy to hirsute in the upper part, margins ciliate; ligule a ciliate membranous rim; blades obliquely erect or spreading, ovate-lanceolate to lanceolate, rarely linear, from a rounded auricled or cordate base, acuminate or gradually tapering to an acute point, 3–7 in. by ⅓–1 in., flat, firm, glabrous, smooth or scantily dotted with small hairs on the upper surface, closely nerved and transversely veined near the base when broad, margins cartilaginous, scabrid to spinulous, often fimbriate at the base; panicle erect or slightly nodding, more or less contracted, lax, 4–8 in. long; axis slender, angular, glabrous or pubescent, particularly at the nodes, angles smooth or scabrid; branches 4–8, rarely more, usually solitary, distant, suberect, 1–5 in. long, simple or shortly and sparingly branched from the base or near it, triangular, very slender to filiform, minutely tomentose at the base, internodes 2–8 lin. long; lower pedicels 2-nate, unequal, upper solitary, the longer 1½–6 lin. long, triangular, scabrid; spikelets oblong-ellipsoid, 2–3 lin. long, turgid, glabrous, smooth, light green, usually obscurely nerved; glumes oblong, rather firm, subcarinate above, with callous, laterally compressed tips, lower shorter by ⅓ or less, 3-nerved, upper equalling the spikelet, 5-nerved; lower floret ♂; valve very similar to the upper glume; pale equal, acute; ⚥ floret equalling the ♂ floret, oblong; valve coriaceous, smooth, shining, straw-coloured or whitish, 5-nerved; pale subauriculate near the base; anthers 1 lin. long. *Kunth, Rév. Gram.* i. 233, *t.* 28; *Enum.* i. 118; *Nees, Agrost. Bras.* 143; *Trin. Pan. Gen.* 188, *and in Mém. Acad. Sc. Pétersb. sér.* vi. iii. 276; *Steud. Syn. Pl. Glum.* i. 75; *Doell in Mart. Fl. Bras.* ii. ii. 228. *P. oryzoides, Sw. Prodr.* 23, *not Arduin.; Kunth, Enum.* i. 129; *Steud. l.c.* 80. *P. balbisianum, Schult. Mant.* ii. 254. *P. numidianum, Hook. Niger Fl.* 560, *not Lam. P. pseudoryzoides, Steud. l.c.* 75. *P. Ridleyi, Hack. in Trans. Linn. Soc. ser.* ii. iii. 400. *P. latifolium, Hook. f. Fl. Brit. Ind.* vii. 39, *not Linn.*

EASTERN REGION: Natal; without precise locality, *Gerrard*, 480!

Also in West Africa, India and tropical America.

23. **P. deustum** (Thunb. Prodr. i. 19); perennial; culms from a very short rhizome or the tops of slender stolons, fascicled with few intravaginal innovation shoots, erect or geniculately ascending, 2–4 ft. long, glabrous or finely pubescent below the nodes, sometimes hirsute towards the panicle, terete, 3–6-noded, subsimple or scantily branched, upper internodes more or less exserted; leaves glabrous or sparingly (rarely densely) hirsute with tubercle-based hairs; sheaths rather firm, terete or subcompressed, striate, nodes pubescent; ligule a narrow membranous softly ciliate rim; blades linear to lanceolate-linear from a contracted and rounded or gradually attenuate base, long tapering to a very fine point, ½–1½ ft. by 4–8 lin. (rarely by 1 in.), flat, smooth except the scabrid margins, midrib rather stout, whitish; panicle erect or nodding, contracted or lax, 3–9 in. long; axis slender, angular, scabrid with fine tubercle-based hairs, often pubescent or subhirsute below the nodes, branches scattered or subopposite or scantily whorled, suberect or spreading, rather distant, up to 4 in. long simple, laxly racemose with 2-nate spikelets or divided almost from the base, filiform, flexuous or strict, angular, very scabrid, sometimes with scattered hairs; pedicels usually 2-nate, unequal, the terminal sometimes up to 6 in. long, filiform, scabrid, glabrous or with few long hairs; spikelets oblong, obtuse, turgid, 2–2½ lin. long, light green, usually with purple or blackish tips, glabrous; lower glume membranous, obtuse or subacute, 1–1½ lin. long, 5–7-nerved, sometimes purple at the base; upper glume firmly membranous, oblong, 2–2½ lin. long, 7-nerved; lower floret ♂; valve ovate-oblong, obtuse, somewhat shorter than the upper glume; pale oblong, obtuse; anthers 1½ lin. long; ⚥ floret exceeding the ♂ and sometimes also the upper glume; narrowly oblong, obscurely and obtusely apiculate, yellowish, smooth, shining; valve coriaceous, 5-nerved; grain obovoid-elliptic, over 1 lin. long; hilum oblong, ⅓ the length of the grain. *Thunb. Fl. Cap. ed. Schult.* 104. *P. unguiculatum, Trin. Pan. Gen.* 187, *and in Mém. Acad. Pétersb. sér.* vi. iii. 275; *Steud. Syn. Pl.*

Glum. i. 75; *Durand & Schinz, Consp. Fl. Afr.* v. 767; *K. Schum. in Engl. Pfl. Ost-Afr. A.* 31, *C.* 102. *P. numidianum, Nees, Fl. Afr. Austr.* 33, *not Lam. P. corymbiferum, Steud l.c.* 76; *Durand & Schinz, l.c.* 744. *P. pubivaginatum, K. Schum. in Engl. l.c. A.* 34, *C.* 102.

SOUTH AFRICA: without precise locality, *Thunberg!* *Zeyher*, 451! 4455! *Mund and Maire! Drège*, 4243!

COAST REGION: Riversdale Div.; near the Vet River, *Gill!* Uitenhage Div.; near the Zwartkops River, *Ecklon & Zeyher!* and without precise locality, *Alexander!* Alexandria Div.; near the Bontjes River in the Zuurberg Range, *Drège!* Bathurst Div.; near Port Alfred, *Hutton*, 55! Albany Div.; in wet places by streamlets near Grahamstown, *MacOwan*, 1311! Fort Beaufort Div.; Kat River Poort, 2000 ft., *Drège!* British Kaffraria, *Cooper*, 3345!

CENTRAL REGION: Somerset Div.; near the Platte River, *Burchell*, 2956! near the Blyde River, *Burchell*, 2977! Graaff Reinet Div.; near the Sunday River, 1500–2000 ft., *Drège!* between Kruid Fontein and Milk River, *Burchell*, 2946!

EASTERN REGION: Pondoland; between the Umtata River and St. Johns River, *Drège!* Natal; by streamlets near Umpumulo, *Buchanan*, 265a! Ubabi, *Sutherland!* near Durban, *Williamson*, 17! 18! and without precise locality, *Gerrard*, 483! *Cooper*, 3342! *Buchanan*, 265!

Also in tropical East Africa.

This grass is very variable with respect to hairiness and the length of the pedicels.

24. **P. maximum** (Jacq. Ic. Pl. Rar. i. 2, t. 13; Collect. i. 76); perennial (sometimes flowering the first year?), tufted; innovation shoots intravaginal; rhizome short, sometimes very stout; culms erect or geniculate, robust, usually tall, up to 10 ft. long, compressed below, glabrous or the lower part more or less hirsute, usually 3–6-noded, subsimple or more or less branched, branches erect, upper internodes exserted; leaves glabrous or softly hairy or coarsely hirsute with tubercle-based hairs; sheaths rather firm, the lower compressed, striate, nodes glabrous, pubescent or bearded; ligules membranous, very short, ciliate and often with a dense beard behind; blades linear to lanceolate-linear from a contracted and rounded or attenuate base, long tapering to a fine (sometimes convolute and filiform) point, $\frac{1}{2}$–2 ft. by 2–8 lin., rarely broader; flat, minutely tomentose at the junction with the blade, midrib rather stout, whitish; panicle erect or nodding, contracted or effuse and lax, decompound, from $\frac{1}{2}$ to over 1 ft. long; axis slender, angular, glabrous, smooth or scaberulous above; lower branches whorled, suberect or spreading, rather distant, up to 6 in. long, divided almost from the base, or undivided for 1–2$\frac{1}{2}$ in., filiform, scaberulous above, smooth below, glabrous except at the often minutely tomentose or pubescent callous base; pedicels fascicled, 3–2-nate or the upper solitary, very unequal, very short to several times longer than the spikelet, capillary, flexuous, scabrid; spikelets oblong, subobtuse or obtuse, somewhat turgid, 1$\frac{1}{4}$–1$\frac{1}{2}$ lin. long, light green, sometimes tinged with purple, glabrous, rarely puberulous; lower glume rounded, $\frac{1}{3}$–$\frac{3}{8}$ lin. long, subhyaline, faintly 3-nerved to nerveless, upper oblong, acute or obtuse, 1$\frac{1}{4}$–1$\frac{1}{2}$ lin. long, membranous, 5-nerved; lower floret ♂, valve very similar to and

very slightly shorter than the upper glume; pale oblong, obtuse; ⚥ floret equalling the ♂ or scarcely shorter, oblong, obtuse; valve 5-nerved, finely transversely rugose; anthers ½–¾ lin. long; grain over ½ lin. long. *Trin. Pan. Gen.* 180, *and in Mém. Acad. Pétersb. sér.* vi. iii. 268; *Nees, Fl. Afr. Austr.* 36; *Hook. Niger Fl.* 560; *Steud. Syn. Pl. Glum.* i. 72; *Oliv. in Trans. Linn. Soc.* xxix. 171; *Baker, Fl. Maurit.* 436; *Engl. Hochgebirgsfl. Trop. Afr.* 119; *Durand & Schinz, Consp. Fl. Afr.* v. 753; *K. Schum. in Engl. Pfl. Ost-Afr. B.* 81, *C.* 103; *Hook. f. Fl. Brit. Ind.* vii. 49. *P. polygamum, Sw. Prodr. Veg. Ind. Occ.* 24. *P. læve, Lam. Illustr.* i. 172. *P. jumentorum, Pers. Syn.* i. 83; *Kunth, Enum.* i. 101; *A. Rich. Tent. Fl. Abyss.* ii. 373; *Peters, Reise Mossamb.* ii. 546. *P. Sorghi, Delile, Fl. Égypte,* 51, *t.* 63, *fig.* 6, *ex Barbey, Herbor. Levant, t.* 8. *P. pamplemoussense, Steud. Syn. Pl. Glum.* i. 71. *P. hirsutissimum, Steud. l.c.* 72. *P. porphyrrhizos, Steud. l.c.* 72. *P. confine, Hochst. ex Durand & Schinz, l.c.* 754. *P. trichoglume, K. Schum. in Engl. Veget. Usambara,* 38 (*name only*).

SOUTH AFRICA: without precise locality, *Drège*, 4275!
COAST REGION: Stellenbosch Div.; Hottentotts Holland, *Ecklon*; Knysna Div.; Plettenberg Bay, *Mund & Maire!* Humansdorp Div.; near the Gamtoos River, *Gill!* Uitenhage Div.; near Uitenhage, *Zeyher!* near the Zwartkops River, *Ecklon & Zeyher!* hill near the mouth of the Zwartkops River, *Drège!* Alexandria Div.; Enon, 1000–2000 ft., *Ecklon! Drège!* Bathurst Div.; Port Alfred, *Hutton!* Albany Div.; near Grahamstown, *MacOwan!* Glenfilling, 1000 ft., *Drège;* Komgha Div.; near the Kei River, below 500 ft., *Drège.*
CENTRAL REGION: Jansenville Div. (?); near the Sunday River, *Drège;* Somerset Div.; near the Blyde River, *Burchell*, 2976! Graaff Reinet Div.; stony hills near Graaff Reinet, 2600 ft., *Bolus*, 674! *Ecklon!* Albert Div.; Cooper, 659! 667!
KALAHARI REGION: Diamond Fields, *Nelson*, 32. Orange Free State; without precise locality, *Hutton!* Transvaal, Houtbosch, *Rehmann*, 5702!
EASTERN REGION: Griqualand East; Kokstad, 5000 ft., *Tyson*, 1410! Natal; common between 2000 and 2500 ft., *Buchanan*, 263! Umpumulo, 2000–2500 ft., *Buchanan*, 264! Hills near the Umlazi River, *Krauss*, 183! Durban, *Drège*, *McKen*, 100! *Buchanan*, 30! by the Umzimkulu River, *Drège*, and between the Umkomanzi River and the Umlazi River, *Drège;* Delagoa Bay, *Forbes!*

25. P. lævifolium (Hack. in Bull. Herb. Boiss. iii. 378); annual; culms fascicled, erect or ascending, 1–2½ ft. long, glabrous, compressed to subterete, 3–5-noded, with flowering branches from most of the nodes; leaves glabrous; sheaths lax, mostly shorter than the internodes, pallid, striate; ligules membranous, up to ½ lin. long, ciliate; blades linear, shortly tapering to a very acute point, 3–8 in. by 2–3½ lin., suberect, flat or folded, light green, smooth; panicle erect, ovate, ½–1 ft. long, very lax, open; axis slender, smooth; branches solitary or 2–3-nate, remote, the longest up to ¼–½ ft. long, repeatedly and very laxly divided from ½–1½ in. above the base, branchlets finely filiform to capillary, scaberulous, the ultimate divisions 2-spiculate; pedicels capillary, 2–4 lin. long, tips clavate; spikelets oblong, obtuse, not compressed, slightly over 1 lin. long, pallid or tinged with purple, quite glabrous; lower glume much broader than long, clasping, very obtuse or subacuminate, $\frac{1}{6}$–$\frac{1}{4}$ lin. long; upper glume thin, subherbaceous, oblong, obtuse, slightly over 1 lin. long, 7–9-nerved; lower floret ♂; valve like the upper glume, 9-nerved; pale subequal to the valve; anthers $\frac{1}{2}$ lin. long; ⚥ floret elliptic-oblong, subobtuse, 1 lin. long, whitish, quite smooth; valve obscurely 5-nerved; grain oblong, obtuse.

KALAHARI REGION: Transvaal; Donkers Hoek, *Rehmann*, 6552! Hooge Veld, between Porter and Trigaards Fontein, *Rehmann*, 6614! Pretoria, at Kudus Poort, *Rehmann*, 4697! Bosh Veld, between Elands River and Klippan, *Rehmann*, 5123.

26. P. meyerianum (Nees, Fl. Afr. Austr. 32); perennial; culms ascending from a decumbent branched rooting base (always?), rather stout, 2–3 ft. long, terete, glabrous, 5- to many-noded, sheathed all along or the upper 1–2 internodes exserted; leaves more or less glaucous; sheaths rather loose, glabrous, except at the pubescent nodes, or with scattered tubercle-based hairs; ligule a fringe of very short hairs; blades linear-lanceolate from a rounded contracted base, long tapering to a fine point, 4–8 in. by 3–6 lin., glabrous except at the minutely villous junction with the sheath; panicle subpyramidal, 3–6 in. long, up to 4 in. wide; axis very slender, angular, glabrous or pubescent below the nodes, angles scaberulous at least in the upper part; branches usually obliquely spreading, scattered or the lower opposite or subopposite, the upper simple or subsimple, racemose; the lower distantly branched almost from the finely tomentose base, more or less secund, filiform, scaberulous, the lowest up to 3 in. long; pedicels solitary or 2-nate, the longer up to 1 lin. long, filiform, scaberulous and usually with 1–3 fine long hairs, tips discoid; spikelets rather crowded, at least towards the tips of the branches and branchlets, ovate-oblong, acute to subacuminate, $1\frac{1}{2}$ lin. long, glabrous, glaucous, contracted at the base into a very short stalk; lower glume minute, truncate or apiculate, hyaline, white or purplish; upper membranous, oblong, acute to subacuminate, faintly 5-nerved; lower floret ♂; valve like the upper glume; pale equal to the valve, oblong; anthers $\frac{1}{2}$ lin. long; ⚥ floret oblong, obtuse or minutely apiculate; valve subcoriaceous, 5-nerved, whitish, very finely pitted; anthers up to $\frac{3}{4}$ lin. long. *Steud. Syn. Pl. Glum.* i. 61; *Durand & Schinz, Consp. Fl. Afr.* v. 754. *Helopus meyerianus, Doell in Mart. Fl. Bras.* ii. ii. 189, *in obs.*

EASTERN REGION: Natal; near the Umzimkulu River, *Drège!* banks of the Tugela River, *Buchanan*, 266!

Also in tropical Arabia. The Abyssinian *P. schimperianum*, Hochst. ex A. Rich., is hardly specifically distinct.

27. P. proliferum (Lam. Encycl. iv. 747) var. *β* **longijubatum** (Stapf); perennial (?) tufted; culms erect or geniculate, very stout, spongy below, about 4 ft. long, glabrous, smooth, many-noded, sheathed all along, or the upper internodes exserted, subsimple; sheaths loose, striate, glabrous; ligules very short, membranous, densely ciliate; blades linear, long tapering to an acute point, up to 1 ft. by 2–4 lin., flat, smooth below, very scabrid above, sometimes with a few fine hairs, midrib whitish; panicle erect or nodding, decompound, narrow or rather open, large, 10–14 in. long; axis slender, sulcate, scabrid, at least upwards; branches subopposite or partly solitary or the lower 3–4-nate at very unequal distances, the longest 6–10 in. long, undivided for 1–2 in. from the base, remotely branched, filiform, subflexuous and, like the very fine branchlets, angular and usually very scabrid; pedicels usually 2-nate, unequal, suberect, the longer up to 2½ lin. long, tips slightly thickened; spikelets oblong to oblong-lanceolate, acute to acuminate, scarcely compressed, 1¼ lin. long, greenish; lower glume hyaline, whitish, broadly clasping at the base, truncate, ¼–⅖ lin. long, obscurely 3–5-nerved; upper glume thinly membranous below, oblong to oblong-lanceolate, acute to subacuminate, 1¼ lin. long, prominently 7–9-nerved; lower floret barren; valve like the upper glume; pale oblong, subacute, 2-keeled, 1 lin. long; ⚥ floret narrowly oblong, acute up to 1 lin. long, smooth, shining, yellowish; valve subcoriaceous, 7-nerved; anthers ½ lin. long; grain over ½ lin. long, white. *P. paludosum, Nees, Fl. Afr. Austr.* 35; *Durand & Schinz, Consp. Fl. Afr.* v. 758, *not Roxb.*

VAR. γ **paludosum** (Stapf); culms subsimple or proliferously branched; blades 2–9 in. by 2–4 lin.; panicle 3–9 in. long, stiff; axis, branches and branchlets stouter, branches at length spreading, the longer 2–4 in. long; pedicels shorter than in the preceding variety; spikelets 1¼–1⅓ lin. long, usually conspicuously acuminate; lower glume obscurely nerved or almost nerveless; upper glume and lower valve 9-nerved; lower floret with or without a pale, rarely ♂. *P. proliferum, Hook. f. Fl. Brit. Ind.* vii. 50 (*as to the Indian specimens and synonyms*). *P. paludosum, Roxb. Fl. Ind.* i. 307; *Trin. Pan. Gen.* 179, *and in Mém. Acad. Pétersb. sér.* vi. iii. 267.

SOUTH AFRICA: Var. γ: without precise locality, *Drège*, 4271!

COAST REGION: Var. β: Komgha Div.; by the Kei River, near Komgha, *Flanagan*, 953! Var. γ: Albany Div.; in cultivated grounds at Coldspring, near Grahamstown (introduced?), *Flanagan*, 761!

EASTERN REGION: Var. β: Natal, near the Umzimkulu River, *Drège!* common near Umpumulo, *Buchanan*, 267! near Durban, *Williamson*, 21!

Var. *longijubatum* occurs also on the River Shire in Nyasaland, whilst var. *paludosum* is common throughout India in marshes and still waters.

A polymorphous species inhabiting India, South Africa, and the Atlantic side of America from Maine to the Argentine; elsewhere very rare and apparently introduced. The typical form occurs only in the Atlantic States of North America, and is distinguished by its annual duration, rather low growth, prolific ramification, and scantier nervation of the spikelets, the upper glume being 7- and the lower only 5-nerved.

28. P. capillare (Linn. Sp. Plant. 58); annual; culms fascicled erect, or ascending, 1–2 ft. high, robust, about 5-noded with flowering branches from some or most of the nodes, sheathed all along, glabrous or hairy below the panicle or nodes; leaves more or less (often very copiously) hirsute or villous, rarely subglabrous; sheaths lax; ligules membranous up to ½ lin. long, ciliate; blades linear to linear-lanceolate, long tapering to a fine point, subrigid to flaccid, margins scabrid; panicle often very large, decompound, lax, contracted, then opening out from the top downwards, up to 1 ft. or more by ¾ ft.; rhachis angular, often sparsely hairy, smooth below, scabrid above; branches solitary, subopposite or 3-nate, or irregularly approximate, filiform, angular, scabrid, undivided for ¼–1 in. from the base, then repeatedly and very laxly divided, the longest up to 1 ft. long; branchlets long, finely filiform to capillary, at length divaricate, scabrid; pedicels very unequal, from ½ lin. to more than ½ in. long, capillary, very scabrid, tips subclavate; spikelets oblong to lanceolate-oblong, acuminate, from less than 1 to 1¼ lin. long, greenish or purplish, glabrous; lower glume broadly ovate, acute, equalling about ½ of the spikelet, 3- to sub-5-nerved; upper glume oblong, acuminate, from less than 1 to 1¼ lin. long, 5- to 7-nerved; lower floret reduced to the valve, which very much resembles the upper glume; ⚥ floret oblong, subacute, ¾–⅞ lin. long, very smooth, shining, yellowish; valve faintly 7-nerved; anthers ⅓ lin. long. *Host, Gram. Austr.* iv. *t.* 16; *Kunth, Enum.* i. 114; *Trin. Pan. Gen.* 203, *and in Mém. Acad. Pétersb. sér.* vi. iii. 291; *Steud. Syn. Pl. Glum.* i. 83.

KALAHARI REGION: Orange Free State; without precise locality, *Cooper*, 3375!

A native of North America; introduced elsewhere.

29. **P. miliare** (Lam. Illustr. i. 173); annual; culms erect or geniculate, 1–3 ft. long, glabrous, 3–5-noded, usually with flowering branches from some of the nodes; leaves glabrous, very rarely more or less hirsute with tubercle-based hairs; sheaths loose, strongly striate, smooth, longer or shorter than the internodes; ligules very short, truncate, ciliolate; blades linear from a usually broader and rounded base, tapering to an acute point, ½ to almost 2 ft. by 2–7 lin., flat, flaccid; panicle erect or nodding, contracted, narrow, decompound, lax or dense, ½–1 ft. long; axis slender, striate, smooth; branches alternate, 2- or 3-nate, the lower rather distant, filiform, angular, scaberulous, the longest 2½ to more than 6 in. long, undivided for ½–2½ in., often nodding, branchlets and pedicels contracted, the latter very unequal, very short or up to ½ in. long, scabrid, tips slightly thickened; spikelets subturgid, ellipsoid or oblong, more or less acute, 1¼–1¾ lin. long, glabrous, green or purplish; lower glume very broadly ovate, clasping, acute or subacuminate, ⅓ to almost ½ the length of the spikelet, about 3-nerved; upper glume oblong, subacuminate, very concave; 1¼–1⅝ lin. long, 11–13-nerved; lower floret barren; valve like the upper glume, but 9-nerved; pale subequal, narrow, 2-keeled; ⚥ floret oblong, acute, 1 lin. long, smooth, shining, pallid or brownish; valve coriaceous, 7-nerved; anthers ½ lin long. *Kunth, Enum.* 104; *Nees, Fl. Afr. Austr.* 39; *Steud. Syn. Pl. Glum.* i. 73; *Durand & Schinz, Consp. Fl. Afr.* v. 754; *Hook. f. Fl. Brit. Ind.* vii. 46.

EASTERN REGION: Natal; between Umzimkulu and Umkomanzi River, *Drège*.

Commonly cultivated all over India, and possibly originated from *P. psilopodium*, Trin., an equally common Indian grass.

30. P. repens (Linn. Sp. Plant. ed. ii. 87); perennial; rhizome long, creeping; innovation shoots extravaginal, often growing into long stolons, or intravaginal; culms erect or ascending, ½–2 ft. long, glabrous, densely and distichously leafy below, many-noded, simple or branched at the base; sheaths generally exceeding the nodes, rather tight, firm, finely striate, glabrous, rarely finely hirsute with tubercle-based hairs; ligule a membranous ciliate rim; blades linear, shortly tapering to a callous point, 2½–6 in. by 1½–3 lin., usually involute, rarely flat, rigid, spreading, glaucous, glabrous or hairy, particularly on the upper surface, margins cartilaginous, smooth or adpressedly spinulous and sparingly tubercled near the base; panicle strict, contracted, 3–8 in. long; branches usually erect, solitary or 2–3-nate, the lower remote and 2–6 in. long, sparingly divided, filiform, often wavy, angular, scaberulous; pedicels usually solitary, the lateral about ½ lin. long, the terminal much longer, scabrid, tips subcupular; spikelets erect, ovate-oblong, acute, 1–1⅜ lin. long, glabrous, pallid; lower glume subhyaline, whitish, very broad, rounded or sometimes shortly subacute, 5–3-nerved to nerveless; upper glume thin, membranous, except the firmer tip, 1–1⅜ lin. long, 9–7-nerved; lower floret ♂; valve like the upper glume; pale subequal to the valve, oblong, obtuse; anthers ½–¾ lin. long; ⚥ floret elliptic, subacuminate or subobtuse, 1 lin. long, white, smooth, shining; valve coriaceous, finely 7-nerved; grain white, ¾ lin. long. *Forsk. Fl. Æg. Arab.* xix.; *Cav. Ic. et Descr.* ii. 6, *t.* 110; *Sibth. Fl. Græc.* i. *t.* 61; *Trin. Pan. Gen.* 182, *and in Mém. Acad. Pétersb. sér.* vi. iii. 270; *Kunth, Enum.* i. 103; *Durand & Schinz, Consp. Fl. Afr.* v. 760; *Hook. f. Fl. Brit. Ind.* vii. 49. *P. arenarium, Brot. Phytogr. Lusit.* i. *t.* 6; *Trin. Pan. Gen.* 181, *and in Mém. Acad. Pétersb. sér.* vi. iii. 269; *Nees, Fl. Afr. Austr.* 37; *Steud. Syn. Pl. Glum.* i. 73. *P. leiogonum, Delile, Fl. Égypte,* 51; *Poir. Encycl.* iv. 284; *Kunth, l.c.* 104; *Steud. l.c.* 79; *Barbey, Herbor. Levant. t.* 8; *Aschers. & Schweinf. Illustr. Fl. Égypte,* 160; *Durand & Schinz, l.c.* 752.

COAST REGION: Clanwilliam Div.; Wupperthal, 1500–2000 ft., *Drège,* 2556! *Ecklon,* Bosch Kloof Valley, *Drège, Ecklon.*

Throughout the Mediterranean Region and India to South China and the Malayan Archipelago; elsewhere introduced (?).

31. P. coloratum (Linn. Mant. i. 30); perennial; culms erect or ascending from a geniculate, often decumbent and branched, base, rather stout, 2 ft. or more long, glabrous, 4–8-noded, internodes more or less exserted; sheaths striate, glabrous, or more or less hirsute with tubercle-based hairs, the upper rather tight; ligule a membranous minutely or obscurely ciliate rim; blades linear to lanceolate-linear from a usually widened and rounded base, tapering to an acute point, 5–8 in. by 2½–4½ lin., flat, suberect, slightly rigid or sometimes flaccid, glabrous or sparsely hairy, glaucous or subglaucous, margins smooth or scaberulous, more or less tubercled towards the base; panicle erect or nodding, lax, 4–9 in. long, up to 6 in. broad when fully expanded; axis very slender, smooth, at least below; branches solitary, opposite or 2–4-nate, distant, the lowest 4–7 in. long, filiform to capillary, straight or flexuous, loosely divided from $\frac{1}{2}$–$2\frac{1}{2}$ in. above the base, scaberulous or smooth below; pedicels solitary or 2-nate, unequal, the longer 1–$1\frac{1}{2}$ lin., with cupular tips; spikelets scattered or in scattered clusters or more or less approximate, oblong, acute, 1–$1\frac{1}{4}$ lin. long, glabrous, green or purple; lower glume very broadly ovate, acute, up to $\frac{1}{2}$ lin. long, 3- to sub-5-nerved; upper glume oblong, subacute, prominently 7-nerved; lower floret ♂; valve like the upper glume, sometimes very slightly longer, 9-nerved; ⚥ floret narrow, oblong, subacute, almost 1 lin. long, yellowish, shining, smooth; valve 7-nerved; anthers $\frac{3}{4}$–$\frac{4}{5}$ lin. long. *Kunth, Enum.* i. 104; *Trin. Pan. Gen.* 182, *and in Mém. Acad. Pétersb. sér.* vi. iii. 270; *Fig. et De Not. in Mem. Acc. Torin. ser* ii. xiv. 355; *Steud. Syn. Pl. Glum.* i. 73; *Aschers. & Schweinf. Ill. Fl. Égypte, Suppl.* 778; *Durand & Schinz, Consp. Fl. Afr.* v. 743 (*in part*), *not Nees.*

KALAHARI REGION: Bechuanaland; Batlapin Country, *Holub!* Transvaal, Apies River, *Nelson*, 36*!

EASTERN REGION: Natal; banks of the Tugela River, 600 ft., *Buchanan*, 262!

Also in Egypt, Nyasaland and on the Lower Zambesi.

The spikelets are very similar to those of *P. repens*, but on the whole smaller. This, together with the mode of growth and the more graceful and ampler ramification of the panicle, makes the distinction of *P. coloratum* from broad-leaved forms of *P. repens* easy. *P. coloratum* of Jacquin (Ic. i. 12), which is often quoted as synonymous with *P. coloratum*, Linn., is identical with *P. virgatum*, Linn., an American plant.

32. P. minus (Stapf); perennial; culms densely tufted, mostly with intravaginal erect innovation shoots on an oblique, generally short, rhizome, erect or geniculate, slender, 1–$1\frac{1}{2}$ ft. long, glabrous, about 3-noded, internodes (at least the upper) exserted; leaves glabrous or more or less beset with scattered fine rigid hairs, rather crowded at the base; sheaths tight, striate, glabrous, rarely finely villous or silky at the nodes, the lowest firm, fugaciously tomentose at the base, persistent; ligule a fine ciliolate or ciliate-membranous rim; blades linear, tapering to an acute point, gradually passing into the sheath, 3–6 in. by 1–2 lin., firm, usually tightly convolute, subfiliform, flexuous, smooth or the margins obscurely scaberulous and with a few tubercle-based bristles near the base; panicle erect, contracted or open, lax, 3–4 in. long, rarely longer; axis subfiliform, smooth or scaberulous; branches solitary, the lower distant or remote, filiform, scaberulous or smooth below, often dull or blackish-green, the longest 2–4 rarely 5 in. long, erect or spreading, scantily divided from $\frac{1}{2}$–$1\frac{1}{4}$ in. from the base; pedicels solitary or 2-nate, unequal, fine, scaberulous, the longer equalling or exceeding the spikelet, tip cupular; spikelets oblong, acute or subacute, scarcely compressed, over 1 to almost $1\frac{1}{2}$ lin. long, green tinged with purple, glabrous; lower glume membranous, very broad, rotundate-ovate to ovate, acute to subacuminate, equalling $\frac{1}{3}$–$\frac{2}{3}$ of the spikelet, 1-nerved with or without 1–2 short side-nerves on each side; upper glume ovate-oblong, acute or subacute, 1 to almost 1½ lin. long, distinctly 7-nerved; lower floret ♂; valve like the upper glume, 9-nerved; pale oblong, obtuse; ⚥ floret elliptic, subacute, 1 lin. long, smooth, shining, yellowish, faintly 7-nerved; anthers ½–⅝ lin. long. *P. coloratum, var. a, Nees in Linnæa,* vii. 274; *var. capense, Nees, Fl. Afr. Austr.* 38; *Durand & Schinz, Consp. Fl. Afr.* v. 743.

VAR. β, **planifolium** (Stapf); taller blades usually flat and less rigid or sometimes flaccid, 1–2 lin. broad, rarely broader; panicle larger, usually laxer, 4–8 in. long, lower branches 4–6 in. long. *P. coloratum, var. γ, in part, Nees, Fl. Afr. Austr.* 38*. *P. coloratum, var. glaucum, Hack. in Engl. Jahrb.* xi. 399; *Durand & Schinz, Consp. Fl. Afr.* v. 743.

COAST REGION: Uitenhage Div.; without precise locality, *Ecklon & Zeyher,* 478! Stutterheim Div.; by the Kabousie River, near Komgha, *Flanagan,* 949! Var. β: Bathurst Div.; Port Alfred, *Hutton!* Albany Div.; in grassy places near Grahamstown, 2000 ft., *MacOwan,* 514! 1306! Queenstown, Div.; Finchams Nek, near Queenstown, 4000 ft., *Galpin,* 2375!

CENTRAL REGION: Richmond Div.; Winter Veld, Limon Fontein and Great Tafelberg, 3000–4000 ft., *Drège!* Var. β: Somerset Div., *Bowker,* 182! Colesberg Div.; Wonderheuvel, 4000–5000 ft., *Drège!*

WESTERN REGION: Little Namaqualand; near Verlaptpraam, by the Orange River, *Drège!*

KALAHARI REGION: Orange Free State; near the Caledon River, *Burke,* 424! Var. β: Griqualand West, Hay Div.; between Griqua Town and Graaff Reinet, cultivated specimen, *Burchell,* seed, 119! near Griqua Town, *Burchell,* 1843! along the Vaal River, *Burchell,* 1779! between Griqua Town and Witte Water, *Burchell,* 1979! and between Witte Water and Riet Fontein, *Burchell,* 2005! sandy places near Kimberley, 4000 ft., *Marloth,* 862! Bechuanaland; stony plains near Kuruman, *Marloth,* 1519! Groot Knil Fontein, 3900 ft., *Marloth,* 992! Transvaal; near Johannesburg, *E.S.C.A. Herb.,* 316!

Nees quotes his *P. coloratum* also from the following localities: Tulbagh Waterfall (Tulbagh Div.), between the Coega River and the Sunday River, and by the Zwartkops River (Uitenhage Div.), and by Kromme River (Humansdorp Div.); but as he has mixed up not less than 3 species under this name, these determinations are doubtful.

The densely-tufted culms and the very narrow blades distinguish this species from *P. coloratum,* Linn.

33. **P. dregeanum** (Nees, Fl. Afr. Austr. 42); perennial, tufted; innovation shoots intravaginal; culms erect, slender, firm, 1 to over 2 ft. long, glabrous, terete, with about 2 exserted nodes; leaves mostly crowded at the base; sheaths tight, striate, ciliate, and usually villous at the junction with the blade, otherwise glabrous or sparingly and softly hairy, the lower very firm, fugaciously tomentose at the base, persistent; ligule a minutely ciliate rim; blades narrowly linear, gradually passing into and often distinctly narrowed towards the sheath, tapering to a fine point, 5–14 in. by 1–2 lin., erect, usually convolute, sometimes flat, rigid, glabrous except the more or less hairy or villous base, scabrid, closely and prominently nerved; panicle erect, more or less contracted, decompound, delicately branched, oblong, 4–8 in. long; axis slender, scaberulous; branches solitary, 2- or 3-nate or irregularly approximate, the longest 2–3½ in. long, and undivided for ½–1½ in., then laxly branched, filiform,

scabrid, branchlets subcapillary or capillary; pedicels solitary or 2-nate, unequal, the longer often up to 3–5 lin. long, capillary, scabrid, tips subcupular; spikelets somewhat obliquely ovoid and obliquely acuminate, slightly over 1 lin. long, greenish or more or less purplish, gaping, prominently nerved; glumes similar, almost boat-shaped, ovate to oblong-ovate, acute or subacuminate; lower about $\frac{1}{2}$–$\frac{3}{4}$ the length of the spikelet, often mucronulate, 5–7-nerved; upper equalling the spikelet, 7-nerved; lower floret ♂; valve very similar to the upper glume, 5-nerved; pale equal to the valve, oblong, subacute, flaps very broad at the base; anthers $\frac{1}{2}$–$\frac{1}{3}$ lin. long; ☿ floret oblong, obtuse, $\frac{3}{4}$ lin. long, smooth, shining, white or yellowish, tips sometimes purplish; valve subcoriaceous, very faintly 5–7-nerved. *Steud. Syn. Pl. Glum.* i. 87; *Durand & Schinz, Consp. Fl. Afr.* v. 748.

SOUTH AFRICA: without precise locality, *Drège*, 4266! 4278! 4279!
KALAHARI REGION: Transvaal; Barberton, Umlomati Valley, 4000 ft., *Galpin*, 1362! Rhenoster Poort, *Nelson*, 14*!
EASTERN REGION: Natal; at the mouth of the Umzimkulu River, *Drège*; between the Umtentu and the Umzimkulu Rivers, *Drège*! near Durban, *Drège*! *M'Ken*, 133! *Williamson*, 25! Ingone, *Sutherland*! and without precise locality, *Buchanan*, 117! 261! *Gerrard*, 477!

Also in Nyasaland.

34. P. natalense (Hochst. in Flora, 1846, 113); perennial; compactly tufted, whole plant glabrous; culms erect, slender, wiry, $\frac{3}{4}$–$1\frac{1}{2}$ ft. long, 1–2-noded, uppermost internode usually long exserted; leaves crowded near the base; sheaths terete, tight, firm, the lowest persistent; ligule an obscurely ciliolate rim; blades filiform, $\frac{1}{2}$ to over 1 ft. long, $\frac{3}{8}$–$\frac{1}{2}$ lin. thick, terete, acute, canaliculate, wiry, flexuous, very rarely partly flat, strongly nerved and sometimes minutely hairy on the upper surface, smooth; panicle erect, lax, contracted or open, 2–6 in. long; axis filiform like the subcapillary or capillary branches, branchlets and pedicels smooth; branches mostly solitary or irregularly approximate, laxly divided from near the base, the longest $\frac{3}{4}$–2 in. long; pedicels solitary or 2-nate, very unequal, the longer 1–4 lin. long, tips subcupular; spikelets turgid, broadly ovate to ellipsoid, obtuse, 1 lin. long, light green, sometimes tinged with purple, glabrous; glumes subequal, subherbaceous, very similar, elliptic, more or less obtuse, 5-nerved; lower floret ♂; valve like the glumes; pale subequal to the valve; anthers $\frac{1}{2}$–$\frac{5}{8}$ lin. long; ☿ floret ovate, obtuse or subacute, $\frac{3}{4}$–$\frac{4}{5}$ lin. long, white, smooth; valve coriaceous, faintly 5-nerved. *Steud. Syn. Pl. Glum.* i. 88; *Durand & Schinz, Consp. Fl. Afr.* v. 756.

KALAHARI REGION: Basutoland; without precise locality, *Cooper*, 920! Transvaal; Houtbosch, *Rehmann*, 5690! Stink Fontein by the Steel Poort River, *Nelson*, 91*! Keeron River, *Nelson*, 47*! between Middelburg and the Crocodile River, *Wilms*, 1666!
EASTERN REGION: Tembuland; Entwanazana, 4000 ft., *Baur*, ! 1152 Natal, Novelo Hills, 7000 ft., *Sutherland*! margins of woods near the Umlazi River, *Krauss*, 188! Inanda, *Wood*, 1592! and without precise locality, *Buchanan*, 73! 258!

35. P. Ecklonii (Nees, Fl. Afr. Austr. 43); perennial; compactly tufted, innovation shoots intravaginal; culms erect, very slender, 1–1½ ft. long, glabrous, 1-noded, 3–4 in. above the base, uppermost internode long exserted; leaves crowded at the base, strongly and closely nerved, more or less hirsute with tubercle-based hairs; sheaths tight, shortly bearded at the nodes and often at the mouth, the lower firm, persistent; ligule a ciliate rim; blades linear, acute, 1½–5 in. by 1½–2 lin., rather firm, flat, margins very scabrid; panicle erect, linear to oblong, rather lax or contracted, 2–4 in. long; axis filiform, like all its divisions glabrous; branches alternate or the lowest 2-nate, suberect, loosely divided almost from the base or from ¼–½ in. above it, ultimate branchlets 4–1-spiculate, capillary, flexuous; pedicels 1–4 lin. long, tips subclavate; spikelets oblong, obtuse, 1¼–1½ lin. long, erect, glabrous, pallid or tinged with purple; glumes herbaceously-membranous, nerves running out into teeth; lower broad, ovate, ⅓–½ the length of the spikelet, 3-nerved and 3-toothed; upper glume oblong, 3–5-toothed, 5-nerved, equalling ⅔ to ⅘ the length of the spikelet; lower floret barren, reduced to the valve which is similar to the upper glume and 1½ lin. long; ⚥ floret oblong, acute, equalling the lower floret; valve subcoriaceous, white or yellowish or purplish towards the minutely hairy, subrostrate tip, 7-nerved; anthers 1 lin. long. *Steud. Syn. Pl. Glum.* i. 87; *Durand & Schinz, Consp. Fl. Afr.* v. 748.

COAST REGION: Albany Div.; Grahamstown, 2000 ft., *McOwan*, 1016! *Flanagan*, 760! and without precise locality, *Williamson!* Stockenstrom Div.; Katriviers Berg, above the wood-zone, *Ecklon;* Kat Berg, 4500–5000 ft., in marshy places, *Drège!* King Williamstown Div.; Amatola Mountains, *Buchanan*, 10! Komgha Div., near Komgha, *Flanagan*, 939!
KALAHARI REGION: Orange Free State; without precise locality, *Cooper*, 917!
EASTERN REGION: Tembuland; Bazeia Mountains, *Baur*, 751! Griqualand East, grassy places at the foot of the Zuur Bergen, 3500 ft., *Tyson*, 1868!

Nearly allied to *P. pectinatum*, Rendle, from Nyasaland which has different leaves and glandular pedicels.

36. P. interruptum (Willd. Sp. Plant. i. 341); perennial; culms ascending from a creeping, rooting base, 3–5 ft. high, stout, very spongy, internodes mostly exserted; leaves glabrous; sheaths loose, striate, transversely veined, the submerged sometimes spreading, flattened and bladeless; ligules membranous, truncate, up to 1 lin. long; blades linear from a scarcely constricted base, long tapering to an acute point, 4–12 in. by 3–6 lin., flat, flaccid, very closely nerved, scaberulous, margins scabrid; panicle erect, spike-like, cylindric, 6–12 in. by 3½–5 lin.; axis stout, sulcate, smooth; branches spirally arranged, very numerous, adpressed, filiform, smooth, up to 2 lin. long, divided from the base or reduced to fascicles of disc-tipped pedicels; spikelets oblong, acute or subacute, sometimes slightly curved, 1½–2 lin. long, glabrous, olive-green with dark tips; lower glume hyaline, almost orbicular, ¼–⅓ the length of the spikelet, finely 5–7-nerved; upper herbaceous-membranous, oblong, prominently 9-nerved; lower floret barren; valve like the upper glume; pale about $\frac{3}{4}$ the length of the valve, hyaline; ⚥ floret, oblong, obtuse, whitish or yellowish, 1–1$\frac{1}{3}$ lin. long; valve chartaceous, obscurely 5-nerved; anthers $\frac{1}{2}$ lin. long. *Kunth, Enum.* i. 87; *Nees, Fl. Afr. Austr.* 51; *Steud. Syn. Pl. Glum.* i. 66; *Durand & Schinz, Consp. Fl. Afr.* v. 751; *Hook. f. Fl. Brit. Ind.* vii. 40. *P. uliginosum, Roth, Nov. Pl. Spec.* 50. *P. inundatum, Kunth, Rév. Gram.* i. 34; *Enum.* i. 88; *Steud. l.c.* 66.

EASTERN REGION: Natal; in stagnant water near Durban, *Drège*, 4709! Durban Flat, *Wood*, 3589!

Throughout tropical Africa and India to South China and Malaya.

This species is usually referred to the section *Hymenachne*, the type of which is the South American *P. Myurus*, Lam., and Bentham even reduced it to *P. Myurus*. But they are, as Sir Joseph Hooker remarks, very different plants, *P. Myurus*, Lam., and its immediate allies, which form the section *Hymenachne*, have very narrow and very acute lanceolate spikelets with a 5-nerved upper glume and a 5-nerved lower valve. The branches (at least the lower) of the very compound panicle are long or very long, and much divided, even in those species where they are contracted into a cylindric false spike.

37. P. typhurum (Stapf); perennial, tufted; culms erect, slender, 2–3 ft. long, glabrous, 3-noded, uppermost internode long exserted, simple; sheaths firm, closely striate, glabrous or more or less hairy, particularly in the upper part, the lowest persistent; ligules firmly membranous, truncate, very short; blades linear, tapering to a very long fine point, 6–12 in. by 1$\frac{1}{2}$–5 lin., erect, rigid, flat or convolute, glabrous or more or less hairy and scaberulous above, very closely and strongly nerved; panicle erect, straight or curved, spike-like, cylindric, 5–10 in. by 3 lin., very dense; axis stout, striate, smooth; branches spirally arranged, very numerous, adpressed, filiform, smooth, up to 3 lin. long, divided from the base or reduced to fascicles of disc-tipped pedicels; spikelets ovoid, turgid, slightly oblique and often curved, acute to subacuminate, 1 lin. long, dull purplish, glabrous or hairy; glumes membranous, lower broadly ovate, acute, $\frac{1}{2}$ the length of the spikelet, 3- to sub-5-nerved; upper ovate, curved, very concave, 7–9-ribbed; lower floret ♂; valve like the upper glume, but less concave and curved; pale about $\frac{3}{4}$ the length of the upper glume, hyaline; anthers $\frac{1}{3}$–$\frac{3}{8}$ lin. long; ⚥ floret ovate-oblong, acute, $\frac{1}{2}$ lin. long, whitish or yellowish; valve chartaceous, obscurely 5-nerved.

KALAHARI REGION: Transvaal; Naboon Fontein, on the Nylstroom River, *Nelson*, 74*!

Also in the Manganja Hills, Nyasaland.

38. P. curvatum (Linn. Syst. Nat. ed. xii. 732); perennial; culms ascending from a decumbent or rambling base, very slender, many-noded, glabrous, internodes exserted; sheaths tight, striate, ciliate along the margins, otherwise glabrous or sparsely hairy; ligule a very narrow, minutely ciliolate rim; blades more or less spreading, linear to linear-lanceolate from a strongly and suddenly constricted base tapering to an acute point, 2–4 in. by 2–3 lin., flat, thin, glabrous or sparsely hairy towards the base, smooth, margins scaberulous; panicle erect, contracted and linear, or open and ovate, 1–3 in. long; axis slender, smooth; branches spirally arranged, rather distant, not very numerous, the lower up to 1½ in. long, loosely divided almost from the base, subcapillary, smooth; lateral pedicels very short and fine, tips discoid; spikelets curved, semi-ovate to suboblong, acute or obtuse, 1–1¼ lin. long, green, strongly nerved; lower glume very minute, broadly ovate to orbicular, nerveless; upper equalling the spikelet, very concave, gibbous below, strongly curved, acute or obtuse, membranous, 9-ribbed; lower floret barren; valve oblong or obovate-oblong, obtuse, straight, equalling the lower glume, herbaceous on the sides and at the tip, 7-nerved; ⚥ floret elliptic-oblong, acute, ¾ lin. long, strongly convex; valve chartaceous, very obscurely 5-nerved; anthers ½ lin. long. *Kunth, Enum.* i. 87; *Nees, Fl. Afr. Austr.* 50 (*excl. syn.*); *Durand & Schinz, Consp. Fl. Afr.* v. 745; *K. Schum. in Engl. Pfl. Ost-Afr. C.* 103; *Hook. f. Fl. Brit. Ind.* vii. 42. *P. curvatum, var. tenellum, Durand & Schinz, l.c.* (*excl. syn. Roxb. & Kunth*). *P. coryophorum, Kunth, Rév. Gram.* i. *t.* 107; *Enum.* i. 88; *Steud. Syn. Pl. Glum.* i. 67; *Durand & Schinz, l.c.* 744.

EASTERN REGION: Natal; between Umzimkulu River and Umkomanzi River, *Drège*, 4252! Coastland, *Sutherland!* near Durban, *Williamson*, 30! and without precise locality, *Gerrard*, 479!

Also in the Mascarene Islands and in South India.

Imperfectly known species.

39. P. mollifolium (Link, Hort. Berol. ii. 215); "culms often branched, up to 2 ft. long; sheaths glabrous, striate; blades long, scarcely more than 3 lin. broad, very finely and softly pubescent; panicle elongate, slender, nodding, branchos scantily divided; spikelets about 1 lin. long, glabrous; glumes acute, lower ⅓ the length of upper, this and the lower valve acute, many-nerved."

Raised in the Berlin Botanic Gardens from seeds stated to have been received from the Cape of Good Hope.

40. P. lycopodioides (Bory ex Nees, Agrost. Bras. 236) is indicated for the Cape Colony in Durand & Schinz, Consp. Fl. Afr. v. 753, and in Bull. Herb. Boiss. iv. App. iii. 14; but this is evidently a slip, as Marloth's specimens quoted under that name are referred to *P. madagascariense*, var. *minus*, Hack. I have not seen *P. lycopodioides* from the African continent.

XV. OPLISMENUS, Linn.

Spikelets ovate-oblong to oblong, awned from the glumes, falling entire from the pedicels, in small clusters or in spicate more or less secund racemes along a common axis. Lower *floret* ♂ or barren, upper ⚥. *Glumes* similar, more or less subequal, herbaceous to membranous, 3–7-nerved, both or at least the lower awned. Lower *valve* resembling the upper glume, 5–9-nerved, muticous or very shortly awned, with or without a hyaline 2-nerved or more or less reduced pale; upper valve chartaceous to coriaceous, 5–7-nerved, muticous with a subequal 2-nerved pale of similar substance. *Lodicules* 2, broadly cuneate, often extremely delicate. *Stamens* 3. *Styles* distinct, very long; stigmas terminally exserted, plumose. *Grain* tightly enclosed by the hardened valve and pale, oblong; hilum basal, oblong; embryo about ½ the length of the grain.

Perennial often weak grasses with slender many-noded branched and ascending culms, the internodes of which have usually a decurrent villous line facing the subtending leaf; ligules thinly membranous, very short, truncate, ciliate; blades flat, thin, often slightly asymetrical; spikelets in small clusters or in spicate often distant racemes, generally secund on a slender triquetrous or 3-winged common axis; the awn of the lower glume by far the longest.

Species about 8, in all the warmer parts of the world.

The genus consists of 2 distinct groups, one, characterized by smooth rather thick and obtuse more or less viscous awns, corresponding to R. Brown's genus *Orthopogon*, and another with capillary scabrid awns (Section *Scabrisetæ*, Schlechtend.). The latter is not represented in South Africa.

1. O. africanus (Beauv. Fl. Owar. ii. 15, t. 68, fig. 1); perennial; culms ascending from an often long decumbent rooting base, 1–1½ ft. high, slightly compressed below; sheaths rather tight, strongly striate, finely villous or ciliate along the outer margin, otherwise glabrous except a transverse villous line at the junction with the blade (in the African specimens), rarely more or less hirsute; ligules up to ½ lin. long; blades lanceolate from a slightly rounded base, acuminate, 2–5 in. by 4–7 lin. (those of the lower leaves and barren shoots often much smaller and ovate-lanceolate), flat, thin, dull green, soft, scantily and minutely hairy to almost velvety below, scaberulous above, at least upwards, margins scabrid; racemes 3–8 on an erect straight or flexuous more or less 3-winged glabrous or finely hairy axis, the lower distant, erect or obliquely spreading, 3–9 lin. long, rarely longer, the upper closer, much shorter or reduced to 2–3-spiculate clusters, the axis usually terminated by a solitary spikelet; rhachis of racemes straight, triangular, dorsally flat, greyish from very minute adpressed hairs, usually fringed with tubercle-based bristles (in the African specimens), tomentose or hispid at the base; pedicels 2-nate or solitary, very short and stout, usually with a few bristles; spikelets ovate-oblong, about 1⅓ lin. long, greyish-green or green, minutely bearded at the base; glumes subequal, 1–1½ lin. long, thinly herbaceous, ovate to elliptic, produced into filiform subobtuse smooth more or less viscous often purplish awns, hairy at least near the margins, rarely quite glabrous, lower 5-nerved, awn 3–6 lin. long, upper 7–5-nerved, awn 1–2 lin. long; lower floret barren, rarely ♂, equalling the spikelet; valve 9–7-nerved obscurely or shortly mucronate; ⚥ floret lanceolate-oblong, acute or obscurely cuspidate, 1–1½ lin. long, whitish, smooth, shining; valve coriaceous, 7-nerved; anthers ¾–⅞ lin. long; grain 1 lin. long. *Beauv. Agrost. t.* xi. *fig.* 3; *Kunth, Enum.* i. 141; *Nees, Fl. Afr. Austr.* 60; *Durand & Schinz, Consp. Fl. Afr.* v.

771. *O. loliaceus, Beauv. Agrost.* 168 (*O. foliaceus erroneously on p.* 54); *Humb. & Bonpl. Nov. Gen. et Spec.* i. 106, *not Lam. O hirtellus, Roem. & Schult. Syst.* ii. 481 (*excl. syn. Mich.*), *not Schult. f. O. brasiliensis, Raddi, Agrost. Bras.* 40. *O. velutinus, Schult. f. Mant.* ii. 271; *Kunth, Enum. l.c. O. oahuensis, Steud. Nomencl. ed.* 2, ii. 260. *Panicum hirtellum, Linn. Amœn. Acad.* v. 391; *Sw. Obs.* 35; *Steud. Syn. Pl. Glum.* i. 44 (*the West Indian plant*). *P. loliaceum, Lam. Encycl.* iv. 743 (*the West Indian plant, not illustr.*). *P. africanum, Poir. Encycl. Suppl.* iv. 275. *P. velutinum, E. Mey. Prin. Fl. Esseq.* 51; *Steud. l.c. P. compositum, var.* 1, *Trin. Spec. Gram. Ic. t.* 188. *P. sylvaticum, var. γ, Trin. Gen. Pan.* 124, *and in Mém. Acad. Pétersb. sér.* vi. iii. 212. *P. sylvaticum, Steud. l.c.* 45 (*in part*), *not Lam. P. compositum, Doell in Mart. Fl. Bras.* ii. ii. 146 (*at least in part*). *Orthopogon sp., R. Br. Prodr.* 194, *in obs. O. hirtellus, Spreng. Syst.* i. 306, *not Nutt. O. loliaceus and O. velutinus, Spreng. l.c. O. africanus, Sweet, Hort. Brit. ed.* i. 448.

VAR. β, **capensis** (Stapf); blades usually smaller; spikelets in small clusters of 4–2 (often 2 or 1 of them reduced to an awn) or the uppermost solitary, rather longer than in the type. *O. capensis, Hochst. in Flora*, 1846, 114. *Panicum Kraussii, Steud. Syn. Pl. Glum.* i. 45; *Durand & Schinz, l.c.* 752.

VAR. γ, **simplex** (Stapf); culms very slender and weak; blades linear-lanceolate to linear, 1–3 in. by 2–3 lin., acute or acutely and often finely acuminate; spikelets 2-nate or the lowest and the upper solitary, rarely 4–3 (of which sometimes 2 or 1 are reduced to an awn), in clusters or very short spike-like racemes; spikelets 2–2¼ lin. long, glabrous; glumes 1¾–2 lin. long. *O. simplex, K. Schum. in Engl. Veget. Usambara* 48 (*name only*).

COAST REGION: Var. β: Albany Div.; in damp woods near Grahamstown, *MacOwan*, 1508! Berg Plaats near Grahamstown, under forest trees, *Flanagan*, 781! Knysna Div.; by streamlets in woods near Knysna, *Krauss*, 91. Var. γ: King Williamstown Div.; Dohne Mountain, in a wooded kloof, 4400 ft., *Galpin*, 2449!

CENTRAL REGION: Var. β: Somerset Div.; Bosch Berg, in very shady woods, *Burchell*, 3159!

KALAHARI REGION: Var. β: Transvaal; Houtbosch, *Rehmann*, 5735!

EASTERN REGION: Natal; between the Umzimkulu and Umkomanzi Rivers, *Drège*, 4336! between Mapumulo and Riet Vlei, 2000–5000 ft., *Buchanan*, 219! and without precise locality, *Cooper*, 3378! Var. γ: Inanda, *Wood*, 1306!

The typical form also in tropical Africa and America and in the Sandwich Islands; var. γ also in tropical Africa and the Mascarene Islands.

The European *O. undulatifolius*, Beauv., very much resembles the var. *capensis*, and differs only in the absence of tubercle-based hairs on the sheaths and the axis of the inflorescence, and in the more numerous nerves of the glumes which are also, on the whole, shorter than in the African form.

XVI. AXONOPUS, (Beauv.) Hook. f.

Spikelets ovate to lanceolate-oblong, mucronate or awned, slightly or conspicuously compressed from the back, falling entire from the solitary, geminate or fascicled (and then very unequal) pedicels of more or less digitate or whorled racemes; lower floret ♂, upper ⚥. *Glumes* unequal, lower smaller, hyaline, 3–1-nerved, mucronate, acute or acuminate, upper equal or subequal to the spikelet, membranous, 5-nerved, mucronulate or muticous, submarginal nerves densely villous or ciliate. *Valves* subequal, lower resembling the upper glume, shortly or scantily villous or glabrous, upper chartaceous, glabrous or scantily ciliate, 5-nerved, produced into a mucro or short awn. *Pales* dissimilar, of the lower floret very short, hyaline, 2-partite, flaps auricled, of the ☿ floret more or less equalling the valve, entire, 2-keeled, flaps very narrow, broadly auricled near the base. *Lodicules* 2, broadly cuneate. *Stamens* 3. *Styles* distinct; stigmas laterally exserted. *Grain* enclosed by the rigid valve and pale, oblong, plano-convex; embryo about ½ the length of the grain; hilum basal, punctiform.

Annual or perennial, of various habit; blades flat or convolute; ligules membranous, ciliate, or reduced to a ciliate rim; panicles consisting of digitate or more or less whorled, slender or stout, often spike-like racemes.

Species 3; one from South Africa and the Mascarene Islands to India and Australia, one in the Indo-Malayan region, the third (*A. paniculatus*, Stapf = *Urochloa paniculata*, Benth.) in tropical Africa.

Axonopus was very vaguely defined by Beauvois, and consisted, according to the author, of 4 species of which only one belongs to the genus as understood by Sir Joseph Hooker (Fl. Brit. Ind. vii. 63). This species was made the type of a new genus *Coridochloa* by Nees in Edinb. New Phil. Journ. xv. 381, in 1833, whilst he founded on the species described below another new genus, *Bluffia*, in Lehm. Ind. Sem. Hort. Hamb. 1835. This remarkable small genus approaches on the one hand the *Trichachne* group of *Digitaria*, and on the other hand, *Arundinella*.

1. **A. semialatus** (Hook. f., Fl. Brit. Ind. vii. 64) var. **Ecklonii** (Stapf); perennial, compactly tufted; culms erect, 1–3 ft. long, glabrous, or more or less hairy, usually 2-noded with the uppermost internode long exserted; sheaths strongly striate, rather tight, more or less hairy, lowest fugaciously tomentose, firm, persistent; ligules very short, truncate; blades linear from a slightly or distinctly narrowed base, acute, 3–12 in. by 2–3 lin., flat, firm, rigid, strongly and closely nerved, softly hirsute with tubercle-based hairs; racemes 2–5, digitate or subdigitate, suberect, usually straight, rather stout and dense, often subsecund, 1½–3 in. long, rarely longer, subsessile or some distinctly peduncled; rhachis slender, angular, hairy, at least below; pedicels mostly 2-nate, angular, scabrid or pubescent, very short, up to 2 lin. long; spikelets ovate-oblong, 2½–3 lin. long, very slightly compressed, light green, usually tinged with purple or quite purplish; lower glume ovate, mucronulate, ½–¾ the length of the spikelet, glabrous or minutely pubescent at the tip; upper glume ovate-oblong, densely villous along the margins; valve of ♂ floret finely woolly above near the margins, of ☿ floret oblong, scantily and finely ciliate, produced into a fine mucro or short awn (up to 1½ lin. long); anthers up to 1¾ lin. long. *Bluffia eckloniana, Nees in Lehm. Ind. Sem. Hort. Hamb.* 1835, *and in Lindl. Introd. Nat. Syst. ed.* ii. 447. *Panicum semialatum, var. ecklonianum, Hack. ex Durand & Schinz, Consp. Fl. Afr.* v. 764.

COAST REGION: Alexandria Div.; Zuurberg Range, 2000–3000 ft., *Drège*; Bathurst Div.; between Blue Krantz and Kaffir Drift, *Burchell*, 3712! Albany

Div.; Grahamstown, *MacOwan!* and without precise locality, *Ecklon & Zeyher*, 871! Stockenstrom Div.; grassy places on the Kat Berg 3000–4000 ft., *Drège!* Philipton, *Ecklon!* Komgha Div.; near the mouth of the Kei River, *Flanagan*, 928!

KALAHARI REGION: Griqualand West, Barkly West Div.; Hebron, on the Vaal River, *Nelson*, 48*!

EASTERN REGION: Tembuland; near Bazeia, 2000–2500 ft., *Baur*, 316! Natal; Umpumulo to Riet Vlei, 2000–5000 ft., *Buchanan*, 180! near Durban, *Krauss*, 55! Inanda, *Wood*, 1593! Drakensberg Range at Polela, 6000–7000 ft., *Evans*, 521! and without precise locality, *Gerrard*, 470!

Also in India.

The typical form, corresponding to R. Brown's *Panicum semialatum*, with narrow usually convolute and glabrous or subglabrous blades, generally much longer and laxer racemes and, on the whole, smaller narrower and nearly always light green spikelets, occurs in Australia, Malaya, India, and the Mascarene Islands. In India, it passes into var. *Ecklonii*, and intermediate forms occur also in Nyasaland. A curious variation is due to the widening of the submarginal nerves of the upper glume into fimbriate and sometimes transversely ribbed and very conspicuous wings. It is common to the type and the variety *Ecklonii*, and may occur in all the spikelets of an inflorescence or only in a limited number, the others being quite normal.

XVII. SETARIA, Beauv.

Spikelets ovate to oblong, falling entire from the pedicel, subtended by 1 to many persistent bristles (modified branchlets), which often form a one-sided involucre (see Section *Ptychophyllum*), subsessile in contracted spike-like or more or less open panicles. *Lower floret* ♂, or reduced to the valve and a more or less arrested pale; upper floret ☿. *Glumes* membranous, lower generally much smaller, usually 3–5-, rarely 1- or 7-nerved, upper usually 5-, sometimes 7-nerved. *Lower valve* more or less exceeding and resembling the upper glume; upper valve chartaceous to coriaceous, 5-nerved. *Pales* subequal to their valves or that of the lower floret more or less arrested, flat, 2-nerved, hyaline in the lower, of the same substance as the valve in the upper floret. *Lodicules* 2, broadly cuneate. *Stamens* 3. *Styles* distinct; stigmas laterally exserted. *Grain* tightly enclosed by the hardened valve and pale, oblong or ellipsoid; hilum basal, punctiform or orbicular; embryo about $\frac{1}{2}$ as long as the grain.

Perennials or annuals of various habit; ligules usually reduced to a ciliate rim, rarely a distinct membrane; panicle mostly cylindric, spike-like, dense, with the solitary or clustered spikelets on very short branches which are more or less produced into bristles beyond the spikelets or divided into a one-sided bristly involucre at their base, or more or less open with elongate branches and more distant spikelets, often with or without subtending bristles in the same inflorescence; bristles always persistent.

Species probably over 40 in the warm regions of the world, some common as weeds in the more temperate parts.

Chamæraphis, R. Br. (1810), which has been united with *Setaria*, Beauv. (1812), and for which priority has been claimed over *Setaria*, is a genus, consisting chiefly of Australian species and well distinguished by the different structure of the spikelets. The species of *Setaria* are often very difficult to distinguish, particularly if the specimens are not complete, as is often the case in those cited below.

Section 1. PTYCHOPHYLLUM (A. Braun under *Panicum*). Blades long, plicately folded when young, at length opening out; panicles almost spike-like or more or less open with elongated branches and crowded or somewhat distant spikelets; bristles solitary, terminating the branches and branchlets and usually also subtending at least a part of the lateral spikelets, or sometimes in fascicles at the base of the branches.

Culms 5–12 ft. long; blades 1½–3 ft. by 1–3½ in.; panicle 1–2 ft. long; lower glume 3-, upper usually 5-nerved; fruiting valve smooth or almost so ... ... (1) **sulcata.**
Culms 2–3 ft. long; blades ¼–1 ft. by ⅓–½ in.; panicle ⅙–½ ft. long, usually spike-like; lower glume 5- to sub-7-, upper 7-nerved; fruiting valve finely and closely wrinkled ... ... (2) **lindenbergiana.**

Section 2. EU-SETARIA (Stapf). Blades not plicately folded when young; panicles usually spike-like and dense or compact, with very short (rarely elongated) branches; bristles often crowded into more or less one-sided involucres subtending solitary or clustered spikelets.

Blades deeply sagittate ... ... (3) **appendiculata.**
Blades not sagittate:
  Perennial; panicles dense, cylindric; bristles with the barbs pointing upwards:
    Spikelets rather turgid; lower glume 5-, upper 7-nerved, the latter ¾ the length of the spikelets or less:
      Blades linear to linear-lanceolate, ⅓–⅔ ft. by 3–4 lin.; tips of bristles slightly thickened, blackish, almost smooth ... ... (4) **nigrirostris.**
      Blades very narrow, more than 1 ft. by 1¼–2 lin., usually folded; bristles slender, not thickened above, scabrid all along ... ... (5) **Gerrardii.**
    Spikelets not or scarcely turgid; lower glume 3-, upper 5-nerved:
      Culms ¾–2 ft. high, 2–3-noded, uppermost internode very slender and longer than the others taken together; panicle ½–2 in. long:
        Blades usually setaceously convolute, 2–6 in. by ½–¾ lin. (flattened out) ... ... (6) **perennis.**
        Blades folded or flat, 3–6 in. by 1–2 lin. (unfolded) ... ... (7) **flabellata.**
      Culms 2–6 ft. high, rather firm, 3-7- (rarely more) noded; panicle 2 in. to more than 1 ft. long:
        Culms very rough for a long distance below the panicle; panicle often subinterrupted; bristles coarse; spikelets about 1 lin. long (8) **rigida.**
        Culms scabrid or puberulous close to the panicle only, otherwise smooth; panicle not interrupted, very dense; bristles slender, often orange-coloured; spikelets 1¼–1½ lin. long ... (9) **aurea.**

Annual; panicles cylindric, often lobed, dense, or lax:
Bristles of panicle with the barbs pointing upwards:
Culms slender, weak; blades 2–6 in. by 1–3 lin.; panicles erect, slender, ½–3 in. by 1½–2 lin.; fruiting valve transversely wrinkled ... ... (10) **imberbis.**
Culms often stout; blades ½–1½ ft. long by 3–10 lin.; panicles often nodding, 1½–12 in. by 4–12 lin.; often lobed; fruiting valve smooth (11) **italica.**
Bristles of panicle with the barbs reversed (12) **verticillata.**

1. S. sulcata (Raddi, Agrost. Bras. 50); perennial; culms erect or ascending, from a short, prostrate and rooting base, stout, 5–12 ft. high, compressed below, sometimes pubescent and scabrid close to the panicle, otherwise usually glabrous and smooth, 5- or more-noded, sheathed almost all along or upper internodes exserted; sheaths long, rather tight, glabrous or hirsute, the lowest strongly compressed, subpersistent; ligule a densely ciliate rim; blades lanceolate to linear-lanceolate from a long and much attenuate or even petioled base, tapering to an acute point, 1½–3 ft. by 1–3½ in., closely plicately folded when young, then opening out (folds very numerous), glabrous or hairy, scabrid above towards the tip; panicle linear or linear-oblong, usually interrupted, 1 to more than 2 ft. long, often nodding; axis angular, glabrous or puberulous, scaberulous above; branches solitary, irregularly approximate or almost whorled, ½–6 in. long, scabrid, spike-like, dense, bearing fascicles of spikelets below and solitary spikelets above, or the lower with similar more or less distant branchlets at the base; bristles solitary, fine, scaberulous, wavy, 1–8 lin. long, terminating the branches and branchlets and at the base of some or most of the lower and middle spikelets; pedicels very short, scabrid, tips subdiscoid; spikelets oblong, acute, 1½ lin. long, glabrous, green or tinged with purple; glumes herbaceous-membranous, very broadly ovate, obtuse or subacute, lower 3-nerved, ⅓ to almost ½ as long as the spikelet, upper 5- to sub-7-nerved, ½–⅔ as long as the spikelet; lower floret barren, rarely ♂, equalling the upper or almost so; valve ovate-oblong, acute, 5- to sub-7-nerved, of the same texture as the glumes; pale slightly shorter than its valve or more or less reduced; ⚥ floret oblong, acuminate, 1½ lin. long, tips often recurved; valve subcoriaceous, 5-nerved, pale or finally brown particularly upwards, smooth or very obscurely wrinkled; anthers ⅔ lin. long. *Schult. f. Mant.* ii. 278. *Panicum sulcatum, Aubl. Guian.* i. 50; *A. Bertol. Excerpt.* 14; *Kunth, Enum.* i. 93; *Trin. Pan Gen.* 132, *and in Mém. Acad. Pétersb. sér.* vi. iii. 220; *Steud. Syn Pl. Glum.* i. 49. *P. flabellatum, Steud. l.c.* 53. *P. racemiferum, Wawra, Oester. bot. Zeitschr.* 1862, 171. *P. megaphyllum, Steud. l.c.* 53. *P. plicatile, Hochst. in Flora,* 1855, 198. *P. excurrens, Nees, Fl. Afr. Austr.* 48; *Steud. l.c.* 49 (*in part*), *not Trin.* *P. nepalense, Steud. l.c.* 49, *and Durand & Schinz, l.c.* 756, *not Spreng.*

EASTERN REGION: Natal; near Durban, *Drège*; Pondoland; between Umtata

River and St. Johns River, 1000–2000 ft., *Drège*; in woods and on rocks along the Umzimkulu River, *Drège*; between the Umtata River and Umgaziana River, *Drège*; and between the Umtenta River and Umzimkulu River, *Drège*.

Throughout tropical Africa and America.

2. **S. lindenbergiana** (Stapf); perennial; culms erect, simple or branched near the base, slender, 2–3 ft. long, glabrous or pubescent and sometimes with long soft hairs near the panicle, 5- or more-noded with the lower internodes short and enclosed when simple; branches few-noded; sheaths tight, keeled above, striate, glabrous or hairy to hirsute, lower firm, persistent, often strongly compressed; ligule a ciliate rim; blades linear or linear-lanceolate from a long attenuate base, tapering to a fine point, 3–12 in. by $1\frac{1}{2}$–6 lin., closely plicately folded when young, at length opening out (folds 8–12), usually glabrous, scaberulous above; panicle linear to oblong, usually almost spike-like, often interrupted or lobed, dense or lax, 2–6 in. long; axis angular, pubescent; branches alternate or irregularly approximate, 2–12 lin. long, filiform, wavy, puberulous, simple or divided from near the base; bristles solitary, fine, scaberulous, wavy, 1–5 lin. long, terminating the branches and branchlets, or sometimes in fascicles at the base of the branches or solitary at the base of the lower spikelets; pedicels very short, tips discoid; spikelets ovate-oblong, acute, 1–$1\frac{1}{4}$ lin. long, glabrous, light green or tinged with purple; glumes membranous, broadly ovate, obtuse, lower 5- to sub-7-nerved, rather less than $\frac{1}{2}$ as long as the spikelet, upper 7-nerved, equalling about $\frac{3}{4}$ the length of the spikelet; lower floret ♂, equalling the spikelet; valve membranous, broadly elliptic, obtuse, 7-nerved; pale subequal to the valve, with narrowly winged keels; ☿ floret elliptic-oblong, acute, equalling the lower floret; valve subcoriaceous, 5-nerved, pallid or purplish, particularly upwards, finely and closely transversely wrinkled; anthers $\frac{1}{2}$ in. long. *Panicum lindenbergianum, Nees, Fl. Afr. Austr.* 47; *Steud. Syn. Pl. Glum.* i. 49; *Durand & Schinz, Consp. Fl. Afr.* v. 753.

SOUTH AFRICA: without exact locality, *Drège*, 2049! 3914!

COAST REGION: Bathurst Div.; between Blue Krantz and "Robber Station," *Burchell*, 3890! between Port Alfred and Kaffir Drift, *Burchell*, 3859! Albany Div.; in shrubberies near Grahamstown, 2000 ft., *MacOwan*, 1289! between Grahamstown and Bothas Berg, *Ecklon*; Lower Albany; Glenfilling, *Drège!* Alexandria Div.; Enon, by the river in woods and thickets, *Drège!* Fort Beaufort Div.; grassy hills near Katriver Poort, 2000 ft., *Drège!* Queenstown Div.; Zwart Kei River, *Ecklon*; British Kaffraria, *Cooper*, 3356! Komgha Div.; borders of woods near Komgha, *Flanagan*, 947!

CENTRAL REGION: Albert Div.; without precise locality, *Cooper*, 3355!

KALAHARI REGION: Orange Free State, without precise locality, *Cooper*, 3370!

EASTERN REGION: Natal; Inanda, *Wood*, 1419! in shady woods by the Umgeni River, *Krauss*, 130!

Also in tropical East Africa.

3. **S. appendiculata** (Stapf); perennial; culms erect, simple, about 1 ft. long, glabrous; densely sheathed below; leaves glabrous; lower sheaths subcompressed; ligules short, ciliate; blades linear-lanceolate from a deeply and acutely sagittate base, with subulate lobes, acumi-

nate, sessile, 4–6 in. by 5 lin., scaberulous, margins subundulate, finely nerved; panicle linear, strict, dense, 3½–5 in. long; axis glabrous; branches obliquely erect, crowded, ½–6 in. long, fine, glabrous; bristles solitary, shorter (rarely longer) than the spikelets, sometimes 0; pedicels pubescent, very short; spikelets 2-ranked, subimbricate, ovate, obtuse, apiculate, 1 lin. long, subgibbous, green variegated with purple; glumes ovate, obtuse, lower ½ as long as the spikelet, very broad, 3-nerved, upper equalling the spikelet, 7–9-nerved; lower floret ♂; valve ovate, apiculate, as long as the spikelet, grooved along the middle, 3–7-nerved; pale equal to the valve; ⚥ floret almost as long as the lower, elliptic, slightly gibbous; valve mucronulate, transversely rugose. *Panicum appendiculatum, Hack. in Bull. Herb. Boiss.* iv. *App.* iii. 13.

WESTERN REGION: Great Namaqualand; by the Onanis River, *Belck*, 63*c*.

Allied to *S. sagittæfolia*, Walp., but differing according to Hackel in the sessile subulate-sagittate blades, the short bristles and less gibbous spikelets.

4. **S. nigrirostris** (Durand & Schinz, Consp. Fl. Afr. v. 774); perennial; rhizome short, oblique, præmorse; culms erect or subgeniculate, slender, ½ to more than 2 ft. long, compressed, pubescent and scabrid just below the panicle and usually more or less tomentose below the nodes, otherwise glabrous and smooth, 1–3-noded, internodes usually enclosed except the uppermost; sheaths tight, firm, strongly striate, glabrous except on the ciliate margins and tomentose nodes, or sparingly hairy, lowest short and fugaciously hairy like the innovation bud-scales, very firm, persistent, dark brown; ligules very short, ciliate; blades linear to linear-lanceolate, often with a gradually narrowed base, tapering to an acute point, 4–9 in. (rarely more) by 3–4 lin., rather firm, flat or involute, glabrous, smooth or margins scaberulous; panicle spike-like, cylindric, stout, rather dense, sometimes interrupted, 1–5 in. by 3–5 lin.; axis angular, finely villous; branches reduced to subsessile clusters of 2–3 spikelets or to a single spikelet, each subtended by a fascicle of 3–4 or, if solitary, by as many as 8 coarse subflexuous scaberulous bristles 2½–4½ lin. long with slightly thickened almost smooth and blackish tips; pedicels stout, very short, tips discoid; spikelets turgid, obliquely ovoid, 1¾–2¼ lin. long, pale with dark tips, glabrous; glumes firmly membranous, lower broadly ovate, acute, about ½ as long as the spikelet or slightly longer, 5-nerved, upper oblong, very concave, subapiculate, almost as long as the spikelet or shorter by ¼–⅕, 7-nerved; lower floret ♂; valve equalling the upper floret, oblong, subapiculate, 5-nerved; pale equalling the valve; ⚥ floret elliptic-oblong, slightly beaked; valve 5-nerved, very convex, finely honey-combed, yellowish, beak purple or blackish; anthers 1–1¼ lin. long. *Panicum nigrirostre, Nees, Fl. Afr. Austr.* 55; *Steud. Syn. Pl. Glum.* i. 50.

SOUTH AFRICA: without precise locality, *Drège!*

COAST REGION: Komgha Div.; Komgha, *Flanagan*, 938! Queenstown Div.; Finchams Nek, near Queenstown, 4000 ft., *Galpin*, 2372! British Kaffraria, *Cooper*, 3341!

KALAHARI REGION: Transvaal; Klerksdorp, near Schoen Spruit, *Nelson*, 51*! Orange Free State; on the Drakensberg, near Harrismith, *Buchanan*, 113!

EASTERN REGION: Tembuland; Bazeia, 2000 ft., *Baur*, 296! Natal; Riet Vlei, 4000 ft., *Buchanan*, 174! near Camperdown, 2000 ft., *Wood*, 4096!

5. **S. Gerrardii** (Stapf); perennial; rhizome short, præmorse, oblique; culms erect, 2½–3½ ft. long, compressed below, villous close to the panicle, otherwise smooth and glabrous, 2–4-noded, internodes usually long exserted, except the lowest; sheaths tight, striate, glabrous except on the ciliate margins and very finely silky nodes, lower compressed, keeled, pallid or purplish, firm, persistent, fugaciously hairy at the base, 2–6 in. long; ligule a very densely and long ciliate rim; blades narrow, linear, tapering to a very long, fine point, over 1 ft. by 1¼–2 lin. (unfolded), usually folded or involute above, flexuous, glaucous, glabrous, margins cartilaginous, scabrid; panicle spike-like, cylindric, dense 1½–3 in. by 3–4 lin.; axis subvillous; branches reduced to subsessile clusters of 2–3 spikelets or to a single spikelet, each spikelet subtended by a fascicle of 3–4 or, if solitary, by 6–8 slender subflexuous bristles, 4–6 lin. long, purple from the middle, scabrid all along; pedicels very short, tips discoid; spikelets obliquely ovoid, 1¼–1½ lin. long, whitish, glabrous; glumes firmly membranous, lower ovate, acute, about half as long as the spikelet, 5-nerved, upper oblong, very concave, subapiculate, shorter than the spikelet by ¼, 7-nerved; lower floret ♂; valve equalling or slightly exceeding the upper floret, oblong, 5-nerved, subapiculate; pale equal to the valve; ⚥ floret elliptic-oblong, slightly beaked, 1½ lin. long; valve very convex, 5-nerved, very finely honey-combed, pallid, sometimes with a dark spot near the tip; anthers almost 1 lin. long.

VAR. **purpurea** (Stapf); panicle more slender, about 4 in. long; bristles fine, 3–4 lin. long; spikelets livid purple; lower glume often 7-, upper often 9-nerved.

KALAHARI REGION: Var. β: Transvaal, Limvuba River, *Nelson*, 42*!

EASTERN REGION: Basutoland; Leribe, *Buchanan*, 224! Natal; without precise locality, *Gerrard*, 681! *Buchanan*, 302!

6. **S. perennis** (Hack. in. Bull. Herb. Boiss. iii. 379); perennial, densely cæspitose; culms erect, very slender, 8–16 in. long, slightly compressed below, 2–3-noded, simple, finely puberulous and scaberulous close to the panicle, otherwise smooth, internodes enclosed or slightly exserted except the uppermost; sheaths terete, tight, glabrous, or sparingly hairy, lowest very firm, closely and strongly striate, persistent; ligule a minutely ciliolate rim; blades erect, very narrow, linear, tapering to a very fine point, usually setaceously convolute, 2–8 in. by ½–¾ lin. (rarely 1½ lin. when expanded), rigid, with scattered very fine spreading hairs; panicle spike-like, cylindric, very dense, ¾–2 in. by 2–3 lin.; axis puberulous; branches reduced to a single spikelet subtended by a subsessile one-sided involucre of 4–6 fine subequal scaberulous bristles, 2–3 lin. long, purplish or yellow above; spikelets oblong, acuminate, apiculate, 1½ lin. long, pallid, glabrous; glumes membranous, lower broadly ovate, acute or subacute, about ½ as long as the spikelet or slightly longer, 3-nerved, upper ovate-oblong, acute or acuminate, shorter than the spikelet by

$\frac{1}{4}$, 5-nerved; lower floret ♂; valve oblong, acuminate or apiculate, $1\frac{1}{2}$ lin. long, 5-nerved; pale slightly shorter than the valve; ⚥ floret slightly shorter than the ♂, elliptic-oblong, apiculate, pallid; valve very finely and closely transversely wrinkled, 5-nerved, tips obscurely 3-toothed or mucronulate; anthers over $\frac{1}{2}$ lin. long.

COAST REGION: Bathurst Div.; Port Alfred, *Hutton!* Komgha Div.; near Komgha, *Flanagan*, 936!

KALAHARI REGION: Transvaal, Pretoria Distr., Kudus Poort, *Rehmann*, 4698; Makapans Mountains at Streyd Poort, *Rehmann*, 5385! Bloemhof on the Vaal River, *Nelson*, 67*! near Lydenburg, *Wilms*, 1768! Apies River, *Nelson*, 28*!

7. S. flabellata (Stapf); perennial; culms fascicled, often with numerous innovation shoots, from long slender simple or branched many-noded rhizomes or stolons; culms usually ascending, slender, $\frac{3}{4}$–2 ft. long, compressed, often ancipitous, scabrid or pubescent close to the panicle, otherwise smooth and glabrous, 2–3-noded, uppermost internode long exserted; leaves crowded at the base; sheaths tight, striate, glabrous, rarely very sparsely hairy, lower strongly compressed, keeled, often flabellate, 1–$2\frac{1}{2}$ in. long, firm, persistent; ligule a very minutely ciliolate rim; blades narrow, linear, tapering to an acute point, 3–6 in. by 1–2 lin. when expanded, flat or folded, usually rather rigid and bright green, glabrous or with scattered fine spreading hairs, particularly below, margins scaberulous; panicle spike-like, cylindric, 1–2 in. (rarely more) by $2\frac{1}{2}$–3 lin., erect, very dense; axis puberulous; branches reduced to a subsessile one-sided involucre, consisting of 4–8 fine scaberulous bristles 2–3 lin. long, yellow or purplish, and subtending 1 perfect and sometimes also 1 or 2 arrested spikelets; spikelets oblong, acute to subobtuse, $1\frac{1}{2}$ lin. long, pallid or tinged with purple, glabrous; glumes firmly membranous, lower broadly ovate, acute or subacute, $\frac{1}{3}$–$\frac{1}{2}$ the length of the spikelet, 3-nerved, upper ovate-oblong, subobtuse, $\frac{1}{2}$–$\frac{3}{4}$ the length of the spikelet, 5-nerved; lower floret ♂; valve oblong, acute to subobtuse or subapiculate, $1\frac{1}{2}$ lin. long, 5-nerved; pale slightly shorter; ⚥ floret elliptic-oblong, equalling the ♂ or almost so; valve strongly convex, often slightly beaked, 5-nerved, very finely and closely transversely wrinkled, yellowish, or tips purplish; anthers 1 lin. long. *S. dasyura*, *Durand & Schinz, Consp. Fl. Afr.* vi. 772 (*excl. part of syn.*), *not Schlechtend.* *Panicum dasyurum*, *Nees, Fl. Afr. Austr.* 56 (*excl. part of syn.*).

COAST REGION: Swellendam Div.; along the Zonder Einde River to the Breede River, on dry hill-slopes, 500–2000 ft., *Zeyher*, 4465! Riversdale Div.; near Zoetemelks River, *Burchell*, 6634! Knysna Div.; near Melville, *Burchell*, 5468! Uitenhage Div.; between Galgebosch and Melk River, *Burchell*, 4767! near the Zwartkops River, in grassy places, *Ecklon*, and without precise locality, *Gill!* Port Elizabeth, *E.S.C.A. Herb.*, 103! Alexandria Div.; near Enon, 1000–2000 ft., *Drège!* near Addo, *Ecklon*. Albany Div.; in woods on the hills of the Bushman River, *Ecklon & Zeyher*, 844! Grahamstown, 2000 ft., *MacOwan*, 1017! 1018! 1310! grassy places near Grahamstown, *MacOwan*, 1305a! 1305b! on the banks of streamlets near Grahamstown, *MacOwan*, 1317! Bathurst Div.; near Port Alfred, *Hutton!* *Burchell*, 4021! and between Port Alfred and Kaffir Drift, *Burchell*, 3839!

KALAHARI REGION: Orange Free State; by the Caledon River, *Burke!* Draai Fontein, *Rehmann*, 3676!
EASTERN REGION: Tembuland; Bazeia, 2000 ft., *Baur*, 1158!

Schlechtendal's *Setaria dasyura* was described from specimens raised from seeds collected in the Nilgherris. It occurs also in Abyssinia and is a form of *S. aurea*, from which *S. flabellata* differs very conspicuously in the formation of long stolons or rhizomes and compressed more or less flabellate fascicles of shoots.

8. **S. rigida** (Stapf); perennial; culms erect, rather stout, over 2½ ft. long, more or less compressed or angular, glabrous, very rough below the panicle; sheaths long, rather tight and firm, smooth, glabrous except at the bearded mouth or along the ciliate margin; ligules truncate, very short, densely ciliolate; blades erect, very narrow, linear, subpungent, ½ to more than 1 ft. long, 1–1½ lin. wide (unfolded), rigid, folded (or involute in the upper part), sparingly hairy towards the base, margins rough; panicle spike-like, cylindric, sometimes subinterrupted, 5–9 in. by 3 lin., coarsely bristly; axis subangular, hairy; branches reduced to subsessile clusters of 5–8 partly arrested spikelets, each of which is subtended by 1 coarse scabrid subflexuous bristle, 3–5 lin. long; pedicels very short, tips cupular; spikelets oblong, acute or subacute, slightly curved, slightly over 1 lin. long, straw-coloured; lower glume ovate or almost round, obtuse or acute, less than ½ lin. long, hyaline, 3-nerved; upper similar, ½–⅔ lin. long; lower floret barren, equalling the upper; valve oblong, membranous, 5-nerved; pale 0; ⚥ floret oblong; valve subcoriaceous, very finely punctate, 5-nerved; anthers over ½ lin. long.

EASTERN REGION: Natal; Umpumulo, 2400 ft., *Buchanan*, 12! 173!

9. **S. aurea** (A. Braun in Flora, 1841, 276); perennial or annual or at least flowering the first year; rhizome short, oblique, covered with the remains of old scales and sheaths, sometimes with subglobose innovation buds; culms suberect or ascending, often geniculate, 2–6 ft. long, usually strongly compressed or even ancipitous below, strongly striate and scabrid or puberulous below the panicle, otherwise glabrous and smooth, 3–7-noded, internodes exserted except the lowest, uppermost usually very long and slender; sheaths striate, glabrous or softly hirsute, lower compressed, keeled, bases often persistent and breaking up into fibres; ligule a shortly and densely ciliate rim; blades linear, long tapering to an acute point, ½–1½ ft. by 1½–4 lin. or rarely 6 lin., flat, rather firm, sometimes rather rigid and more or less involute, glabrous or scantily hairy towards the base, scaberulous or almost smooth; panicle erect, straight or subflexuous, cylindric, 2 in. to more than 1 ft. long, 2½–3 lin. thick (exclusive of the bristles), very dense, very bristly, usually orange-coloured or reddish; axis slender, minutely villous or puberulous; branches reduced to a subsessile one-sided involucre consisting of 6–10 slender scabrid bristles, 2–7 lin. long, yellowish to bright orange or reddish, and subtending usually 1 perfect and 1–2 arrested spikelets; spikelets obliquely ovate to ovate-oblong, subapiculate or obtuse, 1¼–1½ lin.

long, pallid or purplish at the tips, glabrous; glumes very thin, membranous, ovate, acute or subacute, whitish or purplish, lower 3-nerved, $\frac{1}{3}$ as long as the spikelet, upper 5-nerved, $\frac{1}{2}$ as long as the spikelet, nerves faint; lower floret ♂; valve equal or subequal to the spikelet, flat or depressed along the middle, similar to the upper glume; pale subequal to the valve; ☿ floret equalling or slightly exceeding the ♂, plano-convex, oblong, usually minutely apiculate, pallid or purplish upwards; valve coriaceous, transversely wrinkled, 5-nerved; anthers $\frac{1}{2}$–1 lin. long; grain depressed ellipsoid, $\frac{3}{4}$ lin. long. *Walp. Ann.* iii. 721; *Engl. Hochgebirgsfl. Trop. Afr.* 121; *K. Schum. in Engl. Pfl. Ost-Afr. C.* 104. *S. sciuroidea, C. Muell. in Bot. Zeit.* 1861, 316. *S. glauca, Hook. f. Fl. Brit. Ind.* vii. 78 (*partly*); *S. glauca var. elongata, Kunth, Rév. Gram.* ii. *t.* 118; *Durand & Schinz, Consp. Fl. Afr.* v. 773. *Panicum penicillatum, Nees, Fl. Afr. Austr.* 56, *not Agrost. Bras.* 242. *Steud. Syn. Pl. Glum.* i. 50, *not Willd. P. chrysanthum, Steud. Syn. Pl. Glum.* i. 50; *Oliv. in Trans. Linn. Soc.* xxix. 172.

VAR. β, **pallida** (Stapf); culms from the tops of short stolons, 3–4 ft. long, branched below, many-noded, lower 5–6 nodes much shorter than the long, strongly compressed firm sheaths; blades very long and narrow, usually tightly folded; panicles about 2–4 in. long with short pallid bristles. *Panicum dasyurum, Nees, Fl. Afr. Austr.* 56 (*the South African plant*), *not Agrost. Bras., nor Schlechtend.*

COAST REGION: Var. β: Alexandria Div.; Zuurberg Range, near Enon, 1000–2000 ft., *Drège!* Komgha Div.; near Komgha, *Flanagan*, 938!

CENTRAL REGION: Var. β: Graaff Reinet Div.; among shrubs on mountain sides near Graaff Reinet, 3700 ft., *Bolus*, 675!

KALAHARI REGION: Transvaal; Cave Mountains, Tabanas Kraal, *Nelson*, 19*! near Johannesburg, *E.S.C.A. Herb.*, 310!

EASTERN REGION: Natal; Impuzane, *Sutherland!* Nottingham, *Buchanan*, 70! Umpumulo, *Buchanan*, 170! and without precise locality, *Gerrard*, 473! *Buchanan*, 169!

Distributed in slightly different forms throughout tropical Africa and Asia.

The mode of growth of var. *pallida* is rather distinct, resembling that of *S. flabellata.* Certain specimens of *S. aurea* from tropical Africa, however, come rather near it in this respect, and it is possible that this peculiarity is merely due to local conditions. Nees indicates his *Panicum dasyurum* also from the Zwartkops River (Uitenhage Div.) and from Addo (Alexandria Div.).

10. S. imberbis (Roem. & Schult. Syst. ii. 891); annual; culms often copiously fascicled, ascending from a geniculate base, slender, $\frac{1}{2}$–3 ft. long, branched, terete or compressed below, deeply grooved and compressible in the upper part where they are sometimes swollen very finely villous or puberulous and scabrid, otherwise glabrous and smooth, 3–7-noded, internodes slightly exserted or at length slipping out of the sheaths; sheaths striate, glabrous; ligules very short, membranous, truncate, up to $\frac{1}{2}$ lin. long, ciliate; blades narrow, linear, long tapering to an acute point, 2–6 in. by 1–3 lin., flat, flaccid, glabrous (in the African specimens), rarely slightly rigid and involute and with long scattered hairs, finely scaberulous; panicle erect, straight or flexuous, cylindric, slender, $\frac{1}{2}$–3 in. by about 2 lin., dense; axis slender, minutely villous or puberulous; branches reduced to a subsessile one-sided involucre of 3–8 fine scabrid bristles 1½–4 lin. long, pallid below, fulvous or reddish above, and subtending usually 1 perfect and often 1 arrested spikelet; spikelets elliptic-oblong, minutely apiculate or obtuse, 1–1½ lin. long, pallid, or purplish at the tips, glabrous; glumes very thin, membranous, ovate, acute or subacute, whitish, lower 3-nerved, rather less than ½ as long as the spikelet, upper 5-nerved, rather more than ½ as long as the spikelet, nerves green; lower florets equal, lower barren or ♂; valve equalling the spikelet, flat, similar to the upper glume; pale subequal to the valve; ⚥ floret plano-convex, ovate-elliptic, usually minutely apiculate, pallid or purplish at the tips; valve coriaceous, transversely wrinkled, 5-nerved; anthers ¼ lin. long. *Kunth, Enum.* i. 150; *Durand & Schinz, Consp. Fl. Afr.* v. 773. *Panicum imberbe, Poir. Encycl. Suppl.* iv. 272; *Nees, Agrost. Bras.* ii. 237; *Fl. Afr. Austr.* 54; *Trin. Pan. Gen.* 134, *and in Mém. Acad. Pétersb. sér.* vi. iii. 222; *Doell in Mart. Fl. Bras.* ii. ii. 156. *P. glaucum, var. a, Trin. Ic. Gram. Spec. t.* 196, *fig. A.*

COAST REGION: Uitenhage Div.; grassy places near the Zwartkops River, *Ecklon*; Komgha Div.; banks of Key River, *Drège*, 4311! Queenstown Div., Griffithsville near Queenstown, 3500 ft., *Galpin*, 2366!

KALAHARI REGION: Griqualand West, Herbert Div.; near Douglas, *Orpen*, 245! Transvaal; common on open fields near Johannesberg, *E.S.C.A. Herb.*, 313! 317!

EASTERN REGION: Greytown, 3000 ft., *Buchanan*, 177! near the Tugela River, 6000 ft., *Buchanan*, 176! and without precise locality, *Buchanan*, 177c!

Also in the Mascarene Islands and in tropical East Africa; very common throughout tropical America.

11. S. italica (Beauv. Agrost. 51); annual; culms fascicled, erect or geniculate, often stout, terete, 2–5 ft. long, simple or almost so, glabrous, smooth or scabrid below the panicle, 5–7-noded, sheathed all along or some internodes at length exserted; sheaths terete, striate, glabrous except at the finely villous or ciliate margins and a similar transverse line at the junction with the blade; ligule a densely hairy rim; blades linear, very long tapering to a setaceous point, ½–1½ ft. by 3–10 lin., flat, flaccid, scaberulous, particularly above, glabrous, margins scabrid to spinulous; panicle often nodding or flexuous, spike-like, cylindric, stout, often lobed, 1½–12 in. by 4–12 lin., rather dense to compact, shortly or long bristly; axis rather stout, usually tomentose; branches spirally arranged or more or less whorled, very short and branched from the base or the lower ½–¾ in. long, and undivided at the tomentose base; branchlets reduced to dense sessile or subsessile clusters; each spikelet usually subtended by a pair of scabrid green or purplish bristles, 1½–4, rarely up to 8 lin., long, or the bristles more numerous owing to the reduction or suppression of some of the spikelets; spikelets oblong-ellipsoid to globose, obtuse, 1–1¼ lin. long, light green, glabrous; lower glume hyaline, broad, ovate, acute, 1–3-nerved, about ⅓ as long as the spikelet; upper glume thin, membranous, elliptic, concave, 5- to sub-7-nerved, ⅔–¾ as long as the spikelet; lower floret barren; valve similar to the upper glume, dorsally flattened, 5-nerved, equalling the upper (at least when in flower); pale hyaline, more or less arrested or 0; ⚥ floret ellipsoid to globose, obtuse or obscurely apiculate, 1–1¼ lin. long, whitish yellowish or reddish; valve papery to crustaceous, smooth or very obscurely wrinkled, 5-nerved; anthers ⅓ lin. long; grain ellipsoid or subglobose, about ¾ lin. long, whitish. *Kunth, Rév. Gram.* i. 46; *Enum.* i. 153, *Suppl.* 108; *Durand & Schinz, Consp. Fl. Afr.* v. 773; *K. Schum. in Engl. Pfl. Ost-Afr. B.* 73, *C.* 105; *Hook. f. Fl. Brit. Ind.* vii. 78. *Panicum italicum, Linn. Sp. Pl.* 56; *Host, Gram. Austr.* iv. *t.* 14; *Trin. Gram. Spec. Ic. t.* 198; *Pan. Gen.* 135, *and in Mém. Acad. Pétersb. sér.* vi. iii. 223; *Steud. Syn. Pl. Glum.* i. 51; *Koern. & Wern. Handb. Getreidebaues*, i. 259.

CENTRAL REGION: Albert Div.; without precise locality, *Cooper*, 638!

Cultivated from South Europe to Japan, particularly in Northern China, occasionally also elsewhere. Koernicke considers it as descended from *S. viridis*, Linn. As to the copious synonymy of this grass which is very rarely grown in Africa, see Hook f. Fl. Brit. Ind. l.c., and as to the numerous varieties Koernicke & Werner, l.c.

12. **S. verticillata** (Beauv. Agrost. 51); annual; culms erect or ascending from a geniculate base, ½–5 ft. long, usually compressed below, more or less branched, glabrous, smooth, or scabrid below the panicle, 4–9-noded, internodes mostly at length exserted; sheaths thin, rather lax, usually compressed, striate, glabrous or finely hairy upwards; ligules short, truncate, densely ciliate; blades linear or lanceolate-linear from a broad and rounded, or from a narrow base, long tapering to an acute or subsetaceous point, 2–12 in. by 2–6 (rarely up to 12) lin., thin, flat, flaccid, scaberulous, usually finely and scantily hairy; panicle erect or curved, spike-like, cylindric or oblong, dense or rather lax, 1–5 in. long, coarsely bristly; axis scabrid and often pubescent; branches spirally arranged, close, in robust specimens the lower up to 4 lin. long with a distinct scabrid angular rhachis and 2-nate spikelets, otherwise very short or reduced to sessile clusters, each spikelet subtended by a coarse reversely scabrid bristle 2–7 lin. long; spikelets ellipsoid, obtuse, about 1 lin. long, light green, glabrous; lower glume hyaline, broad, ovate, acute, 1- to sub-3-nerved, ⅓–⅖ as long as the spikelet, upper membranous, elliptic, concave, 5–7-nerved, equal to the spikelet or almost so; lower floret barren; valve similar to the upper glume, dorsally flattened, 5–7-nerved; pale hyaline, more or less arrested or 0; ⚥ floret elliptic-oblong, plano-convex, subapiculate or obtuse, almost 1 lin. long, greenish or straw-coloured; valve subcoriaceous, very obscurely wrinkled, 5-nerved; anthers ⅜ lin. long; grain broadly ellipsoid, over ¾ lin. long, white, subtranslucent. *Kunth, Enum.* i. 152; *Reichb. Ic. Fl. Germ.* i. *t.* 47, *fig.* 1465; *Eng. Bot. ed.* iii. *t.* 1694; *Sowerb. Brit. Grass.* 63, *t.* 52; *Durand & Schinz, Consp. Fl. Afr.* v. 774; *Hack. in Bull. Herb. Boiss.* iv. *App.* iii. 16; *Hook. f. Fl. Brit. Ind.* vii. 80. *S. Rottleri, Spreng. Syst. Veg.* i. 304; *Kunth, l.c.* 153. *S. nubica, Link, Hort. Berol.* i. 220; *Kunth, l.c. S. respiciens, Miq. Fl. Ind. Bat.* iii. 467. *Panicum verticillatum, Linn.*

Sp. Pl. ed. ii. 82; *Eng. Bot t.* 874; *Host, Gram. Austr.* ii. *t.* 13; *Nees, Fl. Afr. Austr.* 53; *Trin. Sp. Gram. Ic. t.* 202; *Pan. Gen.* 137, *and in Mém. Acad. Pétersb. sér.* vi. iii. 225; *Steud. Syn. Pl.* i. 52. *P. adhærens, Forsk. Fl. Aegypt.-Arab.* 20; *A. Braun, Ind. Sem. Hort. Berol.* 1871, *App.* 5. *P. respiciens, Hochst. ex A. Rich. Tent. Fl. Abyss.* ii. 379; *Steud. l.c.* *P. Aparine, Steud. l.c.*

SOUTH AFRICA: without precise locality, *Cooper*, 3367!

COAST REGION: Uitenhage Div.; Zwartkops River, *Ecklon.* Albany Div.; cultivated ground near Grahamstown, 2000 ft., *MacOwan*, 1318! Komgha Div.; banks of Key River, below 500 ft., *Drège!*

CENTRAL REGION: Graaff Reinet Div.; in cultivated ground near Graaff Reinet, 2500 ft., *Bolus*, 723! Aliwal North Div.; at the junction of Stormberg Spruit and the Orange River, 4200 ft., *Drège!*

WESTERN REGION: Little Namaqualand; near Verleptpram on the Orange River, *Drège!*

KALAHARI REGION: Transvaal; near Lydenburg, *Wilms*, 1670! *Atherstone!* Zoutpans Berg, *Nelson*, 16!

EASTERN REGION: Tembuland; cultivated land near Bazeia, 2000 ft., *Baur*, 454! Natal; Durban Flat, *Buchanan*, 11! near Durban, *Williamson*, 14! Umpumulo and near the coast, *Buchanan*, 165! and without precise locality, *Gerrard*, 682!

Throughout Africa and India to Malaya, elsewhere (Europe, Australia, America) only as a weed. A. Braun, l.c., distinguishes several subspecies and a considerable number of varieties of this grass. I cannot follow him. There is, however, this remarkable fact that the plant is very uniform where it occurs exclusively as a weed, but very variable in the remainder of its area. The species was originally described by Linnæus from the European plant which thus represents the type. Nearly all the other forms (including some of the South African specimens) come under A. Braun's subspecies *Aparine.*

Imperfectly known species.

13. S. pumila (Schult. Mant. ii. 274); annual; culms ascending, branched, about 1 ft. long, scabrid below the panicle, otherwise smooth; sheaths pubescent; ligules hardly any; blades broad, glabrous, scabrid; panicle spike-like, dense, short, without bristles; branches very short, closely packed; spikelets glabrous; glumes and lower valve obtuse, smooth, green-striped; fruiting valve very finely transversely rugose. *Panicum pumilum, Link, Enum. Hort. Berol.* i. 76.

This description is compiled from Link's description in Enum. Hort. Berol. i. 76 and in Hort. Berol. i. 218, which were drawn up from specimens of unknown origin, raised in the Berlin Botanic Garden. Nees referred to this species a specimen collected between the Umtentu and Umzimkulu Rivers, and gave an amended description in Fl. Afr. Austr. 52, in which he describes, however, the sheaths as hirsute with tubercle-based hairs instead of pubescent, and as narrow instead of broad. He describes further the lower glume as 3-nerved and $\frac{1}{3}$ the length of the spikelet, and the upper as 5-nerved and $\frac{1}{2}$ the length of the spikelet. I have seen neither Link's nor Drège's specimen.

XVIII. **PENNISETUM**, Pers.

Spikelets oblong or lanceolate, solitary or in clusters of 2–4, subtended by and deciduous with sessile or peduncled involucres of naked or plumose bristles (rarely reduced to a solitary bristle in Section *Beckeropsis*), and arranged round the axis of spike-like usually cylindric panicles; lower floret ♂ or barren with or without a pale; upper ⚥. *Glumes* usually small and hyaline, lower sometimes suppressed, upper rarely ½ the length of the spikelet or more and then several- to 7-nerved. *Valves* equal or subequal, membranous to chartaceous, 5–7-nerved, or the lower more or less reduced, thinner, fewer-nerved. *Pales* subequal to the valve and of similar texture, 2-nerved, or more or less reduced in the lower floret. *Lodicules* small, usually in front and outside the pale, or 0. *Stamens* 3. *Styles* distinct, slender or connate. *Grain* enclosed by the slightly changed valve and pale (see also *P. typhoideum*), broadly oblong, slightly dorsally compressed to subglobose; hilum basal, punctiform; embryo large, ½–¾ the length of the grain.

Perennial or annual; culms simple or often profusely branched; blades flat or convolute; ligules usually reduced to a ciliate rim or a fringe of hairs, rarely membranous; panicle spike-like, usually dense, branches very numerous all around the axis, very short, simple with a solitary spikelet, or scantily divided with the spikelets in clusters of 2–5; the solitary spikelets or the clusters subtended by and deciduous with an involucre (very rarely a solitary bristle) of often very numerous and usually unequal scabrid or plumose simple rarely branched bristles.

Species about 40, in most warm countries, particularly in dry regions.

Section 1. PENICILLARIA: Involucre often peduncled; all or at least the innermost bristles plumose; fruiting floret readily deciduous, its valve chartaceous, very smooth and shining below; anther tips penicillate; styles connate ... (1) **typhoideum.**

Section 2. EU-PENNISETUM: Involucre sessile; all or some of the bristles plumose; fruiting floret not readily deciduous, its valve scarcely hardened, membranous; anther tips naked; styles free or more or less connate ... ... (2) **cenchroides.**

Section 3. GYMNOTHRIX: Involucre sessile; spikelets usually solitary, rarely 2–3 in each involucre; bristles never plumose: anther tips usually naked; styles almost free or more or less connate or cohering to ½ their length, rarely higher up.

Lodicules distinct; anther tips naked; styles almost free:
  Culms glabrous, smooth or scaberulous below the panicle; blades up to 1 ft. by 2–4 lin.; panicles up to 1 ft. long, dense:
    Panicles 3½–6 lin. thick; spikelets 2–2½ lin. long, lanceolate; lower floret barren ... (3) **macrourum.**
    Panicles slender, about 3 lin. thick; spikelets 1½–1¾ lin. long, ovate-oblong; lower floret ♂ ... ... (4) **natalense.**
  Culms rough and pubescent or adpressedly hirsute below the panicle; blades very narrow, filiform-convolute, 2–8 in. long; panicles 2–4 in. long, very dense, usually tinged with purple ... (5) **sphacelatum.**
Lodicules 0 or very minute; anther tips minutely penicillate; styles usually connate or coherent to ½ their length ... ... (6) **Thunbergii.**

Section 4. BECKEROPSIS: Involucre reduced to a solitary bristle; spikelets in long-peduncled, spike-like racemes, springing from the upper nodes ... ... (7) **unisetum.**

1. **P. typhoideum** (Rich. in Pers. Syn. i. 72); annual; culms erect, stout, 1 to several feet high, usually terete and simple, 5- or more-noded, hairy to villous below the panicle, otherwise usually glabrous; sheaths terete, glabrous except the bearded nodes and the often villous junction with the blade, rarely hirsute, usually slightly rough, rather shorter than the internodes; ligule a narrow long and densely ciliate rim; blades linear to linear-lanceolate from a rounded base, acute, ½–2 ft. by ⅓–1½ in., flat, more or less rough, glabrous, rarely hirsute; panicle spike-like, cylindric, very dense, 4–8 in. by 5–9 lin. (in the South African specimens) or longer and thicker, often purplish; rhachis stout, villous; branchlets reduced to a peduncled involucrate cluster of 3–1 spikelets; peduncles villous, straight, 1–2½ lin. long, often horizontally spreading or partly deflexed; involucre of very numerous ciliate often purplish bristles about as long as the spikelets; spikelets sessile or shortly pedicelled within the involucre, readily deciduous when ripe, oblong, 2–2½ lin. long, pale or purplish upwards; glumes broadly ovate, obtuse, minute, hyaline, nerveless, ciliate, or larger (the upper to ½ the length of the spikelet), firmer and 3-nerved; florets similar, subequal, lower ♂ or reduced to a minute empty hyaline valve; valves broadly oblong, cuspidate or mucronate, 5–7-nerved, glabrous, ciliate or pubescent towards the margins or the tips; pales broad, oblong, truncate, glabrous, ciliate, or the flaps pubescent below; lodicules 0; anthers 1–1¼ lin. long, tips bearded; styles connate; grain ellipsoid to subglobose, equalling the gaping chartaceous very smooth valve and pale. *Delile, Fl. Égypte,* 17, *t.* 8, *fig.* 3; *Trin. Gram. Pan.* 71; *Pan. Gen.* 96, *and in Mém. Acad. Pétersb. sér.* vi. iii. 184 (*typhoides*); *Steud. Syn. Pl. Glum.* i. 108; *Hook. f. Fl. Brit. Ind.* vii. 82. *P. alopecuroides, Spreng. Syst.* i. 303. *P. Linnæi, Kunth, Rév. Gram.* i. 49. *P. spicatum, Koern. in Koern. & Wern. Handb. Getreidebaues,* i. 284; *Engl. Hochgebirgsfl. Trop. Afr.* 126; *Schweinf. in Bull. Herb. Boiss.* ii. *App.* ii. 25; *Durand & Schinz, Consp. Fl. Afr.* v. 784; *K. Schum. in Engl. Pfl. Ost-Afr. B.* 51–59; *Hack. in Bull. Herb. Boiss.* iv. *App.* iii. 17. *P. nigricans, P. Nigritarum and P. Plukenetii, Durand & Schinz, l.c.* 781, 782. *Penicillaria spicata, Willd. Enum. Hort. Berol.* 1037; *Jacq. Ecl. Gram. t.* 17; *Beauv. Agrost.* 59, *t.* 13, *fig.* 4; *Kunth, Enum.* i. 165; *Suppl.* 120, *t.* 11, *fig.* 1. *P. ciliata, Willd. l.c.* 1037. *P. cylindrica, Roem. & Schult. Syst.* ii. 498. *P. Plukenetii, Link, Hort. Berol.* 221; *Nees, Fl. Afr. Austr.* 72. *P. typhoidea, P. fallax, P. raddiana, Fig. & De Not. Mem. Acad. Tor.* ii. xii. (1852), 371, 372. *P. Nigritarum and P. sieberiana, Schlechtend. in Linnæa,* xxv. 561, 565. *Panicum americanum, Linn. Sp. Pl.* 56. *P. alopecuroides, Thunb. Fl. Cap. ed.* i. 387; *ed. Schult.* 102, *not Linn. Holcus spicatus, Linn. Syst. ed.* x. 1305. *H. racemosus, Forsk. Fl. Aegypt.-Arab.* 175. *Alopecurus indicus, Linn. Syst. Veg. ed.* xiii. *Murr.* 92 (*excl. syn. Pluk.*). *Cenchrus alopecuroides, Thunb. Prod.* 24.

SOUTH AFRICA: without precise locality, *Thunberg!*

COAST REGION: Cape Div.; in Mr. Hesse's garden at Cape Town, brought from Inhambane, *Burchell*, 755!
KALAHARI REGION: Orange Free State; without precise locality, *Cooper*, 3349! Transvaal; Cave Mountains, in Tabana's gardens, *Nelson*, 83*!
EASTERN REGION: Natal; near Durban, *Drège*; and without precise locality, *Cooper*, 3338! Delagoa Bay, *Scott!*

Cultivated in numerous forms in tropical Africa and in India.

Numerous forms of this cereal have been described by A. Braun in Ind. Sem. Hort. Berol. 1855, App. 24; but it is hopeless to attempt to reduce the South African specimens to Braun's species, in the absence of the types. K. Schumann who has had access to them, has given up A. Braun's classification and attempted to break up the species into 2 subspecies: *Willdenowii* and *Plukenetii*, according to the length of the peduncles and the general shape of the spikelets, and to divide the subspecies *Willdenowii* into several varieties. I cannot, however, agree with him, and confine myself in this place to the statement, that all the South African specimens represent more or less the typical form, drawn in Delile's figure quoted above. Those where the lower floret appears reduced to a small and empty valve are evidently starved states.

2. **P. cenchroides** (Rich. in Pers. Syst. i. 72); perennial; culms ascending from a branched, geniculate and often decumbent, many-noded base, 1½–2 ft. long, smooth, glabrous, or scantily hairy, upper internodes more or less exserted, the uppermost much so, and very slender; leaves quite glabrous or sometimes with scattered fine stiff hairs; sheaths tight, the lower persistent, or at length decaying, leaving the internodes naked; ligule a very narrow densely ciliate rim; blades linear, long tapering to a setaceous point, 3–8 in. by 1½–3 lin., usually flat, often flaccid, usually scaberulous above or along the margins; panicle spike-like, cylindric, dense, 1–4 in. by 4–6 lin., pallid or purplish, often flexuous; rhachis finely scaberulous like the very short pedicels; involucres of very numerous bristles, outer bristles fine scabrid, shorter or slightly longer than the spikelets, inner thickened towards the base, ciliate, much longer (about ½ in. long), one usually conspicuously exceeding all the rest; spikelets 3–1 within each involucre, lanceolate–oblong, 2–2½ lin. long, pallid, glabrous; glumes hyaline, ovate, acute or acuminate, usually 1-nerved, upper about 1 lin. long, lower shorter sometimes nerveless; florets equal or subequal, lower ♂ or barren very rarely ⚥; valves ovate-oblong, abruptly mucronate-acuminate, 5-nerved; pales subequal, truncate; lodicules 0; anthers slightly over 1 lin. long, tips acute, naked; styles free nearly from the base. *Beauv. Agrost.* 59, *t.* 13, *fig.* 5; *Nees in Linnæa*, vii. 277, *and Fl. Afr. Austr.* 70; *Kunth, Enum.* i. 162; *Trin. Pan. Gen.* 93, *and in Mém. Acad. Pétersb. sér.* vi. iii. 181; *Steud. Syn. Pl. Glum.* i. 105; *Baker, Fl. Maurit.* 441; *Hook. f. Fl. Brit. Ind.* vii. 88. *P. ciliare, Link, Hort. Berol.* i. 213; *A. Rich. Tent. Fl. Abyss.* ii. 384; *Engl. Hochgebirgsfl. Trop. Afr.* 125; *Durand & Schinz, Consp. Fl. Afr.* v. 778; *Hack. in Bull. Herb. Boiss.* iv. *App.* iii. 16. *Cenchrus ciliaris, Linn. Mant.* 302. *Panicum vulpinum, Willd. Enum. Hort. Berol.* 1031.

SOUTH AFRICA: without precise locality, *Forster! Zeyher*, 1794!
COAST REGION: Swellendam Div.; near Swellendam and above the Vormanns Bosch, *Ecklon!* George Div.; on the Karro by the Gauritz River, *Mund.*

CENTRAL REGION: Beaufort West Div.; near Beaufort by the Gamka River, *Drège*. Graaff Reinet Div.; hillsides near Graaff Reinet, 2500–3400 ft., *Bolus*, 724! 724b! Somerset Div.; banks of Klein Fish River, near Somerset East, 3000 ft., *MacOwan*, 1548! near Somerset East, *Bowker*, 198!

WESTERN REGION: Little Namaqualand; banks of the Orange River, near Verleptpram, *Drège!*

KALAHARI REGION: Griqualand West; Hay Div.; Asbestos Mountains, between the "Kloof Village" and Witte Water, *Burchell*, 2058! 2067! Bechuanaland; on the rocks at Chue Vlei, *Burchell*, 2388! Orange Free State; Jackals Fontein, *Burke!* and without precise locality, *Buchanan*, 266! Transvaal; on the Bosch Veld, between Elands River and Klippan, *Rehmann*, 5114!

Throughout Africa, in Sicily and eastwards as far as North-west India.

3. **P. macrourum** (Trin. Gram. Pan. 64); perennial; culms erect from a short oblique rhizome, rather stout, up to 3 ft. high or more, usually sheathed all along, simple, few-noded, scabrid below the panicle, otherwise smooth; leaves mostly crowded at the base; sheaths long, rounded on the back, firm, persistent, striate, glabrous or somewhat rough or hispid; ligule a fringe of rather long hairs; blades linear, long tapering to a fine or setaceous point, up to 2 ft. by 2–3 lin., firm, usually convolute, glabrous, smooth beneath in the lower part, very rough above and along the margins, nerves strong, prominent above; panicle spike-like, cylindric, erect or nodding, 3–12 in. by 4–6 lin., dense, usually pallid; rhachis scaberulous; pedicels very short or obsolete; involucre of numerous pallid, very slender, scabrid bristles, shorter or slightly longer than the spikelet except one which is 4–7 lin. long and conspicuously stouter; spikelets solitary, lanceolate, acute to acuminate, 2–2$\frac{1}{2}$ lin. long, pallid, glabrous; glumes very small or lower obsolete, hyaline, ovate to rounded, acute to truncate, nerveless or 1-nerved; florets about equal, lower reduced to the valve; valves similar, lanceolate, acute to subulate-acuminate, 5-nerved; lodicules truncate, fleshy, $\frac{1}{6}$–$\frac{1}{8}$ lin. long; anthers 1$\frac{1}{2}$ lin. long, tips naked; styles free or almost so. *Steud. Syn. Pl. Glum.* i. 103; *Durand & Schinz, Consp. Fl. Afr.* v. 780 (*the Cape plant*). *P. hordeiforme, Spreng. Syst.* i. 302 (*in part*); *Durand & Schinz, l.c.* 779. *P. asperum, Schult. f. Mant.* ii. 149 (?). *Cenchrus hordeiformis, Thunb. Prod.* 24. *Panicum hordeiforme, Thunb. Fl. Cap.* i. 388, *and ed. Schult.* 103 (*the Cape plant*). *Gymnothrix cenchroides, Roem. & Schult. Syst.* ii. 499, *and Kunth, Enum.* i. 158 (*the Cape plant*). *G. hordeiformis, Nees in Linnæa*, vii. 276; *Fl. Afr. Austr.* 67. *G. caudata, Schrad. in Goett. Gel. Anz.* 1821, 2073, *and in Schult. Mant.* ii. 284; *Kunth. Rév. Gram.* i. 253, *t.* 38; *Enum.* i. 159; *Suppl.* 112. *Catatheraphora hordeiformis, Steud. in Flora,* 1829, 465.

COAST REGION: Vanrhynsdorp Div.; near Ebenezer, *Drège*. Cape Div.; in swampy spots at the foot of Devils Mountain, *Ecklon*, 973! Table Mountain, *MacGillivray*, 395! *Milne*, 244! by a stream beyond Kamps Bay, *Wolley Dod*, 3136! in meadows below Klein Constantia, *Wolley Dod*, 2042! Paarl Div.; banks of the Berg River, *Drège!* Tulbagh Div.; near the Waterfall, *Ecklon!* New Kloof, 500 ft., *Schlechter*, 9036! Worcester Div.; Hex River East, *Wolley Dod*, 3713! Stellenbosch Div.; wet spots near Lowrys Pass, at the foot of Hottentots

Holland Mountains, *MacOwan, Herb. Austr. Afr.*, 1693! Riversdale Div.; by the Great Vals River, *Burchell*, 6555! Knysna Div.; Vleigt, in water, 600 ft., *Bolus*, 2700! Uitenhage Div.; by the Zwartkops River, 50–500 ft., *Zeyher*, 4468! Albany Div.; wet places near Grahamstown, 2000 ft., *MacOwan*, 1308! Queenstown Div., Shiloh, *Baur*, 918!

CENTRAL REGION: Clanwilliam Div.; Wupperthal, *Wurmb!* Graaff Reinet Div.; by streams near Graaff Reinet, 2500 ft., *Bolus*, 190!

Also in S. Helena and Ascension Island.

4. **P. natalense** (Stapf); perennial; culms branched near the base, over 1½ ft. long, sheathed all along (or the lowest internodes at length naked), firm, smooth and glabrous; leaves quite glabrous, glaucous; sheaths firm, glabrous, finely striate, tight or the uppermost slightly tumid; ligule a fringe of short silky hairs; blades linear, long tapering to a fine or setaceous point, ½ to almost 1 ft. by 1–3 lin., firm, rather rigid, and generally convolute, smooth except the cartilaginous scabrid margins; panicle spike-like, slender, cylindric, 5–8 in. by about 3 lin., pallid; rhachis slender, scaberulous like the very short pedicels; involucres of numerous pallid slender scabrid bristles of unequal length, the longer half as long again as the spikelet, one conspicuously stouter and much longer than the rest; spikelets solitary, ovate-oblong, slightly over 1½ lin. long, pallid, or purple at the tips, glabrous; glumes hyaline, lower very minute, nerveless, upper ovate, acuminate, less than ½ lin. long, 1-nerved; lower floret ♂; valves very similar, broadly ovate-oblong, suddenly and shortly acuminate or the upper mucronate, 5-nerved; lodicules small, but distinct; anthers not quite 1 lin. long, tips acute, naked; styles connate at the very base.

EASTERN REGION: Natal; Umpumulo, *Buchanan*, 172!

In many respects very similar to *P. riparioides*, Hochst. ex Rich., but distinguished by smaller more ovoid spikelets, and long glaucous more rigid leaves with scabrid margins.

5. **P. sphacelatum** (Durand & Schinz, Consp. Fl. Afr. v. 784); perennial; culms on a short oblique rhizome, slender, 1–3 ft. high, about 3-noded, simple, rough and pubescent or adpressedly hirsute below the panicle, otherwise glabrous and smooth, internodes mostly exserted, the uppermost often very long; sheaths firm, lower crowded, short, at least on the innovation shoots, persistent, striate, glabrous, or pubescent along the margins and near the mouth; ligule a fringe of short silky hairs; blades very narrow, linear, filiform-convolute, 2–8 in. by ¾–1½ lin., firm, flexuous, glabrous or hairy towards the base, rough in the upper part; panicles cylindric, very dense, erect or nodding, 2–4 in. by 3–4 lin., pallid or slightly purplish; rhachis slender, scabrid or pubescent like the very short pedicels; involucre of numerous, pallid, slender, scabrid bristles of unequal length, the longer half as long again as the spikelet or longer, one usually longer and stouter than the rest; spikelets solitary, lanceolate, acuminate to subulate-acuminate, 1¾–2½ lin. long, glabrous, pallid, tips usually purplish; glumes very small, lower often suppressed, obtuse or acute, hyaline, nerveless or upper 1-nerved; florets

equal or the lower which is reduced to an empty valve slightly shorter; valves oblong-lanceolate, subulate-acuminate, 5-nerved or the lower 3-nerved; lodicules subquadrate; anthers about 1 lin. long, tips naked; styles connate at the very base. *Gymnothrix sphacelata, Nees, Fl. Afr. Austr.* 68.

VAR. β, **tenuifolium** (Stapf); culms and blades very slender, almost flaccid, quite glabrous. *P. tenuifolium, Hack. in Bull. Herb. Boiss.* iii. 380.

COAST REGION: Queenstown Div.; at the foot of the Stormberg Range between Klass Smits River and Stormberg Spruit, 4000-5000 ft., *Drège!* on mountains at Fincham's Nek, near Queenstown, 3900 ft., *Galpin*, 2368! 2371!

KALAHARI REGION: Orange Free State; near Winberg, *Buchanan*, 249! 250! Var. β: Transvaal; Pretoria Distr.; near Wanderboom Poort, *Rehmann*, 4490! Lydenburg Distr.; Spitzkop Goldmine, *Wilms*, 1699!

EASTERN REGION: Transkei; between Gekau (Gcua) River and Bashee River, 1000-2000 ft., *Drège!* and by the Gekau River, 1000-2000 ft., *Drège!* Natal; Umsinga and at the base of the Biggarsberg Range, *Buchanan*, 107! Reit Vlei, 4000 ft., *Buchanan*, 171! and without precise locality, *Buchanan*, 63! 72! 75!

The variety *tenuifolium* is scarcely more than a very weak state or shade form. The Abyssinian *P. Schimperi*, A. Rich., comes also very near and is perhaps not specifically distinct.

6. **P. Thunbergii** (Kunth, Rév. Gram. i. 50); perennial; culms from a rather slender rhizome, ascending, conspicuously geniculate, slender, compressed below, $\frac{3}{4}$–$1\frac{1}{2}$ ft. long, about 2-noded, smooth, glabrous or pubescent below the panicle; upper internodes exserted, uppermost often long; leaves for the most part crowded at the base, glabrous except a hairy line at the junction of blade and sheath, rarely hairy all over except towards the base of the sheaths; sheaths firm, lower short, persistent, striate, keeled or ultimately rounded, upper long, loose or subtumid; ligule a dense silky fringe of short hairs; blades linear, tapering to an acute point, 2–3 in. (rarely more) by 2–3 lin., keeled and folded or flat or involute, rigid, rather dull green, smooth except towards the tips and along the scaberulous margins; panicle cylindric, very dense, erect, $1\frac{1}{4}$–2 in. by 3–5 lin., purplish; rhachis scabrid; pedicels obsolete; involucres of very numerous and very fine scaberulous bristles, purplish above, of unequal length, the longer sometimes twice as long as the spikelets; spikelets solitary, oblong-lanceolate, $1\frac{1}{2}$–2 lin. long, pallid, glabrous; lower glume suppressed, upper very minute, hyaline, nerveless; lower floret reduced to an ovate acuminate 1–3-nerved valve of about half the length of the ☿ floret and usually with a short fine bristle; fertile valve ovate-oblong, abruptly mucronate, acuminate, 5- to sub-7-nerved; lodicules 0 or very minute; anthers about 1 lin. long, tips minutely bearded; styles almost free, although often more or less cohering. *Kunth, Enum.* i. 164; *Durand & Schinz, Consp. Fl. Afr.* v. 786; *Hackel in Bull. Herb. Boiss.* iv. *App.* iii. 17. *P. alopecuroides, Steud. in Flora*, 1829, 472, *and Syn. Pl. Glum.* i. 103; *Trin. Pan. Gen.* 90, *and in Mém. Acad. Pétersb. sér.* vi. iii. 178, *not of Spreng. P. cenchroides, Nees in Linnæa*, vii. 277, *and Fl. Afr. Austr.* 70 (*in part*), *not Rich. P. purpurascens, Durand & Schinz, Consp. Fl. Afr.* v. 783, *not of Anderss. nor of H. B. K. Cenchrus*

geniculatus, Thunb. Prod. 24. *Panicum geniculatum, Thunb. Fl. Cap.* i. 388, *and ed. Schult.* 103 (*excl. syn. P. hordeiforme*). *Gymnothrix purpurascens, Schrad. in Goett. Gel. Anz.* 1821, 2072; *Kunth, Enum.* i. 159; *Nees in Linnæa*, vii. 277; *and Fl. Afr. Austr.* 68.

VAR. β, **Galpinii** (Stapf); culms erect, up to 2 ft. long, scabrid and pubescent above; basal sheaths usually rather long, strongly compressed, pallid; blades ½–1 ft. by 2–4 lin., usually glaucous, more scabrid than the type; anthers more conspicuously bearded.

COAST REGION: Cape Div.; in ditches and on hillsides near Capetown, *Thunberg! Spielhaus!* moist spots between Table Mountain and Kamps Bay, *Ecklon!* in moist spots at the foot of Devils Mountain, *Ecklon, Bergius;* in swamps in Orange Kloof, *Wolley Dod*, 3225! (intermediate between the type and the variety). Worcester Div.; Hex River Valley, *Wolley Dod*, 3712! George Div.; near George, *Burchell*, 6001! Uitenhage Div.; near Uitenhage, *Herb. Harvey*, 96! Stockenstrom Div.; Kat Berg, 4000 ft., *Drège.* Cathcart Div.; Glencairn, 4800 ft., *Galpin*, 2418! Var. β: Cape Div.; near King's Blockhouse, *Wolley Dod*, 2668! Queenstown Div.; Fincham's Nek, near Queenstown, 3800 ft., *Galpin*, 2368!

CENTRAL REGION: Prince Albert Div.; by the Gamka River at Weltevrede, 2500–3000 ft., *Drège!* Var. β: Somerset Div.; in moist places on the Bosch Berg, 3500 ft., *MacOwan*, 1661! Graaff Reinet Div.; Spitz Kop (Compass Berg), *Bowker*, 48! Sneeuwberg Range, near Houd Constant, 3800 ft., *Bolus*, 1960!

WESTERN REGION: Great Namaqualand; near Keetmanshoop, *Fenchel*, 3; near Aus, *Schinz*, 645, *Schenck*, 78!

KALAHARI REGION: Transvaal; near Lydenburg, *Wilms*, 1698!

EASTERN REGION: Var. β: Tembuland; Bazeia, sandy river margins, 2000–2500 ft., *Baur*, 252! 314! Natal; without precise locality, 4000–5000 ft., *Buchanan*, 85!

The type is also found in Hereroland, German South-west Africa.

The variety *Galpinii* is very distinct from the type in the eastern parts of Cape Colony and in Natal; but in the west, it seems to pass completely into the latter as Captain Wolley Dod's specimens show.

7. **P. unisetum** (Benth. in Journ. Linn. Soc. xix. 47, 49); perennial; rhizome short, præmorse; innovation-buds rather stout, acute, with firm, ovate, striate, ciliate scales; culms 3–12 ft. high, erect, branched, particularly from the upper nodes, or simple below, terete or rarely semiterete in the lower and subtriquetrous in the uppermost internodes, very firm, obscurely striate, glabrous, glaucous, nodes 8 or more, longest internodes ½–1 ft. long; sheaths strong, the lower and middle as long as the internodes or shorter, ultimately spreading, the upper longer, tighter, glabrous except the usually ciliate mouth and the sometimes hairy nodes; ligules very short and ciliate or a fringe of hairs; blades lanceolate-linear from a usually long-narrowed base, or petioled, long tapering to a setaceous point, the longest from ½–1¼ ft. by 3–10 lin., rather firm, glaucous, scabrid above, sometimes shortly and sparingly hairy, margins very rough; racemes on long very slender peduncles, 1 or several from the upper nodes, sometimes very numerous, ½–2 in. long; rhachis angular, scabrid; spikelets subimbricate, oblong or lanceolate-oblong, acute, 1¼–1½ lin. long, often purplish, basal bristles ½–10 lin. long; glumes reduced to minute rounded or truncate nerveless scaberulous scales, lower ⅙–¼, upper ¼–½ lin. long; lower floret reduced to a valve

minutely ciliate below the tip, otherwise scaberulous or almost smooth, firmly membranous, finely 5-nerved, margins obscurely inflexed in the upper third; upper valve similar to the lower, glabrous; pale almost as long as the valve, hyaline, 2-nerved; lodicules broadly cuneiform, fleshy, $\frac{1}{10}$ lin. long; anthers $\frac{3}{4}$–$\frac{7}{8}$ lin. long; grain unknown. *Durand & Schinz, Consp. Fl. Afr.* 786; *Étud. Fl. Congo*, i. 328. *P. dioicum, Engl. Hochgebirgsfl. Trop. Afr.* 122 (♂ *partly*); *Durand & Schinz, Consp. Fl. Afr.* 778 (*partly*). *P. longisetum, K. Schum. in. Engl. Pfl. Ost-Afr. C.* 105 (*excl. most syn.*). *Setaria dioica* ♂, *Hochst. in Flora*, 1841, i. *Intell.* 19 (*name only*). *Gymnothrix uniseta, Nees, Fl. Afr. Austr.* 66. *Beckera dioica*, ♂, *Nees in Linnæa*, xvi. 219. *B. uniseta, Hochst. in Flora*, 1844, 512; *Steud. Syn. Pl. Gram.* i. 118. *B. glabrescens, Steud. l.c.* 117. *Beckeria dioica, Heynh. Nom.* ii. 63.

EASTERN REGION: Natal; near Durban, in coffee gardens, *Drège!* Umpumulo, 1500–2000 ft., *Buchanan*, 220! 221!

This grass is not diœcious as was supposed, whence the names *Pennisetum dioicum, Setaria dioica, Beckera dioica.* The specimens considered as representing the ♀ sex belong to *P. dioicum*, A. Rich. (*Beckera petiolaris*, Hochst.), which is, however, as little diœcious as *P. unisetum*, the supposed ♂ sex of the composite species *P. dioicum*.

XIX. STENOTAPHRUM, Trin.

Spikelets lanceolate to ovate-oblong, sessile, singly or 2–4 on the very short branches of an apparently simple secund spike, more or less sunk in hollows on or adpressed to the anterior face of a dorsally flattened herbaceous continuous or jointed rhachis; lower floret ♂ or reduced to an empty valve; upper ⚥. *Glumes* usually very dissimilar, lower minute, hyaline, upper almost equalling the spikelet, 5–7-nerved, rarely both more or less similar and minute, or the lower nerved like the upper, although smaller. *Valves* equal, lower (outer) firm chartaceous, 3–9-nerved, upper thinner, 5-nerved. *Pales* almost equalling the valves, 2-keeled. *Lodicules* 2, obliquely quadrate, nerved. *Stamens* 3. *Styles* free or almost so, very slender; stigmas long, plumose, laterally exserted. *Grain* broadly elliptic-oblong, slightly compressed from the back, tightly enclosed; hilum punctiform, basal; embryo about $\frac{1}{2}$ the length of the grain.

Creeping or prostrate branched perennials with ascending culms; sheaths strongly compressed, lower more or less flabellate; blades mostly obtuse; spikes terminal and often also lateral from the upper leaves, tough or at length breaking up at the joints.

Species 5, mainly on the shores of the tropical and subtropical seas.

1. S. glabrum (Trin. Fund. Agrost. 176); culms ascending, prostrate or creeping and frequently rooting at the nodes, often very long, strongly compressed, glabrous, smooth, many-noded; leaves glabrous or more or less hairy at the mouth of the sheath or at its junction with the blade, lower crowded at the base of the branches, more or less flabellate, followed often by a pseudo-opposite pair of and

2–4-distant leaves; sheaths strongly compressed, keeled, pallid, soon thrown aside, lower persistent; ligule a fringe of very short hairs; blades linear, obtuse, 1–6 in. by 1–3 lin., first folded, then flat or at length involute, rather firm, glaucous or light green, smooth; false spikes terminal and lateral from the upper leaves, erect, stiff or curved, 1–4 in. long, compressed, glabrous; rhachis linear, entire, 1–2½ lin. broad, with or without transverse depressed lines on the back indicating the joints, often hollowed out in front; branches very short, more or less sunk in the hollows or adpressed to the margin of the rhachis, compressed or angular, often very stout; spikelets solitary from the inner side of the base of the branch, sunk in the adjoining hollow, or 2–5 crowded along the branch, lanceolate-oblong to oblong, acute, 2½ lin. long, pallid; lower glume hyaline, very short, broad, truncate, nerveless, upper ovate-oblong, concave, almost as long as the spikelet, 7-nerved, firmly membranous; lower floret ♂; valve lanceolate-oblong to oblong, acute, equalling the spikelet, chartaceous, faintly 7–9-nerved, dorsally flattened, somewhat rough; anthers 1¼–1½ lin. long; upper floret ⚥; valve similar to the lower, but more acute, firmly membranous, 5-nerved. *Trin. Gram. Pan.* 60; *Pan. Gen.* 102, *and in Mém Acad. Pétersb. sér.* vi. iii. 190 (*in part*); *Nees in Linnæa*, vii. 273; *Steud. Syn. Pl. Glum.* i. 118 (*excl. the two last syn.*); *Doell in Mart. Fl. Bras.* ii. ii. 300, *t.* 39 (*excl. several syn.*); *not Hook. f. Fl. Brit. Ind.* vii. 90. *S. americanum, Schrank, Plant. Rar. Hort. Monac.* 98, *t.* 8; *Kunth, Enum.* i. 138. *S. dimidiatum, Brongn. Bot. Voy. Coq.* 127, *and Durand & Schinz, Consp. Fl. Afr.* v. 787 (*in part*). *S. swartzianum, Nees, Fl. Afr. Austr.* 62 (*var.* α, *and var.* β *in part*), *not Peters, Reise Mossamb. Bot.* 549. *Panicum dimidiatum, Linn. Spec. Pl.* 57 (*partly*), *not Mant.* 323. *Rottbœllia dimidiata, Linn. f. Suppl.* 114; *Thunb. Prod.* 23; *Fl. Cap. ed. Schult.* 118. *R. tripsacoides, Lam. Ill.* i. 205, *t.* 48, *fig.* 1*b*. *R. compressa, Beauv. Agrost. t.* xxi. *fig.* 8, *not Linn. f.*

SOUTH AFRICA: without precise locality, *Thunberg! Bergius!*

COAST REGION: Cape Div.; near Capetown and Table Mountain, *Thunberg! Ecklon*, 17! *Drège! Burchell*, 22! *MacGillivray*, 252! *Schlechter*, 194! Camps Bay, *Burchell*, 322! by the sea shore in Simons Bay, *MacGillivray*, 389! near Klein Constantia, *Wolley Dod*, 1989! Uitenhage Div.; banks of the Zwartkops River, below 100 ft., *Drège! Ecklon & Zeyher! Thunberg!* banks of the Sunday and Coega Rivers, *Drège.* Port Elizabeth Div.; near Port Elizabeth, *E.S.C.A. Herb.*, 117! 194! *Burchell*, 4312! Bathurst Div.; Kowie River, *MacOwan*, 1357!

EASTERN REGION: Natal; Durban Flat, *Buchanan*, 16! and without precise locality, *Gerrard*, 678!

Throughout the hot parts of America, in West Africa, the Sandwich Islands and in Australia.

The form having solitary spikelets sunk in the hollows of the rhachis, on which the species was originally based, is prevalent in America. It occurs in the Cape side by side with the other extreme which has branches, each bearing 3–5 spikelets, and being merely adpressed to the sides of the scarcely hollowed and narrow rhachis; and the two are there connected by a complete series of intermediate stages.

XX. ANTHEPHORA, Schreb.

Spikelets lanceolate to ovate-oblong, subsessile in deciduous squarrose clusters of 4–6, on the very short branches of a cylindric apparently simple spike; lower floret reduced to an empty valve, upper ⚥ or ♂ with or without the rudiments of an ovary. *Glumes* very dissimilar, lower always turned away from the centre of the cluster, coriaceous, oblong to lanceolate, distinctly nerved on the inner side, upper hyaline, setaceously subulate from a small ovate base. *Valves* more or less equal, usually shorter than the lower glume; lower oblong, hyaline, 5–7-nerved; upper similar, but firmer and 3-nerved. *Pale* equalling the valve and of similar substance, 2-nerved. *Lodicules* 0. *Stamens* 3; filaments fleshy at the base when young. *Ovary* lanceolate-oblong; styles free or nearly so, very slender; stigmas plumose, slender, exserted from or near the apex. *Grain* oblong-ellipsoid, enclosed in the spikelet; hilum punctiform; embryo large, equalling about $\frac{1}{2}$ the length of the grain.

Perennial or annual; culms tufted and erect, or ascending from a prostrate base; ligules large scarious; spikelets in subspicate clusters with their respective lower glumes turned outwards so as to form a false involucre, usually with gaps between their bases.

Species 5; in tropical Africa, South Africa, South Persia and tropical America.

Perennial; lower (involucral) glume minutely tubercled and villous outside; lower valve 5–7-nerved ... ... (1) **pubescens.**
Annual; lower (involucral) glume glabrous except at the base; lower valve 3-nerved ... ... (2) **undulatifolia.**

1. **A. pubescens** (Nees, Fl. Afr. Austr. 74); perennial, densely tufted; culms erect or subgeniculate, rather slender, 1–2 ft. high, glabrous, smooth, 2-noded; leaves mainly crowded at the base; basal sheaths about 1 in. long, pallid, persistent, glabrous, sometimes scantily bearded at the mouth, or more or less pubescent to villous all over; upper sheaths 3–5 in. long, usually quite glabrous; ligules oblong, obtuse, up to 3 lin. long; blades linear, long tapering to a subsetaceous point, 2–6 in. by $1\frac{1}{2}$–2 lin., flaccid, flat, glabrous or pubescent above, slightly rough, margins cartilaginous, almost smooth, wavy; false spike cylindric, $2\frac{1}{2}$–4 in. by 2–3 lin., villous; rhachis very slender, obscurely wavy, glabrous; spikelets in subsessile clusters of about 5 which are bearded at the base, $2\frac{1}{2}$–5 lin. long; lower glume lanceolate, setaceously acuminate, $2\frac{1}{2}$–5 lin. long, coriaceous, minutely tubercled and villous outside, 3–4-nerved, upper setaceously subulate from a short ovate base, hyaline, ciliate, $1\frac{1}{2}$–2 lin. long, 1-nerved; lower valve oblong, acute, hyaline, 5–7-nerved, dorsally flattened, finely pubescent, long ciliate from tubercled-based hairs in the upper half; upper valve with a ⚥ or in the 2 inner spikelets (facing the rhachis) with a ♂ flower, oblong, acute, 3-nerved, glabrous; anthers $1\frac{1}{2}$ lin. long. *Durand & Schinz, Consp. Fl. Afr.* v. 732; *Hack. in Bull. Herb. Boiss.* iv. *App.* iii. 12. *Tripsacum pubescens, Lichtenst. ex Nees, l.c. Cenchrus pubescens, Steud. Nomencl. ed.* ii. i. 317.

KALAHARI REGION: Griqualand West; between the Orange River and Laauwwaters Kloof, *Lichtenstein;* between Griqua Town and Witte Watter, *Burchell*, 1984! Basutoland; Leribe, *Buchanan*, 134! Bechuanaland; on the rocks at Chue Vlei, *Burchell*, 2387! Transvaal; near Lydenburg, *Atherstone!*

Also in Hereroland.

Closely allied to *A. Hochstetteri* from tropical North-east Africa, which has, however, longer, more slender spikes and an altogether more straggling habit. Hackel's *A. Schinzii* from Amboland (German South-west Africa) is perhaps only a particularly villous weak state of *A. pubescens.*

2. A. undulatifolia (Hack. in Bull. Herb. Boiss. iv. App. iii. 12); annual, dwarf; culms fascicled, erect, about 3 in. long, sheathed to beyond the middle, slender, glabrous, smooth; sheaths lax, strongly striate, sparsely beset with long fine spreading hairs, particularly towards the margins; ligules membranous, obtuse, 1½ lin. long; blades linear-lanceolate from a rounded base, acute, 1–2 in. by 2½–3 lin., soft, sparsely pubescent, long hairy towards the base, margins cartilaginous, strongly wavy; false spike cylindric, about 1 in. long; rhachis slender, slightly wavy, scaberulous, minutely bearded at the nodes; spikelets in subsessile clusters of 5–4, bearded at the base, 5 lin. long, outer ♂ or one of them reduced to the lower glume, inner 2 ☿; lower glume lanceolate, acutely acuminate, 5 lin. long, coriaceous, silky villous at the base, otherwise glabrous, 5- to many-nerved, contracted at the base so as to form oblong pores with the adjoining outer glumes; upper glume setaceously subulate, ciliolate, over 1 lin. long, 1-nerved; lower valve dorsally flattened, oblong, obtuse or truncate, 2 lin. long, hyaline, 3-nerved, long ciliate above; upper valve similar to the lower in the ♂ (but glabrous), subchartaceous, oblong-lanceolate, acute in the ☿; pale 0 in the ♂, equal to the valve in the ☿; anthers 1¼ lin. long; styles free or almost so, about 1½ lin. long; stigmas 2½ lin. long.

WESTERN REGION: Great Namaqualand, in stony places near Rehoboth, *Fleck!*

The only specimen which I have seen (in Hackel's Herbarium), is a small young plant, which, when collected, was still in connection by means of a lengthened internode with the cluster from which it sprang, and consists of a single flowering and several young leafy shoots, fascicled on and rooting from the first node.

XXI. TRICHOLÆNA, Schrad.

Spikelets oblong, laterally compressed, more or less gaping, panicled, deciduous from capillary pedicels; lower floret usually ♂ or barren, upper ☿. *Glumes* very dissimilar or at least unequal, lower reduced to a minute scale or obsolete, very rarely ½ the length of the spikelet, somewhat remote from the upper; upper glume membranous, emarginate, muticous or finely mucronate or aristate from the sinus, 5-nerved, usually hairy. *Valves* very dissimilar, lower like the upper glume with a hyaline 2-nerved subequal pale, upper much smaller, thinly chartaceous, glabrous, shining, obtuse or subemarginate, obscurely 5-nerved, with an equal 2-nerved pale of similar substance. *Lodicules* 2, very small. *Stamens* 3. *Style* free, slender; stigma densely plumose, laterally exserted. *Grain* oblong-ellipsoid, closely embraced by the valve and pale; hilum basal, punctiform; embryo about ½ the length of the grain.

Perennial, rarely annual; blades linear to setaceous; ligule a fringe of hairs; panicles open or contracted, often much divided, with capillary branchlets and pedicels, the latter with thickened tips; spikelets often completely enveloped by soft shining hairs, rarely pubescent or quite glabrous.

Species 10–12, chiefly in the dry and hot countries of the Old World.

Spikelets pubescent or villous:
  Upper glume and lower valve more or less gibbous at or below the middle, 2–3 lin. long:
    Culms from compact tufts; blades setaceous; panicle narrow, more or less spike-like; tips of pedicels shortly hairy; spikelets silky villous ... ... (1) **setifolia.**
    Culms usually rather loosely fascicled; blades flat, rarely convolute; panicle open (at least temporarily); tips of pedicels with long hairs:
      Spikelets silky villous ... ... (2) **rosea.**
      Spikelets adpressedly and scantily hairy ... (3) **brevipila.**
  Upper glume and lower valve not gibbous, 1 lin. long or rather longer:
    Spikelets loosely villous; lower glume about ½ as long as the spikelet ... ... (4) **capensis.**
    Spikelets pubescent; lower glume almost microscopic (5) **arenaria.**
Spikelets quite glabrous ... ... (6) **glabra.**

1. **T. setifolia** (Stapf); culms in compact tufts, with numerous intravaginal innovation shoots, erect or geniculate, ¾–1½ ft. long, rather slender, glabrous, 2–3-noded, upper internodes more or less exserted; leaves mostly crowded at the base; lower sheaths compressed, keeled, firm, striate, hairy to tomentose, at least towards the base and the margins (hairs spreading or adpressed), at length often glabrescent, persistent, upper glabrous except at the villous nodes; ligule a dense fringe of very short hairs; blades filiform, setaceously convolute, flexuous, 4–8 in. (rarely 12 in.) long, glabrous or hairy; panicle contracted, linear, often spike-like, usually flexuous or nodding, 2–4 in. long, silky; rhachis slender, scabrid or hairy in the upper part and below the nodes; branches solitary or 2-nate, filiform, flexuous, loosely divided almost from the base, erect, like the subcapillary branchlets and pedicels, usually dark and finely hairy, tips slightly thickened with slightly longer hairs; spikelets oblong, 2–2½ lin. wide, very densely villous with shining white or purple hairs, bearded at the base; lower glume oblong, obtuse, less than ½ in. long, hyaline, quite hidden by hairs; upper equalling the spikelet, subchartaceous, oblong, minutely emarginate with a mucro or short awn from the sinus, 5-nerved, densely villous with the hairs increasing in length from the base to beyond the middle, then glabrous towards the apex except along the ciliate margins, hairs usually adpressed and exceeding the spikelet by 1–1½ lin.; lower floret ♂ or barren; valve equal and very similar to the upper glume; pale hairy; ⚥ floret oblong to linear-oblong, 1–1¼ lin. long,

glabrous, shining; valve obtuse, subemarginate, faintly 5-nerved, membranous; anthers $\frac{3}{4}$ lin. long. *T. rosea, Nees, Cat. Sem. Hort. Vratisl.* 1835, *and in Linnæa,* xi. *Lit. Ber.* 129, *and Fl. Afr. Austr.* 17 (*in part*).

COAST REGION: Caledon Div.; Baviaans Kloof, near Genadendal, *Drège!* Uniondale Div.; between Avontuur and the sources of the Keurbooms River in Lange Kloof, *Burchell*, 5038! on a rocky hill near the Groot River in Lange Kloof, *Burchell*, 5018! Uitenhage Div.; without precise locality, *Ecklon & Zeyher*, 835! Alexandria Div.; near Enon, 1000–2000 ft., *Drège!* and in the Zuurberg Range, 2000–3000 ft., *Drège!* Albany Div.; near Grahamstown, 2000 ft., *MacOwan*, 1312! Fish River Heights, *Hutton!* Komgha Div.; near Komgha, *Flanagan*, 946! Queenstown Div.; stony places near Shiloh, 3500 ft., *Baur*, 782! 928! Fincham's Nek, near Queenstown, 4000 ft., *Galpin*, 2377!
CENTRAL REGION: Somerset Div.; without precise locality, *Bowker*, 183! Graaff Reinet Div.; Sneeuwberg Range, 4000 ft., *Bolus*, 1962! Aliwal North Div.; "Cis Garipina," *Zeyher!*
KALAHARI REGION: Griqualand West; Klip Fontein, *Burchell*, 2618! Transvaal; Houtbosch, *Rehmann*, 5722! Johannesburg; common, *E.S.C.A. Herb.*, 306! sources of the Limpopo, *Nelson*, 27*! Pretoria, at Wonderboom Poort, *Rehmann*, 4495!
EASTERN REGION: Tembuland; near Bazeia, 2000 ft., *Baur*, 326! Natal; at the foot of Table Mountain, *Krauss*, 382! by the Umbloti River at Oakfort (Oakford?), *Rehmann*, 8456! Biggarsberg Range, *Rehmann*, 7036! Inanda, *Wood*, 1595! Pietermaritzburg, 2000–3000 ft., *Sutherland*, Umpumulo, *Buchanan*, 298 in part! and without precise locality, *Buchanan*, 86!

The name was proposed by Professor Hackel for Rehmann's specimen, 8456.

2. **T. rosea** (Nees, Cat. Sem. Hort. Vratisl. 1835, and in Linnæa, xi. Lit. Ber. 129, in part); perennial or annual; culms laxly fascicled, rarely densely tufted, usually geniculate, 1–1$\frac{1}{2}$ ft. long, simple or scantily branched below, terete, glabrous or sometimes hirsute from tubercled-based hairs, 3–4-noded; sheaths terete or the lower slightly compressed, glabrous except at the villous nodes or hirsute from usually tubercled-based hairs; blades linear, long tapering to a subsetaceous point, 2–8 in. by 1–4 lin., rarely broader or narrower, flat, often spreading and rather flaccid, rarely convolute, glabrous or sometimes scantily hairy, rough above; panicle oblong to ovoid, lax or contracted, 3–6 in. long, straight or flexuous; rhachis slender, finely scaberulous; branches fascicled or 2-nate, finely filiform, undivided below or laxly branched almost from the base; branchlets and pedicels finely capillary, flexuous, glabrous or more or less hairy, scaberulous, tips thickened with long fine hairs (up to 2 lin. long); spikelets oblong, about 2 lin. long, villous from shining white or purple hairs, shortly bearded at the base; lower glume oblong, obtuse, about $\frac{1}{2}$ lin. long, hyaline, almost hidden by hairs; upper equalling the spikelet, semi-ovate, more or less gibbous below the middle and slightly narrowed into an oblong beak, obtusely 2-lobed or emarginate, with or without a mucro or a short fine awn from the sinus, subchartaceous, often olive-brown, villous with the hairs increasing in length from the base to beyond the middle, then glabrous except the ciliate margins; hairs often springing from minute or sometimes coarse and partly confluent tubercles, adpressed or sometimes spreading, usually exceeding the tips of the glume;

lower floret ♂ or barren; valve equal and very similar to the upper glume; pale more or less hairy; ⚥ floret 1¾ lin. long, elliptic-oblong, obtuse, glabrous; valve emarginate, faintly 5-nerved, membranous; anthers over ¾ lin. long; grain obovoid, brown, ¾ lin. long. *Nees, Fl. Afr. Austr.* 17 (*in part*); *Durand & Schinz, Consp. Fl. Afr.* v. 770; *K. Schum. in Engl. Pfl. Ost-Afr. C.* 104; *Hack. in Bull. Herb. Boiss.* iv. *App.* iii. 15. *T. tonsa, Nees, l.c.* 6; *Durand & Schinz, l.c. T. fragilis, A. Braun in Flora,* 1841, 275 (*name only*); *Durand & Schinz, l.c.* 769. *T. sphacelata, Benth. in Hook. Niger Fl.* 559; *Durand & Schinz, l.c.* 770. *T. grandiflora, Hochst. in Flora,* 1841, i. *Intell.* 19 (*name only*); *A. Rich. Tent. Fl. Abyss.* ii. 445; *Engl. Hochgebirgsfl. Trop. Afr.* 121; *Hack. in Engl. Jahrb.* xi. 400, *and in Bull. Herb. Boiss.* iv. *App.* iii. 15; *Durand & Schinz, l.c.* 769. *T. dregeana, Durand & Schinz, l.c.; Hack. in Bull. Herb. Boiss. l.c. T. ruficoma, Durand & Schinz, l.c.* 770; *Hack. in Bull. Herb. Boiss. l.c. Rhynchelythrum dregeanum, Nees, Fl. Afr. Austr.* 64. *R. ruficoma, Hochst. ex Steud. Syn. Pl. Glum.* i. 120. *Saccharum grandiflorum and S. sphacelatum, Walp. Ann.* iii. 792, 793. *Panicum tonsum, P. roseum, P. sphacelatum, P. insigne, Steud. l.c.* 92. *P. Braunii, Steud. l.c.* 93; *Oliv. in Trans. Linn. Soc.* xxix. 170. *Monachyron roseum and M. tonsum, Parl. Fl. Ital.* i. 131, *in obs.*

SOUTH AFRICA: without precise locality, *Drège*, 4319! 4320! 4321!

COAST REGION: Albany Div.; without precise locality, *Williamson*, 114!

CENTRAL REGION: Graaff Reinet Div.; hillsides near Graaff Reinet, 4000 ft., *Bolus*, 251! Albert Div.; near Aliwal North, *Cooper*, 1388!

WESTERN REGION: Great Namaqualand; Amhub, *Schinz*, 628. Keetmanshoop, *Fenchel*, 4.

KALAHARI REGION: Griqualand West; Herbert Div.; near Douglas, *Orpen*, 251! Hay Div.; Asbestos Mountains, *Burchell*, 2047/1! Klip Fontein, *Burchell*, 2150! Orange Free State; Seven Fonteins, *Burke!* Caledon River, *Burke*, 198! Bechuanaland; Maadji Mountain, *Burchell*, 2363! banks of the Moshowa River between Takun and Molito, *Burchell*, 2314! at Kosi Fontein, 2564! 2578! near Kuruman, *Burchell*, 2184! stony plains near Groot Kuil, *Marloth*, 998! Transvaal; Houtbosch, *Rehmann*, 5723! near Lydenburg, *Wilms*, 1674a!

EASTERN REGION: Griqualand East; near Clydesdale, *Tyson*, 3103! Natal; near Durban, *Plant*, 98! Draakensberge, *Rehmann*, 7175! near Umpumulo and Reit Vlei, common, *Buchanan*, 298 in part! and without precise localities, *Gerrard*, 461! *Buchanan*, 122! Delagoa Bay, *Scott!*

Throughout tropical Africa, in Madagascar and South Arabia.

An extremely variable plant. Nees confounded two different plants under *T. rosea.* I have maintained the name for the species having generally flat leaves, laxer panicles, very fine capillary branchlets and long hairs at the tips of the pedicels, as this is the species of which he distributed seeds from the Breslau Botanic Garden. It is also represented by Drège, 4319 and 4321, which numbers he quotes under *T. rosea.* His *T. tonsa* was supposed to differ in the shorter hairs of the spikelets, weaker culms and more glaucous leaves. These characters are very little marked in Drège, 4320 (which is the type), and fail entirely when tested by the ample material at Kew. The broad-leaved state which is represented by specimens grown in various European gardens, seems to occur spontaneously on road-sides and in cultivation, and is probably truly annual, while another extreme, which seems to form small, dense tufts and has slender culms, and narrow, sometimes convolute leaves, appears to be characteristic of dry and hard ground. According to Nees, the area of *T. rosea* extends towards the south-west as far as "Swellendam." I have not seen any specimens from there nor from various other localities which he quotes. They may be either *T. rosea* or *T. setifolia*.

3. **T. brevipila** (Hack. in Verhandl. Bot. Ver. Prov. Brandenb. xxx. 143); annual; culms geniculate, branched below, slender, 8–10 in. long, terete, glabrous; sheaths subterete, uppermost often widened, finely striate, glabrous or scantily hairy, partly from tubercled-based hairs, minutely villous at the nodes; ligule a dense fringe of short hairs; blades linear, tapering to a fine point, 2–2½ in. by 1–1½ lin., flat, flaccid, scantily pubescent and more or less rough on the upper side; panicle oblong, 2–4 in. long, scanty, lax; rhachis very slender, scaberulous; branches 2-nate or solitary, sparingly branched from near the base, capillary like the pedicels, scaberulous or almost smooth, tips of the latter thickened with a few long fine hairs; spikelets oblong, 2¼–3½ lin. long, olive-brown, minutely hairy, shortly bearded at the base; lower glume oblong, obtuse, ¾ lin. long, glabrous, upper equalling the spikelet, chartaceous, obliquely ovate-lanceolate, slightly gibbous below the middle, obtusely or subacutely 2-toothed with a fine awn from the sinus about as long as the spikelet, shortly sparingly and adpressedly hairy, 5-nerved; lower floret barren; valve equal and very similar to the upper glume; pale linear-lanceolate, slightly shorter than the valve; anthers rudimentary; ⚥ floret slightly over 1 lin. long, elliptic-oblong, obtuse, glabrous; valve emarginate, faintly 5-nerved, membranous. *Durand & Schinz, Consp. Fl. Afr.* v. 769; *Hackel in Bull. Herb. Boiss.* iv. *App.* iii. 15.

WESTERN REGION: Great Namaqualand, Gamochab, *Schinz*, 631!

4. **T. capensis** (Nees, Cat. Sem. Hort. Vratisl. 1835, and in Linnæa xi. Lit. Ber. 130); perennial, very glaucous; culms erect or geniculately ascending from a short rhizome or a suffrutescent base, ¾–1 ft. high, pruinose and very finely silky below the nodes; sheaths glabrous, tight or the uppermost widened, lower exceeding the internodes, whitish, subpersistent; ligule a short fringe of hairs; blades linear, acute, 2–3 in. by 1¼–2 lin., flat or involute, softly pubescent on the upper side, otherwise glabrous; panicle oblong to ovate, lax, open, 3–4 in. long; rhachis slender, smooth; branches solitary, rather distant, 1–2½ in. long, loosely and repeatedly divided from the base, very rarely undivided in the lower half; branchlets and pedicels capillary, smooth, the latter ½–1 lin. long; spikelets oblong, 1 lin. long or slightly longer, villous, pallid or more or less purplish; glumes membranous, lower oblong, obtuse, about ½ lin. long, 1-nerved, loosely villous, upper lanceolate-oblong in profile, almost as long as the spikelet, obtuse or obtusely emarginate, finely 5-nerved, villous all over; lower floret ♂: valve very similar to the upper glume, slightly longer and broader, minutely emarginate and mucronate; pale subequal to the valve, ciliate on the keels; ⚥ floret oblong, obtuse, ¾ lin. long; valve rigidly membranous, shining, obscurely 5-nerved; anthers ½ lin. long. *Nees, Fl. Afr. Austr.* 19; *Durand &*

Schinz, Consp. Fl. Afr. v. 769. *Panicum capense, Lichtenst. in Roem. & Schult. Syst.* ii. 457; *Mant.* ii. 591; *Nees in Linnæa*, vii. 275; *Kunth, Enum.* i. 132; *Trin. Pan. Gen.* 209, *and in Mém. Acad. St. Pétersb. sér.* vi. iii. 297; *Steud. Syn. Pl. Glum.* i. 92. *P. dasyanthum, Lichtenst. ex Nees, Fl. Afr. Austr. l.c.*

CENTRAL REGION: Prince Albert Div.; at Zwartbulletje near the Gamka River, 2500–3000 ft., *Drège*, 899! Calvinia Div.; between Lospers Plaats and Springbok Kuil River, *Zeyher*, 1786!

KALAHARI REGION: Orange Free State; Jackals Fontein, *Burke*! Bechuanaland (?) Jan Bloms Fontein, *Lichtenstein*.

5. **T. arenaria** (Nees, Cat. Sem. Hort. Vratisl. 1835, and in Linnæa, xi. Lit. Ber. 130); perennial; culms branched below, geniculate, slender, rigid, over 1 ft. long, glaucous, rough below the nodes; leaves glaucous; sheaths terete, tight, glabrous except at the often very minutely villous nodes, finely striate, the upper shorter than the internodes; ligule a short fringe of hairs; blades linear, very narrow, tapering to an acute point, 2–3½ in. by 1–1½ lin., rather rigid, involute or convolute, sparingly hairy and scaberulous on the upper side, otherwise glabrous; panicle ovoid, lax, open, 2–5 in. long; rhachis very slender, smooth; branches fascicled or 2-nate, rather distant, filiform, 1–2 in. long, often undivided for ⅓–½ their length from the base, then loosely divided (the longer repeatedly); branchlets and pedicels capillary, smooth, the latter 1–2 lin. long; spikelets oblong, 1–1¼ lin. long, finely pubescent, often purplish; lower glume an extremely minute ciliate scale, upper lanceolate-oblong, equalling the spikelet, membranous, minutely or obscurely emarginate, mucronulate or muticous, finely 5-nerved, pubescent all over; lower floret ♂: valve equal and similar to the upper glume, but broader and more truncate; pale subequal, minutely 2-lobed, ciliolate above, keels glabrous; ⚥ floret elliptic or ovate-oblong, ⅞ lin. long; valve obtuse, rigidly membranous, shining, very faintly 5-nerved; anthers ¾ lin. long; grain elliptic-oblong, ½ lin. long. *Nees, Fl. Afr. Austr.* 20; *Durand & Schinz, Consp. Fl. Afr.* v. 769; *Hack. in Bull. Herb. Boiss.* iv. *App.* iii. 15. *Panicum ammophilum, Steud. Syn. Pl. Glum.* i. 92.

VAR. β, **glauca** (Stapf); spikelets glabrescent towards the base; lower glume very obscure, quite glabrous. *Anthænantia glauca, Hack. in Verhandl. Bot. Ver. Prov. Brandenb.* xxx. 237; *in Bull. Herb. Boiss.* iv. *App.* iii. 13.

WESTERN REGION: Little Namaqualand; sand-hills between Lekkerring and Kaus, *Drège*, 2568! Var. β: Great Namaqualand; Tiras, *Schinz*, 673! sandy plains between Ausis and Kuias, *Schenk*, 80.

6. **T. glabra** (Stapf); perennial; culms densely fascicled, erect or subgeniculate, slender, branched and wiry below, 1–1½ ft. long, glabrous, smooth, 5–7-noded; upper internodes at length exserted; sheaths tight, terete, striate, lower subpersistent, often finely hairy; ligule a dense fringe of fine hairs; blades linear, long tapering to a fine point, 2½–4½ in. by 1–2½ lin., flat or involute, somewhat rigid and glaucous, finely pubescent or glabrous, smooth; panicle oblong or obovoid, open or contracted, lax, 2½–6 in. long; rhachis slender,

smooth; branches filiform, smooth, 2-nate or solitary, the longest 2–4 in. long, very laxly and repeatedly divided from the base, or the strongest simple for ½–1 in.; branchlets and pedicels very delicate, capillary, flexuous, the latter 1–2 lin. long; spikelets elliptic-oblong, 1 lin. long, glabrous, pallid or tinged with purple; lower glume an extremely minute scale, upper oblong almost equalling the spikelet, emarginate, membranous, faintly 5-nerved; lower floret ♂: valve very similar to the upper glume, but slightly longer and broader; pale equal to the valve, glabrous; ☿ floret oblong, ⅔ lin. long; valve rigidly membranous, shining, obscurely 5-nerved; anthers ½–⅝ lin. long.

EASTERN REGION: Natal; sandy valley near the Tugela River, 1000 ft., *Buchanan*, 259! "Holpmaakar Spruit, Potgeiter's Rust" (in Natal ?), *Nelson*, 23*!

XXII. MELINIS, Beauv.

Spikelets linear-oblong, small, laterally compressed, panicled, deciduous from the capillary pedicels; lower floret reduced to an empty valve; upper ☿. *Glumes* very dissimilar, lower reduced to a minute scale, upper membranous, shortly 2-lobed, mucronulate from the sinus, prominently 7-nerved, glabrous, rarely hairy. *Valves* very dissimilar; lower rather like the upper glume, but more deeply lobed and usually with a fine awn from the sinus, 5-nerved; upper smaller, very thin and rigidly membranous, minutely 2-lobed, shining, obscurely 3–1-nerved with an equal, obscurely 2-nerved pale of similar substance. *Lodicules* 2, very delicate. *Stamens* 3. *Styles* free, slender; stigmas plumose, laterally exserted. *Grain* unknown.

Perennial or annual (?); culms ascending from a prostrate, very slender, many-noded, branched base; ligules reduced to a fringe of short hairs; panicle contracted narrow, much divided, with capillary branchlets and pedicels; spikelets usually purple, at least at the tips.

Species 1, in Brazil, tropical and subtropical Africa and Madagascar.

Very closely allied to *Tricholæna*.

1. **M. minutiflora** (Beauv. Agrost. 54, t. xi. f. 4) var. **pilosa** (Stapf); perennial or annual (?); culms 1–3 ft. high, more or less hirsute except at the upper nodes; leaves loosely hirsute to tomentose; sheaths tight, terete, striate, finely tubercled; blades linear to linear-lanceolate, long tapering to a very fine point, 2–6 in. by 2½–4½ lin., rather rigid, flat or involute; panicle linear to linear-oblong, contracted or almost spike-like, 4–8 in. long, erect, stiff or rather flexuous; rhachis slender, angular, smooth below; branches 2-nate or the lower in scanty fascicles, erect, up to 3 in. long, distantly and repeatedly divided from the base; branchlets and pedicels capillary, flexuous, puberulous, the latter very unequal, up to almost 2 lin. long, with or without white stiff fine hairs below the tips; spikelets 1 lin. long, very minutely bearded at the base; lower glume very minute, oblong, obtuse, nerveless, upper equalling the spikelet, linear-oblong, like the lower valve, more or less hairy above the base, hairs white, very fine;

awn of the lower valve very slender, scaberulous, up to 4 lin. long; ⚥ floret $\frac{7}{8}$ lin. long, glabrous; anthers $\frac{1}{2}$ lin. long.

EASTERN REGION: Natal; near Umpumulo hills, 2000–2500 ft., *Buchanan*, 299!

Also in Nyasaland and on Mt. Ruwenzori.

The typical form which is very common in Brazil and occurs also in tropical Africa and Madagascar, only differs in being completely glabrous, and in the occasional absence of awns.

XXIII. ARUNDINELLA, Raddi.

Spikelets small, $2\frac{1}{2}$ lin. to less than 1 lin. long, acute or acuminate, panicled, continuous with or imperfectly jointed on the pedicels; rhachilla disarticulating between the valves, glabrous, not produced beyond the upper floret. *Florets* 2, heteromorphous, lower ♂, rarely barren or ⚥, upper ⚥. *Glumes* persistent, acuminate, membranous, strongly 3–5-nerved, upper usually longer. Lower *valve* thin, equalling the lower glume or slightly longer, subacute or minutely truncate, 3–7 nerved; upper valve terete, rarely dorsally subcompressed, thin, finally cartilaginous, entire or minutely bifid, 3–7-nerved, minutely scaberulous, lobes sometimes produced into fine bristles; awn terminal or from between the lobes, short, kneed or straight, or 0; callus very short, obtuse. *Pales* linear, 2-keeled; flaps more or less auricled. *Lodicules* 2, cuneate, rather fleshy or 0 in the lower floret. *Stamens* 3 or 0 in the lower floret. *Ovary* glabrous, oblong or more or less arrested or 0 in the lower floret; styles distinct; stigmas plumose, laterally exserted. *Grain* oblong to ellipsoid, terete or dorsally subcompressed, tightly embraced by the valve and pale, free; hilum punctiform; embryo large; albumen hard.

Perennial, rarely annual; leaves various; panicles usually rather stiff; branches stiff, divided from the base.

About 25 species, mainly in tropical Asia and America.

1. **A. Ecklonii** (Nees, Fl. Afr. Austr. 80); perennial; culms erect, $2\frac{1}{2}$–4 ft., glabrous, 4–6-noded; sheaths rather tight, glabrous or ciliate along the margins or hirsute; ligules extremely short, ciliolate; blades linear, tapering to a long fine point, from a few inches to more than 1 ft. by 2–4 lin., flat or convolute, rigid, glabrous, margins rough; panicle linear-oblong to oblong, contracted or open, up to 1 ft. long; branches solitary or 2-, rarely 3–4-, nate, the longest up to 3 in. long, stiff, scabrid; spikelets subsecund, 2–$2\frac{1}{2}$ lin. long; lower glume ovate-oblong, mucronate, $1\frac{1}{2}$–2 lin. long, 3-, rarely 4–5-, nerved; upper glume ovate-lanceolate, 2–5 lin. long, 5-nerved; lower floret ♂; valve linear-oblong, subacute, $1\frac{3}{4}$–2 lin. long, sub-7-nerved; pale linear-oblong, $1\frac{1}{2}$ lin. long, keels winged below, scaberulous above; lodicules present; anthers $\frac{3}{4}$ lin. long; ⚥ floret: valve oblong, obscurely bifid, $1\frac{1}{2}$ lin. long, thin, minutely scaberulous; callus minutely 2-bearded; awn fine, column dark brown, $1\frac{1}{2}$ lin. long, bristle whitish, 1 lin. long; pale linear, $1\frac{1}{4}$ lin. long, scaberulous

between the keels; lodicules and stamens as in the ♂ floret. *Steud. Syn. Pl. Glum.* i. 114; *Durand & Schinz, Consp. Fl. Afr.* v. 735.

COAST REGION: George Div.; along the Nuakamma River, *Burchell*, 5106! Uitenhage Div.; Zwartkops River, and Mountains near the Van Stadens River, *Ecklon*; Uitenhage, *Zeyher*! Komgha Div.; banks of the Kei River, below 500 ft., *Drège*! Cathcart Div.; Old Kat Berg Pass, 5200 ft., *Galpin*, 2415! Queenstown Div.; Shiloh, along the river, 3500 ft., *Baur*, 883!

KALAHARI REGION: Transvaal; Johannesburg, *E.S.C.A. Herb.*, 303!

EASTERN REGION: Natal; on the banks of rivulets between Umzimkulu River and Umkomanzi River, *Drège*! Riet Vlei, 4000–5000 ft., *Buchanan*, 275! Umpumulo, near water, 2000 ft., *Buchanan*, 274! and without precise locality, *Gerrard*, 684! 685!

Very closely allied to *A. brasiliensis*, Raddi, a very polymorphic species ranging over tropical Asia and America. (See Hook. fil., Flora of Brit. India, vii. 73.)

A. rigida (Nees, Fl. Afr. Austr. 80) from the Kamies Bergen, Namaqualand, seems to be a hairy state of *A. Ecklonii*.

XXIV. TRICHOPTERYX, Nees.

Spikelets usually 2–18 lin. long, more or less lanceolate to linear, pedicelled, panicled; rhachilla disarticulating between (and less readily below) the valves, glabrous, not produced beyond the upper floret. *Florets* 2, heteromorphous; lower ♂, rarely barren; upper ⚥. *Glumes* persistent, membranous or subcoriaceous, unequal, 3- (very rarely 4–6-) nerved, glabrous or bristly from black glands. Lower *valve* membranous, 3- (very rarely 7-) nerved, with the innermost side-nerves much shorter; upper valve terete, membranous to cartilaginous, shortly bifid with the lobes sometimes produced into bristles, 5–9-nerved; awn from between the lobes, kneed, twisted below. *Pales* membranous, 2-keeled, narrower in the ⚥ floret and channelled between the stout keels. *Lodicules* 2, cuneate, usually very fleshy. *Stamens* 2, rarely 3, or 0 in the lower floret. *Ovary* oblong, glabrous, rudimentary in the ♂ floret; styles distinct, glabrous; stigmas plumose, long, laterally exserted. *Grain* obovoid to linear-oblong, grooved or almost terete, tightly embraced by its valve and pale, free; hilum linear, long; embryo large; albumen very hard.

Perennial, rarely annual grasses of very different habit; leaves more or less rigid; ligules a line of hairs; spikelets slender, from 2 lin. to 1½ in. long, scattered in mostly contracted, narrow panicles.

Species 15, in tropical Africa, extra-tropical South Africa, and in Madagascar.

Spikelets 3–7 lin. long; valve of ♂ floret pubescent, not bearded except at the callus:
  Valve of ♂ floret 3-nerved; culms erect, simple, never suffrutescent:
    Valve of ⚥ floret 7- to sub-5-nerved; callus 2-toothed at the base ... ... (1) **simplex.**
    Valve of ⚥ floret distinctly 9-nerved; callus very acute (2) **flavida.**
  Valve of ♂ floret sub-7-nerved; suffrutescent ... ... (3) **ramosa.**
Spikelets 2–3 lin. long; valve of ♂ long bearded below the middle; culms very slender, many-noded; leaves short, reflexed ... ... (4) **dregeana.**

1. T. simplex (Hack. ex Engl. Hochgebirgsfl. Trop. Afr. 129 in note); densely tufted; culms 2–3 ft. long, glabrous, 2- (rarely 3-) noded; leaves all but 2 or 3 near the base, sheathing the culms to ½ ft.; sheaths firm, tight, glabrous or more or less hispid, bearded at the nodes, the lowest tomentose at the base and finally splitting into rigid fibres; blades narrow linear, tapering to a long setaceous point, ½ ft. by 2 lin. or less, flat or convolute, rigid, hirsute or glabrescent; panicle erect, contracted, 6–10 in. long; rhachis glaucous; branches fascicled or geminate, erect, very unequal, the longest to 2–3 in. long, filiform, sparingly branched, scabrid or almost smooth; spikelets light brown, 5–6 lin. long; glumes glabrous, rarely sparingly bristly, the lower lanceolate-oblong, minutely subtruncate, 2½–3 lin. long, the upper lanceolate, produced into a linear convolute truncate beak, 5–7 lin. long; lower floret ♂: valve lanceolate, acute or acutely acuminate, usually shorter than the upper glume, glabrous or with a very few bristles along the outer nerves; pale linear-oblong, 3½ lin. long; stamens 2, anthers 1½–2 lin. long; ⚥ floret: valve oblong-linear, terete, 2½–3 lin. long (in this and the following species measured from the callus to the base of the awn), pubescent, faintly 7-nerved, shortly bifid; lobes acute; callus 2-toothed, villous, bearded, ½ lin. long; column of awn 4–5 lin. long; bristle 9–15 lin. long; pale lanceolate, acute, 3 lin. long; stamens as in the ♂. *Durand & Schinz, Consp. Fl. Afr.* v. 846. *K. Schum. in Engl. Pfl. Ost-Afr. C.* 109 (*partly*); *T. sp.*, *Benth. in Journ. Linn. Soc.* xix. 98. *Tristachya simplex*, *Nees, Fl. Afr. Austr.* 269; *Steud. Syn. Pl Glum.* i. 238.

VAR. β, **minor** (Stapf); culms 1 ft. long, 1-noded above the base, nodes minutely villous; blades convolute, setaceous, those of the barren shoots as long as the culms, glaucous, glabrous; panicle 3–6 in. long; spikelets 4½–5 lin. long, usually quite glabrous.

VAR. γ, **crinita** (Stapf ex Rendle in Cat. Afr. Pl. Welw. ii. i. 214); culms 1–2½ ft. long, 1-2-noded; blades convolute, subsetaceous, rarely flat, those of the barren shoots 1–1¼ ft. long, glaucous, glabrous, nodes glabrous, rarely silky; panicle 5-6 in. long; spikelets 5–6½ lin. long; lower glume 2–3½ lin. long, like the upper bristly along the nerves; valves of ♂ shorter than the upper glume, acutely acuminate, sub-5-nerved, bristly along the inner side-nerves.

VAR. δ, **sericea** (Stapf); culms 1–2½ ft. long, 2–3-noded; blades convolute, subsetaceous, rarely flat, hirsute or the uppermost glabrous, sheaths glandular-bristly or ultimately glabrescent, nodes bearded; panicle 6 in. long; spikelets 5 lin. long; glumes finely and densely glandular-bristly, the lower 3 lin., the upper 5 lin. long, rather thin; valve of ♂ as long as the upper glume, finely bristly above the middle.

KALAHARI REGION: Transvaal; on the top of Rhenoster Poort in Zoutpansberg Div., *Nelson*, 88*! near Lydenburg, *Wilms*, 1706! Var. *minor:* Orange Free State (?), *Buchanan!* Var. *sericea:* Transvaal; Houtbosch, *Rehmann*, 5663! 5664! Houtbosch Berg, by the Broeder Stroom, *Nelson*, 20*!

EASTERN REGION: Pondoland; between St. Johns River and Umtsicaba, *Drège*. Natal; between Umtentu River and Umzimkulu River, *Drège!* on a bare mountain slope at Umpumulo, *Buchanan*, 294! Var. *crinita:* Natal; Inanda, *Wood*, 1580! margins of ditches at Umpumulo, 2000 ft., *Buchanan*, 293! and without precise locality, *Buchanan*, 65! South Downs in Weenen County, 4000 ft., *Wood*, 4404! *Drège*, 4236, from Port Natal, is probably a state with glabrous spikelets. Its lower branches are nodding and over 3 in. long.

Also in tropical Africa.

I subjoin to this species a number of forms, which though well characterized in some specimens, are on the whole so intricately related to each other that my efforts to separate them satisfactorily as species have so far failed. Very similar forms occur in Madagascar, viz. *T. stipoides*, Hack., and *Stipa madagascariensis*, Baker.

2. T. flavida (Stapf in Kew Bulletin, 1897, 298); cæspitose; culms 1½ ft. long, glabrous or sparingly hairy above, 3–2-noded; leaves mostly basal; sheaths tight, lowest broad, whitish, softly and long villous at the base, not breaking up in fibres, upper glabrous except the shortly bearded nodes; blades narrow, linear, tapering to a long setaceous point, up to 6–8 in. by 1 lin., convolute, rather rigid, minutely puberulous above, glabrous or scantily hairy beneath; panicle 3–6 in. long, narrow, subflexuous; branches geminate or 3–4-nate, filiform, scabrid, scantily branched, the longest up to 2 in. long; spikelets brownish-yellow, 4–5 lin. long, glabrous; lower glume ovate, acute, 2½–3 lin. long; upper lanceolate, truncate, 4–5 lin. long; lower floret ♂; valve lanceolate, acute, sub-5-nerved; pale linear-lanceolate, 3 lin. long; stamens 2; anthers 1 lin. long; ⚥ floret: valve linear-oblong, 2 lin. long, pubescent, 9-nerved, shortly bifid; callus pungent, villous; awn about 1 in. long, fine, kneed at the middle; pale not quite 2 lin. long; stamens as in the ♂.

KALAHARI REGION: Transvaal; Pretoria, *Rehmann*, 4730! Magalies Bergen, at Derde Poort, *Nelson*, 75a!

3. T. ramosa (Stapf in Kew Bulletin, 1897, 298); suffrutescent, much branched, 1½ ft. high; stems very slender, smooth, hard, flowering branches 3-noded; leaves 3–4 at the base of the branches, the other 3 more or less remote; lowest sheaths rather wide and open, upper tight, all glabrous; blades linear, tapering to a very fine point, 3–4 in. by 1–1½ lin., flat or convolute, glabrous, smooth; panicles linear, 2–4 in. long, erect; branches filiform, scantily branched or simple, smooth, the longest up to 1½ in. long, all erect; spikelets pallid, 3–3½ lin. long, glabrous; lower glume lanceolate-oblong, acute, thin, 2–2½ lin. long; upper lanceolate, acute, thin, 3–3½ lin. long; lower floret ♂: valve lanceolate, acute, 7-nerved with the innermost nerves obsolete below the middle; pale lanceolate-oblong, 2¼ lin. long,; stamens 2; ⚥ floret: valve linear-oblong, 2–2½ lin. long, glabrous, 7-nerved, bifid; lobes acute; callus short, obtuse, bearded; awn fine, 4 lin. long, kneed below the middle; pale 2½ lin. long; stamens 2; grain ovoid-oblong, grooved, 1 lin. long; embryo ¾ lin. long.

KALAHARI REGION: Griqualand West, Hay Div.; Klip Fontein, *Burchell*, 2164! 2164/1!

4. T. dregeana (Nees in Lindl. Nat. Syst. ed. ii. 449, Trichopteria by error); culms ascending or erect from a slender creeping rhizome, branched below, 1–2 ft. high, very slender, glabrous, many-noded, internodes exserted; leaves equally distributed over the culms; sheaths tight, glabrous; blades narrowly linear, tapering to an acute point, usually involute, somewhat subulate, reflexed, ¾–1¼ lin. long, glabrous, margins cartilaginous; panicle narrow, 1–3 in. long, flaccid or erect and flexuous, contracted or open; rhachis filiform; branches 2–3-nate, capillary, almost smooth, branched from below the middle, up to 1 in. long, tips of pedicels with long fine hairs; spikelets light brown, 2–3 lin. long; lower glume ovate, subacute or obtuse, 1¼–1½ lin. long, glabrous, thin, upper lanceolate acuminate, acute, 2–2½ lin. long; lower floret ♂: valve like the upper glume, slightly longer; pale linear-oblong, 1½ lin. long; stamens 2, anthers ½ lin. long; ⚥ floret: valve oblong, 1 lin. long, deeply bifid, lobes produced into fine bristles 1½ lin. long, glabrous except a long beard on each side at the middle, sub-7-nerved; callus minute, obtuse, minutely bearded; awn very fine, 3 lin. long, column very short; pale 1½ lin. long; stamens 2. *Nees, Fl. Afr. Austr.* 339. *Danthonia Trichopteryx, Steud. Syn. Pl. Glum.* i. 244; *Durand & Schinz, Consp. Flor. Afr.* v. 855.

KALAHARI REGION: Transvaal, Lydenburg Distr., Spitzkop Goldmine, *Wilms*, 1705!

EASTERN REGION: Natal; common on wet banks, *Buchanan*, 291! between Umtentu River and Umzimkulu River, 500 ft., *Drège!* Zululand; in a swamp at Entumeni, *Wood*, 3995!

Also in the Shire Highlands.

XXV. TRISTACHYA, Nees.

Spikelets 5 lin. to 2 in. long, lanceolate to linear, sessile or shortly pedicelled in clusters of 3 at the tips of the branches of a raceme or panicle; rhachilla disarticulating between the valves and less readily or imperfectly below them, glabrous, not produced beyond the upper floret. *Florets* 2, heteromorphous, lower ♂, upper ⚥. *Glumes* persistent, membranous or subcoriaceous, more or less unequal, 3-nerved, glabrous or bristly from black glands. Lower *valve* membranous, 3-nerved, or 5–9-nerved with some of the nerves much shorter than the others; upper valve convolute, membranous or ultimately cartilaginous, bifid, 7- (rarely 5-) nerved, awned from between the lobes; awn kneed, twisted below. *Pales* membranous, hyaline, 2-keeled. *Lodicules* 2, cuneate, very fleshy. *Stamens* 3. *Ovary* obovoid to oblong, glabrous or the top hairy, rudimentary in the ♂ floret; stigmas plumose, long, laterally exserted. *Grain* obovoid to linear-oblong, grooved, tightly embraced by the valve and pale, free; hilum linear, long; embryo large; albumen very hard.

Perennial, generally coarse grasses, growing mostly in compact tufts; leaves rigid; ligule a line of hairs; triplets of spikelets large and few in racemes, or smaller and more numerous in scantily branched panicles, sometimes deciduous from the jointed peduncles.

Species 11–12, in Africa, Arabia and Beluchistan, and in tropical America.

Glumes glabrous ... ... (1) **Rehmanni.**
Glumes more or less hairy:
 Glumes equal, about 10 lin. long, lower with a submarginal row of large black bristle-bearing glands on each side, otherwise glabrous, upper pubescent ... ... (2) **biseriata.**
 Glumes unequal, 9–12 and 14–18 lin. long respectively, lower densely beset with minute black glands bearing soft spreading hairs along all 3 nerves, upper along the margins only... ... ... (3) **leucothrix.**

1. T. **Rehmanni** (Hack. in Bull. Herb. Boiss. iii. 384); compactly cæspitose; culms ½–3 ft., glabrous, rough, 1-noded at the middle or above; leaves mainly basal, sheathing the culm for 4–6 in., lowest sheaths very firm, rigid, persistent, densely tomentose below, glabrescent above, upper sparsely hirsute or glabrous, nodes villous or glabrescent; blades linear, tapering to a sharp point, 4–6 in. by 1–1½ lin. long, rigid, scabrid above, glabrous or hirsute, triplets of spikelets 1–1¼ in. long, 3–9 in erect contracted racemes; rhachis villous at the lowermost node; peduncles 2–3-nate, up to 1½ in. long, straight or flexuous, glabrous; pedicels 0; lower glume lanceolate, acute, 7–10 lin. long, firm, brown, glabrous, smooth, strongly 3-nerved; upper glume lanceolate, long subulate-acuminate, 12–15 lin. long, glabrous, smooth; lower floret ♂: valve lanceolate, subulate-acuminate, 10–12 lin. long, 5-nerved or 7-nerved above, glabrous, smooth; pale linear-oblong, 2-toothed, 9–10 lin. long, keels narrowly winged; anthers 2½–3 lin. long; ⚥ floret: valve linear-lanceolate, pubescent, 5-nerved, submarginal nerves rather strong, lobes 1-nerved, produced into fine bristles 10 lin. long; callus pungent, 1½ lin. long, villous; awn yellowish, 2½–3 in. long; pale oblong-linear, tip truncate spoon-shaped; top of ovary and styles pubescent. *T. glabra, Stapf in Kew Bulletin*, 1897, 294.

KALAHARI REGION: Transvaal; by the Mooi River, near Potchefstroom, *Nelson*, 31*! Johannesburg, *Barber!* Makapans Range, at Stryd Poort, *Rehmann*, 5383! (5884 according to Hackel).

A more glabrous form in the Shire Highlands.

2. T. **biseriata** (Stapf in Kew Bulletin, 1897, 295); culms slender, 1–1½ ft. long, glabrous, 1-noded; leaves mainly basal, lowest sheaths firm, rigid, persistent, narrow, densely tomentose below, glabrous or glabrescent above, upper glabrescent; nodes villous; blades very narrow, linear, tapering to a very fine point, usually convolute and subsetaceous, up to 9 in. by ¾ lin., glaucous, glabrous, smooth; triplets of spikelets 10–11 lin. long, 6–7 in erect contracted racemes; rhachis villous at the lowermost node; peduncles solitary or in pairs, flexuous, erect, 1–1½ in. long, glabrous; pedicels 0; lower glume lanceolate, acuminate, about 10 lin. long, rather firm, yellowish, strong, 3-nerved, bristly along the margins; upper glume similar, thinner, scantily pubescent, finely nerved; lower floret ♂: valve lanceolate, subulate-acuminate, thin, pubescent; pale obtuse, 7 lin. long, keels ciliate above; ⚥ floret: valve lanceolate-linear 3½ lin. long (from the callus to the base of the awn), pubescent above, finely 7-nerved with a callous transverse line below the awn, lobes 1-nerved produced into fine bristles 4–5 lin. long; callus pungent, ⅝ lin. long, villous; awn yellowish, column 5 lin. long, bristle 15 lin. long; pale linear-oblong, 3–3½ lin. long, tips obtuse, sometimes spoon-shaped; top of ovary and styles villous.

KALAHARI REGION: Basutoland; Leribe, *Buchanan*, 220!

3. T. **leucothrix** (Trin. ex Nees, Agrost. Bras. 460); culms 1–2 ft.

long, glabrous, 2-(rarely 1-) noded; leaves mostly basal; lowest sheaths very firm, rigid, persistent, fulvously tomentose below, glabrescent above, upper rather tight, sparsely hairy or glabrous; nodes villous; blades linear, tapering to a fine point or acute, 6–9 in. by $1\frac{1}{2}$–2 lin., uppermost much shorter, rigid, glabrous and smooth, or sparsely hairy, rarely hispidly villous; triplets of spikelets 1–$1\frac{1}{4}$ in. long, 5–3 in racemes, rarely 2 or solitary; peduncles $\frac{1}{2}$–$1\frac{1}{2}$ in. long, flexuous, glabrous and smooth or hispid above; pedicels very short, stout or hardly any; lower glume lanceolate, acute or subulate-acuminate, 9–12 lin. long, firm, 3-nerved, with long white soft spreading hairs from closely crowded small tubercles along the nerves; upper glume lanceolate, long subulate-acuminate, 14–18 lin. long, scarious, 3-nerved, minutely tubercled and spreadingly hairy along the margin; lower floret ♂: valve similar to the upper glume, but slightly longer, 7-nerved, side-nerves close to the margins with a double row of bristle-bearing glands on each side; pale linear-oblong, 7–8 lin. long; keels narrowly winged below, minutely ciliolate above; anthers $3\frac{1}{2}$ lin.; ⚥ floret: valve lanceolate, $4\frac{1}{2}$ lin. long, bifid to $\frac{1}{3}$ its length, glabrous, 7-nerved, lobes lanceolate, acuminate; callus pungent, $1\frac{1}{2}$ lin. long, villous; awn yellowish, 2–3 in. long, obscurely geniculate; pale linear-oblong, 5–6 lin. long, keels almost smooth; ovary obovoid, villous except at the base; styles pubescent; grain linear-oblong, 3–$3\frac{1}{2}$ lin. long. *Nees in Linnæa*, vii. 345; *Fl. Afr. Austr.* 267; *Kunth, Enum.* i. 308; *Trin. Pan. Gen.* 337; *Drège in Linnæa,* xx. 253; *Steud. Syn. Pl. Glum.* i. 238; *Oliv. in Trans. Linn. Soc.* xxix. 174, *t.* 115, *fig. B*; *Durand & Schinz, Consp. Fl. Afr.* v. 845. *T. monocephala, Hochst. in Flora,* 1846, 120; *Steud. l.c.* 238; *Durand & Schinz, l.c.* 845. *T. hispida, K. Schum. in Engl. Pfl. Ost-Afr. C.* 109. *Avena hispida, Thunb. Prodr.* 22. *Anthistiria hispida, Thunb. Flor. Cap. ed. Schult.* 107. *Danthonia hispida, Spreng. Syst. Veg.* i. 331.

COAST REGION: Swellendam Div.; *Mund.* George Div.; *Mund.* Uniondale Div.; Lange Kloof, *Thunberg!* Humansdorp Div.; Kromme River, *Thunberg!* Port Elizabeth Div.; Krakakamma Mountains, *Ecklon.* Port Elizabeth, *E.S.C.A. Herb.*, 87! Alexandria Div.; Zuur Bergen, *Drège;* Zwart Hoogte, *Burke!* Addo, *Drège.* Albany Div.; near Grahamstown, 2000–3000 ft., *Zeyher*, 4525! *MacOwan,* 1298! Mill River, *Gill!* Karregari River, *Ecklon;* and without precise locality, *Ecklon & Zeyher*, 872! Stockenstrom Div.; Winter Berg near Philippston, *Ecklon.* King Williamstown Div.; near the Kachu (Yellowwood) River, 1500 ft., *Drège.* Komgha Div.; near Komgha, *Flanagan*, 937!

CENTRAL REGION: Somerset Div.; on the plateau at the summit of Bosch Berg, 4800 ft., *MacOwan & Bolus, Herb. Norm. Austr. Afr.* 789! Alexandria Div.; on the rocks of Zwartwater Poort, *Burchell*, 3381!

KALAHARI REGION: Transvaal; near Lydenburg, *Wilms*, 1709! Spitzkop, in Lydenburg district, *Nelson*, 93*!

EASTERN REGION: Tembuland; Bazeia, 2000 ft., *Baur*, 275 partly! Pondoland; between St. Johns River and Umtsikaba River, *Drège!* Natal; on the summit of Table Mountain near Pieter Maritzburg, *Krauss*, 366! Inanda, *Wood,* 1594! Umpumulo and Riet Vlei, *Buchanan*, 237! and without precise locality, *Buchanan*, 59!

Also in tropical Africa.

XXVI. PRIONANTHIUM, Desv.

Spikelets more or less oblong, sessile or subsessile in a spike-like panicle or raceme; rhachilla tough, minutely produced beyond the upper floret. *Florets* 2, ⚥. *Glumes* 2, persistent, subequal, navicular; keels cartilaginous, muricate-pectinate, tubercled or smooth. *Valves* enclosed by and shorter than the glumes, subequal, hyaline, 3-nerved, acute, muticous. *Pales* slightly shorter than the valves, hyaline, 2-nerved. *Lodicules* 2, broadly cuneate or subquadrate, minute, fleshy. *Stamens* 3. *Ovary* glabrous; styles short, distinct; stigmas plumose, laterally exserted. *Grain* enclosed by the scarcely altered valve and pale.

Annual; culms slender; leaves narrow; ligule a line of hairs; panicles spike-like, or reduced to spike-like racemes; spikelets 2- (rarely 3-) nate, or solitary, adpressed, rigid.

Species 3, in the Cape Colony.

Panicle spike-like, not secund, nor 2-ranked :
  Lateral nerves of the glumes close to the keel, membranous margins very broad ... ... (1) **rigidum.**
  Lateral nerves of the glumes more equally distributed over the sides, margins narrow ... ... (2) **Ecklonii.**
Panicle reduced to a 2-ranked secund spike-like raceme ... (3) **pholiuroides.**

1. **P. rigidum** (Desv. Opusc. 65, t. iv. fig. 3); culms tufted, simple, geniculate, up to 8 in. high, glabrous or pubescent above, 2–3-noded, internodes enclosed; sheaths rather loose, pubescent above, glabrous below, the uppermost exceeding the base of the panicle; blades linear, narrow, 1–1¼ in. long, softly pubescent; panicle spike-like, compact, rigid, 1¼ in. long; rhachis pubescent; spikelets imbricate, elliptic-oblong, acute, 1½–1¾ lin. long; glumes ovate-oblong, scantily pubescent or glabrescent, the lower with 2 side-nerves on the outer and 1 on the inner side, the upper 3-nerved, nerves close to the obscurely winged muricate-pectinate keel, membranous wings broad; valves obliquely ovate in profile, finely pubescent; pales pubescent on the back; anthers ¾ lin. long. *Phalaris dentata, Linn. f. Suppl.* 106; *Thunb. Prodr.* 19; *Fl. Cap. ed. Schult.* 106; *Durand & Schinz, Consp. Fl. Afr.* 797 (*partly*). *Phleum dentatum, Pers. Syn.* i. 79. *Chilochloa dentata, Trin. Gram. Unifl. et Sesquifl.* 168.

CENTRAL REGION: Calvinia Div.; Bockland (Bokke Veld), *Thunberg!*

2. **P. Ecklonii** (Stapf); culms tufted, scantily branched below, slender, up to 1 ft. long, glabrous, 3-noded; internodes gradually longer, exserted from the base; sheaths tight, glabrous except near the mouth; blades very narrow, linear to filiform, ⅓–3 in. long, more or less convolute, pubescent; panicle spike-like, rigid, up to 3 in. long; rhachis glabrous; spikelets less imbricate below, oblong, 2–2½ lin. long; glumes linear-oblong, obtuse, pubescent or glabrescent, 7-nerved, nerves rather equally distant, firmly membranous, margins narrow,

keel muricate-pectinate ; valves oblong, glabrous or almost so, sometimes mucronulate ; pales glabrous except at the ciliolate tips ; anthers 1½ lin. long. *Prionachne Ecklonii, Nees in Lindl. Nat. Syst. ed.* ii. 448. *P. dentata, Nees, Fl. Afr. Austr.* 134. *Chilochloa dentata, Trin. Ic. Gram. t.* 73. *Chondrolæna phalaroides, Nees, l.c. C. dentata, Steud. Syn. Pl. Glum.* i. 430 (*partly*). *Lasiochloa pectinata, Trin. ex Pritz. Ic. Bot. Ind. Phalaris dentata, Durand & Schinz, Consp. Fl. Afr.* 797 (*partly*).

COAST REGION: Clanwilliam Div.; Olifants River, *Ecklon.* Piquetberg Div.; near Groene Vallei in the Piquetberg Range, below 1000 ft., *Drège!* Tulbagh Div.; mountains near Tulbagh, *Ecklon.*

3. **P. pholiuroides** (Stapf); annual; quite glabrous; culms tufted, geniculate, ascending, slender, 3–6 in. long, smooth, about 4–5-noded, with flowering branches from most nodes excepting the upper 1 or 2, middle and upper internodes exserted, the uppermost and base of the spike enclosed; sheaths smooth, lower short, purplish, uppermost more or less tumid, ¾–1½ in. long; ligule a line of very short and stiff hairs on a very fine membranous rim; blades linear, tapering to a fine point, ½–2½ in. by ½–¾ lin., more or less spreading, strongly striate, smooth below, scabrid above; panicle reduced to a stiff erect spike-like 2-ranked secund raceme, 1–1½ in. long; rhachis wavy, triquetrous, subscaberulous, back flat, sides hollowed out; pedicels extremely short, stout; spikelets adpressed to the rhachis, several times longer than the internodes, about 3 lin. long; continuation of rhachilla minute; glumes subequal, linear-oblong, subacute, keel stout, smooth or minutely tubercled, side-nerves about 3 on each side, close, prominent, hyaline margins as broad as the herbaceous part at the middle, lower glumes of the successive spikelets imbricate, laterally adpressed to the rhachis, upper much spreading at the time of flowering; valves slightly shorter than the glumes, hyaline except a herbaceous portion near the junction of the 3 nerves, keel scaberulous upwards; pales glabrous; anthers nearly 1½ lin. long.

COAST REGION: Cape Div.; damp hollows in Fish Hook Valley, *Wolley Dod*, 3394!

Very like *Pholiurus pannonicus* in habit and structure, except for the spikelets being strongly compressed from the sides and the glumes being not so thickened.

XXVII. ACHNERIA, Munro (non Beauv.).

Spikelets oblong to elliptic, laterally compressed, in lax or dense, very rarely spike-like panicles; rhachilla disarticulating above the glumes and between the valves, more or less produced beyond the upper floret. *Florets* 2, ⚥. *Glumes* subhyaline, rigid or subcoriaceous, persistent, subequal, 1- (near the base rarely 3-) nerved, keeled. *Valves* membranous, truncate or acute or acuminate, entire or minutely dentate, rarely mucronulate (cf. *A. setifolia*), usually hairy, 5–7- (rarely 3- or 9–11-) nerved; nerves fine, often very faint or obscure. *Pales* narrow, 2-keeled. *Lodicules* 2, minute, cuneate. *Stamens* 3. *Ovary* glabrous; styles distinct; stigmas plumose. *Grain* linear-oblong, grooved in front, enclosed by the unaltered valve and pale, free; hilum short, linear-oblong; embryo $\frac{1}{4}$–$\frac{2}{5}$ the length of the grain.

Perennial, tufted, rarely annual; blades narrow, usually convolute or setaceous; ligule a fringe of fine hairs; panicles usually trichotomously divided and at least temporarily open with fine lengthened branches and branchlets, rarely densely contracted or spike-like; spikelets more or less shining or glistening.

Species 9, all endemic in extra-tropical South Africa.

Achneria and *Pentaschistis* are very closely allied genera, the former being scarcely more than the awnless parallel of the latter. The affinity appears particularly close in *A. setifolia*, where the upper floret is occasionally awned and provided with side bristles, exactly as in *Pentaschistis*, although on a much reduced scale. In *Achneria*, as well as in *Pentaschistis* and a few other genera, tubercular glands occur frequently on the leaves, the divisions of the panicle and the keels of the glumes. They are usually pitted in the dry state owing to the shrinkage of the central portion.

Panicle very dense to compact; branchlets extremely short:
  Panicle compact, spike-like, cylindric, 1–2½ in. long; spikelets 1½–1¾ lin. long ... ... (1) **Ecklonii.**
  Panicle dense, ovoid, 7–10 lin. long, lower branches sometimes spreading, less than ½ in. long; spikelets 1¼ lin. long ... ... (2) **curvifolia.**
Panicle trichotomous, loose and open (at least temporarily); branches as long as or longer than ½ of the panicle, like the branchlets and pedicels finely filiform to capillary:
  Spikelets 1–1¼ lin. long; leaves very small, nearly all basal ... ... (3) **microphylla.**
  Spikelets larger; leaves longer:
    Glumes obtuse to subacute, 1½–2 lin. long:
      Spikelets ovate to elliptic; glumes subacute; valves prominently 9–11-nerved; blades narrow, linear ... ... (4) **aurea.**
      Spikelets linear-oblong; glumes obtuse; valves obscurely 5-nerved; blades finely setaceous ... ... (5) **capensis.**
    Glumes very acute or acutely acuminate:
      Valves much shorter than the glumes; spikelets 1½ lin. long ... ... (6) **capillaris.**
      Valves equalling the glumes or almost so:
        Culms many-noded; panicle large, effuse; spikelets lanceolate-oblong, 1½–1¾ lin. long... ... ... (7) **ampla.**
        Culms few-noded; spikelets oblong, 2–3½ lin. long:
          Panicle small; spikelets 2–2½ lin. long; blades setaceous ... ... (8) **setifolia.**
          Panicle 4–6 in. long and wide; spikelets 2¾–3½ lin. long; blades not setaceous ... ... (9) **hirsuta.**

1. A. Ecklonii (Durand & Schinz, Consp. Fl. Afr. v. 836), perennial; culms ½–1½ ft. long, ascending, wiry, many-noded, glabrous or the lowest internodes villous below; leaf-sheaths rigid, tight, glabrous except the ciliate upper margins or the lowest villous; blades convolute, subulate, $\frac{1}{2}$–$1\frac{1}{2}$ in. long, obtuse or subacute, very rigid, glabrous; panicle spike-like, compact, rarely interrupted, 1–$2\frac{1}{2}$ in. long; rhachis glabrous; branches, branchlets and pedicels extremely short or the lowermost branches up to 1 in. long; spikelets oblong-elliptic, $1\frac{1}{4}$–$1\frac{3}{4}$ lin. long, straw-coloured or silvery and tinged with green and purple, shining, much compressed; glumes acute, 1-nerved, keels scabrid; valves 1–$1\frac{1}{4}$ lin. long, truncate, slightly erose, hairy below the middle, faintly 7-nerved; pales rather shorter than the valves, keels ciliate; anthers 1–$1\frac{1}{4}$ lin. long. *A. assimilis, Durand & Schinz, Consp. Fl. Afr.* v. 836. *Eriachne Ecklonii, Nees, Fl. Afr. Austr.* i. 273; *Steud. Syn. Pl. Glum.* i. 236. *E. assimilis, Steud. Syn. Pl. Glum.* i. 236.

COAST REGION: Cape Div.; Kenilworth, near Capetown, in sandy places, *Bolus*, 7967! Paarl Div.; Klein Drakenstein Mountains, in damp meadows by the Berg River, 400 ft., *Drège!* Tulbagh Div.; Tulbagh Waterfall, *Ecklon.*

2. **A. curvifolia** (Stapf); perennial, densely cæspitose; culms slender, erect or ascending, 3–4 in. long, smooth, glabrous, simple, 3–4-noded, upper internodes shortly exserted; leaves glabrous, smooth; sheaths tight except the tumid uppermost, terete, firm, ciliolate near the mouth, lowest persistent; ligule a ciliate rim; blades setaceous, canaliculate, rather obtuse, up to 1 in. long, very rigid, often strongly curved, margins scaberulous; panicle ovoid, 7–10 lin. long, contracted, dense; rhachis smooth or almost so; branches solitary or 2-nate, filiform, rigid, finely scaberulous or smooth below, lowest up to 4, rarely 5 lin. long and undivided for $\frac{1}{3}$–$\frac{1}{4}$ their length; pedicels very short, stout; spikelets crowded, straw-coloured and tinged with purple, about $1\frac{1}{4}$ lin. long; glumes obliquely ovate to ovate-lanceolate in profile, acute, firmly chartaceous, glabrous, shining, 1-nerved, keels obscurely scabrid, lower scarcely 1 lin. long; valves lanceolate in profile, subacuminate, mucronulate, 1 lin. long, membranous, glabrous, excepting a few minute hairs towards the base, faintly 5-nerved; pales slightly shorter than the valves, obscurely 2-nerved; anthers scarcely $\frac{1}{2}$ lin. long. *Agrostis curvifolia, Hack. in Bull. Herb. Boiss.* iii. 384.

COAST REGION: Malmesbury Div.; Hopefield, *Bachmann!*

This is a typical *Achneria*. Hackel evidently overlooked the second floret. He calls the species peculiar and easy to distinguish from all other species of *Agrostis* by the subcoriaceous glumes and the structure of the valves.

3. **A. microphylla** (Durand & Schinz, Consp. Fl. Afr. v. 836); perennial; compactly cæspitose; culms erect, slender, $\frac{1}{2}$–$\frac{3}{4}$ ft. long, glabrous, smooth, sometimes viscous, 2-noded, uppermost internode long exserted; leaf-sheaths tight, the lowest imbricate to 4 lin. long, broad, strongly striate, ciliate at the mouth and margins; blades lanceolate to linear, acute or subobtuse, up to 6 lin. by 1 lin., the uppermost very short, flat or plicate, thick, firm, glabrous, subglaucous; nerves few, strong, close; margins scabrid; panicle erect, up to 2 in. long and as much broad or broader, very loose; rhachis glabrous,

usually tubercled and viscous; branches 2-nate, filiform, spreading, dichotomous from the middle; pedicels capillary, 1–4 lin. long; spikelets variegated, narrow-ovate, 1–1⅓ lin. long, somewhat shiny; glumes 1-nerved, acute, smooth; valves slightly shorter, ovate-oblong, truncate or minutely 3-dentate, long-hairy to ⅔ their length, faintly 7-nerved; pales as long as their valves, truncate or obscurely 2-toothed, tips ciliolate; anthers ¾ lin. long. *Eriachne microphylla, Nees, Fl. Afr. Austr.* 277; *Steud. Syn. Pl. Glum.* i. 236.

COAST REGION: Queenstown Div.; Storm Berg Range, in rocky places, 3000–4000 ft., *Drège!*

4. A. aurea (Durand & Schinz, Consp. Fl. Afr. v. 836); perennial; culms erect, ½–1 ft. long, glabrous, 3–4-noded, usually sheathed all along; sheaths rather tight or the uppermost subtumid, glabrous except the woolly margins, smooth or scabrid, prominently striate; blades linear, usually convolute, obtuse, the lower up to 6 in. by 1½ lin., the upper very short, hairy to villous above and along the margins, nerves prominent and few; panicle erect, effuse or contracted, oblong, 2–4 in. long, purplish; rhachis compressed, smooth; branches 2-nate, erect or suberect, callous at the base, filiform, smooth, branched from the middle or above, the lowest 1½–2½ in. long; pedicels 1½–5 lin. long; spikelets ovate, 1½–1¾ lin. long, variegated, shining; glumes subacute, 3-nerved at the base; valves slightly shorter than the glumes, ovate, broadly truncate, minutely or obscurely 3-dentate, prominently 11-nerved, shortly hairy between the nerves; pales truncate, tips ciliolate; anthers 1–1¼ lin. long. *Aira aurea, Steud. in Flora,* 1829, 470. *Airopsis aurea, Nees in Linnæa,* vii. 317; *Kunth, Rév. Gram.* ii. 216; *Enum.* i. 294, 525; *Suppl.* 244. *Eriachne aurea, Nees, Fl. Afr. Austr.* 276; *Steud. Syn. Pl. Glum.* i. 236.

VAR. β, **virens** (Stapf); culms 1½ ft. long, slender; leaves subfiliform, 8–9 in. long; panicle up to 6 in. long; lowest branches up to 3 in. long; spikelets light green. *Eriachne aurea, var. virens, Nees, Fl. Afr. Austr.* 276.

COAST REGION: Cape Div.; on the summit of Table Mountain, among shrubs, *Ecklon,* 915! Var. *virens:* Cape Div.; Table Mountain, *Spielhaus!* Worcester Div.; Dutoits Kloof, 1000–3000 ft., *Drège!*

5. A. capensis (Durand & Schinz, Consp. Fl. Afr. v. 836); perennial; stems ascending, copiously branched and fascicled at or above the base; culms very slender, ½–¾ ft. long, glabrous, smooth, the lower internodes very short, enclosed in the sheaths, the upper 3 lengthened, the uppermost exserted; leaf-sheaths tight, glabrous, finely striate; blades very narrow, setaceous or filiform, obtuse, 2–3 in. long, glabrous; panicle erect, oblong, narrow, 1–1½ in. long; rhachis filiform to capillary, glabrous; branches 2-nate, scantily branched, 5–2-spiculate, usually erect, the lowest up to ½ in. long, capillary, minutely scaberulous; pedicels 2–4 lin. long; spikelets linear-oblong, 1¾–2 lin. long, somewhat shiny; glumes linear-lanceolate, obtuse, subchartaceous, yellowish, tips dark, 1-nerved, keels smooth or scaberulous; valves lanceolate, subacute (in profile), faintly 5-

nerved, hairy along the margins and sometimes along the middle-nerve, tips obscurely 3-toothed or entire, ciliolate; pales obtuse, tips ciliolate; anthers $\frac{7}{8}$ lin. long; grain $\frac{3}{4}$–1 lin. long. *Eriachne capensis, Steud. in Flora*, 1829, 470. *E. Steudelii, Nees, Fl. Afr. Austr.* 278; *Syn. Pl. Glum.* i. 236. *Airopsis Steudelii, Nees in Linnæa*, vii. 318; *Kunth, Rév. Gram.* ii. 217; *Enum.* i. 294; *Suppl.* 245.

VAR. β, **firmula** (Stapf); all parts more rigid; panicle 1–1½ in. long, and almost as broad, effuse; spikelets 1½ lin. long; glumes deep purple towards the base; valves minutely hairy.

VAR. γ, **capillacea** (Stapf); culms 1½ ft. long; leaves finely filiform, up to 9 in. long; panicle ovate, 2–3½ in. by 1½–2 in.; effuse, pedicels very long and fine, up to 9 lin. long; spikelets 2 lin. long, pallid.

SOUTH AFRICA: without precise locality, *Drège*, 1673! 1687!

COAST REGION: Clanwilliam Div.; Ceder Bergen, Honig Vallei and Ezelsbank, 3000–5000 ft., *Drège*. Cape Div.; top of Table Mountain, *Ecklon*, 949! near Capetown, *Harvey*, 149! Cape Peninsula, in Orange Kloof, *Wolley Dod*, 2124! Worcester Div.; Dutoits Kloof, 2500–3000 ft., *Drège!* Stellenbosch Div.; on rocks in Lowrys Pass, 1000–2000 ft., *Drège!* Swellendam Div.; Paspas Vallei, *Ecklon*. Var. *firmula*: Riversdale Div.; near Zoetemelks River, *Burchell*, 6645! Var. *capillacea*: Riversdale Div.; on the lower part of the Lange Bergen, near Kampsche Berg, *Burchell*, 6953!

A. pallida (Durand & Schinz, Consp. Fl. Afr. v. 836. *Eriachne pallida*, Nees, Fl. Afr. Austr. 275) is probably a form of *A. ampla* with contracted smaller panicles and slightly smaller spikelets. It was collected by Ecklon in Uitenhage Division on the Zwartkops River. I have not seen it.

6. **A. capillaris** (Stapf in Hook. Icon. Plant. t. 2604); annual; culms geniculate, ascending or suberect, up to 1 ft. long, smooth, glabrous, about 2–3-noded, branched at the base; branches mostly flowering, uppermost internodes at length exserted; leaves loosely villous all over; sheaths rather lax or tumid, the lower beset with rows of sessile or subsessile, finely pitted tubercles; ligule a fringe of hairs; blades linear, acute, ½–2 in. by 1–1½ lin., involute or convolute when dry, margins often beset with stalked pitted tubercles; panicle obovate to pyramidal, 3 in. by 3–4 in., at length very lax; branches mostly 2-nate, repeatedly trichotomously divided with the spikelets approximate at the tips of the ultimate divisions, subdivaricate, filiform to capillary, glabrous or finely hairy in the axils, smooth except some scattered sessile tubercles; pedicels capillary, $\frac{3}{4}$ to almost 2 lin. long; spikelets ovate-oblong, 1½ lin. long, light green, somewhat shining; continuation of rhachilla fine, minute; glumes ovate-lanceolate in profile, acute to subacuminate, hyaline, pubescent, 1-nerved; valves broadly ovate-oblong in profile, obtuse or obscurely 3-lobed, $\frac{3}{4}$ lin. long, membranous, glabrous, 5–7-nerved; nerves very faint, but distinct towards the tips; pales about as long as the valves; anthers $\frac{3}{4}$ lin. long. *Holcus capillaris, Thunb. Fl. Cap. ed.* i. 412; *ed. Schult.* 110 (*excl. diagn.*), *not Prod.* 20. *Sorghum capillare, Roem. & Schult. Syst.* ii. 840. *Andropogon* (?) *capillaris, Kunth, Rév. Gram.* i. 166; *Enum.* i. 510.

SOUTH AFRICA: without precise locality, *Thunberg!*

Thunberg says, "flosculo hermaphrodito mutico, masculo aristato" in Prodr. l.c. and in the diagnosis of the species in Fl. Cap. l.c. This does not agree with his specimen, and I suspect that his diagnosis was originally drawn up from a different plant, probably a true *Holcus*, whilst the description, which follows it, was added later and drawn from the plant which is now in his herbarium under the name of *Holcus capillaris*. This would also explain another discrepancy, namely, that he says "glumis . . . glabris" in the diagnosis, but "glumæ . . . carina subvillosa" in the description.

7. **A. ampla** (Durand & Schinz, Consp. Fl. Afr. v. 836); perennial; culms erect, up to 2 ft. long, glabrous, many-noded; leaf-sheaths tight, the lower imbricate, glabrous except the sometimes ciliate mouth, striate; blades linear, tapering to a fine point, the lower up to 10 in. by 1–2 lin., convolute, glabrous, margins scabrid; panicle erect, usually large, 4–6 in. long and almost as broad; effuse; rhachis slightly compressed, filiform, glabrous, often tubercled; branches hairy near the base, branched from $\frac{1}{3}$ their length, like the branchlets filiform to capillary, divaricate; pedicels flexuous, finely capillary, 2–6 lin. long; spikelets lanceolate-oblong, $1\frac{1}{2}$–$1\frac{3}{4}$ lin. long, pallid or purplish near the base, somewhat shiny; glumes acute, thin, 1-nerved, keels smooth or scaberulous; valves slightly shorter than the glumes, ovate-lanceolate, acute (in profile), sometimes minutely 3-toothed, long, hairy to $\frac{2}{3}$ their length, tips ciliolate, obscurely 3–5-nerved; pales as long as their glumes, or almost so, bifid, tips ciliolate; anthers 1 lin. long; grain 1 lin. long. *Airopsis ampla, Nees ex Steud. Nomencl. ed.* ii. i. 45. *Eriachne ampla, Nees, Fl. Afr. Austr.* 277; *Steud. Syn. Pl. Glum.* i. 236.

COAST REGION: Cape Div.; Table Mountain, lower slope above Fernwood, *Wolley Dod*, 2379! between Wynberg and Constantia, *Burchell*, 817! Muizenberg Vlei, *Wolley Dod*, 2357! marshy ground near Tokay, *Wolley Dod*, 2193! Orange Kloof farm, *Wolley Dod*, 2326. Paarl Div.; Paarl Mountains, 1000–2000 ft., in wet rocky places, *Drège!* Worcester Div.; Dutoits Kloof, 1000–2000 ft., *Drège!* Stellenbosch Div.; Lowrys Pass, 1600 ft., *Schlechter*, 7212! Humansdorp Div.; on the north side of Kromme River, *Burchell*, 4847!

8. **A. setifolia** (Stapf); perennial, densely tufted; culms slender, erect, $\frac{1}{2}$ to over 1 ft. long, glabrous, smooth, 3–4-noded, lower internodes enclosed; leaves crowded towards the base, pubescent all over or glabrous, often with tubercular glands; sheaths tight, conspicuously bearded at the mouth, lower very firm, strongly striate, persistent; ligule a dense fringe of hairs; blades setaceously involute with fine setaceous tips, 3–8 in. long, curved or curled, firm, more or less scaberulous, glaucous; panicle open, very lax and broadly ovate or contracted and obovate, $1\frac{1}{2}$–$2\frac{1}{2}$ in. long, often with tubercular glands; branches 2-nate, trichotomous from $\frac{1}{3}$–$\frac{1}{2}$-way above the base, the longest 1–2 in. long, smooth or tubercled like the filiform axis and the capillary branches and pedicels, glabrous or scantily hairy at the callous axils; pedicels 1–$3\frac{1}{2}$ lin. long; spikelets oblong, about 2 lin. long, straw-coloured or slightly purplish below, somewhat glistening; glumes equal or almost so, oblong-lanceolate in profile, acute to acuminate, subhyaline, 1-nerved, or obscurely 3-nerved at the base, glabrous; keels smooth or tubercled; valves lanceolate in profile, acuminate, $1\frac{3}{4}$ lin. long, finely hairy, faintly 5–7-nerved, tip entire, finely subulate or 3–5-denticulate with a minute mucro, more rarely

with a delicate bristle or awn from the middle; pales glabrous, slightly shorter than the valves; anthers 1–1¼ lin. long; grain ⅔ lin. long. *Holcus setifolius, Thunb. Fl. Cap. ed. Schult.* 110. *Danthonia porosa, Nees, Fl. Afr. Austr.* 283; *Steud. Syn. Pl. Glum.* i. 239; *Durand & Schinz, Consp. Fl. Afr.* 853. *D. circinnata, Steud. Syn. Pl. Glum.* i. 239. *Eriachne tuberculata, Nees, Fl. Afr. Austr.* 274. *Achneria tuberculata, Durand & Schinz, l.c.* 836.

SOUTH AFRICA: without precise locality, *Thunberg!*
COAST REGION: Uitenhage Div. (?), in grassy places on the rocky ridges between Klipplaats River and Zwartkops River, *Ecklon.* Stockenstrom Div.; Kat Berg, 4000–5000 ft., *Drège!* Cathcart Div.; between Windvogel Berg and Zwartkei River, 3000–4000 ft., *Drège!* King Williamstown Div.; Amatola Mountains, *Buchanan*, 28!
CENTRAL REGION: "Cis Garipina," *Zeyher!* Aliwal North Div.; Witte Bergen, on rocks, 7000–8000 ft., *Drège!*

The variability of the valves is very puzzling. They are usually entire or minutely 3-toothed and awnless; but there is, in the upper valve, frequently a tendency towards the formation of an awn, the middle nerve running into a slender and short mucro or a delicate bristle, up to 1 lin. long. In extreme cases, this bristle is replaced by a slender kneed awn, almost as long as the valve, whilst the minute side lobes have each a short very fine bristle at the inner sides. That state is, however, always, as it seems, confined to a small number of spikelets, and does not affect the whole panicle.

9. **A. hirsuta** (Stapf); perennial, densely cæspitose; culms erect, somewhat stout, up to more than 2 ft. long, simple, glabrous, smooth, 2–3-noded, uppermost internode exserted; leaves crowded at the base, softly hairy all over; sheaths tight in the innovation shoots, otherwise lax, very long, lower open and at length flattened, firm, persistent, closely striate, ciliate along the margins, bearded at the mouth; ligule a dense fringe of hairs; blades linear, tapering to a setaceous point, up to almost 2 ft. long, 2 lin. wide at the base when expanded, more or less convolute or involute, firm, flexuous, margins scabrid; panicle pyramidal, open, very lax, 8–10 in. long and wide; branches usually 2-nate, smooth like the filiform axis and the more or less capillary branchlets and pedicels, slightly viscid, glabrous, or sparingly hairy at the callous axils, loosely trichotomous, the longest up to 5 in. long; pedicels 2–7 lin. long; spikelets oblong, 3–3½ lin. long, pallid or purplish below, somewhat glistening; glumes lanceolate in profile, acuminate, subequal, subhyaline, 1-nerved, obscurely pubescent in the upper part, keels smooth; valves linear-oblong in profile, minutely truncate or obtusely 2-lobed and mucronulate, or abruptly subulate-acuminate, 2½ lin. long, finely hairy, rather distinctly 7-nerved; pales as long as the valves, glabrous; anthers 1½ lin. long; grain 1¼ lin. long. *Danthonia hirsuta, Nees, Fl. Afr. Austr.* 282; *Durand & Schinz, Consp. Fl. Afr.* v. 850.

VAR. β, **glabrata** (Stapf); smaller; leaves shorter, glabrous excepting the ciliate margins and bearded mouth of the sheaths; panicles 4 by 4–5 in.; spikelets 2½ lin. long.

COAST REGION: Aliwal North Div.; Witte Bergen Range, 6000 ft., *Drège!*
Var. β: Albany Div.; mountain slopes near Grahamstown, 2000 ft., *MacOwan*, 1295!

XXVIII. AIRA, Linn.

Spikelets small, laterally compressed, loosely panicled; rhachilla disarticulating below and also between the valves, minutely produced beyond the upper floret. *Florets* 2, ⚥. *Glumes* persistent, delicately membranous, subequal, acute, 1-nerved, keeled. *Valves* slightly distant, subequal, shorter than the glumes, lanceolate, acute or acuminate, often 2-toothed, rounded on the back, faintly or very obscurely 5-nerved, awned from below the middle or awnless; awn kneed, twisted below; callus minute, glabrous or minutely hairy. *Pales* slightly shorter than the glumes, narrow, 2-toothed. *Lodicules* 2, delicate. *Stamens* 3; anthers small. *Ovary* glabrous; styles short, distinct; stigmas plumose, laterally exserted near the base. *Grain* more or less adhering to its valve and pale; hilum oblong, minute; embryo suborbicular, $\frac{1}{5}$–$\frac{1}{6}$ the length of the grain.

Small annuals with slender stems and leaves and with delicate, usually open panicles.

About 6 species, mainly in the Mediterranean region, 1 extending to tropical Africa and the Cape, and introduced into various parts of the world.

1. **A. caryophyllea** (Linn. Spec. Pl. 66); culms very slender, 3–9 in. high, geniculate, glabrous, few-noded; leaf-sheaths rather loose, minutely scaberulous; ligules hyaline, lanceolate, up to 4 lin. long; blades usually subsetaceous, convolute, the lowest up to 2 in. long, subscaberulous; panicle erect, ovate or obovate, $1\frac{1}{2}$–3 in. long, effuse or contracted; branches 2–5-nate, branched from or above the middle, dichotomous, capillary, smooth or almost so; pedicels 1–3 lin. long; spikelets gathered together near the tips of the branches, ovate, 1–$1\frac{1}{2}$ lin. long, pallid, somewhat shiny; valves $\frac{5}{8}$–1 lin. long, granular or minutely scaberulous, usually both awned or the lower awnless; awns $1\frac{1}{2}$–2 lin. long; callus glabrous or minutely hairy; anthers $\frac{1}{8}$–$\frac{1}{5}$ lin. long; grain $\frac{1}{2}$ lin. long, slightly grooved. *Fl. Dan.* iii. *t.* 382; *Engl. Bot. t.* 812; *ed.* iii. *t.* 1734; *Host, Gram. Austr.* ii. 44; *Trin. Gram. Gen.* 58; *Nees in Linnæa*, vii. 306; *Kunth, Enum.* i. 289; *Suppl.* 241, *t.* xviii. *fig.* 2; *Nees, Fl. Afr. Austr.* 272; *A. Rich. Tent. Fl. Abyss.* ii. 414; *Steud. Syn. Pl. Glum.* i. 221; *Engl. Hochgebirgsfl. Trop. Afr.* 128; *Durand & Schinz, Consp. Fl. Afr.* v. 834; *Schweinf. in Bull. Herb. Boiss.* ii. *App.* ii. 31.

COAST REGION: Cape Div.; Devils Mountain, *Ecklon*, 946! *Wilms*, 3880! bushes near Noahs-Ark, *Wolley Dod*, 2953! Simons Bay, *Wright!* Paarl Div.; Paarl Mountains, on wet rocks, 1500–2000 ft., *Drège!* Tulbagh Div.; Tulbagh Waterfall, *Ecklon!* Stellenbosch Div.; Hottentotts Holland, *Zeyher*. Caledon Div.; Genadendal, *Zeyher*. Mossel Bay Div.; Little Brak River, *Burchell*, 6184! Uitenhage Div.; Zwartkops River, *Ecklon!* Albany Div.; Grahamstown, *MacOwan*, 1297!

Mediterranean region, Central Europe, Abyssinia and Cameroon Peak; elsewhere introduced.

XXIX. HOLCUS, Linn.

Spikelets in rather dense, oblong or interrupted panicles, laterally compressed, disarticulating from the tips of the pedicels; rhachilla slightly produced beyond the upper floret, disarticulating more or less readily below the valves; joints slender, lower curved and often appendaged. *Florets* 2, lower ⚥, upper usually ♂, sometimes ⚥ or barren. *Glumes* 2, membranous, keeled, acute or acuminate, lower 1-nerved, upper 3-nerved, sometimes awned. *Valves* shorter than the glumes, chartaceous, very obscurely 5–3-nerved, lower awnless, upper awned. *Pales* narrow, 2-keeled. *Lodicules* 2, delicate. *Stamens* 3. *Ovary* glabrous; styles distinct; stigmas plumose, laterally exserted. *Grain* laterally compressed, enclosed by the valve and pale and often adhering to the latter, soft; hilum short; embryo small.

Annual or perennial; blades flat or convolute when dry; panicle usually more or less contracted, sometimes almost spike-like; spikelets deciduous, pallid.

Species about 6; 2 common in Europe, but naturalized in many temperate countries; 1 in South Africa, the rest Mediterranean.

The definition of *Holcus*, as given above, excludes *H. cæspitosus*, Boiss., and *H. grandiflorus*, Boiss. & Reut., which form Bentham's section *Homalachne*, and differ in having persistent glumes and perfectly equal florets, the lower of which is separated from the upper glume by an extremely short joint.

Annual; spikelets $1\frac{1}{2}$ lin. long; upper glume awned ... ... (1) **setiger.**
Perennial; spikelets 2 lin. long; upper glume awnless ... (2) **mollis.**

1. H. setiger (Nees in Linnæa, vii. 278); annual; culms slender, up to 2 ft. long, about 3-noded, glabrous; leaf-sheaths pubescent or glabrous except the nodes, the upper rather tumid; ligules membranous, oblong, pubescent, 1 lin. long; blades linear, acuminate, 1–3 in. by 1–2 lin., flat, flaccid, softly pubescent; panicle contracted, often almost spike-like, 1–$2\frac{1}{2}$ in. long; rhachis scabrid; branches, branchlets and pedicels fine, hairy; spikelets ovate, $1\frac{1}{2}$–2 lin. long, pallid; glumes almost equally long, scabrid, keels pectinate-ciliate, margins ciliolate below the tips, the lower narrow and subulate-acuminate, the upper much broader and awned; awn terminal, straight, 2–3 lin. long; lower floret ⚥, upper ♂ or barren; lower valve obliquely ovate, $\frac{3}{4}$ lin. long, subacute, smooth, shining, glabrous or with a few hairs on the keel, very obscurely 5-nerved; callus with a few long hairs; upper valve small, very thin, awn subterminal, fine, often bent, much shorter than the awn of the upper glume; pale as long as the valve; anthers $\frac{1}{4}$ lin. long. *Nees, Fl. Afr. Austr.* i. 9; *Steud. Syn. Pl. Glum.* i. 15 (*partly*); *Durand & Schinz, Consp. Fl. Afr.* v. 833.

WESTERN REGION: Little Namaqualand; between Pedros Kloof and Lily Fontein, 3000–4000 ft., *Drège.*

COAST REGION: Cape Div.; roadside at Camps Bay, *Wolley Dod*, 3131! Paarl Div.; 1000–2000 ft., *Drège*, 2582! Tulbagh Div.; Great Winter Hoek Mountain, *Zeyher.* Stellenbosch Div.; Somerset West, *Zeyher!* Riversdale Div.; near Zoetemelks River, *Burchell*, 6700!

There is often an imperfect male flower in the axil of the upper glume, consisting of 2 usually reduced stamens, and supported by a 2-keeled pale. (See Grœnland in Bull. Soc. Bot. France, ii. (1855) 172, with fig.)

Closely allied and very similar to the Mediterranean *H. setosus* (Trin.), but differing from it in the slightly smaller and comparatively broader glumes, the longer awn of the lower glume and the very small yellow anthers.

2. H. lanatus (Linn. Sp. Pl. 1048); *sphalmate* **mollis** in Key; perennial, tufted, 2–3 ft. high; culms 3–4-noded, softly hairy, at least below the panicle, rarely quite glabrous; leaf-sheaths reversedly and softly hairy, rarely glabrous, villous at the nodes, the uppermost inflated; ligule membranous, oblong, pubescent, 1 lin. long; blades linear to linear-lanceolate, up to 6 in. by 2–3½ lin., the uppermost very short, flat, softly hairy; panicle erect, oblong, 2–6 in. long, usually contracted; rhachis, branches, branchlets and pedicels hairy; spikelets oblong, 2¼–2½ lin. long, whitish or purplish; glumes almost equally long, mucronate, scabrid, keels pectinate-ciliate, the lower narrower, the upper broader with prominent side-nerves; lower floret ☿, upper ♂: lower valve obliquely lanceolate-oblong, rather more than 1 lin. long, with a few hairs on the keel, very obscurely 5-nerved; callus with a few long hairs; upper valve smaller and thinner, awn shorter than the valve, at length recurved, rather stout; pales as long as their valves; anthers ¾–1 lin. long. *Host, Gram. Austr.* i. 2, *t.* 2; *Beauv. Agrost.* 87, *t.* xvii. *fig.* 10; *Trin. Phalar.* 40; *Steud. Syn. Pl. Glum.* i. 15; *Reichb. Icon. Fl. Germ.* i. *t.* 105, *fig.* 1718–1720.

COAST REGION: Cape Div.; on waste land near Waterford Bridge, *Wolley Dod*, 1824! between Herzog and Retreat, *Wolley Dod*, 2184! King Williamstown Div.; Amatola Mountains, *Buchanan*, 22!

Native of Europe, Siberia, and North Africa, introduced into most temperate regions of both hemispheres.

XXX. ANTHOXANTHUM, Linn.

Spikelets oblong to narrow-lanceolate, slightly laterally compressed; rhachilla disarticulating above the upper glume, not produced beyond the uppermost floret. *Florets* 3, heteromorphous, the lower 2 ♂ or barren, the terminal ☿. *Glumes* persistent, membranous, 1–3-nerved, keeled, acuminate; upper longer. Lower 2 *valves* equal and very similar, oblong, emarginate, membranous, strongly laterally compressed, 5–7-nerved, keeled, hairy, awned, awn of the lower valve short from ⅓–½-way below the tip, of the upper longer, kneed, from near the base, rarely from the middle; terminal valve much shorter than the lower 2, broadly elliptic, very thin, delicately 7–1-nerved. *Pales* of the lower 2 florets, if present, 2-keeled, of the terminal 1-nerved. *Lodicules* 0. *Stamens* 3 in the ♂, 2 in the ☿ florets. *Styles* distinct, long; stigmas long, exserted from the top of the spikelet, plumose. *Grain* ovoid, slightly laterally compressed; hilum punctiform; embryo ¼ the length of the grain.

Perennial or annual; blades flat, usually flaccid; panicle slender, very narrow or spike-like, sometimes reduced to scanty racemes; sweet-scented.

Species 15, belonging to 2 sections: (1) *Eu-Anthoxanthum* with 6 species (4 in the West Mediterranean countries, 1 indigenous in Europe and North-West Asia, but introduced into many parts of the world, and 1 in tropical Africa), and (2) *Ataxia*, with 9 species (3 in South Africa, 5 in tropical Asia, and 1 in Mexico). The section *Ataxia* is characterized by the lowermost valve exceeding, or at least equalling, the subtending lower glume, and by having usually a ♂ flower either in both of the 2 lower florets or only in the lowermost floret. *Hierochloa*, to which the South African species of *Anthoxanthum* have been referred by Nees, possesses lodicules, and has also a somewhat different habit.

Panicle spike-like; culms erect:
  Spikelets very pallid, 4 lin. long; glumes very nearly hyaline, lower 1-nerved ... ... (1) **Ecklonii.**
  Spikelets greener, 3 lin. long; glumes herbaceous between the nerves, lower 3-nerved ... ... (2) **dregeanum.**
Panicle small, oblong, contracted or reduced to a scanty raceme; culms and leaves very weak, fine ... ... (3) **Tongo.**

1. **A. Ecklonii** (Stapf); perennial; culms tufted, erect, simple, 1–3 ft. long, smooth, glabrous or scaberulous, finely striate, about 3-noded; internodes gradually longer from the base, up to 1 ft. or more long; sheaths rather tight, smooth or scabrid, rarely reversedly hispidulous, striate, usually very much shorter than the internodes except the lowest, the basal 1 or 2 bladeless, acute; ligules scarious, white, ovate, obtuse, 1–2 lin. long; blades linear, gradually tapering, acute, up to 3 in. by 1–1½ lin., the upper very short, rather rigid, glabrous, smooth or scabrid, rarely reversedly hispidulous, prominently many-nerved; panicle erect, spike-like, sometimes interrupted near the base, up to 2 in. long, shiny, pallid; rhachis glabrous; pedicels with spreading hairs; spikelets oblong-lanceolate, 3½–4 lin. long, pallid; glumes lanceolate to ovate-lanceolate (in profile), acuminate, scaberulous on the keels, lower 2½–3 lin. long, 1-nerved, hyaline, upper 3½–4 lin. long, 3- (very rarely 1-) nerved, subhyaline; lowest floret ♂ or barren: valve about 2½ lin. long, 5-nerved, with a short straight awn from above the middle; intermediate floret equal and very similar to the preceding, always empty, with a kneed awn 4 lin. long from below the middle; uppermost floret ⚥, 1–1¼ lin. long; valve 5-nerved; anthers 1¼–1½ lin. long; grain ⅞ lin. long. *Anthoxanthum odoratum, Kunth in Ann. Sc. Nat. sér.* 1, xiii. 224 (*not Linn.*). *Ataxia Ecklonis, Nees ex Trin. Phalar.* 31; *Steud. Syn. Pl. Glum.* i. 13. *Hierochloa Ecklonii, Nees, Fl. Afr. Austr.* 7; *Durand & Schinz, Consp. Fl. Afr.* v. 798.

COAST REGION: Stockenstrom Div.; Kat Berg, 4000–5000 ft., in wet places, *Drège!* Queenstown Div.; by the Klipplaats River at Shiloh, *Ecklon.* King Williamstown Div.; Amatola Mountains, *Buchanan*, 47!
EASTERN REGION: Natal; Riet Vlei, 4000–5000 ft., *Buchanan*, 158!

A very closely allied species occurs in Madagascar.

2. **A. dregeanum** (Stapf); perennial; rhizome slender, short; culms erect or ascending, simple, slender, 1½–2½ ft. long, glabrous, finely striate, about 3-noded, uppermost internode longest, up to 1 ft. or more long; sheaths tight or the lower ultimately spreading,

glabrous or very rarely reversedly hairy or scaberulous, striate, usually very much shorter than the internodes, the basal bladeless; ligules scarious, white, oblong, obtuse, 1–2 lin. long; blades linear, gradually tapering, acute, 2–5 in. by 1–1½ lin., the upper very short, glabrous, rarely scantily hairy near the base, prominently nerved; panicle erect, spike-like, slender, sometimes interrupted near the base, up to 1½ in. long; rhachis glabrous; pedicels with spreading hairs; spikelets oblong, 2½–3½ lin. long, shiny; glumes broad-lanceolate to oblong (in profile), shortly acuminate, minutely pilose or glabrous, prominently 3-nerved, herbaceous between the nerves, keels smooth or scaberulous, lower 2–2½ lin., upper 2½–3½ lin. long; lowest floret ♂; valve 2–3 lin. long, 5-nerved, with a very short straight awn from above the middle; intermediate floret equal and very similar to the preceding, ♂, or sometimes empty, with a straight or kneed awn (3–4 lin. long) from below the middle of the valve; uppermost floret ⚥, 1 lin. long, 5-nerved; anthers up to 1¼ lin. long. *Hierochloa dregeana, Nees ex Trin. Phalar.* 37. *H. Dregei, Nees, Fl. Afr. Austr.* 9; *Steud. Syn. Pl. Glum.* i. 14; *Durand & Schinz, Consp. Fl. Afr.* v. 798.

COAST REGION: Cape Div.; by the railway between Claremont and Kenilworth, *Wolley Dod*, 3116! by Raapenberg Vley, *Wolley Dod*, 3112! Paarl Div.; by the Berg River, in the Klein Drakenstein Mountains, *Drège!* Worcester Div.; Hex River Valley, "the Willows," *Wolley Dod*, 3708!

Hierochloa Tongo (Nees, Fl. Afr. Austr. 7), as represented by Drège's specimen from the Cederbergen, Clanwilliam Div. (2558!), is perhaps a weak state of *A. dregeanum*. The spikelets are rather smaller, the glumes thinner and more obscurely nerved, the male flowers sometimes reduced to rudimentary anthers or filaments, and also in most of the spikelets examined the hermaphrodite flower was imperfectly developed. It seems to be specifically distinct from what Nees described as *H. Tongo*, var. *minor* and var. *simplex*, which answer to Trinius's *Ataxia Tongo* and *A. tenuis*.

Wolley Dod's specimens from Raapenberg Vley represent a very vigorous state with barren ovaries bearing 4 or more stigmas of which some are bifid. A similar condition of *Hierochloa australis*, Roem. & Schult., is described by Hackel in Bot. Centralb. viii. (1881), 155.

3. A. Tongo (Stapf); culms ascending, geniculate, very slender to filiform, weak, copiously branched below, ½–1 ft. long, glabrous, smooth, 3–4-noded, uppermost internode by far the longest, up to ¾ ft. long; sheaths rather tight or the upper looser, smooth, with few striations, those of the flowering branches shorter than the internodes; ligules scarious, white, oblong, acute or obtuse, ½–1 lin. long; blades subfiliform to setaceous, flat or folded, up to 3 in. by ¼–½ lin., smooth, glabrous, few-nerved; panicle contracted, narrowly oblong, up to 1 in. long, rather loose or reduced to a scanty raceme; rhachis glabrous; pedicels scaberulous or hairy, up to 1½ lin. long; spikelets oblong, 3–3½ lin. long, shiny; glumes broadly-lanceolate to oblong, acute, hyaline, 3-nerved, glabrous or spreadingly hairy on the back; lower 2¼–2½ lin. long, upper 3–3½ lin. long; lowest floret usually ♂, rarely barren; valve about 2 lin. long, obscurely 3–5-nerved, with a short straight awn from ⅓ way below the tip; intermediate floret almost equal and similar to the preceding, but slightly narrower, empty, with a kneed and twisted awn 4–5 lin. long from near the base of the valve; uppermost floret ⚥, 1 lin. long, faintly 3–1-nerved; anthers 1½ lin. long. *Ataxia Tongo, Trin. Phalar.* 32. *A. tenuis, Trin. l.c. (a very weak state); Steud. Syn. Pl. Glum.* i. 13. *Hierochloa Tongo, vars. minor and simplex, Nees, Fl. Afr. Austr.* 8. *H. tenuis, Durand & Schinz, Consp. Fl. Afr. Austr.* v. 798.

COAST REGION: Cape Div.; Devils Peak, by the second Waterfall, frequent, *Wolley Dod*, 3119! Cape Peninsula, summit of the 12 Apostles, *Wolley Dod*, 3484! in damp clefts of rocks near the top of a hill above Klassenbosch and Orange Kloof, *Wolley Dod*, 2949! above Millers Point, *Wolley Dod*, 2853! Paarl Div.; Paarl Mountains, 2000–3000 ft., *Drège*! Stellenbosch Div.; Lowrys Pass, 4500 ft., *Schlechter*, 7217! Worcester Div.; in Dutoits Kloof at Pieterwetbosch, in shady places, *Drège*!

XXXI. KŒLERIA, Pers.

Spikelets laterally compressed in spike-like panicles; rhachilla glabrous or finely hairy, disarticulating above the glumes and between the valves, produced with or without a rudimentary valve. *Florets* 1–5, ⚥, or the uppermost more or less reduced. *Glumes* 2, persistent, subequal or unequal, subacute to acuminate, keeled, the lower usually 1-nerved or like the upper 3-nerved, margins hyaline. *Valves* exceeding the glumes, acute or obtuse with the margins and tips broadly hyaline, 3–5-nerved; side-nerves usually faint, conniving above, middle nerve percurrent or excurrent into a mucro or a short subterminal awn; callus very minute, glabrous. *Pales* shorter than the valves or almost as long, 2-keeled, 2-toothed, conspicuously hyaline and white. *Lodicules* 2, hyaline. *Stamens* 3. *Ovary* glabrous; styles distinct, very short, stigmas laterally exserted, plumose. *Grain* oblong, laterally compressed, whitish, soft, tightly embraced by the hardened back of the valve; hilum basal, short, obscure; embryo small.

Perennial or annual; blades usually very narrow; ligules hyaline; panicle usually cylindric, often interrupted, glabrous and glistening from the hyaline white margins of the valves and pales, or more or less hairy.

Species 12–15, mainly in Europe, North Africa, and temperate Asia, 1 species almost cosmopolitan.

Perennial; spikelets awnless; valves 3-nerved ... ... (1) **cristata.**
Annual; valves 5-nerved, finely awned or mucronate from below the tips ... ... (2) **phleoides.**

1. K. cristata (Pers. Syn. i. 97); perennial, densely cæspitose; culms erect, rarely geniculately ascending, ½ ft. to more than 1 ft. long, glabrous or villous, pubescent, 1–2-noded, upper internodes very long, exserted; leaves crowded near the base; sheaths rather tight, striate, glabrous or pubescent to villous, the lower more or less persistent; ligule obtuse, minutely ciliolate, rarely more than ½ lin. long, usually produced into lateral auricles; blades linear, acute, from 1 in. to 1 ft. long, to 1 lin. broad, flat, soft and even flaccid, or setaceously convolute and flexuous or rigid, glabrous and smooth below, pubescent above, or ciliate, or pubescent to villous all over; panicle cylindric, often interrupted or lobed, 1–4 in. by 2½–6 lin., dense; branches repeatedly and very shortly branched from the base, like the rhachis pubescent to minutely villous; pedicels very short; spikelets 2–2½ lin. long, 2–3-flowered; glumes glabrous or pubescent, the lower narrow, lanceolate, subacute to acute, 1¼–1¾ lin. long, 1-nerved, the upper much broader, subacute to acutely acuminate, 1½–2¼ lin. long, 3-nerved; valves oblong to lanceolate in profile, subobtuse to acuminate, sometimes mucronulate, 1¾–2 lin. long, glabrous or pubescent; lodicules 2–3-lobed or toothed; anthers ¾–1 lin. long; grain up to 1¼ lin. long. *Flor. Dan. t.* 2223; *Kunth, Enum.* i. 381; *Suppl.* 315; *Reichb. Ic. Fl. Germ.* i. *t.* 93, *fig.* 1668; *A. Rich. Tent. Fl. Abyss.* ii. 431; *Steud. Syn. Pl. Glum.* i. 292; *Engl. Hochgebirgsfl. Trop. Afr.* 134; *Durand & Schinz, Consp. Fl. Afr.* v. 892; *Hook. f. Fl. Brit. Ind.* vii. 308. *K. parviflora, Bertol. in Schult. Mant.* ii. 344. *K. Alopecurus, Nees in Linnæa,* vii. 320; *Kunth, Enum.* i. 384; *Steud. l.c.* 294. *K. capensis, Nees, l.c.* 321; *Kunth, l.c.* 382; *Steud. l.c.* 293; *Durand & Schinz, l.c. Aira cristata, Linn. Sp. Pl.* 63; *Engl. Bot. t.* 648; *Knapp, Gram. Brit. t.* 30. *Alopecurus capensis, Thunb. Prodr.* 19; *Fl. Cap. ed. Schult.* 105. *Dactylis villosa, Thunb. Prodr.* 22; *Fl. Cap. ed. Schult.* 115. *Aira capensis, Steud. in Flora,* 1829, 469. *Poa cristata, Wither. Arr. Brit. Pl.* 51; *Host, Gram. Austr.* ii. 54, *t.* 75. *Airochloa parviflora, Nees, Fl. Afr. Austr.* 425. *A. Alopecurus, Nees, l.c.* 424.

SOUTH AFRICA: without precise locality, *Thunberg!*

COAST REGION: Clanwilliam Div.; Dwars River, *Verreaux!* between Dwars River and Ezelsbank, and in the Onder Bockeveld, 2500–3000 ft., *Drège.* Ceder Bergen, *Ecklon.* Cape Div.; Lion Mountain, *Ecklon,* 945! *Burchell,* 119! Table Mountain and Cape Flats, *Ecklon;* near Constantia Nek, *Wolley Dod,* 2126! sand-bills by Muizenberg Vley, *Wolley Dod,* 3568! Orange Kloof, *Wolley Dod,* 2200! Paarl Div.; Paarl Mountains, 2000–3000 ft., *Drège!* Tulbagh Div.; Great Winter Hoek Mountain, 2000–3000 ft., *Drège!* Tulbagh Waterfall, Tulbagh Kloof, &c., *Ecklon.* Stellenbosch Div.; Hottentots Holland, *Ecklon.* Bredasdorp Div.; sand-dunes near Cape Agulhas, *Ecklon.* Swellendam Div.; hills between Riet Kuil and Grootvaders Hoek, *Zeyher,* 4583! near Swellendam, *Zeyher,* 4594! along the lower part of the Zonder Einde River, 500–2000 ft., *Zeyher,* 4586! Riversdale Div.; near the Zoetemelks River, *Burchell,* 6609! 6691! between Great Vals River and Zoetemelks River, *Burchell,* 6566! Mossel Bay Div.; dry hills on the eastern side of the Gauritz River, *Burchell,* 6432! Knysna Div.; Plettenberg Bay, in pastures, *Bowie,* 275! Uitenhage Div.; Zwartkops River, *Ecklon;* near Uitenhage, *Burchell,* 4246! Port Elizabeth Div.; grassy hills near Addo, *Ecklon.* Albany Div.; grassy places near Grahamstown, 2000 ft., *MacOwan,* 1301! Fort Beaufort Div.; grassy hills near Kat Rivers Poort, 2000–25000 ft., *Drège.* Stockenstrom Div.; Kat Berg, 3000–4000 ft., *Drège!* Komgha Div.; near Komgha, *Flanagan,* 998! Queenstown Div.; Table Mountain, 5500–6000 ft., *Drège.* Shiloh, *Baur,* 949!

CENTRAL REGION: Graaff Reinet Div.; hills near Graaff Reinet, 4500 ft., *Bolus,* 710! Aliwal North Div.; Witte Bergen, in grassy valleys and in gorges, 6000–7000 ft., *Drège!*

WESTERN REGION: Little Namaqualand; Kamies Bergen, *Drège,* Hartebeest River, 4000 ft., *Drège.*

KALAHARI REGION: Orange Free State (?); without locality, *Cooper,* 914 partly!

EASTERN REGION: Tembuland; Bazeia, 2000–2500 ft., *Baur*, 315! Gat Berg, 4000 ft., *Baur*, 1149! Natal, *Buchanan*, 95! 157!

Most temperate regions and in highlands of the tropics.

I have refrained from describing varieties of this extremely variable grass as I find the characters used for this purpose by other authors so unstable, even in specimens from the same locality, and entering into so many more or less distinct combinations as to defy any satisfactory classification.

2. **K. phleoides** (Pers. Syn. i. 97); annual; culms tufted, geniculate, erect or ascending, from a few inches to more than 1 ft. long, smooth, glabrous, 2–3- (rarely 4-) noded, simple or sometimes with a flowering branch from one of the lower nodes; leaves more or less softly hairy to villous; sheaths rather loose or the upper tight, thin; ligules hyaline, delicate, toothed, up to $\frac{1}{2}$ lin. long; blades linear, tapering to an acute point, 1 to more than 6 in. by $\frac{3}{4}$–4 lin., flat, flaccid; panicle spike-like, cylindric, sometimes lobed, $\frac{1}{2}$–3 in. by 3–7 lin., very dense; branches repeatedly and very shortly branched from near the base, glabrous and smooth like the rhachis; pedicels very short, scantily scabrid or glabrous; spikelets ovate-oblong, $1\frac{1}{2}$–$2\frac{1}{2}$ lin. long; florets 3–7; rhachilla very shortly ciliolate, joints short; glumes glabrous or scantily pubescent, or scaberulous in the upper part, lower narrow, lanceolate, acute, 1–$1\frac{1}{4}$ lin. long, 1-nerved, upper much broader, slightly longer, 3-nerved; valves oblong, always entire but easily splitting to the base of the very fine mucro or awn (up to $1\frac{1}{2}$ lin. long) and then 2-toothed, 1–$1\frac{1}{2}$ lin. long, glabrous or subpubescent, often scabrid, prominently 5-nerved, the upper often different, barren, strongly compressed and recurved; anthers $\frac{1}{10}$ lin. long; grain up to $\frac{3}{4}$ lin. long. *Kunth, Enum.* i. 383; *Nees, Fl. Afr. Austr.* 428 (*in part*); *Steud. Syn. Pl. Glum.* i. 294; *Durand & Schinz, Consp. Fl. Afr.* v. 893; *Hook. f. Fl. Brit. Ind.* vii. 309. *Festuca cristata, Linn. Sp. Pl.* 76. *F. phleoides, Vill. Hist. Pl. Dauph.* ii. 95, *t.* 2, *fig.* 7; *Desf. Fl. Atlant.* i. 90, *t.* 23; *Host, Gram. Austr.* iii. *t.* 21. *Poa phleoides, Lam. Ill.* i. 182. *Alopecurus ciliatus, All. Fl. Pedem.* ii. 235.

COAST REGION: Cape Div.; Green Point, common, *Wolley Dod*, 3576! without precise locality, *Pappe*. "Swellendam and George District," *Mund*.

Throughout the Mediterranean Region to North-West India; also in Abyssinia.

Drège's specimens from the Hex River, distributed as *K. phleoides* and quoted by Nees, l.c., under this name are, judging from the examples at Kew, *Trisetum pumilum*, Kunth. The two grasses are, indeed, very similar in appearance, and, no doubt, closely allied. *K. phleoides* may, however, be easily distinguished by the very minutely ciliolate (not long hairy) rhachilla.

XXXII. **TRISETUM**, Pers.

Spikelets usually rather small, 1–4 lin. long, in usually close, often spike-like panicles; rhachilla ciliate or long hairy, very rarely glabrous; disarticulating above the glumes and between the valves, produced into a short bristle beyond the uppermost floret.

Florets 2–6, ⚥, or the uppermost more or less reduced. *Glumes* 2, persistent, equal or more or less unequal, acute, keeled, lower 1- (or like the upper more or less distinctly 3-) nerved, hyaline. *Valves* equalling or exceeding the glumes, membranous with hyaline tips and broad margins, acutely 2-toothed, sometimes with fine short bristles from the teeth, faintly or obscurely 5–3-nerved, awned; awn from the back near the tip, fine, straight or kneed and twisted below; callus minute, more or less hairy or glabrous. *Pales* shorter than the valves, 2-keeled, 2-toothed, hyaline. *Stamens* 3. *Ovary* glabrous; styles distinct, very short; stigmas laterally exserted, plumose. *Grain* oblong, whitish, soft, embraced by the usually slightly hardened back of the valve; hilum basal, short; embryo small.

Perennial, rarely annual; blades flat, usually flaccid; ligules hyaline; panicle usually contracted, often spike-like, rarely open and lax, more or less glistening.

Species 50–60, mainly in the temperate region of the northern hemisphere; and along the Andes to Patagonia; a few in Australia and New Zealand.

1. T. pumilum (Kunth, Rév. Gram. i. 102); annual; culms fascicled, slender, geniculate, ascending or suberect, 2–9 in. long, glabrous or hairy below the lower nodes, smooth, 3–4-noded, simple or branched below, upper internodes exserted; leaves more or less finely and softly hairy; sheaths thin, usually rather lax; ligules very obtuse, dentate, ciliolate, often hairy from the back, up to $\frac{3}{4}$ lin. long; blades linear, tapering to a subacute or acute, subcallous point, $\frac{1}{2}$–2 in. by $\frac{1}{2}$–1 lin., flat, flaccid; panicle spike-like, oblong, $\frac{1}{2}$–$1\frac{1}{2}$ in. long, very dense, sometimes lobed; rhachis very slender, puberulous or glabrous below; branches solitary, erect, the longest up to 6 (rarely to 10) lin. long, closely branched from the base, finely filiform, puberulous to almost villous; lateral pedicels very short; spikelets very crowded, $1\frac{1}{2}$–2 lin. long, pallid; florets 2–4; rhachilla hairy, hairs much exceeding the joints; glumes equal or subequal, ovate-lanceolate, finely subacuminate, 3-nerved, pubescent; valves oblong in profile, acute, very minutely 2-toothed, $1\frac{1}{4}$–$1\frac{1}{2}$ lin. long, hyaline except at the subherbaceous back, 5-nerved; awn very fine, straight, subterminal or from $\frac{1}{4}$ way below the readily splitting tip; pales about $\frac{1}{2}$ the length of the valves; anthers ellipsoid, scarcely $\frac{1}{5}$ lin. long; grain about $\frac{3}{5}$ lin. long. *Kunth, Enum.* i. 297; *Trin. Gram. Suppl.* 15, *and in Mém. Acad. Pétersb. sér.* vi. iv. 16; *Steud. Syn. Pl. Glum.* i. 227; *Durand & Schinz, Consp. Fl. Afr.* v. 840. *Avena pumila, Desf. Fl. Atlant.* i. 103. *Kœleria phleoides, Nees, Fl. Afr. Austr.* 428 (*in part*), *not Pers.*

COAST REGION: Ceres Div.; between Hex River Mountains and the Warm Bokkeveld, 3000–4000 ft., *Drège*, 532!

CENTRAL REGION: Fraserberg Div.; near the Dwaal River, *Burchell*, 1473!

Warmer parts of the Mediterranean region; probably introduced into Cape Colony.

Similar to *Kœleria phleoides* and often confounded with it. (See under that species.)

XXXIII. AVENASTRUM, Jess.

Spikelets usually erect or suberect, rarely nodding, medium-sized (4–7 lin. long, rarely less or more), in nearly always erect, often stiff panicles; rhachilla more or less long-hairy, disarticulating above the glumes and between the valves, produced into a short bristle beyond the uppermost floret or ending with a rudimentary valve. ***Florets*** 3–6, ⚥, or the uppermost more or less reduced. *Glumes* 2, persistent, scarious, more or less unequal, acute or acuminate, more or less distinctly keeled, lower 1- or 3-nerved, upper 3- (very rarely 5- to sub-7-) nerved. *Valves* usually distinctly exserted from the glumes, more or less herbaceous with scarious or hyaline tips, often rather firm, acute or acuminate, bifid, with or without bristles from the lobes, 5–9-nerved, awned; awn dorsal from the middle or slightly above it, kneed and twisted below; callus short or elongate, villous. *Pales* shorter than the valves, 2-keeled, 2-toothed or bifid. *Lodicules* 2, rather large, hyaline. *Stamens* 3. *Ovary* hairy from the middle upwards or at the top only; styles distinct, short; stigmas laterally exserted, plumose. *Grain* oblong, slightly laterally compressed, usually grooved in front, hairy at the top, pallid, rather soft, embraced by the somewhat hardened valve and the pale; hilum linear, up to ½ the length of the grain; embryo small.

Perennial, cæspitose; blades linear, usually narrow, flat or convolute; often setaceous; ligules hyaline; panicle narrow, erect, often stiff, rarely flaccid or expanded.

Species about 45, chiefly in the temperate regions of the northern hemisphere and through the high mountains of tropical Africa to South Africa.

After having worked through the whole material of *Avena* (in the sense of Bentham & Hooker's Genera Plantarum) and *Trisetum*, in the Kew Herbarium, I have come to the conclusion that not only *Trisetum* has a claim to generic rank, but also that the section *Avenastrum*, Koch, which, together with the section *Crithe*, makes up the genus *Avena* of most botanists, should be considered as a distinct genus, so that *Avena* is confined to the section *Crithe*, comprising the common oats.

Spikelets 6–9 lin. long; rhachilla joints 1½–2 lin. long, hairy:
  Spikelets 3–4-flowered, much longer than the glumes:
    Panicle 6–9 in. long, loosely contracted; flaccid; leaves long, flaccid ... ... (1) **longum.**
    Panicle 2–3 in. long, secund, nodding; leaves mostly basal, short, subrigid ... ... (2) **dregeanum.**
  Spikelets 2- or sub-3-flowered, as long as or slightly longer than the glumes; panicle much contracted, dense, stiff or flexuous ... ... (3) **quinquesetum.**
Spikelets 3½–6 lin. long; rhachilla joints ½–1 lin. long:
  Spikelets rather compact, slightly turgid; valves imbricate ... ... (4) **turgidulum.**
  Spikelets more loosely flowered; valves not imbricate (except when quite young), spreading, narrow:
    Panicle very narrow, branches adpressed; valves subcartilaginous below the awn, scaberulous or granulate:

Blades 9–20 in. by 2–2½ lin., lower long attenuated at the base; spikelets 5–6 lin. long; rhachilla joints long hairy ... ... (5) **Dodii.**
Blades 2–9 in. long, very narrow; spikelets 3½–5½ lin. long; rhachilla joints glabrous or scantily and shortly hairy towards the tips ... ... (6) **antarcticum.**
Panicle flexuous; rhachilla long hairy; valves thin, smooth ... ... (7) **caffrum.**

1. **A. longum** (Stapf); culms 2–3 ft. long, glabrous, about 3-noded, sheathed almost all along; leaves 3–6 from near the base, 3 higher up; sheaths rather loose, glabrous, slightly rough; ligule truncate up to 1½ lin.; blades linear, long tapering to a fine point, 6–10 in. by 1½–3 lin., flat or more or less involute, flaccid, glabrous, quite smooth or rough above; panicle contracted, 9–10 in. long, nodding or flexuous; branches fascicled, very unequal, the longest up to 2½ in. long, branched from near the base or simple, filiform, flexuous, scaberulous; spikelets 6–9 lin. long, narrow, 4- to sub-5-flowered; glumes lanceolate, acuminate, the lower 3½–4 lin., the upper 4½–5 lin. long; rhachilla very slender, joints up to 1½ lin. long, with long white hairs; valves long exserted, lanceolate, the lowest 5 lin. long, glabrous, rather firm, smooth, pallid, sometimes purplish below the scarious scaberulous bifid tips, lobes produced into fine bristles 2–3 lin. long; callus subulate, up to ¾ lin. long, hairy; awn from above the middle, fine, column pallid, 3–4 lin. long, bristle 7–8 lin. long; pales 3½ lin. long; keels ciliate; anthers 1¼–1½ lin. long; ovary puberulous below the oblique hispidulous top. *Trisetum antarcticum, Nees in Linnæa,* xx. 254, *not in Linnæa,* vii. 307. *Avena longa, Stapf in Kew Bulletin,* 1897, 292.

Var. β **grande** (Stapf); ligules up to 2½ lin. long; blades up to 5 lin. broad, slightly rough on the upper side; panicle larger, denser; spikelets 8–11 lin. long; glumes and valves proportionally larger.

Coast Region: "Cape of Good Hope," *R. Brown!* Cape Div.; Cape Flats at Doorn Hoogte, *Ecklon & Zeyher*, 1807! 1807 β! Cape Peninsula; Klein Slangkop, *Wolley Dod*, 3004! upper northern slope of Lions Head, *Wolley Dod*, 3571! Var. *grande:* Cape Div.; in grassy rocky places on the sides of Lion Mountain, above Camps Bay, *MacOwan, Herb. Aust.-Afr.* 1793! Cape Peninsula; plentiful above the road beyond Sea Point, *Dod*, 3128! Orange Kloof, *Wolley Dod*, 3168! Smitwinkel, rare, *Wolley Dod*, 3003!

2. **A. dregeanum** (Stapf); stems slightly bulbous at the base, 1 ft. long, glabrous, 1-noded much below the middle; leaves about 4 from near the base, 1 sheathing the culm for the greatest part; sheaths rather tight, glabrous; ligule truncate, ½–⅔ lin. long; blades linear with callous tips, the lower 1½–2½ in. by 1 lin. or less, flat or convolute, rather rigid, subglaucous, hairy above, glabrous beneath; panicle (or raceme) nodding, 2–3½ in. long, very loose, secund; branches paired, 3–1-spiculate, the lowest up to 1 in. long, more or less spreading, filiform, scabrid; spikelets 5–7 lin. long, very loosely 3- to sub-4-flowered; glumes unequal, lanceolate, acuminate, the lower

4, the upper 6 lin. long; rhachilla joints up to $1\frac{1}{2}$ lin. long, with long spreading hairs; valves long exserted, oblong-lanceolate, the lowest 5–6 lin. long, glabrous, pallid, firm, scaberulous below the middle, purplish below the scarious bifid tips, lobes produced into fine bristles, 1–2 lin. long; callus subulate, curved, $\frac{3}{4}$ lin. long, long hairy; awn from the middle, column 3–5 lin. long, bristle 6–8 lin. long; pales linear-lanceolate, 4 lin. long; keels ciliate; anthers $1\frac{1}{2}$–$1\frac{3}{4}$ lin. long; ovary hairy above the middle. *Trisetum barbatum, Nees, Fl. Afr. Austr.* 345, *not Steud. T. dregeanum, Steud. Syn. Pl. Glum.* i. 227; *Durand & Schinz, Consp. Fl. Afr.* v. 838.

WESTERN REGION: Little Namaqualand; Roode Berg and Ezels Kop, 3000–4000 ft., *Drège*, 2625!

3. A. quinquesetum (Stapf); culms $1\frac{1}{2}$–3 ft. long, glabrous, about 2-noded, sheathed to 2–3 in. below the panicle; leaves 4–6 near the base, 2 higher up; lowest sheaths compressed, subcarinate, firm, minutely puberulous, strongly nerved, the upper more terete, glabrous; ligule truncate, up to 1 lin. long; blades linear with callous tips, flat or plicate, rigid, strongly and closely nerved, glabrous; panicle contracted, rather dense, narrow, 5–6 in. long, strict or subflexuous; branches fascicled, very unequal, branched from near the base, adpressed to the rhachis, the longest up to 2 in. long; spikelets 5–6 lin. long, very loosely 2- to sub-3-flowered; glumes narrow-lanceolate, acuminate, the lower 4–5 lin., the upper 5–$6\frac{1}{2}$ lin. long; rhachilla joints very slender, up to $2\frac{1}{2}$ lin. long, with very long white hairs; valves shortly exserted, linear-lanceolate, the lowest 6–$6\frac{1}{2}$ lin. long, glabrous, pallid, rather firm, smooth or almost so, tips scarious, bifid, each lobe produced into 1 or 2 fine bristles, $1\frac{1}{2}$–3 lin. long; callus subulate, up to $\frac{3}{4}$ lin. long, bearded; awn from the middle, column pallid, 4–5 lin. long, bristle 8–9 lin. long; pales $4\frac{1}{2}$ lin. long, keels ciliolate; anthers $1\frac{1}{4}$ lin. long; ovary pubescent near the hispidulous top. *Avena quinqueseta, Steud. in Flora*, 1829, 485; *Kunth, Enum.* i. 305. *Trisetum Steudelii, Nees in Linnæa*, vii. 308; *Fl. Afr. Austr.* 349; *Steud. Syn. Pl. Glum.* i. 228; *Durand & Schinz, Consp. Fl. Afr.* v. 840.

COAST REGION: Cape Div.; without locality, *Harvey*, 295! Table Mountain, *Ecklon*, 929!

4. A. turgidulum (Stapf); culms 1–$2\frac{1}{2}$ ft. long, glabrous, 2–3-noded, upper 2–3 internodes more or less exserted, uppermost finally long exserted; leaves usually few near the base, about 3 higher up; sheaths terete, rather tight, very minutely puberulous or glabrous; ligule truncate, up to $\frac{3}{4}$ lin. long; blades linear, tapering to an acute point, up to 6 in. by $1\frac{1}{2}$ lin., flat or involute, subflaccid or more or less rigid, subglaucous, glabrous, rarely scantily hairy, scaberulous above; panicle contracted, erect or slightly nodding, $\frac{1}{2}$–1 ft. long; branches fascicled, very unequal, the longer up to $1\frac{1}{2}$ in. long, scantily branched or simple, erect or a few spreading, filiform, scabrid; spikelets 4–$5\frac{1}{2}$ lin. long, greenish, compactly 3–4-flowered; rhachilla slender, joints up to 1 lin. long, bearded near the tips; glumes lanceolate, acuminate, the lower $2\frac{1}{2}$–3 lin., the upper 4–$4\frac{1}{2}$ lin. long; valves exserted, oblong-lanceolate, the lowest $3\frac{1}{2}$–4 lin. long, glabrous, light green, slightly purplish below the tips; obscurely granulated, tips scarious, 2-toothed, teeth produced into fine bristles 1–$1\frac{1}{2}$ lin. long; callus bearded, very short; awn from the middle, rather fine, column $2\frac{1}{2}$–3 lin. long, bristle 5–6 lin. long; pales 3 lin. long, keels ciliate; anthers $\frac{1}{2}$–1 lin. long; ovary pubescent from the middle, top hispidulous; grain $1\frac{1}{4}$ lin. long. *Trisetum antarcticum, Nees, Fl. Afr. Austr.* 346 (*partly*). *T. imberbe, Nees, l.c.* 347; *Steud. Syn. Pl. Glum.* i. 228; *Durand & Schinz, Consp. Fl. Afr.* v. 838. *Avena turgidula, Stapf in Kew Bulletin,* 1897, 293.

SOUTH AFRICA: without locality, *Ecklon & Zeyher*, 463!

COAST REGION: Queenstown Div.; Shiloh, 3500 ft., *Drège! Baur*, 776!

CENTRAL REGION: Aliwal North Div.; Leeuwen Spruit, between Kraai River and the Witte Bergen, 4500–5000 ft., *Drège!*

KALAHARI REGION: Transvaal; Pretoria, at Wonderboom Poort, *Rehmann*, 4493!

EASTERN REGION: Transkei; Gcua (Gekau) River, below 1000 ft., *Drège*. Tembuland; Bazeia, 2000 ft., *Baur*, 364! Natal; Umsinga, and foot of the Biggars Berg, *Buchanan*, 100! near Greytown, *Buchanan*, 172! Riet Vlei, 4000–5000 ft., *Buchanan*, 156!

5. **A. Dodii** (Stapf); perennial; culms erect, slender, about 3 ft. high, glabrous, smooth, 3–4-noded, sheathed all along or nearly so, with 1–2 erect intravaginal branches from the lowest nodes; leaves about 3 from near the base, and 3–4 higher up, distant; sheaths narrow, tight or the upper rather loose, lowest 6–9 in. long, involute, glabrous, smooth; ligules oblong, up to 2 lin. long; blades linear, lower tapering from a long attenuate base to a fine point, 9–20 in. by 2–$2\frac{1}{2}$ lin., flat or with involute margins, rigid, more or less glaucous, glabrous, smooth below, strongly striate and scabrid on the upper surface; panicle contracted, 8–12 in. long, narrow, dense, slightly nodding; rhachis smooth; branches fascicled, unequal, divided from the base or nearly so, longest up to 2 in. long, erect, strict, scaberulous or smooth below; pedicels very unequal, lateral very short or scarcely any; spikelets 5–6 lin. long, narrow, erect; florets about 4–5; glumes narrow, linear-lanceolate, shortly aristulate, lower up to 4, upper to 5 lin. long (inclusive of the bristle); rhachilla slender, joints up to 1 lin. long, hairs up to $1\frac{1}{2}$ lin. long; valves distinctly exserted, lanceolate, lowest 3 lin. long (up to the awn), glabrous, light green, rather firm, finely granulate, tips scarious, bifid, lobes produced into fine long bristles; callus short, rather stout, bearded; awn from above the middle, fine, scabrid, pallid, column 3–2 lin. long, bristle 6–7 lin. long; pales $2\frac{1}{2}$ lin. long, keels scabrid; anthers 1 lin. long; ovary puberulous below the suboblique top.

COAST REGION: Cape Div.; wet slopes near Oatlands Point, *Wolley Dod*, 2775!

This differs from *A. longum* in the very long rigid blades, the more crowded smaller spikelets, the shorter hairs of the rhachilla and the small callus. It is in some respects intermediate between it and *A. antarcticum.* This is, however, a smaller plant with a much scantier panicle, shorter blades, and a glabrous or very scantily hairy rhachilla. The African species of *Avenastrum* represent a series of very closely allied forms the definition of which is very difficult and still not quite satisfactory.

6. **A. antarcticum** (Stapf); densely cæspitose with numerous barren shoots; culms $1\frac{1}{2}$–3 ft. long, glabrous, 3-noded, upper 2–3 internodes more or less exserted; leaves few near the base, about 3 ft. higher up, distant; sheaths terete, tight, glabrous or with fine spreading hairs, smooth or scaberulous, the lowest persistent; ligule oblong, up to $1\frac{1}{2}$ lin. long; blades very narrow, linear, tapering to an acute point, usually involute, subsetaceous, the lower 2–9 in. long, subrigid, rarely flaccid, glabrous or hairy, finely nerved, smooth or scaberulous; panicle contracted, linear, erect, stiff or flexuous, 5–7 in. long, lower branches in pairs, unequal, the longer up to 2 in. long and to 8-spiculate, erect or suberect, scaberulous; spikelets $3\frac{1}{2}$–$5\frac{1}{2}$ lin. long, greenish, loosely 3–4-flowered; glumes lanceolate, acuminate, very unequal, the lower 2–$3\frac{1}{2}$ lin., the upper 3–$4\frac{1}{2}$ lin. long; rhachilla joints up to $\frac{1}{2}$–1 lin. long; glabrous or scantily and shortly hairy towards the tips; valves long exserted, lanceolate, acuminate, the lower 3–4 lin., rarely 5 lin. long, glabrous, light green, scaberulous, tips scarious, 2-toothed or bifid, the teeth usually produced into very fine and short bristles; callus very short, minutely bearded; awn from the middle, fine, column pallid, 2–$3\frac{1}{2}$ lin. long, bristle 3–$4\frac{1}{2}$ lin. long, horizontally spreading or reflexed; pales $1\frac{1}{2}$–$3\frac{1}{2}$ lin. long, keels ciliolate or ciliate; anthers $\frac{1}{4}$–1 lin. long; ovary pubescent below the hispidulous top. *Avena antarctica, Thunb. Prodr.* 22; *Fl. Cap. ed. Schult.* 117; *Roem. & Schult. Syst.* ii. 676; *Kunth, Enum.* i. 305. *A. leonina, Steud. in Flora,* 1829, 484; *Kunth, Rév. Gram.* ii. 521, *t.* 175; *Enum.* i. 303; *Trin. Gram. Suppl.* 29. *Trisetum antarcticum, Nees in Linnæa,* vii. 307; *Fl. Afr. Austr.* 346 (*partly*); *Steud. Syn. Pl. Glum.* i. 227; *Durand & Schinz, Consp. Fl. Afr.* v. 838. *Danthonia leonina, Steud. ex Kunth, Enum.* i. 303.

SOUTH AFRICA: without locality, *Drège in Hb. Lübeck!*
COAST REGION: Cape Div.; Lion Mountain, *Ecklon*, 928! Signal Hill near Lion Battery, *Wolley Dod*, 2747! Table Mountain, *Pappe!* near Maitland Station, *Wolley Dod*, 3167! below P.W. Block-house, *Wolley Dod*, 1474! Caledon Div.; Zwart Berg, near the Hot Springs, 1000–2000 ft., *Ecklon & Zeyher*, 4553! Riversdale Div.; near the Zoetemelks River, *Burchell*, 6694! Uitenhage Div.; Zwartkops River, *Ecklon*. Albany Div.; near Grahamstown, *MacOwan*, 1302! Glenfilling, 1000 ft., *Drège*. Komgha Div.; near Komgha, *Flanagan*, 935! Queenstown Div.; Finchams Nek, near Queenstown, 4000 ft., *Galpin*, 2381!

Avena leonina seems to be a shade form with flaccid leaves and slightly larger spikelets.

Avena symphicarpa, Trin. ex Steud. Nom. ed ii. i. 173 (*Trisetum hirtum*, Nees, Fl. Afr. Austr. 350), was described from specimens collected by Ecklon in virgin forests on Olifants Hook near the Boschesman River, and is possibly a state of *A. antarcticum*. I have not seen these specimens, but Drège's example in the Lübeck Herbarium was named "*T. hirtum*," and answers very well the description. Another specimen, named *T. hirtum* by Nees in Linnæa, xx. 254, from the Zwarte Bergen, Caledon Div., is undoubtedly *A. antarcticum*.

7. **A. caffrum** (Stapf); culms 2 ft. long, glabrous, 3–4-noded, sheathed to the base of the panicle; leaves 4–6 near the base, 3 higher up; sheaths tight below, loose or open above, glabrous; ligule oblong, up to $\frac{3}{4}$ lin. long; blades very narrow, subsetaceous, convolute, the lower 5–7 in. long, those of the barren shoots to 1 ft. long, glabrous, strongly and closely few-nerved, margins rough; panicle contracted, about 6 in. long, slightly nodding and subflaccid; rhachis filiform; branches fascicled, very unequal, the longest up to 2 in. long, sparingly branched or simple, finely filiform, flexuous, erect, scaberulous to finely hispidulous; spikelets 4–4$\frac{1}{2}$ lin. long, loosely 3- to sub-4-flowered; rhachilla joints very slender, up to 1 lin. long, with long white hairs; glumes very thin, lanceolate, acuminate, often aristulate, the lower 2–2$\frac{1}{2}$ lin., the upper 2$\frac{1}{2}$–3$\frac{1}{2}$ lin. long; valves exserted, linear-lanceolate, the lowest 3–4 lin. long, glabrous, pallid, rather thin, smooth, faintly nerved, tips scarious, subbifid, produced into short fine bristles; callus $\frac{1}{4}$ lin. long, bearded; awn from above the middle, fine, pallid, column 2$\frac{1}{2}$–2 lin. long, bristle 5 lin. long; pales 3 lin. long, keels minutely ciliolate; anthers up to $\frac{1}{2}$ lin. long; ovary top hispidulous. *Trisetum longifolium, Nees, Fl. Afr. Austr.* 348 (*partly?*). *Avena caffra, Stapf in Kew Bulletin*, 1897, 293.

VAR. (?) **natalensis** (Stapf); internodes exserted, very minutely puberulous below the nodes; leaves up to 2 lin. broad, flat or involute, subflaccid, more or less hairy above; spikelets rather smaller than in the typical form.

CENTRAL REGION: Aliwal North Div.; Witte Bergen, on rocks, 7500 ft., *Drège!*

EASTERN REGION: Var. *natalensis*, Natal: Riet Vlei, 4000–5000 ft., *Buchanan*, 238!

Nees also indicates the species from the dunes near Capetown and from Tulbagh. I have not seen Ecklon's specimens from these localities, but it seems to me very probable that his *Trisetum longifolium* is a mixture of two different species.

The variety is extremely similar to *Avena Rothii, Kew. Bull.* 1897, 292, from Abyssinia, from which it differs in the much smaller number of nodes, and in the valve tips being acute or very obscurely aristulate.

XXXIV. AVENA, Linn.

Spikelets large or very large, 7–20 lin. long, pendulous in open, usually very lax panicles; rhachilla hairy or glabrous, disarticulating above the glumes and between the valves, or only above the glumes, or not at all in cultivated forms, usually terminated by a rudimentary valve. *Florets* 3–5, the lower 1 or 2 (rarely 3) ⚥, the upper reduced, smaller, ♂ or barren or quite rudimentary. *Glumes* 2, persistent, scarious, equal or subequal, acute or acuminate, rounded on the back, 7–11-nerved. *Valves* distinctly shorter or just equalling the glumes, subherbaceous, with rather rigid scarious tips, acute or acuminate, bifid, with or without bristles from the lobes, 5–9-nerved, lower 1–3-awned; awn dorsal from the middle or slightly above it, kneed and twisted below or (in the upper valves) imperfect; callus short, villous (or imperfect and glabrous in cultivated forms). *Pales* shorter than the valves, 2-keeled, 2-toothed or bifid. *Lodicules* 2, rather large, hyaline, entire. *Stamens* 3. *Ovary* densely villous from the base; styles distinct, extremely short or 0; stigmas laterally exserted, plumose. *Grain* oblong, subterete, grooved in front, hairy, pallid, somewhat soft or at least easy to cut, tightly embraced by the hardened valve and the pale; hilum fine, linear, long; embryo small.

Annual; blades linear, flat, flaccid; ligules hyaline or scarious; panicle usually very lax, often secund, with large pendulous spikelets.

Species about 7, indigenous in the Mediterranean region, some of them widely spread as weeds, 1 known only in numerous cultivated forms.

Valves hairy at the base only or quite glabrous; rhachilla wholly tough or tardily disarticulating above the glumes (1) **sativa.**
Valves hairy to or beyond the middle; rhachilla freely disarticulating above the glumes, often also between the valves:
  Spikelets 15–20 lin. long; rhachilla tough and glabrous between the valves ... ... (2) **sterilis.**
  Spikelets 9–14 lin. long; rhachilla villous and disarticulating between the valves:
    Valves shortly bifid ... ... (3) **fatua.**
    Valve lobes produced at the tips into long bristles ... (4) **barbata.**

1. **A. sativa** (Linn. Spec. Pl. 79); culms simple; leaf-sheaths glabrous; ligules shortly ovate, $1\frac{1}{2}$–$2\frac{1}{2}$ lin. long; blades linear or lanceolate-linear, glabrous, scaberulous; panicle open; branches spreading equally all round or contracted and secund; spikelets 9–12 lin. long or longer, usually with a 1-awned floret at the base and 1 or 2 awnless florets above, or with all the florets awnless; rhachilla tough or tardily disarticulating at the base, glabrous or almost so; glumes broad-lanceolate, 7–11-nerved; valves lanceolate, acuminate, shortly bifid or 2–4-toothed, glabrous, rarely with a few scattered hairs, the lowest 7–10 lin. long; awn from the middle or slightly above it, scabrid; anthers 2 lin. long; ovary hairy all over; grain tightly enclosed by its valve and pale, free, silky all over. *Host, Gram. Austr.* ii. *t.* 59; *Kunth, Enum.* i. 301; *Suppl.* 255, *t.* 20, *fig.* 1; *Trin. Gram. Suppl.* 23; *Nees in Linnæa*, vii. 305; *Fl. Afr. Austr.* 351; *Steud. Syn. Pl. Glum.* i. 230; *Benth. & Trin. Med. Pl. t.* 292; *Körn. & Wern. Handb. Getreidebaues*, i. 192; *Sowerby, Grasses Gr. Brit. t.* 110; *Durand & Schinz, Consp. Fl. Afr.* v. 843.

VAR. β **orientalis** (Trin. Gram. Suppl. 23); panicle contracted, secund; rhachilla glabrous; valves glabrous, or the lowest with stiff hairs at the base, awnless or the lowest awned. *Körn. & Wern, Handb. Getreidebaues*, i. 212. *A. orientalis*, *Schreb. Spicil. Fl. Lips.* 52; *Host, Gram. Austr.* iii. *t.* 44; *Kunth, Enum.* i. 302; *Suppl.* 255; *Nees in Linnæa*, vii. 306; *Fl. Afr. Austr.* 352; *Steud. Syn. Pl. Glum.* i. 230; *Sowerby, Grasses Gr. Brit. t.* 111; *Durand & Schinz, Consp. Fl. Afr.* v. 843.

COAST REGION: Cape Div.; foot of Table Mountain, in fields and gardens, *Ecklon*. Tulbagh Div.; Tulbagh Waterfall, *Ecklon*. Var. *orientalis*: Cape Div., with the typical form according to *Nees*.

Widely cultivated in the temperate regions of both hemispheres. Probably of Mediterranean origin.

The occurrence of the variety in the Cape Colony is doubtful. Nees's note at the end of the paragraph on *A. orientalis* is rather in contradiction to the description given in the same place, and points to a confusion with some form of the typical *A. sativa.*

2. **A. sterilis** (Linn. Spec. Pl. ed. ii. 118); culms usually fascicled with few or no barren shoots; sheaths glabrous, rarely the lower hairy; ligules obtuse, 1–2 lin. long; blades linear to lanceolate-linear, up to 1 ft. by 3–9 lin., glabrous, rarely scantily hairy, scabrid; panicle lax; branches spreading equally all round or secund; spikelets 15–20 lin. long, with 2-awned florets at the base and 1–2 (rarely 3) smaller awnless florets above them; rhachilla freely disarticulating above the glumes, tough and glabrous between the valves; glumes broad-lanceolate, acuminate, 9–11-nerved; valves lanceolate, long acute, finely 2-toothed, lowest 10–13 lin. long, usually pallid below and greenish towards the tips, scabrid to hispidulous upwards, with long white or whitish hairs up to the middle, 7-nerved, only the 2 lowest awned; awn from the middle, scabrid to almost villous below, column more or less brown, often pallid, stout, 7–9 lin. long, bristle 1–2 in. long; anthers almost 2 lin. long; grain tightly embraced, free, silky all over. *Thunb. Prod.* 23; *Host, Gram. Austr.* ii. *t.* 57; *Jacq. Icon.* i. *t.* 23; *Kunth, Enum.* i. 303; *Trin. Gram. Suppl.* 24, *and in Mém. Acad. Pétersb. sér.* 6, iv. 26; *Reichb. Icon. Fl. Germ.* i. *t.* 103, *fig.* 1711; *Durand & Schinz, Consp. Fl. Afr.* v. 843.

SOUTH AFRICA: without precise locality, *Thunberg!*

A weed of Mediterranean origin.

3. **A. fatua** (Linn. Spec. Pl. 80); culms solitary or few in a tuft, with few or no barren shoots; leaf-sheaths glabrous or the lower more or less hairy; ligules short, very obtuse, up to 1½ lin. long; blades linear to lanceolate-linear, up to 1 ft. by 6 lin., glabrous or rarely sparsely hairy, scabrid; panicle open or contracted; branches spreading equally all round or more or less erect and subsecund; spikelets 9–11 lin. long, with 2–3-awned florets and with or without a rudimentary, usually minute, awnless floret above them; rhachilla freely disarticulating below and more or less so between the valves, joints between the valves villous; glumes broad-lanceolate, acuminate, 7–9-nerved; valves lanceolate, acute, shortly 2–4-toothed, the lowest 6½–9 lin. long, usually brown below and green towards the tips, scaberulous, with stiff brown hairs to the middle or subglabrous with the exception of the very short callus, 7-nerved, all awned except the rudimentary uppermost; awn from the middle, scabrid, column very dark, 4½–7 lin. long, bristle ¾–1½ in. long; anthers 1–1¼ lin. long; ovary villous all over; grain 3–4 lin. long, tightly embraced, free, silky all over. *Schreb. Beschreib. Gräs.* 109, *t.* 15; *Host, Gram. Austr.* ii. *t.* 58; *Engl. Bot. t.* 2221; *Fl. Dan. t.* 1629; *Kunth, Enum.* i. 302; *Trin. Gram. Suppl.* 24; *Nees in Linnæa,* vii. 306; *Fl. Afr.*

Austr. 352; *Reichb. Icon. Fl. Germ.* i. *t.* 103, *fig.* 1712; *Steud. Syn. Pl. Glum.* i. 230; *Sowerby, Grasses Gr. Brit.* 126, *t.* 109; *Schweinf. in. Bull. Herb. Boiss.* ii. *App.* ii. 31; *Durand & Schinz, Consp. Fl. Afr.* v. 842.

COAST REGION: Cape Div.; Table Mountain and Lion Mountain, *Ecklon*, 50, 52, 925! by the railway near Rondebosch, *Wolley Dod*, 1903!

A widely-spread weed of Mediterranean origin.

4. **A. barbata** (Brot. Fl. Lusit. i. 108); stems erect, usually 2–3 from a dense tuft of barren shoots, glabrous; leaf-sheaths villous to glabrescent; ligules broad-ovate, up to 2 lin. long; blades linear, up to 9 in. by 3 lin., but usually shorter and much narrower, villous to glabrescent, scabrid; panicle or racemes up to $\frac{3}{4}$ ft. long, nodding, usually secund; branches scantily branched or simple, finely filiform, flexuous; spikelets 9–14 lin. long, 2–3-flowered, 2–3-awned, 7–9-nerved; rhachilla freely disarticulating below and between the valves, joints between the valves villous; glumes lanceolate, acuminate; valves narrowly lanceolate, acuminate, bifid; lobes produced into fine bristles, the lowest 6–9 lin. long (excluding the bristles), pallid to brown below the long whitish scarious tips, densely covered with fine stiff whitish or yellowish hairs to the middle, scaberulous, 7-nerved, all awned except the uppermost; awn from the middle, scabrid, column dark; anthers 1–1$\frac{1}{4}$ lin. long; ovary villous all over; grain tightly embraced, free, slender, oblong, 3–4 lin. long, silky. *Coss. et Dur. Expl. Algér.* ii. 112; *Durand & Schinz, Consp. Fl. Afr.* v. 840. *A. hirsuta, Roth, Catal. Bot.* iii. 19; *Kunth, Enum.* i. 302; *Trin. Gram. Suppl.* 25; *Nees, Fl. Afr. Austr.* 352; *Webb, Phyt. Canar.* iii. 400, *t.* 247; *Steud. Syn. Pl. Glum.* i. 230; *Doell in Mart. Fl. Bras.* ii. iii. 99, *t.* 29, *fig.* 1.

COAST REGION: Cape Div.; by the railway near Rondebosch, *Wolley Dod*, 1902! Worcester Div.; Hex River, *Drège*. Mossel Bay Div.; Gauritz River, *Drège*.

Throughout the Mediterranean region, introduced elsewhere.

XXXV. **PENTASCHISTIS**, Stapf.

Spikelets from slightly over 1 lin. to 6 lin. long, very rarely longer, laterally compressed, pedicelled, panicled; rhachilla disarticulating above the glumes and between the valves, continued as a usually very minute bristle beyond the upper floret. *Florets* 2, ⚥, much shorter than the glumes. *Glumes* equal or subequal, lanceolate in profile, acuminate, keeled, hyaline or subhyaline, 1-nerved or closely 3-nerved at the very base. *Valves* membranous, hairy with the hairs seriate between the nerves, or glabrous excepting the always shortly hairy small callus, finely or obscurely 5–9-nerved, 2-lobed, awned from the sinus; lobes with a fine bristle from the inner angle, to which they are usually more or less adnate or pass into, sometimes very small, rarely 3–4-fid with all the divisions bristle-like. *Pales* 2-keeled, 2-toothed, more or less equalling the valves (exclusive of the lobes). *Lodicules* 2, small, cuneate, nerved, usually glabrous.

Stamens 3. *Ovary* oblong, glabrous; styles distinct, very slender; stigmas plumose, laterally exserted. *Grain* oblong, semiterete to subterete, shallowly grooved in front; hilum obscure, linear-oblong, $\frac{1}{4}$–$\frac{1}{3}$ the length of the grain; embryo about $\frac{1}{4}$ the length of the grain.

Perennial, rarely annual; leaves very variable; panicle usually distinctly trichotomous with swollen and often bearded axils, open or contracted, sometimes spike-like; spikelets more or less glistening.

Species over 40, mainly in South, a few in tropical Africa, 1 in Madagascar, and 1 in St. Paul's Island.

Recognized by Nees as a sub-genus (*Ind. Sem. Hort. Bot. Vratisl.* 1835); now separated from *Danthonia* by endemic distribution, habit and reduction of florets to 2.

Section 1. Panicle eglandular; spikelets 3–8 lin. long; blades mostly filiform-convolute, long; perennial.

*Spikelets 4–8 lin. long; basal sheaths not breaking up into fine fibres:
  Panicle obovate when contracted; basal sheaths broad, silky tomentose:
    Spikelets 7–8 lin. long ... ... (1) **aristidoides.**
    Spikelets 5–6 lin. long ... ... (2) **viscidula.**
  Panicle oblong to ovoid; basal sheaths not silky tomentose:
    †Blades rather broad, flat or loosely involute or folded; ligular fringe straight:
      Panicle 4–6 in. long; spikelets brownish-straw-coloured; keel of glumes glabrous and smooth; side-bristles of valves stiff, 2–2½ lin. long ... ... (3) **pallescens.**
      Panicle 2–5 in. long; spikelets silvery, much glistening; keel of glumes ciliate below, scaberulous above; side-bristles of valves very slender, 4–5 lin. long ... (4) **argentea.**
    ††Blades very narrow, filiform-convolute:
      ‡Panicle not spike-like:
        Panicles oblong, 4–10 in. long, usually dense; blades 1 to more than 2 ft. long:
          Sheaths broad; ligular fringe curved:
            Spikelets 5½–6 lin. long; glumes quite smooth; blades smooth ... ... (5) **nutans.**
            Spikelets 4¼–4½ lin. long; glumes scaberulous on the keel; blades more or less scabrid ... ... (6) **tortuosa.**
          Sheaths narrow; blades almost wiry, channelled; ligular fringe straight:
            Sheaths densely woolly or villous in the upper part, lower often subflabellate (7) **eriostoma.**
            Sheaths adpressedly hairy between the nerves, at length glabrescent, or glabrous, not flabellate (8) **juncifolia.**

Panicles oblong, 2–3 in. long, lax, often very scanty; blades ½–1 ft. long, finely setaceously convolute (9) **colorata.**

‡‡Panicle spike-like, compact, ovoid to oblong; leaves quite glabrous, blades setaceously involute, very firm ... (10) **curvifolia.**

**Spikelets 3–4½ lin. long:

Spikelets 3¾–4½ lin. long; basal sheaths at length breaking up into persistent fibres; culms slender, very few-noded; panicle 2–4 in. long, lax and delicately branched, not flaccid:

Blades up to 1 lin. broad when unfolded, flat or filiform-convolute, pubescent above the ligule; awns distinctly exserted from the glumes ... ... (11) **fibrosa.**

Blades setaceous, subterete, folded, quite glabrous; awns scarcely exserted from the glumes ... ... (12) **Tysonii.**

Spikelets 3–4 lin. long; basal sheaths not breaking up into persistent fibres; culms 4- to many-noded; panicle effuse or flaccid:

Panicle 6 in. by 6 in., effuse; awn kneed and twisted below, side-bristles 2–2½ lin. long ... ... (13) **natalensis.**

Panicle 2–4 in., flaccid; awn not kneed nor twisted, fine, flexuous, side-bristles 5–6½ lin. long ... ... (14) **capensis.**

Section 2. Panicle eglandular; ½–2 in. long, very scanty; spikelets 3½–8 lin. long; blades ½–1½ in. long, often spreading, and more or less subulate.

Annual; spikelets 7–8 lin. long; blades soft ... ... (15) **triseta.**

Perennial; spikelets 3½–4½ lin. long; blades rigid:

Axils of panicle long-bearded; blades very firm, conspicuously distichous, spreading, glabrous, very numerous ... ... (16) **acinosa.**

Axils of panicle glabrous; blades erect or suberect, thinner, hirsute ... ... (17) **elegans.**

Section 3. Panicle or leaves or both more or less gland-tubercled, or if eglandular, then the spikelets less than 3 lin. long.

Panicle very slender, acute, about 5 in. long; blades filiform-convolute, up to 1 ft. long, very rough from innumerable minute tubercles ... ... (18) **Lima.**

Panicle obtuse in the contracted state, or if more or less acute, then much smaller and the blades different:

Culms erect, rather stout, 1–2 ft. long; blades ½–¾ ft. by 2–2½ lin., rigid, often flat; panicles 3–6 in. long; spikelets 3–3½ lin. long; perennial:

Panicle contracted, rather dense in the upper part; divisions unequal and rapidly decreasing upwards; axis and lower branches with tubercular glands ... ... (19) **Zeyheri.**

Panicle open, very lax, rather symmetrically dichotomous, all divisions (except the ultimate) long:

Axis and branches of panicle scabrid, without tubercular glands; glumes 3 lin. long, narrow, smooth ... ... (20) **hirsuta.**

Axis and branches of panicle with copious tubercular glands; spikelets clustered at the tips of the branchlets; glumes 3–3½ lin. long, rather broad, scaberulous along the keels ... ... (21) **rupestris.**

Culms usually more or less ascending, or, if erect, then annual or short or the leaves not flat and broad (see No. 36, *P. longipes*); panicles ½–3 in. long; spikelets 1–4 lin. long, very rarely longer:

*Perennial:

†Blades more or less flat or subulate-involute, 1–2½ lin. broad; culms loosely fascicled, ascending from an often many-noded and densely leafy base; spikelets 3–4 lin. long:

Blades rather thick, very rigid, straight, not curling at length, not or scantily tubercled, slightly striate:

Culms many-noded, densely and distichously leafy at the base; sheaths scantily bearded or beardless; ligular fringe inconspicuous; blades 1–3 in. long ... ... (22) **subulifolia.**

Culms 5–6- or more-noded; lower leaves fewer and more distant; sheaths more or less dull purplish, very conspicuously bearded; ligular fringe very conspicuous; blades 2–5 in. long ... ... (23) **leucopogon.**

Blades thinner, less rigid, very closely and prominently striate, flexuous and at length curled, margins fringed with numerous tubercles ... ... (24) **aspera.**

††Blades filiform-convolute or setaceous, or flat; spikelets scarcely 2½ lin. long:

Blades firm, filiform or setaceous; culms densely tufted with numerous short-sheathed innovations; spikelets 2¼–4 lin. long:

Side-bristles 1 on each side of the valve:

Spikelets 3¾–4 lin. long; culms 5–6 lin. long; blades setaceous, glabrous, 1–4 in. long ... (25) **Burchellii.**

Spikelets 2¼–3 lin. long:

Both valves alike:

Blades densely pubescent to villous, 1–1½ in. long, recurved; glumes pubescent ... ... (26) **tomentella.**

Blades glabrous, rarely very scantily hairy, 2–8 in. long, flexuous or circinate; glumes glabrous ... (27) **angustifolia.**

Lower valve with 3 equal fine bristles united at the base, slightly exceeding the glume; upper with a long exserted kneed and twisted awn and exserted side-bristles (28) **heterochæta.**

Side-bristles 3–4 on each side of the valve ... ... (29) **heptamera.**

Blades soft or flaccid, or somewhat rigid and then usually very fine or very short (see also No. 36, *P. longipes*); spikelets 1–2½ lin. long:

‡Panicle very lax, open (at least temporarily) (see also No. 35, *P. Thunbergii*), ¾–3 in. long:

Valves with the lobes gradually passing into bristles; blades not over 2 in. long, hairy ... (30) **Jugorum.**

Valves with the lobes distinct from the bristles:

Culms ¾–1½ ft. long; blades 2–6 in. long, finely filiform or setaceous:

Both valves awned, lobes truncate; blades 2–3½ in. long, glabrous, obscurely striate, margins scantily tubercled ... (31) **filiformis.**

Lower valve usually awnless, lobes acute; blades 3–6 in. long, rather rigid, distinctly striate, not tubercled along the margins, although often with tubercled-based hairs on the back ... (32) **imperfecta.**

Culms 3–4 in. long, many-noded, densely leafy below, sometimes longer and thin, forming loose tufts of weak tangled prostrate or ascending branches; blades ½–1 in. long, flat ... (33) **densifolia.**

‡‡Panicle contracted, usually dense (rarely open in No. 35, *P. Thunbergii*), 1–5 in. long:

Valves glabrous; lobes passing into bristles; awn very short and fine, scarcely kneed; panicle compact ... ... (34) **brachyathera.**

Valves usually hairy; lobes distinct from the bristles, acute; awn distinctly kneed and exserted:

Panicle 1–3 in. long; lowest branches ½–1½ in. long, ultimate divisions and pedicels short; spikelets crowded towards the ends of the branches (35) **Thunbergii.**

Panicle 5 in. long; lowest branches up to 4 in. long; ultimate divisions and pedicels very long and fine ... (36) **longipes.**

**Annual:

Spikelets 2–2½ lin. long; anthers 1 lin. long:

Leaves pubescent all over from tubercle-based hairs, or glabrous; blades filiform or at least involute, without or with a few minute tubercles along the margins; panicle very lax, more or less divaricate ... ... (37) **patula.**

Leaves with seriate, often stalked, conspicuous glands on the sheath-nerves and the blade margins; blades flat; panicle more or less contracted ... ... (38) **euadenia.**

Spikelets 1¼–1½ lin. long; anthers ⅛–⅙ lin. long ... ... (39) **airoides.**

1. P. aristidoides (Stapf); perennial, compactly cæspitose; culms erect, rather stout, 1–2 ft. long, glabrous, smooth, few-noded, internodes enclosed or the upper exserted; sheaths lax, striate, ciliate or villous along the margin, otherwise glabrous except the lowest which are densely tomentose at the base and more or less persistent; ligule a dense fringe of short hairs; blades linear, tapering to an oblique subacute point, filiform-involute, ½–1½ ft. by 1–1¼ lin., flexuous or curved, firm, glabrous, smooth; panicle open, ovate, lax, or contracted, 4–6 in. long, erect or somewhat nodding; branches 2-nate, filiform, smooth, glabrous and purplish like the axis, branchlets and pedicels, laxly trichotomous from ¾–1 in. above the base, lowest 2½–4½ in. long; pedicels very unequal, 1–4 lin. long; spikelets brownish-yellow, purplish at the back below, 6½–7½ lin. long, erect on sometimes nodding branchlets; glumes long and setaceously acuminate, subhyaline, scaberulous or puberulous; body of valves about 3 lin. long, shortly pubescent to villous, 7–9-nerved; lobes short or produced, acute and almost 1 lin. long; pubescent with a fine bristle 4–5 lin. long from the inner angle; awn 12–13 lin. long; kneed at a little below the middle; callus slender to ¾ lin. long; pales equalling the valves, glabrous except at the ciliolate tips; anthers 2 lin. long. *Avena aristidoides*, *Thunb. Prodr.* 22; *Fl. Cap.* 433; *ed. Schult.* 116. *Danthonia trichotoma*, *Nees*, *Fl. Afr. Austr.* 318; *Steud. Syn. Pl. Glum.* i. 242; *Durand & Schinz*, *Consp. Fl. Afr.* v. 855. *D. aristidoides*, *Lehm. ex Nees*, *l.c.*

COAST REGION: Piquetberg Div.; moist places on Piquetberg, 1500–2000 ft., *Drège!* Cape Div.; Table Mountain, 500 ft., *Ecklon*, *Bolus & MacOwan*, *Herb. Norm. Aust. Afr.* 793! Cape Flats near Rondebosch, *Bolus*, 4892! Constantia, *Ecklon*, Simons Bay, *Wright!* Simons Town, *Wolley Dod*, 542! Swellendam Div.; Hottentots Holland, *Ecklon!*

2. **P. viscidula** (Stapf); perennial, tufted; culms erect, slender, 1–1½ ft. long, glabrous, smooth, about 3-noded, sheathed all along or upper and intermediate internodes exserted; lower sheaths rather crowded, lowest scarious, softly villous or tomentose, particularly at the base and along the margins, pubescent upwards, that above it herbaceous, lax, striate, glabrescent or glabrous except at the villous margins, uppermost tight and narrow or tumid, sometimes quite glabrous, all villously bearded; ligule a dense fringe of hairs; blades filiform, canaliculate or involute, tapering to an acute point, 3–5 in. long, firm, flexuous, finely pubescent, particularly on the upper surface, or glabrous on the back, smooth; panicle erect, obovate, contracted, 2¼–5 in. long; branches smooth like the filiform axis and the subcapillary branchlets, bearded at the axils, laxly trichotomous from ½–1 in. above the base, lowest 2–3½ in. long, ultimate divisions and pedicels long, the latter scaberulous towards the tips; spikelets light straw-coloured, faintly tinged with purple, 5–6 lin. long; glumes finely acuminate, subhyaline, smooth or scaberulous upwards, keels smooth or very finely scaberulous; body of valves 2½–2¾ lin. long, finely pubescent near the base, otherwise glabrous, often purplish, 7-nerved; lobes very distinct, narrow, acute, ciliolate, almost or quite from the lateral bristles, these very fine, 3–5 lin. long; awn 8–9 lin. long, kneed below the middle; callus rather slender; pales 2½ lin. long, minutely 2-toothed; anthers 1½ lin. long. *Danthonia viscidula*, *Nees*, *Fl. Afr. Austr.* 303; *Steud. Syn. Pl. Glum.* i. 240; *Durand & Schinz*, *Consp. Fl. Afr.* v. 855.

SOUTH AFRICA: without precise locality, *Zeyher!*

COAST REGION: Clanwilliam Div.; Cederberg Range, near Ezelsbank, 3000–4000 ft., *Drège!*

3. **P. pallescens** (Stapf); perennial, densely tufted; culms erect, stout, 2–2½ ft. long, glabrous, smooth, 4–6-noded, sheathed nearly

all along; leaves hairy all over; sheaths lax, strongly striate, lowest crowded, about 4 in. long, broad, rather pallid, subpersistent, often villous, those above much longer, often slipping from the culm and rolling in, sparingly hairy except along the margins, uppermost subglabrous, all with dense villous beards at the mouth; ligule a dense transverse fringe of hairs; blades linear, long tapering to a filiform point, up to more than 1½ ft. long, 1–3 lin. broad, loosely involute or partly flat, flexuous, coarsely striate, somewhat hairy or glabrous below, smooth or scabrid towards the tips, upper side often glaucous; panicle oblong or ovate, open and lax or contracted, nodding, 4–6 in. long; axis rather stout below, filiform above; branches 2-nate, rarely the lowest 3-nate, like the branchlets and pedicels subflexuous, filiform, smooth, glabrous except the villously bearded axils, loosely trichotomous from 1–1½ in. above the base, lowest 1½–3½ in. long; lateral pedicels short or very short, stout; spikelets brownish straw-coloured, sometimes purple at the base, scarcely shining, 5–6 lin. long, clustered on the tips of the branchlets; glumes lanceolate in profile, long and finely acuminate, subscarious, finely pubescent, at least below, 1-nerved, keel smooth; valves oblong, body 2–2¼ lin. long, loosely hairy, 9-nerved; lobes ovate to lanceolate, acute, ¾–1 lin. long, scaberulous, more or less adnate to a stiff bristle from the inner angle, 2–2½ lin. long; awn 5–6 lin. long, kneed at a little below the middle, rather stout; purple and twisted below the knee; callus short, subacute, minutely villous; pales equalling the body of the valves, 2-toothed, scaberulous upwards; lodicules scantily ciliate; anthers 1½ lin. long; grain 1¼ lin. long. *Danthonia pallescens, Schrad. in Schult. Mant.* ii. 386; *Trin. Gram. Ic. t.* 64; *Gram. Gen.* 72, *and in Mém. Acad. Pétersb. sér.* 6, i. 72, *and* iv. 35; *Nees, Fl. Afr. Austr.* 316 (*excl. syn. Thunb.*); *Steud. Syn. Pl. Glum.* i. 242; *Durand & Schinz, Consp. Fl. Afr.* v. 853 (*excl. syn. Thunb.*). *Pentameris pallescens, Nees in Linnæa,* vii. 312; *Kunth, Enum.* i. 316.

COAST REGION: Clanwilliam Div.; Olifants River, *Ecklon & Zeyher*, 130! Cape Div.; Table Mountain, *Ecklon, Bergius! Spielhaus! MacGillivray! Burchell*, 544! *Harvey*, 150! Devils Peak, by the second waterfall, *Wolley Dod*, 1955! 3491! rocks on Constantia Berg, *Wolley Dod*, 1962! Worcester Div.; in marshy ground in Dutoits Kloof, 3500–4000 ft., *Drège!*

4. **P. argentea** (Stapf); perennial; culms erect or subascending, slender, 1–2 ft. high, glabrous or sometimes silky close to the nodes, smooth, about 4-noded, sheathed all along or intermediate internodes shortly exserted; sheaths somewhat tight or the uppermost tumid, open and exceeding the base of the panicle, lowest silky tomentose at the base, otherwise as those above more or less coarsely villous, particularly on the back and in the upper part, uppermost scantily hairy or finely pubescent, all conspicuously bearded; ligule a fringe of hairs; blades linear, very long tapering to a setaceous point, 6–9 in. by 1–2 lin., flat or loosely involute, rather firm, scantily hairy or glabrescent, strongly nerved and scaberulous on the upper surface, margins scabrid; panicle oblong, erect, or nodding, usually

dense, $2\frac{1}{2}$–5 in. long, silvery; branches 2-nate, smooth like the filiform axis and the subcapillary branchlets and pedicels conspicuously bearded at the axils, repeatedly trichotomous from 3–7 lin. above the base, lowest 1–$2\frac{1}{2}$ in. long, ultimate divisions and pedicels short; spikelets silvery, very shining, sometimes tinged with purple below, crowded, 5–6 lin. long; glumes long and delicately acuminate, hyaline, smooth or finely scaberulous; keels usually shortly ciliate below, scaberulous above; body of valves $2\frac{1}{4}$ lin. long, adpressedly or loosely hairy to or beyond the middle, distinctly 7-nerved; lobes narrow, acute, very distinct, ciliolate, almost free from the very slender side-bristles, these 4–5 lin. long and much exserted; awn 7–9 lin. long, kneed below the middle, slightly twisted; callus rather slender; pales $2\frac{1}{2}$–$2\frac{3}{4}$ lin. long, subentire or finely 2-mucronate; anthers $1\frac{1}{2}$ lin. long.

SOUTH AFRICA: without precise locality, *MacGillivray*, 406!
COAST REGION: Cape Div.; Table Mountain, *Spielhaus!* slopes of Orange Kloof, *Wolley Dod*, 3342! pastures above Simons Bay, frequent, *Milne*, 247!

This is very probably *Danthonia involuta* (Steud. Syn. Pl. Glum. i. 240; Durand & Schinz, Consp. Fl. Afr. v. 851).

5. **P. nutans** (Stapf); perennial, tufted; culms erect, stout, 3 ft. long, glabrous, smooth, about 4-noded, sheathed nearly all along; sheaths lax, firm, strongly striate, lower 4–5 in. long, rather broad, pallid, hairy in the upper part, upper glabrous, margins glabrous, mouth softly hairy; ligule a semi-oval fringe of very short hairs; blades tightly convolute or involute, filiform, tapering to a fine point, up to more than 2 ft. by less than 1 lin. when expanded, rigid, flexuous, glabrous, smooth; panicle oblong, contracted or rather lax, nodding, up to 8 in. long; rhachis rather stout below, filiform above; branches 2-nate, like the branchlets and pedicels flexuous, filiform, smooth and glabrous, unequal, loosely and unequally divided from $\frac{1}{2}$–1 in. above the base, longest up to 4 in. long; pedicels unequal, not under 2 lin. long; spikelets brownish straw-coloured with violet backs, $5\frac{1}{2}$–6 lin. long; glumes lanceolate, long acuminate, scarious, glabrous, smooth, 1- (or lower 3-) nerved at the very base; valves oblong, body $2\frac{1}{2}$ lin. long, rather firm, shortly pubescent all along, 9-nerved; lobes narrow, subacute, scaberulous, $\frac{1}{2}$–$\frac{3}{4}$ lin. long, with a fine bristle $2\frac{1}{2}$–3 lin. long from the inner angle, almost wholly adnate to it; awn 7–9 lin. long, kneed $\frac{1}{3}$ way up, twisted, dark below; callus short, obtuse, minutely villous; pales equalling the body of the valve, 2-mucronulate; lodicules ciliate; anthers $1\frac{3}{4}$ lin. long. *Danthonia nutans, Nees, Fl. Afr. Austr.* 314; *Steud. Syn. Pl. Glum.* i. 241; *Durand & Schinz, Consp. Fl. Afr.* v. 852.

COAST REGION: Caledon Div.; near Genadendal, by the Zonder Einde River, in the sandy river-bed, 500 ft., *Drège!*

6. **P. tortuosa** (Stapf); perennial, tufted; culms fascicled, with intravaginal shoots, erect, rather stout, 2–3 ft. long, glabrous, smooth,

about 4-noded, sheathed nearly all along; sheaths lax, firm, strongly striate, lower crowded, 4–5 in. long, pallid or purplish, sometimes slightly rough, finely hairy in the upper part or towards the margins or the lowermost glabrescent or all quite glabrous, mouth softly bearded; ligule a semi-oval fringe of long soft hairs; blades tightly and finely convolute-filiform, up to more than 1½ ft. long, rigid, flexuous or curled, glabrous, scabrid in the upper part, otherwise smooth or the upper very scabrid all along, sometimes dotted with longitudinal white striæ; panicle oblong, narrow, erect, 6–8 in. long, contracted; branches 2-nate, filiform, like the rhachis and the finely filiform branchlets and pedicels scabrid or scaberulous, glabrous, divided from 6–8 lin. above the base or almost from the base, lowest up to 3 in. long, divisions more or less (or the ultimate very) unequal; pedicels not under 1 lin. long; spikelets brownish straw-coloured, greenish towards the base or tinged with violet, 4¼–4½ lin. long; glumes lanceolate in profile, long and finely acuminate, scarious except on the narrow subherbaceous central part, rather thin, scaberulous in the upper part, 1-, or lower very closely 3-, nerved; keels scaberulous; valves oblong, body 2 lin. long, loosely pubescent all along, white, faintly 7- to sub-9-nerved; lobes narrow, oblong, scaberulous, subacute, up to ¾ lin. long, adnate in part to a fine bristle, 2½ lin. long from the inner angle; awn about 7 lin. long, kneed at ⅓ way up, twisted and dark below; callus short, obtuse, minutely villous; pales equal to the body of the valve, 2-mucronulate; lodicules glabrous; anthers 1 lin. long. *Danthonia tortuosa, Trin. Gram. Ic. t.* 68; *Gram. Gen.* 72, *and in Mém. Acad. Pétersb. sér.* 6, i. 72 (*not Gram. Suppl.* 34, *nor in Mém. Acad. Pétersb. sér.* 6, iv. 35); *Nees, Fl. Afr. Austr.* 313; *Steud. Syn. Pl. Glum.* i. 241; *Durand & Schinz, Consp. Fl. Afr.* v. 855.

SOUTH AFRICA: without precise locality, *Sieber*, 119!

COAST REGION: Cape Div.; top of a Kloof beyond Millers Point, rare, *Wolley Dod*, 2846! "top of Rhodes S. fence," *Wolley Dod*, 3117! 3120! Tulbagh Div.; Winterhoek Berg, 800–5000 ft., *Ecklon.*

Very closely allied to *P. nutans*, but differing in the finer blades which are very scabrid in the upper part, the longer hairs of the ligule, stiffer panicles, smaller spikelets and scaberulous glumes. Trinius describes the spikelets as 3 lin. long, Nees as 3–3¾ lin. long. I found them never less than 4 lin. long in Sieber's specimen.

7. **P. eriostoma** (Stapf); perennial, densely tufted; culms fascicled, often from a woody base, erect, firm, 2 ft. long, glabrous, smooth, very densely leafy and often closely covered below with the remains of the sheaths, 2–3-noded above, sheathed all along or nearly so; sheaths tight, firm, lower crowded, sometimes subflabellate, cobwebby, woolly or villous along the margins, and often all over in the upper part, otherwise glabrous, smooth and shining, not striate, lower very coriaceous, persistent; ligule a dense fringe of hairs; blades filiform, tapering to a setaceous point, canaliculate, wiry, 1–1½ ft. long, very hard, flexuous, finely striate, glabrous and smooth on the lower, minutely and densely scaberulous on the upper

side, margins scabrid; panicle narrow, oblong, 4–6 in. long; axis slender, scaberulous or smooth below; branches mostly 2-nate, unequal, suberect, filiform, usually branched from the base or near it, glabrous; branchlets and pedicels very unequal, subcapillary, minutely scaberulous or almost smooth; spikelets straw-coloured to whitish, 4–5½ lin. long; glumes lanceolate, long and finely acuminate, subhyaline, glabrous, smooth, 1-nerved; valves linear-oblong in profile, body 1½–2 lin. long, long hairy to villous, 7–9-nerved; lobes lanceolate, acute, ⅜–⅝ lin. long, thin, smooth or almost so, with a fine bristle 2–3 lin. long from the inner angle, adnate to it except at the tips; awn 5½–6½ lin. long, kneed below the middle, twisted; callus short, shortly hairy; pales equalling the valves, 2-toothed, glabrous, smooth; lodicules glabrous; anthers 1¼–1½ lin. long. *Danthonia eriostoma, Nees, Fl. Afr. Austr.* 304; *Steud. Syn. Pl. Glum.* i. 241; *Durand & Schinz, Consp. Fl. Afr. Austr.* v. 849.

SOUTH AFRICA: without precise locality, *Thom!*
COAST REGION: Caledon Div.; near Appels Kraal on the Zonder Einde River, *Zeyher*, 4552 in part! Swellendam Div.; Cannaland, *Ecklon & Zeyher*, 123, in St. Petersburg Herb.! Mossel Bay Div.; between the landing-place at Mossel Bay and Cape St. Blaize, *Burchell*, 6268! Little Brak River, 6185! Albany Div.; Bothas Berg, *Ecklon.*

8. **P. juncifolia** (Stapf); perennial, densely tufted; culms erect, firm, 2–3 ft. high, glabrous, smooth, sheathed all along or 1 of the intermediate internodes exserted; sheaths tight, firm, striate, sometimes very minutely hairy between the striæ, at length glabrescent or quite glabrous, lower close, long, coriaceous, persistent; ligule a fringe of hairs; blades very narrow from the base, filiform, canaliculate or convolute, terete, 1 to more than 2 ft. long, very rigid, erect, straight or subflexuous, adpressedly hairy, at length glabrescent or glabrous below, villous to subvillous on the upper surface, smooth; panicle linear to linear-oblong, contracted, dense, 4–10 in. long, erect, stiff; axis slender, stiff, scaberulous or smooth below; branches mostly 2-nate, unequal, erect, branched from the base or undivided for 3–6 (rarely 9) lin.; branchlets and pedicels unequal, filiform, scabrid, lateral pedicels short; spikelets light straw-coloured, 3½–5 lin. long; glumes lanceolate in profile, acutely acuminate, subhyaline to hyaline, smooth, glabrous, 1-nerved or obscurely 3-nerved at the base; keel smooth or scaberulous; valves oblong in profile, body 1½–2 lin. long, long and densely hairy, all over 7-nerved; lobes ovate to lanceolate, acute, ½–⅝ lin. long, with a fine bristle from the inner angle, 2½–3 lin. long, free or more or less adnate to it; awn 5–8 lin. long, kneed at ⅓ way up, twisted below; callus short, minutely hairy; pales equalling the body of the valve, entire or subentire, scantily pubescent, ciliolate above; anthers 1–1½ lin. long; grain 1¼ lin. long. *Danthonia curvifolia, β livida, Nees in Linnæa,* xx. 254.

COAST REGION: Swellendam Div.; Buffeljaghts River, 1000–2000 ft., *Zeyher*, 4545! Riversdale Div.; hills near Zoetemelks River, *Burchell*, 6750! 6761!

9. **P. colorata** (Stapf); perennial, densely tufted; culms fascicled, with intravaginal innovation shoots, branched below, like the branches erect, slender, wiry below, 1–1½ ft. long, glabrous, smooth, few-noded, uppermost node usually near the middle, exserted; sheaths very tight, firm, glabrous, rarely the lower hairy, usually beardless, smooth, often purple, striate, lower crowded, 2–3 in. long, persistent; ligule a dense fringe of very short hairs; blades very finely and setaceously convolute, ½–1 ft. long, rigid, very flexuous or curled, wholly glabrous or scaberulous on the upper side, smooth below; panicle oblong, rather lax, contracted, 2–3 in. long, stiff or somewhat nodding, often very scanty; rhachis filiform; branches 2-nate, finely filiform like the branchlets and pedicels, scaberulous, glabrous, subtrichotomous, or alternately and sparingly divided from 3–6 lin. above the base, often reduced to 3–1 spikelets, the longest 1–2 in. long; pedicels very unequal, down to 1 lin. long; spikelets brownish straw-coloured, purple at the back, 4½–6 lin. long; glumes lanceolate, finely acuminate, glabrous, smooth, scarious except the coloured back, 1-nerved, or lower closely 3-nerved at the base; valves oblong, body 2 lin. long, rigidly pubescent, sub-9-nerved; lobes narrow, rigid, acute, scabrid, up to 1 lin. long, free or more or less adnate to a stiff bristle from the inner angle, 2½ lin. long; awn 6–8 lin. long; kneed at a little below the middle, twisted and dark below; callus acute, minutely bearded; pales equalling the body of the valves, scantily puberulous; lodicules glabrous or subciliate; anthers 1¾ lin. long. *Avena colorata, Steud. in Flora,* 1829, 481. *Danthonia tortuosa, Trin. Gram. Suppl.* 34, *and in Mém. Acad. Pétersb. sér.* 6, iv. 35 (*excl. Syn. Pentameris Thouarsii*), *not* i. 72. *D. tortuosa, var. tenuior, Nees ex Drège in Linnæa,* xx. 254. *D. crispa, Nees, Fl. Afr. Austr.* 310, *excl. var. β*; *Steud. Syn Pl. Glum.* i. 241; *Durand & Schinz, Consp. Fl. Afr.* v. 848. *D. colorata, Steud. l.c. Pentameris tortuosa, Nees in Linnæa,* vii. 311.

VAR β, **polytricha** (Stapf); culms ½–¾ ft. high, very slender, usually sheathed all along; lowest sheaths pallid, glabrous, or almost so, the others purplish, with long white fine spreading hairs at least in the upper part and conspicuously bearded at the mouth; blades glabrous or with hairs as on the upper sheaths at the base, 3–4 in. long; panicles meagre, stiff; spikelets 6–6½ lin. long.

COAST REGION: Cape Div.; Table Mountain, *Ecklon,* 931! 936 in part in Kew Herb.! *MacOwan, Herb. Austr. Afr.* 1697! west slope of Slang Kop, *Wolley Dod,* 3001! near Camps Bay, *Wilms,* 3864! and without precise locality, *Alexander!* Paarl Div.; Great Drakenstein and at the foot of Paarl Mountain, 1000–5000 ft., *Drège!* Swellendam Div.; Puspus Valley, *Zeyher,* 4544! Caledon Div.; near Genadendal, 1500–2500 ft., *Drège,* 1681! Var. β: Cape Div., near Paulsberg and Batseta rocks, *Wolley Dod,* 2952!

The specimens from Cape Division have, on the whole, longer spikelets than the remainder (5–6 against 4½–5 lin.).

10. **P. curvifolia** (Stapf); perennial, compactly cæspitose; culms erect, firm, 1–2 ft. long, smooth, glabrous, 3–4-noded, internodes exserted; leaves crowded at the base, quite glabrous, smooth; sheaths tight, very firm, lower short, imbricate, persistent, slightly striate in the upper part; ligule a fringe of very short hairs; blades

narrow, linear, tapering to an acute or subacute point, up to more than $1\frac{1}{2}$ ft. by 1 lin. when expanded, rarely broader, nearly always setaceously involute, very firm, strongly flexuous or curled; panicle spike-like, oblong or ovate, usually very dense, 1–3 in. by $\frac{1}{2}$–1 in.; all divisions very short, scabrid; spikelets straw-coloured or brownish, $5\frac{1}{2}$–7 lin. long, imbricate; glumes lanceolate in profile, long, setaceously acuminate, scarious, 1-nerved, scaberulous; valves oblong in profile, body $1\frac{1}{2}$ to almost 2 lin. long, loosely hairy all over to villous; lobes $\frac{1}{4}$ lin. long, acute, more or less adnate to a bristle (3–4 lin. long) from the inner angle; awn 6–8 lin. long, kneed and twisted below the middle; pales equalling the valves, keels scabrid, back hairy; lodicules glabrous; anthers $1\frac{1}{4}$ lin. long. *Danthonia curvifolia, Schrad. in Schult. Mant.* ii. 386; *Trin. Gram. Gen.* 70, *and in. Mém. Acad. Pétersb. sér.* 6, i. 70; *Gram. Suppl.* 33, *and in Mém. Acad. Pétersb. sér.* 6, iv. 34 (*in part*); *Nees, Fl. Afr. Austr.* 321 (*excl. Syn. Thunb.*); *Steud. Syn. Pl. Glum.* i. 242; *Durand & Schinz, Consp. Fl. Afr.* v. 848. *D. livida, Trin. Gram. Ic. t.* 50, *and in Mém. Acad. Pétersb. sér.* 6, i. 70. *D. denudata, Nees, Fl. Afr. Austr.* 320; *Steud. l.c.*; *Durand & Schinz, l.c.* 849. *Avena glomerata, Steud. in Flora*, 1829, 483. *Pentameris curvifolia, Nees in Linnæa*, vii. 313; *Kunth, Enum.* i. 317; *Suppl.* 272.

SOUTH AFRICA: without locality, *MacGillivray*, 405! 406! *Harvey*, 274! 280!

COAST REGION: Cape Div.; Cape Flats, *Zeyher*, 1824! stony slopes of Table Mountain, 1500 ft., *MacOwan, Herb. Austr. Afr.* 1695! 1696! *Burchell*, 523! frequent at the summit of Table Mountain, *Milne*, 249! between the Lions Head and Signal Hill, *Wolley Dod*, 3149! Wynberg, *Wilms*, 3868! wood beyond Alphen Bridge, *Wolley Dod*, 3490! Constantia Nek, *Wolley Dod*, 3487! Silvermine swamp, common, *Wolley Dod*, 3389! Simons Bay, *Wright*! Green Point, *Bergius*! Paarl Div.; near French-Hoek or Great Drakenstein, *Drège*! Agter de Paarl, *Drège*. Tulbagh Div.; Tulbagh Waterfall, *Ecklon*! Worcester Div.; Drakenstein Mountains, Bains Kloof, *Rehmann*, 2300! mountains above Worcester, *Rehmann*, 2590! peaty plains near Dutoits Kloof, 3000–4000 ft., *Drège*, 1679! Stellenbosch Div.; Hottentots Holland, *Ecklon & Zeyher*! Caledon Div.; Klein River Berg, *Ecklon*, 321! Bredasdorp Div.; sand-dunes near Cape Agulhas, *Ecklon*. Swellendam Div.; Puspus Valley, *Ecklon*. Riversdale Div.; between Vet River and Krombeks River, *Burchell*, 7171! Uitenhage Div.; mountains near the Van Staadens River, *Ecklon*; near Zwartkops River, *Zeyher*! Albany Div.; near Grahamstown, *MacOwan*, 155! 1286!

11. P. fibrosa (Stapf); perennial, compactly cæspitose; culms erect, slender, 1–2 ft. long, glabrous, smooth, 2-noded, nodes inconspicuous, upper 2 internodes more or less exserted; sheaths very tight, striate, lower crowded, hairy to tomentose, particularly towards the base, then glabrescent, very firm, persistent, finally breaking up into fibres; ligule a dense fringe of hairs; blades very narrow, linear, finely acute, up to more than 1 ft. by 1 lin., flat or more or less finely filiform, firm, striate, glaucous, pubescent above the ligule, otherwise glabrous, smooth except the scaberulous margins; panicle oblong, narrow, contracted, lax, about 4 in. long, erect; branches mostly 2-nate, unequal like the filiform axis, the subcapillary branchlets and the capillary pedicels glabrous, smooth, pallid, lowest up to $1\frac{3}{4}$ in. long, trichotomous from 4–6 lin. above the base, divisions unequal, pedicels not under 2 lin. long; spikelets pallid, straw-coloured, shining, $4–4\frac{1}{2}$ lin. long; glumes lanceolate in profile, long and finely acuminate, hyaline, whitish, glabrous, smooth, 1-keeled; valves lanceolate-oblong, in profile, body $1\frac{1}{2}–1\frac{3}{4}$ lin. long, shortly and rigidly pubescent to beyond the middle, scaberulous above, 7-nerved; lobes lanceolate, very acute, $1–1\frac{1}{4}$ lin. long, partly adnate to or almost free from a fine bristle ($1\frac{1}{2}$ lin. long) from the inner angle; awn about 5 lin. long, distinctly exserted from the glumes, kneed at $\frac{1}{3}$ way up, distinctly twisted below; pales exceeding the body of the valve, $2\frac{1}{2}$ lin. long, 2-mucronulate, scaberulous above; lodicules glabrous; anthers almost $1\frac{1}{2}$ lin. long.

COAST REGION: Albany Div.; on steep mountain slopes near Grahamstown, 2000 ft., *MacOwan*, 1299!

12. P. Tysonii (Stapf); perennial, compactly cæspitose; culms erect, very slender, 1 ft. long, glabrous, smooth, 2-noded, nodes distinct, upper 2 internodes long exserted; sheaths very tight, very narrow, closely striate, lower crowded, hairy then glabrescent, very firm, persistent, breaking up into fibres; ligule a dense fringe of short hairs; blades setaceous, plicate and subterete, up to 8 in. by $\frac{1}{2}$ lin. when unfolded, not striate, perfectly glabrous and smooth; panicle ovate, lax, scanty, more or less open, 2 in. long; branches mostly 2-nate, glabrous like the filiform axis and the capillary branchlets and pedicels, smooth, somewhat purplish, flexuous, longest up to $1\frac{1}{4}$ in. long, sparingly divided from 3–4 lin. above the base, 6–3-spiculate; lateral pedicels $1–1\frac{1}{2}$ lin. long; spikelets $3\frac{3}{4}–4\frac{1}{2}$ lin. long, straw-coloured; glumes lanceolate in profile, acutely acuminate, subhyaline, smooth, glabrous, 1-nerved; valves lanceolate-oblong in profile, body $1\frac{3}{4}$ lin. long, somewhat pubescent with long hairs in nearly all parts, 7-nerved, hairs fine, twisted, glittering; lobes lanceolate, acute, up to 1 lin. long, wholly adnate or almost so to the short lateral bristle; awn scarcely exserted from the glumes, up to 3 lin. long, kneed below the middle, scantily twisted below; pales exceeding the body of the valve, 2-mucronulate, scantily pubescent or at least the keels ciliate; lodicules glabrous; anthers $1\frac{1}{4}$ lin. long; grain $1\frac{1}{4}$ lin. long.

EASTERN REGION: Griqualand East; rocks on the summit of Mount Currie, 7500 ft., *Tyson*, 1312!

13. P. natalensis (Stapf); perennial; culms densely tufted or erect from short stolons, rather slender, 2 ft. high, glabrous, smooth, 4–5-noded, sheathed all along; sheaths tight, striate, softly hairy or the upper glabrous; mouth shortly bearded; ligule a fringe of hairs; blades very long, filiform-convolute, lower up to more than 1 ft. long by $1\frac{1}{2}$ lin. wide at the base, glabrous or scantily hairy towards the base, smooth, margins scaberulous; panicle open, very lax, up to 6 in. by 6 in., somewhat flaccid; branches 2-nate, smooth like the filiform axis and the subcapillary or capillary branchlets and pedicels, glabrous, except the sparingly hairy callous axils, very laxly and spreadingly di- and tri-chotomous from ½ to 1½ in. above the base, lowest over 3 in. long, ultimate divisions and pedicels rather short; lateral pedicels about 1½ lin. long; spikelets very pallid, 3–3¼ lin. long; glumes lanceolate in profile, finely acuminate, hyaline, glabrous, 1-nerved, keel minutely tubercled; valves linear-oblong, body 1½ lin. long, finely and loosely hairy all over, finely 7-nerved; lobes short, acute, adnate to or more or less free from the fine bristle 2–2½ lin. long at the inner side; awn about 6 lin. long, kneed and twisted just below the middle; callus short, acute, minutely villous; pales equalling the valves, 2-toothed, tips ciliolate; anthers 1 lin. long; grain slightly over 1 lin. long.

EASTERN REGION: Natal; Riet Vlei, 5000–6000 ft., *Buchanan*, 283!

14. **P. capensis** (Stapf); perennial; laxly tufted; culms slender, geniculate, ascending from a tuft of leaves and branches springing from the slender decumbent and rooting base or an equally slender rhizome, ½–1½ ft. long, glabrous, smooth, nodes rather many, sheathed nearly all along or upper internodes shortly exserted; leaves glabrous; sheaths rather thin, glabrous, striate, finally more or less loose, often purplish; ligule a ciliolate rim; blades very narrow, linear, very long, tapering to a fine or setaceous point, up to more than 6 in. by 1–1½ lin., usually involute and filiform, at least in the upper part, flaccid, smooth below, densely scaberulous or pruinose on the upper face; panicle lax, flaccid, ovate or oblong, 2–4 in. long; branches 3–2-nate, flexuous like the filiform axis and capillary branchlets and pedicels, glabrous and smooth, or scantily hairy at the callous axils, trichotomously divided from near the base, longest 1½–3 in. long, ultimate divisions 1-spiculate and 5–7 lin. long, or 2-spiculate with the lateral spikelet very shortly pedicelled; spikelets about 3½–4 lin. long, very pallid, somewhat shining; glumes equal, lanceolate, finely and setaceously acuminate, hyaline, almost white, 1-nerved, obscurely 3-nerved at the very base, glabrous, keel smooth; valves much shorter than the glumes, lanceolate in profile, body about 2 lin. long, finely and rather rigidly hairy in the lower part, distinctly 7-nerved, nerves scaberulous; lobes minute, truncate, minutely ciliolate or more often produced, acute, delicately hyaline, up to ½ lin. long with a very fine bristle, 5–6½ lin. long from the inner side; awn 8–9 lin. long, fine, flexuous, scaberulous, not kneed; callus acute, minutely villous; pales equalling the body of the valves, 2-toothed, teeth produced into short mucros or bristles; lodicules glabrous; anthers almost 1 lin. long. *Triraphis capensis, Nees, Fl. Afr. Austr.* 271. *Danthonia radicans, Steud. Syn. Pl. Glum.* i. 243; *Durand & Schinz, Consp. Fl. Afr.* v. 853.

COAST REGION: Cape Div.; Table Mountain, *Spielhaus!* Cape Peninsula, Orange Kloof, *Wolley Dod*, 2122! Worcester Div.; Dutoits Kloof, on rocks in river-beds, 3000–4000 ft., *Drège!* Caledon Div. (?); Vogel Gat, 2000 ft., *Schlechter*, 9520! Riversdale Div.; Lange Bergen, near Kampsche Berg, *Burchell*, 6969!

15. P. triseta (Stapf); annual; culms scantily fascicled, erect or suberect, slender, $\frac{3}{4}$–1 ft. long, glabrous, smooth, few-noded, upper 2 internodes occupying $\frac{3}{4}$ or more of the culm, exserted; sheaths rather lax, thin, striate, ciliate along the margins, lower short, villous in the upper part; ligule a fringe of short hairs; blades linear, acute, 1–1$\frac{1}{2}$ in. by $\frac{1}{2}$–1 lin. (upper very short), more or less spreading, flat, soft (at least the lower), shortly villous, eglandular; panicle flaccid, lax, meagre, sometimes reduced to 2 or 3 spikelets, about 2 in. long; lowest branches 3-2-nate, spreading or nodding, 2-1-spiculate, finely filiform, 4–9 lin. long; spikelets more or less erect at the incurved tips of the pedicels, 7–8 lin. long; glumes equal, very narrow, lanceolate-linear in profile, setaceously acuminate, purplish-brown except at the straw-coloured subhyaline margins and tip, sparingly hairy near the margins and sparingly scaberulous in the upper part, 3-nerved, nerves very close; valves linear in profile, body 3$\frac{1}{2}$ lin. long, thinly chartaceous, finely puberulous towards the tip, obscurely 7-nerved; lobes distinct, with a straight bristle 7–9 lin. long from the inner angle; awn about 15 lin. long, stout below the bend; callus very slender, acute, 1 lin. long, minutely villous; pales equal to the body of the valves, entire, puberulous near the tip; lodicules glabrous. *Avena triseta, Thunb. Prod.* 22; *Fl. Cap.* i. 434; *ed. Schult.* 116. *Danthonia collinita, Nees, Fl. Afr. Austr.* 315; *Steud. Syn. Pl. Glum.* i. 241; *Durand & Schinz, Consp. Fl. Afr. Austr.* 848.

SOUTH AFRICA: without precise locality, *Thunberg!*

COAST REGION: Paarl Div.; Klein Drakenstein Mountains, *Drège!* Clanwilliam Div.; in sandy places near the Olifants River and near Brack Fontein, *Ecklon & Zeyher*, 129!

16. P. acinosa (Stapf); perennial, cæspitose; stems undivided at the base, covered with the persistent leaves or their remnants, then branching and forming fascicles of filiform flowering culms, 4–8 in. long, and slender barren shoots, the former closely and distichously leafy at the base, with 2 long internodes above it, smooth, glabrous, the latter 1–3 in. long, very closely distichously leafy; sheaths tight, pubescent, bearded at the mouth, lower very short, firm, persistent; ligule a fringe of minute hairs; blades linear, tapering from the base to a blunt point, 5–10 lin. by $\frac{3}{4}$–1 lin. at the base, horizontally spreading, very rigid, subpungent, flat or concave or canaliculate, margins scaberulous, otherwise smooth; panicle small, lax, contracted, very meagre, often reduced to very few spikelets, up to 1$\frac{1}{2}$ in. long; branches 2-nate, capillary, 2-1-spiculate, up to $\frac{1}{2}$ in. long, glabrous except at the long-bearded axils; spikelets greenish, tinged with violet, 4–4$\frac{1}{2}$ lin. long; glumes lanceolate in profile, long and setaceously acuminate, subhyaline, glabrous, scaberulous towards the tips, 1-nerved; valves oblong, body 2 lin. long, densely hairy along the margins and more loosely on the back, faintly 9-nerved; lobes distinct, acute, with a fine bristle, 1$\frac{1}{2}$–2$\frac{1}{2}$ lin. long, from the inner angle or more or less adnate to it; awn 5–6 lin. long, kneed below the middle and twisted; pales equal to the valves, scaberulous along the keels and towards the tips; lodicules ciliate; anthers 1¼ lin. long; grain 1 lin. long. *Danthonia sp. Drège, Linnæa*, xx. 254.

VAR. β, **truncatula** (Stapf); barren shoots very short; blades very rigid, setaceously involute, subobtuse; spikelets 4½–5 lin. long; glumes less acuminate, closely 3-nerved. *Danthonia crispa, β truncatula, Nees, Fl. Afr. Austr.* 310.

COAST REGION: Caledon Div.; hills near Appels Kraal, by the Zonder Einde River, *Zeyher*, 4539! Riversdale Div.; on the Kampsche Berg, *Burchell*, 7068! Var. β: Caledon Div.; Bavian Kloof, near Genadendal, 1000–2000 ft., *Drège!*

17. P. elegans (Stapf); perennial, tufted; culms fascicled with short innovation shoots; erect from a loosely branched base, covered with old leaves, slender, ¾–1 ft. long, glabrous, smooth, very closely leafy below, 2-noded above the base, upper 2–3 internodes exserted; leaves crowded at the base, those of the last year persistent on the basal branches; sheaths tight, lowest closely imbricate, short, striate in the upper part, glabrous or very sparingly bristly, particularly at the mouth, upper 2–2½ in. long, very slender, finely striate; glabrous, smooth, beardless; ligule a fringe of minute hairs; blades subulate or linear-subulate, obtusely pointed, ¾–1 in. by ½–¾ lin., suberect, rigid, involute at least in the upper part, bristly from coarse tubercles; panicle scanty, oblong, lax, slightly nodding or flaccid, sometimes secund, 1–1½ in. long; branches 2-nate, scantily scaberulous like the finely filiform axis and the capillary branchlets, obscurely pubescent at the axils, more or less unequal, loosely 4–1-spiculate, lowest ½–1 in. long; pedicels 2–5 lin. long; spikelets brownish, 3½–4¼ lin. long; glumes long and finely acuminate, subhyaline, scaberulous in the upper part, glabrous; body of valves 1½–1¾ lin. long, with scattered adpressed stiff minute hairs; lobes short, ciliolate, passing into short bristles, scarcely more than ½ lin. long; awn rather stout, 7–8 lin. long, kneed at the middle; pales equal to the valve, minutely 2-toothed or entire; anthers slightly over 1 lin. long. *Danthonia elegans, Nees, Fl. Afr. Austr.* 296; *Steud. Syn. Pl. Glum.* i. 240; *Durand & Schinz, Consp. Fl. Afr.* v. 849.

COAST REGION: Caledon Div.; Klein Riversberg Mountains, *Ecklon & Zeyher* 120 in St. Petersburg Herb.!

18. P. Lima (Stapf); perennial; culms erect, slender, over 1 ft. long, glabrous, rough below the panicle and just below the nodes, otherwise smooth, few-noded, sheathed almost all along; sheaths tight, pallid, glabrous except the minutely bearded mouth, smooth below, scaberulous above; ligule a fringe of hairs; blades finely filiform, convolute, setaceously acute, up to 1 ft. long, rigid, subflexuous, glabrous and rough on the lower, adpressedly hairy on the upper side, particularly near the base, sometimes tomentose just above the ligule; panicle narrow, oblong, somewhat lax, erect, about 5 in. long; branches 2- (rarely 3-) nate, like the filiform axis and the subcapillary branchlets and pedicels scabrid and dotted with glandular tubercles, not hairy except at the pubescent axils, distantly trichotomous from 5–10 lin. above the base, longest about 2 in. long, divisions unequal; pedicels down to 1 lin. long; spikelets light straw-coloured, shining, $3\frac{1}{2}$–$4\frac{1}{4}$ lin. long; glumes lanceolate in profile, mucronate-acuminate, hyaline, scaberulous, 1-nerved; keel scabrid; valves linear-oblong in profile, body $1\frac{3}{4}$ lin. long, scantily hairy near the base or almost glabrous, 9-nerved; nerves scaberulous in the upper part; lobes short, ovate, passing into fine bristles, 1–$1\frac{1}{4}$ lin. long; awn 4–5 lin. long, kneed at and twisted below the middle, slender; callus small, obtuse, shortly hairy; lodicules glabrous; anthers $1\frac{1}{2}$–$1\frac{3}{4}$ lin. long. *Danthonia Lima, Nees, Fl. Afr. Austr.* 312; *Steud. Syn. Pl. Glum.* i. 241; *Durand & Schinz, Consp. Fl. Afr.* v. 852.

WESTERN REGION: Little Namaqualand; on the Kamies Bergen between Lily Fontein and Pedros Kloof, 3500–4000 ft., *Drège!*

19. **P. Zeyheri** (Stapf); perennial; culms rather stout, about 2 ft. high, glabrous, more or less beset with small tubercular glands below the panicle, otherwise smooth, many-noded; lower internodes very short, upper 2 up to more than $\frac{1}{2}$ ft. long; sheaths tight, lowest close, firm, glabrous below, shortly and spreadingly hairy to villous upwards, or almost glabrous throughout, persistent, intermediate and upper glabrous, the latter at least with few to many rows of tubercular glands on the nerves towards the mouth all (except the uppermost) distinctly bearded; ligule a fringe of hairs; blades linear, long tapering to a fine point, $\frac{1}{2}$–$\frac{3}{4}$ ft. by 2 lin., flat or loosely involute, very firm, rather rigid, glabrous, smooth below, scantily scaberulous on the upper side, margins scabrid, with a line of tubercular glands below; panicle erect, oblong, contracted, about 3 in. long, rather dense; branches 2-nate, like the filiform axis and the subcapillary branchlets dotted with small glands at least below, finely scaberulous, suberect, trichotomous from 4–7 lin. above the base, lowest up to more than 2 in. long, divisions somewhat unequal, rapidly decreasing upwards; pedicels $\frac{1}{2}$–2 lin. long, scaberulous; spikelets light brownish, 3–$3\frac{1}{2}$ lin. long; glumes acutely acuminate, subhyaline, glabrous, smooth, excepting the glandular keels; body of valves $1\frac{1}{2}$ lin. long or slightly longer, subvillous below, laxly and long hairy at the middle, faintly 7-nerved; lobes short, acute or subtruncate, scaberulous, with a bristle $1\frac{1}{2}$–2 lin. long from the inner side; awn 5–6 lin. long, kneed at the middle; pales subentire, as long as the valves. *Danthonia scabra, var.* a, *Nees, Fl. Afr. Austr.* 287 (*excl. syn.*). *D. viscidula, Drège in Linnæa,* xx. 254, *not Nees.*

COAST REGION; Uitenhage Div.; near the Zwartkops River, *Zeyher!*

20. **P. hirsuta** (Stapf); perennial; culms erect, slender, over $1\frac{1}{2}$ ft. long, glabrous, smooth; about 3–4-noded, upper 2 internodes very long, exserted; sheaths tight, lower short, obscurely striate,

hirsute with tubercule-based spreading hairs, villous at the very base, upper glabrous, scantily tubercled or almost smooth, margins ciliate, mouth conspicuously bearded; ligule a fringe of hairs; blades linear, acute, 3–4 in. by almost 2 lin., rigid, flat or involute, distinctly striate, lower densely beset with tubercle-based hairs, upper like those of the shoots more or less glabrous, scaberulous on both sides or smooth on the back below, margins minutely tubercled; panicle very lax, open, 4 in. by 4 in.; branches 2-nate, scabrid like the filiform axis and the capillary branchlets and pedicels, glabrous except at the hairy swollen axils, spreadingly and loosely trichotomous from $\frac{1}{3}$ to $\frac{1}{2}$ way above the base, lower to $2\frac{1}{2}$ in. long, ultimate divisions and pedicels rather short; spikelets 3 lin. long, straw-coloured, slightly crowded towards the tips of the branchlets; glumes equal, lanceolate in profile, finely acuminate, glabrous, smooth, 1-nerved; valves linear-oblong, body almost $1\frac{1}{2}$ lin. long, obscurely 5-nerved, long and loosely hairy, hairs seriate; lobes minute, acute with a fine bristle, $1\frac{1}{2}$ lin. long from the inner side; callus acute, minutely villous; awn kneed below the middle and twisted, $3\frac{1}{2}$–4 lin. long; pales as long as the valves, 2-toothed; grain slightly over $\frac{1}{2}$ lin. long. *Danthonia scabra, var. hirsuta, Nees, Fl. Afr. Austr.* 287.

COAST REGION: Clanwilliam Div.; near Zwartbast Kraal, between Berg Valley and Lange Valley, below 1000 ft., in sandy places, *Drège!*

21. P. rupestris (Stapf); perennial; culms erect, rather stout, almost 3 ft. long, glabrous, more or less verrucose from small tubercles below the panicle, otherwise smooth, many-noded, scantily branched below; lower internodes very short, intermediate 2–3 in. long, shortly exserted, uppermost much longer, long exserted; sheaths rather tight, striate, basal short, at least the inner tomentose at the very base, otherwise glabrous, the cauline glabrous at the base, shortly hirsute with tubercle-based hairs above and bearded near the mouth, upper quite glabrous with 1–3 rows of tubercular glands; ligule a fringe of hairs; blades linear, long tapering to an acute point, $\frac{1}{2}$–$\frac{3}{4}$ ft. by 2 lin., flat or loosely involute, firm, rather rigid, more or less hairy like the sheaths, or the uppermost glabrous, very scabrid on the upper side or all over near the tips, margins scabrid, not tubercled; panicle erect, 5 in. by 4 in., very lax, subdivaricate; branches 2-nate, like the filiform axis and the very fine branchlets densely dotted with small glands, at least below, scabrid above, bearded at the lower axils, very loosely and divaricately trichotomous from $\frac{1}{2}$–$\frac{3}{4}$ in. above the base, lowest up to 4 in. long, divisions rather equal, very slender and (except the ultimate) long; pedicels very short, rarely 1 lin. long; spikelets in clusters of 3–5 at the tips of the branchlets, light brownish, 3 lin. long; glumes acutely acuminate, subhyaline, glabrous; keels scaberulous, otherwise smooth; body of valves $1\frac{1}{2}$ lin. long, densely villous below, loosely and long hairy at the middle, faintly 7-nerved; lobes short, scaberulous, denticulate, with a bristle $1\frac{1}{4}$ lin. long from the inner side; awn 5 lin. long, kneed at the middle; pales almost as long as the valves, subentire; grain $\frac{7}{8}$ lin. long. *Danthonia rupestris, Nees, Fl. Afr. Austr.* 300; *Steud. Syn. Pl. Glum.* i. 240; *Durand & Schinz, Consp. Fl. Afr.* v. 854.

COAST REGION: Clanwilliam Div.; Ceder Berg Range, on the Blaauwe Berg, 4000 ft., *Drège*, 1682*b*!

Nees describes the branches of the panicle as very rough, but eglandular, and the keels of the glumes as glandular. I find them as described above. Another specimen in Drège's Herbarium (2580) consisting of 3 fragments, base and middle part of a culm and a panicle, agrees exactly with *P. rupestris* in the panicle except for a few occasional minute glands on the keels of the glumes, but the sheaths are glabrous and densely covered with lines of glands and the blade margins are also tubercled. Drège's 2580 is named "*Danthonia glandulosa β*" in the herbarium at Berlin, and the locality given there is "Modder Fonteins Berg."

22. P. subulifolia (Stapf); perennial, loosely fascicled; culms ascending, slender, $\frac{1}{2}$–1 ft. long, glabrous, smooth, very many-noded and densely distichously leafy to $\frac{1}{3}$ or $\frac{1}{2}$ way up, upper 2 or 3 internodes often at length elongated and exserted; leaves softly hairy or glabrous except at the always ciliate margins of the sheaths and their scantily bearded mouth; sheaths tight and closely imbricate except the more or less tumid upper ones, often purplish, smooth or beset with tubercular glands, firm, persistent; ligule a fringe of very short hairs; blades linear-subulate, pungent or almost so, 1–3 in. long, $1\frac{1}{2}$–$2\frac{1}{2}$ lin. broad at the base, almost coriaceous, very rigid with the tubercled or smooth margins involute, closely and finely striate, those of the upper leaves very short; panicle broad, ovate to subglobose, contracted, dense, erect, 1–2 in. long and broad; branches 2-nate, like the filiform axis and the subcapillary branchlets and pedicels more or less minutely tubercled, the latter at the same time scabrid, scantily hairy at the lower swollen axils, otherwise glabrous, trichotomously divided, from $\frac{1}{3}$–$\frac{1}{4}$ way above the base, lower $\frac{3}{4}$–1 in. long, ultimate divisions and pedicels rather short; spikelets crowded, 3–4 lin. long, straw-coloured, tinged with brown and purple, shining; glumes equal or almost so, lanceolate in profile, finely acuminate, subhyaline, 1-nerved, glabrous; keel scabrid or tubercled; valves lanceolate-oblong in profile, body $1\frac{3}{4}$ to almost 2 lin. long, finely and shortly hairy with the hairs adpressed and seriate, finely 7–9-nerved; lobes minute, obtuse, denticulate-ciliate, with a fine bristle $2\frac{1}{2}$ lin. long from the inner side; awn kneed below the middle and twisted, $5\frac{1}{2}$–$6\frac{1}{2}$ lin. long; callus distinct, acute, minutely villous; pales as long as the valves, entire or minutely 2-toothed; anthers 1 lin. long.

COAST REGION: Cape Div.; near Capetown, *Harvey*, 300! northern slopes of Table Mountain, *MacOwan, Herb. Austr. Afr.* 1698! Simons Bay, *Wright!* Vley Ground near Pauls Berg, *Wolley Dod*, 2954! South slope of Slang Kop, *Wolley Dod*, 3296!

Ecklon's 936 from the top of Table Mountain very probably also belongs to this species, although the leaves of the specimen in the Kew Herbarium are setaceous; this may, however, be due to its starved condition, the whole panicle bearing only 6 spikelets. Steudel described it as *Avena papillosa* in Flora, 1829,

ii. 484, whilst Nees quotes "Ecklon No. 936" under *Danthonia scabra* (Fl. Afr. Austr. 287), notwithstanding that it has not "folia scaberrima," as Nees says. Whether Ecklon's specimens from Worcester, Clanwilliam and Uitenhage, enumerated by Nees under *Danthonia scabra*, belong to *P. subulifolia* or to *P. hirsuta* (*Danthonia scabra β hirsuta*, Nees) or to a third species I cannot say.

23 P. leucopogon (Stapf); perennial, loosely tufted; culms fascicled, geniculate, ascending from the centre of the tuft, slender, 1–1½ ft. long, smooth, glabrous, 5- or more-noded; internodes enclosed (except the uppermost) or shortly exserted, upper 3 occupying ¾ or more of the culm; sheaths loose, with ciliate or villous margins and very conspicuously white bearded mouth, or scaberulous or rather smooth, dark purplish, lower short, persistent or subpersistent, upper much longer, usually tumid, the upper somewhat glabrous; ligule a dense conspicuous fringe of hairs; blades linear, tapering to an obtuse somewhat callous point, 2–5 in. long, 1–1½ lin. broad at the base, flat or with the margins involute, rigid, rather thick, spreading, deep green, glabrous, scabrid on the upper side, smooth below or scaberulous towards the tip; panicle broadly ovate, contracted, dense, 1½–3 in. long; branches 2-nate, strict, erect or subobliquely erect, filiform like the axis and branchlets and pedicels, scabrid, without tubercles, pubescent at the swollen nodes, lowest 1–2¼ in. long, trichotomous from 3–9 lin. above the base; ultimate divisions and pedicels very short; spikelets very crowded at the tips of the branchlets, greenish white below, brownish in the upper part, somewhat shining, 3–3½ lin. long; glumes lanceolate in profile, setaceously acuminate, subhyaline, scaberulous or smooth, glabrous or minutely puberulous, 1-nerved; keel scabrid, sometimes minutely tubercled; valves oblong in profile, body 1½ lin. long, loosely hairy all over, very thin, white, faintly sub-7-nerved; lobes distinct, short, broad, with a fine bristle about 2 lin. long from the inner angle; awn 6–7 lin. long, kneed at ⅓ or ¼ way up, twisted; callus subacute, minutely villous; pales as long as the valves, finely puberulous at the 2-toothed tip; anthers 1¼ lin. long.

South Africa: without precise locality, *Zeyher!*
Coast Region: Cape Div.; sandy places between Groot and Klein Slang Kloof, *Wolley Dod*, 3228!

There is little doubt that this is *Danthonia angulata*, Nees, Fl. Afr. Austr. 313.

24. P. aspera (Stapf); perennial; culms fascicled, rather slender, ascending from a long decumbent and branched slender base, ¾–1½ ft. long, glabrous, smooth, very many-noded, densely leafy to or beyond the middle, upper 2 or 3 internodes often elongate and exserted; leaves softly hairy to villous or glabrescent, more or less beset with tubercular glands, sometimes viscid; sheaths tight, closely imbricate (except the upper ones), margins ciliate, mouth conspicuously bearded; ligule a fringe of hairs; blades linear, tapering almost from the base to a fine callous point, 1–6 in. by 1–2 lin., very firm, rigid or at length flexuous or curled, flat, prominently striate on both sides, sometimes scaberulous on the upper side, margins tubercled; panicle ovoid to oblong, contracted, dense, erect, 1½–3 in. long; branches 2–4-nate, smooth like the filiform axis and the capillary branchlets and pedicels, or the latter scabrid, glabrous or scantily hairy at the callous axils, rarely tubercled, trichotomously divided from the base or the lower from above it; ultimate divisions and pedicels rather short; spikelets crowded, 3–3½ lin. long, light greenish or slightly tinged with purple, scarcely shining; glumes equal or almost so, lanceolate in profile, acuminate, subhyaline, 1-nerved, glabrous; keel smooth or sparingly and minutely tubercled; valves much shorter than the glumes, linear-oblong in profile, body 1½–1⅔ lin. long, finely hairy, with the hairs seriate, finely 7-nerved; lobes minute, truncate, denticulate, with a fine bristle, 1½–2 lin. long, from the inner side; awn kneed near the middle, twisted below; callus short, villous; pales as long as the valves, obscurely 2-toothed; anthers 1¼ lin. long. *Holcus asper, Thunb. Prodr.* 20; *Fl. Cap. ed. Schult.* 111. *Sorghum asperum, Roem. & Schult. Syst.* ii. 839. *Avena muricata, Spreng. Neue Entdeck.* i. 247. *A. capensis* (?), *Steud. in Flora,* 1829, ii. 481, *not of Spreng. Andropogon* (?) *asper, Kunth, Rév. Gram.* i. 166. *Pentameris papillosa, Nees in Linnæa,* vii. 311 (*excl. syn. Schrad.*). *Danthonia papillosa, Trin. Spec. Gram. Ic. t.* 66; *Gram. Suppl.* 33, *and in Mém. Acad. Pétersb. sér.* 6, iv. 34 (*not sér.* 6, i. 71); *Nees, Fl. Afr. Austr.* 292; *Steud. Syn. Pl. Glum.* i. 239; *Durand & Schinz, Consp. Fl. Afr.* v. 853.

COAST REGION: Cape Div.; Table Mountain, *Thunberg! Ecklon,* 935! *Harvey! Burchell,* 555! *MacGillivray,* 396! *Bolus!* Devils Mountain, *Bergius!* west slopes of Lions Head, *Wolley Dod,* 3570! rocks over Millers Point, *Wolley Dod,* 2776! rocky slopes of Beacon Hill, *Wolley Dod,* 2831! Pauls Berg slope, *Wolley Dod,* 2955!

25. P. Burchellii (Stapf); perennial, densely tufted; culms erect, very slender, 5–6 in. long, glabrous, smooth, 2–3-noded, upper 1–2 internodes by far the longest, exserted; leaves mostly basal, glabrous except at the ciliate sheath-margins; sheaths very tight, lower very short, firm, broad, imbricate, persistent, striate; ligule a dense fringe of short hairs; blades setaceous, fine, subacute, 1–4 in. long, tightly involute from the base, rigid, curved, smooth, purplish; panicle small, contracted or open and lax, 1–1½ in. long, erect; branches like the subcapillary axis, the capillary branchlets and pedicels smooth or very minutely tubercled, glabrous except at the hairy swollen axils, loosely 5–2-spiculate, undivided for 3–4 lin. above the base, lowest ⅓–½ in. long; pedicels 2–3 lin. long; spikelets straw-coloured, tinged with brown and violet, 3¾–4 lin. long, glistening; glumes lanceolate in profile, long and finely acuminate, glabrous, subhyaline, 1-nerved; keel minutely tubercled; valves linear-oblong, body over 1½ lin. long, finely and shortly hairy all over, hairs seriate, rather rigid; lobes short, subacute, with a fine bristle, 2–2½ lin. long, from the inner side, adnate to it; awn 5–6 lin. long, kneed at and twisted below the middle; callus short, minutely bearded; pales as long as the valves, minutely 2-toothed, tips ciliolate; anthers 1 lin. long.

COAST REGION: Mossel Bay Div.; between the landing-place at Mossel Bay and Cape Saint Blaize, *Burchell*, 6270!

26. P. tomentella (Stapf); perennial, densely cæspitose; culms erect, very slender, up to 1 ft. long, glabrous, smooth, 2–3-noded; upper 2 internodes very long, exserted; leaves softly hairy to villous all over, glaucous; sheaths tight, lower very short, whitish, thin, remotely striate; ligule a fringe of hairs; blades linear, subacute, 1–1½ in. by ¾ lin., usually setaceously involute, rigid, recurved; panicle erect, ovate, contracted, 1–1½ in. long; branches 2-nate, like the very fine rhachis and the capillary branchlets and pedicels scaberulous or very minutely tubercled, glabrous except at the pubescent swollen axils, ½–1 in. long, closely and sparingly trichotomous from 3–1 lin. above the base; ultimate divisions and pedicels short; spikelets whitish below, straw-coloured above, 2½ lin. long, glistening; glumes broad-lanceolate in profile, subacuminate, hyaline, minutely pubescent, 1-nerved; keel minutely tubercled; valves oblong in profile, body 1 lin. long or slightly longer, glabrous except at the base and along the lower margins, faintly 7–9-nerved, nerves raised in the upper part; lobes almost wholly reduced to a fine bristle not exceeding the glumes; awn 2½ lin. long, kneed and twisted below the middle; callus very short, sparingly bearded; pales as long as the valves, 2-toothed; anthers slightly over 1 lin. long.

WESTERN REGION: Little Namaqualand; Modder Fonteins Berg, *Drège!*

Distributed by Drège as "*Pentaschistis glandulosa, β speciosa*," but not taken up in Nees, Fl. Afr. Austr. "*Danthonia glandulosa, β speciosa*," Nees, Fl. Afr. Austr. 289, is *Pentaschistis euadenia*, Stapf, an annual.

27. P. angustifolia (Stapf); perennial, densely tufted; culms erect or subascending, simple or branched and fascicled below with numerous crowded innovation-shoots, ¾–1½ ft. long, glabrous, smooth, 4- to many-noded, closely sheathed below, upper 2–3 internodes more or less exserted and very much longer than the rest; sheaths tight or the uppermost tumid, bearded or beardless at the mouth, otherwise glabrous, rarely softly pubescent, striate, lower firm, persistent; ligule a fringe of hairs; blades usually filiform, tapering to a fine point, 2–8 in. by ½–1 lin. when expanded, firm, tightly involute, very rarely flat, flexuous or curled, glabrous, or scaberulous towards the tips; panicle ovate, open or contracted, 1½–3 in. long; branches 2-nate, like the filiform axis and the capillary branchlets and pedicels smooth or minutely tubercled, finely bearded at the axils, trichotomous from ¼–⅓ way above the base, lowest 1–2 in. long; ultimate divisions and pedicels rather short; spikelets rather crowded towards the ends of the branches, 2½–3 lin. long, light green or tinged with violet and brown; glumes acuminate, subhyaline, glabrous, smooth, keel scaberulous; body of valves 1¼–1½ lin. long, finely and more or less adpressedly hairy near the base, rarely higher up, finely 7–9-nerved; lobes distinct, usually acute or produced into an acumen or a minute bristle, free or more or less adnate to the lateral bristles, these 1½–2 lin. long, equalling the spikelets or slightly exserted; pales as long as the valves, minutely 2-toothed; anthers 1 lin. long. *Danthonia angustifolia, Nees, Fl. Afr. Austr.* 302; *Steud. Syn. Pl. Glum.* i. 240; *Durand & Schinz, Consp. Fl. Afr.* v. 847.

VAR. β, **micrathera** (Stapf); intermediate and upper sheaths usually dotted with seriate glands; blades very scabrid at least upwards; panicle 3–4 in. long; all the divisions somewhat longer and looser than in the type; lobes of valve obtuse, almost wholly adnate to the bristles, these 1–1½ lin. long; awn 4–5 lin. long. *Danthonia angustifolia, var. micrathera, Nees, Fl. Afr. Austr.* 302.

VAR. γ, **cirrhulosa** (Stapf); sheaths and sometimes also the blades more or less hairy; spikelets 2 lin. long; lobes of valves very short. *Danthonia cirrhulosa, Nees, Fl. Afr. Austr.* 309; *Steud. Syn. Pl. Glum.* i. 241; *Durand & Schinz, Consp. Fl. Afr.* v. 848.

VAR. δ, **albescens** (Stapf); leaves quite glabrous; sheaths (except the lowest) dotted with seriate patelliform glands in the upper part; blades very rigid; panicle very pallid and dense; spikelets 2 to almost 2½ lin. long; valves more or less suddenly contracted at the level of the insertion of the awn, not lobed; side-bristles very fine, 2–3 lin. long, long exserted.

SOUTH AFRICA: without precise locality, *Boivin!* *MacOwan*, 1304!

COAST REGION: Uitenhage Div.; by the Zwartkops River, *Zeyher*, 4531! Port Elizabeth Div.; Port Elizabeth, *E.S.C.A., Herb.* 99! 119! Alexandria Div.; Zuur Berg Range (Zuur Veld), *Gill!* near Addo, *Drège*. Bathurst Div.; Port Alfred, *Hutton!* near Theopolis between Riet Fontein and the source of the Kasuga River, *Burchell*, 4150! Albany Div.; near Grahamstown, 2000 ft., *MacOwan*, 1296! *Glass*, 769! Var. β: Queenstown Div.; Table Mountain, 5000–6000 ft., *Drège!* Var. γ: Swellendam Div.; Puspus Valley, *Ecklon & Zeyher*, 201, in St. Petersb. Herb.! Riversdale Div.; near the Zoetemelks River, *Burchell*, 6685! 6762! between the Great Vals River and Zoetemelks River, *Burchell*, 6562! Mossel Bay Div.; between Mossel Bay and Zout River, *Burchell*, 6328! Var. δ: Cape Div.; near Upper North, *Wolley Dod*, 2064! Robertson or Caledon Div.; Bosjes Veld or near Caledon, *Thom!* (intermediate between var. γ and var. δ, glabrous).

CENTRAL REGION: Var. β: Graaff Reinet; on the eastern side of Cave Mountain, near Graaff Reinet, 4000 ft., *Bolus*, 700!

28. P. heterochæta (Stapf); perennial, tufted with numerous crowded innovation shoots; culms erect, slender, ¾ ft. high, glabrous, smooth, about 2-noded, uppermost 1–2 internodes very long, exserted; sheaths tight, firm, lower broad, short, glabrous, except along the pubescent margins, with or without tubercular glands, upper long, purplish, finely pubescent, mouth naked or indistinctly bearded; ligule a fringe of hairs; blades setaceously involute, minutely acute, 1–2 in. long, rarely longer, rather rigid, curved or subflexuous, glabrous below, pubescent on the upper side, smooth or scaberulous towards the tip, margins with tubercular glands; panicle broad, ovoid, lax, rather open or contracted, about 2 in. long; branches 2-nate, like the filiform axis and the capillary branchlets and pedicels dark purplish, smooth or the pedicels scaberulous with scattered tubercles, glabrous except at the scantily hairy swollen axils, scantily trichotomously divided from 2–5 lin. above the base, lowest up to 1½ in. long; ultimate divisions and pedicels contracted, rather short; spikelets 2¼–3 lin. long, straw-coloured, shining; glumes lanceolate in profile, finely acuminate, subhyaline, glabrous, smooth except on the scaberulous keel; valves linear-oblong in profile, body 1¼ to almost 1½ lin. long, sparingly and minutely hairy below and along the lower margins, faintly 7-nerved, linear with 3 equal fine bristles slightly exceeding the glumes and united at the base, upper with a middle awn, about 4–4½ lin. long (kneed near the middle, twisted below) and with 2 lateral bristles about 2–2½ lin. long and hardly any side-lobes; pales as long as the valves, minutely 2-toothed or entire; anthers over 1 lin. long; grain ¾ lin. long. *Danthonia viscidula (b), Nees in Drège & Meyer, Zwei Pflanzengeogr. Docum.* 178, *not in Fl. Afr. Austr.*

CENTRAL REGION: Calvinia Div.; Onder Bokke Veld, between Grasberg River and Waterval River, 2500–3000 ft., *Drège!*

29. P. heptamera (Stapf); perennial, branched at or above the base; culms fascicled, erect or subascending, slender, about 1 ft. long, glabrous, smooth, many-noded, lower internodes short, enclosed, uppermost by far the longest, exserted; sheaths tight, lower closely imbricate, firm, persistent, sometimes purple, all glabrous except the ciliolate upper margins; ligule a fringe of very short hairs; blades very narrow, setaceously involute, subobtuse, 1–1¾ in. long, rigid, glabrous, smooth, glaucous; panicle contracted, dense, oblong, about 2 in. long; branches 2-nate, like the filiform axis and the very fine branchlets and pedicels glabrous, smooth, swollen at the base, trichotomously divided nearly or quite from the base, longest up to 1 in. long; pedicels ½–2 lin. long; spikelets about 3½ lin. long, pallid, shining; glumes lanceolate, acuminate, subhyaline, 1-nerved, glabrous, lower slightly longer; valves much shorter than the glumes, lanceolate-oblong in profile, body 1½ lin. long, finely hairy all over, delicate, faintly 5- (or near the tips 7–9-) nerved; lobes divided to the base into 3–4 very fine scaberulous bristles, those nearest the middle awn 5–6 lin. long, the others gradually shorter; middle awn about 10 lin. long, very fine, kneed at ⅓ way above the base and twisted below; callus obscure, minutely hairy; pales as long as the valves, glabrous; anthers 1 lin. long or slightly longer; grain linear-oblong, terete or subterete, ⅞ lin. long. *Danthonia heptamera, Nees, Fl. Afr. Austr.* 309; *Steud. Syn. Pl. Glum.* i. 241; *Durand & Schinz, Consp. Fl. Afr.* v. 850.

COAST REGION: Uitenhage Div.; on the downs near the mouth of the Zwartkops River, *Ecklon & Zeyher*, 659! Alexandria Div.: primeval forests of Olifants Hoek, *Ecklon & Zeyher*, 126, in St. Petersburg Herb.!

30. P. Jugorum (Stapf); perennial, densely tufted; culms erect, very slender, over ½ ft. long, smooth, glabrous, 2–3-noded; internodes enclosed except the long exserted uppermost one; lower sheaths short, loosely striate, finely hairy, upper much longer, tight, glabrescent, sparingly dotted with seriate glands, obscurely bearded; ligule a fringe of very short hairs; blades linear, very narrow, acute, $1-1\frac{1}{2}$ in. by $\frac{1}{2}$ lin., usually involute, lower hairy like the sheaths, upper glabrous, margins lined with small glands; panicle erect, ovoid, open, very lax or contracted, $1-1\frac{1}{4}$ in. long; branches like the finely filiform axis and the capillary branchlets dotted with scattered tubercles, otherwise smooth, obscurely bearded at the axils, very scantily trichotomous from 2–1 lin. above the base, sometimes divaricate, lowest up to 10 lin. long; branchlets and pedicels rather long, the latter scaberulous; spikelets $1\frac{3}{4}-2$ lin. long, straw-coloured; glumes acutely acuminate, subhyaline, glabrous, smooth; valves oblong, body 1 lin. long, loosely hairy all over, very obscurely 7-nerved; lobes more or less gradually passing into fine bristles, $\frac{7}{8}-1$ lin. long; awn $3\frac{1}{2}$ lin. long, kneed below the middle; pales scantily hairy, as long as the valves; grain $\frac{2}{3}$ lin. long. *Danthonia glandulosa, var. minor, Nees, Fl. Afr. Austr.* 289 (*partly*).

CENTRAL REGION: Aliwal North Div.; on the Witte Bergen, 7000–8000 ft., *Drège!*

31. P. filiformis (Stapf); perennial, densely tufted; culms erect, filiform, about $\frac{3}{4}-1$ ft. long, glabrous, smooth, about 2-noded, uppermost internode occupying more than $\frac{1}{2}$ of the culm, exserted; leaves glabrous excepting a long scanty beard at the mouth of the sheaths; sheaths tight or lower at length loosened, rather thin, finely striate, lower short, pallid; ligule a fringe of minute hairs; blades very finely filiform, acute, lower 3–4 in. long, upper much shorter, weak, flexuous, glabrous, smooth, margins sometimes with very minute tubercular glands; panicle ovate to oblong, lax, open or contracted, $1-2\frac{1}{2}$ in. long, erect; branches 2-nate, like the finely filiform axis and the capillary branchlets and pedicels dotted with very minute tubercles or almost smooth, glabrous except at the minutely bearded swollen axils, remotely trichotomously divided from 2–8 lin. above the base, lower $\frac{3}{4}-2$ in. long; ultimate divisions and pedicels rather long; spikelets greenish-straw-coloured, tinged with brown or purple, $1\frac{1}{4}-2\frac{1}{4}$ lin. long; glumes lanceolate in profile, acuminate and finely mucronate, subhyaline, 1-nerved, glabrous, smooth, keel of lower sometimes with very minute tubercles; valves oblong in profile, body 1 lin. long, loosely and long hairy, often pinkish; lobes distinct, truncate, denticulate-ciliate, with a fine bristle from the inner side, $\frac{3}{4}-1$ lin. long, adnate to it; awn very fine, $2\frac{1}{2}-3$ lin. long, kneed below the middle and twisted; callus rather slender, acute, minutely villous; pales as long as the valves, minutely 2-toothed; anthers 1 lin. long. *Danthonia filiformis, Nees, Fl. Afr. Austr.* 293; *Steud. Syn. Pl. Glum.* i. 240; *Durand & Schinz, Consp. Fl. Afr.* v. 850.

WESTERN REGION: Little Namaqualand; Kamies Bergen, at Ezels Fontein, between Rood Berg and Ezels Kop, 4000–5000 ft., *Drège*, 2584! near Modder Fonteins Berg, 4000–5000 ft., *Drège*, 2577!

32. P. imperfecta (Stapf); perennial, tufted; culms ascending from a sometimes decumbent and branched base, or suberect, very slender, 1–$1\frac{1}{2}$ ft. high, smooth, glabrous, lowest internodes rather numerous, very short, enclosed, uppermost 2–3 occupying by far the greater part of the culm, exserted; sheaths tight, finely striate, hairy with fine spreading hairs all over or along the margins or the lowest glabrescent to glabrous, the upper sometimes with a row of minute glands, distinctly bearded; ligule a fringe of hairs; blades linear, very narrow, tapering to a setaceous point, often very fine, setaceously convolute, 2–6 in. long, up to 1 lin. broad when expanded, somewhat rigid, erect, curved or flexuous, hairy like the sheaths or glabrous, smooth or scaberulous in the upper part; panicle ovoid, open, very lax or contracted and oblong, 2–3 in. long; branches 2-geminate, like the finely filiform axis and the capillary branchlets and pedicels scaberulous, distinctly bearded at the axils, lowest up to $1\frac{1}{2}$ in. long, loosely and repeatedly dichotomous from 2–6 lin. above the base, often divaricate; ultimate divisions and pedicels long; spikelets 2 lin. long or slightly more, yellowish, tinged with purple on the back; glumes acutely acuminate, hyaline, glabrous, smooth, keel scaberulous; valves oblong, body $1\frac{1}{2}$ lin. long, finely and sparingly hairy to or beyond the middle, faintly 7-nerved, both alike or lower usually awnless, acuminate, or with 2–3 minute teeth near the tips, upper 2-lobed, awned; lobes acute, scaberulous, $\frac{1}{4}$ lin. long, with a fine bristle, up to $\frac{1}{2}$ lin. long, from the inner angle, adnate to it to the middle; awn 3 lin. long, obscurely kneed at the middle or almost straight; pales minutely 2-toothed, $1\frac{1}{4}$ lin. long; anthers 1–$1\frac{1}{4}$ lin. long.

COAST REGION: Cape Div.; Tokay Mountain, *Wolley Dod*, 407! eastern slopes of Constantia Berg, 2000 ft., *Schlechter*, 449!

33. P. densifolia (Stapf); perennial, tufted; culms erect or suberect, very slender, 3–4 in. long, glabrous, smooth, rather many-noded, densely leafy below, upper 2 internodes by far the longest, uppermost exserted; sheaths tight, glabrous or minutely villous, sparingly bearded at the mouth, striate, lower very short, closely imbricate, not tubercled, upper long and tubercled in the other part; ligule a fringe of very minute hairs or quite obscure; blades linear, tapering to an acute point, $\frac{1}{2}$–1 in. by $\frac{1}{2}$–$\frac{3}{4}$ lin. (upper very small), flat, rather rigid, spreading or flexuous, glabrous, quite smooth except at the conspicuously tubercled margins; panicle small, contracted, oblong, $\frac{3}{4}$–1 in. long; branches 2-nate, like the axis and the branchlets very finely filiform, dark purple, densely tubercled, scantily trichotomous, 6–3-spiculate, lowest 4–7 lin. long; pedicels capillary, scaberulous, 1 lin. long or more; spikelets straw-coloured, about 2 lin. long, scarcely shining; glumes lanceolate in profile, finely acuminate, subhyaline to subherbaceous, glabrous, 1-nerved, keel scaberulous; valves linear-oblong, body 1 lin. long, laxly hairy, obscurely 5–7-nerved; lobes distinct, acute, with a fine bristle, $\frac{1}{2}$ lin. long, from the inner side; awn 3 lin. long, kneed below the middle and twisted; pale as long as the valve, 2-fid; anthers scarcely ¾ lin. long; grain ½ lin. long. *Danthonia densifolia*, α, *Nees, Fl. Afr. Austr.* 291; *Steud. Syn. Pl. Glum.* i. 239; *Durand & Schinz, Consp. Fl. Afr.* v. 849 (*in part*).

VAR. β, **intricata** (Stapf); forming loose tufts of numerous weak, tangled, prostrate or ascending, densely leafy branches; flowering culms ¼–¾ ft. long, weak, geniculate; sheaths very thin, pallid, withering; blades soft, flaccid, the lower often twisting or curling; panicle lax, divaricate, glands very minute. *Danthonia densifolia*, β, *Nees, Fl. Afr. Austr.* 291.

SOUTH AFRICA: without precise locality, *Zeyher*, in Lübeck Herb.!

COAST REGION: Var. β: Paarl Div.; in clefts of rocks near the waterfall of Simons Berg, 2000 ft., *Drège!* Worcester Div.; mountains above Worcester, *Rehmann*, 2586!

CENTRAL REGION: Worcester Div.; Constable, 3000–3500 ft., *Drège.*

Although the habit of var. β is very different from that of the type, I have little doubt that it is only a peculiar state of it, due to the conditions of the habitat. There is practically no difference in the spikelets.

34. **P. brachyathera** (Stapf); perennial; culms erect, slender, ½ ft. long, glabrous, smooth, 1-noded at the middle, uppermost internode long exserted; leaves crowded at the base; sheaths tight or the upper slightly tumid, lower very finely villous, upper glabrescent to glabrous, all with numerous glands on the nerves, at least in the upper part, sparingly bearded; ligule a line of hairs; blades subulate-involute, tapering to a fine pungent point, up to 1½–2 in. long, rigid, lower finely villous, upper glabrescent or glabrous, at least on the back, margins and prominent nerves near the base glandular; panicle contracted, very dense, oblong, 1–1¼ in. long; branches 2-nate, hairy at the axils, like the filiform axis and the subcapillary branchlets sparingly tubercled, closely branched from 1–2 lin. above the base, lowest up to 10 lin. long; ultimate divisions and pedicels short, latter scaberulous; spikelets very crowded, straw-coloured, 2 lin. long; glumes finely acuminate, subhyaline, glabrous, obscurely scaberulous, keels scabrid and minutely tubercled; body of valves scarcely 1 lin. long, glabrous except the small callus, finely 7-nerved; lobes very short, passing into short bristles (½ lin. long); awn 2–1½ lin. long, shortly exserted from the spikelet, very fine, scarcely kneed; pales ¾ lin. long, obscurely emarginate; anthers ¾ lin. long. *Danthonia papillosa, Nees in Drège & Meyer, Zwei Pflanzengeogr. Docum.* 178, *not of Schrader, nor of Nees in Fl. Afr. Austr.*

WESTERN REGION: Little Namaqualand; between Pedros Kloof and Lily Fontein, 3000–4000 ft., *Drège!*

35. **P. Thunbergii** (Stapf); perennial, tufted; culms fascicled, very slender, geniculate, ascending, ½–1½ ft. long, glabrous, smooth or scaberulous close below the panicle, 3–5- (rarely more) noded; upper 2 internodes usually occupying more than ½ of the culm, exserted; leaves softly hairy to villous, or glabrous, sometimes minutely tubercled; sheaths tight, rather thin and withering, conspicuously bearded at the mouth; ligule a fringe of hairs; blades very narrow, linear, tapering to a filiform point, 1–4 in. by ½–1 lin., rarely

broader, usually filiform-involute, rarely flat, flaccid or flexuous, finely striate; panicle erect, more or less ovate, contracted, dense, or sometimes open and almost divaricate, 1–3 in. long; branches 2-nate, like the filiform axis and the capillary branchlets and pedicels scaberulous, glabrous except at the more or less hairy swollen axils, sometimes minutely tubercled, trichotomously divided from some distance above the base, lowest $\frac{1}{2}$–$1\frac{1}{2}$ in. long; ultimate divisions and pedicels short; spikelets crowded towards the ends of the branches, 2–$2\frac{1}{2}$ lin. long, light green or tinged with violet, scarcely shining; glumes equal, lanceolate in profile, finely acuminate, subhyaline, 1-nerved, glabrous, keels finely scabrid; valves much shorter than the glumes, oblong in profile, body 1 lin. long, finely hairy from almost villous to subglabrous (hairs soft, seriate), faintly 5- to sub-9-nerved; lobes $\frac{1}{4}$ to almost $\frac{1}{2}$ lin. long, more or less acute, with a very fine bristle, 1–$1\frac{1}{2}$ lin. long, from the inner side and more or less adnate to it, rarely free; awn about $3\frac{1}{2}$ lin. long, fine, kneed at $\frac{1}{3}$ way up, twisted below; callus short, acute, minutely hairy; pales as long as the body of the valves, 2-toothed; anthers $\frac{3}{4}$ lin. long; grain slightly over $\frac{1}{2}$ lin. long. *Avena pallida, Thunb. Prodr.* 22; *Fl. Cap.* i. 435; *ed. Schult.* 117 (*in part*). *A. aristidoides, Steud. in Flora*, 1829, ii. 481, *not of Thunb. Trisetum nudum, Pers. Syn.* i. 97. *Danthonia Thunbergii, Kunth, Rév. Gram.* i. 107, ii. 523, *t.* 176; *Enum. Pl.* i. 314 (*excl. Thunberg's syns.*); *Durand & Schinz, Consp. Fl. Afr.* v. 855 (*excl. some syns.*). *D. papillosa, Schrad. in Schult. Mant.* ii. 385, *not of Trin. D. micrantha, Trin. Gram. Gen.* 71, *and in Mém. Acad. Pétersb. sér.* 6, i. 71; *Kunth, Enum. Pl.* 314. *D. villosa, Steud. ex Trin. Gram. Suppl.* 33, *and in Mém. Acad. Pétersb. sér.* 6, iv. 34 (*in part*); *Nees, Fl. Afr. Austr.* 294; *Steud. Syn. Pl. Glum.* i. 240.

VAR. β, **ebarbata** (Stapf); leaves glabrous, beardless; blades 2–6 in. by 1 lin. *Danthonia procumbens, Nees ex Drège in Linnæa*, xx. 254.

VAR. γ, **brevifolia** (Stapf); 5–8 in. high; leaves glabrous, beardless; blades 3–9 lin. long; $\frac{1}{2}$ to almost 1 lin. broad, rigid, spreading; panicle dense, up to 1 in. long. *Danthonia procumbens, var.* (?), *Nees ex Drège in Linnæa*, xx. 254. *D. curvifolia, var. livida, Nees, l.c.*

VAR. δ, **bulbothrix** (Stapf); all sheaths or at least the lower bearded, finely hirsute with tubercle-based partly deciduous hairs, upper sometimes glabrescent or glabrous and beardless; blades 2–6 in. by $\frac{3}{4}$–1 lin., hairy like the sheaths or the hairs not tubercle-based, or the upper glabrous, but rough on the upper side; panicle dense, $1\frac{1}{2}$–2 in. long; valves very sparingly and shortly hairy near the base.

SOUTH AFRICA: without precise locality, *Mund & Maire!* *Harvey*, 286! *Zeyher!* var. β, *Thunberg!* var. δ, *Herb. Caley!* *Harvey*, 300! 330!

COAST REGION: Cape Div.; Table Mountain, *Ecklon*, 939! *MacGillivray*, 397! *Milne*, 242! sandy plains near Capetown, *Ecklon*, 938! between Raspenberg Vley and Watchhouse, *Wolley Dod*, 3494! Lions Head, towards Kamps Bay, *Wolley Dod*, 3132! common all over the Lions Head, *Wolley Dod*, 3572! common between Rondebosch and Newlands, *Wolley Dod*, 3558! Wynberg Hill, *Wolley Dod*, 1818! Fish Hook Valley, *Wolley Dod*, 3401! Herschel Lane, *Wolley Dod*, 1821! hedges near Claremont, *Wolley Dod*, 1822! Simons Bay, *Wright!* Paarl Div.; Paarl Berg, 1000–2000 ft., *Drège!* Tulbagh Div.; Tulbagh Waterfall, *Ecklon & Zeyher*, 119! Worcester Div.; near Worcester, *Zeyher!* Var. β: Swellendam Div.; dry hills along the lower part of the Zonder Einde River, 500–2000 ft., *Zeyher*, 4536! Riversdale Div.; between Little Vet River and

Kampsche Berg, *Burchell*, 6878! wet places near Riversdale, 300 ft., *Schlechter*, 1948! Var. γ: Caledon Div.; on the Kenko River, between Riet Kuil and Hemel en Aarde, below 1000 ft., *Zeyher*, 4535! Swellendam Div.; on the Buffeljaghts River, *Zeyher*, 4545! Var. δ: Cape Div.; eastern side of the Lions Rump, *Burchell*, 116! Stellenbosch Div.; Hottentots Holland, near Somerset, *Ecklon & Zeyher*, 128!

There are 3 distinct species, named *Avena pallida*, in Thunberg's Herbarium; the specimens are marked α, *a*, β, γ, δ; of which α, *a*, and β correspond to my variety *ebarbata;* but they do not agree with Thunberg's description of *Avena pallida*, nor of *Avena pallida*, var. β, which are said to possess villous leaves and hirsute sheaths respectively. The type of Thunberg's *Avena pallida* was very probably the form figured and described by Kunth as *Danthonia Thunbergii*. Var. δ *bulbothrix* is possibly *D. buekeana*, Nees, Fl. Afr. Austr. 297. To *P. Thunbergii* belong very probably also *Danthonia barbata* (Nees, Fl. Afr. Austr. 301; Steud. Syn. Pl. Glum. i. 240; Durand & Schinz, Consp. Fl. Afr. v. 847) from the Olifants River, Clanwilliam Division, and from various places in Tulbagh Division, and *Danthonia propinqua* (Nees, l.c. 299; Steud. l.c.; Durand & Schinz, l.c. 853) from Tulbagh Division. Nees referred *Danthonia micrantha*, Trin., which was described from starved cultivated specimens to *Danthonia villosa*, i.e. *Pentaschistis Thunbergii*, and himself used the same name, *D. micrantha*, for a species distinct from Trinius's *D. micrantha*, which he diagnosed, at the same time, from some examples collected by Ecklon near Constantia, Cape Division, and by Mund in the Swellendam District. However, I cannot find anything in his description to separate them from *P. Thunbergii*, except perhaps the slightly smaller size of the spikelets, which are stated to be 1½–1¾ lin. long.

36. P. longipes (Stapf); perennial; culms erect, scantily branched near the base with the branches intravaginal and erect, about 2 ft. long, glabrous, smooth, except for some hairs below the panicle, 5-noded, internodes gradually increasing upwards, intermediate exserted, upper enclosed; lowest 4–5 leaves crowded at the base; sheaths somewhat tight or the upper rather lax, strongly striate, lowest short, 1–1½ in. long, spreadingly hairy with tubercle-based hairs in the upper part, glabrous below, upper gradually longer (uppermost 6 in. long) and less hairy to glabrous except along the pubescent margins and with seriate patelliform glands along the middle nerve, all bearded at the mouth; ligule a fringe of hairs; blades linear, long tapering to a fine point 3–6 in. by 1–1½ lin., flat, soft or somewhat rigid, strongly and closely striate, finely hairy or the lower almost hirsute all over, margins lined with patelliform glands in the lower part; panicle oblong, contracted, 5 in. long; branches 2-nate, finely scaberulous like the filiform axis and capillary branchlets and pedicels, bearded at the axils, repeatedly trichotomous from 1 in. above the base, lowest up to 4 in. long with the lowest spikelets about 2½ in. from the base; ultimate divisions and pedicels very long and very fine; spikelets 2¼–2½ lin. long, brownish-yellow, tinged with violet on the back; glumes acutely acuminate, subhyaline, glabrous, smooth, keel finely scaberulous and minutely glandular; valves oblong, body 1¼ lin. long, scantily and very shortly hairy below the middle; lobes distinct, almost wholly adnate to the lateral bristles, these about 1–1½ lin. long; awn 3½ lin. long, kneed at ⅓ way from the base; pales slightly over 1 lin. long, shortly 2-fid; anthers scarcely 1 lin. long.

COAST REGION: Albany Div.; Albany Plains, *Bowie!*

37. P. patula (Stapf); annual; culms fascicled, very slender, $\frac{1}{2}$–$\frac{3}{4}$ ft. long, glabrous, smooth, 2–3-noded, uppermost internodes by far the longest, exserted; sheaths tight, thin, lower short, pubescent, striate, closely beset with tubercle-based hairs except at the base, upper long glabrous, smooth or almost so; ligule a fringe of short hairs; blades very narrow, linear, tapering to an acute or subacute point, 1–1$\frac{1}{2}$ in. long, usually filiform-convolute, rather soft, pubescent like the sheaths or the upper glabrous; panicle open, very lax, more or less divaricate, 1$\frac{1}{4}$–2 in. long and broad, erect; branches 2-nate, smooth, like the finely filiform axis and the capillary branchlets and pedicels obscurely pubescent at the axils, very loosely and divaricately dichotomous from 2–5 lin. above the base, lowest $\frac{3}{4}$–1$\frac{1}{2}$ in. long; ultimate divisions and pedicels rather long; spikelets straw-coloured, 2$\frac{1}{4}$–2$\frac{1}{2}$ lin. long; glumes setaceously acuminate, hyaline, glabrous, smooth; body of valves 1$\frac{1}{4}$ lin. long, loosely hairy to or beyond $\frac{1}{2}$ way, very faintly 5-nerved; lobes distinct, acute, with a fine bristle $\frac{1}{2}$–1 lin. long from the inner side and usually more or less adnate to it; awn about 4 lin. long, kneed below $\frac{1}{2}$ way up, very fine; pales as long as the valves, minutely 2-toothed; anthers $\frac{1}{2}$–$\frac{5}{8}$ lin. long; grain $\frac{2}{3}$ lin. long. *Danthonia patula, Nees, Fl. Afr. Austr.* 285; *Steud. Syn. Pl. Glum.* i. 239; *Durand & Schinz, Consp. Fl. Afr. Austr.* 853.

VAR. β, **glabrata** (Stapf); leaves glabrous or the sheaths slightly hairy below the more conspicuously bearded mouth; blades more rigid, less involute, sometimes minutely tubercled along the margins. *Danthonia patula, Nees in Linnæa,* xx. 254.

VAR. γ, **acuta** (Stapf); leaves as in var. β, but rather larger with a line of rigid bristles at the junction of blade and sheath; lowest sheaths pubescent; divisions of panicle stiffer and coarser than in the type; spikelets light olive-brown; glumes acute, scarcely acuminate, less hyaline.

SOUTH AFRICA: Var. γ: without precise locality, *Drège*, 2587!

COAST REGION: Vanrhynsdorp Div.; sand-hills near Ebenezer, below 500 ft., *Drège!* Var. β: Stellenbosch Div.; Hottentots Holland, 1000–3000 ft., *Zeyher!* Swellendam Div.; lower part of the Zonder Einde River, 500–2000 ft., *Zeyher*, 4534!

38. P. euadenia (Stapf); annual; culms fascicled, erect, or geniculate and ascending, slender, $\frac{1}{2}$–1 ft. long, glabrous, smooth, 2–3-noded, upper 2 internodes by far the longest, exserted; sheaths usually tumid, or the intermediate tight, pallid, glabrous or pubescent, usually dotted with numerous seriate often stalked tubercular glands except at the base, margins ciliate, mouth scantily bearded; ligule a fringe of hairs; blades linear, long tapering to a fine point, 1–3$\frac{1}{2}$ in. by 1–1$\frac{1}{2}$ lin., flaccid, flat, glabrous, margins fringed with often stalked tubercular glands; panicle oblong to ovoid, rather lax, contracted or somewhat open, 1$\frac{1}{2}$–3$\frac{1}{2}$ in. long; branchlets 2–3-nate, very sparingly tubercled, like the filiform axis and the capillary branchlets and pedicels, otherwise smooth, pubescent at the axils, laxly trichotomous from 2–6 lin. above the base, longest $\frac{3}{4}$–2 in. long; ultimate divisions and pedicels usually rather long; spikelets light greenish-straw-coloured, 2 lin. long; glumes rather broad, finely acuminate, glabrous, subhyaline, keels scaberulous; body of valves 1½ lin. long, very sparingly hairy below, very minutely scaberulous above, faintly 5–7-nerved; lobes obscure passing into fine bristles, ½–1 lin. long; awn 3–4 lin. long, kneed below ½ way, very fine; pales as long as the valves, 2-toothed; anthers 1 lin. long. *Danthonia glandulosa, var. speciosa, Nees, Fl. Afr. Austr.* 289 (*excl. syn. in part*). *D. glandulosa, Steud. Syn. Pl. Glum.* i. 239, *not Schrad.*

WESTERN REGION: Little Namaqualand; Kamies Bergen, between Pedros Kloof and Lily Fontein, 3000–4000 ft., *Drège!* Modder Fonteins Berg, 4500 ft., *Drège*, 2581!

Another species from Modder Fonteins Berg, collected by Drège, was also issued as "*Pentaschistis glandulosa. β speciosa*"; it is *P. tomentella.* What *Danthonia glandulosa* (Schrad. ex Schult. Mant. ii. 385) is I do not venture to say. It is described as a perennial, about 1½ ft. high, whilst Nees quotes Schrader's original specimen in the herbarium at Berlin apparently under his variety *β speciosa.* This is probably due to a printer's error, and was meant to stand under variety *α.* Nees's variety *α*, however, comprises certainly at least 2 species, one of which is Steudel's "*Danthonia heteropla*" (Syn. Pl. Glum. i. 239; Durand & Schinz, Consp. Fl. Afr. v. 850), whilst another is here described above as *Pentaschistis Jugorum*, Stapf. I have not seen the original of Steudel's *Danthonia heteropla*, but I suspect that it is *Pentaschistis patula.*

39. P. airoides (Stapf); annual; culms fascicled, erect or geniculate and ascending, very slender, 2–6 in. long, glabrous, smooth, 2–3-noded, uppermost internodes by far the longest; leaves softly hairy to villous all over or glabrous; sheaths thin, lower short, at length loose, upper tight or tumid, with or without small tubercular glands, bearded or beardless; ligule a fringe of short hairs; blades linear, tapering to an acute point, ½–1 in. by ½–1 lin., flat or involute, subflaccid or at least soft, margins smooth, rarely minutely tubercled to tubercular-pectinate; panicle lax, open or contracted, ½–1½ in. long and broad; branches 2-nate, like the very fine axis and the capillary branchlets and pedicels smooth or scantily and minutely tubercled, glabrous, loosely and often divaricately trichotomous from 1–3 lin. above the base, longest ¼–1 in. long; ultimate divisions and pedicels rather short; spikelets straw-coloured, usually tinged with purple, 1¼–1½ lin. long; glumes acuminate to acute, smooth, glabrous, subhyaline; body of valves ¾–1 lin. long, shortly hairy or glabrous, callus hairy; lobes obscure, passing into fine bristles, ½–1½ lin. long; anthers ellipsoid, ⅛–⅙ lin. long; grain slightly over ½ lin. long. *Pentameris airoides, Nees in Sem. Hort. Bot. Vratisl.* 1834, *and in Linnæa,* x. *Litt.-Ber.* 118. *Danthonia airoides, Nees, Fl. Afr. Austr.* 284; *Steud. Syn. Pl. Glum.* i. 239; *Durand & Schinz, Consp. Fl. Afr.* v. 847. *D. cyatophora, Nees, Fl. Afr. Austr.* 286; *Steud. l.c.* 239; *Durand & Schinz, l.c.* 849. *D. webbiana, Steud. Syn. Pl. Glum.* i. 240 (*from the description*); *Durand & Schinz, Consp. Fl. Afr.* v. 855.

COAST REGION: Cape Div.; north slope of Lions Head, *Wolley Dod*, 3095! Albany Div.; Bothas Berg, 2200 ft., *MacOwan*, 1291! Fort Beaufort Div.; Kat River, *Zeyher!*

CENTRAL REGION: Fraserburg Div.; between Quagga Fontein and Dwaal River, *Burchell*, 1440! by the Dwaal River, *Burchell*, 1472.

WESTERN REGION: Little Namaqualand; Silver Fontein, near Ookiep, 2000 ft., *Drège!* Vanrhynsdorp Div.; Karee Bergen, 1000 ft., *Schlechter*, 8176!

According to Nees (Sem. Hort. Bot. Vratisl. l.c.), the original from which the species was described, was collected on the Zwartkops River; he does, however, not quote this locality in Fl. Afr. Austr. On the other hand, he indicates it here from various localities in "Swellendam," from which district I have not seen it.

XXXVI. PENTAMERIS, Beauv.

Spikelets 7–12 lin. long, rarely shorter or longer, compressed, pedicelled, panicled or racemose; rhachilla disarticulating above the glumes and between the valves, continued as a short slender bristle beyond the upper floret. *Florets* 2, ⚥, much shorter than the glumes. *Glumes* equal, lanceolate in profile, acuminate, keeled, hyaline to scarious, 1-nerved or very closely 3-nerved at the very base. *Valves* membranous, more or less hairy 7–9- (rarely 11-) nerved, 2-lobed, awned from the sinus; lobes with a fine bristle from the inner side, usually more or less adnate to it; callus very short. *Pales* 2-keeled, more or less equalling the valves, 2-fid or 2-toothed. *Lodicules* 2, cuneate, nerved, glabrous or ciliate. *Stamens* 3. *Ovary* obovoid, with a very dense deciduous tomentum, consisting of branched hairs at the top; styles distinct, short or very short; stigmas plumose, laterally exserted. *Grain* (known only in *P. Thuarii*) globose-ellipsoid, truncate at the top, enclosed by the slightly hardened valve and pale; pericarp crustaceous, granulate; seed free, except along the hilum; hilum linear, half the length of the grain or longer, obscure on the outside; embryo small.

Perennial; culms fascicled, from a usually woody or suffrutescent base, densely leafy below or almost all along; blades rigid or wiry, usually long and filiform; panicle open or contracted, lax, usually scanty.

Species 5, in the Cape Colony.

The structure of the ovary is so alike in the 5 species of this genus that it is very probable that they agree also in the peculiarities of the ripe fruit which is known only in *P. Thuarii*. The glumes and valves are essentially like those of *Pentaschistis*, and the branching of the panicle is also similar, although usually much scantier. On the other hand, there is in *Pentaschistis* no approach to the characteristic structure of the ovary and the fruit of *Pentameris*.

Valves and pales very broad; lobes of former short, broad, dentate or abruptly subulate-acuminate ... (1) **Thuarii.**

Valves and pales narrow; lobes of former long, narrow, acute, adnate for $\frac{1}{2}$–$\frac{4}{5}$ of their length to the lateral bristles, or passing into them:

Blades filiform, $\frac{1}{3}$–1 ft. long, flexuous:

Panicle very lax, more or less open, 6–7 in. by 5–7 in.; spikelets about 10 lin. long; lobes of valve half as long as the body ... ... (2) **longiglumis.**

Panicle narrow, contracted:

Spikelets 8–12 lin. long; awns 9–11 lin. long; lobes of valve about $\frac{1}{3}$ as long as the body ... ... (3) **speciosa.**

Spikelets 6½–7½ lin. long; awns about 6 lin. long; lobes of valves half as long as the body ... ... (4) **dregeana.**

Blades squarrose, subulate, 2–3½ in. long, very hard and rigid, with obtuse tips ... ... (5) **squarrosa.**

1. **P. Thuarii** (Beauv. Agrost. 93, t. 18, fig. 8); culms erect, very firm, in dense and tight fascicles from a woody, sometimes branched and suffrutescent, base, which is densely covered by the remnants of old sheaths, 1–2 ft. high, glabrous, smooth, densely leafy to ⅓–½ way up, upper 2–3 internodes usually exserted; sheaths tight, firm, glabrous, except at the often villous or ciliate margins, rarely loosely hairy on the back, lower coriaceous, persistent, about 2 in. long, uppermost 3–3½ in. long; ligule a dense fringe of hairs, about 1 lin. long; blades very narrow, linear, usually convolute-filiform from a short suddenly contracted blackish base, 6–12 in. by 1 lin., rigid, subflexuous, striate, glabrous and smooth below, shortly hairy on the upper side; panicle obovoid, lax, open or contracted, 2–4 in. long, erect or slightly nodding; axis filiform like all the divisions, scaberulous to scabrid; branches 3–2-nate or solitary, finely filiform, obliquely erect, sparingly divided from ⅓–½ way above the base or simple, 6–1-spiculate, longest up to 2 in. long; branchlets and pedicels subcapillary, unequal, latter 3–12 lin. long; spikelets erect, whitish or straw-coloured, 7–9 lin. long, shining; glumes very long and finely acuminate, hyaline to subhyaline, glabrous, smooth; valves broad, oblong in profile, body 1½ lin. long, pubescent, 9–11-nerved; lobes broad, scaberulous, acute or often suddenly contracted into a setaceous acumen, or 2-toothed with a bristle 3–3½ lin. long from the inner angle, adnate to ½ way; awn 7–9 lin. long, kneed between ½ and ⅓ way up, column rather stout; pales very broad, as long as the valve, hairy below, obtusely or acutely 2-toothed; anthers 1¾ lin. long; grain broadly ellipsoid, truncate, 1½ lin. long, deep brown, granular; hilum almost ½ the length of the grain; seed whitish. *Roem. & Schult. Syst.* ii. 693. *P. Thouarsii, Kunth, Rév. Gram.* i. *t.* 66; *Enum.* i. 315 (*excl. syn.*); *Suppl.* 270. *Danthonia Thuarii, Desv. Opusc.* 99 (*excl. syn.*); *Durand & Schinz, Consp. Fl. Afr.* v. 856. *D. Thouarsii, Nees, Fl. Afr. Austr.* 337; *Steud. Syn. Pl. Glum.* i. 243; *Durand & Schinz, Consp. Fl. Afr.* v. 854 (*excl. syn.*).

VAR. β, **Burchelli** (Stapf); culms stout, 3 ft. high, simple for more than 1 ft., then with a few erect, barren, intravaginal branches and simple again above them; blades up to 2 lin. broad; panicle very effuse, up to more than 6 in long, longest branches up to 6 in. long.

COAST REGION: Worcester Div.; rocky mountain ridges near Dutoits Kloof, 2000–3000 ft., *Drège!* Caledon Div.; rocky places near Genadendal, 2000–3000 ft., *Drège!* Swellendam Div.; in a wooded gorge on Zuurbraak Mountain, 1000 ft., *Galpin*, 4845! and without precise locality, *Zeyher!* var. β: Riversdale Div.; lower part of the Lange Bergen, near Kampsche Berg, *Burchell*, 6964!

The specimen, representing the variety β, has the appearance of a robust annual; but the spikelets are exactly as in the type. Burchell says that specimens grown at Fulham from the seeds of his no. 6964 attained a height of 6 ft.

2. **P. longiglumis** (Stapf); culms erect, slender, firm, in dense and tight fascicles from a slender woody base, $1\frac{1}{2}$–2 ft. high, glabrous, smooth, densely leafy below, sheathed all along; sheaths tight except the uppermost, firm, striate, glabrous or more or less sparsely hairy and ciliate near the mouth, lower 1–$1\frac{1}{2}$ in. long, uppermost 6–9 in. long, almost reaching to or beyond the base of the panicle; ligule a dense fringe of hairs; blades finely filiform-convolute from a broader base, 6–9 in. long, lower up to 2 lin. broad at the base, rigid, flexuous, glabrous and smooth below, finely puberulous or scaberulous on the upper side, margins scaberulous; panicle very lax, open, 6–7 in. long; rhachis filiform, smooth, bearded at the lower axils, otherwise smooth; branches 2–3-nate or solitary, very loosely di- to tri-chotomous or alternately divided from $\frac{1}{2}$ to $1\frac{1}{2}$ in. above the base, longest 4–5 in. long, filiform, flexuous, glabrous, smooth; branchlets and pedicels unequal, subcapillary, smooth, the latter 2–18 lin. long; spikelets slightly nodding, straw-coloured, about 10 lin. long; glumes setaceously acuminate, scarious, shining, smooth, glabrous; valves oblong, body $1\frac{3}{4}$ lin. long, densely hairy, 9-nerved; lobes large, lanceolate, acute, scaberulous, as long as the body of the valve or almost so, with a bristle 3–4 lin. long, from the inner side, adnate to it for $\frac{3}{4}$ its length; awn about 7 lin. long, kneed at $\frac{1}{3}$ way up; pales as long as the valve (inclusive of the lobes), densely pubescent, 2-toothed; lodicules ciliate; anthers 2 lin. long. *Danthonia longiglumis*, *Nees*, *Fl. Afr. Austr.* 306; *Steud. Syn. Pl. Glum.* i. 241; *Durand & Schinz, Consp. Fl. Afr.* v. 852.

COAST REGION: Cape Div.; Table Mountain, *Burchell*, 542! 598! *Spielhaus! Milne*, 246!

3. **P. speciosa** (Stapf); culms erect, very firm and stiff, woody below, fascicled, simple or usually branched above the base, with the branches in tight fascicles, $1\frac{1}{2}$ to more than 2 ft. long, glabrous, smooth, closely and very tightly sheathed beyond the middle or all along; sheaths coriaceous, extremely tight, smooth, often shining, not striate, glabrous except a fugacious dense flake of wool at the mouth; ligule a dense fringe of hairs; blades filiform, finely pointed or subpungent, wiry, plicate, terete, $\frac{1}{2}$–1 ft., long, very hard, flexuous or straight, glabrous and very smooth outside, densely and minutely puberulous or tomentose inside; panicle narrow, linear to oblong, usually contracted, 2–6 in. long; rhachis very slender, very scabrid like the unequal filiform or subcapillary branches, branchlets and pedicels; lower branches 2-nate or fascicled, longest up to $2\frac{1}{2}$ in. long, scantily and loosely branched from near the base or undivided for 3–9 lin., 6–1-spiculate; spikelets straw-coloured, 8–12 (rarely to 15) lin. long, erect; glumes long and setaceously acuminate, scarious except at the hyaline tips, glabrous, smooth; valves linear-oblong, body $2\frac{1}{2}$–3 lin. long, pubescent to or beyond the middle, 7-nerved; lobes lanceolate, minutely and unequally bifid, scaberulous, about 1–$1\frac{1}{4}$ lin. long, outer tooth acute, hyaline,

inner produced into a fine bristle, 5–6 lin. long; awn 9–11 lin. long, kneed at or below the middle; pales much exceeding the insertion of the awns, 4½–5 lin. long, deeply bifid, hairy; anthers 3 lin. long. *Danthonia speciosa, Lehm. ex Nees, Fl. Afr. Austr.* 307; *Steud. Syn. Pl. Glum.* i. 241; *Durand & Schinz, Consp. Fl. Afr.* v. 854.

SOUTH AFRICA: without precise locality, *Mund & Maire!*
COAST REGION: Cape Div.; summit of Table Mountain, *Bolus!* all over the Lower Plateau of Table Mountain, *Wolley Dod*, 3300! 1686! slopes of Elsie Peak, *Wolley Dod*, 2957! Ceres Div.; Cold Bokke Veld, 5000 ft., *Schlechter*, 8917! Worcester Div.; Dutoits Kloof, 3000–4000 ft., *Drège!* Caledon Div.; rocky slopes of mountains above Genadendal, *Drège*; by the Zonder Einde River, near Appels Kraal, *Zeyher*, 4546! Riversdale Div.; lower part of the Lange Bergen near Kampsche Berg, *Burchell*, 7010!

4. **P. dregeana** (Stapf); culms erect, simple or branched from a woody base, with the branches in tight fascicles, 1½ ft. long, glabrous, or the short basal internodes hairy, smooth, densely and tightly sheathed to ⅓ way, upper 3 internodes exserted; sheaths very tight and firm, striate, long and loosely villous, particularly along the margins, finally glabrescent on the back, bearded at the mouth; ligule a short fringe of hairs; blades filiform-convolute, terete, gradually tapering to an acute or blunt point, 4–6 in. long, rigid, flexuous or curved to circinnate, long and usually sparsely hairy below (at least towards the base), pubescent or puberulous to scaberulous on the upper side, margins scabrid; panicle oblong narrow, 2–2½ in. long, erect, contracted; rhachis scabrid or smooth below; branches fascicled to 2-nate, unequal, erect, longest up to 1 in. long, scantily branched from the base or undivided for 2–3 lin., 4–1-spiculate; branchlets and pedicels unequal, finely filiform, scaberulous, the latter, when short, clavate; spikelets straw-coloured, 6½–7½ lin. long; glumes acutely acuminate, finely scaberulous or puberulous, subhyaline, scaberulous; valves oblong, body 2 lin. long, pubescent to villous, 7-nerved; lobes lanceolate, acute, about 1 lin. long, densely and very minutely pubescent with a fine bristle, 2 lin. long, from the inner side, adnate to it to ¾ their length; awn 6 lin. long, kneed at ⅓ way up; pales exceeding the insertion of the awn, up to 3 lin. long, 2-toothed, pubescent. *Danthonia distichophylla, Nees, Fl. Afr. Austr.* 305 (*in part*), *not Lehm.*

COAST REGION: Paarl Div.; rocks of Paarl Mountain, 2000–5000 ft. (or Worcester Div.; Dutoits Kloof?), *Drège!*

Lehmann's description of *Danthonia distichophylla* (Pugill. iii. 41) does not at all agree with this plant. It refers probably to a *Pentaschistis*.

5. **P. squarrosa** (Stapf); culms ascending, 1½–2 ft. high, fascicled, from a woody base, scantily branched below, very densely leafy to ⅓–½ way, then slender and 2-noded, upper 3 internodes long (uppermost 6–7 in.), exserted, glabrous, smooth; lower sheaths very numerous, short, close, tight, firm, persistent, striate, pubescent to tomentose or villous, particularly towards the mouth and at the junction with the blade, brownish or purplish, upper 3–5 in. long, very tight, glabrous, smooth; ligule a fringe of hairs; blades

squarrose, subulate-involute from a broad and almost flat base, with callous obtuse tips, 2–3½ in. by 1½–2 lin. at the base, very firm and rigid, closely and strongly striate, glabrous, very scabrid near the tips, margins scabrid all along; panicle contracted, sublinear, strict, 3½–4 in. long; branches 2-nate, 4–2-spiculate from 3–6 lin. above the base or 1-spiculate, filiform like the axis, smooth, pubescent at the axils; pedicels very unequal, 2–12 lin. long, finely filiform, smooth; spikelets 8–9 lin. long, straw-coloured; glumes long acuminate, scarious, glabrous, smooth; valves linear-oblong in profile, body 3 lin. long, villous all along, 7–9-nerved; lobes very narrow, about ½–¾ lin. long, passing into a bristle about 3 lin. long; awn 9–10 lin. long, kneed at the middle, column rather stout; pales 3½ lin. long, narrow, very hairy all over; lodicules ciliate.

COAST REGION: Caledon Div.; Nieuw Kloof in Houw Hoek Mountains, *Burchell*, 8076!

XXXVII. DANTHONIA, DC.

Spikelets small to large, laterally compressed, pedicelled, panicled, very rarely sessile or subsessile in distichous spikes; rhachilla disarticulating above the glumes and between the valves, usually ending with a rudimentary valve. *Florets* 3 to many, ⚥, the uppermost reduced, exceeded by the glumes or more or less equalling them. *Glumes* equal or subequal, more or less lanceolate in profile, usually acute or acuminate, keeled, hyaline to scarious or subherbaceous, 3–9- (rarely 1-) nerved, often with transverse veins. *Valves* firmly membranous to chartaceous, hairy, often with the hairs partly gathered in variously arranged tufts, very rarely almost glabrous, 7–11- (rarely 5-) nerved, 2-lobed, awned from the sinus or sometimes the lobes reduced to minute teeth and the awn to a mucro; lobes free, rarely more or less adnate to the awn, acute or obtuse, with or without a bristle from the tip; callus small or rather long and acute, hairy. *Pales* 2-keeled, usually entire or almost so, mostly exceeding the insertion of the awn of the valve. *Lodicules* 2, usually large or ciliate, many-nerved, or sometimes small, fleshy, glabrous. *Stamens* 3. *Ovary* oblong or obovoid, glabrous; styles distinct, slender, long or short and then sometimes with the stigmatic hairs decurrent on the inner side and joining over the top of the ovary; stigmas plumose, laterally exserted. *Grain* oblong, obovoid or ellipsoid, usually semiterete, rarely almost terete; hilum obscure, oblong, short; embryo ⅓ to ½ the length of the grain.

Perennial, rarely annual; leaves very variable; panicle with usually solitary, alternately and closely divided branches, often very dense, sometimes compactly capitate, rarely transformed into a distichous spike.

Species 65–70, mostly in the temperate regions of the Southern Hemisphere, particularly in South Africa, Australia and New Zealand.

Spikelets 1½–2 in. long; stigmatic hairs decurrent on the inner side of the styles and joining over the top of the ovary:

Blades filiform, tapering to a long setaceous point, ½–1 ft. long ... ... (1) **macrantha.**
Blades linear, tapering almost from the base to the involute filiform obtuse tips, 2–3 in. long ... ... (2) **brachyphylla.**
Spikelets smaller: styles naked below, distinct:
Lower sheaths more or less covered with a coat of wool; panicles usually capitate:
Panicle loosely contracted; branches with few (or 1) spikelets on filiform very unequal pedicels; spikelets 7½–9½ lin. long ... ... (3) **zeyheriana.**
Panicle compact, spike-like or capitate:
Spikelets 10–12 lin. long in large heads, 2–2½ in. long; tips of glumes usually suddenly contracted and mucronate (4) **macrocephala.**
Spikelets smaller:
Spikelets 6–8½ lin. long; glumes narrow, long and acutely acuminate; awn with 2 very loose twists below ... ... (5) **lanata.**
Spikelets 4–7 lin. long; glumes rather broad, shortly acuminate or mucronate; awn straight or imperfectly twisted and bent below ... ... (6) **lupulina.**
Lower sheaths not covered with a coat of wool; panicles never capitate:
*Spikelets about 6 lin., rarely to 10 lin. long; rather coarse grasses with rigid to wiry blades:
Inflorescence a true, contracted panicle, ⅓ to more than 1 ft. long:
Spikelets 5½–6½ lin. long; valves loosely villous all over, hairs not in tufts; lower sheaths very minutely pubescent or tomentose and pruinose ... ... (7) **elephantina.**
Spikelets 6, rarely 10 lin. long; valves with distinct tufts of hairs; sheaths smooth and glabrous, except sometimes near the mouth or at the very base:
Sheaths woolly at the mouth and the lower enclosing a dense tuft of hairs at the very base; hair tufts of valves 3 on each side at the base of the lobes and equal to them ... ... (8) **papposa.**
Sheaths quite glabrous or almost so; hair tufts not arranged as in the preceding section:
Culms and leaves very robust; blades tightly involute or partly flat, up to more than 3 ft. long; tufts of hairs long, gathered in a dense straight trans-

verse fringe at or near the middle of the body of the valve ... ... (9) **cincta.**

Culms and leaves more slender, firm; blades filiform, terete, or subterete, canaliculate, often wiry, ½ to over 2 ft. long:

Glumes 1-, or lower sub-3-nerved; tufts of hairs 3 on each side in an oblique transverse fringe near the base, much shorter than the valve; lobes almost wholly adnate to the straight awn ... (10) **Macowanii.**

Glumes 5–3-nerved; valves with a submarginal fringe of hairs on each side below the middle; lobes free from the loosely twisted awn:

Blades wiry, up to more than 2 ft. long; lobes of valves mucronate ... (11) **dura.**

Blades finely filiform, flexuous, ½–1½ ft. long; lobes of valves acute with a slender bristle from the side (12) **stricta.**

Inflorescence a distichous linear spike; keels of pales approximate, smooth... (13) **disticha.**

**Spikelets less than 6 lin. long, or if so long, then in small scanty panicles and the plant dwarf, or the spikelets turgid and awnless (21):

†Spikelets not turgid, nor cleistogamous, distinctly awned (except in No. 20):

Dwarf; culms 1–2-noded; panicles small, scanty, often reduced to a few spikelets; spikelets 5–6 lin. long, with mucronate valve-lobes or 2–4 lin. long with hyaline obtuse side-lobes:

Spikelets 5–6 lin. long; valve-lobes mucronate; blades subulate, plicate, very rigid, obtuse or mucronulate, ½–1 in. long ... ... (14) **pumila.**

Spikelets 2–4 lin. long; valve-lobes obtuse, hyaline:

Perennial; blades setaceous to subulate, canaliculate, rigid, ½–2 in. long; spikelets 2½–4 lin. long; valve-lobes ciliolate, obtuse ... ... (15) **purpurea.**

Annual; blades finely setaceous, soft, ⅓–1 in. long; spikelets 2–3 lin. long; valve-lobes truncate, not ciliate ... (16) **tenella.**

Not dwarf, 4- or more-noded; panicles almost spike-like; spikelets 2–4½ lin. long, or if dwarf and fewer-noded (starved specimens of No. 17), then the valve-lobes not hyaline and obtuse:

Spikelets 2–2½ lin. long, awned; glumes 3–5-nerved; valve-lobes long, setaceously acuminate ... ... (17) **curva.**

Spikelets 3½–4½ lin. long and awned, at least from the lower florets, or 2 lin. long and awnless; glumes prominently 5–7-nerved; valve-lobes not long setaceously acuminate;

Culms and leaves more or less puberulous to tomentose; spikelets 3½–4½ lin. long; awns, at least of the lower deeply bifid valves, distinct; anthers 1 lin. long:

Suffrutescent; lower sheaths coriaceous (18) **suffrutescens.**

Herbaceous; lower sheaths thin ... (19) **glauca.**

Culms and leaves quite glabrous; spikelets 2 lin. long; awns reduced to a mucro from the sinus of the 2-toothed valves; anthers ⅓–¼ lin. long ... (20) **inermis.**

††Spikelets turgid, 4–6 lin. long, awnless; cleistogamous, in scanty racemes or lowest branches 2–3-spiculate ... (21) **decumbens.**

1. **D. macrantha** (Schrad. in Schult. Mant. ii. 385); perennial, tufted; culms erect or shortly ascending, stout below, slender above, more or less branched at some distance above the base (branches fascicled), densely leafy to the middle or towards it, 2–3 ft. long, glabrous, smooth, lower internodes very short, upper 3 or 4 long, exserted; leaves glabrous; sheaths firm, striate, smooth, lower usually very numerous and close, upper much narrower, often loose; ligule a fringe of short hairs; blades tightly filiform-involute or canaliculate, rarely loosely involute or partly flat, tapering to a long setaceous

point, $\frac{1}{2}$–1 ft. by 1–2 lin. (when expanded), firm, flexuous, smooth, except at the scaberulous margins; panicle contracted, ovate to oblong, 3–7 in. long, erect or slightly nodding; axis angular below, striate above; branches solitary, alternately divided from near the base, smooth like the very unequal branchlets and pedicels, filiform; pedicels $\frac{1}{4}$–1 in. long; spikelets straw-coloured, $1\frac{1}{2}$–2 in long, erect; florets (fertile) 2, much shorter than the glumes; rhachilla long continued with a rudimentary floret; glumes unequal to subequal, lower shorter, lanceolate, long and finely acuminate, scarious, smooth, glabrous, lower 5-, upper 3-nerved or both with 1–2 short additional nerves, transversely veined; valves spindle-shaped in profile, body 4–$4\frac{1}{2}$ lin. long, firm, at length cartilaginous, 6–9-nerved, smooth and glabrous below, increasingly villous from the middle; lobes narrow, acute, up to $1\frac{1}{2}$ lin. long, almost wholly adnate to the side-bristles, these 5–6 lin. long, very loosely twisted; awn $\frac{3}{4}$–1 in. long, kneed at or above $\frac{1}{3}$ way up, stout, tightly twisted, almost black below the bend; callus slender, pungent, densely villous; pales as long as the valves, minutely 2-toothed, glabrous, cartilaginous below; lodicules obovate-cuneate, glabrous, many-nerved, $\frac{1}{2}$–$\frac{3}{4}$ lin. long; anthers $3\frac{1}{2}$ lin. long; ovary glabrous; stigmas sessile, stigmatic hairs joining over the top of the stigma; grain linear-oblong, subterete; hilum more than $\frac{1}{2}$ the length of the grain. *Trin. Gram. Ic. t.* 63; *Gram. Gen.* 71, *and in Mém. Acad. Pétersb. sér.* 6, i. 71; *Gram. Suppl.* 34, *and in Mém. Acad. Pétersb. sér.* 6, iv. 35; *Nees, Fl. Afr. Austr.* 319; *Steud. Syn. Pl. Glum.* i. 242; *Durand & Schinz, Consp. Fl. Afr.* v. 852. *D. juncea, Trin. Gram. Suppl. l.c.*; *Steud. l.c.* 243; *Durand & Schinz, l.c.* 851. *Avena macrocalycina, Steud. in Flora,* 1829, 482. *Pentameris macrantha, Nees in Linnæa,* vii. 316; *Kunth, Enum.* i. 316. *Pentaschistis* (*by error*) *macrantha, Nees, Fl. Afr. Austr. l.c.*

COAST REGION: Piquetberg Div.; Piquetberg, 2000 ft., *Drège.* Cape Div.; Table Mountain, *Drège!* *Ecklon,* 932! *Ecklon & Zeyher!* 132! among rocks on the north side of Table Mountain, *MacOwan, Herb. Aust. Afr.* 1699! near Capetown, *Spielhaus!* Farmer Peck Valley, *Wolley Dod,* 2808! Devils Peak, *Bergius!* Cape Flats near Doorn Hoogte, *Ecklon & Zeyher,* 1825! Cape Peninsula, low hills of South Kommetje, *Wolley Dod,* 1573! near Witzenberg, Vogel Vlei and Constantia, *Ecklon.* Malmesbury Div.; Groene Kloof, *Ecklon.* Caledon Div.; Klyn Riversberg Range, *Ecklon.* Mossel Bay Div.; hills near the landing-place at Mossel Bay, *Burchell,* 6292! Riversdale Div.; hills near the Zoetemelks River, *Burchell,* 6752!

2. **D. brachyphylla** (Stapf); perennial; culms subascending from a short slender rhizome, covered with tight hard scales, very densely leafy at the base or up to the middle, 1–$1\frac{1}{2}$ ft. long, smooth, glabrous, lowest internodes numerous, very short, upper 3–4 slender, 3–5 in. long, the intermediate or also the uppermost more or less exserted; leaves glabrous; sheaths smooth, lower very tight, close, short, striate, persistent, upper herbaceous, lax, uppermost tumid, and usually exceeding the base of the panicle; ligule a fringe of short hairs; blades linear, tapering almost from the base to the involute filiform scabrid obtuse tips, 2–3 in. by $1\frac{1}{2}$–2 lin. at the

base, very rigid and firm, flat below, closely and strongly striate, margins scabrid; panicle ovate to oblong, contracted, erect or slightly nodding, $2\frac{1}{2}$–4 in. long; axis angular or compressed, striate; branches solitary, alternately and very sparingly divided from near the base, 4–2-spiculate, or 1-spiculate, filiform like the very unequal branchlets and pedicels, smooth; pedicels 1–10 lin. long; spikelets straw-coloured, $1\frac{1}{2}$–$1\frac{2}{3}$ in. long, erect; florets (fertile) 2, much shorter than the glumes; rhachilla long continued, glabrous with a rudimentary floret; glumes equal or subequal, lanceolate, long and finely acuminate, scarious, smooth, glabrous, lower 5–7-, upper 3-nerved, or both with 1 or 2 short additional nerves at the base, transversely veined; valves spindle-shaped in profile; body 3 lin. long, firm, villous all along, 9-nerved; lobes narrow, subobtuse, 2–$2\frac{1}{4}$ lin. long, scaberulous, ciliolate, villous below, wholly adnate to the loosely twisted and flexuous side-bristles, these 5–8 lin. long; awn about 1 lin. long, kneed at $\frac{1}{3}$ way up, stout, brown, tightly twisted below; callus slender, subpungent, densely villous; pales very narrow, about 5 lin. long, tips puberulous; lodicules glabrous or ciliolate, $\frac{1}{2}$–$\frac{2}{3}$ lin. long; ovary glabrous, stigmas sessile, stigmatic hairs joining over the top of the ovary.

SOUTH AFRICA: without precise locality, *Zeyher*, 1826 δ!

3. **D. zeyheriana** (Steud. Syn. Pl. Glum. i. 244); perennial, compactly tufted; culms erect, slender or rather stout, $1\frac{1}{2}$ ft. long, glabrous, smooth, 2-noded, upper 2 internodes usually exserted, rarely sheathed all along; leaves crowded at the base; basal sheaths densely woolly tomentose, usually at length breaking up into fibres, or the inner glabrous except at the base and near the mouth, 1–2 in. long, upper quite glabrous, or with a woolly beard at the mouth; ligule a dense villous fringe; blades linear, filiform-involute, or flat below, acute to subobtuse, 4–8 in. by 1–2 lin. when expanded, erect, straight or slightly curved, very firm, scantily hairy on the back, at least when young, smooth, whitish above, strongly striate; panicle oblong to ovate, erect, loosely contracted, 3–4 in. long; axis more or less angular, smooth, striate; branches fascicled or 2-nate and very unequal, or solitary and sparingly divided from or near the base, bearing few to 1 spikelet, filiform, glabrous like the branchlets and pedicels except at the hairy axils, smooth; pedicels 1–8 lin. long; spikelets $7\frac{1}{2}$–$9\frac{1}{2}$ lin. long, erect; rhachilla glabrous, joints about $\frac{1}{2}$ lin. long; florets 4–5, the uppermost rudimentary; glumes lanceolate, narrow, long and acutely acuminate, equal, subherbaceous on the back, otherwise scarious, smooth or scaberulous near the tips, 3–5-nerved below, scantily transversely veined; valves oblong-lanceolate in profile, body 2–$2\frac{1}{4}$ lin. long, 9-nerved, densely villous to the middle, hairs in rows between the nerves terminating in a transverse fringe of long tufts, glabrous above the middle; lobes long lanceolate, subulate- or setaceous-acuminate, up to $3\frac{1}{2}$ lin. long, scaberulous, shortly adnate to the base of the awn; awn flat, up to 6 in. long, kneed at $\frac{1}{3}$ way up, with 2 loose twists below; callus up

to $\frac{1}{2}$ lin. long, bearded; pales oblong, minutely 2-toothed, 3 lin. long, hairy at or below the middle; keels densely and rigidly ciliolate; lodicules 1 lin. long, tips ciliolate; anthers $2\frac{1}{2}$ lin. long; ovary glabrous; styles distinct. *Durand & Schinz, Consp. Fl. Afr.* v. 855.

VAR. β, **trichostachya** (Stapf); upper sheaths woolly; spikelets 9–12 lin. long; glumes long and finely hairy or almost woolly; lobes of valves very narrow, finely pubescent.

COAST REGION: Swellendam Div.; mountains near Puspus Valley, *Zeyher*, 4555! George Div.; on Post Berg (Cradock Berg) near George, *Burchell*, 5933! Var. β: Cape Div.; between Slang Kop and Red Hill, frequent, *Wolley Dod*, 3002! low hills south of Kommetje, *Wolley Dod*, 1572!

The typical form is perhaps identical with *Danthonia decora*, Nees, Fl. Afr. Austr. 332, which I have not seen.

4. D. macrocephala (Stapf); perennial, compactly tufted; culms rather slender, erect, 1 ft. long, glabrous, smooth, 1–2-noded, upper internodes shortly exserted; leaves crowded at the base; basal sheaths very densely woolly tomentose, 2–4 in. long, firm, persistent, the following very broad and tumid, herbaceous, glabrous, or with a woolly beard at the mouth, strongly striate; ligule a dense villous fringe; blades tightly filiform-involute, wiry, rather stout, obtuse, lower about $\frac{1}{2}$ ft. long, uppermost very short, glabrous, smooth; panicle capitate, compact, globose-ovoid, $2–2\frac{1}{2}$ in. long; axis smooth, striate; branches very closely divided from the base; branchlets and pedicels very short; spikelets imbricate, 10–12 lin. long; florets 4–5, uppermost much reduced; rhachilla glabrous, joints short; glumes lanceolate, gradually acuminate or usually with the tips suddenly contracted and mucronate, scarious, glabrous, scaberulous upwards, 5- to sub-7-nerved, side-nerves short; valves oblong-lanceolate in profile; body $2\frac{1}{2}–3$ lin. long, 9-nerved, densely villous to $\frac{2}{5}$ the way from the base with the hairs in rows between nerves, abruptly ending in a dense tranverse fringe about $2\frac{1}{2}$ lin. long, otherwise glabrous; lobes lanceolate, acutely acuminate or mucronate, over $2\frac{1}{2}$ lin. long, scaberulous, hyaline, adnate to the awn at the base; awn flat, 5–6 lin. long with 2 very loose twists in the lower $\frac{1}{2}$; callus short, bearded; pales obovate to oblong, obtuse, $3\frac{1}{2}–3\frac{3}{4}$ lin. long, long hairy near the middle; keels densely and rigidly ciliate; lodicules obovate, $\frac{3}{4}$ lin. long, ciliate; grain obovoid-oblong, $1\frac{1}{4}$ lin. long.

SOUTH AFRICA: without precise locality, but probably from Caledon, Tulbagh, or Clanwilliam Division, *Thom!*

5. D. lanata (Schrad. in Schult. Mant. ii. 386); perennial, compactly tufted; culms slender, erect, $\frac{1}{2}–1\frac{1}{4}$ ft. long, glabrous, smooth, 1–2-noded, upper 2 internodes exserted; leaves crowded at the base; basal sheaths very densely woolly tomentose, $1\frac{1}{2}–2$ in. long, firm, persistent, the following glabrous or with a flake of wool at the mouth, striate, uppermost 1 or 2 tumid, herbaceous; ligule a villous fringe; blades linear, tightly or loosely filiform-convolute, subpungent, 3–7 in.

by $\frac{1}{2}$ lin. when expanded, very rigid, more or less flexuous or curved, glabrous, or villous just above the ligule, smooth; panicle capitate or subcapitate, ovoid to subglobose, very dense, about $1\frac{1}{2}$ in. long; axis striate; branches fascicled, scantily bearded at the base, very closely divided from the base; branchlets and pedicels very short; spikelets imbricate, 6–$8\frac{1}{2}$ lin. long; florets 3–4, uppermost more or less reduced or at least smaller; rhachilla glabrous, joints short; glumes equal, narrow lanceolate, acutely acuminate, subherbaceous and often purple on the back, otherwise scarious, glabrous, scaberulous upwards, usually closely 3- to sub-5-nerved, or sometimes 1-nerved; valves oblong-lanceolate in profile, body $2\frac{1}{2}$ lin. long, 9-nerved, densely villous to $\frac{1}{3}$ or $\frac{2}{5}$ way from the base, with the hairs in rows between the nerves, abruptly ending in a dense transverse fringe about 2 lin. long, otherwise glabrous; lobes subulate-lanceolate, 3 lin. long, scaberulous, hyaline, adnate to the awn for $\frac{1}{3}$–$\frac{1}{4}$ of its length; awn flat, 5–6 lin. long, with 2 very loose twists in the lower $\frac{1}{3}$; callus short, bearded; pales oblong, obtuse, $3\frac{1}{2}$–$3\frac{3}{4}$ lin. long, long hairy near the middle; keels densely and rigidly ciliate; lodicules obovate, cuneate, long ciliate, $\frac{3}{8}$ lin. long; anthers 2–$1\frac{1}{2}$ lin. long; ovary glabrous; styles slender, distinct. *Nees in Linnæa*, vii. 316; *Fl. Afr. Austr.* 329; *Trin. Gram. Ic. t.* 62; *Gram. Gen.* 70, *and in Mém. Acad. Pétersb. sér.* 6, i. 71, *not in Gram. Suppl.* 33, *nor in Mém. Acad. Pétersb. sér.* 6, iv. 34; *Kunth, Enum.* i. 314; *Steud. Syn. Pl. Glum.* i. 242; *Durand & Schinz, Consp. Fl. Afr.* v. 851. *D. rufa, Nees, l.c.* 330; *Steud. l.c.* 243; *Durand & Schinz, l.c.* 854. *Avena lanata, Schrad. in Goett. Gelehrte Anzeig.* 2075. *A. lupulina, Steud. in Flora*, 1829, 486, *not of Thunb.*

VAR. major (Nees, l.c.); panicles somewhat laxer, spikelets 10–11 lin. long; florets 5–7.

COAST REGION: Clanwilliam Div.; Ceder Berg Range, on Blaauw Berg, *Drège*, 2559; Cape Div.; Table Mountain, *Ecklon*, 924! *Ecklon & Zeyher*, 135 in St. Petersburg Herb.! *Wolley Dod*, 1688! *Spielhaus!* near Constantia Nek, *Wolley Dod*, 1678! behind Hout Bay Hotel, *Wolley Dod*, 3169! near Constantia, and at Doorn Hoogte, *Ecklon*. Tulbagh Div.; Witzenberg, Vogel Vley and near Tulbagh Waterfall, *Ecklon*. Stellenbosch Div.; Hottentots Holland, moist places on the summits of the hills, *Ecklon!* Swellendam Div.; near Puspus Vlei or on the Keureboom River, *Zeyher*, 4541! Caledon Div.; in stony places near Genadendal, *Drège*, 1664! mountains of Baviaans Kloof near Genadendal, *Burchell*, 7629! Uitenhage Div.; Steenbocks Flats to the north of Winterhoek Mountains, *Ecklon*. Var. β: Cape Div.; Vley beyond Smitwinkel, *Wolley Dod*, 2956! Swellendam Div.; without precise locality, *Maire & Mund!*

An imperfect specimen, collected in the Orange Free State by Buchanan (164), belongs, perhaps, also to this species; its spikelets are only 5–$6\frac{1}{2}$ lin. long.

6. D. lupulina (Roem. & Schult. Syst. ii. 690); perennial, compactly cæspitose; culms slender, erect or suberect, $\frac{1}{2}$–$1\frac{1}{4}$ ft. long, glabrous; smooth, 1–2-noded, upper 2 internodes exserted; leaves crowded at the base; basal sheaths densely woolly tomentose, about 2 in. long, firm, persistent, the upper glabrous or with a woolly beard at the mouth, striate, uppermost 2 tumid, herbaceous; ligule a villous fringe; blades filiform, involute, subacute or subobtuse, 3–6 in. long, very rigid, subflexuous, glabrous or villous just above the ligule, smooth; panicle capitate or spike-like, globose to oblong, 1–1½ in. long; axis striate, smooth, sometimes puberulous; branches scantily bearded at the axils, very closely divided from the base; branchlets and pedicels very short; spikelets imbricate, 4–7 lin. long; florets 4–5, uppermost rudimentary; rhachilla glabrous, joints very short; glumes lanceolate, rather broad, acuminate in profile or mucronate, 3-nerved, scaberulous upwards, herbaceous on the back and sometimes purplish, otherwise scarious; valves lanceolate-oblong in profile, body 2½ lin. long, 9-nerved, densely villous to about the middle with the hairs between the nerves in rows abruptly ending in a somewhat rigid dense transverse fringe, about 1½ lin. long, otherwise glabrous; lobes short, ovate, acute, about 1 lin. long, often almost wholly adnate to the awn, scaberulous or puberulous; awn 2–4 lin. long, flat, straight or imperfectly twisted below; callus short, villous; pales obovate, obtuse, 2–2¾ lin. long, long hairy near the middle; keels densely and rigidly ciliolate; lodicules obovate or cuneate, ciliate, ⅜ lin. long; anthers 1–1½ lin. long; ovary glabrous; styles distinct, slender. *Nees in Linnæa*, vii. 315; *Fl. Afr. Austr.* 330; *Kunth, Enum.* i. 315; *Steud. Syn. Pl. Glum.* i. 243; *Durand & Schinz, Consp. Fl. Afr.* v. 852. *D. lanata, Trin. Gram. Suppl.* 33, *and in Mém. Acad. Pétersb. sér.* 6, iv. 34, *and (partly) Gram. Gen.* 71, *and in Mém. Acad. Pétersb. sér.* 6, i. 71. *D. coronata, Trin. Gram. Gen.* 70, *and in Mém. Acad. Pétersb. sér.* 6, i. 70; *Kunth, l.c.* 314. *Avena lupulina, Thunb. Prod.* 23; *Fl. Cap.* i. 438; *ed. Schult.* 118.

SOUTH AFRICA: without precise locality, *Thunberg! Herb. Harvey*, 272! *Bergius!*

COAST REGION: Cape Div.; sandy plains between Tyger Berg and Simons Mountain, *Drège!* Cape Flats near Doorn Hoogte, *Ecklon.* Table Mountain, *Spielhaus!* Paarl Div.; Klein Drakenstein Mountains, *Drège!* Paarl Mountain, *Drège!* Tulbagh Div.; near Tulbagh Waterfall, *Ecklon; Ecklon & Zeyher*, 203 in St. Petersburg Herb.! Winterhoek Mountain, 5000 ft., *Ecklon; Ecklon & Zeyher*, 136 in St. Petersburg Herb.! Worcester Div.; Dutoits Kloof, in moist rocky places, 1500–3000 ft., *Drège.* Stellenbosch Div.; near Stellenbosch, *Harvey! Alexander!* Hottentots Holland, *Ecklon!* Swellendam Div.; Puspus Valley, Voormans Bosch and Keurbooms River, *Ecklon & Zeyher.*

7. **D. elephantina** (Nees, Fl. Afr. Austr. 334); perennial, tufted; culms erect, very robust, fascicled at the base with 1 or 2 intravaginal innovation shoots, 2–3 ft. high, glabrous, smooth, about 3-noded, sheathed right up to the panicle, or one of the intermediate nodes shortly exserted; basal sheaths usually crowded, very close, coriaceous, ¼–½ ft. long, obscurely striate, very minutely pubescent or tomentose and at the same time pruinose from waxy granules, at length glabrescent, upper looser, pubescent or glabrous, slightly scaberulous, uppermost tumid, pubescent and scaberulous; ligule a dense fringe of short hairs; blades loosely filiform-involute below, very tightly involute or canaliculate and very long and finely attenuate in the upper part, acute, 1 to more than 2 ft. long, up to 2½ lin.

broad when expanded, very hard, flexuous, glabrous, scabrid along the margins and on the back in the upper part, upper surface whitish, strongly and closely striate; panicle erect, oblong, very dense, 4–7 in. long; axis stout, striate, angular, glabrous; branches solitary, lowest up to 3 in. long, abundantly and closely divided from the base, branchlets and pedicels filiform to subcapillary, scaberulous to scabrid; pedicels ½–2 lin. long; spikelets densely crowded, straw-coloured, 5½–6½ lin. long; florets 3–4, the uppermost rudimentary; rhachilla glabrous, joints ½–3¼ lin. long; glumes lanceolate, long acuminate, hyaline, pubescent, 1-nerved, keel smooth; valves oblong in profile, body 1¾ lin. long, villous all over, 7–9-nerved; lobes narrow, about 1–1½ lin. long, scaberulous and ciliolate, gradually passing into a mucro or short fine bristle, as long as the lobe or shorter; awn 4–6 lin. long, kneed at the middle, loosely twisted below, bristle flattened; callus obtuse, very short, bearded; pales linear-oblong, obtuse, 1¾ lin. long, hairy below, keels stout, densely ciliolate above; lodicules ciliate; anthers almost 1½ lin. long; ovary glabrous, styles slender, distinct. *Drège in Linnæa*, xx. 254; *Steud. Syn. Pl. Glum.* i. 243; *Durand & Schinz, Consp. Fl. Afr.* v. 849. *Avena elephantina, Thunb. Prod.* 23; *Fl. Cap.* i. 437; *ed. Schult.* 117; *Kunth, Enum.* i. 305.

SOUTH AFRICA: without precise locality, *Thom!*

COAST REGION: Vanrhynsdorp Div.; Gift Berg, *Drège;* Malmesbury Div.; Swartland, *Thunberg!* between Eikenboom and Riebeck Castle, *Drège!* Tulbagh Div.; Tulbagh Valley, Winterhoek, &c., *Ecklon; Ecklon & Zeyher*, 138 in St. Petersburg Herb.! Worcester Div.; between Driekoppen and Hex River, *Drège.* Swellendam Div.; lower part of Zonder Einde River, *Zeyher*, 4549!

8. **D. papposa** (Nees, Fl. Afr. Austr. 333, in part); perennial, densely tufted; culms rather stout, erect, about 1½ ft. long, glabrous, smooth, 2–3-noded, upper internodes exserted; leaves crowded at the base; sheaths woolly at the mouth, otherwise glabrous, very smooth, tight, striate or sulcate, basal very close, sometimes numerous and subflabellate, 2–3 in. long, broad, often shining as if lacquered, enclosing a dense tuft of hairs at the very base, uppermost often tumid; ligule a woolly fringe; blades filiform-involute, tapering to a fine point, rather stout, 4 in. to almost 1 ft. long, very hard and rigid, glabrous, smooth except at the scabrid margins; panicle erect, narrow, oblong, very dense, almost spike-like, 5–6 in. long; axis rather stout, striate, subangular, glabrous; branches solitary, lowest up to 3 in. long, closely and frequently divided from the base; branchlets and pedicels filiform, scaberulous to scabrid; pedicels 1–3 lin. long; spikelets very crowded, straw-coloured, sometimes tinged with purple, 6½–7½ lin. long; florets 3, the uppermost rudimentary; rhachilla glabrous, joints up to ¾ lin. long; glumes subequal, lanceolate, acuminate, scarious, smooth, glabrous, closely and strongly 3-nerved, scantily transversely veined; body of valves 2 lin. long, rather firm, 9-nerved, loosely pubescent on the back, with 3 submarginal tufts of long hairs on each side between the middle and the base of the lobes; lobes subulate-lanceolate,

acute, 2½ lin. long, equalled by the hair tufts; awn about 6–7 lin. long, kneed at the middle, stout, loosely twisted below, bristle rather flat; callus over ½ lin. long, bearded; pales linear-oblong, 3¼ lin. long, scantily hairy, keels ciliolate; lodicules glabrous; anthers 2–2¼ lin. long; ovary glabrous; styles distinct, slender. *Steud. Syn. Pl. Glum.* i. 243; *Durand & Schinz, Consp. Fl. Afr.* v. 853.

COAST REGION: Riversdale Div.; between Little Vet River and Kampsche Berg, *Burchell*, 6902! Uitenhage Div.; by the Zwartkops River, *Ecklon & Zeyher*, 469! *Burchell*, 4430!

Nees also refers to this species Drège's specimens from the Kromme River (Humansdorp Div.) and Enon (Alexandria Div.) which I have not seen.

9. **D. cincta** (Nees, Fl. Afr. Austr. 332); perennial, densely tufted; culms erect, very robust, fascicled with long, very stout innovation shoots, several feet high, up to 3 lin. thick below, solid, glabrous, smooth, sheathed all along, or the uppermost internode exserted and then to more than 1½ ft. long; sheaths tight or the uppermost tumid, very firm, striate, glabrous or ciliate along the margins below, lower very broad, coriaceous, up to more than ½ ft. long, uppermost ¾–1 ft. long; ligule a dense fringe of hairs; blades very long, linear, tapering to a fine pungent point, up to more than 3 ft. by 2–2½ lin. when expanded, usually tightly involute, sometimes partly flat, very rigid and firm, straight, glaucous, very smooth, hairy to densely villous above the ligule, otherwise glabrous, closely and strongly nerved on the upper surface; panicle erect, almost spike-like, lanceolate or linear-oblong, very dense, up to more than 1 ft. by 1–3 in., shining, almost silvery; axis rather stout, more or less angular, glabrous or finely hairy, smooth; branches solitary, divided from the very base, up to 5 in. long, erect; branchlets fascicled, densely contracted, filiform, scaberulous; pedicels very unequal, ½–4 lin. long; spikelets 6–7 lin. long, white, tinged with purple then straw-coloured; florets 3–4, distant, with the uppermost rudimentary; rhachilla glabrous, joints ½ lin. long; glumes narrow, lanceolate, acuminate, subhyaline, glabrous, smooth, 1-nerved or shortly 3-nerved; valves oblong-lanceolate in profile, body 2½ lin. long, distinctly 7-nerved, with a transverse fringe of tufts of long hairs (2½ lin. long) at the middle, otherwise glabrous; lobes narrow, subulate-lanceolate, very acute, 1½ lin. long, hyaline, scaberulous; awn 2–3 lin. long, slightly kneed and twisted at ¼ the way from the base, flat; callus slender, shortly villous; pales linear-oblong, 2½ lin. long, keels densely and rigidly ciliolate; lodicules glabrous; styles distinct, slender. *Steud. Syn. Pl. Glum.* i. 243; *Durand & Schinz, Consp. Fl. Afr.* v. 848.

COAST REGION: Cape Div.; near Constantia and Doorn Hoogte, *Ecklon*; Muizenberg Vley, *Wolley Dod*, 2356! near the convict establishment at Tokay, *Wolley Dod*, 2192! Tulbagh Div.; near Tulbagh Waterfall, *Ecklon*! Ceres Div.; on rocks at the foot of the mountains, near Ceres, 1600 ft., *Bolus*, 7460! Uitenhage Div.; by the Zwartkops River, *Ecklon & Zeyher*, 979! 137 and 204 (in St. Petersburg Herb.)! between Van Stadensberg Range and Bethelsdorp, *Drège*! between Leadmine River and Van Stadens River, *Burchell*, 4638!

Albany Div.; wet places on the mountains around Howisons Poort near Grahamstown, 1800 ft., *MacOwan & Bolus, Herb. Norm. Aust. Afr.* 792!

10. D. Macowanii (Stapf); perennial, tufted; culms erect, rather slender, over 2 ft. long, smooth, glabrous, all the nodes close to the base, uppermost internode occupying almost the whole of the culm, long exserted; leaves glabrous, smooth; sheaths firm, strongly striate, basal 5–6 in. long, slightly compressed in the upper part, uppermost up to 1½ ft. long, purplish, tight, terete; ligule a dense, short, villous fringe; blades filiform, tightly involute or canaliculate, subpungent, lower over 1½ ft., uppermost ½–¾ ft. long, very rigid, straight, glaucous, scarcely striate; panicle erect, contracted, sublinear, 8–10 in. long; axis filiform, terete, smooth, glabrous; branches solitary, erect or suberect, lowest to ½ ft. long, divided from 1 in. or less or the upper from close above the base, like the alternate branchlets filiform, glabrous, smooth; ultimate divisions and pedicels more or less fascicled, very unequal, latter from 1–6 lin. long, the longer almost capillary; spikelets close, straw-coloured, 7 lin. long, shining; florets 4, uppermost 1 or 2 rudimentary or at least reduced in size; rhachilla glabrous, joints ½ lin. long; glumes narrow lanceolate, finely acuminate in profile, scarious, glabrous, smooth, lower 3-nerved near the base; valves linear-lanceolate in profile, 4 lin. long, including the almost completely adnate lobes, rather firm, purple on the back, subhyaline in the upper half and along the margins as far as the fringe, 7–9-nerved with the nerves prominent below, with an oblique transverse fringe of 3 tufts of hairs on each side near the base and towards the margins (about 1½ lin. long), otherwise glabrous; awn a straight scaberulous bristle up to 3 lin. long; callus ½–⅔ lin. long, with a beard about 1½ lin. long; pales linear, as long as the valve, keels ciliolate; lodicules oblique, ciliate, over 1 lin. long; anthers 1 lin. long; ovary glabrous; styles distinct, short.

CENTRAL REGION: Somerset Div.; on the banks of streamlets near the summit of Bosch Berg, and rarely in the bed of the Little Fish River, 4800 ft., *MacOwan*, 4800 ft., 1986!

11. D. dura (Stapf); perennial, tufted, with numerous innovation shoots; culms erect, firm, about 3 ft. long, glabrous, smooth, 3-noded, upper internodes ½ ft. long, exserted; leaves crowded at the base; sheaths tight, very obscurely striate, glabrous, usually beardless, lowest firmly scarious, pallid, up to ½ ft. long; ligule a dense villous fringe; blades filiform, much narrower at the base than the sheath, wiry, canaliculate, subacute, over 2 ft. long, very firm, flexuous, glabrous, smooth; panicles linear to oblong-linear, nodding, 6–7 in. long, loosely contracted; axis subcompressed or subangular, smooth at least below; branches solitary, alternately divided from near the base, or the lowest from ⅓–⅔ in. above it and up to 2 in. long and 10–12-spiculate, filiform like the branchlets and pedicels, subflexuous, compressed or angular, scabrid at least above, glabrous, or almost so; pedicels very unequal, from 1–8 lin. long, tips usually slightly pubescent; spikelets erect, straw-coloured, about 9 lin. long; florets about 5, rather close, uppermost reduced; rhachilla glabrous; glumes equal, lanceolate, acutely acuminate, scarious, glabrous, lower 5-, upper 3- nerved; valves linear-oblong in profile, body 2 lin. long, rather firm, smooth, 9-nerved, with a single short submarginal fringe of hairs on each side below the middle; lobes ovate-lanceolate, decurrent, over 1 lin. long, hyaline, scaberulous, mucronate; awn 5 lin. long, kneed at $\frac{1}{3}$ way up, very loosely twisted, broad, dark below the bend; callus short, bearded; pales oblong, almost 3 lin. long, keels densely and rigidly ciliolate; lodicules obovate, long ciliate; grain oblong, almost $1\frac{1}{2}$ lin. long. *Chætobromus stricta, var. β, Nees, Fl. Afr. Austr.* 342.

WESTERN REGION: Little Namaqualand; Kamies Bergen, between Pedros Kloof and Lily Fontein, 3000–4000 ft., *Drège!*

Drège indicates *C. stricta*, var *β*, also from Modderfonteins Berg (Little Namaqualand, the Zuurberg Range (Alexandria Div.), but the latter locality is extremely doubtful.

12. D. stricta (Schrad. in Schult. Mant. ii. 383); perennial, densely cæspitose with numerous innovation shoots; culms erect, rarely ascending and geniculate, slender, 1–2 ft. high, glabrous, smooth, 1–2-noded, uppermost 2 internodes very long, exserted; leaves crowded at the base; sheaths tight, striate, glabrous, bearded or beardless at the mouth, lowest short, very firm, persistent; ligule a fringe of usually long hairs; blades finely filiform, canaliculate, subacute, $\frac{1}{2}$–$1\frac{1}{4}$ ft. long, firm, flexuous, glabrous, smooth, light green; panicle linear to oblong-linear, erect or slightly nodding, 3–6 in. long, contracted, rather dense; axis compressed or angular, scabrid and pubescent or puberulous; branches solitary, sparingly and alternately divided from near the base, like the branchlets and pedicels filiform, straight, compressed or angular, scabrid and pubescent, usually 5–2-spiculate (rarely 6–11-spiculate); pedicels very unequal, $\frac{1}{2}$–5 lin. long, tips shortly bearded; spikelets erect, straw-coloured, sometimes tinged with brown or purple, 6–10 lin. long; florets 5–7, close, uppermost 1–2 reduced; rhachilla glabrous; glumes equal or subequal, linear-lanceolate in profile, lower slightly longer, acute or acuminate, glabrous, smooth or finely granular, 3–5-nerved, sometimes with 1–2 short additional nerves; valves closely imbricate at first, at length rolling in and then oblong-linear in profile; body about 2 lin. long, rather firm, smooth, faintly 9-nerved with a submarginal row of 4–5 tufts of hairs on each side below the middle and 1 more inside close to the base; lobes ovate to ovate-lanceolate, acute, decurrent, 1–$1\frac{1}{2}$ lin. long, hyaline, scaberulous, almost wholly adnate to the fine side-bristles, these $1\frac{1}{2}$–$2\frac{1}{2}$ lin. long; awn 3–$4\frac{1}{2}$ lin. long, kneed at or below the middle, very loosely twisted, broad, brown below; callus acute, villous; pales oblong, $2\frac{1}{4}$–$2\frac{3}{4}$ lin. long, keels densely and rigidly ciliolate; lodicules obovate to cuneate, $\frac{1}{2}$ lin., long ciliate; anthers $1\frac{1}{4}$–$1\frac{1}{2}$ lin. long; ovary glabrous; styles distinct, short. *Trin. Gram. Suppl.* 35, *and* *in Mém. Acad. Pétersb. sér.* 6, iv. 36; *Steud. Syn. Pl. Glum.* i. 243; *Durand & Schinz, Consp. Fl. Afr.* v. 854. *D. fascicularis, Steud. l.c.; Durand & Schinz, l.c.* 850. *Avena hexantha, Steud. in Flora*, 1829, 487. *Pentameris stricta, Nees in Linnæa*, vii. 310; *Kunth, Enum.* i. 317. *Chætobromus strictus, Nees, Fl. Afr. Austr.* 341. *C. fascicularis, Nees, l.c.*

SOUTH AFRICA: without precise locality, *Mund & Maire!* *Herb. Harvey*, 267! 285! *Drège*, 1686!

COAST REGION: Clanwilliam Div.; Ceder Bergen, between Wupperthal and Ezelsbank, 3500–4000 ft., *Drège*, 2575c! Cape Div.; Table Mountain and Devils Mountain, *Ecklon!* Lions Head over Sea Point, *Wolley Dod*, 3574! Constantia Nek, *Wolley Dod*, 2123! 2254! Paarl Div.; rocks of Paarl Berg, 1500–3000 ft., *Drège!* Tulbagh Div.; Great Winter Berg, 1000–2000 ft., *Drège!* Worcester Div.; Dutoits Kloof, 2000–2500 ft., *Drège*. Caledon Div.; Koude River, 500 ft., *Schlechter*, 9600! Onrust River, 2500 ft., *Schlechter*, 9499! by the Zonder Einde River near Appels Kraal, *Zeyher*, 4552! Nieuw Kloof in Houw Hoek Mountains, *Burchell*, 8056! Riversdale Div.; near the Zoetemelks River, *Burchell*, 6693/2! 6688! 6782! between the Little Vet River and Kampsche Berg, *Burchell*, 6884! Mossel Bay Div.; Little Brak River, *Burchell*, 6187!

CENTRAL REGION: Beaufort West Div.; Nieuw Veld Mountains, *Drège*, 679!

The eastern specimens (from Caledon Div. to Mossel Bay Div.) have, on the whole, smaller spikelets (6–7 lin. long) than the western and northern.

13. **D. disticha** (Nees, Fl. Afr. Austr. 335); perennial, densely tufted with numerous innovation shoots; culms slender, erect, or geniculate, 1–1½ ft. long, smooth, glabrous, about 2–3-noded, internodes exserted or the uppermost sheathed up to the panicle; leaves crowded at the base; sheaths very tight, firm, obscurely striate, glabrous except at the sometimes bearded mouth, lower very hard, persistent; ligule a fringe of minute hairs; blades finely filiform, canaliculate, acute or subacute, lower ½ to more than 1 ft. long, rarely all short, rigid, strongly flexuous, densely scabrid and whitish on the upper side, and sometimes hairy at the base, ciliate along the margins below, otherwise glabrous and smooth; panicle reduced to an erect linear dense distichous spike, 1–3½ lin. long; rhachis compressed, slightly wavy, angles ciliate; spikelets sessile or subsessile with a rudimentary shortly bearded pedicel, imbricate (the two ranks at first contiguous, then divergent), 5½–6½ lin. long, the lowest rudimentary; florets 5–3, very close, uppermost rudimentary; rhachilla glabrous; glumes about equal, lanceolate, acute to acuminate, strongly compressed, glabrous, smooth, sides hyaline or subhyaline, nerves 3, close; body of valves about 2½ lin. long, rather firm, 9–11-nerved, with a submarginal tuft of stiff hairs somewhat above the base and a submarginal line of similar hairs from the tuft to the base of the lobes, otherwise smooth and glabrous; lobes subulate-lanceolate, almost 2 lin. long, ciliate, excurrent into a bristle of equal length; awn 6–8 lin. long, kneed at ⅓ way from the base, loosely twisted below; bristle rather flat; callus acute, bearded, ½ lin. long; pales almost equalling the lobes, over 4 lin. long; keels finely scaberulous; lodicules long ciliate; anthers 2½–3 lin. long;

ovary glabrous; styles distinct. *Drège in Linnæa*, xx. 254; *Steud. Syn. Pl. Glum.* i. 243; *Durand & Schinz, Consp. Fl. Afr.* v. 849.

SOUTH AFRICA: without precise locality, *Boivin!*

COAST REGION: Swellendam Div.; on mountain ridges along the lower part of the Zonder Einde River, 500–2000 ft., *Zeyher*, 4550! Riversdale Div.; near the Zoetemelks River, *Burchell*, 6693! Uitenhage Div.; grass-fields near the Zwartkops River, *Ecklon & Zeyher*, 453! Alexandria Div.; near Addo, *Ecklon!* Lower Albany; Glenfilling, 500–1000 ft., *Drège.* Albany Div.; Howisons Poort near Grahamstown, *Glass*, 7700! near Grahamstown, *Burke!* Bothasberg Range, 2700 ft., *MacOwan*, 1313! Queenstown Div.; Storm Bergen, 6000 ft., *Drège!*

CENTRAL REGION: Somerset Div.; between the Zuurberg Range and Klein Bruintjes Hoogte, 2000–2500 ft., *Drège!* Graaff Reinet Div.; mountain slopes near Graaff Reinet, 4000–6000 ft., *Bolus*, 672! Oude Berg, 3500 ft., *Drège!* Albert Div.; rocks by the Stormberg Spruit, 5000–5500 ft., *Drège!* Wodehouse Div.; Zuur Poort, *Zeyher*, 139! Aliwal North Div.; rocky summit of the Witte Bergen, 7500 ft., *Drège.*

KALAHARI REGION: Orange Free State; Witte Bergen, near Harrismith, *Buchanan*, 260!

14. **D. pumila** (Nees, Fl. Afr. Austr. 323); perennial, with few short innovation shoots; culms very slender, erect or ascending, 2–4 in. long, greyish pubescent, at least in the upper part, 2-noded, internodes enclosed or upper 2 exserted; leaves mainly basal, distichous in the innovation shoots; sheaths finely puberulous, lower bearded at the mouth, closely imbricate, very short, more or less coriaceous, at length glabrescent, upper very narrow and tight; ligule a fringe of minute hairs; blades subulate, plicate, obtuse or mucronulate, $\frac{1}{2}$–1 in. by $\frac{3}{4}$ lin. when unfolded, very rigid, strongly striate on the back and puberulous in the grooves, densely puberulous above; panicle reduced to an erect or somewhat nodding raceme of about 5 spikelets, $\frac{3}{4}$–1 in. long; rhachis and very short pedicels greyish pubescent to tomentose; spikelets rather close, 5–6 lin. long; florets 6–7, uppermost rudimentary; rhachilla joints very short, glabrous or the uppermost minutely hairy; glumes equal, lanceolate, acute, finely pubescent, 5–7-nerved, scarious, or herbaceous below; valves oblong in profile, body 2 lin. long, olive-brown or purplish, scantily hairy on the back with 1–2 submarginal tufts of snow-white hairs at the middle of each side and 4 smaller tufts in a transverse row between them and the sinus, 9-nerved, outer 3 nerves very marked in the upper part; lobes oblong, almost 1 lin. long, scaberulous, minutely ciliolate, mucronulate; awn 3–3$\frac{1}{2}$ lin. long, flat, kneed, with 2 loose twists below the middle; callus short with a beard on each side; pales elliptic-oblong, over 2 lin. long, flaps scantily and minutely hairy, keels rigidly and densely ciliolate; lodicules fleshy, broadly cuneate; anthers 1 lin. long; grain elliptic, $\frac{5}{8}$ lin. long. *Steud. Syn. Pl. Glum.* i. 242; *Durand & Schinz, Consp. Fl. Afr.* v. 853.

WESTERN REGION: Little Namaqualand; Karroo near the mouth of the Orange River, below 600 ft., *Drège*, 2562!

15. **D. purpurea** (Beauv. Agrost. 160); perennial, very densely cæspitose with numerous very short densely leafy innovation shoots;

culms shortly ascending, very slender, from 2–9 in. long, glabrous, smooth, densely leafy at the very base, 2-noded, internodes exserted, often purplish; sheaths tight, strongly striate, rather firm, glabrous except for the ciliate upper margins and a few coarse hairs at the mouth; ligule a fringe of short hairs; blades setaceous-subulate, canaliculate with a callous point, $\frac{1}{2}$–2 in. long, rigid, the basal more or less curved, glaucous, scaberulous towards the tip, with spreading long white papilliform hairs; panicle small, contracted, ovoid to subglobose, $\frac{1}{3}$–$\frac{3}{4}$ in. long; rhachis and all the divisions scaberulous, purplish; branches fascicled, unequal, filiform, very scantily divided from the base or 1-spiculate; pedicels $\frac{1}{2}$–2 lin. long; spikelets $2\frac{1}{2}$–4 lin. long, crowded; florets about 5, uppermost rudimentary; rhachilla glabrous, joints short; glumes lanceolate, acute, herbaceous and usually purple on the back with thin whitish margins, glabrous, very minutely scaberulous in the upper part, closely 5-nerved; valves oblong, body $\frac{3}{4}$–1 lin. long, with a transverse fringe of tufts of hairs below the insertion of the awn and several submarginal tufts below it, obscurely 9-nerved; lobes oblong, obtuse, 1 lin. long, each with 3 distinct parallel nerves below the middle, hyaline and ciliolate above; awn about $2\frac{1}{2}$ lin. long, slightly twisted below the middle; callus short, shortly hairy; panicles oblong or panduriform, $1\frac{1}{2}$ lin. long, hairy below the middle, keels scabrid; lodicules fleshy, scantily ciliate; anthers $\frac{7}{8}$ lin. long; grain obovoid, $\frac{1}{2}$ lin. long. *Roem. & Schult. Syst.* ii. 690; *Kunth, Enum.* i. 314; *Nees, Fl. Afr. Austr.* 325; *Steud. Syn. Pl. Glum.* i. 242; *Durand & Schinz, Consp. Fl. Afr.* v. 853. *D. purpurea β setosa, Nees ex Drège in Linnæa,* xx. 254. *D. setosa, Nees, l.c.*; *Steud. l.c.*; *Durand & Schinz, l.c.* 854. *Avena purpurea, Thunb. Prod.* 23; *Fl. Cap. ed. Schult.* 118; *Linn. f. Suppl.* 112.

COAST REGION: Queenstown Div.; summits of Sterkstroom Mountains, *MacOwan Herb. Austr. Afr.*, 1694! Stormberg Range, on rocks near Zuurplats, 5000 ft., *Drège!* Stormberg Mountains, 6000–7000 ft., *Zeyher!* Table Mountain, 6500 ft., *Drège!*

CENTRAL REGION: Sutherland Div.; Roggeveld, *Thunberg!* Beaufort West Div.; Nieuw Veld Mountains, *Drège!* Graaff Reinet Div.; Sneeuwberg Range, near Graaff Reinet, 5000 ft., *Drège! Bolus,* 520! Aliwal North Div.; Wittebergen, 7500 ft., *Drège!*

16. D. tenella (Nees, Fl. Afr. Austr. 324); annual, dwarf, tufted; culms fascicled, very slender, erect or ascending, 1–5 in. long, glabrous, smooth, 1–2-noded, internodes exserted; sheaths tight, thin, lower scantily beset with fine stiff spreading hairs, or glabrous like the upper except for a few hairs at the mouth; ligule a fringe of short hairs; blades fine, setaceously involute, acute, $\frac{1}{3}$–1 in. long, soft, curved or straight, slightly scaberulous towards the tips, beset with fine spreading tubercle-based hairs; panicle very small, ovate, loosely contracted, often reduced to 4–3 spikelets; branches and pedicels finely filiform, flexuous, scaberulous, pedicels unequal, up to $2\frac{1}{2}$ lin. long; spikelets light green, 2–3 lin. long; florets 4–5, uppermost rudimentary; rhachilla glabrous, joints short; glumes equal, linear-oblong in profile, acute to subobtuse, subherbaceous except the hyaline margins, glabrous, smooth, 5-nerved; valves broadly oblong in profile, body $\frac{1}{2}$–$\frac{5}{8}$ lin. long, with a transverse fringe of tufts of hairs below the insertion of the awn and several submarginal tufts; lobes oblong, very obtuse, $\frac{1}{2}$–$\frac{5}{8}$ lin. long, each with 3 distinct parallel nerves, hyaline, ciliate; awn 1–1$\frac{1}{2}$ lin. long, slightly twisted and bent below the middle; callus short, shortly hairy; pales oblong, truncate, $\frac{7}{8}$ lin. long, keels scaberulous; lodicules fleshy, broad, cuneate, ciliolate; anthers 1 lin. long or almost so; ovary glabrous; styles distinct, short. *Steud. Syn. Pl. Glum.* i. 242; *Durand & Schinz, Consp. Fl. Afr.* v. 854.

South Africa: without precise locality, *Thom!*

Western Region: Little Namaqualand; between Buffels (Kousies) River and Silver Fontein, 2000 ft., *Drège!*

17. **D. curva** (Nees, Fl. Afr. Austr. 328); perennial, tufted, often stoloniferous; culms very slender, often ascending from a decumbent base, $\frac{1}{3}$–1$\frac{1}{2}$ ft. long, glabrous, smooth, 4–5-noded, upper internodes mostly exserted; leaves glabrous except at the bearded mouth of the sheaths; sheaths tight, striate; ligule a fringe of hairs; blades very narrow, linear, tapering to an acute point, 1–6 in. by $\frac{1}{2}$–1 lin., usually setaceously involute, rarely flat, subflaccid or flexuous, scaberulous upwards; panicle more or less spike-like, ovate to cylindric, usually lobed or interrupted, $\frac{1}{2}$ to more than 2 in. long; rhachis scaberulous, angular; branches solitary, longest not more than $\frac{1}{2}$ the length of the panicle, lower undivided for 1–3 lin. or like the rest closely divided from the base, branchlets contracted, more or less secund, like the pedicels very short, scaberulous; spikelets light green, very crowded, 2–2$\frac{1}{2}$ lin. long; florets about 5, uppermost quite rudimentary; rhachilla glabrous, joints short; glumes equal or subequal, broad lanceolate, acute, herbaceous on the back, closely 3–5-nerved, keels scabrid; valves oblong in profile, body $\frac{2}{3}$–$\frac{3}{4}$ lin. long, 9-nerved, hairy along the margins to beyond the middle or on the back almost to the base of the awn, hairs in dense or loose (usually interrupted) rows between the nerves, often ending in tufts of short hairs; lobes ovate-lanceolate, setaceously acuminate, about $\frac{3}{4}$ lin. long, adnate for $\frac{3}{4}$ their length to the awn, at length usually free; awn at first straight or slightly bent, then twisted below the bend, up to 1$\frac{1}{2}$ lin. long; callus very short, minutely bearded; pales oblong, 1$\frac{1}{2}$ lin. long, hairy below, keels scabrid; lodicules minute, cuneate, subglabrous; anthers $\frac{3}{4}$–$\frac{7}{8}$ lin. long; ovary glabrous; styles distinct, slender. *Steud. Syn. Pl. Glum.* i. 242; *Durand & Schinz, Consp. Fl. Afr.* v. 848. *D. Bachmanni, Hack. in Bull. Herb. Boiss.* iii. 385.

Var. β, **elongata** (Stapf); blades to 1 ft. long by 1$\frac{1}{2}$ lin.; panicle to more than 3 in. long, lower and intermediate branches 1$\frac{1}{2}$–2 in. long, divided from $\frac{1}{4}$–$\frac{1}{2}$ in. above the base, obliquely erect or almost nodding; spikelets 2$\frac{1}{2}$–3 lin. long.

South Africa: without precise locality, *Boivin!*

Coast Region: Malmesbury Div.; Mooresbury near Hopefield, *Bachmann*, 1018! Cape Div.; Devils Mountain, *Verreaux!* Worcester Div.; on the

Boschjes Veld, at the Doorn River, near Moord Kuil, *Drège*. Mossel Bay Div.; dry hills on the east side of the Gauritz River, *Burchell*, 6451! George Div.; Lange Kloof, *Ecklon*. Humansdorp Div.; Karroo by the Gamtoos River, *Gill!* Uitenhage Div.; by the Zwartkops River, *Ecklon & Zeyher*, 134! *Zeyher*, 4529! near Uitenhage, *Burchell*, 4236! Albany Div.; near Grahamstown, 2000 ft., *MacOwan*, 1303; King Williamstown Div.; Amatola Mountains, *Buchanan*, 26! Var. β: Albany Div.; Brand Kraal near Grahamstown, 2000 ft., *MacOwan*, 1283!

WESTERN REGION: Little Namaqualand; near the mouth of the Orange River, *Drège!* between the Buffels River and Silver Fontein, 2000 ft., *Drège!*

Var. β is perhaps nothing but a luxuriant state of *D. curva*.

18. D. suffrutescens (Stapf); perennial, suffrutescent, with a woody much branched base and the stout basal branches closely covered with imbricate scales or remnants of old sheaths; flowering culms rather slender, firm, geniculate-ascending or suberect, about 1 ft. long, finely puberulous or scaberulous in the upper part or towards the nodes, glabrous and smooth below, 3–4-noded, intermediate internodes usually exserted; lowest sheaths and scales coriaceous, persistent, tomentose at the very base, otherwise very minutely puberulous, upper culm sheaths subherbaceous or papery, loose or tumid, from very minutely puberulous (on the nerves) to almost glabrous, uppermost reaching or exceeding the base of the panicle; sheaths of the rigid slender innovation shoots tight, glaucous; ligule a dense fringe of short hairs; blades linear, tapering to an acute point, those of the culm leaves 1–1½ in. by 1½ lin., flat or the tips involute, finely striate, very minutely and densely puberulous or tomentose on both sides, those of the lower leaves and the innovation shoots very short, up to 5 lin. long, very rigid, involute, pungent or the lowest quite suppressed; panicle spike-like linear, erect or nodding, interrupted, 1–4 lin. long; axis slender, minutely scaberulous or puberulous; branches erect, a few in a fascicle or solitary, very unequal, divided from the base or the longer undivided for $\frac{1}{4}$–$\frac{1}{3}$ of their length, filiform, scaberulous; branchlets and pedicels short to very short; spikelets 4–4½ lin. long, straw-coloured, crowded; florets 3–4, the uppermost reduced or quite rudimentary; glumes lanceolate in profile, acutely acuminate, scarious, scaberulous, 6–7-nerved, lower somewhat shorter; valves oblong in profile, body 1 lin. long, membranous to subchartaceous, 9-nerved, with dense rows of very short hairs between the nerves to the base of the lobes or slightly beyond, the second or third row ending in a tuft of long hairs; lobes ovate-lanceolate, acute or mucronate-acuminate, 1–1½ lin. long, scaberulous, tips ciliolate; awn erect, 1½–2 lin. long with 1 or 2 loose twists below the middle; callus very slender, pungent, villous; pales linear-oblong, 1½ lin. long, keels scabrid; lodicules cuneate, glabrous, fleshy, small; anthers 1 lin. long; ovary glabrous; styles slender, almost as long as the slender dense stigmas.

CENTRAL REGION: Carnarvon Div.; at Buffels Bout, *Burchell*, 1607! Prieska Div.; between Modder Fontein and Keikams Poort, *Burchell*, 1612/6!

The spikelets are very like those of *D. glauca*, but they are slightly larger and more acuminate.

19. D. glauca (Nees, Fl. Afr. Austr. 327); perennial; culms from a short decumbent base, covered with papery or membranous coarsely villous scales or sheaths, strongly and often repeatedly geniculate, ½–1 ft. long, finely greyish tomentose, 3–4-noded, intermediate nodes exserted; lower sheaths thin, wide, short, very finely puberulous and also with numerous long rigid fine often tubercle-based hairs, upper loose or subtumid, finely greyish tomentose, uppermost very wide from a narrow base, embracing the base of the panicle; ligule a dense fringe of short hairs; blades linear, acute, subpungent, 1–2 in. long, 2 lin. broad, loosely involute or almost flat at the base, rigid, finely greyish tomentose all over; panicle erect, straight, spike-like, 3–4 in. long; rhachis slender, puberulous and scabrid; branches erect, a few in fascicles, unequal, closely divided from the base, filiform and scabrid like the very short branchlets and pedicels; spikelets crowded, 3½ lin. long, straw-coloured or brownish to purplish; florets 3, rarely 4, uppermost much reduced or quite rudimentary; glumes subequal, lanceolate to linear-oblong in profile, subacute to acute or acuminate, herbaceous on the back, strongly 7-nerved, scabrid, lower slightly shorter; valves oblong in profile, body ¾ lin. long, membranous to subchartaceous, often with a dark purplish spot below the insertion of the awn, 9-nerved with dense rows of very short hairs between the nerves to the base of the lobes, the second and third row or all 3 outer rows ending with a fine tuft of hairs about as long as the lobes; lobes ovate-lanceolate, acute, 1 lin. long, scaberulous, tips ciliate or ciliolate; awn erect or slightly bent, 1–1½ lin. long, with 1 or 2 loose twists below the middle; callus very slender, pungent, villous; pales linear-oblong, obtuse, 1½ lin. long, keels scabrid; lodicules cuneate, glabrous, rather fleshy, small; anthers about 1 lin. long; ovary glabrous; styles distinct, slender. *Steud. Syn. Pl. Glum.* i. 242; *Durand & Schinz, Consp. Fl. Afr.* v. 850.

WESTERN REGION: Little Namaqualand; sandy hills near Kuigunjels, near the mouth of the Orange River, 200 ft., *Drège*, 2536!

20. D. inermis (Stapf); perennial (?); culms geniculate, very slender, about 1 ft. long, glabrous, smooth, 2-noded, internodes exserted; leaves glabrous; sheaths tight, lower very obscurely scaberulous, upper smooth; ligule a fringe of very fine hairs; blades fine, setaceously convolute, acute, 3–4 in. long, smooth; panicle contracted, linear, 1–2½ in. long, subflexuous; rhachis filiform, smooth below; branches solitary, 2-nate or a few in fascicles, lowest up to 1 in. long, divided from the base or nearly so, finely filiform, scaberulous; branchlets and pedicels short to very short, scabrid; spikelets clustered along the branches, greenish, 2 lin. long; florets 4–6, very close, uppermost rudimentary; glumes oblong-lanceolate, acute, subherbaceous except at the margins, sometimes tinged with purple on the sides, glabrous, scaberulous above, prominently 5-nerved; valves oblong-lanceolate in profile, broadly rounded on the back, 1¼ lin. long, shortly 2-toothed, mucronate from the sinus, densely hairy up to the middle on the back and ciliate along the margins to $\frac{3}{4}$ the way up, otherwise glabrous, 9-nerved, tip hyaline; callus short, bearded; pales elliptic-oblong, slightly over 1 lin. long, hairy below, keels scaberulous; lodicules minute, ciliate; anthers oblong-linear, $\frac{1}{5}$–$\frac{1}{4}$ lin. long; grain ellipsoid, $\frac{3}{8}$ lin. long.

COAST REGION: Port Elizabeth Div.; near Port Elizabeth, *E.S.C.A. Herb.*, 178!

21. D. decumbens (DC. Fl. Franç. iii. 33); perennial, tufted; culms erect, slender, 1–2 ft. high, glabrous, smooth, about 2-noded, upper internodes exserted; leaves glabrous except for a more or less distinct beard at the mouth of the sheaths, or scantily beset with very fine stiff hairs; sheaths tight, striate, lower persistent; ligule a fringe of short hairs; blades linear, acute, 3–8 in. by 1–1$\frac{1}{2}$ lin., flat or involute, striate, scaberulous towards the tip and along the margins; panicle usually reduced to a raceme or the lowest branches 2–3-spiculate, often scanty, linear, 1–2 in. long; rhachis and branches or pedicels filiform, angular, scabrid, the latter very unequal, 1–9 lin. long; spikelets turgid, ovoid, awnless, 4–6 lin. long, greenish or tinged with purple; florets cleistogamous, imbricate, 5–7, very close, uppermost rudimentary; glumes lanceolate in profile, acute, scarious, often tinged with purple on the sides, smooth glabrous, sub-5- to sub-7-nerved in the lower part, nerves joining and connected by a few transverse veins; valves oblong in profile, minutely 2-toothed, mucronulate from the sinus, broadly rounded on the back, 3 lin. long, rather firm, tips often purplish, otherwise light green with a distinct or obscure submarginal fringe of short hairs below the middle, otherwise glabrous, smooth; callus very short, minutely bearded; pales broad elliptic-oblong, 2$\frac{1}{4}$ lin. long, flaps narrow, keels slightly winged, densely and rigidly ciliolate; lodicules 0; anthers ellipsoid, $\frac{1}{10}$–$\frac{1}{8}$ lin. long; ovary glabrous; styles short, distinct, stigmas very delicate; grain strongly dorsally compressed, ellipsoid, 1 lin. by $\frac{2}{3}$ lin. *Trin. Gram. Gen.* 67, *and in Mém. Acad. Pétersb. sér.* 6, i. 67; *Kunth, Enum.* i. 311; *Suppl.* 265, *t.* xxi. *fig.* 2. *Festuca decumbens, Linn. Sp. Pl.* 75; *Fl. Dan. t.* 162; *Leers, Fl. Herborn.* 34, *t.* vii. *fig.* 5. *Sieglingia decumbens, Bernh. Syst. Verz. Erf.* 20, 44. *Poa decumbens, Schrad. Fl. Germ.* 305; *Engl. Bot. t.* 792; *Host, Gram. Austr.* ii. *t.* 72; *Knapp, Gram. Brit. t.* 59. *Melica decumbens, Weber, Spicil. Goett.* 3. *M. rigida, Wibel, Prim. Fl. Werthem.* 117. *Bromus decumbens, Koel. Descr. Gram.* 242. *Triodia decumbens, Beauv. Agrost.* 179, *t.* xv. *fig.* 9; *Steud. Syn. Pl. Glum.* i. 249; *Durand & Schinz, Consp. Fl. Afr.* v. 877.

COAST REGION: King Williamstown Div.; Amatola Mountains, *Buchanan*, 49!
A native of Europe and North Africa; elsewhere introduced.

Imperfectly known species.

22. D. calycina (Roem. & Schult. Syst. ii. 691; Kunth, Enum. i.

315; Durand & Schinz, Consp. Fl. Afr. v. 848), based on *Avena calycina* (Lam. Ill. i. 200; Poir. Encycl. Suppl. i. 540), was referred to *D. macrantha* by Nees, Fl. Afr. Austr. 319; but the very insufficient description points rather to a species of *Pentameris.*

23. D. holciformis (Nees, Fl. Afr. Austr. 326); perennial; culms simple, slender, $\frac{3}{4}$ ft. long; leaves glabrous except for a dense tomentose beard at the mouth of the sheath; blades setaceously convolute, curved, lower up to 4 in. long, upper very short, somewhat rigid; panicle narrow, slender, loose, 2–2$\frac{1}{2}$ in. long; rhachis and branches flexuous, slightly viscid; branches 2-nate or solitary, bearing few spikelets from the middle; florets 2; glumes equal, acute or mucronulate, glabrous, smooth, 1-nerved; valves 1$\frac{1}{2}$–1$\frac{3}{4}$ lin. long, deeply bifid, densely villous, 7–9-nerved in the lower, 3-nerved in the upper part; lobes lanceolate, mucronate, pubescent, ciliolate; awn scarcely 1 lin. long, flat, pallid, with a single twist at the middle. *Steud. Syn. Pl. Glum.* i. 242; *Durand & Schinz, Consp. Fl. Afr.* v. 851.

COAST REGION: Caledon Div.; by the Palmiet River, in a place called Gretzaget, and between the Palmiet River and the Steenbrazen River, *Ecklon.*

This species was placed by Nees between *Danthonia purpurea,* Beauv., and *D. glauca,* Nees.

24. D. Kuhlii (Steud. Syn. Pl. Glum. i. 240); cæspitose; culms geniculate, $\frac{1}{3}$–1 ft. long, sheathed almost all along or the uppermost internode exserted; sheaths pubescent, striate, long bearded at the mouth; blades flat or convolute, 2–5 in. by scarcely 1 lin., the uppermost exceeding the panicle; panicle ovate, dense, 1–1$\frac{1}{2}$ in. long, branches divided from near the base; branchlets and pedicels capillary, short to very short; spikelets scarcely 1 lin. long; glumes ovate, acute; valves shorter than the glumes, silky or subglabrous; lateral bristles twice, awn four times the length of the valve. *Durand & Schinz, Consp. Fl. Afr.* v. 851.

COAST REGION: Cape Div. (?), without precise locality, *Kuhl.*

Probably a species of *Pentaschistis.*

25. D. obtusifolia (Hochst. in Flora, 1846, 120); perennial; culms fascicled, almost woody at the base, rigid, 1 ft. long, densely covered with old sheaths below, then distichously leafy; sheaths purplish, striate, bearded at the mouth; ligule a fringe of hairs; blades convolute, obtuse, 2–3 in. long, rigid, glabrous, finely punctate, sometimes scabrid at the tips; panicle raceme-like, loose, 2 in. long; pedicels twice the length of the florets; spikelets 2-flowered; glumes subequal, acute, quite glabrous, twice the length of the florets; valves villous, 7-nerved, lateral bristles slightly longer than the valve, awn 3 times the length of the valve. *D. rigida, Steud. Syn. Pl. Glum.* i. 243; *Durand & Schinz, Consp. Fl. Afr.* v. 854. *Avena rigida, Steud. in Flora,* 1829, 482.

SOUTH AFRICA: without precise locality, *Ludwig.*

Hochstetter and Steudel place this plant near *Danthonia lanata*, Schrad.

XXXVIII. CHÆTOBROMUS, Nees (partly).

Spikelets laterally compressed, pedicelled, panicled, deciduous with the bearded or plumose upper part of the pedicel; rhachilla tough or almost so above the glumes, readily disarticulating between the valves. *Florets* 3–4, ⚥, exceeded by the glumes. *Glumes* equal, lanceolate in profile, keeled, subherbaceous on the back, otherwise scarious, prominently and closely nerved. *Valves* membranous, 7–9-nerved, slightly heteromorphous, lowest glabrous, with an obscure, glabrous callus, and usually also with a reduced awn or without bristles, following valves slightly larger or smaller, pubescent, deeply 2-lobed with a kneed awn from the sinus, a bristle from the inner side of each lobe, and a slender pungent villous callus. *Pales* 2-keeled, more or less equalling the valves. *Lodicules* 2, cuneate, glabrous. *Stamens* 3. *Ovary* glabrous, styles distinct, short; stigmas plumose, laterally exserted. *Grain* almost spindle-shaped or linear-oblong, very slightly compressed from the sides, grooved in front, tightly enclosed by the slightly altered valve and pale; embryo short; hilum linear, long.

Perennial, tufted; blades more or less flat, soft; panicle contracted, narrow, sometimes very meagre; spikelets deciduous with a part of the pedicel and disarticulating between the valves.

Species 3, in South-western South Africa.

Blades usually silky or silvery below from adpressed hairs; spikelets 5–6 lin. long; lowest valve suddenly contracted into a short bristle ... ... (1) **involucratus.**
Blades glabrous below; spikelets 6–8 lin. long; all the valves with a kneed awn:
Culms erect; blades 2–5 in. long; spikelets 6–7½ lin. long; lowest valve without side-bristles ... (2) **dregeanus.**
Culms ascending from a sometimes decumbent base; blades 6–10 in. long; spikelets 7½–8 lin. long; all the valves with side-bristles ... ... (3) **Schraderi.**

1. **C. involucratus** (Nees, Fl. Afr. Austr. 344, excl. syn.); tufted; culms geniculate, ascending or suberect, slender, ½ to 1 ft. long, glabrous, smooth, 2–3-noded, upper internodes exserted; sheaths rather tight, striate, lowest scarious, whitish, glabrous, following herbaceous, glabrous or adpressedly hairy, particularly in the upper part; ligule a fringe of short hairs; blades linear, tapering to an acute point, 2–4 in. by 1–2 lin., soft, flat, or those of the innovation-shoots filiform-involute, densely silky or silvery from adpressed hairs or glabrescent; panicle linear to linear-oblong, erect, 1½–2 in. long; rhachis filiform; branches solitary, ½ in. long, filiform, flexuous, scaberulous, 6–1-spiculate from near the base; pedicels unequal, deciduous part 1–1½ lin. long, bearded below, scarcely exceeding the beard; spikelets light yellow or greyish-green, 5–6 lin. long; florets 3; glumes acutely acuminate, scaberulous in the upper part and

along the keel, lower much broader, sub-9-nerved, upper 5-nerved, nerves scabrid; valves lanceolate-oblong in profile, 7- to sub-9-nerved, lowest 2 lin. long, glabrous, suddenly contracted into a rather flexuous bristle, about 2½ lin. long, upper 1¼ lin. long (exclusive of the callus), loosely pubescent, 2-lobed; lobes acute, ciliolate, ½ lin. long, adnate to a fine bristle (up to 2 lin. long) from the inner side; awn 4–5 lin. long, kneed at ⅓ way up; callus ½ lin. long; pales as long as the valves; anthers 1¼ lin. long. *Danthonia involucrata, Steud. Syn. Pl. Glum.* i. 244; *not Schrad., nor Trin. Gen. Gram. Suppl.* 33, *and in Mém. Acad. Pétersb. sér.* 6, iv. 34.

WESTERN REGION: Little Namaqualand; near the mouth of the Orange River, below 600 ft., *Drège!*

2. C. dregeanus (Nees, Fl. Afr. Austr. 343); densely tufted; culms erect, slender, 1–1½ ft. long, glabrous, smooth, 3-noded, middle and upper internodes (at least ultimately) exserted; sheaths tight or the upper rather loose, striate in the upper part, smooth, glabrous, lowest pallid, scarious or firm and persistent; ligule a fringe of hairs; blades linear, acute, 2–5 in. by 1–2 lin., flat, or those of the innovation-shoots filiform-convolute, soft, glabrous and smooth below, very finely pubescent or scaberulous and whitish on the upper side; panicle contracted, oblong to linear, erect or nodding, 1–4½ in. long; rhachis slender, smooth and angular below; branches solitary, up to 2 in. long, angular, divided from the base; branchlets more or less fascicled, very unequal, terete, scaberulous or smooth, or the branches much reduced, 5–2-spiculate from the very base; pedicels unequal, deciduous part 1½ lin. long, bearded below or plumose nearly all along, more or less exceeded by the hairs; spikelets 6–7½ lin. long, light yellow or purplish-green; florets 3; glumes acute or acuminate, puberulous or scaberulous (at least the lower and particularly on the outer side), keels scaberulous to scabrid, lower slightly broader, 5–10-nerved, upper 5-nerved, nerves scabrid; valves lanceolate-oblong in profile, 2-lobed, lowest 2 lin. long, smooth with a twisted and more or less kneed awn, 6–7 lin. long from between the short and obtuse lobes, without side-bristles, following 2–2¼ lin. long (exclusive of the callus), very laxly pubescent or somewhat scabrid, more deeply lobed with an awn (as in the lowest valve) from between the acute ciliolate lobes, and with 2 fine, loosely twisted, flexuous or spreading bristles 3–4 lin. long and adnate to them; callus to ¾ lin. long; pales 2½ lin. long; anthers 1½–2 lin. long; grain 1½ lin. long. *Danthonia dregeana, Steud. Syn. Pl. Glum.* i. 244; *Durand & Schinz, Consp. Fl. Afr.* v. 849.

CENTRAL REGION; Clanwilliam Div.; Wupperthal, *Wurmb.*
WESTERN REGION: Little Namaqualand; between Pedros Kloof and Lily Fontein, 3000–4000 ft., *Drège!* between Silver Fontein, Koper Berg and Kaus Mountain, 2000–3000 ft., *Drège!*

3. C. Schraderi (Stapf); culms ascending from a geniculate, sometimes decumbent and branched base, 2 ft. long, glabrous, smooth, 6–12-noded, lower internodes ½–1½ in. long, slender, upper 3 much longer, stouter, shortly exserted; leaves glabrous except at the sometimes scantily bearded mouth of the sheaths; sheaths rather loose, lower more or less slipping from the culm and rolling in; ligule a fringe of minute hairs; blades linear, long tapering to an acute point, 6–10 in. by 1½–2½ in., flat, striate, smooth or obscurely scaberulous and glaucous from a waxy covering on the upper surface; panicle ovate to oblong, contracted, 3½–4 in. long, erect or slightly nodding; rhachis angular, smooth; branches solitary, lower distant, very unequally divided from the base or nearly so, longest about 2 in. long, obliquely erect, filiform, glabrous, smooth below; branchlets and pedicels contracted, scabrid; pedicels very unequal, deciduous part 1½–2 lin. long, bearded to ½ way, hairs equalling the joint; spikelets 7½–8 lin. long, light brownish-green, erect; florets 3–4; glumes acute or mucronulate, scabrid on the keel and nerves, lower much broader, 8–9-nerved, scaberulous, upper 3–4-nerved; valves oblong-lanceolate in profile, 2-lobed, lobes ovate, produced into a fine bristle (about 2½ lin. long); awn 6–7 lin. long, kneed at or below the middle, very slender; lowest valve 2½–2¾ lin. long, glabrous, those above 2 lin. long (excluding the callus), loosely pubescent, with a callus up to 1 lin. long; pales 2½–2¾ lin. long; anthers 2½ lin. long.

COAST REGION: Cape Div.; Paarden Island, *Wolley Dod*, 3078!

This is probably identical with Schrader's *Avena involucrata*, in Goett. Gel. Anzeig. 1821, iii. 2075 (*Danthonia involucrata*, Schrad. in Schult. Mant. ii. 383; Kunth, Enum. i. 317; Trin. Gram. Gen. 71, and in Mém. Pétersb. sér. 6, i. 71. *Pentameris involucrata*, Nees in Linnæa, vii. 310).

Imperfectly known species.

4. C. interceptus (Nees, Fl. Afr. Austr. 342); culms 1½ ft. long, simple; sheaths glabrous excepting a short beard at the mouth; panicle oblong, 2½ in. long; branches alternate, scabrid; branchlets 2-spiculate; pedicels densely covered with white erect hairs increasing in length upwards and exceeding the pedicels; spikelets 5 lin. long; glumes lanceolate, acuminate, scabrid, 5-nerved; florets 2; valves 2 lin. long, glabrous, smooth, faintly 7-nerved, lateral bristles 1½ lin. long, awn 4½–5 lin. long, kneed at ⅓ of its length. *Danthonia intercepta, Steud. Syn. Pl. Glum.* i. 243; *Durand & Schinz, Consp. Fl. Afr.* v. 851.

COAST REGION: "Cape District, Groote Post." Collector not mentioned.

Similar to *C. involucratus*, Nees, according to the author, but differing in the glabrous valves.

XXXIX. ARUNDO, Linn.

Spikelets 2–7-flowered, laterally compressed, in very compound panicles, rhachilla disarticulating above the glumes and between the valves, joints short, glabrous; florets hermaphrodite, the uppermost reduced. *Glumes* equal, broadly lanceolate, shortly acuminate, keeled, membranous, 3–5-nerved. *Valves* more or less equalling the glumes, ovate to lanceolate-ovate, acuminate, finely bifid or entire, long hairy below, 5–9-nerved, 3 nerves more or less percurrent or excurrent, the rest short, the middle nerve often produced into a short, fine bristle; callus short, shortly bearded. *Pales* slightly exceeding ½ the length of the valve, 2-keeled. *Lodicules* 2, obovate, nerved, glabrous. *Stamens* 3. *Ovary* glabrous; styles distinct, almost as long as the laterally exserted plumose stigmas. *Grain* obovoid-oblong, broad, loosely enclosed by the valve and pale; hilum basal, punctiform; embryo occupying almost wholly one side of the grain.

Perennial with creeping rhizomes, extravaginal shoots and very tall and stout culms; leaves rather evenly distributed over the culms; sheaths slightly exceeding the internodes; blades long, broad, flat, ligules very short, membranous; panicles large, much compound; spikelets hairy.

Species 1, all over the Mediterranean regions to the Himalaya; introduced into America and South Africa.

1. **A. Donax** (Linn. Sp. Pl. 81); glabrous; culms erect, 6–15 ft. long, smooth, hollow, very many-noded, simple or scantily branched, internodes slightly exceeded by the sheaths, these very tight, firm, smooth; blades linear-lanceolate from a broad base, long tapering to a very fine point, more or less drooping, 1–2 ft. by 1–2 in., smooth; panicles erect, 1–2 ft. long; branches scaberulous, erect or drooping; spikelets 4–5½ lin. long, light brown; glumes glabrous; valves 3–5 lin. long; hairs 2½–4 lin. long; anthers 1¼ lin. long; grain 1 lin. by almost ½ lin. long. *Host, Gram. Austr.* iv. 22, *t.* 38; *Kunth, Enum.* i. 246; *Suppl.* 189; *Reichb. Ic. Fl. Germ.* i. *t.* 109; *Steud. Syn. Pl. Glum.* i. 193; *Durand & Schinz, Consp. Fl. Afr.* v. 874; *Hook. f. Fl. Brit. Ind.* vii. 302. *A. sativa, Lam. Fl. Franç.* iii. 616. *Donax arundinaceus, Beauv. Agrost. t.* 16, *fig.* 4, *Expl.* 11; *Nees, Fl. Afr. Austr.* 357. *Scolochloa Donax, Gaud. Fl. Helv.* i. 202. *S. arundinacea, Mert. & Koch in Roehling, Deutschl. Fl.* i. 530.

SOUTH AFRICA: without precise locality; *Drège!*
COAST REGION: Riversdale Div.; Lange Bergen (?), *Mund & Maire.* Uitenhage Div.; by the Zwartkops River, near P. Maré's Farm, *Ecklon.*
Introduced.

Imperfectly known species.

2. **A. webbiana** (Steud. Syn. Pl. Glum. i. 194), described from specimens collected by Webb in the Cape Colony and preserved in the Lenormand Herbarium, is almost certainly a species of *Danthonia.*

XL. PHRAGMITES, Trin. (partly).

Spikelets loosely 3–10-flowered, awnless, in large panicles; rhachilla disarticulating above the first and between the following valves, slender, glabrous, joints very short; lowest flower male or abortive, the following hermaphrodite, the uppermost florets reduced. *Glumes* thin, unequal to subequal, lanceolate, acute, more or less rounded on the back, 3-nerved, or the lowest sometimes sub-5-nerved. *Valves* heteromorphous, the lowest linear-lanceolate, much longer than the subtending glume, otherwise of a similar structure, quite glabrous, persistent, the following valves very thin, linear, long and more or less caudate-acuminate, 3-nerved, middle nerve percurrent, side-nerves fine, short, callus long, slender, with very long silky hairs. *Pales* linear-oblong, about $\frac{1}{2}$ as long as the valves, 2-keeled. *Lodicules* 2 (or sometimes 3 in the lowest floret) obovate, 2–3-nerved, glabrous. *Stamens* 3, or 2 in the lowest floret. *Ovary* glabrous, in the lowest flower rudimentary or quite suppressed; styles distinct, rather short; stigmas laterally exserted, densely plumose. *Grain* loosely enclosed by the valve and pale, free, oblong, semiterete, hilum oblong, short basal; embryo about $\frac{1}{2}$ as long as the grain.

Perennials with a creeping rhizome, extravaginal innovation shoots, and tall sheathed annual or perennial culms; blades flat; ligule a narrow, ciliate, membranous rim; panicle lax, usually very large and much compound; spikelets conspicuously silky from the long callus hairs.

Species 1 (or 2, very closely allied), almost cosmopolitan.

1. **P. communis** (Trin. Fund. Agrost. 134); culms erect, 4–10 ft. long, sometimes much taller or dwarfed, annual, many-noded, usually sheathed all along; sheaths embracing each other, tight, terete, smooth, glabrous, the lowest with reduced blades or bladeless, firm; blades linear to lanceolate, tapering to a setaceous or subulate point, very variable in length and width (up to $1\frac{1}{2}$ ft. long and $\frac{1}{2}$–$1\frac{1}{2}$ in. broad in the South African specimens), firm, more or less glaucous below, glabrous, smooth or slightly rough towards the base, margins smooth or scabrid; panicle oblong to ovate-oblong, erect or nodding, more or less secund, dense or rather lax, $\frac{1}{2}$–$1\frac{1}{2}$ ft. long (in the South African specimens), brownish-purple or brownish-yellow, branched, fascicled or the upper solitary, the longest $\frac{1}{2}$–$\frac{3}{4}$ ft., loosely and repeatedly branched, like the rhachis angular and scabrid, ultimate branchlets more or less terete, filiform; lateral pedicels usually 1–2 lin. long, the terminal up to 4 lin. long; spikelets 6–8 lin. long (in the South African specimens); glumes $1\frac{1}{2}$–$2\frac{1}{2}$ lin. and 2–$3\frac{1}{2}$ lin. long respectively; lower valve oblong-linear, like the following valves 2–$6\frac{1}{2}$ lin. long, callus to $\frac{3}{4}$ lin., hairs to 3 lin.; anthers about 1 lin. long; grain $\frac{3}{4}$ lin. long. *Kunth, Enum.* i. 251; *Suppl.* 193; *Fl. Dan. t.* 2464; *Reichb. Ic. Fl. Germ.* i. 108; *Nees, Fl. Afr. Austr.* 355; *Steud. Syn. Pl. Glum.* i. 195; *Boiss. Fl. Or.* v. 563; *Durand & Schinz, Consp. Fl. Afr.* v. 876; *Hook. f. Fl. Brit. Ind.* vii. 303. *P. mauritiana, Kunth, Rév. Gram.* i. 277, *t.* 50. *P. barbata, Burch. Trav.* ii. 271. *P. capensis, Nees, l.c.* 356. *Arundo Phragmites, Linn. Sp. Pl.* 81; *Host, Gram. Austr.* iv. 23, *t.* 39; *Engl. Bot. t.* 401; *Knapp, Gram. Brit. t.* 95. *A. vulgaris, Lam. Fl. Franç.* iii. 615.

SOUTH AFRICA: without precise locality, *Mund & Maire.*

COAST REGION: Cape Div.; Muizenberg Vley, *Wolley Dod*, 2592! Worcester Div.; between Genadendal and Tulbagh, *Burchell*, 8642! The Straat, *Drège.* Uitenhage Div.; Zwartkops River, *Ecklon & Zeyher!* Albany Div.; near streams, Grahamstown, *MacOwan*, 774! Queenstown Div.; Shiloh, *Baur*, 923!

CENTRAL REGION: Somerset Div.; near Somerset East, *Cooper*, 1506! Graaff Reinet Div.; banks of the Sunday River, *Burchell*, 2863!

KALAHARI REGION: Hopetown Div.; banks of the Orange River, *Burchell*, 2667/2! Prieska Div.; banks of the Orange River, *Burchell*, 1636! Orange Free State, *Cooper*, 3355! Basutoland; Leribe, *Buchanan*, 122! Transvaal, near Lydenburg, *Wilms*, 1711!

EASTERN REGION: Natal; common through the colony, *Buchanan*, 273! at the mouth of the Umzimkulu River, *Drège!* near Durban, *Cooper*, 3354! *Drège!* Inanda, Umhlanga Valley, *Wood*, 1335! *Sutherland!*

Almost cosmopolitan.

Imperfectly known species.

2. **P. nudus** (Nees, Fl. Afr. Austr. 356) was described from very imperfect, young specimens, collected by Drège, without indication of locality. Durand & Schinz in Consp. Fl. Afr. v. 876 refer it as var. *nuda* to the preceding species. Nees says that it is mainly characterized by the very long hairs of the ligule.

3. **P. Xenochloa** (Trin. ex Steud. Syn. Pl. Glum. i. 196. *Xenochloa arundinacea*, Lichtenst. in Roem. & Schult. Syst. ii. 501). The description is quite insufficient to ascertain the genus; but it is possibly a *Danthonia*. It was collected by Lichtenstein in Bechuanaland.

XLI. LAGURUS, Linn.

Spikelets laterally compressed, in compact spike-like softly villous panicles; rhachilla disarticulating above the glumes, more or less continued beyond the floret, terminating with or without a rudimentary valve. *Floret* 1 (very rarely 2), ⚥ shorter than the glumes. *Glumes* subequal, very narrow, gradually attenuate into a bristle, hyaline, plumose. *Valve* membranous, long acuminate, 5-nerved, tips 2-setose, finely awned from the back; callus small, minutely hairy. *Pale* 2-nerved, somewhat shorter than the valve. *Lodicules* 2, hyaline. *Stamens* 3. *Ovary* glabrous; styles distinct, short; stigmas plumose, laterally exserted. *Grain* linear-oblong, laterally compressed, soft, tightly enclosed by the slightly hardened valve; hilum basal, linear-oblong, very short, embryo small.

Annual; blades flat, soft; ligules membranous; panicle spike-like, compact, oblong to almost globose, softly villous from the plumose glumes.

Species 1, in the Mediterranean countries.

1. **L. ovatus** (Linn. Spec. Pl. 81); culms fascicled, suberect or geniculate, ascending, from a few inches to more than 1 ft. long, very finely tomentose or pubescent (at least below the panicle), few-noded; upper internodes exserted, uppermost very long and slender; leaves finely tomentose or villous all over; sheaths loose or the upper tumid, lower membranous, white; ligules obtuse, decurrent, hyaline, pubescent; blades linear to linear-lanceolate, acute, very variable in size, from 1–7 in. by 1–7 lin., flat, soft, margins often wavy; panicle oblong or ovate to globose, compact, $\frac{1}{2}$–$2\frac{1}{2}$ in. long; branches, branchlets and pedicels puberulous or scaberulous; spike-

lets imbricate, 2½–3 lin. long (exclusive of the bristles); continuation of rhachilla pubescent; glumes linear-lanceolate, white, plumose from near the base to the tips of the bristles; valve linear-lanceolate, long and finely acuminate in profile, 2–3 lin. long, glabrous, awned from the middle; awn fine, 4–8 lin. long; anthers ¾ lin. long; grain 1¼ lin. long. *Schreb. Beschr. Graes.* 143, *t.* 19, *fig.* 3; *Host, Gram. Austr.* ii. *t.* 46; *Sibth. & Sm. Fl. Græc. t.* 90; *Engl. Bot. t.* 1334; *Beauv. Agrost.* 35, *t.* viii. *fig.* 12; *Kunth, Enum.* i. 295; *Suppl.* 247; *Trin. Agrost.* 251, *and in Mém. Acad. Pétersb. sér.* 6, v. 272; *Steud. Syn. Pl. Glum.* i. 183; *Durand & Schinz, Consp. Fl. Afr.* v. 832.

COAST REGION: Cape Div.; Herrschel Lane, in a gravel pit, *Wolley Dod*, 1840! King Williamstown Div.; Amatola Mountains, *Buchanan*, 27!

Evidently introduced.

XLII. POLYPOGON, Desf.

Spikelets 1-flowered, in dense spike-like, often lobed, panicles, disarticulating from the pedicels; rhachilla disarticulating below the valve, not produced; flower hermaphrodite. *Glumes* 2, subequal, awned from the entire or 2-lobed tips. *Valve* shorter than the glumes, very thin to hyaline, glabrous, truncate, faintly or obscurely 5-nerved, the lateral nerves shortly excurrent or evanescent within the hyaline tips; awn very fine, usually deciduous, subterminal, sometimes reduced to a mucro or 0. *Pale* 2-keeled, slightly shorter than the valve, rarely much shorter. *Lodicules* 2, delicate. *Stamens* 3. *Ovary* glabrous; styles distinct, very short; stigmas laterally exserted, loosely plumose. *Grain* oblong, subterete or slightly grooved, enclosed by the unaltered valve and pale; hilum short; embryo small.

Annual or subperennial; blades flat, uppermost sheath often tumid; spikelets 1–2 lin. long, often hidden by the numerous awns.

Species about 8; 4 in America, 1 in Africa, the others widely dispersed over the warm regions of both hemispheres.

Awn of glumes 2–3 lin. long; valve minutely 2–4-mucronulate, very obscurely nerved; awn as long as the valves or shorter or 0 ... ... (1) **monspeliensis.**

Awn of glumes finely capillary, up to 1 in. long; valve shortly 2-lobed, outer nerves excurrent into fine short bristles; awn very fine, up to 5 lin. long (2) **tenuis.**

1. **P. monspeliensis** (Desf. Fl. Atlant. i. 67); annual, tufted; culms erect or geniculately ascending, ½–2 ft. long, simple, glabrous, smooth, 3–4-noded, nodes exserted or enclosed; sheaths loose, often tumid, glabrous, smooth; ligules scarious, oblong, denticulate, ciliolate, 1½–3 lin. long; blades linear, tapering to a callous point, 2–6 in. by 1–3 lin., flaccid to subrigid, flat, scabrid; panicle spike-like, cylindric, sometimes lobed or interrupted below, 1–5 in. long, light green or straw-coloured; branches with numerous branchlets from the base, scaberulous; lateral pedicels extremely short, disarticulating near the base; spikelets 1 lin. long; glumes subequal, linear or oblanceolate-oblong, shortly 2-lobed or emarginate, scaberulous,

ciliate or ciliolate; awn 2–3 lin. long; valve ½ lin. long, minutely 2–4-mucronulate, nerves very obscure, awn as long as the valve or shorter or 0; pale 2-toothed or 2-mucronulate, almost as long as the valve; anthers ⅕–¼ lin. long; grain oblong, ½ lin. long, subterete or terete, obscurely grooved. *Beauv. Agrost.* 17, *t.* vi. *fig.* 8; *Kunth, Enum.* i. 232; *Suppl.* 181, *t.* xiii. *fig.* 7; *Nees, Fl. Afr. Austr.* 143; *Reichenb. Ic. Fl. Germ.* i. *t.* 31, *fig.* 1416; *A. Rich. Tent. Fl. Abyss.* ii. 402; *Steud. Syn. Pl. Glum.* i. 184; *Doell in Mart. Fl. Bras.* ii. iii. 44, *t.* 126. *P. monspeliensis var. capensis, Steud. in Flora,* 1829, 466. *P. polysetus, Steud. l.c.* 467. *Phleum crinitum, Schreb. Beschr. Graes.* i. 151, *t.* 20, *fig.* 3; *Sibth. & Sm. Fl. Græc.* i. 46, *t.* 62. *Alopecurus monspeliensis, Linn. Sp. Pl.* 61; *Thunb. Prodr.* 19. *A. paniceus, Linn. l.c. ed.* ii. 90. *Agrostis panicea, Ait. Hort. Kew,* i. 94; *Engl. Bot. t.* 1704.

SOUTH AFRICA: without precise locality, *Thunberg!* *Miller!*

COAST REGION: Vanrhynsdorp Div.; Ebenezer, below 100 ft., *Drège!* Cape Div.; Simons Bay, *MacGillivray,* 401! *Milne,* 241! Table Mountain, Doorn Hoogte, Hosacks Platz, and Van Kamps Bay, *Ecklon,* near Constantia, *Ecklon & Zeyher!* shore near Muizenberg, *Wolley Dod,* 2085! Tulbagh Div.; Tulbagh Waterfall, *Ecklon.* Swellendam Div.; Buffeljagts River, *Gill!* Mossel Bay Div.; between Duyker River and Gauritz River, *Burchell,* 6385! Uitenhage Div.; Zwartkops River, *Ecklon.* Port Elizabeth Div.; Port Elizabeth, *Drège, E.S.C.A. Herb.* 147! Albany Div.; Brand Kraal near Grahamstown, *MacOwan,* 1281! King Williamstown Div.; Amatola Mountains, *Buchanan,* 13! Queenstown Div.; Zwart Kei River, *Baur,* 988!

CENTRAL REGION: Aberdeen Div.; Camdeboo Mountains, *Drège.* Graaff Reinet Div.; banks of the Sunday River, near Graaff Reinet, *Bolus,* 187!

WESTERN REGION: Little Namaqualand; banks of the Orange River, near Verleptpram, *Drège!*

KALAHARI REGION: Griqualand West; Griqua Town, *Burchell,* 1910! Orange Free State, *Cooper,* 3350! 3351!

EASTERN REGION: Natal, *Cooper,* 3520!

Very common throughout the Mediterranean region to India, introduced into most warm countries.

All the Cape specimens which I have seen, with the exception of MacGillivray's from Simon's Bay, have the glumes more distinctly 2-lobed than is the case in the Mediterranean *P. monspeliensis,* but at the same time less than in *P. maritimus.* They are also more ciliate, as Nees has already remarked. The general habit, however, is on the whole that of typical *P. monspeliensis.*

2. P. tenuis (Brongn. in. Duperr. Voy. Coq. Bot. 22); annual; culms erect or geniculately ascending, ½–2 ft. long, simple, rarely branched above the base, glabrous, smooth, 3- (rarely 4–5-) noded; sheaths rather loose, the upper often tumid, glabrous, smooth; ligules scarious, oblong, denticulate, ciliolate, ½–4 lin. long; blades linear, tapering to an acute point, 3–8 in. by 1–2½ lin., flat or involute, flaccid to rigid, scaberulous; panicle spike-like, often interrupted below, 2–9 in. long, pallid, silky plumose, branches with numerous branchlets from the base, lateral pedicels very short, disarticulating near the base; spikelets ¾ to almost 1 lin. long; glumes subequal, linear-oblong, minutely or obscurely 2-lobed, scaberulous, ciliolate, awn finely capillary, up to 1 in. long; valve shortly 2-lobed, outer side-nerves excurrent into short fine bristles, the inner into obscure mucros, nerves almost evanescent below; awn very fine, up to 5 lin. long; pale 2-toothed, $\frac{4}{5}$ lin. long; anthers $\frac{1}{5}$–$\frac{1}{4}$ lin. long; grain oblong, not quite $\frac{1}{2}$ lin. long, terete, finely channelled. *P. Adscensionis, Trin. Agrost.* 257; *Steud. Syn. Pl. Glum.* i. 183; *Durand & Schinz, Consp. Fl. Afr.* v. 825. *P. strictus, Nees in Linnæa,* vii. 297; *Fl. Afr. Austr.* 145; *Kunth, Enum.* i. 234; *Trin. Agrost.* 255; *Steud. l.c.* 183; *Durand & Schinz, l.c.* 826.

COAST REGION: Cape Div.; Cape sand-dunes, *Zeyher*, 1805! Cape Flats, *Burke!* Dorn Hoogte, *Ecklon*, Liesbeck River, *Bergius*, road-side towards Vyges Kraal, *Wolley Dod*, 2210! and without precise locality, *Harvey*, 296! Tulbagh Div.; Witzen Berg, *Ecklon*, Vogel Vallei, *Ecklon*. Caledon Div.; near Caledon, *Thom*, 852! Uitenhage Div.; ditches near the Zwartkops River, *Ecklon!* Port Elizabeth Div.; Port Elizabeth, *E.S.C.A. Herb.* 175!

CENTRAL REGION: Aliwal North Div.; in a cave on the Witte Bergen, 6000 ft., *Drège!*

Drège's specimens from the Witte Bergen represent a very starved state, described by Nees as var. *spelæus*.

Also in Ascension Island.

XLIII. AGROSTIS, Linn.

Spikelets from less than 1 lin. to $2\frac{1}{2}$ lin. long, panicled; rhachilla not continued beyond the floret or produced into a minute point or short delicate bristle, glabrous or shortly hairy. *Floret* 1, ⚥, shorter than the glumes. *Glumes* equal or subequal, usually lanceolate and acute, rarely oblong and obtuse, awnless (except in *A. polypogonoides*), membranous, usually 1-nerved, keeled. *Valve* broadly oblong, delicately membranous, glabrous or hairy, usually truncate, 5- (rarely 3-) nerved, awned from the back or awnless; side-nerves evanescent below, often excurrent into fine mucros or bristles; callus very small, glabrous or minutely hairy or rarely bearded. *Pale* delicate, hyaline, usually shorter than the valve or very short or obsolete, 2-nerved or nerveless. *Lodicules* 2, lanceolate, delicately hyaline. *Stamens* 3. *Ovary* glabrous; styles distinct, very short; stigmas plumose, laterally exserted. *Grain* free, enclosed in the scarcely altered floret, oblong, more or less dorsally compressed, grooved in front, rarely subterete; embryo short; hilum punctiform, basal.

Annuals or perennials, of varying habit; blades usually flat, often flaccid; ligules membranous; panicle usually much divided, often delicate, more or less effuse or contracted, rarely spike-like; branches and branchlets fine to very fine; spikelets usually very numerous, much gaping, at least temporarily.

Species numerous, all over the world, but mostly in the temperate regions.

Valve 5-nerved, usually awned; awn sometimes very short, rarely 0 (in *A. verticillata*):
  Glumes awnless:
    Spikelets $\frac{1}{2}$–1 lin. long:
      Panicles spike-like, dense; spikelets $\frac{1}{2}$–$\frac{3}{4}$ lin. long; glumes usually obtuse, mucronulate or emarginate:

Panicles very slender; spikelets $\frac{1}{2}$ lin. long; valve with a minute terminal awn; culms 2-noded ... ... (1) **griquensis.**
Panicle stouter, interrupted or lobed; spikelets $\frac{3}{4}$ lin. long; valve awnless; culms many-noded... ... ... (2) **verticillata.**
Panicles very lax, effuse or contracted; spikelets $\frac{3}{4}$–1 lin. long; glumes acute ... (3) **bergiana.**
Spikelets about $2\frac{1}{2}$ lin. long:
Panicle spike-like, lobed; rhachilla not continued; valve glabrous; pale minute, nerveless ... ... (4) **natalensis.**
Panicle often with long capillary branches; rhachilla shortly continued, hairy, often bearing a bristle; valve scantily pubescent: pale 2-nerved, almost equalling the valve ... ... (5) **barbuligera.**
Glumes mucronate or finely awned ... ... (6) **polypogonoides.**
Valve 3-nerved, muticous or mucronate, usually hairy, at least along the side-nerves ... ... (7) **lachnantha.**

1. **A. griquensis** (Stapf in Kew Bulletin, 1897, 290); annual, glabrous; culms geniculate-ascending, up to 1 ft. long, 2-noded; leaf-sheaths smooth; ligules 1 lin. long, acute; blades very narrow, linear, acute, 1–2 in. by $\frac{1}{3}$–$\frac{1}{2}$ lin., flat, green, smooth or subscaberulous above; panicle erect, spike-like, very narrow, sometimes interrupted, 1–$1\frac{1}{2}$ in. by 2 lin.; branches fascicled, very unequal, the longest up to 4 lin. long, hispidulous, branched from the base, adpressed to the rhachis; pedicels very short; spikelets $\frac{1}{2}$ lin. long or slightly more, greenish, rhachilla not produced; glumes subequal, oblong, obtuse or subemarginate, sometimes minutely mucronate, scaberulous, margins ciliolate, keels scabrid; valve broadly oblong, not quite $\frac{1}{2}$ lin. long, truncate, minutely denticulate or ciliolate, smooth, very faintly 5-nerved; callus glabrous; awn terminal, $\frac{1}{5}$–$\frac{1}{4}$ lin. long or less; pale shorter by $\frac{1}{3}$ than the valve, hyaline, nerveless or almost so; anthers $\frac{1}{8}$ lin. long, obtuse; grain narrow-oblong, deeply grooved, $\frac{1}{4}$ lin. long, tightly embraced by the delicate valve and pale.

KALAHARI REGION: Griqualand West, Hay Div.; at Griqua Town, *Burchell*, 1863!

2. **A verticillata** (Vill. Prosp. Fl. Dauph. 16); perennial, tufted, 1–2 ft. high; culms geniculate, ascending or erect, sometimes rooting from the lowest nodes, glabrous like the whole plant, many-noded, internodes usually enclosed or shortly exserted; sheaths rather loose, smooth, strongly striate; ligules scarious, 1–$1\frac{1}{2}$ lin. long, ciliolate; blades linear, tapering to an acute point, 2–5 in. by 1–3 lin., glaucous, flat, flaccid or subrigid, often spreading at right angles, smooth below, scabrid on the upper surface, finely but prominently nerved; panicle erect, more or less spike-like, interrupted or lobed, 1–3 in. long, very dense; branches fascicled, unequal, longest to $\frac{1}{2}$ in. long, rarely longer, scabrid, branched from the base, straight; pedicels usually very short; spikelets greenish, $\frac{3}{4}$ lin. long; rhachilla not or very obscurely produced; glumes subequal, oblong, subacute

or obtuse, minutely mucronulate or emarginate, scaberulous, keels scabrid, margins minutely ciliolate; valve broadly oblong, $\frac{2}{5}$ lin. long, truncate, minutely denticulate or ciliolate, smooth, very faintly 5-nerved, awnless; pale almost as long as the valve, hyaline, obscurely 2-nerved, obtuse, tip ciliolate; anthers apiculate, $\frac{1}{3}$–$\frac{1}{5}$ lin. long; grain broadly obovoid, terete, $\frac{3}{8}$ lin. long. *Trin. Gram. Unifl. & Sesquifl.* 195; *Gram. Ic. t.* 36; *Agrost.* 358, *and in Mém. Acad. Pétersb. sér.* 6, ii. 370; *Reichb. Ic. Fl. Germ.* i. *t.* 35, *fig.* 1435; *Steud. Syn. Pl. Glum.* i. 169; *Durand & Schinz, Consp. Fl. Afr.* v. 829. *A. alba, var. schimperiana, Engl. Hochgebirgsfl. Trop. Afr.* 128.

COAST REGION: Cape Div.; Wynberg Park, *Wolley Dod*, 1816! roadside beyond Black River, *Wolley Dod*, 2200! Muizenberg Vlei, *Wolley Dod*, 2594! Cape Flats, *MacOwan, Herb. Aust. Afr.*, 1794! *MacOwan*, 3157.

Throughout the Mediterranean countries, and as a weed occasionally in other parts of the temperate zones.

3. **A. bergiana** (Trin. Gram. Unifl. & Sesquifl. 203); annual, glabrous; culms erect or geniculate-ascending, very slender, smooth or subscaberulous below the nodes, 3–4-noded; leaf-sheaths minutely scaberulous; ligules 1–2$\frac{1}{2}$ lin. long; blades very narrow, linear, tapering to a fine point, 2–6 in. by $\frac{1}{3}$–1 lin., flat, flaccid, minutely scaberulous; panicle very lax, effuse or contracted, 3–9 in. long, erect or nodding; branchlets in fascicles of 2–6, capillary, scaberulous, the longest 3–6 in. long, remotely and repeatedly branched; pedicels usually longer than the spikelets, the latter light green or purplish, $\frac{3}{4}$–1 lin. long; rhachilla not produced; glumes subequal, lanceolate, acute or acuminate, keels scabrid; valve oblong, truncate, 2–4-mucronate, $\frac{3}{5}$–$\frac{4}{5}$ lin. long, glabrous, 5-nerved; callus minutely bearded; awn fine, straight, from the middle or below, exceeding the valve or shorter; pale $\frac{1}{2}$–$\frac{3}{4}$ lin. long; anthers $\frac{1}{2}$ lin. long, apiculate; grain oblong, $\frac{1}{2}$ lin. long. *Nees in Linnæa*, vii. 296; *Fl. Afr. Austr.* 150; *Kunth, Enum.* i. 221; *Trin. Agrost.* 363; *Steud. Syn. Pl. Glum.* i. 171; *Durand & Schinz, Consp. Fl. Afr.* 826. *A. stolonifera, Thunb. Prodr.* 19. *A. capensis, Steud. in Flora*, 1829, 467. *A. Ecklonis, Trin. l.c.* 364; *Steud. l.c.* 171; *Durand & Schinz, l.c.* 827.

VAR. β, **læviuscula** (Stapf); leaves almost smooth; blades glaucescent, the upper spreading at right angles; panicle straight, very divaricate, longest branches 1 in. long, rarely longer; spikelets a little over 1 lin. long; pale as long as or slightly longer than the valve.

COAST REGION: Clanwilliam Div.; Olifants River, *Ecklon.* Cape Div.; hills near Capetown, *Thunberg!* Table Mountain, in moist places, *Ecklon*, 943! *Burchell*, 637! near Cape Town, *Burchell*, 492! *Drège! Harvey*, 148! Devils Peak, below the Waterfall, *Wolley Dod*, 2294! Lions Head, *Wolley Dod*, 2345! Cape Peninsula, watercourse close to Kirstenbosch, *Wolley Dod*, 2384! Muizenburg Vlei, *Wolley Dod*, 2348! old road to Constantia Nek, *Wolley Dod*, 2086! Tulbagh Div.; Tulbagh Kloof, Tulbagh Valley, and Winterhoek, *Ecklon.* Swellendam Div.; in the forest of Grootvaders Bosch, *Burchell*, 7236! Knysna Div.; Plettenberg Bay, *Mund!* Albany Div.; stony places near Bothas Hill, 2000 ft., *MacOwan*, 1321!

EASTERN REGION: Natal; Durban Flats, *Buchanan!* and inland, 2000 ft., *Buchanan!*

KALAHARI REGION: Var. β; Orange Free State, *Buchanan*, 17!

Also in St. Helena (*Burchell*, 39!).

4. **A. natalensis** (Stapf in Kew Bulletin, 1897, 290); perennial, tufted, 2–3 ft. high, glabrous; culms erect, 5–6-noded, smooth; leaf-sheaths rather tight, smooth; ligule $\frac{1}{2}$–$\frac{3}{4}$ lin. long; blades narrowly linear, tapering to a fine point, 3–7 in. by $\frac{1}{2}$–1 lin., flat, flaccid, minutely asperulous or almost smooth; panicle erect, spike-like, lobed, 4–7 in., branches in distant, very dense, oblong fascicles, very unequal, up to 1$\frac{1}{2}$ lin. long, branched from the base, or the longest some way above it, scaberulous; pedicels very short; spikelets light green, 2$\frac{1}{2}$ lin. long, narrow; rhachilla not produced; glumes subequal, linear-oblong, mucronate, scaberulous, keels scabrid; valve oblong, truncate, 4-mucronate, 1 lin. long, smooth, 5-nerved at the base, 4-nerved above; callus minutely hairy in front; awn straight, tapering towards both ends, scabrid, from above the base, 1 lin. long; pale subquadrate, hyaline, denticulate, $\frac{1}{4}$–$\frac{1}{3}$ lin. long; anthers $\frac{1}{2}$ lin. long, apiculate; grain oblong, dorsally compressed, grooved, $\frac{7}{8}$ lin. long.

EASTERN REGION: Natal; Umpumulo, 2000 ft., *Buchanan*, 159!

Allied to *A. Elliotii* (Hack.) from Madagascar, but differing in having somewhat larger spikelets, shorter and broader valve and truncate pale.

5. **A. barbuligera** (Stapf); perennial, quite glabrous, compactly tufted; culms erect or suberect, and more or less geniculate, slender, 1$\frac{1}{2}$–2 ft. long, smooth, 1–2-noded, uppermost node much below the middle; upper internode very long, exserted; leaves crowded at the base; lower sheaths subflabellate, 1–2 in. long, finally breaking up into fibres, upper tight, smooth, striate; ligules scarious, oblong, obtuse, 1 lin. long; blades linear, tapering to an acute point, 3–5 in. by 1–1$\frac{1}{2}$ lin., flat, subrigid, striate, smooth, except along the scaberulous margins, upper side whitish when young; panicle open, flexuous, lax, about $\frac{1}{2}$ lin. long; axis filiform, smooth; branches 2–4-nate, capillary, very flexuous, more or less spreading, purplish, smooth, lower 2–4 in. long, undivided for $\frac{1}{2}$–2 in. from the base; branchlets and pedicels contracted, very fine, the latter 1$\frac{1}{2}$–3 lin. long, scaberulous; spikelets more or less tinged with purple, 2$\frac{1}{4}$–2$\frac{1}{2}$ lin. long; rhachilla slightly continued, continuation $\frac{1}{4}$–$\frac{1}{3}$ lin. long, sometimes bearing a very fine scaberulous bristle 1–1$\frac{1}{2}$ lin. long, hairy, hairs up to $\frac{1}{2}$ lin. long; glumes lanceolate, acute, keels scaberulous; upper slightly shorter; valve awned, oblong, truncate, 1$\frac{1}{2}$–2 lin. long, scantily pubescent, at least below the middle, 5-nerved, side-nerves excurrent, inner into mucros, outer into minute bristles up to $\frac{1}{3}$ lin. long; awn from near the base, 3$\frac{1}{2}$–4 lin. long, very fine, somewhat flexuous; callus small, minutely bearded; pale 1$\frac{1}{4}$–1$\frac{3}{4}$ lin. long, glabrous, keels excurrent into very short fine bristles; anthers 1$\frac{1}{4}$ lin. long; grain linear-oblong, almost 1 lin. long.

COAST REGION: King Williamstown Div.; Amatola Mountains, *Buchanan*, 24!

CENTRAL REGION: Somerset Div.; on the higher rocks of Bosch Berg, 4300 ft., *MacOwan*, 2189!

Allied to *A. continuata*, Stapf, and *A. Mannii*, Stapf (*Deyeuxia Mannii*, Hook. f.); the continuation of the rhachilla is in both species very similar to that in *A. barbuligera*, except for being glabrous in *A. continuata*. *A. Mannii*, which is more nearly allied, differs in the mode of growth and in having smaller spikelets and shorter panicle branches.

6. **A. polypogonoides** (Stapf); perennial, tufted; culms erect, about 3 ft. long, glabrous, smooth or slightly rough below the nodes, 3–4-noded, internodes exserted except the lowest; sheaths rather tight, striate, glabrous, smooth or somewhat rough, lower pallid; ligules oblong, up to 4 lin. long, scarious, nerved; blades linear, long tapering to an acute point, 8–12 in. by 1–1½ lin., flat, glabrous, glaucous, scabrid, upper rather rigid; panicle narrow, oblong, 5–7 in. by ¾–1 in., contracted, erect; rhachis very slender, smooth; branches unequal, in fascicles of 5 or less, longest up to 2 in. long, undivided for 2–9 lin., finely filiform, scaberulous; branchlets and pedicels scabrid, the latter with thickened tips, the lateral about ½ lin. long; spikelets pallid, 1½–1¾ lin. long; rhachilla not produced beyond the floret; glumes subequal, lanceolate, acute, produced into fine fragile scabrid bristles (up to 1½ lin. long), scaberulous, keel scabrid; valve oblong in profile, minutely 4-toothed, ⅞ lin. long, hairy on the sides and across below the tip, 5-nerved, side-nerves shortly excurrent, very faint below; awn capillary, scabrid, about 2 lin. long; callus glabrous; pale almost ¾ lin. long, 2-toothed; anthers ⅜ lin. long.

COAST REGION: Cape Div.; Muizenburg Vlei, *Wolley Dod*, 2349!

This species comes very near to *Polypogon*, the only difference being in the habit, and the absence of a terminal notch in the persistent glumes.

7. **A. lachnantha** (Nees in Ind. Sem. Hort. Vratisl. 1834, and in Linnæa, x. 115); perennial or annual, glabrous; culms erect or geniculate-ascending, ½–2 ft. long, smooth or scaberulous below the nodes, 2–4-noded; leaf-sheaths minutely scaberulous; ligule 1–2½ lin. long; blades linear, tapering to an acute point, 1½–8 in. by 1–2 lin., green or subglaucous, flat, flaccid, scaberulous on both sides; panicle contracted, narrow, 2–12 in. long, erect, branches very unequal, in distant fascicles, the longest up to 4 in., or all very short, capillary, erect or flexuous, scabrid, branched from the base or the longer ones simple for ½–1 in.; pedicels mostly as long as or shorter than the spikelets, the latter light green, ¾–1¼ lin. long, shining; rhachilla not produced; glumes subequal, lanceolate, acutely subacuminate, keels rather stout, scabrid; valve oblong, truncate, ⅝–1 lin. long, hairy, often only along the side-nerves, rarely glabrous, 3-nerved, sometimes mucronate; callus scantily bearded; pale ½–¾ lin. long; anthers ¼ lin. long; grain oblong, ½ lin. long. *Steud. Syn. Pl. Glum.* i. 173; *Durand & Schinz, Consp. Fl. Afr.* v. 828. *A. Neesii, Trin. Agrost.* 361; *Steud. l.c.* 170; *Durand & Schinz, l.c.* 828. *A. vestita, Hochst. ex A. Rich. Tent. Fl. Abyss.* ii. 401; *Steud.*

l.c. 173; *Schweinf. Beitr. Fl. Aethiop.* 297; *Engl. Hochgebirgsfl. Trop. Afr.* 128; *Durand & Schinz, l.c.* 829. *A. dregeana, Steud. l.c.* 173; *Durand & Schinz, l.c.* 827 (*from the description*). *Podosæmum lachnanthum, Nees, Fl. Afr. Austr.* 148. *P. angustum, Nees, l.c.* 147.

COAST REGION: Cape Div.; Newlands, in a dry pond, *Wolley Dod*, 3559! Liesbeck River, *Bergius*; about the ponds at Salt River, *Burchell*, 681! Tulbagh Div.; Tulbagh Waterfall, *Ecklon*. Uitenhage Div.; in stagnant water by the Zwartkops River, *Ecklon*; between Sunday River and Coega River, *Drège*. Albany Div.; everywhere on the banks of streams, *MacOwan*, 1846! Cathcart Div.; between Windvogel Mountain and Zwart Kei River, 3000–3500 ft., *Drège!* Queenstown Div.; near Shiloh, 3500 ft., *Baur*, 1135!

CENTRAL REGION: Graaff Reinet Div.; on the Sneeuw Bergen, 3300 ft., *Bolus*, 1966! Aliwal North Div.; Witte Bergen, in a cave, 6000 ft., *Drège!*

WESTERN REGION: Little Namaqualand; between Buffels (Kousies) River, and Pedros Kloof, 2000–3000 ft., *Drège!*

KALAHARI REGION: Griqualand West, Hay Div.; at Griqua Town, *Burchell*, 1913! 1944! Orange Free State; Drakens Bergen, near Harrismith, *Buchanan*, 119!

EASTERN REGION: Natal; Umpumulo, 2000 ft., common by streamlets, *Buchanan*, 280! Riet Vlei, 4000–5000 ft., *Buchanan*, 286!

Also in Abyssinia.

A very variable plant, so far as size and habit are concerned. Drège's specimens from the Witte Bergen were collected in a cave, and represent an extreme shade form. They have been described by Nees (Fl. Afr. Austr. 148) as *Podosæmum lachnanthum*, var. *humile*.

A. gymnostyla, Steud. Syn. Pl. Glum. i. 170 (*Podosæmum gymnostylum*, Nees, Ind. Sem. Hort. Vratisl. 1850, and in Linnæa, xxiv. 236. *Mühlenbergia gymnostyla*, Walp. Ann. iii. 753), is indicated by Steudel from the Cape of Good Hope; but it was described from a cultivated specimen of unknown origin, and Fenzl in Ind. Sem. Hort. Vindob. 1850, says it is identical with *Cinna mexicana*, Beauv.

XLIV. CALAMAGROSTIS, Roth.

Spikelets very narrow, lanceolate, acuminate, in contracted much divided panicles; rhachilla disarticulating above the glumes, not or very shortly continued beyond the floret. *Floret* 1, ☿, much shorter than the glumes. *Glumes* equal or subequal, very narrow, linear-lanceolate, acuminate, membranous, keeled, lower 1-, upper 3-nerved. *Valve* narrow, lanceolate in profile, membranous, glabrous, more or less shortly bifid, 5–3-nerved with a fine short dorsal, rarely subterminal, awn; callus small, long hairy, hairs usually much exceeding the valve. *Pale* 2-nerved, as long as the valve or somewhat shorter. *Lodicules* 2, hyaline. *Anthers* 3. *Ovary* glabrous; styles distinct, short; stigmas plumose, laterally exserted. *Grain* enclosed by the hardly changed valve and pale, free, subterete; hilum basal, small; embryo small.

Perennial, usually rather robust; blades long, linear, flat; ligules scarious; panicle more or less contracted, narrow, rather dense, with much divided branches, and short branchlets and pedicels; florets surrounded by long fine hairs.

Species few, in the temperate regions of the northern hemisphere; 1 also in South Africa.

1. **C. epigeios** (Roth, Tent. Fl. Germ. i. 34) var. **capensis** (Stapf); culms tufted on a creeping rhizome, erect, rather stout, 2–3 ft. high, glabrous, smooth except close to the panicle, about 3-noded, internodes (at least the upper) finally exserted; sheaths striate, firm, without hairs but slightly rough, lower tight, upper more or less loose or tumid; ligules oblong, $1\frac{1}{2}$–$2\frac{1}{2}$ lin. long; blades linear, very long tapering to a setaceous point, up to more than 1 ft. by $1\frac{1}{2}$–$2\frac{1}{2}$ lin., somewhat firm, flat, closely striate, without hairs but more or less scabrid, at least towards the tips and along the margins; panicle linear to oblong-linear, somewhat spike-like, lobed, or interrupted below, 5–6 in. long, erect; rhachis terete, scabrid; branches fascicled, erect or the lower suberect, closely and repeatedly branched, lowest almost to 2 in. long; branchlets and pedicels filiform, very short, scabrid; spikelets pallid, straw-coloured, somewhat shining, very crowded, 3–$3\frac{1}{4}$ lin. long; glumes subequal, very narrow, scaberulous in the upper part and along the keels, lower slightly longer, linear-subulate, upper linear-lanceolate; rhachilla continued beyond the floret, as a very delicate bristle up to $\frac{1}{2}$ lin. long, scantily hairy; valve linear-lanceolate in profile, $1\frac{1}{4}$ lin. long, shortly bifid (lobes minutely 2-mucronate), 3-nerved, side-nerves percurrent; awn from $\frac{1}{3}$ to $\frac{1}{2}$ way above the base, $1\frac{1}{2}$–$1\frac{3}{4}$ lin. long; callus hairs 2 lin. long; pale $\frac{1}{2}$–$\frac{5}{8}$ lin. long; anthers $\frac{5}{8}$–$\frac{3}{4}$ lin. long. *C. epigeios, Nees, Fl. Afr. Austr.* 152, *not Roth; Durand & Schinz, Consp. Fl. Afr.* v. 831 (*partly*).

COAST REGION: Swellendam Div.; on mountains, *Krebs!*

CENTRAL REGION: Aliwal North Div.; at the foot of the Witte Bergen, 4500–5000 ft., *Drège!*

KALAHARI REGION: Griqualand West, Hay Div.; at Griqua Town, *Burchell*, 1912!

The South African specimens differ very slightly from the European form in the dense panicle, longer spikelets, the deeper insertion of the awn, the comparatively shorter callus hairs and the hairy continuation of the rhachilla. The apex of the rhachilla is often discernible as a very minute point at the base of the pale in the European specimens; but I have never seen it produced and hairy as in the South African variety.

XLV. ARISTIDA, Linn.

Spikelets 1-flowered, narrow, panicled, rhachilla disarticulating above the glume, not produced. *Glumes* usually persistent, narrow, 1–3-nerved, muticous or mucronate, awnless. *Valve* convolute, cylindric or oblong-cylindric, 3-nerved, awned, rather rigid, tips gradually tapering or minutely bilobed, sometimes jointed at or above the middle; callus villous, shortly bearded, usually pungent; awn nearly always 3-partite from the base or above the simple base (stipitate) very rarely simple, continuous with the valve or disarticulating from it or deciduous with a portion of the valve, foot straight or twisted, bristles plumose or the lateral or all naked. *Pale* small oblong, 2-nerved or nerveless. *Lodicules* 2, finely nerved. *Stamens* 3. *Ovary* glabrous; styles distinct, short; stigmas plumose, laterally exserted. *Grain* slender, cylindric or oblong-cylindric, terete, sometimes grooved, tightly embraced by the valve; hilum linear, almost as long as the grain; embryo short or long.

Annual or more often perennial, tufted, usually with more or less wiry culms; blades narrow, usually convolute; ligule usually a line of very short hairs; panicle varying from spike-like to effuse.

Numerous species in the dry and warm regions of both hemispheres.

Section 1. CHÆTARIA. Awns continuous with the valve, or articulated just below their branching point, but not deciduous, glabrous; glumes 1-nerved.

*Awns perfectly continuous with the valve:
Annual (at least in the South African specimens); valve linear in profile, compressed above, not beaked; anthers ½ lin. long ... ... (1) **Adscensionis.**
Perennial; valve more or less cylindrical, narrowed into a more or less distinct and usually twisted beak; anthers 1–3 lin. long:
Panicle contracted, narrow, often spike-like:
Culms glabrous; panicle 2–6 in. long:
Glumes equal or subequal, shortly mucronate from the minutely 2-toothed tips; awns suberect, 10–15 lin. long ... ... (2) **æquiglumis.**
Glumes unequal; awns more or less divaricate:
Culms terete; spikelets 5–6 lin. long; valve exserted from the shortly mucronate glumes; awn 5–7 lin. long ... ... (3) **angustata.**
Culms distinctly compressed below; spikelets 3½–4½ lin. long; valve not exserted from the long acuminate aristulate glumes; awn 7–15 lin. long ... ... (4) **junciformis.**
Culms with fugacious adpressed wool below the nodes, rather stout; panicle 1–1½ ft. long ... ... (5) **Sciurus.**
Panicle open, lax; branches 3–6 in. long:
Blades 8–12 in. long; branchlets of panicle capillary, flexuous; spikelets 5–6 lin. long, secund, nodding; awns up to 1 in. long ... ... (6) **Burkei.**
Blades 1–4 in. long; spikelets 3–4 lin. long, 1–3 at the tips of short divaricate branchlets; awns 4 lin. long ... (7) **bipartita.**
**Awns articulated with the valve just below their branching point, not deciduous; perennial:
Sheaths naked at the mouth: panicle simply spike-like or with 1–2 peduncled additional false spikes at the base ... ... (8) **congesta.**
Sheaths bearded at the mouth; panicle composed of (often long) peduncled false spikes, ovate to oblong ... ... (9) **barbicollis.**

Section 2. ARTHRATHERUM. Awns disarticulating from the valve, distinctly stipitate, glabrous; glumes 1-nerved.

Panicle usually contracted; branchlets and pedicels filiform, short; spikelets up to 10 lin. long (exclusive of the awns); glumes aristulate; foot of awn 8–20 lin. long ... ... (10) **sieberiana.**

Panicle effuse or contracted; branchlets and pedicels very fine, long; spikelets 5–8 lin. long (exclusive of the awns); glumes not aristulate; foot of awn 1–6 lin. long:
  Ligule a line of very short hairs; panicle 3–6 in. long; lower branches 1½–3 in. long; foot of awn 1–2½ lin. long ... ... (11) **vestita.**
  Ligule a more or less woolly fringe; panicle up to 1 or 1½ ft. long, very effuse; lowest branches 4–8 in. long; foot of awn 2½–6 lin. long:
    Glumes subequal; foot of awn 2½ lin. long... (12) **spectabilis.**
    Glumes very unequal; foot of awn 2½–6 lin. long ... ... (13) **stipoides.**

Section 3. STIPAGROSTIS. Awns nearly always disarticulating from the valve or together with the upper part of the valve (see No. 14. *A. sericans*), all or only the middle bristle plumose; glumes 3-nerved.

Awn continuous with the valve; the whole plant more or less hairy ... ... (14) **sericans.**
Awn disarticulating:
  *Valve gradually passing into the awn:
    Glumes firm, linear-oblong, emarginate; sheaths long bearded at the mouth and the nodes ... ... (15) **ciliata.**
    Glumes linear-lanceolate, acuminate, tips minutely truncate, hyaline, convolute:
      All bristles of the awn plumose:
        Awn tardily disarticulating with the tip of the valve; densely cæspitose: blades to 1 ft. long or longer ... ... (16) **capensis.**
        Awn readily deciduous with the upper half of the valve:
          Suffrutescent; spikelets 6–7 lin. long; middle bristle 9–12 lin. long ... ... (17) **namaquensis.**
          Not suffrutescent; spikelets up to 4 lin. long; middle bristle 3–4 lin. long ... ... (18) **proxima.**
      Side-bristles of awn sparingly and adpressedly ciliate; blades subulate, pungent, 1–3 in. long; culms many-noded, simple from a scantily branched base ... ... (19) **lutescens.**
  **Valve minutely and obtusely 2-lobed; awn from the sinus:
    Lower glume rather longer than upper, compactly tufted, with 1-noded very slender culms, 1–12 in. high:
      Perennial; panicles 1–6 in. long, exserted from the leaf-tufts ... ... (20) **obtusa.**
      Annual, about 1 in. high, inclusive of the very reduced panicles ... ... (21) **subacaulis.**
    Lower glume shorter than upper:
      Panicle narrow, rather lax, but not spike-like:
        Culms 1½–2 ft. long, 4-noded; blades up to ½ ft. long; panicle

3-8 in. long; awn plumose from below the branching point ... (22) **uniplumis.**
Culms ½ ft. long, 1-noded; blades ½-3 in. long; panicle 2-3 in. long; middle awn plumose in the upper half ... ... (23) **dregeana.**
Panicle spike-like:
Culms up to ½ ft. long; internodes very unequal; a pair of sub-opposite leaves above the middle; panicle hairy ... ... (24) **geminifolia.**
Suffrutescent; internodes very many, subequal or upper gradually longer; panicle glabrous ... ... (25) **brevifolia.**

1. A. Adscensionis (Linn. Sp. Pl. 82); annual or occasionally perennial with an oblique rhizome, glabrous; culms tufted, geniculately ascending, slender, from a few inches to 2 ft. long, usually branched from one or several of the lower nodes, smooth, upper internodes long exserted; sheaths tight, rather firm, smooth; ligule a line of short hairs; blades very narrow, linear, tapering to a very fine point, 1-9 in. long, up to 1 lin. broad, convolute, rarely flat, smooth below, scabrid above and on the margins; panicle linear, spike-like, usually interrupted or oblong and more or less lax, rigid or flaccid; branches single or 2-nate, unequal, branched from the base or simple to the middle, erect or nodding or flexuous, filiform, scabrid, lateral pedicels short; spikelets 3-5 lin. long, often purplish; glumes linear to linear-lanceolate, acute or subobtuse, 1-nerved, the lower 2-3½ lin., the upper 3-4½ lin. long, sometimes mucronate; valve linear, laterally compressed, as long as the upper glume or slightly longer, rarely shorter, scabrid along the keel and the outer nerve, otherwise smooth or scabrid, particularly below the straight tip; callus ¼ lin. long; awns 6-9 lin. long, rarely shorter (down to 4 lin.) or longer (up to 1 in.), diverging, continuous with the valve, the lateral somewhat shorter; pale obtuse, less than ½ lin. long; lodicules similar to the pale, 3-5-nerved, ⅓-½ lin. long; anthers ½ lin. long; grain almost as long as the valve. *Brongn. in Duperr. Voy. Coq.* 13; *Kunth, Enum.* i. 190; *Trin. & Rupr. Stip.* 138; *Steud. Syn. Pl. Glum.* i. 139; *Baker, Fl. Maurit.* 450; *Durand & Schinz, Consp. Fl. Afr.* 799; *Hook. f. Flor. Brit. Ind.* vii. 224. *A. paniculata, Forsk. Fl. Aegypt.-Arab.* 25 (?). *A. americana, Linn. in. Amoen. Acad.* v. 393, *not Swartz.* *A. gigantea, Linn. f. Suppl.* 113. *A. depressa, Retz. Obs.* iv. 22. *A. cœrulescens, Desf. Fl. Atlant.* i. 109, *t.* 21, *fig.* 2; *Schum. & Thonn. Beskr. Guin. Pl.* 47; *Trin. Spec. Gram. Ic. t.* 313. *A. interrupta, Cav. Ic.* v. 45, *t.* 471, *fig.* 2. *A. elatior, Cav. Ic.* vi. 65, *t.* 589, *fig.* 1. *A. canariensis, Willd. Enum.* 99. *A. humilis, A. bromoides, and A. coarctata, H.B.K. Nov. Gen. et. Spec.* i. 121, 122. *A. divaricata, Jacq. Ecl. Gram.* 7, *t.* 6, *not Willd.* *A. mauritiana, Kunth, Rév. Gram.* i. 61; 265, *t.* 44; *Trin. & Rupr. Stip.* 139. *A. setacea, Trin. Gram. Gen.* 84, *not Retz.* *A. nigrescens, Presl, Reliq. Haenk.* i. 223. *A. cognata and A. dispersa, Trin. &*

Rupr. Stip. 127 *and* 129. *A. festucoides, Hochst. & Steud.* (*not Poir.*) *ex Trin. & Rupr. l.c.* 129. *A. laxa, Willd.* (*not Cav.*) *ex Trin. & Rupr. l.c.* 130. *A. vulgaris, Trin. & Rupr. l.c.* 131–136 (*excl. vars. æthiopica, senegalensis, Ehrenbergii, spicigera*). *A. nutans, Ehrenb. & Hempr. ex Trin. & Rupr. l.c.* 135. *A. pusilla, Trin. & Rupr. l.c.* 140; *Steud. Syn. Pl. Glum.* i. 139; *Durand & Schinz, Consp. Fl. Afr.* v. 807. *A. swartziana, A. maritima, A. nana, A. arabica, A. tenuiflora, A. modatica, A. simplicissima, Steud. Syn. Pl. Glum.* i. 137–139. *A. chætophylla and A. Teneriffæ, Steud. l.c.* 420. *A. macrochloa, Hochst. in Flora,* 1855, 200. *A. vulpioides, Hance in Ann. Sc. Nat. sér.* 5, v. 251. *A. Heymanni, Regel in Act. Hort. Petrop.* vii. (1880) 649. *Chætaria ascensionis, C. canariensis, C. cærulescens, C. depressa, C. elatior, C. gigantea, and C. interrupta, Beauv. Agrost.* 30. *C. humilis, C. bromoides and C. coarctata, Roem. & Schult. Syst.* ii. 396. *C. nana, Nees ex Steud. Nom. ed.* 2, i. 340. *C. mauritiana, Nees, Fl. Afr. Austr.* 188.

COAST REGION: Uitenhage Div.; near Uitenhage, *Zeyher!* Steenboks Flats, *Ecklon & Zeyher!*

CENTRAL REGION: Ceres Div.; at Yuk River, near Yuk River Hoogte, *Burchell,* 1266! Somerset Div.; *Bowker,* 149! Graaff Reinet Div.; stony hills near the Sunday River, 1500–2000 ft., *Drège!* mountain sides near Graaff Reinet, 2900 ft., *Bolus,* 678! Calvinia Div.; between Lospers Plaats and Springbok Kuil, *Zeyher,* 1817! Albert Div., *Cooper,* 778! 1364!

WESTERN REGION: Little Namaqualand; between Holgat River and Orange River, *Drège!* between Buffels River and Silver Fontein, 2000 ft., *Drège.*

KALAHARI REGION: Griqualand West, Hay Div.; between the Kloof Village and Witte Water, *Burchell,* 2083! between Witte Water and Riet Fontein, *Burchell,* 2006! Orange Free State; between Kimberley and Bloemfontein, *Buchanan,* 283!

Common in most dry and hot countries.

Drège's specimens from Little Namaqualand represent Nees's *Chætaria mauritiana* var. *nana* (*Aristida nana,* Steud.). I consider them only as very dwarfed individuals of this species.

2. **A. æquiglumis** (Hack. in Bull. Herb. Boiss. iii. 381); perennial, densely cæspitose, glabrous; culms very slender, terete, erect, simple, wiry, 1–1¼ ft. long, smooth, about 3–4-noded, internodes exserted; sheaths very tight throughout, smooth; ligule a minutely ciliolate rim; blades setaceously convolute, fine, 2–4 in. long, firm, strongly curved, smooth below, minutely puberulous on the upper surface, margins scabrid; panicle linear-oblong, contracted, or somewhat open and lax, 3–4 in. long; rhachis filiform, smooth or subscaberulous; branches remotely 2- or 3-nate, or the upper solitary, erect or oblique, longest up to more than 2 in. long and sparsely divided from the middle, the others almost simple, subcapillary, scaberulous; branches 1–3-spiculate; lateral pedicels very short; spikelets straw-coloured or tinged with purple, 4–5 lin. long; glumes equal or subequal, lanceolate-linear, shortly mucronate from the minutely 2-toothed tips, 1-nerved; valve linear-convolute, 3½ lin.

long, produced into an often slightly exserted and more or less twisted almost solid scabrid beak, minutely scaberulous below it, violet; awns continuous with the valve, capillary, suberect, scaberulous, about 10–15 lin. long; pale oblong, acute, slightly over $\frac{1}{2}$ lin. long; lodicules about as long as the pale.

KALAHARI REGION: Transvaal; Pretoria Distr., at Kudus Poort, *Rehmann*, 4696! Klip Spruit, *Nelson*, 96*!

3. **A. angustata** (Stapf); perennial, densely cæspitose, light green or glaucous; culms very slender, terete, erect, simple or very scantily branched, wiry, 1–1$\frac{1}{2}$ ft. long, about 3-noded, internodes exserted; sheaths very tight, often slightly widened at the base, the lower sometimes woolly; blades setaceously convolute, rather fine, up to $\frac{3}{4}$ ft. long, curved or flexuous, smooth below, densely pubescent or hispidulous and almost white above; panicle very narrow, often spike-like, 2–4 in. long; rhachis straight or flexuous; branches solitary, the longest up to 1$\frac{1}{4}$ in. long, scantily branched; branchlets filiform; lateral pedicels very short; spikelets yellowish, 4$\frac{1}{2}$–5$\frac{1}{2}$ lin. long; glumes unequal, oblong-linear, usually shortly mucronate, the lower 2–3$\frac{1}{2}$ lin. long, acute, the upper about 4 lin. long, acute or minutely truncate; valve linear, convolute, produced into a short or usually very short beak (when of sufficient length to be slightly exserted then more or less twisted), scaberulous below the beak; callus $\frac{1}{4}$ lin. long; awns continuous with the valve, divaricate, 5–7 lin. long, fine; pale hyaline, $\frac{2}{3}$ lin. long; lodicules 5-nerved, $\frac{7}{8}$ lin. long; anthers 1$\frac{1}{2}$–2 lin. long, not apiculate.

COAST REGION: Cape Div.; between Newlands and Clairmont, *Wolley Dod*, 2387! Sand Road near Clairmont, *Wolley Dod*, 2388! Tulbagh Div.: Tulbagh, 600 ft., *Schlechter*, 7509! Worcester Div.; mountains above Worcester, *Rehmann*, 2582! 2587! 2667! Cathcart Div.; Windvogel Mountain, 3500 ft., *Baur*, 1115!

KALAHARI REGION: Transvaal; Magalies Berg, *E.S.C.A. Herb.* 304! Hol Fontein, *Nelson*, 79*! Houtbosch, *Rehmann*, 5667! Bosch Veld, between the Elands River and Klippan, *Rehmann*, 5112!

4. **A. junciformis** (Trin. & Rupr. Stip. 143); perennial, loosely cæspitose, sometimes stoloniferous; rhizome slender, oblique or creeping, covered like the innovation-buds with imbricate short ovate acute or pungent scales; culms fascicled, erect, 1–1$\frac{1}{2}$ ft. long, slender, usually branched, glabrous, compressed below, wiry, 3–4-noded, internodes exserted, culm-nodes slightly swollen and sometimes protruding above the obscure sheath-nodes; sheaths tight, the lower keeled, often slipping from the stem and rolling in, glabrous or villous; blades very narrow, gradually passing into the sheaths, subsetaceous, acute, up to 1 ft. long, rigid, convolute or folded below, curved or flexuous, smooth below, minutely scaberulous above; panicle narrow, 2–6 in. long, erect or nodding; rhachis angular; branches fascicled, erect, flexuous or nodding, the lowest $\frac{1}{2}$–2 in. long, often undivided for 1 in. or more, filiform to capillary,

scaberulous; lateral pedicels short or almost 0; spikelets light green to yellowish, 2–3 near the tips of the branchlets, $3\frac{1}{2}$–$4\frac{1}{2}$ lin. long; glumes unequal, thin, acute or subacute, 1-nerved, nerve excurrent into a fine bristle, 1–$1\frac{1}{2}$ lin. long, the lower lanceolate, 2–$2\frac{1}{2}$ lin. long, the upper sublinear; valve linear, convolute, produced into a short slightly twisted beak, as long as the upper glume, smooth or scaberulous below the beak; callus obtuse; awns continuous with the valve, very fine, 7–15 lin. long; pale $\frac{3}{8}$ lin. long; lodicules delicate, lanceolate, over $\frac{1}{2}$ lin. long; anthers 1 lin. long, not apiculate. *Steud. Syn. Pl. Glum.* i. 140; *Durand & Schinz, Consp. Fl. Afr.* v. 804.

KALAHARI REGION: Orange Free State, *Buchanan*, 67!

EASTERN REGION: Natal; coast land to 1000 ft., *Sutherland!* Mohlamba Range, 5000–6000 ft., *Sutherland!* West Town, on the Mooi River, *Rehmann*, 7342; near Durban, *Williamson*, 34! *Plant*, 61! and without precise locality, *Buchanan*, 1! 289! *Drège*, 4349 in the Lübeck Herb.!

5. **A. Sciurus** (Stapf); culms rather stout, over 3 ft. long, with a fugacious snow-white adpressed woolly indumentum below the nodes; sheaths long, tight, glabrous except for some wool near the mouth, striate; ligule a flake of fine wool; blades linear, acute, involute or setaceously convolute above, over 1 ft. long, 2 lin. broad near the base, smooth below, asperulous above; panicle contracted, dense, 1–$1\frac{1}{2}$ ft. by 1–$1\frac{1}{2}$ in., erect; rhachis rather stout, smooth; branches fascicled, the lowest up to $\frac{1}{2}$ ft. long, erect, remotely and repeatedly branched; branchlets filiform to capillary, scaberulous; lateral pedicels short; spikelets 2–3 at the tips of the branchlets, yellowish, 5–6 lin. long, very slender; glumes very unequal, rather thin, minutely truncate or 2-toothed (when expanded), the lower lanceolate-oblong, acute, $2\frac{1}{2}$ lin. long, the upper linear-oblong, 5–6 lin. long; valve linear, 4–5 lin. long, obscurely beaked, beak straight; callus very short; awns continuous with the valve, fine, scaberulous, 7–8 lin. long; pale $\frac{2}{3}$ lin. long; lodicules few-nerved, $\frac{3}{4}$ lin. long; anthers 3 lin. long, cells minutely apiculate.

KALAHARI REGION: Transvaal; Zebedelis Kraal, near the Inkumpi River, *Nelson*, 26*!

EASTERN REGION: Natal; without precise locality, *Gerrard*, 471!

Similar to *A. setacea*, Retz., and *A. multicaulis*, Baker, but differing from both in the woolly, rather stout, not wiry and apparently simple culms and in the woolly ligules.

6. **A. Burkei** (Stapf); perennial, tufted, glabrous; culms rather slender, erect, 2 ft. long, wiry, smooth, 2-noded, sheathed from the lower node to the panicle; sheaths firm, tight; ligule a line of very short hairs or the lowest a tuft of wool; blades coarsely setaceous, involute, $\frac{1}{2}$–1 ft. long, firm, curved or flexuous, the uppermost exceeding the panicle, smooth below; panicle about $\frac{3}{4}$ ft. by $\frac{1}{2}$ ft., very lax; rhachis smooth; branches distant, 2–3-nate, up to 6 in. long, scantily and remotely branched, filiform, scaberulous or smooth below; branchlets capillary, flexuous; pedicels as long as the spike-

lets or shorter; spikelets yellowish, secund, often nodding, 5–6 lin. long; glumes rather firm, rounded at the back, obtuse or emarginate, not mucronate, the upper twice as long as the lower; valve linear, produced into a short slightly twisted beak, smooth, somewhat exceeding the upper glume; callus $\frac{1}{2}$ lin. long; awns continuous with the valve, up to 1 in. long, fine; pale broad, $\frac{1}{2}$ lin. long, 2-nerved; lodicules $\frac{2}{3}$ lin. long; anthers 2 lin. long, cells apiculate.

KALAHARI REGION: Orange Free State; near the Vaal River, *Burke*, 165! near Hoopstad, *Grindon Herb.!*

7. **A. bipartita** (Rupr. & Trin. Stip. 144); perennial, light green to glaucous; rhizome short, oblique with compact tufts of short barren shoots and culms, these erect or ascending, 1–2 ft long, simple, terete or compressed below, smooth, glabrous, rarely puberulous below the nodes, about 3-noded; basal sheaths short, compressed, firm, persistent, whitish, the upper widened and loose in the upper part, at length open, smooth or bearded at the mouth; blades very narrow, linear, acute, 1–4 in. by 1 lin., rigid, curved, folded, more rarely flat, smooth below, scabrid above; panicle effuse, 5–6 in. by 3–5 in., very lax; rhachis straight or flexuous; branches solitary, distant, spreading, the lower 3–4 in. long, 2- (rarely 3–4-) partite close to the base, very scantily and remotely branched; branchlets divaricate, 1–3-spiculate at the tips, filiform, straight or flexuous, scabrid; lateral pedicels very short; spikelets 3–4 lin long, sometimes purplish; glumes subequal, linear-lanceolate, abruptly and shortly mucronate or the upper emarginate; valve linear, not or obscurely beaked, as long as the glumes or slightly shorter, smooth or finely scaberulous above, purplish; callus $\frac{1}{4}$ lin. long; awns continuous with the valve, subequal, divaricate, 4 lin. long; pale $\frac{1}{2}$ lin. long, shortly 2-nerved; lodicules up to $\frac{3}{4}$ lin. long, 3-nerved; anthers 1–2 lin. long. *Steud. Syn. Pl. Glum.* i. 140; *Durand & Schinz, Consp. Fl. Afr.* v. 801. *Chætaria bipartita, Nees, Fl. Afr. Austr.* 187.

SOUTH AFRICA: without precise locality, *Zeyher*, 1810!
COAST REGION: Fort Beaufort Div.; Kat River, *Drège!*
CENTRAL REGION: Somerset Div.; *Bowker*, 166!
KALAHARI REGION: Orange Free State; Vaal River, *Burke*, 430! Bloemfontein, *Rehmann*, 3736! Basutoland; Leribe, *Buchanan*, 125!
EASTERN REGION: Natal; Biggars Berg, *Rehmann*, 7102!

8. **A. congesta** (Roem. et Schult. Syst. ii. 401); perennial, tufted, light green or glaucous, glabrous; culms slender, rather wiry, erect or geniculately ascending, compressed below, $\frac{1}{4}$–2 ft. long, simple or branched from some of the lower nodes, 3–4-noded; sheaths tight, smooth; blades usually very narrow, linear, acute, 1–6 in. by 1 lin., rarely larger, usually folded or convolute, rigid, curved, rarely flat, smooth below, scabrid to hispidulous above; panicle spike-like, often interrupted, with 1–2 shortly peduncled, more or less spreading lateral pseudo-spikes, 2–6 in. long; pedicels very short; spikelets densely crowded, $3\frac{1}{2}$–4 lin., rarely up to 5 lin. long; glumes keeled, keels smooth or almost so, the lower lanceolate, gradually passing

into a long mucro, 3 lin. long, the upper linear, emarginate, long mucronate, 3½–5 lin. long; valve linear, produced into a short twisted beak, usually slightly shorter than the upper glume, minutely scaberulous above; callus ½ lin. long; awns jointed with the valves, but not disarticulating, diverging, fine, 5–7 lin. long; pale not quite ½ lin. long, nerveless or almost so; lodicules ½–⅝ lin. long, 5–6-nerved; anthers ⅜ lin. long; grain up to 1¾ lin. long, deeply grooved. *Kunth, Rév. Gram.* ii. *t.* 172; *Enum.* i. 195; *Suppl.* 152; *Trin. & Rupr. Stip.* 153; *Steud. Syn. Pl. Glum.* i. 142; *Durand & Schinz, Consp. Fl. Afr.* v. 802; *Hack. in Bull. Herb. Boiss.* iv. *App.* iii. 18. *A. coarctata, Lichtenst. ex Roem. & Schult. l.c.* 401. *Chætaria congesta, Nees, Fl. Afr. Austr.* 189.

SOUTH AFRICA: without precise locality, *Zeyher*, 1818!

COAST REGION: Uitenhage Div.; Steenboks Flats, *Ecklon & Zeyher!* Komgha Div.; Kei River, 500 ft., *Drège.* Queenstown Div.; Shiloh, *Baur*, 55! Wodehouse Div.; Indwe, *Baur*, 71!

CENTRAL REGION: Somerset Div.; *Bowker*, 134! 148! Graaff Reinet Div.; near Graaff Reinet, 2900 ft., *Bolus*, 679! Colesberg Div.; near Colesberg, in crevices of rocks, 4500 ft., *Drège! Shaw*, 16! Albert Div.; without precise locality, 4500 ft., *Drège.*

WESTERN REGION: Great Namaqualand; Gubub, *Schinz*, 657! Warmbad, *Wandres*, 19!

KALAHARI REGION: Prieska Div.; near Keikams Poort, *Burchell*, 1612! Griqualand West, Hay Div.; Witte Water, *Lichtenstein*, between Griquatown and Witte Water, *Burchell*, 1983! between Kloof Village and Witte Water, *Burchell*, 2084/1! Herbert Div.; St. Clair, near Douglas, *Orpen*, 254! Kimberley; near Dutoits Pan, *Tuck!* Orange Free State; Great Vet River, *Burke*, 210! Bloemfontein, *Rehmann*, 3739! Basutoland; Leribe, *Buchanan*, 149! Bechuanaland; *Marloth*, 696! Transvaal; near Lydenburg, *Atherstone!* Bosch Veld, by the Elands River, *Rehmann*, 4999! Pretoria Div., at Kameel Poort, *Nelson*, 64*!

EASTERN REGION: Natal, *Buchanan*, 124!

9. **A. barbicollis** (Trin. & Rupr. Stip. 152); perennial, tufted, light green to glaucous, glabrous except at the mouths of the sheaths; culms slender, rather wiry, more or less compressed below, geniculately ascending or suberect, ½–1½ ft. long, simple or scantily branched from some of the lower nodes, smooth, 2–3-noded; sheaths tight, smooth; ligule a dense line of short hairs passing into beards or a ring of long hairs at the mouths of the sheaths; blades usually very narrow, linear, acute, 1–6 in. by ¾–1 lin., folded; convolute, curved, rigid or flat and then often twisted or curled, smooth below, scabrid above; panicle ovate to oblong, 2–6 in. long; rhachis straight or flexuous, smooth; branches solitary, distant, filiform, spreading, flexuous or straight, scaberulous, dense, spike-like from ¼–1½ in. above the base; pedicels very short; spikelets 3½ lin. long; glumes keeled, the lower lanceolate, shortly mucronate, 2 lin. long, keels smooth or scabrid, the upper linear, emarginate, mucronate, 3½ lin. long; valve linear, produced into a short, stout, tightly twisted beak, somewhat shorter than the upper glume, minutely scaberulous below the beak; callus less than ½ lin. long; awns jointed with the valve, not disarticulating, fine, 5–9 lin. long; pale, lodicules, stamens and grain

as in. *A. congesta*. *Steud. Syn. Pl. Glum.* i. 141; *Durand & Schinz, Consp. Fl. Afr.* v. 800. *Chætaria Forskolii, Nees, Fl. Afr. Austr.* 188.

COAST REGION: Cape Div.; Cape Flats, near Claremont, *Schlechter*, 492! Uitenhage Div., *Zeyher! Ecklon.* Alexandria Div.; Enon, 1500–2000 ft., *Drège*, 388! Bathurst Div.; Port Alfred, *Hutton!* Fort Beaufort Div.; between the Kunab River and the Kat River, *Drège;* between Hermanns Kraal and Beaufort Castle, *Drège.* Komgha Div.; on the Kei River, 500 ft., *Drège.* Queenstown Div.; plains near Queenstown, 3500 ft., *Galpin*, 2355!

CENTRAL REGION: Prince Albert Div.; by the Gamka River, near Weltevrede, 2500–3000 ft., *Drège!* Graaff Reinet Div.; near the Sunday River, 2000 ft., *Drège!* near Graaff Reinet, 2900 ft., *Bolus*, 677!

KALAHARI REGION: Transvaal; Lydenburg, *Atherstone!* Orange Free State; without precise locality, *Buchanan*, 57!

EASTERN REGION: Natal; near Durban, *Williamson!* near Tugela, 4000 ft., *Buchanan*, 290! *Wood*, 3588! near Colenso, 3000 ft., *Wood*, 4418! Umsinga and base of Biggars Berg, *Buchanan*, 90! and without precise locality, *Gerrard!*

Very close to *A. congesta*, but the branches of the panicle are more numerous and longer, the spikelets a little larger, and the mouth of the sheaths is distinctly bearded, the beards sometimes uniting into a ring at the junction of the blade and the sheath.

10. A. sieberiana (Trin. in Spreng. Neue Entdeck. ii. 61); perennial, glabrous; rhizome short, slender, oblique; culms fascicled, erect and simple or branched and geniculate, ½–3 ft. long, wiry, smooth, about 3-noded; culm-nodes slightly swollen and often protruding above the annular dark sheath-nodes; sheaths tight, the lower at length slipping from the stem and rolling in, smooth; blades narrow, linear, acute, 4–6 in. by 1 lin., convolute, smooth below, scaberulous to hispidulous above; panicle usually contracted, narrow, 6–8 in. long, rarely open and secund; rhachis straight or somewhat flexuous; branches solitary, 2-nate or fascicled, all or some branched from the base, some undivided for ½–1½ in., the longest 2½ in. long; branchlets short, filiform, scaberulous; lateral pedicels short; spikelets pallid or purplish, very narrow, up to 10 lin. long; glumes unequal, very narrow, the lower sublinear, about 5 lin. long, passing into a fine bristle, the upper linear, bifid, with a bristle 3–4 lin. long from the sinus, lobes fine; valve linear, smooth or almost so, 4–5 lin. long; callus slender, 1½ lin. long; awn stipitate, foot slender, twisted, 8–12 lin. long; bristles fine, 2–2½ lin. long; pale broad, over ½ lin. long; grain very slender, subcylindric, 3–3½ lin. long. *Kunth, Enum.* i. 191; *Suppl.* 147; *Trin. & Rupr. Stip.* 160; *Steud. Syn. Pl. Glum.* i. 143; *Boiss. Fl. Or.* v. 492; *Durand & Schinz, Consp. Fl. Afr.* v. 808. *Chætaria sieberiana, Roem. & Schult. Syst.* ii. *Mant.* iii. 578.

KALAHARI REGION: Transvaal; Klip Fontein, *Nelson*, 103*! Bosch Veld, between Elands River and Klippan, *Rehmann*, 5111!

EASTERN REGION: Delagoa Bay, *Mrs. Monteiro!*

Also in Kordofan and in Southern Palestine.

A variety with a much longer foot to the awn, occurs in Nubia, and another with a still longer foot, in German South-west Africa (*A. stipitata*, Hack.).

11. A. vestita (Thunb. Prod. Cap. 19); perennial, light green to glaucous; rhizome very short, with dense tufts of barren shoots and culms, the latter erect, 1–2 ft. high, simple, 2–1-noded, terete, wiry, glabrous, smooth; sheaths tight, smooth, scarcely striate, glabrous or the lower more or less covered with a very fugacious wool; ligule a line of very short hairs; blades convolute-setaceous from a few inches to more than 1 ft. by scarcely 1 lin. when expanded, rigid, curved or flexuous, glabrous, smooth below, scabrid to hispidulous above; panicle effuse or contracted, 3–6 in. by 2–5 in.; rhachis strict or flexuous; lower branches 2–3-nate, $1\frac{1}{2}$–3 in. long, usually spreading, sparingly and remotely branched; branchlets very flexuous, filiform to capillary, scaberulous; pedicels very fine, the longest equalling the spikelets; spikelets often secund, nodding, yellowish, rarely purplish, 5–6 lin. long; glumes rather firm, rounded at the back, obtuse or more or less 2-toothed, the lower linear-oblong, about $\frac{1}{2}$ the length of the upper or less, this narrow lanceolate-linear, 5–6 lin. long; valve linear, $4\frac{1}{2}$–$5\frac{1}{2}$ lin. long, not beaked, scaberulous from the middle or almost smooth; callus $\frac{1}{2}$ lin. long; awn disarticulating from the valve, stipitate, foot 1–$2\frac{1}{2}$ lin. long, twisted, bristles divaricate or the lateral upright, $\frac{2}{3}$–1 in. long; pale broad, $\frac{1}{2}$ lin. long; lodicules $\frac{2}{3}$–$\frac{3}{4}$ lin. long, finely nerved; anthers $2\frac{1}{2}$ lin. long; grain very slender, 3 lin. long. *Thunb. Fl. Cap. ed. Schult.* 104; *Nees in Linnæa*, vii. 287; *Kunth, Enum.* i. 197; *Trin. & Rupr. Stip.* 157; *Steud. Syn. Pl. Glum.* i. 142; *Durand & Schinz, Consp. Fl. Afr.* v. 810. *A. Hystrix, Thunb. Prodr.* 19; *Fl. Cap. ed.* i. 394; *ed Schult.* 104, *not Linn. A. diffusa, Trin. Gram. Gen.* 86. *A. lanuginosa, Burch. Trav.* ii. 226. *Chætaria vestita, Beauv. Agrost.* 30. *Arthratherum Hystrix, Nees in Linnæa*, vii. 287. *A. lanuginosum, Burch. l.c.* 612 (*index*). *A. vestitum, Nees, Fl. Afr. Austr.* 174.

VAR. β, **parviflora** (Trin. & Rupr. l.c. 158); culms branched with several usually naked wiry internodes below; blades finer than in the type, straight; panicle contracted, very narrow; spikelets 5 lin. long; valve $3\frac{1}{2}$ lin. long, smooth.

VAR. γ, **schraderiana** (Trin. & Rupr. l.c. 158); spikelets 7–8 lin. long; valve 6–7 lin., foot of awn 1–2 lin., bristles $1\frac{1}{4}$–$1\frac{3}{8}$ in.

SOUTH AFRICA: without precise locality, *Thunberg! Boivin! Zeyher*, 1811! 1808! 447! Var. γ: *Zeyher*, 4405!

COAST REGION: Cape Div.; Table Mountain, *Ecklon*, 976! Swellendam Div.; between Breede River and Zonder Einde River, *Burchell*, 7491! Uitenhage Div.; near the Zwartkops River, 50–500 ft., *Zeyher*, 4504! *Ecklon*; between Sunday River and Koega River, *Drège*. Alexandria Div.; in virgin forests by the Bushmans River, Olifantshoek Mountains, *Ecklon*, Addo, *Ecklon*. Cathcart Div.; Windvogel Berg, 3000–4000 ft., *Drège*. Queenstown Div.; Engotini, near Shiloh, 3500 ft., *Baur*, 964! between the Klipplats River and the Zwart Kei River, *Drège*, Finchams Nek, near Queenstown, 4000 ft., *Galpin*, 2383! Var. γ: Cape Div.; Capetown, *Harvey*, 158! Port Elizabeth Div.; near Port Elizabeth, *E.S.C.A. Herb.* 121!

CENTRAL REGION: Prince Albert Div.; near Zwartbulletje, by the Gamka River, 2500 ft., *Drège!* near Klaarstroom, 2500–3000 ft., *Drège*. Aberdeen Div.; Camdebo, 2500–3000 ft., *Drège*. Somerset Div.; Little Fish River, 2000–2500 ft., *Drège*. Blyde River, *Burchell*, 2978! Graaff Reinet Div.; hillsides near Graaff Reinet, 2500–2700 ft., *Bolus*, 459! Colesberg Div.; Colesberg, *Shaw*, 22! Albert Div.; near Gaatje in the Klein Buffel Valei, 4500–5000 ft., *Drège*. Burghersdorp, *Cooper*, 3372!

WESTERN REGION: Var. β: Little Namaqualand; Silver Fontein, near Ookiep, 2000–3000 ft., *Drège!*

KALAHARI REGION: Griqualand West, Hay Div.; Griqua Town, *Burchell*, 1842! 1917! Asbestos Mountains, at the Kloof Village, *Burchell*, 2038! Basutoland; Leribe, *Buchanan*, 124! Transvaal; near Lydenburg, *Atherstone!* Potgeiters Rust, near the Nylstroom, *Nelson*, 55*! Bosh Veld, between Elands River and Klippan, *Rehmann*, 5124! Apies River, *Nelson*, 5*!

Also in Hereroland.

Trinius and Ruprecht, l.c., distinguished 7 varieties (*diffusa, densa, eckloniana, schraderiana, parviflora, brevistipitata* and *pseudo-hystrix*); of these, only the 2 above described appear to me fairly well characterized.

Thunberg, in describing his *A. Hystrix* as having 2 equal glumes, was misled by the frequent splitting of the lower glume into 2 equal halves, whilst the upper glume is in most of the spikelets of his specimens so tightly rolled round the floret that it appears as the lower part of it.

12. **A. spectabilis** (Hack. in Bull. Herb. Boiss. iii. 380); perennial, cæspitose; culms rather stout, over 4 ft. long, terete, simple, glabrous, smooth, about 5-noded, sheathed nearly all along; sheaths long, tight, glabrous, excepting sometimes a flake of wool at the mouth, lowest imbricate, firm, persistent, with fugacious white adpressed wool near the base; ligule a fringe of soft, almost woolly cilia; blades long linear, tapering to a setaceous point, up to 2 ft. by 2–2½ lin. (near the base), firm, involute, glaucous and scaberulous on the upper surface, smooth below, glabrous; panicle oblong or ovate, lax, open, nodding, up to 1 ft. long; rhachis very slender, smooth; branches distantly 2-nate or solitary, lowest up to 8 in. long, obliquely spreading, remotely and repeatedly branched, filiform, scaberulous; branchlets and pedicels capillary, flexuous, the latter as long as the spikelets or considerably longer; spikelets yellowish, 5–6 lin. long; glumes subequal, linear-lanceolate, about 5 lin. long; acute, rather firm, except the delicate hyaline tips and narrow margins, 1-nerved, smooth; valve linear-convolute, produced, 4 lin. long, not beaked, finely scaberulous; callus slender, shortly bearded, ⅓ lin. long; awn disarticulating from the valve, stipitate; foot 2½ lin. long, twisted; bristles fine, scaberulous, spreading, the middle up to 1¼ in., the lateral 10 lin. long; pale ovate, ⅖ lin. long; lodicules ovate-oblong, nerved, ⅖ lin. long; anthers 2½–2¾ lin. long, cells minutely apiculate.

KALAHARI REGION: Transvaal; Pretoria Div., at Kudus Poort, *Rehmann*, 4695!

13. **A. stipoides** (Lam. Illustr. i. 157, not R. Br.) var. **meridionalis** (Stapf); perennial, compactly tufted with numerous innovation shoots; culms erect, 3 ft. or more long, simple or sparingly branched, terete, glabrous, smooth, 2–3-noded; sheaths tight or loose, the lowest more or less slipping from the culm, glabrous, smooth, often black at the mouth, shiny; ligule a dense line of long soft hairs, surrounding the mouth of the sheath like a flake of wool; blades narrow, linear, up to 1½ ft. by 1–2½ lin., setaceously convolute, flexuous, subglaucous, smooth below, scabrid above; panicle effuse,

very lax, 1–1½ ft. by ½–1 ft.; rhachis strict or flexuous, smooth; branches 2–3-nate, spreading, the lower 4–8 in. long, repeatedly and remotely branched; branchlets filiform to capillary, flexuous; pedicels capillary, as long as the spikelets or longer; spikelets very scattered, yellowish, 4–5 lin. long; glumes very unequal, rather firm, rounded at the back, the lower linear-oblong, obtuse or emarginate, about ½–⅓ as long as the upper, the latter lanceolate-linear, 2-toothed; valve linear, not beaked, ⅓ shorter than the upper glume, smooth, mottled with purple; callus ½–¾ lin. long; awn disarticulating, stipitate, foot 2½–6 lin. long, slender, twisted, bristles up to 1 in. long, fine; pale broad, over ½ lin. long; lodicules finely nerved; anthers 2 lin. long; grain fusiform, 3–4 lin. long.

KALAHARI REGION: Griqualand West; Hunernest Kloof, *Rehmann*, 3386! Orange Free State, *Buchanan*, 56! 68! between Kimberley and Bloemfontein, *Buchanan*, 281! Bechuanaland; between Kuruman and Matlareen River, *Burchell*, 2188!

Also in Northern Bechuanaland.

The typical form is loosely tufted with very few innovation shoots, and perhaps occasionally annual. It has less convolute or flat leaves, longer spikelets (6–7½ lin. long) and awns (foot 10–15 lin.; bristles 1¾–2 in. long). It inhabits tropical Africa from the Senegal and Kordofan to German East Africa. The following synonyms are referable to it:—*A. stipiformis, Poir. Encycl. Suppl.* i. 452; *Kunth, Enum.* i. 194; *Durand & Schinz, Consp. Fl. Afr.* v. 809. *A. amplissima, Trin. & Rupr. Stip.* 155; *Steud. Syn. Pl. Glum.* i. 142; *Durand & Schinz, l.c.* 800. *A. gracillima, Oliv. in Trans. Linn. Soc.* xxix. 173, *t.* 114, *f.* 1; *Durand & Schinz, l.c.* 803. *Chætaria Lamarckii, Roem. & Schult. Syst.* ii. 393. *Arthratherum comosum, Gay ex Kunth, l.c.*

14. A. sericans (Hack. in Bull. Herb. Boiss. iii. 381); perennial, densely tufted; culms very slender, erect, 1 ft. long, wiry, softly hairy, 1-noded, internodes exserted; sheaths very tight, striate, hairy all over except at the nodes; blades finely setaceous, convolute, acute, up to 8 in. long, glaucous, hairy below, hispidulous above; panicle spike-like, rather dense, narrow, up to 3 in. long, erect; rhachis hairy; branches solitary or 2–3-nate, adpressed, the lowest ½–1 in. long, simple, 1-spiculate or very sparingly branched; branchlets and pedicels hairy or scabrid; spikelets yellowish or purplish, 6–7 lin. long; glumes lanceolate to linear, acuminate, hairy or glabrous, 3-nerved, the lower 5–6 lin. long, acute, the upper 6–7 lin. with hyaline tips; valve linear, produced into a short twisted hairy beak, 3½–4 lin. long, purple, smooth; callus ½ lin. long; awns continuous with the valve, the middle one 4–5 lin. long, all plumose except at the very tips; pale over 1 lin. long, obtuse, nerveless; lodicules finely nerved, over ½ lin. long; anthers 2 lin. long.

KALAHARI REGION: Transvaal; Hooge Veld, at Standarton, *Rehmann*, 6793! Pretoria, near Apies Poort, *Rehmann*, 4046!

15. A. ciliata (Desf. in Schrad. Neues Journ. iii. 255); perennial, compactly cæspitose, with numerous, usually short innovation shoots; culms geniculate, slender, 1–2 ft. long, glabrous, smooth, 2–3-noded, internodes exserted; sheaths crowded at the base, the lowest broad, whitish, firm, persistent, woolly near the margins below, the upper tight, long bearded at the nodes and at the mouths; blades coarsely setaceous, convolute, those of the innovation shoots sometimes very short, recurved or like those of the culms up to ½ ft. long, rigid, glabrous, smooth below, minutely hairy above; panicle narrow, oblong-linear, usually contracted and strict, 4–6 in. long; rhachis smooth; branches erect, solitary, often bifid near to or branched from the base; branchlets filiform, few spiculate, tips clavate; lateral pedicels short; spikelets linear-oblong, 5–6 lin. long, straw-coloured or purplish; glumes subequal, linear-oblong, emarginate, firm, 3-nerved, usually glabrous; valve cylindric, 5–6 lin. long, narrowed from below the middle into and jointed with a beak, beak conical below, filiform and straight or twisted above; callus 1 lin. long; awns deciduous with the beak, divaricate, plumose from the base, the side-bristles very fine, naked; pale ½ lin. long, nerveless; lodicules few-nerved, ⅝ lin. long; anthers 2½–3 lin. long; grain cylindric, truncate, 1½ lin. long; hilum as long as the grain. *Delile, Ill. Fl. Égypte*, 31, *t.* 13, *fig.* 3; *Kunth, Enum.* i. 195; *Suppl.* 150; *Trin. & Rupr. Stip.* 163; *Hack. in Bull. Herb. Boiss.* iv. *App.* iii. 17; *Steud. Syn. Pl. Glum.* i. 143; *Boiss. Fl. Or.* v. 494. *A. plumosa, Desf. Fl. Atlant.* i. 109, *not Linn. A. Schimperi, Hochst. & Steud. ex Trin. & Rupr. l.c.* 164. *A.* (?) *piligenu, Burch. Trav.* i. 288 (*name only*). *A. piligera, Burch. ex Schult. Mant.* ii. 478. *A.* (?) *centrifuga, Burch. Trav.* i. 266. *Arthratherum ciliatum, Nees in Linnæa*, vii. 289; *Jaub. & Spach, Ill. Pl. Or. t.* 334. *A. Schimperi, Nees, Fl. Afr. Austr.* 178. *Schistachne ciliata, Fig. & Not. in Mem. Acc. Torin. sér.* 2, xii. (1852) 252.

COAST REGION: Clanwilliam Div.; Wind Hoek, 400 ft., *Schlechter*, 8338! Cape Div.; near Capetown, *Schlechter*, 37!

CENTRAL REGION: Calvinia Div.; between Lospers Plaats and Springbok Kuil River, 2000–3000 ft., *Zeyher*, 1812! Prince Albert Div.; Jackals Fontein, *Burke*, 22! Gamka River, *Mund & Maire!* Somerset Div.; *Bowker!* Fraserburg Div.; between Patrys Fontein and Great Brak River, *Burchell*, 1521!

WESTERN REGION: Great Namaqualand; Angra Paquena, *Schinz*, 672! 668; *Schenck*, 1/a; between Angra Paquena and Guos, *Schinz*, 669; Aus, *Schinz*, 667; between Ausis and Kuias, *Schenck*, 218; Anib Plains, between Aus and the Orange River, *Schenck*, 326! Little Namaqualand, near the mouth of the Orange River, below 600 ft., *Drège*, 2548! Bitterwater, 2500 ft., *Drège.*

KALAHARI REGION: Prieska Div.; between Modder Fontein and Keikams Poort, *Burchell*, 1612/4!

Hackel describes (in Bull. Herb. Boiss. iv. App. iii. 18) two varieties, viz. var. *tricholæna* with the back of the upper glume rigidly ciliate above the middle, and var. *villosa*, with the sheaths and blades woolly. To the former he refers specimens collected by Schinz at Lüderitzhafen (672, see above), and by Schenck between Nama and Tschirub (10), to the latter, Schenck's 327 from the Anib Plains.

Also in Nubia and throughout the southern part of the Mediterranean region.

Steudel (Syn. Pl. Glum. i. 144) indicates *A. concinna*, Sond., from the Cape of Good Hope, quoting "*Arthratherum uniplume*, Drège, herb. ex parte." I have not seen any specimen of *A. concinna*, Sond., which is synonymous with *A. papposa*, Trin. & Rupr., from the Cape; but there is, in Drège's herbarium at Lübeck, a specimen marked "*Aristida concinna*, Sonder," communicated by Drège in 1827, and numbered 901. This, however, is typical *A. ciliata*, Desf. *A. pungens*, Desf., indicated by Hackel in Bull. Herb. Boiss. iv. App. iii. 19, as collected by Dr. Schenck near Aus in Great Namaqualand, is probably also *A. ciliata*.

16. A. capensis (Thunb. Prod. 19); perennial, compactly cæspitose, glabrous; culms slender, erect, 1–2 ft. long, wiry, smooth, 1-noded; internodes shortly exserted; sheaths firm, very tight, smooth or obscurely striate, sometimes purplish; blades finely setaceous, convolute, acute, up to 1 ft. or more long, firm, flexuous, smooth below, hispidulous above; panicle erect or nodding, contracted or effuse, 6–10 in. long; rhachis smooth; branches solitary, 2- (rarely 3-) partite near the base, the lowest 4–6 in. long, very remotely and sparingly branched, finely filiform, smooth; pedicels clavate-tipped, the lateral ½ as long as the spikelets; spikelets suberect or nodding, brownish or purplish, 6–8 lin. long; glumes subequal, linear-lanceolate, acuminate, minutely truncate, 3-nerved; valve cylindric, smooth, 2½–3 lin. long, greyish; callus ¾–1 lin. long; awn stipitate, tardily disarticulating from the valve; foot 4–6 lin. long, slightly twisted, scantily plumose or glabrous; bristles all plumose except at the tips, the middle one 8–12 lin. long, the side-ones shorter; pale ⅛ lin. long; lodicules scarcely ¾ lin. long, finely nerved; anthers up to 3 lin. long; grain cylindric, 2 lin. long, very finely grooved; hilum as long as the grain. *Thunb. Fl Cap. ed. Schult.* 105; *Kunth, Rév. Gram.* ii. *t.* 171; *Enum.* i. 195; *Suppl.* 151; *Trin. & Rupr. Stip.* 178; *Steud. Syn. Pl. Glum.* i. 145; *Durand & Schinz, Consp. Fl. Afr.* v. 801. *Avena capensis, Linn. f. Suppl.* 112. *Arthratherum capense, Nees in Linnæa*, vii. 288; *Fl. Afr. Austr.* 176. *Chætaria capensis, Beauv. Agrost.* 30.

VAR. β, **Zeyheri** (Trin. & Rupr. Stip. 179); panicle contracted or rather loose, often secund; lowest branches 2–4 in. long; glumes more unequal than in the type; foot of awn 1–1½ lin. long, hairy. *A. Zeyheri, Steud. Nomencl. ed.* ii. i. 132. *Arthratherum Zeyheri, Nees, Fl. Afr. Austr.* 177.

VAR. γ, **macropus** (Trin. & Rupr. Stip. 179); panicle as in the type; spikelets yellowish; foot of awn 4½–5 lin. long, glabrous; bristles densely plumose to the very tips.

VAR. δ, **barbata** (Stapf); culms sheathed all along; sheaths long bearded at their mouths; blades very long, curled, overtopping the panicle; spikelets 8–10 lin. long; foot of awn 1 lin. long, hairy; bristles densely plumose except at the tips.

SOUTH AFRICA: without precise locality, *Thunberg! Bergius!* Var. β, *Harvey*, 324! 298!

COAST REGION: Clanwilliam Div.; Olifants River, *Ecklon.* Malmesbury Div.; Mamre, *Baur*, 1176! Cape Div.; Table Mountain, *Ecklon*, 977! Kloof between Lions Head and Table Mountain, *Burchell*, 272! near Capetown, *Burchell*, 890! Cape Flats, near Claremont, *Schlechter*, 557! Tulbagh Div.; Tulbagh Kloof, *Ecklon & Zeyher!* Worcester Div.; Bains Kloof, 1000 ft., *Schlechter*, 9105! Caledon Div.; Zonder Einde River, 500–1000 ft., *Drège!* between Bot River and Zwart Berg, *Ecklon*, mountain slopes near Genadendal, *Bolus*, 7432! Uitenhage Div.; Uitenhage, *Bowie!* Var. β: Cape Div.; Simons Bay, *MacGillivray*, 391! *Milne*, 255! near Tokay, *Wolley Dod*, 1969! 2190! near Constantia Nek, *Wolley Dod*, 2235! Farmers Peak Valley, *Wolley Dod*, 2361! Constantia, and Doorn Hoogte, *Ecklon.* Tulbagh Div.; Witzenberg, Vogel Vallei, Tulbagh Kloof and Tulbagh Vallei, *Ecklon.* Worcester Div.; mountains near Worcester, *Rehmann*, 2669! Caledon Div.; Houw Hoek, 2000 ft., *Schlechter*, 7366! Var. γ: Paarl Div.; Paarl Mountains, 1000–2000 ft., *Drège!* Var. δ: Uitenhage Div.; on the downs between the Koega and the Zwartkops Rivers, *Ecklon & Zeyher*, 502! *Zeyher*, 4501! Bathurst Div.; Port Alfred, *Hutton*, 11a!

WESTERN REGION: VAR. β: Little Namaqualand; Kamies Bergen, *Ecklon!* near Kaus Mountain, 2000–3000 ft., and in the plains between Kuil and Modder Fontein, 3500 ft., *Drège*. Vanrhynsdorp Div.; Karee Bergen, 1500 ft., *Schlechter*, 8213! near Ebenezer, *Drège!*

Arthratherum capense β, Nees, Fl. Afr. Austr. 176, from Tulbagh and Paarl Mountains belongs probably to var. β, which, however, represents hardly more than individual variations.

17. **A. namaquensis** (Trin. ex Steud. Nomencl. ed. ii. i. 131); suffrutescent; rhizome long creeping, stoloniferous, innovation-buds covered with densely imbricate sheaths, the latter glabrous except on the woolly lower margins, and bearing tiny spine-like blades; culms fascicled, ascending or prostrate, woody below, simple or more often with fascicles of erect branches from the lower or the middle nodes, several feet long, glabrous, smooth; sheaths very tight, firm, pallid, glabrous, and smooth or hairy, longer or slightly shorter than the internodes; blades setaceous or subulate, convolute, 1–8 in. long, when short, then very rigid and pungent, glaucous, glabrous, smooth below, hispidulous above; panicle narrow, linear, more or less contracted, 3–8 in. long; rhachis straight or subflexuous, smooth; branches solitary, erect or suberect, bifid or sparingly branched from near the base, the lowest 1–2 in. long, scabrid and filiform like the branchlets; lateral pedicels very much shorter than the spikelets, the latter yellowish, 6–7 lin. long; glumes rather firm, lanceolate to linear, acuminate, 3-nerved, glabrous, tips minutely truncate or 2-toothed, involute, the lower shorter; valve subcylindric, 3–4½ lin. long, glabrous, smooth, produced into and articulated with a straight or slightly twisted beak, articulated at or below the middle; callus ½ lin. long; bristles plumose to the very tips, the middle 9–12 lin. long; pale broad, ¾ lin. long, nerveless; lodicules ½ lin. long, few-nerved; anthers 2¼ lin. long. *Trin. & Rupr. Stip.* 174; *Steud. Syn. Pl. Glum.* i. 145; *Durand & Schinz, Consp. Fl. Afr.* v. 805; *Hack. in Bull. Herb. Boiss.* iv. *App.* iii. 19. *A.* (?) *fruticans, Burch. Trav.* i. 492. *Arthratherum namaquense, Nees, Fl. Afr. Austr.* 185. "*Aristida capensis* α, *Thunb. MS.*"

SOUTH AFRICA: without precise locality, *Thunberg!*

COAST REGION: Uitenhage Div.; Kromme River, 3000 ft., *Drège*. Queenstown Div.; Klipplaats River, *Drège*.

CENTRAL REGION: Calvinia Div.; between Lospers Plaats and Springbok Kuil River, *Zeyher*, 1814! Worcester Div.; on the Karoo, at "Maggis Fontein" (probably Mutjies Fontein), *Rehmann*, 2910! Prince Albert Div.; Bitterwater, near the Gamka River, 2500 ft., *Drège*; by the Gamka River, *Mund & Maire!* Graaff Reinet Div.; on the banks of dried-up streams in the Sneeuwberg Range, near Riviertje, 3700 ft., *Bolus*, 1981! Victoria West. Div.; Nieuwe Veld, near Brak River, 3000–4000 ft., *Drège!*

WESTERN REGION: Great Namaqualand, Rehoboth, *Fleck*, 21a! Warmbad, *Wandres*, 28! Little Namaqualand, Vanrhynsdorp Div.; Bushmans Karoo, *Drège!* between New Fontein and Plat Klip, 3000–3500 ft., *Drège*, 2040! and without precise locality, *Ecklon & Zeyher!*

KALAHARI REGION: Griqualand West; in Leeuwenkuil Valley at Griqua Town, *Burchell*, 1885!

18. **A. proxima** (Steud. Syn. Pl. Glum. i. 145); suffrutescent, branched from the base; culms ½–1 ft. long, 1–2-noded, glaucous,

woolly below the nodes, otherwise pubescent, internodes shortly exserted; sheaths tight, pubescent, nodes densely tomentose or woolly; blades convolute, filiform, subpungent, 1–3 in. long, rigid, glaucous, glabrous, scaberulous above; panicle narrow, rather loose, 3–4 in. long; branches 2-nate, erect, simple or the lowest branched above the middle, 2 in. long, 2–4-spiculate, woolly pubescent; pedicels 1½–3 lin. long, tips subclavate, woolly; spikelets 4–4½ lin. long, pallid; glumes lanceolate, minutely truncate or obscurely 2-toothed, glabrous, 3-nerved, the lower over 3 lin. long; valve oblong, 1½–2 lin. long, disarticulating just above the middle; callus 1 lin. long; awn deciduous with the upper half of the valve, bristles subequal, plumose all along, 3–4 lin. long; pale broader than long, ⅖ lin. long, truncate, nerveless; lodicules very obtuse, ⅓ lin long; anthers almost 2 lin. long. *Durand & Schinz, Consp. Fl. Afr.* v. 807.

CENTRAL REGION: Aliwal North; between Kraai River and Witte Bergen, *Zeyher!*

19. A. lutescens (Trin. & Rupr. Stip. 173); rhizome creeping, covered like the innovation-buds and the culm bases with strong whitish scale-like sheaths, the lowest scales woolly, the upper glabrous with mucro-like blades; culms fascicled, erect, slender, 1–2 ft. long, smooth, many-noded, enclosed except the uppermost; leaves crowded towards the base, glaucous; sheaths very tight, glabrous, smooth; blades subulate, convolute, pungent, 1–3 in. long, very rigid, spreading, glabrous and smooth below, puberulous above; panicle ovate to pyramidal, erect, ½ ft. by 2–3 in., very lax; rhachis smooth; branches 2–3-nate or solitary and bifid near the base, remotely and very scantily branched, finely filiform, flexuous, smooth, the lowest up to 3 in. long, 5–6-spiculate, the upper 3–2-spiculate; lateral pedicels shorter than the spikelets, these very scattered, often nodding, 6–7 lin. long, light green or yellowish; glumes lanceolate, acuminate, minutely truncate or 2-toothed, glabrous, 3-nerved, the lower 5–5½ lin. long, the upper 6–7 lin.; valve subcylindric, smooth, purplish-grey, 8 lin. long, produced into and articulated with a slightly twisted beak, articulation above the middle; callus 1 lin. long; middle bristle about 9 lin. long, plumose all along; side-bristles somewhat shorter, finer, glabrous below, scantily and adpressedly ciliate above; pale ¾ lin. long; lodicules ½ lin. long; anthers 2–2½ lin. long. *Steud. Syn. Pl. Glum.* i. 145; *Durand & Schinz, Consp. Fl. Afr.* v. 804; *Hack. in Bull. Herb. Boiss.* iv. *App.* iii. 18. *Arthratherum lutescens, Nees, Fl. Afr. Austr.* i. 179.

WESTERN REGION: Little Namaqualand; hills near the mouth of the Orange River, below 600 ft., *Drège*, 727!

Also in Hereroland.

Very near *A. namaquensis* from which it differs mainly in the non-suffrutescent habit, the very loose and open panicle and the almost naked side awns. *A. Marlothii*, Hack. in Engl. Jahrb. xi. 400, is a variety with bearded nodes and panicles occurring at Walfish Bay.

20. A. obtusa (Del. Fl. Égypte, i. 175, t. 13. fig. 2); perennial, compactly cæspitose with numerous innovation shoots; culms from a few inches to 1 ft. long, very slender, 1-noded, geniculate, glabrous, smooth, internodes exserted; lower sheaths short, firm, persistent, glabrous or woolly near the margin, upper tight; ligule a line of short hairs often passing into a long spreading beard at the mouth; blades setaceous, convolute, $\frac{1}{2}$–3 in. long, rigid, curved or flexuous, glaucous, glabrous, smooth beneath, minutely villous above; panicle very narrow, contracted, 1–6 in. long; branches solitary or 2-nate, erect, branched from the base or simple for $\frac{1}{6}$–$\frac{1}{2}$ in., filiform, scaberulous near the tips; lateral pedicels short; spikelets $3\frac{1}{2}$–$4\frac{1}{2}$ lin. long, pallid; glumes equal or the lower slightly longer, the lower lanceolate, obtuse, 3-nerved, tip hyaline, the upper narrower, 1- to sub-3-nerved; valve oblong-cylindric (broadly obovate when expanded), broadly emarginate, 1 lin. long, smooth; callus $\frac{1}{2}$ lin. long; awn from the sinus, stipitate, foot straight or slightly twisted, very fine, $2\frac{1}{2}$–3 lin. long; middle bristle up to 1 in. long, dark, plumose above the middle; side-bristles very fine, divaricate, up to $\frac{1}{2}$ in. long, naked; pale nerveless, broad, $\frac{2}{5}$ lin. long; lodicules nerveless, slightly longer than the pale; anthers 2 lin. long; grain subobovate, terete, almost 1 lin. long, whitish. *Roem. & Schult. Mant.* ii. 212; *Trin. & Rupr. Stip.* 167; *Steud. Syn. Pl. Glum.* i. 144; *Boiss. Flor. Or.* v. 494; *Durand & Schinz, Consp. Fl. Afr.* v. 805; *Hack. in Bull. Herb. Boiss.* iv. *App.* iii. 19. *Stipagrostis* (?) *obtusa, Nees in Linnæa,* vii. 293; *Kunth, Enum. Pl.* i. 198; *Suppl.* 154. *S. capensis, Nees, l.c.* 291; *Fl. Afr. Austr.* 171; *Kunth, l.c.* i. 197. *Arthratherum obtusum, Nees, Gram. Afr. Austr.* 179; *Jaub. & Spach, Ill. Pl. Or.* iv. *t.* 338.

COAST REGION: Clanwilliam Div.; Wind Hoek, 400 ft., *Schlechter*, 8343!
CENTRAL REGION: Calvinia Div.; between Lospers Plaats and Springbok Kuil River, *Zeyher*, 1815! Fraserburg Div.; between Patrys Fontein and Great Brak River, *Burchell*, 1520! between Great Riet River and Stink Fontein, *Burchell*, 1392! Carnarvon Div.; at the northern exit of Karee Bergen Poort, near Carnarvon, *Burchell*, 1556! Beaufort West Div.; Nieuw Veld, *Ecklon*, near Bock Poort, 3500–4500 ft., *Drège!*
WESTERN REGION: Great Namaqualand; Aus, *Schinz*, 659! Little Namaqualand, in sand, between Holgat River and the Orange River, 1000–1500 ft., *Drège*, 2542! Kamies Bergen, *Drège*.
KALAHARI REGION: Griqualand West, Hay Div.; at Griqua Town, *Burchell*, 196/1! Kimberley Diamond fields, *Tuck!* Orange Free State, at Olifants Fontein, *Rehmann*, 3525!

Also in Hereroland, and in the deserts of North Africa and Arabia.

21. **A. subacaulis** (Steud. Nomencl. ed. ii. i. 132); annual, in small compact tufts, about 1 in. high; culms very slender and short, 1-noded, minutely hairy, sheathed all along; sheaths loose, the lowest bladeless, submembranous, the upper strongly striate, scabrid above, margins broad, membranous; ligules passing into small beards at the mouth; blades setaceously convolute, acute, 2–5 lin. long, rigid, strongly striate, asperulous on both sides; panicle contracted, few spiculate, $\frac{1}{2}$ in. long; rhachis like the usually simple fascicled branches hispidulous; spikelets $4\frac{1}{2}$–$5\frac{1}{2}$ lin. long, whitish; glumes linear-oblong or linear-lanceolate, long acuminate, 3-nerved, mem-

branous, scaberulous above, tips hyaline, the lower 4½–5½ lin., the upper 4 lin. long; valve as in *A. obtusa*, 1¼ lin. long; callus ½ lin. long; awn stipitate, foot 1–1½ lin. long; middle bristle 10–12 lin. long, purplish, plumose above the middle, scaberulous below; side-bristles about ½ the length, very fine; pale not quite ½ lin. long; lodicules ½ lin. long; anthers 2 lin. long; grain obliquely oblong, 1 lin. long, whitish. *Trin. & Rupr. Stip.* 171; *Steud. Syn. Pl. Glum.* i. 144; *Durand & Schinz, Consp. Fl. Afr.* v. 809; *Hack. in Bull. Herb. Boiss.* iv. *App.* iii. 19. *Arthratherum subacaule, Nees, Fl. Afr. Austr.* 180.

WESTERN REGION: Great Namaqualand, Lüderitzhafen, *Schinz*, 665, 666; *Schenck*, 32. Little Namaqualand; on the banks of the Orange River, near Verleptpram, 300 ft., *Drège!*

Also in Hereroland.

22. A. uniplumis (Lichtenst. in Roem. & Schult. Syst. ii. 401); perennial, tufted; culms very slender, erect or geniculate, 1½–2 ft. long, simple or branched, about 4-noded, wiry, glabrous, smooth, internodes exserted; sheaths very tight, the upper quite glabrous, the lower long, bearded at the mouths, smooth; blades setaceously convolute, fine, up to ½ ft. long, light green to glaucous, smooth below, minutely scaberulous above, margins scabrid; panicle narrow, linear to oblong, contracted or open, 3–8 in. long; rhachis strict or flexuous; branches geminate or fascicled, simple for ¼–½ in., then sparingly and remotely branched; branchlets capillary, flexuous, smooth, or minutely scaberulous above; pedicels as long as the spikelets or longer, tips clavate; spikelets erect or nodding, 5–6 lin. long, straw-coloured or purplish; glumes linear-lanceolate, acuminate, 3-nerved, thin, the lower slightly shorter and broader; valve cylindric, subemarginate, scarcely 2 lin. long, purplish, smooth; callus almost ½ lin. long; awn disarticulating from the valve, stipitate, foot twisted 2½–3½ lin. long, plumose below the knee, middle bristle plumose all along or naked below, 8–10 lin. long, side-bristles very fine, shorter, glabrous; pale truncate ½ lin. long; lodicules ½ lin. long, few-nerved; anthers 2 lin. long; grain oblong-cylindric, 1 lin. long, whitish, finely grooved. *Kunth, Enum.* i. 195; *Trin. & Rupr. Stip.* 172; *Steud. Syn. Pl. Glum.* i. 144; *Durand & Schinz, Consp. Fl. Afr.* v. 809.

CENTRAL REGION: Prince Albert Div.; by the Gamka River, *Lichtenstein.* Colesberg Div.; rocks of the Table Mountains near Colesberg, 4500 ft., *Drège.*

KALAHARI REGION: Griqualand West; Kimberley, *Rehmann*, 3470! Eitalers Fontein, *Rehmann*, 3346! Orange Free State; between Kimberley and Bloemfontein, *Buchanan*, 291! Bloemfontein, *Rehmann*, 3722! and without precise locality, *Hutton! Buchanan*, 69! Transvaal; Bosch Veld, between Eland River and Klippan, *Rehmann*, 5113! near Lydenburg, *Atherstone!* Transvaal Plains, 4000 ft., *McLea*, 140!

WESTERN REGION: Great Namaqualand; Byzondermeid, *Schinz*, 660; Great Fish River, *Fleck*, 289a.

23. A. dregeana (Trin. & Rupr. Stip. 169); perennial, branching from the base, up to ½ ft. long; culms very slender, erect or geniculate, 1-noded, glabrous, smooth, internodes exserted; sheaths very tight, glabrous, smooth, the lowest broad, pallid, firm, persistent;

blades setaceous, convolute, subacute, $\frac{1}{2}$–2 in. long, rigid, glabrous and smooth below, pubescent above; panicle erect or nodding, rather lax, more or less secund, 2–3 in. long; rhachis smooth; branches 2-nate, 1–2-spiculate, the lowest $\frac{1}{2}$–$\frac{3}{4}$ in. long, erect or suberect, filiform, smooth; spikelets 6 lin. long, yellowish or purple; glumes linear-lanceolate, subacuminate, glabrous, 3-nerved, the lower very slightly shorter, acute, the upper emarginate, mucronate; valve oblong-cylindric, minutely and obtusely bilobed, glabrous, 1$\frac{1}{2}$ lin. long; callus $\frac{3}{4}$ lin. long; awn from the sinus of the valve, tardily disarticulating, stipitate, foot slightly twisted, slender like the middle bristle, purple, smooth, plumose above the middle, to 1$\frac{1}{2}$ in. long, side-bristles much finer, up to 8 lin. long, glabrous, smooth; pale very broad, $\frac{3}{4}$ lin. long; lodicules scarcely $\frac{1}{2}$ lin. long; anthers 1$\frac{1}{2}$ lin. long. *Steud. Syn. Pl. Glum.* i. 144; *Durand & Schinz, Consp. Fl. Afr.* v. 803; *Hack. in Bull. Herb. Boiss.* iv. *App.* iii. 18. *Stipagrostis dregeana, Nees, Fl. Afr. Austr.* 172.

WESTERN REGION: Great Namaqualand; Aus, *Schinz*, 659; sandy plains between Ausis and Kuias, *Schenck*, 219! Little Namaqualand; near the mouth of the Orange River, below 600 ft., *Drège!*

Also in Hereroland.

24. **A. geminifolia** (Trin. & Rupr. Stip. 169); perennial, branched from the base, almost suffrutescent, $\frac{1}{2}$ ft. long or less; culms erect or pendulous, slender, glabrous or hairy below the panicle, rather few-noded, internodes very unequal, the lowest short, sheathed, the following long exserted (2–4 in.), the next very short, usually enclosed, the last slightly longer and shortly exserted, innovations intravaginal with a similar alternation of short and long internodes; leaves crowded near the base, the uppermost usually subopposite; sheaths tight, mouths bearded, margins ciliate or woolly, nodes glabrous or bearded; blades subulate, involute, subpungent, 1–6 lin. long, rigid, more or less spreading, rarely up to 1$\frac{1}{4}$ in. long and subflexuous, smooth and glabrous below, pubescent above; panicle usually nodding, very short, spike-like, often secund, 1–1$\frac{1}{2}$ in. long; branchlets very short, branched from the base, hairy; spikelets crowded, purple or yellowish, 5 lin. long; glumes unequal, linear-lanceolate, acuminate, 3-nerved, hirsute, the lower 4 lin., the upper 5 lin. long, tips bifid, convolute; valve linear-oblong, minutely bilobed, 1$\frac{1}{4}$ lin. long, whitish or purplish, smooth; callus over $\frac{1}{2}$ lin. long; awn from the sinus, tardily disarticulating, foot 1 lin. long, slightly twisted; middle bristle up to 10 lin. long, gradually plumose from a glabrous base; side-bristles very fine, shorter, glabrous; pale emarginate, 2-nerved, $\frac{1}{2}$ lin. long; lodicules $\frac{1}{2}$ lin. long; anthers 2 lin. long. *Steud. Syn. Pl. Glum.* i. 144 (*errore geminiflora*); *Durand & Schinz, Consp. Fl. Afr.* v. 803. *Stipagrostis geminifolia, Nees, Fl. Afr. Austr.* 173.

WESTERN REGION: Little Namaqualand; near the mouth of the Orange River, *Drège;* between Kook Fontein and Holgat River, 1000–2000 ft., *Drège!*

25. **A. brevifolia** (Steud. Nomencl. ed. ii. i. 130); suffrutescent, $\frac{1}{2}$–1$\frac{1}{2}$ ft. long, branched, woody below; culms erect or ascending,

slender, many-noded, minutely puberulous, gland-dotted, lower and middle internodes short, the upper gradually longer, exserted; sheaths tight, gland-dotted, covered with evanescent wool near the mouth, margins and nodes, or more or less glabrate, those of the thick cylindric innovations broad with rudimentary blades; blades of the culm leaves subulate, convolute, rarely flat, subpungent, ½–2 in. long, rigid, spreading, glaucous, asperulous below, hispidulous or pubescent above, densely striate; panicle contracted, linear, somewhat spike-like, 2–4 in. long, erect, sometimes reduced to a raceme; rhachis smooth; branches short, branched from the base, 3–2-spiculate or simple, smooth; spikelets 6½–7 lin. long, light green or straw-coloured; glumes equal, lanceolate, acuminate, minutely truncate, prominently 3-nerved, the lower 5, the upper 6½–7 lin. long; valve linear-oblong, about 2½ lin. long, smooth, articulated below the minutely and obtusely 2-lobed tip; callus 1 lin. long, pungent; awn stipitate, foot very slender, slightly twisted, 2 lin. long; middle bristle glabrous at the base, then plumose all along, up to 14 lin. long; side-bristles very fine, glabrous, 6 lin. long; pale very broad, nerveless, ⅓ lin. long; lodicules ½ lin. long; anthers 3 lin. long. *Trin. & Rupr. Stip.* 170; *Hack. in Bull. Herb. Boiss.* iv. *App.* iii 17. *Arthratherum brevifolium, Nees, Fl. Afr. Austr.* 183.

CENTRAL REGION: Calvinia Div.; near Kamos and Springbock Kuil, *Zeyher*, 1813! Prince Albert Div.; between the Dwyka River and Zwartbulletje, *Drège!*

WESTERN REGION: Great Namaqualand, in the southern part, *Fleck*, 290a. Little Namaqualand; Silver Fontein, 2000–3000 ft., *Drège!* in the Karroo, Lat. South, 29°, *Drège!*

Imperfectly known species.

26. A. hochstetteriana (Beck ex Hack. in Verh. Bot. Ver. Prov. Brandenb. xxx. 144); perennial; blades glaucous, convolute, glabrous, scabrid; panicle reduced to a spike-like raceme, 2–2½ in. long; spikelets subsessile, 7½–8 lin. long; glumes subequal, long acuminate, the lower shortly hirsute; valve 3½ lin. long, bearded (where ?), disarticulating at the middle; awn stipitate, foot 4½ lin. long; middle bristle 3–3½ lin. long, the lower ¼ naked, then plumose except at the tip; side-bristles very fine, glabrous.

WESTERN REGION: Little Namaqualand (?), "Buschmansland," *Wyley*.

Also in Hereroland.

Evidently a marked species, but of doubtful affinity.

XLVI. STIPA, Linn.

Spikelets 1-flowered, narrow, paniculate; rhachilla disarticulating above the glumes, not produced. *Glumes* usually persistent, narrow, 1–3-nerved, muticous or mucronate. *Valve* convolute, cylindric or oblong-cylindric, 5–7- (rarely 3-) nerved, rather rigid, tip gradually tapering or minutely 2-lobed; callus more or less bearded, usually pungent; awn simple, continuous with or jointed on the valve, bent or geniculate, twisted below, plumose or naked above the knee. *Pale*

2-keeled or 2-nerved, almost as long as the valve or much shorter. *Lodicules* usually 3, the posterior smaller or suppressed. *Stamens* 3, rarely fewer. *Ovary* glabrous; styles distinct, short; stigmas plumose. *Grain* slender, cylindric or oblong-cylindric, terete or subterete, sometimes grooved, tightly embraced by the hardened valve and the pale; hilum linear, almost as long as the grain; embryo rather small.

Perennial, rarely annual; leaves often convolute, rarely flat; ligules membranous; panicle from spike-like to effuse.

Numerous species, principally in the drier and warm regions of both hemispheres.

Annual; blades fine; panicle spike-like; awns long, intertwisted ... ... (1) **tortilis.**
Perennial; blades rather broad and flat; panicle contracted or open, large; awns short, slightly twisted at the base only ... ... (2) **dregeana.**

1. **S. tortilis** (Desf. Fl. Atlant. i. 99, t. 31, fig. 1); annual, tufted; culms erect or geniculate, very slender, from a few inches to 1½ ft. long, with a few intravaginal branches from the lower nodes, 1–3-noded, glabrous or reversedly hairy, internodes enclosed or the uppermost exserted; sheaths rather loose, glabrous, smooth, the uppermost often tumid and embracing the base of the panicle; ligules ciliolate, up to ¼ lin. long; blades linear, tapering to a fine point, 2–7 in. by 1 lin., convolute or flat, glabrous below, more or less hairy above or hairy all over; panicle spike-like, very narrow, 1–6 in. long, straight; branches fascicled, very unequal, scaberulous, simple or almost so, up to 4 lin. long; pedicels shorter than the spikelets, these very narrow, 6–9 lin. long, pallid, glistening; glumes very narrow, linear-lanceolate to linear, acuminate, whitish, hyaline, 1-nerved (with 2 fine short side-nerves at the base), the lower slightly shorter; valve cylindric, tightly convolute, callous tipped, 2 lin. long, pubescent, 5–7-nerved; callus 1 lin. long, minutely bearded; awn disarticulating, 2½–3 in. long, geniculate at the middle, strongly twisted below, hispidulous; pale broadly oblong, obtuse, ¾ lin. or more; lodicules 2, oblong-lanceolate, up to ¾ lin. long; anthers 1¼–1½ lin. long, scantily ciliolate; grain subcylindric, 1½ lin. long. *Kunth, Enum.* i. 180; *Trin. & Rupr. Stip.* 64; *Steud. Syn. Pl. Glum.* i. 130; *Boiss. Fl. Or.* v. 500; *Durand & Schinz, Consp. Fl. Afr.* v. 812. *S. humilis, Brot. Fl. Lusit.* i. 86, *non Cav. S. seminuda, Vahl ex Hornem. Hort. Hafn.* i. 76. *S. paleacea, Sibth. & Sm. Fl. Græc. t.* 86, *non Vahl. S. capensis, Thunb. Prodr.* 19, *and Fl. Cap. ed. Schult.* 106; *Nees, Fl. Afr. Austr.* i. 170; *Trin. & Rupr. l.c.* 63; *Steud. Syn. Pl. Glum.* i. 129.

SOUTH AFRICA: without precise locality, *Thunberg!*
COAST REGION: Vanrhynsdorp Div.; banks of the Oliphants River, near Ebenezer, *Drège!* George Div.; Gauritz River, *Ecklon.*

Common throughout the Mediterranean region.

2. **S. dregeana** (Steud. Syn. Pl. Glum. i. 132); perennial, glabrous; culms fascicled from a short præmorse rhizome, erect, 2–4 ft. high,

3-noded, smooth, internodes enclosed or more or less exserted; lowest sheaths much reduced, firm, scale-like, the following very tight, long, the uppermost sometimes tumid, slightly rough, striate; ligules obtuse, erose, up to 2 lin. long; blades linear from a broad or slightly narrowed base, tapering to a very long fine point, up to 1½ ft. by 3–5 lin., flat, rather firm, smooth below, slightly rough above, closely and very finely many-nerved; panicle erect or slightly nodding, oblong, contracted, rarely open and pyramidal, ½–1 ft. long; branches fascicled or 3–2-nate, very unequal, the longest ¼–½ ft. long, undivided to ⅓ of their length or more, filiform, smooth; pedicels scabrid, the lateral shorter than the spikelets, the latter light green, 2½–3½ lin. long; glumes subequal, 3-nerved, subhyaline above, glabrous, the lower lanceolate, acuminate, the upper lanceolate-oblong, acute or subacuminate; valve oblong-cylindric, convolute, obscurely bilobed, 2–2½ lin. long, shortly hairy all over, 5-nerved; callus minute, obtuse, minutely hairy; awn not disarticulating, 5–6 lin. long, scabrid, slightly twisted below, bent 1–2 lin. above the base; pale almost as long as the valve, obtuse, hairy on the back; lodicules 3, oblong, obtuse, the posterior smaller; anthers 1½ lin. long, naked; grain oblong, cylindric, 1½ lin. long. *Durand & Schinz, Consp. Fl. Afr.* v. 811. *Lasiagrostis capensis, Nees, Fl. Afr. Austr.* 167; *Trin. & Rupr. Stip.* 88.

VAR. β, **elongata** (Stapf); panicle flaccid, larger, up to 1½ ft. long, more effuse; lower branches usually 2-nate or solitary, 4–8 in. long, undivided for 2–6 in. above the base; spikelets fewer and more scattered. *S. elongata, Steud. Syn. Pl. Glum.* i. 132; *Durand & Schinz, Consp. Fl. Afr.* v. 811. *Lasiagrostis elongata, Nees, Fl. Afr. Austr.* 168. *L. capensis, var. elongata, Trin. & Rupr. Stip.* 89, 185.

SOUTH AFRICA: without precise locality, var. *elongata*; *Harvey*, 299! *Mund & Maire!*

COAST REGION: Stellenbosch Div.; Hottentots Holland, *Mund & Maire!* Swellendam Div. ? at Kenko River, between Riet Kuil and Hemel en Aarde, *Zeyher*, 4499! Mossel Bay Div.; near the Gauritz River, in woods, *Mund & Maire, Ecklon.* Uitenhage Div.; by the Zwartkops River, *Ecklon & Zeyher*, 770! Alexandria Div.; in the forests of Addo, *Ecklon & Zeyher*, 770! in sunny, grassy places, *Drège.* Albany Div.; Grahamstown, *Drège*, in thornbush near Burnt Kraal, 2200 ft., *MacOwan*, 1276! Var. *elongata:* Cape Div.; Cape Peninsula, Kloof on the western side of the 12 Apostles, *Wolley Dod*, 1201! Devils Peak, by the waterfall, *Wolley Dod*, 2301! Swellendam Div.; in the forest at Grootvaders Bosch, *Burchell*, 7238! Port Elizabeth Div.; Kakakama Mountains, in virgin forest, *Drège!* King Williamstown Div.; Dohne Peak, in forest, 4400 ft., *Galpin*, 2455! Komgha Div.; near Komgha, *Flanagan*, 903!

CENTRAL REGION: Somerset Div.; between the Zuureberg Range and Klein Bruintjes Hoogte, 2000–2500 ft., *Drège!* at Blyde River, *Burchell*, 2983! Graaff Reinet Div.; shady places near Graaff Reinet, 2500 ft., *Bolus*, 707! along the Sunday River, north of Monkey Ford, *Burchell*, 2861! Var. *elongata:* Somerset Div.; in forests at the foot of Bosch Berg, 3000 ft., *MacOwan*, 1520!

EASTERN REGION: Natal; Riet Vlei, in bush, 5000 ft., *Buchanan*, 239!

Var. *elongata* is probably an extreme shade form.

Doubtful species.

3. S. parvula (Nees, Fl. Afr. Austr. 169); annual (?); culms tufted, erect, 3–4 in. long, branched from the base, glabrous; sheaths tight, purplish below, glabrous except the bearded mouth; ligules obsolete;

blades setaceous, canaliculate-complicate, 1 in. long, glabrous; panicle straight, contracted, secund, 1 in. long; lower branches 2-nate, 2–3-spiculate, scabrid; spikelets bright violet, shining, glabrous, almost 2 lin. long; glumes 1-nerved, the lower entire, mucronate, 1 lin. long, the upper 2-toothed, mucronate from the sinus, 2 lin. long; valve glabrous, 1½ lin. long; callus bearded, beard 1 lin. long, hairs elastic, somewhat spreading; awn disarticulating, purple, scabrid, over 2 lin. long, twisted to ⅓ of the way from the base, gently recurved; anthers naked. *Trin. & Rupr. Stip.* 53; *Steud. Syn. Pl. Glum.* i. 129; *Durand & Schinz, Consp. Fl. Afr.* v. 812.

WESTERN REGION: Little Namaqualand; near the mouth of the Orange River, *Drège*.

I suspect this is not a *Stipa*, but a simple-awned *Aristida*, allied to *A. uniseta*, Stapf.

XLVII. ORYZOPSIS, Michx.

Spikelets ovoid or lanceolate, usually awned, loosely panicled; rhachilla disarticulating above the glumes, not continued beyond the floret. *Floret* 1, ⚥, shorter than the glumes. *Glumes* equal or subequal, persistent, usually acute or subacuminate, thin, 3–9-nerved, side-nerves often very short. *Valve* convolute, more or less flattened from the back, ovate to obovate or lanceolate, 5- (very rarely 3-) nerved, rather rigidly membranous to coriaceous, tips obscurely 2-lobed; callus very short and broad, obtuse; awn (if present) from the minute sinus of the tip, jointed on the valve and easily deciduous, bristle-like, naked. *Pale* almost as long as the valve or rather shorter, 2-nerved. *Lodicules* 2, rarely 3, hyaline. *Stamens* 3. *Ovary* glabrous; styles distinct, short, stigmas plumose, laterally exserted. *Grain* tightly enclosed by the hardened valve and pale, oblong, ovoid or ellipsoid or obovoid, terete; embryo small; hilum filiform, shorter than the grain, often obscure.

Perennial, tufted; blades linear, usually long, flat, flaccid; ligules membranous; panicles effuse or more or less contracted, always very lax, often nodding.

Species about 16, in the northern hemisphere.

1. **O. miliacea** (Richter, Pl. Europ. 33); perennial, loosely tufted; rhizome short, thick, innovation buds short, extravaginal, covered with silky scales; culms erect, slender or more or less robust, 1½–3 ft. long, smooth, glabrous, 4–5-noded, internodes more or less exserted; sheaths tight or somewhat lax, finely striate, glabrous excepting the lowest bladeless ones (innovation scales), smooth, lower at length dry and scarious; ligules very short, truncate or the uppermost oblong and up to 1½ lin. long; blades linear, long tapering to a setaceous point, up to 1 ft. long, 1½–4 lin. broad, flat and flaccid, or rolling up when dry, more or less glaucous, glabrous or finely hairy on the upper side, smooth beneath, scaberulous above or all over in the upper part; panicle large, oblong to linear, open or contracted, more or less nodding, ½ to more than 1 ft. long; axis very slender, terete; branches few to very many in distant semiwhorls, finely

filiform to capillary, scaberulous, lower 2–6 in. long, often undivided to ½ their length from the base; branchlets few, subracemose, contracted; pedicels very unequal, the lateral usually very short; spikelets oblong-lanceolate, greenish or tinged with purple, 1½–1¾ lin. long; glumes acuminate, rather broad when expanded, 3- to sub-5-nerved, side-nerves very short; valve obovate to oblong, 1 lin. long, smooth, glabrous, rigidly membranous, whitish, 3-nerved; awn a very fine and caducous flexuous bristle, 1½–2 lin. long; lodicules 2, oblanceolate; anthers ¾ lin. long, tips very minutely penicillate; grain obovoid-oblong, over ¾ lin. long, terete. *Durand & Schinz, Consp. Fl. Afr.* v. 812. *Agrostis miliacea, Linn. Sp. Plant.* 61; *Host, Gram. Austr.* iii. *t.* 45; *Reichb. Ic. Fl. Germ.* i. *t.* 45, *fig.* 1459. *Milium multiflorum, Cav. Descr.* 36. *M. arundinaceum, Sibth. & Smith, Fl. Græc. Prod.* i. 45; *Fl. Græc.* i. *t.* 66. *Piptatherum multiflorum, Beauv. Agrost.* 18; *Kunth, Enum.* i. 177. *P. miliaceum, Coss. Pl. Crit. fasc.* 3, 129. *Urachne parviflora, Trin. Fund. Agrost.* 110; *Stip.* 10, *and in Mém. Acad. Pétersb. sér.* 6, v. 10; *Steud. Syn. Pl. Glum.* i. 121.

KALAHARI REGION: Orange Free State, without precise locality, *Buchanan*, 18!

A native of the Mediterranean region, elsewhere introduced.

XLVIII. PEROTIS, Ait.

Spikelets very small, narrow, sessile or subsessile on the continuous axis of a spike or a lax spike-like raceme, jointed on and falling entire from the axis or the rudimentary pedicels; rhachilla not continued beyond the floret. *Floret* 1, ⚥, much shorter than the glumes. *Glumes* equal, linear or linear-lanceolate, rigidly membranous, 1-nerved, passing into capillary awns. *Valve* lanceolate, acute, delicately hyaline, 1-nerved. *Pale* very minute, hyaline nerveless. *Lodicules* 2, broad, cuneate. *Stamens* 3. *Styles* distinct, short; stigmas plumose, laterally exserted. *Grain* cylindric, slender, exserted from the unchanged floret and enclosed with it in the glumes; embryo about ⅓ the length of the grain; hilum punctiform, basal.

Annual or subperennial; culms leafy; blades usually broad, rigid and ciliate; ligules hyaline; spikes or pseudo-spikes slender, crinite from the long capillary awns.

Species 2 or 3, in the tropics of the Old World and in subtropical Australia.

1. **P. latifolia** (Ait. Hort. Kew. i. 85); culms fascicled, geniculate, suberect, ascending often from a few inches to 1½ in. long, smooth, glabrous, many-noded, lower internodes short, not or slightly exserted, uppermost 1 or 2 by far the longest, long exserted; leaves rather numerous in the lower ⅓–¼ of the culm; sheaths thin, striate, smooth; ligules very delicate, short, ciliolate; blades linear-lanceolate to ovate-lanceolate from a clasping broad base, acute or acuminate, ¾–3 in. by 1½–4 lin., flat or somewhat wavy, glaucous, margins rigidly ciliate or fimbriate, rarely smooth; spike slender, rigid or flexuous, 3–8 in.

long, rather dense; axis smooth, terete; spikelets about 1 lin. long, linear-lanceolate; glumes scaberulous, 3–10 lin. long, very fine, flexuous, often purplish; valve less than $\frac{1}{2}$ lin. long; pale under $\frac{1}{4}$ lin. long; tip finely ciliolate; anthers $\frac{1}{4}$ lin. long; grain almost 1 lin. long. *Beauv. Agrost.* 172, *t.* iv. *fig.* 9 (*P. laxifolia* (*by error*), 6); *Kunth, Rév. Gram.* i. 357, *t.* 92; *Enum.* i. 470; *Suppl.* 380; *Hook. Niger Fl.* 569; *Steud. Syn. Pl. Glum.* i. 186; *Anderss. in Peters, Reise Mossamb.* 561; *Oliv. in Trans. Linn. Soc.* xxix. 176; *Durand & Schinz, Consp. Fl. Afr.* v. 734; *Hook. f. Fl. Brit. Ind.* vii. 98; *not Thunb. P. scabra, Willd. ex Trin. Gram. Unifl. et Sesquifl.* 172. *Agrostid.* ii. 21, *and in Mém. Acad. Pétersb. sér.* 6, vi. 267; *Steud. l.c.* 186; *Durand & Schinz, l.c.* 734. *P. hordeiformis, Nees ex Steud. Nomencl. ed.* ii. ii. 306; *Fl. Afr. Austr.* 139; *Trin. Agrostid.* ii. 21, *and in Mém. Acad. Pétersb. sér.* 6, vi. 267; *Steud. Syn. Pl. Glum.* i. 186. *Saccharum spicatum, Linn. Sp. Plant.* 54; *not of Thunb. Prodr.* 20. *Agrostis spicæformis, Linn. fil. Suppl.* 108.

KALAHARI REGION: Transvaal; Bosch Veld, at Klippan, *Rehmann*, 5370!

EASTERN REGION: Natal, near Durban, below 500 ft., *Drège!* valleys between Tugela and Umpumulo, 1000 ft., *Buchanan*, 168! and without precise locality, *Gerrard*, 687! Delagoa Bay, *Forbes!*

Throughout tropical Africa and Asia.

Saccharum spicatum of Thunberg's Prodromus, 20, from Table Mountain, referred by him to *Perotis latifolia* in Prodromus, 192 (and in Fl. Cap. ed. i. 416; ed. Schult. 111), is *Imperata arundinacea*, Cyr., according to Nees; but there is no specimen of it in Thunberg's herbarium.

XLIX. TRAGUS, Haller.

Spikelets sessile, in deciduous clusters of 2–4 on the filiform continuous axis of a cylindric, spike-like panicle; rhachilla tough, not continued beyond the floret. *Floret* 1, ⚥, somewhat shorter than the upper glume. *Glumes* very dissimilar, lower facing the rhachis, minute, hyaline, or suppressed, upper 5-ribbed or 5-nerved, membranous between the hispid or spine-hooked ribs or nerves, exceeding the valve. *Valve* lanceolate or lanceolate-oblong, membranous, 3-nerved. *Pale* as long as the valve, 2-nerved. *Lodicules* 2, broad, cuneate, fleshy. *Stamens* 3. *Styles* distinct, very slender; stigmas narrow, plumose, terminally exserted. *Grain* enclosed by the valve and pale, oblong to ellipsoid, slightly compressed from the back; embryo about $\frac{1}{3}$ the length of the grain; hilum punctiform, basal.

Annual or perennial; culms erect, ascending or decumbent; blades linear, rather rigid with cartilaginous spinulously ciliate margins; ligules reduced to a delicate ciliate rim; panicles cylindric, slender; all the spikelets of a cluster fertile, or often 1 more or less reduced.

Species 2; one in South Africa, the other throughout the warm parts of both hemispheres.

Culms more or less erect, 2–3-noded, with the upper internodes long exserted; upper glume subacuminate, hispid or spinulously ciliate along the nerves; anthers 1 lin. long ... ... (1) **major.**

Culms ascending or decumbent, 3–5-noded, with the uppermost internodes usually enclosed; upper glume acute, with rows of hooked spines on the stout ribs, completely enveloping the floret; anthers $\frac{1}{6}$–$\frac{1}{5}$ lin. long ... ... (2) racemosus.

1. **T. major** (Stapf); perennial; culms erect or suberect from a short slender rhizome or from short rooting runners, sometimes densely tufted with short innovation-shoots, slender, $\frac{1}{2}$–$1\frac{1}{2}$ ft. high, glabrous, smooth, 2–3-noded, internodes exserted, uppermost by far the longest (3–9 in. long) and at length long exserted; leaves crowded at the base; lower sheaths short, broad, very pallid, persistent, upper tight or the uppermost subtumid; blades linear from an almost clasping base, tapering to a very acute point, 1–3 in. by 1–2 lin., flat, rigid, glaucous, closely striate; panicle 1–3 in. long, slender; axis pubescent or glabrous below; branches close or lowest more or less distant, very short, 2–4-spiculate, continued beyond the uppermost spikelet; spikelets about 2 lin. long; lower glumes minute, delicately hyaline, oblong-lanceolate, acute or notched, up to $\frac{3}{4}$ lin. long, nerveless; upper glume lanceolate, subacuminate, prominently 5-nerved or 5-ribbed, thin between the hispid or spinulously ciliate nerves; valve lanceolate, subacute, $1\frac{1}{2}$ lin. long, 3-nerved, rigidly membranous to subhyaline, smooth, glabrous; pale equal to the valve, obtuse or 3-toothed, 2-nerved; lodicules broad, cuneate, fleshy, $\frac{1}{6}$ lin. long; anthers linear, 1 lin. long or almost so. *T. racemosus, var. major, Hack. in Engl. Jahrb.* xi. 397; *Durand & Schinz, Consp. Fl. Afr.* v. 733. *T. occidentalis, Nees, Fl. Afr. Austr.* 72 *(in part), not Agrost. Bras.* 286. *T. berteroanus, Durand & Schinz, l.c. (in part), not T. berteronianus, Schult. Mant.* ii. 205.

COAST REGION: Cape Div.; above Sea Point, *Wolley Dod*, 1088! Uitenhage Div.; Steenboks Flats, *Ecklon & Zeyher!* Bathurst Div.; Port Alfred, *Hutton*, 9*a!* Queenstown Div.; Shiloh, 3500 ft., *Baur*, 944!

CENTRAL REGION: Somerset Div.; by the Fish River, *Baur!* and without precise locality, *Bowker*, 154! Albert Div.; without precise locality, *Cooper*, 777! Aliwal North Div.; at the confluence of Stormberg Spruit and the Orange River, 4200 ft., *Drège!*

KALAHARI REGION: Griqualand West; Du'oits Pan, *Tuck!* between Griqua Town and Witte Water, *Burchell*, 1977! foot of the Asbestos Mountains, between Kloof Village and Witte Water, *Burchell*, 2094! near Griqua Town, *Burchell*, 1877! Bechuanaland; sandy plains near Kuruman, 4000 ft., *Marloth*, 1514!

T. occidentalis, Nees, Agrost. Bras. 286, is only a small state of *T. racemosus*, All., and has been described from Brazilian specimens. It occurs, however, also in the Old World, and one at least of Ecklon's specimens, enumerated by Nees in Fl. Afr. Austr. l.c., under the name of *T. occidentalis*, belongs to it. Of Drège's specimens, quoted in the same place, I have seen only that from the confluence of the Stormberg Spruit and Orange River, which is *T. major*. The others are from Weltevrede and Rhinosterkop (Beaufort West Div.), and from the Gekau River (Transkei Div.).

2. **T. racemosus** (All. Fl. Pedem. ii. 241); annual or subperennial (?); culms fascicled, simple or branched with the branches often fascicled and densely leafy, geniculate, ascending often from a decumbent base or wholly decumbent, slender, from a few inches to 1 ft. long, glabrous or pubescent near the panicle, smooth, 3–5-noded,

intermediate internodes exserted, uppermost 1 or 2 usually enclosed, and from less than 1 to 4 in. long; lowest sheaths short, broad, pallid, the following more or less herbaceous, rather loose, uppermost tumid, usually embracing the base of the panicle; blades linear to lanceolate, acute, $\frac{1}{3}$–2 in. by 1–2 lin., flat or wavy, rigid, very glaucous, closely striate; panicle 1–5 in. long, slender; axis straight or slightly wavy, pubescent; branches very close or the lowest distant, very short, 2–3-spiculate, sometimes minutely continued beyond the uppermost spikelets; spikelets facing each other when paired, $1\frac{1}{2}$–2 lin. long, one of a cluster often reduced; lower glume very minute, up to $\frac{1}{4}$ lin. long, hyaline, ciliolate or quite suppressed; upper glume slightly curved, involute, completely enveloping the floret, strongly 5-ribbed, thin between the ribs, these with rows of stout hooked spines; valve lanceolate-oblong, apiculate or mucronulate, 1–$1\frac{1}{2}$ lin. long, thinly membranous, very minutely pubescent, faintly 3-nerved; pale subacute, obscurely 2-nerved; anthers ellipsoid, $\frac{1}{6}$–$\frac{1}{8}$ lin. long; grain oblong to obovoid-ellipsoid, subterete, $\frac{1}{2}$ lin. long. *Beauv. Agrost.* 23, *t.* vi. *fig.* 13; *Nees, Agrost. Bras.* 287; *Fl. Afr. Austr.* 73; *Doell in Mart. Fl. Bras.* ii. ii. 122; *Hack. in Engl. Jahrb.* xi. 397; *Durand & Schinz, Consp. Fl. Afr.* v. 733; *Hook. f. Fl. Brit. Ind.* vii. 97. *T. berteronianus, Schult. Mant.* ii. 205. *T. berteroanus, Durand & Schinz, l.c.* 733. *T. occidentalis, Nees, ll.cc.* 286, *and* 72 (*for the greatest part*) *respectively*. *T. brevicaulis, Boiss. Diagn. Pl. Or. Sér.* 1, xiii. 44. *Cenchrus racemosus, Linn. Spec. Pl.* 1049; *Schreb. Beschreib. Graes.* 45, *t.* iv. *Lappago racemosa, Honck. Syn. Pl. Germ.* i. 440; *Host, Gram. Austr.* i. *t.* 36; *Sibth. Fl. Graec.* ii. *t.* 101; *Kunth, Rév. Gram. t.* 120; *Enum.* i. 170; *Suppl.* 124; *Reichb. Ic. Fl. Germ.* i. *t.* 30; *Steud. Syn. Pl. Glum.* i. 112. *L. aliena, Spreng. Neue Entdeck.* iii. 15; *Steud. l.c.* 112. *L. phleoides and L. decipiens, Fig. & De Not. in Act. Acad. Torin.* 1854, 360 *and* 387, *t.* 38, *fig.* 1–12, *and t.* 37, *fig.* 1–12.

COAST REGION: Uitenhage Div.; without precise locality, *Zeyher!* Albany Div.; Bothas Hill, *MacOwan*, 571! Queenstown Div.; plains near Queenstown, 3500 ft., *Galpin*, 2350! Wodehouse Div.; Indwe, 3000 ft., *Baur*, 979!

CENTRAL REGION: Graaff Reinet Div.; stony places near Graaff Reinet, 2500 ft., *Bolus*, 554! Albert Div.; without precise locality, *Cooper*, 640!

KALAHARI REGION: Griqualand West, Hay Div.; Griqua Town, *Burchell*, 1938! between Griqua Town and Witte Water, *Burchell*, 1973/2! sandy places near Kimberley, 4000 ft., *Marloth*, 746! 858! Orange Free State, near Boshoff, *Barber!*

EASTERN REGION: Natal; banks of Tugela River near Colenso, 3000 ft., *Wood*, 4417! banks of lower Tugela River, 600 ft., *Buchanan*, 175! and without precise locality, *Gerrard*, 673! Delagoa Bay, *Forbes!*

Throughout most warm regions.

The figure in Beauvois, Agrost. l.c., represents the anthers as linear and as long as in *T. major*; but this is evidently an error. I never found them otherwise than as described above, and so they are figured by all authors cited, except Beauvois.

L. SPOROBOLUS, R. Br.

Spikelets usually very small, variously panicled, continuous on the pedicels; rhachilla more or less readily disarticulating above the glumes, not continued, or very rarely produced into a bristle. *Floret* 1, ⚥. *Glumes* 2, delicately membranous, lower usually smaller, nerveless, upper 1-nerved, falling away one after the other. *Valve* more or less resembling the upper glume, 1-nerved or more or less distinctly 3-nerved. *Pale* usually almost as long as the valve, 2-nerved, folded between the nerves, often split by the maturing grain. *Lodicules* 2, small, broadly cuneate, glabrous, thin. *Stamens* 3, rarely 2. *Ovary* glabrous; styles short, distinct, terminal; stigmas plumose or subaspergilliform. *Grain* free, falling out or retained and dehiscing; pericarp thin, usually swelling in water, rigid, dehiscing, or the inner layers mucilaginous when wetted, and adherent, or the whole pericarp adnate and indistinct; hilum small, punctiform, basal; embryo rather large.

Annuals or perennials of various habit; ligules reduced to a ciliate or ciliolate rim; panicles contracted to spike-like, or more or less open, sometimes extremely lax; spikelets mostly $\frac{1}{2}$–1 lin. long.

Species 60–70, chiefly in the tropical and subtropical regions of both hemispheres.

Section 1. EU-SPOROBOLUS (Stapf). Rhachilla long, not produced.

Culms from a few inches to 1 ft. long, very slender, few-noded; uppermost node usually much below the middle; all the blades, or at least the culm-blades, small; panicle ovate, lax, often divaricate (except no. 1. *S. albicans*); spikelets less than 1 lin. long, rarely 1 lin. (no. 1. *S. albicans*):
  Valve 3-nerved (often very faintly, or 1-nerved in no. 4. *S. festivus*), exceeding the glumes:
    Seed mucilaginous in water:
      Panicle contracted, somewhat spike-like; blades subpungent ... ... (1) **albicans.**
      Panicle effuse; blades obtuse or truncate (2) **tenellus.**
    Seed not mucilaginous; pericarp swelling in water:
      Rhizome much branched; branches distinct; blades subulate; panicle dichotomously divaricate, very lax ... ... (3) **acinifolius.**
      Compactly cæspitose; culms with a coat of fibres at the base; blades setaceous; panicle denser than in the preceding ... ... (4) **festivus var. stuppeus.**
  Valve 1-nerved, about as long as the upper glume:
    Culms and barren shoots in small fascicles; blades denticulate-fimbriate, $\frac{1}{3}$–1 in. long (5) **discosporus.**
    Culms and barren shoots from an oblique or creeping rhizome; blade margins smooth:
      Rhizome creeping; barren shoots very short, cylindric, distinct ... ... (6) **Ludwigii.**
      Rhizome oblique; barren shoots much longer in compact tufts ... ... (7) **ioclados.**
Culms $\frac{2}{3}$–2 ft. long, 2–1-noded, slender; uppermost node usually much below the middle; blades 3–10 in. long,

usually setaceously convolute; panicle ovate or spike-like, rather dense; spikelets 1½–2¼ lin. long ... (8) **centrifugus.**

Culms 2–4 ft. long, rather robust, 2–4-noded, much more leafy than in the preceding species; blades ½–1 ft. long; panicle narrow, oblong, ½–1½ ft. long, with the lower branches obliquely spreading and 1–3 in. long, or spike-like:

Upper glume equalling or slightly exceeding the valve:

Panicle at length rather lax; blades 3–4 lin. broad, mostly flat ... ... (9) **Rehmanni.**

Spikelets crowded on the branchlets, often secund; blades usually much narrower and convolute ... ... (10) **fimbriatus.**

Upper glume slightly exceeding ½ of the valve; panicle narrowly oblong or spike-like ... ... (11) **indicus.**

Culms very many-noded, from a creeping stoloniferous rhizome; leaves often apparently subopposite; blades mostly short, subpungent; panicle more or less compact, spike-like ... ... (12) **pungens.**

Section 2. CHÆTORHACHIA (Stapf). Rhachilla produced into a bristle half as long to almost as long as the floret.

Only species ... ... (13) **subtilis.**

1. **S. albicans** (Nees, Fl. Afr. Austr. 154); perennial, tufted; culms 1–6 in. long, erect, glabrous, smooth, 1–2-noded, sheathed almost all along; basal leaves densely imbricate; sheaths glabrous, smooth, the lower short, broad, the upper long, very tight; ligule a ciliolate rim; blades lanceolate to linear, usually involute and subpungent, the lowest 2–3 lin. by 1 lin. (when expanded), the upper up to 1½ in. broad, rigid, glabrous, smooth below, densely papillose and almost white above; panicle spike-like, 1½–2 in. long, erect; branches solitary, adpressed to the rhachis, up to 9 lin. long, filiform, laxly branched, smooth; pedicels capillary, as long as the spikelets or longer; spikelets pallid, rather more than 1 lin. long; glumes unequal, hyaline, the lower oblong, obtuse, sometimes irregularly toothed, nerveless, equalling ½ of the spikelet or almost so, the upper lanceolate-oblong, obtuse to subacute, 1-nerved, equalling ⅘–⅚ of the spikelet; valve oblong, subobtuse, 3-nerved; pale slightly shorter; stamens 3; anthers ½ lin. long; grain globose-ellipsoid, very slightly compressed, rounded at the back, ½ lin. long, pericarp delicate; testa mucilaginous; albumen opaque. *Durand & Schinz, Consp. Fl. Afr.* v. 818. *Vilfa albicans, Trin. Agrostid.* 79; *Steud. Syn. Pl. Glum.* i. 161.

COAST REGION: Queenstown Div.; between Table Mountain and Wildschuts Berg, 4000 ft., *Drège!*

2. **S. tenellus** (Kunth, Enum. i. 215); perennial, densely cæspitose; rhizome much branched; branches short, densely covered with the imbricate persistent sheath-bases and sheaths; culms erect, very slender, 2–6 in. long, glabrous, smooth, 1-noded; sheaths of the barren shoots and basal leaves short, broad, closely imbricate, firm, glabrous, smooth, of the culms narrow, tight, scarcely striate; ligule

a minutely ciliolate rim; blades ovate or lanceolate to linear, obtuse to truncate, the lower 2–6 lin. by ½–¾ lin., flat or folded, very rigid, spreading, glaucous, glabrous, smooth or scaberulous above, margins cartilaginous, smooth; panicle erect, ovate, 1–2 in. by 1–1½ in. very lax; branches solitary, obliquely erect, capillary, smooth, branched from the middle; branchlets divaricate, usually simple, 2–4-spiculate; pedicels very fine, longer than the spikelets, the latter subsecund, purplish, ⅞ lin. long; glumes obtuse, ovate-oblong, the lower equalling rather more than ⅓ of the spikelet, nerveless, hyaline, the upper slightly exceeding ½ of the spikelet, 1-nerved; valve broadly oblong, subobtuse in profile, 3-nerved; pale equalling the valve; stamens 3; anthers ½ lin. long; grain orbicular-elliptic, ⅝ lin. long, half as thick as broad, rounded at the back; testa mucilaginous. *S. brevifolius, Nees, Fl. Afr. Austr.* 160; *Hack. in Engl. Jahrb.* xi. 401; *Durand & Schinz, Consp. Fl. Afr.* v. 819. *Ehrharta tenella, Spreng. Tent. Suppl. Syst. Veg.* 11. *Vilfa brevifolia, Nees in Linnæa*, vii. 294; *Trin. Agrostid.* 66; *Steud. Syn. Gram.* 154.

CENTRAL REGION: Beaufort West Div.? Beaufort, *Zeyher!* Graaff Reinet Div.; by the Sunday River, 1500–2000 ft., *Drège;* and without precise locality, *Ecklon.* Colesberg Div.; Carolus Poort, *Burchell*, 2756!

KALAHARI REGION: Griqualand West; at Kimberley, in sandy depressions, 3900 ft., *Marloth*, 884! Orange Free State; between Kimberley and Bloemfontein, *Buchanan*, 286! Bechuanaland; stony plains near Kuruman, 4000 ft., *Marloth*, 1517!

The plant collected by Marloth at Kimberley is Hackel's variety *major*. The only difference indicated by the author is in the longer leaves.

3. **S. acinifolius** (Stapf); perennial, cæspitose; rhizome slender, branched, with numerous short barren shoots; culms subascending, slender, ¾–1 ft. long, glabrous, smooth, 2–3-noded, uppermost internode much longer than ½ of the culm; leaves of the barren shoots and those at the base of the culms densely crowded, distichous; lower sheaths firm, persistent, prominently striate, glabrous, smooth, culm sheaths rather tight, hardly constricted at the inconspicuous nodes; ligule a ciliate rim; blades subulate, involute, or more or less open near the base, acute, ⅓–1 in. long, very rigid, often pungent, glabrous, densely papillose and almost white above; panicle erect, ovate to oblong-ovate, 3–5 in. by 2–3 in., very lax, divaricate; branches solitary, filiform to capillary, dichotomously branched, smooth; pedicels 3–5 lin. long; spikelets oblong, light purplish, ⅝–⅞ lin. long; glumes subequal, hyaline, broadly oblong or elliptic, obtuse, nerveless, the lower equalling ½, the upper ¾ of the spikelet; valve oblong, obtuse, delicate, hyaline above, 3-nerved, all the nerves evanescent above the middle; pale slightly shorter; stamens 3, anthers ½ lin. long; grain shortly pyriform, ⅜ lin. long, light brown, compressed, back rounded, pericarp thin; testa not mucilaginous; albumen opaque.

KALAHARI REGION; Griqualand West, Hay Div.; at Griqua Town, *Burchell*, 1846!

4. S. festivus (Hochst. ex A. Rich. Tent. Fl. Abyss. ii. 398) var. **stuppeus** (Stapf); perennial, compactly tufted; culms erect or geniculate-ascending, slender, ½–1 ft. long, glabrous, smooth, 2–3-noded; upper sheaths glabrous except at the ciliate margins, smooth, the lower short, firm, at length breaking up into numerous persistent fibres, about ½ in. long, the inner covered with pallid tow-like hairs; ligules a minutely ciliolate or almost woolly rim; blades narrowly linear, tapering to an acute point, usually setaceously convolute, 1–2 in. long, rarely more, by 1 lin. (when expanded), glabrous, smooth; panicle oblong to ovate, erect, 2–4 in. by 1–1½ in., lax; rhachis straight; branches solitary, or irregularly fascicled, at length spreading, filiform to capillary, repeatedly branched from near the base; secondary branchlets flexuous, capillary; pedicels extremely fine, smooth, 2–3 times the length of the spikelets, rarely longer; spikelets oblong, rather obtuse, purplish, ¾ lin. long or rather less; glumes hyaline, acute or acuminate, minutely denticulate, the lower oblong, nerveless, almost ½ the length of the spikelet, the upper ovate, nerveless or faintly 1-nerved, about ½ the length of the spikelet; valve oblong, obtuse or subacute in profile, 1- to sub-3-nerved; pale slightly shorter; stamens 3; anthers ⅜ lin. long; grain globose-ellipsoid, ¼ lin. long, pericarp swelling and bursting in water; seed free, compressed, obtusely quadrangular; albumen glassy.

KALAHARI REGION: Transvaal; Nylstroom River district at Klip Fontein, Springbok Vlatke, *Nelson*, 274!
EASTERN REGION: Natal; Umsinga and at the base of Biggars Berg, *Buchanan*, 96!

Also in Nyasaland and Jur.

The type, which is known from Abyssinia, Eritrea and Somaliland, only differs in the absence of the peculiar fibrous coating of the culm-bases, the very lax, divaricate panicles and slightly larger spikelets with comparatively shorter glumes. Another variety with similar fibrous tufts to those of var. *stuppeus*, but without the tow-like tomentum, with culms usually taller, larger panicles and smaller spikelets (slightly exceeding ½ lin. long), occurs in most parts of tropical Africa, from the Niger and Nubia to the Zambesi.

5. S. discosporus (Nees, Fl. Afr. Austr. 158); perennial, tufted, with short barren shoots from the base of the culms; the latter 4–8 in. long, slender, glabrous, smooth, 1–3-noded; sheaths tight or the upper slightly tumid, glabrous, particularly those of the barren shoots deeply grooved; ligule a line of short hairs; blades broad or narrowly lanceolate from a broad base, finely acuminate, the lower ⅓–1 in. by 1–2 lin., the upper much reduced, rigid, glabrous, smooth; margins cartilaginous, spinulously fimbriate; nerves numerous, very close; panicle erect, oblong to ovate, lax, 1–2 in. by ½–1¼ in.; branches in whorls or semiwhorls of 3–9, rarely solitary or in pairs, spreading at right angles, capillary, smooth, usually simple, 6–3-spiculate from the middle; lateral pedicels very short; spikelets secund, drooping, ½ lin. long, dark purplish or olive-green; glumes very dissimilar, the lower equalling ½ the length of the spikelet, ovate, acute, hyaline, nerveless, the upper as long as the valve and like it, but firmer, oblong-ovate, acute, 1-nerved; pale as long as the valve; stamens 3; anthers about ⅛ lin. long, elliptic; grain lenticular, ½ lin. long, very flat with subacute edges. *Durand & Schinz, Consp. Fl. Afr.* v. 820. *S. blepharíphyllus, A. Rich. Tent. Fl. Abyss.* ii. 398; *Durand & Schinz, l.c.* 818. *Triachyrium discosporum, Steud. Syn. Pl. Glum.* i. 176. *T. adoënse, Hochst. ex A. Rich. l.c.* 398.

CENTRAL REGION: Colesberg Div.; between Plettenbergs Beacon and "Flat Station," *Burchell*, 2750! Colesberg, *Shaw*, 7! Aliwal North Div.; on a rocky plateau at Kraai River, 4500 ft., *Drège!*

KALAHARI REGION: Orange Free State; Thaba Unchu, *Burke!*

Also in Abyssinia.

Similar to *S. coromandelianus*, Kunth, but distinguished by the lenticular grain, &c.

6. S. Ludwigii (Hochst. in Flora, 1846, 118); perennial; rhizome creeping, covered with the persistent imbricate sheath-bases; barren shoots short, stout; culms erect, very slender, ½–1 ft. long, 1–3-noded, glabrous, smooth, uppermost internode long exserted; sheaths of the barren shoots and the culm-bases closely imbricate, short, broad, firm, glabrous, smooth, of the culm-leaves tight, narrow, obscurely striate; ligule a line of short hairs; blades lanceolate, subobtuse, the lower 3–6 lin. by ¾–1 lin., the uppermost rudimentary, flat, rigid, often curled, glabrous, smooth, or minutely puberulous below; margins cartilaginous, smooth; panicle erect, ovate, lax, 1–3 in. by ¾–1½ in.; branches 5–2-nate or solitary, oblique-erect, subcapillary, smooth, branched from the middle, or simple; branchlets 2–4-spiculate; lateral pedicels very short; spikelets close, secund, dark olive-green, ¾ lin. long; glumes very unequal, obtuse, the lower ovate, ¼ lin. long, hyaline, nerveless, the upper oblong, 1-nerved; valve like the upper glume, but slightly shorter; pale slightly shorter than the valve; stamens 3; anthers ½ lin. long. *Durand & Schinz, Consp. Fl. Afr.* v. 821. *Vilfa Ludwigii, Steud. Syn. Pl. Glum.* i. 155.

CENTRAL REGION: Graaff Reinet Div.; on stony heights by the Sunday River, 1500–2000 ft., *Drège!*

KALAHARI REGION: South-western Transvaal?, Batlapin Territory, *Holub!*

Drège's specimens cited above have been issued as "*S. brevifolius*, N. ab. E."

7. S. ioclados (Nees, Fl. Afr. Austr. 161); perennial; rhizome oblique, densely beset with barren shoots and the basal portions of dead shoots; culms geniculate-ascending, 1–1½ ft. long, glabrous, smooth, 2-noded; lower sheaths broad, firm, pallid, glabrous, smooth; ligule a ciliolate rim; blades linear, tapering to a long fine point, 3–6 in. by 1½ lin. (when expanded), flat or setaceously convolute, subglaucous, glabrous, smooth below, scaberulous above; panicle erect, ovate, 4–6 in. by 2–3 in., very lax; lower branches in whorls of 7–4, the upper whorled or scattered, ultimately spreading, remotely branched, up to 2 in. long, filiform to capillary, straight or flexuous, smooth; lateral pedicels extremely short; spikelets dark olive-grey, oblong-lanceolate, ⅝–⅞ lin. long; glumes very unequal, the lower hyaline, oblong, obtuse or subobtuse, equalling $\frac{1}{3}$ of the spikelet or less, the upper equalling the valve or very slightly longer, oblong, acute or subacute in profile, 1-nerved; valve very like the upper glume; pale slightly shorter; stamens 3; anthers $\frac{1}{2}$ lin. long; grain obovoid, truncate, obtusely quadrangular, slightly compressed, $\frac{1}{2}$ lin. long; pericarp thin. *Durand & Schinz, Consp. Fl. Afr.* v. 821. *Vilfa ioclados, Nees ex Trin. Agrostid.* 43; *Steud. Syn. Pl. Glum.* i. 156.

COAST REGION: Uitenhage Div.; Springbok Flats, by the Zwartkops River, *Ecklon!*

CENTRAL REGION: Prince Albert Div.; Gamka Poort, *Mund & Maire!* Calvinia Div.; Uien Vallei, 2000 ft., *Drège.* Richmond Div.; Winter Veld, between Nieuwjaars Fontein and Ezels Fontein, 3000–4000 ft., *Drège!* Colesberg Div.; Wonderhuivel, 4500 ft., *Drège.*

The Indian "*S. ioclados*" (Hook. f. Fl. Brit. Ind. vii. 249) has the upper glume constantly shorter than the valve.

8. **S. centrifugus** (Nees, Fl. Afr. Austr. 158); perennial, compactly tufted; culms erect, rather slender, $\frac{3}{4}$–2 ft. long, glabrous, smooth, usually 2, rarely 1-noded; lower sheaths very firm, persistent, $\frac{1}{2}$–$1\frac{1}{2}$ in. long, glabrous except the usually long-ciliate margins, sometimes hairy all over, smooth, finely striate, upper tight, the uppermost up to $\frac{3}{4}$ ft. long; ligule a very minutely ciliolate rim; blades linear, usually very narrow, tapering to an acute point, 3–10 in. by $\frac{1}{2}$–2 lin., involute, often setaceous, particularly those of the barren shoots, rarely flat, firm, more or less glaucous, glabrous, except the often serrulate-fimbriate lower margins, rarely scantily hairy, smooth below, subscaberulous above, margins rough or tubercled; panicle erect, ovate or ovate-oblong, 1–4 in. by $\frac{3}{4}$–2 in., usually rather dense; lower branches in whorls of 8–5, obliquely erect or spreading, filiform, smooth, branched from the middle or above it; branchlets contracted; lateral pedicels very short; spikelets rather crowded towards the tips of the branches, dark olive-grey, $1\frac{1}{2}$–$2\frac{1}{4}$ lin. long; glumes unequal, the lower lanceolate-acuminate or acute, $\frac{2}{3}$–$\frac{3}{4}$ the length of the spikelet, rarely longer, 1-nerved or nerveless, the upper broad-oblong-lanceolate, acutely acuminate, somewhat longer than the valve, 1-nerved, rarely with 2 obscure side-nerves; valve very similar, sometimes with 2–4 short obscure side-nerves; pale equalling the valve; stamens 3; anthers $\frac{3}{4}$ lin. long. *Durand & Schinz, Consp. Fl. Afr.* v. 819. *Vilfa centrifuga, Trin. Agrostid.* 35; *Steud. Syn. Pl. Glum.* i. 154.

VAR. β, **angustus** (Nees, l.c. 159); leaves setaceous; panicle linear or spike-like; branches very short, branched from the base or nearly so.

COAST REGION: Cathcart Div.; between Windvogel Mountain and Zwart Kei River, *Drège!*

CENTRAL REGION: Aliwal North Div.; on the Witte Bergen, 4000–6000 ft., *Drège!*

KALAHARI REGION: Orange Free State; on the Drakens Bergen, near Harrismith, *Buchanan*, 117!

EASTERN REGION: Tembuland; Bazeia, *Baur*, 555! Griqualand East, grassy slopes near Kokstad, 5100 ft., *Tyson*, 1473! Natal; near Durban, *Williamson*, 3!

Durban Flats, *Buchanan*, 42! 64! on bare hills at Umpumulo, 2400–2800 ft., *Buchanan*, 297! Umsinga and base of Biggars Berg, *Buchanan*, 92! Var. β: Transkei; between Gekau (Gcua or Geuu) River and Bashee River, *Drège*. Natal; Inanda, *Rehmann*, 8254! *Wood*, 1578! Zululand, 1000 ft., *Buchanan*, 300!

The type also in Nyasaland.

9. S. Rehmanni (Hack. in Bull. Herb. Boiss. iii. 383); perennial; culms rather robust, geniculate, more or less compressed below, 2–4 ft. long, glabrous, smooth, 3–4-noded; leaf-sheaths rather tight, the lower slipping from the culms and rolling in or folding, broad, glabrous and smooth, or tubercled and hispid; ligules reduced to a ciliate rim; blades linear, tapering to a long fine point, 6–10 in. by 3–4 lin., flat or almost so, glaucous, glabrous or tubercled and hispid, primary nerves distant; panicle erect, narrow, oblong, 1–1¼ ft. by 2 in., ultimately rather lax; rhachis smooth; branches solitary or often irregularly crowded, obliquely erect or at length spreading, filiform, smooth or almost so, 1–3 in. long, loosely and repeatedly branched, the lowest branchlets up to 9 lin. long; lateral pedicels extremely short; spikelets olive-grey, lanceolate-oblong, ¾ to almost 1 lin. long; glumes unequal, the lower oblong, subacute, nerveless, about ½ the length of the spikelet, the upper lanceolate-oblong, acute, about ⅘ the length of the spikelet, 1-nerved; valve like the upper glume, but longer; pale almost equalling the valve; stamens 3; anthers ½ lin. long.

KALAHARI REGION: Transvaal; Bosch Veld, at Klippan, *Rehmann*, 5373! EASTERN REGION: Natal; Durban Flats, *Buchanan*, 6! at the borders of woods near the Umlazi River, *Krauss*, 7 partly! Tugela River, 600–1000 ft., *Buchanan*, 245! 246!

10. S. fimbriatus (Nees, Fl. Afr. Austr. 156); perennial, densely tufted; culms usually geniculate, 2–3 ft. long, smooth, glabrous, 2–4-noded; sheaths glabrous except the sometimes ciliate or fimbriate margins, smooth, firm, the lowest pallid, more or less compressed and subcarinate; ligule a ciliate rim; blades linear, tapering to a long setaceous point, 5–10 in. by 1–2 lin., flat or folded with the margins rolled in, glabrous, rarely with long fine spreading hairs near the base, smooth or scaberulous; panicle erect, subflexuous or nodding, 8–12 in. by 1–2 in. (when open); branches solitary, irregularly crowded, 1–3 in. long, flat, at length more or less spreading, filiform, repeatedly branched from the base, lower secondary branchlets up to 9 lin. long, smooth or almost so; lateral pedicels very short; spikelets greyish-green, ⅞–1 lin. long, crowded or rather lax; glumes unequal, lanceolate, acute or acuminate, the lower hyaline, equalling about ½ the length of the spikelet, the upper as long as the valve or slightly longer, 1-nerved; valve ovate-lanceolate, acute or acuminate, 1-nerved; stamens 3; anthers ½ lin. long; grain obovoid, truncate, quadrangular, very slightly compressed, ⅜ lin. long; pericarp delicate. *Durand & Schinz, Consp. Fl. Afr.* v. 820. *Vilfa fimbriata, Trin. Agrostid.* 47; *Steud. Syn. Pl. Glum.* i. 156.

COAST REGION: Uitenhage Div.; Springbok Flats by the Zwartkops River, *Ecklon*.

CENTRAL REGION: Graaff Reinet Div.; stony hills near Graaff Reinet, 2600 ft., *Bolus*, 555! Somerset Div.; near Somerset East, *Bowker!* Colesberg Div.; Slingers Fontein, *Drège!*

KALAHARI REGION: Griqualand West, Herbert Div.; St. Clair, Douglas, *Orpen*, 248! Hay Div.; Griqua Town, *Burchell*, 1943/1! between Griqua Town and Witte Water, *Burchell*, 1982! 1982/2! Orange Free State, Winberg, *Buchanan*, 242! and without precise locality, *Buchanan*, 273! 277! Transvaal; near Lydenburg, *Atherstone!* Natal; near Durban, below 500 ft., *Drège!* and without precise locality, *Gerrard*, 602!

S. Marlothii (Hack. in Engl. Jahrb. xi. 401) seems to be a state of this with obtuse glumes and valves. Hackel's elaborate description agrees otherwise point for point with *S. fimbriatus*. *S. Marlothii* was collected by Marloth, near Koo Fontein, south-east of Kuruman, Bechuanaland.

11. **S. indicus** (R. Br. Prodr. 170); perennial, tufted; culms erect, 2–3 ft. long, glabrous, smooth, usually 2-noded below the middle, sheathed all along or the upper internodes exserted; leaves mostly crowded near the base, often numerous; sheaths glabrous except at the often ciliate margins, smooth, the lowest sometimes compressed, short, broad, pallid, the upper tight; ligule a minutely ciliolate rim; blades linear, long tapering to a fine point, 4 to almost 12 in. by 1–1½ lin., usually convolute, glabrous, smooth; panicle erect, spike-like, slender, often interrupted below; branches solitary, often irregularly crowded, very short and adpressed to the rhachis or the lowest up to 1 in. long, filiform, smooth or scaberulous; lateral pedicels very short; spikelets dark olive-green, crowded, 1 lin. long; glumes unequal, the lower oblong or elliptic, obtuse, often denticulate, about ⅓ the length of the spikelet, nerveless, the upper ovate-oblong, acute or subacute, about ⅝ the length of the spikelet, sometimes 1-nerved; valve lanceolate-oblong, acute or acuminate, 1-nerved; pale scarcely shorter; stamens 3; anthers ½ lin. long; grain ellipsoid, truncate, quadrangular, slightly compressed, ½ lin. long, brown; pericarp thin. *Kunth, Enum.* i. 211; *Durand & Schinz, Consp. Fl. Afr.* v. 820 (*partly*); *K. Schum. in Engl. Pfl. Ost-Afr. C.* 107 (*partly*). *S. indicus, var. capensis, Engl. Hochgebirgsfl. Trop. Afr.* 127. *S. capensis, Kunth, Enum.* i. 212; *Hochst. ex A. Rich. Tent. Fl. Abyss.* ii. 395; *Durand & Schinz, l.c.* 819. *S. ruppeliana, Hochst. ex Steud. Syn. Pl. Glum.* i. 160. *Panicum caudatum, Thunb. Prod.* 19. *Agrostis spicata, Thunb. Prod.* 19; *Fl. Cap. ed. Schult.* 106. *A. capensis, Willd. Sp. Pl.* i. 372. *A. africana, Poir. Encyc. Suppl.* i. 254. *Vilfa capensis, Beauv. Agrost.* 16; *Trin. Spec. Gram. Ic. t.* 56; *Agrostid.* 72; *Nees in Linnæa*, vii. 293; *Steud. Syn. Pl. Glum.* i. 160. *V. ruppeliana, Steud. l.c. Vilfa indica, Steud. Syn. Pl. Glum.* i. 162; *Baker, Fl. Maurit.* 449.

VAR. β, **laxus** (Stapf); usually more robust; panicle looser, ½ to 1½ ft. long; branches more distant, longer, more or less spreading; spikelets often secund. *S. pyramidalis, Nees, Fl. Afr. Austr.* 155, *not Beauv.* *S. natalensis, Durand & Schinz, Consp. Fl. Afr.* v. 822. *Vilfa natalensis, Steud. Syn. Pl. Glum.* i. 154. *Vilfa capensis, var. laxa, Nees, l.c.*

SOUTH AFRICA: without precise locality, *Thunberg!* *Mund & Maire!* *Zeyher*, 4497!

COAST REGION: Cape Div.; wet rocky places on Table Mountain and Devils Mountain, *Ecklon*, 941! *Drège!* Lion Mountain, *Drège*; Devils Mountain, *Ecklon*, 942! near Duiker Vallei below Tyger Mountain, and by the vineyards near Classenbosch, *Ecklon!* by the railway near Rondebosch, *Wolley Dod*, 3341! Tulbagh Div.; Tulbagh Waterfall, *Ecklon*. Stellenbosch Div.; Hottentots Holland, *Ecklon*. Caledon Div.; near Caledon and Genadendal, *Ecklon*. Robertson Div.; Boschjes Veld, *Thom*, 1077! Riversdale Div.; hills near the Zoetemelks River, *Burchell*, seed 777! Port Elizabeth Div.; Port Elizabeth, *E.S.C.A. Herb.*, 196! Uitenhage Div.; Zwartkops River, *Ecklon*. Albany Div.; near Grahamstown, 2000 ft., *MacOwan*, 1309! Var. β: Cape Div.; Devils Mountain, *Ecklon*. Knysna Div.; Plettenbergs Bay, *Mund & Maire!* near Melville, *Burchell*, 5472! Uitenhage Div.; Uitenhage, *Harvey*, 101! Port Elizabeth Div.; near the Leadmine, *Burchell*, 4486! Bathurst Div.; Kasuga River, *MacOwan*, 1015! Komgha Div.; near Komgha, *Flanagan*, 910!

EASTERN REGION: Natal, *Gerrard*, 688! Var. β: Tembuland; Bazeia, 2000 ft., *Baur*, 405! Natal, "all over the colony," *Buchanan*, 243! near Durban, *Williamson*, 2! 101! at the borders of woods near the Umlazi River, *Krauss*, 7 partly! between the Unzimkulu and Umkomanzi Rivers, *Drège!*

Also in German East Africa and Abyssinia, in the Mascarene Islands and St. Helena, and in Australia.

The Indian form of *S. indicus* is identical with the American *S. tenacissimus*, Beauv., which Doell maintains as a distinct species. It has smaller spikelets and smaller anthers. *S. elongatus*, R. Br. (not Durand & Schinz, l.c. 820), which has also often been reduced to *S. indicus*, and which is indicated by various authors as indigenous in Africa, is a different diandrous species the area of which extends from Australia and New Caledonia to China and Japan.

12. S. pungens (Kunth, Rév. Gram. i. 68); perennial; rhizome often long creeping, stoloniferous; stolons emitting fascicled or solitary ascending culms, these 2–12 in. long, glabrous, very many-noded, sheathed nearly all along, internodes alternately very short and long, hence the leaves appear opposite; culm-sheaths rather tight, slightly compressed, glabrous or sometimes ciliate along the margins and bearded at the mouth, smooth; ligule a ciliolate rim; blades subulate-involute, often pungent, rarely flat towards the base, from $\frac{1}{3}$–4 in. long, rigid, firm, closely and strongly nerved, glabrous or scantily long-hairy above, margins scaberulous; panicle spike-like, cylindric, compact, rarely somewhat loosened, $\frac{1}{2}$–3 in. long; branches short, branched from the base, scaberulous; pedicels very short; spikelets light to dark olive-green, 1–1$\frac{1}{2}$ lin. long; glumes lanceolate, acute or acuminate, keels acute, scaberulous above, the lower equalling $\frac{1}{2}$–$\frac{4}{5}$ of the upper, the latter as long as the valve or slightly longer and like it 1-nerved; pale slightly shorter; stamens 3; anthers $\frac{3}{4}$–1 lin.; grain ellipsoid, $\frac{2}{5}$ lin. long, light brown; pericarp thin. *Kunth, Enum.* i. 210; *Coss. & Durieu, Expl. Scient. Algér.* ii. 62. *S. littoralis, Kunth, Rév. Gram.* i. 68; *Enum.* i. 213; *Durand & Schinz, Consp. Fl. Afr.* v. 821. *S. virginicus, Kunth, Rév. Gram.* i. 67; *Enum.* i. 210; *Nees, Fl. Afr. Austr.* 153; *Durand & Schinz, l.c.* 824. *S. Matrella, Nees, l.c.* 152; *Durand & Schinz, l.c.* 822. *S. arenarius, Duval-Jouve in Bull. Soc. Bot. France*, xvi. 294; *Durand & Schinz, l.c.* 818. *Agrostis arenaria, Gouan, Ill. & Obs. Bot.* 3. *A. pungens, Schreb. Beschreib. Graes.* ii. 46, *t.* 27, *fig.* 3;

Cav. Icon. & Descr. ii. *t.* 111; *Reichenb. Ic. Fl. Germ.* i. *t.* xxxvi. *fig.* 1437. *A. barbata, Pers. Syn.* i. 75. *A. littoralis, Lam. Illustr.* i. 161. *A. virginica, Linn. Sp. Pl.* 63. *Vilfa pungens, Beauv. Agrost.* 16; *Trin. Spec. Gram. Icon. t.* 47; *Agrostid.* 49; *Steud. Syn. Pl. Glum.* i. 157. *Vilfa virginica, Beauv. l.c.*; *Trin. Spec. Gram. Icon. t.* 48; *Agrostid.* 48; *Steud. Syn. Pl. Glum.* i. 157; *Doell in Mart. Fl. Bras.* ii. iii. 30, *t.* viii. *fig.* 1. *Vilfa barbata, Beauv. l.c.* *V. littoralis, Beauv. l.c.*; *Steud. l.c.* 162. *V. murina, Sieb. ex Steud. l.c.* 157. *Podosæmum pungens, Link, Hort. Berol.* i. 84. *P. virginicum, Link. l.c.* 85. *Calotheca sabulosa, Steud. in Flora,* 1829, 488. *Phalaris disticha, Forsk. Fl. Ægypt.-Arab.* 17 (?). *Zoysia pungens, Nees in Linnæa,* vii. 299. *Crypsis maritima, Munro ex MacOwan in Cape Monthly Mag. new series,* iii. (1871), *Suppl.* 7.

SOUTH AFRICA: without precise locality, *Boivin! Mund & Maire! Zeyher!*

COAST REGION: Cape Div.; Green Point, *Ecklon!* in saline soil near Doorn Hoogte and between Tyger Berg and Simons Berg, 100–200 ft., *Drège!* Muizenberg Vley, *Wolley Dod,* 925! Tulbagh Div.; Tulbagh Waterfall, *Ecklon.* Uitenhage Div.; *Zeyher!* mouth of the Zwartkops River and mountains on the Van Stadens River, *Ecklon.* Port Elizabeth Div.; Port Elizabeth, *Drège!* sand-hills along the coast, *E.S.C.A. Herb.,* 250! Bathurst Div.; near Port Alfred, *Burchell,* 4032! mouth of the Kasuga and Kowie Rivers, *MacOwan,* 710!

WESTERN REGION: Little Namaqualand; in sandy saline plains at the mouth of the Orange River, *Drège!*

EASTERN REGION: Natal; at the mouth of the Umzimkulu River, *Drège!* sand-dunes around Durban Bay, *Krauss,* 67!

A very variable litoral plant of most warm countries. The specimens from Bathurst Division and from Natal are rather different in habit from the western, approaching the form common in the Mediterranean region which originally was understood under *S. pungens.*

13. S. subtilis (Kunth, Rév. Gram. ii. t. 124); perennial, densely cæspitose, sometimes stoloniferous, erect, 1–1½ ft. long, smooth, glabrous, about 4-noded, internodes enclosed or shortly exserted; leaves glabrous, smooth; sheaths very tight, more or less bearded at the mouth; ligule a minutely ciliolate rim; blades very narrow, setaceously convolute, acute, 2–5 in. long, striate; panicle embraced at the base by the uppermost sheath, ovoid to oblong, open, very lax, 2–4 in. long, much branched; branches and branchlets capillary with very long and fine hairs from the axils; pedicels very variable in length (from ½–1½ lin. long in the Natal, up to 5 lin. long in the Madagascar specimens); spikelets lanceolate, acute, ¾ lin. long; rhachilla produced into a fine bristle half as long to almost as long as the floret; glumes subequal, lanceolate in profile, acute, ½–⅝ lin. long, lower 1-, upper 1–3-nerved; valve ovate-lanceolate in profile, ¾ lin. long, 3-nerved, lateral nerves evanescent above the middle; pale as long as the valve or very slightly longer; keels very fine, percurrent or evanescent below the subciliolate tips; anthers ⅜–½ lin. long; grain oblong, ⅖ lin. by ⅙–⅕ lin., subterete, finely striate; pericarp adnate to the seed, indistinct; embryo not quite ⅕ the length of the grain. *Kunth, Enum.* i. 215; *Suppl.* 171; *Krauss in Flora,* 1846, 118; *Durand*

& Schinz, Consp. Fl. Afr. v. 824. *Vilfa subtilis, Trin. Agrostid.* 66, *and in Mém. Acad. Pétersb. sér.* 6, v. *part.* ii. 88; *Steud. Syn. Pl. Glum.* i. 159.

EASTERN REGION: Natal; grassy flats between Umlazi River and Durban Bay, *Krauss*, 212!

Also in Madagascar.

The Natal plant differs from the Madagascar specimens which I have seen (Hildebrandt, 4906; Baron, 672 and 4092) in the much shorter ramifications of the panicle. The presence of a bristle-like continuation of the rhachilla is unique in the genus; as the structure of the spikelet is, however, otherwise essentially that of *Sporobolus*, it does not seem expedient to separate this species from that genus.

LI. POGONARTHRIA, Stapf.

Spikelets laterally compressed, subsessile, more or less imbricate, secund on the irregularly spirally arranged branches of a panicle; rhachilla disarticulating above the glumes and between the valves, tips of the joints ciliate. *Florets* 2–8, ⚥. *Glumes* rigidly membranous, 1-nerved. *Valves* oblong, rigidly membranous, acuminate, quite glabrous, 3-nerved; side-nerves evanescent above the middle. *Pales* 2-keeled, slightly shorter than the valves. *Lodicules* 2, minute, delicate. *Stamens* 3. *Ovary* glabrous; styles distinct; stigmas plumose. *Grain* tightly embraced by the scarcely altered valve and pale, linear-oblong, obtusely triquetrous or oval in cross section; embryo less than ½ the length of the grain; hilum basal, punctiform.

Perennial, stiff; blades rigid, usually convolute; ligule a fringe of cilia; panicles straight, with spreading more or less curved branches in irregular spirals; spikelets secund, crowded, livid, purplish or dark grey.

Species 1, in tropical South-east Africa, and in extra-tropical South Africa.

Hackel, who described the only species of this genus under *Leptochloa*, has already remarked, that it differs considerably from all other species of *Leptochloa*. The differences exist mainly in the coarse rigid habit and in the structure of the spikelets, the glumes and valves of which are more rigidly membranous, livid purplish or dark grey, and quite glabrous, whilst the tips of the rhachilla joints are ciliate; the valves resemble more those of *Eragrostis* than of *Leptochloa*, and the affinity of the genus lies most certainly with the former.

1. P. falcata (Rendle in Cat. Afr. Pl. Welw. ii. 232); perennial, cæspitose, quite glabrous except at the mouth of the sheath; culms strictly erect or subgeniculate, 1–2½ ft. long, terete, smooth, about 3-noded, internodes exserted; sheaths tight, terete, smooth, mouth bearded; ligule a fringe of minute cilia; blades linear, setaceously attenuated, 4–8 in. by 1–2 lin., flat or more often convolute, rigid, subglaucous, quite smooth, striate; panicle linear, 4–10 in. by ½–2 in., usually straight; rhachis sulcate, scaberulous; branches often 2–5 close together, more or less spreading, usually curved, up to 1 in. long, flat on the back, wavy, simple, bearing spikelets from the base, scabrid; spikelets 1½–3 lin. long, livid, purplish or dark grey; rhachilla joints up to ¼ lin. long; glumes lanceolate to lanceolate-oblong, reddish subacuminate, scaberulous, lower ⅜–½ lin. long, upper ¾–1 lin. long valves lanceolate in profile, oblong when expanded, acutely acuminate or mucronulate, 1 lin. long; callus very minute, obtuse, glabrous; pales 1 lin. long; keels scabrid; lodicules $\frac{1}{8}$ lin. long; anthers $\frac{3}{8}$–$\frac{1}{2}$ lin. long; grain linear-oblong, oval in cross section, $\frac{3}{8}$ lin. long. *Stapf in Hook. Ic. Plant. t.* 2610. *Leptochloa falcata, Hack. in Bull. Herb. Boiss.* iii. 386, *and* iv. *App.* iii. 21. *Eragrostis sp., Nees in Linnæa,* xx. 255. *E. Marlothii, Hack. in Engl. Jahrb.* xi. 404 *(from the description).*

KALAHARI REGION: Griqualand West, Herbert Div.; St. Clair, Douglas, *Orpen*, 256! Hay Div.; at the foot of the Asbestos Mountains, between the Kloof Village and Witte Water, *Burchell*, 2101! Orange Free State, Olifants Fontein, *Rehmann*, 3514! rocky and grassy hills by the Great Vet River and Little Vet River, 4000–5000 ft., *Zeyher*, 1840! *Burke*, 209! Bloemfontein, *Rehmann*, 3753! Basutoland; Leribe, *Buchanan*, 128! Transvaal; Bosch Veld, between Eland River and Klippan, *Rehmann*, 5118! near Lydenburg, *Atherstone*, 72!

EASTERN REGION: Natal; Tugela River, 600 ft., *Buchanan*, 242! at Umlaas Drift, *Wood*, 1910!

Also in tropical South Africa as far as the Zambesi.

LII. DIPLACHNE, Beauv.

Spikelets shortly pedicelled or subsessile, somewhat distant or remote on the simple slender branches of a panicle; rhachilla disarticulating above the glumes and between the valves, glabrous. *Florets* 2–10, ⚥, or the uppermost reduced. *Glumes* unequal or subequal, membranous, 1-nerved, keeled, persistent. *Valves* oblong to linear-oblong, 2-toothed or minutely notched, rarely quite entire, muticous or mucronulate from the sinus, very rarely shortly awned from below the apex, membranous, 3-nerved, usually finely ciliate in the lower part of the nerves or sometimes quite glabrous; side-nerves percurrent or almost (or sometimes very shortly) excurrent. *Pales* 2-keeled, shorter than the valves. *Lodicules* 2, cuneate, fleshy, nerved. *Stamens* 3. *Ovary* glabrous; styles distinct, slender; stigmas plumose, laterally exserted. *Grain* enclosed by the slightly altered valve and pale, oblong to obovoid-oblong, dorsally compressed, sometimes quite flat, rarely terete; embryo equalling $\frac{1}{3}$–$\frac{1}{2}$ the length of the grain; hilum punctiform, basal.

Mostly perennial, tufted, somewhat coarse grasses; blades long, narrow, flat or involute; ligules membranous, sometimes reduced to a rim; panicles consisting of slender, usually long, simple, loosely spike-like and more or less distant branches; spikelets light or olive-green, often tinged with purple and dark.

Species about 12, mainly in the warm regions of the Old World and in North America.

The genus, as usually understood, consists of heterogeneous elements. The description given here applies to about a dozen species which group round *D. fascicularis*, Beauv., on which the genus was founded; the definition is, however, not quite satisfactory, as some of the American species and the species of the *D. serotina* group require further examination.

Culms erect or ascending, 1–5 ft. long, usually 3–5-noded; blades 3–9 in. long, flaccid or subrigid, never pungent:

Spikelets 5–10-flowered, light or olive-green; valves awnless; grain much compressed:
Blades more or less flaccid; ligules very short; valves obtuse, tips broad, hyaline ... ... (1) Eleusine.
Blades more or less rigid; ligules up to 2½ lin. long; valves usually minutely emarginate, often with a minute tooth on each side and mucronulate ... ... (2) fusca.
Spikelets 2–3-flowered, reddish; valves shortly awned from below the tips; grain terete ... (3) biflora.
Culms densely fascicled with numerous innovation shoots from a descending rhizome, distichously leafy all along; blades ½–2 in. long, convolute-subulate, pungent, very rigid ... ... (4) paucinervis.

1. **D. Eleusine** (Nees, Fl. Afr. Austr. 255); perennial, glabrous; culms tufted, 1–2 ft. long, geniculately ascending, terete, smooth, simple or branched below, 3-noded, internodes exserted; sheaths rather tight and firm, smooth, the lower keeled; ligules membranous, very short, truncate, denticulate, ciliate; blades linear, tapering to a fine point, 4–9 in. by 1–2 lin., more or less flaccid, scabrid on both sides or rather smooth below; panicle narrow, consisting of 2–8 erect, distant spikes or spike-like racemes; rhachis angular, finely scaberulous or almost smooth; branches 1–4 in. long, flexuous; spikelets unilateral, imbricate or 2-seriate, subsessile, 2–3 lin. long, 5–8-flowered, light green; glumes lanceolate in profile, obtuse or subacute, 1¼ and 1½ lin. long respectively, whitish, keel green; valves oblong, very obtuse, entire, up to ½ lin. long, tips broad, hyaline, side-nerves finely silky, evanescent below the tips; pales obtuse; anthers not quite ½ lin. long; grain elliptic, flat, ¾–⅞ lin. by ½ lin. *Uralepis Eleusine, Steud. Syn. Pl. Glum.* i. 248. *Triodia Eleusine, Durand & Schinz, Consp. Fl. Afr.* v. 877.

COAST REGION: Fort Beaufort Div.; Kat River Poort, 2000 ft., *Drège.* Komgha Div.; near Komgha, *Flanagan*, 942!
KALAHARI REGION: Basutoland; Leribe, *Buchanan*, 232!
EASTERN REGION: Tembuland, between Bashee River and Morley, 1000–2000 ft., *Drège.* Transkei, near the Gekau (Gcua or Geuu) River, below 1000 ft., *Drège!* Natal; banks of Tugela River, 700 ft., *Buchanan*, 217!

2. **D. fusca** (Beauv. Agrost. 163); perennial, glabrous; culms tufted, stout, geniculately ascending or erect, often branched from the lower nodes, 3–5 ft. long, terete, smooth, 3–4-noded, or many-noded when branched, internodes enclosed except the uppermost or shortly exserted; sheaths smooth, almost shining or the upper rough, the basal whitish, slightly compressed, bluntly keeled; ligules hyaline, oblong, acute, up to 2½ lin. long; blades very narrow, linear, tapering to a fine often subpungent point, 3–6 in. by 1–1½ lin. when expanded, folded or convolute or sometimes flat, rather rigid, rough on both sides, rarely almost smooth below; panicle erect, straight or slightly nodding, obovate-oblong to linear, contracted or open; rhachis slender, angular, rough; branches scattered or 2–3 close together, often more or less flexuous, the

longest 3–5 in., usually racemose; pedicels short; spikelets distant by half their length or more, narrow, oblong, 3–5 lin. long, 5–10-flowered, usually dark olive-grey, rarely light or whitish; glumes lanceolate to oblong, obtuse or acute, often obscurely mucronate, the lower about 1 lin. long, the upper 1½–2 lin.; valves oblong, tips broad, entire or minutely emarginate, and with a tooth on one or both sides, middle and side-nerves excurrent into a short or obscure mucro, or only the former, side-nerves silkily ciliate below; callus hardly any; pales minutely 2-toothed, flaps hairy along the keels; anthers ¾ lin. long; grain oblong, dorsally compressed, up to 1 lin. long; embryo almost ½ the length of the grain. *Aschers. & Schweinf, Ill. Fl. Égypte*, 171; *Durand & Schinz, Consp. Fl. Afr.* v. 878; *Hack. in Bull. Herb. Boiss.* iv. *App.* iii. 25; *Hook. f. in Fl. Brit. Ind.* vii. 329. *D. livida, Nees, Fl. Afr. Austr.* 254. *D. capensis, Nees, l.c.* 256. *D. alba, Hochst. in Flora*, 1842, i. *Beibl.* 134; *Durand & Schinz, l.c.* 878. *D. pallida, Hack. in Bull. Herb. Boiss.* iii. 387. *Festuca fusca, Linn. Sp. Pl. ed.* ii. 109; *Del. Fl. d'Égypte*, 24, *t.* xi. *fig.* 1. *Bromus polystachyus, Forsk. Fl. Ægypt.-Arab.* 23. *Leptochloa fusca, Kunth, Rév. Gram.* i. 91; *Enum.* i. 271; *Suppl.* 223. *Tridens capensis, Nees in Linnæa*, vii. 324. *Uralepis fusca, Steud. Syn. Pl. Glum.* i. 247. *U. livida, Steud. l.c.* 248. *U.* (?) *capensis, Kunth, Enum.* 319. *U. alba, Steud. l.c.* 248. *Triodia livida, Durand & Schinz, Consp. Fl. Afr.* v. 877. *T. capensis, Durand & Schinz, l.c.*

COAST REGION: Cape Div.; Cape Flats, in moist sandy places, *MacOwan, Herb. Austr. Afr.* 1795! and at Doorn Hoogte, *Ecklon!* near Muizenberg Vley, *Wolley Dod*, 2354! 2355! about the ponds near Cape Town and at Salt River, *Burchell*, 678! Mossel Bay Div.; in a dry channel of an arm of the Gauritz River, *Burchell*, 6473! Oudtshoorn Div.; Gamkas Poort, *Mund & Maire!* Uitenhage Div.; Uitenhage, *Zeyher!* banks of the Zwartkops River, *Ecklon.* Albany Div.; in running streams near Grahamstown, 2000 ft., *MacOwan*, 1315! and in wet places near Grahamstown, *MacOwan*, 1315b! King Williamstown Div.; Amatola Mountains, *Buchanan*, 36! Knysna Div.; Plettenberg Bay, *Mund & Maire!* Komgha Div.; near Komgha, *Flanagan*, 952! Queenstown Div.; between Table Mountain and Zwart Kei River, 4000–4500 ft., *Drège.*

CENTRAL REGION: Prince Albert Div.; near Weltevrede, 2500–3000 ft., *Drège!* Somerset Div.; at Blyde River, *Burchell*, 2963! between the Zuur Berg Range and Klein Bruintjes Hoogte, 2000–2400 ft., *Drège.* Graaff Reinet Div.; Sneeuwberg, on the banks of rivulets, 3700 ft., *Bolus*, 1970! near Graaff Reinet, *Ecklon.*

WESTERN REGION: Little Namaqualand; between Pedros Kloof and Lily Fontein, 3000–4000 ft., *Drège!*

KALAHARI REGION: Griqualand West, Herbert Div.; along the Vaal River, *Burchell*, 1778! and Lower Campbell, *Burchell*, 1808! Basutoland; Leribe, *Buchanan*, 135! Bechuanaland; at Kosi Fontein, *Burchell*, 2555! Transvaal; Bosch Veld, at Klippan, *Rehmann*, 5371! by the Sand River, at Olifants Poort, *Nelson*, 85*!

Widely spread throughout the warm regions of the Old World, mainly near the water.

Nees distinguishes 4 varieties of his *D. capensis*, mainly by the colour of the spikelets, and the more or less contracted state of the panicle. They appear to me only to be slight variations and different grades of development. Hackel's

D. pallida, founded on Rehmann, 5371, represents a state with strictly erect culms and pale spikelets. Similar erect states occur in India, and particularly in Australia, along with the ordinary form.

3. **D. biflora** (Hack. in Bull. Herb. Boiss. iii. 387); perennial, almost glabrous; culms tufted on a short oblique rhizome, erect, 1–2 ft. long, terete, simple, slender, rough below the nodes, about 3-noded, internodes usually enclosed except the uppermost; leaves crowded near the base; sheaths tight, terete, scaberulous or smooth, firm, the lowest reduced to bladeless scales; ligule a membranous, ciliolate rim; blades linear, tapering to an acute point, 3–7 in. by 2–3 lin., rigidly erect, flat or convolute, with scattered stiff hairs, particularly near the base, rough on both sides, glaucous; panicle contracted, obovate to linear-oblong, 2–6 in. long, glaucous, purplish; rhachis scabrid, angular; branches simple, solitary or paired, subflexuous, bearing spikelets from the base or almost so, lowest up to 2 in. long; spikelets 2–3-flowered, subsecund, 2-ranked, lower slightly exceeding the internodes, upper closer, shortly but distinctly pedicelled, about 2½ lin. long; glumes subequal, lanceolate, acute, 2–2¼ lin. long, minutely scabrid, margins and tips hyaline; valves up to 2¼ lin. long, entire, acute or very minutely 2-toothed, very shortly awned from below the readily splitting pruinose tips, nerves silky-ciliate to the middle (at least in the lower floret); callus minute, acute, bearded; pales obtuse, not quite 2 lin. long, keels scabrid; anthers 1 lin. long; grain oblong-linear, terete, 1 lin. long.

VAR. β, **Buchanani** (Stapf); spikelets rather more distant, 4 lin. long; all parts proportionally larger and at the same time more slender.

KALAHARI REGION: Transvaal; Makapans Mountains, at Streyd Poort, *Rehmann*, 5386! Var. β: Basutoland; Leribe, *Buchanan*, 219!
EASTERN REGION: Natal; mountain slopes near Umpumulo, 2500 ft., *Buchanan*, 282!

Rather different from the other species of the genus, and perhaps not a true *Diplachne*.

4. **D. paucinervis** (Stapf ex Rendle in Cat. Afr. Pl. Welw. ii. 232); perennial, tufted; culms densely fascicled with numerous innovation shoots from a descending rhizome (covered with papery scales), distichously leafy all along, slender, ½–1 ft. long, pubescent below the panicle, otherwise glabrous, smooth, many-noded, internodes all enclosed or sometimes the intermediate and upper shortly exserted; leaves glaucous, finely hairy to glabrous; sheaths very tight, striate, scantily bearded at the mouth, lower short, firm, persistent, lowest bladeless or with minute mucro-like blades; ligule a minute, ciliolate rim; blades convolute-subulate from a usually almost flat base, spinously pungent, ½–2 in. by 1–1½ lin. (at the base), rigid to very rigid, scabrid along the margins and towards the tips; panicle linear to obovate-oblong, contracted, dense, erect, 1–2 in. long, light green; rhachis angular, scabrid, very finely puberulous and sometimes also with scattered long hairs; branches solitary or geminate, adpressed

or obliquely erect, subsecund, bearing spikelets from the base or almost so, angular, scabrid and finely puberulous, lowest $\frac{1}{3}$–1 in. long; pedicels very short; spikelets closely imbricate, lanceolate, acute, 2–3$\frac{1}{2}$ lin. long, closely 3–7-flowered; glumes subequal, lanceolate, acute, membranous, margins and tips hyaline, keel smooth or scaberulous, upper 1$\frac{1}{2}$–2 lin. long, lowest slightly shorter; valves oblong to lanceolate in profile, acute, entire or very minutely 2-toothed, mucronulate from or close to the tip, rigidly membranous, smooth, 3-nerved, side-nerves and keel silky-ciliate, the former to the middle or beyond, the latter scarcely to the middle, scabrid above, or upper valves almost glabrous; pales oblong, obtuse, 1–1$\frac{1}{4}$ lin. long, keels scabrid, flaps glabrous or hairy; lodicules cuneate, fleshy, small; anthers $\frac{1}{2}$ lin. long. *Dactylis paucinervis, Nees, Fl. Afr. Austr.* 429; *Steud. Syn. Pl. Glum.* i. 297; *Durand & Schinz, Consp. Fl. Afr.* 904.

WESTERN REGION: Little Namaqualand; near Henkries, *Atherstone*, 7! Vanrhynsdorp Div.; at Strand Fontein, near the mouth of the Olifants River, *Drège!*

Also in Damaraland and tropical Bechuanaland.

This is evidently very closely allied to *D. pungens*, Hack. (Bull. Herb. Boiss. iv. App. iii. 25), from Hereroland.

LIII. ERAGROSTIS, Beauv.

Spikelets usually strongly laterally compressed, pedicelled in open or contracted panicles, rarely sessile in simple or compound spikes, very rarely articulate on the pedicels; rhachilla disarticulating above the glumes and between the valves or tough and persistent, glabrous, sometimes more or less scaberulous, very rarely minutely hairy. *Florets* 2 to many, ⚥ or the uppermost reduced. *Glumes* unequal or equal, usually membranous, 1-nerved, or the upper sometimes 3-nerved, keeled, persistent or deciduous. *Valves* more or less imbricate, ovate to lanceolate, acute or obtuse, entire, muticous, membranous to chartaceous, 3-nerved, glabrous, very rarely minutely pubescent; side-nerves short or almost percurrent. *Pales* equal to the valves or slightly shorter, membranous, 2-keeled, deciduous or persistent on the rhachilla. *Lodicules* 2, small, cuneate, more or less fleshy. *Stamens* 3, rarely 2. *Ovary* glabrous; styles distinct; stigmas plumose, laterally exserted. *Grain* enclosed by the scarcely altered valve and pale and deciduous with them, or more commonly falling with the deciduous valve, leaving the more or less persistent pale behind, oblong to obovoid or globose, round or very obtusely triquetrous or quadrangular in cross section; pericarp thin, sometimes slightly swelling or separating; embryo often $\frac{1}{2}$ as long as the grain (or sometimes longer); hilum punctiform, basal.

Perennial or annual, of very varying habit; blades narrow; ligule reduced to a fringe of usually minute hairs: panicles lax to effuse or contracted to spike-like, or transformed into simple or compound spikes; spikelets usually more or less olive-green or olive-grey, breaking up variously, very rarely deciduous as a whole.

Species very numerous in the warm parts of the world.

Section 1. PTEROESSA, Doell. Spikelets linear to oblong or ovate, often several times longer than broad, with glabrous (very rarely with pubescent) valves, variously breaking up, but not into false fruits; rhachilla usually persistent.

A. *Leptostachyæ.* Panicle usually delicate with fine divisions, flexuous or nodding, or sometimes rigid; spikelets lanceolate to linear, ½–1 lin. broad, few- to many-flowered; valves thinly membranous (except in No. 16. *E. dura*), ½–1 lin. long, rarely longer (No. 2. *E. cæsia*):

*Perennial, or if annual, then the spikelets 3–11-flowered and very shortly pedicelled:

Creeping in sand; culms short, distichously leafy, fascicled on long slender rhizomes or stolons ... (1) **glabrata.**

Not creeping; culms densely tufted with numerous innovation-shoots, usually more or less erect, simple and few-noded, sometimes branched and then often more-noded:

†Glumes slightly unequal to equal:

Valves 1½–1¾ lin. long ... ... (2) **cæsia.**

Valves about 1 lin. long or shorter:

‡Spikelets linear, 4–13- (rarely 3-) flowered, 2–7 lin. long, rarely shorter:

Valves 1 lin. long or slightly longer ... (3) **curvula.**

Valves ⅔–1 lin. long:

§Spikelets 4–13-flowered, more or less dark grey:

‖Valves not variegated:

Perennial; valves acute or obtuse, not truncate:

Culms usually repeatedly geniculate, branched; valves tightly imbricate, obtuse ... (4) **lehmanniana.**

Culms simple, or if sparingly branched below, then the spikelets rather broader and the valves looser and more acute:

Sheaths bearded, lower not or indistinctly compressed ... (5) **chloromelas.**

Sheaths beardless or almost so, lower distinctly compressed and keeled:

Blades finely filiform below, capillary above, canaliculate, 1 ft. long or longer; panicle at length effuse; divisions smooth or the upper scaberulous; pedicels long ... (6) **nebulosa.**

Blades loosely convolute or involute, about 6 in. long; panicle 8–9 lin. long; branches flaccid or very flexuous, all divisions densely scaberulous; pedicels short ... ... (7) **margaritacea.**

Annual; valves very obtuse, truncate or subemarginate... ... (8) **porosa.**

||||Valves purple or violet with yellowish tips ... ... (9) **bicolor.**

§§Spikelets 3–6-flowered, pallid:

Spikelets about 1 lin. broad, on the spreading or pendulous branchlets of an open pyramidal panicle with naked axils ... ... (10) **Poa.**

Spikelets $\frac{2}{3}$–$\frac{3}{4}$ lin. broad, more scattered on the usually contracted branchlets of an oblong panicle with fine virgate branches and bearded axils ... ... (11) **Wilmsii.**

‡‡Spikelets lanceolate or oblong, 2–4-flowered, $\frac{3}{4}$–2 lin. long:

Glumes and valves thin to very thin; spikelets lanceolate:

Culms simple or almost so, 2–5-noded; branches of panicles more or less whorled:

Panicle ovoid, 1½–2 in. by 1–1½ in., somewhat dense and rigid; pedicels short; valves acute... ... (12) **sporoboloides.**

Panicle narrow, oblong to obovoid-oblong, 2–6 in. by 1–2 in., very lax; divisions capillary, flexuous; pedicels 1 lin. long or much longer; valves obtuse or subobtuse:

Panicle long exserted from the uppermost sheath ... ... (13) **Atherstonei.**

Panicle shortly exserted from or enclosed at the base in the uppermost sheath ... ... (14) **micrantha.**

Culms branched, 5–6-noded; branches solitary or the lowest subopposite... (15) **Burchellii.**

Glumes and valves firm, almost chartaceous, shining; spikelets oblong ... (16) **dura.**

††Glumes very unequal, lower often a minute scale, or suppressed:

Lower sheaths strongly compressed; spikelets 3–6 lin. by 1 lin. ... ... (17) **plana.**

Lower sheaths not compressed; spikelets 2–4 lin. by $\frac{2}{3}$–$\frac{3}{4}$ lin. ... ... (18) **heteromera.**

Annual; spikelets 1 lin. long, 2–1-, rarely 3-, flowered on long and fine pedicels... ... ... (19) **biflora.

B. *Megastachyæ.* Panicle more or less rigid (except sometimes in No. 28. *E. gangetica*); spikelets ovate or oblong to linear, 1–1½ lin. broad, rarely broader, usually many-flowered; valves rigidly membranous to subchartaceous, $\frac{3}{4}$–1½ lin. long:

*Perennial (see also No. 34. *E. barbinodis*):

Blades more or less subulate, pungent; culms leafless and nodeless above the base:

Spikelets in dense, distant clusters along the stout axis of a panicle ... ... (20) **cyperoides.**

Spikelets loosely racemose on the spine-like branches of a panicle, usually deflexed ... (21) **spinosa.**

Blades neither subulate nor pungent; culms 1- or more-noded and more or less leafy above the base:

Panicle very broad and lax, divaricate; spikelets scattered over the periphery ... ... (22) **patentissima.**
Panicle contracted, narrow; spikelets much closer:
Rhachilla breaking up; valves rather loosely imbricate, very acute ... ... (23) **denudata.**
Rhachilla persistent or subpersistent; valves closely imbricate, usually less acute:
Spikelets reddish-brown; stamens 2 ... (24) **Chapelieri.**
Spikelets not reddish-brown; stamens 3:
Basal sheaths tomentose-woolly below ... (25) **sclerantha.**
Basal sheaths glabrous or with scattered spreading hairs:
Valves ovate to ovate-oblong, 1 lin. long, side-nerves rather inconspicuous; anthers $\frac{1}{2}$–$\frac{3}{8}$ lin. long:
Spikelets ovate-oblong to oblong, $2\frac{1}{2}$–4 lin., olive-green to leaden-grey... (26) **chalcantha.**
Spikelets linear, 4–5 lin., yellowish-green ... ... (27) **pallens.**
Valves oblong, $\frac{3}{4}$–1 lin. long, side-nerves distinct, prominent; anthers $\frac{1}{4}$–$\frac{3}{8}$ lin. long:
Panicle often nodding, oblong, usually contracted; spikelets up to 8 lin., 8–30-flowered ... ... (28) **gangetica.**
Panicle erect, very slender, with the short distant or crowded branches adpressed to the axis; spikelets $2\frac{1}{2}$–4 lin., 5–15-flowered:
Not stoloniferous; culms erect or shortly ascending, 2–3-noded; spikelets $2\frac{1}{2}$–4 lin. by 1–$1\frac{1}{4}$ lin. (29) **elatior.**
Stoloniferous; culms ascending from a creeping or decumbent base, many-noded; spikelets $1\frac{1}{2}$–$3\frac{1}{2}$ lin. by $\frac{3}{8}$–$\frac{4}{5}$ lin.... ... (30) **sarmentosa.**

**Annual:
Leaves usually covered with very minute, soft, gland-tipped hairs; pedicels with a viscous ring at the middle; grains excavated in front (31) **annulata.**
Leaves, pedicels, and grains not as in 31:
Nodes glabrous, 3–4; valves 1–$1\frac{1}{4}$ lin. long:
Leaves usually glandular along the blade-margins and sheath-keel; panicle dense or rather lax, 2–6 in. long; valves obliquely ovate in profile, obtuse or subobtuse; grain globose ... ... (32) **major.**
Leaves eglandular; panicle dense, $1\frac{1}{2}$–2 in. long; valves oblong, acute or mucronulate; grain oblong ... ... (33) **procumbens.**
Nodes bearded, 5–6; valves $\frac{3}{4}$ lin. long ... (34) **barbinodis.**

Section 2. PLATYSTACHYA, Benth. Spikelets orbicular to ovate-oblong in outline, obtuse, not more than twice as long as broad, with glabrous valves, variously breaking up but not into false fruits, very rarely falling entire (No. 35. *E. superba*); rhachilla usually disarticulating.

Spikelets articulate with the pedicels, deciduous, flat (35) **superba.**
Spikelets not articulate with the pedicels, breaking up, smaller:

*Perennial, densely tufted with innovation-shoots; culms simple:
Rhachilla persistent; spikelets 2–6 lin.; valves 1¼–1½ lin. long ... ... (36) **brizoides.**
Rhachilla readily disarticulating; spikelets 1½–4 lin.; valves ¾–1 lin. long:
Sheaths of innovation-shoots covered more or less with fugacious wool; spikelets crowded on very short pedicels or subsessile:
Valves broadly truncate, their broad upper halves closely imbricate, sides rather flat (37) **truncata.**
Valves rounded-obtuse, loosely imbricate, sides rather convex ... ... (38) **bergiana.**
Sheaths of innovation-shoots not covered with fugacious wool; spikelets more or less nodding on capillary branchlets and pedicels ... ... (39) **obtusa.**
**Biennial or annual, without or with very few feeble barren shoots; culms simple or branched:
Annual; culms simple; valves obtuse ... ... (40) **brizantha.**
Biennial; culms branched; valves acute ... (41) **echinochloidea.**

Section 3. LAPPULA, Stapf. Spikelets oblong, with long rigid tubercle-based cilia along the side-nerves of the valves.]

Only species ... ... (42) **Lappula.**

Section 4. CATACLASTOS, Doell. Spikelets of various shape, small to very small, with glabrous valves, breaking up into false fruits, consisting of the grain enclosed by its valve and pale.

Panicle large, thyrsoid, very lax; spikelets linear, on long fine pedicels ... ... (43) **aspera.**
Panicle linear to oblong, usually contracted; spikelets very shortly pedicelled:
Perennial; valves ¾ lin. long ... ... (44) **gummiflua.**
Annual; valves ½ lin. long or shorter:
Keels of pales long and rigidly ciliate; stamens 3 (45) **ciliaris.**
Keels of pales smooth; stamens 2 ... ... (46) **namaquensis.**

1. E. glabrata (Nees in Linnæa vii. 332); perennial; culms in fascicles on the slender branches of a long creeping slender rhizome, or from similar stolons, with barren branches from the lower sheaths, rising 2–4 in. above the ground, sheathed all along, with 5–7 blade-bearing leaves to each flowering culm, quite glabrous; sheaths loose, striate, lowest pallid, bladeless or with minute blades, those above often purplish or like the upper green; ligule a fringe of very minute hairs; blades subulate-linear, subacute to almost pungent, 1–1½ in. by 1½ lin., rarely up to 3 in. by 2 lin., flat with the margins only involute or subulate-convolute, rigid, smooth, dull green below, paler above; panicle scarcely exserted from the leaves, ovoid to obovoid-oblong, contracted, very dense, ½–1 in. long; axis and branches glabrous, smooth, the latter divided almost from the base; pedicels ½–1 lin. long, compressed, smooth; spikelets densely crowded, linear-oblong to oblong, acute, 2½–3½ lin. by 1–1¼ lin., densely 4–9-flowered, dark olive-grey; rhachilla tardily disarticulating; glumes very unequal, lanceolate to ovate-lanceolate in profile, acute, 1-nerved, rather firmly membranous, lower $\frac{1}{2}$ lin. long, upper 1 lin. long; valves ovate-oblong in profile, obtuse or subobtuse, 1 lin. long, smooth, firmly membranous, side-nerves fine; pales equal to the valves, keels scaberulous; anthers $\frac{1}{2}$ lin. long; grain ellipsoid, $\frac{1}{2}$ lin. by over $\frac{1}{4}$ lin., reddish brown. *Nees in Linnæa*, vii. 344; *Fl. Afr. Austr.* 394; *Steud. Syn. Pl. Glum.* i. 271. *Poa glabrata, Kunth, Enum.* i. 344.

COAST REGION: Cape Div.; Cape Flats, *Krauss*, 61; sand-hills along the shore, *MacGillivray*, 404! wet places above Simons Bay, frequent, *Milne*, 253! sands on Paarden Island, *Wolley Dod*, 3182! without precise locality, *Herb. Harvey*, 279!

2. **E. cæsia** (Stapf); perennial, densely tufted; culms erect, slender, compressed, simple, $\frac{3}{4}$–$1\frac{1}{2}$ ft. long, glabrous, smooth, 1-noded at or below the middle, internodes shortly exserted, or both or the upper alone enclosed; lower sheaths crowded, almost flabellate, strongly compressed and keeled, often pinkish with white margins, upper tight or widening upwards, all quite glabrous and smooth except at the scantily bearded mouth; ligule a fringe of minute hairs; blades tightly convolute, finely setaceous, 3–10 in. long, flexuous, rather firm, glabrous, smooth; panicle nodding, contracted, more or less linear, 3–8 in. long; axis filiform, smooth; branches solitary, rather distant, lowest often enclosed at the base in the uppermost sheath, finely filiform, compressed, smooth, divided from the base or some distance above it; branchlets distant, simply racemose or the lower again divided and then up to $\frac{3}{4}$ in. long, usually adpressed to the branches; pedicels short to very short; spikelets rather crowded, lanceolate, acute, $2\frac{1}{2}$–$5\frac{1}{2}$ lin. by 1 lin., grey, closely 3–5-flowered; rhachilla disarticulating, smooth; glumes unequal, deciduous, linear-oblong in profile, acute or subobtuse, subhyaline, 1-nerved or (particularly the lower) nerveless, lower $\frac{3}{4}$–$\frac{4}{5}$ lin. long, upper over 1 lin. long; valves lanceolate, acute or sometimes mucronulate, $1\frac{1}{2}$–$1\frac{3}{4}$ lin. long, thin, smooth except on the scaberulous acute keels; pales 1 lin. long, keels narrowly winged, scaberulous; anthers $\frac{3}{4}$ lin. long.

COAST REGION: Cathcart Div.; Glencairn, 4800 ft., *Galpin*, 2414!
EASTERN REGION: Natal; Riet Vley, 4000–5000 ft., *Buchanan*, 240!

3. **E. curvula** (Nees, Fl. Afr. Austr. 397, excl. var β); perennial, very densely tufted, with numerous closely packed innovation-shoots; culms erect or geniculate, usually slender, simple, 1–2 ft. high, glabrous, smooth, 2–3-noded, internodes usually exserted, uppermost very long; lower sheaths crowded, short, firm, strongly striate, tomentose at the base, gradually less hairy to glabrous upwards, persistent, upper tight, glabrous or rarely hairy, smooth; ligule a fringe of short hairs; blades narrow, linear, long tapering and usually capillary in the upper part, 3 in. to more than 1 ft. long, 1–$1\frac{1}{2}$ lin. wide at the base when expanded, more or less filiform-involute or convolute, at least in the upper part, flexuous, somewhat firm, glabrous, very rarely hairy, scabrid on the upper side and all

over towards the tips, otherwise smooth; panicle open or contracted, erect or more or less nodding, 3–10 in. long; axis filiform, more or less angular, smooth, at least below; branches solitary, unequally distant or partly subverticillate, first erect, then more or less spreading, finely filiform, flexuous, smooth or almost so, glabrous or sometimes with a few fine hairs at the axils, lower divided from 3–6 lin. above the base; branchlets rather loose, usually contracted, simple or the lowest again divided, smooth, rarely the ultimate divisions scaberulous; pedicels unequal, lateral usually short, rarely up to 2 lin. long; spikelets linear-oblong to oblong, 2–3 lin. by 1 lin., loosely 3–6- (rarely to 8-) flowered, usually dark olive-grey; rhachilla subpersistent, then disarticulating, more or less very minutely hairy; glumes more or less unequal, lanceolate to oblong, acute to subobtuse, thinly membranous to almost hyaline, 1-nerved, or sometimes nerveless, keel if present scaberulous, upper up to 1 lin. long, lower slightly shorter; valves lanceolate-oblong in profile, obtuse or subobtuse, 1 lin. long or slightly longer, membranous, scaberulous above the middle, tips usually hyaline and white, side-nerves fine; pales equal to the valves, obtuse, keels fine, smooth or scaberulous above; anthers $\frac{3}{8}$–$\frac{1}{2}$ lin. long; grain subellipsoid, obtusely quadrangular, $\frac{1}{2}$ lin. long, brown; embryo large. *Steud. Syn. Pl. Glum.* i. 271; *Durand & Schinz, Consp. Fl. Afr.* v. 882. *E. filiformis, Nees in Linnæa*, vii. 330; *Fl. Afr. Austr.* 396; *Trin. Gram. Suppl.* 75, *and in Mém. Acad. Pétersb. sér.* 6, iv. 76; *Durand & Schinz, l.c.* 883, *not of Link. E. thunbergiana, Steud. l.c.* 271. *Poa filiformis, Thunb. Prod.* 21; *Fl. Cap.* 420, *ed. Schult.* 112; *in Mém. Nat. Mosc.* iii. 44, *t.* 4; *Kunth, Enum.* i. 345. *P. curvula, Schrad. in Goett. Gelehrt. Anzeig.* 1821, 2073; *and in Schult. Mant.* ii. 308; *Kunth, l.c. P. capensis, Steud. in Flora*, 1829, 488.

VAR. β, **conferta** (Nees, Fl. Afr. Austr. 398); on the whole taller and more robust; panicle contracted, dense, with the branches more or less verticillate and divided from the base, divisions more often scabrid than in the type; spikelets usually crowded, linear to linear-oblong, up to 5 lin. long and to 13-flowered, light olive-green to dark olive-grey.

VAR. γ, **valida** (Stapf); culms usually robust, tall, 3–4-noded; sheaths glabrous and smooth or more or less hairy from often tubercle-based hairs; blades up to more than 2 ft. by 2–3 lin.; panicle $\frac{1}{2}$–1 ft., contracted or open; axis smooth or scabrid; branches 3–6 in. long, flexuous, much divided from the base or simple for as much as 1 in.; spikelets linear to linear-oblong, $3\frac{1}{2}$–$5\frac{1}{2}$ lin. long, 7–13-flowered; glumes and valves very slightly larger than in the type.

SOUTH AFRICA: without precise locality, *Thunberg! Harvey*, 289! 116! *MacOwan*, 2191! Var. β: *Mund! Zeyher*, 1806!

COAST REGION: Cape Div.; near Capetown, *Burchell*, 443! north slopes of the Lions Head, *Wolley Dod*, 3096! Table Mountain, *Ecklon*, 950! 952! *Milne*, 246! Devils Peak, *Wilms*, 3883! by the railway near Rondebosch, *Wolley Dod*, 3345! Paarl Div.; Paarl Mountains, 1000–2000 ft., *Drège!* (approaching var. β). Tulbagh Div.; Piquetberg Road, 400 ft. *Schlechter*, 7838! Riversdale Div.; by the Zoetemelks River, *Burchell*, 6632 partly! near the Great Vals River, *Burchell*, 6545! Albany Div.; at the foot of Bothas Berg, 2000 ft., *MacOwan*, 1278! Komgha Div.; near Komgha, *Flanagan*, 948! Stutterheim Div.; grassy hills near the Kabousie River, 2000 ft., *Flanagan*, 913! Var. β: Worcester Div.; Brand Vley, *Rehmann*, 2420! Hex River East, *Wolley Dod*, 3704! "Swellendam and George," *Mund & Maire!* (intermediate between the type and var. β).

Swellendam Div. ? at Kenko River, between Riet Kuil and Hemil en Aarde, below 1000 ft., *Zeyher*, 1806! Knysna Div.; Plettenberg Bay, *Mund & Maire!* Uitenhage Div.; near Uitenhage, *Ecklon & Zeyher*, 452! *Harvey*, 100! Port Elizabeth Div.; near Port Elizabeth, *E.S.C.A. Herb.* 133! Alexandria Div.; Zuur Veld, *Gill!* Albany Div.; Albany Plains, *Bowie!* Komgha Div.; by the Kei River, *Drège!* "Kaffraria," *Baur*, 276! (with the typical form).

CENTRAL REGION: Prince Albert Div.; near Klaarstroom, on the Great Zwarte Bergen, 2000–3000 ft., *Drège!* Sutherland Div.; by the Great Riet River, *Burchell*, 1383! (both approaching var. β). Var. β: Somerset Div.; at Biesjes Fontein near Loots Kloof, 2800 ft., *MacOwan*, 1610! near Somerset East, *Bowker*, 148! Graaff Reinet Div.; near Graaff Reinet, 2600 ft., *Bolus*, 555! Colesberg Div.; near Colesberg, *Shaw*, 8! 10! Wodehouse Div.; Indwe, *Baur*, 69!

EASTERN REGION: Natal; Umsinga and base of the Biggars Berg, *Buchanan*, 93! Var. γ: near Durban, *Plant*, 57! *Gerrard & MacKen*, 35! (approaching the type); Umpumulo, 2000 ft., *Buchanan*, 248! 249a! very common at Riet Vley, 4000 ft., *Buchanan*, 250; without precise locality, *Buchanan*, 78! 249! *Gerrard*, 675! (approaching the type). Var. γ: Natal; Berea, *Wood*, 5940! Umhlanga, *Wood*, 6060! Van Reenans Pass, *Wood*, 7224! Pieter Maritzburg, *Wood*, 7229!

KALAHARI REGION: Orange Free State; without precise locality, *Hutton!* Transvaal; near Lydenberg, *Wilms*, 1714!

Nees quotes numerous other localities, partly under *E. filiformis*, partly under *E. curvula*, from which I have not seen any specimen. As they come, however, entirely within the area indicated by the localities cited above, and as I am not certain as to the variety under which I should have to enumerate them, I preferred omitting them altogether. Drège's specimen from Paarl Mountain, mentioned above, was distributed as *Eragrostis capillifolia*, Nees. Nees described this plant from Paarl Mountain, and does not mention any other locality; but the description in Nees, Fl. Afr. Aust. 403, does not agree with Drège's specimen, named *E. capillifolia* in the Kew Herbarium, as Nees described his plant as having hairy lower and scabrid upper internodes, a pyramidal panicle with a very scabrid axis and similar spreading branches. Nevertheless, I am convinced that this species as well as *E. subulata*, Nees (l.c. 399), from the Gauritz River and the Lange Kloof which I have also not seen, are but states of *E. curvula*.

4. **E. lehmanniana** (Nees, Fl. Afr. Austr. 402); perennial, tufted; culms usually repeatedly geniculate; sometimes prostrate at the base and rooting from the nodes, branched, with the branches often fascicled, slender, 1 to more than 2 ft. long, glabrous or smooth, rather wiry, more or less compressed, 4-noded, internodes exserted, uppermost the longest; lower sheaths short, close, firm, softly hairy to villous at the very base, persistent, upper tight, glabrous, long bearded at the mouth; ligule a fringe of short hairs; blades very narrow linear, tapering to a fine rigid point, usually filiform convolute, sometimes subulate and pungent, 2–6 in. by $\frac{3}{4}$–1 lin., rarely longer; very rigid to subflexuous, glabrous, smooth below, scaberulous on the upper side; panicle open, ovoid to oblong, lax, 3–6 in. by 1–3 in., erect, or slightly nodding; axis filiform, smooth; lower branches 2-nate or sometimes whorled, or all solitary, unequally distant, spreading, at least ultimately, finely filiform, straight and rather rigid or subflexuous, glabrous or with hairs in the axils, longest 1–2$\frac{1}{2}$ in. long, undivided for $\frac{1}{6}$–$\frac{1}{2}$ in. from the base, then loosely and (at length) subdivaricately branched, smooth or the ultimate divisions scaberulous; pedicels fine, the lateral rarely over 1 lin. long; spikelets linear, acute, 2–4 lin. by $\frac{1}{2}$ lin., closely 4–13-flowered, dark olive-grey; rhachilla subpersistent, then dis-

articulating, very slender, flexuous, smooth or almost so, joints $\frac{1}{2}$ lin. long; glumes more or less unequal, lanceolate-oblong to oblong in profile, subacute to obtuse, very thin to subhyaline, scaberulous, 1-nerved or almost nerveless, margins minutely serrulate, lower about $\frac{2}{5}$ lin., upper $\frac{3}{4}$ lin. long; valves oblong in profile, rounded on the back, closely imbricate with the tips adpressed, obtuse, scarcely $\frac{1}{8}$ lin. long, membranous, pallid towards the base, minutely scaberulous above the middle, side-nerves faint, sometimes whitish, keel obscure below, more distinct and scaberulous near the tip; pales equal to the valves, keels fine and scaberulous; anthers $\frac{3}{8}$–$\frac{2}{5}$ lin. long; grain obovoid-subellipsoid, obtusely quandrangular, over $\frac{1}{4}$ lin. by $\frac{1}{5}$ lin., light brown. *Steud. Syn. Pl. Glum.* i. 271; *Durand & Schinz, Consp. Fl. Afr.* v. 884.

VAR. β, **ampla** (Stapf); culm up to 3 ft. long, stout, more or less hairy below the nodes; lower sheaths long, compressed and more or less keeled; leaves scantily hairy, hairs very fine and spreading; blades up to 8 in. long, setaceous to capillary and flexuous in the upper part; panicle 7 in. by 6 in.

CENTRAL REGION: Graaff Reinet Div.; in shady places near Graaff Reinet, 2700 ft., *Bolus*, 557! Colesberg Div.; near Colesberg, *Shaw*, 9! Albert Div.; in rocky and stony places, *Drège.*

KALAHARI REGION: Griqualand West, Herbert Div.; St. Clair, Douglas, *Orpen*, 257! Hay Div.; Griqua Town, *Burchell*, 1943/2. Orange Free State; between Kimberley and Bloemfontein, *Buchanan*, 280! and without precise locality, *Buchanan*, 267! Transvaal; near Lydenberg, *Atherstone!* Var. β: Basutoland; Leribe, *Buchanan*, 141!

EASTERN REGION: Natal, without precise locality, *Buchanan*, 277!

5. E. chloromelas (Steud. Syn. Pl. Glum. i. 271); perennial, very densely tufted with closely packed innovation-shoots; culms erect or geniculate, slender, simple, very rarely branched above the base, $\frac{1}{2}$–$1\frac{1}{2}$ ft. long, subcompressed, glabrous, or very rarely scantily hairy, smooth, usually 2-noded, internodes exserted, uppermost very long; lower sheaths crowded, very short, firm, adpressedly hairy to tomentose at the very base or quite glabrous, persistent, upper tight, glabrous, or with few fine scattered hairs, long-bearded at the mouth; ligule a fringe of short hairs; blades very narrow, filiform-convolute, capillary above, flexuous, 3–6 in. long, rarely longer, $\frac{1}{2}$–1 lin. broad when expanded, somewhat rigid, glaucous, glabrous or scantily hairy, scaberulous or scabrid on the upper face and all over towards the tips, otherwise smooth; panicle open, ovoid or pyramidal, lax, 2–8 in. long, erect, rather rigid; axis filiform, smooth; lower branches in whorls of 5–3, or 2-nate, rarely all solitary, spreading, finely filiform, straight or subflexuous, glabrous or sometimes with a few fine hairs at the axils, longest $1\frac{1}{2}$–4 in. long, undivided for $\frac{1}{3}$–1 in. from the base, then very loosely and at length divaricately branched, smooth, or the ultimate divisions scaberulous; pedicels capillary, the lateral 1–3 lin. long; spikelets scattered, linear, acute, 2–4 lin. by $\frac{1}{2}$–$\frac{3}{4}$ lin., loosely 5–13-flowered, dark olive-grey to slate-grey; rhachilla subpersistent, then disarticulating, very slender, flexuous, smooth, or almost so, joints up to $\frac{3}{8}$ lin. long; glumes unequal, deciduous, lanceolate to lanceolate-oblong in profile, thinly acute or subacute, membranous or hyaline, 1-nerved, scaberulous on the nerve, lower $\frac{1}{2}$–$\frac{3}{4}$ lin., upper $\frac{3}{4}$–1 lin. long; valves obliquely oblong in profile, subacute to acute, $\frac{3}{4}$–$\frac{7}{8}$ lin. long, membranous, smooth, side-nerves fine; pales equal to the valves, obtuse, keels fine, smooth or nearly so; anthers about $\frac{2}{5}$ lin. long; grain oblong-ellipsoid, obtusely quadrangular, $\frac{1}{2}$ lin. long, brown; embryo large. *Durand & Schinz, Consp. Fl. Afr.* v. 881. *E. atrovirens, Nees, Fl. Afr. Austr.* 400, *not of Trin.*

COAST REGION: Cathcart Div.; between Kat Berg and Klipplaats River, *Drège!* Queenstown Div.; plains near Queenstown, 3500 ft., *Galpin*, 2354!
CENTRAL REGION: Graaff Reinet Div.; by the Sunday River, *Drège!* Somerset Div.; near Somerset East, *Bowker*, 153! Wodehouse Div.; Indwe, *Baur*, 72!
KALAHARI REGION: Orange Free State; without precise locality, *Cooper*, 3348! Transvaal; Houtbosch, *Rehmann*, 5689! 5697! near Potschefstroom, *Nelson*, 89*!
EASTERN REGION: Natal, near Ladysmith, *Rehmann*, 7130! 7134! Umhlanga, *Wood*, 6058! near Van Reenens Pass, 5000–6000 ft., *Wood*, 7221!

6. E. nebulosa (Stapf); perennial, densely tufted on a short oblique rhizome; culms erect, rather slender, stiff, 2–3 ft. long, glabrous, smooth, about 3-noded, internodes long, exserted, nodes slightly marked; lower sheaths compressed, more or less keeled, 4–5 in. long, very firm, usually scarcely striate, quite glabrous, very smooth, often shining, upper tight, terete, scantily bearded at the mouth or quite glabrous; ligule a fringe of very minute hairs; blades very narrow, finely filiform and canaliculate below, capillary in the upper part, 1 ft. long or longer, conspicuously narrower than the sheath at their junction, rather rigid below, very flexuous above, glabrous, smooth or scaberulous towards the tips; panicle erect or nodding, large, at length open and very lax, $\frac{3}{4}$–1$\frac{1}{4}$ ft. long and almost as wide; axis filiform, terete, smooth; branches 3–2-nate, or partly solitary, at length spreading, the longest 4–6 in. long and undivided for 1–2 in. from the base, then like the rest distantly branched, finely filiform, glabrous, smooth or scaberulous; branchlets again scantily and loosely divided, up to 1$\frac{1}{2}$ in. long, like the often long pedicels capillary and very flexuous; spikelets linear, acute, 2–3 lin. by $\frac{1}{3}$–$\frac{2}{3}$ lin., loosely 4–10-flowered, olive-grey; rhachilla subpersistent, very slender, smooth; glumes subequal, lanceolate, acute in profile, $\frac{2}{3}$–$\frac{3}{4}$ lin. long, delicate, 1-nerved, keel scaberulous; valves lanceolate-oblong in profile, acute to subacuminate, $\frac{5}{6}$ lin. long, membranous, smooth, slightly shining, side-nerves faint and short; pales equalling the valves, keels smooth or almost so; anthers $\frac{1}{3}$–$\frac{1}{2}$ lin. long; grain oblong, $\frac{2}{5}$ lin. by $\frac{1}{6}$ lin., brown.

COAST REGION: Swellendam Div.; right bank of the Zonder Einde River, *Burchell*, 7521! Riversdale Div.; Zoetemelks River, *Burchell*, 6632 in part!
KALAHARI REGION: Orange Free State; Hoopstad, *Herb. Grindon!* and without precise locality, *Buchanan*, 34! Basutoland; between Harrismith and Leribe, *Buchanan*, 209! Transvaal; marshy places in the Magalies Bergen, *E.S.C.A. Herb.*, 315!
EASTERN REGION: Natal; on the Drakensberg Range, near Newcastle, *Buchanan*, 196! De Beers Pass, *Wood*, 5992! Mooi River, 3000–4000 ft., *Wood*, 7325!

7. **E. margaritacea** (Stapf); perennial; culms erect, rather slender, firm, simple, 2–3 ft. long, glabrous, smooth, 4-noded, internodes long exserted; sheaths tight, glabrous, striate, lowest compressed and keeled, tomentose at the very base; ligule a fringe of minute hairs; blades narrow, linear, long tapering, up to more than 6 in. by 1–1½ lin., when expanded, usually convolute or involute, keeled below, smooth on the lower, densely scaberulous on the fine prominent nerves on the upper side, often reddish; panicle oblong, erect or somewhat flaccid, 8–9 in. long; axis filiform, terete below, or more or less angular throughout; branches in rather distant whorls of 3 or the lowest of 5–7, suberect or rather spreading, flaccid or at least flexuous, capillary almost from the base, like all the divisions very scaberulous all over, the lowest with hairs in their axils, distantly divided from ⅓–½ in. above the base, the longest to more than 3 in. long; branchlets contracted, very fine and flexuous, 4–1-spiculate; pedicels up to 1 lin. long; spikelets contracted in brush-like fascicles, linear, 5–7 lin. by ½ lin., loosely 7–13-flowered, pearl-grey, often tinged with purple; rhachilla very slender, smooth, subpersistent; glumes deciduous, lanceolate to oblong in profile, acute or subobtuse, very delicate, 1-nerved to nerveless, lower ½ lin. long, upper ⅔ lin. long; valves lanceolate-oblong, obtuse, ⅔–¾ lin. long, membranous except at the hyaline sometimes whitish tips, smooth, slightly shining, side-nerves short, fine; pales subequal to the valves, keels fine, scaberulous upwards; anthers ⅓ lin. long; grain ellipsoid-oblong, ⅓ lin. by ⅕ lin., brown.

KALAHARI REGION: Orange Free State; between Kimberley and Bloemfontein, *Buchanan*, 282! Transvaal; Bosch Veld, at Klippan, *Rehmann*, 5372!

8. **E. porosa** (Nees, Fl. Afr. Austr. 401); annual or subperennial (?), tufted; culms geniculate, simple, or with leafy branches from the lower nodes about 1 ft. long, glabrous, smooth, usually with a ring of depressed glands below each node, more or less viscous, 3-noded, internodes exserted; leaves scantily beset with tubercle-based spreading hairs, or the upper quite glabrous; sheaths striate, smooth, bearded at the mouth; ligule a fringe of short hairs; blades linear, tapering to a fine point, 2–3 in. by 1–1½ lin., soft, smooth on the back, scabrid on the upper side and along the margins; panicle erect, ovoid to oblong, 4–6 in. long, open or contracted; axis filiform, smooth; lower branches in whorls of 5–7, upper alternate or irregularly approximate, obliquely erect or more or less spreading, finely filiform, flexuous, glabrous below, lower loosely divided from ½–¾ in. above the base; branchlets contracted or more or less spreading, scabrid, the lowest again divided and up to ½ in. long; pedicels very short; spikelets linear to linear-oblong, 1½–2½ lin. by ¾–1 lin., crowded or rather scattered, olive-green, 3–11-flowered; glumes unequal, deciduous, oblong-lanceolate, subacute, delicate, 1-nerved, keel scaberulous, margins minutely serrulate, lower scarcely ½ lin. long, upper ⅔–¾ lin. long; rhachilla subpersistent, smooth or subscaberulous; valves oblong in profile, very obtuse, truncate or subemarginate,

$\frac{2}{3}$ lin. long or slightly longer, smooth or almost so, side-nerves fine, close to the margin; pales equal to the valves, keels scaberulous above; anthers $\frac{1}{4}$–$\frac{3}{8}$ lin. long; grain ellipsoid, $\frac{1}{4}$–$\frac{3}{8}$ lin. by $\frac{1}{5}$–$\frac{1}{6}$ lin., pallid, translucent. *Steud. Syn. Pl. Glum.* i. 271; *Durand & Schinz, Consp. Fl. Afr.* v. 888. *E. emarginata, Hack. in Verhandl. Bot. Ver. Brandenb.* xxx. 238; *Durand & Schinz, Consp. Fl. Afr.* v. 883.

VAR. β, **parvula** (Stapf); dwarf, $2\frac{1}{2}$–3 in. high; culms 1–2-noded; panicle contracted, 1–$1\frac{1}{2}$ in. long; spikelets dark olive-green, $1\frac{1}{2}$ lin. long, 3–6-flowered; glumes equal, $\frac{1}{2}$–$\frac{2}{3}$ lin. long; valves slightly broader than in the type, up to $\frac{3}{4}$ lin. long.

CENTRAL REGION: Albert Div.; by the Gamka River, near Weltevrede, 2500–3000 ft., *Drège!* Aberdeen Div.; Camdeboo plains, *Drège.*
WESTERN REGION: Great Namaqualand, between Ausis and Kuias, *Schenck,* 82, near Keetmanshoop, *Schinz.*
KALAHARI REGION: Var. β, Orange Free State? Jackals Fontein, *Burke!*

9. **E. bicolor** (Nees, Fl. Austr. 407); perennial, densely tufted; culms erect, simple, slender, terete, 1–$1\frac{1}{2}$ ft. long, glabrous, smooth, about 3-noded, upper 2 internodes exserted; sheaths glabrous, striate, lowest pallid, about 2 in. long, shortly tomentose at the insertion; ligule a fringe of minute hairs; blades narrow, linear, tapering to a setaceous point or setaceously filiform almost from the base, 4–8 in. by 1 lin. when flat, glabrous, rather firm, flexuous, smooth below, densely scaberulous on the prominent nerves on the whitish upper side; panicle ovoid to lanceolate, erect or nodding, 3–5 in. long, lax, open or more or less contracted; axis angular, glabrous, smooth, very slender; branches 2-nate or solitary, suberect or more or less spreading, distantly divided from $\frac{1}{3}$–1 in. above the base, subcapillary, flexuous or finally rigid, subcompressed, smooth or scaberulous, glabrous at the axils, the longest up to $2\frac{1}{2}$ in. long; branchlets capillary, contracted or divaricate, few- to 1-spiculate; pedicels very fine, scaberulous, the lateral $\frac{1}{2}$–2 lin. long; spikelets oblong to linear, acute, $1\frac{1}{2}$–$2\frac{1}{2}$ lin. by $\frac{1}{2}$–$\frac{2}{3}$ lin., loosely 4–9-flowered, variegated; rhachilla very slender, scaberulous, disarticulating; glumes subpersistent oblong-lanceolate, acute to subobtuse, 1-nerved, subhyaline, lower $\frac{1}{2}$ lin. long, or almost so, upper $\frac{2}{3}$ lin. long; valves broadly oblong in profile, obtuse or subobtuse, somewhat turgid, $\frac{3}{4}$ lin. long, smooth, slightly shining, usually purple with yellowish tips, side-nerves stout, though not very distinct, short; pales equal or subequal to the valves, keels fine, scaberulous; anthers almost $\frac{1}{2}$ lin. long. *Steud. Syn. Pl. Glum.* i. 272; *Durand & Schinz, Consp. Fl. Afr.* v. 880.

CENTRAL REGION: Richmond Div.; near Styl Kloof, 4000–5000 ft., *Drège!* Colesberg Div.; near Wonderheuvel, *Drège*; Albert Div.; near Leeuwen Fontein, *Drège!*
KALAHARI REGION: Griqualand West, Hay Div.; between Witte Water and Riet Fontein, *Burchell,* 2001! Herbert Div.; along the Vaal River, *Burchell,* 1780!

10. **E. Poa** (Stapf); perennial, densely tufted; culms erect, simple, rather slender, terete, up to 2 ft. high, glabrous, smooth, about 3-noded, internodes exserted; lowest sheaths persistent, firm,

tomentose at the base, and, like the following, scantily hairy in the upper part, upper glabrous, all striate, beardless; ligule a fringe of minute hairs; blades convolute, filiform, tapering to a setaceous point, erect, $\frac{1}{2}$–1 ft. long, rigid, or flexuous in the upper part, glabrous, smooth below, scaberulous on the stout equal nerves of the upper side; panicle erect, ovoid or pyramidal, rather stiff, open, lax, 5–8 in. by 3–6 in.; axis filiform, terete, smooth; branches more or less in whorls of 4–3, horizontally spreading or almost so with the tips often nodding, lowest 3–4 in. long, loosely divided from $\frac{1}{4}$–$\frac{3}{4}$ in. above the base, finely filiform, glabrous, scaberulous above; branchlets simple, capillary, 3–5 lin. long (or lowest again divided and up to 9 lin. long), mostly pendulous, ultimate divisions 3–2-spiculate; pedicels $\frac{1}{2}$–$\frac{3}{4}$ lin. long; spikelets clustered on the branchlets and towards the tips of the branches, light olive-green, $1\frac{1}{2}$–$2\frac{1}{2}$ lin. by 1 lin., loosely 5–6-flowered; rhachilla disarticulating; glumes subequal or lower somewhat shorter, lanceolate in profile, acute, up to more than 1 lin. long, very delicate, 1-nerved or the lower nerveless, keel scaberulous; valves obliquely oblong in profile, subobtuse, $\frac{7}{8}$–1 lin. long, thinly membranous, keel scaberulous, side-nerves faint; pales equal to the valves, keels fine, scaberulous; anthers $\frac{1}{2}$ lin. long; grain oblong-ellipsoid, $\frac{2}{3}$–$\frac{3}{4}$ lin. by $\frac{1}{5}$–$\frac{1}{3}$ lin., light brown.

COAST REGION: Caledon Div.; moist grounds in the neighbourhood of Caledon, *Thom*, 350!

KALAHARI REGION: Bechuanaland; between Kosi Fontein and Kneghts Fontein, *Burchell*, 2606!

Thom's specimen is imperfect and may possibly belong to another species.

11. E. Wilmsii (Stapf); perennial, densely tufted; culms geniculate, suberect, simple, about $1\frac{1}{2}$ ft. high, smooth, glabrous, 2-noded, internodes exserted; lower sheaths compressed, keeled, loosely striate, glabrous except at the bearded mouth, rather firm, herbaceous except at the base, uppermost tight, terete; ligule a fringe of very short hairs; blades narrow, linear, long tapering to a fine point, 6–7 in. by 1–$1\frac{1}{2}$ lin. when flattened, folded, keeled below, with the margins involute, or convolute in the upper part, glabrous, smooth below, scaberulous to scabrid on the upper surface along the nerves, primary side-nerves about 4 on each side, stout; panicle erect, oblong, open, lax, up to 10 in. by 3 in.; axis smooth, terete; branches scattered subopposite or irregularly approximate, virgate, obliquely spreading, rather stiff, distantly divided from about $\frac{1}{2}$ in. above the base, bearded at the axils; branchlets spreading or usually contracted, flexuous, subcapillary to capillary, smooth or almost so, the lower again divided, up to $1\frac{1}{2}$ in. long, the others loosely 3–1-spiculate; pedicels capillary, lateral 1–3 lin. long; spikelets lanceolate to linear-lanceolate, acute, $1\frac{1}{2}$–$2\frac{1}{2}$ lin. by $\frac{2}{3}$–$\frac{3}{4}$ lin., straw-coloured or light olive-green, loosely 3–6-flowered; rhachilla subpersistent, at length disarticulating; glumes lanceolate in profile, acute to acuminate, thin, 1-nerved, lower $\frac{1}{2}$–$\frac{2}{3}$ lin., upper $\frac{3}{4}$ lin. long; valves lanceolate in profile, acute, $\frac{5}{6}$–1 lin. long; keel scaberulous, side-nerves fine, tips hyaline; pales distinctly shorter than or almost equal to the valves, keels fine and scaberulous; anthers $\frac{1}{8}$ lin. long.

KALAHARI REGION: Transvaal; near Pretoria, *Wilms*, 1713a!

12. E. sporoboloides (Stapf); perennial, compactly tufted; culms geniculate, suberect, slender, terete, wiry, not quite 1 ft. long, 2-noded, internodes exserted; sheaths very tight, firm, striate, lower sparingly and finely hairy or glabrescent, bearded at the mouth, upper glabrous, lowest persistent; ligule a fringe of minute hairs; blades setaceous, wiry, tightly convolute, very flexuous, 3–6 in. long, firm, glabrous or the lower sparingly hairy, striate, smooth on the back, scaberulous along the prominent nerves of the upper side; panicle erect, ovoid, $1\frac{1}{2}$–2 in. by 1–$1\frac{1}{2}$ in., rather rigid, dense or lax; axis terete, smooth; branches in whorls of 5–3, or the upper subopposite or solitary, obliquely erect, loosely divided from $\frac{1}{6}$–$\frac{1}{3}$ in. above the base, subcapillary, rather stiff, scaberulous; branchlets short, simply racemose, 4–1-spiculate or the lower again divided, divaricate or subsecund and more or less contracted, particularly toward the tips; pedicels short, scaberulous; spikelets lanceolate $\frac{3}{4}$–$1\frac{1}{2}$ lin. long, dark olive-grey, 2–4-flowered; glumes lanceolate, acute or acuminate in profile, hyaline, 1-nerved or nerveless, lower scarcely $\frac{1}{2}$ lin., upper $\frac{2}{3}$ lin. long; valves ovate-lanceolate, subacute to acutely acuminate in profile, $\frac{3}{4}$–$\frac{7}{8}$ lin. long, membranous, smooth or almost so; pales equal or subequal to the valves, keels fine and almost smooth; anthers $\frac{2}{5}$ lin. long.

KALAHARI REGION: Transvaal; Houtbosch, *Rehmann*, 5686! 5695!

13. E. Atherstonei (Stapf); perennial; culms erect, simple or with 1–2 branches from the upper nodes, over 2 ft. long, wiry, glabrous, smooth, about 3-noded, lowest lengthened, internodes about $\frac{1}{2}$ ft. long, like the upper long exserted, with a ring of more or less distinct glands below the nodes; sheaths tight, glabrous, smooth, substriate, intermediate 1–$1\frac{1}{2}$ in. long; ligule a fringe of minute hairs; blades narrow, linear, tapering to a fine point, 6–8 in. by 1–$1\frac{1}{2}$ lin., more or less involute or convolute, rather firm, scaberulous on the upper side, otherwise smooth; panicle oblong, erect, lax, 5–6 in. by 1–$1\frac{1}{2}$ lin.; axis filiform, smooth; branches whorled, the lower in whorls of 4–7, obliquely erect or slightly spreading, loosely divided from near the base, longest 1–$1\frac{1}{2}$ in. long, capillary or subcapillary, straw-coloured, smooth, bearded at the axils; branchlets capillary, subdivaricate, short, 3–1-spiculate, smooth or subscaberulous; lateral pedicels about 1 lin. long; spikelets scattered, lanceolate, 1–2 lin. long, grey, 2–4-flowered; rhachilla subpersistent; glumes subequal, lanceolate in profile, acute or acuminate, very delicate, faintly 1-nerved, up to 1 lin. long, or the lower slightly shorter; valves oblong, subobtuse, $\frac{7}{8}$ lin. long, thinly membranous, smooth or almost so, side-nerves short, faint; pales equal to the valves, keels scaberulous and fine; anthers $\frac{3}{8}$ lin. long; grain pallid, translucent, oblong-ellipsoid, over $\frac{1}{4}$ lin. long.

KALAHARI REGION: Transvaal; near Lydenburg, *Atherstone!*

14. E. micrantha (Hack. in Bull. Herb. Boiss. iii. 389); perennial, densely tufted; culms ascending or erect, simple, slender, 1–1½ ft. long, glabrous, smooth, sometimes with a ring of obscure glands below the nodes, terete, 3–5-noded, internodes exserted; sheaths tight, glabrous except at the often scantily bearded mouth, striate, lower usually short; ligule a fringe of short hairs; blades very narrow, linear, tapering to a setaceous point, flexuous, 3–5 in. by ½ lin., rather soft, glabrous, smooth on the back, scaberulous on the face; panicle erect, obovoid-oblong to oblong, 2–6 in. by 1–2 in., lax, shortly exserted from or enclosed at the base in the uppermost sheath; axis filiform, smooth; branches verticillate, obliquely spreading, subcapillary to capillary, glabrous or scaberulous, at least in the upper part, the longest 1–2 (rarely to 3) in. long, loosely and repeatedly divided from 2–4 lin. above the base; branchlets rather long and divaricate, or more or less contracted; pedicels very fine, long, divisions often pinkish; spikelets lanceolate, 1–1½ lin. long, loosely 3–4-flowered; rhachilla subpersistent, at length disarticulating, scaberulous; glumes lanceolate in profile, acute or subacute, delicate, smooth except on the scaberulous keels, the lower sometimes nerveless, ½ lin. long or scarcely so, upper ⅔ lin. long; valves narrow-oblong to oblong-lanceolate in profile, subobtuse, ¾ lin. long, membranous, smooth, except near the tips, side-nerves faint, short; pales more or less equal to the valves, keels smooth or scaberulous; anthers ⅖ lin. long.

KALAHARI REGION: Orange Free State; near Draai Fontein, *Rehmann*, 3645! Basutoland; near Leribe, *Buchanan*, 126!

15. E. Burchellii (Stapf); perennial, tufted; culms erect or ascending, firm, slender, branched from the lower and intermediate nodes, 1½ ft. high, glabrous, smooth, 5–6-noded, internodes exserted; sheaths glabrous, finely striate, firm, tight at first, then loosened or thrown aside, the lowest tomentose at the very base; ligule a fringe of soft short hairs; blades linear, tapering to an acute point, 3–6 in. by 1½–2 lin., firm, setaceously convolute and rigid when young, at length flat and subflaccid, glabrous or with a few scattered fugacious hairs, smooth below, scaberulous or pruinose on the upperside; panicle pyramidal or oblong, erect, stiff, 3–4 in. by 1–2 in., lax, open; axis filiform, smooth; branches solitary or the lowest subopposite, obliquely erect or spreading, loosely and often divaricately divided from a few lines above the base, or simply racemose, filiform, smooth or scaberulous in the upper part, the longest ¾–2 in. long; lowest branchlets again divided, or like the rest simple, 4–1-spiculate, spreading or adpressed, all divisions yellowish; pedicels short; spikelets scattered, lanceolate to oblong, 1–2 lin. long, dark olive-grey, 2–4-flowered; rhachilla disarticulating; glumes lanceolate, subequal in profile, delicate, finely 1-nerved or nerveless, lower about ½ lin., upper ⅔ lin. long; valves ovate-oblong, obtuse, ¾–⅚ lin. long, very thin, smooth or scaberulous towards the margins, keel smooth, side-nerves faint; pales equal to the valves, keels fine, almost smooth; anthers ¼–⅓ lin. long; grain obovoid-ellipsoid, ⅓ lin. by ⅙–⅕ lin.

KALAHARI REGION: Carnarvon Div.; at Buffels Bout, *Burchell*, 1606! Prieska Div.; between Modder Fontein and Keikams Poort, *Burchell*, 1612/11!

16. E. dura (Stapf); perennial, densely tufted; culms erect or suberect, very firm, terete, simple, up to 2½ ft. long, glabrous, smooth, 2-noded, internodes exserted, all long; lower sheaths firmly papery or scarious, striate, perfectly glabrous and smooth, persistent, enclosing very slender fugaciously woolly innovation-shoots, upper loose, subcoriaceous, very finely striate, separating at length at the very base; ligule a fringe of minute hairs; blades narrow, linear, tapering to a fine setaceous point, over ½ ft. by 1½ lin. (at the base), canaliculate, or almost flat below, rigid, smooth on the back, scaberulous on the prominent nerves of the upper surface, those of the innovations finely setaceous; panicle erect, rather stiff, open, lax, up to 7 in. by 2 in.; rhachis filiform, terete, smooth; branches oblique erect, solitary or the lowest in whorls of 3, very loosely divided from the base or a few lines above it, filiform, smooth or scaberulous in the upper part, longest 2–3 in. long; branchlets 2–4 lin. long, more or less adpressed, 3–1-spiculate, finely filiform; pedicels very short; spikelets oblong, 1½ to more than 2 lin. long, 2–4-flowered, light brown, shining; rhachilla disarticulating; glumes subequal, ovate-oblong, acute to acuminate in profile, about ½ lin. long, firm on the back, keel scaberulous; valves ovate-oblong, subobtuse in profile, $\frac{5}{6}$–$\frac{7}{8}$ lin. long, almost chartaceous, keel stout and smooth or almost so, side-nerves inconspicuous; pales equal to the valves, back broad, truncate, keels prominent and obtusely scaberulous.

KALAHARI REGION: Bechuanaland; between Kuruman and Matlareen River, *Burchell*, 2190!

17. E. plana (Nees, Fl. Afr. Austr. 390); perennial, densely tufted; culms erect or suberect, strongly compressed, 2–3 ft. long, glabrous, smooth, 3-noded, upper internodes long, usually more or less exserted; leaves crowded, and almost flabellate at the base, striate, glabrous; lower sheaths strongly compressed, keeled, pallid; ligule a dense fringe of short hairs; blades very narrow, linear, long tapering to a setaceous point, tightly folded, flexuous, 3 to more than 12 in. long, closely striate, smooth on the lower, scabrid and whitish on the upper side; panicle narrow, linear to oblong, nodding, ½–1 ft. long; axis angular, smooth; branches solitary, very unequally distant, erect or slightly spreading, subflexuous or somewhat nodding, longest 1–4 in. long, finely filiform, more or less triquetrous, smooth or scabrid along the angles, remotely divided from near the base with the lower branchlets 3–2-spiculate, or all simply racemose; pedicels up to 1½ lin. long; spikelets linear, 3–6 lin. by 1 lin., olive-green to olive-grey, loosely 7–15-flowered; rhachilla subpersistent; glumes very unequal, lanceolate to oblong, acute or obtuse, pallid, 1-nerved, lower about ¼ lin. long, upper ½ lin. long; valves somewhat spreading, obliquely oblong in profile, folded, acute or subacute, 1–1¼ lin. long, keel smooth, like the side-nerves prominent, rigid, almost straight; pales equal to the valves, keels curved, scaberulous above; anthers

$\frac{2}{3}$–$\frac{3}{4}$ lin. long; grain oblong, $\frac{2}{3}$ lin. by $\frac{1}{3}$ lin., reddish-brown. *Steud. Syn. Pl. Glum.* i. 270; *Durand & Schinz, Consp. Fl. Afr. Austr.* 888.

COAST REGION: King Williamstown Div.; by the Yellowwood River, *Drège.* Queenstown Div.; plains near Queenstown, 3500 ft., *Galpin*, 2359!

KALAHARI REGION: Orange Free State; between Harrismith and Leribe, *Buchanan*, 216! Transvaal; near Lydenburg, *Wilms*, 1715!

EASTERN REGION: Transkei; between the Gekau (Gcua or Geuu) River and Bashee River, 1500–2000 ft., *Drège!* Natal; Berea, *Wood*, 5928! 5937! Mooi River, *Wood*, 7320! Riet Vlei, *Buchanan*, 247! near Durban, *Williamson*, 54! and without precise locality, *Buchanan*, 244!

18. E. heteromera (Stapf); perennial; culms geniculate-ascending, stout, simple, over 2 ft. long, glabrous, smooth, 3-noded, internodes (except the lowest) exserted; sheaths quite glabrous except at the usually bearded mouth, smooth, more or less coarsely striate, the lower not compressed, often purplish; ligule a dense fringe of minute hairs; blades linear, tapering to a long fine point, 6–8 in. by $1\frac{1}{2}$–2 lin. long, flat or more or less involute, rather soft, glabrous, smooth on the lower, scaberulous on the upper side, midrib rather stout below, primary side-nerves 4–5, prominent; panicle oblong, nodding, 10–12 in. long; axis angular, striate or sulcate, glabrous; branches somewhat irregularly arranged, in false whorls or 2–4-nate or solitary, suberect, flexuous to flaccid, unequal, divided from near the base or undivided for 1 in. or more, capillary, scaberulous, the longest 4–6 in. long; branchlets somewhat distant, short, contracted, 3–1-spiculate, very fine; lateral pedicels very short; spikelets linear, acute, 2–4 lin. by $\frac{2}{3}$–$\frac{3}{4}$ lin., olive-green, loosely 4–12-flowered; rhachilla subpersistent, sparingly scaberulous; glumes very unequal, lower a minute scale or quite suppressed, rarely over $\frac{1}{3}$ lin. long, upper lanceolate to oblong, subacute, $\frac{1}{2}$–1 lin. long, hyaline, 1-nerved; valves obliquely oblong, obtuse, $\frac{3}{4}$–1 lin. long, keel scabrid and prominent like the side-nerves, rigid, almost straight; pales equal to the valves, keels curved and scabrid; anthers $\frac{2}{3}$–$\frac{3}{4}$ lin. long. *Sporobolus fimbriatus, Nees in Drège, Zwei Pflanzengeogr. Docum.* 223 (*b. only*). *Poa filiformis, Krauss in Flora,* 1846, 121, *not Thunb.*

EASTERN REGION: Natal; near Durban, *Drège!* by the Umlazi River, and near Pietermaritzburg, *Krauss*, 43! by the Tugela River, 600–1000 ft., *Buchanan*, 241! 245a!

19. E. biflora (Hack. in Bull. Herb. Boiss. iii. 390); annual, tufted; culms erect, or ascending, very slender, 8 in. to almost 2 ft. long, glabrous, smooth, 2–3-noded, internodes more or less exserted (at least ultimately); sheaths tight, more or less compressed, glabrous, smooth; ligule a fringe of minute hairs; blades linear, tapering to a fine point, 2–10 in. by $\frac{1}{2}$–1 lin., flat, very flaccid, green, glabrous or scantily hairy, scabrid; panicle oblong, very loose and open, 4–10 in. long; axis smooth, filiform; branches 3–2-nate, rarely solitary, spreading, capillary, scabrid, glabrous, loosely divided from near the base, lowest branchlets divided again, and 3- or more-spiculate; pedicels very long and fine; spikelets minute, elliptic to lanceolate,

1 lin. long, 2- or sometimes 1- (very rarely 3-) flowered, green, more or less tinged with purple or almost wholly purple; rhachilla very slender, persistent, continued and scaberulous beyond the upper floret; glumes subequal, linear-lanceolate, $\frac{1}{2}$ to almost $\frac{3}{4}$ lin. long, 1-nerved, keel scaberulous; valves oblong to lanceolate in profile, acute to subobtuse, $\frac{2}{3}$ lin. long, very thin, glabrous, side-nerves faint; pales persistent equalling the valves, truncate, keels fine and smooth; anthers $\frac{1}{6}$–$\frac{1}{4}$ lin. long; grain subglobose, about $\frac{2}{5}$ lin. in diam., granulate, excavated above the punctiform hilum, pit oblong, pallid; embryo $\frac{1}{2}$ the length of the grain; pericarp slightly swelling in water. *Hack. in Bull. Herb. Boiss.* iv. *App.* iii. 26.

WESTERN REGION: Great Namaqualand; without precise locality, *Fleck*, 281a.
KALAHARI REGION: Orange Free State; near Bloemfontein, *Rehmann*, 3759! and without precise locality, *Buchanan*, 270! Transvaal; by the Vaal River, *Nelson*, 149! Bosch Veld, at Klippan, *Rehmann*, 5364! under trees, on the plains near Rustenberg, 3500 ft., *McLea*, 124!

20. E. cyperoides (Beauv. Agrost. 162); perennial; culms erect or shortly ascending from a creeping rhizome, branched and densely sheathed at the base, then quite leafless and nodeless, stout to very stout, terete, 1–1$\frac{1}{2}$ ft. long, glabrous, very closely striate, glaucous; sheaths subcoriaceous, rather wide, glabrous, closely striate, lowest blade-less and pungent, or with a rudimentary pungent blade, those of the rhizome scale-like, scarious, more or less silky hairy; ligule a short dense fringe of hairs; blades subulate-convolute from a rather broad base, pungent, gradually increasing upwards, longest to more than 3 in. by 4 lin. (at the base), very rigid, glabrous and smooth below, very finely white tomentose on the upper side; panicle consisting of one or, usually, several very distant branches bearing dense clusters of spikelets from the base or almost so, the panicle rarely reduced to a terminal cluster, 4–9 in. long; axis not differentiated from the culm below, naked, and more or less pungent at the top or ending with a cluster of spikelets; branches $\frac{1}{2}$–2 in. long, erect or spreading, stout, rigid, densely covered with clustered spikelets, finely tomentose or puberulous like the very short branchlets and pedicels; spikelets oblong, 3–4 lin. by 1$\frac{1}{2}$ to almost 2$\frac{1}{2}$ lin., light green, sometimes slightly tinged with purple, closely 4–9-flowered; rhachilla tardily disarticulating, more or less minutely puberulous; glumes lanceolate-oblong, subobtuse or acute in profile, firmly membranous, keel scaberulous, lower about 1 lin. long, upper slightly longer; valves ovate-oblong to ovate, subacute to obtuse in profile, about 1$\frac{1}{2}$ lin. long, chartaceous to subcoriaceous, very finely puberulous towards the margins, keel almost smooth, side-nerves strong; pales equal to the valves, keels stout and densely ciliolate; anthers 1 lin. long; grain oblong, elliptic in cross section, $\frac{5}{8}$ lin. by $\frac{1}{3}$ lin., brown. *Roem. & Schult. Syst.* ii. 577; *Nees in Linnæa*, vii. 328; *Durand & Schinz, Consp. Fl. Afr.* v. 882. *Poa cyperoides, Thunb. Prodr.* 22; *Fl. Cap.* i. 424; *ed. Schult.* 113; *Kunth, Enum.* i. 345. *Brizopyrum cyperoides, Nees, Fl. Afr. Austr.* 374. *E. enodis, Hack. in Verhandl. Bot. Ver. Brandenb.* xxx. 148.

SOUTH AFRICA: without precise locality, *Herb. Harvey*, 332! *Drège*, 2540!

COAST REGION: Clanwilliam Div.; Lamberts Bay, *Schlechter*, 8546! Cape Div.; in dunes and loose sand near Capetown, *Thunberg!* Green Point, *Ecklon*; Table Bay and Paarden Island, *Drège!* sand hills of Paarden Island, *Wolley Dod*, 3181! in sand at the mouth of Eerste River, False Bay, *MacOwan, Herb. Austr. Afr.*, 1767!

WESTERN REGION: Great Namaqualand; Nautilus Point near Angra Pequena (ex *Hackel*). Little Namaqualand; near the mouth of the Orange River, below 600 ft., *Drège!*

Also in tropical German South-West Africa.

In some of the spikelets of Harvey's specimen I observed that the valves are much broader than usual, and have 1–2 additional side-nerves.

21. E. spinosa (Trin. in Gram. Gen. 416, and in Mém. Acad. Pétersb. sér. 6, i. 416); perennial, creeping with a long slender woody rhizome or stolon, branching near the surface of the soil and above it; culms fascicled, sheathed at the base, stout, hard, terete or compressed, 1–5 in. long (to the base of the panicle), glabrous, smooth, node-less above the base, overtopped by the sheaths or more or less exserted from them; sheaths subcoriaceous to scarious, persistent, glabrous except the often finely silky base, faintly striate, at first very tight, then loose, lowest thinner, short, more or less blade-less and more silky; ligule a fringe of short hairs; blades linear, flat, subulate-convolute above the base, pungent, gradually increasing upwards, from a few lines to $1\frac{1}{2}$ in. by $1\frac{1}{4}$ lin., usually very hard and rigid, spreading, glabrous and smooth below, minutely papillose on the upper side; panicle oblong, narrow, very rigid, open, about $\frac{1}{2}$ ft. long; axis not differentiated from the culm, ending with a pungent tip, terete below or more or less angular all along, sulcate above the branches, and marked with more or less obsolete transverse ridges between them, glaucous; branches solitary, distant or some subopposite, horizontally spreading, very rigid, pungent, from a few lines to 2 in. long, somewhat compressed or quite flat and up to 1 lin. broad, smooth, glabrous, glaucous, simply racemose with 1–9-spikelets or with very short spinous 1-spiculate branchlets in the lower part; pedicels filiform, smooth, 1 lin. long; spikelets lanceolate to linear, acute, 3–9 lin. long by $1\frac{1}{2}$–2 lin., pallid, sometimes tinged with purple, deflexed, loosely 3–18-flowered; rhachilla disarticulating, smooth or almost so; glumes subequal, lanceolate, acute or subobtuse in profile, about 1 lin. long, hyaline, 1-nerved; valves oblong, acute or subacute in profile, $1\frac{1}{2}$ lin. long, sides convex, firmly membranous, finely scaberulous; side-nerves almost percurrent, almost straight, subprominent, keel scaberulous; pales equal to the valves; keels strong, scaberulous; anthers 1 lin. long. *Nees, Fl. Afr. Austr.* 382; *Steud. Syn. Pl Glum.* i. 270; *Hack. in Engl. Jahrb.* xi. 406; *Durand & Schinz, Consp. Fl. Afr.* v. 889. *Festuca spinosa, Linn. f. Suppl.* 111. *Poa spinosa, Thunb. Prodr.* 22; *Fl. Cap.* i. 425; *ed. Schult.* 114; *Kunth, Rév. Gram.* ii. 551, *t.* 190; *Enum.* 334; *Suppl.* 289.

COAST REGION: Vanrhynsdorp Div.; dry plains near the Olifants River, 400–800 ft., *Drège!* Clanwilliam Div.; Lange Kloof, 600 ft., *Schlechter*, 8038!

Swellendam Div.; without precise locality, *Mund & Maire!* Mossel Bay Div.; by the Gauritz River, *Ecklon;* in the dry channel of an arm of the Gauritz River, *Burchell*, 6174!

CENTRAL REGION: Calvinia Div.; Bokke Veld and Hantam, *Thunberg!* Ceres Div.; between Little Doorn River and Great Doorn River, *Burchell*, 1205!

WESTERN REGION: Little Namaqualand; by the Buffels (Kousies) River, *Ecklon & Zeyher*, 1836! between Kook Fontein and Holgat River, covering vast tracts, *Drège!* plains between Kosies Mountain and Waggas, *Drège*, 2539! dunes at the mouth of the Orange River, *Drège.*

The rhachilla of the spikelets is rather fragile; at the same time, the valves separate easily at the base whilst the pales remain attached to the rhachilla joints. The whole spikelets break up therefore very readily, leaving the glumes or, as they are also more or less deciduous, the naked recurved pedicels. This is the "Vogelstruis" grass (ostrich-grass) of the colonists.

22. E. patentissima (Hack. in Bull. Herb. Boiss. iii. 391); perennial; culms tufted, shortly ascending, subcompressed, $\frac{1}{3}$–1 ft. long, glabrous, smooth, few-noded, sheathed almost to the base of the panicle; sheaths lax, particularly the lower, glabrous except at the scantily bearded mouth, striate; ligule a ciliolate rim; blades linear to lanceolate-linear, tapering almost from the base to a fine point, 2–3 in. long, 1–2$\frac{1}{2}$ lin. wide at the base, flat or involute, more or less hairy on the upper side, otherwise glabrous, smooth below, scabrid in the upper part; panicle erect, ovate-orbicular in outline, divaricate, effuse, very loose, up to 8 in. long; axis glabrous, smooth, subangular; branches solitary, rarely 2-nate, subdistichous, lower almost $\frac{3}{4}$ the length of the panicle, obliquely erect, divided from near the base; branchlets long divaricate, 2–1-spiculate, filiform to subcapillary, angular, scabrid; pedicels $\frac{1}{2}$–1$\frac{1}{4}$ in. long; spikelets oblong, compressed, 3–4 lin. by 1$\frac{1}{4}$–1$\frac{1}{2}$ lin, 6–9-flowered, light green, tinged with purple; rhachilla persistent, smooth; glumes subequal, lanceolate in profile, acuminate, almost 1$\frac{1}{2}$ lin. long, herbaceous-membranous, 1-nerved; valves ovate-lanceolate in profile, acuminate, 1$\frac{1}{2}$ lin. long, firmly membranous; lateral nerves somewhat prominent, keels scaberulous; pales somewhat shorter than the glumes, strongly curved, keels stout and spinulously ciliolate; anthers $\frac{2}{5}$ lin. long.

KALAHARI REGION: Transvaal; Houtbosch, *Rehmann*, 5684.

EASTERN REGION: Natal; hill tops at Umpumulo, 2700–2800 ft., *Buchanan*, 278!

23. E. denudata (Hack. in Bull. Herb. Boiss. iii. 392); perennial, tufted; culms erect, simple, rather slender, $\frac{1}{2}$–2 ft. long, smooth, 1–2-noded, uppermost internode very long and long exserted; leaves mainly crowded at the base; upper sheaths tight, glabrous, except for a few long hairs at the mouth, lowest short, strongly striate, persistent, tomentose at the broader base, the outer at length glabrescent; ligule a short ciliate rim; blades narrow, linear, tapering to a setaceous point, convolute-filiform, 3–6 in. long, rather rigid, glabrous or with few scattered fine hairs on the margins, closely striate, margins scaberulous; panicle erect, linear-oblong, 3–4 in. long; axis glabrous, angular, scaberulous, at least in the upper part; branches solitary, short, spreading more or less at right angles, 6–2-spiculate from the base, simple or almost so, angular, scaberulous; pedicels extremely short, puberulous; spikelets more or less spreading at right angles, linear to linear-lanceolate, strongly compressed, 4–8 lin. by 1½–2 lin., olive-green, somewhat loosely 7–20-flowered; rhachilla persistent, scaberulous; glumes subequal, ovate, acute, 1 lin. long, 1-nerved, keel scaberulous; valves slightly imbricate, obliquely ovate in profile, acutely and shortly acuminate or acute, obscurely mucronate, 1½ lin. long, rather firm, very smooth, side-nerves obscure in reflected light, keel scaberulous above; pales almost equalling the valves, keels scaberulous and in the upper part narrowly winged; anthers 1 lin. long.

KALAHARI REGION: Griqualand West, Hay Div.; between Griqua Town and Witte Water, *Burchell*, 1981! between Witte Water and Riet Fontein, *Burchell*, 2010! Herbert Div.; near St. Clair, Douglas, *Orpen*, 252! Bechuanaland; at Kosi Fontein, *Burchell*, 2579! between Kosi Fontein and Knegts Fontein, *Burchell*, 2605! Orange Free State; between Kimberley and Bloemfontein, *Buchanan*, 279! near the Caledon River, *Burke!* Basutoland, Leribe, *Buchanan*, 132! Transvaal; Bosch Veld, near Klippan, *Rehmann*, 5360!

24. E. Chapelieri (*Chapellieri*, by error, Nees, Fl. Afr. Austr. 392); perennial, tufted; culms erect or suberect, simple, somewhat stout, 2–3 ft. high, glabrous, 3–4-noded, intermediate and upper internodes exserted, uppermost very long; sheaths glabrous except at the more or less bearded mouth, or the lower scantily hairy, striate, lower very firm, persistent; ligule a narrow long hairy rim; blades very narrow, linear, tapering to a fine point, usually involute or convolute, 3–10 in. long, 1 lin. broad at the base when expanded, rigid, closely striate, more or less hairy towards the base on the upper side, glabrous and smooth underneath; panicle erect, contracted, very narrow, 3–9 in. long; axis slender, striate; branches erect, more or less adpressed to the axis, lowest solitary, 2–5 in. long, undivided for some distance, then (like the upper part of the axis) bearing fascicles of shortly pedicelled or subsessile spikelets on short branchlets crowded towards the tips, or all branches very short, and then the panicle resembling an interrupted false spike; ultimate divisions and pedicels scabrid; spikelets linear, much compressed, 3–8 lin. by 1¼–1½ lin., reddish-brown, 7–20-flowered; rhachilla persistent, glabrous, smooth, joints very short; glumes equal or more or less unequal, deciduous, lanceolate, about 1 lin. long, 1-nerved or upper sub-3-nerved, membranous, keels scaberulous above; valves broad, obliquely ovate in profile, shortly subacuminate or acute, 1 lin. long or very slightly longer, rather firm, deciduous from the base upwards, keel scaberulous above, side-nerves strong; pales slightly shorter than the valves, persistent, keels stout and rigidly ciliolate; stamens 2; anthers about ⅙–⅓ lin.; grain short, ellipsoid, laterally compressed, ¼–⅓ lin. long, whitish, subtranslucent. *Steud. Syn. Pl. Glum.* i. 271; *Anderss. in Peters, Reise Mossamb. Bot.* 560; *Durand*

& Schinz, Consp. Fl. Afr. v. 881. *Poa Chapelieri, Kunth, Rév. Gram.* ii. 543, *t.* 186; *Enum.* i. 336; *Suppl.* 293.

EASTERN REGION: Natal; Umpumulo, 2000 ft., *Buchanan*, 254a! near Durban, *Williamson*, 58! Valley of the Umlazi River, *Drège!* without precise locality, *Plant*, 59! *Gerrard*, 481! Delagoa Bay, *Kuntze*, 219!

Also in tropical Africa and the Mascarene Islands.

When the spikelets are very crowded, the lower florets are often more or less reduced and barren.

25. **E. sclerantha** (Nees, Fl. Afr. Austr. 388); perennial; culms erect, straight, from an oblique rhizome, up to 1 ft. long, sheathed all along or the uppermost internode exserted, few-noded; leaves crowded at the base; basal sheaths numerous, woolly-tomentose at the base, otherwise like the upper glabrous and smooth, distichously imbricate; ligule a dense fringe of very short hairs; blades linear, long tapering to a fine point, 4–7 in. by 1–2 lin., rather rigid, glabrous or adpressedly hairy, striate; panicle narrow, contracted, oblong-cuneate, $2\frac{1}{2}$–6 in. long; axis subflexuous, filiform, smooth; branches solitary, sometimes irregularly approximate, lower erect, upper more or less spreading, longest up to 2 in. long, filiform, flexuous, smooth, lower repeatedly divided from near the base; pedicels $\frac{1}{3}$–$\frac{3}{4}$ lin. long; spikelets ovate, much compressed, 2–$2\frac{1}{2}$ lin. by 1–$1\frac{1}{2}$ lin., closely 6–8-flowered, olive-green; rhachilla persistent; glumes unequal, ovate-lanceolate, acute or subacute, 1-nerved, upper longer, 1 lin. long, keels scabrid; valves oblong-lanceolate, subacute, $1\frac{1}{4}$–$1\frac{1}{2}$ lin. long, rather firmly membranous, side-nerves pallid and inconspicuous, keel stout and scaberulous towards the tip; pales 1 lin. long, keels scabrid; anthers $\frac{1}{2}$–$\frac{2}{3}$ lin long. *Steud. Syn. Pl. Glum.* i. 270; *Durand & Schinz, Consp. Fl. Afr.* v. 889.

COAST REGION: King Williamstown Div.; between Yellowwood (Kachu) River and Zandplaat, 1000–2000 ft., *Drège*, 4328!
KALAHARI REGION: Transvaal; Houtbosch, *Rehmann*, 5699!

26. **E. chalcantha** (Trin. Gram. Gen. 401, and in Mém. Acad. Pétersb. sér. 6, i. 401); perennial, densely cæspitose; culms erect, straight, $\frac{1}{2}$–1 ft. long, glabrous, smooth, 1-noded, uppermost internode occupying $\frac{3}{4}$ or more of the culm and long exserted; leaves crowded at the base, more or less beset with spreading often tubercle-based hairs, rarely quite glabrous; sheaths striate, tight, bearded at the mouth or not, lowest firm, persistent; ligule a dense fringe of very short hairs; blades linear, tapering to a fine often subcallous point, 1–4 in. by 1–2 lin., flat or more or less involute or convolute, particularly in the upper part, rigid, smooth or scaberulous on the upper side, obscurely striate above, conspicuously so below; panicle ovate to oblong, 1–3 in. long, more or less contracted; axis smooth below, compressed and scabrid along the angles above; branches solitary, spreading, 3–9 lin. long, racemosely 6–2-spiculate or shorter and reduced to a single spikelet, filiform, more or less angular, scabrid; pedicels often puberulous, very short; spikelets ovate-oblong

to oblong, obtuse, somewhat turgid, $2\frac{1}{2}$–4 lin. by $1\frac{1}{4}$–$1\frac{3}{4}$ lin., closely 7–15-flowered, olive-green to almost leaden-grey; rhachilla persistent, smooth, joints very short; glumes unequal, ovate, obtuse to subacute, upper longer, about $\frac{3}{4}$ lin. long, keels scabrid; valves broadly and obliquely ovate, obtuse to subacute, 1 lin. long, membranous, side-nerves more or less inconspicuous, keel scaberulous near the tip; pales $\frac{3}{4}$ lin. long, keels spinulously scabrid; anthers about $\frac{1}{2}$–$\frac{2}{3}$ lin. long; grain subglobose to almost cubic, less than $\frac{1}{4}$ lin. long, brown; embryo very large. *Nees, Fl. Afr. Austr.* 389; *Drège in Linnæa*, xx. 255; *Steud. Syn. Pl. Glum.* i. 270; *Durand & Schinz, Consp. Fl. Afr.* v. 881. *E. racemosa, Steud. Syn. Pl. Glum.* i. 271; *Durand & Schinz, Consp. Fl. Afr.* v. 889. *Poa racemosa, Thunb. Prodr.* 21; *Fl. Cap.* i. 422; *ed. Schult.* 113; *Kunth, Enum.* i. 344. *Poa chalcantha, Kunth, Enum.* i. 339.

SOUTH AFRICA: without precise locality, *Thunberg!* *Drège*, 4329!

COAST REGION: Swellendam Div.; Puspus Valley, *Ecklon & Zeyher!* Riversdale Div.: hills near Zoetemelks River, *Burchell*, 6755! between Little Vet River and Kampsche Berg, *Burchell*, 6885! Uitenhage Div.; Vanstadensberg Range, *Zeyher!* Peddie Div.; Fredricksburgh, on the Golana River, *Gill!* Alexandria Div.; Zuurberg Range, 2000–3000 ft., *Drège!* Albany Div.; near Grahamstown, *MacOwan*, 153! Komgha Div.; near Komgha, *Flanagan*, 912! between Zandplaat and Komgha, 2000–3000 ft., *Drège*. Queenstown Div.; Finchams Nek near Queenstown, 3900 ft., *Galpin*, 2376! Shiloh, 3500 ft., *Baur*, 910! Stockenstrom Div.; near Philipton and on the Winterberg Range, *Ecklon*, on Kat River Berg, *Ecklon*.

KALAHARI REGION: Orange Free State; by the Caledon River, near Komissie Drift, 4000 ft., *Zeyher*, 1835! *Burke*, 114! and without precise locality, *Cooper*, 3363! Transvaal; Mooi River near Potchefstroom, *Nelson*, 62*! Wonderboomport, near Pretoria, *Rehmann*, 4496! Houtbosch, *Rehmann*, 5688! 5685! Bosch Veld, at Menaars Farm, *Rehmann*, 4859! Apies River, *Nelson*, 98*! Blauw Bank, *Nelson*, 24*! near Lydenberg, *Wilms*, 1718! Apies Poort near Pretoria, *Rehmann*, 4043!

EASTERN REGION: Tembuland, Tabase near Bazeia, *Baur*, 320! Pondoland (Faku's Territory), *Sutherland!* Natal; Durban Bay, *Krauss*, 295! Durban Flats, *Buchanan*, 46! Umpumulo, common, 2000 ft., *Buchanan*, 256! Rovelo Hills, 7000 ft., *Sutherland!* De Beers Pass, *Wood*, 5995! Pieter Maritzburg, *Wood*, 7230! Drakensberg Range, near Van Reenens Pass, 5000–6000 ft., *Wood*, 7223!

27. **E. pallens** (Hack. in Bull. Herb. Boiss. iii. 392); perennial; culms erect, simple, 2–3 ft. high, glabrous, smooth, about 5-noded, internodes exserted, except the lowest; sheaths tight, glabrous or sometimes bearded at the mouth, finely striate, firm, lowest short, tomentose at the very base and persistent; ligule a shortly ciliate rim; blades linear, tapering to a setaceous point, 4–8 in. by $1\frac{1}{2}$–2 lin. (when expanded), more or less convolute, rigid, suberect, glabrous, smooth; panicle erect, oblong, rather dense, 5–6 in. long; axis slender, smooth; branches solitary, irregularly approximate or the lowest somewhat remote and up to 2 in. long, branched from the base; branchlets very short, 3–1-spiculate, scabrid; spikelets subsessile, very crowded, linear, compressed, 4–5 lin. by 1 lin., closely 10–14-flowered, yellowish-green; rhachilla persistent, slightly scaberulous, joints very short; glumes subequal, ovate, obtuse,

½ lin. long, keel scabrid; valves broadly and obliquely ovate-oblong in profile, more or less obtuse, 1 lin. long, firm, very smooth, yellowish below, dull green in the upper part, side-nerves rather inconspicuous in reflected light; pales persistent, subequal to the valves, rather broad, keels stout and scabrid; anthers ½ lin. long.

KALAHARI REGION: Transvaal; between Elands River and Klippan, *Rehmann*, 5116!

28. E. gangetica (Steud. Syn. Pl. Glum. i. 266); perennial, tufted; culms geniculate and suberect or erect, rather stout, simple or branching below, 1–3 ft. long, glabrous, smooth, about 4-noded, upper internodes exserted; leaves few at the base of each culm; sheaths glabrous except at the often bearded mouth, smooth, firm, upper tight, lowest persistent; ligule a very minutely ciliolate rim; blades linear, tapering to an acute or setaceous point, 2–6 in. by 1½–2 lin., flat or more often involute, rather rigid, glabrous, smooth below, densely scaberulous and whitish above; panicle oblong, 2–6 in. long, rarely shorter or longer, generally contracted, often nodding, or the lower branches spreading; axis smooth below; branches solitary, rather distant, lower up to 4 in. long, more or less flexuous, filiform, scaberulous, the lowest undivided for ½–1 in. or like the rest divided from near the base, their branchlets usually short and bearing shortly pedicelled crowded spikelets; spikelets linear-oblong to linear, up to 8 lin. by 1 lin., sometimes rather flexuous, usually olive-grey to leaden-grey, 8–30-flowered; rhachilla persistent; glumes subequal, ovate-oblong, acute, up to 1 lin. long, 1-nerved, deciduous; valves obliquely oblong, acute to subobtuse, ⅘–1¼ lin. long, side-nerves slender, prominent, keel scaberulous above; pales deciduous, slightly shorter than the valves, keels scabrid; anthers ⅔–⅜ lin. long; grain oblong, ⅜ lin. by ⅙–⅕ lin., brown. *E. luzoniensis, Steud. Syn. Pl. Glum.* i. 266. *E. elegantula, Stapf in Hook. f. Fl. Brit. Ind.* vii. 318, *not of Nees. Poa gangetica, Roxb. Fl. Ind.* i. 340.

EASTERN REGION: Natal; near Durban, *Williamson*, 61! moist sandy soil, near Pinetown, *Buchanan*, 115! Umpumulo, 2000 ft., *Buchanan*, 254! Clairmont, *Wood*, 7262! and without precise locality, *Gerrard*, 672!

Tropical Africa and throughout tropical Asia.

29. E. elatior (Stapf); perennial, densely cæspitose, not stoloniferous; culms erect or shortly ascending from an oblique short rhizome, simple or scantily branched below, ½ to more than 1 ft. long, glabrous, smooth, 2–3-noded, internodes exserted, upper much longer than the lower; sheaths tight, striate, very smooth and glabrous except at the rarely bearded mouth, lowest short, firm, persistent; ligule a fringe of very minute hairs, sometimes mixed with long ones at both ends; blades linear, tapering to a setaceous point, 1½ in. to more than ½ ft. by 1–2 lin., rigid to flaccid, flat or more or less involute, green, glabrous or with long hairs close above the ligule, smooth or scabrid in the upper part, finely and closely striate;

panicle erect, very slender, contracted, 2–4½ in. long; axis slender, smooth or almost so; branches erect, adpressed to the axis, solitary, lowest rarely more than 1 in. long, usually short and like the uppermost part of the panicle racemose from the base, 6–2-spiculate, scabrid; pedicels short; spikelets often distant on the lower branches, close towards the top, 2½–4 lin. by 1–1¼ lin., dark olive-green, closely 5–11-flowered; rhachilla subpersistent, joints very short; glumes lanceolate-oblong, subacute, equal, ¾–1 lin. long or the lower shorter, 1-nerved, membranous, keels scabrid; valves obliquely oblong, subacute to acute, 1 lin. long, membranous, side-nerves prominent, keel scaberulous; pales slightly shorter than the valves, keels strong and spinulously ciliate; anthers ⅜ lin. long. *E. sarmentosa, Nees, Fl. Afr. Austr.* 391 (*in part*); *Drège in Linnæa*, xx. 255.

VAR. β, **Burchellii** (Stapf); somewhat taller and firmer; blades subglaucous, more scabrid than in the type; spikelets on the whole more distant and on longer pedicels, up to 6 lin. long, paler; valves over 1–1¼ lin. long; anthers ⅜ lin. long; grain oblong, ½ lin. by ¼ lin., reddish-brown, opaque.

COAST REGION: Cape Div.; Cape Flats, near Doorn Hoogte, *Zeyher*, 1839 Paarl Div.; by the Berg River near Paarl, *Drège!* Var. β: Swellendam Div.; right bank of the Zonder Einde River, *Burchell*, 7516!

It appears from the material at Kew, as well as from the specimens in Drège's own Herbarium at Lübeck, that two species, namely *E. sarmentosa* and *E. elatior*, have been distributed by Drège as "*E. sarmentosa*, Nees, var. *pumila*." The particular specimen, however, to which the original number 1666 is attached in the Lübeck collection, is *E. sarmentosa*. The locality given for 1666 in the Berlin Herbarium is, according to a kind communication from Dr. H. Harms, "am grossen Bergfluss, Sandboden, i. Höhe."

30. E. sarmentosa (Trin. Gram. Gen. 398, and in Mém. Acad. Pétersb. sér. 6, i. 398); perennial, loosely cæspitose, stoloniferous; culms ascending from a creeping or decumbent base, branched below, ½ to more than 1 ft. long, glabrous, smooth, many-noded, internodes exserted; sheaths tight or more or less loose, striate, very smooth and glabrous except at the sometimes bearded mouth, short, lowest subpersistent; ligule a fringe of very minute hairs mixed with long ones at both ends; blades linear to linear-lanceolate, acute, ¾–2 in. by 1–1½ lin., flat or involute, rigid, glaucous or subglaucous, glabrous or with long hairs close above the ligule, finely and closely striate; panicle erect, very narrowly linear, contracted, dense, 1–4 in. long; axis slender, smooth or almost so; branches erect, adpressed to the axis, solitary, lowest up to more than 1 in. long, undivided for some distance, then bearing (like the upper part of the axis) fascicles of shortly pedicelled or subsessile spikelets on short branchlets crowded towards the tips, or all the branches very short and then the panicle resembling a dense interrupted or lobed false spike; ultimate divisions and pedicels scabrid; spikelets linear, 1½–3½ lin. by ⅜–⅘ lin., greyish-brown or purplish, closely 7–15-flowered; rhachilla subpersistent, at length breaking up from the top towards the base, almost smooth, joints very short; glumes unequal, deciduous, lanceolate-oblong, acute or subacute, membranous, 1-nerved, upper somewhat longer, up to more than $\frac{1}{2}$ lin. long; valves obliquely ovate-oblong in profile, subacute, $\frac{3}{4}$ lin. long, rather firm, side-nerves straight, prominent, keel scaberulous above; pales about $\frac{1}{2}$ lin. long, keels scabrid; stamens 3; anthers $\frac{1}{4}$ lin. long, ellipsoid; grain elliptic-ovoid, over $\frac{1}{4}$ lin. long, brown, opaque. *Nees in Linnæa*, vii. 330; *Fl. Afr. Austr.* 391; *Trin. Gram. Suppl.* 70, *and in Mém. Acad. Pétersb. sér.* 6, iv. 71; *Steud. Syn. Pl. Glum.* i. 270; *Durand & Schinz, Consp. Fl. Afr.* v. 889. *Poa sarmentosa*, *Thunb. Prodr.* 21; *Fl. Cap.* i. 422; *ed. Schult.* 113; *in Mém. Nat. Mosc.* iii. 45, *t.* 5; *Steud. in Flora*, 1829, 488; *Kunth, Enum.* i. 344. *P. racemosa* (?), *Steud. in Flora, l.c., not of Thunb.*

SOUTH AFRICA: without precise locality, *Zeyher*, 1836!

COAST REGION: Cape Div.; swampy places at the foot of Devils Mountain, *Ecklon*, 954! in wet places on Table Mountain and by streamlets between Table Mountain and the Lions Head, *Ecklon;* by the Salt River, *Ecklon;* in the sand of the "Lang'sche" dunes, *Ecklon;* Cape Flats near Doorn Hoogte, *Zeyher*, 1938! roadside beyond Black River, *Wolley Dod*, 2201! sandy places near Claremont, *Schlechter*, 152! Paarl Div.; sandy places by the Berg River near Paarl, *Drège*, 1666!

See note on p. 618, under *E. elatior*.

31. E. annulata (Rendle in Journ. Bot. 1891, 72); annual, tufted; culms geniculate, suberect, slender, terete, $\frac{1}{2}$–$\frac{3}{4}$ ft. long, glabrous or scantily hairy, with a viscous ring below each node, 2–3-noded, simple or with flowering branches from the lower and intermediate nodes, internodes at length exserted; leaves loosely covered with very minute soft gland-tipped hairs, glaucous; sheaths with long fine spreading usually tubercle-based hairs, particularly towards the bearded nodes, sometimes more or less viscous, striate, the lower compressed and keeled in the upper part; ligule a fringe of minute hairs; blades linear, tapering to a fine point, $1\frac{1}{2}$–3 in. by 1–$1\frac{1}{2}$ lin., spreading, flat or involute, margins gland-tubercled; panicle ovoid to oblong, erect, rather stiff, 2–4 in. by $1\frac{1}{2}$–$2\frac{1}{2}$ in., open, lax; axis terete or subangular, striate, scantily glandular-pubescent or glabrous; branches solitary or subopposite, spreading, usually stiff, simply racemose or the lowest branchlets 2-spiculate, finely filiform to capillary, slightly rough above; pedicels spreading at right angles or almost so, capillary, 1–$3\frac{1}{2}$ lin. long, with a viscous ring at the middle; spikelets scattered, elliptic-oblong to linear, $2\frac{1}{2}$–4 lin. by $\frac{3}{4}$–1 lin., whitish or tinged with purple, 7–16-flowered; rhachilla persistent; glumes ovate, subacute to obtuse, 1-nerved or the lower nerveless, keels scaberulous, lower $\frac{1}{2}$–$\frac{2}{3}$ lin., upper about $\frac{3}{4}$ lin. long; valves oblong, obtuse, $\frac{3}{4}$–1 lin. long, rigidly membranous, keel and the almost percurrent side-nerves rigid, prominent, usually pallid, keel scaberulous near the tip; pales subequal to the valves, keels strong, scabrid; anthers $\frac{1}{2}$ lin. long or almost so; grain ellipsoid, almost $\frac{1}{3}$ lin. by over $\frac{1}{6}$ lin., hollowed out in front, pit oblong.

WESTERN REGION: Little Namaqualand, near Ookiep, *Scully!*

KALAHARI REGION: Griqualand West, Herbert Div.; St. Clair, Douglas, *Orpen*, 214!

32. E. major (Host, Gram. Austr. iv. 14, t. 24); annual, tufted; culms geniculate-ascending or suberect, usually stout and branched below, ½–2 ft. long, glabrous, smooth, 3–4-noded, internodes more or less exserted; sheaths loose, strongly striate, keeled in the upper part, often glandular, particularly on the keel and the nerves above, glabrous or scantily hairy, bearded; ligule a fringe of short hairs; blades linear or lanceolate-linear, long tapering to a fine point, 2–6 in. by 1½–4 lin., flat, more or less flaccid, light green or subglaucous, glabrous or very scantily hairy, smooth below, scaberulous above, usually glandular along the margins; panicle oblong to ovate-oblong, stiff, 2 to more than 6 in. long, dense or rather lax; axis terete, smooth; branches subsolitary, spreading, stiff or flexuous, lowest up to 3½ in. long or all short, branched from near the base; lateral pedicels ½–1½ lin. long, all the divisions filiform, angular, scabrid; spikelets linear to ovate-oblong, 2–6 lin. long by 1 to almost 2 lin., subflexuous if very long, light or dark olive-green, few- to 50-flowered; rhachilla persistent; glumes subequal, ovate oblong, subobtuse to acute, ¾ or almost 1 lin. long, 1- (or the upper 3-) nerved, keels scabrid, margins minutely serrulate; valves broadly and obliquely ovate in profile, obtuse or subobtuse, 1–1¼ lin. long, side-nerves prominent, strong; pales persistent, somewhat shorter than the valves, broad, keels scabrid or ciliolate; anthers oblong, about $\frac{1}{5}$–$\frac{1}{4}$ lin. long; grain globose, brown, loose within the turgid valves, $\frac{1}{3}$–$\frac{1}{4}$ lin. in diameter. *Stapf in Hook. f. Fl. Brit. Ind.* vii. 320. *E. megastachya, Link, Enum. Hort. Berol.* i. 87; *Kunth, Enum.* i. 333; *Reichb. Ic. Fl. Germ.* i. *t.* 91, *fig.* 1662; *Nees, Fl. Afr. Austr.* 387; *Schweinf. in Bull. Herb. Boiss.* ii. *App.* ii. 40. *E. poæoides, Trin. Gram. Gen.* 404, *and in Mém. Acad. Pétersb. sér.* 6, i. 404; *Steud. Syn. Pl. Glum.* i. 263. *E. multiflora, Aschers. & Schweinf. Beitr. Fl. Aethiop.* 299, 310; *Durand & Schinz, Consp. Fl. Afr.* v. 885; *Hack. in Bull. Herb. Boiss.* iv. *App.* iii. 26. *E. vulgaris, var. megastachya, Coss. & Dur. Fl. Alger.* 147. *Briza Eragrostis, Linn. Spec. Pl.* 70. *Poa multiflora, Forsk. Fl. Ægypt.-Arab.* 21. *P. megastachya, Koel. Descr. Gram.* 181; *Kunth, Enum.* i. 333; *A. Rich. Tent. Fl. Abyss.* ii. 426. *P. Eragrostis, Cav. Ic.* i. 63, *t.* 92; *Sibth. Fl. Græc. t.* 73.

COAST REGION: Queenstown Div.; near Shiloh, 3500 ft., *Baur*, 869! plains near Queenstown, *Galpin*, 2352! Bathurst Div.; Port Alfred, *Hutton*, 38a!

CENTRAL REGION; Graaff Reinet Div.; in cultivated ground near Graaff Reinet, 2500 ft., *Bolus*, 683! Hanover Div.; near the Zeekoe River, 5000 ft., *Drège!*

KALAHARI REGION: Orange Free State; near Bloemfontein, *Rehmann*, 3755! Transvaal; near Lydenburg, *Atherstone!*

EASTERN REGION: Tembuland; in gardens at Bazeia, *Baur*, 460! Natal; by the Tugela River, 600–1000 ft., *Buchanan*, 253! and without precise locality, *Gerrard*, 472!

Probably introduced; a native of the Mediterranean regions and India.

This is probably the plant, enumerated as *E. poæoides*, Beauv., by Hackel in Bull. Herb. Boiss. iv. App. iii. 26, from Aus, in Great Namaqualand, *Schinz*, 617!

33. E. procumbens (Nees, Fl. Afr. Austr. 386); annual, tufted;

culms geniculate, ascending, branched, ½ ft. long, glabrous, smooth, about 3-noded, internodes shortly (or the uppermost long) exserted; sheaths loose or the uppermost tumid, striate, eglandular, scantily bearded at the mouth, otherwise glabrous; ligule a fringe of short hairs; blades linear, tapering to an acute point, 1–2 in. by 1–1½ lin., flat or involute, glabrous, smooth, eglandular; panicle oblong, dense or almost spike-like, 1½–2 in. long; axis terete, smooth; branches subsolitary, irregularly approximate, short or reduced to subsessile fascicles of spikelets; pedicels very short, all the divisions filiform, smooth or almost so; spikelets ovate-oblong to linear, 2–3½ lin. by 1–1¼ lin. long, light green, 5–15-flowered; rhachilla persistent; glumes somewhat unequal, ovate-oblong, acute, upper ¾ lin. long, 1-nerved, keels scabrid, margins minutely serrulate; valves oblong in profile, acute or minutely mucronate, 1 lin. long, side-nerves prominent, strong; pales persistent, somewhat shorter than the valves, keels scabrid; anthers ellipsoid, $\frac{1}{8}$–$\frac{1}{7}$ lin. long; grain oblong, ⅜ lin. by ⅙ lin., reddish-brown. *Steud. Syn. Pl. Glum.* i. 270; *Durand & Schinz, Consp. Fl. Afr.* v. 888.

CENTRAL REGION: Aberdeen Div.; Camdeboo, on the flats and by the river near Camdeboo Mountains, 2000–3000 ft., *Drège!*

Very closely allied to *E. major*, but with smooth, eglandular leaves, narrower and acute valves, smaller anthers and oblong, not globose, grains.

34. **E. barbinodis** (Hack. in Bull. Herb. Boiss. iii. 390); annual (?); culms ascending, geniculate, simple or with few leafy branches from the intermediate nodes, from less than 1 to more than 2 ft. long, scantily hairy below the nodes, and sometimes resinous, 5–6-noded, internodes exserted; sheaths tight, bearded at the nodes and mouth, otherwise glabrous or beset with very fine often tubercle-based hairs; ligule a short-ciliate rim; blades linear to lanceolate-linear, acute to pungent, ¾–4 in. by 2–2½ lin. more or less spreading, flat or convolute, rigid, glaucous, with scattered fine tubercle-based hairs, scaberulous or finely puberulous besides, finely striate; panicle erect, ovate, open, 2–6 in. long, rather lax; axis scaberulous; branches solitary or lower 3–5-nate and undivided for ¼–⅕ of their length, then branched, scabrid; branchlets divaricate, bearing few or rather many spikelets; lateral pedicels short; spikelets linear, slightly compressed, 3½ lin. long by ¾ lin., 6–8-flowered, greyish-green; rhachilla subpersistent, at length breaking up from the top towards the base, somewhat scabrid; glumes unequal, ovate-lanceolate in profile, acute, lower over ½ lin., upper over ¾ lin. long, membranous, 1-nerved, margins minutely serrulate; valves oblong, obtuse to subacute, about ¾ lin. long, membranous, scabrid from papillæ, particularly along the slightly prominent side-nerves, keels rather obtuse, scabrid, margins minutely serrulate; pales equalling the valves, keels scaberulous; anthers ⅜–½ lin. long; grain elliptic, slightly compressed from the back, ⅜ lin. long, brown; pericarp slightly swelling in water.

KALAHARI REGION: Transvaal; Bosch Veld, at Klippan, *Rehmann*, 5362! 5364.

35. E. superba (Peyr. in Sitz. Ber. Acad. Wien, Math.-Nat. Cl. xxviii. 584); perennial, densely cæspitose with intravaginal innovations, glabrous; culms erect or geniculately ascending, 2–3 ft. long, rather stout, smooth, 2-noded, internodes exserted, uppermost very long; sheaths smooth, bearded at the mouth, lowest crowded, broad at the base, keeled, persistent, upper terete, tight; ligule a fringe of short hairs; blades linear, long tapering to an acute point, 2–8 in. or more by 1–3 lin., firm, more or less rigid, upper often spreading, usually more or less involute or convolute, rarely quite flat, smooth below, scaberulous on the upper side; panicle narrow, linear or oblong, often interrupted below, erect, 4–10 in. long; axis usually straight, smooth, terete below, angular above; branches distant, erect or suberect, solitary, filiform, usually simply racemose, $\frac{1}{2}$–3 in. long, 1–10-spiculate, rarely branched; pedicels very unequal, lateral, $\frac{1}{2}$–3 lin. long; spikelets articulated with the pedicels, deciduous, rather distant or clustered towards the tips of the branches or branchlets, strongly compressed from the side, suborbicular, ovate or ovate-oblong, 3–8 lin. by $2\frac{1}{2}$–$4\frac{1}{2}$ lin., straw-coloured, rarely more or less purplish, 7–37-flowered; glumes subequal, ovate-lanceolate to lanceolate in profile, acute or mucronate, $1\frac{1}{2}$–$2\frac{1}{2}$ lin. long, 1-nerved, firmly membranous, strongly keeled; valves obliquely oblong, subacute, $1\frac{3}{4}$–$2\frac{3}{4}$ lin. long, acutely keeled, subchartaceous, side-nerves prominent and often green; pales 2-fid, keels winged, wings narrowed upwards, produced into obtuse auricles at the base, ciliolate; anthers 1 lin. long; grain oblong, 1 lin. long; pericarp loose; seed truncate at both ends, subquadrangular, brown. *Hack. in Engl. Jahrb.* xi. 405; *Durand & Schinz, Consp. Fl. Afr.* 890; *Hack. in Bull. Herb. Boiss.* iv. *App.* iii. 27. *E. elata, Munro ex Ficalho & Hiern in Trans. Linn. Soc. ser.* 2, ii. 32; *Durand & Schinz, l.c.* 883.

SOUTH AFRICA: without precise locality, *Zeyher*, 1834!

KALAHARI REGION: Hopetown Div.; by the Orange River at "Amaryllis Station," near Hope Town, *Burchell*, 2659! Griqualand West, Herbert Div.; between Spuigslang Fontein and the Vaal River, *Burchell*, 1709! at Blaauwbosch Drift on the Vaal River, *Burchell*, 1735! and Kimberley Div.; near Dutoits Pan, *Tuck*, 13! Orange Free State; Olifants Fontein, *Rehmann*, 3524! Caledon River, *Burke*, 428! and without precise locality, *Hutton!* Bechuanaland; between Hamapery and Kosi Fontein, *Burchell*, 2546! Transvaal; Bosch Veld, between Elands River and Klippan, *Rehmann*, 5115! plains near Rustenburg, *McLea*, 113! near Lydenburg, *Atherstone!* near Johannesburg, *E.S.C.A. Herb.* 314! and without precise locality, *McLea*, 111!

EASTERN REGION: Natal; Weenen County, 5000 ft., *Wood*, 4416! at 3500 ft., *Wood*, 3587! banks of the Tugela River, *Buchanan*, 255! and without precise locality, *Gerrard*, 468! Zululand, *Wood*, 7307! Delagoa Bay, *Forbes!*

Also in tropical Africa.

36. E. brizoides (Nees in Linnæa, vii. 328); perennial, compactly tufted with intravaginal innovation-shoots; culms erect or geniculate-ascending, slender, firm, from $\frac{3}{4}$ to more than 2 ft. long, glabrous, smooth, 1–3-noded, internodes exserted, uppermost occupying from $\frac{1}{2}$ to the length of the culm; sheaths tight, glabrous except the bearded mouth, lowest short, crowded, persistent; ligule a fringe of minute hairs; blades narrow, linear, tapering to a fine point, 2–4 in. (rarely to 8 in.) by 1–1½ lin., usually more or less convolute or involute and rigid, glabrous, smooth below, scaberulous above, striate; panicle linear to oblong, contracted, 2–4 in. long, erect or nodding; axis filiform, flexuous, smooth; branches solitary, filiform, angular, smooth, racemosely 6–2-spiculate; pedicels very short; spikelets crowded or sometimes more distant, ovate to ovate-oblong or suborbicular, strongly compressed, 2–6 lin. by 1½–3½ lin., densely 5–40-flowered, straw-coloured, usually tinged with dull purple; rhachilla persistent, rather stout, joints very short; glumes and valves similar, closely imbricate, rigidly membranous to subchartaceous, obtuse, back broad, keeled; glumes subequal, oblong in profile, 1¼–1½ lin. long, lower 1-nerved; valves obliquely elliptic-oblong in profile, 1½–1¾ lin long, side-nerves prominent, keel adpressedly ciliate; pales subequal to the valves, broad, keels very densely and minutely ciliolate; anthers ¾ lin. long; grain narrowly oblong, not quite 1 lin. by ½ lin.; pericarp slightly swelling in water. *Nees, Fl. Afr. Austr.* 384; *Steud. Syn. Pl. Glum.* i. 270; *Durand & Schinz, Consp. Fl. Afr.* v. 880; *Hack. in Bull. Herb. Boiss.* iv. *App.* iii. 26. *E. capensis, Trin. Gram. Gen.* 400, *and in Mém. Acad. Pétersb. sér.* 6, i. 400. *Briza capensis, Thunb. Prodr.* 21; *Fl. Cap.* i. 419; *ed. Schult.* 112 *Poa brizoides, Linn. f. Suppl.* 110; *Kunth, Enum.* i. 327. *Megastachya brizoides, Roem. & Schult. Mant.* ii. 329.

SOUTH AFRICA: without precise locality, *Bergius!* *Zeyher*, 1832! 1833!

COAST REGION: Clanwilliam Div.; by the Olifants River and at Brack Fontein, *Ecklon.* Cape Div.; near Capetown, *Drège*, 2! Devils Mountain, *Ecklon*, 959! north slopes of the Lions Head, *Wolley Dod*, 3575! between Capetown and Table Mountain, *Burchell*, 21! 919! Table Mountain, eastern slopes near Tokay, *Ecklon*, Tokay Plantation, *Wolley Dod*, 1965! Cape Flats, near Doorn Hoogte, *Ecklon;* near Rondebosch, *Burchell*, 201! Sandown Road, *Wolley Dod*, 2494! Dido Valley, *Wolley Dod*, 2826! between Groot and Klein Slang Kop, *Wolley Dod*, 3297! Railway between Kenilworth and Claremont, *Wolley Dod*, 3115! 3565! above Oatlands, *Wolley Dod*, 2951! Tulbagh Div.; Tulbagh Kloof, *Ecklon!* Paarl Div.; sandy plains near Paarl Berg, *Drège.* Stellenbosch Div.; Hottentots Holland, *Ecklon.* Caledon Div.; near Genadendal, *Drège.* Swellendam Div.; on mountain ridges along the lower part of the Zonder Einde River, *Zeyher*, 4562! Riversdale Div.; between Great Vals River and Zoetemelks River, *Burchell*, 6564! near the Zoetemelks River, *Burchell*, 6695! between the Gauritz River and the Great Vals River, *Burchell*, 6504! Uniondale Div.; between Apies River and Roode Krantz River, *Burchell*, 4964! Bathurst Div.; between Blue Krantz and Kaffir Drift Military Post, *Burchell*, 3708! near Port Alfred, *Hutton*, 44a! Port Elizabeth Div.; at Krakakamma, *Burchell*, 4537! Humansdorp Div.; near Kromme River, *Krauss*, 365! Uitenhage Div.; by the Zwartkops River, *Ecklon & Zeyher!* between the Coega River and Sunday River, *Ecklon;* Olifantshoek Mountain by the Bushman River, *Ecklon;* among shrubs in the Vanstaadens Berg Range, *Ecklon & Zeyher*, 287! Albany Div.; between Bushman River and Karega River, *Ecklon;* and without precise locality, *MacOwan*, 153! Alexandria Div.; hills near Addo, *Ecklon!* Bathurst Div.; *Port Alfred*, 44a! Zwart Hoogte, *Burke!* Stockenstrom Div.; Winterberg Mountains, near Philipton, *Ecklon.* Cathcart Div.; Glencairn, 5000 ft., *Galpin*, 2419a!

CENTRAL REGION: Jansenville Div.; hills near Klein Bruintjes Hoogte, 3000 ft., *Drège.* Graaff Reinet Div.; mountain tops near Graaff Reinet 4500 ft., *Bolus*, 779! Aliwal North Div.; sandy plains at the foot of the Witte Bergen, 5000 ft., *Drège.*

KALAHARI REGION: Orange Free State; without precise locality, *Cooper*, 913! 3337! Transvaal; Klip Drift, by the Vaal River, *Nelson*, 78*! Spitzkop Goldmine, *Wilms*, 1720a! Apies Poort, *Rehmann*, 4042!

EASTERN REGION: Transkei; between Gekau (Gcua or Geuu) River and Bashee River, 1000–2000 ft., *Drège!* Tembuland; near Bazeia, *Baur*, 299! 300! Natal; Riet Vlei, 4000–5000 ft., *Buchanan*, 251! Durban Flats, *Buchanan*, 27! near Durban, *Williamson*, 59! Coastland, *Sutherland!* throughout Natal, *Krauss*, 365! Inanda, *Wood*, 993! Mooi River, *Wood*, 4068! Berea, *Wood*, 5934! Van Reenens Pass, *Wood*, 7222! 7245, and without precise locality, *Buchanan*, 102! 252! *Plant*, 62!

37. **E. truncata** (Hack. in Engl. Jahrb. xi. 405); perennial; branched at the base, branches very short, in compact fascicles of short densely leafy innovation-shoots and flowering culms; culms geniculate or erect, slender, $\frac{1}{2}$–1 ft. long, glabrous, smooth, or slightly viscous, 2–3-noded, internodes exserted; sheaths of the barren shoots closely imbricate, very short, broad, strongly striate, covered more or less with wool, persistent, sheaths of culm-leaves very tight, glabrous, smooth; ligule a dense, very short, woolly fringe; blades of the barren shoots linear, subobtuse, or with a callous point, $\frac{1}{3}$–1$\frac{1}{2}$ in. by $\frac{1}{2}$–1$\frac{1}{4}$ lin., flat or folded, rigid, usually obliquely erect, sometimes spreading and curved, glaucous, smooth below, strongly nerved and densely scabrid above; culm-blades acute or subacute, up to 3 in. long; panicle oblong or ovoid-oblong, erect, dense or somewhat lax, 1–2$\frac{1}{2}$ in. long; axis slender, angular, smooth, or slightly rough; branches solitary, obliquely erect or spreading, up to 1 in. long, divided from the base or from 2–3 lin. above it, filiform, angular, scaberulous; pedicels very short; spikelets crowded, more or less secund, strongly compressed, ovate to oblong, 2–3 lin. by 1–1$\frac{1}{2}$ lin., 5–12-flowered, pallid or slightly tinged with purple; rhachilla flexuous, disarticulating; glumes subequal, oblong in profile, acute, 1 lin. long, somewhat rigidly membranous, keel stout and scabrid; valves very broad, elliptic-oblong in profile, truncate, 1 lin. long, finely granular, somewhat rigidly membranous except at the very narrow hyaline margins and a hyaline area along the keel, side-nerves strong, prominent, keel slightly rough; pales equal to the valves, keels acute and scabrid; anthers $\frac{1}{2}$–$\frac{2}{3}$ lin. long.

CENTRAL REGION: Graaff Reinet Div.; near Wagenpads Berg, on the southern side, *Burchell*, 2834!

KALAHARI REGION: Griqualand West, Hay Div.; at the foot of the Asbestos Mountains between the Kloof Village and Witte Water, *Burchell*, 2076! and Kimberley Div.; Dutoits Pan, *Tuck!* Philipstown Div.; between "Bare Station" and "Gnu Halt," *Burchell*, 2687! Basutoland; Leribe, *Buchanan*, 130! Bechuanaland; stony places near Kachun Fontein, 3900 ft., *Marloth*, 1023!

38. **E. bergiana** (Trin. in Bull. Scient. Acad. St. Pétersb. i. 70); perennial; branched at the base, branches creeping, and often rooting, up to 4 in. long, emitting numerous very short and imbricate curved barren branches and one flowering culm from near the apex; flowering culm geniculate-ascending or suberect, slender, 4–6 in. long, glabrous, smooth or viscous, 1-noded, internodes exserted, upper by far the longest; sheaths of the barren branches closely imbricate, very short

and broad, strongly striate, covered with a partly fugacious wool towards the base; sheaths of the flowering culms very tight, glabrous, smooth; ligule a dense very short woolly fringe; blades of the barren branches linear, obtuse, 2–4 lin. by ½–¾ lin., flat or folded, very rigid, spreading and somewhat curved, glaucous, smooth below, strongly nerved and densely scabrid above; culm-blades ½–1½ in. by 1 lin., less rigid; panicle ovoid or oblong, erect, usually dense, 1–2 in. long; axis slender, angular, smooth or slightly rough; branches solitary, obliquely erect or spreading, up to 1 in. long, divided from the base or from 2–3 lin. above it, filiform, angular, scaberulous; branchlets and pedicels very short; spikelets crowded, broad, elliptic to oblong, obtuse, strongly compressed, 2–4 lin. by 1½–2 lin., 5–12-flowered, more or less tinged with dull purple; rhachilla flexuous, disarticulating; glumes subequal, oblong, subobtuse, 1 lin. long, rigidly membranous except at the narrow hyaline margins, lower 1-nerved, keels rough; valves broad, obliquely ovate, very obtuse, 1 lin. long, rigidly membranous to subchartaceous, except at the very narrow hyaline margins, granular-scabrid, side-nerves strong, prominent, keel scabrid; pales equal to the valves, keels stout and scaberulous; anthers ½ lin. long. *Trin. Gram. Suppl.* 71, *and in Mém. Acad. Pétersb. sér.* 6, iv. 72; *Durand & Schinz, Consp. Fl. Afr.* v. 880. *E. striata, Nees, Fl. Afr. Austr.* 385; *Steud. Syn. Pl. Glum.* i. 270; *Durand & Schinz, l.c.* 890. *E. stricta, Nees, ex Durand & Schinz, l.c.* 880 (*by error*). *Poa striata, Thunb. Prod.* 22; *Fl. Cap.* i. 421; *ed. Schult.* 113; *Kunth, Enum.* i. 345. *P. bergiana, Kunth, Rév. Gram.* ii. 549, *t.* 189; *Enum.* i. 334; *Suppl.* 290. *P. floccosa, Lehm. Nov. Stirp. Pug.* iii. 39; *Nees in Linnæa*, vii. 328; *Kunth, Enum.* i. 363.

COAST REGION: Riversdale Div.; banks of the Gauritz River, *Thunberg! Ecklon.* Oudtshorn Div.; "Cannaland," *Thunberg!*

CENTRAL REGION: Prince Albert Div.; by the Gamka River, *Bergius.* Beaufort West Div.; near Beaufort, *Ecklon;* between Beaufort and Rhinoster Kop in the Nieuw Veld Mountains, *Drège!* Somerset Div.; near Somerset East, *Bowker*, 80! Graaff Reinet Div.; by the Sunday River, 1500–2000 ft., *Drège!* near Graaff Reinet, 2500 ft., *Bolus*, 552! higher parts of Compass Berg (Spitz Kop), *Bolus*, 1288! Middelburg Div.; between Wolve Kop and Middelburg, *Burchell*, 2789! Colesberg Div.; near Wonderheuvel, *Drège.*

39. **E. obtusa** (Munro ex Ficalho & Hiern in Trans. Linn. Soc. ser. 2, ii. 32); perennial, densely cæspitose, with numerous intravaginal innovation-shoots; culms usually repeatedly geniculate, wiry, ½–1½ ft. long, smooth, glabrous, 2–4-noded, internodes exserted; sheaths tight, glabrous except at the bearded mouth, very rarely finely ciliate along the margins, lowest very firm, persistent; ligule a line of short hairs; blades linear, long tapering to a subsetaceous point, 2–5 in. by ¾–1½ lin., usually involute or convolute, rather rigid, glabrous, very rarely with scanty fine and spreading hairs, smooth or finely scaberulous above; panicle erect, ovoid to oblong, 2–4 in. by ¾–3 in., rather lax; axis straight, smooth; branches solitary, rarely subopposite, rather distant, obliquely spreading, straight, filiform,

smooth, divided from near the base or 3–4 lin. above it; branchlets straight below, capillary and flexuous above; pedicels capillary, $\frac{1}{2}$–2 lin. long; spikelets loose or rather crowded on the branchlets, more or less nodding, broadly ovate-oblong or oblong, obtuse or almost orbicular, strongly laterally compressed, whitish or usually more or less grey, $1\frac{1}{2}$–$2\frac{1}{2}$ lin. by $1\frac{1}{2}$–2 lin., closely 8–20-flowered; rhachilla disarticulating; glumes and valves very similar, broad, boat-shaped, membranous, keeled, subacute; lower glume 1-nerved, $\frac{1}{2}$–$\frac{2}{3}$ lin. long, upper 3-nerved, $\frac{3}{4}$ lin. long; valves obliquely truncate at the base, 1 lin. long, keel stouter upwards, smooth, side-nerves prominent; pales subequal to the valves, truncate, keels rigidly ciliolate; anthers $\frac{1}{2}$ lin. long; grain dorsally compressed, broadly elliptic or suborbicular in outline, plano-convex, $\frac{1}{2}$ lin. long, brown, smooth. *Durand & Schinz, Consp. Fl. Afr.* v. 886. *Briza geniculata, Thunb. Prodr.* 21; *Fl. Cap. ed. Schult.* 112; *Kunth, Enum.* i. 372; *Steud. Syn. Pl. Glum.* i. 282. *B. nigra, Burch. Trav.* i. 537 (*name only*).

SOUTH AFRICA: without precise locality, *Thunberg!*

COAST REGION: Riversdale Div.; between the Gauritz River and Great Vals River, *Burchell*, 6527! Uitenhage Div.; between the Zwartkops River and Sunday River, *Ecklon & Zeyher!* Steenbok Flats, to the north of the Winterhoek Mountains, *Ecklon.* Albany Div.; summit of Bothas Berg, 2300 ft., *MacOwan*, 1277! Alexandria Div.; Sam Tees Vlakte near Enon, *Drège.* Bathurst Div.; Port Alfred, *Hutton*, 7c! Fort Beaufort Div.; Fort Beaufort, *Ecklon;* Kat River, *Baur*, 1063. Stockenstrom Div.; Winterberg, near Philipton, *Ecklon.*

CENTRAL REGION: Prince Albert Div.; near Zwartbulletje by the Gamka River, 2500–3000 ft., *Drège.* Jansenville Div.; between Sunday River and the Zuurberg Range, *Drège!* Somerset Div.; near Somerset East, *Bowker*, 148! Graaff Reinet Div.; stony hills near Graaff Reinet, 2500 ft., *Bolus*, 368! Albert Div. (Nieuwe Hantem), 4500–5000 ft., *Drège*, and without precise locality, *Cooper*, 657! Richmond Div.; by the Zeekoe River, 5000 ft., *Drège.* Colesberg Div.; at Carolus Poort, *Burchell*, 2752! near Colesberg, *Shaw*, 2!

KALAHARI REGION: Griqualand West, Hay Div.; between Griqua Town and Witte Water, *Burchell*, 1978! Orange Free State; Caledon River, *Burke!* Transvaal; Derde Poort, near Pretoria, *Nelson*, 22*!

Also in tropical Africa, according to Ficalho and Hiern.

40. E. brizantha (Nees, Fl. Afr. Austr. 411); annual, with very few feeble barren shoots; culms fascicled, geniculate-ascending, $\frac{1}{3}$–$\frac{1}{2}$ ft. long, glabrous, smooth or somewhat rough and compressed, at least towards the panicle, 1-noded, internodes exserted; lowest sheaths very short, broad, hairy at the very base, like the remainder ciliate along the margins, otherwise glabrous or scantily puberulous, bearded at the mouth; ligule a fringe of short hairs; blades linear, tapering to a fine point, 1–2 in. long, $1\frac{1}{2}$ lin. wide at the base, flat or more or less involute or convolute above, somewhat rigid, glaucous, scabrid except on the back towards the base, scantily hairy on the upper side or quite glabrous; panicle consisting of 5–8 somewhat distant dense short false spikes, up to $2\frac{1}{2}$ in. long, straight; axis smooth or slightly rough; branches up to 7 lin. long, filiform, stiff, spreading, bearing dense clusters of subsessile spikelets; spikelets ovate or broad-oblong, obtuse, 1½–2 lin. by 1–1¼ lin., compressed, densely 8–17-flowered; rhachilla disarticulating; glumes and valves similar, membranous, whitish, becoming dark towards the tip, glabrous, smooth; glumes subequal, ovate, acute, ½ lin. long, lower sometimes 1-nerved; valves very broad-ovate in profile, obtuse, inflated, boat-shaped, ⅓ lin. long, side-nerves prominent; pales subequal to the valves, broad, very obtuse, keels scabrid above; anthers ½ lin. long. *Steud. Syn. Pl. Glum.* i. 272; *Durand & Schinz, Consp. Fl. Afr.* v. 880; *Hack. in Bull. Herb. Boiss.* iv. *App.* iii. 26.

WESTERN REGION: Little Namaqualand; by the Orange River, near Verleptpram, *Drège!* Great Namaqualand; between Aus and Kuias, *Schenck*, 221; near Gobachab, between Aus and the Orange River, *Schenck*, 337! *Pohle*, 17!

Also in the tropical part of German South-west Africa.

41. E. echinochloidea (Stapf); biennial (?), without innovation-shoots (?); culms geniculate, suberect from a slender short rhizome, branched, about 1½ ft. long, rather firm, glaucous, glabrous, smooth, 3–4-noded, internodes exserted, branches sometimes fascicled; sheaths glabrous, loosely striate, upper tight, lower at length loosened, rather firm, more or less persistent; ligule a fringe of minute hairs; blades linear, tapering to a fine point, 1½–3 in. by 1–1½ lin., flat or involute, rigid, often spreading, glaucous, glabrous and smooth below, scaberulous and sometimes finely hairy in the upper part; panicle ovate-oblong to oblong, sometimes secund, about 2 in. long; axis smooth, angular towards the nodes; branches 5–7, spreading, triquetrous, more or less secund and spike-like from densely clustered spikelets, the lowest up to 1 in. long; pedicels very short; spikelets ovate, subacute, strongly compressed, 2 lin. by 1¼–1½ lin., whitish, 7–13-flowered, often tinged with greyish-purple; rhachilla disarticulating; glumes oblong to ovate in profile, acute or mucronate-acuminate, ¾–1 lin. long, 3-nerved; valves very broad-ovate in profile, acute, boat-shaped, 1 lin. long, hyaline margins broad below, side-nerves prominent; pales ¾ lin. long, curved; keels scabrid above and with a large decurrent tooth-like appendage at the middle; anthers ⅖ lin. long; grain oblong, ⅖ lin. long, brown, smooth; pericarp loose.

KALAHARI REGION: Orange Free State; between Kimberley and Bloemfontein, *Buchanan*, 284!

42. E. Lappula (Nees, Fl. Afr. Austr. 412); perennial, densely cæspitose; culms erect or geniculate, firm, rather stout, 2–3 ft. long, glabrous, smooth, about 3-noded, upper internodes (sometimes also the lower) exserted; lower sheaths firm, strongly striate, pubescent or glabrous, usually bearded at the mouth, persistent, upper tight, glabrous, smooth; ligule a very minutely ciliolate rim; blades narrow, linear, filiform-convolute, tapering to a fine point, 6–12 in. long, 1½ lin. broad when expanded, rigid, flexuous, glabrous or sparsely hairy particularly towards the base, smooth on the back, scaberulous or smooth on the face, strongly striate; panicle erect or nodding, narrow, linear to lanceolate, contracted, dense, sometimes spike-like, 6–8 in. long; axis filiform, smooth; branches solitary, sometimes 2–3-nate or irregularly approximate, adpressed, lower up to 4 in. long, undivided for 1–1½ in. or like the others divided from near the base; branchlets somewhat distant, adpressed, simple or again divided; lateral pedicels very short; all the divisions finely filiform, angular, smooth or scaberulous; spikelets oblong, 1½–4 lin. long, brownish or purplish, loosely 4–17-flowered; rhachilla persistent; glumes subequal, lanceolate, acute, 1 lin. long, thin, deciduous, keel scabrid; valves somewhat spreading, stiff, lanceolate in profile, acute, 1 lin. long or slightly longer, membranous, side-nerves prominent, like the keels rigidly ciliate, with the cilia tubercle-based (or rarely with the keels glabrous); pales subequal to the valves, keels tubercled, long and rigidly ciliate from the tubercles; anthers over ½ lin. long. *Steud. Syn. Pl. Glum.* i. 272; *Durand & Schinz, Consp. Fl. Afr.* v. 884.

VAR. β, **divaricata** (Stapf); panicle open, broadly oblong or ovate; branches and branchlets more or less divaricate; pedicels up to 3 lin. long.

KALAHARI REGION: Transvaal; Klip Spruit, *Nelson*, 58*! Var. β: Bechuanaland; Pellat Plains between Matlareen River and Takun, *Burchell*, 2199!

EASTERN REGION: Natal; near Durban, *Drège! Plant*, 63! *Gerrard*, 33! 475! *Williamson*, 57! *Wood*, 6047! *Rehmann*, 8630! Berea, *Wood*, 5938! Zululand, *Buchanan*, 301!

43. E. aspera (Nees, Fl. Afr. Austr. 408); annual; culms scantily fascicled, erect or suberect, ½–¾ ft. long (excluding the panicle), glabrous, smooth, simple, 2–3-noded, internodes usually enclosed; sheaths keeled, glabrous, except at the bearded mouth, or sparingly hairy, hairs fine, tubercle-based; ligule a fringe of long hairs; blades linear, tapering to a long setaceous point, 4 in. to more than 1 ft. by 2–3 lin., flat, flaccid, scabrid on both sides, glabrous; panicle large, very lax and open, thyrsiform, oblong to obovate-oblong, 8–20 in. long; axis terete, filiform, smooth below; branches whorled or irregularly approximate, finely filiform to capillary, scabrid, bearded at the callous base, loosely and repeatedly divided from near the base, longest up to 6 in. long; pedicels very long and fine; spikelets scattered, linear, obtuse, 2½–4 lin. by ⅔–¾ lin., pallid or tinged with purple, loosely 4–16-flowered; rhachilla very slender, breaking up; glumes subequal, oblong, obtuse, ½ lin. long, 1-nerved; valves obliquely ovate-oblong, truncate, ⅔–¾ lin. long, thin, side-nerves prominent, strong; pales equal to the valves and falling with them, obtuse, keels scabrid; anthers about ⅙ lin. long; grain globose, about ⅕ lin. in diameter, brown, loose in the somewhat turgid florets. *Steud. Syn. Pl. Glum.* i. 272; *Engl. Hochgebirgsfl. Trop. Afr.* 133; *Schweinf. in Bull. Herb. Boiss.* ii. *App.* ii. 38; *Durand & Schinz, Consp. Fl. Afr.* v. 879; *Stapf in Hook. f. Fl. Brit. Ind.* vii. 314. *E. laxiflora, Schrad. in Linnæa*, xii. 451 (*name only*); *Durand & Schinz, Consp. Fl. Afr.* v. 884. *Poa aspera, Jacq. Hort. Bot. Vindob.* iii. *t.* 56; *Kunth, Enum.* i. 332 (*excl. var.* β); *A. Rich. Tent. Fl. Abyss.* ii. 427.

COAST REGION: Cape Div.; near Capetown, *Spielhaus!*
EASTERN REGION: Natal; slopes of Tugela, 600–1000 ft., *Buchanan*, 257! in coffee plantations, near Durban, *Drège*, 4285!

Southern India, tropical Africa, and Mascarene Islands.

44. **E. gummiflua** (Nees, Fl. Afr. Austr. 393); perennial, compactly cæspitose; culms firm, erect, slender or somewhat stout, 1–2 ft. long, glabrous, smooth, usually very viscous below the nodes, 3–5-noded, internodes exserted; sheaths tight, striate, glabrous except at the bearded mouth, more or less viscous, lowest persistent; ligule a dense fringe of minute hairs; blades narrow, linear, tapering to a setaceous point, usually filiform-convolute, 4–10 in. long, 1–2 lin. wide at the base when expanded, rigid, glabrous, smooth on the back, densely scabrid along the projecting nerves on the upper side; panicle linear or narrow-oblong, erect, 6–10 in. long; rhachis angular, smooth; branches solitary, or irregularly approximate and subverticillate, erect or obliquely erect, short or the lower up to 3 in. long, angular, smooth, divided from the base; branchlets mostly very short, subsecund, with the spikelets in dense clusters; pedicels very short; spikelets oblong, obtuse, up to 2 lin. long, purplish or light brown, rigid, loosely 5–9-flowered; rhachilla disarticulating; glumes subequal, oblong-lanceolate, acute, about ½ lin. long, strongly keeled, keel scaberulous; valves oblong in profile, obtuse, ⅔ lin. long, rigidly membranous, side-nerves very prominent, keel scaberulous above; pales subequal to the valves, and falling with them, keels scabrid above; stamens 3; anthers ⅖ lin. long; grain oblong-ellipsoid, ⅕–¼ lin. long, brown, smooth. *Steud. Syn. Pl. Glum.* i. 271; *Durand & Schinz, Consp. Fl. Afr.* v. 883.

COAST REGION: Uitenhage Div.; by the Zwartkops River, *Zeyher*, 1841! Winterhoek Mountain, *Krauss*, 59! Humansdorp Div.; between Galgebosch and Melk River, *Burchell*, 4772! Uniondale Div.; in Lange Kloof, between Wagenbooms River and Apies River, *Burchell*, 4939! Komgha Div.; banks of Kei River, below 500 ft., *Drège!* Queenstown Div.; by the Zwart Kei River, 4000 ft., *Drège!*
KALAHARI REGION: Basutoland; near Leribe, *Buchanan*, 140! Transvaal; Bosch Veld, between Elands River and Klippan, *Rehmann*, 5117! Rustenberg Plains, *McLea*, 126! Dwars River, *Nelson*, 34*!
EASTERN REGION: Transkei; banks of the Bashee River, *Drège*. Natal; without precise locality, *Gerrard*, 680!

45. **E. ciliaris** (Link, Hort. Berol. i. 192); annual or subperennial (?), tufted; culms geniculate, ascending, often from a procumbent base, slender, ½–2 ft. long, glabrous, smooth, simple or branched below, about 3-noded, internodes exserted; sheaths striate, tight, glabrous or scantily hairy, bearded with long hairs at the mouth; ligule a fringe of short hairs; blades linear, tapering to a fine point, 3–6 in. by 1–2 lin., usually involute, somewhat stiff and spreading, glabrous, or with scattered fine long hairs, scaberulous; panicle spike-like, more or less lobed or interrupted, dense to very dense, 2–6 in. long; axis scabrid; branches adpressed, usually all very short or the lowest up to 1 in. long, divided from the base;

pedicels very short; spikelets crowded, ovate, strongly compressed, 1 to almost 2 lin. long, loosely 6–12-flowered, pallid, sometimes purplish; rhachilla breaking up; glumes oblong-lanceolate, acute, $\frac{2}{5}$ to almost $\frac{1}{2}$ lin. long, 1-nerved, keel scabrid; valves oblong in profile, subtruncate and mucronulate, spreading, about $\frac{1}{2}$ lin. long, thin, side-nerves prominent, keel scabrid; pales equal to the valves and falling with them, keels of pale very long and rigidly ciliate; anthers $\frac{1}{5}$–$\frac{1}{6}$ lin. long; grain elongate-ovoid, $\frac{1}{4}$ lin. long, brown. *Trin. Gram. Gen.* 397, *and in Mém. Acad. Pétersb. sér.* 6, i. 397; *Nees, Fl. Afr. Austr.* 413; *Steud. Syn. Pl. Glum.* i. 265; *Anderss. in Peters, Reise Mossamb. Bot.* 558; *Baker, Fl. Maurit.* 456; *Durand & Schinz, Consp. Fl. Afr.* v. 881; *Stapf in Hook. f. Fl. Brit. Ind.* vii. 314. *E. lepida, Hochst. ex A. Rich. Tent. Fl. Abyss.* ii. 424. *E. pulchella, Parl. in Hook. Niger Fl.* 188. *Poa ciliaris, Linn. Sp. Pl.* 102; *Jacq. Ic. Pl. Rar.* ii. *t.* 304; *Kunth, Enum.* i. 337. *A. Rich. l.c.* 423. *Megastachya ciliaris, Beauv. Agrost.* 167; *Roem. & Schult. Syst.* ii. 592.

EASTERN REGION: Natal; common near the coast, *Buchanan*, 160! margins of woods near the Umlazi River, *Krauss*, 349! near Durban, *McKen*, 124! Durban Flats, *Buchanan*, 38! Berea, *Wood*, 5926! between Umzimkulu River and Umkomanzi River, *Drège*, 4270! and without precise locality, *Gerrard*, 601! 482! Delagoa Bay, *Wilms*, 1691!

Common throughout tropical Africa and America, and in North India.

46. E. namaquensis (Nees, Ind. Sem. Vratisl. 1835, and in Linnæa, Lit.-Ber. xi. 125); annual or subperennial (?), tufted; culms erect, slender, $\frac{1}{2}$–1 ft. long, glabrous, smooth, 1-noded, simple, upper internode by far the longest, long exserted; sheaths striate, glabrous, tight, lowest more or less compressed and keeled; blades linear, tapering to a fine point, 2–4 in. by 1–1$\frac{1}{2}$ lin., flat, flaccid, glabrous, smooth; panicle tightly contracted and linear or more or less open and oblong, 4–6 in. long, erect; axis smooth, terete; branches solitary or 2–3-nate or irregularly approximate, erect or obliquely spreading, rather loosely and repeatedly divided from near the base, all divisions subcapillary, glabrous, smooth or almost so; lateral pedicels usually very short; spikelets crowded or more or less scattered, elliptic, obtuse, 1 lin. long, about 5–6-flowered, light purplish or brownish; rhachilla disarticulating; glumes subequal, broad, oblong, obtuse, emarginate, $\frac{1}{4}$ lin. long, hyaline, 1-nerved, persistent; valves oblong in profile, obtuse, $\frac{2}{5}$ lin. long, hyaline, side-nerves prominent, smooth, like the keel; pales subequal to the valves, keels smooth, falling with the valves; stamens 2; anthers almost $\frac{1}{4}$ lin. long; grain oblong, $\frac{1}{4}$ lin. by $\frac{1}{8}$ lin., brown, smooth. *Nees ex Schrader in Linnæa*, xii. 452, *and Fl. Afr. Austr.* 408; *Steud. Syn. Pl. Glum.* i. 272. *E. interrupta, var. namaquensis, Durand & Schinz, Consp. Fl. Afr.* v. 884 (*the South African plant*).

VAR. β, **robusta** (Stapf); culms stout, up to 3 ft. high, 3-noded, simple or branched below; sheaths long, exceeding the internodes (except the uppermost), usually slipping from the stem and rolling inwards in the upper part; panicle $\frac{3}{4}$–1$\frac{1}{4}$ ft. long, usually contracted, dense; branches more numerous, often very long; anthers $\frac{1}{8}$ lin. long or almost so.

WESTERN REGION: Little Namaqualand; sand hills on the right bank of the Orange River, near Verleptpram, *Drège!*

KALAHARI REGION: Var. β: Griqualand West, Herbert Div.; near St. Clair, Douglas, *Orpen*, 247!

EASTERN REGION: Var. β: Natal; by streamlets, at 1000 ft., without precise locality, *Buchanan*, 276!

The typical form occurs also in Angola (Huilla, by the Cunene River, *Newton*); the variety β seems to range all over tropical Africa; both may be easily distinguished from *E. interrupta*, Beauv. (which is very similar in habit), by the perfectly smooth keels of the pales.

Imperfectly known species.

47. **E. homomalla** (Nees, Fl. Afr. Austr. 406); annual; culms fascicled, ascending from a decumbent base, simple, 2-noded, angular; sheaths much shorter than the internodes, glabrous; ligule a fringe of hairs; blades linear, tapering to an acute point, 2–2½ in. by 1 lin., flat, scabrid; panicle dense, 1½–2 in. long; branches solitary or gathered in groups of 2–3, rigid, those of one side erect, of the other side spreading, triquetrous, divided almost from the base; pedicels short or very short; spikelets linear, 1½–2 lin. by ⅓ lin., purplish, 8–10-flowered; glumes unequal, pallid, oblong, 1-nerved; valve ovate, obtuse, membranous, nerves subexcurrent, keel scaberulous; pales shorter than the valves by ⅓, keels serrulate *Steud. Syn. Pl. Glum.* i. 272; *Durand & Schinz, Consp. Fl. Afr.* v 884.

COAST REGION: Vanrhynsdorp Div.; in the sand near Ebenezer, *Drège.*

48. **E. hornemanniana** (Nees, Fl. Afr. Austr. 395); perennial; culms with numerous barren branches near the middle; sheaths long-bearded at the mouth; blades setaceously convolute; panicle decompound, contracted; branches solitary, straight, naked at the base, divided from below the middle; pedicels short; spikelets oblong, glaucous-green, glabrous; glumes unequal, keel scabrid; valves ovate, subacute, lateral nerves very prominent; pales slightly shorter than the valves, keels rigidly ciliolate. *Steud. Syn. Pl. Glum.* i. 271; *Durand & Schinz, Consp. Fl. Afr.* v. 884.

EASTERN REGION: Natal; between Umzimkulu and Umkomanzi Rivers, *Drège.*

49. **E. planiculmis** (Nees, Fl. Afr. Austr. 391); perennial; culms branched at the base, erect, straight, compressed, glabrous; sheaths smooth; ligule a minute fringe of cilia; blades tapering to a fine filiform point, compressed, canaliculate; panicle decompound, contracted, narrow; branches solitary, rigid, almost naked at the base, divided from near the base; branchlets short, fascicled, adpressed, subcontiguous; pedicels short; spikelets linear, compressed, glabrous, grey, shining, loosely 4–8-flowered; glumes subequal, obtuse, 3-nerved. *Steud. Syn. Pl. Glum.* i. 270; *Durand & Schinz, Consp. Fl. Afr.* v. 888.

COAST REGION: Alexandria Div.; wet places on the Quaggas Flats, *Ecklon.*

LIV. DESMOSTACHYA, Stapf.

Spikelets linear, strongly laterally compressed, closely imbricate, alternate, sessile or subsessile on, and falling entire from, the slender rhachis of secund more or less distinctly 2-ranked spikes which are crowded into long narrow spike-like panicles; rhachilla tough. *Florets* numerous, ☿, rather loose. *Glumes* very unequal, membranous, 1-nerved, keeled. *Valves* ovate, acute or subacute, entire, muticous, rigidly membranous, 3-nerved, acutely keeled, glabrous, side-nerves evanescent upwards. *Pales* slightly shorter than the valves, 2-keeled. *Lodicules* 2, rather large, asymmetric, hyaline, nerved at the base. *Stamens* 3. *Ovary* glabrous; styles distinct, slender; stigmas plumose, laterally exserted. *Grain* loosely enclosed by the scarcely altered valve and pale, obliquely ovoid, obtusely triquetrous; pericarp thin, adnate to the seed; embryo about $\frac{1}{3}$ the length of the grain; hilum small, basal, punctiform.

Perennial, branched at the base; branches covered with leathery sheaths at or above the base and with a tuft of coarse leaves; panicle spike-like, often interrupted below; branches (spikes) more or less spreading, irregularly approximate or spirally arranged on a stiff axis, persistent; spikelets on the lower side of, and often at right angles to, the rhachis, closely packed, light straw-coloured or tinged with brown or purple, often very many-flowered.

Species 1, from Egypt to India, and southwards to East Tropical Africa.

This genus was erroneously supposed to occur within the area of the Flora Capensis when the Key was drawn up; but although not actually represented in the South African collections at Kew, it is very probable that it will be found in the Eastern Region. The only species of this genus *D. bipinnata*, Stapf, was originally described as *Uniola bipinnata* by Linnæus. Subsequently, it has been redescribed in, or referred to, at least 5 other genera, viz. *Poa, Briza, Cynosurus, Eragrostis* and *Leptochloa*. Hochstetter pointed out the affinity to the latter genus in Flora, 1855, 422. My recent researches in *Eragrostis* have convinced me that it is one of the links which connect *Eragrosteæ* and *Chlorideæ* (especially the *Leptochloa* group), and I have (following Sir Joseph Hooker's example in the case of *Myriostachys*) now separated *Desmostachya* generically from *Eragrostis*, where, under the name of *E. cynosuroides*, Beauv., it represented a separate section, *Desmostachya* in Hook. f. Fl. Brit. Ind.

LV. SPARTINA, Schreb.

Spikelets 1-flowered, laterally compressed, narrow, densely imbricate or distant by half their length, alternately biseriate, sessile, unilateral on a triquetrous excurrent rhachis, tardily disarticulating at the base; rhachilla tough, not produced. *Floret* hermaphrodite. *Glumes* 2, narrow, unequal, 1–5-nerved, keeled. *Valve* as long as or shorter than the upper glume, oblong (linear-oblong in profile), 1-nerved, very thin except along the keel. *Pale* exceeding the valve, finely 2-nerved, very thin. *Lodicules* 0. *Stamens* 3; filaments very long. *Ovary* glabrous, acute; styles connate at the base, very long; stigmas narrow, densely plumose, terminally exserted. *Grain* enclosed by the little changed valve and pale, free, oblong; embryo narrow, as long as the grain; hilum basal, small.

Perennial, cæspitose or creeping, usually tall and coarse; leaves rigid; spikelets in an interrupted or continuous spike or raceme, erect or oblique; rhachis more or less produced beyond the uppermost spikelet.

Species about 7–8, mostly maritime on the coasts of the Atlantic.

1. S. stricta (Roth, Catal. Bot. iii. 9); glabrous; rhizome creeping, stoloniferous; culms ½–2 ft. long, very slender, sheathed all along, uppermost internode long; sheaths imbricate, the lowest bladeless, short, whitish, smooth and persistent like the others; ligule a ciliate rim; blades articulated with the sheaths, linear, tapering to a pungent point, convolute or involute, 2–6 in. by 2–3 lin. when expanded, very rigid, smooth; spikes 1–3, erect, 2–6 in. long, yellowish; common rhachis short; special rhachis terminating in a mucro not or slightly exceeding the uppermost spikelet; spikelets 6–7 lin. long, imbricate, those of the same side 3–4 lin. distant, pubescent; lower glume very narrow, linear, 4½–6½ lin. long, finely 1–2-nerved, the upper lanceolate, acute, 6–7 lin. long, 3–5-nerved, nerves very close; valve subacute or obtuse, 5–5½ lin. long, tips hyaline; pale 5½–6 lin. long; anthers 3 lin. long; grain about 3–4 lin. long. *Kunth, Enum.* i. 278; *Suppl.* 232; *Reichb. Ic. Fl. Germ.* i. *t.* 25, *fig.* 1401; *Steud. Syn. Pl. Glum.* i. 215; *Engl. Bot. ed.* iii. *t.* 1687; *Sowerby, Brit. Grass. t.* 140; *Durand & Schinz, Consp. Fl. Afr.* v. 858. *S. capensis, Nees, Fl. Afr. Austr.* 260; *Durand & Schinz, l.c.* 857. *Dactylis stricta, Ait. Hort. Kew.* i. 104; *Engl. Bot. t.* 380. *Limnetis pungens, Rich. in Pers. Syn.* i. 72; *Host, Gram. Austr.* iv. *t.* 66.

COAST REGION: Port Elizabeth Div.; along the strand of Algoa Bay, near Cape Recife and Port Elizabeth, *Ecklon & Zeyher*, 662! in salt marshes at the mouth of the Zwartkops River, *Drège! MacOwan & Bolus, Herb. Norm. Austr. Afr.*, 791.

Atlantic coast of Europe from Holland and South England to the Strait of Gibraltar and in the Bay of Friaul in the Adriatic.

LVI. CYNODON, Pers.

Spikelets 1-flowered, small, laterally compressed, sessile, imbricate, alternately 2-seriate and unilateral on a slender keeled rhachis; rhachilla disarticulating above the glumes, produced, or not, beyond the valve. *Floret* hermaphrodite. *Glumes* narrow, keeled, acute or subulate-mucronate, the upper usually deciduous with the valve, the lower subpersistent. *Valve* exceeding the glumes, navicular, firmly membranous, 3-nerved, awnless, keel ciliate. *Pale* somewhat shorter than the valve, 2-keeled. *Lodicules* 2, minute, obovate-cuneate, glabrous. *Stamens* 3. *Ovary* glabrous; styles distinct, slightly shorter than the plumose styles. *Grain* oblong, subterete; embryo about ⅓ the length of the grain; hilum linear, ⅔ the length of the grain.

Perennial; stems creeping, rooting at the nodes and emitting from them fascicles of barren shoots and flowering culms; spikes 2–6 in terminal umbels.

Species 2; 1 in extra-tropical South Africa, the other almost cosmopolitan.

Culms many-noded with the leaves mostly crowded at the base; ligule a ciliate rim; rhachilla produced ... ... (1) **Dactylon.**
Culms 2–3-noded; ligule membranous; rhachilla not produced ... ... (2) **incompletus.**

1. **C. Dactylon** (Pers. Syn. i. 85); culms from a few inches to 1 ft. long, slender, glabrous, smooth, many-noded, the lower internodes very short, enclosed, the upper 3–4 much longer, more or less exserted; leaves usually conspicuously distichous in the barren shoots and at the base of the culms; sheaths tight, glabrous or hairy, often bearded at the mouth; ligule a very fine ciliate rim; blades linear, finely acute to pungent, $\frac{1}{2}$–6 in. by 1–$1\frac{1}{2}$ lin., very rigid to flaccid, folded or convolute or flat, more or less glaucous, glabrous or hairy, smooth below, scaberulous above; spikes 2–6, straight, $\frac{1}{2}$–$2\frac{1}{2}$ in. long; rhachis pubescent at the base, keel and margins scabrid or the keel smooth; spikelets light green or purplish, $\frac{7}{8}$–$1\frac{1}{8}$ lin. long; rhachilla produced, very slender, equalling $\frac{1}{2}$ the length of the spikelet; glumes lanceolate, acute to subulate-mucronate, the lower $\frac{1}{2}$–$\frac{3}{4}$ lin. long, the upper usually slightly longer, keels scabrid or smooth; valve obliquely oblong to semi-ovate, subobtuse or minutely apiculate, about 1 lin. long, keel ciliate; keels of pale scaberulous; anthers oblong, $\frac{1}{2}$ lin. long; grain $\frac{1}{2}$ lin. long. *Kunth, Enum.* i. 259; *Suppl.* 203, *t.* 16, *fig.* 1; *Reichb. Ic. Fl. Germ. t.* 26, *fig.* 1404; *Nees, Fl. Afr. Austr.* 241; *Steud. Syn. Pl. Glum.* i. 212; *Engl. Hochgebirgsfl. Trop. Afr.* 132; *Durand & Schinz, Consp. Fl. Afr.* v. 856; *Engl. Pfl. Ost-Afr. A.* 11, 79; *B.* 78; *C.* 110. *C. linearis, Willd. Enum. Hort. Berol.* 90. *C. stellatus, Willd. l.c.*; *Kunth, l.c.* 260. *C. pascuus, Nees, Agrost. Bras.* 425; *Kunth, l.c.* 259; *Nees, Fl. Afr. Austr.* 243; *Steud. Syn. Pl. Glum.* i. 212; *Durand & Schinz, l.c.* 857. *C. glabratus, Steud. l.c.*; *Durand & Schinz, l.c.* *Panicum Dactylon, Linn. Sp. Pl.* 58; *Thunb. Prodr.* i. 19; *Fl. Cap. ed. Schult.* 103; *Host, Gram. Austr.* ii. 15, *t.* 18; *Engl. Bot. t.* 850; *Knapp, Gram. Brit. t.* 13. *Dactylon officinale, Vill. Hist. Pl. Dauph.* ii. 69; *Aschers. & Schweinf. Ill. Fl. Égypte*, 170. *Digitaria Dactylon, Scop. Fl. Carn. ed.* ii. 53. *D. stolonifera, Schrad. Fl. Germ.* i. 165, *t.* 3, *fig.* 9; *Steud. in Flora*, 1829, ii. 468.

SOUTH AFRICA: without precise locality, *Thunberg!*

COAST REGION: Vanrhynsdorp Div.; near Ebenezer, below 100 ft., *Drège!* Cape Div.; by streamlets and ditches, near Capetown, *Ecklon*, 967! sandy places near Green Point, *Ecklon*, 764! Platt Klipp, near Capetown, *Wilms*, 3874! roadside towards Constantia Nek, *Wolley Dod*, 2473! Paarl Div.; near Paarl, in wet places, *Drège*. Tulbagh Div.; Tulbagh Waterfall, *Ecklon*. Uitenhage Div.; grassy places by the Zwartkops River, *Ecklon*; near Uitenhage, *Burchell*, 4229! Port Elizabeth Div.; on the dunes near Port Elizabeth, below 100 ft., *Drège! E.S.C.A. Herb.* 93! Queenstown Div.; plains near Queenstown, 3500 ft., *Galpin*, 2353!

CENTRAL REGION: Somerset Div.; among shrubs, near Somerset East, *MacOwan*, 2119.

WESTERN REGION: Little Namaqualand, on the right bank of the Orange River, near Verleptpram, below 500 ft., *Drège!*

KALAHARI REGION: Orange Free State; between Kimberley and Bloemfontein, *Buchanan*, 289! Kanon Fontein, *Rehmann*, 3549! Bechuanaland;

Kosi Fontein, *Burchell*, 2570/1! Transvaal; Houtbosch, *Rehmann*, 5713! near Lydenburg, *Wilms*, 1701!

EASTERN REGION: Transkei; near the Gekau (Gcua or Geuu) River, below 1000 ft., *Drège!* Natal; Durban Flats, *Buchanan*, 12! 34! Berea, *Wood*, 5930! and without precise locality, *Buchanan*, 200!

Almost cosmopolitan.

Extremely variable in habit according to the station. *C. pascuus*, Nees, is, in my opinion, only a shade form of *C. Dactylon*.

2. C. incompletus (Nees in Linnæa, vii. 301); culms slender, 2–7 in. long, 2–3-noded, glabrous, smooth; sheaths glabrous or hairy to villous, bearded at the mouth, the lower loose; ligules membranous, truncate, up to $\frac{1}{2}$ lin. long; blades linear, very shortly acute or subobtuse, 1–2 in. by 1–$1\frac{1}{2}$ lin., rather flaccid, glaucous, rough on both sides, glabrous or hairy to villous; spikes 3–5, straight, $\frac{3}{4}$–$1\frac{1}{4}$ lin. long; spikelets $1\frac{1}{4}$–$1\frac{1}{2}$ lin. long; rhachilla not produced; glumes ovate to lanceolate, acute, the lower $\frac{1}{4}$–$\frac{2}{5}$ lin. long, the upper $\frac{1}{3}$–$\frac{3}{4}$ lin. long; valve $1\frac{1}{4}$–$1\frac{1}{2}$ lin. long, keel narrowly or obscurely winged, rigidly ciliolate, keels of the pale stout, very close, scabrid above; anthers rather more than $\frac{1}{2}$ lin. long; otherwise like the preceding species. *Fl. Afr. Austr.* 243; *Kunth, Enum.* i. 260; *Steud. Syn. Pl. Glum.* i. 213; *Durand & Schinz, Consp. Fl. Afr.* v. 857. *C. notatus*, *Nees in Linnæa*, vii. 302; *Fl. Afr. Austr.* 244; *Kunth, l.c.* 260; *Steud. l.c.* 213; *Durand & Schinz, l.c.*

COAST REGION: Cape Div.? Zeekoe River, *Ecklon*. Riversdale Div.; by the Gauritz River, and on the Lange Kloof Mountains, *Ecklon*. Uitenhage Div.; Steenbok Flats, north of Winterhoek Bergen, *Ecklon, Zeyher*. "Kaffraria," 2000 ft., *Baur*, 267!

CENTRAL REGION: Beaufort West Div.; in the Gouph, *Drège*. Aberdeen Div.; Camdeboo Plains, 3000 ft., *Drège*. Colesberg Div.; Wonderhuivel, *Drège!* Colesberg, *Shaw*, 11!

WESTERN REGION: Little Namaqualand; banks of the Orange River, *Drège*.

KALAHARI REGION: Transvaal; near Lydenburg, *Atherstone!*

LVII. MICROCHLOA, R. Br.

Spikelets 1–2-flowered, small, sessile, crowded, unilateral on a flattened rhachis, alternately 2-seriate from near the margins of the rhachis, or in a single row; rhachilla disarticulating above the glumes, more or less produced. *Floret* hermaphrodite, or if 2, the lower hermaphrodite, the upper male or indicated by an empty valve. *Glumes* 2, persistent or (particularly the upper) deciduous, strongly 1-nerved, flattened from the back or keeled, subequal. *Valve* shorter than the glumes, delicate, white, minutely or obscurely mucronulate or emarginate, 3-nerved, densely hairy along the nerves (if 2, the upper glabrous); callus small, acute, hairy. *Pale* slightly shorter than the valve or almost equal, 2-keeled. *Lodicules* 2, cuneate, glabrous, thin, faintly nerved. *Stamens* 3. *Ovary* glabrous, (quite suppressed in the upper floret); styles distinct; stigmas plumose, laterally exserted. *Grain* oblong, terete, triquetrous or compressed, embraced by the unchanged valve and pale, free; hilum punctiform; embryo equalling $\frac{1}{4}$–$\frac{1}{2}$ the length of the grain.

Perennial, rarely annual, sometimes densely tufted; leaves narrow, often subsetaceous; ligule reduced to a minutely ciliolate rim; spikes solitary, terminal (in the African species) or 2–4 in a terminal umbel, straight or curved.

Species 7; 1 widely distributed throughout the tropics, 3 in Africa, 3 in Australia.

Spikelets strictly 1-flowered:
- Spikelets 1–1½ lin. long; valve abruptly and shortly acuminate, like the pale long ciliate along the nerves ... ... (1) **setacea.**
- Spikelets 1¾–2¼ lin. long; valve minutely cuspidate, like the pale ciliate along the nerves except at the glabrous tips ... ... (2) **caffra.**

Spikelets with 2 valves; the lower with a hermaphrodite, the upper with a male flower or empty ... ... (3) **altera.**

1. **M. setacea** (R. Br. Prodr. 208); perennial, cæspitose or annual; culms geniculate-erect, very slender, branched or almost simple, from 2 in. to more than 1 ft. long, compressed below, 1–3-noded, glabrous, smooth, uppermost internode by far the longest; leaves crowded at the base; sheaths tight or slightly tumid, compressed, keeled, glabrous, smooth, the basal persistent, breaking up into fibres; blades subsetaceous with an acute or callous point, ½–1½ in. long, plicate, firm, the lower often curved, glabrous, rarely sparingly hairy, smooth, margins rough; spike solitary, 1–6 in. long, very slender, usually curved, often purplish; rhachis minutely ciliate; spikelets 1-flowered, dorsally compressed, in a single row, 1–1½ lin. long, glabrous; glumes lanceolate-oblong, acute to cuspidate-acuminate, the lower slightly longer, asymmetric; valve abruptly and shortly acuminate, sometimes mucronulate, up to ⅞ lin. long, very densely hairy along the nerves; pale ciliate on the nerves; anthers ¼–⅓ lin. long; grain dorsally compressed, over ½ lin. long. *Beauv. Agrost.* 115, *t.* 20, *fig.* 8; *H. B. & K. Nov. Gen. & Spec.* i. 84, *t.* 22; *Kunth, Enum.* i. 258, *Suppl.* 201; *Nees, Fl. Afr. Austr.* 247; *Steud. Syn. Pl. Glum.* i. 202; *Doell in Mart. Fl. Bras.* ii. iii. 76, *t.* 21, *fig.* 2; *Oliv. in Trans. Linn. Soc.* xxix. 173; *Engl. Hochgebirgsfl. Trop. Afr.* 131; *Durand & Schinz, Consp. Fl. Afr.* v. 856; *Hook. f. Fl. Brit. Ind.* vii. 283. *M. abyssinica, Hochst. ex A. Rich. Tent. Fl. Abyss.* ii. 404; *Steud. Syn. Pl. Glum.* i. 202; *Schweinf. in Bull. Herb. Boiss.* ii. *App.* ii. 31. *Rottboellia setacea, Roxb. Pl. Corom.* ii. 17, *t.* 132.

KALAHARI REGION: Transvaal; Houtbosch, *Rehmann*, 5725!

Throughout the tropics.

2. **M. caffra** (Nees, Fl. Afr. Austr. 246); perennial, compactly cæspitose; culms erect or geniculate-erect, very slender, simple, ½–1 ft. long, compressed below, 1–2-noded, glabrous, smooth, internodes exserted, the uppermost by far the longest; leaves crowded at the base; sheaths tight or the uppermost subtumid, glabrous or ciliate

at the mouth, smooth, the lowest persistent, breaking up into fibres; blades subsetaceous, with an acute or callous point, $\frac{1}{2}$–6 in. long, folded, firm, often curved, glabrous or scantily hairy near the base, smooth, margins rough; spike solitary, 2–$3\frac{1}{2}$ in. long, usually curved, often purple, margins of the rhachis ciliolate; spikelets 1-flowered, divergent and biseriate or imbricate and more or less uniseriate, slightly dorsally compressed, $1\frac{3}{4}$–$2\frac{1}{4}$ lin. long, glabrous; glumes lanceolate-oblong, acute or the upper acuminate, the lower asymmetric, slightly longer; valve minutely cuspidate, $1\frac{1}{2}$ lin. long, densely hairy along the nerves except at the very tip; keels of pale scabrid, ciliolate above the middle; anthers not quite 1 lin. long; grain terete, over $\frac{1}{2}$ lin. long. *Steud. Syn. Pl. Glum.* i. 202; *Durand & Schinz, Consp. Fl. Afr.* v. 856.

SOUTH AFRICA: without locality, *Boivin! Harvey! Bolus!*

COAST REGION: Tulbagh Div.; near Tulbagh Waterfall, *Ecklon & Zeyher!* Uitenhage Div.; near the Zwartkops River, *Zeyher*, 946! Stokenstrom Div.; Kat Berg, in grassy places, 3000–4000 ft, *Drège! Zeyher*, 4518! Cathcart Div.; Blesbok Flats, near Windvogel Berg, 3000 ft., *Drège.* Albany Div.; Brand Kraal, near Grahamstown, 2200 ft., *MacOwan*, 1272!

CENTRAL REGION: Somerset Div.; Bosch Berg, *Burchell*, 3185!

KALAHARI REGION: Transvaal; Incomate (King George's) River, *Nelson*, 13!

EASTERN REGION: Tembuland; Tabaze, near Bazeia, 2000–2500 ft., *Baur*, 322! Natal; Riet Vlei, 4000–5000 ft., *Buchanan*, 162! Weenen County, South Downs, 4000 ft., *Wood*, 4403! Benvie, 3000–4000 ft., *Wood*, 6007! Pieter Maritzburg, *Wood*, 7226!

3. **M. altera** (Stapf; *Harpechloa altera*, Rendle in Trans. Linn. Soc. ser. 2, Bot. iv. 57) var. **Nelsonii** (Stapf); perennial, densely cæspitose; culms erect, very slender, simple, 8–10 in. long, compressed below, 2-noded, glabrous or woolly, upper 2 internodes very long, exserted; leaves mainly crowded at the base; sheaths tight or scantily woolly, the basal compressed, keeled, very narrow, persistent, at length breaking up into fibres; blades setaceous, folded, acute, scarcely distinct from the sheaths, 3–6 in. long, glabrous, smooth; spike solitary, $\frac{3}{4}$–1 in. long, usually straight; rhachis glabrous; spikelets 2-flowered, dorsally and obliquely compressed, uniseriate or biseriate, $2\frac{1}{4}$ lin. long, glabrous, brown; glumes lanceolate-oblong, obtuse, the lower slightly longer and asymmetric, the upper firmer; lower floret hermaphrodite; valve minutely bilobed, $1\frac{1}{2}$ lin. long, ciliate along the nerves; pale glabrous, keels very finely scabrid above; upper floret barren, slightly smaller; valve and pale glabrous, more delicate, the latter often reduced or quite suppressed; grain 1 lin. long.

KALAHARI REGION: Transvaal; Incomate (King George's) River, *Nelson*, 14!

EASTERN REGION: Natal; Riet Vlei, 4000–5000 ft., *Buchanan*, 163! Zululand, 2000–3000 ft., *Wood*, 7304!

The type of the species, only known from the Shire Highlands, differs mainly in the more robust habit, broader basal sheaths, longer and curved spikes, and in the upper floret being (always ?) male.

LVIII. CTENIUM, Panz.

Spikelets of 3–4 florets, sessile, compactly crowded, unilateral, alternately biseriate along the midrib of the flattened rhachis; rhachilla disarticulating above the glumes, continuous between the valves, the lower 2 florets barren or the second male, the third hermaphrodite, the fourth male or barren, or quite rudimentary. *Glumes* unequal, the lower persistent, keeled, thin, 1-nerved, the upper much longer, oblong to lanceolate, flattened or rounded on the back, firm, 2–3-nerved, with a stiff awn from the middle. *Valves* oblong in profile, obtuse, 3-nerved, awned just below the tips, ciliate along the nerves or the uppermost glabrous, white, thin. *Pales* slightly shorter, 2-keeled or 2-nerved. *Lodicules* 2, quadrate-cuneate, delicate, faintly nerved. *Stamens* 3 in the hermaphrodite, 2 in the male florets. *Ovary* glabrous; styles distinct, stigmas slender, long, laterally exserted. *Grain* free, embraced by the unchanged valve and pale, oblong; embryo up to ½ the length of the grain; hilum basal, punctiform.

Perennial, densely tufted, rarely annual; leaves narrow, flat or convolute; spikes terminal, solitary or in umbels of 2–3, usually curved; spikelets prettily pectinate and awned.

Species about 9, in Africa and America.

1. **C. concinnum** (Nees, Fl. Afr. Austr. 237); perennial, densely tufted; culms erect, 1½–2 ft. long, villous or pubescent below the spike, 2-noded, upper 2 internodes very long, at length more or less exserted; leaves mostly crowded at the base; sheaths tight or the upper subtumid, striate, glabrous, smooth, the basal ones compressed, persistent; ligule extremely short, minutely ciliolate; blades narrow, linear, acute, the basal up to 1 ft. by ½–1 lin., flat or setaceously convolute, smooth below, scaberulous above and along the margins; spikes solitary, rarely paired, olive-grey, 3–10 in. by 2–2½ lin.; spikelets 3–3½ lin.; lower glume ovate, acuminate, about 1½ lin. long, keel coarsely scabrid; upper glume broadly lanceolate, acuminate, 3–3¼ lin. long, scabrid to hispidulous, tubercled on the nerves, 2-nerved, middle nerve emitting an obliquely erect awn, not produced beyond it or faintly so or percurrent and even excurrent, side-nerve percurrent or excurrent, awn scarcely exceeding the glume; lowest valve barren, more or less cuspidate or apiculate, ciliate, 1½ lin. long, awn 2–4 lin. long; second valve slightly longer, narrower, the cilia prolonged into a beard above the middle with a rudimentary pale and 2 perfect or imperfect stamens, awn 3–3½ lin. long; third valve like the second, but more delicate, shorter awned, with a 2-keeled glabrous pale, and a hermaphrodite flower; fourth valve glabrous, delicate, 1½ lin. long, with a broad 2-nerved delicate pale and 2 stamens; lodicules up to ⅓ lin. long; anthers of the hermaphrodite flower 1½ lin. long, those of the male usually shorter, anther-cells acute; styles very short; stigmas 1½ lin. long.

EASTERN REGION: Pondoland; between St. Johns River and Umtsikaba River, below 1000 ft., *Drège!* between Umtentu River and Umzimkulu River, *Drège.* Natal; Umpumulo, 2000 ft., *Buchanan*, 179!

LIX. HARPECHLOA, Kunth.

Spikelets of 3–4 florets, sessile, crowded, unilateral, alternately biseriate along the midrib of a flattened rhachis; rhachilla disarticulating above the glumes, continuous between the valves, the lower floret hermaphrodite, the following 1 or 2 male, the uppermost barren, rudimentary. *Glumes* unequal, the lower persistent, keeled, very thin, 1-nerved, the upper much longer, oblong, flattened on the back, 2–3-nerved, firm. *Hermaphrodite floret* about equalling the upper glume; valve folded, obliquely oblong in profile, obtuse, white, thin, 3-nerved, densely ciliate along the nerves; callus obscure; pale slightly shorter, 2-keeled; lodicules cuneate, fleshy, almost 3-winged; stamens 3; ovary glabrous; styles distinct; stigmas slender, plumose, laterally exserted. *Upper florets* crowded in a club-shaped body, not exceeding the hermaphrodite floret, enveloped by the valve of the lower male floret; valves 2-nerved or with a trace of the middle nerve near the apex, ciliolate or glabrous; pales 2-nerved; stamens 3 or 0; ovary usually quite suppressed. *Grain* free, embraced by the unchanged valve and pale, oblong, obtusely triquetrous; embryo $\frac{1}{2}$ the length of the grain; hilum punctiform, basal.

Perennial, densely cæspitose; leaves firm, folded or convolute above, more or less curved; spikes terminal, solitary, rarely geminate, dark olive-grey.

Species 1, endemic.

1. H. capensis (Kunth, Rév. Gram. i. 92); culms erect, $\frac{3}{4}$–2 ft. long, compressed below, 2-noded, woolly below the spike, otherwise usually glabrous, upper 2 internodes very long, exserted; leaves mostly crowded at the base; basal sheaths imbricate, firm, persistent, compressed, keeled, hairy or glabrescent, bearded at the mouth, striate, the uppermost slightly tumid, glabrous; ligule a ciliolate rim; blades linear, acute to obtuse, 3–10 in. by $\frac{1}{2}$–$1\frac{1}{2}$ lin. when flattened out, glaucous, glabrous or hairy to woolly, smooth; spikes 1–$2\frac{1}{2}$ in. long to 4 lin. broad; rhachis ciliate or woolly; spikelets 3–$3\frac{1}{2}$ lin. long; lower glume ovate, acute to obtuse, $1\frac{1}{2}$–$1\frac{3}{4}$ lin. long, upper 3–$3\frac{1}{2}$ lin. long; valve of the hermaphrodite floret 3–$3\frac{1}{2}$ lin. long; pale ciliate along the margins and hairy near the tip, keels scaberulous; lodicules $\frac{1}{3}$ lin. long; anthers $1\frac{1}{2}$–$1\frac{3}{4}$ lin. long; upper valves and anthers smaller; grain $1\frac{1}{2}$ lin. long. *Kunth, Enum.* i. 274; *Suppl.* 225; *Steud. Syn. Pl. Glum.* i. 203; *Durand & Schinz, Consp. Fl. Afr.* v. 858. *Melica Falx, Linn. f. Suppl.* 109. *Cynosurus falcatus, Thunb. Prodr.* 23. *Chloris falcata, Sw. in Berl. Ges. Naturf. Freunde, Neu. Schr.* iii. 160, *t.* 1, *fig.* 1; *Thunb. Fl. Cap. ed.* i. 408; *ed. Schult.* 109. *Dactyloctenium falcatum, Willd. Enum. Hort. Berol.* 1030. *Campulosus hirsutus, Desv. in Nouv. Bull. Soc. Philom.* ii. (1810) 189. *C. falcatus, Beauv. Agrost.* 157. *Campuloa hirsuta,*

Desv. in Journ. Bot. iii. (1813) 69. *Eleusine falcata, Spreng. Syst.* i. 349.

SOUTH AFRICA: without precise locality, *Thunberg!*

COAST REGION: Swellendam Div.; *Mund.* Mossel Bay Div.; between Duyker River and Gauritz River, *Burchell,* 6402! Uniondale Div.; Lange Kloof, *Ecklon.* Uitenhage Div.; Van Stadensberg Range, nearest to Galgebosch, *Burchell,* 4729! Port Elizabeth Div.; Port Elizabeth, *E.S.C.A. Herb.,* 95! Alexandria Div.; between Sunday River and Bushmans River, *Ecklon;* near Addo, *Drège;* northern slopes of the Zuur Berg Range, 2500–3000 ft., *Drège!* Albany Div.; mountains near Grahamstown, 2200–2400 ft., *MacOwan,* 1293! *Burke!* Stockenstrom Div.; above Philipton near the sources of the Kat River, *Zeyher!* Kat Berg, 3000–4000 ft., *Drège.* Queenstown Div.; Finchams Nek, near Queenstown, 3900 ft., *Galpin,* 2370! British Caffraria, *Cooper,* 3357!

CENTRAL REGION: Somerset Div.; plateau on the summit of Bosch Berg, near Somerset East, 4800 ft., *MacOwan & Bolus, Herb. Norm. Aust.-Afr.,* 790! Graaff Reinet Div.; top of Kaudeveld Mountain, 5000 ft., *Bolus,* 1373!

KALAHARI REGION: Orange Free State, *Cooper,* 916! Transvaal; Lydenburg District, Spitzkop Goldmine, *Wilms,* 1662!

EASTERN REGION: Tembuland; near Bazeia, *Baur,* 283! Natal; near Pieter Maritzburg, 1000–2000 ft., *Krauss,* 445! Riet Vlei, 4000–5000 ft., *Buchanan,* 178! Mooi River, 4000 ft., *Wood,* 7317! and without precise locality, *Buchanan,* 57! 93!

LX. CHLORIS, Swartz.

Spikelets of 2–4 florets, sessile, crowded, unilateral, 2-seriate on a slender rhachis; rhachilla disarticulating above the glumes, tough between the valves, more or less produced; lowest floret hermaphrodite, the second male or barren, the following, if present, barren, often minute. *Glumes* 2, persistent, narrow, keeled, acute and mucronate, very thin, or broad, and the upper obtuse, more or less bilobed and rounded on the back. *Hermaphrodite floret:* valve narrow or broad, 3-nerved, acute or obtuse, minutely 2-toothed, usually awned from below the apex, often ciliate; pale almost equalling the valve, 2-keeled; lodicules 2, minute, delicate, glabrous; stamens 3; ovary glabrous, styles distinct, short; stigmas laterally exserted. *Male floret:* valve and pale as in the hermaphrodite flower, but smaller and glabrous. *Rudimentary florets* glabrous, awned or awnless, small to very small, usually without a trace of a pale. *Grain* oblong, triquetrous; embryo rather large; hilum punctiform, basal.

Perennial or annual; leaves flat or folded; spikes solitary or several to many in terminal umbels or short racemes, erect or stellately spreading.

Species 40–45, in the tropical and subtropical regions of both hemispheres.

Glumes narrow, lanceolate, acute or mucronate, keeled:
- Valves long and finely awned:
  - Rudimentary floret minute, oblong on a long rhachilla joint ... ... (1) **pycnothrix.**
  - Rudimentary floret cuneate on a rather short rhachilla joint, more than ½ the length of the hermaphrodite floret (sometimes with a minute second rudimentary floret) ... ... (2) **virgata.**
- Awns as long as the valve or little longer; second floret usually male on a short rhachilla joint ... (3) **gayana.**

Upper glume very broad, shortly bilobed, mucronate, rounded or almost flat on the back; valves mucronate (4) **petræa.**

1. **C. pycnothrix** (Trin. Gram. Unifl. 234); perennial or annual (flowering the first year ?), ½–1 ft. high; stems prostrate, emitting tufts of barren shoots and culms from the rooting nodes; culms geniculately-ascending, 2–3-noded, more or less sulcate below, glabrous, upper internodes exserted; leaves conspicuously distichous; basal sheaths much compressed, keeled, short, uppermost usually subtumid, all glabrous, smooth; ligules membranous, up to ¼ lin. long, ciliolate; blades linear, obtuse, 1½–2 in. by 1½–2 lin., rarely longer, flat, glaucous, glabrous, smooth or scaberulous above, margins rough; spikes 3–9, sessile or some shortly peduncled, umbelled or subumbelled, suberect, at length usually spreading, pallid or purplish, 1½–3 in. long; rhachis scabrid; spikelets 2-awned, ¼ lin. long; rhachilla joints between the valves ⅔ the length of the lower valve, fine, rhachilla not produced; glumes very narrow, lanceolate, acuminate, the lower ½–⅔ lin. long, the upper 1–1¼ lin., keels very scabrid; lower valve linear-oblong in profile, acute, minutely 2-toothed, 1¼ lin. long, glabrous, keel and tip scaberulous; callus minutely bearded, awn very fine, 6–8 lin. long; pale glabrous, keels scabrid; anthers ¼ lin. long; grain linear-oblong, ⅞ lin. long; upper valve rudimentary, empty, ⅓ lin. long, awn 2–3 lin. long. *Nees, Agrost. Brasil.* 423; *Kunth, Enum.* i. 266; *Anderss. in Peters, Reise Mossamb. Bot.* 556. *C. beyrichiana, Kunth, Rév. Gram.* 89 *and* 289, *t.* 56; *Enum* i. 265; *Steud. Syn. Pl. Glum.* i. 205. *C. leptostachya, Hochst. ex A. Rich. Tent. Fl. Abyss.* ii. 407; *Steud. Syn. Pl. Glum.* i. 206; *Schweinf. in Bull. Herb. Boiss.* ii. *App.* ii. 32; *Durand & Schinz, Consp. Fl. Afr.* v. 861. *C. intermedia, A. Rich. l.c.* 407; *Steud. l.c. C. leptostachya, var. intermedia, Durand & Schinz, l.c. C. radiata, Durand & Schinz, Consp. Fl. Afr.* v. 862, *non Swartz.*

EASTERN REGION: Natal; Umpumulo, 1800 ft., *Buchanan*, 185! Inanda, *Wood*, 1590!

Also in tropical Africa, and Eastern Brazil and Paraguay.

2. **C. virgata** (Swartz, Fl. Ind. Occ. i. 203); perennial or annual (flowering the first year ?), 1–3 ft. high; culms erect or geniculately-ascending or prostrate below, rooting and emitting fascicles of barren shoots from the nodes, 3–5-noded, more or less compressed below, glabrous, smooth, internodes exserted; sheaths glabrous, rarely sparingly hairy, smooth, the lower much compressed, keeled; ligules membranous, very short, very minutely ciliolate; blades linear, gradually tapering to an acute point, 1–4 in. by 1–1¾ lin., flat or folded, sometimes flaccid, glaucous, glabrous, rarely sparingly hairy, smooth below, scaberulous above, margins rough; spikes 6–15 or more, suberect, sessile, whitish-green or purplish, 1–2½ in. long, straight; rhachis pubescent or villous at the base, scabrid; spikelets 2- (rarely sub-3-) flowered, 2-awned, almost 2 lin. long; rhachilla-joint between the valves rather long, terminal joint very minute; glumes narrow, lanceolate, hyaline, mucronate, keels scabrid, the lower 1–1¼ lin. long, the upper almost 2 lin.; lower valve obliquely oblong, acute or obscurely 2-toothed, 1¼ lin. long, whitish or almost

black when mature, ciliate along the marginal nerves and bearded below the tip, finely grooved on the faces, keel glabrous or minutely ciliate below the middle; awn 5–8 lin. long, straight; pale glabrous; anthers $\frac{1}{5}$–$\frac{1}{4}$ lin. long; grain linear-oblong, obtusely triquetrous, $\frac{3}{4}$ lin. long; upper valve (or valves) quite empty, obliquely cuneate in profile, 1 lin. or less long, awn from below the tip, 3–6 lin. long. *Doell in Mart. Fl. Bras.* ii. iii. 65. *C. compressa, DC. Cat. Hort. Monsp.* 1813, 94; *Nees in Linnæa,* vii. 300; *Fl. Afr. Austr.* 240; *Steud. Syn. Pl. Glum.* i. 204; *Anderss. in Peters, Reise Mossamb. Bot.* 556. *C. meccana, Hochst. ex Steud. Syn. Pl. Glum.* i. 205; *Boiss. Fl. Or.* v. 554. *C. multiradiata, Hochst. in Flora,* 1855, 204; *Durand & Schinz, Consp. Fl. Afr.* vi. 861. *C. barbata, var. meccana, Aschers. & Schweinf. Ill. Fl. Égypte,* 170; *Schweinf. in Bull. Herb. Boiss.* ii. *App.* ii. 32.

VAR. β, **elegans** (Stapf); spikes up to 3 in. long; spikelets usually sub-3-flowered; lower valve conspicuously gibbous, $1\frac{1}{2}$ lin. long, more deeply grooved on the faces, keel glabrous or ciliate to, and bearded at the middle. *C. elegans, H.B.K., Nov. Gen. & Spec.* i. 166, *t.* 49; *Kunth, Enum.* i. 264; *Steud. Syn. Pl. Glum.* i. 204. *C. polydactyla, Jacq. Ecl. Gram.* 12, *t.* 9 (*non Swartz*). *C. alba, Presl, Rel. Haenk.* 289. *C. compressa, Oliv. in Trans. Linn. Soc.* xxix. 174, *non DC. C. brachystachys, Anderss. in Peters, Reise Mossamb.* 556.

SOUTH AFRICA: without precise locality, *Zeyher,* 1822!

COAST REGION: Uitenhage Div.; Steenbok Flats, northern side of Winterhoeks Bergen, *Ecklon;* near Uitenhage, *Zeyher! MacOwan,* 2190! Komgha Div.; banks of Kei River, *Drège!* Queenstown Div.; Shiloh, *Baur,* 781! plains of Queenstown, 3500 ft., *Galpin,* 2363!

CENTRAL REGION: Aberdeen Div.; Camdeboo Mountains, 2500–3000 ft., *Drège.* Graaff Reinet Div.; by the Sunday River, north of Monkey Ford, *Burchell,* 2862! Colesberg Div.; near Colesberg, *Shaw,* 4! Richmond Div.; Winterveld, *Ecklon, Drège,* 843! Albert Div.; *Cooper,* 674!

KALAHARI REGION: Griqualand West, Herbert Div.; St. Clair, Douglas, *Orpen,* 249! Albania, between the Orange River and Vaal River, 4500 ft., *Bolus,* 1968! Orange Free State; Olifants Fontein, *Rehmann,* 3538! near the Caledon River, *Burke,* 429! Thabunchu, *Burke!* Transvaal; near Lydenburg, *Atherstone!*

EASTERN REGION: Transkei; between Gekau (Gcua or Geuu) River, and Bashee River, 1000–2000 ft., *Drège!* Natal; Berea, *Wood,* 5948! river banks at Tugela, 600–1000 ft., *Buchanan,* 186! Van Reenens Pass, 5500 ft., *Wood,* 5990! Var. β: Natal; Inanda, *Wood,* 687!

Widely spread through the tropics of both hemispheres.

3. **C. gayana** (Kunth, Rév. Gram. i. 89, ii. 293, t. 58); perennial or annual, 2–4 ft. high; culms erect or geniculately-ascending, or prostrate at the base, simple or branched, often emitting fascicles of barren shoots or short runners from the lower nodes, often robust, 3–9-noded, compressed below, glabrous, smooth, upper internodes usually exserted; sheaths glabrous or sparingly hairy near the mouth, smooth, the lower strongly compressed, keeled, keels sometimes scabrid, the uppermost sometimes tumid; ligules membranous, very short, long-hairy; blades linear, long-tapering to a fine point, $\frac{1}{2}$ to more than 1 ft. by 3–4 lin. when expanded, flat or folded, glabrous or hirsute near the base, green, smooth below, rough above and on the margins; spikes 6–15, umbelled, sessile, suberect, rarely

spreading, 2½–4 in. long, greenish or brownish; rhachis scabrid; spikelets 1½ lin. long, 3–4-flowered, shortly 2-awned, glumes very unequal, the lower ovate-lanceolate, acute, subhyaline, ½–¾ lin. long, the upper oblong, obtuse, mucronate, 1–1½ lin. long, firmer, scaberulous; lowest valve oblong, subobtuse or acute, minutely 2-toothed, ciliolate along the marginal nerves and shortly bearded below the tips or only finely bearded (in the South African species) or almost glabrous, with a (sometimes minutely hairy) groove on each face; awn as long or slightly longer than the valve, straight; callus minutely bearded; pale glabrous, keels scabrid; anthers ¾ lin. long; second valve with a male flower, like the preceding, but glabrous, 1 lin. long, awn 1 lin. long or less; third (and fourth) valve rudimentary, cuneate in profile, empty, awnless. *Kunth, Enum.* i. 267; *Suppl.* 216; *Nees, Fl. Afr. Austr.* 240; *Steud. Syn. Pl. Glum.* i. 207; *Oliv. in Trans. Linn. Soc.* xxix. 174; *Durand & Schinz, Consp. Fl. Afr.* v. 861. *C. abyssinica, Hochst. ex A. Rich. Tent. Fl. Abyss.* ii. 406; *Steud. l.c.*; *Engl. Hochgebirgsfl. Trop. Afr.* 132; *Schweinf. in Bull. Herb. Boiss.* ii. *App.* ii. 32; *Durand & Schinz, Consp. l.c.* 860. *C. glabrata, Anderss. in Peters, Reise Mossamb. Bot.* ii. 557.

COAST REGION: Komgha Div.; banks of the Kei River, below 500 ft., *Drège!* near Komgha, *Flanagan*, 904! Queenstown Div.; Gwatyn, 2900 ft., *Galpin*, 2050!

KALAHARI REGION: Transvaal; Houtbosch, *Rehmann*, 5717!

EASTERN REGION: Transkei, between Kei River and Bashee River, *Drège*. Natal; Umlazi River, *Drège*. Umaduana, *Sutherland!* near Durban, *Williamson*, 43! Umpumulo, *Buchanan*, 188!

Also in tropical Africa.

4. C. petræa (Thunb. Prodr. 20); perennial, densely tufted; culms erect, or suberect, 1–2 ft. long, 2-noded, compressed below, glabrous, smooth, internodes long-exserted; leaves crowded at the base in a fan-like manner; sheaths strongly compressed, keeled, glabrous, smooth, except on the scabrid keels; ligule a ciliate rim; blades linear, acute or subobtuse, 2–8 in. by ½–3 lin. when expanded, usually folded, glabrous, glaucous, smooth; spikes 3–8, sessile, 2–4 in. long, suberect, brown, straight or gently curved; rhachis pubescent at the base, scabrid; spikelets about 1 lin. long, 2-flowered; rhachilla-joint between the valves very short, terminal joint a bristle half as long as the upper valve or longer; lower glume oblong, subobtuse, not quite ¾ lin. long, compressed; upper glume broadly oblong, shortly and obtusely 2-lobed, about 1 lin. long, rounded or flat on the back, scaberulous, mucronate; lower valve obliquely ovate-oblong in profile, very obtuse, emarginate, curved-mucronate, ciliate along the side-nerves (except towards the base) and along the keel slightly beyond the middle, brown, with an elliptic-oblong subacute pale; anthers ¾ lin. long; upper valve obliquely cuneate in profile, ¾–1 lin. long, glabrous, faintly nerved, with a delicate nerveless pale, and subtending a male flower. *Thunb. Fl. Cap. ed. Schult.* 109;

Steud. Syn. Pl. Glum. i. 207 (*partly*); *Hack. in Engl. Jahrb.* xi. 403; *Durand & Schinz, Consp. Fl. Afr.* v. 862 (*partly*), *non Swartz. Andropogon muticum, Houtt. Handl.* xiii. 579, *t.* 93, *fig.* 3, *non Linn. A. capense, Houtt. Linn. Pfl. Syst.* xii. *t.* 93, *fig.* 3. *Schultesia petræa, Spreng. Pug.* ii. 17 (*partly*). *Eustachya petræa, Nees in Linnæa*, vii. 299; *Fl. Afr. Austr.* 248; *Roem. & Schult. Syst.* ii. 613; *Kunth, Enum.* i. 262 (*partly*), *non Desv.*

SOUTH AFRICA: without precise locality, *Thunberg!*

COAST REGION: Caledon Div.; Genadendal, 500–1000 ft., *Drège.* Riversdale Div.; between Zoetemelks River and Little Vet River, *Burchell*, 6841! Uitenhage Div.; between Koega River and Sunday River, *Drège.* Klein Winterhoek Berg, *Drège*, and without precise locality, *Ecklon & Zeyher*, 43! *Zeyher!* Alexandria Div.; near Addo, *Ecklon*, between Hoffmanns Kloof and Drie Fontein, 1000–2000 ft., *Drège!* Albany Div.; Grahamstown, *MacOwan!* Bothas Berg, 2000 ft., *MacOwan*, 493! Queenstown Div.; between Kat Berg and Klipplaats River, 3000–4000 ft., *Drège!* on mountains at Finchams Nek, near Queenstown, 3900 ft., *Galpin*, 2369!

CENTRAL REGION: Graaff Reinet Div.; stony hills near Graaff Reinet, 2700 ft., *Bolus*, 369! Albert Div.; *Cooper*, 664!

KALAHARI REGION: Griqualand West, Hay Div.; Griqua Town, *Burchell*, 1951! plains at the foot of the Asbestos Mountains, *Burchell*, 2096! Orange Free State, *Buchanan*, 268! Bechuanaland, *Marloth.*

EASTERN REGION: Tembuland; Bazeia, 2000 ft., *Baur*, 449! Natal; Umpumulo, *Buchanan*, 189! Pieter Maritzburg, 1000–2000 ft., *Wood*, 7234!

Also in tropical Africa.

Very closely allied to the South American *C. bahiensis*, Steud., but less to *C. swartziana*, Doell (*C. petræa*, Swartz), with which it has often been confused.

LXI. ELEUSINE, Gaertn.

Spikelets 3–6-flowered, laterally compressed, densely imbricate, alternately biseriate, unilateral, sessile on a flattened rhachis, the uppermost terminal, perfect; rhachilla disarticulating above the glumes and between the valves, or tough, produced, sometimes terminating with a rudimentary valve. *Florets* ☿. *Glumes* 2, subequal, persistent, obtuse or obscurely mucronate, membranous, strongly keeled, 3–5-nerved, the lateral nerves close to the keel, the lower shorter, with the keel crested. *Valves* very similar, 3-nerved near the base; lateral nerves submarginal above, with 1–2 short additional nerves close to the keel. *Pales* slightly shorter than the valves, 2-keeled, keels winged. *Lodicules* 2, minute, cuneate. *Stamens* 3. *Ovary* glabrous; styles slender from a broadened base, distinct; stigmas plumose, laterally exserted. *Grain* broadly-oblong to globose, broadly grooved; pericarp loose, delicate, breaking up irregularly or almost circumscissile; seed finely striate; embryo suborbicular, basal; hilum punctiform, basal.

Annual or perennial; leaves long, flat or folded, flaccid or firm; spikes in interrupted spikes or the upper or all in a terminal umbel, straight, suberect, spreading or deflexed; spikelets glabrous.

Species 6 in tropical Africa and Asia; 1 widely spread through the tropics.

Spikes slender, straight; rhachilla disarticulating at least above the glumes; valves lanceolate, oblong in profile, acute ... ... (1) **indica.**
Spikes thick, often curved; rhachilla tough; valves obliquely ovate in profile, obtuse ... ... (2) **coracana.**

1. **E. indica** (Gaertn. Fruct. i. 8); annual; culms erect or geniculate-erect, from a few inches to 2 ft. long, slender or stout, compressed, 2–3-noded, glabrous, smooth, upper internodes exserted; leaves often numerous, crowded near the base and conspicuously distichous; sheaths compressed, pallid, glabrous except at the often ciliate margins, striate; ligules thin, membranous, short, long-fimbriate; blades linear, long tapering to an acute point, ½ to more than 1 ft. by 1½–3 lin., flat or folded, sometimes flaccid, glabrous, rarely sparingly hairy below, smooth; spikes rather slender, straight, 1–7 in. long, sessile, 2–14 in a terminal umbel, usually with 1–2 (rarely to 7) additional spikes ¼–3 in. below it; rhachis pubescent to villous at the base, otherwise glabrous, smooth; spikelets 1½–2 lin. long, 3–6-flowered, disarticulating above the glumes and very tardily or tough between the valves; glumes and valves ovate (lanceolate-oblong in profile), acute, the latter about 2 lin. long; anthers ⅜ lin. long; grain oblong; seed heart-shaped in cross section, ½ lin. long, dark reddish-brown, obliquely striate; embryo small. *Lam. Ill.* i. 203, *t.* 48, *fig.* 3; *Kunth, Enum.* i. 272; *Suppl.* 224, *t.* 16, *fig.* 4; *Trin. Spec. Gram. Ic. t.* 71; *Nees, Fl. Afr. Austr.* 251; *Steud. Syn. Pl. Glum.* i. 211; *Anderss. in Peters, Reise Mossamb. Bot.* 558; *Schweinf. in Bull. Herb. Boiss.* ii. *App.* ii. 36; *Durand & Schinz, Consp. Fl. Afr.* v. 866. *Cynosurus indicus, Linn. Sp. Pl.* 72.

COAST REGION: Cape Div.: common near Capetown, by the roadsides, *MacOwan, Herb. Aust.-Afr.*, 1566! sandy roadside between Lansdown Road and Kenilworth, *Wolley Dod*, 2448! in ditches near Claremont Station, *Wolley Dod*, 2383! Albany Div.; Grahamstown, *MacOwan*, 1357! Komgha Div.; banks of the Kei River, *Drège*, 4295!

KALAHARI REGION: Transvaal; Johannesburg, *E.S.C.A. Herb.*, 307!

EASTERN REGION: Tembuland; Bazeia, a weed in gardens, *Baur*, 408! Natal; Durban Flats, *Buchanan*, 14! 33! Berea, *Wood*, 5996! from the coast to Umpumulo, in neglected gardens, *Buchanan*, 181! and without precise locality, *Gerrard*, 694! *Cooper*, 3361! Delagoa Bay, *Forbes!*

Tropics of the Old World; introduced (?) in the New World.

2. **E. coracana** (Gaertn. Fruct. i. 8, t. i. fig. 11); very like *E. indica*, but more robust, up to 5 ft. high; spikelets more crowded, to as thick as a finger, often curved; rhachilla tough; glumes and valves broader, ovate in profile, obtuse; grain globose, ½–¾ lin. in diam., usually dark reddish-brown, finely striate, striæ curved. *Lam. Ill.* 203, *t.* 28, *fig.* 1; *Schreb. Beschreib. Gräs.* ii. *t.* 35; *Trin. Sp. Gram. Ic. t.* 70; *Kunth, Enum.* i. 273; *Suppl.* 225; *Nees, Fl. Afr. Austr.* 251; *Steud. Syn. Pl. Glum.* i. 211; *Anderss. in Peters, Reise Mossamb. Bot.* 558; *Schweinf. in Bull. Herb. Boiss.* ii. *App.* ii. 36; *Durand & Schinz, Consp. Fl. Afr.* v. 866. *E. cerealis, Salisb. Prod.* 19. *E. sphærosperma, Stokes, Bot. Mat. Med.* i. 149. *E. stricta, Roxb.*

Fl. Ind. i. 344; *Kunth, l.c.* 273; *Steud. l.c.* 211. *E. Tocussa, Fresen. in Mus. Senckenb.* ii. (1837) 141; *A. Rich. Tent. Fl. Abyss.* ii. 411; *Steud. l.c.*; *Durand & Schinz, Consp. Fl. Afr.* v. 866.

EASTERN REGION: Natal; in a coffee plantation, near Durban, *Drège*, 4294! *Williamson*, 42! Tugela, *Buchanan*, 182! and without precise locality, *Gerrard*, 469! Zululand, *Wood*, 3869! Delagoa Bay, *Forbes*!

Cultivated by the natives as a cereal and for making beer.

Grown in many parts of tropical Africa, tropical Arabia, and throughout India; originated very probably from *E. indica.* The figure in Gaertner represents the seed as smooth (not striate); there is little doubt that it refers to a variety which is grown in India and in Southern Arabia, distinguished by smooth, whitish seeds. This I have not seen from any part of Africa.

LXII. DACTYLOCTENIUM, Willd.

Spikelets 3–5-flowered, laterally compressed, densely imbricate, biseriate, sessile, unilateral on a flattened rhachis, the uppermost reduced; rhachilla tardily disarticulating above the glumes, tough between the valves. *Florets* ⚥, the uppermost rudimentary. *Glumes* 2, unequal, strongly keeled, the lower ovate, acute, thin, persistent, the upper elliptic-oblong in profile, obtuse, mucronate or awned, firm, deciduous. *Valves* ovate, subacuminate, 3-nerved, mucronate or awned, deciduous with the grains. *Pales* about as long as the valves, 2-keeled, subpersistent. *Lodicules* 2, cuneate, minute. *Stamens* 3. *Ovary* glabrous; styles distinct, very long, subterminally exserted. *Grain* subglobose, slightly laterally compressed, not grooved or hollowed, rugose or punctate; pericarp very delicate, irregularly breaking away; embryo scarcely equalling $\frac{1}{2}$ the length of the grain; hilum basal, punctiform.

Annual or perennial; leaves flat, subflaccid; spikes in umbels of 2–6, erect or stellately spreading; tips of the rhachis barren, mucroniform, usually curved.

Species 3; 1 widely spread throughout the tropics.

1. **D. ægyptiacum** (Willd. Enum. Pl. Hort. Berol. 1029); annual, 1–1$\frac{1}{2}$ ft. high; stems sometimes prostrate, rooting from the proliferously branched nodes; culms geniculately ascending, compressed, 2–3-noded, glabrous, smooth, internodes exserted; sheaths striate, the lower whitish, keeled above, glabrous, or scantily hispid; ligules membranous, very short, scantily ciliolate; blades linear, tapering to a fine point, 1–5 in. by 1–2 lin., flat, subflaccid, glaucous, glabrous or hispid or hispidly ciliate, hairs tubercle-based; spikes 2–6, rarely solitary, $\frac{1}{2}$–2 in. long, light or dark olive-grey; rhachis keeled, scabrid; spikelets 3–5-flowered, spreading at right angles, up to 1$\frac{1}{2}$ lin. long, glabrous; lower glume about $\frac{3}{4}$ lin. long, the upper cuspidately mucronate or awned; awn curved, sometimes exceeding the glume; valves 1$\frac{1}{4}$–1$\frac{1}{2}$ lin. long, mucronate or awned; anthers about $\frac{1}{2}$–$\frac{3}{4}$ lin. long; grain $\frac{1}{2}$–$\frac{3}{4}$ lin. long, very rugose, reddish. *Beauv. Agrost. t.* xv. *fig.* 2; *Kunth, Enum.* i. 261; *Suppl.* 204 *A. Rich. Tent. Fl. Abyss.* ii. 406; *Steud. Syn. Pl. Glum.* i. 2

Baker, Fl. Maurit. 452; *Schweinf. in Bull. Herb. Boiss.* ii. *App.* ii. 34; *Durand & Schinz, Consp. Fl. Afr.* v. 868. *D. mucronatum, Willd., l.c.; Trin. Sp. Gram. Ic. t.* 69; *Nees, Fl. Afr. Austr.* 150; *Steud. Syn. Pl. Glum.* 212; *Anderss. in Peters, Reise Mossamb. Bot.* 555. *D. Figarii, De Not. Catal. Sem. Hort. Genuens.* 1847, *and in Ann. Sc. Nat. sér.* 3, ix. (1848) 325; *Steud, l.c. D. australe, Steud. l.c.; Durand & Schinz, l.c.* 869. *Cynosurus ægyptius, Linn. Sp. Pl.* 72. *Cenchrus ægyptiacus, Beauv. Agrost.* 157. *Rabdochloa mucronata, Beauv. l.c.* 176. *Eleusine ægyptiaca, Desf. Fl. Atl.* i. 85; *Hook. f. Fl. Ind.* vii. 295. *E. cruciata, Lam. Illustr.* i. 203, *t.* 48, *fig.* 2. *E. mucronata, Stokes, Mat. Med.* i. 150. *E. pectinata, Moench, Meth. Suppl.* 68. *E. prostrata, Spreng. Syst.* i. 350.

COAST REGION: Bathurst Div.; Port Alfred, *Hutton!* Komgha Div.; near the mouth of the Kei River, *Flanagan*, 969!

KALAHARI REGION: Transvaal; Bosch Veld, at Klippan, *Rehmann*, 5357!

EASTERN REGION: Pondoland; on the shore, between Umtsikaba River and Umtentu River, *Drège*, 4310! Natal; near Durban, *Williamson*, 38! *Plant*, 85! Durban Flats, *Buchanan*, 36! Berea, *Wood*, 5929! Delagoa Bay, *Forbes!*

Widely spread throughout the tropical and subtropical regions.

LXIII. LOPHACME, Stapf.

Spikelets laterally compressed, subsessile, somewhat distant on the long slender simple subdigitate branches of a panicle; rhachilla slender, glabrous, disarticulating above the glumes and between the 2 lowest valves, tough above. *Florets* about 6, the lowest 2 ⚥, shorter than the contiguous glumes, the following gradually reduced, barren, embracing each other and forming a tuft of awns. *Glumes* unequal, narrow, membranous, 1-nerved, keeled persistent. Fertile *valves* linear-lanceolate in profile, 2-toothed, membranous, 3-nerved, with the side-nerves evanescent above, finely awned from between the teeth; callus minutely hairy; barren valves entire, glabrous, passing into fine awns, without a callus. *Pales* very narrow, 2-keeled, slightly shorter than the valves. *Lodicules* 2, very minute, cuneate, hyaline. *Stamens* 3. *Ovary* glabrous; styles distinct, short; stigmas loosely plumose, laterally exserted. *Grain* unknown.

Perennial (?); blades flat; ligule a ciliate rim; panicle subdigitate, of very slender somewhat flexuous spike-like racemes.

Species 1, in Transvaal.

1. **L. digitata** (Stapf in Hook. Icon. Pl. t. 2611); culms slender, erect, over 1 ft. long, glabrous, smooth, uppermost internodes over ½ ft. long, exserted; culm-leaves (only known) glabrous; sheaths tight, terete, smooth, uppermost 5–5½ in. long; blades linear, acute, lower up to 1½ in. by 1¼ lin., flat, subglaucous, smooth, uppermost very short or obsolete; panicle 3½–5 in. long; axis filiform, less than 1 in. long, scaberulous, purplish; branches 3–4½ in. long, finely filiform, subflexuous, scaberulous, purplish, bearing subsessile spikelets from the base or little distance above it; spikelets narrow, reddish, 2½–3 lin. long (exclusive of the awns), lowest distant, upper

rather close; glumes linear-lanceolate, acute, glabrous, lower somewhat shorter; fertile valves white above, purple below, finely pubescent, 1½–2 lin. long, teeth slender; awn scabrid below, very fine, 3–4 lin. long, erect; cluster of barren valves distinctly overtopped by the fertile (except for the awns); pales 1½ lin. long; anthers unknown; stigmas orange-coloured.

KALAHARI REGION: Transvaal; Rhenoster Poort, *Nelson*, 32*!

LXIV. **LEPTOCARYDION**, Hochst. ex Benth.

Spikelets 4–9-flowered, laterally compressed, sessile or subsessile, secund, biseriate, close, on a very slender rhachis; rhachilla disarticulating above the glumes and between the valves. *Florets* ⚥, the uppermost reduced. *Glumes* subequal, lanceolate, acuminate, 1-nerved, keeled. *Valves* oblong, truncate, minutely 4-toothed (teeth hyaline), thin, 3-nerved, margins inflexed, nerves ciliate, the middle nerve excurrent into a fine bristle, the side-nerves not excurrent; callus slender, acute, bearded. *Pales* linear-oblong, slightly shorter than the valves, 2-keeled. *Lodicules* 2, cuneate, delicate. *Stamens* 3; anthers minute. *Ovary* glabrous; styles distinct, slender; stigmas laterally exserted, very slender, plumose. *Grain* linear, obtusely triquetrous, tightly embraced by the scarcely changed valve and pale, free; embryo less than ½ the length of the grain; hilum basal, punctiform.

Annual, culms tufted, many-noded; blades linear to oblong-lanceolate; ligule hyaline, very short or obscure; panicle spike-like, dense; branches erect, simple, or with adpressed branchlets.

Species 3, in Africa.

Allied to *Triraphis*, but differing in the non-excurrent side-nerves and the sessile unilateral spikelets.

1. **L. Vulpiastrum** (Stapf); culms erect or ascending, 2–4 ft. long, simple or sometimes branched (branches intravaginal), many-noded, internodes shortly exserted, glabrous, smooth; sheaths tight, glabrous, smooth or somewhat rough, striate; ligule up to ¼ lin. long, truncate, ciliolate, soon evanescent; blades lanceolate-oblong from a rounded abruptly constricted base, acute, 1–3 in. by 3–6 lin., flat or involute, smooth or finely scaberulous below, glaucescent, finely many-nerved, primary nerves about 7 on each side; panicle spike-like, 2–8 in. by ½–¾ in., pallid or faintly purplish, very dense; branches up to 1 in. long, branched from the villous base; branchlets 5–1-spiculate, up to 3 lin. long; spikelets crowded, adpressed, 5–9-flowered, up to 3 lin. long; rhachilla very slender; glumes reddish, subhyaline, the lower lanceolate, acuminate, mucronate, about 1¼ lin. long, the upper linear-oblong, about 1¾ lin. long; valves 1¼ lin. long, pubescent below the middle, long and finely ciliate along the side-nerves; anthers ⅕ lin. long, ovate; grain linear, obtusely triquetrous, less than ½ lin. by less than ⅛ lin. *Rabdochloa Vulpiastrum, De Not. Cat. Sem. Hort. Genuens.* 1852, *and in Ann.*

Sc. Nat. sér. 3, xix. (1853), 372. *Leptochloa plumosa, Anderss. in Peters, Reise Mossamb. Bot.* 557. *Triodia Vulpiastrum, K. Schum. in Engl. Pfl. Ost-Afr. C.* 113. *Triodia plumosa, Benth. in Journ. Linn. Soc.* xix. 110, *and in Benth. & Hook. f. Gen. Plant.* iii. 1176 (*partly*); *Durand & Schinz, Consp. Fl. Afr.* v. 877 (*partly*).

EASTERN REGION: Natal; banks of the Tugela River and its tributaries, 600–1000 ft., *Buchanan*, 187!

Also in tropical East Africa as far north as Usambara.

Closely allied to *L. alopecuroides* (*Diplachne alopecuroides*, Hochst. ex Steud.) from Abyssinia; but this is a smaller plant with very slender culms and smaller leaves, having only 2–3 primary nerves on each side.

LXV. CROSSOTROPIS, Stapf.

Spikelets laterally compressed, subsessile, more or less distinctly 2-ranked on the rigid simple branches of a panicle; rhachilla slender, disarticulating above the glumes and between the valves. *Florets* 3–9, ⚥ or the uppermost more or less reduced, equalling the glumes or slightly exserted, or overtopped by the awn-like tips of the glumes. *Glumes* subequal or equal, narrow, membranous, strongly 1-nerved, keeled, persistent. *Valves* somewhat distant, linear-oblong in profile, shortly 2-lobed, mucronate or shortly awned from the sinus, membranous, 3-nerved, side-nerves submarginal, subpercurrent, rigidly ciliate, margins inflexed; callus small, hairy. *Pales* narrow, 2-keeled, slightly shorter than the valves. *Lodicules* 2, cuneate, small. *Stamens* 3. *Ovary* glabrous; styles distinct, very slender; stigmas plumose, laterally exserted. *Grain* oblong, strongly compressed from the back, concave or flat, enclosed by the slightly altered valve and pale; embryo about ½ the length of the grain; hilum basal, punctiform.

Annual or perennial; blades usually flat; ligules hyaline; panicle contracted and narrow, or open with the branches spreading at right angles; spikelets rather close to very distant, the uppermost terminal.

Species 3, in Africa and Arabia.

1. **C. grandiglumis** (Rendle in Cat. Afr. Pl. Welw. ii. i. 226); perennial; culms tufted, erect or geniculate, ½–1½ ft. long, glabrous, smooth, terete or slightly compressed, 2–3-noded, upper internodes finally exserted; leaves crowded near the base; sheaths glabrous or very rarely with scattered fine long spreading hairs, rather firm, striate, the upper scabrid, the lower smooth; ligules truncate, up to ¾ lin. long; blades linear, shortly tapering to an acute (often subpungent) point, 1–2½ in. by 1½–2½ lin., flat or subulately convolute, glabrous, scabrid all over; panicle 4–6 in. by 4–9 lin. when ripe, straight; rhachis angular, scabrid or hispidulous; branches simple, singly or 2–3 close together, straight, 2–5 in. long, at first erect, at length spreading at right angles, hispidulous, villous at the base; spikelets 3–5-flowered, distant by more than their own length, adpressed, shortly pedicelled, 3½–5 lin. long; rhachilla minutely

pilose; glumes lanceolate, subulate-acuminate, scaberulous, $3\frac{1}{2}$–5 lin. long; valves oblong, shortly bilobed, mucronate, up to 2 lin. long, side-nerves rigidly ciliate; pales truncate, finely pubescent on the back, keels scabrid; anthers up to $\frac{1}{2}$ lin. long; grain narrowly-oblong, flat, over 1 lin. *Stapf in Hook. Icon. Pl. t.* 2609. *Leptochloa grandiglumis, Nees, Fl. Afr. Austr.* 252; *Steud. Syn. Pl. Glum.* i. 210. *Diplachne grandiglumis, Hack. in Engl. Jahrb.* xi. 404; *Durand & Schinz, Consp. Fl. Afr.* v. 878.

CENTRAL REGION: Albert Div. (New Hantam); in stony places, 4500–5000 ft., *Drège.* Aliwal North Div.; between the Witte Bergen and Krai River, 4500–5000 ft., *Drège!*

KALAHARI REGION: Griqualand West, Hay Div.; between Klip Fontein and Kneghts Fontein, *Burchell,* 2167! Orange Free State; Caledon River, *Burke,* 228! by the Caledon River, at Komissie Drift, *Zeyher,* 1844! near Winberg, *Buchanan,* 246! Bechuanaland; near Groot Kuil, *Marloth,* 989! Transvaal; plains near Rustenberg, *MacLea,* 125! Buys Kop, near the Nylstrom, *Nelson,* 99*!

EASTERN REGION: Natal; sandy valley, near Tugela, 1000 ft., *Buchanan,* 279!

LXVI. TRIRAPHIS, R. Br. (partly).

Spikelets 5–15-flowered, laterally compressed, pedicelled, panicled; rhachilla disarticulating above the glumes and between the valves. *Florets* ☿, the uppermost gradually reduced. *Glumes* subequal, lanceolate to linear-lanceolate, acuminate or truncate, or minutely 2-toothed and aristulate, 1-nerved, keeled, thin. *Valves* oblong, 3-lobed, thin, 3-nerved, 3-awned, the middle lobe more or less bifid, awned from the sinus, the side lobes shorter, entire, asymmetric, awned from the inner side, margins inflexed, nerves ciliate, particularly the lateral; awns fine, scabrid, often longer than the valves; callus slender, acute, bearded. *Pales* linear or linear-oblong, somewhat shorter than the valves. *Lodicules* 2, cuneate, delicate, minute. *Stamens* 3. *Ovary* glabrous; styles distinct, slender; stigmas laterally exserted, very slender, plumose. *Grain* tightly embraced by the scarcely changed valve and pale, linear, terete or obtusely triquetrous; embryo short; hilum basal, punctiform.

Annual or perennial; blades narrow, linear; ligule a ciliate membranous rim; panicle contracted, spike-like, or open, much branched; spikelets distinctly pedicelled.

Species 8 in Africa, 1 (*T. mollis,* R. Br.) in Australia.

Perennial:
- Culms many-noded, profusely branched ... ... (1) **ramosissima.**
- Culms about 3-noded, simple ... ... (2) **Rehmanni.**

Annual:
- Valves about $1\frac{1}{2}$ lin. long; anthers $\frac{3}{4}$–1 lin. long:
  - Culms 5–7-noded, glabrous; middle awn about twice as long as the valve or almost so ... ... (3) **Elliotii.**
  - Culms 1–2-noded, glabrous or hairy; middle awn about as long as the valve:

Culms with scattered long spreading hairs; panicle rather lax, often flexuous; glumes smooth ... ... (4) **Fleckii.**
Culms glabrous, scaberulous or sometimes scantily hairy; panicle dense, straight; glumes scaberulous ... (5) **purpurea.**
Valves ¾ lin. long; anthers $\frac{1}{8}$–$\frac{1}{10}$ lin. long; a dwarf grass ... ... (6) **nana.**

1. T. ramosissima (Hack. in Verhandl. Bot. Ver. Brandenb. xxx. (1888) 237); perennial, glabrous; culms wiry, straight, many-noded, profusely branched, particularly from the middle nodes, terete, smooth, branches solitary or in fascicles of 2–3, with the lower nodes very short and sheathed by subpersistent bud-bearing scales, internodes exserted; sheaths tight, terete, at length open and breaking away at the base or withering; blades very narrow, linear, setaceously convolute, tapering to a very fine point, 2–4 in. by ½ lin., scaberulous above, smooth below; panicle linear-oblong, 2–2½ in. long, subcontracted; branches subcapillary, flexuous, 3–5-nate, 5–1-spiculate, up to ½ in. long; lateral pedicels up to 1 lin. long; spikelets 5–7-flowered, 3½–4 lin. long, light purplish; rhachilla glabrous; glumes lanceolate, hyaline, mucronate, the lower about 1–1¼ lin., the upper up to 1½ lin. long, with bifid tips; valves up to 1½ lin. long; middle awn up to 2½ lin. long, side-awns half as long or longer; keels of the pale scaberulous; anthers 1 lin. long. *Hack. in Bull. Herb. Boiss.* iv. *App.* iii. 23, 24; *Durand & Schinz, Consp. Fl. Afr.* 872.

WESTERN REGION: Great Namaqualand; between Ausis and Kuias, *Schenck*, 83!

2. T. Rehmanni (Hack. in Bull. Herb. Boiss. iii. 388); perennial, compactly cæspitose, glabrous; rhizome short, oblique; culms erect or geniculate, 2 ft. long, very firm, terete, striate, smooth, about 3-noded, internodes exserted; sheaths tight, firm, smooth, striate, the basal reddish or purplish-brown, persistent, the lowest reduced to short acute bladeless scales; blades linear, narrow, tapering to a setaceous point, usually tightly convolute, 4–8 in. by 1–1½ lin. (when expanded), firm, smooth, coarsely striate; panicle 2–12 in. long, contracted or open, and then 2–4 in. broad, erect or slightly nodding; rhachis smooth; branches solitary or fascicled, closely or loosely branched from the base or almost so, smooth, filiform, straight or flexuous; pedicels ½–1½ lin. long; spikelets subsecund, crowded, 4–8-flowered, 3–5 lin. long; rhachilla glabrous; glumes linear-oblong, erose, minutely mucronate, the lower 1–2 lin., the upper 1½–2¼ lin. long, often with a fine lateral nerve on one or both sides; valves oblong (when expanded), not quite 2 lin. long; awns stiff, middle awn ½–1½ lin. long, side-awns up to 1 lin. long, or mere mucros; pale glabrous or the flaps hairy, keels scaberulous; anthers ¾ lin. long; grain linear, terete, ⅞ lin. by ⅙ lin. *Diplachne andropogonoides, Nees, Fl. Afr. Austr.* 258. *Avena andropogonoides, Steud.*

in Flora, 1829, 486; *Kunth, Enum.* i. 307. *Trisetum andropogonoides, Steud. Syn. Pl. Glum.* i. 228; *Durand & Schinz, Consp. Fl. Afr.* v. 837.

SOUTH AFRICA: without precise locality, *Zeyher*, 1829!

COAST REGION: Uitenhage Div.; stony places in the channel of the Zwartkops River, *Ecklon & Zeyher*, 947! between Galebosch and Melk River, *Burchell*, 4786! Alexandria Div.; in shrubberies, in Olivenhout Kloof near Enon, 500–1000 ft., *Drège!*

KALAHARI REGION: Bechuanaland; between Kuruman and Matlareen River, *Burchell*, 2189! Orange Free State; Draai Fontein, *Rehmann*, 3622! near Winberg, *Buchanan*, 245! between Kimberley and Bloemfontein, *Buchanan*, 278 bis! Caledon River, *Burke*, 426! Transvaal; Zoutpansberg District, at Rhenoster Poort, *Nelson*, 35*!

Burke's specimen has rather larger spikelets on short stiff branches; otherwise it is not distinct.

3. **T. Elliotii** (Rendle in Journ. Bot. 1891, 73); annual (?); culms fascicled, erect, $\frac{3}{4}$–1$\frac{1}{4}$ ft. long, simple, smooth, glabrous, 5–7-noded, internodes exserted; sheaths tight, smooth, glabrous; ligule a minute ciliolate rim; blades linear, tapering to a fine setaceous point, 3–4 in. by $\frac{3}{4}$–1 lin., flat, glabrous, smooth; panicle dense, somewhat spike-like, interrupted below, subovate to oblong, 1–2 in. by $\frac{3}{4}$–$\frac{4}{5}$ lin., purplish; lowest branches up to $\frac{3}{4}$ in. long; lateral pedicels very short; spikelets 4, rarely 5 lin. long; florets up to 11; rhachilla glabrous, joints up to $\frac{1}{5}$ lin. long; glumes lanceolate in profile, lower acute, submucronate, 1 lin. long, upper emarginate with a short bristle from the sinus, up to 1$\frac{1}{2}$ lin. long; valves oblong, over 1$\frac{1}{2}$ lin. long, purple, hyaline teeth minute (but distinct) at the base of the middle awn; middle awn 2–2$\frac{1}{2}$ lin. long, side-awns 1–1$\frac{3}{4}$ lin. long; pales 1$\frac{1}{2}$ lin. long, keels scaberulous; anthers 1 lin. long. *Hack. in Bull. Herb. Boiss.* iv. *App.* iii. 23, 24.

WESTERN REGION: Little Namaqualand, Ookiep, *Scully!*

Also in Damaraland.

4. **T. Fleckii** (Hack. in Bull. Herb. Boiss. iv. App. iii. 23, 24); annual; culms fascicled, $\frac{1}{2}$–1 ft. high, geniculate, sometimes from a decumbent base, or suberect, terete, striate, with scattered long spreading hairs, about 2-noded, upper internodes at length more or less exserted; sheaths rather tight, terete, striate, with long spreading tubercle-based hairs; ligule a ciliate rim; blades linear, tapering to a fine setaceous point, 2–5 in. by $\frac{1}{2}$–1 lin., flat, subflaccid, sparsely hairy below, otherwise glabrous, scaberulous; panicle linear-oblong, contracted or more or less open, lax and subsecund, 4–6 in. long; rhachis scabrid; branches distinctly 2-nate or solitary, erect or spreading, filiform, straight or flexuous, scabrid, sparingly divided or simple, bearing spikelets from the base or the upper from somewhat above it or towards the tips only; lateral pedicels $\frac{1}{2}$–1 lin. long; spikelets linear-oblong, more or less purple, 4–7 lin. long; florets 9–16; rhachilla glabrous; glumes

linear-oblong, mucronate from the more or less 2-toothed tips, 1-nerved, glabrous, smooth, lower slightly over 1 lin., upper 1½ lin. long; valves linear-oblong, 1½ lin. long, middle lobe split to ⅓–½ way down, side-nerves ciliate all along; middle awn about as long as the valve, side-awns half as long; pales slightly shorter than the valves, keels scabrid; anthers almost 1 lin. long.

WESTERN REGION: Great Namaqualand; Rehoboth, *Fleck!*
KALAHARI REGION: Griqualand West, Herbert Div.; St. Clair, Douglas, *Orpen*, 246!

Also in Angola.

5. **T. purpurea** (Hack. in Verh. Bot. Ver. Prov. Bradenb. xxx. (1888) 146); annual; culms fascicled, geniculately-ascending, 3-6 in. long, terete or compressed, glabrous or sparsely hairy, scaberulous, usually 1-noded, upper internodes exserted; sheaths rather tight or at length loosened, terete or subcompressed, striate, sparsely hairy, hairs long, spreading, tubercle-based; ligule a short membranous ciliolate rim; blades linear, tapering to a very fine point, 1¼–2 in. by ½–¾ lin., glabrous or sparsely hairy below, scabrous; panicle linear to ovate-oblong, 1¼–2½ in. long, dense, sometimes spike-like; rhachis straight, angular, scabrid; branches in alternately approximate and distant pairs or the upper solitary, ½–¾ in. long, erect or obliquely erect, filiform, scabrid, bearing spikelets from the base; spikelets crowded, linear, 3–5 lin. long, purplish; florets 5–11; rhachilla glabrous; glumes oblong or linear-oblong, mucronulate from the truncate or obscurely 2-toothed tips, 1-nerved, glabrous, scaberulous, lower almost 1 lin., upper over 1 lin. long; valves oblong, not quite 1½ lin. long, side-nerves ciliate all along, middle lobe split to ½ way down; middle awn as long as the valve, side-awns scarcely exceeding the middle lobe; callus silky-bearded; pales almost equalling the valve, keels scabrid; anthers ¾ lin. long. *Durand & Schinz, Consp. Fl. Afr.* v. 872; *Hack. in Bull. Herb. Boiss.* iv. *App.* iii. 23, 24.

WESTERN REGION: Great Namaqualand; Gubub, to the south of Aus, *Pohle!* sandy plains near Gobachab, between Aus and the Orange River, *Schenck*, 279!

6. **T. nana** (Hack. in Engl. Jahrb. xi. 403); annual, scarcely more than 1 in. high, the whole plant reddish; culms fascicled, straight above the subbasal node, terete, striate, finely scaberulous; peduncle long exserted; sheaths rather wide, short, striate, with scattered fine stiff tubercle-based hairs; blades linear-lanceolate, acuminate, 4 lin. by 1 lin. (when expanded), rigid, curved, folded, striate, scaberulous; panicle spike-like, ovate to oblong, 3–4 lin. long; branches 6–1-spiculate, scaberulous, erect, up to 2 lin. long; pedicels up to ¾ lin. long; spikelets 5–7-flowered, up to 2 lin. long; glumes acutely keeled, mucronulate, straw-coloured, the lower lanceolate, up to ¾ lin. long, the upper oblong, obtuse, up to ⅞ lin. long; valves oblong, ¾ lin. long; middle awn as long as the valve,

side-awns somewhat longer than the middle lobe; pale glabrous, keels scabrid; anthers $\frac{1}{10}$–$\frac{1}{8}$ lin. long; grain linear, terete, $\frac{1}{2}$ lin. by $\frac{1}{12}$ lin. *Hack. in Bull. Herb. Boiss.* iv. *App.* iii. 24. *Diplachne nana, Nees, Fl. Afr. Austr.* 259 (*also Boiss. Fl. Or.* v. 562; *Ascherson & Schweinf. Ill. Fl. Egypte*, 171 ?).

WESTERN REGION: Little Namaqualand; bank of the Orange River near Verlaptpram, *Drège*, 2593!
KALAHARI REGION: Bechuanaland, *Marloth*, 1513.

I am not quite certain whether the Nubian plant, referred by Boissier to *Diplachne nana*, Nees, is identical with this. A specimen of it, which I have seen (collected by Th. Bent on the Nubian coast in 21° N. Lat.), has very much larger panicles, and also the individual spikelets and florets are longer.

LXVII. ENNEAPOGON, Desv.

Spikelets 3-flowered, paniculate; rhachilla disarticulating above the glumes, minutely scaberulous or almost smooth. Lowest *floret* ⚥, the intermediate male or barren, the uppermost rudimentary, minute. *Glumes* 2, persistent, membranous, acute or obtuse or minutely truncate, 3–5- or sub-7- nerved. Hermaphrodite *floret*: valve very broad, rounded on the back, rather firm, more or less villous, 9-nerved, 9-awned; awns subulate, equal or subequal, plumose, ciliate or scaberulous; callus minute, short; pale oblong, 2-keeled, exceeding the valve; lodicules 2, minute, cuneate, fleshy; stamens 3; ovary glabrous; styles distinct, short; stigmas laterally exserted, loosely plumose. Second *floret* like the lower, but the valve about $\frac{1}{2}$ as long, glabrous, the ovary rudimentary or suppressed. Uppermost *floret* reduced to a tuft of minute awns. *Grain* oblong, dorsally more or less compressed; hilum punctiform, subbasal; embryo large, occupying $\frac{3}{4}$ or more of the front.

Perennial, rarely subannual; blades usually narrow, often convolute; ligules reduced to a line of hairs; panicle contracted, more or less spike-like, elegantly bristly-plumose from the numerous awns.

Species about 6 in the dry and warm regions of the Old World and in Australia, 1 in Western North America.

Awns distinctly plumose or ciliate:
  All parts softly glandular-pubescent; nodes villous; culms not wiry:
    Culms 2–6 in. long; leaves crowded from more or less compact tufts; blades fine, usually setaceously convolute; side-nerves of the glumes evanescent above ... ... (1) **brachystachyus.**
    Culms 1–3 ft. long, fascicled; blades 1–3 lin. broad, flat or convolute; the 2 side-nerves of the upper glume percurrent or shortly excurrent ... ... (2) **mollis.**
  Almost glabrous; culms very wiry, 1–2 ft. ... (3) **scoparius.**
Awns scaberulous along the margins ... ... (4) **scaber.**

1. **E. brachystachyus** (Stapf); perennial, often compactly cæspitose, all parts finely glandular-pubescent, rarely subglabrous;

culms fascicled, geniculately ascending, 2–6 in. long, slender, often with a bulbous thickening at the base, 2–4-noded, simple or sparingly branched below, internodes mostly exserted; leaves mostly near the base; sheaths tight or those at the base of the branches loose, finely striate, nodes pubescent to villous; blades very narrow, linear, finely attenuated, 1–5 in. long, usually setaceously convolute, sometimes more or less scabrid; panicle spike-like, $\frac{1}{2}$–$1\frac{1}{2}$ in. by 3–5 lin., dense, light to dark grey; spikelets $1\frac{3}{4}$–2 lin. long; glumes subequal, oblong, obtuse or emarginate, scantily pubescent, thin, usually 5- (rarely 3- or 7-) nerved, side-nerves evanescent above; lower valve $\frac{3}{4}$ lin. long, shortly villous; awns about 1–$1\frac{1}{2}$ lin. long, shortly plumose to or beyond the middle; pale 1 lin. long, keels scabrid; anthers ellipsoid-oblong, $\frac{1}{8}$–$\frac{1}{6}$ lin. long; grain almost $\frac{1}{2}$ lin. long. *Pappophorum phleoides, Trin. in Spreng. Neue Entdeck.* ii. 73; *Gram. Gen.* 91, *and in Mém. Acad. Pétersb. sér.* 6, i. (1831) 91; *Kunth, Enum.* i. 254; *not Cav., nor Steud. P. brachystachyum, Jaub. & Spach in Ann. Sc. Nat. sér.* 3, xiv. (1850) 365; *Illustr.* iv. 34; *Steud. Syn. Pl. Glum.* i. 200; *Durand & Schinz, Consp. Fl. Afr.* v. 870; *Hook. f. Fl. Brit. Ind.* vii. 302. *P. figarianum, Fig. & De Not. in Mem. Acc. Torin. ser.* 2, xii. (1852) 254. *P. bulbosum, Fig. & De Not. l.c. P. vincentianum, Schmidt, Beitr. Fl. Cap. Verd. Ins.* 144; *Durand & Schinz, l.c.* 871. *P. nanum, Steud. l.c. P. senegalense, Steud. l.c.* 199; *Durand & Schinz, l.c.* 871.

VAR. β, **macranthera** (Stapf); spikelets 2–$2\frac{1}{2}$ lin. long; glumes less obtuse, or subacute; anthers linear, $\frac{3}{8}$ lin. long.

SOUTH AFRICA: without precise locality, *Zeyher*, 1795!

KALAHARI REGION: Orange Free State; between Kimberley and Bloemfontein, *Buchanan*, 290! Var. β: Griqualand West; Dutoits Pan, *Tuck!* plains at the foot of the Asbestos Mountains, *Burchell*, 2078!

The typical form in the Cape Verdes and in Senegambia, and from North Africa to North-West India; a variety with slightly larger anthers in Central Asia (*Pappophorum boreale*, Griseb.).

2. E. mollis (Lehm. Pugill. iii. 40); perennial (sometimes flowering in the first year?), more or less glandular-pubescent or villous all over; culms fascicled, erect or geniculately-ascending, 1–3 ft. long, pubescent to finely villous, 2–5-noded, simple or sparingly branched below, internodes more or less exserted; sheaths rather tight, finely striate, nodes villous; blades linear to lanceolate-linear, long tapering to a fine point, 3–8 in. by 1–3 lin., flat or convolute, rigid or subflaccid, scaberulous above and along the margins; panicle spike-like, often interrupted and lobed below, 1–6 in. long; lowest branches sometimes up to 1 in. long; spikelets crowded, $1\frac{1}{2}$–$2\frac{1}{2}$ lin. long; glumes unequal, scantily and finely hairy, thin, greyish, tips usually dark, the lower ovate, subacute or minutely truncate, rarely 3–2-nerved, the upper somewhat longer, oblong, truncate, 3-nerved, nerves percurrent or very shortly excurrent; lower valve $\frac{3}{4}$ lin. long, shortly villous; awns $1\frac{1}{2}$–2 lin. long, plumose beyond the middle; pale 1–$1\frac{1}{4}$ lin. long, keels ciliolate; anthers $\frac{5}{8}$–$\frac{3}{4}$ lin. long; grain not quite $\frac{1}{2}$ lin. by $\frac{1}{4}$ lin. *Nees in Linnæa*, vii.

304; *Fl. Afr. Austr.* 233. *Pappophorum cenchroides, Licht. ex Roem. & Schult. Syst.* ii. 616; *Trin. Gram. Gen.* 92, *and in Mém. Acad. Pétersb. sér.* 6, i. (1830) 92; *Kunth, Enum.* i. 254; *Steud. Syn. Pl. Glum.* i. 199; *Durand & Schinz, Consp. Fl. Afr.* v. 870 *P. molle, Kunth, l.c.* 255; *Steud. l.c.*; *Schweinf. in Bull. Herb. Boiss.* ii. *App.* ii. 36; *Durand & Schinz, l.c. P. abyssinicum, Hochst. in Flora,* 1855, 202; *Durand & Schinz, Consp. Fl. Afr.* v. 869. *P. robustum, Hook. f. Fl. Brit. Ind.* vii. 302.

CENTRAL REGION: Beaufort West Div., *Ecklon.*
WESTERN REGION: Great Namaqualand; Arasab plains, between Aus and the Orange River, *Schinz,* 282, Naiams, *Schinz,* 633, Gamochab, *Schinz,* 632.
KALAHARI REGION: Griqualand West, Herbert Div.; St. Clair, Douglas, *Orpen,* 253! right bank of the Vaal River, at Blaauwbosch Drift, *Burchell,* 1743! between the Vaal River and Leeuwenkuil, *Lichtenstein.* Bechuanaland; on rocks at Chue Vley, *Burchell,* 2384!

Also in Eastern tropical Africa, in tropical Arabia and the Punjab.

The specimens distributed by Drège and by Zeyher (1820) as *E. mollis* are *E. scaber.* Where the original specimens of Lehmann's *E. mollis* are, I do not know; but from his description, it is almost certain that he meant the plant described here as *E. mollis.* Lichtenstein's specimens of *P. cenchroides* are also unknown to me, and their identity with *E. mollis* is assumed from the very short description and the locality, which coincides with that of Burchell's 1743. Nees records *E. mollis* also from the Sunday River in Jansenville Div., and the Springbok Flats in Uitenhage Div.

3. **E. scoparius** (Stapf); perennial; culms fascicled on a very short rhizome, erect from a subbulbous base or slightly geniculate, wiry, very slender, 1–2 ft. long, more or less branched, glabrous or finely pubescent below the nodes, few- to 6-noded, internodes usually exserted; sheaths very tight, glabrous or finely pubescent, at least at the nodes, the bases of the lowermost persistent; blades very narrow, linear, finely attenuated, usually setaceously convolute, 2–5 in. long, rigid, glabrous, rarely pubescent, smooth; panicle spike-like, ovoid to cylindric, ½–3 in. by 3–6 lin., greyish; all branches usually very short; spikelets crowded, 1¾–2 lin. long; glumes subequal, very thin, pubescent, subacute, sometimes minutely mucronate, 3-nerved, side-nerves evanescent above, sometimes sub-5 to 7-nerved, the additional nerves very short, lower glume ovate, 1¼–1½ lin. long, the upper oblong, 1¾–2 lin. long; lower valve not quite 1 lin. long, villous; awns slightly unequal, up to 1½ lin. long, plumose beyond the middle; keels of the pale ciliolate; anthers over ½ lin. long.

CENTRAL REGION: Graaff Reinet Div.; stony hills near Graaff Reinet, 2600 ft., *Bolus,* 691! Colesberg Div.; Colesberg, *Shaw!* Phillipstown Div.; by the Orange River near Vissers Drift, *Burchell,* 2674! Albert Div., *Cooper,* 651!
KALAHARI REGION: Orange Free State, *Buchanan,* 248! 271! Transvaal; Bosch Veld, Klippan, *Rehmann,* 5361!

Also in the tropical part of Transvaal, and in Abyssinia (*Schimper,* 2235).

4. **E. scaber** (Lehm. Pug. iii. 41); perennial, glandular-pubescent all over and often scabrid besides; culms tufted, erect or geniculately-ascending, often zigzag, ½–1 ft. long, rather slender, branched below,

3–5-noded, internodes usually enclosed except the uppermost; sheaths rather loose, the lower whitish, usually scabrid, nodes villous; blades linear, tapering to a fine point, 2–4½ in. by 1–2 lin., flat or more often convolute, rather rigid, scabrid on both sides; panicle spike-like, cylindric, sometimes interrupted, or oblong to ovate and rather lax, 1½–3 in. by ½–¾ in., light to dark grey; lowest branches up to 1½ in. long, branched from the base or simple in the lowest ½ in.; spikelets 2–2½ lin. long, crowded or rather lax; glumes subequal, ovate, subacute or minutely truncate, scantily hairy, 7–9-nerved, nerves prominent, 2 of the side ones sometimes percurrent in the upper glume; lower valve ¾–1 lin. long, densely villous; awns 1–1½ lin. long, rather broad at the base, scaberulous; pale 1½ lin. long, keels ciliolate; anthers ¾–1 lin. long; grain not quite ½ lin. by ¼ lin. *Nees in Linnæa*, vii. 304; *Fl. Afr. Austr.* 234. *Pappophorum scabrum, Kunth, Enum.* i. 255; *Steud. Syn. Pl. Glum.* i. 199; *Durand & Schinz, Consp. Fl. Afr.* v. 870; *Hack. in Bull. Herb. Boiss.* iv. *App.* iii. 22.

COAST REGION: Knysna Div.; Plettenberg Bay, *Mund.* Queenstown Div.; between Kat Berg and Klipplaat River, 3000–4000 ft., *Drège.*

CENTRAL REGION: Calvinia Div.; between Lospers Plaats and Springbok Kuil River, 2000–3000 ft., *Zeyher*, 1820! Beaufort West Div.; on stony heights and by the Gamka River at Zwartbulletje, 2500–3000 ft., *Drège!* Graaff Reinet Div.; stony hills near Graaff Reinet, 2700 ft., *Bolus*, 556! Sutherland Div.; at Yuk River, near Yuk River Hoogte, *Burchell*, 1270! 1271! Fraserburg Div.; Dwaal River, *Burchell*, 1471! Carnarvon Div.; at the northern exit of Karree Bergen Poort, near Carnarvon, *Burchell*, 1557! Richmond Div.; Zeekoe River, 5000 ft., *Drège.*

WESTERN REGION: Great Namaqualand; Kukaus, near Aus, *Schenck*, 117! Aus, *Schinz*, 634! *Steingrover*, 40. Little Namaqualand; Silver Fontein, near Ookiep, 2000–3000 ft., *Drège!*

KALAHARI REGION: Griqualand West, Hay Div.; on the Asbestos Mountains, *Burchell*, 2052! Herbert Div.; St. Clair, *Orpen!*

Also in tropical South-west Africa, Morocco, and Algiers.

LXVIII. **SCHMIDTIA**, Steud.

Spikelets closely 4–6-flowered; rhachilla disarticulating above the glumes and between the valves, joints extremely short, glabrous. *Florets* hermaphrodite, except the rudimentary uppermost one. *Glumes* 2, persistent, membranous, acute or subobtuse, 9–11-nerved. *Valves* broad, rounded on the back, rather firm and villous below, 9-nerved, 5 of the nerves excurrent into straight subulate scabrid awns, the 4 alternate ones into very thin lanceolate muticous lobes; callus slender, minute, bearded. *Pales* oblong, 2-keeled, rather longer than the body of the valves. *Lodicules* 2, minute, cuneate, fleshy. *Stamens* 3. *Ovary* glabrous; styles distinct, slender; stigmas laterally exserted; loosely plumose. *Grain* oblong, dorsally compressed; hilum punctiform, subbasal; embryo large, occupying ¾ or more of the back of the grain.

Perennial or annual (?), more or less glandular-pubescent; blades rather rigid, flat or convolute; ligule a line of hairs; panicle contracted, narrow, oblong or spike-like; spikelets turgid, many-bristled.

Species 3–4, closely allied, in tropical and South Africa.

1. **S. bulbosa** (Stapf); perennial, with numerous villous conical innovation-buds at the base, (base hence more or less bulbous); culms simple or scantily branched, erect or shortly ascending, up to 2 ft. long, 5–8-noded, villous or pubescent below the nodes, internodes mostly enclosed or subexserted; leaves villous or pubescent or glabrescent; sheaths striate, the upper tight; blades linear, long and finely attenuated, 2½–4 in. by 2–3 lin., usually convolute at least above the middle, firm, rather rigid; panicle linear to oblong, 2–4 in. by 6–12 lin., somewhat loosely contracted; branches up to 1 in. long, 3–7-spiculate; spikelets 4–5 lin. long; glumes pubescent, nerves prominent, the lower 3–4 lin. long, ovate, the upper 4–5 lin., oblong; body of valves 1½–2 lin. long, linear-lanceolate, lobes about 1½ lin. long; awns 3–4 lin. long; flaps of pales long silky; anthers 1–1½ lin. long; grain ⅞ lin. long. *S. quinqueseta, Ficalho and Hiern, in Trans. Linn. Soc. ser.* 2, *Bot.* ii. 31 (*partly*); *Durand & Schinz, Consp. Fl. Afr.* v. 871 (*partly*).

KALAHARI REGION: Griqualand West; Hüuernest Kloof, *Rehmann*, 3401! St. Clair, Douglas, *Orpen*, 196! Orange Free State, *Buchanan*, 52! Bechuanaland; Chooi Desert, near "Giraffe Station," *Burchell*, 2361! Transvaal; on the plains, *McLea*, 136! Bosch Veld, at Klippan, *Rehmann*, 5365! near Lydenburg, *Atherstone!*

S. quinqueseta, Benth., (according to Ficalho and Hiern, as "illustrated by specimens which have been distributed from the Polytechnic School of Lisbon as part of the herbarium of D. A. R. Fereira,") is a grass, much branched above the base, the branches being intravaginal and erect, the culms very slender and glabrous, the blades rather short (1½–2 in.), stiff, and like the sheaths clothed with spreading gland-tipped hairs.

Hackel (Bull. Herb. Boiss. iv. App. iii. 22) indicates *S. pappophoroides*, Steud., and *S. quinqueseta*, Benth., from various places in Great Namaqualand; but, in the absence of the specimens, to which he refers, I must leave open the question how far they belong to *S. bulbosa* or to one of the other 2 species, if the specific distinction of these can be maintained at all.

LXIX. POTAMOPHILA, R. Br.

Spikelets laterally slightly compressed, on short or rather long pedicels, panicled; rhachilla jointed above the rudimentary basal glumes. *Florets* 3; lower 2 reduced to minute empty valves, uppermost ⚥ or unisexual with the organs of the other sex reduced. *Glumes* reduced to very minute rounded or truncate scales, or to an obscure entire or bilobed hyaline rim. Empty *valves* very small, hyaline, nerveless, subulate or elliptic and rounded or lobed; fertile valve membranous, 5-nerved, awnless, nerves raised, sometimes slightly winged. *Pale* 3-nerved; otherwise similar to the fertile valve. *Lodicules* 2, finely nerved. *Stamens* 6. *Styles* distinct; stigmas feathery. *Grain* obovate, compressed, crowned by the thickened bases of the styles, enclosed by the unaltered glume and pale, free.

Rather tall aquatic grasses; blades flat; ligules membranous; panicle effuse or contracted.

Species 3; 1 in South Africa, 1 in Madagascar, and 1 in New South Wales.

1. **P. prehensilis** (Benth. in Journ. Linn. Soc. xix. 55); stems several feet high, branched, many-noded, slightly compressed or terete, glabrous, smooth; internodes up to 4 in. long; branches spreading at a right angle or almost so; leaf-sheaths rather tight, shorter than, or as long as the internodes, slightly compressed and more or less keeled, strongly striate, scabrid in the uppermost part of the keel, hairy near the mouth or glabrous, except the sometimes minutely villous nodes; ligule membranous, oblong or truncate, $\frac{1}{2}$–1 lin. long, pubescent; blades linear-lanceolate from a very short contracted base, acute, 3–5 in. by 3–6 lin., flat, flaccid, glaucous, sparingly hairy and minutely scabrid on both sides, margins and midrib very rough from minute reversed spines; primary nerves 3–4 on each side; panicle terminal, 4–6 in. long, open, very lax, rigid; branches spreading, fine, up to 4 in. long, the lowest usually paired, sparingly branched, compressed and angular, scaberulous or smooth below, branchlets 2–3-spiculate; pedicels 1–7 lin. long; spikelets lanceolate-oblong, acute, 3–4 lin. long, slightly twisted, pallid; glumes extremely minute truncate hyaline scales; empty valves subulate, $\frac{1}{2}$–$\frac{3}{4}$ lin. long, nerveless; fertile valve tightly clasping the similar pale with the inflexed margins, nerves raised, slightly winged, scaberulous; lodicules ovate; anthers 3 lin. long; styles $\frac{1}{2}$ lin. long; stigma exserted near the base of the valve, 1–1$\frac{1}{2}$ lin. long; grain unknown. *Maltebrunia prehensilis, Nees, Fl. Afr. Austr.* 194; *Durand & Schinz, Consp. Fl. Afr.* v. 788. *Oryza prehensilis, Steud. Syn. Pl. Glum.* i. 3.

EASTERN REGION: Natal; near the mouth of the Umzimkulu River, in copses and woods, *Drège!* at Umbilo Waterfall, *Rehmann*, 8156! Umpumulo, to 2000 ft., common in bush, *Buchanan*, 288! Inanda, *Wood*, 1305!

LXX. LEERSIA, Sw.

Spikelets laterally compressed, very shortly pedicelled, panicled; rhachilla jointed above the rudimentary glumes. *Floret* 1, ⚥. *Glumes* reduced to an obscure hyaline entire or 2-lobed rim. *Valve* 5-nerved, subcartilaginous, awnless, keel and margins rigidly ciliate. *Pale* narrow, 3-nerved, subcartilaginous except at the hyaline margins, grooved along the outer nerves and tightly clasped by the inflexed margins of the valve, keel rigidly ciliate. *Lodicules* 2, fleshy, finely nerved. *Stamens* 6, 3 or 1. *Styles* distinct; stigmas feathery. *Grain* ovate or oblong, compressed, embraced by the valve and the pale, free; embryo short.

Perennial; leaves narrow; panicle usually flaccid with very slender branches.

Species 6–7, mostly in the tropics and the subtropical regions of both hemispheres.

1. **L. hexandra** (Sw. Prod. Veg. Ind. Occ. 21); perennial; rhizome creeping, stoloniferous; innovation-buds ovoid, subacute,

scales smooth, striate; stems erect from a prostrate or ascending base, rooting from the lower nodes, 2–4 ft. high, simple or very sparingly branched, usually slender and weak, many-noded, smooth, firmly striate; uppermost internode longest (up to ½ ft., rarely to 1 ft., long); sheaths rather tight or the lower looser and ultimately spreading, terete, usually shorter than the internodes, the uppermost longest, reaching to or almost to the panicle, glabrous, slightly scabrid or smooth, except the villous nodes; ligule short, obliquely truncate or bilobed, firmly membranous; blades narrowly linear, tapering to a fine point, 3–6 in. by 1½–4 lin., glaucous, usually subrigid, very slightly scabrid; panicle 2–4 in. long, erect or more or less flaccid and nodding, narrow; branches suberect, simple, up to 1½ in. long, filiform, flexuous, angular, slightly scabrid or smooth; spikelets often closely imbricate, subsecund and laterally concavo-convex, obliquely oblong, 1½–2 lin. long, sometimes purplish; sides of valve scabrid or smooth; stamens 6; anthers 1–1¼ lin. long. *Kunth, Enum.* i. 6; *Trin. in Mém. Acad. Petrop. sér.* 6, v. (1839) 172; *Steud. Syn. Pl. Glum.* i. 2; *Durand & Schinz, Consp. Fl. Afr.* 789; *Étud. Fl. Congo*, i. 329. *L. australis, R. Br. Prod.* 210. *L. mexicana, Kunth in H. B. & K. Nov. Gen. et Spec.* i. 195; *Kunth, Rév. Gram.* 179, *t.* 1; *Nees, Fl. Afr. Austr.* 193; *Krauss in Flora*, 1846, 121. *L. luzonensis, Presl, Rel. Haenk.* i. 207. *L. elongata, Petit-Thouars in Herb. Willd. ex Presl, l.c.*; *Trin. in Mém Acad. Pétersb. sér.* 6, v. (1839) 172. *L. parviflora, Desv. Opusc.* 61. *L. mauritanica, Salzm. ex Trin. l c.* 174. *L. triniana, Sieb. ex Trin. l.c.* *L. abyssinica, Hochst. ex A. Rich. Tent. Fl. Abyss.* ii. 356. *L. ægyptiaca, Fig. & De Not. in Mem. Acc. Torin. ser.* 2, xiv. (1853), 317. *L. ferox, Fig. & De Not. l.c.* 319. *L. griffithiana, C. Muell. in Bot. Zeit.* 1856, 345. *L. capensis, C. Muell. l.c.* *Asprella hexandra, A. australis and A. mexicana, Roem. & Schult. Syst.* ii. 267. *A. purpurea, Boj. Hort. Maur.* 376 (*name only*). *Oryza hexandra, Doell in Mart. Fl. Bras.* ii. ii. 10. *O. australis, A. Br. ex Schweinf. Beitr. Fl. Aethiop.* 300; *Aschers. & Schweinf. Ill. Fl. Égypte*, 167.

COAST REGION: Robertson Div.; Bosjes Veld, *Thom!* George Div.; Plettenberg Bay, *Mund.* Uitenhage Div.; Zwartkops River, *Ecklon & Zeyher*, 510! in the bed of the Vanstaadens River, *MacOwan*, 2188! Lower Albany; near Glenfilling, 500–1000 ft., *Drège!* Komgha Div.; in the valley of the Key River, 100 ft., *Drège.*

KALAHARI REGION: Orange Free State; near Winberg, *Buchanan*, 251! Transvaal; near Lydenburg, *Wilms*, 1703!

EASTERN REGION: Tembuland; near Bazeia, 2000 ft., *Baur*, 594! Natal; Durban Flats, *Buchanan*, 22! 75! 96! Berea, 100 ft., *Wood*, 5944! by the Umlazi River, *Krauss*, 9! by the Tugela River, 600 ft., and at Umpumulo, 2000 ft., *Buchanan*, 281!

Widely spread through the tropical and subtropical regions.

LXXI. EHRHARTA, Thunb.

Spikelets laterally compressed, panicled or racemed, sometimes solitary, pedicelled; rhachilla disarticulating below the valves, more or less obscurely produced. *Florets* 3; lower 2 reduced to empty valves, uppermost ⚥. *Glumes* persistent, membranous. *Valves* 3, heteromorphous; the lower 2 empty, usually exceeding the glumes, more or less cartilaginous, often bearded, and the upper with a callous appendage at the base, awnless or awned; the uppermost fertile, smaller, thinner, awnless, sometimes with a knob-like appendage at the base forming a hinge with the appendage of the upper empty valve. *Pale* narrow, keeled, finely 2-nerved, nerves very close. *Lodicules* 2. *Stamens* 6 or 3, very rarely 1. *Styles* distinct, short; stigmas plumose or brush-like, exserted above the base. *Grain* elliptic, much compressed; hilum a fine line almost as long as the grain; embryo about $\frac{1}{5}$ of the grain.

Perennials or annuals of very varied habit, sometimes bulbous at the base or suffrutescent; blades flat or convolute, sometimes much reduced or suppressed; ligules membranous, usually short or reduced to a narrow rim; panicle or racemes sometimes very scanty or even reduced to solitary spikelets.

Species 25 in South Africa, 1 of them also in a slightly different form in East Africa, tropical Arabia, the Mascarene Islands and India (here probably introduced).

Empty valves subulate-caudate, similar in outline, equal or unequal, 2–4 times the length of the glumes, upper not appendaged at the base:
- Flaccid annuals; lower empty valve exceeding the fertile; upper empty valve not pitted at the base:
  - Spikelets small; empty valves subequal, fertile valve 1–1¼ lin. long, equalling or slightly exceeding the glumes; stamens 3 ... ... (1) **triandra.**
  - Spikelets long; upper empty valve longer than the lower, fertile valve 2½–2¾ lin. long, much exceeding the glumes; stamens 6 ... ... (2) **longiflora.**
- Perennials with crowded basal leaves; lower empty valve distinctly shorter than the fertile, upper pitted at the base:
  - Leaves lanceolate to linear-lanceolate, 2–5 lin. broad, when flat ... ... (3) **dura.**
  - Leaves filiform ... ... (4) **microlæna.**

Empty valves not subulate-caudate, less than twice the length of the glume, or if subulate-caudate, the culm with a bulbous thickening at the base or the empty valves very dissimilar:
- Culms bulbous at the base, simple, 2–3-noded, with tufts of basal leaves; empty valves subequal, upper not appendaged at the base, though sometimes with a prominent ridge:
  - Lower empty valve widening above the middle, smooth, or like the upper coarsely rugose from projecting transverse lamellæ; blades usually flat, broad:
    - Spikelets 3 lin. long ... ... (5) **bulbosa.**
    - Spikelets 4–6 lin. long ... ... (6) **capensis.**
  - Lower empty valve almost equally wide throughout, like the upper smooth or more or less transversely rugose; blades usually convolute, rigid ... ... (7) **longifolia.**
- Culms not bulbous, or if subbulbous at the base, then branched and 4–6-noded, or the empty valves very unequal:

Empty valves very dissimilar; lower minute, ⅓ the length of the upper or less, 3–5-ribbed or nerved, hyaline between the ribs or nerves:
Culms coarse, almost woody below; blades very rigid; spikelets in 5–12-spiculate racemes:
Glumes about ⅓ the length of the spikelet ... ... (8) **rupestris.**
Glumes ¾–⅝ the length of the spikelet ... (9) **setacea.**
Culms and blades fine to very fine; spikelets in 2–9-spiculate racemes, or solitary on delicate peduncles:
Glumes scarcely ⅓ the length of the spikelet, subquadrate:
Spikelets 2½–3 lin. long, in 2–9-spiculate racemes ... ... (10) **tricostata.**
Spikelets 2 lin. long, solitary on delicate rather long peduncles ... (11) **Dodii.**
Glumes equalling the spikelet, linear-lanceolate to linear-oblong ... ... (12) **uniflora.**
Empty valves similar, or lower conspicuously shorter and narrower, but then more or less like the upper in substance:
Spikelets 1–2 lin., rarely to 2½ lin. long; empty valves obtuse or truncate, not mucronate, rarely subacuminate, and then not exceeding 1½ lin. in length:
Empty valves very similar and subequal, distinctly exserted from the glumes:
Perennial; spikelets 1½–2½ lin. long; stamens 6 ... ... (13) **erecta.**
Annual; spikelets 1 lin. long; stamens 3 ... ... (14) **delicatula.**
Empty valves unequal, lower about ½ the length of the upper and very narrow, upper about equalling the glumes:
Perennial with basal tufts of villous leaves; valves glabrous ... ... (15) **melicoides.**
Annual; empty valves hairy .. (16) **brevifolia.**
Spikelets 2–7½ lin. long:
*Spikelets 2–4 lin. long; glumes and empty valves subequal; upper empty and fertile valves hinged together by an ear-shaped appendage on a pivot-like knob:
Annual; empty valves hairy, equal, subcaudate ... ... (17) **pusilla.**
Perennial:
Not suffrutescent; blades flat or convolute:
Empty valves long hairy (18) **calycina.**
Empty valves glabrous:
Spikelets erect, more or less adpressed to the rhachis of a stiff spike-like panicle or raceme, 3–4 lin. long ... (19) **subspicata.**
Spikelets nodding on the flexuous rhachis

of an often scanty raceme, 2½–3 lin. long ... ... (20) **Rehmannii.**

Suffrutescent; blades very much reduced or suppressed:

Culms stout, up to 2½ lin. thick at the base; sheaths broad, leathery, usually quite bladeless; glumes usually shorter than the valves ... (21) **ramosa.**

Culms slender; sheaths narrow, bladeless or sometimes bearing blades, up to 1½ in. long and scarcely ½ lin. broad; glumes slightly exceeding the valves ... (22) **aphylla.**

**Spikelets 4½–7½ lin. long; glumes and villous empty valves subequal or the latter more or less exserted; upper empty valve not appendaged at the base, not hinged to the fertile:

Lower nodes reversedly hairy; keels of the empty valves long hairy ... ... (23) **barbinodis.**

Nodes glabrous; empty valves hairy all over:

Innovation-buds (or young stolons) stout, densely covered with villous scales; culms erect, stout or slender, simple or branched; spikelets 4½–6 lin. long ... ... (24) **gigantea.**

Innovation-buds rather slender, only the basal scales fugaciously tomentose; culm ascending from an often decumbent and rooting base, slender, with long erect branches from the base; spikelets 5–7½ lin. long ... (25) **villosa.**

1. **E. triandra** (Nees ex Trin. Phalar. 15); annual; culms usually simple, weak, ½–1 ft. long, glabrous, smooth, about 3-noded; sheaths rather loose, glabrous or scantily hairy above; ligules obtuse, ½ lin. long; blades lanceolate-linear, tapering to a long fine point, up to 5 in. by 4 lin., flat, flaccid, softly hairy or glabrescent, margins wavy, scabrid; panicle ovate or narrow-oblong, up to almost 6 in. long, flaccid; rhachis smooth; branches distant, whorled, the lowest up to 2 in. long, spreading at right angles or deflexed, flexuous, branched from the base, scabrid; pedicels capillary, up to 3 lin. long; spikelets pallid, 4–6 lin. long; glumes ovate, acuminate, 1¼ lin. long, delicate, ciliolate, 5-nerved, the upper slightly longer and broader; empty valves subequal, lanceolate, narrowed into a long purplish scabrid beak or awn, 5-nerved, transversely rugose above the base from short strongly projecting lamellæ, margins and keels

ciliate, base beardless; fertile valve ovate, subacute in profile, 1–1¼ lin. long, glabrous; pale ciliolate near the base; lodicules glabrous; stamens 3, anthers ⅝ lin.; stigma brush-like. *Nees, Fl. Afr. Austr.* 221; *Steud. Syn. Pl. Glum.* i. 5; *Durand & Schinz, Consp. Fl. Afr.* v. 794.

COAST REGION: Vanrhynsdorp Div.; Ebenezer, dry hills, below 300 ft., *Drège*, 523! Tulbagh Div.; between Roodesand and Hex River, 1000–1500 ft., *Drège*! Paarl Div.; near Paarl, in copses, 500–1000 ft., *Drège*. Worcester Div.; Hex River Valley, *Wolley Dod*, 3700!

2. E. longiflora (Sm. Pl. Ic. Ined. t. 32); culms geniculate-ascending or erect, usually simple, weak, 1–2 ft. long, glabrous, smooth, 3–4-noded, uppermost internode up to 9 in. long; sheaths rather loose, glabrous, smooth; ligules truncate, up to 1 lin. long; blades linear from a clasping fimbriate base, 3–6 in. by ½–5 lin., flaccid, scaberulous or smooth, usually softly hairy or glabrescent, margins wavy, scabrid; panicle narrow, up to ½ ft. long, flaccid, loose; rhachis smooth, branches in distant semi-whorls, very unequal, the longest up to 2 in., simple or sparingly branched, filiform to capillary, flexuous, glabrous below, or like the capillary pedicels spreadingly hairy or scabrid; pedicels up to 4 lin. long; spikelets oblong-lanceolate, more or less secund, often nodding, 3½ lin. long; lower glume lanceolate, acuminate, 1–1¼ lin., 5-nerved; upper glume oblong, cuspidate, 1¾–2 lin. long, 7-nerved; empty valves lanceolate-oblong, produced into a beak or a scabrid subulate awn of varying length (1–5 lin.), 7-nerved, the lower 3 lin. long, with a minute beard in front and 2 small tufts of hairs at the back of the base, scabrid above and often transversely rugose below, the upper up to 3½ lin. long, with a pair of inconspicuous bearded ridges at the base; fertile valve oblong, subacute, 2½–2¾ lin. long, 7-nerved, glabrous, not appendaged; lodicules ciliate; stamens 6; anthers ¾ lin. long; stigmas brush-like. *Thunb. Prodr.* 192; *Fl. Cap. ed. Schult.* i. 336; *Sw. in Trans. Linn. Soc.* vi. 56, *t.* 4, *fig.* 7; *Schrad. in Schult. f. Syst.* vii. 1377; *Kunth, Enum.* i. 14; *Trin. Phalar.* 14; *Nees, Fl. Afr. Austr.* 219; *Steud. Syn. Pl. Glum.* i. 5; *Durand & Schinz, Consp. Fl. Afr.* v. 792. *E. longiflora, vars. ecklonianа, urvilleana, longiseta, Nees, l.c.* 220. *E. Banksii, J. F. Gmel. Syst.* 549 (*excl. ic. cit.*). *E. aristata, Thunb. Prodr.* 66. *E. urvilleana, Kunth, Rév. Gram.* i. 9, 189, *t.* 6; *Brongn. in Duperr. Voy. Coq. Bot. t.* 24; *Kunth, Enum.* i. 14. *E. ecklonianа, Schrad. in Schult. f. l.c.* 1376; *Nees in Linnæa,* vii. 338; *Kunth, Enum.* i. 14; *Steud. l.c.* 6; *Durand & Schinz, l.c.* 791. *E. longiseta, Schrad. in Goett. Gel. Anz.* iii. (1821) 2078, *and in Schult. f. l.c.* 1377; *Kunth, Enum.* i. 15. *E. longifolia, Durand & Schinz, l.c.* 792 (*partly, not Schrad.*). *E. avenacea, Willd. ex Schult. f. l.c.* 1378 (*from the description*).

SOUTH AFRICA: without precise locality, *Thunberg*!

COAST REGION: Clanwilliam Div.; near Clanwilliam, 350 ft., *Schlechter*, 8591! Brak Fontein and by the Olifants River, in sand, *Drège*. Cape Div.; foot of Table and Devils Mountains, in shady places, *Ecklon*, *Pappe* 13!

near Oatlands Point, *Wolley Dod*, 2958! Piers Road, Wynberg, *Wolley Dod*, 3230! precipices by the second waterfall on Devils Peak, *Wolley Dod*, 3455! near Klein Constantia, *Wolley Dod*, 2069! near Cape Town, *Alexander!* Tulbagh Div.; Tulbagh Waterfall, *Ecklon.* Swellendam Div.; Zonder Einde River, *Zeyher!* Riversdale Div.; Gauritz River, *Ecklon.* Mossel Bay Div.; Attaquas Kloof, *Gill!*

Introduced into St. Helena and Bourbon.

3. **E. dura** (Nees ex Trin. Phalar. 13); perennial, densely tufted; culms erect, simple, 1–2 ft. long, glabrous, smooth, 2-noded; leaves crowded at the base; lowest sheaths very firm, compressed or open, persistent, glabrous, very rarely pubescent, strongly striate, the upper tight or the uppermost tumid; ligules ovate, obtuse, up to 1½ lin. long; blades of the basal and culm-leaves linear-lanceolate to lanceolate from a long and narrow or broad and clasping base, acute, 4–8 in. by 2–5 lin., rather thick and firm, flat or convolute, glaucous, glabrous, smooth or rough, margins sometimes wavy, upper blades linear, convolute, short; panicle oblong to sublinear, erect or nodding, 3–5 in. long, rather loose; rhachis smooth; branches in distant semi-whorls or pairs, unequal, up to 2½ in. long, the lowest branched, the others simple, 3–1-spiculate, filiform, glabrous, tips thickened; pedicels up to 6 lin. long; spikelets lanceolate-oblong, strongly compressed, 5–6 lin. long, greenish; glumes ovate-oblong, scarious, obscurely 5-nerved, glabrous, the lower 1½–2 lin. long, obtuse, the upper 2½ lin., sometimes emarginate; lower empty valve linear-oblong, gradually or abruptly contracted into a short bristle, 3–4 lin. long, prominently 5–7-nerved, scaberulous above, bearded at the base in front and behind; upper empty valve lanceolate, gradually narrowed into an awn, 5 lin. long, 7–9-nerved, scaberulous, base with a short beard and a tympanum-like pit on each side; awn 2–4 lin. long; fertile valve linear-oblong, acute, 5 lin. long, glabrous except for a beard at the base and the ciliate tip, 7–9-nerved; lodicules glabrous, 1 lin. long; stamens 6; anthers 2½ lin. long; stigmas plumose. *Nees, Fl. Afr. Austr.* 218; *Steud. Syn. Pl. Glum.* i. 5; *Durand & Schinz, Consp. Fl. Afr.* v. 791.

COAST REGION: Tulbagh Div.; Tulbagh Waterfall, *Ecklon.* Worcester Div.; Dutoits Kloof, 3000 ft., on grassy rocks, *Drège!* Swellendam Div.; on a mountain near Puspus Valley, *Ecklon & Zeyher*, 4513! Caledon Div.; Houw Hoek Mountains, Nieuw Kloof, *Burchell*, 8058! Mossel Bay Div.; Attaquas Kloof, *Gill!* Uitenhage Div.; Witteklip Mountain, *MacOwan*, 2126!

4. **E. microlæna** (Nees ex Trin. Phalar. 13); perennial, densely tufted; culms erect, slender, simple, 2 ft. long, glabrous, smooth, 2- (rarely 3-) noded; leaves crowded at the base; lower sheaths compressed and keeled, or open, firm, persistent, strongly striate, glabrous, sometimes scaberulous; ligules oblong-lanceolate, up to 2 lin. long; blades setaceous, canaliculate, acute, the lower to ½ ft. long or more, the upper much shorter, glabrous, smooth or scaberulous; panicle erect or nodding, 4–6 in. long, subsecund, often reduced to a raceme; rhachis filiform, scaberulous except at the base; branches 4–2-nate or solitary, usually simple, 2–1- (rarely 6–3-) spiculate, capillary, scabrid;

pedicels up to 6 lin. long, tips thickened; spikelets oblong-lanceolate, 5–7 lin. long; glumes ovate-oblong, obtuse, glabrous, 5–7-nerved, the lower 1½ lin. long, the upper 2½–3 lin.; lower empty valve oblong-lanceolate, shortly awned from the gradually narrowed or truncate tip, 3–4 lin. long, prominently 5–7-nerved, scaberulous above, obscurely bearded at the base; upper empty valve lanceolate, 5–7 lin. long, gradually narrowed into a rigid awn, prominently 5–7-nerved, scaberulous above, base glabrous or obscurely bearded with a linear or oblong pit on each side, awn up to 7 lin. long; fertile valve linear-oblong, obliquely truncate or subacute, 5–6 lin. long, faintly 7-nerved, glabrous except at the ciliate tips; lodicules 1 lin. long, glabrous; stamens 6; anthers 3 lin. long; stigmas plumose. *Nees, Fl. Afr. Austr.* 217; *Steud. Syn. Pl. Glum.* i. 5; *Durand & Schinz, Consp. Fl. Afr.* v. 793.

COAST REGION: Tulbagh Div.; Tulbagh Waterfall, *Ecklon*. Worcester Div.; Dutoits Kloof, in bogs, 3000–4000 ft., *Drège!* Swellendam Div., *Mund*. Riversdale Div.; between Vet River and Krombecks River, *Burchell*, 7184!

5. **E. bulbosa** (Sm. Pl. Ic. Ined. sub t. 33); perennial; rhizome short, slender; culms erect, 1–1¼ ft. long, glabrous, smooth, 2–3-noded, lowest internode swollen (normally?) into a depressed subglobose tuber 3 lin. in diam., covered by the villous sheath-bases; basal leaves 3–4; sheaths rather loose to tumid, glabrous, smooth; ligule almost reduced to a line of short hairs; blades linear, gradually tapering, 3 in. by 1 lin., glabrous, smooth or scaberulous, rather rigid, the uppermost very small or suppressed; panicle narrow, linear, nodding, 2–4 in. long, secund, almost reduced to a raceme; rhachis very slender, smooth; branches in semi-whorls or pairs or solitary, erect or spreading; pedicels subcapillary, smooth, 2–6 lin. long; spikelets elliptic-oblong, 3 lin. long; glumes subequal, lanceolate-oblong, acute or subobtuse, 2–2½ lin. long, 5–7-nerved, glabrous, purplish; empty valves oblong or the lower obovate-oblong, truncate, mucronate, 3 lin. long, 5–7-nerved, strongly transversely rugose, pubescent, the lower bearded at the base in front and behind, the upper with a very narrow curved 2-bearded base; fertile valve obliquely oblong, truncate, smooth or obscurely rugose, 7-nerved; lodicules ciliate; stamens 6; anthers 1½–1¾ lin. long; stigmas plumose. *Lam. Illustr. t.* 263, *fig.* 3, *a–g.*; *Thunb. Prodr.* 188; *Fl. Cap. ed. Schult.* i. 336; *Sw. in Trans. Linn. Soc.* vi. 60, *t.* 4, *fig.* 9; *Kunth, Enum.* i. 12; *Trin. Phalar.* 19; *Steud. Syn. Pl. Glum.* i. 6; *Durand & Schinz, Consp. Fl. Afr.* v. 791. *E. Trochera, Schrad. in Goett. Gel. Anz.* iii. (1821) 2077; *Schult. f. Syst.* vii. 1368; *Nees, Fl. Afr. Austr.* 200. *E. æmula, Schrad. l.c.*; *Schult. f. l.c.* 1367; *Kunth, l.c.* 11; *Nees, l.c.* 199; *Steud. l.c.* 6; *Durand & Schinz, l.c.* 790. *Trochera striata, Rich. in Rozier, Journ. Phys.* xiii. (1779) 225, *t.* 3; *Beauv. Agrost.* 62, *t.* 12, *fig.* 3. *T. spicata, Rich. ex Poir. Encyc. Suppl.* ii. 542.

COAST REGION: Cape Div.; Table Mountain, *Ecklon*, 907! Lion Mountain, *Pappe!* lower slopes of Lion Mountain, near Sea Point, *Wolley Dod*, 3532!

near Cape Town, *Harvey!* Constantia, *Ecklon.* Tulbagh Div.; Tulbagh Waterfall, *Ecklon.* Stellenbosch Div.; Hottentots Holland, *Ecklon.* Caledon Div.; Zwart Berg by the Steenboks River, *Ecklon.*

6. **E. capensis** (Thunb. in Vet. Acad. Handl. Stockh. 1779, 216, t. 8); perennial; culms tufted, erect, sometimes geniculate, 1–2 ft. long, ovoid-bulbous at the base, glabrous, smooth, 2-noded; basal leaves 5–8; sheaths loose, glabrous, smooth, the lowest pallid, more or less open; ligules very short, ciliolate; blades linear, from a clasping (often fimbriate) base, gradually tapering to a fine point, 3–8 in. by 2–3½ lin., flat or convolute, rigid or subflaccid, rather thick, glabrous or hairy, margins cartilaginous, often wavy; panicle erect or slightly flexuous, 2–6 in. long, sometimes secund; rhachis glabrous; branches usually paired or solitary, up to 1½ in. long, suberect, scantily branched or simple, filiform, flexuous, glabrous; pedicels 2–8 lin. long; spikelets oblong, 5–6½ lin. long;. glumes subequal, ovate, acute or acuminate, 2–3 lin. long, glabrous, often purplish, 5–6½ lin. long, 5–7-nerved; empty valves equal, lower ovate-oblong, much compressed, obtuse or subacute, usually mucronate, 7–9-nerved, obscurely rugose below, glabrous or pubescent above on the sides, ciliate along the margins and the straight or concave keel, bearded at the base in front and behind; upper empty valve linear, sometimes curved, usually mucronate, strongly transversely rugose, glabrous or pubescent, ciliate along the margins and the convex keel, contracted base very short and bearded, beards 1½–3 lin. long; fertile valve linear-oblong, obtuse, 4–4½ lin. long, glabrous, smooth, 7-nerved; pale ciliolate above; lodicules ciliate, 1 lin. long; stamens 6; anthers 3–3½ lin. long; stigmas plumose. *Thunb. Prodr.* 192; *Fl. Cap. ed. Schult.* i. 335. *E. Mnemateia, Linn. f. Suppl.* 209; *Thunb. Prodr.* 66; *Durand & Schinz, Consp. Fl. Afr.* v. 793. *E. mnematea, Sw. in Trans. Linn. Soc.* vi. 44, *t.* iii. *fig.* 1; *Schrad. in Schult. f. Syst.* vii. 1366; *Nees, Fl. Afr. Austr.* 198. *E. Mnemateja, Trin. Phalar.* 21; *Steud. Syn. Pl. Glum.* i. 7. *E. nutans, Lam. Encyc.* ii. 346 (*excl. Syn. Trocherea*); *Illustr.* ii. 397, *t.* 263, *fig.* 2. *E. cartilaginea, Sm. Pl. Ic. Ined. sub t.* 33.

SOUTH AFRICA: without precise locality, *Thunberg!*
COAST REGION: Malmesbury Div.; Eikenboom, 1000–5000 ft., *Drège.* Cape Div.; Capetown, *Harvey! Alexander!* Constantia, *Wolley Dod,* 2125! Paarl Div.; Paarl Mountains, 1000–1500 ft., *Drège!* Tulbagh Div.; Tulbagh Waterfall, 1200 ft., *Schlechter,* 9060! Tulbagh Waterfall, Tulbagh Kloof, Winterhoek Mountain, &c., 800–5000 ft., *Ecklon.* Worcester Div.; on decaying wood, *Ecklon.* Stellenbosch Div.; between Hottentots Holland Kloof and Hau Hoek, *Ecklon!* Caledon Div.; Zonder Einde River, *Drège.* Swellendam Div.; Puspus Valley, *Ecklon.* Hills between Riet Kuil and Grootvaders Hoek, *Ecklon & Zeyher,* 4512! Riversdale Div.; near Zoetemelks River, *Burchell,* 6686! between Great Vals River and Zoetemelks River, *Burchell,* 6567!

7. **E. longifolia** (Schrad. in Goett. Gel. Anz. iii. (1821) 2077, and in Schult. f. Syst. vii. 1368); perennial, tufted, simple from a more or less tuberous base, 1½–2 ft. long, glabrous, smooth, 2–3-noded; basal leaves 3–4; sheaths loose or tight or the lowest open, firm,

glabrous, striate; ligule a line of hairs; blades linear from a clasping and sometimes ciliate base, long tapering to a fine point, up to ½ ft. by 2 lin., the uppermost very short, convolute to subsetaceous or involute below the middle, firm, glabrous, smooth or scaberulous above; panicle linear to linear-oblong, erect, loose or rather dense, usually subsecund, 4–5 in. long; rhachis glabrous; branches in semi-whorls or pairs or solitary, unequal, up to ¾ in. long, rarely longer, erect or spreading, 3–1- (rarely to 7-) spiculate, subcapillary, glabrous; pedicels up to 6 lin. long; spikelets oblong, 3½–4 lin. long; glumes subequal, oblong-lanceolate, acuminate, 3–3½ lin. long, glabrous, 5-nerved, often purplish; lower empty valve linear-oblong, truncate, apiculate, exceeding the lower glume, scabrid to hispidulous above and on the keel, transversely rugose or not, bearded at the base in front and behind; upper empty valve similar, slightly longer and narrower, usually mucronate, transversely rugose, rarely smooth, base shortly contracted with a callous projecting ridge and a long beard on each side; fertile valve linear-oblong, 3¼ lin. long, truncate, apiculate, glabrous, 5-nerved, with a knob on each side of the base; lodicules ciliate, ½–⅔ lin. long; stamens 6; anthers 1¾–2½ lin. long; stigmas plumose. *Trin. Phalar.* 21; *Steud. Syn. Pl. Glum.* i. 6; *Durand & Schinz, Consp. Fl. Afr.* v. 792. *E. Ottonis, Kunth, Rév. Gram.* i. 9 (*name only*); *Enum.* i. 12; *Suppl.* 10, *t.* 2, *fig.* 2; *Trin. l.c.* 20; *Nees, Fl. Afr. Austr.* 201; *Steud. l.c.* *E. varicosa, Nees ex Trin. l.c.* 19; *Nees, l.c.* 200; *Steud. l.c.*; *Durand & Schinz, l.c.* 794.

VAR. β, **robusta** (Stapf); culms stout, up to 1½ lin. thick and 4 ft. long, 3–4-noded; blades very rigid, convolute or flat below, up to more than 1 ft. long; panicle up to 10 in. long, more compound; spikelets 4–5½ lin. long; glumes ¼–⅓ shorter than the empty valves, the latter transversely rugose, mucronate and hispidulous above.

COAST REGION: Cape Div.; Table Mountain, *Ecklon*, *Bergius!* Simons Bay, *Wright*, 196! Tulbagh Div.; Tulbagh Waterfall, *Ecklon*, *Schlechter*, 9056! Stellenbosch Div.; Hottentots Holland, *Ecklon*; *Alexander!* Caledon Div.; Genadendal, 3500–4000 ft., *Drège!* Var. β: Cape Div.; top of Tokay Plantation, *Wolley Dod*, 1958! lower slopes of Orange Kloof, *Wolley Dod*, 3343! Albany Div.; near Grahamstown, 2000 ft., *MacOwan*, 1294!

8. **E. rupestris** (Nees ex Trin. Phalar. 25); perennial; culms ascending from a decumbent base, very slender, almost woody, simple or almost so in the lower part, divided into short aggregate branches above, lower internodes pruinose; leaves crowded, imbricate in the upper branches; sheaths compressed, keeled, very firm, pruinose or almost rough, persistent; ligules very short, ½ lin. long, woolly-ciliate; blades folded, subulate, obtusely keeled, pungent, 1–1½ in. by 1 lin. (when expanded), rigid, prominently nerved, densely puberulous on the upper surface, pruinose or almost rough below; raceme erect, rigid, about 1 in. long; rhachis wavy, glabrous; pedicels erect, 1–1½ lin. long, subscaberulous; spikelets 7–5, adpressed to the rhachis, broad-oblong, 2–2½ lin. long; glumes subequal, ovate-oblong, very obtuse, equalling about ⅓ of the spikelet, thin, finely pruinose, very shortly 5–9-nerved; empty valves very unequal,

the lower very small, oblong, obtuse, mucronulate, thin, almost 1 lin. long, with a stout middle nerve and 2–3 very short side-nerves below, upper oblong, very obtuse, coriaceous, obscurely 5–7-ribbed, finely tubercled along the ribs, pruinose, base quite plain; fertile valve similar, somewhat shorter; lodicules denticulate, glabrous, $\frac{3}{4}$ lin. long; stamens 6; anthers 1 lin. long; stigmas plumose. *Nees, Fl. Afr. Austr.* 227; *Steud. Syn. Pl. Glum.* i. 7; *Durand & Schinz, Consp. Fl. Afr.* v. 792.

COAST REGION: Caledon Div.; Genadendal, 3000–4000 ft., *Drège!* Swellendam Div.; on a mountain peak near Swellendam, *Burchell,* 7419!

9. **E. setacea** (Nees, Fl. Afr. Austr. 228); perennial; culms tufted, ascending or suberect, 1–1½ ft. long, slender, almost woody, smooth, striate, very many-noded, simple or bearing long branches near the base, usually with 1 or several shorter branches higher up, branches contracted, internodes dark; leaves of the young branches closely imbricate; sheaths subterete, tight, firm, glabrous or scantily bearded at the mouth, smooth, striate, exceeding the internodes except the uppermost, the lower persistent, throwing off their blades; ligules reduced to a minutely ciliolate rim; blades setaceously subulate from a linear base, involute or convolute above, with a callous base, very rigid, striate, glabrous, smooth below, scabrid above; raceme erect, straight, usually 1–2 in. long, 6–12-spiculate; rhachis very slender, smooth; pedicels filiform, smooth, erect, 1–4 lin. long, tips thickened; spikelets oblong, 2½–3 lin. long, sometimes secund, erect; glumes subequal, ovate-oblong or oblong, very obtuse, $\frac{3}{4}$–$\frac{4}{5}$ the length of the spikelet, scarious, glabrous, smooth, shortly 5–9-nerved, often purple; barren valves very unequal, lower minute, ovate-oblong, obtuse, $\frac{1}{2}$–$\frac{3}{4}$ lin. long, scarious, strongly 3–5-ribbed, upper oblong, very obtuse, narrowed below, coriaceous, 7-ribbed, ribs tubercled, rough, base quite plain; fertile valve similar to the preceding, somewhat smaller; lodicules obliquely obovate, denticulate, $\frac{3}{4}$ lin. long; stamens 6; anthers 1 lin. long. *Steud. Syn. Pl. Glum.* i. 7; *Durand & Schinz, Consp. Fl. Afr.* v. 793.

VAR. **scabra** (Stapf); leaves very rough, sheaths distinctly bearded at the mouth; spikelets up to 3½ lin. long; intermediate and upper fertile valves narrower, granular, scarcely ribbed.

COAST REGION: Cape Div.; near Capetown, *Harvey,* 135! 136! by a stream at the head of Waai Vley, *Wolley Dod,* 3334! on Table Mountain, 3000 ft., *Drège!* Tulbagh Div.; Witzenberg and Vogel Vley, *Ecklon.* Var. β: Swellendam Div.; on a mountain peak near Swellendam, *Burchell,* 7312!

10. **E. tricostata** (Stapf); perennial; culms tufted or fascicled, ascending or suberect, branched, filiform, ½–1½ ft. long, glabrous, smooth, 6- or more-noded, upper intermediate internodes usually shorter than the preceding ones, branches erect; sheaths very tight, terete, glabrous, smooth, very lowest longer than the internodes, those above mostly much shorter excepting some of the upper: ligule a minute ciliolate rim; blades linear, long tapering to a subsetaceous

point, flat or finely setaceously convolute, $1\frac{1}{2}$–$2\frac{1}{2}$ in. long, $\frac{1}{2}$–1 lin. wide at the base when expanded, somewhat rigid, glabrous, quite smooth, uppermost very short and fine; spikelets 2–9, in erect slender racemes (up to $1\frac{1}{4}$ in. long), shortly pedicelled, adpressed to the filiform smooth somewhat wavy axis, linear-oblong, $2\frac{1}{2}$–3 lin. long, glabrous; glumes subquadrate, equal, scarcely a quarter as long as the spikelet, subherbaceous and obscurely 5–7-nerved at the very base, otherwise membranous, white; lower barren valve ovate, strongly concave, $\frac{1}{2}$ lin. long, strongly 3-ribbed, otherwise delicately hyaline; upper linear obtuse, equalling the spikelet, coriaceous, light green, 7-ribbed, ribs fine and minutely tubercled; fertile valve very similar to the upper barren valve, but broader, slightly shorter and less obtuse, 7–9-ribbed; pale over $1\frac{1}{2}$ lin. long, distinctly 2-nerved; lodicules obliquely ovate, denticulate, over $\frac{3}{4}$ lin. long; stamens 6; anthers 1 lin. long; stigmas densely plumose, sub-sessile.

COAST REGION: Paarl Div.; French Hoek, 2400 ft., *Schlechter*, 9292!

11. **E. Dodii** (Stapf); perennial; culms tufted, geniculately-ascending from a subdecumbent base, branched, very finely filiform, about 6–8 in. long, glabrous, smooth, many-noded, lower internodes several times as long as the upper; branches rather contracted; sheaths very tight, terete, glabrous, minutely rough, lower usually shorter than the internodes, upper longer except the uppermost; ligules short, truncate, ciliate; blades very fine, setaceously convolute, 1–$1\frac{1}{2}$ in. (rarely 3 in.) by $\frac{1}{3}$–$\frac{2}{3}$ lin. when expanded, glabrous, smooth or obscurely rough, finely striate and scantily scabrid on the upper surface; spikelets solitary, terminal, linear-oblong, 2 lin. long, glabrous; glumes subquadrate, equal, scarcely equalling $\frac{1}{3}$ of the length of the spikelet, herbaceous and 5–7-nerved near the base, otherwise hyaline, usually purplish; lower barren valve ovate, hyaline, 5–7-nerved, scarcely exceeding the glumes; upper linear-oblong, obtuse, equalling the spikelet, coriaceous in the upper part, light green, 7-ribbed, ribs fine and minutely tubercled; fertile valve very similar to the upper barren valve, but slightly shorter; pale $1\frac{1}{2}$ lin. long, 2-nerved below; lodicules obliquely obovate, many-nerved, $\frac{1}{2}$ lin. long; stamens 6; anthers $1\frac{1}{4}$ lin. long.

COAST REGION: Cape Div.; rocks on Constantia Berg, *Wolley Dod*, 1961!

12. **E. uniflora** (Burch. Trav. i. 57; name only); perennial; culms tufted, filiform, very weak, ascending from a decumbent base, simple in the lower part, abundantly branched above, over 1 lin. long, smooth, glabrous, many-noded, lower internodes often several times longer than the upper; leaves quite glabrous, smooth; sheaths tight, terete, shorter than the internodes (except sometimes the upper); ligules very short, truncate, ciliate; blades linear, acute, very narrow, setaceously convolute, $\frac{1}{2}$–$1\frac{1}{2}$ in. by almost 1 lin. at the base when expanded, striate; spikelets solitary or in 2–3-spiculate racemes, terminal on very fine smooth peduncles, scarcely exserted

from the uppermost sheath, oblong, 2½–3½ lin. long, light green, glabrous; glumes spreading, linear-lanceolate to linear-oblong, equal, as long as the spikelet, prominently 5–7-nerved, smooth, thin; lower barren valve ovate, ¾ lin. long, smooth, 5-ribbed; upper linear-oblong, obtuse, as long as or slightly shorter than the glumes, coriaceous, finely 7-nerved, rough and minutely tubercled in the upper part; fertile valve similar to the upper barren valve, but shorter; pale over 1½ lin. long, obscurely 2-nerved at the very base; lodicules obliquely ovate, denticulate, glabrous, 1 lin. long; stamens 6; anthers 1 lin. long.

COAST REGION: Cape Div.; Cape Flats near Rondebosch, *Burchell*, 182!

13. E. erecta (Lam. Encyc. ii. 347); perennial; culms tufted, geniculate-ascending from a procumbent often copiously branched base, slender, weak or wiry below, up to 2 ft. long, glabrous, very rarely reversedly pubescent below the nodes, smooth, 5–6-noded; sheaths tight, the lowest ultimately slipping from the culm, glabrous, rarely finely hairy, smooth; ligules obtuse or truncate, up to 3 lin. long; blades linear from a clasping often fimbriate base, gradually tapering, 2–6 in. by 1½–5 lin., flat, flaccid, glabrous, rarely hairy, smooth or subscaberulous, margins often wavy, scabrid or ciliate; panicle erect or nodding, narrow, 2–8 in. long, loose, sometimes reduced to a raceme; branches distant, the lowest 2–3-nate, very unequal, the longest up to 1–3 in., erect or spreading, simple or sparingly branched, filiform to almost capillary, flexuous, glabrous or scaberulous above; pedicels capillary, scaberulous to puberulous, up to 4 lin. or more long; spikelets light green, oblong, 1¼–2 lin., very rarely 2½ lin. long; glumes ovate, obtuse or apiculate, 5-nerved, the lowest ⅝–1 lin., the upper 1–1¼ lin. long; empty valves oblong, obtuse or truncate, smooth and shiny or scaberulous, transversely rugose, faintly 5-nerved, the lower 1¼–1½ lin., the upper longer by ⅙–¼, with a pair of obscure beardless ridges at the base; fertile valve oblong, obtuse to subacute, 1¼–1¾ lin. long, glabrous, smooth, obscurely 5–7-nerved; lodicules usually glabrous; stamens 6; anthers ½–⅝ lin. long; stigmas brush-like; grain 1 lin. long. *Lam. Illustr.* ii. 397, *t.* 263, *fig.* 1, *a–e*. *E. panicea, Sm. Pl. Ic. Ined. t.* 9; *Thunb. Prodr.* 188; *Fl. Cap. ed. Schult.* i. 335; *Sw. in Trans. Linn. Soc.* vi. 47, *t.* 3, *fig.* 2; *Beauv. Agrost.* 61, *t.* 12, *fig.* 2; *Schrad. in Schult. f. Syst.* vii. 1370; *Trin. Phalar.* 18; *Nees, Fl. Afr. Austr.* 225; *Steud. Syn. Pl. Glum.* i. 6; *Durand & Schinz, Consp. Fl. Afr.* v. 793. *E. paniciformis, Nees ex Trin. Phalar.* 18; *Fl. Afr. Austr.* 226 (*excl. var. β*); *Steud. l.c.* 6; *Durand & Schinz, l.c.* 793. *Panicum deflexum, Guss. in Ten. Fl. Nap.* v. 320. *Trochera panicea, H. Baill. Hist. Pl.* xii. 171, *fig.* 313, 314.

VAR. β, **natalensis** (Stapf); culms usually reversedly pubescent below the nodes; sheaths and blades pubescent; branches of the panicle and the pedicels often densely and minutely pubescent; spikelets 2–2½ lin. long; empty valves scabrid and puberulous; fertile valve obscurely bearded at the base; stamens 3 (always?).

COAST REGION: Cape Div.; Table Mountain, as a weed on roadsides and in gardens, *Ecklon*, 918! *Mac Gillivray*, 398! Kloof between the Lions Head and Table Mountain, *Burchell*, 256! Wynberg Park, *Wolley Dod*, 1886! Wynberg, *Wilms*, 3857! in ditches in Sand Road, Claremont, *Wolley Dod*, 2384! Paarl Div.; Paarl Mountains, among rocks, 2000–3000 ft., *Drège*, 522! Worcester Div.; hills on the Hex River, *Drège*. Swellendam Div.; near Grootvaders Bosch, *Zeyher*, 4515! Riversdale Div.; between Kochmans Kloof and Gauritz River, *Ecklon*. Mossel Bay Div.; dry hills on the eastern side of Gauritz River, *Burchell*, 6453/2! George Div.; George, in forests, *Burchell*, 6049! Uitenhage Div.; under bushes by the Zwartkops River, *Ecklon & Zeyher*, 186! Addo, *Ecklon*. Port Elizabeth Div.; near Port Elizabeth, on sand hills, *E.S.C.A. Herb.*, 190! Alexandria Div.: Enon, in shady woods, 1500–2000 ft., *Drège*. Komgha Div.; near Komgha, *Flanagan*, 911! King Williamstown Div.; in woods by the Kachu (Yellowwood) River, 2000 ft., *Drège!*

CENTRAL REGION: Prince Albert Div.; Great Zwart Bergen, near Klaarstroom, 2000–3000 ft., *Drège!* Somerset Div.; between the Zuurberg Range and Klein Bruintjes River, by shady brooks, 2000–3000 ft., *Drège!* Bosch Berg, *Burchell*, 3244! Graaff Reinet Div.; Klein Visch River, *Ecklon* (?); Albert Div.; Stormberg Range, in gorges, 5000–6000 ft., *Drège!* Aliwal North Div.; on the Witte Bergen, *Drège*.

KALAHARI REGION: Orange Free State; Bloemfontein, *Rehmann*, 3837!

EASTERN REGION: Tembuland; Bazeia, *Baur*, 411! Var. β, Natal; Clairmont, 50 ft., *Wood*, 7260! on the Drakens Bergen, near Newcastle, *Buchanan*, 177! Umsinga and base of Biggars Berg, *Buchanan*, 94! Riet Vlei, *Buchanan*, 287!

In a slightly different form also in tropical East Africa as far as Abyssinia, and (probably introduced) in tropical Arabia, the Mascarene Islands, and India.

14. **E. delicatula** (Stapf in Kew Bulletin, 1897, 288); annual; culms tufted, geniculate, simple or branched below, up to 1½ ft. long, delicate, glabrous or minutely and reversedly hairy, 2–3-noded; sheaths rather tight or loose, the lower finally slipping from the culms, glabrous or hairy; ligules truncate, short; blades linear, tapering to a fine point, 1–4½ in. by 1–3½ lin., flat, flaccid, scaberulous, minutely hairy or glabrescent; panicle narrow, 1–6 in. long, subsecund, loose; rhachis filiform, glabrous; branches in distant semi-whorls, spreading, very unequal, up to 2 in. long, simple or sparingly branched, capillary, flexuous, minutely scaberulous; pedicels of very variable length, up to 6 lin.; spikelets often nodding, 1–1½ lin. long, light green or purplish; glumes lanceolate, acuminate, 3-nerved, glabrous or somewhat pilose, the lower ¾ lin., the upper 1 lin. long; empty valves narrowly oblong or obovate-oblong, subacuminate, 5-nerved, transversely rugose, scaberulous, beardless, the upper slightly longer and with 2 semicircular shield-shaped appendages at the base; fertile valve equalling the upper glume, elliptic-oblong, minutely truncate or subacute, scaberulous above, 5-nerved; lodicules glabrous; stamens 3; anthers ⅙ lin. long; stigmas brush-like. *E. panicea, var. cuspidata, Nees, Fl. Afr. Austr.* 225 (*var. mucronata*, 226, *by error*), *partly*.

WESTERN REGION: Little Namaqualand, Vanrhynsdorp Div.; near Mieren Kasteel, among shrubs, *Drège*, 508!

COAST REGION: Tulbagh Div.; Roodesand, *Drège!* Worcester Div.; Hex River Valley, *Wolley Dod*, 3701! 3703!

Nees, l.c. 226, quotes "*E. panicea* var. *mucronata*" also from the following localities: Elleboog Fontein (Little Namaqualand), Ebenezer (Vanrhynsdorp

Div.) and Slangenhuivel (Tulbagh Div.). I have not seen specimens from these places, but from the description on pp. 225 and 226, it appears that at least some of the specimens referred to belong to *E. delicatula.*

15. **E. melicoides** (Thunb. Prodr. 192); perennial; rhizome very short, the fascicled culms and barren shoots densely clustered, with an almost bulbous thickening at the base; culms erect, slender, simple, 1–2 ft. long, glabrous, smooth, 1-noded at or below the middle; sheaths reversedly villous, the lower and outermost rather firm, open, straw-coloured, glabrescent, the others tighter, glaucous; ligules short, truncate; blades linear, narrow, finely acute, up to 6 in. by $\frac{3}{4}$–$1\frac{1}{2}$ lin., involute or flat, glaucous, villous or glabrous, strongly nerved; panicle ovate to oblong, open or contracted, very loose, up to $\frac{1}{2}$ ft. long; rhachis very slender, smooth; branches usually paired, up to $2\frac{1}{2}$ in. long, spreading or erect, filiform, the lower sparingly branched, smooth; pedicels very unequal, up to 5 lin. long, capillary; spikelets very scattered, oblong, greenish or purplish, $1\frac{1}{2}$–$1\frac{3}{4}$ lin. long, quite glabrous and smooth; glumes subequal, oblong, subacute, 5-nerved; empty valves unequal, the lower narrow-oblong, obtuse, 1 lin. long, 3-nerved, the upper obliquely oblong, equalling the upper glume, 5–7-nerved, with 2 orbicular-reniform appendages at the base; fertile valve like the upper empty valve, not appendaged; pale obscurely 2-nerved; lodicules glabrous; stamens 6; anthers 1 lin. long; stigmas brush-like. *Thunb. Fl. Cap. ed. Schult.* i. 335; *Sw. in Trans. Linn. Soc.* vi. 51, *t.* 3, *fig.* 4; *Schrad. in Goett. Gel. Anz.* iii. 2078, *and in Schult. f. Syst.* vii. 1371; *Kunth, Enum. Pl.* i. 13; *Trin. Phalar.* 23; *Nees, Fl. Afr. Austr.* 202; *Steud. Syn. Pl. Glum.* i. 6; *Durand & Schinz, Consp. Fl. Afr.* v. 792. *Melica capensis, Thunb. Prodr.* 21.

SOUTH AFRICA: without precise locality, *Thunberg!*

COAST REGION: Cape Div.; Lion Mountain, *Ecklon;* above Sea Point, *Wolley Dod,* 1608! Swellendam Div.; by the Kenko River, between Riet Kuil and Hemel-en-Aarde, below 1000 ft., *Zeyher,* 4516!

WESTERN REGION: Little Namaqualand; Kamies Bergen, between Lily Fontein and Pedros Kloof, 3000–4000 ft., *Drège!*

16. **E. brevifolia** (Schrad. in Goett. Gel. Anz. iii. 2077, and in Schult. f. Syst. vii. 1371); annual; culms geniculate, ascending or erect, simple or branched, up to 1 ft. long, weak, very slender, glabrous, smooth, 3-noded; sheaths loose or the upper tighter, glabrous, smooth; ligules very short, truncate; blades linear, gradually tapering, acute, up to 2 in. by $\frac{1}{2}$–1 lin., very flaccid, glabrous, margins minutely scabrid; panicle linear, very slender, up to 3 in. long, flaccid; rhachis filiform, flexuous, glabrous; branches in distant semiwhorls, very unequal, up to 1 in. long, capillary, scaberulous; pedicels up to 4 lin. long; spikelets greenish, erect, $1\frac{1}{2}$ lin. long; glumes subequal, lanceolate, acute, very thin, 5–7-nerved; empty valves very unequal, the lower linear-oblong, obtuse, $\frac{3}{4}$ lin. long, 1-nerved, shortly bearded at the base, loosely pubescent, the upper oblong, obtuse, almost $1\frac{1}{2}$ lin. long, muticous or mucronate, 5-nerved, loosely pubescent, with 2 unequally lobed appendages at

the base; fertile valve elliptic-oblong, subobtuse, 1 lin. long, loosely pubescent, 1- (or very obscurely 3–5-) nerved; pale nerveless; lodicules glabrous; stamen usually 1, rarely 3; anthers ¼ lin. long; stigmas brushlike; grain ¾ lin. long. *Kunth, Enum. Pl.* i. 13; *Trin. Phalar.* 23; *Nees, Fl. Afr. Austr.* 204; *Steud. Syn. Pl. Glum.* i. 7; *Durand & Schinz, Consp. Fl. Afr.* v. 791. *E. calycina, β, MS. in Herb. Thunberg.*

VAR. β, **cuspidata** (Nees, l.c. 205); more robust; leaves glabrous or minutely puberulous to villous; spikelets up to 2 lin. long; glumes sometimes rather pilose; the upper empty valve minutely cuspidate; stamen 1. *E. brevifolia*, var. *adscendens*, Nees, l.c.

SOUTH AFRICA: without precise locality, *Thunberg!*

COAST REGION: Malmesbury Div.; between Groene Kloof and Saldanha Bay, below 500 ft., *Drège!* Zwartland and Paarde Berg, *Ecklon.* Uniondale Div.; Lange Kloof, *Bergius.* Var. β: Vanrhynsdorp Div.; sand hills near Ebenezer, below 500 ft., *Drège!*

WESTERN REGION: Var. β: Little Namaqualand; near the mouth of the Orange River, below 600 ft., *Drège*, 2563!

A glabrous and robust state of this variety was collected by Drummond in the Swan River district, West Australia; it has probably been accidentally introduced there.

17. **E. pusilla** (Nees ex Trin. Phalar. 22); culms tufted, ⅓–1 ft. long, geniculate, ascending, slender, glabrous, 2–3-noded; sheaths lax or the upper tumid, glabrous, smooth; ligules truncate, up to ½ lin. long; blades linear, acute or mucronate, ½–1½ in. by 1–1¼ lin., subrigid, softly hairy above, smooth or scaberulous below; panicle narrow, oblong, ½–2 in. long, sometimes reduced to a raceme, flaccid; rhachis flexuous, filiform; branches 2–3-nate, short, flexuous, capillary, smooth, 1–3-spiculate; pedicels subclavate, scaberulous; spikelets secund, nodding; glumes subequal, lanceolate, acuminate, 2–3 lin. long, membranous, finely 7-nerved, glabrous, margins and back white hyaline; empty valves subequal, lanceolate, acuminate, 2–3 lin. long, produced into a scaberulous purplish awn, up to 1 lin. long, 3-nerved, loosely villous, the lower minutely bearded at the base, the upper with ear-shaped appendages; fertile valve oblong, subobtuse, 2–2½ lin. long, delicately 7-nerved, scantily hairy; lodicules glabrous; stamens 6; anthers ⅜ lin. long; stigmas brushlike. *Nees, Fl. Afr. Austr.* 223; *Steud. Syn. Pl. Glum.* i. 6; *Durand & Schinz, Consp. Fl. Afr.* v. 793.

CENTRAL REGION: Fraserburg Div.; between Klein Quaggas Fontein and Dwaal River, *Burchell*, 1460!

WESTERN REGION: Little Namaqualand; between Pedros Kloof and Lily Fontein, 3000–4000 ft., *Drège!* between Buffels (Kousies) River and Silver Fontein, near Ookiep, *Drège!* Hazenkraals River, 2000 ft., *Drège!* Kamies Bergen, *Ecklon.* Mouth of the Orange River, *Ecklon.*

18. **E. calycina** (Sm. Pl. Ic. Ined. t. 33); perennial, tufted, rarely stoloniferous; culms usually geniculate, slender, simple or scantily branched, 1–2 ft. long, smooth, glabrous, very rarely minutely villous below, 4–6-noded; sheaths glabrous, rarely reversedly

pubescent, rather tight; ligules very short, truncate, denticulate, ciliate; blades linear from a clasping, often denticulate and ciliate base, long and gradually tapering or shortly acute, rarely subobtuse, 1–4 in. by 1–3 lin., flaccid or rigid, flat or involute to setaceous, glaucous, scaberulous, glabrous or hairy, margins sometimes wavy, scabrid; panicle very narrow, nodding, 3–9 in. long, subsecund; rhachis flexuous, smooth; branches in distant semiwhorls, very unequal, the longest rarely more than 1 in. long, simple or scantily branched, spreading or erect, subcapillary, flexuous, smooth; spikelets pallid, rarely purplish, oblong, 2½–3 lin. long; glumes subequal, narrow-oblong, acute or subobtuse, 7-nerved; empty valves unequal, loosely villous, the lower very narrow, linear-oblong, acute, as long as the lower glume or shorter, sub-5-nerved, shortly bearded at the base in front, the upper oblong, obtuse, mucronate (mucro up to ¾ lin. long), as long as the upper glume or longer, 5-nerved, with 2 large semilunar appendages at the base, beardless; fertile valve oblong, obtuse, slightly shorter than the upper empty valve, glabrous or scantily hairy, obscurely 5–7-nerved; lodicules glabrous; stamens 6; anthers 1½ lin. long; stigmas brush-like; grain 1½ lin. long. *Thunb. Prodr.* 188; *Fl. Cap. ed. Schult.* i. 335; *Sw. in Trans. Linn. Soc.* vi. 53, *t.* 4, *fig.* 5; *Schrad. in Schult. f. Syst.* vii. 1372; *Kunth, Enum. Pl.* i. 13; *Nees in Linnæa,* vii. 336; *Trin. Phalar.* 24; *Steud. Syn. Pl. Glum.* i. 7; *Durand & Schinz, Consp. Fl. Afr.* v. 791. *E. geniculata, Thunb. Prodr.* 192; *Fl. Cap. ed. Schult.* 336; *Sw. l.c.* 55, *t.* 4, *fig.* 6; *Schrad. l.c., and in Schult. f. l.c.* 1373; *Kunth, l.c.; Steud. l.c.; Durand & Schinz, l.c.* 792. *E. paniculata, Poir. Encyc. Suppl.* ii. 542 (*by error*). *E. adscendens, Schrad. l.c. E. laxiflora, Schrad. l.c.* 2078, *and in Schult. f. l.c.* 1373; *Kunth, l.c.; Steud. l.c.; Durand & Schinz, l.c.* 792. *E. auriculata, Steud. in Flora,* 1829, 491. *E. ovata, Nees in Linnæa,* vii. 336; *Fl. Afr. Austr.* 203; *Kunth, l.c.; Steud. Syn. Pl. Glum.* i. 7; *Durand & Schinz, l.c.* 793. *E. undulata, Nees ex Trin. Phalar.* 24; *Nees, Fl. Afr. Austr.* 208. *Aira capensis, Linn. f. Suppl.* 108. *Melica geniculata, Thunb. Prodr.* 21. *M. festucoides, Lichtenst. ex Trin. Phalar.* 24. *Trochera calycina, Beauv. Agrost.* 62, *t.* 12, *fig.* 4.

VAR. β, **versicolor** (Stapf); densely cæspitose; culms erect, usually 1½–3 ft. long, rigid; leaves mostly convolute, rigid, up to 1 ft. long; spikelets 3–4½ lin. long, acute, the lower empty valve slightly exceeding the lower glume; tips of the upper empty valve almost acute, mucronate, exceeding the upper glume by 1 lin. or more. *E. versicolor, Schrad. in Goett. Gel. Anz.* iii. (1821) 2078, *and in Schult. f. Syst.* vii. 1374; *Kunth, Enum. Pl.* i. 14; *Trin. Phalar.* 17; *Nees, Fl. Afr. Austr.* 212; *Steud. Syn. Pl. Glum.* i. 6; *Durand & Schinz, Consp. Fl. Afr.* v. 794. *E. stricta, Nees ex Trin. l.c.; Nees, l.c.* 214; *Steud. l.c.; Durand & Schinz, l.c.*

SOUTH AFRICA: without precise locality, *Thunberg! Mund & Maire!*

COAST REGION: Vanrhynsdorp Div.; Ebenezer, by the Olifants River, *Drège.* Clanwilliam Div.; Wupperthal, 1500–2000 ft., *Drège.* Malmesbury Div.; Riebeks Kasteel, *Drège.* Cape Div.; Lion Mountain, *Ecklon,* 909! Kloof between the Lions Head and Table Mountain, *Burchell,* 256! common on rocks of Lions Head, *Wolley Dod,* 3528! (a state with setaceous leaves); saddle of

Lions Head, *Wolley Dod*, 3101! Capetown, *Harvey*, 138! 139! 140! Cape Flats at Doorn Hoogte, *Zeyher*, 1845! near Sea Point, *MacOwan, Herb. Aust. Afr.*, 1792! Simons Bay, *Wright*, 192! by Raapenberg Vley, *Wolley Dod*, 2752! common all over Simons Town Hill, *Wolley Dod*, 3005! Paarl Div.; Paarl, 500–1000 ft., *Drège*. Tulbagh Div.; Tulbagh Waterfall, *Drège*. Worcester Div.; Hex River Valley, 1500 ft., *Drège*; *Wolley Dod*, 3715! Swellendam Div; hills near Riet Kuil and Grootvaders Hoek, 2000–3000 ft., *Zeyher*, 4574! Riversdale Div.; hills near Zoetemelks River, *Burchell*, 6778! Mossel Bay Div.; Gauritz River, *Ecklon*. Uitenhage Div.; Uitenhage, *Zeyher!* Zwartkops River and near Addo, *Ecklon*. Port Elizabeth Div.; Port Elizabeth, *E.S.C.A. Herb.*, 102! Bathurst Div.; between Blue Krantz and Kaffir Drift Military Post, *Burchell*, 3693! Albany Div.; near Grahamstown, *MacOwan*, 1284a! King Williamstown Div.; Amatola Mountains, *Buchanan*, 48! Var. β: Uitenhage Div.; Grass Ridge, *Ecklon & Zeyher*.

CENTRAL REGION: Prince Albert Div.; Gamka River, *Mund & Maire!* Var. β: Graaff Reinet Div.; Graaff Reinet, 2900–4300 ft., *Bolus*, 676! 706! Fraserburg Div.; between Karee River and Klein Quaggas Fontein, *Burchell*, 1414! Carnarvon Div.; Klip Fontein, *Burchell*, 1526!

WESTERN REGION: Vanrhynsdorp Div.; Karee Bergen, 1500 ft., *Schlechter*, 8212! Little Namaqualand; between Kous Mountain, Silver Fontein and Koper Berg, 2000–3000 ft., *Drège!* between Kousies River and Orange River, *Ecklon*. Var. β: Little Namaqualand; between Pedros Kloof and Lily Fontein, 3000–4000 ft., *Drège*, 2567!

KALAHARI REGION: Basutoland; Leribe, *Buchanan*, 218!

EASTERN REGION: Natal; sand-dunes near the mouth of the Umlazi River, *Krauss*, 414; Clairmont, 50 ft., *Wood*, 7261!

A very polymorphic species of which Nees distinguishes 6 varieties and several subvarieties; but the characters used by him are so uncertain that I find it useless to retain his subdivisions.

19. **E. subspicata** (Stapf); perennial; culms geniculate, ascending or suberect from and fascicled on a very slender rhizome, 2 ft. high, slender, glabrous, smooth, 5–8-noded, simple or branched at the base, lower internodes short and enclosed, the others exserted; sheaths rather tight, except the uppermost, glabrous, striate, lower scarcely persistent; ligule a finely ciliolate rim; blades linear from a clasping ciliate base, tapering to an acute point, 1–5 in. by $2\frac{1}{2}$–$3\frac{1}{2}$ in., flat or almost so, rigid, glabrous, closely striate, margins smooth or nearly so; panicle linear, contracted, spike-like, erect, 3–4 in. long; rhachis slender, smooth, somewhat wavy; branches (or pedicels) in semiwhorls, unequal, erect, the longest $\frac{1}{2}$–$\frac{3}{4}$ in. long, and up to 9-spiculate, filiform, usually scaberulous or minutely hispid; pedicels very unequal, lateral short; spikelets oblong, 3–4 lin. long, light green; glumes subequal, oblong, acute, glabrous, 5-nerved, as long or almost as long as the spikelet, light green on the back; margins whitish, tips sometimes purplish; barren valves very similar, equal or subequal, linear-oblong, truncate, finely rough along the margins, otherwise transversely rugose almost all along, light green except the often purplish tips, 7- (or sub-7-) nerved, lower bearded above the base, upper with a large semilunar appendage on each side of the shortly contracted base and a short beard above it; fertile valve oblong, truncate, slightly shorter than the barren valves, glabrous, finely granular, 5-nerved; pale closely 2-nerved; lodicules obliquely obovate, broad, up to 1 lin. long; stamens 6; anthers $1\frac{1}{2}$ lin. long.

COAST REGION: Cape Div.; without precise locality, *Harvey*, 318! by a stream from Retreat to Muizenberg Vley, *Wolley Dod*, 3519! Riversdale Div.; by the Zoetemelks River, *Burchell*, 6712! George Div.; on the Post Berg (Cradock Berg) near George, *Burchell*, 5974!

20. E. Rehmannii (Stapf in Kew Bulletin, 1897, 288); perennial, herbaceous; culms geniculately ascending from a sometimes prostrate and often copiously branched base, 1–2 ft. long, slender, glabrous, rarely minutely villous below; flowering branches 6–8-noded; sheaths tight or the lower at length loosened, glabrous or reversedly pubescent to finely villous, often scantily and very finely bearded at the mouth; ligules very short, ciliolate; blades linear, tapering to an acute point, 1–4 in. by 1–3 lin., usually flat, often spreading, glabrous and smooth, or scaberulous to finely villous above; panicle erect, very narrow, 1–4 in. long, subsecund, reduced to a scanty raceme; rhachis filiform, flexuous, glabrous, smooth below; pedicels solitary or 2-nate, subcapillary, glabrous, smooth or almost so, unequal, up to 4 lin. long, flexuous; spikelets nodding, oblong, 2½–3 lin. long, pallid; glumes equal, lanceolate-oblong, acute to mucronate-acuminate, glabrous, 5–7-nerved, slightly shorter than the valves or equalling them; barren valves oblong, obtuse or truncate, lower slightly shorter, narrower, 5–7-nerved, with a callous ridge and a very minute beard at the base, upper 7-nerved, transversely rugose in the upper part, with a large semilunar appendage and a short beard on each side of the base; fertile valve subobliquely oblong, truncate, intermediate in length between the two preceding valves, glabrous, 7-nerved with a knob on each side of the base; lodicules glabrous; stamens 6; anthers 1 lin. long. *E. aphylla, Nees, Fl. Afr. Austr.* 207 (*in part*). *E. ramosa, var. Rehmannii, Hack. MS.*

VAR. β, **filiformis** (Stapf); culms filiform; blades very narrow, mostly finely setaceous; panicle reduced to 1–2 spikelets.

COAST REGION: Cape Div.; without precise locality, *Harvey*, 329! 335! Table Mountain, *Ecklon*, 914 partly! Cape Peninsula, in a clearing near Newlands Reservoir, *Wolley Dod*, 3121! and in a wood above Newlands House, *Wolley Dod*, 2385! common by the second waterfall on Devils Mountain, *Wolley Dod*, 3118! George Div.; Outeniqua Mountains, Montagu Pass, *Rehmann*, 74! Var. β: Cape Div.; top of Constantia Berg, *Wolley Dod*, 3477! Stellenbosch Div.; Lowrys Pass, 1500 ft., *Schlechter*, 7285! Caledon Div.; Houw Hoek Mountains, 3000 ft., *Schlechter*, 9417!

E. aphylla, var. *filiformis*, Nees, from the Klein River Bergen (Paarl Div.) is probably identical with Schlechter's plant from Lowrys Pass.

21. E. ramosa (Thunb. Prodr. 192); suffrutescent, 1–4 ft. high, much branched; culms stout, firm, more or less woody, up to 2½ lin. thick, green, glabrous, smooth, many-noded, middle internodes often up to 6 in. long; branches solitary or fascicled, flowering branches 2–4-noded; leaves almost wholly reduced to broad dry more or less leathery glabrous persistent finally spreading sheaths, rarely with a minute setaceous blade; ligules very short, obtuse, glabrous; panicle erect, very narrow, linear, 2–4 in. long, often reduced to a raceme; rhachis filiform, glabrous, scabrid; branches solitary or paired or fascicled, unequal, the longest ½–1 in. long, 3–1-spiculate; pedicels puberulous, scabrid, 1–3 lin. long; spikelets erect, oblong, 3–3½ lin. long; glumes equal, usually distinctly shorter than the spikelet, lanceolate-oblong, acute, glabrous, pallid, 5- to sub-9-nerved; barren valves very similar, oblong, obtuse or truncate with usually purplish tips, scabrid, lower slightly shorter, narrower, 7-nerved, with a callous ridge and a minute beard at the base, upper often minutely cuspidate, 7- to sub-11-nerved, sometimes obscurely transversely rugose with a large semilunar appendage and a short beard on each side of the base; fertile valve subobliquely oblong, truncate, intermediate in length between the 2 preceding valves, glabrous, 7–9-nerved, with a knob on each side of the base; lodicules glabrous; stamens 6; stigmas short, plumose. *Thunb. Fl. Cap. ed. Schult.* 335; *Sw. in Trans. Linn. Soc.* vi. 49, *t.* 3, *fig.* 3; *Schrad. in Goett. Gel. Anz.* iii. (1821) 2077, *and in Schult. f. Syst.* vii. 1370; *Kunth, Enum. Pl.* i. 12; *Trin. Phalar.* 25, *and in Mém. Acad. Pétersb. sér.* 6, v. 71 (*in part*); *Nees, Fl. Afr. Austr.* 205 (*in part*); *Steud. Syn. Pl. Glum.* i. 7; *Durand & Schinz, Consp. Fl. Afr.* v. 793. *E. digyna, Thunb. Prodr.* 192.

COAST REGION: Worcester Div.; Dutoits Kloof, 3000–4000 ft., *Drège!* Swellendam Div.; on a peak near Swellendam, *Burchell*, 7312! Riversdale Div.; Zoetemelks River, *Thunberg!* lower part of the Lange Bergen, near Kampsche Berg, *Burchell*, 7011! Knysna Div.; Plettenberg Bay, *Mund & Maire!* Uitenhage Div.; by a rivulet between Leadmine River and Van Stadens River, *Burchell*, 4648!

22. E. aphylla (Schrad. in Goett. Gel. Anz. iii. (1821) 2077, and in Schult. f. Syst. vii. 1369); suffrutescent, 1–2½ ft. high, very much branched; culms slender, firm, woody below, glaucous or light green, glabrous, very smooth, middle internodes often 6 in. long; branches solitary, sometimes fascicled, long, the flowering 2–3-noded and very slender; leaves usually reduced to narrow short or long dry rather firm glabrous persistent sheaths, tight in the flowering branches, often slipping from the culms, but erect in the lower parts, rarely bearing distinct blades; ligules very short, obtuse, glabrous; blades (if present) short, setaceous or linear, flat, up to 1½ in. by scarcely ½ lin., glabrous; panicle erect, straight or flexuous, narrow, more or less linear, often secund, 1–3 in. long, nearly always reduced to a raceme; rhachis filiform, usually very fine, glabrous, scaberulous; pedicels fascicled or geminate or the upper solitary, finely filiform, thickened upwards, densely puberulous, unequal, up to 4 lin. long, often flexuous; spikelets often nodding, oblong, 3–4 lin. long, pallid; glumes equal, lanceolate, acute, glabrous, 5-nerved, slightly exceeding the valves; barren valves oblong, obtuse or truncate, scaberulous upwards, lower slightly shorter, narrower, 7-nerved, with a callous ridge and a minute beard at the base, upper often minutely cuspidate, 7- to sub-9-nerved, often obscurely transversely rugose with a large semilunar appendage and a short beard on each side of the base; fertile valve subobliquely oblong, truncate, intermediate in length between the 2 preceding valves, glabrous, 7-nerved with a knob on each side of the base; lodicules glabrous; stamens 6; anthers $1\frac{1}{2}$ lin. long; stigmas short, plumose. *Nees, Fl. Afr. Austr.* 207 (*in part*); *Steud. Syn. Pl. Glum.* i. 7 (*in part*); *Durand & Schinz, Consp. Fl. Afr. Austr.* v. 791 (*in part*). *E. ramosa, Trin. Phalar.* 25, *and in. Mém. Acad. Pétersb. sér.* 6, v. 71 (*in part*); *Nees, Fl. Afr. Austr.* 205 (*in part*).

VAR. β, **fasciculata** (Stapf); scarcely $\frac{1}{2}$ ft. high; branches fascicled, very numerous, ascending, divided, ultimate divisions (flowering branches) about 2 in. long; sheaths broad, short, open, bladeless or almost so; panicles almost reduced to racemes, about 1 in. long; rhachis and pedicels smooth; spikelets $2\frac{1}{4}$–$2\frac{1}{2}$ lin. long.

COAST REGION: Vanrhynsdorp Div.; Giftberg, 1500–2500 ft., *Drège!* Cape Div.; Table Mountain, *Ecklon*, 914 partly! *Drège! Milne*, 248! *Burchell*, 543! *MacGillivray*, 392! *MacOwan, Herb. Aust. Afr.*, 1692! Simons Bay, *Wright!* Paarl Div.; Paarl Mountains, 1000–1500 ft., *Drège!* Tulbagh Div.; Tulbagh Waterfall, *Ecklon & Zeyher!* Worcester Div.; Bains Kloof, 2800 ft., *Schlechter*, 9180! Caledon Div.; Genadendal, to the top of the mountains, *Drège!* Alexandria Div.; Zuurberg Range, between Enon and Drei Fontein, 2000–3000 ft., *Drège!* Var. β: Caledon Div.; tops of mountains of Baviaans Kloof, near Genadendal, *Burchell*, 7725!

The variety β *fasciculata* seems to represent a modified state of *E. aphylla* or of *E. ramosa*, probably due to the conditions of habitat. I have placed it with *E. aphylla* on account of the slender culms and the comparatively long glumes; but the sheaths are those of *E. ramosa*. The smooth rhachis and pedicels distinguish it from typical *E. aphylla*, as well as *E. ramosa*.

Ecklon, 914, is represented in the Kew Herbarium by 2 different plants: *Ehrharta aphylla* and *E. Rehmannii*. Although Schrader's description of *E. aphylla* clearly shows that he meant the suffrutescent bladeless species, Nees, evidently having the other form before him, was misled into taking this for Schrader's *E. aphylla*. Other specimens which he enumerates under *E. ramosa*, but which I have not seen, may be *E. ramosa*, *E. Rehmannii* or *E. aphylla*; they are :—

COAST REGION: Clanwilliam Div.; Ezelsfontein, *Drège.* Tulbagh Div.; Winterhoek Berg, 800–5000 ft., *Ecklon.* Caledon Div.; Caledon, *Ecklon.* Zwart Berg, *Drège.* Swellendam Div.; Cannaland, *Ecklon.* Riversdale Div.; Kochmanns Kloof to Gauritz River, *Ecklon.* Uitenhage Div.; Van Stadens Berg, 3000 ft., *Drège.*
WESTERN REGION: Little Namaqualand; Rood Berg, 4500–5000 ft., *Drège.*

23. **E. barbinodis** (Nees ex Trin. Phalar. 20); perennial; culms geniculate, ascending, $1\frac{1}{2}$–2 ft. long, glabrous, smooth, many-noded, lower internodes short, uppermost up to 1 ft. long; sheaths tight, glabrous except the reversedly villous nodes of the lower; ligules extremely short, ciliolate; blades linear, subacute or gradually tapering to a fine point, up to $1\frac{1}{2}$ in. by $1\frac{1}{2}$ lin., rather firm, flat, glabrous or scantily hairy, margins scabrid; panicle narrow, linear, erect, 3 in. long, usually reduced to a raceme, loose; rhachis glabrous; branches in rather distant semiwhorls or pairs, unequal, up to $\frac{1}{2}$ in. long, usually 1-spiculate, capillary, flexuous, glabrous; spikelets ovate-oblong, $4\frac{1}{2}$–6 lin. long; glumes ovate to lanceolate, acuminate, glabrous, sometimes purple, the lower $2\frac{1}{2}$–$3\frac{1}{2}$ lin. long, 5–7-nerved, the upper $3\frac{1}{2}$–4 lin., 7–9-nerved; lower empty valve linear-oblong, obliquely truncate, mucronate, $3\frac{1}{2}$–$4\frac{1}{2}$ lin. long, strongly compressed,

5–7-nerved, bearded at the base in front and behind, keel long-ciliate, margins ciliolate; upper empty valve similar, but longer, obscurely transversely rugose, scantily hairy below, obscurely ridged and bearded on each side of the base; fertile valve linear-oblong, truncate, almost as long as the preceding valve, glabrous, 7-nerved; lodicules glabrous, 1 lin. long; stamens 6; anthers 2 lin. long. *Nees, Fl. Afr. Austr.* 215; *Steud. Syn. Pl. Glum.* i. 6; *Durand & Schinz, Consp. Fl. Afr. Austr.* v. 791.

COAST REGION: Vanrhynsdorp Div.; near Ebenezer, among shrubs, 300 ft., *Drège!* Karee Bergen, 1500 ft., *Schlechter*, 8281!

WESTERN REGION: Little Namaqualand; Kamies Bergen, Vallei Fontein, and near the mouth of the Orange River, *Ecklon*. Kaus Mountain, 2500 ft., Goedemans Kraal, and near Rustbank, *Drège*.

24. **E. gigantea** (Thunb. Prodr. 192); perennial, stoloniferous; young stolons stout, densely covered with villous scales; culms erect, stout, simple, up to 6 ft. high, terete, green, glabrous, except below, about 3-noded, internodes exserted, base covered with about 6 tightly adpressed imbricate firm tomentose ultimately glabrescent bladeless sheaths; blade-bearing sheaths tight, terete, lowest finely tomentose, following glabrous, long (uppermost over 1 ft.); ligules very short, ciliate; blades linear, from an almost clasping base, acute, up to 4 in. by 3 lin., firm, involute, glabrous, smooth; panicle linear, erect, 1 ft. long; rhachis slender, flexuous in the upper part, smooth below, scaberulous above; branches in rather distant contracted fascicles, longest up to 2 in. long and up to 6-spiculate, others much shorter and few- to 1-spiculate, subcapillary, very flexuous, scaberulous, purplish-black; lateral pedicels 3–4 lin. long; spikelets nodding, ovate-oblong, 5¼–6 lin. long; glumes subequal, oblong, subacute, 3½–4 lin. long, scarious, sometimes purplish, glabrous, except at the ciliolate margins, lower 1- to sub-3-nerved, upper 3- to sub-5-nerved; barren valves similar, coriaceous, long villous and ciliate along the margins, awned, awn 2 lin. long and scabrid; lower 4¼–4½ lin. long, linear, 3- to sub-5-nerved, bearded at the base in front and behind, upper 5¼–6 lin. long, linear-oblong, 5- to sub-7-nerved, bearded on each side above the shortly contracted base; fertile valve oblong, truncate, apiculate, 5 lin. long, subcoriaceous, 7-nerved, keel ciliate above; pale 4½ lin. long, closely 2-nerved; lodicules broad, 2-lobed, glabrous, 1 lin. long; stamens 6; anthers 3 lin. long. *Thunb. Fl. Cap. ed. Schult.* 336; *Sw. in Trans. Linn. Soc.* vi. 58, *t.* 4, *fig.* 8; *Schrad. in Goett. Gel. Anz.* iii. (1821) 2079, *and in Schult. f. Syst.* vii. 1375; *Kunth, Enum. Pl.* i. 14; *Trin. Phalar.* 16, *and in Mém. Acad. Pétersb. sér.* 6, v. 62. *Melica gigantea, Thunb. Prodr.* 21; *Willd. Spec. Pl.* i. 382. *Aira villosa, Linn. f. Suppl.* 109.

VAR. β, **Neesii** (Stapf); culms shorter, more slender, 4–5-noded; panicles up to 6 in. long; spikelets 4½–5½ lin. long; fertile valve sometimes scantily hairy on the sides. *E. gigantea, Nees, Fl. Afr. Austr.* 216; *Steud. Syn. Pl. Glum.* i. 5; *Durand & Schinz, Consp. Fl. Afr.* v. 792 (*excl. syn.*).

Var. γ, stenophylla (Stapf); culms subbulbous and densely covered with imbricate bladeless sheaths at the base as in the type, 2–3 ft. high, slender, hard, 4–6-noded, branched from or above the base; branches long, erect, like the main culm; lower sheaths purplish; blades very narrow, setaceously convolute, 1–6 in. long; spikelets 4½–5½ lin. long; glumes few-nerved, upper 3–4 lin. long.

South Africa: without precise locality, *Thunberg!*

Coast Region: Var. β: Malmesbury Div.; near Riebecks Castle, among shrubs, 500–1000 ft., *Drège!* Var. γ: Tulbagh Div.; Tulbagh Waterfall, 1200 ft., *Schlechter*, 9058! Mossel Bay Div.; on dry hills near the Gauritz River, *Burchell*, 6456! between Great Brak River and Little Brak River, *Burchell*, 6160! Uitenhage Div.; on the sand-hills near the Zwartkops River, *Ecklon & Zeyher*, 167!

Nees indicates "*Ehrharta gigantea*" also from Vanrhynsdorp Div.; near the Olifants River and Brackfontein (*Ecklon*); on the Giftberg, 2000–2500 ft., Clanwilliam Div.; in the Ceder Bergen, near Ezelsbank, 3500 ft. (*Drège*); Cape Div., near Capetown (*Ecklon*); and Stellenbosch Div.; Hottentots Holland (*Ecklon*). The description of the spikelets of "*Aira villosa*" by Linnæus f. is altogether faulty and misleading; but a note in Thunberg's herbarium leaves no doubt that he meant *Ehrharta gigantea*, Thunb.

25. E. villosa (Schult. f. Syst. vii. 1374); perennial; innovation-buds rather slender, only the basal scales fugaciously tomentose; culms geniculate, ascending from an often decumbent and rooting base, up to 3 ft. long, slender, hard, perfectly smooth and glabrous, 5–8-noded, branched below, branches long and like the main culm; sheaths rather firm, lowest bladeless, puberulous or tomentose, glabrescent, at length opening and withering away, thereby laying bare the slender internodes, the upper sheaths tight, often slipping from the culms and rolling in, quite glabrous, smooth; ligule a ciliolate rim; blades linear from a clasping ciliate base, acute, spreading, 2–3 in. by 2–3 lin., flat or involute, rigid, subpungent, smooth below, scabrid above and along the margins; panicle linear, erect, 3–4 in. long, usually reduced to a subsecund raceme; rhachis filiform, smooth or scaberulous in the upper part; branches or pedicels finely filiform, lower fascicled, the others 2-nate or solitary, flexuous, 2–3 lin. long, rarely longer, smooth or scaberulous; spikelets often nodding, ovate-oblong, 5–6 lin. long; glumes subequal, oblong, acute, 4–5½ lin. long, scarious, pallid, rarely purple, glabrous except at the ciliolate margins, 5–7-nerved, side-nerves short or long; barren valves similar, subequal, linear-oblong, coriaceous, ciliate along the margins, long and densely villous all over, mucronate, mucro up to 1 lin. long, scabrid, lower slightly shorter and narrower, 5-nerved, bearded at the base in front and behind, upper 7-nerved, bearded on each side above the shortly contracted base; fertile valve linear-oblong, truncate, apiculate, about intermediate between the barren valves, subcoriaceous, 7-nerved, villous; pale, lodicules and anthers as in *E. gigantea*. *Nees in Linnæa*, vii. 338; *Fl. Afr. Austr.* 213; *Trin. Phalar.* 16, *and in Mém. Acad. Pétersb. sér.* 6, v. 62; *Steud. Syn. Pl. Glum.* i. 5; *Durand & Schinz, Consp. Fl. Afr.* v. 794. *E. gigantea, Steud. in Flora*, 1829, 491.

Var. β, maxima (Stapf); culms stouter; sheaths somewhat looser; panicle longer and much more compound; spikelets 6–7½ lin. long; glumes 5½–7 lin. long with 7–9 long nerves.

SOUTH AFRICA: without precise locality, *Thom*, 80! Var. β, *Ecklon & Zeyher*, 409! *Thom!*

COAST REGION: Cape Div.; Camps Bay, *Ecklon! Burchell*, 332! Simons Bay, *Alexander!* sandy slopes at Upper North Battery and common in deep sand all above Simons Town, *Wolley Dod*, 2827! very common in sand on the beach beyond the Mill, *Wolley Dod*, 3229! Var. β: Port Elizabeth Div.; common near Port Elizabeth, *E.S.C.A. Herb.*, 94!

CENTRAL REGION: "Somerset," *Bowker!*

E. villosa and *E. gigantea* are very closely allied, the difference being mainly in the mode of growth and the relative length of the glumes and valves. How far the former is influenced by the conditions of soil will have to be investigated on the spot. Nees also refers to *E. villosa* specimens collected by Ecklon at Tulbagh Waterfall, Cape Agulhas, and near the Zwartkops River.

LXXII. PHALARIS, Linn.

Spikelets laterally compressed, in contracted more or less spike-like panicles; rhachilla disarticulating above the glumes, not or obscurely produced beyond the terminal floret. *Florets* 3, the lower 2 minute, rudimentary, the uppermost ⚥, enclosed by the glumes. *Glumes* subequal, boat-shaped, keeled, keel often winged. *Empty valves* very small, subulate to lanceolate, membranous, with a callous base, or 1 or both reduced to a minute callous scale; fertile valve thin, ultimately rigid, 5-nerved, awnless, ovate, acute. *Pale* almost as long as the valve, 2-nerved (sometimes obscurely). *Lodicules* 2, hyaline. *Stamens* 3. *Styles* long, distinct; stigmas plumose, exserted from the top of the spikelet. *Grain* much compressed, ovate, free, enclosed by the valve and pale; hilum oblong, short; embryo equalling ¼ the length of the grain.

Annuals or perennials; leaves flat; panicle terminal, stiff, spike-like, subcapitate or interrupted and lobed; pedicels very short.

Species 10, mainly natives of the Mediterranean region, but widely dispersed as weeds; 1 species in the boreal region and in South Africa, and another from California to Chile.

Annual; keels of glumes conspicuously winged ... ... (1) **minor.**
Perennial; keels of glumes not or very obscurely winged ... (2) **arundinacea.**

1. **P. minor** (Retz. Obs. iii. 8); annual; culms tufted, erect or ascending, geniculate, 1–3 ft. long, glabrous, finely striate, 4–7-noded; internodes gradually longer from the base, up to 5 in. long; sheaths shorter than the internodes, the lower tight, the uppermost more or less inflated, striate; ligule scarious, white, obtuse, 1–3 lin. long; blades linear, gradually tapering, 2–6 in. by 1½–3 lin., flaccid, glabrous, smooth or almost so, margins slightly rough; panicle spike-like from subglobose to cylindric, up to 2½ in. long, compact; rhachis and branches glabrous; spikelets obliquely elliptic, 2–2½ lin. long; glumes subequal, 3-nerved, acute, glabrous, white, nerves green, keel serrulate, suddenly contracted below the apex; lower empty valve a minute callous scale; upper somewhat subulate, firmly

membranous, hairy with a callous base, up to $\frac{3}{4}$ lin. long; fertile valve scantily silky; pale ciliate on the back, obscurely 2-nerved or almost nerveless; anthers $\frac{1}{2}$–$\frac{5}{8}$ lin. long; grain $\frac{7}{8}$ lin. long. *Kunth, Enum. Pl.* i. 32; *Trin. Spec. Gram.* i. *t.* 79; *Phalar.* 8; *Nees, Fl. Afr. Austr.* 5; *Steud. Syn. Pl. Glum.* i. 11; *Durand & Schinz, Consp. Fl. Afr.* 795. *P. aquatica, Ait. Hort. Kew. ed.* 1, i. 86 (*not Linn.*); *Thunb. Prodr.* 19; *Trin. Gram. Pan.* 254; *Host, Gram. Austr.* ii. *t.* 39; *Reichb. Fl. Germ. Ic.* i. *t.* 52, *fig.* 1493. *P. capensis, Thunb. Prodr.* 19, *and Fl. Cap. ed. Schult.* i. 106. *P. bulbosa, Desf. Fl. Atl.* i. 55 (*not Linn.*). *P. ambigua, Fig. & De Not. in Mem. Acc. Torin. ser.* 2, xii. 326.

SOUTH AFRICA: without precise locality, *Thunberg!*
COAST REGION: Vanrhynsdorp Div.; Ebenezer, *Drège!* Malmesbury Div.; between Groene Kloof and Saldanha Bay, *Drège!* Cape Div.; near Capetown, *Ecklon*, 975! Mossel Bay Div.; on dry hills on the east side of the Gauritz River, *Burchell*, 6436! Uitenhage Div.; sandy places near Uitenhage, *MacOwan*, 1565! Port Elizabeth Div.; common along the coast, *E.S.C.A. Herb.*, 233! Albany Div.; in cultivated ground near Grahamstown, *MacOwan*, 1565!
EASTERN REGION: Tembuland; Gatberg, 4000 ft., *Baur*, 1148!

A native of the Mediterranean countries; introduced in many other parts of the world.

2. **P. arundinacea** (Linn. Sp. Pl. 55); perennial; rhizome short, præmorse, stoloniferous; stolons with firm scarious sheathing scales; culms erect from a creeping or ascending base, 2–4 ft. long, rooting at the lower nodes, simple or very sparingly branched, firm, glabrous, finely striate, 5–7- or more-noded, internodes gradually longer from the base, up to 1 ft. long; sheaths glabrous, smooth, strongly striate, lower tight, longer than the internodes, upper looser, shorter; ligule scarious, white, obtuse, 1$\frac{1}{2}$–3 lin. long; blades linear to linear-lanceolate, long tapering, $\frac{1}{2}$–1 ft. by 3–8 lin., rigid, glaucous, glabrous, smooth, many-nerved; panicle erect, sometimes nodding, contracted, lobed or spike-like, up to 8 in. long; branches very short, adpressed to the rhachis, or longer (to 1$\frac{1}{2}$ in.) and more or less spreading, copiously and densely branched, glabrous, smooth or scabrid; spikelets ovate-lanceolate, 2$\frac{1}{2}$–3 lin. long; glumes subequal, whitish-green or purplish, acute, 3-nerved, keel not or very obscurely winged, minutely serrulate, nerves raised; empty valves subequal, lanceolate to subulate, obscurely 1-nerved or nerveless, hairy with a callous base, $\frac{3}{4}$ lin. long; fertile valve scantily silky, 1$\frac{1}{2}$ lin. long; pale ciliate on the back; lodicules obliquely ovate-lanceolate; anthers 1$\frac{1}{2}$ lin. long. *Fl. Dan. t.* 259; *Host, Gram. Austr.* ii. *t.* 33; *Schrad. Fl. Germ.* i. 180, *t.* 6, *fig.* 5; *Engl. Bot. t.* 2160, *fig.* 2; *Kunth, Enum. Pl.* i. 33; *Trin. Phalar.* 11; *Steud. Syn. Pl. Glum.* i. 11; *Baill. Hist. Pl.* xii. 169. *P. cæsia, Nees, Fl. Afr. Austr.* 6; *Steud. Syn. Pl. Glum*, i. 10; *Durand & Schinz, Consp. Fl. Afr.* 795.

COAST REGION: Uitenhage Div.; Olifants Hoek Mountain, by the Bosjesman River, in virgin forests (*Ecklon & Zeyher?*); Bedford Div.; Bedford, *Hutton!* Fort Beaufort Div.; by the Kat River, *Zeyher!* by the Great Fish River, *Bowie!* between the Fish River and Fort Beaufort, in copses, 1000 ft., *Drège!* Cathcart Div.; Glencairn, 4800 ft., *Galpin*, 2416! British Caffraria, *Cooper*, 286!

CENTRAL REGION: Somerset Div.; Bosch Berg, on the banks of ditches, 2500 ft., *MacOwan*, 1640! Albert Div.; Mooi Plaats, 4500–5000 ft., *Drège!* Aliwal North Div.; at the foot of the Witte Bergen, 4500–5000 ft., *Drège!*
EASTERN REGION: by the Mooi River, 4000 ft., *Wood*, 4097! Van Reenens Pass, *Wood*, 7215 partly!

Also in Europe.

P. aquatica, Thunb. Prod. 19, is perhaps this plant. It is not mentioned in his Flora Capensis, nor is it in his herbarium.

LXXIII. MELICA, Linn.

Spikelets in spike- or raceme-like or open panicles, laterally or dorsally compressed, or subterete, jointed (sometimes imperfectly) on their pedicels or continuous with them; rhachilla tardily disarticulating above the glumes, readily between the fertile valves. Lower 1 or 2 (rarely 3) *florets* ⚥, the following 2–3 barren, small, embracing each other and forming a clavate or oblong body. *Glumes* 2, membranous, hyaline or scarious, obtuse or acute, 3–5-nerved, or the upper 7-nerved. Fertile *valves* firmly membranous except at the hyaline margins and tips, awnless, 7–9-nerved, nerves evanescent below the tips; callus minute, obtuse. *Pales* shorter than the valves, 2-keeled. *Lodicules* 2, small, truncate, quite connate. *Stamens* 3. *Ovary* glabrous; styles distinct, short; stigmas laterally exserted, finely plumose. *Grain* enclosed by the more or less hardened (chartaceous) valve and the pale, free, oblong, semi-terete to subterete; hilum a fine line as long as the grain; embryo small.

Perennial; blades flat or convolute; ligules hyaline; panicles open, spike-like or almost reduced to a raceme, many- to few-spiculate, often secund; spikelets more or less scarious, often vividly coloured, nodding on capillary pedicels, the tips of which are usually strongly incurved.

About 40 species, mainly in the northern temperate zone, a few in the temperate regions of the southern hemisphere.

Valves hairy only along the sides:
  Leaves softly hairy all over; fertile floret 1, shorter than the upper glume ... ... (1) **Bolusii.**
  Leaves glabrous; fertile florets 2, equalling or exceeding the upper glume:
    Culms 2–3 ft. long, simple or scantily branched above the base; leaves very scabrid:
      Spikelets 3–5 lin. long; lower glume ovate, upper oblong, subacuminate, rather narrow ... ... (2) **racemosa.**
      Spikelets 2½–3 lin. long; glumes very broad, subacute or subcuspidate ... (3) **ovalis.**
    Culms 3–6 in. long, profusely branched, in dense fascicles from a long prostrate very slender base; leaves scaberulous; spikelets 2½–3 lin. long; upper glume acuminate ... (4) **pumila.**
Valves hairy all over:
  Culms profusely branched, in dense fascicles from a long prostrate base, up to ½ ft. long; spikelets up to 4 lin. long ... ... (5) **Neesii.**
  Culms simple, suberect or ascending from a short base; spikelets 5½–7 lin. long ... (6) **decumbens.**

1. **M. Bolusii** (Stapf); culms scantily fascicled, ascending from a weak prostrate very slender base, up to 1 ft. long or more, glabrous, scaberulous below the panicle, otherwise smooth, rather many-noded, internodes (except the uppermost) enclosed; leaves softly and spreadingly hairy all over; sheaths tight, striate, rather thin; ligules up to 1 lin. long; blades linear, tapering to an acute point, 2–4 in. by 1–1½ lin., flat, flaccid, or somewhat stiff, glaucous, strongly striate; panicle linear, 8–12-spiculate, secund; branches up to 7 lin. long, about as long as the internodes, 3–1-spiculate, erect, subcapillary, supported by ovate or oblong hyaline bracts, tips of pedicels minutely villous; spikelets pallid, 3 lin. long, with 1 fertile floret; glumes unequal, the lower broadly ovate, acute, 2 lin. long, subhyaline, the upper narrowly oblong, subacuminate, firmer, often light purplish, 3 lin. long, hyaline margins very narrow; valves shorter than the upper glume, fertile valve oblong, obtuse, almost smooth on the back, densely hairy along the sides, hairs 1–1¼ lin. long; body of barren valves clavate, glabrous; anthers almost ½ lin. long.

CENTRAL REGION: Graaff Reinet Div.; on rocks, below the summit of the Compass Berg, 8300–8500 ft., *Bolus*, 1985!

2. **M. racemosa** (Thunb. Prodr. 21); culms ascending from a usually long very slender wiry sometimes procumbent base, simple or branched below, 2–3 ft. long, glabrous, smooth or scabrid below the panicle, many-noded, internodes (except the uppermost) mostly enclosed; leaves glabrous, scabrid, rather crowded above the base, the lowest more or less reduced; sheaths tight, striate; ligules up to 1 lin. long; blades linear, tapering to a fine point, 3–6 in. long, ½–1½ lin. wide when expanded, usually convolute, subglaucous; panicle very narrow, 3–9 in. long, erect or nodding, more or less secund; branches mostly solitary, distant, erect or suberect, branched or more often simple, filiform to capillary, flexuous, often much shorter than the internodes; pedicels ½–1½ lin. long, tips thickened, pubescent; spikelets pallid, rarely slightly purplish, 3–5 lin. long; fertile flowers 2; glumes unequal or subequal, 5-nerved, the lower hyaline, ovate, acute, 3–3½ lin. long, the upper oblong, firmer, acuminate, nerves rather close and prominent, slightly scabrid, 3½–4½ lin. long; valves slightly exceeding or equalling the glumes, the fertile oblong, obtuse or minutely truncate or emarginate, 7–9-nerved, nerves rather prominent, scabrid, sides only hairy; hairs 2 lin. long; body of barren valves clavate, glabrous, scaberulous; anthers 1 lin. long; grain 1–1¼ lin. long, semiterete. *Thunb. Fl. Cap. ed.* 1, 417; *ed. Schult.* 111; *Kunth, Enum.* i. 378; *Durand & Schinz, Consp. Fl. Afr.* v. 898. *M. Caffrorum, Schrad. in Goett. Gel. Anz.* iii. (1821) 2073, *and in Schult. Mant.* ii. 296; *Kunth, Enum. Pl.* i. 376; *Nees in Linnæa*, vii. 327, *and Fl. Afr. Austr.* 418; *Steud. Syn. Pl. Glum.* i. 289; *Durand & Schinz, Consp. Fl. Afr.* v. 896.

COAST REGION: Cape Div.; between Lions Head and Table Mountain, *Burchell*, 251! near Lion Battery, *Wolley Dod*, 3107! East side of Table Mountain, near Tokay, *Ecklon*. Capetown, *Harvey*, 137! in and about Camps

Bay, *Wolley Dod*, 3106! lower slopes of Smithwinkel Bay, *Wolley Dod*, 2977! Riversdale Div.; near the Zoetemelks River, *Burchell*, 6631! Uitenhage Div.; amongst shrubs on sand hills near the Zwartkops River, *Ecklon & Zeyher*, 253! in virgin forest on Olifants Hoek, *Ecklon*. Port Elizabeth Div.; near Port Elizabeth, *Ecklon*, *E.S.C.A. Herb.*, 132! Alexandria Div.; near Addo, in woods and grassy places, 1000–2000 ft., *Drège!* Albany Div.; near Grahamstown, 1000–2000 ft., *Drège!* in the scrub near Brand Kraal, 2000 ft., *MacOwan*, 1282! in woods near the Kowie River, *MacOwan*, 705! Albany, *Zeyher!* King Williamstown Div.; Amatola Mountains, *Buchanan*, 53! Komgha Div.; near Komgha, *Flanagan*, 432! Stockenstrom Div.; in mountain woods, by the sources of the Kat River, near Philipton, *Ecklon*. Queenstown Div.; on rocks of Table Mountain, 5000–6000 ft., *Drège*.

CENTRAL REGION: Middelburg Div.; between Compass Berg and Rhinoster Bergen, 4500–5000 ft., *Drège*.

KALAHARI REGION: Basutoland, *Cooper*, 3358!

EASTERN REGION: Tembuland; Bazeia, at Klip Krantz, 2000 ft., *Baur*, 410! Griqualand East; mountain sides around Clydesdale, *Tyson*, 2863! Natal; Berea, near Durban, by the roadside, *Wood*, 3926! Merebank, near Durban, *Wood*, 7258! Vernon, *Buchanan*, 155! Weenen County, 4000 ft., *Wood*, 3590!

Very like the Mediterranean *M. ciliata*, but differing in the rough leaves, the less acuminate upper glume and the more obtuse valves.

3. **M. ovalis** (Nees, Fl. Afr. Austr. 417); culms ascending from a very slender, wiry base, simple, 2 ft. long, glabrous or slightly scabrid below the panicle, otherwise smooth, many-noded, internodes (except the lower) exserted; leaves glabrous, scabrid; sheaths rather tight, striate; ligule up to ¾ lin. long; blades linear, acute, 2–3 in. by ½ lin., convolute, rather rigid, erect; panicle linear, 2–5 in. long, erect or nodding; branches simple, distant, 7–1-spiculate, the longest up to ¾ in. long, subcapillary, flexuous, the lower supported by a hyaline ovate bract; pedicels ½–2 lin. long, tips slightly thickened, pubescent; spikelets pallid, ovoid to ellipsoid, 2½–3¼ lin. long; fertile florets 2; glumes scarious, 5-nerved, subacute or obscurely cuspidate, the lower very broadly ovate, 2–2½ lin. long, the upper broadly oblong, 2½–3 lin. long; valves slightly or scarcely exceeding the glumes, the fertile broadly oblong, obtuse, truncate or emarginate, up to 3¼ lin. long, 7-nerved, scaberulous, sides long and densely hairy, hyaline tips short, broad, nerves somewhat prominent; body of barren valves clavate, scaberulous and sparingly and shortly hairy; anthers ½ lin. long. *Steud. Syn. Pl. Glum.* i. 289; *Durand & Schinz, Consp. Fl. Afr.* v. 899.

COAST REGION; Queenstown Div.; Stormberg Range, 5000–6000 ft., *Drège!*

4. **M. pumila** (Stapf); culms much branched, densely fascicled from a long very slender prostrate base, 3–6 in. long, glabrous, rather many-noded, internodes enclosed except the uppermost; leaves glabrous, finely scaberulous; sheaths tight, striate; ligules ½ lin. long; blades linear, acute, very narrow, 1–1½ in. long, setaceously convolute, rather rigid; panicle reduced to a loose 4–6-spiculate secund raceme, 1–1½ lin. long; pedicels erect, tips curved; spikelets nodding, 2½–3 lin. long, pallid; fertile florets 2; glumes subequal, both scarious, the lower ovate, acute, up to 2 lin. long, the upper oblong,

acuminate, slightly longer; valves equalling or slightly exceeding the glumes, the fertile oblong, subobtuse, almost smooth on the back, hairy along the sides; hairs 1 lin. or slightly more long, innermost side-nerves very faint, body of barren valves clavate to oblong, glabrous; anthers over ½ lin. long.

CENTRAL REGION: Beaufort West Div.; near Weltevrede, *Drège*, 752 in part!

Two different plants have been distributed by Drège under the name of *Melica Caffrorum* γ *decumbens* (a). One, which is represented in Bentham's herbarium and in Drège's own herbarium, now in Lübeck, is *M. Neesii*, whilst the other, represented in the Hookerian herbarium, is *M. pumila.* Nees quotes under his var. *decumbens* of *M. Caffrorum* two other specimens from the mountains between Ganze Fontein and Bock Poort (Beaufort West Div.) and from Stylkloof (Richmond Div.); but, not having seen either, I do not know if they belong to *M. pumila* or *M. Neesii. M. pumila* differs from *M. racemosa* mainly in the habit, the dwarfed stature, the short scarcely rough leaves and the much smaller spikelets.

5. **M. Neesii** (Stapf); culms much branched, densely fascicled from a long slender wiry prostrate base, ½ ft. long, glabrous, rather many-noded, internodes enclosed except the uppermost; leaves glabrous, very scabrid all over; sheaths very tight, striate; ligules 1 lin. long; blades tightly convolute, setaceous to subulate, acute, 1–2 in. long, very rigid; panicle reduced to a loose 4–7-spiculate secund raceme (or the lowest branch 2–3-spiculate) 1–2 in. long, tips of slender pedicels thickened, pubescent; spikelets nodding, 4 lin. long, pallid; fertile florets 2; glumes very unequal, both subhyaline, the lower ovate, acute, 2–2½ lin. long, the upper oblong, subacuminate, 4 lin. long; valves shorter than the glumes, the fertile oblong, obtuse or subemarginate, 7-nerved, hairy all over; body of barren valves clavate, hairy; anthers ½ lin. long. *M. Caffrorum, var. decumbens, Nees, Fl. Afr. Austr.* 418 (*in part?*).

CENTRAL REGION: Beaufort West Div.; near Weltevrede, *Drège*, 752 in part!

Very distinct from *M. racemosa* in the habit and in the valves, being hairy all over. (See note under *M. pumila.*)

6. **M. decumbens** (Thunb. Prodr. 21); culms tufted, suberect or ascending from a usually short (sometimes slender) base, ½–1½ ft. long, simple, scabrid, many-noded, internodes enclosed, except the uppermost; leaves somewhat crowded above the base, glabrous, very scabrid, green to glaucous; sheaths rather tight, striate; ligules up to 1 lin. long; blades linear, acute, 1–3 in. by 1½ lin., usually convolute, rather firm and rigid, sometimes almost pungent, strongly striate; panicle linear, secund, 1½–4 in. long; branches erect, 4–1-spiculate, usually as long as the internodes or longer, filiform; pedicels ½–2 lin. long, tips thickened, villous; spikelets 5½–7 lin. long, purplish; the florets hidden by long hairs, fertile florets usually 2; glumes unequal, ovate-oblong to oblong, purple, rarely pallid, firmly membranous, the lower broader, acute or subacute, 5-nerved, 3–4 lin.

long, the upper acuminate, 7-nerved, 5½–7 lin. long; valves exceeded by the upper glume, the fertile oblong, obtuse, 5 lin. long, scabrid and strongly 7-nerved, hairy all over except at the hyaline tips; hairs very long; body of barren valves clavate, hairy; anthers ¾–1 lin. long. *Thunb. Fl. Cap. ed. Schult.* 111; *Kunth, Enum. Pl.* i. 378, *non Web. M. dendroides, Lehm. Pugill.* iii. 39; *Nees in Linnæa,* vii. 327; *Fl. Afr. Austr.* 419; *Kunth, Enum. Pl.* i. 378; *Steud. Syn. Pl. Glum.* i. 289; *Durand & Schinz, Consp. Fl. Afr.* v. 898.

SOUTH AFRICA: without precise locality, *Thunberg!*
COAST REGION: Fort Beaufort Div.; Fort Beaufort, *Ecklon.* Queenstown Div.; Bontebock Flats and near the Klipplaats River, *Ecklon.* Finchams Nek, near Queenstown, 3800 ft., *Galpin,* 2373! at the foot of the Stormbergen, between Klass Smits River and Stormberg Spruit, 4000–5000 ft., *Drège!*
CENTRAL REGION: Somerset Div.; near Somerset, *Bowker,* 180! Graaff Reinet Div.; near Graaff Reinet, 3200 ft., *Bolus,* 681! Colesberg Div.; near Colesberg, *Shaw,* 12! and 18! Albert Div. (?); "Cis-Garipina," banks of the Orange River, 5000–6000 ft., *Zeyher!*
KALAHARI REGION: Orange Free State, *Hutton!*

This is according to Hutton the "dronk grass" of the Boers, so-called on account of the effect it has on cattle and horses, making them stagger as if intoxicated. Oxen are even said to die from eating it.

Melica festucoides (Lichtenst. in Roem. & Schult. Syst. ii. 530) probably does not belong to this genus. Kunth suggests *Ehrharta.*

LXXIV. LAMARCKIA, Moench.

Spikelets dimorphic, fascicled, in dense unilateral panicles; fascicles deciduous, consisting usually of 5 spikelets, 2 of them awned, 3 awnless, elongated, many-valved, barren; one of the awned spikelets imperfect and paired with the lowest awnless spikelet, the other fertile between the other 2 awnless spikelets. *Fertile spikelet:* rhachilla slender, tough; florets 2, lower ⚥, upper barren; glumes subequal, narrow, linear-lanceolate, acuminate, 1-nerved, keeled; lower valve ovate-oblong (linear-oblong in profile), rounded on the back, thin, finely 5-nerved, awned from below the apex; awn fine, straight, callus obscure; upper valve empty, minute, delicate, 1-nerved, awned at or below the middle; pale narrow, 2-keeled; lodicules 2, minute, delicate, nerveless; stamens 3; anthers minute; ovary obtuse, glabrous; styles distinct, very short; stigmas very slender, barbellate, terminally exserted. *Barren awned spikelet* similar, smaller, with a single valve, like the upper valve in the fertile spikelet. *Awnless spikelets* long; rhachilla tough, somewhat pilose, wavy, thickened at the nodes; glumes rather broader than in the awned spikelet and more or less asymmetric; valves empty, broad, truncate, hyaline, 3-nerved, the upper imbricate. *Grain* tightly embraced by the valve and pale, adhering to the latter, linear-oblong, grooved; embryo oblong, short; hilum linear, ¼ the length of the grain.

Annual, small, glabrous; culms numerous; leaves flat; panicle very elegant, glistening; fascicles of spikelets ebracteate, jointed on the branches, at length deciduous, nodding; fertile spikelet quite hidden by the awnless spikelets.

Species 1, Mediterranean, occasionally found as an alien in other countries.

1. **L. aurea** (Moench, Meth. 201); culms $\frac{1}{4}$–$\frac{3}{4}$ ft. long, usually with 1 or more intravaginal branches near the base, slender, smooth, few-noded, internodes enclosed except the uppermost; leaves rather crowded near the base; sheaths lax, thin, smooth; ligules membranous, hyaline, oblong, acute, up to 4 lin. long; blades linear, tapering to an acute or setaceous point, 2–5 in. by 2–4 lin., smooth; panicle oblong to linear, 1–3 in. by $\frac{1}{2}$–1 in., light green or straw-coloured, rarely purplish or variegated; branches solitary or the lowest 2-nate, flexuous, nodding; branchlets glabrous below, hispidulous above the joint as are the short pedicels; fertile spikelet about 2 lin. long; rhachilla joints glabrous, the basal $\frac{1}{2}$ lin., the upper up to $\frac{3}{4}$ lin. long; glumes slightly exceeding the valves; lower valve almost $1\frac{1}{2}$ lin. long; awn 3–4 lin. long; anthers about $\frac{1}{4}$ lin. long; grain $\frac{7}{8}$ lin. long; barren spikelets 2–4 lin. long; valves white, 1 lin. long or less. *Kunth, Enum.* i. 389; *A. Rich. Tent. Fl. Abyss.* ii. 432; *Steud. Syn. Pl. Glum.* i. 300; *Durand & Schinz, Consp. Fl. Afr.* v. 906. *Cynosurus aureus, Linn. Sp. Plant.* 73; *Host, Gram. Austr.* iii. *t.* 4; *Sibth. & Sm. Fl. Græc.* i. *t.* 79. *Chrysurus aureus, Beauv. Agrost.* 123. *C. cynosuroides, Pers. Syn.* i. 80; *Viv. Fl. Lyb.* 4.

COAST REGION: Cape Div.; Simons Bay, *Wright!* among houses by the new reservoir at Simons Town, *Wolley Dod*, 2950! Introduced.

Mediterranean countries from the Canaries to the Punjab and Abyssinia.

LXXV. CYNOSURUS, Linn.

Spikelets dimorphic, laterally compressed, shortly pedicelled or subsessile, fascicled, fascicles gathered in usually dense spike-like unilateral panicles, the terminal spikelet of each fascicle fertile, the lateral barren, more or less concealing the former; rhachilla readily disarticulating and short-jointed in the fertile, tough in the barren spikelets. *Fertile spikelets*; florets 1–5, ☿ or the uppermost rudimentary; glumes equal or almost so, subulate to lanceolate, 1-nerved, more or less hyaline, slightly longer or shorter than the florets; valves oblong to lanceolate in profile, rounded on the back, minutely or obscurely 2-toothed, mucronate or awned from close below the tip, membranous, 5-nerved; callus small, obtuse, glabrous; pales subequal to the valves, 2-keeled, 2-toothed; lodicules 2, small, oblong; stamens 3; ovary glabrous; styles distinct, short; stigmas loosely plumose, laterally exserted. *Grain* oblong, rounded on the back, grooved in front, more or less adhering to the valve and pale; hilum linear, short; embryo small. *Barren spikelets* consisting of more or less numerous persistent undifferentiated bracts, the lower narrow subulate to lanceolate, somewhat distant and spreading, upper broader to ovate, closer to imbricate, all or at least the lower passing into bristle-like awns.

Annual or perennial, scantily fascicled to cæspitose; blades flat, more or less flaccid; ligule hyaline; panicle usually very dense, rarely somewhat loose, spike-like or capitate, secund; sterile and fertile spikelets very dissimilar, the former placed outside and more or less concealing the latter.

Species 2 or 3, in the Mediterranean countries, 1 all over Europe.

1. **C. echinatus** (Linn. Sp. Plant. 72); annual; culms fascicled or solitary, erect or geniculate and ascending, from a few inches to 2 ft. long, glabrous, 2–6-noded, internodes more or less exserted; leaves glabrous; sheaths thin, loosely striate, tight or the upper usually looser to tumid; ligules hyaline, oblong, obtuse, up to 5 lin. long; blades linear to lanceolate-linear from a very oblique base, tapering to an acute point, 2–6 in. by 1½–5 lin., flat, thin, flaccid, more or less scaberulous to scabrid in the upper part; panicle very dense, spike-like, secund, obliquely globose to oblong, ½–2½ in. long; axis sulcate, smooth; branches and branchlets very short, much contracted, granular or scaberulous, branchlets often bracteate; spikelets fascicled, heteromorphous, the terminal fertile, the lateral barren, more or less concealing the former; fertile spikelets more or less cuneate, 4–6 lin. long, 1–5-flowered; rhachilla readily disarticulating, produced as a short bristle or ending with a rudimentary floret; glumes equal, subulate-lanceolate, long and finely acuminate, 4–6 lin. long, hyaline, white, 1-nerved, scaberulous on the nerve and margins; valves lanceolate-oblong in profile, minutely 2-toothed or entire, 2½–3½ lin. long, membranous, hyaline and white below, somewhat firmer, light greenish, scabrid or puberulous above, 5-nerved; awn from close to the tip, fine, straight, scabrid, 3–8 lin. long, often purple; pales almost equalling the valve, keels nearly smooth; anthers 1¼–1½ lin. long; grain oblong, 1½ lin. long, dorsally compressed, grooved; barren spikelets broad-obovate, 3½–5 lin. long; rhachilla tough, granular-scabrid; glumes and valves not differentiated, like the supporting bracts subulate-lanceolate, long attenuate into scabrid often purplish bristles, more or less spreading, rigidly membranous, light green or purplish, decreasing and closer towards the top, longest 3–10 lin. long (including the awns). *Host, Gram. Austr.* ii. *t.* 95; *Engl. Bot. t.* 1333; *Kunth, Enum.* i. 388; *Steud. Syn. Pl. Glum.* i. 299; *Durand & Schinz, Consp. Fl. Afr.* v. 905. *C. coloratus, Lehm. ex Steud. Nomencl. ed.* ii. i. 465; *Nees, Fl. Afr. Austr.* 439; *Boiss. Fl. Or.* v. 571; *Durand & Schinz, l.c. Chrysurus echinatus, Beauv. Agrost.* 123.

COAST REGION: "Swellendam District," *Mund.*

Mediterranean countries, elsewhere introduced.

LXXVI. FINGERHUTHIA, Nees ex Lehm.

Spikelets strongly laterally compressed, in compact spike-like panicles, jointed on and deciduous from the pedicels; rhachilla tough. *Florets* 1, ⚥, or if more, then the uppermost ♂ or rudimentary. *Glumes* 2, subequal, narrow, thin, complicate, 1-nerved, keeled, shortly awned or mucronate. *Valves* oblong to lanceolate, mucronate, rather firm, 7–5- (rarely 3-) nerved, the upper smaller. *Pales* slightly shorter than the valves, ovate-oblong, 2-keeled, flaps broad. *Lodicules* 2, cuneate. *Stamens* 3. *Ovary* glabrous, slightly constricted below the apex (at least after fecundation); styles distinct, rather long; stigmas very slender, finely plumose, subterminally exserted. *Grain* unknown.

Perennial, cæspitose; innovation shoots intravaginal; blades narrowly linear; ligule a dense line of silky hairs; panicle compact, spike-like; the lowest spikelets barren, consisting of a few empty glumes.

Species 2 in South Africa; 1 of which is also found on the Afghan-Indian frontier.

Glumes long and softly ciliate along the keels and upper margins, finely awned ... ... (1) **africana.**
Glumes rigidly ciliolate along the keels, mucronate ... (2) **sesleriæformis.**

1. F. africana (Lehm. Ind. Sem. Hort. Hamb. 1834; and in Linnæa, x. Litt.-Ber. 112); densely tufted, glabrous; culms slender, about 1 ft. long, erect, simple, smooth, 2-noded, internodes exserted; sheaths glabrous, smooth, tight, the lower short, firm, whitish, persistent; blades linear, acute, 3–6 in. by 1–2 lin., flat or usually convolute, smooth below, finely scaberulous above, striate; panicle ellipsoid to cylindric, ½–1½ in. by 4–6 lin., sometimes purplish; spikelets 2–2½ lin. long, 2–4-flowered; glumes linear-lanceolate in profile, 1½–2 lin. long, long and softly ciliate along the keel and upper margins; awn fine, 1–1½ lin. long; valves ovate-lanceolate in profile, abruptly mucronate, 2 lin. long, rather firm except at the hyaline margin and obtuse tip, grooved below parallel to the keel, finely hairy near the margin below the middle, 7–9-nerved, side-nerves conniving and joining the middle nerve below the tip, the inner evanescent below, the middle one excurrent into a mucro; lodicules narrow, cuneate, ⅛ lin. long; anthers 1–1¼ lin. long; styles ¼ lin. long; stigmas 1 lin. long. *Benth. in Hook. Icon. Pl.* xiv. 54, *t.* 1373; *Durand & Schinz, Consp. Fl. Afr.* v. 873; *Aitch. in Journ. Linn. Soc.* xix. 193. *F. capensis, Lehm. ex Nees, Fl. Afr. Austr.* 136. *F. ciliata, Nees, l.c. F. afghanica, Boiss. Fl. Or.* v. 569; *Hook. f. Fl. Brit. Ind.* vii. 306.

SOUTH AFRICA: without precise locality, *Zeyher*, 1784!

COAST REGION: Vanrhynsdorp Div.; Attys, 4000 ft., *Schlechter*, 8333! Riversdale Div.; in the Karoo, near the Gauritz River, below 1000 ft., *Ecklon*. Oudtshoorn Div.; Cango, *Mund & Maire*.

CENTRAL REGION: Prince Albert Div.; Jackal Fontein, *Burke*, 23! Zwartbulletje, near the Gamka River, 2500–3000 ft., *Drège!* Graaff Reinet Div.; mountain sides near Graaff Reinet, 2800 ft., *Bolus*, 673! Ceres Div.; at Yuk River, near Yuk River Hoogte, *Burchell*, 1269! Fraserburg Div.; between Karree River and Klein Quaggas Fontein, *Burchell*, 1408!

WESTERN REGION: Little Namaqualand; on rocks near Noagas, North of Kaus Mountains, 1500–2000 ft., *Drège*.

KALAHARI REGION: Griqualand West; Herbert Div., on the right bank of the Vaal River at Blaauwbosch Drift, *Burchell*, 1745! and between the Vaal

River and Lower Campbell, *Burchell*, 1788! Hay Div., at Griqua Town, *Burchell*, 1876! 1961/2! plains between Griqua Town and Witte Water, *Burchell*, 1980! between Witte Water and Riet Fontein, *Burchell*, 2009! and plains at the foot of the Asbestos Mountains, *Burchell*, 2086! Basutoland; Leribe, *Buchanan*, 129!

Also in tropical Transvaal and on the Afghan-Indian frontier.

2. F. sesleriæformis (Nees, Fl. Afr. Austr. 138); culms very densely tufted on a short oblique rhizome, glabrous, rather robust, 1–2 ft. long, erect, smooth, 2-noded, internodes long exserted; sheaths tight, smooth, the lower short, very firm, pallid; blades linear, tapering to a subsetaceous point, 4–8 in. long, $1\frac{1}{2}$–$2\frac{1}{2}$ lin. wide when expanded, convolute, rarely flat, rigid, glaucous, smooth below, finely scaberulous above, margins rough; panicle ellipsoid to cylindric, $\frac{1}{2}$–$2\frac{1}{2}$ in. by 5–6 lin., sometimes purplish; spikelets 2–4-flowered, $2\frac{1}{2}$–3 lin. long; glumes lanceolate in profile, mucronate-acuminate, $1\frac{3}{4}$–2 lin. long, rigidly ciliolate along the keel; valves oblong-lanceolate in profile, mucronate-acuminate or the upper emucronate, about 2 lin. long, rather firm except at the narrow hyaline margins, glabrous or scantily and minutely hairy below towards the margins, 5–3-nerved, side-nerves rather close, more or less prominent (or the inner evanescent below the middle or quite suppressed) joining the middle nerve below the tip; lodicules obliquely cuneate, $\frac{1}{4}$ lin. long; anthers $1\frac{1}{4}$ lin. long; styles $\frac{1}{2}$ lin. long; stigmas $\frac{1}{2}$ lin. long. *Durand & Schinz, Consp. Fl. Afr.* v. 873.

SOUTH AFRICA: without precise locality, *Zeyher*, 1785!

COAST REGION: Albany Div.; Glenfilling, by a brook, below 1000 ft., *Drège!* in swampy places near Grahamstown, 2200 ft., *MacOwan*, 1319! Queenstown Div.; Zwart Kei Flats, 4000 ft., *Drège!* plains near Queenstown, 3500 ft., *Galpin*, 2356!

CENTRAL REGION: Beaufort West Div.; Nieuw Veld, near Bok Poort, 3500–4000 ft., *Drège*. Richmond Div.; Winter Veld, between Nieuwjaars Fontein and Ezels Fontein, 3000–4000 ft., *Drège*. Colesberg Div.; Wonderhuivel, 4500 ft., *Drège*.

KALAHARI REGION: Orange Free State, *Hutton!* Seven Fontein, *Burke*, 206! Winberg, *Buchanan*, 255! Transvaal; Hooge Veld, between Trigards Fontein and Standerton, *Rehmann*, 6751!

EASTERN REGION: Natal; Weenen County, South Downs, 4000 ft., *Wood*, 4405!

LXXVII. SCHISMUS, Beauv.

Spikelets laterally compressed, panicled; rhachilla disarticulating above the glumes and between the valves, slender, glabrous. *Florets* 5–7, ⚥, the uppermost reduced. *Glumes* much longer than the single valves, acute, herbaceous on the back, with white subhyaline margins, subequal, the lower broader, 5–7-nerved, the upper 3–5-nerved. *Valves* obliquely obovate to oblanceolate in profile, 2-lobed or 2-fid, with or without a mucro or a minute awn from the sinus, thin, hairy below, rounded on the back, nerves 7–9, prominent, the lateral evanescent and obscurely anastomosing below the hyaline tips; callus small. *Pales* spathulate, 2-keeled below the broad top, longer or shorter than the valves. *Lodicules* 2, cuneate, nerved, ciliate. *Stamens* 3. *Ovary* subglobose, glabrous; styles distinct, slender, about as long as the narrowly and densely plumose laterally exserted stigmas. *Grain* loosely embraced by the unaltered valve and pale, oblong to obovate, trigonous to plano-convex; hilum small, elliptic, basal; embryo less than $\frac{1}{2}$ the length of the grain.

Annual, very rarely subperennial, rather small; blades very narrow, often subsetaceous; ligule reduced to a line of hairs; panicle contracted, or at least narrow; spikelets conspicuously nerved, only the upper florets exserted from the glumes.

Species about 4, in the Mediterranean Region from the Canaries to India, and in South Africa.

Anthers $\frac{1}{8}$–$\frac{1}{5}$ lin. long, elliptic; valves without or with a mucro, not or slightly exceeding the lobes:
- Spikelets narrow; glumes thin, acute or acuminate; annual, very rarely subperennial ... ... (1) **fasciculatus.**
- Spikelets slightly turgid; glumes somewhat firm, subobtuse; perennial ... ... (2) **kœlerioides.**

Anthers $\frac{1}{2}$ lin. long, linear; glumes somewhat firm; valves minutely awned ... ... (3) **aristulatus.**

1. **S. fasciculatus** (Beauv. Agrost. 74, 177); annual, tufted; culms often very numerous, geniculately ascending, 1–8 in. long, very slender, smooth, 2-noded, internodes enclosed or exserted; leaves glabrous or very sparingly hairy, often equalling or overtopping the culms; lowermost sheaths scarious, pallid, prominently few-nerved, the upper very tight, all bearded at the mouth; ligule a ciliate rim; blades very narrow, linear, acute, 1–3 in. long, up to $\frac{2}{3}$ lin. broad, flat or setaceously convolute, filiform, flaccid or somewhat rigid and curved, scaberulous or smooth; panicle contracted, oblong, 5–14 lin. by 2–7 lin., dense, sometimes much reduced; spikelets narrow, $2\frac{1}{2}$–$3\frac{1}{2}$ lin. long, greenish, 6–8-flowered; glumes with rather broad and marked white margins, the lower oblong-lanceolate, acute, 5–7-nerved, $1\frac{1}{2}$–2 lin. long, the upper lanceolate, acute to subacuminate, 3–5-nerved, slightly longer; valves obliquely obovate-oblong in profile, $\frac{7}{8}$–$1\frac{1}{8}$ lin. long, shortly and subobtusely 2-lobed, minutely mucronate or emucronate, densely hairy all over below the broad hyaline tip or subglabrous except at the sides; lower pales exceeding their valves; anthers $\frac{1}{8}$–$\frac{1}{6}$ lin. long; grain globose-obovoid to obovoid, $\frac{3}{8}$ lin. long. *Trin. Fund. Agrost.* 148. *S. marginatus, Beauv. Agrost. in the index* 177, *and in the Expl. Pl. et Fig.* 10, *t.* 15, *fig.* 4; *Nees in Linnæa*, vii. 323; *Fl. Afr. Austr.* 421; *Kunth, Enum.* i. 385; *Suppl.* 318, *t.* 28, *fig.* 2; *Steud. Syn. Pl. Glum.* i. 295. *S. brevifolius, Nees, Fl. Afr. Austr.* 422; *Steud. Syn. Pl. Glum.* i. 295. *S. calycinus, Duval-Jouve in Billot, Adnot.* 285–290; *Coss. in Coss. & Dur. Expl. Scient. Algér. Phan.* 138; *Hack. in Oest. Bot. Zeit.* 1878, 189; *Aschers. in Oest. Bot. Zeit.* 1878, 254; *Durand & Schinz, Consp. Fl. Afr.* v. 907; *non C. Koch. Festuca calycina, Loefl. It. Hisp.* 116. *F. barbata, Linn. Amoen. Acad.* iii. 400. *Kœleria calycina, DC. Fl. Franç.* vi. 271. *Electra calycina, Panz. in Muench. Denkschr.* iv. (1813) 299, *cum ic.*

VAR. β, **tenuis** (Stapf); culms and blades very fine; panicles linear, usually very narrow, laxer than in the type, sometimes reduced to 3–6-spiculate racemes; spikelets very narrow; glumes narrower and more acuminate. *S. tenuis, Steud. Syn. Pl. Glum.* i. 295. *S. scaberrimus, Nees, Fl. Afr. Austr.* 423 (*in part*). *S. calycinus, var. tenuis, Durand & Schinz, Consp. Fl. Afr.* v. 907. *Hemisacris gonatodes, Steud. in Flora,* 1829, 490.

VAR. γ, **flaccidus** (Stapf); culms 4–10 in. long; blades up to 6 in. long, flaccid, usually flat, up to 1 lin. broad; panicles flaccid, flexuous, linear, 1–3 in. long, lax; lower branchlets sometimes up to 1 in. long, all finely filiform, erect or slightly spreading; spikelets as in the preceding variety or rather longer.

VAR. δ, **perennans** (Stapf); dwarf, 1½–4 in. long; culms and innovation shoots from a slender rooting rhizome-like base; blades filiform, canaliculate, 1½–4 in. long, curved, subrigid; panicles contracted, oblong to ovate, ⅓–⅔ in. long, dense; spikelets as in the type, but smaller in all parts.

COAST REGION: Worcester Div.; roadside near Hex River East Station, *Wolley Dod,* 3705! Var. β: Vanrhynsdorp Div.; Ebenezer, *Drège!* Uitenhage Div.; near Uitenhage, *Zeyher!* Var. γ: Swellendam Div.; mountain ridges along the lower part of the Zonder Einde River, *Zeyher,* 4579!

CENTRAL REGION: Beaufort West Div.; on the Nieuw Veld, near Bok Poort, 3500–4500 ft., *Drège!* Somerset Div.; near Somerset, *Bowker!* Graaff Reinet Div.; by the Sunday River, near Graaff Reinet, 2500 ft., *Bolus,* 461 partly! 738! Var. β: Somerset Div.; Biesjes Fontein, near Loots Kloof, 2800 ft., *MacOwan,* 1611! Var. γ: Somerset Div.; in woods at the foot of the Bosch Berg, 3000 ft., *MacOwan,* 1495! Graaff Reinet Div.; common on sandy ground on river banks near Graaff Reinet, 2500 ft., *Bolus,* 461 (with the type)! Fraserburg Div.; on the rocky hill at Dwaal River Poort, *Burchell,* 1483! Ceres Div.; at Ongeluks River, *Burchell,* 1227! Var. δ: Colesberg Div.; Colesberg, *Shaw,* 13!

KALAHARI REGION: Griqualand West, *Mrs. Barber!*

WESTERN REGION: Var. β: Little Namaqualand; near the Haazenkraals River, 2000–2500 ft., *Drège!* under shrubs near Mieren Kasteel, 1000 ft., *Drège!* Vanrhynsdorp Div.; Zout River, 450 ft., *Schlechter,* 8107!

I have very little doubt that these varieties are (with perhaps the exception of the last) entirely dependent on the conditions of the habitat, and also that *S. ovalis* (Nees, Fl. Afr. Austr. 421) and *S. scaberrimus* (Nees, l.c. 423), so far as the latter does not come under var. *tenuis,* represent only states of *S. fasciculatus* in a broad sense.

The localities given by Nees for them are:—

S. ovalis—COAST REGION: Robertson Div.; between Kochmans Kloof and Gauritz River, along with the type, *Ecklon.*

S. scaberrimus—CENTRAL REGION: Beaufort West Div.; between Gauze Fontein and Bok Poort, 3500 ft., *Drège.* WESTERN REGION: Little Namaqualand; Kamies Bergen, 3500 ft., *Drège.*

S. scaberrimus, var. *nanus*—CENTRAL REGION: Richmond Div; near the Zeekoe River, 5000 ft., *Drège.* WESTERN REGION: Little Namaqualand; near Kuigunjels on the Orange River, *Drège.*

2. S. kœlerioides (Stapf); subperennial; innovation shoots and culms in dense tufts, the latter 6–9 in. long, erect or suberect, very slender, smooth, 2–3-noded, internodes exserted; leaves glabrous; lowermost sheaths rather firm, pallid, closely striate, the upper very tight; cilia of ligule over ½ lin. long; blades fine, setaceously involute, 2–3 in. long, somewhat rigid, smooth, about 7-nerved; panicles contracted, linear, subflexuous, 1–1¼ in. long; branches erect, the longest up to ½ in., racemose or almost so; lateral pedicels very short; spikelets scarcely more than 2 lin. long, somewhat turgid, greenish or variegated with purple, 4–6-flowered; upper florets not or slightly exserted; glumes rather firm, scarious margins rather broad, not very conspicuous, with a hyaline glistening edge, the lower oblong, subobtuse to acute, 7-nerved, almost 2 lin. long, the upper oblanceolate, acute, 5-nerved, 2 lin. long; valves obliquely oblong in profile, over 1 lin. long, shortly and acutely bifid, mucronate, hairy all over below the broad hyaline sometimes purple tip; hairs long, very fine, acute; callus shortly bearded; pale equalling the valves; anthers about $\frac{1}{6}$ lin. long; grain obovate-oblong, over $\frac{3}{8}$ lin. long.

COAST REGION: Uitenhage Div.; near the mouth of the Zwartkops River, *Ecklon & Zeyher*, 411!

3. **S. aristulatus** (Stapf); annual, in small dwarf tufts, scarcely more than $1\frac{1}{2}$ in. high; culms sheathed all along, overtopped by the glabrous leaves; basal sheaths thinly scarious, glistening, few-nerved with the blades reduced to bristles, the upper very tight, suddenly contracted at the mouth; ligule a ciliolate rim; blades finely setaceous, canaliculate, 1 in. long, curved, rigid, smooth, 3-nerved; spikelets 2–3 in a short raceme, shortly pedicelled, 2 lin. long or slightly more; upper florets very shortly exserted; glumes subequal, rather firm, scarious margins rather broad, but not very conspicuous, the lower oblong, acute to subobtuse, 7-nerved, the upper broadly oblanceolate-oblong, acute, 5-nerved; valves obliquely oblong in profile, $1\frac{1}{4}$–$1\frac{1}{2}$ lin. long, shortly and subacutely bifid, with a fine short bristle from the sinus (bristle twice to three times the length of the lobes) hairy all over below the very broad hyaline tip; hairs very fine, acute; pales equalling the valves; anthers $\frac{1}{2}$ lin. long.

CENTRAL REGION: Sutherland Div.; between Kuilen Berg and Great Riet River, *Burchell*, 1353!

The specimens represent probably a dwarfed state; but they differ from all other species of *Schismus* in the comparatively long linear anthers, and in the presence of a distinct (though very fine) awn.

LXXVIII. DACTYLIS, Linn.

Spikelets much laterally compressed, very shortly pedicelled in dense compound often secund clusters of a panicle or a false spike; rhachilla very tardily disarticulating above the glumes and between the valves or quite tough. *Florets* 3–7, ⚥, or the uppermost rudimentary, exserted from the glumes. *Glumes* rigid, more or less hyaline, strongly keeled, subequal or the lower shorter, 1–3-nerved. *Valves* oblong, rigid, subherbaceous, mucronate or shortly awned, 5-nerved, keeled, keel ciliate; callus 0 or obscure. *Pales* slightly shorter than the valves, 2-keeled. *Lodicules* 2, bilobed. *Stamens* 3. *Ovary* glabrous; styles distinct; stigmas plumose, laterally exserted. *Grain* enclosed by the valve and pale, oblong, strongly convex on the back, grooved in front, somewhat soft; embryo rather small; hilum punctiform, basal.

Perennial, of varying habit; blades flat; ligules scarious; clusters of spikelets compact in an interrupted or uninterrupted false spike, or at the end of the branches or branchlets of a panicle.

Species about 3 (or 1, very polymorphic) in the temperate regions of the Old World; introduced elsewhere.

1. D. glomerata (Linn. Sp. Plant. i. 71); tufted; rhizome short, oblique; culms shortly ascending, geniculate or erect, rather robust, 1 to several feet high, simple, glabrous, smooth or slightly rough below the panicle, 2–3-noded; internodes more or less exserted; leaves mostly crowded at the base, glabrous; sheaths tight, striate, lower more or less compressed and keeled (particularly those of the innovation shoots), pallid, subpersistent or breaking up into fibres, upper terete, rough or almost smooth; ligules oblong, acute, up to 3 lin. long; blades linear, tapering to an acute point, $\frac{1}{3}$–1 ft. long or longer, $1\frac{1}{2}$–4 lin. broad, flat, flaccid, rough on the upper side and along the margins, striate; panicle erect, 1 in. to more than $\frac{1}{2}$ ft. long; clusters crowded into a dense (usually lobed) terminal false spike with or without 1–4 (rarely more) distant branches below it; these erect or spreading, straight or flexuous, terminated by similar clusters or groups of clusters; axis, branches and branchlets scabrid or pubescent in the upper part, the latter often minutely hispid or ciliate; spikelets oblong, $2\frac{1}{2}$–4 lin. long, light green, often concave on the inner side; glumes 1–2 lin. long, glabrous to pubescent; valves about $2\frac{1}{2}$ lin. long, tips obtuse, sides scaberulous to pubescent, mucro or awn up to 1 lin. long; keels of pales ciliate; anthers $1\frac{1}{4}$–$1\frac{1}{2}$ lin. long; grain 1–$1\frac{1}{4}$ lin. long. *Beauv. Agrost.* 85, *t.* 17, *fig.* 5; *Fl. Dan. t.* 743; *Host, Gram. Austr.* ii. 67, *t.* 94; *Engl. Bot. t.* 335; *Kunth, Enum.* i. 386; *Suppl.* 320, *t.* 29, *fig.* 1; *Reichb. Ic. Fl. Germ.* i. *t.* 59, *fig.* 1523; *Steud. Syn. Pl. Glum.* i. 297; *Durand & Schinz, Consp. Fl. Afr.* v. 904.

COAST REGION: Cape Div.; Herschel Lane, *Wolley Dod*, 1909!

A native of the temperate regions of the Old World; introduced elsewhere.

LXXIX. STIBURUS, Stapf.

(TRIPHLEBIA, Stapf.)

Spikelets laterally compressed, subsessile or shortly pedicelled, in spike-like cylindric panicles; rhachilla disarticulating above the glumes and between the valves. *Florets* 4–5, ⚥, the uppermost reduced, shortly exserted from the glumes. *Glumes* equal or subequal, membranous, lanceolate, caudate-acuminate, 1-nerved. *Valves* very similar to the glumes, but 3-nerved; callus very minute. *Pales* shorter than the valves, 2-keeled. *Lodicules* 2, minute, hyaline, cuneate. *Stamens* 3. *Ovary* glabrous; styles short, distinct; stigmas plumose, laterally exserted. *Grain* enclosed by the hardly changed valve and pale, free, oblong, terete; embryo short; hilum basal, punctiform.

Perennial, tufted; blades very narrow, usually subsetaceous, long; ligule a ciliate rim; panicle cylindric, dense, usually dark purple, greyish-villous.

Species 1, in extra-tropical South Africa and in tropical Transvaal.

1. **S. alopecuroides** (Stapf); densely cæspitose; culms erect, $\frac{1}{3}$–$1\frac{1}{2}$ ft. long, glabrous, simple, or branched at the base; leaves all basal, with scattered fine spreading hairs all over to glabrous; sheaths crowded, rather firm, pallid, smooth, persistent; blades usually setaceous or filiform, very acute, 3 to more than 12 in. long, sometimes flat and then up to 1 lin. broad, rather rigid; panicle $\frac{3}{4}$–3 in. by 3–$3\frac{1}{2}$ lin., sometimes interrupted at the base; branches solitary, adpressed to the rhachis; lowest $\frac{1}{3}$ to almost 1 in. long, divided from the base or nearly so, smooth; pedicels unequal, mostly very short; spikelets about 2 lin. long, densely crowded, usually dark purple; glumes, valves and pales equally villous from fine greyish hairs; glumes about $1\frac{1}{2}$ lin. long, tips firm, subulate; valves very slightly shorter, often mucronulate; anthers $\frac{2}{3}$ lin. long; grain $\frac{7}{8}$–1 lin. long, reddish-brown. *Lasiochloa alopecuroides, Hack. in Bull. Herb. Boiss.* iii. 393. *Triphlebia alopecuroides, Stapf in Hook. Icon. Pl. t.* 2612. *Kœleria Gerrardi, Munro ex Benth. in Benth. & Hook. Gen. Pl.* iii. 1184 (*name only*).

KALAHARI REGION: Orange Free State; without precise locality, *Cooper*, 723! 3352! Transvaal; Spitzkop Goldmine, *Wilms*, 1697! Lymklip Spruit, *Nelson*, 52*! Steelport River, *Nelson*, 12*! Houtbosch Berg, *Nelson*, 82*!

EASTERN REGION: Pondoland; Fakus Territory, *Sutherland!* Griqualand East; grassy rocky places on the summit of Malowe Mountain, 6000 ft., *Tyson*, 2773! *MacOwan & Bolus, Herb. Norm. Aust. Afr.*, 1217! summit of Mount Currie, near Kokstad, 7500 ft., *Tyson*, 1311! Natal; De Beers Pass, *Wood*, 5993! Noods Berg, *Wood*, 884! Kar Kloof, *Rehmann*, 7361! Umpumulo to Riet Vlei, *Buchanan*, 166! 167! and without precise locality, *Buchanan*, 32! *Gerrard*, 474!

LXXIX. bis. LASIOCHLOA, Kunth.

Spikelets 2-4-flowered, laterally more or less compressed, subsessile in a spike-like panicle; rhachilla tardily disarticulating above the glumes and between the valves; florets hermaphrodite or the uppermost gradually reduced, not exserted from the glumes. *Glumes* subequal, firmly membranous except at the hyaline margins, acute or acuminate or sometimes subulate-caudate, usually asymmetric, 5-nerved. *Valves* navicular, obtuse or acute, mucronulate, firmly membranous except at the thin margins and tips, with a submarginal line of acute hairs on each side, 7–9-nerved, side-nerves conniving above, evanescent below; callus obscure, glabrous. *Pales* shorter than the valves, 2-keeled. *Lodicules* 2, truncate-cuneate, scantily ciliolate from articulated hairs. *Stamens* 3. *Ovary* glabrous; styles short, distinct; stigmas laterally exserted, plumose. *Grain* enclosed by the hardly changed valve and pale, free, oblong, dorsally compressed; embryo small; hilum basal, minute.

Perennial or annual; innovations intravaginal; leaves flaccid or rigid and convolute; ligule a hairy rim; panicles dense, spike-like, often lobed; spikelets mostly hispid from tubercle-based hairs.

Species 3 or 4, endemic.

Annual; glumes long caudate-acuminate ... ... (1) **ciliaris.**
Perennial; glumes acutely or cuspidately acuminate or acute:
Glumes acuminate, coarsely hispid ... ... (2) **longifolia.**
Glumes acute, scantily and finely hispid or almost glabrous ... ... (3) **obtusifolia.**

1. **L. ciliaris** (Kunth, Rév. Gram. ii. 555, t. 192); annual, tufted, 6–8 in. high; culms geniculately ascending, slender, 3–4-noded, glabrous, middle internodes exserted, the uppermost 1–2 usually enclosed; sheaths tight or the uppermost subtumid, glabrous and smooth or hispid from fine tubercle-based hairs, bearded at the mouth, striate; ligule a long ciliate rim; blades linear, tapering to an acute point, 1–2½ in. by 1–1½ lin., rather flaccid, flat or convolute, hairy to glabrescent, subglaucous; panicle ovate-oblong to oblong-linear, sometimes lobed, ½–1½ in. by 4–5 lin., usually embraced at the base by the 2 uppermost sheaths; branches branched from the base, up to ½ in. long, rarely longer; pedicels extremely short, up to 1 lin. long; spikelets broadly obovate, greenish, 2–4-flowered; florets much exceeded by the glumes, these ovate-lanceolate, mucronate-acuminate, 2 lin. long, mucros mostly curved, ¾–1½ lin. long; valves oblong-lanceolate in profile, acute, obtuse or subemarginate, mucronate, 1½ lin. long, cilia short, rigid; anthers ¾–1 lin. long; grain over ½ lin. by ⅓ lin. *Kunth, Enum.* i. 387; *Nees, Fl. Afr. Austr.* 432; *Durand & Schinz, Consp. Fl. Afr.* v. 902 (*all in part*). *L. adscendens, Kunth, l.c.* 388; *Nees, l.c.* 435. *L. hispida, var. longifolia, Nees, l.c.* 433 (*in part*). *L. utriculosa, Drège in Linnæa,* xx. 255, *non Nees. Alopecurus echinatus, Thunb. Prodr.* 19; *Fl. Cap. ed.* i. 398; *ed. Schult.* 105. *Dactylis adscendens, Schrad. in Schult. Syst. Mant.* ii. 351; *Nees in Linnæa,* vii. 323; *Steud. Syn. Pl. Glum.* i. 299. *D. ciliaris, Nees in Linnæa,* vii. 322, *not Thunb. Hystringium acuminatum, Trin. ex Steud. Nomencl. ed.* ii. ii. 11.

SOUTH AFRICA: without precise locality, *Thunberg!*

COAST REGION: Vanrhynsdorp Div.; Ebenezer, below 100 ft., *Drège!* Malmesbury Div.; between Greene Kloof and Saldanha Bay, 500 ft., *Drège!* Cape Div.: Lion Mountain, *Ecklon,* 957 in part! near Lion Battery, *Wolley Dod,* 3109! Wynberg Park, *Wolley Dod,* 1815! Swellendam Div.; mountain ridges along the lower part of Zonder Einde River, *Zeyher,* 4588!

Nees indicates *L. ciliaris* also from Stellenbosch and George Div., and several more of the specimens quoted under *L. hispida* and *L. adscendens* also very likely belong here.

2. **L. longifolia** (Kunth, Rév. Gram. ii. 557. t. 193); perennial, cæspitose, with numerous short- or long-leaved innovation shoots; culms geniculate or suberect, 1–1½ ft. long, glabrous, smooth, often purple, almost wiry, 3–4-noded, internodes often long exserted; sheaths tight, glabrous, rarely scantily hairy, conspicuously bearded at the mouth, often all round, the lowermost pallid, subpersistent; ligules a hairy rim; blades linear, narrow, tapering to an acute point, 6–12 in. by ¾–1 lin. (when expanded), setaceously convolute, flexuous, glabrous or sometimes more or less hairy, hairs fine, spread-

ing from minute or obscure tubercles; panicle ovoid to oblong-cylindric, lobed or interrupted, 1–4 in. by 6–12 lin.; rhachis and branches smooth below, the latter branched from the base or the lowest undivided for ½–2 lin. and to 1 in. long; pedicels very short; spikelets obovate, greenish or purplish, 2–3-flowered; florets slightly exceeded by the glumes, these ovate-lanceolate in profile, cuspidate-acuminate, 1½–2 lin. long, hispid, tips tightly complicate, rigid, usually purplish or brown; valves lanceolate-oblong in profile, acute to obtuse, often mucronulate, over ½ lin. long, cilia short, rigid; anthers ¾ lin. long; grain not quite ½ lin. by less than ¼ lin. *Kunth, Enum.* i. 387; *Suppl.* 321; *Durand & Schinz, Consp. Fl. Afr.* v. 903. *L. hispida, var. longifolia, Nees, Fl. Afr. Austr.* 433 (*excl. subvar. longiglumis*). *Dactylis longifolia, Schrad. in. Schult. Mant.* ii. 351; *Nees in Linnæa*, vii. 322; *Steud. Syn. Pl. Glum.* i. 298.

VAR. β, **hispida** (Stapf); culms ½–⅔ ft. high, rarely taller, often repeatedly geniculate; leaves glabrous or hirsute; blades short, 1–4 in. long, rarely longer, fine or coarsely setaceous, the upper often spreading; panicle ½–1½ in. long, rarely more than slightly lobed. *L. hirta, Kunth, Rév. Gram.* ii. 559, *t.* 194; *Enum.* i. 387; *Nees, Fl. Afr. Austr.* 434; *Durand & Schinz, Consp. Fl. Afr.* v. 903. *L. hispida, Kunth, Enum.* i. 388; *Nees, Fl. Afr. Austr.* 432. *Dactylis hispida, Thunb. Prodr.* i. 22; *Fl. Cap. ed. Schult.* 115; *Steud. in Flora*, 1829, 490. *D. hirta, Schrad. in Schult. Mant.* ii. 350; *Nees in Linnæa*, vii. 322; *Steud. Syn. Pl Glum.* i. 298. *Festuca melangæa, Spreng. Syst.* i. 352. *Lappago setiformis, Spreng., L. setacea, Spreng. & Phleum subulatum, Spreng. ex Steud. Nomencl. ed.* ii. ii. 11, 12.

VAR. γ, **pallens** (Stapf); culms ½–1¼ ft. long, overtopped by the leaves; blades as in the type or rather finer; panicles often lobed or subinterrupted, pallid, 1–2 in. long; tips of the glumes long, finely acuminate, hyaline, pallid or purplish; tips of the valves also longer, acute, hyaline. *L. hispida, var. longifolia, Nees, Fl. Afr. Austr.* 433 (*partly*).

SOUTH AFRICA: without precise locality, *Harvey*, 258!

COAST REGION: Cape Div.; Cape Flats, *Rehmann*, 1770! *MacOwan & Bolus, Herb. Norm. Aust. Afr.*, 956! Herschel Lane, *Wolley Dod*, 1820! rocks to the west of Lions Head, *Wolley Dod*, 3573! plain above Simons Bay, *Milne*, 254! *MacGillivray*, 403! Paarl Div.; Paarl Mountains, *Drège!* Var. β: Malmesbury Div.; Swartland, *Thunberg!* Cape Div.; by a brook above Simons Bay, *Milne*, 259! by the roadside, near Tokay, *Wolley Dod*, 1970! Lion Mountain, *Burchell*, 118! *Pappe!* Stellenbosch Div.; Hottentots Holland, *Ecklon*, 956! Worcester Div.; *Zeyher!* Mossel Bay Div.; near the landing-place at Mossel Bay, *Burchell*, 6318! Uitenhage Div.; near the Zwartkops River, *Zeyher*, 4591! Uitenhage, *Alexander!* Albany Div.; near Grahamstown, *Glass*, 773! sandy places at Brand Kraal, near Grahamstown, 2000 ft., *MacOwan*, 1279! Port Elizabeth Div.; Port Elizabeth, *E.S.C.A. Herb.*, 101! Var. γ: Vanrhynsdorp Div.; near Ebenezer, below 100 ft., *Drège!* Cape Div.; Cape Peninsula, on sandhills, *Wolley Dod*, 1881! foot of Lion Mountain, *Pappe!* Capetown, *Harvey*, 146! Humansdorp Div.; on the Kromme River Heights, *Bolus*, 2698! Bathurst Div.; near Theopolis, between Riet Fontein and the shore, *Burchell*, 4082!

WESTERN REGION: Var. β: Little Namaqualand; between Silver Fontein, Koper Berg and Kaus Mountain, *Drège!* Var. γ: Little Namaqualand; Kamies Bergen, between Pedros Kloof and Lily Fontein, *Drège!*

The varieties, admitted here, represent probably only two rather marked states due, as it seems, to particular conditions of the habitat.

3. **L. obtusifolia** (Nees, Fl. Afr. Austr. 430); perennial, tufted: emitting long runners and numerous innovation shoots; culms 4–8 in.

long, erect, glabrous, smooth, 2-noded, internodes exserted; leaves glabrous; sheaths tight, smooth, striate, sometimes scantily bearded at the mouths, the basal short, whitish; ligule a ciliolate rim; blades very narrow, linear, with an obtuse callous point, 1–4 in. long, rigid, convolute, those of the culms spreading or curved, smooth, dark green; panicle spike-like, scarcely lobed, dense, ovate-oblong to oblong, ½–1 in. by 4–7 lin.; rhachis and branches scabrid, the latter very short, branched almost from the base; pedicels extremely short; spikelets broadly truncate-obovate, 2 lin. long, 3–4-flowered, variegated; florets equalling the glumes or slightly shorter; glumes ovate, acute, 5-nerved, rather firm, dull purplish except at the whitish margins, hispid from scanty fine tubercle-based hairs, rarely glabrescent; valves oblong-lanceolate in profile, subacute or submucronate, not quite 2 lin. long, cilia short and rigid; anthers 1 lin. long; grain not quite ½ lin. long. *Durand & Schinz, Consp. Fl. Afr.* v. 903. *L. ovata, Nees, l.c.* 431; *Durand & Schinz, l.c. Dactylis obtusifolia, Steud. Syn. Pl. Glum.* i. 298. *D. ovata, Steud. l.c.*

SOUTH AFRICA: without precise locality, *Herb. Harvey*, 314!
COAST REGION: Uitenhage Div.; in dry places and on limestone, near the Zwartkops River, *Drège.*

Imperfectly known species.

4. **L. utriculosa** (Nees, Fl. Afr. Austr. 436); annual; culms fascicled, about 3 in. long, repeatedly geniculate, ascending, glabrous, sheathed all along; sheaths tight, glabrous or scantily hirsute, the basal ones wider, more hirsute; ligules very short, truncate, ciliate; blades narrow, linear, 1–3 in. long, ultimately convolute, finely hirsute from tubercle-based hairs; panicle oblong, dense, ¾ in. long, sheathed at the base; branchlets very short, mostly 3–4-spiculate; spikelets ovate, 1¼–1½ lin. long, greenish, 3-flowered; glumes equal, rather rigid, subulate-acuminate with the acumen recurved, scabrid and purplish, coarsely and long hispid on the back below the middle, exceeding the florets by almost ½ of their length; valves thin, pallid, setaceously mucronate, pubescent below with a submarginal line of fascicled clavate-tipped cilia, faintly 5-nerved in the upper part; pales 3–4-toothed; lodicules linear, glabrous, emarginate. *Drège in Drège & Meyer, Zwei Pflanzengeogr. Docum.* 197; *Durand & Schinz, Consp. Fl. Afr.* v. 904. *L. ustuloides, Drège in Drège & Meyer, Zwei Pflanzengeogr. Docum.* 92. *Dactylis utriculosa, Steud. Syn. Pl. Glum.* i. 299.

WESTERN REGION: Little Namaqualand; on rocks in gorges near Noagas, *Drège.*

This description points rather to *Brizopyrum*, with the exception of the glumes so conspicuously exceeding the florets.

LXXX. UROCHLÆNA, Nees.

Spikelets few- to 7-flowered, in very short spike-like or capituliform panicles, laterally slightly compressed; rhachilla tough; florets

⚥, the upper gradually reduced. *Glumes* subequal, ovate-oblong, rigidly aristate-acuminate, rounded on the back, membranous, 5-nerved. *Valves* similar, 7-nerved, side-nerves conniving below the tip, awns shorter. *Pales* linear-oblong, almost equalling the valves (exclusive of the awns), 2-keeled. *Lodicules* minute, 2, cuneate, emarginate. *Stamens* 3. *Ovary* obtuse, glabrous; styles distinct, very short; stigmas very slender, delicately plumose, laterally exserted. *Grain* enclosed by the little altered valve and pale, free, oblong, dorsally compressed, convex on the back, obscurely concave in the front; embryo elliptic, equalling about $\frac{1}{3}$ of the grain; hilum elliptic, minute, basal.

Annual, dwarf, glabrous, much branched from the base; panicles compact, ovoid, small, embraced at the base by the uppermost sheath and deciduous with it, terminal, rarely with an additional somewhat remote peduncled cluster of spikelets below it; branches very short, 4–1-spiculate; spikelets sessile or subsessile, oblong to obovate, hispid, those at the base of the lower branches 1-flowered or barren, consisting of 2–4 empty glumes.

Species 1, endemic.

Nearest allied to *Lasiochloa.*

1. U. pusilla (Nees, Fl. Afr. Austr. 438); culms 3–4 in. long, very slender, smooth, about 3-noded, disarticulating below the uppermost node; middle internodes exserted, the uppermost one short, quite enclosed, the one preceding it longest; sheaths loose, scantily bearded at the mouth, smooth, striate; ligule a line of long and short hairs; blades linear, narrow, acute, $\frac{3}{4}$–1 in. long; panicles 2–4 lin. long; spikelets up to 2 lin. long; glumes $1\frac{1}{2}$–$1\frac{3}{4}$ lin. long, margins finely ciliate, otherwise very scantily hairy or the lower with a submarginal row of tubercle-based hairs on the exterior side; awns 3–5 lin. long, attenuated into bristles; valves $1\frac{1}{2}$ lin. long, finely hairy below along the middle nerve and both (or the exterior) margins, tips of hairs clavate, awns 2 lin. long or shorter; anthers 1 lin. long; grain $\frac{3}{4}$–$\frac{7}{8}$ lin. long. *Steud. Syn. Pl. Glum.* i. 299; *Benth. in Hook. Icon. Pl.* xiv. 46, *t.* 1363, *fig. B*; *Durand & Schinz, Consp. Fl. Afr.* v. 873.

WESTERN REGION: Vanrhynsdorp Div.; sand-hills near Ebenezer, *Drège!*

LXXXI. BRIZOPYRUM, Nees (in part, not of other authors).

Spikelets 4–9-flowered, more or less laterally compressed, subsessile in contracted panicles or 2-ranked spikes; rhachilla tardily disarticulating above the glumes and between the valves. *Florets* ⚥, the uppermost reduced, exserted from the glumes. *Glumes* keeled, mucronulate, firmly membranous or subcoriaceous except at the scarious margins, 4–5- (rarely to 7-) nerved, the lower often somewhat asymmetric. *Valves* navicular, obtuse or mucronulate, firmly membranous or subcoriaceous, minutely villous above the base or ciliate along 2–3 short lines from clavate or acute hairs, rarely glabrous, 7-nerved, lateral nerves conniving above, evanescent below or altogether invisible except by transmitted light; callus obscure, glabrous. *Pales* shorter than the valves, 2-keeled, keels sometimes narrowly winged. *Lodicules* 2, truncate-cuneate, nerved, ciliate, sometimes also papillose. *Stamens* 3. *Ovary* obovoid, obtuse, glabrous; styles distinct, very short; stigma laterally exserted, plumose. *Grain* enclosed by the scarcely changed valve and pale, free, oblong, semiterete; hilum basal, minute; embryo small.

Perennial or annual; innovations intravaginal; blades flaccid or rigid, and convolute; ligules a hairy or ciliate rim; panicles spike-like, or partly or wholly transformed into dense 2-ranked spikes; glumes glabrous or hispid from tubercle-based hairs.

Species 5, in extra-tropical South Africa; 1 also in St. Helena.

- Spikelets in spike-like panicles, all round the axis; valves with submarginal lines of cilia, otherwise glabrous or with a short dorsal line of hairs, rarely uniformly hairy at the base; anthers $\frac{1}{8}$ lin. long or shorter (except in 3. *B. glomeratum*):
  - Valves with acute cilia or bristles; spikelets broad, ovate to elliptic, $1\frac{1}{4}$–2 lin. long; glumes ovate, subacute to subacuminate:
    - Leaves very finely hirsute all over; spikelets $1\frac{1}{4}$–$1\frac{1}{2}$ lin. long, turgid; lower glume and valves with stiff tubercle-based bristles; annual ... ... (1) **ciliare.**
    - Leaves glabrous (except sometimes a beard at the mouth); spikelets about 2 lin. long, not turgid; glumes glabrous; valves with short fine rigid cilia:
      - Uppermost sheath usually exceeding the base of the panicle; anthers $\frac{1}{5}$ lin. long ... ... (2) **obliterum.**
      - Uppermost sheath remote from the panicle; anthers up to almost 1 lin. long ... ... (3) **glomeratum.**
  - Valves with clavate-tipped cilia along the keel and near the margins in the lower part; spikelets oblong, 2–$2\frac{1}{2}$ lin. long; glumes ovate-lanceolate, acutely acuminate ... (4) **acutiflorum.**
- Spikelets throughout in 2-ranked spikes, or the lower clustered on very short branches; valves finely and uniformly villous near the base, rarely glabrous, hairs clavate; anthers $\frac{3}{4}$ lin. or more long:
  - Glumes and valves subcoriaceous, very obscurely nerved; spikelets often clustered near the base; leaves rigid ... ... (5) **capense.**
  - Glumes and valves thinner; nerves slightly prominent in the upper part; spikelets strictly 2-ranked; leaves glaucous, not rigid:
    - Spikelets loosely imbricate, $3\frac{1}{2}$–4 lin. long; leaves long, flaccid ... ... (6) **alternans.**
    - Spikelets very tightly imbricate, 2–$2\frac{1}{2}$ lin. long; leaves short ... ... (7) **brachystachyum.**

1. **B. ciliare** (Stapf in Hook. Icon. Pl. t. 2602); annual, tufted; culms geniculate, suberect or ascending, very slender, 4–6 in. long,

glabrous, smooth, 2–3-noded, sheathed all along or 1–2 intermediate internodes slightly exserted; leaves very finely hirsute all over; sheaths tight, lower often purplish; ligule a minute rim; blades linear, very narrow, 1–3 in. by $\frac{1}{2}$ lin., often convolute, rather flaccid, uppermost much exceeding the panicle; panicle spike-like, oblong, very dense, sometimes lobed, 4–9 lin. by 2–3 lin.; rhachis and branches terete, smooth, the latter very short; pedicels up to $\frac{1}{2}$ lin. long; spikelets very crowded, very broad, ovate, turgid, $1\frac{1}{4}$–$1\frac{1}{2}$ lin. long, 3–6-flowered; glumes broad, ovate, acute to subacute, up to 1 lin. long, faintly 5-nerved, white margins very broad; lower glume with stiff tubercle-based bristles on the herbaceous back, upper almost glabrous; valves ovate in profile, subobtuse, 1 lin. long, with a submarginal row of stiff tubercle-based acute bristles, otherwise glabrous, rather firm, tips somewhat compressed and purple, nerves obscure; pales broad, obtuse; keels scaberulous; lodicules minute, scantily and minutely ciliolate; anthers $\frac{1}{2}$ lin. long. *Dactylis ciliaris, Thunb. Prodr.* 22; *Fl. Cap. ed.* i. 429; *ed. Schult.* 115 (*not Nees in Linnæa*, vii. 322).

SOUTH AFRICA: without precise locality, *Thunberg!*

A very distinct species, quite different from *Lasiochloa ciliaris* (p. 698), which Kunth (*Rév. Gram.* ii. 556) took to be Thunberg's *Dactylis ciliaris* on the authority of a specimen so named in the Berlin Herbarium. He repeated the error later (*Enum.* i. 387) and was followed by Nees (*Fl. Afr. Austr.* 432).

2. **B. obliterum** (Stapf); perennial, tufted with numerous closely sheathed innovation shoots 2–6 in. long; culms geniculately ascending or prostrate, slender, glabrous, smooth, 2–3-noded, internodes exserted except the uppermost; sheaths rather tight, glabrous, except at the often bearded mouth, striate, the lower short, whitish, the upper 1 or 2 usually exceeding the base of the panicle; ligule a ciliolate rim; blades very narrow, more or less setaceous with a subobtuse callous point, $\frac{1}{2}$–$1\frac{1}{2}$ in. long, glabrous or rough below the tips, dark green, often spreading; panicle spike-like, linear-oblong, $\frac{1}{2}$–1 in. by 2–3 lin.; rhachis and branches terete, scaberulous, the latter very short; pedicels up to $\frac{1}{2}$ lin. long, rarely more; spikelets crowded, elliptic, up to 2 lin. long, greenish, 3–6-flowered; glumes ovate, acute or subacute, prominently 3–5-nerved, not quite $1\frac{1}{2}$ lin. long, glabrous, rather firm except the broad whitish margins; keels scabrid; valves shortly ovate-lanceolate in profile, subacute or mucronulate, slightly over 1 lin. long, rather firm, with a short submarginal line or tuft of short rigid pointed cilia on each side, otherwise glabrous; nerves rather distinct above the middle; pales broad, obtuse or emarginate, 1 lin. long; keels scabrid; lodicules scantily papillose; anthers $\frac{1}{5}$ lin. long; grain $\frac{2}{5}$ by $\frac{1}{4}$ lin. *Demazeria oblitera, Hemsl. in Chall. Exped. Bot.* i. ii. 90, *t.* 51, *fig.* 1–8; *Durand & Schinz, Consp. Fl. Afr.* v. 900.

VAR. β, **erectum** (Stapf); culms erect or suberect, 6–7 in. long; blades $2\frac{1}{2}$–$4\frac{1}{2}$ in. long.

COAST REGION: Swellendam Div.; near Swellendam, *Zeyher*, 4589! Var. β: Mossel Bay Div.; between Great Brak River and Little Brak River, *Burchell*, 6159!

Also in St. Helena.

This species is intermediate between *Lasiochloa* and *Brizopyrum*. I have placed it in the latter genus on account of the florets being exserted from the glumes.

Burchell's specimens have more the habit of *B. glomeratum*, but the spikelets are exactly as in *B. obliterum*.

3. **B. glomeratum** (Stapf in Hook. Icon. Pl. t. 2603); perennial, tufted, with numerous closely-sheathed intravaginal innovation shoots, up to 1 ft. high; culms geniculate, suberect, very slender, glabrous, 2–4-noded, internodes exserted; sheaths tight, quite glabrous, lower short, whitish, rather firm; ligule a minutely ciliolate rim; blades subsetaceous with an obtuse callous point, 2 to more than 6 in. long, glabrous; panicle spike-like, linear-oblong, slightly lobed or subinterrupted, 1–1¼ in. by 3–4 lin.; rhachis and branches terete, scaberulous to hispidulous, the latter very short like the pedicels; spikelets very crowded, broad, ovate to elliptic, about 2 lin. long, greenish, 3–6-flowered; glumes ovate, acute to subacuminate, about 1½ lin. long, glabrous, 3–5-nerved, nerves rather close to the scabrid keel, faint, margins very broad, white; valves ovate in profile, usually abruptly mucronate, slightly over 1–1½ lin. long, rather firm, with lines of fine rigid pointed hairs along the margins and keel or hairy all over to the middle, nerves fine; pales broad, obtuse, 1 lin. long, keels ciliate below, scabrid above; anthers almost 1 lin. long. *Poa glomerata, Thunb. Prodr.* 22; *Fl. Cap. ed.* i. 423; *ed. Schult.* 113; *Kunth, Enum.* i. 363.

SOUTH AFRICA: without precise locality, *Thunberg*!

Nees, who had not seen Thunberg's type of *Poa glomerata*, referred it (*Fl. Afr. Austr.* 376) erroneously to *Tetrachne Dregei*—a mistake which passed into other works (e.g. *Steud. Syn. Pl. Glum.* i. 299 and *Durand & Schinz, Consp. Fl. Afr.* v. 865).

4. **B. acutiflorum** (Nees, Fl. Afr. Austr. 371); perennial, tufted with short-leaved innovation shoots, glabrous; culms slender, geniculate, suberect, ½–1 ft. long, smooth, 4–5-noded, lower internodes exserted, the uppermost enclosed; sheaths tight, striate, the uppermost exceeding the base of the panicle; ligule a ciliolate rim; blades narrow, linear, acute, the middle ones longest, up to 6 in. by 1 lin. when expanded, often setaceously convolute, rather rigid, smooth below, scaberulous above; panicle spike-like, cylindric, dense, ¾–2 in. by 2–3 lin., greenish to straw-coloured, with some smaller additional clusters from the sides of the uppermost 1 or 2 internodes; the rhachis and the very short branches scabrid; pedicels to ½ lin. long; spikelets imbricate, oblong, acute, 2–2½ lin. long, 5–8-flowered; glumes ovate-lanceolate, acutely acuminate, 1½ lin. long or slightly more, glabrous, 4–5-nerved, nerves close to the keel, sides thinly membranous; valves lanceolate in profile, mucronate-acuminate, nerves rather distinct above the middle, with a short line of clavate cilia along the keel and near the margins from the base to

$\frac{1}{2}$ or $\frac{1}{3}$ way up the valve; pales broad, 1 lin. long; keels obscurely winged, scaberulous; lodicules papillose and with a few cilia; anthers $\frac{1}{8}$–$\frac{1}{5}$ lin. long; grain over $\frac{1}{2}$ lin. by $\frac{1}{3}$ lin. *Steud. Syn. Pl. Glum.* i. 281. *Desmazeria acutiflora, Durand & Schinz, Consp. Fl. Afr.* v. 900.

VAR. β, **capillare** (Nees, l.c.); culms 3–12 in. long; blades capillary, with scattered spreading long hairs.

COAST REGION: Vanrhynsdorp Div.; sand-hills near Ebenezer, below 500 ft., *Drège!* Piquetberg Div.; Piquetberg Range, near Groen Vallei, below 1000 ft., *Drège!* Var. β: Cape Div.; at Green Point and on the shore beyond Sea Point, common, *Wolley Dod*, 3569!

5. B. capense (Trin. in Mém. Acad. St. Pétersb. sér. 6, iv. (1838), and in Gram. Gen. Suppl. 54); perennial, usually quite glabrous; innovation shoots from the axils of short firm basal scales, erect or growing into stolons, densely covered at the base with short broad strongly striate scales; culms tufted, 1–2 ft. long, erect, subgeniculate, firm, smooth, 3–4-noded, internodes at length exserted; leaves glabrous, rarely sparingly to copiously hirsute from tubercle-based hairs; sheaths tight, smooth, sometimes bearded at the mouth; ligule a very dense line of fine hairs; blades very narrow, linear, acute, convolute, 3–10 in. by 1 lin. when expanded, firm, flexuous, smooth or obscurely scaberulous and strongly nerved above; panicle spike-like, very shortly branched below, reduced to a bifarious spike above or throughout; rhachilla triquetrous, often wavy, angles scabrid; spikelets imbricate, all in one plane or the two ranks conniving in front, 3–4$\frac{1}{2}$ lin. long, ovate, 5–9-flowered, pallid or variegated with purple; glumes ovate to ovate-lanceolate in profile, acuminate, 2–3 lin. long, firmly chartaceous, 5- to sub-7-nerved, glabrous or with a few tubercle-based bristles, the scarious margins broad, finely scaberulous; valves obliquely oblong or subovate in profile, shortly or obscurely and obtusely acuminate, usually finely tomentose near the base, with clavate-tipped hairs, rarely quite glabrous, subcoriaceous except at the narrow scarious margins, nerves scarcely visible except in transmitted light; pales 2-toothed, 2$\frac{1}{2}$ lin. long; flaps glabrous or hairy; keels winged, ciliolate; lodicules ciliate; anthers 1 lin. long. *Nees, Fl. Afr. Austr.* 372 (*excl. var.* β); *Steud. Syn. Pl. Glum.* i. 281. *Cynosurus Uniolæ, Linn. f. Suppl.* 110; *Thunb. Prodr.* 23; *Fl. Cap. ed. Schult.* 119; *Nees in Linnæa*, vii. 327. *C. paniculatus, Thunb. l.c.; Kunth, Enum.* i. 389; *Durand & Schinz, Consp. Fl. Afr.* v. 906. *Triticum capense, Spreng. Pugill.* ii. 23. *Briza imbricata, Steud. in Flora*, 1829, 489. *B. Uniolæ, Nees ex Steud. Syn. Pl. Glum.* i. 283. *Poa Uniolæ, Schrad. in Goett. Gel. Anz.* 1821, 2074, *and in Schult. Mant.* ii. 312; *Kunth, Enum.* i. 341. *P. papillosa, Schrad, l.c., and in Schult. Mant. l.c.; Kunth, l.c. Uniola capensis, Trin. in Mém. Acad. St. Pétersb. sér.* 6, i. (1831) 360, *and in Gram. Gen.* 360.

VAR. β, **villosum** (Stapf); leaves densely villous all along; blades up to more than 1 lin. broad; panicle reduced to a one-sided, 2-ranked, usually simple spike; glumes hairy; hairs fine, long, flexuous.

SOUTH AFRICA: without precise locality, *Thunberg!*

COAST REGION: Clanwilliam Div.; Cederberg Range, on sandy flats, near Ezelsbank, *Drège.* Cape Div.; Table Mountain, *Ecklon*, 960! *MacOwan & Bolus, Herb. Norm. Aust. Afr.*, 794! *MacGillivray*, 394! *Milne*, 245! between Capetown and Table Mountain, *Burchell*, 29! near Capetown, *Harvey*, 143! *Burchell*, 491! Dido Valley, *Wolley Dod*, 2829! 2824! Lion Mountain, *Drège*, 33! Devils Mountain, *Ecklon*, *Wilms*, 3875! 3877! near Constantia, *Ecklon.* Simons Bay, *Wright!* Paarl Div.; in moist meadows by the Berg River, near Paarl, *Drège!* Paarl Mountain, 1000–1500 ft., *Drège.* Tulbagh Div.; top of Winterhoek Mountain, 5000 ft., *Ecklon*, Tulbagh Waterfall, *Ecklon.* Worcester Div.; Hex River Valley, at Els Kloof, *Wolley Dod*, 3711! "Swellendam and George Distr.," *Mund & Maire!* Swellendam Div.; Zonder Einde River, *Zeyher!* Riversdale Div.; Great Vals River, *Burchell*, 6546! between Great Vals River and Zoetemelks River, *Burchell*, 6563! Oudtshorn Div.; Cango, *Mund & Maire!* Mossel Bay Div.; between Duyker River and Gauritz River, *Burchell*, 6403! Uitenhage Div.; Witteklip Mountain, *MacOwan*, 2074! Koegas Kop, *Zeyher!* Port Elizabeth Div.; Port Elizabeth, *E.S.C.A. Herb.*, 195! George Div.; Zuur Berg, 1500 ft., *Bolus*, 2701! Var. β: Cape Div.; near Lion Battery, *Wolley Dod*, 3110!

Burchell's 6546, Bolus' 2701, and Zeyher's specimens from Koegas Kop, have short spreading blades (1–2 in. long) and a slightly different habitat, but they are so connected with the typical long-leaved form by C. Wright's and one of Mund & Maire's specimens, that I consider them only as a form occurring in dry stations.

6. **B. alternans** (Nees, Fl. Afr. Austr. 369); perennial; innovation shoots long-leaved; culms simple, ascending, up to 1 ft. long, slender, about 4-noded, internodes shortly exserted; leaves more or less spreadingly villous all over with fine tubercle-based hairs; sheaths tight except the lower, striate; ligule a line of long stiff hairs; blades linear with an acute callous point, $\frac{1}{2}$–1 ft. by 1 lin., very flaccid, flat, prominently nerved; panicle reduced to a rather lax spike, up to 2 in. long, erect; rhachis triquetrous; angles narrowly winged, wings produced into fine subsetaceous scaberulous lobes below and nearly opposite to each spikelet; spikelets alternately biseriate, ovate, $3\frac{1}{2}$–4 lin. long, pallid, 4-flowered; glumes subequal, lanceolate in profile, acutely or cuspidately acuminate, 3–$3\frac{1}{2}$ lin. long, outer sides of the lower copiously hirsute with 2 lateral nerves and broad membranous margins, the inner subglabrous, with short obscure lateral nerves and almost wholly membranous, the upper hirsute on both sides, 5-nerved; valves obliquely oblong in profile, subacute to almost obtuse, the lowest $2\frac{1}{2}$–$2\frac{3}{4}$ lin. long, with rather broad membranous margins in the upper part, finely tomentose from very short clavate-tipped hairs near the base, nerves prominent in the upper part, keel obscurely winged; pales subobtuse, $2\frac{1}{2}$ lin. long, keels scaberulous; lodicules ciliate; grain not quite $\frac{1}{2}$ lin. by $\frac{1}{3}$ lin. *Steud. Syn. Pl. Glum.* i. 281. *Desmazeria alternans, Durand & Schinz, Consp. Fl. Afr.* v. 900.

COAST REGION: Worcester Div.; Dutoits Kloof, 1000–2000 ft., *Drège!*

The narrow lobes on the angles of the rhachis have been erroneously described as bracts by Nees. I suspect they are the product of some abnormal condition.

7. **B. brachystachyum** (Stapf); perennial, tufted, with numerous short-leaved innovation shoots; culms very slender, geniculately ascending or prostrate at the base, 6–8 in. long, branched from the base, glabrous, smooth, 4–5-noded, internodes exserted; sheaths rather thin, the lower short, somewhat looser, pallid, strongly striate, hirsute from tubercle-based hairs, the upper tight, glabrescent or glabrous, smooth; ligule a line of hairs; blades linear, obtusely pointed, $\frac{1}{2}$–$1\frac{1}{2}$ in. by 1 lin., more or less involute, glaucous, strongly nerved, hirsute all over or the upper glabrescent; panicle reduced to a spike, oblong, 4–7 lin. by 3–4 lin., compact, 2-ranked, secund; rhachis wavy, angles scabrid; spikelets imbricate, 2–$2\frac{1}{2}$ lin. long, 5–6-flowered, pallid or variegated with purple; glumes shortly ovate-lanceolate in profile, mucronate-acuminate, $1\frac{3}{4}$ lin. long, the lower hispid on the outer side, almost glabrous on the inner side, 4–5-nerved, upper hispid on both sides, 5-nerved, nerves distinct; valves obliquely oblong, obscurely and obtusely acuminate, $1\frac{1}{2}$ lin. long, firm, usually purplish above with a narrow yellow margin, lower finely tomentose from the middle downwards with clavate-tipped minute hairs, upper glabrescent to glabrous, nerves slightly prominent above; pales slightly tomentose on the back, 1 lin. long; keels obscurely winged; lodicules ciliate; anthers 1 lin. long; grain $\frac{1}{2}$ lin. by $\frac{1}{3}$ lin. *B. capense, var. brachystachyum, Nees, Fl. Afr. Austr.* 373.

COAST REGION: Worcester Div.; on moor-like ground, in Dutoits Kloof, 2500–3500 ft., *Drège!*

Very distinct from *B. capense* by having much weaker shoots and culms, thinner sheaths, short glaucous scarcely rigid blades and smaller spikes and spikelets. I have referred Schrader's *Poa papillosa* to *B. capense*, and not to *B. brachystachyum*, because Schrader describes his plant as having erect culms and rigid filiform leaves.

LXXXII. BRIZA, Linn.

Spikelets many-flowered, laterally compressed, panicled; rhachilla disarticulating above the glumes and between the valves; florets hermaphrodite, the upper gradually reduced. *Glumes* scarious or firmly membranous, keeled or boat-shaped or saccate with the back rounded, persistent, 3- to sub-7-nerved, subequal. *Valves* close, firmly membranous with scarious margins or almost wholly scarious, keeled or boat-shaped or saccate with the back rounded, obtuse, acute, subacuminate or subaristate, 7–9-nerved, outer 3 or all the side-nerves spreading from a common base, rarely 5-nerved, with the side-nerves distant at the base. *Pales* broad, shorter than the valves, 2-keeled, keels often winged. *Lodicules* 2, obliquely ovate, hyaline, fleshy at the base. *Stamens* 3. *Ovary* glabrous; styles short, distinct; stigmas very slender, loosely plumose, laterally exserted. *Grain* tightly embraced by the hardened back of the valve and the pale, usually adherent to the latter, concavo-convex to plano-convex, usually dorsally compressed; hilum basal, small, elliptic, oblong or linear; embryo small.

Annual or perennial; blades flat and rather broad, or convolute and narrow; ligules hyaline; panicle effuse with capillary branchlets and pedicels and nodding spikelets, sometimes reduced to a raceme, or straight, contracted or almost spiciform.

Species about 11; 4 mainly in the Mediterranean region, of which 2 have been introduced into various temperate countries, 1 all over temperate Europe and Asia, the rest in South America.

Spikelets 5–8 lin. long ... ... (1) **maxima.**
Spikelets 1½–2 lin. long ... ... (2) **minor.**

1. **B. maxima** (Linn. Sp. Pl. 70); annual, glabrous (except the spikelets); culms fascicled, geniculate, 1–2 ft. long, smooth, 2–4-noded; internodes exserted, at least ultimately; sheaths rather loose, smooth, the lower thin and distantly nerved; ligules oblong, up to 1–2½ lin. long; blades linear, acute or long tapering to a fine point, usually 3–6 in. by 2–3 lin., sometimes much shorter and narrower, flat, flaccid, slightly scabrid or almost smooth; panicle oblong, nodding, secund, lax, sparingly branched or reduced to a 6–2-spiculate raceme or a solitary spikelet, 1–3 in. long; rhachis filiform, flexuous, scabrid or smooth below; branches distant, usually solitary, simple, bifid or distantly branched from near the base, subcapillary, scabrid, the lowest up to 2 in. long; pedicels 3–10 lin. long, smooth above; spikelets nodding, ovate, 5–8 lin. by 4–6 lin., 7–17-flowered, scarious, straw-coloured to purple; glumes very broadly-ovate, obtuse, sub-7- to sub-9-nerved, lower about 2½ lin. long, upper slightly longer; valves very close, very broadly cordate-ovate, shortly and obtusely acuminate, 3–3½ lin. long, lower glabrous or finely pubescent near the base from clavate-tipped hairs, upper silky-pubescent, all 7–9-nerved, side-nerves joining at the base, hyaline margins broad; pales broadly obovate, almost 2 lin. long, winged, wings densely ciliolate; lodicules up to ½ lin. long; anthers up to more than 1 lin. long in the lower flowers, much smaller in the upper; grain rotundate-obovate, convexo-concave, over 1 lin. by ¾–1 lin. *Berg. Descr. Pl. Cap. B. Spei*, 12; *Host, Gram. Austr.* ii. *t.* 30; *Sibth. & Sm. Fl. Græc.* i. *t.* 76; *Schrad. in Goett. Gel. Anz.* 1821, 2074; *Trin. in Mém. Acad. St. Pétersb. sér.* 6, i. (1831) 362, *and in Gram. Gen.* 362; *Nees in Linnæa*, vii. 326; *Kunth, Enum.* i. 371; *Nees, Fl. Afr. Austr.* 415; *Steud. Syn. Pl. Glum.* i. 283; *Durand & Schinz, Consp. Fl. Afr.* v. 899.

South Africa: without precise locality, *Thunberg!*

Coast Region: Cape Div.; near Capetown, *Drège! Burchell*, 95! Sea Point, *Wilms*, 3894! Table Mountain, *Ecklon, Milne*, 243! *MacGillivray*, 399! *Mudd!* Constantia, *Ecklon*, by the railway, near Rondebosch, *Wolley Dod*, 1825! Simons Bay, *MacGillivray! Wright!* Paarl Div.; in moist meadows, near Paarl, 400–1000 ft., *Drège!* Tulbagh Div.; near Tulbagh Waterfall, Tulbagh Kloof, &c., *Ecklon & Zeyher!* Stellenbosch Div.; Hottentots Holland, *Ecklon.* Riversdale Div.; Great Vals River, *Burchell*, 6541! between Zoetemelks River and Little Vet River, *Burchell*, 6840! Port Elizabeth Div.; along the coast, *E.S.C.A. Herb.*, 148! Albany Div.; near Grahamstown, *MacOwan! Glass!* Bathurst Div.; Port Alfred, *Hutton!*

Mediterranean region; introduced into the Cape and Australia.

2. **B. minor** (Linn. Sp. Pl. 70); annual, glabrous; culms tufted, geniculate, $\frac{1}{2}$–2 ft. long, smooth or somewhat rough above, 2–3-noded; internodes exserted, at least ultimately; sheaths loose, smooth, the lower thin, striate; ligules oblong, 2–3 lin. long; blades linear to lanceolate-linear, acute, 2–8 in. by $1\frac{1}{2}$–$4\frac{1}{2}$ lin., flat, flaccid, more or less scabrid or almost smooth; panicle broadly obovate, 2–4 in. long and almost as broad, erect, lax, rather divaricate; rhachis slender, straight; branches geminate, distantly and repeatedly tri- or di-chotomously branched, scabrid, filiform to capillary, the lowest up to 3 in. long; pedicels 6–2 lin. long, finely capillary, smooth above; spikelets triangular to ovate, very obtuse, often broader than long, $1\frac{1}{2}$–2 lin. long, 4–7-flowered, nodding, green, rarely purplish below; glumes thinly scarious, horizontally spreading, subequal, obtuse or subacute, 3-nerved, 1–$1\frac{1}{4}$ lin. long; valves very close, very broadly cordate-ovate, very obtuse with the tips often inflexed, very gibbous below, 1–$1\frac{1}{4}$ lin. long, glabrous, 7-nerved, the side-nerves joining at the base, hyaline margins very broad; pales elliptic, scarcely $\frac{3}{4}$ lin. long, finely winged, wings very minutely ciliolate; lodicules up to $\frac{1}{4}$ lin. long; anthers almost $\frac{1}{4}$ lin. long in the lower florets, much smaller in the upper; grain shortly oblong, truncate, convexo-concave or subtriquetrous, broadly grooved, $\frac{3}{8}$ lin. long. *Berg. Descr. Pl. Cap. B. Spei*, 13; *Host, Gram. Austr.* ii. *t.* 28; *Sibth. & Sm. Fl. Græc.* i. *t.* 74; *Engl. Bot. t.* 1316; *Trin. in Mém. Acad. St. Pétersb. ser.* 6, i. (1831) 362, *and in Gram. Gen.* 362; *Kunth, Enum.* i. 372; *Reichb. Ic. Fl. Germ.* i. *t.* 92; *Steud. Syn. Pl. Glum.* i. 282; *Baker, Fl. Maurit.* 457; *Durand & Schinz, Consp. Fl. Afr.* v. 900. *B. deltoidea, Burm. Fl. Cap. Prodr.* 3. *B. virens, Linn. Sp. Pl. ed.* ii. 103; *Steud. in Flora*, 1829, 490; *Kunth, Enum.* i. 372; *Nees in Linnæa*, vii. 326; *Fl. Afr. Austr.* 414 (*non Trin.*).

SOUTH AFRICA: without precise locality, *Thunberg!*

COAST REGION: Clanwilliam Div.; by the Olifants River, *Ecklon.* Vanrhynsdorp Div.; Gift Berg, 1500–2000 ft., *Drège.* Cape Div.; common near Capetown, *Pappe!* Table Mountain, *Ecklon; Mudd!* Devils Mountain and Somerset, *Ecklon.* Simons Bay, *MacGillivray*, 400! *Wright!* near Westerford, *Wolley Dod*, 1904! Paarl Div.; Paarl Mountains, by streamlets, 1000–1500 ft., *Drège.* Tulbagh Div.; Tulbagh Kloof, *Ecklon & Zeyher!* Riversdale Div.; Great Vals River, *Burchell*, 6540! hills near the Zoetemelks River, *Burchell*, 6751! Uitenhage Div.; cultivated ground, by the Zwartkops River, *Ecklon.* Albany Div.; near Coldspring, *Flanagan*, 777!

Mediterranean regions; introduced into many parts of the world.

Imperfectly known species.

3. **B. spicata** (Burm. Fl. Cap. Prodr. 3, not Sibth. & Sm.) is possibly *Brizopyrum spicatum.*

LXXXIII. TETRACHNE, Nees.

Spikelets laterally compressed, awnless, densely imbricate, biseriate, sessile, unilateral on the flattened rhachis of a compound spike, jointed on and falling entire from it, uppermost terminal; rhachilla tough or tardily disarticulating between the fertile valves. *Florets* 5–6, lowest 2 reduced to barren valves, remainder ⚥ or the uppermost reduced. *Glumes* thin, acute, 1-nerved, keeled, keel narrowly winged. Empty *valves* very similar to the glumes; fertile valves rather firm, subobtuse, 5-nerved, keeled, keel winged. *Pales* slightly shorter than the valves, 2-keeled, keels winged. *Lodicules* 2, cuneate, fleshy, nerved. *Stamens* 3. *Ovary* glabrous; styles distinct, much shorter than the laterally exserted plumose styles. *Grain* loosely enclosed by the valve and pale, oblong, subterete; embryo large; hilum punctiform, basal.

Perennial, tufted; blades convolute, rigid; spikes arranged in an uninterrupted spike.

Species 1, endemic.

1. **T. Dregei** (Nees, Fl. Afr. Austr. 376); barren and flowering shoots crowded on a very short oblique rhizome; culms erect-geniculate, 1–1½ ft. long, terete, glabrous, smooth, 2–3-noded, internodes more or less exserted, or enclosed except the uppermost; sheaths tight, firm, glabrous except at the finely ciliate margins and mouths, smooth; ligule a dense fringe of hairs; blades narrow, linear, usually setaceously convolute, 2–5 in. by ½ lin. (when expanded), rigid, glabrous, smooth except at the scabrid margins, glaucous; compound spike 2–8 in. long, erect, edges of rhachis scabrid; partial spikes ½–1 in. long, erect or suberect, straight or slightly curved; secondary rhachis glabrous, ciliate or hairy along the margins; spikelets about 2½ lin. long, light to dark olive-grey; glumes and empty valves 1½ to almost 2 lin. long; fertile valves 2 lin. long; anthers ¾–1 lin. long; grain rather over 1 lin. long; embryo ⅔ lin. long. *Steud. Syn. Pl. Glum.* i. 299; *Durand & Schinz, Consp. Fl. Afr.* v. 865. *Poa glomerata, Thunb. Prodr.* 22; *Fl. Cap. ed. Schult.* i. 113; *Kunth, Enum.* i. 363.

COAST REGION: Queenstown Div.; by the Klipplaat River, near Shiloh, 3500 ft., *Drège!* Zwart Kei River, 4000 ft., *Drège.*

CENTRAL REGION: Colesberg Div.; Wonderheuvel, 4000–5000 ft., *Drège.* Colesberg, *Shaw,* 17! Graaff Reinet Div.; Sneeuw Berg Range, 3800 ft., *Bolus,* 1802! between Compass Berg and Rhenoster Berg, 5000–6000 ft., *Drège.* Albert Div., *Cooper,* 1368! "Cis Gariepina," *Zeyher!* Aliwal North Div.; Witte Bergen, 4500–5000 ft., *Drège.*

KALAHARI REGION: Hopetown Div.; by the Orange River, near Hopetown, *Burchell,* 2667!

LXXXIV. ENTOPLOCAMIA, Stapf.

Spikelets strongly laterally compressed, mucronate, sessile, solitary or clustered on the angular or flattened rhachis of a simple or compound spike, jointed on and falling entire from it, uppermost terminal; rhachilla tough, joints extremely short. *Florets* few to 20, lowest 2 reduced to barren valves, the next ⚥, the uppermost reduced. *Glumes* thin, acute, lower strongly 3-, upper 5-nerved. Empty *valves* intermediate in shape and texture between the glumes and fertile valves, 6–8-nerved; fertile valves with stoutly mucronate subrecurved tips, thin below, prominently 9–11-nerved above, keeled; callus 0. *Pales* as long as the valves, 2-toothed, 2-keeled, keels winged with a tuft of long delicate wool from the inner base of the broad flaps. *Lodicules* 0. *Ovary* glabrous; styles connate at the very base, very slender; stigmas terminally exserted, very narrow, plumose. *Grain* loosely enclosed by the valve and pale, laterally flattened; pericarp thin, loose; embryo ½ as long as the grain or less; hilum punctiform, basal.

Perennial; blades linear, flat or involute; ligule a hairy rim; spikelets solitary or in clusters or secondary spikes of few to 9 spikelets on the rhachis of a simple or compound spike, the lowest of a cluster often smaller or barren.

Species 1, in Hereroland and Namaqualand.

1. **E. aristulata** (Stapf); tufted; culms erect, 1–2 ft. long, glabrous, smooth, 2–3-noded; internodes exserted; sheaths glabrous except for a beard at the mouth, smooth, lowest keeled; blades linear, tapering to a setaceous point, 3–6 in. by 1–1½ lin. long, flat or rolled in when dry, glaucous, finely puberulous or scaberulous on both sides, or with scattered long hairs towards the base; spike 2–3 in. long; longest branches up to 8–9 lin. long, triquetrous, broad, more or less secund, like the rhachis scabrid to hispid along the angles; spikelets ovate, 4½–8 lin. long, 4–5 lin. broad; glumes lanceolate, lower 2 lin. long, upper broader and over 2½ lin. long, keels scabrid, margins ciliolate; fertile valves broadly and obliquely ovate, up to 4 lin. long, whitish below, dark grey and punctate upwards, nerves curved and conniving above, stout, margins very narrow and ciliolate above, broad and ciliate at the base; pales oblong, white and hyaline except at the green tips, keel-wings ciliate, flaps long ciliate along the margins, almost glabrous on the faces, with a tuft of long wool at the inner base, hairs up to 2 lin. long; filaments 4–4½ lin. long; anthers 2–2½ lin. long; styles over 3 lin. long; stigmas almost 2 lin. long; grain broad, oblong, truncate, more than 1 lin. long; seed brown, obscurely punctate. *Tetrachne aristulata, Hack. & Rendle in Journ. Bot.* 1891, 72.

WESTERN REGION: Little Namaqualand; Ookiep, *Scully!*

Also in Hereroland (*Dinter*, 10!).

LXXXVI. POA, Linn.

Spikelets mostly 2–6-flowered, in loose or close (rarely in spike-like) panicles; rhachilla disarticulating above the glumes and between the valves, glabrous or scantily and minutely hairy; flowers hermaphrodite or the upper imperfect. *Glumes* thin, membranous, keeled, acute or obtuse, 1–3-nerved. *Valves* membranous, sometimes rather firm, obtuse or acute, 5–7-nerved; callus small, obtuse, often with a tuft of long wool. *Pales* shorter than the valves, 2-keeled. *Lodicules* 2, more or less 2-lobed. *Stamens* 3. *Ovary* glabrous, styles short, free; stigmas plumose, laterally exserted. *Grain* ovoid, oblong or linear, often grooved, free or adherent to the pale; hilum punctiform, basal; embryo small.

Annual or perennial; blades flat and flaccid or convolute and more or less rigid; ligules hyaline; panicles open, often effuse, rarely contracted, spike-like; spikelets rather small, awnless.

Numerous species in the temperate regions, particularly of the northern hemisphere, few in the tropics.

Perennial; anthers 1 lin. or more long:
- Lowest sheaths thickened at the base, forming a more or less distinct bulb; innovation blades usually finely filiform; spikelets very often viviparous ... ... (1) **bulbosa.**
- Lowest sheaths not thickened at the base; blades not finely filiform:
  - Valves connected by long, often copious wool:
    - Spikelets $2\frac{1}{2}$–3 lin. long, elegantly variegated, 2–6 on the 2-nate smooth branches of a very lax and scanty panicle ... ... (2) **Atherstonei.**
    - Spikelets $1\frac{1}{2}$–$2\frac{1}{2}$ lin. long, not or scarcely variegated, 3 to many on the 3–6-nate, more or less scaberulous branches of a rather compound panicle:
      - Culms many-noded, strongly compressed; valves minutely 2-toothed ... ... (3) **bidentata.**
      - Culms terete or subcompressed below, about 3-noded; valves entire ... ... (4) **trivialis.**
  - Valves glabrous or almost so, without connecting wool; basal sheaths ultimately breaking up into persistent fibres ... (5) **binata.**

Annual or subperennial; anthers less than $\frac{1}{2}$ lin. long (6) **annua.**

1. P. bulbosa (Linn. Sp. Pl. 70); perennial, densely cæspitose, 4–16 in. high, glabrous; culms erect, terete, smooth, 2–3-noded; upper internodes exserted; leaves crowded near the base, lowest sheaths usually with a bulbous thickening at the base, rather loose in the upper part, remainder tighter; ligules oblong or ovate, $1\frac{1}{2}$–2 lin. long; blades linear, acute, those of the barren shoots often finely filiform, 2–$2\frac{3}{4}$ in. long, upper much shorter, up to 2 lin. broad; panicle oblong to ovate, contracted, lobed, $\frac{4}{5}$–2 in. long; the lower branches geminate, almost smooth, branchlets short, scaberulous; pedicels short or very short; spikelets ovate-oblong or oblong, light green, $2\frac{1}{4}$–3 lin. long, 4–7-flowered, often viviparous; lower glume ovate, acute, 1–$1\frac{1}{2}$ lin. long, 3-nerved, rarely 1-nerved, margins broadly hyaline, keel asperulous; upper glume somewhat larger and more acuminate, 3-nerved; valves oblong, acute, $1\frac{1}{2}$–$1\frac{3}{4}$ lin. long, upper $\frac{1}{4}$–$\frac{1}{3}$ hyaline, keel and outer side-nerves silkily hairy to the middle, connecting wool scanty or 0, nerves obscure; pales slightly shorter than the valves; keels scaberulous; anthers $\frac{2}{3}$–1 lin. long. *Host, Gram. Austr.* ii. *t.* 65; *Engl. Bot. t.* 1071; *Kunth, Enum.*

i. 352; *Nees, Fl. Afr. Austr.* 380; *Reichb. Ic. Fl. Germ.* i. *t.* 81; *Steud. Syn. Pl. Glum.* i. 250; *Durand & Schinz, Consp. Fl. Afr.* v. 908.

WESTERN REGION: Little Namaqualand; Kamies Bergen, by a dried-up streamlet near Kuil, 3500 ft., *Drège.*

Temperate regions of Europe, Asia, and North Africa.

The description is drawn up from European and Asiatic specimens, as I have not seen Drège's plant.

2. **P. Atherstonei** (Stapf); culms smooth, slender; sheaths glabrous, striate, purplish above; ligule ovate, 1 lin. long; blades narrow, linear, acute, up to 2 in. long, glabrous, smooth; panicle very lax, 2½ in. long; rhachis very slender, smooth; branches geminate, distant, flexuous, often deflexed, finely filiform, up to 1¼ in. long, quite smooth, 2–6-spiculate near the tips; lateral pedicels very short, tips thickened, smooth; spikelets variegated, 2½–3 lin. long, 3-flowered; glumes thin, lower narrow, oblong when expanded, acute, 1-nerved, up to 1¾ lin. long, upper broad, oblong, subacute, 3-nerved, up to 2 lin. long, almost wholly purple; valves rather thin, oblong, acute to subacute, up to 2½ lin. long, silky pubescent below along the keel and outer nerves, with scanty short wool from the base, inner side-nerves very faint, tips hyaline; pale of the lower valves ⅓–¼ shorter than these, 2-toothed, keels scabrid.

CENTRAL REGION: Graaff Reinet Div.; summit of Compass Berg, *Atherstone,* 46!

Only one specimen is known. It is 9 in. high, inclusive of the panicle, and comprises only 2 culm internodes, the lower of which is much exceeded by its sheath. It represents evidently only the upper part of a culm.

3. **P. bidentata** (Stapf); perennial, glabrous; culms ascending from a sometimes prostrate and branched base, 1–2 ft. long, strongly compressed, smooth, many-noded; internodes enclosed except the uppermost; leaves numerous; lower sheaths embracing each other, keeled, striate, like the upper rather tight; ligules a narrow membranous margin, up to ⅓ lin. long; blades linear, acute, lower and middle 6–9 in. by 1½–2 lin., uppermost much shorter, all rather firm, smooth except towards the tips; panicle 2½–3½ in. long, lax, flexuous, nodding; rhachis smooth, very slender; branches 5–3-nate, longest up to 1½ in. long and naked for ½–¾ in., simple or almost so, capillary, very flexuous, scaberulous above; lateral pedicels ½–1 lin. long; spikelets distant almost by their own length, 2–2½ lin. long, pallid, about 4-flowered; glumes rather firm, the lower ovate when expanded, acute, 3- or sub-3-nerved, 1¼ lin. long, upper oblong, acute, 1½ lin. long, 3-nerved; valves linear-oblong in profile, over 1½ lin. long, minutely 2-toothed, sometimes with an obscure mucro in the sinus, firm except at the very tips, nerves distinct, silkily pubescent below along the keel and the outer nerves and with long wool from the base; pales obtuse, almost equalling the valves, keels scabrid; anthers almost 1 lin. long; grain oblong, grooved, ¾ lin. long.

SOUTH AFRICA: without precise locality, *Zeyher!*

The mode of growth is similar to that of *P. compressa;* the panicle, however, is quite different. The spikelets are as in certain forms of *P. pratensis* or of *P. nemoralis*, but the tips of the valves are constantly notched.

4. **P. trivialis** (Linn. Sp. Pl. 67); perennial, loosely tufted, glabrous; culms geniculately ascending, usually from a short arched rooting base, 1½–2 ft. long, terete or subcompressed below, scabrid above, rarely smooth all along, about 3-noded; internodes exserted; sheaths somewhat loose, striate, smooth or rough, the lower thin; ligules ovate-oblong, 2–5 lin. long; blades linear, acute, 2–5 in. by 1–3 lin., usually flat and flaccid, scaberulous; panicle oblong to ovate or pyramidal when open, erect or slightly nodding, 3–7 in. long; rhachis usually smooth below; branches in distant semiwhorls of 4–6 (mostly of 5), unequal, filiform, scaberulous, the longest up to 3 in. long and undivided often for more than ½ their length, distantly or closely branched, ultimate branchlets closely 6–2-spiculate; lateral pedicels very short; spikelets green or purplish, ovate to oblong, acute, ½–2 lin. long, 3–4-flowered; glumes subequal, rather firm, lower narrow, oblong, acute, 1⅛–1⅓ lin. long, 1-nerved; keel scaberulous, upper glume ovate and acuminate, 1½ lin. long, 3-nerved; side-nerves prominent, keel very scabrid; valves oblong, acute, rather firm, lower 1½ lin. long, pubescent along the keel to the middle, otherwise glabrous, side-nerves rather prominent; callus with a small tuft of very long wool; pales 1¼ lin. long, 2-toothed, keels finely and very densely scabrid; anthers 1 lin. long; grain ½ lin. long, grooved. *Host, Gram. Austr.* ii. 45, *t.* 62; *Engl Bot. t.* 1072; *Trin. in Mém. Acad. Pétersb. sér.* 6, i. (1831) 380, *and in Gram. Gen.* 380; *Kunth, Enum.* i. 352; *Reichb. Ic. Fl. Germ.* i. *t.* 89; *Steud. Syn. Pl. Glum.* i. 251; *Durand & Schinz, Consp. Fl. Afr.* v. 909.

EASTERN REGION: Natal; without precise locality, *Buchanan*, 33!

Probably introduced; a native of the temperate regions of the Old World.

5. **P. binata** (Nees, Fl. Afr. Austr. 378); perennial, compactly cæspitose, glabrous; culms erect, ½–1½ ft. high, rarely taller, more or less compressed below, usually 2-noded, internodes exserted; leaves mostly near the base; sheaths tight, smooth, the lower more or less keeled, ultimately breaking up into persistent fibres; ligules ovate, ½–1½ lin. long; blades linear, acute, 2–6 in. by 1–1½ lin., flat, with the tips often complicate and curved, smooth or finely scaberulous above; panicle ovoid or pyramidal when open, 2½–6 in. long, erect or nodding, lax; rhachis smooth, very slender; branches distant, geminate, finely filiform, flexuous or wavy, up to 2 in. long, undivided to the middle or beyond it, smooth or scaberulous above; branchlets contracted; lateral pedicels very short, scaberulous; spikelets crowded on the tips of the branches, ovate-oblong, 2–2½ lin. long, closely 3–5-flowered, pallid, rarely variegated with purple; glumes rather unequal, lower ovate to oblong when expanded, acute,

1–1½ lin. long, 1-nerved; keel scaberulous, upper ovate, acuminate, 1¼–2 lin. long, 3-nerved, side-nerves usually very short; valves oblong, subacute to acute, lower 2 lin. long, glabrous or minutely pubescent below along the keel and the outer nerves, not connected by wool; tips hyaline; nerves prominent; pales 1¾ lin. long, keels scabrid; anthers 1–1¼ lin. long. *Steud. Syn. Pl. Glum.* i. 257; *Durand & Schinz, Consp. Fl. Afr.* v. 908.

COAST REGION: Albany Div.; Howisons Poort, near Grahamstown, coming up after a fire, 2000 ft., *MacOwan*, 1292! *Zeyher*, 4598! on cultivated land, near Grahamstown, *Glass*, 775! Albany, *Ecklon & Zeyher*, 870! Cathcart Div.; between Kat Berg and Klipplaat River, 4000–5000 ft., *Drège*. Komgha Div.; near Komgha, *Flanagan*, 934! Queenstown Div.; Table Mountain, 6000–7000 ft., *Drège!*

KALAHARI REGION: Orange Free State; on the Drakens Bergen, near Harrismith, *Buchanan*, 116!

EASTERN REGION: Tembuland; mountains near Bazeia, 4000 ft., *Baur*, 526! Natal; on mountains about 100 miles inland, 4000–6000 ft., *Sutherland!* Riet Vlei, 4000–5000 ft., *Buchanan*, 284! near the Mooi River, 3000–4000 ft., *Wood*, 7326! *Mason*, 68! and without precise locality, *Buchanan*, 285!

Very closely allied to the Abyssinian *P. simensis*, Hochst. ex A. Rich., which differs mainly in having narrower, more acute valves and longer ligules. Both species belong to the group of *P. polycolea*, Stapf.

6. **P. annua** (Linn. Sp. Pl. 68); annual or subperennial, tufted, glabrous; culms erect from a geniculate and often rooting base, from a few inches to 1 ft. long, slightly compressed below, smooth; internodes exserted or enclosed; sheaths rather loose, slightly compressed, smooth; ligules oblong to ovate, up to 1½ in. broad; blades linear, acute, ⅘–1½ in. (rarely longer) by 1–1½ lin., flat, flaccid; margins scaberulous; panicle rather stiff, ovate, lax, up to 3½ in. long, often subsecund; the lower branches 2- (rarely 3–5-) nate or solitary, spreading, ultimately often deflexed, up to 1⅔ in. long, branched from the middle, smooth; spikelets more or less crowded, oblong-ovate or ovate, green or sometimes purplish, 2–2½ lin. long, 3–7-flowered; lower glume lanceolate, acute, 1–1¼ lin. long, 1-nerved to sub-3-nerved, upper slightly longer, ovate when expanded, 3-nerved; valves oblong, obtuse or subacute, 1½ lin. long; margins and tips broadly hyaline; nerves slightly prominent, silkily ciliate below along the keel and the outer side-nerves, rarely glabrescent, without connecting wool (in the type); pales slightly shorter than the valves, keels ciliate; anthers ⅜ lin. long; grain oblong, ½–1 lin. long. *Host, Gram. Austr.* ii. *t.* 64; *Fl. Dan. t.* 1686; *Engl. Bot. t.* 1141; *Reichb. Ic. Fl. Germ.* i. *t.* 82; *Nees, Fl. Afr. Austr.* 378; *Steud. Syn. Pl. Glum.* i. 250; *Durand & Schinz, Consp. Fl. Afr.* v. 907; *Hook. f. Fl. Brit. Ind.* vii. 345.

COAST REGION: Cape Div.; near Capetown, *Ecklon*, 935! Devils Peak, *Wilms*, 3895! Rondebosch, *Drège;* Table Mountain, to the top, *Bergius!* near the Orange Kloof River, *Wolley Dod*, 2624! near Alpen Bridge, *Wolley Dod*, 2073! Albany Div.; near Grahamstown, *Ecklon*.

EASTERN REGION: Natal, *Cooper*, 3362!

Throughout Europe and temperate Asia, introduced into most other temperate regions, rare in the tropics.

P. squarrosa, Lichtenst in Roem. et Schult. Syst. ii. 553 and *P. tenuiflora*, Lichtenst. l.c. 566, are evidently species of *Eragrostis*.

LXXXVII. ATROPIS, Griseb.

Spikelets 2- or 3- to many-flowered, narrow, laterally compressed or subcylindric; rhachilla disarticulating above the glumes and between the valves, glabrous, produced. *Glumes* broad, obtuse, rounded on the back, unequal, 1–3-nerved. *Valves* oblong, obtuse, rounded on the back, rather firm below the hyaline tips, 5-nerved, nerves usually obscure, except in transmitted light; callus minute. *Pales* almost as long as the valves, 2-keeled. *Lodicules* 2, ovate, hyaline, *Stamens* 3. *Ovary* glabrous; styles very short, distinct; stigmas laterally exserted, delicately plumose. *Grain* tightly embraced by the hardened valve and pale, oblong, almost semiterete, subconcave in front; hilum basal, punctiform; embryo small.

Perennial; blades linear, flat, plicate or convolute; ligules hyaline; panicles open or contracted; branches usually spreading, often naked for a considerable distance; branchlets and pedicels adpressed; spikelets close, awnless.

Species about 12, mainly in the temperate regions of the northern hemisphere. Most of them very closely allied.

Very closely allied to *Poa*, from which it chiefly differs in the glumes and valves being not keeled.

Leaves 1–1½ lin. broad when expanded, more or less plicate; panicles ovate to subdeltoid, with rather coarse and stiff more or less spreading branches ... ... (1) **Borreri.**
Leaves up to ⅔ lin. broad, when expanded, setaceously convolute; panicles linear with erect finely filiform to capillary branches ... ... (2) **angusta.**

1. **A. Borreri** (Stapf); culms fascicled with the intravaginal innovation shoots, ½–1½ ft. long, suberect or geniculate below, smooth, terete, somewhat stout, 1-noded near the base, uppermost internode long exserted; leaves glabrous, glaucous, somewhat thick; sheaths smooth, lower rather broad, pallid; ligules ovate, to almost 1 lin. long; blades linear with an obtuse callous hooded point, 1–3 in. (in the Cape specimens) by 1–1½ lin., folded or more or less open, rigid, smooth below, scaberulous above; panicle erect, ovate to subdeltoid, 1–4 in. long; rhachis straight, scabrid except quite below; branches very unequal, in fascicles of 2–5, stiff, coarsely filiform, very scabrid, the shortest 2–4-spiculate almost from the base, the longest up to 1½ in. long, branched from near the base or bare for ¼–¾ in., obliquely spreading; lateral pedicels hardly any or up to ½ lin. long, stout; spikelets more or less adpressed to the branches, linear-oblong to subcylindric, 1½–3 lin. long (in the South African specimens), 2–4- (rarely to 7-) flowered, pallid; glumes ovate, lower about ½ lin. long, hyaline except along the nerve, upper ¾–1 lin. long, 3-nerved, hyaline, margins broad; valves obliquely oblong in profile, rather more than 1 lin. long, minutely hairy at the very

base; pales minutely 2-toothed, keels ciliolate; anthers $\frac{3}{8}$–$\frac{1}{3}$ lin. long; grain about $\frac{1}{2}$ lin. long. *A. distans, var. permixta, Coss. in Coss. & Dur. Expl. Sc. Alger. Phan.* 140; *Durand & Schinz, Consp. Fl. Afr.* v. 911. *Glyceria Borreri, Babingt. in Engl. Bot. Suppl. t.* 2797; *Crépin, Man. Fl. Belg. ed.* 2, 347. *G. Neesii, Steud. Syn. Pl. Glum.* i. 286. *Poa Borreri, Hook. & Arn. Brit. Fl. ed.* viii. 549. *Sclerochloa arenaria, Durand & Schinz, l.c. S. arenaria, var.* β*, *Nees, Fl. Afr. Austr.* 381 (*excl. syn.*). *S. multiculmis, subsp. Borreri, Babingt. in Engl. Bot. ed.* 3, xi. 105, *t.* 1756.

COAST REGION: "Cape District," Windvalley, on the dunes and in sandy fields, *Ecklon.* Beaufort West Div.; Nieuwefeld, near Ganze Fontein, 3500 ft., *Drège!*

WESTERN REGION: Little Namaqualand; Kamies Bergen, near Ezelfontein, 3500 ft., *Drège!*

A native of Western Europe, along the coast of the Atlantic and on the western shores of the Mediterranean; introduced (?) into North America.

I am not quite certain if Gussone's *Glyceria permixta* from Sicily is the same plant as Cosson assumes, as it is described as having fine, short, flaccid, subinvolute leaves.

2. **A. angusta** (Stapf); densely tufted; culms fascicled with numerous long-leaved intravaginal innovation shoots, 1 ft. long, erect, smooth, 1-noded near the base, terete above, much compressed and slender below the node, sheathed all along; leaves mostly crowded near the base, glabrous, glaucous; sheaths smooth, lower pallid or ultimately brown, persistent, uppermost lax, exceeding the base of the panicle, embraced below by the preceding two sheaths; ligules up to 1 lin. long, ovate; blades very narrow, linear, acute, 3–4$\frac{1}{2}$ in. by $\frac{3}{4}$ lin. (when expanded), setaceously involute, subrigid, smooth, margins rough; panicle erect, linear, contracted (at least when in flower), 4–8 in. long; rhachis slender, subflexuous above; branches very unequal, in distant fascicles of about 5, finely filiform to capillary, scaberulous, the shortest 1–3-spiculate, the longest up to 3 in. long, bare for $\frac{1}{2}$–$\frac{3}{4}$ in.; lateral pedicels $\frac{1}{2}$–2 lin. long, adpressed; spikelets distant by about their own length, linear to linear-oblong, 2–2$\frac{1}{2}$ lin. long, loosely 3–5-flowered, pallid; glumes oblong, with broad hyaline margins, the lower over $\frac{1}{2}$ lin. long, 1- to sub-3-nerved, the upper up to 1 lin. long, 3-nerved; valves obliquely-oblong in profile, about 1 lin. long with a purple mark below the tips, minutely pubescent below along the middle nerve and the outer side-nerves; pales minutely 2-toothed, keels shortly ciliate; anthers $\frac{1}{2}$ lin. long. *Sclerochloa angusta, Nees, Fl. Afr. Austr.* 381; *Durand & Schinz, Consp. Fl. Afr.* v. 911.

COAST REGION: Uitenhage Div.; in saline places by the Zwartkops River, *Ecklon!*

The description is drawn from a specimen in the Lübeck Herbarium which is accompanied by a loose label with the inscription "728. *Poa halophila*, Sched. Nieufeld. m. Drège, 1827." So closely does this specimen agree with Nees' description of *Sclerochloa augusta* that undoubtedly the label has been misplaced and belongs to *A. Borreri.* I presume that the specimen in question is Ecklon's from the Zwartkops River.

LXXXVIII. **SCLEROPOA**, Griseb.

Spikelets narrow, laterally compressed or almost cylindric when closed, in stiff panicles; rhachilla very tardily disarticulating above the glumes and between the valves. *Florets* few to many, ⚥, uppermost reduced, long exserted from the glumes. *Glumes* more or less unequal, herbaceous, oblong, acute or obtuse, 1–3-nerved, stoutly keeled, persistent. *Valves* oblong, obtuse, rounded on the back or slightly keeled, firm, herbaceous except at the very tips, 5-nerved; outer side-nerves more distinct than the faint inner; callus obscure. *Pales* almost as long as the valves, 2-keeled. *Lodicules* 2, ovate, hyaline. *Stamens* 3. *Ovary* glabrous; styles distinct, very short; stigmas laterally exserted, plumose. *Grain* very tightly embraced by the hardened valve and pale, oblong, almost semiterete, subconcave in front with a small soft terminal cap; embryo small; hilum basal, punctiform.

Annual short grasses; blades linear, flat or involute; ligules hyaline; panicles short, very stiff, often rather dense, with short, rigid, 2-ranked, more or less spreading branches; spikelets awnless.

Species 3, in the Mediterranean countries from the Canaries to Persia and in Western and Southern Central Europe.

1. **S. rigida** (Griseb. Spic. Fl. Rum. ii. 431); culms geniculate, ascending or suberect, from a few inches to 1 ft. long, glabrous, smooth, 3–5-noded, internodes more or less exserted; leaves glabrous; sheaths tight, terete, thin, striate; ligules oblong, obtuse, up to 1½ lin. long; blades linear, acute, 1–4 in. by 1–1½ lin., flat, scaberulous on the upper side, smooth below; panicle linear-oblong to ovate, rather dense, 1–3½ in. long; rhachis and all the divisions of the panicle stiff, acutely triquetrous, scabrid to puberulous; branches solitary or 2-nate, alternate on the front of the rhachis, erect, then spreading, lowest ⅙–1 in. long (exclusive of the terminal spikelet), straight, usually racemose, 2–10-spiculate with the spikelets much longer than the internodes and therefore apparently fascicled, more rarely repeatedly divided or reduced to a solitary spikelet; lateral pedicels short, stout; spikelets subcylindric with imbricate valves when closed, or linear, flat, with spreading valves when open, 2–6 lin. long, 4–12-flowered; rhachilla scaberulous; glumes acutely keeled or rounded on the back, ¾–1 lin. long, lower slightly shorter, stoutly 1-nerved, upper 3-nerved, side-nerves usually faint; valves linear-oblong in profile, rounded on the back, 1–1¼ lin. long, smooth, nerves fine, inner side-nerves faint, tips usually mucronulate; keels of pales scabrid; anthers ⅓ lin. long. *Durand & Schinz, Consp. Fl. Afr.* v. 921. *Poa rigida, Linn. Amœn. Acad.* iv. 265; *Engl. Bot. t.* 1371; *Host, Gram. Austr.* ii. *t.* 74. *Glyceria rigida, Sm. Engl. Fl.* i. 119. *Sclerochloa rigida, Panz. in Denkschr. Bot. Regensb.* i. ii. 180; *Link, Hort. Berol.* i. 150. *Festuca rigida, Kunth, Rév. Gram.* i. 129; *Enum.* i. 392; *Suppl.* 327; *Steud. Syn. Pl. Glum.* i. 302.

COAST REGION: Cape Div.; Constantia Road, near Silverlea, *Wolley Dod*, 2039! also in Newlands Woods, according to *Wolley Dod*.

A native of South and Western Europe, North Africa and Western Asia; elsewhere introduced.

LXXXIX. FESTUCA, Linn.

Spikelets laterally compressed (at least after flowering), panicled; rhachilla disarticulating above the glumes and between the valves. *Florets* 2 to many, ⚥ (the uppermost usually reduced), distinctly exserted from the glumes. *Glumes* subequal, rarely conspicuously unequal, subherbaceous or subscarious, more or less keeled, 1–3-(rarely the upper 5-) nerved. *Valves* lanceolate, muticous, mucronate or awned, herbaceous, rounded on the back or keeled towards the tip, 5–7-nerved; mucro or awn from the tip or very close to it, straight; callus small, glabrous or almost so. *Pales* 2-keeled, more or less 2-toothed. *Lodicules* 2, unequally 2-lobed or entire, hyaline. *Stamens* 3. *Ovary* glabrous or the top minutely hairy; styles distinct, very short, terminal or subterminal; stigmas plumose, laterally exserted. *Grain* tightly enclosed by the slightly hardened valve and pale, free or more or less adhering to the pale or to both, oblong, dorsally convex, grooved or concave in front; hilum linear, long; embryo small.

Perennial, tufted; sheaths open or more or less closed; ligules scarious; blades flat, folded or convolute, often setaceous; panicles more or less compound, contracted or open, sometimes effuse.

Numerous species, often very difficult to discriminate, in the temperate regions, particularly of the northern hemisphere, and a few in the high mountains of the tropics.

Florets hermaphrodite:
- Blades permanently folded, filiform or setaceous; panicle contracted, flexuous; spikelets 4–6 lin. long; valves mucronate or shortly awned ... ... (1) **caprina.**
- Blades convolute in bud, then flat or involute when dry:
  - Top of ovary glabrous; panicle almost reduced to a raceme; ligules extremely short ... ... (2) **vulpioides.**
  - Top of ovary hairy; ligules 1–2 lin. long:
    - Panicle compound, more or less nodding; branches long, flexuous, divided from about the middle; spikelets 6–8 lin. long; valves finely but distinctly nerved ... ... (3) **costata.**
    - Panicle subcorymbose, at length more or less divaricate; branches long, straight, with 4–1 spikelets near the tips; spikelets 4–5 lin. long; valves 5-ribbed ... ... (4) **longipes.**

Diœcious; panicles usually dense and almost spike-like; spikelets 4–5 lin. long; culms and young shoots tightly coated at the base by the fibres of the old sheaths ... (5) **scabra.**

1. F. caprina (Nees, Fl. Afr. Austr. 443); densely cæspitose with intravaginal innovation shoots; culms erect, slender, about 1½ ft. high, smooth, 2-noded, uppermost node at or below the middle; sheaths closed at the base only, tight, smooth, thin, lowest subpersistent; ligule very short, more or less obtusely auricled; blades filiform, acute, 4–8 in. long, folded, flexuous, 5–7-nerved, smooth on the back, scabrid on the face; panicle contracted, narrow, flexuous, 2–6 in. long; rhachis angular, scabrid or smooth below; branches usually solitary, lower remote, flexuous or straight, finely filiform, scaberulous, very sparingly branched from near the base, or the lowest up to 1½–2 in. long, and undivided for ½–1 in.; lateral pedicels short to very short; spikelets elliptic-oblong to oblong, 4–6 lin. long, 3–8-flowered, greenish; glumes equal or subequal, lanceolate to oblong-lanceolate, acute, 1½–2½ lin. long, herbaceous-membranous, scaberulous, lower 1-, upper 3-, nerved; valves linear-lanceolate in profile, acute, usually produced into a short fine bristle, smooth or very finely scaberulous near the tip; nerves 5, very faint; keels of pales scaberulous; lodicules ovate-lanceolate; anthers up to 1½ lin. long; top of ovary minutely hispid. *Steud. Syn. Pl. Glum.* i. 313; *Durand & Schinz, Consp. Fl. Afr.* v. 913. *F. costata, var. longiseta, Nees, Fl. Afr. Austr.* 447 (*in part*).

VAR. β, **irrasa** (Stapf); blades shorter and more rigid, 3–6 in. long; valves densely pubescent.

VAR. γ, **macra** (Stapf); blades very finely filiform, rough, 3–5-nerved; panicle very lax and flexuous; branchlets very fine, 2–5-spiculate; pedicels rather long; spikelets slightly smaller and with longer awns.

SOUTH AFRICA: without precise locality, *Drège*, 3920!

COAST REGION: Queenstown Div.; Table Mountain, 5000–6000 ft., *Drège!* Storm Bergen, 5000–6000 ft., *Drège!* King Williamstown Div.; Amatola Mountains, *Buchanan*, 39! Var. β: Albany Div.; Howisons Poort, near Grahamstown, *Flanagan*, 94! King Williamstown Div.; Amatola Mountains, *Buchanan*, 37! 38! 41!

KALAHARI REGION: Orange Free State; without precise locality, *Buchanan*, 51! Var. γ: Orange Free State; on the Witte Bergen, near Harrismith, *Buchanan*, 262!

2. F. **vulpioides** (Steud. Syn. Pl. Glum. i. 305); perennial; culms shortly ascending, subgeniculate, 1½–2 ft. long, smooth, 3–4-noded, internodes (except the lowest) exserted; leaves glabrous; lower sheaths somewhat tight, subpersistent, others loose, all smooth; ligules very short or reduced to a membranous rim; blades linear, acute, 4–6 in. by 2–2½ lin., flat, scaberulous towards the tips and along the margins; panicle very narrow, almost reduced to a raceme, straight or flexuous, 6–7 in. long; rhachis like the branches angular or ancipitous, scaberulous or smooth below; branches 2-nate except the upper, unequal, 1- (or the lower 2-) spiculate and up to 1½ in. long, flexuous, erect; lateral pedicels (if present) short to very short; spikelets oblong, 7–10 lin. long, light green, 5–9-flowered; glumes subulate to lanceolate in profile, 1¾–3 lin. long, equal, 3-nerved, or the lower shorter and 1-nerved, nerves prominent; valves lanceolate in profile, acutely acuminate or produced into a very short awn, 4–5½ lin. long, firmly herbaceous, smooth or scaberulous toward the tip, inconspicuously 5- to sub-7-nerved; keels of pales stout, crested, scaberulous; lodicules unequally 2-lobed, denticulate;

anthers 1½ lin. long; ovary quite glabrous. *Durand & Schinz, Consp. Fl. Afr.* v. 921. *Vulpia megastachya, Nees, Fl. Afr. Austr.* 441.

SOUTH AFRICA: without precise locality, *Drège!*
COAST REGION: King Williamstown Div.; Amatola Mountains, *Buchanan*, 45!

3. **F. costata** (Nees, Fl. Afr. Austr. 447, excl. var.); perennial; densely tufted; innovation shoots intravaginal; culms erect, robust, up to 3 ft. long, smooth, about 2-noded, internodes exserted; leaves glabrous; basal sheaths crowded, the outer coriaceous, brown, at length breaking up into fibres, persistent, upper somewhat lax, striate to sulcate, green, smooth, uppermost more than ½ ft. long; ligules ovate, obtuse, 1½–2 lin. long; blades convolute in bud, then flat, or involute when dry, much narrower than the sheath, linear, tapering to a very acute point, lower up to more than 1 ft. by 1½–3 lin., firm, smooth on both sides or scabrid above and along the margins, prominently ribbed on the upper side, with a continuous sclerenchymatous layer on the back; panicle contracted or open, lax, angular, or ancipitous above; branches 2- or 3-nate, lower very remote, filiform, flexuous, erect or spreading, angular or ancipitous, smooth or scabrid, lowest 4–6 in. long, usually undivided to the middle or beyond, then loosely divided; branchlets contracted, fine; pedicels very unequal, 1–6 lin. long, often very flexuous; spikelets erect or nodding, elliptic, 6–8 lin. long, green, tinged with purple or violet, very loosely 3–7-flowered; rhachilla very slender; joints up to 1 lin. long; glumes subequal or lower distinctly shorter, 2½–3 lin. long, upper 3–4 lin. long, both lanceolate, acute, 3-nerved, smooth; valves lanceolate in profile, acute or minutely mucronate, 4–5½ lin. long, rather firm, scaberulous, distinctly 5-nerved; keels of pale scaberulous; lodicules lanceolate, over ¾ lin. long; anthers 2–3 lin. long; top of ovary pubescent. *Steud. Syn. Pl. Glum.* i. 313; *Durand & Schinz, Consp. Fl. Afr.* v. 913.

VAR. β, **longiseta** (Nees, l.c., in part); blades 4–8 in. long, rigid, convolute; panicle scanty; valves produced into a fine bristle, 1–1½ lin. long.

COAST REGION: Albany Div., *Ecklon & Zeyher*, 874! Queenstown Div.; mountains on the Klipplaat River, *Ecklon*. Stockenstrom Div.; on the Winter Berg, near Philipton, *Ecklon!* Kat Berg, 4000–5000 ft., *Drège!* Cathcart Div.; Windvogel Berg, 5000 ft., *Drège!* Var. β: Stockenstrom Div.; Kat Berg, 4000–5000 ft., *Drège!*
EASTERN REGION: Natal; Riet Vlei, 4000–5000 ft., *Buchanan*, 234! Benvie, Kar Kloof, 3000–4000 ft., *Wood*, 6005! and without precise locality, *Buchanan*, 53!

Also in Nyasaland.

4. **F. longipes** (Stapf); tufted; innovation shoots extravaginal; culms shortly ascending, sometimes geniculate, about ½ ft. high (exclusive of the panicle), usually scabrid below the panicle, 1–2-noded, internodes enclosed or very shortly exserted; leaves glabrous; sheaths mostly rather loose, prominently striate, smooth or slightly rough; ligules ovate, rounded, 1–1½ lin. long; blades linear, tapering to a fine point, 4–6 in. by 2–3 lin., convolute in bud, then flat,

or involute when dry, green or glaucous on the upper side, smooth except on the scabrid margins or rough above, 9–13-nerved, nerves very prominent above; panicle large, subcorymbose, over ½ ft. long, base enclosed in uppermost sheath or very shortly exserted; rhachis straight, angular, or acutely ancipitous above, very scabrid; branches 2–3- (rarely 4-) nate, distant, 4–6 in. long, straight, at first suberect, then spreading, spinulously scabrid along the angles, with 4–1 spikelets near the tips; pedicels very unequal, very short, up to 8 lin. long; spikelets 3–5-flowered, oblong to elliptic, 4–5 lin. long, light green, erect; glumes unequal to subequal, lanceolate in profile, about 2 lin. long, acute, lower smaller, upper broader, margins scarious; valves lanceolate in profile, acute, 3½ lin. long, herbaceous-chartaceous, stoutly 5-ribbed, scaberulous, at least towards the rigid tips, keels stout, scabrid; lodicules unequally 2-lobed, ½ lin. long; anthers 1½ lin. long; ovary top puberulous. *F. costata, var. fascicularis, Nees, Fl. Afr. Austr.* 447.

COAST REGION: Albany Div.; on stony slopes, near Grahamstown, 2000 ft, *MacOwan*, 1323! Komgha Div.; valleys near Komgha, *Flanagan*, 898! King Williamstown Div.; Amatola Mountains, *Buchanan*, 35! Stockenstrom Div.; on the Winter Berg, *Ecklon*.
EASTERN REGION: Tembuland; Bazeia, 2000 ft., *Baur*, 363!

5. **F. scabra** (Vahl, Symb. Bot. ii. 21); diœcious; compactly cæspitose; innovation shoots intravaginal; culms erect, ½–3 ft. long, rather robust, smooth, very rarely scaberulous, 1–2-noded, uppermost internode long exserted; leaves crowded at the base; basal sheaths broad, firm, open to the base, breaking up at length and forming dense tight coats of fibres, smooth; upper sheaths rather tight, gradually longer, striate, smooth, rarely scabrid to very scabrid; ligules ovate, obtuse, up to 1½ lin. long; blades convolute in bud, then open or involute when dry, linear, acute, 3–8 in. by 1½–3 lin., firm, usually rather rigid, 9–13-nerved, smooth below, scabrid above or rarely very scabrid all over, nerves prominent on the upper side; panicle very narrow, dense, often spike-like, sometimes interrupted, straight or slightly nodding, 3–8 in. long; rhachis terete, striate and smooth below, angular and scabrid to hispidulous above; branches 2-nate or solitary, sparingly to abundantly and very closely divided from the base or rarely undivided for a few lines; branchlets and pedicels very short, scabrid; spikelets densely crowded, oblong, 4–5 lin. long, green, often tinged with purple, or glaucescent, closely 4–6-flowered; florets unisexual with the rudiments of the other sex; glumes unequal to subequal, linear-lanceolate in profile, acute to subobtuse, firmly herbaceous except at the hyaline margins, smooth or scaberulous, lower usually shorter, 1¼–2½ lin. long, 1-nerved, upper 1¾–3 lin. long, 3-nerved; valves oblong, acute or subacute, sometimes mucronulate, 2¾–3½ lin. long, scaberulous at least in the upper part, distinctly 5-nerved; pales minutely 2-toothed or entire, keels stout, scabrid; lodicules ovate, acute; anthers 1½ lin. long on fine long filaments in the ♂, very minute and empty and on very short delicate filaments in the ♀; ovary obovoid, very densely

villous from the middle upwards, with rather short loosely plumose stigmas in the ♀, quite rudimentary in the ♂; grain oblong, dorsally compressed, subconcave in front, 2½ lin. long; embryo short; hilum filiform, $\frac{3}{4}$–$\frac{4}{5}$ the length of the grain. *Kunth, Enum.* i. 409; *Nees in Linnæa*, vii. 323; *Fl. Afr. Austr.* 444; *Steud. Syn. Pl. Glum.* i. 313; *Durand & Schinz, Consp. Fl. Afr.* v. 918. *F. aspera, Poir. Encycl. Suppl.* ii. 635, *not of Labill. F. neesiana, Steud. Nomencl. ed.* ii. i. 631. *Dactylis lævis, Thunb. Prodr.* 22; *Fl. Cap.* 426; *ed. Schult.* 114 (♀). *D. serrata, Thunb. Prodr.* 22; *Fl. Cap.* 428; *ed. Schult.* 115 (♂). *Lasiochloa lævis and L. serrata, Kunth, Enum.* i. 388.

SOUTH AFRICA: without precise locality, *Thunberg!*

COAST REGION: Cape Div.; near Capetown, *Ecklon; Harvey*, 167! *Burchell*, 459! Table Mountain, *Bergius*; in a plantation, near Disa Gorge, *Wolley Dod*, 2158! Lions Head, *Ecklon, Wilms*, 3871! common below a plantation near Wynberg, *Wolley Dod*, 3301! near Whitemash Mark, Red Hill, *Wolley Dod*, 2979! slopes towards Smithwinkel Bay, *Wolley Dod*, 2976! Batsata Rock and Pauls Berg, *Wolley Dod*, 2948! Simons Bay, *Wright!* Paarl Div.; Paarl Mountains, 2000 ft., *Drège!* Tulbagh Div.; Tulbagh Kloof, Winterhoek, &c., *Ecklon*. Worcester Div.; Hex River East, *Wolley Dod*, 3706! Hex River Valley at Els Kloof, *Wolley Dod*, 3707! Stellenbosch Div.; Hottentots Holland Mountains, *Ecklon*. Caledon Div.; Bavians Kloof near Genadendal, 1000–2000 ft., *Drège!* near Caledon, *Thom*, 922! on the Zwart Berg, near the Hot Springs, 1000–2000 ft., *Zeyher*, 4595! Swellendam Div.; Puspus Valley, near Swellendam, *Zeyher*, 4594! *Ecklon*; Duivenhoeks River, *Gill!* Riversdale Div.; near the Zoetemelks River, *Burchell*, 6683! Mossel Bay Div.; Little Brak River, *Burchell*, 6186! George Div.; between George and Malgat River, *Burchell*, 6085! and without precise locality, *Mund & Maire*. Uitenhage Div.; by the Zwartkops River, *Ecklon*. Albany Div.; plains of Albany, *Bowie*, 350! stony ledges of Bothas Berg, 2200 ft., *MacOwan*, 1274! near Grahamstown, *MacOwan!* Komgha Div.; Komgha, *Flanagan*, 944! Queenstown Div.; near Shiloh, 3500 ft., *Drège*.

CENTRAL REGION: Graaff Reinet Div.; near Graaff Reinet, 3200 ft., *Bolus*, 682! Aliwal North Div.; Witte Bergen, 7000–7500 ft., *Drège*.

WESTERN REGION: Little Namaqualand; Rood Berg and Ezels Kop, 4000–5000 ft., *Drège!*

KALAHARI REGION: Orange Free State, without precise locality, *Cooper*, 914!

EASTERN REGION: Tembuland; Tabase near Bazeia, 2000 ft., *Baur*, 312! 763! Griqualand East; hills near Kokstad, 5000 ft., *Tyson*, 1789! Natal; on the Drakensberg Range, near Newcastle, *Buchanan*, 185!

XC. VULPIA, Gmel.

Spikelets laterally compressed after flowering, on short clavate pedicels in usually more or less secund and spike- or raceme-like panicles; rhachilla slender, disarticulating above the glumes and between the fertile valves. *Florets* 5–7, long exserted from the glumes, ⚥, except the reduced upper ones, or the lowest ⚥ and the rest reduced to empty valves. *Glumes* very unequal, lower very minute or obsolete, or like the upper subulate to subulate-lanceolate, but much shorter, 1- (or the upper 3-) nerved. *Valves* subulate-lanceolate, passing into an awn, rounded on the back, faintly 5-nerved; awn straight, often long; callus small, obtusely glabrous. *Pales* 2-keeled, entire or minutely 2-toothed. *Lodicules* 2, hyaline, unequally lobed. *Stamens* 1–3; filaments very short; anthers usually enclosed in the floret during flowering or permanently. *Ovary* glabrous (in the South African species) or minutely hispid at the top; stigmas sessile, plumose, permanently enclosed in the floret, or shortly exserted at the base. *Grain* linear, strongly compressed from the back, concave in front, more or less adhering to the pale or also to the valve; embryo small; hilum filiform, long.

Annual or perennial, slender grasses; blades linear, very narrow, usually convolute or involute, at least when dry; panicles contracted, narrow, usually more or less secund, with short clavate pedicels; spikelets subcylindric and acuminate, when young, then opening out, laterally compressed and broader upwards; flowers often cleistogamous.

Species about 20, mostly in the Mediterranean region and the adjacent countries; the 2 species described below have been introduced into many parts of the world.

Uppermost internode usually wholly enclosed in the sheath; panicle 2–10 in. long; lower glume reduced to a minute scale or up to ¾ lin. long and subulate, nerveless or like upper 1-nerved ... ... (1) **Myuros.**

Uppermost internode long exserted; panicle 1–3 in. long; both glumes subulate-lanceolate, lower 1–2 lin. long and 1-nerved, upper 2–3½ lin. long and strongly 3-nerved ... ... (2) **bromoides.**

1. **V. Myuros** (Gmel. Fl. Bad. i. 8); annual, tufted; culms slender, geniculate, ascending or suberect, ½–1½ ft. high, glabrous, smooth, 2- (sometimes 3-) noded, uppermost internode 2½–6 in. long, usually wholly enclosed in the uppermost sheath; sheaths (particularly the upper) rather loose, smooth, glabrous; ligules very short, often obtusely auricled; blades linear, tapering to a very acute point, 1–6 in. by ½–1 lin., flat or involute when dry, or setaceous, flaccid to subrigid, finely and prominently few-nerved, puberulous or scabrid on the upper surface, otherwise glabrous and smooth; panicle spike-like, erect or nodding and flexuous, narrow, subsecund or secund or facing all sides, 2–10 in. long,; rhachis filiform, acutely triquetrous, like the branches scabrid along the angles or smooth below; branches fascicled or 2-nate and very unequal, or solitary (lowest often very remote), racemose from the base or the upper reduced to a solitary spikelet, adpressed or lowest slightly nodding; lateral pedicels about ½ lin. long, smooth; spikelets rather close or the lowest of the lower branches remote, 3½–5 lin. long (exclusive of the awns), loosely 3–6-flowered; rhachilla joints up to ¾ lin. long; lower glume reduced to a minute scale (particularly in the lateral spikelets) or like the upper subulate, but much shorter (up to ¾ lin. long), nerveless or 1-nerved, upper 1½–2½ lin. long, acute, setaceously acuminate, 1-nerved; valves linear-lanceolate, acuminate in profile, 2–3½ lin. long, faintly 5-nerved, scabrid, sometimes ciliate in the upper part; awn 3–10 lin. long, fine, scabrid; stamen 1; anther ⅙–⅖ lin. long; grain 1½–2 lin. long. *Boiss. Fl. Or.* v. 628. *V. Pseudo-myurus*, *Reichb. Fl. Germ. Excurs.* i. 37; *Ic. Fl. Germ.* i. *t.* 60, *fig.* 1525. *V. bromoides*, *Celak. in Kern. Sched. ad Fl. Austr.-Hung. Exsicc.*

no. 1082. *V. bromoides, var. rigida, Nees, Fl. Afr. Austr.* 441. *Festuca Myurus, Linn. Sp. Plant.* 74; *Leers, Fl. Herb.* 33, *t.* 3, *fig.* 5; *Host, Gram. Austr.* ii. 66, *t.* 93; *Engl. Bot. t.* 1412; *Kunth, Enum.* i. 396; *Steud. Syn. Pl. Glum.* i. 303; *Durand & Schinz, Consp. Fl. Afr.* v. 916 (*excl. syn. Festuca bromoides*).

SOUTH AFRICA: without precise locality, *Drège*, 2058!

COAST REGION: Clanwilliam Div.; Clanwilliam, 350 ft., *Schlechter*, 8587! Vanrhynsdorp Div.; Ebenezer, 500 ft., *Drège*. Cape Div.; Table Mountain, *Ecklon*, 972 in Kew Herb. partly! near cultivated ground at Kenilworth, 100 ft., *Bolus*, 7923! Uitenhage Div.; by the Zwartkops River, *Ecklon & Zeyher*, in the Lübeck Herb.!

CENTRAL REGION: Somerset Div.; between Zuurberg Range and Klein Bruntjes Hoogte, 2000–2500 ft., *Drège!*

WESTERN REGION: Little Namaqualand; near Kuil, between Pedros Kloof and Lily Fontein, 3500 ft., *Drège*.

2. **V. bromoides** (S. F. Gray, Nat. Arr. Brit. Pl. ii. 124); annual, tufted; culms geniculate, ascending or suberect, very slender, ½–1½ ft. high, glabrous, smooth, 2–3-noded, uppermost internode usually occupying more than ½ of the culm, long exserted; sheaths somewhat loose, glabrous, smooth; ligules extremely short; blades linear tapering to a very acute point, 1 to more than 6 in. by ½–1 lin., flat or involute or convolute when dry, flaccid to subrigid, finely and prominently few-nerved, scabrid to puberulous on the upper surface, otherwise glabrous and smooth; panicle erect or nodding and flexuous, very narrow, more or less secund, 1–3 in. long, rarely longer, sometimes reduced to a raceme; rhachis filiform, acutely triquetrous like the branches, scabrid along the angles or smooth below; branches fascicled, 2-nate or solitary, racemose or the lowest again divided below, uppermost, or sometimes all reduced to a single spikelet, erect or nodding; lateral pedicels about 1 lin. long, smooth or almost so; spikelets 4–6 lin. long (exclusive of the awns), loosely 4–7-flowered; glumes subulate-lanceolate, acute or acuminate, lower 1–2 lin. long, 1-nerved, upper 2–3½ lin. long, strongly 3-nerved; valves lanceolate-linear, acuminate in profile, 2½–4 lin. long, faintly 5-nerved, scaberulous at least above; awn as long as the valve or longer, up to 6 lin. long, very fine; stamen 1; anther $\frac{1}{6}$–$\frac{1}{3}$ lin. (rarely ½ lin.) long; grain about 2–2½ lin. long. *Link, Hort. Berol.* i. 147, ii. 271; *Reichb. Fl. Germ. Excurs.* 37; *Ic. Fl. Germ.* i. *t.* 60, *fig.* 1529; *Nees in Linnæa*, vii. 324; *Fl. Afr. Austr.* 440 (*excl. var.* δ). *V. sciuroides, Reichb. Fl. Germ. Excurs.* 37; *Boiss. Fl. Or.* v. 628. *V. sp., Gmel. Fl. Bad.* i. 9. *Festuca bromoides, Linn. Sp. Plant.* 75; *Thunb. Prodr.* 22; *Fl. Cap.* 426; *ed. Schult.* 114; *Engl. Bot. t.* 1411; *Kunth, Enum.* i. 396; *Steud. Syn Pl. Glum.* i. 303. *F. sciuroides, Roth, Bot. Abhandl.* 43; *Catal. Bot.* ii. 11; *Durand & Schinz, Consp. Fl. Afr.* v. 919.

COAST REGION: Malmesbury Div.; Zwartland, *Thunberg*. Cape Div.; among shrubs and by the roadside on Table Mountain, *Ecklon*, 972 partly! Lion Mountain, *Pappe!* mouth of tunnel, Orange Kloof, *Wolley Dod*, 2127! by Kaapenberg Vley, *Wolley Dod*, 2751! Old road to Constantia Nek, *Wolley Dod*, 3486! Tulbagh Div.; near Tulbagh Waterfall, *Ecklon*. Paarl Div.; by the

Berg River near Paarl, *Drège!* Stellenbosch Div.; Hottentots Holland, *Ecklon.* Swellendam Div.; hills by the Zonder Einde River, *Ecklon;* in the forest at Grootvaders Bosch, *Burchell,* 7237! Riversdale Div.; near the Zoetemelks River, *Burchell,* 6633! George Div.; Lange Kloof Mountains, *Ecklon.* Uitenhage Div.; by the Zwartkops River, *Ecklon!* near Uitenhage, *Burchell,* 4223! Albany Div.; Brand Kraal, near Grahamstown, *MacOwan,* 1275!

Throughout the Mediterranean region and in Central and Western Europe, on Cameroon Peak, and in Abyssinia; elsewhere introduced.

I found the number of stamens constantly 1 in the South African specimens as well as in the European.

XCI. BROMUS, Linn.

Spikelets laterally compressed, at least after flowering, variously panicled; rhachilla disarticulating above the glumes and between the valves. *Florets* usually numerous, ⚥, the uppermost reduced. *Glumes* more or less unequal, acute to acuminate, persistent, lower 1–7-, upper 3–9-, nerved and sometimes mucronate or aristulate. *Valves* lanceolate or broadly oblong, rounded on the back or keeled, 5–13-nerved, usually awned; awn terminal (rarely 3 or 0) or somewhat distant from the often 2-toothed tip, straight or recurved and then often loosely twisted below, not kneed. *Pales* entire or bifid; keels usually rigidly ciliolate or ciliate. *Lodicules* 2, oblong or lanceolate, entire or lobed. *Stamens* 3, rarely 2. *Ovary* obovoid with an often large, villous, 3- or 2-lobed terminal appendage; styles short, lateral on the appendage; stigmas plumose, laterally exserted, or in the cleistogamic species permanently enclosed. *Grain* linear to linear-oblong, convexo-concave and usually adherent to the valve and pale, or at least to the latter; hilum filiform, long; embryo small.

Annual or perennial, of very varying habit; blades linear, flat, often flaccid; ligules membranous, hyaline; panicle contracted, often very dense, or open and even effuse, or reduced to a raceme; spikelets rather large, erect or pendulous, from ovoid to linear-oblong or linear-cuneate; awns very long or short, sometimes reduced to a mucro or 0.

Species rather numerous, mostly in the temperate regions of the northern hemisphere and of South America, several in the high mountains of the tropics.

Section 1. SERRAFALCUS (Koch). Annual; spikelets ovate-lanceolate to oblong; lower glume 3–5- (rarely 1-) nerved; upper glume 5–9-nerved; valves 7–9-nerved; internodes of rhachilla more or less clavate above.

Panicle contracted, dense; branches shorter than the spikelets ... ... (1) **molliformis.**
Panicle loose, open or contracted; lower branches much longer than the spikelets, flexuous:
  Panicle ample, erect, contracted at first, then more or less effuse with very long flexuous or nodding branches; spikelets 5–9 lin. long, rarely longer; valves about 3½ lin. long; anthers 1 to almost 2 lin. long ... ... (2) **arvensis.**
  Panicle more or less secund; longest branches not exceeding 2 in.; spikelets 6–12 lin. long; valves 3½–6 lin. long; anthers ¼–⅜ lin. long:
    Spikelets ovate-oblong; florets close; valves obliquely obovate-oblong in profile, tips

short, obtuse or subobtuse; awns of lower valves usually much shorter than those of the upper ... ... (3) commutatus.
Spikelets oblong; florets rather loose; valves obliquely oblong to linear-oblong in profile; tips acute, entire or 2-toothed; awns nearly equal ... ... (4) patulus.

Section 2. STENOBROMUS (Griseb.). Annual; spikelets narrow, more or less cuneiform; lower glume 1- (rarely 3-) nerved, upper 3–5-nerved; valves 5–7-nerved; internodes of rhachilla slender.

Only species in South Africa ... ... (5) maximus.

Section 3. FESTUCARIA (Coss. & Dur.). Perennial, often tall; lower glume 1–3-nerved, upper 3–5-nerved; valves 5–7- (very rarely 9-) nerved; awn terminal, subterminal or 0.

Lower glume 1- (rarely 3-), upper 3-nerved; anthers 1–1½ lin. long ... ... (6) leptoclados.
Lower glume 3-, upper usually 5-nerved; anthers about 3 lin. long:
Spikelets glabrous; glumes 5–7½ lin. long; valves 7½–9 lin. long, prominently nerved ... ... (7) natalensis.
Spikelets pubescent to tomentose or villous; glumes 3½–6 lin. long; valves 6–7 lin. long; nerves not prominent:
Florets rather loose; valves linear-lanceolate in profile, narrow ... ... (8) speciosus.
Florets close; valves lanceolate-cuneate, rather broad above ... ... (9) firmior.

Section 4. CERATOCHLOA. Glumes and valves strongly compressed, acutely keeled, many-nerved, the latter usually 13-nerved.

Only species in South Africa ... ... (10) unioloides.

1. **B. molliformis** (Lloyd, Fl. Loire-Inf. 315); annual; culms geniculate, ascending or suberect, 1–2 ft. high, glabrous, smooth, 3–4-noded, lower and intermediate internodes usually enclosed; leaves more or less spreadingly hairy or the blades subglabrous; sheaths tight, thin, strongly striate; ligules hyaline, delicate, ½–1 lin. long, denticulate; blades linear, acute, 3–5 in. by 2–3 lin., flat, strongly striate, scarcely flaccid; panicle oblong to linear-oblong, contracted, erect, rather dense, 2–3 in. long; axis striate, scabrid to scaberulous or almost smooth below; branches 3–6-nate, nearly always 1-spiculate, unequal, 2–4 lin. long, rarely longer, scabrid to coarsely pubescent, erect; spikelets oblong, greyish-green, 6–10 lin. by 2½–3 lin., erect, crowded, closely 9–13-flowered; rhachilla joints ½ lin. long, subclavate, stout, scabrid; glumes unequal, oblong in profile, rather broad, acute, herbaceous except at the narrow membranous margins, pubescent to villous; lower 2½–3 lin. long, 5- to sub-5-nerved, upper 3–4 lin. long, 7-nerved; valves rather narrow, oblong, about 4 lin. long, pubescent to villous, 7-nerved; margins almost straight to ¾ of the length from the base, then gradually curved or turned in at a distinct obtuse angle to the broad 2-toothed hyaline tips; awn from close to the tip, fine, scabrid, straight or finally divaricate, 2½–3½ lin. long; callus very

short; pales somewhat shorter than the valves, keels distinctly ciliate; stamens 3; anthers ½ lin. long; ovary glabrous below the appendage, front lobe of appendage entire, hind lobe very minute; grain linear, about as long as the pale, strongly convexo-concave. *B. divaricatus, Lloyd, l.c.* 314, *not of Rhode. B. confertus, Boreau, Fl. Centr. France, ed.* ii. 586. *Serrafalcus lloydianus, Gren. & Godr. Fl. de France,* iii. 591.

COAST REGION: Cape Div.; by the railway near Rondebosch, *Wolley Dod,* 1826!

A native of the Western Mediterranean region and West Europe.

2. **B. arvensis** (Linn. Sp. Plant. i. 77); annual; culms geniculate, ascending or suberect, rather slender, 1–2 ft. long, glabrous, smooth, 3–5-noded, upper internodes exserted, uppermost usually very long; leaves more or less shortly and spreadingly hairy all over; sheaths tight, thin, striate; ligules hyaline, obtuse, 1–1½ lin. long; blades linear, acute, 2–6 in. by 1–3 lin., flat, soft, strongly striate; panicle erect or slightly nodding, at first contracted, then opening out, sometimes effuse, very lax, 4–9 in. long and about as wide; axis terete, slender, smooth, at least below; branches in distant semiwhorls of 3–8, usually very unequal, finely filiform, flexuous, scabrid, the longest often 3–5 in. long, usually bare to beyond the middle, 8–1-spiculate; spikelets ellipsoid to oblong, 5–9 lin. long, glabrous, green or variegated, 5–10-flowered; rhachilla joints ½–¾ lin. long, stout, scaberulous; glumes unequal, lanceolate to oblong, acute, lower about 2½ lin. long, 3-nerved, upper much broader, about 3 lin. long, 5–7-nerved, nerves strong; valves obliquely obovate-oblong to oblanceolate, about 3½ lin. long, smooth or nearly so, distinctly 7-nerved, margins straight to the middle or beyond, then turned at an obtuse angle or a distinct curve to the broad minutely 2-toothed tip, membranous, whitish; pale slightly shorter than the valves, keels loosely and rigidly ciliate; stamens 3; anthers 1 to almost 2 lin. long; grain linear-oblong, strongly convexo-concave, about 3 lin. long, adhering to the valve and pale. *Fl. Dan. t.* 2527; *Engl. Bot. t.* 1984; *Host, Gram. Austr.* i. *t.* 14; *Kunth, Enum.* i. 417; *Reichb. Ic. Fl. Germ.* i. *t.* 74, *fig.* 1587; *Knapp, Gram. Brit. t.* 82; *Sowerby, Brit. Grass. t.* 107; *Nees, Fl. Afr. Austr.* 452; *Steud. Syn. Pl. Glum.* i. 324; *Durand & Schinz, Consp. Fl. Afr.* v. 922. *Serrafalcus arvensis, Godr. Fl. Lorr. ed.* i. iii. 185.

COAST REGION: Vanrhynsdorp Div.; near Ebenezer, 500 ft., *Drège.*

A native of Europe.

3. **B. commutatus** (Schrad. Fl. Germ. i. 353); annual; culms fascicled, geniculate, ascending or suberect, slender to stout, 1 to more than 2 ft. long, glabrous, smooth, 2–4-noded, upper internodes exserted, uppermost very long; leaves usually loosely hairy; sheaths tight, striate; ligules hyaline, delicate, toothed, about 1 lin. long; blades linear, tapering to an acute point, 2–6 in. by ½–3 lin., flat, flaccid; panicle usually contracted, narrow, suberect or nodding and

more or less secund, 3–6 in. long; axis terete, scaberulous or smooth below; lower branchlets 2–4- (or in robust specimens to 6-) nate, very unequal, 1–2- (rarely 3- or 4-) spiculate, erect or nodding, finely filiform, flexuous, scaberulous to scabrid, the longest up to 2 (rarely 3) in. long; lateral pedicels 2–8 lin. long; spikelets ovate-oblong, light green, 6–12 lin. long, closely 5–12-flowered; rhachilla joints scarcely clavate, almost smooth, ½ lin. long; glumes unequal, oblong, acute or subobtuse, herbaceous except at the narrow subhyaline margins, smooth or scaberulous on the nerves, lower 2½–3 lin. long, 3–5-nerved, upper much broader, 3½–4½ lin. long, 7–9-nerved; valves obliquely obovate-oblong in profile, 3½–5 lin. long, 7–9-nerved, scaberulous, rarely pubescent, rather firm; margins almost straight for ⅔ or ¾ their length, then strongly curved towards the obtuse or subobtuse apex, whitish, scarcely shining; awn permanently straight, scaberulous, usually shorter to almost obsolete in the lower, gradually longer (to 5 lin.) in the upper valves, from ½–¾ lin. below the tip; pales 3½–4 lin. long, obtuse, keels distantly ciliate; anthers elliptic, ⅜–¾ lin. long; grain 3–3½ lin. long, tightly adhering to valve and pale, strongly convexo-concave. *Gmel. Fl. Bad.* iv. 72, *t.* 4; *Kunth, Enum.* i. 414; *Reichb. Ic. Fl. Germ.* i. *t.* 74, *fig.* 1589; *Steud. Syn. Pl. Glum.* i. 325; *Fl. Dan. t.* 2526; *Durand & Schinz, Consp. Fl. Afr.* v. 923.

CENTRAL REGION: Graaff Reinet Div.; in cultivated ground in the Sneeuwberg Range, 4500 ft., *Bolus*, 2057! (introduced).

Throughout Europe and in North Africa, usually as a weed.

4. **B. patulus** (Mert. & Koch in Röhl. Deutschl. Fl. i. 685); annual; culms fascicled or tufted, geniculate, shortly ascending or suberect, slender to stout, 1–1½ ft. high, glabrous, smooth or scaberulous to pubescent below the panicle and the nodes, 2–4-noded, upper internodes exserted, uppermost very long; leaves finely hairy to villous; sheaths rather tight, thin, strongly striate; ligules delicate, ¾ to almost 2 lin. long, fimbriate-toothed; blades linear, tapering to an acute point, 2–4 in. by 1½–3 lin. (in the South African specimens), flat, flaccid; panicle usually more or less nodding, 3–5 in. long (in the South African specimens), broadly ovoid when quite open; axis terete, striate, scabrid; branches in rather distant semiwhorls, lower 3–6-nate, 2–1- (rarely 3-) spiculate, nodding, finely filiform, very flexuous, scabrid; lateral pedicels 2 to more than 6 lin. long; spikelets oblong, light green or purplish, 8–12 lin. long, 7–10-flowered; rhachilla joints clavate, smooth or almost so, up to 1 lin. long; glumes unequal, oblong, acute, lower 3–4 lin. long, 1–3-nerved, upper broader, 4–5½ lin. long, 5–7-nerved, scaberulous on the nerves; valves obliquely oblong, 4–6 lin. long (in the South African specimens), 7- to sub-9-nerved, scaberulous or scabrid at least towards the tips, rarely quite smooth, widest at the middle or slightly above it, margins gently curved from this point towards the tip and white, shining and hyaline, straight towards the base; awn from 1–1½ lin. below the entire or denticulate or 2-toothed hyaline tips, straight to

recurved, 5–7 lin. long, fine; pales 4–4½ lin. long, keels distantly ciliate; anthers elliptic, ¼–½ lin. long; stigmas 1 lin. long; grain 4–5 lin. long, tightly adhering to valve and pale, strongly convexo-concave. *Kunth, Enum.* i. 415; *Reichb. Ic. Fl. Germ.* i. *t.* 74, *fig.* 1588; *Steud. Syn. Pl. Glum.* i. 325; *Stapf in Hook. f. Fl. Brit. Ind.* vii. 361. *B. japonicus, Thunb. Fl. Jap.* 52, *t.* 11; *Kunth, l.c.* 418; *Steud. l.c.* 326. *B. multiflorus, Host, Gram. Austr.* i. 11, *not of Smith. B. pectinatus, Nees, Fl. Afr. Austr.* 452 (*in part*), *not of Thunb.*

VAR. β, **pectinatus** (Stapf); taller, stouter, up to more than 2 ft. high; blades up to 8 in. by 3 lin.; panicles up to 8 in. long; lowest branches sometimes 3–4-spiculate; spikelets 1–1⅓ in. long, all parts proportionally longer; valves up to 8 lin. long. *B. pectinatus, Thunb. Prodr.* 22; *Fl. Cap.* i. 432; *ed. Schult.* 116; *Kunth, Enum.* i. 422; *Nees. Fl. Afr. Austr.* 452 (*in part*); *Steud. Syn. Pl. Glum.* i. 321; *Durand & Schinz, Consp. Fl. Afr.* v. 926. *B. erectus, Durand & Schinz, l.c.* 923 (*in part*), *not of Huds. B. erectus, var. æquifolius, Nees, l.c.* 453.

VAR. γ, **vestitus** (Stapf): habit of the type or of var. *pectinatus*, or sometimes rather dwarf; spikelets ½–1 in. long; glumes and valves narrower than in the type, pubescent to villous; valves 5–6 lin. long. *B. vestitus, Schrad. in Goett. Gel. Anz.* iii. (1821) 2074; *Schult. Mant.* ii. 356; *Nees in Linnæa*, vii. 319; *Fl. Afr. Austr.* 451; *Kunth, Enum.* i. 415; *Steud. Syn. Pl. Glum.* i. 326; *Durand & Schinz, Consp. Fl. Afr.* v. 928. *B. mollis, Thunb. Prodr.* 22; *Fl. Cap.* i. 431; *ed. Schult.* 116, *not of Linn. B. capensis, Steud. in Flora*, 1829, 491. *B. anatolicus, Boiss. & Heldr. in Boiss. Diagn. sér.* i. xiii. 63.

SOUTH AFRICA: Var. γ: without precise locality, *Drège*, 2525!

COAST REGION: Mossel Bay Div.; on dry hills on the eastern side of the Gauritz River, *Burchell*, 6455! in a dry channel of an arm of the Gauritz River, *Burchell*, 6499! Port Elizabeth Div.; Port Elizabeth, *E.S.C.A. Herb.*, 104! Var. β: Malmesbury Div.; near Groene Kloof, and in the sand of Saldanha Bay, *Thunberg!* Cape Div.; roadside towards High Constantia, *Wolley Dod*, 3489! Paarden Island, *Wolley Dod*, 3209! Queenstown Div.; Table Mountain, 5000–6000 ft., *Drège!* Var. γ: Malmesbury Div.; Groene Kloof and Saldanha Bay, *Thunberg!* Cape Div.; near Green Point, *Ecklon, Wilms*, 3887! in cultivated ground near Capetown, *Ecklon*, 197! 970! on the shore at Reed Valley, *Ecklon*; on Paarden Island, *Ecklon!* Lion Mountain, above Sea Point, *Wolley Dod*, 1606! roadside towards High Constantia, *Wolley Dod*, 3489! by Hout Bay Hotel, *Wolley Dod*, 3207! Paarl Div.; Great Draakenstein Mountains, *Drège!* Worcester Div.; by the Hex River, 1000–2000 ft., *Drège*, 2524! Swellendam Div.; Hottentots Holland, *Ecklon*. Uitenhage Div.; in cultivated ground by the Zwartkops River, *Ecklon!* Albany Div.; among shrubs near Grahamstown, 2000 ft., *Bolus*, 1280!

CENTRAL REGION: Carnarvon Div.; Klip Fontein, *Burchell*, 1527! Graaff Reinet Div.; in cultivated ground, near Graaff Reinet, 3200 ft, *Bolus*, 680! Colesberg Div.; near Colesberg, *Shaw*, 1! Var. β: Somerset Div.; near Somerset East, *Bowker!* Var. γ: Sutherland Div.; Klein Roggeveld, *Burchell*, 1295! Clanwilliam Div.; Wupperthal, *Wurmb*. Beaufort West Div.; Nieuw Veld, 3000–4000 ft., *Drège*. Somerset Div.; by the Little Fish River and Great Fish River, 2000–3000 ft., *Drège*.

WESTERN REGION: Little Namaqualand; Kamies Bergen, between Pedros Kloof and Lily Fontein, 3000–4000 ft., *Drège!* Vanrhynsdorp Div.; Karree Bergen, 1500 ft., *Schlechter*, 8235!

The typical form is common in the Eastern parts of the Mediterranean region and in the Western Himalaya, elsewhere (Europe, China, Japan, South Africa) evidently introduced. Forms very similar to the var. *pectinatus* occur in

Afghanistan and in North-West India, whilst specimens agreeing exactly with samples from South Africa have been collected in Asia Minor, the Sinai Peninsula, Afghanistan, and St. Helena.

Steudel suggests that *B. hirtus*, Lichtenst. ex Roem. & Schult. Syst. ii. 654, belongs to *B. patulus*, var. *vestitus*.

5. **B. maximus** (Desf. Fl. Atlant. i. 95, t. 26); annual; culms erect or subascending, geniculate, 1–2 ft. high, glabrous or pubescent to finely villous in the upper part, 5–7-noded, leafy all along, upper internodes usually exserted; leaves scantily to densely hairy all over or almost glabrous; sheaths tight, thin, strongly striate; ligules hyaline, delicate, 2–2½ lin. long, fimbriate-toothed; blades linear, tapering to an acute point, 4–12 in. by 2–4 lin., flat, subflaccid, dark green, margins scaberulous; panicles erect or slightly nodding, more or less contracted, 4–9 in. long (inclusive of the awns); axis terete, striate, scaberulous to hispidulous; branches 3–6-nate, unequal, 2–1- (rarely 3-) spiculate, longest 2–3 in. long, filiform, flexuous, scabrid to minutely hispid or villous; lateral pedicels short, rather stout; spikelets cuneate (when open), 1¼–1¾ in. long, green, 5–7-flowered; rhachilla slender, scabrid, joints 2–3 lin. long; glumes subulate to lanceolate-subulate in profile, finely acuminate, green along the nerves, otherwise scarious, lower 1- to sub-3-nerved, 7-10 lin. long, upper broader, 3-nerved, 11–14 lin. long, keels scabrid; valves linear-lanceolate, acuminate, about 1 in. long, 7-nerved, scabrid, upper margins and the finely bifid tips hyaline, white; awn permanently straight, scabrid, 2–2½ in. long; callus ½ lin. long; pales narrow, about 7 lin. long, keels remotely rigidly ciliolate; stamens 3 (in the South African specimens) or 2; anthers ¾–1½ lin. long; ovary pubescent just below the appendage; grain linear, about 7 lin. long, strongly convexo-concave. *Kunth, Enum.* i. 419; *Engl. Bot. t.* 2820; *ed.* iii. *t.* 1798; *Reichb. Ic. Fl. Germ.* i. *t.* 73, *fig.* 1585; *Steud. Syn. Pl. Glum.* i. 319; *Sowerby, Brit. Grass. t.* 101; *Durand & Schinz, Consp. Fl. Afr.* v. 925. *B. rigidus, Boiss. Fl. Or.* v. 649, *not* (?) *of Roth.*

COAST REGION: Cape Div.; by the railway at Rondebosch, *Wolley Dod*, 3153! by wet rocks at Hout Bay Fisheries, *Wolley Dod*, 3208! Introduced.

A native of the Mediterranean countries and Western Europe.

6. **B. leptoclados** (Nees, Fl. Afr. Austr. 453); perennial; culms erect or suberect from a short oblique rhizome, rather stout, to more than 2 ft. high, glabrous or pubescent, particularly close to the nodes, about 4-noded, uppermost internode to more than 1 ft. long, long exserted; leaves usually scantily and spreadingly hairy or almost glabrous; sheaths tight, striate or sulcate, lowest thin, subpersistent, not breaking up into fibres; ligules hyaline, 1–2 lin. long; blades linear, tapering to an acute point, 6–10 in. by 2–5 lin., flat, subflaccid, dull or sometimes subglaucous, strongly striate, scaberulous to scabrid or almost smooth below, margins scabrid, midrib prominent below, whitish; panicle 6–9 in. long, erect, very lax; axis

slender, scabrid, striate; branches 3–2-nate, filiform, scabrid, very flexuous, up to 5 in. long, undivided to about the middle, then very scantily divided, branchlets 3–1-spiculate; pedicels very unequal, lateral often much shorter than the spikelets; spikelets linear-oblong, 8–15 lin. long, light green, more or less erect, very loosely 5–10-flowered; rhachilla very slender, joints scabrid, up to 1¾ lin. long; glumes unequal, lanceolate to oblong-lanceolate in profile, acute to subacute, subscarious, lower 2½–4 lin. long, 1-nerved, upper broader, 3½–4½ lin. long, 3-nerved, nerves strong and scabrid; valves oblong-linear in profile, 4½–6 lin. long, scaberulous or scabrid on the nerves, prominently 7-nerved, margins almost straight to or beyond the middle, then very gradually curved towards the short very minutely 2-toothed tip, very narrow; awn very close to the tip, fine, straight, scaberulous, 2–3 lin. long; pales 4–5 lin. long, keels rigidly ciliolate; stamens 3; anthers about 1–1½ lin. long; ovary pubescent close below the appendage; grain (when immature) linear, strongly convexo-concave. *Steud. Syn. Pl. Glum.* i. 321; *Durand & Schinz, Consp. Fl. Afr.* v. 924.

COAST REGION: Cape Div.; Cape Peninsula, close to the waterfall, *Wolley Dod*, 2302! bottom of the Kloof at the west side of the Lions Head, *Wolley Dod*, 2786! Swellendam Div.; in the forest at Grootvaders Bosch, *Burchell*, 7249! Bedford Div.; near Bedford, *Hutton!*

CENTRAL REGION: Somerset Div.; woods on Bosch Berg, 4000 ft., *MacOwan*, 1665! Aliwal North Div.; by the Kraai River, 4000–5000 ft., *Drège*.

KALAHARI REGION: Basutoland, near Leribe, *Buchanan*, 222!

EASTERN REGION: Natal; Riet Vlei, 5000 ft., *Buchanan*, 236!

Closely allied to the European *B. asper*, Murr., and *B. scabridus*, Hook. f., from Cameroon Peak, and particularly to *B. cognatus*, Steud., from Abyssinia.

7. **B. natalensis** (Stapf); perennial; culms erect, slender above, 1–3 ft. long, glabrous, smooth, about 4-noded, upper internodes very long, exserted; sheaths tight, glabrous or nearly so, striate; ligules hyaline, ½–¾ lin. long; blades linear, over 4 in. by 1½ lin., flat, firm, strongly striate, very scantily hairy or glabrous; panicle erect, up to 7 in. long, sometimes very scanty; axis terete, smooth; branches 2–6-nate, filiform, 1-spiculate, purplish, smooth, or almost so, flexuous, longest up to 3 in. long; spikelets lanceolate-oblong, broad when quite open, glabrous, purplish, 1¼–1½ in. long, rather loosely 7–9-flowered; rhachilla joints scabrid, 1½ lin. long; glumes slightly unequal, lanceolate in profile, long acuminate, glabrous, lower 5–6½ lin. long, strongly 3-nerved, upper 6½–7½ lin. long, strongly 5-nerved; valves lanceolate-linear, acuminate, 7½–9 lin. long, firm, glabrous, smooth or very sparingly scabrid on the nerves, purplish, 7–9-nerved, outer side-nerves stout, margins gradually curved above, almost straight towards the base, narrow and membranous towards the slender tips; awn strictly terminal, 4–5 lin. long, straight, scabrid; callus short, obtuse; pales 6–7 lin. long, keels loosely and rigidly ciliolate; anthers 3 lin. long.

VAR. β, **lasiophilus** (Stapf); culms slender, scarcely 1½ ft. high; leaves shortly hairy all over; blades up to 3 in. by 1 lin.; panicle very scanty, about 6-spiculate, secund; axis and branchlets scabrid to hispidulous; spikelets light green or tinged with purple; nerves of glumes and valves less prominent than in the type.

EASTERN REGION: Natal; without precise locality, *Buchanan*, 58! Var. β: Natal; Weenen County, South Downs, 4000 ft., *Wood*, 4406!

8. **B. speciosus** (Nees, Fl. Afr. Austr. 454); perennial; culms erect, rather slender, 2 ft. high, glabrous, smooth, 3–4-noded, uppermost internodes very long, long exserted; leaves pubescent to subvillous except the glabrous 1–2 uppermost sheaths; sheaths tight, strongly striate; ligules hyaline, very short; blades linear, acute with a callous point, 2–3 in. by 1 lin., flat or involute above, erect, firm, strongly striate, uppermost very short; panicle erect or somewhat nodding, very lax, scanty, subsecund, 3–5 in. long; axis terete, glabrous, smooth, very slender; branches 2-nate, finely filiform, smooth, purplish, 1-spiculate, lowest 1–2 in. long, very flexuous; spikelets linear-oblong, 1¼–2¼ in. long, loosely 7–8-flowered, purplish; rhachilla joints 1½–2 lin. long, scabrid; glumes unequal, oblong-lanceolate in profile, acute, purple except at the narrow margin and tip, very scantily pubescent, lower 3½–4½ lin. long, 3-nerved, upper much broader, 5–6 lin. long, 5-nerved; valves linear-lanceolate in profile, about 7 lin. long, rather firm, densely pubescent, central part purple, margins narrow, 7-nerved, very gently curved; awn straight, scabrid, 1½–3 lin. long, from close to the short minutely 2-toothed tip; callus very short, obtuse; pales equalling the valves, keels densely and spreadingly ciliate; anthers 3–3½ lin. long. *Steud. Syn. Pl. Glum.* i. 321; *Durand & Schinz, Consp. Fl. Afr.* v. 927.

COAST REGION: Queenstown Div.; by the Klipplaat River, *Ecklon*; on the Storm Bergen, *Drège*. King Williamstown Div.; Amatola Mountains, *Buchanan!*

Nees's description of the type of his *B. speciosus* fits very well the plant from the Amatola Mountains from which the above description is drawn up, with this exception, that he describes the branches of the panicle as scabrid and the culms as thick as a hen's quill.

9. **B. firmior** (Stapf); perennial; culms erect, stout, 2–3 ft. high, glabrous, rough below the panicle, 3–4-noded; uppermost internode long exserted; leaves densely and rigidly villous all over, except the glabrous uppermost sheath; sheaths tight, strongly striate to sulcate; ligule hyaline, about 2 lin. long; blades linear, acute with a callous point, 3–6 in. by 2–3 lin., erect, firm, strongly and very closely striate; panicle erect or somewhat nodding, very lax, subsecund, about 6 in. long; axis terete, scabrid and pubescent; branches 4–6-nate, filiform, very flexuous or nodding, densely hispidulous, 1-spiculate, lowest up to 3 in. long; spikelets linear-oblong, greyish-green, 1–1¼ in. long, rather densely 7–11-flowered; rhachilla joints about 1 lin. long, scabrid to hispidulous; glumes unequal, pubescent, lower subulate to lanceolate in profile, 3½–4½ lin. long, 1–3-nerved, upper oblong-lanceolate, acute or acuminate, 5 lin. long, 3–5-nerved; valves lanceolate-cuneate in profile, about 6 lin.

long, rather firm, pubescent to tomentose or villous, 7-nerved, margins almost straight to the middle or somewhat beyond it, then bent at an obtuse angle or distinctly curved towards the acute tip; awn straight, scabrid to hispidulous, 2–2½ lin. long, strictly terminal; callus very short, obtuse; pales equalling the valves, pubescent, keels densely and spreadingly ciliate. *B. speciosus, var. firmior, Nees, Fl. Afr. Austr.* 454.

VAR. β, **leiorhachis** (Stapf); uppermost internode to more than 1 ft. long, smooth; panicle up to 8 in. long; axis smooth; branches scaberulous to scabrid, longest up to 4 in. long and 2-spiculate; spikelets 1–1½ in. long; valves tinged with purple in the upper part, but greyish from the dense tomentum, 6–8 lin. long; awn from close below the tip; anthers 6–8 lin. long.

COAST REGION: Queenstown Div.; on the Storm Bergen, 5000–6000 ft., *Drège!* Var. β: King Williamstown Div.; Amatola Mountains, *Buchanan*, 42!
KALAHARI REGION: Var. β: Orange Free State; Witte Bergen, near Harrismith, *Buchanan*, 265!

10. B. unioloides (H. B. K. Nov. Gen. i. 151); annual; culms fascicled, erect or suberect, slender to stout, 1–2 ft. long, glabrous, smooth, 2–3-noded, uppermost internodes long exserted; lower sheaths thin, pallid, finely tomentose to spreadingly villous, upper green, prominently striate, glabrescent or glabrous and smooth; ligules ovate, obtuse, 1–2 lin. long; blades linear, tapering to an acute point, 3–8 in. by 1–4 lin., flat, more or less flaccid, scaberulous or scabrid on both sides and along the margins, glabrous or softly hairy to villous; panicle narrow, usually nodding, scantily divided or reduced to a raceme, from a few inches to almost 1 ft. long; axis terete, striate, smooth below, angular and scaberulous above; lower branches remote, 2–3-nate, 2–4- (rarely 5-) spiculate, or all 1-spiculate, filiform, angular, scabrid, longest to 2–3 in. long; pedicels very unequal, lateral usually extremely short; spikelets erect or suberect, 8–15 lin. long, lanceolate-oblong to oblong, strongly compressed, light green to glaucous, 4–10-flowered; rhachilla very fragile; joints stout, up to 1¼ lin. long, scaberulous on the outer side; florets permanently closed, cleistogamous; glumes lanceolate, acute or acuminate, firm except at the narrow white hyaline margins, glabrous, smooth except on the scaberulous keels, lower 4½–6½ lin. long, 5–7-nerved, upper 5½–8 lin. long, 7–9-nerved, nerves prominent; valves lanceolate in profile, usually very minutely 2-toothed, mucronate or shortly awned, 6–8 lin. long, acutely keeled, herbaceous-chartaceous, green, whitish below, scabrid, about 13-nerved, margins straight to ⅓ their length from the base, then hyaline, white, nerves prominent; pales 3–5½ lin. long, folded between the crested rigidly ciliolate keels; filaments short, extremely delicate, clavate-tipped; anthers ellipsoid, usually about ¼ lin. long, permanently enclosed in the floret; ovary top with a large 3-lobed villous appendage; stigmas sessile, short, slender, loosely plumose; grain strongly compressed, linear-oblong, deeply grooved in front, tightly adhering to the valve and pale. *Kunth, Enum.* i. 415; *Nees in Linnæa*, vii. 319; *Steud. Syn. Pl. Glum.* i. 326; *Durand & Schinz, Consp. Fl. Afr.*

v. 928. *B. Willdenowii, Kunth, Rév. Gram.* i. 134; *Enum.* i. 416. *Ceratochloa unioloides, Beauv. Agrost. Expl. planch.* 11, *t.* 15, *fig.* 7; *Roem. & Schult. Syst.* ii. 596; *Nees, Fl. Afr. Austr.* 449. *Festuca unioloides, Willd. Hort. Berol.* i. 3, *t.* 3; *Enum. Hort. Berol.* 115.

COAST REGION: Cape Div.; in gardens near Capetown, *Ecklon!* near Rondebosch Station, *Wolley Dod*, 2156! 3566! Port Elizabeth Div.; Port Elizabeth, *E.S.C.A. Herb.*, 186! Albany Div.; very common throughout the division, *MacOwan*, 1510! Queenstown Div.; plains near Queenstown, 3500 ft., *Galpin*, 2362!

KALAHARI REGION: Transvaal; near Pretoria, *Wilms*, 1714a! near Lydenburg, *Wilms*, 1712!

EASTERN REGION: Natal; Hermannsburg Station, 3000 ft., *Buchanan*, 235!

Throughout America (probably a native of South America); introduced into South Europe, India, South Africa, Tristan d'Acunha, &c.; often grown for fodder.

B. laxiflorus (Spreng. ex Steud. Nomencl. Bot. i. 120; Durand & Schinz, Consp. Fl. Afr. Austr. v. 928) undescribed, is a name referring to one of Zeyher's grasses.

XCII. BRACHYPODIUM, Beauv. (in part).

Spikelets at first cylindric, then laterally compressed, in a simple raceme or false spike with very short pedicels, with the flattened side to the axis; rhachilla glabrous, disarticulating above the glumes and between the valves. *Florets* 5 to many, ⚥, the uppermost more or less reduced. *Glumes* more or less unequal, firm, strongly 3–7-nerved, lower often slightly asymmetric. *Valves* oblong to oblong-lanceolate, usually narrowed into a straight awn, or mucronate, rounded on the back, closely imbricate at first, then more or less diverging and rolling inwards, 7-nerved, nerves prominent in the upper part, faint below; callus very short, obtuse. *Pales* oblong, rather broad, very obtuse or truncate, slightly shorter than the valve, 2-keeled, keels rigidly ciliate. *Lodicules* 2, lanceolate, usually ciliolate. *Stamens* 3, rarely 2. *Ovary* with a villous appendage at the top; styles laterally inserted on the appendage, very short; stigmas plumose, laterally exserted. *Grain* linear or linear-oblong, convexo-concave, adhering more or less to the pale; embryo small; hilum filiform, long.

Perennial or annual; blades flat or setaceously involute; racemes terminal, joints of axis more or less hollowed out (at least the lower) on the side facing the spikelets; spikelets usually few (sometimes 1), erect or spreading, rather large.

Species about 15, often very difficult to discriminate, mainly in the temperate regions of the northern hemisphere.

Annual; anthers $\frac{1}{4}$–$\frac{1}{3}$ lin. long ... ... (1) **distachyum.**
Perennial; anthers $1\frac{1}{2}$–2 lin. long:
Loosely branched or simple; culms leafy for more than half their length; blades linear; racemes more or less flexuous ... ... (2) **flexum.**
Densely cæspitose; leaves crowded at the base; blades linear-lanceolate; racemes straight ... (3) **Bolusii.**

1. **B. distachyum** (Beauv. Agrost. 155); annual; culms fascicled, or simple or branched below, often very strongly and repeatedly

geniculate, ascending, ½–1½ ft. long, glabrous or sparingly hairy below the nodes, smooth or slightly rough in the upper part, 3–4-noded, internodes exserted, uppermost ultimately becoming by far the longest; sheaths rather tight, herbaceous, striate, spreadingly hairy or finely pubescent or glabrescent, spreadingly ciliate along the upper margins, finely tomentose at the nodes; ligules very obtuse, finely pubescent to villous, up to ¾ lin. long; blades linear, tapering to an acute point, 2–4 in. by 1½–2 lin., flat, slightly rigid to subflaccid, scabrid and more or less spreadingly hairy all over; false spike erect, straight, 2-ranked, 4–1- (rarely to 6-) spiculate, up to 3 in. long; rhachis slightly rough except at the scabrid margins, striate; pedicels stout, very short or obsolete; spikelets erect, adpressed, ½ to more than 1 in. long, glabrous or sparingly hairy, 6–12-flowered; glumes subulate-lanceolate, very acute, lower 2½–3 lin. long, 4- to sub-6-nerved, upper about 3½ lin. long, 7-nerved, nerves very prominent; valves lanceolate-acuminate, 4–4½ lin. long, firmly chartaceous, scaberulous above, 7-nerved; awn 4–9 lin. long, very short or absent in the lowest florets; pales about 3½ lin. long, keels very rigidly ciliate in the upper part; stamens 3 (in the South African specimens) or 2; anthers oblong to ellipsoid, ¼–⅓ lin. long; grain oblong-linear, convexo-concave, about 3 lin. long. *Sturm, Deutschl. Fl. Heft*, 86, *t.* 10; *Durand & Schinz, Consp. Fl. Afr.* v. 929. *Bromus distachyos, Linn. Amœn.* iv. 304; *Host, Gram. Austr.* i. *t.* 20. *B. ciliatus, Lam. Fl. Franç.* iii. 609. *Festuca ciliata, Gouan, Hort. Monsp.* 48 *and* 547. *F. monostachya, Poir. Voy. Barb.* ii. 98; *Desf. Fl. Atl.* i. 92, *t.* 24, *fig.* 2. *F. distachya, Willd. Enum. Hort. Berol.* i. 118; *Steud. Syn. Pl. Glum.* i. 317. *Triticum ciliatum, DC. Fl. Franç.* iii. 85; *Kunth, Enum.* i. 447. *Trachynia distachya, Link, Hort. Berol.* i. 43. *Reichb. Ic. Fl. Germ.* i. *t.* 14, *fig.* 1368; *Nees, Fl. Afr. Austr.* 458.

COAST REGION: Cape Div.; near Capetown, *Ecklon;* roadside beyond Alphen Bridge, *Wolley Dod,* 3488! shore between Sea Point and Camps Bay, and common in grassy places all over Signal Hill, *Wolley Dod,* 3105! by the railway between Maitland and Bridge, *Wolley Dod,* 3226! George Div.; in the Karroo by the Gauritz River, *Ecklon.*

CENTRAL REGION: Aberdeen Div.; in the Camdeboo Mountains, *Drège!*

A native of the Mediterranean region.

2. **B. flexum** (Nees, Fl. Afr. Austr. 456); perennial; culms often abundantly branched below, ascending from a geniculate (often decumbent and rooting) base, very slender, 1½ to more than 2 ft. long, glabrous or scantily hairy just below the nodes, smooth, 5–7-noded, intermediate and upper internodes exserted, uppermost the longest; sheaths tight, striate, spreadingly hairy along the margins, otherwise very sparingly hairy or glabrous, excepting a finely tomentose line at the junction with the blade and the densely tomentose or villous (rarely finely pubescent) nodes, lowest sheaths finally thrown aside; ligules extremely short, up to ½ lin. long, truncate, brownish; blades linear, long tapering to a very fine point, 2–5 in. by 1½–3 lin., flat, glaucous, subrigid to almost flaccid, spreadingly hairy on the upper side, less

so or glabrous and scabrid or scaberulous on the lower side; false spike erect or nodding, usually flexuous, 2-ranked, 9–1-spiculate, up to 4 in. long; rhachis very slender, strongly compressed, striate, slightly rough or smooth except along the scabrid margins; pedicels very short, up to $\frac{3}{4}$ lin. long, finely puberulous to almost glabrous; spikelets usually more or less spreading, $\frac{1}{2}$–$1\frac{1}{2}$ in. long, glabrous, very rarely sparingly pubescent, 7–16-flowered; glumes acute to acuminate, lower subulate to subulate-lanceolate, $1\frac{1}{2}$ to almost 3 lin. long, 4–5- (rarely 3-) nerved, upper lanceolate to oblong, $2\frac{1}{2}$ to more than $3\frac{1}{2}$ lin. long, 7-nerved, nerves very prominent; valves lanceolate, acuminate, gradually narrowed into the awn, 3–4 lin. long, rather firm, finely scaberulous, rarely subpubescent above, 7-nerved; awn up to $3\frac{1}{2}$ lin. long, shorter in the lower florets; pales $3\frac{1}{2}$ to almost 4 lin. long, rigidly ciliate in the upper part; anthers over $1\frac{1}{2}$ lin. long, linear. *Durand & Schinz, Consp. Fl. Afr.* v. 929 (*the Cape plant*). *Festuca flexa, Steud. Syn. Pl. Glum.* i. 316.

VAR. β, **trachycladum** (Stapf); culms very rough from very minute reversed hairs; blades more rigid and hispid than in the type; false spike rigid; axis scabrid; spikelets erect; glumes and valves more or less pubescent.

VAR. γ, **simplex** (Stapf); culms simple, up to more than 2 ft. high, erect, rough from minute reversed hairs to reversedly hirsute, rarely smooth towards the panicle; panicle up to $\frac{1}{2}$ ft. long, flexuous; spikelets erect or spreading, up to $1\frac{3}{4}$ in. long, and to 18-flowered, glabrous, scaberulous; glumes and valves by about $\frac{1}{4}$–$\frac{1}{3}$ longer than in the type.

VAR. δ, **tenue** (Stapf); culms as in the type, but very slender; leaves very glaucous, scantily hairy to glabrous; false spike and spikelets very slender; florets loose and very narrow after flowering.

COAST REGION: Cape Div.; near Capetown, *Burchell*, 472! Swellendam Div.; in the forest at Grootvaders Bosch, *Burchell*, 7230! Uitenhage Div.; near Strand Fontein and Matjes Fontein, below 500 ft., *Drège!* and without precise locality, *Ecklon & Zeyher*, 563! Var. β: Cape Div.; Old road to Constantia Nek, *Wolley Dod*, 2385!

CENTRAL REGION: Somerset Div.; in woods at the foot of Bosch Berg, 2800 ft., *MacOwan*, 1495! 1521!

KALAHARI REGION: Orange Free State; on the Witte Bergen near Harrismith, *Buchanan*, 261! 264! Transvaal; Houtbosch, *Rehmann*, 5734!

EASTERN REGION: Var. γ: Natal; Umsinga and base of Biggars Berg, *Buchanan*, 104! on the Drakensberg Range, near Newcastle, *Buchanan*, 198! and without precise locality, *Buchanan*, 60! 61! Var. δ: Natal; Umpumulo, 2400 ft., *Buchanan*, 233! and without precise locality, *Buchanan*, 79!

B. flexum resembles very much *B. sylvaticum*, Beauv., which, however, differs in the more produced ligules, the (on the whole) more numerous nerves of the glumes, and the constantly simple culms. *B. fontanesianum*, Nees (Fl. Afr. Austr. 457), described from a specimen collected by Ecklon near Tulbagh, is very probably only a state of *B. flexum*.

3. B. Bolusii (Stapf); perennial, very densely cæspitose; culms slender, erect, $\frac{1}{2}$–$1\frac{1}{2}$ ft. long, glabrous, smooth or finely puberulous below the nodes, 2–3-noded, uppermost internode occupying more than $\frac{1}{2}$ the length of the culm, long exserted; leaves crowded at the base, lower sheaths compressed, subflabellate, keeled, whitish, striate, glabrous or scantily hairy, subpersistent, upper terete, finely and spreadingly hairy; ligules very short, truncate; blades linear-lanceolate to linear, acute, 2–$3\frac{1}{2}$ (rarely to 6) in. by 2–3 lin., flat, rather rigid, glaucous, hispid all over, strongly striate; spikelets solitary or 2–3 in an erect straight false spike, oblong, up to more than $\frac{3}{4}$ in. long, more than 6-flowered, glabrous, very shortly pedicelled; glumes lanceolate to oblong, acute or mucronulate, firm, rigid, lower up to 3 lin. long, 4-nerved, upper $3\frac{1}{2}$ lin. long, 7-nerved, scaberulous; pales slightly shorter than the valves, keels rigidly ciliolate; grain linear-oblong, 3 lin. long.

CENTRAL REGION: Graaff Reinet Div.; rocky places on Compass Berg, 8500 ft., about 100–200 ft. below the summit, *Bolus*, 1986!

XCIII. LOLIUM, Linn.

Spikelets usually more or less compressed, 2-ranked, alternate, sessile in the hollows of the rhachis of a simple spike; rhachilla glabrous, disarticulating above the glumes and between the valves. *Florets* 3–11, or sometimes more, ☿ or the uppermost reduced. *Glumes* of terminal spikelets equal and similar, lower suppressed in the lateral spikelets, upper linear to oblong, obtuse to acute, flat or slightly rounded, coriaceous, prominently 7–9-nerved. *Valves* oblong, rounded on the back, subobtuse, minutely 2-toothed (or acute in profile), more or less chartaceous except at the short hyaline tips, glabrous, 5-nerved, muticous or awned; awn a straight bristle from close to the tip. *Pales* equalling the valves or nearly so, 2-keeled, keels more or less crested. *Stamens* 3. *Lodicules* 2, lanceolate with a lateral tooth. *Ovary* glabrous, truncate; styles distinct, very short; stigmas laterally exserted, plumose. *Grain* elliptic-oblong to linear-oblong, tightly enclosed by the valve and pale, adhering to both; embryo short; hilum linear, almost as long as the grain.

Annual or perennial; blades linear, flat; ligules hyaline; spikes terminal; spikelets more or less erect, 2-ranked, with the (upper) glume opposite the hollow of the rhachis.

Species 6–8 in the temperate regions of Europe, Asia, and North Africa, elsewhere introduced.

Florets turgid, the uppermost equalled or exceeded by the glume ... ... (1) **temulentum.**
Florets not turgid:
  Rather tall; spikes $\frac{1}{2}$–1 ft. long; axis rather slender; glume much shorter than the spikelet (2) **multiflorum.**
  Short; spikes $\frac{1}{4}$–$\frac{1}{2}$ ft. long; axis stout and stiff; glume equalling the spikelet ... ... (3) **rigidum,** var. **rottbœllioides.**

1. **L. temulentum** (Linn. Sp. Plant. 83); annual; culms fascicled or solitary, erect, rather stout, straight, 1–$1\frac{1}{2}$ ft. long, glabrous, rough, at least in the upper part, 2–4-noded, uppermost internode usually long exserted; leaves glabrous; sheaths rather tight, strongly striate; ligules very short, truncate; blades linear, acute, $\frac{1}{2}$–1 ft. by 2–3 lin., flat, rough all over or only on the upper side and along the margins; spikes erect, $\frac{1}{4}$–1 ft. long, of rather numerous (10–20) spikelets; axis rather stout, slightly rough or smooth on the back; spikelets

about the length of the internodes, or the lower shorter and more distant, the upper more approximate, laterally compressed, 5–9 lin. long, 5–9-flowered; florets turgid; upper glume oblong, equalling or somewhat exceeding the uppermost floret, subobtuse to acute, very rigid, flat, glabrous, smooth, 7–9-nerved; valves elliptic-oblong, obtuse or subobtuse, muticous or usually awned, 3–3½ lin. long, rounded on the back, herbaceous-chartaceous, smooth, 5-nerved, tips rather broad, hyaline, obscurely 2-toothed; awn subterminal, straight, rather stout at the base, scaberulous, up to 8 lin. long; pales broad, keels green, stout; anthers over 1 lin. long; grain elliptic-oblong, semiterete, grooved in front, slightly shorter than the pale. *Fl. Dan. t.* 160; *Host, Gram. Austr.* i. *t.* 26; *Engl. Bot. t.* 1124; *Knapp, Gram. Brit.* 101; *Nees in Linnæa*, vii. 304; *Fl. Afr. Austr.* 364; *Kunth, Enum.* i. 437; *Reichb. Ic. Fl. Germ.* i. *t.* 5, *figs.* 1342-4; *Steud. Syn. Pl. Glum.* i. 340; *Durand & Schinz, Consp. Fl. Afr.* v. 933.

COAST REGION: Malmesbury Div.; near Riebeeks Castle, 500–1000 ft., *Drège*. Cape Div.; cultivated ground near Muizenberg Vlei, *Wolley Dod*, 3567! Klein Constantia, *Wolley Dod*, 2067! by the roadside near Doorn Hoogte, *Wolley Dod*, 3474! woods above Newlands Avenue, *Wolley Dod*, 2131! Stellenbosch Div.; near Stellenbosch, in cornfields, *Ecklon*. Queenstown Div.; Shiloh, 3500 ft., *Baur*, 1140!

CENTRAL REGION: Calvinia Div.; Onder Bokke Veld near Bok Fontein, 2500 ft., *Drège!*

KALAHARI REGION: GriqualandWest, Hay Div.; near Griqua Town, *Burchell*, 1936! Transvaal; Matebe Valley, *Holub!*

EASTERN REGION: Natal, *Buchanan*, 50!

2. **L. multiflorum** (Lam. Fl. Franç. iii. 621); annual; culms fascicled, geniculate, ascending or erect, 1½–3 ft. long, glabrous, smooth or slightly rough in the upper part, 4–5-noded, internodes at length more or less exserted, simple or branched below; leaves glabrous; sheaths striate, smooth, lower sometimes purplish; ligule very short, truncate from an auricled base; blades linear, long tapering to a very slender point, 4–8 in. by 1–2½ lin., flat, somewhat rigid, scabrid on the upper side and along the margins; spikes erect, ½–1 ft. long, of numerous (12–30) spikelets; axis rather slender, smooth except at the scabrid margins; internodes (except the lowest) distinctly shorter than the spikelets; spikelets strongly compressed from the side, elliptic-oblong, 6–8 lin. long, obliquely erect, 9–11-flowered, uppermost floret long exserted from the glume; upper glume narrow, oblong, subobtuse, somewhat rounded on the back, equalling the contiguous floret or nearly so, strongly 7-nerved; valves awned or muticous, oblong, 2½–3 lin. long, subherbaceous-chartaceous, light green, or tinged with purple above, 5-nerved, smooth; awn (when present) straight, very fine, 2½–4 lin. long, close to the short hyaline minutely 2-toothed tip; keels of pales green, crested, scabrid; anthers 1½–2 lin. long; grain linear-oblong, semiterete, 1½ lin. long, deeply channelled in front, adhering to valve and pale. *Kunth, Enum.* i. 436; *Reichb. Ic. Fl. Germ.* i. *t.* 5, *fig.* 1345; *Steud. Syn. Pl. Glum.* i. 340; *Durand & Schinz, Consp. Fl. Afr.* v. 933.

COAST REGION: Cape Div.; Rondebosch Station, *Wolley Dod*, 3560! cultivated field near the Salt River, *Wolley Dod*, 3562!

A native of Central Europe and the Mediterranean countries.

3. **L. rigidum** (Gaudin, Agrost. Helv. i. 334) var. **rottbœllioides** (Heldr. ex Boiss. Fl. Or. v. 680); annual; culms fascicled or solitary, erect or ascending, $\frac{1}{3}$–$\frac{2}{3}$ ft. long, glabrous, smooth, sheathed all along or nearly so, 2–3-noded; leaves glabrous, smooth; sheaths striate, herbaceous, sometimes purplish; ligules very short, truncate, often produced into an obtuse auricle on one side; blades linear, tapering to an acute point, $2\frac{1}{2}$–5 in. by $1\frac{1}{2}$–2 lin., flat or involute, rather firm, erect or spreading; spike straight or curved, $2\frac{1}{2}$ to more than 6 in. long, of 15–20 spikelets, rather stout and stiff; rhachis smooth, striate, deeply hollowed out; internodes (except the lowest) shorter than the spikelets; spikelets 4–6 lin. long, 3–5-flowered, very slightly gaping when in flower; rhachilla glabrous, much compressed; upper glume oblong-linear, obtuse or subobtuse, slightly exceeding or equalling the uppermost floret, very rigid, flat, glabrous, smooth, prominently 7-nerved; valves oblong, obtuse to acute, awnless or the uppermost finely mucronate, $2\frac{1}{2}$ lin. long, flat on the back, subchartaceous, whitish or light green, smooth, 5-nerved, tip hyaline; keels thinly crested or winged, scaberulous; anthers $\frac{1}{2}$–$\frac{3}{4}$ lin. long; grain elliptic-oblong, strongly compressed from the back, scarcely 2 lin. long. *Lolium lepturoides, Boiss. Diagn. ser.* i. *fasc.* 13, 67. *L. subulatum, Vis. Fl. Dalm.* i. 90, *t.* 3. *Rottbœllia loliacea, Bory & Chaub. Fl. Pelop.* 9, *t.* 3, *fig.* 2. *Crypturus loliaceus, Link in Linnæa,* xvii. 387.

COAST REGION: Cape Div.; Cape Peninsula, near Maitland Station, *Wolley Dod*, 3324! waste ground near Westerford Bridge, *Wolley Dod*, 1823!

The variety is otherwise only known from Greece, Dalmatia, and a few places in the Levant. It differs from the type mainly in the shorter, stouter growth, the relatively longer upper glume, fewer florets and smaller anthers.

XCIV. LEPTURUS, R. Br.

Spikelets sessile, solitary, more or less immersed in the hollows of the rhachis of a simple, terminal, more or less articulate, straight or incurved spike; rhachilla very slender, glabrous. *Florets* 1 or 2, lower ⚥, upper usually rudimentary, shorter than the upper glume. *Glumes* of terminal spikelet equal and similar, lower glume suppressed in the lateral, or (if present) minute, hyaline, upper narrow, very rigid, acute to subulate-acuminate, prominently 5–7-nerved. *Valves* lanceolate to oblong, acute, hyaline, glabrous, finely 3-nerved; callus obsolete. *Pales* equalling the valves, faintly 2-nerved. *Lodicules* 2, fleshy below, hyaline above. *Stamens* 3. *Ovary* glabrous, truncate; styles distinct, very short; stigmas laterally exserted, plumose. *Grain* oblong, dorsally more or less compressed, free in the unaltered floret and enclosed with it by the hollow of the rhachis and the glume; hilum linear, rather short; embryo short.

Small, annual or creeping perennial grasses; blades flat or involute; ligules membranous, short; spike slender, more or less cylindric, straight or curved, disarticulating when ripe or almost tough; lower glume spreading during flowering, then adpressed.

Species 3, in the warm coast regions of the Old World.

1. **L. cylindricus** (Trin. Fund. Agrost. 123); annual, $\frac{1}{3}$–1 ft. high; culms fascicled and branched or simple, geniculate, ascending or suberect, glabrous, smooth, 3–5- (rarely more-) noded. usually sheathed all along or intermediate internodes shortly exserted; leaves glabrous; sheaths smooth, striate, tight; ligules very short, hyaline, truncate, denticulate, slightly produced on one side; blades narrow, linear, tapering to a fine point, 2–6 in. by 1 lin., flat or involute, soft or somewhat rigid, finely or obscurely scaberulous above and along the margins; spike slender, cylindric, erect, straight or slightly curved, 2–8 in. long; rhachis dull green, slightly striate, smooth, subarticulate, tough, margins of hollows and articulation lines whitish; spikelets alternate, 1 to each joint or sometimes 2 subopposite, always longer than the joint, 2–3$\frac{1}{2}$ lin. long, 1-flowered; lower glume 0; upper lanceolate-linear, very acute, thick, very hard, prominently 5–7-nerved, smooth; valve lanceolate-oblong, acute, 2–2$\frac{1}{2}$ lin. long, hyaline, 3-nerved; pale delicately 2-nerved, hyaline, minutely 2-toothed; anthers linear, about 1 lin. long; grain oblong, elliptic in cross section or semiterete, 1$\frac{1}{4}$ lin. long. *Webb & Berth. Hist. Nat. Canar.* iii. iii. 424. *L. subulatus, Kunth, Rév. Gram.* i. 151; *Enum.* i. 462; *Steud. Syn. Pl. Glum.* i. 357. *Rottbœllia cylindrica, Willd. Sp. Plant.* i. 464 *(in part)*. *R. subulata, Savi, Due Cent.* 35, *and in Nuovo Giorn. Pis.* 1806, 230, *figs.* 4 *and* 8. *Monerma subulata, Nees in Linnæa,* vii. 280; *Fl. Afr. Austr.* 126; *and Gen.* i. *t.* 88; *Durand & Schinz, Consp Fl. Afr.* v. 931. *M. cylindrica, Coss. in Coss. & Dur. Expl. Scient. Alg. Phan.* 214; *Durand & Schinz, l.c. Ophiurus cylindricus, Beauv. Agrost.* 168; *Link, Hort. Berol.* ii. 171. *O. subulatus, Link, l.c.* i. 3; *Reichb. Ic. Fl. Germ.* i. *t.* 3, *fig.* 1335.

COAST REGION: Cape Div.; stony places on Table Mountain, *Ecklon,* 979! Kirstenbosch, *Wolley Dod,* 2244! common about Capetown, *Pappe!* Tulbagh Div.; near Tulbagh Waterfall, *Zeyher!* Uniondale Div.; Kammanassie Mountains, near Avontuur, *Bolus,* 2697! Uitenhage Div.; by the Zwartkops River, near Uitenhage, *Drège!* near Uitenhage, *Alexander!*

Also in the Mediterranean countries; introduced in Australia and North America.

XCV. OROPETIUM, Trin.

Spikelets sessile, solitary, more or less immersed in the hollows of the rhachis of a simple distichous terminal spike; rhachilla very slender, glabrous, disarticulating above the glumes and between the valves. *Florets* 1–4, ⚥ or uppermost rudimentary, shorter than the upper glume. *Glumes* of terminal spikelet equal and similar, lower glume of the lateral spikelets suppressed or (if present) more or less reduced, sometimes split in 2, upper very narrow, rigid, except at the hyaline margins, subacute to acuminate, 1–3-nerved, nerves often more or less confluent into a broad rigid midrib. *Valves* oblong or lanceolate in profile, elliptic to broadly oblong when expanded, distinctly or obscurely keeled, at least in the upper part, truncate or minutely 3-toothed, hyaline, glabrous, hairy along the nerves, finely 3-nerved; callus bearded. *Pales* equalling the valves, faintly 2-nerved. *Lodicules* 2, minute, cuneate. *Stamens* 3. *Ovary* glabrous; styles slender, distinct; stigmas laterally exserted. *Grain* oblong, subterete or slightly compressed from the sides; embryo about $\frac{1}{3}$ the length of the grain; hilum punctiform, basal.

Dwarf densely tufted perennials, rarely annuals; blades setaceous; ligule hyaline, ciliate, short; spikes very slender, straight or gradually curved; spikelets closely or loosely 2-ranked on the opposite or subopposite sides of the rhachis.

Species 6, in the dry regions of India, Algeria, and South Africa.

The renewed examination of this genus has led me to the conclusion that its proper place is in *Chlorideæ* near *Microchloa*, notwithstanding the superficial resemblance to *Lepturus*. It includes also the North African genus *Kralikia*, Coss. (*Kralikiella*, Coss. & Dur.; *Arcangelina*, O. Kuntze).

1. O. capense (Stapf); perennial, densely tufted, dwarf, 1–2$\frac{1}{2}$ in. high; culms slender, smooth, 1–2-noded, sheathed all along; leaves crowded at the base; lower sheaths flabellate, glabrous, firm, strongly striate, persistent; ligule extremely short, hyaline; blades very narrow, linear, or usually setaceously folded, subacute, $\frac{1}{2}$–1$\frac{1}{4}$ in. by $\frac{3}{8}$–$\frac{1}{2}$ lin. (when unfolded), rigid, with a few very fine spreading hairs or quite glabrous, scaberulous on the upper side, smooth below; spike $\frac{1}{2}$–$\frac{3}{4}$ in. long, erect, straight, very slender, with the base enclosed in the uppermost sheath; rhachis slender, quadrangular, sides striate; spikelets 1–1$\frac{1}{2}$ lin. long, 1-flowered; rhachilla continued as a short glabrous bristle; lower glume of lateral spikelets suppressed, upper lanceolate-subulate, acute or acuminate, 1$\frac{1}{4}$–1$\frac{1}{2}$ lin. long, rigid, except at the hyaline margins, 3-nerved; valve lanceolate in profile, 3-toothed, 1 lin. long, hyaline, 3-nerved, nerves percurrent or very shortly excurrent; anthers linear, $\frac{1}{4}$ lin. long; grain linear-oblong, subterete.

CENTRAL REGION: Somerset Div.; near Somerset East, *MacOwan!*
KALAHARI REGION: Griqualand West; on the Asbestos Mountains, *Burchell*, 2057! plains at the foot of the Asbestos Mountains between Kloof village and Witte Water, *Burchell*, 2091!

Also in German South-West Africa, Herercland (Dinter).

XCVI. AGROPYRUM, Gaertn.

Spikelets more or less laterally compressed, alternate, distichous, sessile at the hollows of, and with the flattened side to, the tough or disarticulating rhachis of a simple spike; rhachilla disarticulating above the glumes and between the valves or almost tough. *Florets* 3 to many, ☿ or the uppermost reduced, exceeding the glumes. *Glumes* equal or subequal, lanceolate, linear or oblong, rigid, prominently 5- to many-nerved, usually somewhat asymmetrical. *Valves* lanceolate to oblong in profile, obtuse and sometimes 3-toothed or mucronate, or acuminate or produced into an awn, more or less coriaceous, 5- (rarely 7–9-) nerved, nerves faint or partly evanescent below; callus very obtuse and short. *Pales* somewhat shorter than the valves, acutely 2-keeled or keels somewhat winged, ciliate. *Lodicules* 2, rather large, oblique or unequally lobed, densely ciliate. *Stamens* 3. *Ovary* obovoid with a large villous top; styles hardly any; stigmas delicately plumose. *Grain* tightly enclosed and adhering to the valve and pale or at least to the latter, linear-oblong, plano- or slightly concavo-convex; embryo short; hilum filiform as long as the grain.

Perennials with an often long creeping rhizome; blades flat or convolute; ligule scarious; spikelets usually long and slender, rigid, rarely stout (in the South African species); spikelets adpressed to the rhachis, usually rather large.

Species 30–35, mainly in the temperate region of Europe and Asia, 1 or 2 widely spread weeds.

1. A. distichum (Beauv. Agrost. 146); perennial; culms ascending from a long creeping slender branched rhizome, 1–1½ ft. high, stout, glabrous, smooth, closely sheathed all along, many-noded; sheaths rather loose or the upper tumid, smooth, thick, somewhat spongy, lower at length becoming scarious and withering away; ligules membranous, obtuse, 1½ lin. long; blades linear, subulate-convolute, subpungent, ½ to more than 1 ft. by as much as 3½ lin. (when expanded), thick, rigid, smooth on the back, prominently and closely striate above, very minutely and densely tomentose above; spike erect, straight, very stout, 3–6 in. long, of 3–16 spikelets; rhachis stout, very fragile; internodes much shorter than the spikelets, ¼–½ in. long or the lowest up to ¾ in., smooth; spikelets more or less elliptic in outline when open, stout, 1–1½ in. long, adpressed to the rhachis, glabrous, 6–11-flowered; rhachilla-joints stout, cuneate, puberulous; glumes equal, lanceolate to oblong, more or less asymmetrical (the side facing the axis much narrower), obtuse, minutely notched, 10–11 lin. long, coriaceous, smooth, glabrous, up to 15-nerved, keeled; valves ovate-lanceolate in profile, 3-toothed or mucronate with 2 lateral teeth, lowest 9–10 lin. long, subcoriaceous, smooth, 7–9-nerved, keeled in the upper part; pales elliptic, minutely truncate, 6½–7½ lin. by almost 3 lin. (when flat), very finely velvety all over, keels broadly winged, wings densely ciliolate; anthers 3½ lin. long. *Roem. & Schult. Syst.* ii. 756. *A. junceum, Durand & Schinz, Consp. Fl. Afr.* v. 935. *Triticum junceum, var. macrostachyum, Nees, Fl. Afr. Austr.* 366. *T. distichum, Thunb. Prodr.* 23; *Fl. Cap.* i. 440; *ed. Schult.* i. 119; *Kunth, Enum.* i. 448.

COAST REGION: Cape Div.; shore at Simons Town, towards Fish Hoek, *Wolley Dod*, 2061! Knysna Div.; between Groene Valley and Zwart Valley, *Burchell*, 5668! Uitenhage Div.; at the mouth of Van Stadens River, *MacOwan*,

703! in rather salt, marshy spots near the mouth of the Zwartkops River, *Ecklon & Zeyher*, 657!

Very nearly allied to *A. junceum*, Beauv., but differing in the larger spikelets, broader, 7–9- (not 5-) nerved valves and broader pales.

XCVII. HORDEUM, Linn.

Spikelets in groups of 3 at the nodes of the disarticulating (except in cultivated forms) rhachis of a dense spike; rhachilla tough, produced as a bristle beyond the floret, sometimes bearing a rudimentary valve. *Floret* 1, ⚥ in the intermediate spikelet, ♂ or barren (⚥ sometimes in cultivated races) in the lateral. *Glumes* subulate or bristle-like or narrowly linear and awned, rigid, 1–3-nerved, persistent, more or less collateral and together forming a kind of involucre. *Valves* lanceolate, subinvolute, rounded on the back, gradually produced into a bristle-like awn, 5-nerved. *Pale* about as long as the valve (exclusive of the awn), 2-keeled, narrowed, deeply folded between the keels. *Lodicules* 2, lanceolate, ciliate or ciliolate, hyaline. *Stamens* 3. *Ovary* villous at the top; styles very short, terminal; stigmas laterally exserted, plumose; grain tightly enclosed and usually adhering to the valve and pale, ovoid-oblong or narrow-oblong, rounded on the back, grooved in front; embryo short; hilum filiform, very long.

Perennial or annual; blades flat; ligules hyaline; spike cylindric, dense, usually very bristly from the awns and bristle-like glumes, usually very fragile in the wild species.

Species about 12, mainly in the temperate regions of the northern hemisphere.

Perennial; all the glumes alike, subulate-aristate ... (1) **secalinum**.
Annual; glumes of the intermediate spikelet and usually the inner of the lateral spikelets narrow, linear-lanceolate, ciliate ... ... (2) **murinum**.

1. **H. secalinum** (Schreb. Spicil. Fl. Lips. 148); perennial, tufted; culms slender, erect or geniculate-ascending, 1–2 ft. long, glabrous, smooth, 4-noded, internodes exserted; sheaths rather tight, thin, striate, glabrous, or the lower hairy, at length withering away or breaking up into fine fibres; ligule very short, truncate, hyaline; blades linear, long tapering to a fine point, 2–10 in. by 1½–3 lin., flat, slightly involute, somewhat stiff to flaccid, subglaucous, glabrous, or sparingly hairy, closely striate and scabrid on the upper side; spike erect, straight or flexuous, 1–3 in. long, light green to glaucous, bristle often purple; rhachis fragile, flattened, margins scabrid; spikelets 3 at each node, intermediate sessile, ⚥, lateral shortly pedicelled, ♂ or barren; glumes equal, reduced to straight scabrid bristles, slightly channelled on the inner side, 6–9 lin. long, those of the intermediate spikelet close to the floret, of the lateral more or less distant from it; valve of ⚥ floret lanceolate, subinvolute, gradually narrowed into the awn, 3½–4½ lin. long, herbaceous-chartaceous, smooth or scabrid in the upper part, 5-nerved; awn exceeding the glumes; pales narrow, keels scaberulous; anthers over

2 lin. long; grain more or less adhering to the valve and pale, oblong, 2 lin. long, semiterete, deeply grooved in front; embryo ½ lin. long; ♂ or barren spikelets similar to the ⚥, but smaller and with the awn exceeded by the glumes. *Host, Gram. Austr.* i. 26, *t.* 33; *Trin. Ic. Gram.* i. *t.* 3; *Durand & Schinz, Consp. Fl. Afr.* v. 943. *H. pratense, Huds. Fl. Angl. ed.* ii. 56; *Engl. Bot. t.* 409; *Steud. Syn. Pl. Glum.* i. 352; *Kunth, Enum.* i. 455; *Reichb. Ic. Fl. Germ.* i. *t.* 11, *fig.* 1363; *Knapp, Gram. Brit.* 105. *H. murinum β, Linn. Sp. Plant.* 85. *H. capense, Thunb. Prodr.* 23; *Fl. Cap.* i. 441; *ed. Schult.* i. 119; *Nees in Linnæa,* vi. 305, *and Fl. Afr. Austr.* 362; *Kunth, Enum.* i. 456; *Steud. Syn. Pl. Glum.* i. 353; *Durand & Schinz, Consp. Fl. Afr.* v. 941.

COAST REGION: Malmesbury Div.; Saldanha Bay and Zwartland, *Thunberg!* Cape Div.; between Table Mountain and Devils Mountain, *Bergius.* Paarden Island, *Wolley Dod,* 3165! on the shore of Table Bay, *Wolley Dod,* 3227! by the Raapenberg Vley, *Wolley Dod,* 3113! Swellendam Div.; mountain ridges along the lower part of the Zonder Einde River, *Zeyher,* 4558! Queenstown Div.; between Table Mountain and Zwart Kei River, 4000 ft., *Drège.*

CENTRAL REGION: Graaff Reinet Div.; in a valley at Houd Constant Waterfall, in the Sneeuwberg Range, 3500 ft., *Bolus,* 1965! banks of the Sunday River, 2600 ft., *Bolus,* 1965! Victoria West Div.; between Brak River and Uitvlugt, 3000–4000 ft., *Drège.* Aliwal North Div.; by the Kraai River, *Drège!* Albert Div.; without precise locality, *Cooper,* 646!

WESTERN REGION: Little Namaqualand; between Pedros Kloof and Lily Fontein, 3000–4000 ft., *Drège.*

KALAHARI REGION: Orange Free State, *Hutton!*

Temperate regions of the northern hemisphere; probably introduced into the Cape Colony.

2. H. murinum (Linn. Sp. Plant. 85, excl. var. *β*); annual, tufted; culms erect or usually geniculate-ascending, ½–1½ ft. long, glabrous, smooth, about 4-noded, sheathed all along or some of the internodes exserted; lower sheaths membranous, whitish, striate, glabrous or scantily hairy, upper herbaceous, laxer, uppermost usually very wide, tumid; ligules hyaline, truncate, denticulate, about ½ lin. long; blades linear, tapering to an acute point, produced at the base into sometimes long, semilunar membranous auricles, up to 8 in. by as much as 4 lin., the upper often very short, flat, flaccid, dull green, loosely hairy on both sides; spikes erect or nodding, 1½–3 in. long, light green; rhachis fragile, flattened, margins scabrid; spikelets 3 at each node, intermediate sessile, lateral shortly pedicelled, ♂ or barren; glumes of the intermediate spikelet equal, very narrow, lanceolate-linear, produced into a stiff scabrid awn, 4–5 lin. long (with the awn up to 1 in. long), 1-nerved, rigid, scabrid on the back, glabrous or hairy on the inner side, green in the middle, margins cartilaginous, rigidly ciliate; valve somewhat distant, lanceolate, compressed from the back, subinvolute, about 5 lin. long, herbaceous-chartaceous, glabrous or sparingly hairy on the upper (inner) side, smooth except at the nerves close to the tip, 5-nerved, awn up to more than 1 in. long; pales narrow, keels smooth or almost so, hairy on the upper (inner) side, sometimes also on the back above; anthers ¼–1 lin.

long; grain adhering to the valve and pale, 2–2½ lin. long, deeply channelled; ♂ or barren spikelets very similar to the ⚥, but glumes usually unequal, outer reduced to a scabrid bristle and usually longer than the inner (1–1½ in. long) which is ciliate only on the inner side or not at all; valve up to 8 lin. long and like the pale usually more hairy than in the ⚥. *Fl. Dan. t.* 629; *Engl. Bot. t.* 1971; *Host, Gram. Austr.* i. *t.* 32; *Kunth, Enum.* i. 456; *Nees in Linnæa*, vii. 304, *and Fl. Afr. Austr.* 363; *Steud. Syn. Pl. Glum.* i. 352; *Durand & Schinz, Consp. Fl. Afr.* v. 941.

COAST REGION: Clanwilliam Div.; near Clanwilliam, 300 ft., *Schlechter*, 8595! Cape Div.; near Capetown, *Ecklon*, 974! Hout Bay Fisheries, *Wolley Dod*, 1561! Mossel Bay Div.; by the Gauritz River, *Burchell*, 6483! Albany Div.; by the roadsides near Grahamstown, 1800–2000 ft., *MacOwan*, 1285!

WESTERN REGION: Little Namaqualand; between Pedros Kloof and Lily Fontein, 3000–4000 ft., *Drège!*

A very widely spread weed of Mediterranean origin.

XCVIII. OLYRA, Linn.

Spikelets unisexual, heteromorphous, both sexes in the same panicle or sometimes in different panicles of the same or very rarely of different individuals; rhachilla disarticulating below the valve. ♂ *spikelets* smaller than the ♀, in mixed panicles below them and much more numerous, 1-flowered. *Glumes* suppressed, indicated by a minute, sometimes obscurely 2-lobed, rim just below the articulation of the rhachilla. *Valve* lanceolate to almost subulate in profile, acuminate or caudate-awned, membranous, 3-nerved. *Pale* shorter than the valve, 2-nerved. *Lodicules* 3, or 0 (?). *Stamens* 3. *Ovary* 0. ♀ *spikelets* terminal, or terminal and lateral on the branches or branchlets of the panicle, above the ♂, 1-flowered. *Glumes* subequal and similar, ovate to lanceolate, acuminate or caudate-awned (particularly the lower), papery or membranous, 3–9-nerved, transversely veined, persistent. *Valve* elliptic to lanceolate, awnless, subcoriaceous to hard, cartilaginous, 5-nerved (or apparently nerveless). *Pale* similar to the valve in shape and substance, 2-nerved. *Lodicules* 3, truncate-cuneate. *Stamens* 0. *Ovary* glabrous, acuminate; style simple below or beyond the middle; stigmas plumose, terminally exserted. *Grain* tightly enclosed by the hardened valve and pale, biconvex; embryo short; hilum linear, almost as long as the grain.

Branched, tall or dwarf perennials; blades convolute in bud, then flat, often broad and asymmetric, shortly petioled, transversely veined; ligules papery, very short; panicles terminal, sometimes with 1 or few axillary additional ones, rarely all axillary, often decompound.

Species about 20, in tropical America, 1 of them extending to tropical Africa, the Mascarene Islands and Zululand.

1. **O. latifolia** (Linn. Amœn. Acad. v. 408); perennial; culms erect, branched 3–8 ft. high, glabrous, very smooth, rarely more or less hairy near the nodes, many-noded, sheathed all along or the upper

internodes shortly exserted; sheaths tight, firm, striate, keeled in the upper part, glabrous except at the ciliate upper margins, or hairy; ligules very short, truncate, broader than the short hairy petioles; blades lanceolate-oblong or ovate-oblong, asymmetric, particularly at the rounded base, conspicuously acuminate, 4–7 in. by 1–2¾ in., flat, glaucous, glabrous except quite at the base, rarely sparingly hairy, slightly rough with about 11–19 primary nerves and a prominent whitish midrib; panicle erect, terminal or with a lateral one from the uppermost leaf axil, 3–6 in. long, rather contracted or subpyramidal; axis slender, angular, pubescent; branches solitary or sometimes fascicled, subflexuous or straight, angular, pubescent; pedicels adpressed, of the ♂ spikelets filiform, of the ♀ clavate; ♂ spikelets lateral, linear-lanceolate, acuminate, awned from the valve, glabrous; glumes quite rudimentary forming an inconspicuous slightly 2-lobed puberulous rim at the tip of the pedicels; valve herbaceous-membranous, gradually passing into an awn of about equal length, 1¾–2 lin. long, 3-nerved; lodicules 3, cuneate, thin; anthers over 1 lin. long; ♀ spikelets solitary and terminal on the tips of the branches, or 2–3, rarely more, below the terminal spikelet, oblong-ovoid, turgid, awned from the lower glume; glumes ovate, elliptic, membranous, strongly nerved and veined, lower acuminate or shortly caudate-acuminate, about 4 lin. long, 5–7-nerved, upper produced into a flexuous scabrid awn of equal or greater length, 7–9-nerved; valve ovate-elliptic, obtuse, 2–3 lin. long, white or greyish, cartilaginous, very hard, shining; styles connate beyond the middle; grain oblong. *Linn. Sp. Plant. ed.* ii. 1379; *Doell in Mart. Fl. Bras.* ii. ii. 316. *O. paniculata*, *Sw. Prodr.* 21; *Obs.* 347; *Trin. Panic. Gen.* 23, *and in Mém. Acad. Pétersb. sér.* 6, iii. (1835) 111; *Ic. Gram. t.* 346; *Kunth, Enum.* i. 69; *Steud. Syn. Pl. Glum.* i. 35. *O. brevifolia*, *Schum. in Schum. & Thonn. Beskr. Guin. Pl.* 402. *O. guineensis*, *Steud. l.c.* 37; *Durand & Schinz, Consp. Fl. Afr.* v. 788.

EASTERN REGION: Zululand; in Ungoya Forest, *Wood*, 3856! Ongor, *Gerrard & McKen*, 2014!

Tropical America and Africa and in the Mascarene Islands.

XCIX. **ARUNDINARIA**, Mich.

Spikelets often long, more or less compressed or subcylindric, variously panicled or racemose; rhachilla tardily disarticulating above the glumes and below the valves. *Florets* 1 to many, ⚥ or the uppermost reduced. *Glumes* 2 or 1, unequal, herbaceous to membranous, variously nerved. *Valves* usually similar to the glumes, but larger, subobtuse, acute, acuminate or mucronate, rounded on the back, membranous-herbaceous, many-noded. *Lodicules* 3, ovate or lanceolate, ciliate. *Stamens* 3. *Ovary* glabrous or often hairy at the top; styles 2–3, connate at the base; stigmas loosely plumose; laterally exserted. *Grain* ovoid to oblong, enclosed by the valve and pale, grooved in front; embryo short; hilum long, filiform.

Erect or climbing shrubs; culms slender; nodes usually prominent, internodes rather short; branches fascicled at the nodes; blades articulate on the sheath, shortly petioled, distinctly transversely veined; panicles or racemes sometimes leafy, terminal on leafy culms or on separate culms; spikelets often partially enclosed in bract-like sheaths.

About 50 species, in tropical and subtropical Asia and America and in temperate Eastern Asia.

1. **A. tesselata** (Munro in Trans. Linn. Soc. xxvi. 31); shrubby or arborescent, 5–20 ft. high; old culms over 4 lin. thick below, fistulous like the very slender crowded leaf-bearing branches, terete, very smooth, internodes of the branches usually less than 2 in. long; sheaths at the base of the branches bladeless or almost so, scarious to subscarious, striate, smooth, the rest coriaceous, very tight, slightly striate, ciliate along the upper and outer margins and fugaciously fimbriate at the mouth, otherwise glabrous; ligules short, obtuse or sometimes produced and up to 2 lin. long; fully developed blades 3–4 at the tips of the branches, lanceolate to linear-lanceolate from a rounded or attenuated base, shortly acute or long attenuate to a very fine point, $2\frac{1}{2}$–5 in. by $4\frac{1}{2}$–7 lin., coriaceous, somewhat glaucous, very sparingly pubescent at the very base and spinulously ciliate along the margins, at least when young, otherwise quite glabrous, primary side-nerves 3 (rarely 4) on each side, transverse veins close and usually very conspicuous. *Kew Report*, 1878, 47. *Nastus tesselatus*, *Nees*, *Fl. Afr. Austr.* 463; *Steud. Syn. Pl. Glum.* i. 333; *Durand & Schinz, Consp. Fl. Afr.* v. 944.

COAST REGION: Queenstown Div.; Table Mountain, 6000–7000 ft., *Drège*. Stockenstrom Div.; Kat Berg, 4000–5000 ft., *Drège!* Winter Berg, *Ecklon*.
CENTRAL REGION: Aliwal North Div.; on the Witte Bergen, in rocky moist valleys, 5000–6000 ft., *Drège!*
KALAHARI REGION: Basutoland, *Cooper*, 922!
EASTERN REGION: Natal; edge of bush at Olivers Hoek Pass, 4500 ft., *Wood*, 3668! shady places near Van Reenens Pass, 8000 ft., *Schlechter*, 6997!

The position of this Bamboo in *Arundinaria* is somewhat hypothetical as the flowers are unknown. It flowers probably like many other Bamboos at very long intervals. No opportunity of collecting flowering specimens should therefore be missed. It is the so-called Mountain (Berg) Bamboo of the Drakensbergen, where it grows according to Bowker to "an unlimited extent, mostly on the northern slopes . . . and on the most exposed sites." The Bamboo Mountains (Bamboes Berg), on the border of Albert and Cradock Division, have most probably derived their name from this species. The canes are much used by the natives for house-building, fences, spear-handles, &c.

C. **BAMBUSA**, Schreb.

Spikelets in sessile clusters on spike-like branchlets of usually large leafless or sometimes leafy compound panicles; rhachilla disarticulating above the glumes and below the valves. *Florets* 2 to many, ⚥ or the uppermost ♂, or the lower sometimes barren.

Glumes gradually passing into the valves and like these rigidly membranous to subcoriaceous, muticous or mucronate-acuminate, upper larger. *Valves* ovate-lanceolate, many-nerved. *Pales* equal to the valves, prominently 2-keeled, keels ciliate or not. *Lodicules* 3, membranous, generally obtuse, ciliate. *Stamens* 6; filaments free. *Ovary* oblong or obovoid, hairy at the top; style simple or 2–3-fid; stigmas short, plumose or hairy. *Grain* oblong or linear-oblong, acute or obtuse, grooved; pericarp thin, adherent; embryo conspicuous.

Arborescent or shrubby, rarely climbing, sometimes thorny; culms from a thick rhizome, usually cæspitose, sometimes stoloniferous; blades medium-sized, rarely large, linear or oblong-lanceolate, acuminate, shortly petioled; panicles often very compound.

Species about 50, in tropical Asia, 1 in North-East Australia; elsewhere only cultivated or as escapes.

1. **B. Balcooa** (Roxb. Hort. Beng. 25; Fl. Ind. ii. 196); tall, stout, cæspitose; culms 50–70 ft. high, 3–6 in. in diam., branched from the base, with the lowest shoots long and leafless; nodes swollen with a whitish ring above them, hairy below, internodes 8–18 in. long; lowest sheaths short, broad, densely and adpressedly hairy on the upper side, ciliate along the margins, bearing short triangular auricled imperfect blades, following sheaths 10–14 in. by 8–10 in. (when expanded), almost glabrous except at the ciliate margins, truncate, bearing imperfect blades, 6–8 in. by 3–4 in., closely hairy, rounded at the base and then again decurrent on the sheath; upper sheaths striate, adpressedly white-hairy, truncate, sometimes with a few stiff deciduous bristles; ligules membranous, broadly triangular; blades oblong-lanceolate, rounded or subcordate at the base, tapering to a twisted scabrid setaceous point, glabrous above, pallid beneath and hairy when young, particularly towards the base, scabrid-ciliate along the margins, midrib prominent, primary side-nerves 7–11; panicle large, compound; rhachis pubescent or scurfy, striate; branches long; clusters of few to many spikelets, bracteate; spikelets ovoid-lanceolate, $3\frac{1}{2}$–8 lin. by $2\frac{1}{2}$–$3\frac{1}{2}$ lin., up to 7-flowered, greenish, straw-coloured or brownish; glumes 2–0, very concave, ovate, acute, many-nerved, if 2, the lower smaller, upper intermediate between it and the lowest valve, both with a small or rudimentary bud in the axil; valves ovate-lanceolate to ovate in profile, acute or mucronate, up to 4 lin. long, herbaceous-subcoriaceous, many-nerved, margins ciliate above; pale acutely 2-keeled, keels long ciliate, back 5-nerved between the keels, flaps about 3-nerved; lodicules oblong to obovate, 3–5-nerved, ciliate, up to 1 lin. long; stamens hardly exserted; anthers $2\frac{1}{2}$ lin. long, glabrous; ovary hairy, attenuated into the villous style, this nearly 2 lin. long; stigmas 3, slender. *Munro in Trans. Linn. Soc.* xxvi. 100; *Gamble in Ann. Bot. Gard. Calcutta,* vii. 42. *B. capensis, Rupr. Bamb.* 54, *t.* 12, *fig.* 54; *Steud. Syn. Pl. Glum.* i. 330; *Durand & Schinz, Consp.*

Fl. Afr. v. 945. *B. vulgaris, Nees, Fl. Afr. Austr.* 462 (*excl. syn.*), *not of Schrad*; *Durand & Schinz, l.c.*

SOUTH AFRICA: without precise locality, *Mund! Zeyher!*
COAST REGION: Tulbagh Div.; in a garden at Winter Hoek, *Burchell*, 1045/3! "George District," *Ecklon*. Uitenhage Div.; by the Zwartkops River near Paul Mare's Estate, *Ecklon*.

A native of lower Bengal and Assam.

ADDENDA AND CORRIGENDA.

1. **Xyris decipiens** (N. E. Brown). This species is not a member of the South African Flora. All Curror's plants were collected in Angola, and the locality "south of the Tropic" (given on p. 4) is an error.

8. **Xyris filiformis** (Lam.). For this name substitute **X. straminea** (Nilss.), given in the synonymy. The description on p. 7 is entirely drawn up from *X. straminea*. The true *X. filiformis*, Lam., according to analyses and drawings made from the type-specimen at Paris, differs in having spikes of a darker colour and the keels of the lateral sepals being minutely ciliate on the apical part.

34a. **Restio multicurvus** (N. E. Brown); plant about $\frac{3}{4}$–1 ft. high; stems cæspitose, slender, branching at the base, irregularly much curved backwards and forwards, ochreous, smooth, not shining; sheaths 3–5 lin. long, closely convolute in the lower half, ending in a rigid subulate subpungent point about as long as the sheathing part, dark brown; male spikelets terminal, solitary or occasionally 2 together, $2\frac{1}{2}$–3 lin. long, lanceolate, about equally acute at both ends, dark brown; bracts 2 lin. long, $\frac{3}{4}$ lin. broad, boat-shaped, shortly mucronate, straw-coloured in the lower half, dark brown at the apex, with slightly paler margins; perianth-segments submembranous, straw-coloured; outer lateral segments $1\frac{1}{2}$ lin. long, densely ciliate with brown hairs on the keel; inner segments $1\frac{1}{4}$ lin. long, oblong, obtuse; anthers narrowly oblong, apiculate; female spikelets not seen.

COAST REGION: Riversdale Div.; on the mountain sides of Garcias Pass, 1000 ft., *Galpin*, 4785!

Allied to *R. curviramis*, Kunth, but easily distinguished by its usually solitary (not racemose) spikelets and much more curved stems. The internodes are often curved nearly into a semicircle.

39a. **Restio scaberulus** (N. E. Brown); stems terete, densely much branched, tuberculate-scabrid, about 1 lin. diam.; branchlets several from each node, again branching, filiform, erect or ascending, tuberculate-scabrid; sheaths of the main stem about $\frac{3}{4}$ in. long, closely convolute in the lower half, ending in a filiform awn as long as the convolute part, scabrid, brown, with white membranous margins, produced on each side of the awn into a lanceolate acuminate point nearly as long as the awn; sheaths of the branchlets similar, but smaller; male spikelets 2–6 in a lax linear terminal spike, sessile, 4–6-flowered; spathes like the sheaths but smaller; bracts $1\frac{1}{2}$ lin. long, boat-shaped, acuminate into a fine awn-like point, brown, with membranous whitish margins; outer perianth-segments about 1 lin. long, $\frac{1}{6}$ lin. broad, boat-shaped, acute, light brown along the keel of the lateral segments, otherwise transparent yellowish-brown; keels minutely ciliate; inner segments rather shorter, oblong, acutely bifid at the apex, hyaline; stamens 3; anthers oblong; female spikelets solitary, rarely 2 together, terminal, $1\frac{1}{2}$ lin. long, ovoid, becoming broadly turbinate in fruit, 1-flowered; spathes and bracts like those of the male spikelets but rather smaller; outer perianth-segments about $1\frac{1}{6}$ lin. long, $\frac{1}{2}$ lin. broad, much larger than those of the male flower, boat-shaped, acute, not ciliate on the rather obtuse keels; inner segments 1 lin. long, $\frac{1}{2}$ lin. broad, ovate, subacute, somewhat rigid, whitish, transparent; ovary compressed-trigonous, 2-celled; styles 3; capsule $\frac{3}{4}$ lin. long, $1\frac{1}{3}$ lin. broad, compressed, somewhat obreniform, 2-celled; seeds trigonous, smooth, whitish.

COAST REGION: Riversdale Div.; on the mountains of Garcias Pass, 2000 ft., *Galpin*, 4789!

This much resembles *R. leptoclados*, Mast., but differs in its more scabrid stems, very different sheaths, and in floral structure.

66a. Restio strictus (N. E. Brown); barren stems 3–5 in. high, branching, rather dense, less than $\frac{1}{2}$ lin. thick, terete, ochreous; leaves 2–3 lin. long, terete-subulate, subpungent, with a convolute brown sheath 1–2 lin. long, glabrous; flowering stems $1–1\frac{1}{2}$ ft. high, simple, $\frac{3}{4}$ lin. thick, terete, smooth, ochreous; sheaths of young stems closely convolute, basal part 6–7 lin. long, coriaceous, brown, with membranous white margins, produced at the apex into a fine brown awn, $\frac{1}{2}$ in. long, with an obliquely lanceolate acuminate membranous lobe nearly as long on each side of it, these lobes have usually fallen away or worn off from the older stems; male and female spikelets similar, about 4 in a cluster at the apex of the stem, each about 7 lin. long, $2\frac{1}{2}$ lin. thick, not compressed; male spikelets oblanceolate, obtuse, narrowed towards the base; bracts 4 lin. long, $1–1\frac{1}{4}$ lin. broad, lanceolate, acuminate, concave, glabrous, light brown, membranous along the margins; outer perianth-segments about $2\frac{1}{4}$ lin. long, acute, rather rigid, yellowish-brown; lateral boat-shaped, ciliate along the keel; inner segments rather shorter, membranous, lanceolate, obtuse; anthers linear-oblong, apiculate; female spikelets lanceolate, acute; bracts and perianth-segments as in the male; ovary suborbicular, compressed-trigonous, glabrous; styles free and plumose to the base; staminodes 3.

COAST REGION: George Div.; Cradock Berg, near George, 3000 ft., *Galpin*, 4795!

This bears some resemblance to *R. Sieberi*, Kunth, and *R. bigeminus*, Nees, but is readily distinguished by its very different sheaths and stouter leaves, besides other characters.

70b. Restio foliosus (N. E. Brown); stems probably several feet in height, $1\frac{1}{2}$–2 lin. thick, terete, smooth, ochreous or light brownish, furnished with tufts of short leafy branchlets at all the nodes nearly or quite up to the panicle; sheaths $1\frac{1}{4}$–$1\frac{1}{2}$ in. long, closely convolute, acute, and with a terminal awn 2–3 lin. long, brown, with white membranous margins; leafy barren branchlets in erect dense tufts at the nodes, 2–5 in. long, very slender, the stoutest not more than $\frac{1}{3}$ lin. thick; leaves 2–$2\frac{1}{2}$ lin. long, subulate, very finely pointed, their sheaths produced into an acutely bifid membrane more than half the length of the leaf; male panicle about 1 ft. long, moderately lax; its main branchlets 2–$4\frac{1}{2}$ in. long, very slender, flattened, $\frac{1}{5}$–$\frac{1}{2}$ lin. broad, apparently ascending, narrowly paniculately branched; female panicle $3\frac{1}{2}$–4 in. long, $\frac{3}{4}$ in. in diam., linear-oblong, moderately compact, its main branchlets $\frac{1}{2}$–$1\frac{1}{2}$ in. long, erect, flattened; male spikelets numerous, 2–$3\frac{1}{2}$ lin. long, $1\frac{1}{2}$ lin. in diam., cylindric-oblong, obtuse, tapering to the base, 5–9-flowered; bracts $1\frac{1}{2}$ lin. long, $\frac{2}{3}$–$\frac{3}{4}$ lin. broad, oblong, obtuse, shortly mucronate, concave, brown, narrowly white margined; outer perianth-segments $1\frac{1}{5}$ lin. long, scarcely $\frac{1}{3}$ lin. broad, linear, obtuse, not ciliate on the keels of the lateral segments, thin, light brown; inner perianth-segments slightly shorter than the outer, linear-lanceolate, obtuse, membranous, light brownish; anthers linear, apiculate; female spikelets 1-flowered, 2 lin. long, 1 lin. in diam., trigonous; inner bracts much shorter than the flower, $1\frac{1}{2}$ lin. long, 1 lin. broad, elliptic or elliptic-oblong, obtusely rounded at the apex, mucronulate, brown, with very narrow membranous white margins; perianth 2 lin. long; segments subequal, lanceolate, acute, concave, coriaceous, brown; ovary globose-trigonous, 3-celled; styles 3, densely plumose to the base.

COAST REGION: Riversdale Div.; on the mountain sides of Garcias Pass, 1000 ft., *Galpin*, 4783, ♂ and ♀!

Allied to *R. giganteus*, N. E. Brown (70a, see p. 755), and *R. Rhodocoma*, Mast.

70c. Restio comosus (N. E. Brown); stems about 7 ft. high, stout, terete, very finely corrugated (from shrinkage?), dull greenish; sheaths $1\frac{1}{2}$–2 in. long, acuminate or awned at the apex, brown, more or less torn in the specimens seen; barren branches numerous, in tufts at the nodes, 5–12 in. long, $\frac{1}{4}$–$\frac{1}{2}$ lin. thick at the base, very much divided into fine capillary branchlets; leaves 2–4 lin. long, capillary, very acute, with a brown bifid membranous sheath, 1–$1\frac{1}{2}$ lin. long; male spikelets either in a terminal panicle about 8 in. long destitute of leafy branchlets, or borne on the basal parts of the uppermost tufts of leafy branches (the latter probably an abnormal condition), 3–4 lin. long, $1\frac{1}{4}$–$1\frac{1}{2}$ lin. in diam., oblanceolate, tapering to the base, several-flowered, bright brown; bracts $1\frac{1}{2}$–2 lin. long, oblong, obtuse or subacute, deeply concave, entire, glabrous; outer perianth-segments $1\frac{1}{2}$ lin. long, $\frac{1}{3}$ lin. broad, linear-oblong, acute, glabrous, not ciliate on the keels of the lateral pair;

inner segments slightly shorter than the outer, more membranous, oblong-lanceolate, obtuse; anthers linear-oblong, apiculate.

COAST REGION: Knysna Div.; on a river-bank at "The Glebe," Knysna, 800 ft., *Galpin*, 4784!

Allied to *R. foliosus*, N. E. Brown, but the leafy branchlets and leaves are longer and much more slender, quite capillary, and the bracts of the spikelets are not white-margined.

9a. **Elegia fusca** (N. E. Brown); stems about 16–18 in. high, simple, terete, smooth, not shining, glabrous, pallid or tinged with smoky-brown; leaf-sheaths not seen, very deciduous, leaving a narrow dark-brown basal ring; female inflorescence terminal, narrow, 1½–2½ in. long, erect; spathes 5–6, erect, convolute, imbricate, ¾–1 in. long, ovate-oblong, probably acute, but apex broken in the specimens seen, smoky-brown, glabrous; flowers under each spathe 2–6, subsessile or in a short spike about 4–6 lin. long, including the peduncle; perianth-segments subequal, 1½ lin. long, glabrous, dark chestnut-brown; outer linear-oblong, subacute, keeled; inner oblong-lanceolate, obtuse, thin, not keeled; ovary slightly compressed, lanceolate-oblong, 1-celled; stigmas 2; male inflorescence not seen.

COAST REGION: George Div.; on Cradock Berg, near George, 2500 ft., *Galpin*, 4802 ♀!

A very distinct species, somewhat resembling *E. juncea*, Linn., but with a much smaller inflorescence.

18a. **Elegia gracilis** (N. E. Brown); stems dimorphic; barren stems 6–7 in. (or more?) high, slender, sparingly branched; sheaths very small, about 2 lin. long or less and nearly as broad, elliptic-ovate, slightly mucronulate, dull brown, spreading, persistent; fertile stems 12–17 in. high, scarcely ½ lin. thick, simple, erect, straight, terete, glabrous, pallid; sheaths 6–8 lin. long, very closely convolute, obtuse, minutely apiculate, persistent, light brown in the lower half, darker brown in the more membranous upper part; inflorescence 1½–2¼ in. long, narrow; spathes 6–7, laxly imbricate or the lowest distant from the rest, about 7 lin. long, 2½–3 lin. broad, ovate, acute, glabrous, convolute below, persistent, golden-brown, darker in the upper part; male flowers very small, in small compact panicles about as long as the spathes; bracts about 2 lin. long, linear-lanceolate, acute; bracteoles about ½ lin. long, orbicular, very obtuse; perianth-segments subequal, broadly elliptic, concave, obtuse; female flowers not seen.

COAST REGION: Caledon Div.; Houw Hoek Mountains, 1500 ft., *Galpin*, 4800!

Allied to *E. obtusiflora*, Mast., but very much more slender.

24a. **Elegia Galpinii** (N. E. Brown); stems apparently tall, simple, about 1¼ lin. thick, terete, opaque, light brown; sheaths not seen, distant, deciduous, leaving a dark brown ring; inflorescence a narrow interrupted erect densely-flowered panicle, 6–8 in. long in the male, $3\frac{1}{2}$–$4\frac{1}{2}$ in. long in the female; spathes (only a part of an upper one seen) 1 in. or more long, oblong, expanded, brown, very deciduous, leaving a dark brown ring; branches of the panicle oblong, very densely covered with very dark brown flowers, erect, glabrous; bracts deciduous; bracteoles persistent, $\frac{1}{2}$–$\frac{2}{3}$ lin. diam., orbicular, concave, very obtuse, more or less lacerated at the margin; perianth-segments of the male flowers unequal, oblong, obtuse, concave; outer scarcely $\frac{1}{2}$ lin. long, inner $\frac{3}{4}$ lin. long; of the female flowers subequal, orbicular, concave, very obtuse; fruit subglobose, 1-celled, about 1 lin. diam., slightly rugose, blackish-brown.

COAST REGION: Riversdale Div.; on the mountains of Garcias Pass, 1200 ft., *Galpin*, 4803!

A most distinct species, allied to *E. nuda*, Kunth, and *E. elongata*, Mast., but very much larger than either.

2. **Thamnochortus giganteus** (Kunth). The discovery of ♀ plants makes it necessary to transfer this species to *Restio*, in which, from its affinity to *R. Rhodocoma*, it will become **70a, Restio giganteus** (N. E. Brown). To the description and distribution are to be added:—Female inflorescence a narrow erect panicle, 14–22 in. long, $1\frac{1}{2}$ in. diam.; branches erect, 1–4 in. long, sparingly branched in their lower part; spikelets $\frac{1}{4}$ in. long, closely placed, some of the lower on the branches and their divisions pedicellate, the rest sessile, 1-flowered; bracts imbricate, the innermost $1\frac{1}{2}$ lin. long, $\frac{3}{4}$ lin. broad, oblong, obtuse, apiculate, the rest gradually smaller, dark brown; outer perianth-segments $2\frac{1}{2}$ lin. long, $\frac{3}{4}$ lin. broad, all alike, lanceolate, acute, concave, rounded (not keeled) on the back, glabrous, chestnut-brown; inner segments 2 lin. long, $1\frac{1}{4}$–$1\frac{1}{3}$ lin. broad, ovate, acute, flattened or slightly concave, chestnut-brown; ovary 3-angled, 3-celled; styles 3, recurved over or upon the angles of the ovary, densely plumose, especially at the base.

COAST REGION: Knysna Div.; on a hillside, "The Glebe," Knysna, 800 ft., *Galpin*, 4781, ♂ and ♀! "Witte Els Bosch," Zitzikama, 500 ft., *Galpin*, 4782, ♂ and ♀!

From the time of Kunth until now only the male plant of this species has been known to authors, but in 1897 Mr. Galpin collected the female plant, from which it is clear (on account of its 3-celled ovary) that it must be removed from the genus *Thamnochortus* and placed under *Restio*, where its alliance is with *R. Rhodocoma*, Mast., *R. foliosus*, N. E. Brown, and *R. comosus*, N. E. Brown. Mr. Galpin found both sexes growing together in two localities, and states that the plant is 10 ft. in height.

12a. **Hypolæna decipiens** (N. E. Brown); stems 1–$1\frac{1}{2}$ ft. high, $\frac{1}{2}$–$\frac{3}{4}$ lin. thick, branching from the base upwards; branches slender, erect, dull ochreous, minutely papillate-rugulose; sheaths 3–6 lin. long, closely convolute, mucronate, brown, with a broad white membranous border at the apex; male spikelets in a linear panicle $\frac{3}{4}$–$2\frac{1}{2}$ in. long, erect, $2\frac{1}{2}$ lin. long, $\frac{1}{2}$ lin. diam., lanceolate, acute, 1-flowered; spathes $2\frac{1}{2}$ lin. long, 1 lin. broad when expanded, ovate-lanceolate, acute, brown, with white membranous margins, glabrous; bracts 2, membranous, light brown on the back, about $1\frac{1}{2}$ lin. long, one linear-spathulate, subacute or obtuse, lacerate-ciliate, the other boat-shaped, bifid, hairy on the back; outer perianth-segments about $1\frac{3}{4}$ lin. long, subequal, lanceolate, acute, light yellowish-brown; inner segments shorter, lanceolate, acute, very membranous, transparent; anthers linear, apiculate; female spikelets not seen.

COAST REGION: Swellendam Div.; Zuurbraak Mountain, 4000 ft., *Galpin*, 4792!

Similar to *H. Burchellii*, Mast., but differing in its 1-flowered spikelets, less membranous spathes, and acute perianth-segments.

4a. **Willdenovia simplex** (N. E. Brown); stems slender, straight, terete, slightly sulcate, glabrous, ochreous; leaf-sheaths 10–11 lin. long, closely convolute, acute, brown; female inflorescence of 1 (or 2?) erect 1-flowered spikelets; spathes 2, erect, $1–1\frac{1}{3}$ in. long, 2 lin. broad, lanceolate, long-acuminate, expanded, thinly coriaceous, inclined to split, but not membranous at the apex; bracts several, outer about 8 lin. long, lanceolate, acuminate into an arista, innermost shorter, ovate; perianth 0; disk or stalk of the ovary corrugated, lobulate, white; young fruit subcylindric, truncate at the apex, with transverse irregular impressed lines, or somewhat rugulose in the lower half, not punctate nor pitted; styles 2.

COAST REGION: Riversdale Div.; mountains of Garcias Pass, 1000 ft., *Galpin*!

Allied to *W. sulcata*, Mast., but differing in its leaf-sheaths, spathes, bracts, and especially by the absence of a perianth and non-pitted fruit. I have not seen the male plant. The specimens are 14 in. long, and show no traces of branching, but appear to be only the upper parts of the plant.

6a. **Willdenovia fraterna** (N. E. Brown); stems terete, branching, very minutely and densely pitted, glabrous, ochreous; leaf-sheaths $1–1\frac{1}{4}$ in. long when entire, closely convolute, acuminate; lower part coriaceous, persistent, at first ochreous with brown margins, afterwards becoming dark olive-brown, upper part membranous, deciduous; male inflorescence not seen; female spikelets 1–2, terminal, erect, 1-flowered; spathes usually 2, somewhat convolute, but the margins not meeting, $1\frac{1}{4}–1\frac{1}{2}$ in. long, $\frac{1}{3}$ in. broad, lanceolate, acuminate into a stiff awn, coriaceous, shining, brown and very minutely pitted in the lower half, greyish and densely marked with purple-brown in the upper half, narrowly membranous at the margin; bracts several, 9–11 lin. long, lanceolate, acuminate into an awn, brown, with narrow white membranous margins; perianth-segments about $\frac{3}{4}$ lin. long and broad, subquadrate, bicuspidate at the apex, with a minute apiculus in the notch and slightly toothed on the sides; disk or stalk of the ovary 3-lobed; lobes oblong, emarginate at the apex with rounded lobules and with a transverse apical groove, giving the appearance of superposed lobes; young ovary turbinate, pitted and dark brown in the apical part; styles 2.

COAST REGION: Knysna Div.; on a hillside near "The Glebe," 800 ft., *Galpin*, 4830!

Allied to *W. lucæana*, Kunth, but easily distinguished by its different perianth-segments and pitted fruit.

6b. Willdenovia Galpinii (N. E. Brown); stems erect, attaining a height of 4 ft., laxly branching, terete, smooth, ochreous, minutely whitish-punctate; leaf-sheaths 1 in. long, closely convolute, ochreous or brown, the apical part 4–5 lin. long, membranous, white, with a dark brown central stripe, which is broadest in the male plant; male inflorescence 1½–1¾ in. long, composed of about 3 or 4 erect dense compound spikes, 4–6 lin. long; spathes 3–4, erect, expanded, deciduous, 1–1¾ in. long, 3–5 lin. broad, lanceolate, acuminate, more or less aristate, brown, coriaceous, somewhat shining and minutely punctate below, membranous above, with whitish margins; bracts about 4 lin. long, linear-subulate, membranous, brownish; flowers pedicellate; perianth-segments about 1½ lin. long, very slender, linear, filiform, more or less contorted; female spikelets 2–3 in a spike 1½–2¼ in. long, 1-flowered, erect; spathes 1¼–1⅓ in. long, like those of the male plant, but more convolute and having broad membranous white margins; bracts several, 6–9 lin. long, lanceolate, acuminate; perianth-segments membranous, ⅔–¾ lin. long, ¾ lin. broad, subquadrate-obovate, very obtuse; styles 2, deciduous; fruit about 2¼ lin. long, 1½ lin. thick, cylindric, subtruncate, thinly pitted, brown at the apex, pallid at the sides, seated on a stout fleshy lobulate-furrowed stalk.

COAST REGION: Riversdale Div.; mountains of Garcias Pass, 1000 ft., *Galpin*, 4831, ♂! 4824, ♀! 4825, ♀ monstrosity!

Allied to *W. lucæana*, Kunth, but readily distinguished by its broadly white-margined sheaths and spathes.

7a. Willdenovia peninsularis (N. E. Brown); stems erect, fastigiately much branched, terete, smooth, ochreous, very minutely punctulate; leaf-sheaths 6–8 lin. long, with a stout subulate point, 3–4 lin. long, closely convolute, dull brown, minutely punctulate, alike in both sexes; male inflorescence about ¾ in. long, spike-like, composed of about 3 erect spikes 3–4 lin. long; spathes about 3, expanded, 7–10 lin. long, 2–4 lin. broad, ovate or lanceolate, acuminate, shortly aristate, lower part light shining brown, upper third membranous, white with a broad brown stripe; bracts 3–3½ lin. long, very slender, linear; flowers pedicellate; perianth-segments 1–1¼ lin. long, filiform; female spikelets solitary, terminal, 1-flowered, erect; spathes 8–9 lin. long, ¼ in. broad, expanded, similar to those of the male plant; bracts several, 6–8 lin. long, 2½–4 lin. broad, ovate or oblong-lanceolate, acuminate, mucronate or shortly aristate; perianth-segments about 1 lin. long, ¾–1¼ lin. broad, all subquadrate or the outer 3 broadly ovate, all emarginate, brown with paler margins in the fruiting state; styles 2; fruit 3½ lin. long, 2¼ lin. thick, barrel-shaped, truncate, apiculate with the remains of the style-base, smooth and shining, not punctate dark olive-brown.

COAST REGION: Cape Div.; on hills of the Cape Peninsula, near Kommetjes, 500 ft., *Galpin*, 4832, ♂! 4822, ♀!

Allied to *W. teres*, Thunb., but differing in the sheaths, spathes, quadrate perianth-segments and smooth (not pitted) fruit.

8a. Willdenovia decipiens (N. E. Brown); plant 5 ft. high; stems terete, minutely punctate, glabrous, ochreous; leaf-sheaths broken, probably $1\frac{1}{2}$ in. or more long, closely convolute; basal part 5–6 lin. long, coriaceous, persistent, more or less bordered or stained all over with brown, upper part thin, submembranous, soon breaking away, greyish or light brownish; inflorescence of both sexes solitary, terminal, erect; spathes 2–3, similar in both sexes, outer varying from $1\frac{3}{4}$–$2\frac{3}{4}$ in. long, inner shorter, lanceolate, convolute into a subulate point, dark brown, shining, minutely punctate; male inflorescence about 2 in. long, consisting of several ovoid sessile many-flowered spikes, scattered along one or more flexuous axes; bracts 3–4 lin. long, $\frac{1}{2}$ lin. broad, lanceolate, tapering to an awn-like point, membranous, white; flowers very shortly pedicellate; perianth-segments $1\frac{1}{2}$–2 lin. long, $\frac{1}{3}$ lin. broad, lanceolate, tapering into an awn-like point, membranous, white; female flower solitary; bracts about 9 lin. long, 2–$2\frac{1}{2}$ lin. broad, ovate-lanceolate, convolute at the apex into a subulate point, very light ochreous-brown; perianth-segments 1 lin. long, $1\frac{1}{2}$ lin. broad, suborbicular-ovate, very obtusely rounded at the apex, which is notched and with a minute tooth in the notch in the 3 outer, but entire in the inner segments, brown; disk or stalk of the ovary $1\frac{1}{2}$–2 lin. long, deeply grooved lengthwise, whitish; styles short and stout, about $1\frac{1}{3}$ lin. long; fruit 3 lin. long, 2 lin. thick, cylindric, obtuse, smooth to the eye but with a very minutely roughened surface when seen under a lens, blackish-brown.

COAST REGION: Riversdale Div.; mountains of Garcias Pass, 1000 ft., *Galpin*, 4828!

Allied to *W. fimbriata*, Kunth, but differing in its smooth fruit and longer spathes.

2a. Eleocharis Schlechteri (C. B. Clarke); stems tufted, $\frac{1}{2}$–$1\frac{1}{4}$ in. long, rigid; a very slender short rhizome sometimes present; uppermost sheath firm, entire, and shortly oblique at the top; spikelet $\frac{1}{4}$ by $\frac{1}{6}$–$\frac{1}{8}$ in., brown, bearing 8–16 nuts, lanceolate at the top; glumes elliptic, obtuse, scarious-margined; hypogynous bristles 7, as long as the nut, retrorse-scabrous, cinnamon-brown; style long, 2-branched, style-base depressed, bulbous, small; nut obovoid, biconvex, much narrowed at the top, brown finally black, smooth.

COAST REGION: Caledon Div.; Onrust River, *Schlechter*, 9484!

The nut and setæ are as of *E. capitata*, R. Br., to which *E. Schlechteri* is allied, but differs by its pointed fewer-flowered spikelet.

4. Eleocharis Lepta (C. B. Clarke); stems 12–15 in. long, thread-like, quadrangular; uppermost sheath firm, truncate, shortly produced on one side; spikelet $\frac{1}{4}$ by $\frac{1}{8}$ in., chestnut-brown, many-flowered; glumes ovate; tip obtuse, scarious-edged; hypogynous bristles 4, about as long as the nut, white, patently papillose-scabrous; style as long as the nut, with 3 linear branches; style-base narrow conic.

COAST REGION: Cape Peninsula; Vaarsche Vley, *Wolley Dod*, 3541!

This may hereafter, when the ripe nut is obtained, be united to one of the numerous American species of this group; there is nothing at all like it in South Africa.

7. Scirpus verrucosulus (Steud.).

VAR. β, Pterocaryon (C. B. Clarke); nut margined all the way round by a thin scarious or greenish wing.

COAST REGION: Cape Peninsula; Fish Hook Valley, *Wolley Dod*, 3382 partly! 3383!

This is a striking variety, but Major Wolley Dod collected it growing with typical *S. verrucosulus*. Such a wing to the nut is almost unknown in *Scirpus*; very broad wings occur on the nuts of several species of *Fimbristylis*, and then are often not present in all the spikelets on the same stem.

17a. Ficinia minutiflora (C. B. Clarke); nearly glabrous; rhizome very slender; stems cæspitose, 2–6 in. long, setaceous; uppermost sheath entire, bearing a setaceous leaf often as long as the stem; head 1, apparently lateral, of 1–5 spikelets, $\frac{1}{12}$–$\frac{1}{8}$ in. in diam.; bract as though a continuation of the stem, 1 in. long; spikelets $\frac{1}{12}$ in. long, few-flowered, white or pallid; glumes ovate, obtuse, the central nerve rarely minutely excurrent.

COAST REGION: Caledon Div.; Houw Hoek Mountains, 2500 ft., *Galpin*, 4834! *Schlechter*, 9409!

50. Ficinia secunda.

VAR. β, maxima (C. B. Clarke); larger in all parts; stems up to 2 ft. long; spikelets $\frac{1}{4}$–$\frac{1}{3}$ in. long, turgid, hardly compressed.

COAST REGION: Cape Peninsula; Slang Kop, *Wolley Dod*, 3231!

2. Ecklonea solitaria (C. B. Clarke); stems tufted, 4–12 in. long; leaves 3–8 by $\frac{1}{8}$–$\frac{1}{6}$ in.; spikelets all solitary, far apart, on peduncles (sometimes carrying 2 spikelets) or subsessile, from several axils, some basal short-peduncled or sessile often added, $\frac{1}{3}$–$\frac{1}{2}$ in. long, narrowly-oblong or linear, greenish in fruit; glumes distant, so that the nuts in fruit are completely superposed; hypogynous bristles broadly ligulate, densely margined by long spreading hair.

COAST REGION: Cape Div.; Rondebosch Camp Ground, *Wolley Dod*, 3348!

The structure of the nut, style, and bristles is exactly that of *E. capensis*, Steud.; the species is very distinct in other respects.

6a. Tetraria lucida (C. B. Clarke); rhizome $\frac{1}{5}$ in. in diam., horizontal, clothed by short broad red scales; stems approximate, more than 2 ft. long, rather stout, terete, carrying 3–5 spikelets close to the top; leaves none, i.e. the uppermost basal sheath produced on one side, lanceolate, 2 by $\frac{1}{4}$ in., shining, chestnut-red; lowest bract $\frac{3}{4}$–$\frac{1}{12}$ in., brown-scarious, slightly overtopping the inflorescence; spikelets $\frac{1}{2}$ in. long; lowest flower male, 3-androus; nut large, obovoid, black-brown, crowned by the long pyramidal style-base.

COAST REGION: Ceres Div.; Prince Alfred, 3000 ft., *Schlechter*, 9987!

8a. Tetraria ferruginea (C. B. Clarke); rhizome horizontal, $\frac{1}{3}$–$\frac{1}{6}$ in. in diam.; stems close together, 2 ft. long, slender, without nodes except close to the base; basal sheaths ferruginous-brown, almost viscid; leaves 1 ft. long, filiform; panicle 1–2 in. long with 12–44 spikelets; lowest bract overtopping the inflorescence, at base ferruginous-brown and sheathing the stem; spikelets subsolitary, with 5–8 glumes.

CENTRAL REGION: Ceres Div.; Cold Bokkeveld, at Elands Fontein, 5000 ft., *Schlechter*, 10,018!

The spikelets, very young, are similar to those of *T. Burmanni*, C. B. Clarke.

3. **Chrysithrix Dodii** (C. B. Clarke); rhizome horizontal, $\frac{1}{20}$–$\frac{1}{16}$ in. in diam., clothed by lanceolate dirty-straw-coloured scales; stems 12–20 in. long, terete to the top, finely striate, slender throughout, $\frac{1}{20}$–$\frac{1}{16}$ in. in diam. at the top; leaves $\frac{3}{4}$ the length of the stem, $\frac{1}{16}$–$\frac{1}{12}$ in. in diam., terete, finely striate; head $\frac{1}{2}$ by $\frac{1}{8}$–$\frac{1}{6}$ in.

COAST REGION: Cape Peninsula; lower plateau above Skeleton Ravine, *Wolley Dod*, 3550! Silvermine Valley, *Wolley Dod*, 3549! George Div.; Cradock Berg, 5000 ft., *Galpin*, 4840!

This resembles closely, but is much slenderer in every part than *C. capensis*, Linn., var. *subteres*, C. B. Clarke, which Mr. Galpin got at the same time and place.

XXVIIA. POAGROSTIS, Stapf.

Spikelets very small, laterally compressed, pedicelled, panicled; rhachilla disarticulating above the glumes, obscurely produced beyond the base of the floret. *Floret* 1, ⚥, slightly shorter than the glumes. *Glumes* persistent, equal, lanceolate in profile, acute, membranous, closely 3-nerved at the very base, middle nerve percurrent or almost so. *Valve* entire, muticous, delicately membranous, finely silky villous, faintly 7-nerved; callus obscure, glabrous. *Pale* faintly 2-nerved below the middle, subequal to the valve. *Lodicules* 2, small, cuneate, glabrous. *Stamens* 3. *Ovary* oblong, glabrous; styles distinct, short; stigmas plumose, laterally exserted. *Grain* oblong, dorsally slightly compressed, elliptic in transverse section; pericarp subcrustaceous; embryo about $\frac{1}{4}$–$\frac{1}{3}$ the length of the grain; hilum obscure.

A very delicate, loosely tufted, much branched dwarf perennial; blades flat, thin; ligule a fringe of hairs; panicle very scanty and lax, sometimes reduced to 3–1 spikelets.

Species 1, on Table Mountain.

The affinity of *Poagrostis* seems to lie with *Achneria*, of which it might be said to represent a much reduced modification. The 2 essential points in which it differs from this genus are the solitary florets and the crustaceous pericarp.

1. **P. pusilla** (Stapf); culms ascending, very delicate, much branched, a few inches high, glabrous, smooth, up to 9-noded, often strongly bent at the uppermost node; sheaths thin, striate, with long fine spreading hairs near the mouth and the margins, upper tight, lower loose and whitish; ligule a fringe of fine hairs; blades linear, tapering to an acute point, $\frac{1}{2}$–1 in. by $\frac{1}{2}$–$\frac{3}{4}$ lin., flat, thin,

glabrous or very sparingly hairy, few-nerved; panicle very scanty, open or contracted; branches (pedicels) 2-nate, finely capillary, glabrous, smooth, up to 4 lin. long; spikelets about 1 lin. long, light greenish; glumes delicately herbaceous and green on the back with white margins, lower somewhat broader; valve oblong in profile, subacute, hairs very fine and more or less tortuous, inner side-nerves joining the middle nerve below the hyaline tip; pale sparingly hairy; tip ciliolate; anthers $\frac{1}{2}$ lin. long; grain brown, almost $\frac{3}{4}$ lin. long. *Colpodium (?) pusillum, Nees, Fl. Afr. Austr.* 149; *Steud. Syn. Pl. Glum.* i. 177; *Durand & Schinz, Consp. Fl. Afr.* v. 910. *Agrostis umbellulata, Trin. Agrost.* 370, *and in Mém. Acad. Petérsb. sér.* 6, vi. ii. 392; *Steud. l.c.* 173; *Durand & Schinz, l.c.* 829.

COAST REGION: Cape Div.; Table Mountain, in shady, rocky places, 3000 ft., *Drège!* and in marshy places at the same altitude, *Schlechter*, 115!

8. **Pentaschistis juncifolia** (Stapf). The synonym *Danthonia curvifolia β livida* founded on *Zeyher* 4545, is cited also under *P. Thunbergii*, as there are two specimens so numbered in the Kew Herbarium, and it is not certain to which of the two plants Nees gave his name.

25a. **Pentaschistis patuliflora** (Rendle in Journ. Bot. 1899, 381); perennial, densely tufted; culms erect, simple, fascicled below with numerous crowded innovation shoots, 10–16 in. long, glabrous, smooth, few- to many-noded, tightly sheathed below, uppermost internode exserted and much longer than the rest; leaves mostly basal, glabrous except at the sometimes shortly bearded mouth of the sheath; sheaths tight, firm, lower persistent, striate; ligule a fringe of short hairs; blades filiform, wiry, mucronulate to subobtuse, up to 5 in. long, tightly involute, flexuous or curled, smooth; panicle ovate in outline, open, 2–2$\frac{1}{2}$ in. long, up to 2 in. wide, erect; branches 2-nate, like the filiform axis and the capillary branchlets minutely tubercled, dark purple, finely bearded at the axils, trichotomous from 3–5 lin. above the base, lowest branches up to 1$\frac{1}{2}$ in. long; ultimate divisions and pedicels rather long, the latter usually quite smooth; spikelets scattered, 2$\frac{1}{2}$ lin. long, light brown, deep purple below; glumes acuminate, subhyaline, glabrous, smooth, keel minutely tubercled, upper slightly shorter; valves linear-oblong, body 1–1$\frac{1}{4}$ lin. long, glabrous or almost so; lobes short, subacute with a fine adnate bristle ($\frac{3}{4}$–1 lin. long) from the inner side; awn 4–5 lin. long, kneed just above the middle, twisted below; callus short, minutely bearded; panicles up to 1$\frac{1}{2}$ lin. long, minutely 2-toothed; anthers 1 lin. long.

COAST REGION: Caledon Div.; Genadendal, *Schlechter*, 10,286!

Intermediate between *P. Burchelli* and *P. angustifolia*.

35. **Pentaschistis Thunbergii** (Stapf). See note under *P. juncifolia* above.

3. Agrostis bergiana (Trin.).

Add to synonymy: *A. aristulifera, Rendle in Journ. Bot.* 1899, 381.

Add to distribution: COAST REGION: Bredasdorp Div. (?); by the Koude River, *Schlechter*, 9596!

5a. Agrostis Schlechteri (Rendle in Journ. Bot. 1899, 380); perennial, glabrous, tufted; culms erect, slender, 1½–2½ ft. long, smooth, 3–4-noded, upper internodes exserted, uppermost very long; leaves crowded near the base; sheaths somewhat rough, lowest short, uppermost very long and tight; ligules scarious, oblong, finely ciliolate, 2 lin. long; blades linear, acute, 2–4 in. by 1 lin., convolute or more or less flat, scaberulous on both sides, glaucous; panicle elliptic or ovate in outline, lax, 4–5 in. by 2–3 in.; axis filiform, scaberulous; branches 2–3-nate, finely filiform, obliquely erect, flexuous, the longest 2–3 in. long, undivided for 1–1½ in.; branchlets and pedicels scaberulous, the latter rather short; spikelets more or less tinged with purple, 1½ lin. long; rhachilla continued, hairy, up to 1 lin. long, sometimes bearing a fine bristle; glumes lanceolate, acute, keels scaberulous; upper slightly shorter; valve awned, oblong, truncate, 1 lin. long, loosely hairy, 5-nerved, side-nerves excurrent (inner into mucros, outer into minute bristles); awn from below the middle, 2–2½ lin. long, very fine, somewhat flexuous; callus small; minutely bearded; pale ¾–1 lin. long, glabrous, keels excurrent into minute fine mucros; anthers ⅞ lin. long; grain slightly over ½ lin. long.

COAST REGION: Paarl Div.; French Hoek, *Schlechter*, 10,274!

Very closely allied to *A. barbuligera*, Stapf, but with distinctly smaller spikelets and more scaberulous leaves.

1. Tragus major (Stapf). This is **T. koelerioides**, Aschers. in Verhandl. bot. Ver. Brandenb. xx. (1878) p. xxx.; the latter name has priority.

16. Ehrharta brevifolia, var. **cuspidata**, Nees.

Add to synonymy: *E. Schlechteri, Rendle in Journ. Bot.* 1899, 380.

Add to distribution: WESTERN REGION: Little Namaqualand; in Vanrhynsdorp Div., at Zout River, *Schlechter*, 8133!

17. Ehrharta pusilla, var. **inæquiglumis** (Rendle in Journ. Bot. 1899, 380); barren valves distinctly unequal, the short weak awn of the lower scarcely exceeding the outer glume.

COAST REGION: Clanwilliam Div.; Windhoek, *Schlechter*, 8361!

Probably only an accidental variation of the type.

1a. Urochlæna major (Rendle in Journ. Bot. 1899, 382); culms 4–6 in. long, very slender, ascending, smooth, about 4-noded, internodes (except the uppermost) exserted; sheaths smooth, striate, naked or with a few long fine hairs at the mouth, the uppermost tumid, with large obtuse hyaline auricles; ligule a line of hairs; blades linear with a minute callous tip, 1–1½ in. by 1 lin., scaberulous upwards, with very fine scattered hairs; heads obovoid, up to 7 lin. by 5 lin.; spikelets about 3-flowered, 2–2½ lin. long (exclusive of awns), finely bearded at the base, green with purplish awns; glumes lanceolate, long acuminate in profile (ovate-oblong when flattened out), 2½–3 lin. long, asymmetrically 4-nerved; outer margin with a row of long tubercle-based hairs, otherwise glabrous or almost so; awn flexuous, purple, up to 5 lin. long; valves 2–2½ lin. long, finely hairy on the sides near the base (with clavate-tipped hairs) and with tubercle-based hairs near the margin in the upper part; awn 1½–2½ lin. long; anthers (young) 1 lin. long.

COAST REGION: Clanwilliam Div.; Hoek, at the foot of the Koude Berg, *Schlechter*, 8699!

Very closely allied to *U. pusilla*, Nees, and possibly only a vigorous state of it.

LXXXIXA. PSEUDOBROMUS, K. Schum.

Spikelets lanceolate, scarcely compressed, pedicelled, panicled; rhachilla disarticulating above the glumes, produced beyond the base of the floret into a fine bristle. *Floret* 1, ⚥, exserted from the glumes. *Glumes* persistent, slightly unequal, lanceolate to oblong, 1–3-nerved. *Valve* lanceolate in profile, rounded on the back, involute, passing into a straight awn or with the awn subterminal, subherbaceous, 3-nerved; callus very short, obtuse, glabrous. *Pale* equalling the valve, 2-keeled, entire. *Lodicules* 2, hyaline, unequally 2-lobed. *Stamens* 3. *Ovary* oblong, pubescent at the top; styles distinct, very short; stigmas slender, plumose, laterally exserted. *Grain* unknown.

Perennial; blades flat, broad-linear, flaccid; ligules membranous; panicle large, lax, nodding.

Species 2, 1 in the Transvaal, the other on Mount Kilimanjaro.

Pseudobromus has very much the habit of *Festuca gigantea*, Vill., and is, in my opinion, closely allied to *Festuca*, in spite of the solitary florets and 3-nerved valves. The presence of a rudimentary valve at the tip of the rhachilla points to the derivation from a more than 1-flowered type, and if the suppression of the second pair of side-nerves were not so complete, this grass would have to be described as a species of *Festuca* of the *F. gigantea* group.

1. **P. africanus** (Stapf); quite glabrous with the exception of the base of the lowest sheaths; culms erect or shortly ascending, simple, up to 2 ft. long, sheathed all along, about 3-noded; sheaths rather loose, strongly striate, the lowest subpersistent; ligules ovate-oblong, 2–2½ lin. long, denticulate; blades linear-lanceolate, tapering to a long setaceous point, to more than 1 ft. by 8 lin., flaccid, green, paler below, scaberulous on the lower side, almost smooth on the upper, margins very scabrid; panicle ovate to oblong, very loose, almost 1 ft. long, nodding; axis angular, smooth below, scaberulous above; branches more or less 3-nate, scantily and remotely divided

from $\frac{1}{2}$–$\frac{1}{3}$ their length above the base, filiform, flexuous, scaberulous to scabrid, the longest 3 in. long or longer; lateral pedicels about 2 lin. long; spikelets lanceolate, 3–4 lin. long, green; continuation of rhachilla glabrous, scaberulous, almost half the length of the spikelets with a minute rudimentary valve at the upper end; glumes lanceolate, acuminate, herbaceous-membranous except at the hyaline margins, lower much narrower, 1-nerved, 2–2$\frac{1}{2}$ lin. long, upper 3-nerved, 2$\frac{1}{2}$–3 lin. long; valve lanceolate in profile, oblong (when expanded), passing into the awn, 3–4 lin. long, herbaceous-membranous except along the narrow hyaline margins, middle nerve stout and prominent towards the tip; awn 7–8 lin. long; anthers 2–2$\frac{1}{2}$ lin. long. *Brachyelytrum africanum, Hack. in Bull. Herb. Boiss.* iii. 382.

KALAHARI REGION: Transvaal; Houtbosch, *Rehmann*, 5732!

Very closely allied to *P. silvaticus*, K. Schum. in Engl. Pfl. Ost-Afr. C. 108, which has broader, less acute glumes and subterminal awns.

XCVIA. SECALE, Linn.

Spikelets laterally compressed, sessile, distichously imbricate in a dense cylindric spike; axis of the spike readily disarticulating except in cultivated forms, compressed, each joint of the axis falling with the spikelet entire; rhachilla continued into a short bristle above the base of the upper floret, otherwise scarcely perceptible. *Florets* 2, ☿, more or less exserted from the glumes. *Glumes* equal or slightly unequal, very narrow, linear-subulate, strongly compressed and keeled. *Valves* oblong, lanceolate in profile, strongly compressed, passing into a straight awn, firmly membranous, 5-nerved, keeled; callus 0. *Pales* 2-keeled, 2-toothed, almost as long as the valves, narrow. *Lodicules* 2, hyaline, ciliate. *Stamens* 3. *Ovary* obovoid, top densely tomentose; styles distinct, very short; stigmas plumose, laterally exserted. *Grain* oblong, subterete, grooved in front, enclosed by the valve and pale, free or almost so; embryo about $\frac{1}{3}$ the length of the grain.

Perennial or annual; blades flat, flaccid; ligules membranous; spikes very fragile.

Species 4; 2 in the Mediterranean countries and in South Eastern Europe and Turkestan, 1 in South Africa, 1 only known as a cereal.

1. S. africanum (Stapf in Hook. Icon. Pl. t. 2601); culms slender, over $\frac{1}{2}$ ft. long, glabrous, smooth, upper internodes exserted; upper leaves glabrous, smooth; sheaths tight; ligules very short, obtuse; blades linear, narrow, up to 6 in. long; spike linear, very dense, 2$\frac{1}{2}$–3 in. by 2$\frac{1}{3}$–3 lin.; rhachis fragile, edges of joints densely villous; spikelets 5–6 lin. long (exclusive of the awns), closely imbricate; glumes linear, gradually narrowed to a fine point or produced into a very short bristle, upper about as long as the valves, lower usually somewhat shorter, keels finely scabrid; valves lanceolate-oblong, produced into a fine scabrid straight awn, 3–3$\frac{1}{2}$ lin. long, scabrid on the sides; nerves distinct, green upwards, keel very minutely spinulous-scabrid; keels of pale scabrid. *S. cereale, Thunb. Prodr.* 23; *Fl. Cap. ed.* i. 440; *ed. Schult.* 118; *Durand & Schinz, Consp. Fl. Afr.* v. 937, *not of Linn.*

CENTRAL REGION: Sutherland Div.; Rogge Veld, *Thunberg!*

Thunberg says in his *Travels*, ed. 3, ii. 168:—"These (Rogge Velds) . . . have been so named from a kind of rye, which grows wild here in abundance near the bushes." Burchell, however, says (*Travels*, i. 256):—"I saw none of the wild rye which has been said to be so abundant as to give the name to this district; but this might be owing to the season of the year." *S. africanum* differs from *S. cereale*, Linn., and *S. montanum*, Guss., in the smaller spikelets, slightly unequal glumes, more conspicuously nerved and scabrid valves, and in all the keels being very minutely spinulous or scabrid.

INDEX.

[SYNONYMS ARE PRINTED IN *italics.*]

CORRIGENDA.

PAGE

11, line 18, for *L.* read *C.*
13 ,, 8 ,, *Dokumente* read *Documente*
40 ,, 35 ,, Hook. *read* Horkel.
42 ,, 30 ,, *maritinum* read *maritimum*
51 ,, 19 ,, R. Schum. *read* K. Schum.
141 ,, 20 ,, **virgata** read **cephalotes**
155 ,, 11 ,, compressed *read* impressed
216 ,, 11 ,, *verrucosula* read *verruculosa*
251 ,, 5 from the bottom; dele *Kunth, Enum.* ii. 256
258 ,, 27, *for* accuminate *read* acuminate
259 ,, 6 from the bottom, *for* natura *read* natural
299 ,, 1, for 1873 read 1783
304 ,, 6 ,, cental *read* central
316 ,, 4 from the bottom, for **Leptocarydium** read **Leptocarydion**
318 ,, 17, for LXXIX. **Lasiochloa** read LXXIX. bis. **Lasiochloa**
323 ,, 13 *for* loosely, hairy to ½ lin. long, *read* loosely hairy, hairs up to ½ lin. long
344 ,, 20 *for* oints *read* joints
417 ,, 9 ,, *Prin.* read *Prim.*
439 ,, 1 ,, 2–4-distant *read* 2–4 distant
464 ,, 24 ,, **mollis** read **lanatus**
465 ,, 37 ,, kneed, from near the base *read* kneed from near the base
476 ,, 8, dele ft.
478 ,, 38, for *Benth. & Trin.* read *Bentl. & Trim.*
529 ,, 35 ,, lin. *read* in.
548 ,, 33 ,, ½ lin. *read* ½ ft.
575 ,, 8 from the bottom, *for* in. *read* ft.
622 ,, 2 from the bottom, *before* the *insert* $\frac{4}{5}$
642 ,; 12 *for* vi. *read* v.
654 ,, 4 from the bottom, *after* 1–2 ft. *insert* long
706 ,, 23 *for* habitat *read* habit
717 ,, 3 from bottom, for *augusta* read *angusta*
741 ,, 8 *for* noded. usually *read* noded, usually

Printed in the United States
By Bookmasters